T0398215

Springer Series in
CHEMICAL PHYSICS

Series Editors: A. W. Castleman, Jr.   J. P. Toennies   K. Yamanouchi   W. Zinth

The purpose of this series is to provide comprehensive up-to-date monographs in both well established disciplines and emerging research areas within the broad fields of chemical physics and physical chemistry. The books deal with both fundamental science and applications, and may have either a theoretical or an experimental emphasis. They are aimed primarily at researchers and graduate students in chemical physics and related fields.

Please view available titles in *Springer Series in Chemical Physics* on series homepage http://www.springer.com/series/676

Paul Corkum
Sandro De Silvestri
Keith A. Nelson
Eberhard Riedle
Robert W. Schoenlein
(Eds.)

Ultrafast Phenomena XVI

Proceedings
of the 16th International Conference,
Palazzo dei Congressi Stresa, Italy,
June 9–13, 2008

With 667 Figures

 Springer

Paul Corkum
Steacie Institute for Molecular Sciences
National Research Council Canada
100 Sussex Drive, Rm 2063
Ottawa, ON K1A 0R6, Canada
E-Mail: paul.corkum@nrc-cnrc.gc.ca

Professor Eberhard Riedle
Ludwig-Maximilians-Universität München
Sektion Physik
Oettingenstrasse 67
80538 München, Germany
E-Mail: eberhard.riedle@physik.uni-muenchen.de

Professor Sandro De Silvestri
Politecnico di Milano
Dipto. Fisica
Piazza Leonardo da Vinci, 32
20133 Milano, Italy
E-Mail: sandro.desilvestri@fisi.polimi.it

Professor Robert W. Schoenlein
Lawrence Berkeley National Laboratory
Materials Sciences Division
1 Cyclotron Road
Berkeley, CA 94720, USA
E-Mail: rwschoenlein@lbl.gov

Professor Keith A. Nelson
Massachusetts Institute of Technology (MIT)
77 Massachusetts Ave
Cambridge, MA 02139, USA
E-Mail: kanelson@mit.edu

Series Editors:
Professor A.W. Castleman, Jr.
Department of Chemistry, The Pennsylvania State University
152 Davey Laboratory, University Park, PA 16802, USA

Professor J.P. Toennies
Max-Planck-Institut für Strömungsforschung
Bunsenstrasse 10, 37073 Göttingen, Germany

Professor K. Yamanouchi
University of Tokyo, Department of Chemistry
Hongo 7-3-1, 113-0033 Tokyo, Japan

Professor W. Zinth
Universität München, Institut für Medizinische Optik
Öttingerstr. 67, 80538 München, Germany

ISSN 0172-6218
ISBN 978-3-540-95945-8   e-ISBN 978-3-540-95946-5
DOI 10.1007/978-3-540-95946-5
Springer Heidelberg Dordrecht London New York

Library of Congress Control Number: 2009926069

Cover design: SPi Publisher Services

Printed on acid-free paper

Springer is part of Springer Science+Business Media (www.springer.com)

Preface

This volume is a compilation of research papers presented at the Sixteenth International Conference on Ultrafast Phenomena held at the Palazzo dei Congressi Stresa, Italy, from June 9 to 13, 2008. The Ultrafast Phenomena conferences are held every two years and are the premier international forum for discussion of the latest results in ultrafast science. These meeting bring together researchers spanning numerous fields of science and engineering to deliberate the latest advances in ultrafast optics and their applications in science and engineering. The conferences and associated published proceedings effectively disseminate the latest scientific advances using ultrashort coherent pulses of light. More than 370 papers were presented at Ultrafast Phenomena XVI. Significant progress in creating every shorter pulses of light was reported, now extending below 100 attoseconds, with new developments in high harmonic generation and frequency comb metrology. Multidimensional spectroscopy is rapidly evolving to provide new insights into quantum coherence and interactions in complex systems. Dramatic advances in time-resolved electron and x-ray diffraction and spectroscopy provide detailed information on atomic and electronic structural dynamic in molecular systems and crystalline solids. These examples are but a small subset of the research summaries gathered in this volume, which provides a valuable synopsis of the recent advances and impact of ultrafast technology in illuminating fundamental processes in physics, chemistry, and biology. There were 434 attendees at the meeting, more than a third of which were graduate and postdoctoral students. Increased student attendance energized the proceedings, and discussions were further enhanced by the beautiful lakeside setting and natural alpine surroundings in Stresa.

Many people and organizations made invaluable contributions to the success of the conference. The international program committee reviewed over 550 submissions and organized the scientific program. The local organizing committee arranged many of the logistics for the meeting including the stimulating venue at Stresa, on the shore of Lake Maggiore. Staff of the European Physical Society deserve thanks for making the conference arrangements and running a smooth and efficient meeting. We thank the Optical Society of America, European Science Foundation, European Cooperation in Scientific and Technical Research (COST P14), Politecnico di Milano, LASERLAB-Europe (a consortium of 17 European Laser Infrastructures), and European Extreme Light Infrastructure for their support. Generous support for Ph.D. student participation was provided by: European DYNA - ESF Network, Coherent Corp., Newport Corp., Kapteyn-Murnane Labs, and Quantronix Corp. We are particularly grateful to André Wobst for coordinating the conference submissions and evaluations, and for his help in bringing this volume together.

Politecnico di Milano, Italy
Steacie Institute, Ottawa, Canada
MIT, Cambridge, USA
LMU, Munich, Germany
LBNL, Berkeley USA
September 2008

Sandro De Silvestri
Paul Corkum
Keith A. Nelson
Eberhard Riedle
Robert W. Schoenlein

Contents

Part II: Ultrafast X-ray and Electron Science

Part III: Correlated Electron Systems, Magnetization and Spin Dynamics

Part IV: Physics - Condensed Phase and Low Dimensional Systems

Part V: Chemistry - Condensed Phase

Part VI: Chemistry - Advanced Spectroscopy, Molecular Control, Hydrogen Bonding, Liquids and Interfaces

Part VII: Biological Systems, Molecular Light Harvesting and Charge-Transfer Complexes

XX

Part VIII: THz Science and Technology, Nano-Optics and Plasmonics

Part IX: Novel Pulsed Sources: oscillators, amplifiers, nonlinear mixing

XXV

Part X: Frequency Combs and Waveform Synthesis

Part XI: Optics, Optoelectronics, Measurement, Diagnostics and Instrumentation

Part XII: Applications of Ultrashort Pulses

XXX

Attosecond and High-Order Harmonic Generation and Measurement, Atomic and Molecular Physics

Sub-100-as soft x-ray pulses

E. Goulielmakis[1], M. Schultze[1], M. Hofstetter[2], M. Uiberacker[2], J. Gagnon[1], V. Yakovlev[2], U. Kleineberg[2], F. Krausz[1,2]

[1]Max-Planck-Pnstitut für Quantenoptik, Hans-Kopfermann-Strasse 1, D-85748 Garching, Germany
[2] Department für Physik, Ludwig-Maximilians-Universität, am Coulombwall 1, Germany
elgo@mpq.mpg.de

Abstract: We demonstrate the generation of powerful sub-100-as soft x-ray pulses by means of 1.5-cycle waveform-controlled laser fields. Our new tool opens the door for exploring electronic processes on a time scale approaching the atomic unit.

Introduction

Attosecond pulses are emitted when energetic electron wave packets, created by the interaction of intense laser fields with atoms, recollide with the atomic core and radiate soft x-rays. Spectral filtering of radiation emerging from a single recollision event comprises the cornerstone of isolated attosecond pulse technology [1],[2]. Indeed, light wave packets emitted within the most intense half-cycle of a few-cycle laser pulse have allowed for generation of powerful attosecond pulses in the sub 200-as regime [3]. The duration of these pulses has been limited by the spectral width of the emitted soft x-ray continuum, which is inextricably related to the intensity contrast between adjacent half-cycles of the driving pulse. Combined with polarization gating, few-cycle, phase-controlled laser pulses have permitted the generation of even shorter bursts of radiation [4], though at the expense of the photon yield, which is highly crucial for a number of experiments.

Here, we demonstrate a new regime of attosecond pulse generation where waveform-controlled pulses, comprising merely 1.5 field oscillations [5], are employed to drive soft x-ray emission from atoms. Due to the high nonlinearity of the tunnelling process, the interaction is virtually restricted to within a single optical-cycle. This enables emission from only two electron trajectories, over a significant fraction (>50 %) of the emitted bandwidth. Adjusting the phase setting to enhance the contribution of a single, most-powerful electron trajectory, we generate a soft x-ray supercontinuum that extends over more than ~28 eV in FWHM. From the temporal characterization of the emitted soft x-rays, pulses were ascertained to have duration of 80±5 attoseconds.

Experimental Methods

The soft x-rays are generated by gently focusing (f=600 mm) sub-4-fs laser pulses into a quasi-static gas cell filled with Neon at a pressure of ~ 350 mbar. The setup is discussed in detail in [2,3]. A quasi-monolithic double-mirror assembly, consisting of a Mo/Si multilayer mirror and an outer concentric perforated mirror is placed approximately 1.5 m downstream and is used to focus the soft x-rays and the laser beam into a second neon target. The multilayer mirror - consisting of a stack of four Mo/Si quarter wave layers - combined with Zr foils allows for a bandwidth greater than 30 eV. The delay between the laser and soft-x-ray pulses can be adjusted with nanometer precision. Electrons set free by the soft x-ray pulse, along the laser polarization, are collected by a time-of-flight (TOF) electron spectrometer. The resolution of the spectrometer is better than ~0.5 eV in the range 40-90 eV.

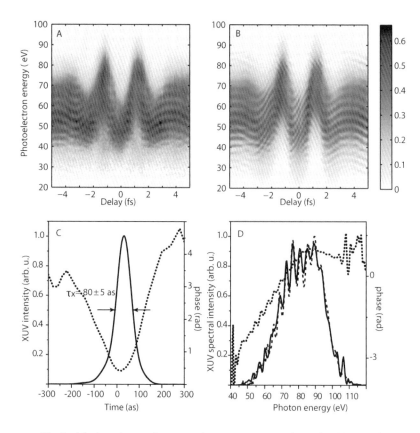

Fig.1. (a) Atomic-transient-recorder spectrogram of a sub-100-as soft x-ray pulse. (b) Reconstructed spectrogram after 10^3 iterations of a FROG-based algorithm [9]. (c) Temporal intensity profile and phase of the pulse retrieved from the measured spectrogram depicted in (a). (d) Retrieved (solid line) and directly measured (dashed line) spectrum. The retrieved spectral phase is depicted as a thin dash.

Results and Discussion

Figure 1.a shows a spectrogram recorded from our soft-x-ray attosecond pulse. The basic principles of the characterization technique have been put forward in [6] and realized for the first time in [1]. In order to guarantee a sufficient resolution, our streaking field is adjusted to produce spectral shifts comparable to the original bandwidth of the photoemission. We analyze our spectrogram utilizing an a version of the FROG technique optimized for attosecond streaking [8,9]. The algorithm converges after $\sim 10^3$ -10^4 iterations. The reconstructed spectrogram is depicted in Fig1.b, while the retrieved temporal intensity and phase of the 80 attosecond (in FWHM) soft x-ray pulse are depicted in figure 1.c. Figure 1.d shows the spectral phase as well as the retrieved spectrum. A soft x-ray spectrum measured directly is also shown for comparison.

The quasi monocycle nature of our driving pulses has played a key role in (i) enabling the generation of sub-100-as pulses and (ii) enhancing the photon yield by giving rise to favourable phase matching conditions in the generating medium. Indeed, utilizing our yield-calibrated soft x-ray camera, we can safely estimate an energy of ~0.5 nJ/pulse within the bandwidth of the attosecond pulse at the high harmonic source. This corresponds to generation efficiency on the order of 10^{-6}.

Conclusions

We have generated powerful sub-100-as pulses in the soft x-ray regime. These pulses, precisely synchronized to the generating laser field, will offer a dramatic improvement in temporally resolved experiments, and hold promise for the real time tracking of ultrafast dynamics on the time scale of electron correlation [10].

[1]. R. Kienberger, E. Goulielmakis, M. Uiberacker, A. Baltuska, V. Yakovlev, F. Bammer, A. Scrinzi, T. Westerwalbesloh, U. Kleineberg, U. Heinzmann, M. Drescher, and F. Krausz, "Atomic transient recorder, Nature **427**, 817 (2004).

[2]. E. Goulielmakis, M. Uiberacker, R. Kienberger, A. Baltuska, V. Yakovlev, A. Scrinzi, T. Westerwalbesloh, U. Kleineberg, U. Heinzmann, M. Drescher, and F. Krausz, "Direct measurement of light waves," Science 317, 1267 (2004).

[3] M. Schultze , E. Goulielmakis, M. Uiberacker , M. Hofstetter, J. Kim J, Kim D, Krausz F, Kleineberg U., Powerful "170-attosecond XUV pulses generated with few-cycle laser pulses and broadband multilayer optics", New. J. Phys 9, 243 (2007)

[4]. G. Sansone, E. Benedetti, F. Calegari, C. Vozzi, L. Avaldi, R. Flammini, L. Poletto, P. Villoresi, C. Altucci, R. Velotta, S. Stagira, S. De Silvestri, and M. Nisoli, "Isolated single-cycle attosecond pulses,"Science **314**, 443 (2006).

[5]. A. L. Cavalieri, E. Goulielmakis, B. Horvath, W. Helml, M. Schultze, M. Fiess, V. Pervak, L. Veisz, V. S. Yakovlev, M. Uiberacker, A. Apolonski, F. Krausz, and R. Kienberger, New J. Phys., 9, 242 (2007).

[6]. J. Itatani, F. Quere, G. Yudin, M. Ivanov, F. Krausz, and P. Corkum, "Attosecond streak camera", Phys. Rev Lett., 88, 173903 (2002)
[8]. Y. Mairesse, and F. Quere, "Frequency-resolved optical gating for complete reconstruction of attosecond bursts," Phys. Rev. A 71, 011401 (2005)

[9] J. Gagnon et al., "The accurate FROG characterization of attosecond pulses from streaking measurements", Appl. Phys. B (published online, doi: 10.1007/s00340-008-3063-x)

[10] J. Breidbach, and L. Cederbaum, "Universal attosecond response to the removal of an electron," Phys. Rev. Lett. **94**, 033901 (2005) 974.

Generation of High-order Harmonics with a Near-IR Self-phase-stabilized Parametric Source

C. Vozzi[1], F. Calegari[1], F. Frassetto[2], E. Benedetti[1], M. Nisoli[1], G. Sansone[1], L. Poletto[2], P. Villoresi[2], and S. Stagira[1]

[1] National Laboratory for Ultrafast and Ultraintense Optical Science - INFM - CNR Dipartimento di Fisica, Politecnico di Milano, Piazza Leonardo da Vinci 32, 20133 Milano, Italy
E-mail: caterina.vozzi@polimi.it
[2] Laboratory for Ultraviolet and X-Ray Optical Research - INFM - CNR, DEI, Universit di Padova, Padova, Italy

Abstract. We generated high-order harmonics with self-phase-stabilized near-IR pulses produced by a parametric source. We observed a significant cutoff extension with respect to 800-nm driving pulses at comparable peak intensity.

An increasing interest has recently been devoted to the development of near-infrared self-phase stabilized sources for driving high-order harmonic generation (HHG) process. This research is triggered by the possibility of pushing harmonic emission toward soft X-ray region and efficiently generating isolated attosecond pulses exceeding 100 eV photon energy [1,2]. Moreover scaling of HHG efficiency and phase-matching condition with driving pulse central wavelength is still a subject of intense investigation [3,4]. We describe here high-order harmonics generated with a self-phase-stabilized high-energy parametric source at 1.5 μm [5].

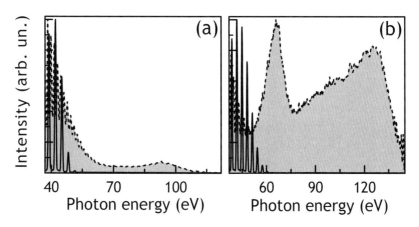

Fig. 1. Harmonic spectra generated in krypton (a) and argon (b) with 800-nm driving pulse (solid curve) and 1.5 μm driving pulse (dashed curve).

The IR source is developed starting from an amplified Ti:sapphire laser system (60 fs, 10 mJ, 800 nm, 10 Hz). A broadband supercontinuum is generated by filamentation in a krypton-filled gas cell and a phase stable seed is then produced by difference frequency (DF) of the supercontinuum spectral components. This seed is then amplified through a two-stage near-IR optical parametric amplifier (OPA) up to 1.2 mJ with a nearly transform limited pulse duration of 18 fs. Carrier-envelope-phase stability of this source has been demonstrated [5]. In order to produce high-order harmonic radiation we focused the 1.5-μm driving pulses on a synchronized gas jet operating at 10 Hz. Observation of harmonic emission was done by means of a soft x-ray spectrometer and a micro-channel plate coupled to a phosphor screen and a CCD camera.

We compared the harmonic radiation generated by the 1.5-μm parametric source with harmonic spectra obtained with standard 800-nm driving pulses. For this purpose, a small portion of our Ti:sapphire laser system was used for generating harmonics with the same experimental setup; the energy of the 800-nm pulses was reduced in order to get comparable pulse peak intensities in the focal region for both sources.

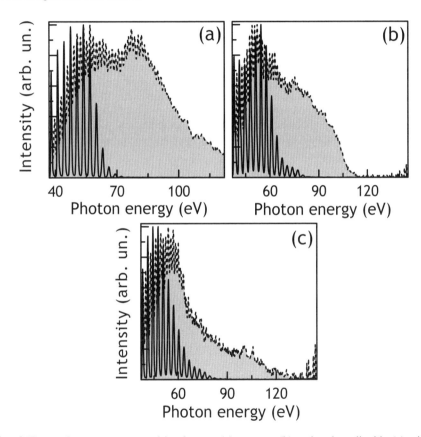

Fig. 2. Harmonic spectra generated in nitrogen (a), oxygen (b) and carbon dioxide (c) with 800-nm driving pulse (solid curve) and 1.5 m driving pulse (dashed curve).

Fig. 1 shows a comparison between normalized harmonic spectra generated by the 1.5-μm DF source (dashed curve) and by the 800-nm laser (solid curve) in krypton (a) and argon (b). For both gases peak intensity of the DF pulse in the focus was about 2×10^{14} W/cm^2, whereas the peak intensity of the 800-nm driving pulse at the focus was estimated in about 2.5×10^{14} W/cm^2. A noticeable spectral extension of harmonic radiation was measured for the 1.5-μm driving pulses with respect to the 800-nm source. As can be seen from Fig. 2, cutoff energy was 114 eV in krypton and 140 eV in argon for the 1.5-μm driving pulses; these values have to be compared to 54 eV for krypton and 64 eV for argon in the case of 800-nm driving pulses. The observed cutoff energy are in good agreement with the prediction of the three step model for argon. In the case of krypton, we observed a difference between experimental cutoff energy and value predicted by the cutoff law; this difference could be ascribed to ionization saturation. It is worth noting that the cutoff photon energy in argon for the 1.5-μm driving pulses represents the instrumental limit of our spectrometer in the present configuration. Aside from the cutoff energy increase, harmonic spectra generated by the 1.5-μm driving source present interesting spectral feature. Indeed for both generating media the harmonic spectrum is continuum above 60 eV and presents deep modulations depending on the harmonic photon energy. These spectral structure seem not be related to absorption in the generating medium. We also experimentally investigated the behavior of the harmonic yield as a function of the driving wavelength. As predicted by Tate and coworkers the emission yield decreases with increasing the driving wavelength [3]. Our measurements show that the emission yield in the case of 1.5-μm driving pulses is about two order of magnitude lower than the one observed with 800-nm driving pulses. Indeed a proper estimation of the yield scaling law with driving wavelength should take into account macroscopic effects like phase matching condition. Since a careful characterization of driving sources and harmonic generation condition is crucial for addressing this issue, these aspects will be the object of further investigation.

We also extended our study to harmonic generation in molecular gases. Fig. 2 shows a comparison between normalized harmonic spectra generated by the 1.5-μm DF source (dashed curve) and by the 800-nm laser (solid curve) in nitrogen (a), oxygen (b) and carbon dioxide (c). In this case peak intensity of the DF pulse in the focus was about 2×10^{14} W/cm^2, whereas the peak intensity of the 800-nm driving pulse at the focus was estimated in about 3.5×10^{14} W/cm^2. Even in this case, harmonics generated by the 1.5-μm DF source show a considerable increase of the cutoff frequency.

In conclusion HHG with a high-energy self-phase-stabilized near-IR parametric source has been demonstrated. The experimental results show a dramatic extension of the cutoff photon energy with respect to a standard Ti:sapphire laser source. Moreover the parametric source exploited in this experiment is characterized by a self stabilization of the carrier-envelope phase. This feature has been proved to be essential for the generation of isolated and reproducible attosecond pulses; thus our experimental result represents a step forward the generation of single attosecond pulses with photon energies exceeding 100 eV.

Acknowledgements. This work was partially supported by the European Union within contracts RII3-CT-2003-506350 (Laserlab Europe) and MRTN-CT-2003-505138 (XTRA). We acknowledge the financial support from the Italian MIUR PRIN project n. 2006027381.

1   V. S. Yakovlev, M. Ivanov, and F. Krausz, "Enanched phase-matching condition for generation of soft X-ray harmonics and attosecond pulses in atomic gases", Opt. Exp. 15, 15351 (2007).
2   H. Merdji, T. Auguste, W. Boutu, J-P. Caumes, B. Carr, T. Pfeifer, A. Jullien, D. Neumark, S. Leone, Opt. Lett. 32, 3134 (2007).
3   I. J. Tate, T. Auguste, H. G. Muller, P. Salires, P. Agostini and L. F. DiMauro, "Scaling of wave-packet dynamics in an intense midinfrared field", Phys. Rev. Lett. 98, 139011 (2007).
4   P. Colosimo, G. Doumy, C. I. Blaga, J. Wheeler, C. Hauri, F. Catoire, J. Tate, R. Chirla, A. M. March, G. G. Paulus, H. G. Muller, P. Agostini and L. F. DiMauro, Nature Physics 4, 386 (2008).
5   C. Vozzi, F. Calegari, E. Benedetti, S. Gasilov, G. Sansone, G. Cerullo, M. Nisoli, S. De Silvestri, and S. Stagira, "Millijoule-level phase-stabilized few-optical-cycle infrared parametric source", Opt. Lett. 32, 2957 (2007).

Quasi-Phase-Matched High-Order Harmonic Generation in the Soft-X-ray Regime

J. Seres[1,2], V. S. Yakovlev[3], E. Seres[1,2], Ch. Streli[4], P. Wobrauschek[4], F. Krausz[3,5], Ch. Spielmann[1,6]

[1] Physikalisches Institut EP1, Universität Würzburg, D-97074 Würzburg, Germany
E-mail: josef.seres@physik.uni-wuerzburg.de

[2] Institut für Photonik, Technische Universität Wien, A-1040 Wien, Austria

[3] Dept. für Physik, Ludwig-Maximilians-Universität München, D-85748 Garching, Germany

[4] Atominstitut, Technische Universität Wien, A-1020 Wien, Austria

[5] Max-Planck-Institut für Quantenoptik, D-85748 Garching, Germany

[6] Institute for Optics and Quantumelectronics, FS University Jena, D-06943 Jena, Germany

Abstract. We realized quasi-phase-matched generation of soft x-rays emitted from two gas jets. The harmonic signal has been enhanced in a broad range (from 250 eV to 600 eV) and up to two orders of magnitude around 400 eV.

Introduction

High-order harmonic (HH) generation from atoms illuminated by short laser pulses is a suitable way to generate coherent, well collimated electromagnetic radiation in the extreme ultraviolet [1] and soft-x-ray [2-8] region. The maximum achievable photon energy (so called cut-off) is proportional to the laser intensity. However, the observed low conversion efficiency, especially at higher photon energies is a serious limitation for several applications. HHG is a coherent process requiring for phase-matching between the fundamental and harmonic beam. The higher laser intensity, which is necessary for short wavelength generation, results in higher ionization rates. The free electrons cause a dramatic decrease of the phase-matching length, reducing the efficient medium length to a few microns or less.

Two approaches are known to extend the phase-matching length namely non-adiabatic self-phase-matching (NSPM) and quasi-phase-matching (QPM). NSPM was proposed theoretically in the last decade [9], however experimental realization was only possible recently [4-6] due to the availability of the suitable laser delivering pulses with a few millijoules energy and duration around 10 fs [10]. Nevertheless, NSPM has been recently the only technique, which allowed the generation of coherent radiation up to few keVs [5,6] with suitable photon flux for the first spectroscopic applications [6,11]. On the other hand, severe effort has put in the development of different QPM schemes. Enhancement of the photon yield has been demonstrated up to the carbon K edge (~4 nm) using Argon-gas-filled hollow fiber with modulated inner diameter [4] or by periodic beating between fiber propagation modes [7]. Counter-propagating pulse modulate the laser intensity along the axis with a period fulfilling the condition for QPM. With this method a substantial enhancement of the harmonic yield has been observed for lower order harmonics [12].

In this work we present a new implementation of QPM by using two successive gas jets. The multi jet configuration in combination of NSPM is feasible for increasing the harmonic conversion efficiency in an energy range between 250-600 eV (~2-5 nm),

9

by coherent superposition of harmonic radiation generated in these successive sources as traversed by the same laser beam. Under optimized conditions, such as pressure and gas jet separation, the enhancement exceeds one order of magnitude, especially at higher photon energies between 400 and 600 eV.

Experimental Set-up

In the recent experiments, the fundamental laser pulses are provided by a two-stage Ti:sapphire amplifier system running at 1 kHz repetition rate and delivering pulses with energy of 2 mJ and pulse duration of 15 fs. Details about the laser system can be found elsewhere [10]. The output pulses are spectrally broadened in filament in an Argon gas filled (0.6 bar) tube and compressed with a set of chirped mirrors. The output pulses have an energy of approx. 1.5 mJ at a center wavelength of 750 nm. However, the pulses are not perfectly re-compressed. A short 5 fs long pulse with an energy of 1.0 mJ is superimposed on a 15-fs-long pedestal containing 0.5 mJ. These pulses are focused with a silver coated mirror (focal length of 500 mm) into the gas jets. The beam diameter at the focal spot is estimated to $55\pm5\mu m$ (FWHM) implying an averaged peak intensity of about $3x10^{15}$ W/cm^2 corresponding to a HH cut-off energy of about 0.6 keV [2]. The interaction between the laser beam and the helium gas takes place within a pair of nickel tubes. Their axes are aligned perpendicularly to the laser beam. The overall effective gas jet length L is 0.8 mm and estimated from the 0.2 mm inner diameter of two tubes and the 0.1 mm wall thickness. The separation between the two jets was set between 0.4 mm and 3.5 mm. The distance is measured from the middle of the jets thus 0.4 mm is the smallest distance when the two tubes are meeting.

In the experiments [8], we examined the 200-600 eV part of the generated HH spectrum. In this range the number of photons was high enough for an accurate measurement, even after filtering out the laser light and the lower order harmonics with a 200-nm-thick Aluminum and a 200-nm-thick Titanium foil. This filter combination attenuates also the soft x-ray flux resulting less than 0.1 detected photon/ pulse. Such a low count rate is a prerequisite for a correct measurement with an energy dispersive x-ray spectrograph (EDX, Si(Li) Oxford).

Results and Discussion

To proof our predictions for QPM with two successive sources we measured the generated HH spectra for different jet distances and gas backing pressures. The harmonic spectra as a function of the jet separation are shown in Fig. 1a. The jet separation was varied from 0.4 mm to 3.5 mm keeping backing pressure fixed at 80 mbar. The harmonic intensity shows pronounced maxima for certain gas jet separations implying constructive interference between the harmonic signals generated in the two parts of the source. A very clear enhancement of the harmonic yield is observable at 300 eV for a source distance of 1.7 mm. The optimum photon yield is four times higher compared to single source with the same length, which is in good agreement with theoretical prediction for quasi-phase matching. For the spectral range between 500 and 550 eV two distinct maxima could be identified for source distances of 1.3 mm and 2.4 mm.

The HHG yield as function of the photon energy is plotted for the shortest (0.4mm) and optimum (1.7 mm) jet separation in Fig. 1 b. Note for this graph, we have not corrected for the exponential roll-off. The observed photon energy enhancement at about 400 eV is in reasonable agreement with our calculations. At this photon energy, we expect a vanishing signal for a single long jet and a reasonable signal for the double jet configuration. For a comparison of the single and double jet configuration we have plotted the contrast, i.e. the ratio of photon yields for the two configurations. The enhancement can be observed in the whole spectral range from 200 to 600 eV and is up to two orders of magnitude in the range of 400 eV.

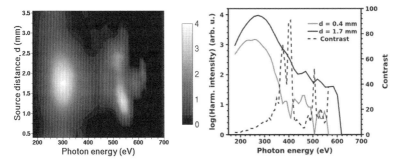

Fig. 1. a) Spectra of soft-x-ray harmonics for successive sources with different source separation. The rapid roll-off of the yield with photon energy has been corrected by the exponential factor. b) Harmonic spectra recorded at two different source distances and the calculated contrast of the photon yield between them.

Conclusions

Coherent addition of harmonics originating from subsequent sources has been proposed and demonstrated experimentally. It offers - by individual adjustment of the source separation, gas density and position of focus - enough degrees of freedom necessary adapting the system parameters to the changing conditions for optimum harmonic growth as the few-cycle driving laser pulses propagating through the medium. Although several technical challenges are left to be solved, the presented route to the development of a powerful coherent kilo-electron-volt-scale source is most promising and justifies further efforts.

Acknowledgements. This study has been sponsored by the Austrian Science Fund (grants No. F016 P02, P03 and grant No. Z63), DFG grant SP 687/1-3, DFG Cluster of Excellence Munich Center for Advanced Photonics – MAP (www.munich-photonics.de), and BMBF grant 06WU266I.

1   E. Constant, et al., Phys. Rev Lett. 82, 166, 1999.

2   Ch. Spielmann, et al., Science 278, 661, 1997

3   E. A. Gibson, et al., Science 302, 95, 2003.

4   E. Seres, J. Seres, Kraus F. and Ch. Spielmann, Phys. Rev. Lett. 92, 163002, 2004.

5   J. Seres, et al., Nature 433, 59, 2005.

6   E. Seres, J. Seres and Ch Spielmann, Appl. Phys. Lett. 89, 181919, 2006.

7   M. Zepf, et al., Phys. Rev. Lett. 99, 143901, 2007.

8   J. Seres, et al, Nature Phys. 3, 878, 2007.

9   G. Tempea, et al., Phys. Rev. Lett. 84, 4329, 2000.

10 J. Seres, et al., Opt. Lett. 28, 1832, 2003.

11 E. Seres and C. Spielmann, Appl. Phys. Lett. 91, 2789732, 2007.

12 X. Zhang, et al., Nature Phys. 3, 270, 2007.

Phase Matching and Quasi-Phase Matching of Extreme High-Order Harmonic Generation

Tenio Popmintchev, Ming-Chang Chen, Oren Cohen, Margaret M. Murnane, and Henry C. Kapteyn

JILA, University of Colorado at Boulder and NIST, 440 UCB, Boulder, CO 80309-0440, USA
E-mail: popmintchev@jila.colorado.edu

Abstract. We present new schemes for phase matching and quasi-phase matching extreme high-order harmonic generation. Using longer wavelength driving light at 1.3 μm, we extend full phase matching of high harmonic generation in argon and neon to 100 and 200 eV respectively, well beyond the phase-matching limit for 0.8 μm. Also, we show that multiple weak quasi-CW waves can induce phase modulated structures in the high-harmonic generation process that can be used for highly efficient quasi-phase matching.

High-order harmonic generation (HHG) is a useful tabletop source of ultrafast, coherent radiation in the extreme ultraviolet region of the spectrum. A major limitation to date, however, is the rapid decrease in the conversion efficiency at extremely high-orders. This is due to the difficulty of phase matching the high harmonic conversion process. Specifically, it is not possible to fully phase match the HHG process at photon energies above ~130 eV using a 0.8 μm driving laser in any HHG geometry. Also, quasi-phase matching (QPM) techniques that have been demonstrated in HHG are limited to the case of long coherence lengths L_c. At keV energies, however, L_c is typically in the micron range.

Here, we discuss new phase matching and quasi-phase matching schemes for enhancing extremely high-order harmonics [1-4]. First, we show that phase matching can be significantly extended to high photon energies by driving the HHG process with mid-IR light. Experimentally, using a 1.3 μm driving laser and pressure-tuned phase matching in a hollow waveguide, full phase matching in Ar gas was extended from ~45 eV (0.8 μm driver) to ~100 eV [4], while full phase matching in Ne was extended from ~100 eV (0.8 μm driver) to ~200 eV. Secondly, we show theoretically that highly efficient quasi-phase matching of high harmonic generation can be induced by multiple weak quasi-cw waves.

In any HHG geometry, the phase mismatch is a sum of the contributions from the pressure-dependent neutral atom and free electron dispersions, as well as the pressure-independent geometric dispersion. The balance between the pressure-dependent terms determines the "critical" ionization limit above which full phase matching is not possible, which is 4% (1.5%) for Ar and 1% (0.4%) for Ne for 0.8μm (1.3μm) driving lasers. When combined with the ADK tunneling ionization model, the phase matching relation predicts *the phase matching cutoff photon energy* as a function of driving laser wavelength λ_L (see Fig. 1(a)). This phase matching cutoff corresponds to the maximum photon energy that is emitted before the ionization level in the medium exceeds the critical ionization, and therefore represents a limit for efficient upconversion. Since the single-atom HHG cutoff scales linearly with the laser intensity and quadratically with the laser wavelength, lower laser intensities are required to generate a given harmonic when longer laser wavelength is used. Therefore, the ionization level in the medium becomes significantly lower, allowing

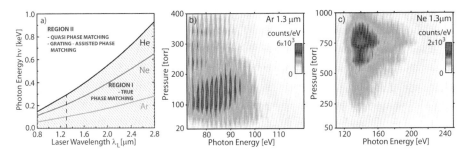

Fig. 1. (a) Theoretically predicted maximum photon energies that can be fully phase-matched as a function of the wavelength. (Region I) Phase-matching in Ar, Ne, or He is possible up to ~1 keV, at ionization levels below critical. (Region II) Quasi-phase matching or grating assisted phase matching techniques can be implemented at higher energies. (b, c) Experimentally observed high harmonic generation in Ar (200 nm Zr filter) and Ne (400 nm Ag filter) as a function of gas pressure, demonstrating bright emission due to phase matching at higher pressures, extending to 100 eV and 200 eV respectively. These phase matching cutoffs are ~2x higher than for λ_L=0.8 μm, and agree well with the theoretical prediction (a).

phase matching to extend to higher photon energies (Fig. 1(a)) in spite of the decreasing critical ionization with increasing λ_L. In our experiments, high energy (up to 5.5 mJ), 35 fs pulses at λ_L=1.3 μm were focused into hollow waveguides of 250 μm and 400 μm inner diameter, 2 cm length, filled with Ar and Ne respectively. Full phase matching of the HHG emission was observed up to 100 eV (Fig. 1(b) [4]) and 200 eV (Fig. 1(c)) in Ar and Ne respectively, in good agreement with the theoretical predictions shown in Fig. 1(a). Because phase matching was achieved at higher pressures (~5x higher in Ar and ~8x in Ne comparing the optimal pressures for HHG near the phase matching cutoffs at λ_L=1.3 vs 0.8 μm), the unfavorable scaling of the single atom yield ($\sim\lambda^{-5.5}$) was significantly mitigated [4]. At high laser intensities corresponding to ionization levels exceeding several times the critical ionization, the HHG emission extends further (200 eV in Ar and above the carbon absorption edge, 285 eV, in Ne) than the phase matching cutoff. However, the HHG flux above and below the phase matching cutoff is much reduced compared with the emission under the optimal phase matching conditions.

Next, we discuss a scheme to optically-induce a highly efficient QPM structure. A distinctive property of HHG is that it is phase shifted relative to the driving laser. This extra phase is acquired by the electron during its "boomerang" oscillatory motion between ionization and recollision. This phase is also in general very large, reaching hundreds of radians, and is proportional to the intensity of the driving laser I_L. Thus, inducing a shallow sinusoidal modulation in I_L along its propagation leads to a sinusoidally-modulated phase shift in the HHG emission. A convenient way to induce such a modulation is by interfering the driving laser with a weak quasi-CW beam that propagates in a different direction. In this case, it is straightforward to control the period and amplitude of this phase modulation. The periodicity of the phase modulation corresponds to the periodicity of the intensity grating, and thus can be controlled, for example, by changing the propagation direction of the quasi-CW beam or its wavelength. Importantly, periodic structures with short periodicity can be created using quasi-CW counterpropagating beams, opening up the possibility for QPM of HHG into the keV region of the spectrum. The amplitude of the phase modulation is determined by the amplitude of the grating, and therefore can be controlled by the intensity of the counterpropagating beam. Remarkably, an

13

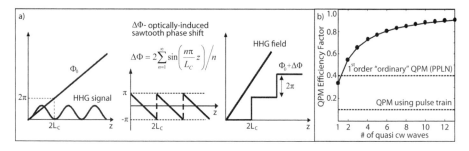

Fig. 2. Full correction of phase mismatch in HHG. (a-left) phase mismatch in the nonlinear medium. The phase shift between the driving laser and HHG vs propagation distance grows linearly, leading to oscillations in the output. (a-center) a saw-tooth wave phase structure for correcting the phase mismatch. (a-right) The combination of the medium phase mismatch and the sawtooth phase shift results in a stairstep function phase, which leads to a linear growth of the HHG field. This linear increase corresponds to full phase matching. (b) The QPM efficiency factor (conversion efficiency normalized to full phase matching) vs the number of quasi-CW waves. The QPM efficiency factor with a *single* Fourier component is already more than three times the QPM efficiency factor predicted while using trains of counterpropagating pulses. The conversion efficiency with two quasi-CW waves is better than that of the "most optimal" QPM where the polarization is flipped every coherence length (e.g. PPLN).

extremely weak quasi-CW field is sufficient to induce a significant phase-shift [1-3]. Employing this effect, complex phase structures can be induced by using multiple quasi-CW weak beams. It is possible to consider each quasi-CW beam as contributing a single Fourier component to the phase structure. Thus, a complex phase structure can be induced by interfering multiple quasi-CW waves with the driving pulse. The generated field at the output of the nonlinear medium ($z=L$), of a given harmonic at a given time in the pulse, is proportional to $\int \exp[i(\Phi_0(z) + \Delta\Phi(z))]$, where $\Phi_0(z) = \pi z / L_c$, L_c is the coherence length, and $\Delta\Phi(z)$ is the phase structure induced by the multiple weak beams. Figure 2(a) shows qualitatively that a saw-tooth phase structure can completely correct any phase-mismatch, making true phase matching of the HHG process possible under a wide variety of conditions. In practice, the conversion efficiency can routinely exceed that obtainable using standard QPM techniques. The QPM efficiency factor of a truncated saw-tooth wave, as a function of the number of quasi-CW waves used to create the phase grating, is shown in Fig. 2(b).

In summary, we showed new schemes for phase matching and quasi-phase matching of extremely high-order harmonics. We extended the full phase matching cutoffs of the HHG process 2x in both Ar and Ne using mid-infrared driving light. Also, we proposed a scheme for highly efficient quasi-phase matching using phase modulated structures induced by multiple weak quasi-CW waves.

The authors gratefully acknowledge support from the NSF ERC for EUV Science and Technology and from the U.S. Department of Energy.

1   X. Zhang, A. L. Lytle, T. Popmintchev, X. Zhou, H. C. Kapteyn, M. M. Murnane, and O. Cohen, in Nature Physics 3, 270, 2007.

2   O. Cohen, X. Zhang, A. L. Lytle, T. Popmintchev, M. M. Murnane, H. C. Kapteyn, in Physical Review Letters 99, 53902, 2007.

3   H. Kapteyn, O. Cohen, I. Christov, and M. Murnane, in Science 317, 775, 2007.

4   T. Popmintchev, M.-C. Chen, O. Cohen, M. E. Grisham, J. J. Rocca, M. M. Murnane, and H. C. Kapteyn, in CLEO Postdeadline CPDA9, 2008.

Comparison of Parallel and Perpendicular Polarized Counterpropagating Light for Quasi-Phase-Matching High Harmonic Generation

T. Robinson[1], K. O'Keeffe[1], M. Landreman[2], B. Dromey[3], M. Zepf[3] and S.M. Hooker[1]

[1] Department of Physics, Clarendon Laboratory, University of Oxford, Parks Road, Oxford, OX1 3PU, UK
E-mail: t.robinson1@physics.ox.ac.uk
[2] Department of Physics, Massachusetts Institute of Technology, Cambridge, Massachusetts 02139, USA
[3] Department of Physics and Astronomy, Queens University Belfast, BT7 1NN, UK

Abstract. The effect of the polarization of counterpropagating pulses on suppression of high harmonic generation is investigated. The results agree well with simple models of harmonic suppression and have application to quasi-phase-matching of harmonics.

High harmonic generation (HHG) is an attractive source for coherent XUV radiation. However, the phase mismatch between harmonics generated at different points in the medium limits the conversion efficiency which can be achieved. It is possible to overcome the problem of phase mismatch using the technique of quasi-phase-matching (QPM), in which HHG is suppressed in regions where the macroscopic harmonic wave would be out of phase with the local harmonic emission. Peatross et al. [1] suggested that a relatively weak parallel polarized counterpropagating laser pulse could be used to suppress harmonic production. The first demonstration of QPM with a train of pulses was recently reported by Zhang et al.[2] - who showed that a train of counterpropagating pulses could enhance the harmonic signal generated in a hollow waveguide by more than an order of magnitude.

In this experiment we further investigate the effect of the polarization of a counterpropagating beam (CPB) on the generation of harmonics in a gas cell [3]. We demonstrate for the first time that it is possible to suppress HHG using perpendicularly polarized light, and therefore QPM schemes should also be achievable using perpendicularly polarized light as well as parallel polarized light. For parallel polarization, harmonic suppression results from phase scrambling of local harmonic emission. This scrambling is caused by direct and intensity-dependent phase modulation of the driver beam in the presence of the CPB. For direct phase modulation a small shift in the phase of the fundamental results in a large shift in the phase of harmonics of sufficiently high order. This fast modulation prevents these harmonics from achieving any significant strength. The intensity-dependent phase modulation is caused by the variation of the harmonic phase with intensity. However, the phase of the induced atomic dipole also depends on the relevant electron trajectory [4]. For perpendicular polarization harmonic generation is extinguished due to the ellipticity that results when the driver beam and CPB overlap.

Experimental Methods

The laser system used in this experiment was a Ti:Sapphire CPA system operating at a 10 Hz repetition rate. A split from this beam was focused to an intensity of 2.5 x 10^{15} W/cm^2 in an Ar gas cell to generate high-harmonics. The remainder of the laser beam was used as the CPB. We have generated trains of ultrashort pulses using two methods. In the first method a pulse train consisting of up to 128 evenly spaced pulses could be produced by passing the CPB though a series of seven birefringent crystals [5]. In the second method a stretched pulse is passed through a waveplate and linear polarizer to produce a train of ultrashort pulses [6,7]. By using an acousto-optic programmable dispersive filter to add specific phase profiles to the stretched pulse, pulse trains with controllable, nonuniform spacing may be produced. This technique has applications for optimizing QPM in waveguides where there is longitudinal variation of the coherence length.

In this experiment the first method for producing pulse trains was used. A train of 16 pulses with the minimum possible spacing was chosen. This gave a roughly square wave pulse with a 4 ps duration. The peak intensity of each pulse in the CPB was 4 x 10^{13} W/cm^2 which is insufficient to cause significant ionization. The experimental setup is shown in Figure 1. The HHG spectrum was recorded as a function of the delay between the driving laser pulse and the CPB, as shown in Figure 2.

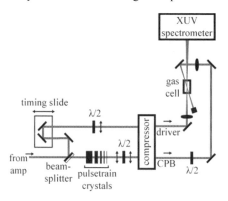

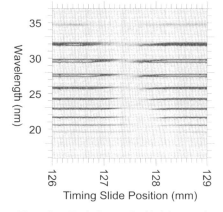

Fig. 1. Experimental setup. The CPB is delayed relative to the driver beam using a computer-controlled timing slide. The polarization of the CPB is changed by adjusting a half-wave plate.

Fig. 2. Variation of high-harmonic spectrum as a function of timing slide position (collision point). The position of the suppression varies with harmonic order, due to longitudinal variation of the driver intensity.

Results and Discussion

A series of scans were made to compare the extinction of HHG by a parallel and perpendicularly polarized CPB. The HHG signal was observed to decrease dramatically at certain timing slide positions corresponding to collision of the CPB with the driver beam in the generating region of the gas. Experiments were performed to determine the dependence of the harmonic extinction on the intensity of the CPB for both parallel and perpendicularly polarized light. Harmonic spectra were recorded as a function of the pulse energy in the CPB. Figure 3. plots the CPB-driver intensity ratio needed to reduce the harmonic signal by 50%. The predicted CPB-intensity ratios resulting from direct phase modulation, indirect phase modulation and position-dependent ellipticity are also shown. It is clear that much higher intensities in the CPB are required for a given reduction in harmonic signal for perpendicular polarization than for parallel polarization. For the perpendicular polarization scheme the data accord very well with that predicted by theory. In the case of parallel polarization the measured intensity required for harmonic extinction agrees well with that predicted for the direct phase modulation effect. The data also indicates that it is the short trajectory which is dominant in this case.

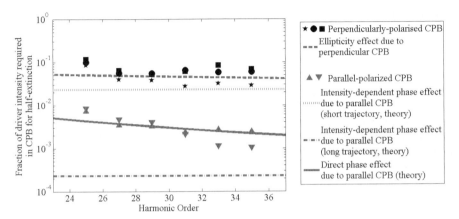

Fig. 3. Measured fraction of the intensity in the CPB to that of the driving pulse, for half-extinction of the harmonic signal for several runs. Theoretical predictions also shown.

Conclusions

In conclusion we have investigated the effect of polarization on suppression of HHG by a counterpropagating train of pulses. Although the perpendicular polarization scheme has the disadvantage that a higher intensity CPB is required to suppress harmonics it offers the practical advantage that the CPB can easily be prevented from propagating back into the laser system, which is particularly important where waveguides are used. Further, we have investigated different techniques for generating pulse trains. Chirped pulse trains offer further scope for control and enhanced output from quasi-phase-matched HHG.

Acknowledgements: This work was supported by EPSRC grant EP/C005449.

1   J. Peatross, S. Voronov, and I. Prokopovich, Opt. Express 1, 114 (1997).

2   X. Zhang et al., Nat. Phys. 3, 270 (2007).

3   M. Landreman et al., J. Opt. Soc. Am. B 24, 2421 (2007).

4   M. Lewenstein, P. Salieres, and A. L'Huillier, Phys. Rev. A 52, 4747 (1995).

5   B. Dromey et al., Appl. Opt. 46 5142 (2007).

6   T. Robinson et al., Optics Letters 32 2203 (2007).

7   R. Yano and H. Gotoh, Jpn. J. Appl. Phys., **44** 8470 (2005).

Enhanced Harmonic Generation in Gas Jets with Expanding Clusters

Bonggu Shim, Xiaohui Gao, Xiaoming Wang, Todd Ditmire, and Mike Downer

Department of Physics, University of Texas at Austin, Austin, TX 78712, USA
E-mail: downer@physics.utexas.edu

Abstract: We report femtosecond-time-resolved enhancement and anisotropy of third-harmonic generation in highly ionized clustered argon at intensities exceeding 10^{15} W/cm^2. Results suggest a path to phase-match high-order harmonic generation in dense plasmas with ultrahigh intensity.

Introduction

A growing theoretical literature [1-3] suggests that solid-density atom clusters embedded in background monomer gas can markedly enhance the efficiency of harmonic generation (HG) at high ionization level (average $n_e > 10^{18}$ cm^{-3}) and high intensity ($I > 10^{15}$ W/cm^2), where highest-order harmonics can be achieved, through at least 3 mechanisms: 1) phase-matching, by compensating normal dispersion of monomer plasma with strong anomalous dispersion of ionized clusters [1]; 2) recombination of photo-electrons with ions other than the parent in a large cluster [2]; 3) Mie resonant enhancement of nth-order cluster susceptibility $\chi^{(n)}$ by collective oscillation of internal cluster electrons [3]. While the last two mechanisms are expected to be most effective for low-order harmonics [3,4], the first can enhance HG up to very high orders (n > 100) if the mass fraction and internal electron density of the clusters can be controlled and optimized [1]. Quasi-phase-matching using corrugated hollow capillaries [5] has been very successful in extending high-order HG (HHG) in monomer gases to high ionization fraction, but is limited to moderate intensity (mid 10^{14} W/cm^2) and plasma density ($n_e \sim$ mid 10^{17} cm^{-3}). Clustered plasmas potentially enable phase-matching at significantly higher I and n_e [1], but so far, experimental studies of HHG in clustered gases [6] have also used only $I <$ mid 10^{14} W/cm^2 and low ionization level. Thus, despite modest improvements over monomer gas, they have not accessed the high-n_e, high-intensity regime in which the unique benefits of clusters are expected [1-3]. Here we report strong enhancement of low-order HG when drive pulse intensity exceeds 10^{15} W/cm^2 in a clustered argon jet that was multiply pre-ionized to average density $n_e > 10^{18}$ cm^{-3} by an intense heating pulse $\sim$ 300 fs earlier [7]. Numerical modelling, combined with independent jet characterization, shows that delayed HG enhancement results from a combination of transient, partial phase-matching (mechanism #1) and Mie resonance (mechanism #3) as the clusters expand, while its strong transient polarization anisotropy resolves the anisotropy of expanding clusters [8] with fs resolution. Analysis also shows that perfect phase-matching of both low- and high-order HG can be achieved with a $\sim$3-fold increase in cluster mass fraction f, potentially achievable in a cryogenic jet.

Experiments

For experiments, 800 nm, 100 fs, $\sim$ 0.1 J pulses from a 10 Hz Ti:S laser system were split into pump and probe beams. Clusters were formed in a room temperature pulsed supersonic gas jet (750 µm orifice, 11° half expansion angle) backed with 600 psi argon. We measured average cluster radius $\bar{r}_c = 20$ nm, total atomic density $n_{tot} = 10^{18}$ cm^{-3} (clusters plus monomers), and cluster mass fraction $f = 0.25 \pm 0.03$ by combining time-integrated Rayleigh scatter and Mach-Zehnder interferometry with femtosecond pump-probe frequency-domain interferometry (FDI). In the last measurement, illustrated in Fig. 1, the negative refractive index contribution of monomer plasma (used to determine monomer partial density n_m) appears immediately after the pump ($\sim 4 \times 10^{15}$ W/cm^2) ionizes the medium, whereas the positive contribution of clusters (used to determine the number density n_c of clusters) appears only after clusters expand to a Mie resonance condition in < 300 fs, enabling separation of n_m from n_c in the time domain. While $f \sim 1$ is often assumed, our much lower value $f = 0.25 \pm 0.03$ thus determined is consistent with the simulations of [9,10] for room-temperature argon jets.

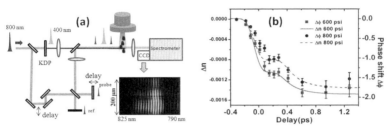

Fig. 1. (a) Experimental schematic for FDI. (b) Time evolution of phase shift and refractive index with theoretical fits.

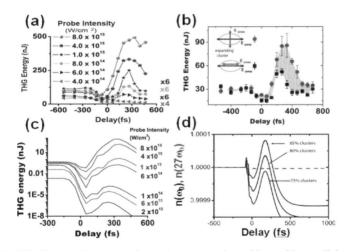

Fig. 2. (a) THG from probe vs Δt, for various probe intensities, with parallel pump-probe polarizations. (b) THG anisotropy with parallel (filled squares) or perpendicular (filled circles) polarizations. (c) Calculated time-dependent THG by probe pulses of indicated intensities, for gas jet with parameters given in Fig. 1. (d) Calculated time-dependent refractive index at fundamental (solid curves) and 27th harmonic (dashed) for same plasma conditions as in panel (c), except that cluster fraction f is increased to 0.75, 0.80 or 0.85.

Figure 2a illustrates results of third-harmonic generation (THG) by an 800 nm probe pulse polarized parallel to the pump, as a function of pump-probe delay Δt. For $I_{probe} > 10^{15}$ W/cm^2 a sharp enhancement at $\Delta t \sim 300$ fs is observed, coincident with partial recovery in refractive index shown in Fig. 1b, and scales as ~ I^3_{probe}. When probe polarization was turned perpendicular to the pump, the delayed THG peak doubled in amplitude (see Fig. 2(b)), consistent with faster cluster elongation along the pump polarization [8], creating stiffer ion density gradient and stronger nonlinearity perpendicular to the pump [3]. HG is evidently uniquely capable of time-resolving cluster expansion anisotropy, which yielded no signature in time-resolved FDI (Fig. 1) or absorption (not shown). Isotropic THG was observed for $\Delta t < 0$ (unexpanded clusters) and $\Delta t > 600$ fs (approaching uniform plasma). Fig. 2(c) presents calculated time-resolved TH response, using the uniform cluster density model [11] and measured linear optical properties (Fig. 1) to calculate cluster ionization, heating and hydrodynamics and TH coherence length, and an empirical model of $\chi^{(3)}$ based on collective oscillations of free electrons inside clusters of nonuniform ion density [3]. The calculations correctly reproduce the temporal evolution and growth with I_{probe} of the delayed THG peak. Further modelling details are described in Ref. [7].

Conclusion and future directions

Because of the complete characterization of, and control over, clustered plasma parameters combined with accurate modelling (Fig. 2c), we can use our THG results to predict quantitatively how to phase-match *high-order* HG in a dense, highly ionized clustered plasma with high drive pulse intensity. To this end, Fig. 2d shows calculated refractive index at fundamental (ω_0, solid curves) and 27th harmonic ($27\omega_0$, dashed curve) for the same conditions as the calculations in Fig. 2c, except that cluster fraction is increased to $0.75 < f < 0.85$. Perfect phase-matching is achieved at $\Delta t \sim 200$ fs. Similar predictive curves are generated for up to relativistic drive pulse intensities ($I_{probe} \sim 10^{18}$ W/cm^2), up to 8-fold-ionized Ar, and other atomic species. In nearly all cases, the fundamental requirement and challenge is to achieve $f > 0.5$, suggesting cryogenic supersonic jets.

1   T. Tajima, Y. Kishimoto and M. C. Downer, Phys. Plasmas **6**, 3759 (1999); J. W. G. Tisch, Phys. Rev. A **62**, 041802 (R) (2000).

2   V. Véniard, R. Taïeb and A. Maquet, Phys. Rev. A **65**, 013202 (2001).

3   B. N. Breizman, A. V. Arefiev, and M. V. Fomyts'kyi, Phys. Plasmas **12**, 056706 (2005); T. M. Antonsen, Jr., T. Tagauchi, A. Gupta, J. Palastro and H. M. Milchberg, Phys. Plasmas **12**, 056703 (2005).

4   M. Kundu, S. V. Popruzhenko, and D. Bauer, Phys. Rev. A **76**, 033201 (2007) and references therein.

5   E. A. Gibson, A. Paul, N. Wagner, R. Tobey, S. Backus, I.P. Christov, M. M. Murnane and H. C. Kapteyn, Phys. Rev. Lett. **92**, 033001 (2004) and references therein.

6   T. Donnelly, T. Ditmre, K. Neuman, M. D. Perry, and R. W. Falcone, Phys. Rev. Lett. **76**, 2472 (1996); C. Vozzi, M. Nisoli, J-P. Caumes, G. Sansone, S. Stagira, S. De Silvestri, Appl. Phys. Lett. **86**, 111121 (2005); C.-H. Pai, C-C Kuo, M-W Lin, J. Sang S.-Y. Chen and J.-Y. Lin, Opt. Lett. **31**, 984 (2006).

7   B. Shim, G. Hays, R. Zgadzaj, T. Ditmire, and M. C. Downer, Phys. Rev. Lett. **98**,123902 (2007).

8   V. Kumarappan, M. Krishnamurthy, and D. Mathur, Phys. Rev. Lett. **87**, 085005 (2001).

9   F. Dorchies, F. Blasco, T. Caillaud, J. Stevefelt, C. Stenz, A. S. Boldarev, and V. A. Gasilov, Phys. Rev. A **68**, 023201 (2003).

10 A. S. Boldarev, V. A. Gasilov, A. Ya. Faenov, Y. Fukuda and K. Yamakawa, Rev. Sci. Instrum. **77**, 083112 (2006).

Observation of Elliptically Polarized High Harmonic Emission from Molecules Driven by Linearly Polarized Light

Xibin Zhou, Robynne Lock, Nick Wagner, Wen Li, Henry C. Kapteyn, and Margaret M. Murnane

JILA and Department of Physics, University of Colorado
Boulder, CO 80309-0440
E-mail:xibin.zhou@colorado.edu

Abstract. Using a metal mirror polarizer, we perform an accurate polarimetry of high harmonic emission from aligned molecules. Surprisingly, we find that harmonic emission from N_2 can be strongly elliptically polarized even when driven by linearly polarized lasers.

High-order harmonic generation (HHG) results from the extreme distortion of an electron wave function in an atom or molecule in the presence of a strong laser field. In the semi-classical picture of HHG, an atom undergoes tunnel ionization, followed by oscillation of the free electron in the laser field. Some of these electrons will recombine with their parent ions a fraction of an optical cycle following ionization, liberating their excess energy as an extreme ultraviolet (EUV) photon. In the case of HHG from molecules, the intensity and phase of the EUV light can encode structural information - particularly if the molecular sample is aligned [1-3].

However, developing a comprehensive understanding of HHG from molecules has proven challenging to date. Because the time dependent Schrödinger equation cannot be implemented for molecules other than H_2, intensive effort has been devoted to extending the Lewenstein model to correct for limitations of the strong field and single active electron approximations. These theories need to be tested carefully with experiments. In past work, we measured the angle-dependent intensity and phase of HHG from CO_2 molecules [2]. This allowed us to unambiguously observe a reversal in phase of π of the high order harmonic emission resulting from molecular-scale quantum interferences between the molecular electronic wave function and the recolliding electron as it recombines with the molecule. This result is consistent with a two-charge center model of CO_2.

In this work we study HHG from N_2 molecules, and find two unexpected results that are not in agreement with previous work. First, we find that the phase of HHG from N_2 *does not* depend on orientation of the molecule, and that there is *no* angle-dependent intensity minimum for harmonic orders 15 - 31. This suggests that any angular dependence of HHG emission from N_2 cannot be explained by a simple two center interference model (as is the case for CO_2). Second, we find the relative phase between the parallel and orthogonal components of HHG strongly depends on the harmonic order, and that *elliptically polarized* HHG beams can be generated from N_2 even when driven by linearly polarized lasers. These findings cannot be explained by the two center interference model or even the strong field approximation (SFA) [3], and will be very useful to benchmark new and more complete theories of molecules in strong fields. Moreover, our results present a straightforward and efficient way to generate circular polarized harmonic beams for applications in molecular and materials science. Our findings differ from previously published results [4] because we achieve stronger molecular alignment (confirmed by the observation of a strong rotation of the HHG ellipse) and we achieve a better signal-to-noise ratio (allowing us to detect the perpendicular component of HHG for the first time).

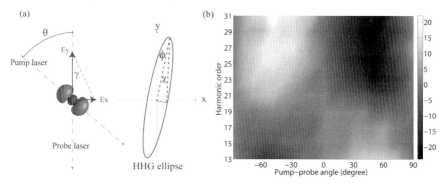

Fig. 1 (a) Schematic of the experiment. θ and φ are positive for clockwise rotation and negative for counter clockwise rotation (b) Color map of the HHG ellipse orientation angle for different harmonic orders and molecular orientations.

In the first experiment, we scanned the time delay between a pump pulse that aligns the N_2 molecules, and a probe pulse with parallel polarization that generates harmonics from the molecular sample with a certain angular distribution. In this geometry, the molecular angular distribution is cylindrically symmetric around the laser polarization, so only the HHG component parallel to the probe laser polarization direction is observable. We monitored the HHG emission as a function of orientation, and observed no angle-dependent intensity minimum in the harmonic emission for orders 15 - 31. We also interfered harmonic emission from aligned molecules with that from randomly-oriented molecules, and saw no obvious orientation dependent phase shift.

In the second experiment, to investigate HHG polarization as a function of molecular orientation, we varied the angle between the aligning pump laser and the HHG-generating probe pulse, while fixing the time delay such that the molecules are strongly aligned (see Fig. 1(a)). Since the cylindrical symmetry around the probe laser is broken, a HHG component perpendicular to the probe laser polarization direction should be observable. In general, the complex amplitude of $E_y(n,\theta)=A_y(n,\theta)\exp(i\Phi_y(n,\theta))$ and $E_x(n,\theta)=A_x(n,\theta)\exp(i\Phi_x(n,\theta))$ component of HHG should depend on the harmonic order n and angle θ between the molecular axis and probe laser polarization (**y** direction in the lab frame). Using a pair of metal mirror as a polarizer, we can measure the polarization state of the harmonic emission by rotating the pump and probe laser polarization together [4]. Figure 1(b) plots a color map of the HHG ellipse orientation angle for different harmonic orders and molecular orientation. We can also measure the ellipticity ε of the HHG ellipse, which is defined as $\varepsilon = \sqrt{I_{min}/I_{max}}$, where I_{min} and I_{max}, are the minimum and maximum intensity during the polarization scan. We find that HHG from aligned N_2 molecules can be strongly elliptical polarized (with ellipticities up to 0.35 for certain harmonics). The measured ellipticity ε for different harmonic orders and orientation angles θ is plotted as a color map in Fig. 2 (a), while Fig.2 (b) shows a lineout of ε for θ=60 degrees.

From the measured ellipticity and the orientation angle of the HHG ellipse, we can extract the phase difference $\delta=\Phi_x-\Phi_y$ and ratio A_x/A_y between the HHG polarization components, using the formulae $\sin(2\chi)=\sin(2\gamma)\sin(\delta)$ and $\tan(2\phi)=\tan(2\gamma)\sin(\delta)$. The extracted δ at θ=60° for different harmonics is shown in Fig.2 (c). We see that the relative phase difference gradually changes from 0.8 π to 0.2 π from harmonic 15 to harmonic 29. This relative phase difference across 8 harmonic orders is not likely to arise from a two-center interference effect, which should exhibit a sharp phase jump depending on the molecular orientation angle. The intrinsic phase accumulated by the electron in continuum should not contribute either, because the ionization and acceleration steps are common for the parallel and orthogonal components of HHG.

The reason we observe elliptical polarized harmonic emission in contrast with previously published results [4] is that we achieve a stronger molecular alignment in our sample (confirmed from the strong rotation of the HHG ellipse), and we also achieve a higher signal-to-noise ratio that allows us to detect the small nonzero harmonic component I_{min} corresponding to the signature of elliptically-polarized HHG. Previous measurements observed a nonzero component E_x orthogonal to the probe polarization direction, when the angle between the pump and probe is between 0° and 90°, but the HHG emission was still measured to be linearly polarized for N_2, O_2 and CO_2, which implies I_{min} =0 thus ε≈0. To explain the polarization behavior of HHG from N_2, a π phase jump in the E_y component of HHG at 23rd harmonic was assumed. Our measurements do not show any phase discontinuity between the two polarization components, making such a π phase jump unlikely.

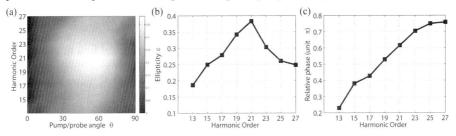

Fig. 2 (a) Color map of the ellipticity ε for different harmonic orders and molecular orientation angles θ. (b) Ellipticity ε for θ=60 as a function of HHG order. (c) Calculated phase difference δ between parallel and orthogonal HHG components for θ=60 as a function of HHG order.

In summary, we uncovered two unexpected features in HHG from N_2 molecules, which have broad implications for the theory of molecules in strong fields. First we found that the phase of HHG from N_2 *does not* depend on the molecular alignment angle and no interference induced π phase shift exists. This suggests that the two center interference model does not work well for N_2. Second, we found that *elliptically polarized harmonics* can be emitted by molecules driven by linearly polarized light. This effect will need to be explained by theories that go beyond strong field approximation or single active electron approximation, and may reflect nontrivial structure or dynamics in the molecules, or contributions to the HHG emission from occupied molecular orbital other than HOMO. Finally, our results present a straightforward and efficient way to generate circular polarized harmonic beams for applications in molecular and materials science.

1   T. Kanai, et al., Nature **435**, 470 (2005).

2   X. Zhou et al., Phys. Rev. Lett. **100**, 073902 (2008)

3   M. Lein et al, Phys. Rev. Lett. **88**, 183903 (2002).

4   J. Levesque et al., Phys. Rev. Lett. **99**, 243001(2007).

Polarization-Resolved Pump-Probe Spectroscopy with High Order Harmonics

E. Mével[1], Y. Mairesse[1], S. Haessler[2], B. Fabre[1], J. Higuet[1], W. Boutu[2], P. Breger[2], E. Constant[1], D. Descamps[1], S. Petit[1], D. Shafir[3], H. Deleon[3], N. Dudovich[3] and P. Salières[2]

[1] CELIA, Université Bordeaux 1, UMR 5107 (CNRS, Bordeaux 1, CEA), 351 Cours de la Libération, 33405 Talence Cedex, France
[2] CEA-Saclay, DSM, Service des Photons, Atomes et Molécules, 91191 Gif-sur-Yvette, France
[3] Dept of Physics of Complex Systems, Weizmann Institute of Science, Rehovot 76100, Israel
E-mail address: mevel@celia.u-bordeaux1.fr

Abstract. High Harmonic generation can be used as a probe of the emitting medium with attosecond and Angström resolutions. We show that polarization-resolved pump-probe spectroscopy with high harmonics improves the detection sensitivity of rotationally excited molecules.

Introduction

During high harmonic generation from molecules, an electron wavepacket tunnels out from the highest occupied molecular orbital (HOMO) into the continuum, is accelerated by the laser field, is driven back to the vicinity of the parent ion and interferes with the part of the HOMO remained bound [1, 2]. This process is periodic and leads to high order harmonic emission that encodes the interference between the recolliding electron wavepacket and the HOMO with both an attosecond and an Angström resolution. Thus the harmonic signal is a very accurate probe of molecular orbital. This property was recently used to perform the tomographic reconstruction of the HOMO of N_2 [3].

Recent experiments have shown that the high harmonic signal could be used to encode rotational [3, 4] and vibrational [5, 6] molecular dynamics. The extension to more complex processes such as electronic excitations and photochemical dynamics will face the problem of the contrast in the measurements. Here, we show that polarization-resolved pump-probe spectroscopy can be used to increase the contrast in the detection of the harmonic emission from excited molecules [7]. We consider the particular case of rotational wavepackets, but the technique could be used with any excitation process leading to an anisotropy of the generating medium. Recent measurements achieved with an improved detection dynamics show that harmonics generated in N_2 are elliptically polarized, in contrast to previous studies [7, 8].

Experimental Methods

In our experiment, rotational wavepackets are excited by focusing an IR fs laser pulse in a molecular gas jet. Excitation of a coherent superposition of rotational states results in a molecular alignment at characteristic revival times. Since harmonic generation is sensitive to the alignment of the molecule with respect to the laser polarization, such revivals can be probed by monitoring the high harmonic signal produced in the rotationally excited gas sample.

In previous pump-probe studies, the revivals of molecular alignment were detected by measuring the total harmonic intensity. In an isotropic medium, the high harmonics generated are polarized parallel to the laser linear polarization. But when an anisotropy is introduced, such as a particular molecular alignment, an orthogonal polarization component can be produced [7, 8]. Thus, measuring the orthogonal polarization of the harmonic light provides a sensitive way of probing any anisotropy of the generating medium.

The experiment was performed on two laser systems: the 1kHz Ti:Sapph. Aurore laser facility from CELIA providing 35 fs, 9 mJ pulses at 800 nm, and the 20Hz Ti:Sapph. Luca laser facility from CEA which delivers 50 fs, 50 mJ pulses. Pulses with a fraction of these energies were focused by a 500 mm lens into a 1kHz pulsed gas jet with a backing pressure of 2 bars. The harmonic spectrum is analyzed by an XUV spectrometer consisting in a grazing incidence grating and a dual MCP in front of a phosphor screen. A silver mirror, placed at 45° between the grating and the MCP acts as a polarizer with an extinction ratio of about 30.

Results and Discussion

Here, we set the pump (aligning pulse)-probe (generating pulse) delay close to 4.1 ps alignment revival of N_2 to maximize the degree of molecular alignment. For 15 different alignment orientations, one measures the harmonic intensity as a function of the polarization angle of the generating pulse. It allows us to measure accurately small variation in the direction of the harmonic polarization angle. Remarkable features, such as sign changes in the orthogonal component of the harmonic field are detected. When the generating pulse is P-polarized with respect to the polarizer (minimum reflectivity), and the molecules aligned at about 40° from P-polarization, we observe a maximum harmonic orthogonal component along S-polarization. This provides an optimum sensitivity for selective detection of aligned molecules.

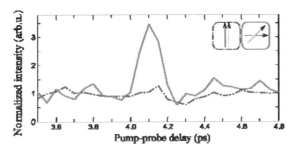

Fig. 1. Evolution of the harmonic 21 yield in N_2 as a function of the delay in the "conventional configuration" (dot-dash) and in the polarization resolved-configuration (full line).

To illustrate the improved sensitivity detection when using the polarization-resolved spectroscopy, we compare the above configuration with the "conventional situation" where both the aligning and generating pulses are parallel and S-polarized (thus equivalent to what would be obtained without any polarizer). It can be clearly seen on figure 1 that a factor 4 increase in the contrast detection of alignment revivals can be obtained with polarization-resolved measurements.

We have applied this method to demonstrate that a signal generated by excited molecules can be extracted from a strong background generated by a carrier gas. For this purpose, we use a 50/50 mixture of Ar and N_2 in the gas target. As harmonic generation in Argon is about three times more efficient as in nitrogen, the total yield is strongly dominated by the argon contribution. With the "conventional configuration" the signature of alignment revivals is hardly distinguishable in the harmonic signal while with the polarization-resolved technique, revivals of molecular alignment can be detected with a contrast of 3.5 [7].

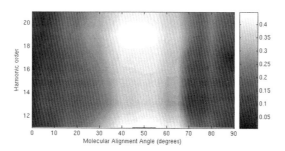

Fig. 2. Harmonic ellipticity (colour bar) generated in aligned N_2

We are currently taking harmonic polarimetry one step forward by measuring the harmonic ellipticity. Accurate measurements performed at CELIA with an improved detection dynamics clearly show a strong ellipticity of the harmonic field generated in aligned N_2 by a linearly polarized laser field (see figure 2).

Conclusions

We have shown that polarization-resolved pump-probe spectroscopy can be used to accurately measure the state of polarization of harmonic from rotationally excited molecules. We observe remarkable features, such as sign changes in the orthogonal component of the harmonic field. These features should be further analyzed, since it was recently predicted to contain signatures of multielectron effects [9].

In pump-probe scans, we have demonstrated an enhancement by a factor 4 of the contrast in the detection of alignment revival in pure N_2. We have also shown that polarization-resolved measurements could be used to extract the signal of excited nitrogen molecules from strong background generated by argon atoms.

Extending this technique to other types of excitation is straightforward, as soon as the excited medium presents some anisotropy. This is an important step towards the investigation of molecular dynamics initiated by low-cross section processes.

Finally, harmonic ellipticity is observed appealing for more refined theoretical models taking into account multi-electron effects.

1   P.B. Corkum, Phys. Rev. Lett. **71**, 1994, 1993.
2   K. J. Schafer *et al.*, Phys. Rev. Lett. **70**, 1599, 1993.
3   J. Itatani *et al.,* Nature **432**, 867, 2004.
4   C. Vozzi *et al.,* Phys. Rev. Lett. **95**, 153902, 2005.
5   S. Baker *et al.*, Science **312**, 424, 2006.
6   N. Wagner *et al.,* Proc. Natl. Acad. Sci. **103**, 13279, 2006.
7   Y. Mairesse *et al.*, New J. Phys. **10**, 025028, 2008.
8   J. Levesque *et al.,* Phys. Rev. Lett. **98**, 183903, 2007.
9   Z. Zhao, J. Yuan, and T. Brabec, Phys. Rev. A **76**, 031404(R), 2007.

Study of quantum-path interferences in the high harmonic generation process.

Amelle Zaïr[1], Mirko Holler[1], Florian Schapper[1], Lukas Gallmann[1], Adam Wyatt[2], Antoine Monmayrant[2] Thierry Auguste[4], Jean Pascal-Caumes[4], Ian Walmsley[2], Eric Cormier[3], Pascal Salières[4], Ursula Keller[1]

[1]Physics Department, ETH Zurich, CH-8093 Zurich, Switzerland

[2]Clarendon Laboratory, Parks Road, Oxford 3PU, UK

[3]CELIA, CNRS-CEA-Université Bordeaux 1, 351 cours de la libération, 33405 Talence, France

[4]Services des Photons Atomes et Molécules, CEA-Saclay, 91191 Gif-sur-Yvette, France
  E-mail: zair@phys.ethz.ch

Abstract. We studied the intensity dependence of the high-order harmonics generated in different atomic gases. We found experimental conditions allowing the observation of quantum-path interferences in all gas species studied. Interference modulations can be observed in neon over a wide range of intensities due to its high ionization threshold. The period of these modulations could indicate that longer electron trajectories contribute to the measured emission.

Introduction

Since many years, quantum mechanical models have been developed to describe high-order harmonic generation in atomic and molecular gases [1-4]. In the case of multiple electron trajectories involved in the XUV harmonic emission, these models predicted quantum-path interferences that result in harmonic yield modulations as a function of laser peak intensity. This intensity dependence results from the fact that the phase of these quantum paths is roughly proportional to the ponderomotive energy (and thereby to the laser peak intensity) times the electron excursion time in the continuum along the corresponding trajectories. Previously, it was shown that mainly two trajectories contribute to the generation of plateau harmonics, referred to as 'short' and 'long' trajectories and that those two quantum paths merge in the cutoff region where only one trajectory subsists. Experimental studies demonstrated the possibility of selecting a single trajectory by means of phase matching conditions [5, 6]. Therefore, the two first quantum paths have been experimentally studied separately providing information on the phase acquired by the electronic wave packet following each trajectory, but no interference pattern between these two first trajectories (first order quantum-path interferences) or more than two (higher-order quantum-path interferences) has been reported so far. With our experimental setup, we were able to find experimental conditions that enabled us to observe the quantum-path interferences (QPI) in argon for the first time [7]. Here, we report a systematic study of QPI in different gases (xenon, argon, neon) and highlight how the measurement can be affected by strong ionization inducing blue shift and limiting the contrast and the number of fringes in the interference pattern (xenon). In the particular case of neon, we are able to observe QPI over a large intensity range without being limited by the ionization threshold, allowing a detailed study of these interferences.

Experimental Methods

Our experiments are based on a laser system delivering 30 fs infrared (IR) pulses at a repetition rate of 1 kHz with maximum pulse energy of 3 mJ. The laser beam enters a vacuum chamber and is focused by a spherical mirror (ROC = 500 mm for the xenon and argon target, ROC = 250 mm for the neon target) into a pulsed gas jet. The jet is movable along the laser propagation direction to control which trajectory is involved in the high-order harmonic generation process: if the jet is positioned before (after) the laser focus, the short trajectory is (the two trajectories are) phase matched. A polariser and a half-wave plate are placed in the IR beam for fine control of the laser peak intensity in the jet. Far-field spatial filtering with 6 mrad acceptance was used before the XUV spectrometer to select different transverse areas (from on- to off-axis positions). This allows to select and control the relative trajectory contributions seen by the XUV spectrometer in order to maximize the interference contrast.

Results and Discussion

For all interaction media used in our experiment we find a generic behavior: when selecting the short trajectory by positioning the jet after the laser focus, the harmonics are narrow, due to the small frequency chirp related to the short trajectory, and their amplitude increases monotonically with the laser peak intensity since only one trajectory is involved in the process. This behavior does not depend on the far field spatial selection: from on-axis to off-axis filtering, the harmonic signal simply decreases by roughly one order of magnitude. This behavior changes when the jet is placed before the laser focus where phase matching conditions allow contributions from longer trajectories. When positioning the far field spatial filter on the off-axis position, and increasing the laser peak intensity, a clear broadening of the plateau harmonic spectral width is observed and modulations of the harmonic amplitude occur. The broadening is due to a large frequency chirp of the long trajectory and the modulations are a signature of QPI. In the cutoff region of the spectra no significant broadening and no modulations are observed since only one trajectory exists. Fig. 1 shows the intensity dependence of experimental harmonic spectra generated in neon where short and long trajectories contribute to the emission and the off-axis spatial filtering is applied. The laser peak intensity was varied from $0.5 \times 10^{14} \, W/cm^2$ to $8 \times 10^{14} \, W/cm^2$. In neon and for the laser peak intensity scale used here, the ionization does limit the observation of the QPI to about five modulations of the harmonic yield. Similar spectra are observed in the case of xenon and argon, but the saturation effects induce a blue-shift of the spectra (case of xenon) and limit the intensity range over which QPI is clearly observed (in argon three modulation periods are seen). In argon, the average periodicity of the modulation ($\approx 0.3 \times 10^{14} \, W/cm^2$) implies that the short and the long trajectories are responsible for the interference pattern and that we measured QPI of first order [7]. In the case of neon, Fig. 2 shows a lineout at the harmonic 33 central frequency (blue curve) compared to the result of a simulation (green curve) where the single-atom response was calculated with an SFA model and the macroscopic response with a propagation code including far-field spatial filtering corresponding to the experimental conditions. The average periodicity of the modulation ($\approx 0.8 \times 10^{14} \, W/cm^2$) is well reproduced by our calculation. In addition the parabolic shape of the interference observed in the experiments with spatial far field filtering is in a good agreement with our simulation highlighting the importance of the spatial selection for studying the QPI.

Conclusions

In conclusion, we were able to find general experimental conditions that allow us to study quantum-path interferences in a variety of high-order harmonic generation media. In the case of neon, our experimental data shows signature of many QPI. In the future, our experimental setup could be used as an interferometrically sensitive measurement of the atomic dipole. In addition, our data demonstrates that by finely tuning the laser intensity, an attosecond control on the electron trajectories is performed, which might become exploitable in future experiments on molecules and clusters.

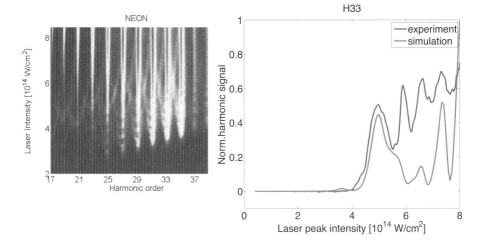

Fig.1. Experimental harmonic spectra versus laser peak intensity generated in Neon. The jet was placed before the focus and the far field spatial filtering was applied. The harmonic amplitude is modulated at the central harmonic frequency and the spectral width increases with the laser peak intensity due to the harmonic frequency chirp.

Fig.2. 33rd neon harmonic amplitude as a function of the laser peak intensity, corresponding in blue curve to a vertical lineout in the harmonic spectra reported in Fig. 1 at the harmonic central frenquency and green curve to our SFA calculation. The average periodicity of the signal $\approx 0.8 \times 10^{14} W/cm^2$ might indicate the contribution of higher-order trajectories to the harmonic emission.

Acknowledgements. This research was supported by the NCCR Quantum Photonics (NCCR QP), research instrument of the Swiss National Science Foundation (SNSF), the Research Councils UK through the UK Attoscience Consortium (EPSRC GR/S24015/01), the French Agence Nationale de la Recherche (ANR-05-BLAN-0295-01, ATTO-SCIENCE), COST-STSM-P14-01910, the European Commission through the RTN XTRA (MRTN-CT-2003-505138) and the M3PEC-UBx1 computer center funded by the Conseil Régional d'Aquitaine.

1 P. B. Corkum, Phys. Rev. Lett 71,1994, 1993.
2 K. Schafer et al., Phys. Rev. Lett. 70, 1599, 1993.
3 M. Lewenstein, P. Salières, A. L'Huillier, Phys. Rev. A 52, 4747, 1995.
4 M. B. Gaarde et al., Phys. Rev. A 59, 1367, 1999.
5 P. Salières et al., Science 292, 902, 2001.
6 G. Sansone et al., Phys. Rev. A 73, 053408, 2006.
7 A. Zaïr et al., Phys. Rev. Lett. 100, 143902, 2008.

Interference Patterns in the Wavelength Dependence of High-Harmonic Generation

Kenichi L. Ishikawa[1,2], Klaus Schiessl[3], Emil Persson[3], and Joachim Burgdörfer[3]

[1] Computational Science Research Program, RIKEN, Hirosawa 2-1, Wako, Saitama 351-0198, Japan
E-mail: ishiken@riken.jp
[2] PRESTO (Precursory Research for Embryonic Science and Technology), Japan Science and Technology Agency, Honcho 4-1-8, Kawaguchi-shi, Saitama 332-0012, Japan
[3] Institute for Theoretical Physics, Vienna University of Technology, Wiedner Hauptstraße 8-10, A–1040 Vienna, Austria, EU

Abstract. We investigate the dependence of the intensity of radiation due to high-harmonic generation (HHG) as a function of the wavelength λ of a few-cycle driver field. Superimposed on a smooth power-law dependence observed previously, strong and rapid fluctuations on a fine λ scale are observed. The origin of these fluctuations can be identified in terms of quantum path interferences with several orbits significantly contributing.

Introduction

High harmonic generation (HHG) represents a versatile and highly successful avenue towards an ultrashort coherent light source covering a wavelength range from the vacuum ultraviolet to the soft X-ray region. The fundamental wavelength λ used in most of existing HHG experiments is in the near-visible range (~ 800 nm). The cutoff law $E_c = I_p + 3.17U_p$, where I_p denotes the ionization potential and $U_p (\propto \lambda^2)$ the pon-deromotive energy, suggests that a longer fundamental wavelength is advantageous to extend the cutoff to a higher photon energy. There is an increasing interest in the development of high-power mid-infrared ($\sim 2~\mu$m) laser systems, and the dependence of the HHG yield on λ has become an issue of major interest. It has been commonly accepted that the spreading of the returning wavepacket would result in a λ^{-3} dependence [1]. Recently, however, Tate *et al.* [2] have reported a more rapidly decreasing HHG yield $\propto \lambda^{-5} \sim \lambda^{-6}$, calculated with the time-dependent Schrödinger equation (TDSE) for Ar and a strong-field approximation (SFA) for He. This surprising finding based on a limited number of data points motivated us to explore the λ-dependence in more detail.

Numerical Model

We investigate the HHG for H and Ar on the level of single-atom response, solving the TDSE,

$$i\frac{\partial}{\partial t}\psi(\mathbf{r},t) = \left[-\frac{1}{2}\nabla^2 + V_{\text{eff}}(r) + zF(t)\right]\psi(\mathbf{r},t), \qquad (1)$$

where $F(t)$ denotes the laser electric field, and $V_{\text{eff}}(r)$ the atomic potential within the single-active electron approximation. We employ two complementary methods to solve Eq. (1) in order to establish reliable and consistent results. The first method is the alternating direction implicit (Peaceman-Rachford) method [3] with a uniform grid spacing. In the second method, the TDSE is integrated by means of the pseudo-spectral method

[4] with a non-uniform mesh point distribution. We adopt the laser parameters used in Ref. [2], with a fixed peak intensity of 1.6×10^{14} W/cm^2, and an envelope function corresponding to a 8-cycle flat-top sine pulse with a half-cycle turn-on and turn-off.

Results and Discussion

The HHG yield ΔY (defined as radiated energy per unit time, integrated from 20 to 50 eV) calculated on a coarse mesh in λ with a spacing of 50 nm (Fig. 1(a)) falls off with a power law, $\Delta Y \propto \lambda^{-x}$ ($x \approx 4.8 - 5.5$) for H and Ar, in qualitative agreement with Ref. [2]. The two alternative integration algorithms employed in this work agree well with each other. A closer look at Fig. 1(a) reveals the remarkable feature that the harmonic yield does not vary smoothly with λ as anticipated in the previous work, but strongly fluctuates. Slight change in fundamental wavelength may lead to variations of the yield by a factor of 2 to 6 (Fig. 1(a)(b)). Such oscillations are largely independent of the atomic species and obviously the result of interference effects.

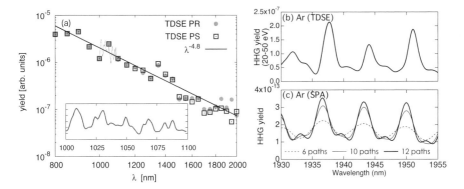

Fig. 1. (a) Harmonic yield (20 to 50 eV) for H as a function of λ calculated on a coarse mesh with $\Delta\lambda = 50$ nm. •: Peaceman-Rachford method; □: pseudo-spectral method, solid line: fit $\Delta Y \propto \lambda^{-x}$. The inset provides a zoom near 1 μm. (b) TDSE- and (c) SPA-calculated variations in the harmonic yield (20 to 50 eV) for Ar in a narrow range of λ.

We apply the saddle-point analysis (SPA) [1], to identify the origin of the interference structures. In this model, the time-dependent dipole moment $d(t)$ is expressed as a sum over paths P that start at the moment of tunnel ionization, evolve in the laser field and recombine upon rescattering at the core. When including up to $10 - 12$ returning paths, the SPA calculation can reproduce the modulation depth and frequency of the oscillations reasonably well, thus unambiguously establishing the quantum path interference as the origin of the fluctuations (Fig. 1(c)). Remarkably, for the present λ-dependence the frequently discussed short and long trajectories are insufficient to account for the oscillations. Note that the spectral width of the few-cycle driving field exceeds, on a wavelength scale, the period $\delta\lambda$ of the modulation. As long as the few-cycle pulse permits the generation of a set of a few quantum paths in subsequent half-cycles, the overall temporal characteristics of the driver pulse is of minor importance.

The modulation period $\delta\lambda$ is a function of λ itself; ca. 20 nm near 1 μm and approaches 6 nm near 2 μm. On the other hand, expressed by the channel-closing parameter $R = (I_p + U_p)/\hbar\omega$ [6], the peak spacing is characterized by $\delta R = 1$ (Fig. 2(b)), regardless of the wavelength region and the atomic species. Expressed by λ, the λ-dependence strongly depends on the peak intensity (Fig. 2(a)). On the other hand, expressed by R, not only the peak positions but also the detailed structure of the λ-dependence is surprisingly robust against the intensity variation (Fig. 2(b)).

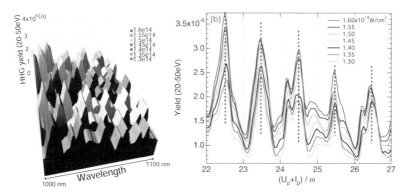

Fig. 2. (a) TDSE-calculated harmonic yield (20 to 50 eV) for H as a function of λ and for different values of peak intensity indicated in the figure. (b) Fluctuations of the harmonic yield as a function of the channel-closing parameter R for H at various driver intensities.

Conclusions

We have found that the fundamental wavelength dependence of HHG in the single-atom response features surprisingly strong oscillations on fine λ scales with modulation periods as small as 6 nm in the mid-infrared regime near $\lambda = 2\,\mu$m. According to SPA, this rapid variation is the consequence of the interference of several rescattering trajectories with long excursion times, confirming the significance of multiple returns of the electron wavepacket [2]. On a large λ scale, apart from the rapid oscillation, our TDSE results show that the HHG yield at constant intensity decreases as λ^{-x} with $x \approx 4.8 - 5.5$ for H and Ar, which is close to the scaling reported in Ref. [2].

Acknowledgements. The work was supported by the Austrian "Fonds zur Förderung der wissenschaftlichen Forschung", under grant no. FWF-SFB016 "ADLIS". K.S. also aknowledges support by the IMPRS-APS program of the MPQ (Germany). K.L.I. gratefully acknowledges financial support by the PRESTO program of the Japan Science and Technology Agency and by the Ministry of Education, Culture, Sports, Science, and Technology of Japan, Grant No. 19686006.

1 M. Lewenstein *et al.*, Phys. Rev. A, **49**, 2117, 1994.
2 J. Tate *et al.*, Phys. Rev. Lett, **98**, 013901, 2007.
3 K. C. Kulander, K. J. Schafer, and J. L. Krause, in Atoms in intense laser fields, Edited by M. Gavrila, 247, Academic, New York, 1992.
4 X.-M. Tong and S. Chu, Chem. Phys. **217**, 119, 1997.
5 K. Schiessl, K. L. Ishikawa, E. Persson, and J. Burgdörfer, Phys. Rev. Lett. **99**, 253903, 2007; J. Mod. Opt., in press.
6 D. B. Milosevic and W. Becker, Phys. Rev. A **66**, 063417, 2002.

Generation of Polarization-Shaped Ultraviolet Femtosecond Pulses

Reimer Selle[1], Patrick Nuernberger[1,2], Florian Langhojer[1,2], Frank Dimler[1,2], Susanne Fechner[1], Gustav Gerber[1], and Tobias Brixner[1,2]

[1] Physikalisches Institut, Universität Würzburg, Am Hubland, 97074 Würzburg, Germany
[2] Institut für Physikalische Chemie, Universität Würzburg, Am Hubland, 97074 Würzburg, Germany
E-mail: brixner@phys-chemie.uni-wuerzburg.de

Abstract. We demonstrate the generation and characterization of polarization-shaped femtosecond laser pulses in the ultraviolet. Polarization-shaped near-infrared pulses are frequency-converted in an interferometrically stable setup comprising two perpendicularly oriented nonlinear crystals.

Optical control of diverse quantum systems has become possible on an ultrafast time scale by shaped femtosecond laser pulses. Although major absorption bands of many organic molecules lie in the ultraviolet (UV), experimental concepts with shaped UV pulses are rare. In indirect UV pulse shaping, a modulated visible or near-infrared (NIR) laser pulse is transferred to the UV in a frequency-upconversion process, either by frequency-doubling or by sum-frequency-mixing [1-3]. In contrast to this, there are a few experiments where the UV pulses are shaped directly [4-6]. However, the pulse shapers employed therein so far do not allow an ultrafast modulation of the polarization state of the UV pulse, thereby excluding optical control scenarios that exploit the three-dimensional (3D) temporal response of the molecular system.

This issue can be addressed by femtosecond polarization pulse shaping [7], as demonstrated for the NIR. Up to now, interferometrically stable polarization shaping is performed only with two-layer liquid-crystal displays (LCDs), which absorb in the UV. Thus, direct UV polarization shaping is not possible with such devices.

We present femtosecond polarization shaping in the UV at a central wavelength of 400 nm by indirect UV pulse shaping employing two perpendicularly oriented nonlinear crystals. As a further novelty, energy losses of the NIR polarization pulse shaper are reduced by integrating volume phase holographic (VPH) gratings [8].

Since conventional gratings have different efficiencies for s- and p-polarized light, usually adequate polarization-dependent attenuators are employed to ensure that different polarization-shaped pulses have the same energy. We have designed a pulse shaper with VPH gratings [8], which exhibit efficiencies close to unity for both polarization directions over the whole employed spectrum (center wavelength 800 nm, 12 nm FWHM). The advantages are a significantly increased energy throughput of 30% in total and the omission of polarization-dependent attenuators. By placing a carefully adjusted Berek compensator as the last element of the pulse shaper, mixing effects of the two shaped polarization components are reduced.

Frequency doubling of polarization-shaped pulses is not feasible directly, since a nonlinear crystal generally acts as a polarizer. In analogy to NIR polarization shaping with two perpendicular LCD layers, the generation of polarization-shaped 400 nm pulses is realized in an interferometrically stable way by two 100 μm BBO (beta barium borate) crystals at the type-I phase-matching angle of 29.2°, but oriented perpendicularly to each other. For instance, a pulse with a linear polarization tilted by

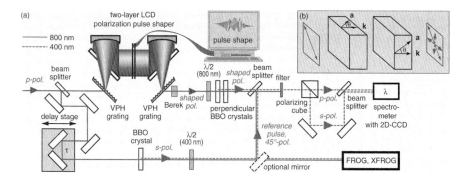

Fig. 1. (a) Setup for the generation and characterization of polarization-shaped UV pulses. The polarization-shaped NIR pulse is frequency-doubled in two perpendicular BBO crystals and combined with the frequency-doubled reference pulse. The two orthogonal polarization components are analyzed separately. The reference pulse passes a delay stage to adjust the temporal delay between shaped and reference pulse. The reference pulse is characterized via FROG and XFROG measurements. (b) Sketch illustrating frequency-doubling of polarization-shaped NIR pulses (propagating along **k**): one component of the fundamental is doubled in the first, the other component in the second crystal (optical axes indicated by arrow **a**)

$45°$ relative to the ordinary polarization directions of the crystals can be decomposed into two orthogonal components, whose second harmonic (SH) is generated in the first and second crystal, respectively [Fig. 1(b)]. The frequency-dependent dispersion in the crystals has to be considered, too, and is precompensated with the pulse shaper.

The setup used to generate and characterize the polarization-shaped UV pulses is sketched in Fig. 1(a). The major fraction of the NIR pulse from the titanium:sapphire amplifier (800 nm, 80 fs, pulse energy up to 1 mJ) is polarization-shaped, while a small fraction is split off in advance, frequency-doubled in a further BBO crystal, and used as 400 nm reference pulse. The electric field of the reference pulse is determined by cross-correlation frequency-resolved optical gating (XFROG) with an 800 nm pulse, which itself is characterized by a FROG measurement first [9]. The polarization-shaped UV pulse is fully characterized by dual-channel spectral interferometry (SI) [7] with the known reference pulse, whose polarization is rotated by $45°$ with a half-wave plate to provide reference pulses for both polarization directions. Shaped and reference pulses are collinearly combined, followed by a setup to separate *s*- and *p*-polarized components so that their spectra can be analyzed separately but simultaneously with an imaging spectrometer. The optical elements (including the spectrometer) that the two components impinge upon have polarization- and wavelength-dependent transmission and reflection properties, which are taken into account to accurately describe the pulse shape after the BBO crystals.

To demonstrate the generation of polarization-shaped UV pulses with our setup, pulse sequences are exemplarily created and characterized. Different linear spectral phases applied to the LCD layers lead to two temporally separated NIR subpulses with perpendicular polarizations, also generating two orthogonally polarized UV subpulses with our setup. Such a UV pulse sequence is shown in Fig. 2, with linear phase coefficients of -500 fs and $+500$ fs applied to component 1 and 2, respectively. The spectral and temporal intensities are plotted in Figs. 2(a) and 2(b), and the temporal evolution of the electric field is visualized in a quasi-3D representation [Fig. 2(c)]. It is noteworthy that the subpulse at -500 fs is slightly elliptical, due to a temporally coincident, but much weaker subpulse of the other polarization,

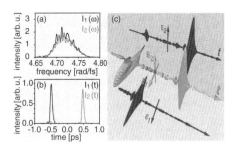
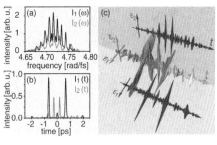

Fig. 2. Polarization-shaped UV double pulse: (a) spectral and (b) temporal intensity, (c) quasi-3D plot of the electric field.

Fig. 3. Polarization-shaped UV pulse train: (a) spectral and (b) temporal intensity, (c) quasi-3D plot of the electric field.

originating from mixing effects between the two polarization components.

The pulse sequence of Fig. 2 has a pulse energy of 500 nJ. While in this case the spectra of the two polarization components are almost identical and reflect the spectrum of the SH of an unshaped laser pulse, one has to keep in mind that frequency doubling of shaped UV pulses generally leads to an amplitude modulation of the SH [1]. Therefore, although the pulse energy and the spectrum of the polarization-shaped NIR pulses are mutually independent of the applied phases, they vary considerably in dependence on the phase for the polarization-shaped UV pulses.

Imposing sinusoidal spectral phase patterns with a pulse shaper results in pulse trains whose temporal subpulse spacing τ is adjustable via the periodicity of the sinusoidal phase. When these pulse trains are frequency-doubled, τ is directly reflected in the spectral modulation of the SH [1]. The pulse sequence of Fig. 3 is generated by applying two sinusoidal phases with τ_1=670 fs and τ_2=470 fs, respectively, to the two LCD layers of the pulse shaper. In the quasi-3D plot [Fig. 3(c)], the periodic appearance of several subpulses is evident. Some subpulses are elliptically rather than linearly polarized due to temporally coinciding subpulses in the orthogonal polarization direction.

In summary, an interferometrically stable pulse shaper setup comprising VPH gratings and two nonlinear crystals has been realized, with which the generation of polarization-shaped femtosecond pulses in the UV has been successfully demonstrated. This technique has many applications, e.g. for manipulation and control of stereochemical reactions and for near-field control in the vicinity of metal nanostructures, also in combination with adaptive femtosecond quantum control.

1   M. Hacker, R. Netz, M.Roth, G. Stobrawa, T. Feurer, and R. Sauerbrey, Appl. Phys. B **73**, 273, 2001.
2   C. Schriever, S. Lochbrunner, M. Optiz, and E. Riedle, Opt. Lett. **31**, 543, 2006.
3   P. Nuernberger, G. Vogt, R. Selle, S. Fechner, T. Brixner, and G. Gerber, Appl. Phys. B **88**, 519, 2007.
4   M. Hacker, G. Stobrawa, R. Sauerbrey, T. Buckup, M. Motzkus, M. Wildenhain, and A. Gehner, Appl. Phys. B **76**, 711, 2003.
5   M. Roth, M. Mehendale, A. Bartelt, and H. Rabitz, Appl. Phys. B **80**, 441, 2005.
6   B. J. Pearson and T. C. Weinacht, Opt. Express **15**, 4385, 2007.
7   T. Brixner and G. Gerber, Opt. Lett. **26**, 557, 2001.
8   I. K. Baldry, J. Bland-Hawthorn, and J. G. Robertson, Publ. Astron. Soc. Pac. **116**, 403, 2004.
9   R.Trebino, Frequency-Resolved Optical Gating, Kluwer, Norwell 2000.

All-Optical Quasi-Phase Matching and Quantum Path Selection of High-Order Harmonic Generation at 140 eV Using Counterpropagating Light

A.L. Lytle, X. Zhang, P. Arpin, O. Cohen, M. M. Murnane, and H. C. Kapteyn

JILA, University of Colorado at Boulder, Boulder CO 80309, USA
E-mail: lytle@colorado.edu

Abstract. We extend all-optical quasi-phase matching of high harmonic generation to 140 eV, where conventional phase matching is not possible. We also demonstrate, and present a model for, selective enhancement of a single quantum trajectory.

Introduction

High-order harmonic generation (HHG) is a unique source of femtosecond to attosecond duration x-ray light that has made possible new applications in ultrafast spectroscopy of atoms, molecules, and materials, as well as enabling new coherent imaging and attosecond dynamics measurements. The main obstacle to generating bright harmonics at wavelengths shorter than $\sim$10 nm has been the phase mismatch present in the harmonic generation process due to high levels of ionization at the intensities required to generate these short wavelengths. Beyond a critical fraction of ionization, η_{cr}, the dispersion of the free electron plasma is too large to be compensated using conventional pressure- and geometry-optimization techniques. Periodic correction of the phase mismatch, known as quasi-phase matching (QPM), is one alternative technique for enhancing the brightness of HHG.

In past work [2,3], we demonstrated the use of counterpropagating light in a hollow waveguide geometry as a flexible and experimentally practical way to both probe the in-situ coherence length of the high-harmonic generation process, and to periodically compensate the phase mismatch. This work resulted in a selective enhancement of HHG through all-optical QPM at photon energies around 65 eV by >100x. Independent influence of the two separate quantum pathways that contribute to harmonic emission was also demonstrated. In the present work, we extend all-optical QPM to higher photon energies of 140-150 eV, which are necessarily generated at ionization levels greater than η_{cr}, and therefore cannot be phase matched conventionally. We also present experimental evidence and a model for a significant difference in phase mismatch between the "long" and "short" quantum trajectories, due to ionization effects during propagation in the waveguide. This effect leads to a mechanism for selectively enhancing either trajectory independently through the proper implementation of all-optical QPM.

Experimental Methods

The output from a 1 kHz Ti:sapphire laser amplifier is split into two beams - a 25 fs, $\sim$1 mJ pump pulse to generate harmonics, and a pair of $\sim$750 fs, $\sim$200 μJ counterpropagating pulses. The beams are coupled into a hollow waveguide, filled with 100-200 torr helium, from opposite directions. The counterpropagating beam is injected into the waveguide using a mirror with a central hole, through which the lower divergence HHG beam can pass. The counterpropagating pulse train is formed by propagating the beam through a separate, grating based compressor, whose configuration determines the pulse durations. In the frequency dispersed plane of the compressor, half the spectral intensity is directed to a delay line, so that the time delay between the two pulses may be independently adjusted. By independent control of the separation and width of two counterpropagating pulses, the maximum possible enhancement from each pulse may be obtained through suppression of the harmonic emission from an out-of-phase coherence zone, one that could otherwise cause destructive interference.

Results and Discussion

First, we show enhancement of photon energies near 140 eV, generated in helium, by >100x using two counterpropagating pulses (see Fig. 1(a)) [4]. Measurement of the coherence length [2] allows determination of the ionization level at which the harmonics were generated: $\eta \approx 1.7\%$ – significantly higher than the critical ionization for helium, $\eta_{cr} = 0.5\%$. The maximum enhancement for a single pulse is also shown to be in agreement with predictions by quantitative analysis [5]. The emission optimized through all-optical QPM is also shown to be >40x brighter than that obtained by optimizing the pressure of the gas, i.e., the most conservative measure of the enhancement factor. This extension of all-optical QPM to a different gas, different ionization level, and higher photon energies shows it to be a robust technique with great promise for increasing the brightness of harmonics at higher photon energies.

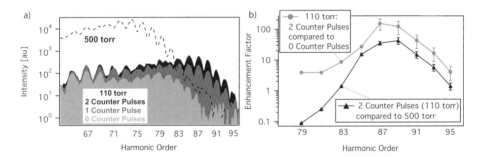

Fig. 1. (a) Harmonic spectra from helium, in the presence of no (light gray), one (dark gray), and two (black) counterpropagating pulses. The dashed curve shows the maximum possible signal obtainable in the same geometry without QPM. (b) Enhancement of several harmonic orders near the 89th, showing selective enhancement using all-optical QPM (gray circles) and enhancement using QPM compared to the maximum signal possible without QPM (black squares).

We also present a model [6], which describes how ionization loss during the formation of a plasma in a hollow waveguide leads to a significantly different phase accumulated by the so-called "long" and "short" trajectories: the two quantum electronic pathways which contribute to a given harmonic. Propagation effects, including time-dependent self-phase modulation and energy loss in the beam, cause a varying intrinsic phase accumulated by the electron wavefunction during the rescattering process. Because the electron spends a longer time in the continuum for the long trajectory than the short, there will be a growing difference in the phase mismatch between emission from the two trajectories with propagation distance (see Figs 2(a) and (b)). A different phase mismatch allows selective enhancement of a single trajectory by varying the separation of the counterpropagating pulses used for all-optical QPM. Fig. 2(c) shows harmonic spectra from helium, where either the long or short trajectory emission has been selected through adjusting the separation of the counterpropagating pulses. This technique can be used for control of HHG on sub-femtosecond time-scales and to increase both the brightness and coherence of HHG.

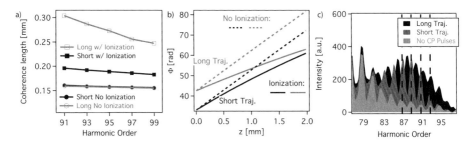

Fig. 2. (a) Phase of long (gray) and short (black) trajectories for the 95th harmonic as a function of z, considering time dependent ionization and preformed plasma contributions to dispersion (dashed lines), and excluding time dependent ionization but including a preformed plasma contributions (solid lines) such that the plasma densities are the same at the peak of the laser field. (b) Coherence length at the peak of the laser pulse as a function of harmonic order for both cases. (c) Selective enhancement of the long (black) and short (dark gray) quantum trajectories using counterpropagating pulse trains containing two pulses with different separations.

Acknowledgements. We gratefully acknowledge support for this work from the NSF Engineering Research Center on Extreme Ultraviolet Science and Technology (award EEC-0310717).

1   A. Paul, R. A. Bartels, R. Tobey, H. Green, S. Weiman, I. P. Christov, M. M. Murnane, H. C. Kapteyn, and S. Backus, Nature 421, 51, 2003.
2   A. L. Lytle, X. Zhang, J. Peatross, M. M. Murnane, H. C. Kapteyn, and O. Cohen, Physical Review Letters 98, 123904, 2007.
3   X. Zhang, A. L. Lytle, T. Popmintchev, X. Zhou, M. M. Murnane, H. C. Kapteyn, and O. Cohen, Nature Physics 4, 270, 2007.
4   A. L. Lytle, X. Zhang, P. Arpin, O. Cohen, M. M. Murnane, and H. C. Kapteyn, Optics Letters 33. 174, 2008.
5   O. Cohen, A. L. Lytle, X. Zhang, M. M. Murnane, and H. C. Kapteyn, Optics Letters 32, 2975 2007.
6   X. Zhang, A. L. Lytle, O. Cohen, M. M. Murnane, and H. C. Kapteyn, New Journal of Physics 10, 025021, 2008.

Ultrafast Molecular and Materials Dynamics probed by Coherent X-Rays

Margaret M. Murnane and Henry C. Kapteyn

University of Colorado at Boulder, 440 UCB, Boulder, CO 80309, USA
E-mail: murnane@jila.colorado.edu

Abstract. Ultrafast short-wavelength light is ideal as a probe of complex, highly excited, systems. We resolve for the first time the decay of core-excited atoms adsorbed onto a surface, and x-ray driven molecular dissociation.

Extreme nonlinear optical techniques make it possible to upconvert visible laser light into coherent x-rays using the process of high-order harmonic generation (HHG). In HHG, the most loosely bound electron is ripped from an atom or molecule by a strong laser field. Once free, the electron follows a trajectory controlled by the laser field, first moving away from the parent ion and then reversing its motion as the laser field oscillates in time. Upon returning to the vicinity of its parent ion, this electron has some probability of recombining with it and giving up its excess kinetic energy as a high-energy photon. Several remarkable scientific and technological opportunities have emerged as a result of this strong field process, including controlling electron rescattering in order to manipulate electrons on attosecond timescales, using the rescattering electrons as a probe of molecular structure and dynamics, and using the ultrashort duration of the generated x-rays as a probe of complex electron dynamics in molecules and materials [1]. Here we present new results that probe complex, highly-excited, dynamics in molecules and materials [1-5].

In the first experiment (see Fig. 1), we report the first direct measurement of core-level relaxation dynamics on a surface-adsorbate system [2]. The coupling between electronic states of an adsorbate and the surface on which it resides is fundamental to the understanding of many surface interactions. However, the coupling of highly excited adsorbate states is an area that has been explored only indirectly to-date. By comparing laser-assisted photoemission from a substrate with a delayed Auger decay process from an adsorbate, we measure the lifetime of a 4d core hole in Xenon on Pt(111) to be 7.1 ± 1.1 fs.

In a second series of experiments (see Fig. 2), we explored x-ray driven molecular dynamics for the first time [3]. The direct observation of molecular dynamics initiated by x-rays has been hindered to date by the lack of bright femtosecond sources of short wavelength light. In this work we used 43 eV soft-x-ray high harmonic beams to photoionize N_2 and O_2 molecule, creating highly excited ions. A strong IR pulse was then used to probe the ultrafast electronic and nuclear dynamics as the molecule explodes. In N_2, we found that significant fragmentation occurs through an electron shakeup process, in which a second electron is simultaneously excited during the soft-x-ray photoionization process. During fragmentation, the molecular potential seen by the electron changes rapidly from nearly spherically symmetric, to a two-center molecular potential. In O_2, the 43 eV beam is sufficiently energetic to access autoionizing states, allowing us to explore why autoionization is suppressed in O_2 until the dissociation has progressed to large distances (≈ 30 Å). Our work captures in real time and with Å resolution the influence of ionizing radiation on a range of molecular systems, probing dynamics that are inaccessible using other techniques.

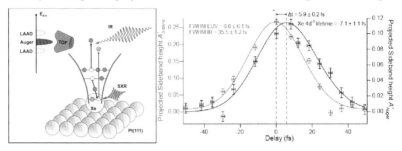

Fig. 1. Experimental (Left) Ultrafast 91 eV soft x-ray (SXR) and IR pulses are focused onto a Xe/Pt(111) surface, The SXR pulse creates a core hole by ejecting an electron from the Xe N4 (4d) shell. This hole is then filled of the by an O2,3 (5p) shell electron followed by ejection of a secondary (Auger) electron from the O2,3 shell. In the presence of the IR beam on the surface, sidebands appear in the photoelectron spectrum of both Xe and Pt. By comparing the sideband amplitude as a function of time delay between the SXR and IR pulses for both Pt and Xe, we extract the Xe 4d core-hole lifetime. (Right) Laser-assisted sideband height vs time delay for the Pt d-band and Xe Auger photoelectrons. The sideband height for the Pt d-band (red curve) corresponds to a cross correlation between the SXR and the IR pulses. In contrast, the Auger sideband height (black curve) is clearly shifted by 5.9 ± 0.2 fs with respect to time-zero, and is also broadened with respect to the Pt d-band curve. The shift and broadening are due to the inherent lifetime of the Xe 4d core-hole on the Pt surface. From these curves we extract a Xe/Pt 4d core-hole lifetime of 7.1 ± 1.1 fs

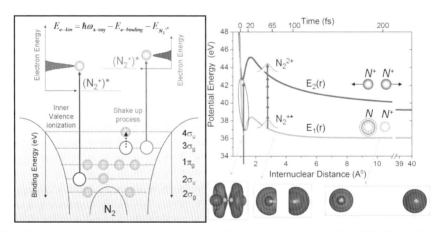

Fig. 2. X-ray driven molecular dynamics. (Left) Schematic of the formation of highly excited N2+ through an inner-valence (i.v.) ionization process (blue), or an electron shakeup process (red) accompanying outer-valence (o.v.) ionization. (Right) A soft-x-ray pump pulse photoionizes N_2 to a highly excited N_2^{+*} shakeup state (orange curve, labeled $E_1(r)$). The probe IR pulse further ionizes N_2^{+*} to the final N_2^{2+} ground state, shown in green and labeled $E_2(r)$. A schematic of the dynamically changing wave function for N_2^{+*} state is also shown at different internuclear separations

In conclusion, many exciting applications of coherent x-rays and attosecond electron recollisons have been demonstrated to date, including high-resolution coherent x-ray imaging [5], femtosecond holography for studying nano-thermal transport, real-time observation of x-ray driven molecular dynamics [3], ultrasensitive molecular spectroscopies and imaging [4], as well as capturing the motion of electrons in atoms, molecules and materials [2].

Acknowledgements. Numerous students and other collaborators have contributed to this work, most principally in the work herein Luis Miaja-Avila, Guido Saathoff, Etienne Gagnon, and Arvinder Sandhu. This work was funded by the National Science Foundation Physics Frontiers Centers program, the Department of Energy Division of Chemical Sciences, Geosciences, and Biosciences. This work made use of facilities provided by the National Science Foundation Engineering Research Center for Extreme Ultraviolet Science and Technology.

1 H.C. Kapteyn, O. Cohen, I. P. Christov, M. M. Murnane Science **317**, 775, 2007.
2 L. Miaja-Avila, G. Saathoff, S. Mathias, J. Yin, C. La-o-vorakiat, M. Bauer, M. Aeschlimann, M. Murnane, H. Kapteyn, "Direct measurement of core-level relaxation dynamics on a surface-adsorbate system", Physical Review Letters, to be published, 2008.
3 E. Gagnon, P. Ranitovic, A. Paul, C. L. Cocke, M. M. Murnane, H. C. Kapteyn, and A. S. Sandhu, Science **317**, 1374, 2007.
4 X. Zhou, R. Lock, W. Li, N. Wagner, M. M. Murnane, H. C. Kapteyn, Physical Review Letters **100**, 073902, 2008.
5 R. Sandberg, C. Song, P. Wachulak, D. Raymondson, A. Paul, B. Amirbekian, E. Lee, A. Sakdinawat, C. La-O-Vorakiat, M. Marconi, C. Menoni, M. Murnane, J. Rocca, H. Kapteyn, J. Miao, Proc. Nat. Acad. Sci. **105**, 24, 2008.

Plasma-Blue-Shift Spectral Shear Interferometry for Characterization of Ultimately Short Optical Pulses

A. Verhoef[1], A. Mitrofanov[1], A.M. Zheltikov[2], and A. Baltuška[1]

[1] Photonics Institute, Vienna University of Technology,
Gusshausstrasse 27-387, A-1040, Vienna, Austria
E-mail: averhoef@mail.tuwien.ac.at
[2] Physics Department, International Laser Center, M.V. Lomonosov Moscow State University,
Vorobyevy gory, 119992 Moscow, Russia
E-mail: zheltikov@phys.msu.ru

Abstract. We introduce a bandwidth-unlimited, dispersion- and shear- self-calibrated, timing-jitter-free pulse measurement technique based on a quasi-linear temporal phase modulation in a gas weakly ionized by a long pump-pulse. Results of a 5-fs pulse characterization are reported.

Measurement of ultrashort ultraweak laser pulses with over-an-octave wide spectra poses a well-known challenge because of the spectral overlap of harmonic orders. The problem of the interference between the nonlinear signal and the fundamental spectrum can be mitigated in both frequency-resolved optical gating (FROG) and spectral shearing interferometry (SPIDER) [1] techniques by using a non-collinear geometry of sum-frequency generation. However, the suitable phase-matching bandwidth for frequency conversion is ensured as a rule by using low conversion efficiency several-μm-thick nonlinear crystals. Additional issues that become increasingly important as the pulse duration shrinks are the minimum added dispersion of the pulse characterization apparatus [2] and the time delay accuracy/jitter between the interfering pulse replicas in SPIDER [3].

To address these issues we introduce a new SPIDER variant based on a controlled blue shift of one test pulse replica in the field of a long weakly ionizing pump pulse while preserving another, reference, replica of the test pulse. We use this technique, dubbed i(onization)-SPIDER, to characterize weak (150-nJ) 5-fs pulses around 800 nm derived from a chirped-mirror-compressed output of a Kr-filled 100-μm capillary. The pump field was a 120-μJ ~70-fs pulse at 390 nm. For weakly ionized gases the instantaneous frequency ω of the phase-modulated test pulse is shifted by Ω [4]:

$$\Omega(\omega) = \frac{e^2 L}{2c(\omega)\varepsilon_0 m_e \omega} \frac{dN_e(t)}{dt},$$ (1)

where m_e and e are the electron mass and charge; c, speed of light; ε_0, electric permittivity of free space; L the interaction length; and N_e is free electron concentration. Figure 1 presents an experimentally observed blue shift of a 5-fs pulse and a numerical simulation for our experimental conditions. The build-up of N_e was calculated using the cycle-dynamics formalism of Ivanov and Yudin [5]. In our experiments, the ionization mechanism was deliberately set to be multiphoton (Keldysh γ=4–5) to achieve a smooth quasi-linear (within a 30-fs time window) temporal dependence of N_e.

i-SPIDER has several advantages compared to traditional SPIDER arrangements in which both pulse replicas are shifted to their new respective central frequencies: (I) in the absence of the temporal overlap with the pump pulse, i-SPIDER turns into a linear spectral interferometer for characterization of the dispersion of the apparatus including, if relevant, the contribution of the plasma dispersion. The group delay of the test pulse then can be obtained as $\tau(\omega) = \left(\varphi_{nl}(\omega + \Omega(\omega)) - \varphi_{linear}(\omega)\right)/\Omega(\omega)$, where φ_{nl} and φ_{linear} are the spectral phases retrieved from an i-SPIDER and a linear interferograms, respectively, using a conventional Takeda algorithm [1]; (II) both the linear (no temporal pump-test pulse overlap) and nonlinear (pump-test pulse overlap) signals are equally strong, because no frequency conversion is involved in rendering a nonlinear interferogram. Therefore the minimum pulse intensity of the test pulse to perform its complete phase characterization is dictated by the spectrometer sensitivity; (III) since one (reference) replica of the pulse remains unshifted, the amount of spectral shear is directly computed from the pump-on and pump-off interferograms.

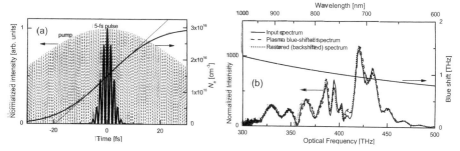

Fig. 1. Spectral blue-shifting of a weak 5-fs 780-nm laser field as the result of a quasi-linear refractive index change of a Kr gas target that is being ionized by a 70-fs 390-nm pump field. (a) Numerical simulations, showing normalized instantaneous intensities of the blue pump pulse (Peak intensity 2.5×10^{13} W/cm2) and a 5-fs test pulse and the evolution of the plasma density. The linear slope indicates a constant ionization rate required to produce a 0.9 THz blue shift at 800 nm for a 5-mm long Kr target at the pressure of 2 bars. (b) Experimental spectra and the calculated magnitude of the blue shift, $\Omega(\omega)$. The back-shifted trace was obtained by subtracting $\Omega(\omega)$ from the optical frequencies of the measured blue-shifted spectrum. The maximum spectral shift due to cross-phase modulation in neutral atoms [6] between the 390-nm pump field and the 800-nm probe field was several times smaller than the ionization blue shift.

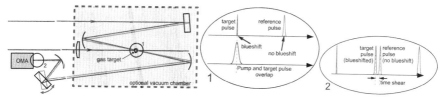

Fig. 2. Experimental i-SPIDER setup and the explanation of pulse sequence. The required time separation between the pulse pair creating a spectral interferogram is obtained by a slight angular detuning of the recombining beamsplitter. The first and the fourth pulse in the sequence have a ps time separation (unresolvable with a spectrometer) and contribute to an unmodulated background of an interferogram.

The experimental i-SPIDER setup is depicted in Fig. 2. Because of the extreme ease of detecting spectral interference of even very weak beams, uncoated flat parallel fused silica plates were used as splitting and recombining beamsplitters. This arrangement satisfies the most stringent timing jitter requirements put forward in Ref.[3] for the case of a single-cycle pulse SPIDER. The summary of i-SPIDER retrieval of a 5.3-fs test pulse is provided in Fig. 3.

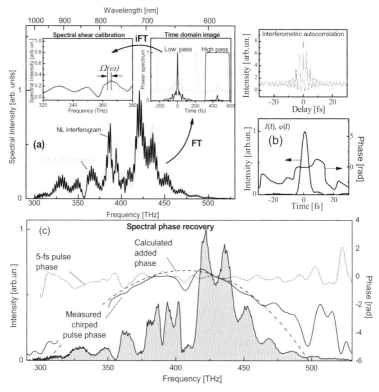

Fig. 3. Retrieved i-SPIDER data for a 5.3-fs test pulse. (a) Input interferograms and the phase retrieval routine. (b) Retrieved pulse and comparison with an independently measured autocorrelation for a 5-fs pulse case. (c) Spectral phases retrieved by i-SPIDER for a nearly TL pulse (red curve) and a pulse transmitted through 0.3 mm of fused silica and 1 m of air. (i)FT, (inverse) Fourier Transform The iFT of the low-pass-filtered temporal images serves as spectral shear calibration, whereas the iFT of the high-pass parts recovers $\varphi_{\text{linear}}(\omega)$ and $\varphi_{\text{nl}}(\omega)$.

An i-SPIDER apparatus can be readily integrated in the beamlines used for many higher-field physics applications which typically already include all the hardware necessary for an i-SPIDER measurement.

Acknowledgements. This work is supported by the Austrian Science Fund (FWF), grants U33-N16 and F1619-N08.

1. Ch. Iaconis and I. Walmsley, IEEE J. Quantum Electron. **35**, 501 (1999).
2. P. Baum, S. Lochbrunner, and E. Riedle, Opt. Lett. **29**, 210 (2004).
3. J. R. Birge, R. Ell, and F. X. Kärtner, Opt. Lett. **31**, 2063-2065 (2006).
4. S.P. Le Blanc and R. Sauerbrey, J. Opt.Soc. Am. B **13**, 72, (1996).
5. G.L. Yudin and M.Yu. Ivanov, Phys. Rev. A **64**, 013409 (2001).
6. S.P. Le Blanc, R. Sauerbrey, S.C. Rae, and K. Burnett, J. Opt. Soc. Am. B **10**, 1801 (1993).

Spatially resolved Ar* and Ar+* imaging as a diagnostic for capillary based high harmonic generation

R.T. Chapman[1], E.T.F Rogers[2], C.A. Froud[2], J. Grant-Jacob[2], M. Praeger[3],
S.L. Stebbings[1], J.G. Frey[1], W.S. Brocklesby[2]

[1] School of Chemistry, University of Southampton, SO17 1BJ.
[2] Optoelectronics Research Centre, University of Southampton, SO17 1BJ
[3] School of Physics and Astronomy, University of Southampton, SO17 1BJ.

Abstract. Spectrally resolved imaging of Ar/Ar+ created by high harmonic generation is demonstrated, and used as a diagnostic of capillary geometry on XUV generation efficiency.

Introduction

Coherent XUV radiation can be generated by the highly non-linear interaction between a gas target and high intensity ultrafast laser pulses using the high harmonic generation (HHG) process. Guiding the fundamental laser field inside a hollow capillary waveguide [1] improves phase matching and extends the potential interaction length. However, propagation of an intense pulse within a capillary waveguide filled with ionizable gas is complex, as the pulse creates a plasma, which in turn strongly affects the propagation. Previous work [2] has used emission from the excited gas to study propagation of ns pulses in capillary guides. In this work we demonstrate spectrally-resolved imaging of the plasma created by intense fs pulses within a capillary during an HHG experiment. The spectral & spatial resolution is used to separate contribution from ions and neutral species, and is an effective diagnostic for the local pressure and modal intensity variations along the waveguide.

Experimental

The laser radiation at 800nm, from a 1 kHz Ti:sapphire chirped pulse amplifier system producing 1 mJ, 35 fs pulses is focused into a 150 m hollow capillary waveguide. The capillary is 70 mm long with two radial 300 m holes drilled 20 mm from each end. Gas flows through the two holes to define a central region of constant gas pressure, with almost linear pressure gradients from the holes to the ends of the capillary, which were maintained at a pressure of 10-5 mbar. The resulting pressure variation along the capillary was calculated using computational fluid dynamics. The capillary was imaged at right angles to the laser propagation direction onto a CCD camera using a lens (fl. 5cm). A short pass filter was used to reject scattered laser light. In addition, 420nm and 488nm narrow band pass interference filters were used to image via transitions arising from excited neutral argon and argon ions respectively.

Results and Discussion

Figure 1 shows an example of a spectrum observed and cross-sections through the images of the capillary taken with the 420nm and 488nm band pass filters, showing the

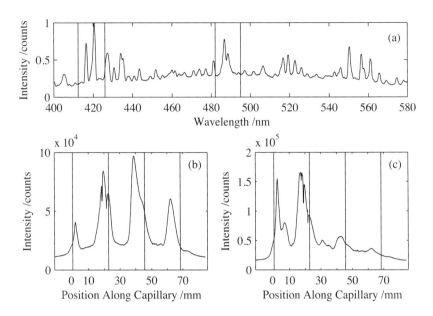

Figure 1: (a)Spectrum of excited Ar and Ar+ measured from capillary during HHG. The black lines indicate 10 − 90 % filter transmissions. (bottom) Section through images of the capillary at emission wavelengths corresponding to (b) argon neutral species and (c) argon ion species. The vertical lines indicate the expected peak positions of the EH11/EH12 mode beats.

intensity of the neutral (b) and ionized (c) argon species respectively. Both these images show a series of intense maxima as a function of length along the capillary. These arise from mode beating between EH11 and higher order modes (principally EH12) launched into the capillary. The intensity of the emission from the argon ion species is significantly reduced as the beam propagates along the capillary. Plasma generation in the capillary is modeled by calculating, firstly, the launched mode distribution via the overlap integrals with the incident laser focal spot. This intensity distribution at the capillary entrance is used to calculate the ionization via ADK theory, and the loss due to ionization calculated as a function of radius. The resulting intensity profile is then decomposed into a new set of capillary modes which are propagated a short distance, allowing for propagation loss differences between modes, and the process repeated. The results of this calculation for a range of pressures equivalent to those measured experimentally are shown in Figure 2a. These figures show the intensity distribution at 488nm as a function of the pressure at the inlet. The model shows the calculated ionization level for sum over four capillary modes. In addition to this comparison the model can predict the laser transmission through the capillary as a function of pressure. Agreement between the calculated and measured transmission vs. pressure is good, demonstrating the validity of the model.

Figure 2 Argon ion emission intensity along the capillary as a function of pressure from experimental data (left), and from modelling (right) Dotted lines show positions of gas inlet holes. (Laser propagates from left to right)

The peaks in emission at the beat positions predicted from mode propagation and

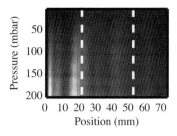 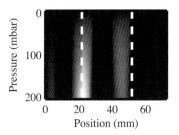

Figure 2: Argon ion emission intensity along the capillary as a function of pressure from experimental data (left), and from modelling (right) Dotted lines show positions of gas inlet holes. (Laser propagates from left to right)

loss are accurate. Further adjustment within the model of the laser wavefront shape at the capillary entrance has shown even better agreement, shifting the overall position of the beat pattern by several mm along the capillary. The significant decrease in emission intensity as the laser propagates along the capillary seen in the data is mirrored by a similar decrease in the calculated ionization level along the capillary. The highly nonlinear ionization process is very sensitive to small reductions in the intensity caused by loss due to ionization and capillary loss. In contrast, emission from excited neutral Ar, shown in Figure 1(b), does not show a similar decrease along the capillary. The excitation of neutral argon by processes such as multiphoton absorption is less strongly nonlinear than ADK ionization, and so will show less variation with intensity. Thus measurement of the total emission intensity is insensitive to the variations that affect the HHG process directly, i.e. the level of ionization. The beat period does not vary significantly along the capillary length, indicating that nonlinear mode coupling out of the EH11 and EH12 modes is not strong even at the highest intensities and pressures. In the 20-30nm XUV region absorption by Ar along the capillary is significant, so the presence of most of the ionization (and thus XUV generation) at the capillary entrance implies that altering the capillary design could increase the amount of XUV at the output.

Conclusions

Spectrally resolved Ar/Ar+ imaging has been observed during HHG in a capillary. The observed images are a useful diagnostic for several aspects of ionization important for capillary-based XUV generation, including the distribution of the laser power between the capillary modes, the effect of losses on ionization along the capillary and the non-linear absorption and propagation inside the capillary. Extension of the modelling using a multimode nonlinear Schroedinger equation, including polarization effects and high-order dispersion as well as wavelength-dependent mode coupling is in progress.

1 Durfee III C.G., Rundquist A.R., Backus S., Herne C., Murnane M.M., Kapteyn H.C., Phase Matching of High-Order Harmonics in Hollow Waveguides Phys. Rev. Lett. 83, 2187 (1999)
2 Pfeifer T., Downer M.C., Direct experimental observation of periodic intensity modulation along a straight hollow-core optical waveguide J. Opt. Am. B. 24, 1025 (2007)

Internal Momentum State Mapping using High Harmonic Radiation

Xinhua Xie[1], A. Scrinzi[1], M. Wickenhauser[1], A. Baltuška[1], I. Barth[2], and M. Kitzler[1]

[1] Photonics Institute, Vienna University of Technology, Austria, EU
[2] Institute for Physical and Theoretical Chemistry, Free University of Berlin, Germany, EU
E-mail: markus.kitzler@tuwien.ac.at

Abstract. We numerically demonstrate so far undescribed features in ionization and high harmonic generation from bound states with non-vanishing electronic angular momentum. The states' modified response to a strong laser pulse can be exploited for the production of near-circularly polarized isolated attosecond XUV/X-ray pulses.

1. Introduction

The way how gas atoms or molecules respond to a strong laser field depends on their internal state. This fact has been exploited in a range of recent experiments, where insight into the bound electron structure, nuclear dynamics, and even electronic dynamics was gained from measurements of photon spectra produced by high-order harmonic generation (HHG) from aligned molecules, e.g. [1, 2]. Here we demonstrate a conceptually reversed approach: instead of extracting information about the bound state from the HH radiation, we prepare the molecular state in order to control the properties of the HH emission. Specifically we consider electronic ring current states, which can be excited in a suitable molecule or atom by (weak) circularly polarized π pulses [3].

We present numerical solutions of the time-dependent Schrödinger equation in two spatial dimensions for a single active electron in a ring-shaped model potential. The initial ring-current states that we use are are constructed by superposition of two equally populated, degenerate real states of m-fold angular symmetry ($|m| \geq 1$), which leads to a ring-current wave function $\Psi_\pm(\rho, \varphi) = 1/\sqrt{2}\, f(\rho)e^{\pm i|m|\varphi}$ (in 2D polar coordinates). Because only the phase of the wave function depends on φ the electron density $|\Psi_\pm|^2$ of a ring-current state is constant along the angular direction.

3. Results and Discussion

We first investigate electron detachment from a molecule in a ring-current state. Fig. 1 shows snapshots of the electron density in a linearly polarized laser pulse at a wavelength of 1600 nm and pulse duration of 10 fs FWHM (gaussian intensity envelope). The insets of Figs. 1(a) and 1(b) show the electron density for the degenerate states used to construct the ring-current state. The nodes in the electron density are imprinted also in the detached part of the wave function, as expected for molecular tunnel ionization [4]. When the two states are superimposed to form the ring-current state, the probability density inside the molecule becomes radially constant; any nodal structure is absent, cf. inset of Fig. 1(c). Ionization of this ring-current state by the same linearly polarized field as in panels (a) and (b) results in the emittance of an electronic wave packet strongly rotated off the polarization axis in counter-clockwise direction, i.e. along the direction of the ring-current, see Fig. 1(c). Correspondingly, ionization from a right-handed ring-current state results in wave packet emission into the opposite, clockwise, direction (not shown).

At first glance this *asymmetric* emission from a rotationally *symmetric* electron density distribution could be surprising. A simple classical picture for the asymmetry observed here is that during ionization internal angular momentum is imparted to the detached electron. In more accurate quantum mechanical terms, the effect of asymmetric emission is due to the interference of two wave functions emitted from the two initial state components, which have opposite parity with respect to reflections about the x-axis, cf. Fig. 1(a) with Fig. 1(b).

Now we turn to a novel application of $|m| \geq 1$ states for the generation of high harmonic pulses. Elliptically (circularly) polarized harmonics cannot be produced by simply using elliptical (circular) driving polarization, because this leads to the suppression of recollision and a dramatic drop in photon yield [5]. In a range of schemes this problem is dealt with by meticulously shaping the polarization of the drive pulse, see e.g. [6] and references therein. Here we demonstrate the reversed approach and use linear drive laser polarization, but prepare the molecules in ring-current states instead.

Fig. 2(a) shows the high harmonic intensity spectrum of two orthogonal polarization components. They are comparable, especially in the cut-off around 1.9 a.u. with a ratio in intensity of 1:4. The two components are out of phase by about $\pi/2$ over a broad spectral range (dots in Fig. 2(a)). This means that those harmonics correspond to near-circularly polarized XUV radiation, with a ratio of $\sim 1:2$ between the largest and the smallest field component, cf. filtered out attosecond pulse in Fig. 2(b).

A simple calculation in the frame of the strong field approximation (SFA) elucidates the mechanism of production of circularly polarized HH radiation. The polarization state of the harmonic radiation is determined by the re-combination step [5]. Evaluating the bound-free dipole matrix element $\langle \Psi_m | \vec{r} | \vec{k}(t) \rangle$ using plane wave approximation for the recolliding electron $|\vec{k}(t)\rangle$ and a Bessel expansion for the ring-current state $\Psi_m = 1/\sqrt{2} f(\rho) e^{im\varphi}$ one can easily show that, as long as the re-scattering wave length, k, is comparable to the extension of the initial bound state, $\Delta\rho$, such that $2\pi/k \sim \Delta\rho$, the x and y components of the harmonic radiation at frequency $\omega(k)$ are (almost) equally strong and exhibit a relative phase of $\pi/2$. This demonstrates that creation of circularly (elliptically) polarized harmonics is a generic effect, which appears for any system with $|m| \geq 1$ and harmonic frequencies in the range $\omega \sim 2\pi^2/\Delta\rho^2 + I_p$.

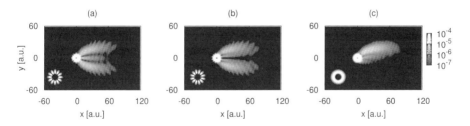

Figure 1: Snapshots of the probability density of an electron wave packet as it is detached from a ring-current state by strong-field ionization in a linearly polarized laser field with peak intensity 5.6×10^{13} W/cm^2. (a) Ionization from one of the degenerate states, (b) the other state, (c) the left-handed ring-current state.

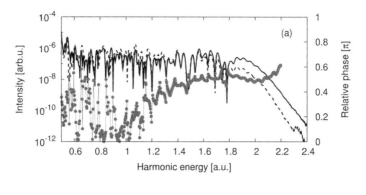

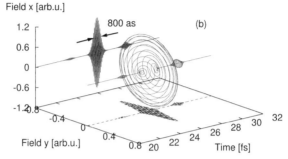

Figure 2: (a) HH spectra of I_x (full line) and I_y (dashed line) and their relative phase (dots) from a ring-current state with $m = +6$. Pulse parameters as in Fig. 1(a)-(c). (b) Nearly circularly polarized electric field of an attosecond pulse obtained by filtering the highest harmonics of the spectra shown in (a).

4. Conclusions

In conclusion we have demonstrated that for ring-current initial states tunneling wave packets receive an initial momentum kick, a so far unexplored mechanism, and that in HHG the internal state is transferred to the polarization state of the emitted photons, which leads to the production of novel, spatially and temporally coherent, nearly circularly polarized attosecond XUV/X ray pulses. These pulses might open the door to tabletop-scale imaging and spectroscopy of ferromagnetic materials on extreme time-scales.

Acknowledgements. Fruitful discussions with S. Gräfe and O. D. Mücke are gratefully acknowledged.

1   R. Torres *et al.*, in Phys. Rev. Lett. **98**, 203007, 2007.
2   J. Levesque *et al.*, in Phys. Rev. Lett. **98**, 183903, 2007.
3   I. Barth *et al.*, in J. Am. Chem. Soc. **128**, 7043, 2006.
4   X. M. Tong *et al.*, in Phys. Rev. A **66**, 033402, 2002.
5   P. B. Corkum and F. Krausz, in Nature Physics **3**, 381, 2007.
6   D. B. Milošević *et al.*, in Phys. Rev. A **61**, 063403, 2000.

Attosecond control of electron localization in one- and two-color dissociative ionization of H_2 and D_2

G. Sansone[1], F. Kelkensberg[2], M.F. Kling[3], W.K. Siu[2], O. Ghafur[2], P. Johnsson[2], S. Zherebtsov[3], I. Znakovskaya[3], T. Uphues[3], E. Benedetti[1], F. Ferrari[1], F. Lépine[4], M. Swoboda[5], T. Remetter[5], A. L'Huillier[5], M. Nisoli[1], M.J.J. Vrakking[2]

[1] National Laboratory for Ultrafast and Ultraintense Optical Science CNR Istituto Nazionale per la Fisica della Materia, Department of Physics, Politecnico, Piazza Leonardo da Vinci 32,20133, Italy
[2] FOM-Institute AMOLF, Kruislaan 407, 1098 SJ Amsterdam, The Netherlands
[3] Max-Planck Insitut für Quantenoptik, Hans-Kopfermann Strasse 1, D-85748 Garching, Germany
[4] Université Lyon 1; CNRS; LASIM, UMR 5579, 43 bvd. du 11 novembre 1918, F-69622 Villeurbanne, France
[5] Department of Physics, Lund University, P.O. Box 118, SE-221 00 Lund, Sweden
Email: m.vrakking@amolf.nl

Abstract. We report experiments where an attosecond pulse launches a wavepacket on the dissociative state of D_2^+, and a few-cycle IR pulse localizes the electron on one ionic fragment with attosecond sensitivity to the XUV-IR delay.

Introduction

Observation and control of electron dynamics on attosecond timescales are two of the main driving forces behind the recent emergence of attosecond science. Using attosecond laser pulses, experiments can be conceived where electron dynamics is initiated on attosecond timescales. Subsequently, the electrons are allowed to evolve and are probed (again, on an attosecond timescale), providing new insights into the fundamentals of photo-excitation, as well as the role of electron-electron correlations and electron-nuclear energy transfer on these ultrafast timescales.

A technical breakthrough that has been very influential and that has greatly improved the ability to generate attosecond laser pulses in a controlled manner, was the development of amplified femtosecond laser systems with a controlled carrier-envelope-phase (CEP) in 2003 [1]. This enabled the controlled generation of isolated attosecond laser pulses with a pulse duration down to the present record of 130 attoseconds [2]. In addition, CEP-stable pulses have been used to control atomic ionization and the localization of electrons in D_2^+ molecules [3].

One-color experiment

Here we present results from two new experiments performed in Milano where one-color localization under the influence of a few-cycle IR pulse is extended to H_2^+ molecules, and where electron localization is accomplished by exposing D_2 and H_2 molecules to a sequence of an isolated attosecond XUV pulse and a few-cycle IR pulse. Figure 1 shows an example from the former, one-color experiment, recorded for linearly polarized, phase-stabilized 5-fs light pulses. The asymmetry in the ejection of D^+ from D_2 is shown as a function of the energy of the fragments and the CEP of the few-cycle IR laser. Such asymmetry is defined as $(N_{up}-N_{down})/(N_{up}+N_{down})$, where Nup

51

and N_{down} are the number of D^+ ions emitted in the up and down directions, respectively, along the laser polarization axis. Importantly, in a significant improvement over the results that were previously reported in ref. [3], the energy dependence of the asymmetry parameter can clearly be observed in the measurements. A detailed analysis of the results, including an absolute calibration of the CEP by means of stereo-ATI measurements on Xe that were simultaneously performed, is currently under way.

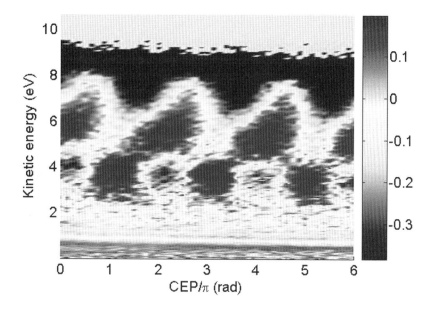

Fig. 1. Asymmetry parameter $(N_{up}-N_{down})/(N_{up}+N_{down})$ for the ejection of D^+ in one-color dissociative ionization of D_2 by a few-cycle IR pulse

Two-color experiment

In the two-color experiment an isolated attosecond laser pulse ionizes D_2 and part of the ionized molecules dissociate. Isolated attosecond pulses are produced by means of high-harmonic generation in Krypton using the polarization gating technique [2,4]. In the photo-ionization by the attosecond pulse the repulsive $2p\sigma_u^+$ state is populated when the photon-energy of the pulse is high enough. Interaction of this dissociating wave packet with a moderately strong IR field localizes the electron on the upper or lower D^+ ion. By varying the delay between the XUV pulse and the few-cycle IR pulse the asymmetry in the ejection of D^+ ions can be controlled with attosecond time resolution. The experiment may be viewed as a first example of the observation of attosecond time-resolved electron dynamics in molecular physics. Similar to the previous, one-color experiment, the asymmetry shows a strong dependence on the kinetic energy of the D^+ fragment.

A model solving the 1D time-dependent Schrödinger equation (TDSE) was used to confirm the interpretation and to give insight into the origin of the observed energy dependence. In the model a nuclear wave packet is projected on the repulsive $2p\sigma_u^+$ state and propagated on the bound $1s\sigma_g^+$ and the repulsive $2p\sigma_u^+$ state of the molecular ion in the presence of a few-cycle IR field. The IR field couples the two states, which generates a coherent superposition where the electron can be localized on the upper or the lower ion. As in the experiment, the timing of the IR field determines the electron localization. The kinetic energy dependence of the electron localization, which is also observed in the calculations, originates from the initial spread of the wave packet.

Electron transfer processes are ubiquitous in chemistry. The present combination of one- and two-color experiments, where electron localization is first controlled on attosecond timescales using the controlled waveform of a few-cycle IR laser pulse and then using the controlled delay between an isolated attosecond pulse and a few-cycle IR laser pulse, are first examples of strong-field control of chemical processes on attosecond timescales and of the direct observation of attosecond electron dynamics in molecules, paving the way towards attempts to observe and control electron transfer in more complicated molecules, such as molecules of biological interest.

1   A. Baltuška, Th. Udem, M. Uiberacker, M. Hentschel, E. Goulielmakis, Ch. Gohle, R. Holzwarth, V. S. Yakovlev, A. Scrinzi, T. W. Hänsch and F. Krausz, "Attosecond control of electronic processes by intense light fields", Nature 421, 611 (2003).

2   G. Sansone, E. Benedetti, F. Calegari, C. Vozzi, L. Avaldi, R. Flammini, L. Poletto, P. Villoresi, C. Altucci, R. Velotta, S. Stagira, S. D.Silvestri, and M. Nisoli, Isolated Single-Cycle Attosecond Pulses, Science 314, 443–446 (2006).

3   M. F. Kling, Ch. Siedschlag, A. J. Verhoef, J. I. Khan, M. Schultze, Th. Uphues, Y. Ni, M. Uiberacker, M. Drescher, F. Krausz and M. J. J. Vrakking, "Control of Electron Localization in Molecular Dissociation", Science 312, 246 (2006).

4   I. J. Sola, E. Mével, L.Elouga, E.Constant, V.Strelkov, L.Poletto, P. Villoresi, E. Benedetti, J.-P. Caumes, S. Stagira, C. Vozzi, G. Sansone and M. Nisoli, "Controlling attosecond electron dynamics by phase-stabilized polarization gating", Nature Physics 2, 319 (2006).

Simultaneous Description of Electron and Nuclear Dynamics: A Quantum Approach for Multi-Electron Systems

Philipp von den Hoff, Dorothee Geppert, and Regina de Vivie-Riedle

Department Chemie und Biochemie, Ludwig-Maximilians-Universität München, 81377 München, Germany
E-mail: Regina.de_Vivie-Riedle@cup.uni-muenchen.de

Abstract. A new and efficient approach to describe molecular electron and nuclear dynamics simultaneously is presented. The method is tested by the photodissociation of D_2^+ and allows for a successive extension to multi-electron systems.

Introduction

The fast development of laser techniques opens the door for new control scenarios. The generation of pulses with a duration of a few femtoseconds or even attoseconds [1] pushes the frontiers of coherent control to the dynamics of the much faster electrons. One challenging idea is the control of chemical reactions by guiding an electronic wavepacket to the wanted position in the molecule. However, in many reactions the nuclear motion is too slow to respond directly to the dynamics of an electronic wavepacket and the two dynamics must be synchronized. In a photochemical reaction a light pulse induces fast nuclear motion. At the same time a coherent superposition of electronic states can create an electronic wavepacket whose time evolution is coupled to the simultaneous motion of the nuclei. A nuclear event like the dissociation of a molecule or a level crossing can serve as defined end of the electron dynamics.

Highly accurate methods have been developed to describe the electronic motion in many-electron systems [2-5], however, a simultaneous quantum mechanical treatment of the nuclear dynamics is computationally demanding and not yet realized. We developed an efficient approach, which exploits quantum dynamics and quantum chemistry in a new way to simultaneously describe the dynamics of the nuclei and the electrons. A first test for our method is the photodissociation of D_2^+. In this reaction the localization of the electron at either one of the D-atoms can be controlled via the carrier-envelope phase of an ultrashort laser pulse. It allows us to check our new method against more rigorous theories [6,7] and experiments [8]. In both cases, we achieved excellent agreement. The detailed analysis of the electron motion after different ionization events reveals the underlying complex dynamics, which are hidden in the experiment. The interplay between the carrier-envelope phase and electron control is elucidated.

Our method is based on the highly developed electronic structure theory and allows for a successive extension to multi-electron system simultaneously enabling a quantum dynamical description of the nuclear motion.

Results and Discussion

For the photodissociation of D_2^+ the theoretical set up assumes the experimental conditions as described by Kling et al. [8]. We start our simulations after the recollision excitation of the electron. The computing procedure for our ansatz follows the idea of the Born-Oppenheimer approach vice versa. With the quantum chemistry package Molpro [9] the potential energy surfaces and the respective transition dipole moments involved are calculated. The timedependent Schrödinger equation for the nuclei is solved numerically on a grid with an appropriate propagator. Subsequently, the expectation value of the nuclear distance $\langle X_{tot}|R|X_{tot}\rangle_R$ as well as the interference term $\langle X_i(R,t)|X_j(R,t)\rangle_R$ is calculated as a function of time, where the total nuclear wavefunction X_{tot} is spanned by the nuclear wavefunctions X_i and X_j of the electronic states $|i\rangle$ and $|j\rangle$ coupled through the short light pulse. These calculations are followed by new quantum chemical computations at the determined nuclear geometries, which now satisfy the requirement of temporal equidistance instead of spatial equidistance. From these quantum chemical data we obtain the form and the eigenenergies of the molecular spinorbitals, needed to analyze the time dependence. The phase $\phi(i/j)$ of the electronic wavefunctions $|i\rangle$ and $|j\rangle$ is calculated recursively and determines their time evolution. From these results the electronic density can be calculated as functions of time t, the electron coordinate r and the nuclear coordinate R. For many-electron systems the electronic wavefunction is represented by a single or a linear combination of Slater determinants (eq.1), which are set up in the basis of molecular spin orbitals χ_i. The most common case of single electron excitation leads to the mixing of two electronic states that differ by one spin orbital.

$$|i\rangle = |\chi_1 \chi_2 \cdots \chi_{n-1} \chi_a\rangle$$
$$|j\rangle = |\chi_1 \chi_2 \cdots \chi_{n-1} \chi_r\rangle \qquad (1)$$

The linear combination of the electronic states $|i\rangle$ and $|j\rangle$ produces a temporal evolution of the electronic density, which is given by the differing orbitals χ_a and χ_r and oscillates with the energy difference ΔE of the two electronic states. The overall electron density ρ can be calculated in laboratory coordinates r'.

$$\rho(r',t;R) = \sum_{i=1}^{n-1} |\chi_i|^2 + a_1(t)^2 |\chi_a|^2 + a_2(t)^2 |\chi_r|^2$$
$$+ 2\mathrm{Re}\left\{ \langle X_j(R,t)|X_i(R,t)\rangle_R \, \chi_{r,t=0}^*(r';R)\chi_{r,t=0}(r';R)e^{-i\Delta Et} \right\} \qquad (2)$$

This electron density gives the probability to find any electron at a specific point in space. If the Slater determinants differ by two spin orbitals, i.e. there are two active electrons, the ansatz can be extended to calculate the electron pair density, which will be treated in future works.

For our test model D_2^+ the induced electron oscillation is much too fast for the nuclei to react. The overall dynamics results from a competition between the influence of the light pulse, the dynamics of the linear combination, which oscillates with ΔE and the interference term of the nuclear wavefunctions, which determines the "degree of state mixture". Exemplarily we show in Fig. 1 the complex landscape of the expectation value of an electron after ionization by one particular maximum of the few cycle pulse for various CEP values. This single maximum event is not observable in the experiment.

By averaging the asymmetry $\left(A = P_{left\,D} - P_{right\,D}/P_{left\,D} + P_{right\,D} \right)$ over five extrema of the electric field the final electron localization was determined as a function of the CEP.

The resulting simple periodic structure is in excellent qualitative and quantitative agreement with the experimental data. Comparison with the single maximum result (Fig. 1) clearly reveals that the periodic dependence is due to multi maxima ionization.

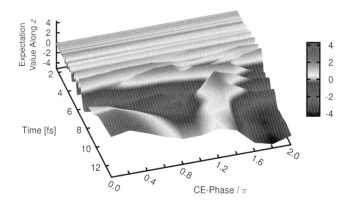

Fig. 1. Expectation value of the electron along the nuclear bond axis z as a function of time and the carrier envelope phase.

Conclusions

We propose a new ansatz to calculate the dynamics of electrons coupled to nuclear dynamics. The new method was tested successfully for the photodissociation of D_2^+. The advantage of our ansatz originates from the exploitation of the highly developed quantum chemical methods and nuclear quantum dynamics. Therefore it can be extended straightforwardly to multi-electron systems and multi-configurational theory. First calculations were performed for the photodissociation of CO^+ molecule, which affords the treatment of 13 electrons of which five actively participate in the electron dynamics.

Acknowledgements. The authors like to acknowledge the DFG for financial support through the Munich Center of Advanced Photonics (MAP).

1 M. Hentschel, R. Kienberger, C. Spielmann, G.A. Reider, N. Milosevic, T. Brabec, P. Corkum, U. Heinzmann, M. Drescher and F. Krausz Nature 414, 509 (2001).

2 T. Klamroth Phys. Rev. B 68, 245421 (2003).

3 N. Rohringer, A. Gordon and R. Santra Phys. Rev. A 74, 043420 (2006)..

4 J. Zanghellini, M. Kitzler, T. Brabec and A. Scrinzi J. Phys. B 37, 763 (2004).

5 A. I. Kule, J. Breidbach and L. S. Cederbaum J. Chem. Phys. 123, 044111 (2005).

6 J. Levesque, S. Chelkowski and A. Bandrauk J. Mod. Opt. 50, 497 (2003)..

7 X. M. Tong and C. D. Lin Phys. Rev. Lett. 98, 123002 (2007).

8   M. F. Kling, C. Siedschlag, A. J. Verhoef, J. I. Khan, M. Schultze, T.Uphues, Y. Ni, M. Uiberacker, M. Drescher, F. Krausz and M. J. J. Vrakking Science 312, 246 (2006)

9 H. J. Werner, P. J. Knowles, R. Lindh, F. R. Manby, M. Schütz et al. (2006) Molpro, version 2006.1, a package of ab initio programs.

Attosecond Photoelectron Spectroscopy of Electron Tunneling in Dissociating Hydrogen Molecular Ion

Stefanie Gräfe[1,2] , Volker Engel[3] , Misha Yu. Ivanov[1]

[1]Steacie Institute for Molecular Sciences, National Research Council, 100 Sussex Drive, Ottawa ON K1A0R6
[2]Institute for Theoretical Physics, Vienna Technical University, Wiedner Hauptstr. 8-10/E136, A-1040 Wien, Austria
[3]Institute for Physical Chemistry, University Würzburg, Am Hubland, 97074 Würzburg, Germany
Email: Steffi.Graefe@nrc.ca

Abstract: We demonstrate the potential of intense-field pump-probe (attosecond XUV) photoelectron spectroscopy to monitor coupled nuclear-electronic tunneling dynamics between the two protons during dissociative ionization of the hydrogen molecular ion.

Time-resolved pump-probe experiments are very versatile tools allowing us to monitor fundamental dynamic processes in atoms, molecules and clusters. Physically, time-resolved experiments require techniques and (ultra) short light pulses which enable us to make 'snapshots' on the same time scale as the dynamics to be observed. Applying femtosecond laser pulses, nuclear dynamics can be monitored, opening the door to the new field of 'femtochemistry', making the real-time observation of chemical bond breaking possible [1]. Recently, isolated pulses as short as 130 attoseconds have been generated [2]. This allows observing processes on the sub-femtosecond timescale, thus making possible the real-time observation of electronic dynamics.

We show that pump-probe ionization spectroscopy, a standard tool in femtosecond spectroscopy, can be applied to resolve processes on the attosecond timescale. Ionization of the neutral hydrogen molecule and subsequent excitation and dissociation of the molecular ion are induced by an intense few-cycle infrared pulse, which acts as a pump. A time delayed attosecond XUV pulse ionizes H_2^+, therefore probing the coupled nuclear-electronic dynamics. Although attosecond pulses are spectrally very broad (several eV), it was shown that information about the electronic state can be extracted [3,4]. We demonstrate that information about electronic motion can be extracted by monitoring asymmetries in the photoelectron distribution. We give a theoretical description for calculating time-resolved spectra in the presence of a strong laser field.

We consider a setup as shown in Fig.1: an intense few-cycle linearly polarized IR pump pulse ionizes H_2, producing the hydrogen molecular ion ($^2\Sigma_g^+$) and a correlated free electron. The free electron follows the optical oscillations of the driving field and recollides with the parent ion, inducing excitation onto the dissociative electronic state $^2\Sigma_u^+$. During the dissociation the bound electron tunnels between the two nuclei and – controlled by the carrier-envelope phase of the driving IR field – localizes on one of the nuclei [5-9].

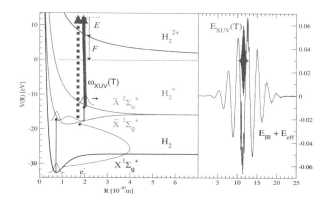

Figure 1: Excitation scheme in H_2^+. After strong field ionization, a vibrational wavepacket in the ground state of the molecular ion (H_2^+) is created. Following the optical oscillations of the IR field (800 nm, 5fs, 1×10^{14} W/cm^2), the electron may recollide and, together with the driving IR field, induces population transfer (recollisional excitation) into the *ungerade* state. The coherent wavepacket dynamics in H_2^+ is monitored by an attosecond XUV probe pulse (15 nm, 350 as), inducing a wavepacket in H_2^{++}. Measuring the photoelectron kinetic energy distribution E as a function of the XUV arrival time T gives information about the nuclear and electronic dynamics in the probed state in H_2^+.

The XUV probe pulse ionizes the system, transferring the electron during its tunneling motion into the continuum. The electron keeps the momentum and upon ionization will be released preferably in the direction of its momentum. Due to the vector potential of the intense IR field, the final momentum of the electron we will measure is distorted and shows the streaking features [10]. Calculating the photoelectron distribution on a detector placed to the left and to the right of the polarization axis shows asymmetries in the electron distribution: for certain delay times and energies, more electrons fly to the right (left) detector, depending on the direction the electron was tunneling to. Additionally, the photoelectron kinetic energy distribution as a function of delay time directly reflects the nuclear motion [11] so that we besides temporal resolution of the electron tunneling dynamics as well obtain spatial information of the nuclear dynamics.

To model the ionization of H_2, occurring in several ionization bursts centered around the maxima of the IR field (800 nm, 5fs, 1×10^{14} W/cm^2), we apply our recently developed approach based on a time-dependent interaction potential tracking the correlated motion of two electrons and a nuclear coordinate in an intense laser field [8]. We then model the molecular ion, confining H_2^+ to the two potential surfaces $^2\Sigma_g^+$ und $^2\Sigma_u^+$, applying the Born-Oppenheimer approximation. Ionization by the weak attosecond XUV is treated perturbatively.

As the two electronic states $^2\Sigma_g^+$ and $^2\Sigma_u^+$ have different symmetry they coherently add up to continuum states with the same energy but different symmetry. Those electronic states with even symmetry will only have a nonzero overlap with the odd continuum functions, and vice versa. This symmetry consideration is true not only in the plane-wave approximation for continuum states. Note, that these transition dipoles are $\pi/2$ out of phase. Using the transition dipoles, the ionic wave function for positive and negative momenta p_E are calculated separately. The different symmetry of the transition dipoles induces interferences in the ionic wave function. These interferences can be seen in the angular distribution, i.e. in one dimension in asymmetries in the left/right direction. Then, the photoelectron spectra on the left and on the right detector can be calculated, as well as the difference between them – the asymmetry. These asymmetry is displayed in Figure 2, showing rich structure which can be directly related to the electron tunneling dynamics.

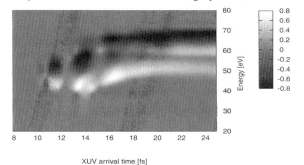

XUV arrival time [fs]

Figure 2: Asymmetric distribution of photoelectrons as a function of the XUV (15 nm, 350 as) arrival time T probing the nuclear and electronic dynamics in H_2^+: Positive values denote more electrons flying to the left detector, for negative values more electrons fly to the right detector, respectively.

1. Gruebele, M., A. H. Zewail. "Femtosecond wave packet spectroscopy: Coherences, the potential, and structural determination." *J. Chem. Phys.* **98**, 883 (1993).

2. Sansone, G., E. Bendetti, F. Calegari, C. Vozzi, L. Avaldi. "Isolated single-cycle attosecond pulses." *Science* **314**, 443 (2006).

3. Yudin, G. L., A. D. Bandrauk, P. B. Corkum. "Chirped Attosecond Photoelectron Spectroscopy." *Phys. Rev. Lett.* **96**, 063002 (2006).

4. Smirnova, O., A. S. Mouritzen, S. Patchkovskii, M. Yu. Ivanov. "Coulomb-laser coupling in laser-assisted photoionization and molecular tomography." *J. Phys. B* **40**, F197 (2007).

5. Kling, M. F., Ch. Siedschlag, A. J. Verhoef, J. I. Khan, M. Schultze, Th. Uphues, Y. Ni, M. Uiberacker, M. Drescher, F. Krausz, M. J. J. Vrakking. "Control of electron localization in molecular dissociation." *Science* **312**, 246 (2006).

6. Haljan, P., M. Yu. Ivanov, P. B. Corkum. "Laser Control of Electron Localization in Molecules and Double Quantum Wells." *Laser Phys.* **7**, 839 (1997).

7. Tong, X. M., S. Watahiki, K. Hino, N. Toshima. "Numerical observation of the rescattering wave packet in laser-atom interactions." *Phys. Rev. Lett.* **99**, 093001 (2007).

8. Gräfe, S., M. Yu. Ivanov. "Effective fields in laser-driven electron recollision and charge localization." *Phys. Rev. Lett.* **99**, 163603 (2007).

9. He, F., C. Ruiz, A. Becker. "Control of electron excitation and localization in the dissociation of H_2^+ and its isotopes using two sequential ultrashort laser pulses."*Phys. Rev. Lett.* **99**, 083002 (2007).

10. Itatani, J., F. Quéré, G. L. Yudin, M. Yu. Ivanov, F. Krausz, P. B. Corkum. "Attosecond Streak Camera."*Phys. Rev. Lett.* **88**, 173903 (2002).

11. Braun, M., C. Meier, V. Engel. "The reflection of predissociation dynamics in pump/probe photoelectron distributions." *J. Chem. Phys.* **105**, 530 (1996).

Acknowledgement: S. G. is funded by the Deutsche Akademie der Naturforscher Leopoldina, award No. BMBF-LPD 9901/8-139. M.I. acknowledges support of the NSERC SRO grant and the Bessel prize of the A.v. Humboldt foundation.

Attosecond angular streaking: an ideal technique to measure an electron tunneling time?

Petrissa Eckle[1], Adrian Pfeiffer[1], Claudio Cirelli[1], André Staudte[2], Reinhard Dörner[3], Harm Geert Muller[4], Markus Büttiker[5], Ursula Keller[1]

[1] Physics Department, ETH Zurich, 8093 Zürich, Switzerland
[2] Steacie Institute for Molecular Sciences, National Research Council of Canada, Ottawa, Ontario K1A 0R6, Canada
[3] Institut für Kernphysik, Johann Wolfgang Goethe Universität, 60438 Frankfurt am Main, Germany
[4] FOM Institute for Atomic and Molecular Physics, Kruislaan 407, 1098 SJ Amsterdam, The Netherlands
[5] Physics Department, University of Geneva, 1211 Geneva, Switzerland

Abstract. We used attosecond angular streaking to measure attosecond ionization dynamics in the non-adiabatic tunneling regime of helium using slightly elliptically polarized 5.9 fs pulses with a peak intensity ranging from 2.3 to 3.5 x 10^{14} W/cm^2 (corresponding to a Keldysh parameter variation of 1.45 to 1.17). With our technique we could demonstrate intensity-independent "instantaneous" ionization with an accuracy of 50 as. Numerical simulations based on the time-dependent Schrödinger equation confirm such ionization behavior with no distinct electron wave packets. This implies that we would not expect a tunneling time or multi-photon ionization delay in the ionization dynamics.

Introduction

The tunneling process is one of the most fundamental quantum phenomena. Since the early days it was asked whether tunneling takes a real time or is instantaneous[1]. A number of approaches have been applied to discuss this question: the two principal ones are the Wigner-Eisenbud-Smith time delay[2] and the Buttiker-Landauer traversal time for tunneling[3]. In the Wigner-Eisenbud approach one considers a wave packet and follows its peak. In the classical allowed region this leads to the familiar group velocity. In the tunneling regime the Wigner-Eisenbud time can be much shorter than the propagation of light through the same distance, and if taken at face value gives superluminal tunneling times. This is rather similar to superluminal group velocities discussed in regions of anomalous dispersion. Such a short time delay can in an experiment effectively appear as zero. Several objections are possible against the Wigner-Eisenbud approach: first, there is no conservation law in physics for peaks of wave packets, second the fact that we obtain superluminal answers makes it obvious that we are not discussing a causal process. The answer of Brillouin and Sommerfeld to superluminal velocities in regions of negative dispersion was to characterize pulse propagation by a signal velocity, which is always limited by the velocity of light[4]. The Buttiker-Landauer traversal time for tunneling is similarly an effort to obtain a more obvious physical answer for the speed of the tunneling process.

We have demonstrated a new technique, termed attosecond angular streaking[5] that allows us to investigate sub-100-attosecond ionization dynamics of helium atoms by an intense near infrared pulse in the non-adiabatic tunneling regime[6]. The helium atom was ionized with a slightly elliptically polarized pulse with a pulse duration in the two optical cycle regime at a center wavelength of 725 nm. We wanted to address the question if the ionization rate really follows the electric field instantaneously or not. Assuming there would be a real time associated with this tunneling ionization process then we should be able to measure such a *delay time* Δt_D with a possible delay between the maximum electric field, which induces the ionization process, and the moment of appearance of the electron after ionization. With a small ellipticity in the angular streaking field we were able to determine "time zero" with very high accuracy for the first time. We define "time zero" as the time or angular position of the maximum electric field that induces tunneling ionization. As we will show we do not detect an intensity-dependent delay time Δt_D with an accuracy of 50 as.

In tunneling ionization, we can distinguish between two regimes with the Keldysh parameter $\gamma = \omega_0 \sqrt{2 I_p / I}$, where I_p is the ionization potential and I the intensity of the infrared field. For $\gamma \gg 1$ the ionization is expected to be dominated by multi-photon ionization (MPI) and for $\gamma \ll 1$ by tunneling ionization. The intermediate regime was referred to as the regime of non-adiabatic tunneling[6]. A simple tunneling time was derived for this regime[6], which is similar to the *traversal time for tunneling* Δt_T of Ref. 3 given by $\gamma = \omega_0 \Delta t_T$ (in atomic units), where ω_0 is the center laser angular frequency.

Note, that this traversal time for tunneling is not the same as the tunneling delay time Δt_D discussed above. We have experimentally explored this regime with a Keldysh parameter γ ranging from 1.17 to 1.45, which would correspond to a traversal time for tunneling between 450 to 560 as, well within our measurement accuracy.

Experimental part

Here we present a technically very simple method for achieving absolute time measurement with angular streaking: (i) no measurement and stabilization of the carrier-envelope offset phase is necessary, and (ii) a small well characterized and selected ellipticity, that is always present for broadband short pulses, does not distort the experiment but rather allows for very accurate "time zero" calibration. The main idea is that the main axis of the E-field ellipse provides a fixed reference point in time. The comparison of the main ellipse axis to the momentum distribution peak of the ionized He atoms then directly yields the delay time Δt_D.

For our experiment we have used linearly polarized 5.5 fs pulses produced by a Ti:Sapphire based laser system and two-stage filament compression[7]. A broadband $\lambda/4$-wave plate was used to generate elliptical polarization. Attosecond angular streaking is very well described by a semi-classical model where an ADK rate is used for the tunnel ionization followed by classical propagation in the angular streaking field[5]. Fig. 1a shows the measured streaking angle of the ionized He atoms as a function of ellipticity. This semi-classical model describes the measurements very well (with an accuracy of 50 as) and a systematic offset can be explained by the residual Coulomb potential effects during the streaking process. For Fig. 1b the ellipticity was fixed at $\varepsilon = 0.88$ and the ionization time was compared to the maximum position of the electric field. The ionization time was determined by the measured streaking angle minus the propagation within the infrared field (which is close to 90 degrees as shown in Fig. 1a). We varied the peak intensity from 2.4 to 3.3 x 10^{14} W/cm^2. Fig. 1b clearly shows that we do not observe any significant intensity dependence on the delay time Δt_D and that a possible Δt_D would be smaller than 50 as.

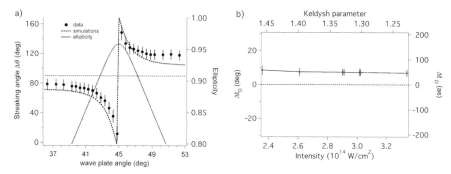

Fig. 1: (a) Streaking angle as a function of $\lambda/4$-wave plate angle (i.e. ellipticity). The semiclassical simulations (dashed line) reproduce the data with an accuracy of ≈ 7.5 degrees corresponding to 50 as. The laser peak intensity is constant over the ellipticity scan at $\approx 2 \times 10^{14}$ W/cm^2. The systematic offset can be partially explained by the residual Coulomb interaction with the escaping electron, which is $\approx 5°$. (b) Measured delay time Δt_D.

Conclusion and Outlook

With our measurement we can put an upper limit of 50 as on a tunneling delay time Δt_D. Under the current condition the Buttiker-Landauer traversal time Δt_T was predicted to range between 450 to 560 as. In the Buttiker-Landauer approach the tunneling process is probed by an oscillating barrier. This leads to the distinction of an adiabatic regime, where the tunneling barrier oscillates slowly compared with the tunneling process and a regime where the oscillation is fast. A traversal time for tunneling is obtained by considering the crossover between these two regimes. For linearly polarized light the Buttiker-Landauer time is directly related with the adiabaticity parameter of Keldysh. In the present experiment the Keldysh parameter varies only between 1.17 and 1.45, which is a too narrow range to observe a crossover. Still there is the interesting question, whether there are other features in the data of the experiment, which could be used to identify a crossover. The main point we wish to make is that these two approaches consider entirely different aspects of the tunneling process. As a consequence a finite traversal time is not in contradiction to zero delay time statement based on a Wigner-Eisenbud phase delay analysis.

1   L. A. MacColl, Phys. Rev. **40**, 621, (1932)

2   E. P. Wigner, Phys. Rev. **98**, 145, (1955)

3   M. Büttiker and R. Landauer, Phys. Rev. Lett. **49**, 1739 (1982).

4   A. Sommerfeld, Physikalische Zeitschrift **8**, 805 (1908).

5   P. Eckle et al., Nature Physics, Published online: 30 May 2008, doi:10.1038/nphys982

6   G. L. Yudin and M. Y. Ivanov, Phys. Rev. A **64**, 013409 (2001).

7   C. P. Hauri et al., Appl. Phys. B **79**, 673 (2004).

Probing Dynamics in Polyatomic Molecules Using High Harmonic Generation: the Role of Ionization Continua

Wen Li[1], Xibin Zhou[1], Robynne Lock[1], Serguei Patchkovskii[2], Olga Smirnova[2], Albert Stolow[2], Margaret Murnane[1] and Henry Kapteyn[1]

[1] JILA and Dept. of Physics, University of Colorado, Boulder, CO, 80309, USA
  E-mail: wli@jila.colorado.edu
[2] Steacie Institute of Molecular Sciences, National Research Council, Ottawa, ON, K1A0R6, Canada

Abstract. We show that to understand harmonic generation from molecules undergoing a large change in configuration, multiple ionization continua must be considered. After exciting large amplitude vibrations in an N_2O_4 dimer, bright bursts of harmonics are emitted at the outer turning point of the vibration.

Introduction

An exciting frontier in strong-field science is to exploit electron recollisions as a probe of molecular structure and dynamics. In high harmonic generation (HHG), an electron is plucked from a molecule, accelerated in the laser field, before recombining with the same molecule a fraction of an optical cycle later while emitting a coherent x-ray. The energy gained by the electron in the laser field can reach tens to hundreds of eV, corresponding to a characteristic electron deBroglie wavelength of ~1Å. X-ray harmonics from molecules are thus very sensitive to the orientation, structure, and dynamic motion of the electrons and atoms in a molecule. To date, this technique has been used to map the static valence electron orbital in a rotationally-excited diatomic molecule, and to observe small-amplitude vibrational dynamics in molecules [1,2].

Vibrational excitation can modulate the nuclear configuration of molecules and thus dynamically change the electronic wave function. Because HHG is mainly an electronic process, more information about the internal structure of a molecule is expected by observing HHG from vibrationally excited molecules. In this work, we observe a very large modulation in the HHG yield from a vibrationally excited molecule, N_2O_4. We explain this large modulation as a result of the large amplitude oscillation of the dimer. At the inner turning point, tunnel ionization favors the first electronically excited state of the ion, which is "dark" for harmonic emission. At the outer point, the "bright" ground state dominates. This work uncovers new understanding of how molecular structure and dynamics can influence HHG.

Experimental Methods

In this experiment, ultrashort laser pulses from a Ti: Sapphire laser system (~30fs, 4mJ, 1 KHz) are split into pump (25%) and probe (75%) beams. The pump beam is focused onto a N_2O_4 gas jet at an incident intensity of 5×10^{13} Wcm^{-2} to impulsively excite a vibrational wavepacket in the sample by impulsive stimulated Raman scattering (ISRS). The probe beam is focused to ~2×10^{14} Wcm^{-2}, and generates high harmonics from these molecules. The relative timing between the pump and probe is controlled by an optical delay stage. The generated harmonics are spectrally separated using an EUV grating, and imaged onto a EUV-sensitive CCD camera (Andor Technology) at each time delay.

Results and Discussion

Fig. 1(a) shows raw harmonic intensity data for orders 19–27, where a time-dependent oscillation of the HHG yield is immediately apparent. Fig. 1(b) shows the integrated intensity of the 21st harmonic as a function of time. The time dependence of the HHG emission for all orders shows several common characteristics: a sudden drop in yield when initially impulsively excited at time-zero, followed by subsequent strong oscillations imposed on a slowly rising baseline. The strong suppression of HHG at time-zero is likely a result of direct interference of the pump and probe pulses, while the drop and subsequent recovery of the HHG is likely due to an induced rotational alignment followed by dephasing. The origin of the strong oscillations becomes apparent if we apply a discrete Fourier transformation to the data. The signal has one clear frequency, at 255 cm^{-1} (T=130 fs) (see Fig. 1(b) inset) corresponding to the N-N stretch mode of the N_2O_4. Given the ISRS excitation mechanism, the outer turning point should occur 1/4 period after the pump pulse. This timing was confirmed by accurately locating the initial peak in HHG emission at 170±10 fs—in good agreement with T+T/4 for the N-N stretch mode of N_2O_4.

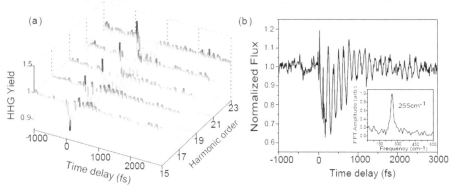

Fig. 1. (a) Harmonic yield vs. pump-probe delay (b) Lineout of the 21st harmonic. The inset shows the Fourier transform.

Using the three-step model, we model ionization and recombination in N_2O_4. For the ionization step, we first performed an *ab initio* calculation of both the neutral and cation ground states. The results (Fig. 2(a)) show that for N-N bond lengths below 1.7Å, structural factors favor ionization to the B2g first excited state of the ion. At longer bond lengths, ionization to the Ag ground state of the ion is favored.

Next we consider the recombination step. First we calculate the Dyson orbital for ionization from the ground state neutral to Ag /B2g cation states. The amplitude of the ionization/recombination process within a given ionization channel is determined by the 1-particle dipole matrix element between the Dyson orbital and the continuum wave function at a given energy. In Fig. 2(b), we show predictions for the Ag ion state. For low order harmonics (H19) the HHG yield is higher for the longer N-N bond lengths. The calculation correctly predicts the harmonic order dependence of the modulation depth, which is larger for lower order harmonics. The recombination dipoles for B2g cation states are zero at all bond lengths due to symmetry. Therefore, even though B2g states compete with Ag states at the inner turning point, they do not emit harmonics and thus the total harmonic yield is reduce at short bond lengths.

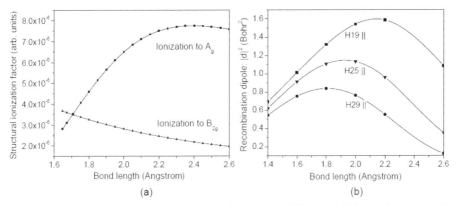

Fig. 2. (a) Structural Ionization factor calculated at different N-N bond lengths (b) Recombination dipoles calculated at different bond length for harmonic 19, 25 and 29.

Conclusions

In conclusion, we observe large modulated HHG signals from the vibrationally excited N_2O_4, where harmonics are emitted primarily from the outer turning point of the vibration. We explain this observation as due to the cation electronic structure that dynamically changes and influences both the ionization and recombination steps. Thus, to understand HHG from molecules during a chemical reaction or other large change in configuration, multiple ionization continua must be considered.

Acknowledgements. The authors gratefully acknowledge support from the Department of Energy and from the National Science Foundation.

1 J. Itatani, D. Zeidler, J. Levesque, D. M. Villeneuve, and P. B. Corkum, Phys. Rev. Lett., Vol. 94, 123902, 2005.
2 N. L. Wagner, A. Wuest, I. P. Christov, T. Popmintchev, X. B. Zhou, M. M.   Murnane, H. C. Kapteyn, Proc. Natl. Acad. Sci., Vol. 103, 13279, 2006.

High harmonic generation from multiple molecular orbitals of N_2

Markus Gühr[1], Brian K. McFarland[1], Joseph P. Farrell[1] and Philip H. Bucksbaum[1]

[1] Stanford PULSE Institute, Via Pueblo Mall, Stanford CA 94305, USA
E-mail: mguehr@stanford.edu

Abstract. We observe the contribution of the HOMO and HOMO-1 orbitals in high harmonics from N_2 and discuss the harmonic modulation in the rotational revivals.

Introduction

High harmonic generation (HHG) proceeds in three steps: (1), A part of the electron wave function tunnels out of the valence orbital; (2) The liberated electron wave packet accelerates in the laser field; (3) The electron wave packet coherently recombines with the initially ionized orbital [1]. For molecules, the highest occupied molecular orbital (HOMO) is generally thought to be responsible for ionization and recombination. The spectrum has been transformed to produce an image of the HOMO's σ_g electronic structure [2]. We have obtained new experimental evidence that other orbitals apart from the HOMO, namely the HOMO-1 with its π_u symmetry, contribute to HHG. We reproduce the experimental features with semi-classical simulations of the recombination process on the HOMO and HOMO-1. This opens the route to imaging coherent superpositions of electronic orbitals.

Experimental Methods

An ultrashort laser pulse from a Ti:Sapphire amplifier interacts with the anisotropic polarizability of N_2 molecules in a supersonic jet and creates a rotational wave packet that results in molecular alignment at wavepacket revivals [3]. We concentrate on the half rotational revival in Fig 1. At 4.1 ps after the alignment pulse, the molecular ensemble is aligned along the pulse polarization with a prolate distribution, followed by an oblate distribution at 4.3 ps. A second laser pulse with a variable intensity from 1.6 to 2.3×10^{14} W/cm^2 generates high harmonics in the aligned molecules. Since the signal we attribute to the HOMO-1 is most pronounced under 90 degrees relative polarization of alignment and generation pulse, we concentrate on this configuration. The HHG beam passes through a thin Al filter and aperture which reject the fundamental as well as harmonics from long electron trajectories. The remaining harmonics are dispersed by a flat-field grating and detected by a MCP-phosphor screen unit. The HHG signal as a function of delay between alignment and harmonic generating pulse is given in Fig. 1 for a selection of harmonics from 15 to 39.

Results and Discussion

Figure 1 b) shows the harmonic signal for perpendicular polarizations of alignment and high harmonic generating laser pulses for a harmonic generation intensity of 2.3×10^{14} W/cm^2. As seen in the inset Fig. 1a), the molecular axes are preferentially perpendicular with respect to the harmonic generating polarization at 4.1 ps and a decrease of the signal at harmonic 15 is observed compared to the unaligned case prior to 3.6 ps. At 4.3 ps the molecular axes are partially parallel to the harmonic generating polarization and an increase of the harmonic radiation compared to the unaligned case is observed on harmonic 15. However, as the harmonic number increases above number 25, a peak grows out of the minimum at 4.1 ps. At the 39th harmonic the peak is most pronounced and the temporal modulation compared to harmonic 15 is inverted. For lower intensities of 1.9 and 1.6×10^{14} W/cm² (c and d), the single peak at 4.1 ps is visible, however at lower harmonics. The peak at 4.1 ps is always most pronounced in the cutoff region.

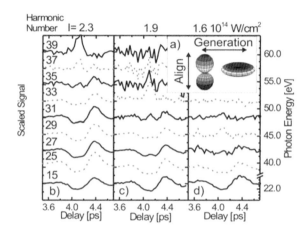

Fig. 1. a) At 4.1 ps a prolate distribution of molecular axes is created standing orthogonal to the generation polarization, whereas at 4.3 ps an oblate ensemble is generated, which has axes partwise parallel to the generation polarization. The alignment polarization is perpendicular to the generation polarization. b)-d) For harmonic 15, a modulation by a minimum at 4.1 ps followed by a maximum at 4.3 ps is visible. At higher harmonics, the signal modulation is reversed. We attribute the peak visible at 4.1 ps for the higher harmonics to ionization and recombination to the HOMO-1.

We interpret the data in terms of tunnel ionization and recombination for the molecular orbitals in Fig. 2 a) and b). For tunnel ionization the magnitude of the molecular wave function far away from the nuclei and in direction of the generation polarization is important. Thus, the σ_g HOMO orbital ionizes more easily if the generation polarization is parallel to the internuclear axis as compared to the perpendicular case. Also the recombination dipole is larger if the molecular axis is aligned with the generation polarization, since the recombination dipole has a larger amplitude in the long direction of the orbital. For the HOMO, ionization and recombination maximize the signal for molecules standing parallel to the HHG polarization and minimize the signal in the perpendicular configuration explaining the observed experimental trend in harmonics 15-25 in Fig. 1 b)-d).

The peak growing out of the minimum at harmonic 25-39 can be explained by reversed conditions for effective and ineffective HHG. The HOMO-1 fulfills exactly those requirements. Tunnelling ionization perpendicular to the molecular axis is higher for the HOMO-1 compared to the HOMO because it extends farther out (Fig. 2b). The ionization and recombination in the direction of the internuclear axis gives rise to a vanishing dipole, because of the opposite sign of the wave function on either side of the axis. In contrast, the recombination dipole perpendicular to the axis is strong.

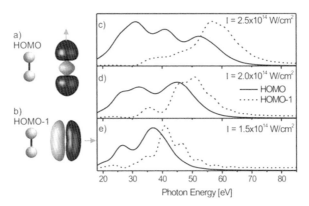

Fig. 2. a) -b) Shape of the HOMO and HOMO-1 with respect to internuclear axis. Directions of strong ionization and recombination are indicated by the arrow. c)-e) Simulation of the harmonic spectra for different intensities by recombination of a free electron wave with the HOMO (solid) and HOMO-1 (dashed).

To support our ideas we simulated the harmonic spectra using the recombination of a free electron wave with the orbitals shown in Fig. 2a) and b), assuming the prolate molecular distribution at 4.1 ps. The recombination of the accelerated free electron wave to the HOMO yields the black spectrum in Fig. 2 c)-d), whereas the HOMO-1 yields the grey spectrum. Destructive and constructive interferences during recombination modulate the spectra. Thus the HOMO dominates in the low harmonics, whereas the cutoff region is dominated by the HOMO-1. This trend is consistent for all intensities in agreement with the experiment.

Conclusions

The harmonic signal of aligned N_2 molecules shows an inversion for harmonics in the cutoff. The features are attributed to the HOMO-1 by qualitative arguments for tunnelling ionization and simulations of the recombination process.

Acknowledgements. We would like to thank H. Merdji and O. Smirnova for discussions. This program was funded by the US Deprtment of Energy within the Stanford PULSE Institute, M.G. thanks the Humboldt Foundation for a research grant.

1   P. Corkum, Phys. Rev. Lett. **71**, 1994-1997, 1993 and J. L. Krause et al., Phys. Rev. Lett. **68**, 3535-3538, 1992
2   J. Itatani et al., Nature **432**, 867-871, 2004
3   H. Stapelfeldt and T. Seideman, Rev. Mod. Phys. **75**, 543-557, 2003

Ultrafast Multiphoton Crystallography

Marina Gertsvolf[1,2], Hubert Jean-Ruel[2], Pattathil P. Rajeev[2], Dennis Klug[2], David M. Rayner[2] and Paul B. Corkum[1,2]

[1] University of Ottawa, Ottawa, Ontario, K1N 6N5, Canada
[2] National Research Council of Canada, Ottawa, Ontario, K1A 0R6, Canada
   E-mail: marina.gertsvolf@nrc.ca

Abstract. We show that non-resonant multiphoton ionization of dielectric crystals depends on the alignment of the laser field to the crystal lattice. Through absorption measurements we probe the local symmetry non-invasively, anywhere inside the sample.

Introduction

For intense laser pulses, multiphoton ionization (MPI) rates in molecules depend on the shape of the highest occupied molecular orbital and on its alignment with respect to the electric field [1]. Experiments in the gas phase with aligned N_2 and O_2 molecules showed differences of up to 50% and more in the ionization yield depending on the orientation of the linearly polarized laser electric field [2]. Transparent dielectrics have band gaps comparable to the ionization potential of such molecules and therefore should exhibit similar MPI behavior. If the MPI rate depends on lattice structure and its symmetry, then we have a structural probe of crystalline dielectrics that can follow melting with femtosecond precision.

We realize multiphoton crystallography by measuring the alignment dependence of the transmitted light. We show that the MPI probability depends on crystal symmetry. Since MPI confines the interaction to a small region of the laser focus [3] this implies that we have a highly local probe of crystalline structure, anywhere in 3- dimensional space. By using an optically contacted interface between fused and crystalline quartz, we confirm the local nature of the interaction. If multiphoton crystallography is to serve as a probe of crystalline structure, then it must not significantly contribute to structural changes itself. Our calculations [4] show that the temperature rise induced by the probe is proportional to the pulse duration and is only 12K for a 5fs pulse at 10% absorption. We show experimentally that probing quartz structure using 45fs pulse does not affect the crystal.

Results and Discussion

All of the crystals we have studied were z-cut to avoid birefringence. We used linearly polarized 50fs, 800nm laser pulses with a repetition rate of 40Hz to 400Hz. A microscope objective with numerical aperture of 0.25 focused the laser beam. This tight focusing allows us to reach high intensities without propagation being influenced by low order effects such as self-focusing or self-phase modulation. In our experiment the pulse energies used were in the range of 20nJ to 300nJ to give intensities between 0.5×10^{13} W cm^{-2} to 7×10^{13} W cm^{-2} if the focus were in vacuum.

The crystals were placed on a translation stage to provide new material at every laser shot if required. A half wave plate (HWP) was placed in the incident laser beam to control its polarization orientation relative to the lattice. All transmitted light was collected by an integrating sphere and the signal was acquired independently for every laser shot. We measured the alignment dependence of the transmission by rotating the HWP. Signal modulation frequencies were obtained by discrete Fourier transform (DFT) analysis. The measured DFT spectrum is presented in Figure 1. (top). All crystal samples show strong DFT features, several orders of magnitude over background noise. α-Quartz (α-SiO$_2$) and sapphire (α-Al$_2$O$_3$) both show modulation with π and $\pi/3$ rotational periods. The $\pi/3$ corresponds to their trigonal lattice system. Lithium fluoride (LiF) has a cubic unit cell and shows strong $\pi/2$ periodicity. It is clear that the absorption of intense femtosecond pulses measures the symmetry of the lattice (Figure 1. (bottom)). The π modulation in α-SiO$_2$ and α-Al$_2$O$_3$ is due to birefringence experienced by the peripheral beams. Notably this modulation is abcent from non-birefringent LiF.

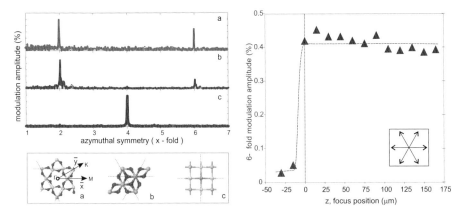

Fig. 1.(left) (a) α-Quartz; (b) Sapphire; (c) LiF. (top) Non-linear transmission modulation spectrum (DFT) for different glasses. The vertical scale is 0 – 0.75% for (a,b) abd 0 – 1.5% for (c). (bottom) Crystal structures (z-cut view) [5]. (right) Absorption measurements in optically bonded fused silica and quartz sample. 6-fold modulation intensities as a function of focusing position: negative values – in fused silica; positive – in quartz. Pulse energy 90nJ.

While crystal structures can be measured in many ways, measuring it with μm^3 spatial precision inside a material is unique. We use an optically bonded sample of fused and crystal SiO$_2$ and repeat the transmission experiments for different positions of the focus relative to fused silica - quartz interface. The magnitude of 6-fold peak Figure 1. (right) is zero before the interface when the absorption occurs in amorphous fused silica. When the focus approaches the interface, the crystalline structure shapes the absorption and the 6-fold peak rises sharply until the focus is fully inside the crystal, where it saturates.

All of the results presented so far were obtained using a fresh spot for every laser shot. We analyzed the transmission for quartz as a function of laser electric field angle, for different number of laser shots. The experiment was done at 20% absorption level using 45fs pulses, estimated to cause material heating by only 156K with every laser shot. We measured almost no changes in the magnitude of the 6-fold symmetry peak which implies that the crystal structure remains unchanged.

For crystals the reduced mass varies along different directions in the lattice and its

70

value can be derived from the electronic band structure. In α-quartz (Fig. 1(a) bottom) we calculated numerically the electronic energy levels and densities of states along different directions in the crystal and derived the reduces mass, using parabolic fitting. The band structure was obtained using the pseudopotential plane-wave method with the ABINIT code [7,8]. Trouiller-Martins-type [9], generalized gradient approximation pseudopotentials were employed with a Perdew-Burke-Ernzerhof exchange-correlation functional [10]. We found that for transitions between the HOMO and the LUMO, the reduced mass value changes from 0.93 m_e along ΓM, to 2.15 m_e along ΓK directions. For tunneling ionization the minimal band gap ΔE corresponds to Γ symmetry point and therefore is direction independant. Therefore, we expect the ionization to be the most probable along the ΓM direction where the reduced mass is the smallest. In our transmission experiments we found that absorption is at its maximum when the laser polarization is parallel to ΓM or x-axis, and at its minimum when the laser field is along ΓK (at 30deg to ΓM), in full agreement with the model.

Conclusions

Over the past few years research has shown that many aspects of multiphoton physics are sensitive to molecular structure. This presents exciting new possibilities for molecular imaging. We have shown that at least some of the structural sensitivity of strong field molecular science transfers to solids. We think strong field science also offers unusual new opportunities in material science.

We have shown that with a pulse as long as 45fs pulses with the multiphoton absorption we can test the structure of quartz non-distractively. A few cycle pulse would be less intrusive generating only about 10^{19} carriers/cm^3. At this pulse duration multiphoton crystallography can be applicable to biological molecules in condensed phase where in-vivo structural measurements are critically important. Finally, structural changes inside bulk (e.g. melting by superheating or non-thermal melting) can be measured by femtosecond infrared light in a pump-probe configuration. Multi-photon absorption and crystallography is an alternative to ultra-fast x-ray diffraction [6] offering $1\mu m^3$ spatial resolution.

MG acknowledges the travel support from the Ontario Centres of Excellence.

1   X. M. Tong, Z. X. Zhao, and C. D. Lin, Phys. Rev. A 66, 033402 (2002).
2   I. V. Litvinyuk, K. F. Lee, P. W. Dooley, D. M. Rayner, D. M. Villeneuve, and P. B. Corkum, Phys. Rev. Lett. 90, 233003 (2003).
3   W. Denk, J. H. Strickler, and W. W. Webb, Science 248, 73 (1990).
4   D. M. Rayner, A. Naumov, and P. B. Corkum, Opt. Exp. 13, 3208 (2005).
5   URL www.crystallography.net.; URL www.jmol.com.
6   C. W. Siders, A. Cavalleri, K. Sokolowski-Tinten, C. Toth, T. Guo, M. Kammler, M. Horn von Hoegen, K. R. Wilson, D. von der Linde, and C. P. J. Barty, Science 286, 1340 (1999).
7   X. Gonze, J. M. Beukena, R. Caracasa, F. Detrauxa, M. Fuchsa, G. M. Rignanesea, L. Sindica, M. Verstraetea, G. Zerahb, F. Jolletb, Comput. Materials Science 25, 478 (2002).
8   URL www.abinit.org.; URL http://www.abinit.org/.
9   N. Troullier and J. L. Martins, Phys. Rev. B 43, 1993 (1991).
10 J. P. Perdew, K. Burke, and M. Ernzerhof, Phys. Rev. Lett. 77, 3865 (1996).

Direct Measurement of Angle-Dependent Single Photon Ionization of N_2 and CO_2

I. Thomann[1], R. Lock[1], V. Sharma[1], E. Gagnon[1], S.T. Pratt[2], H. C. Kapteyn[1], M. M. Murnane[1], and W. Li[1]

[1] JILA and Department of Physics, University of Colorado and NIST, Boulder, Colorado 80309, USA
E-mail: wli@jila.colorado.edu
[2] Chemistry division, Argonne National Laboratory, Argonne, IL, 60439, USA

Abstract. We present a novel method for determining the angular dependence of molecular photoionization, by measuring time-dependent ionization yields from transiently aligned molecules. This method allows us to map the angular dependence of nondissociative single-photon ionization for molecules (N_2 and CO_2) for the first time.

Introduction

Knowledge of the angular dependence of molecular ionization cross sections is fundamental to basic molecular physics as well as being relevant to strong-field physics. However, the ability to directly measure the angular dependence of ionization in free molecules has been very limited. By combining intense femtosecond laser pulses to transiently align a molecular sample, with few-femtosecond high harmonic extreme-ultraviolet (EUV) pulses, a new capability arises for directly measuring the angular dependence of molecular single-photon ionization. We present the first experimental determination of the angular dependence of nondissociative single photon ionization. We use femtosecond EUV pulses to measure the yield of single-photon non-dissociative molecular ionization as a function of time (angle) as a molecule is undergoing field-free alignment. We present results for N_2 and CO_2.

Experimental Methods

A schematic of our experiment is shown in Fig. 1(a). The output of a Ti:Sapphire amplifier (2 mJ, 35 fs) is split into three beams. Approximately 1mJ of energy is used to generate high harmonics in an Argon filled hollow waveguide. Using two Mo/Si multilayer mirrors and Al filters, we preferentially select EUV light centered at ~ 43 eV (27th harmonic). The two remaining IR beams pass through variable delay lines. The first infrared (IR) beam is stretched to 140 fs and is used for impulsive excitation of the molecules to create a rotational wavepacket. This field-free alignment method leads to transient molecular alignment revival structures, determined by the rotational constant of the molecule. As a probe, we use either the second IR beam to strong-field ionize the sample, or the EUV beam to single-photon ionize the molecules. The pump and probe beams have the same polarization, and are focused into a molecular supersonic gas jet in a noncollinear geometry. The ions and electrons created through ionization are detected using a Coltrims (Cold target recoil ion momentum spectroscopy) chamber [1], which allows reconstruction of full momentum vectors of all charged particles resulting from the ionization events.

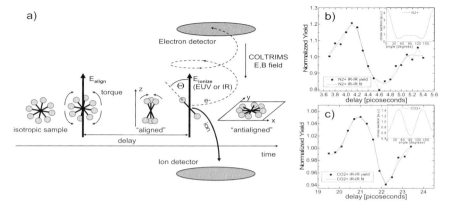

Fig. 1. (a) The alignment laser impulsively torques a molecular sample. Around the half-revival time, the sample changes from aligned to anti-aligned in time. A delayed probe pulse ionizes the sample, and ions and electrons are detected. (b), (c) Measured ion yield (symbols) and fit (lines) as a function of time delay between IR alignment and IR probe pulses. (b) N_2^+, (c) CO_2^+. Insets show the angular dependence of photoionization cross-sections extracted from data.

Results and Discussion

The ionization yield for N_2 and CO_2, as a function of time delay between the IR alignment pulse and the IR probe pulse, in the region around the 1/2 revival, are shown in Figs. 1(b) and (c). During this revival, the molecular axes distribution changes from aligned to antialigned. Information on the angular dependence of the molecular ionization cross section is obtained from the variation in ion yield with time delay. In the case of an ionization cross section that is peaked when the molecular axis is aligned parallel to the polarization of the ionizing laser, the yield rises for aligned samples and drops for antialigned samples.

To determine angular variations of the ionization cross section, the ion yield versus delay τ is fitted to $Yield(\tau) = \int A(\theta, \tau)\sigma(\theta)\sin(\theta)d\theta$. Here, $A(\theta, \tau)$ is the calculated [2] angular probability distribution of the molecules. We expand the angular dependence of the ionization cross section in terms of Legendre polynomials, $\sigma(\theta) = C(1 + \beta P_2(\cos(\theta)) + \gamma P_4(\cos(\theta)) + ...)$.

To test our method, we first studied strong-field ionization using an IR probe. Figures 1 (b) and (c) show the yield of N_2^+ and CO_2^+, respectively. For N_2, we find a cross-section strongly peaked in the case where the molecular axis is parallel to the electric field (inset Fig. 1(b)) and is suppressed nearly to zero away from zero angle. For CO_2, it is strongly suppressed around $0°$, peaks around $45°$, and is again suppressed for angles around $90°$ (inset Fig. 1(c)). All these features in both N_2^+ and CO_2^+ are in good qualitative agreement with recent work [3].

Next we study, for the first time to our knowledge, the ionization dipole moments for nondissociative single-photon ionization (N_2^+, CO_2^+) and also for a dissociative ionization channel ($O^+ + CO$) in which the axial recoil approximation is not fulfilled. These channels could not be measured using previous methods, which were limited to dissociative ionization fulfilling the axial recoil approximation (i.e. the dissociation into the ionic state is faster than the molecular rotation). We find that both N_2^+ and CO_2^+ show angle-dependent photoionization clearly peaked at 90° with respect to the molecular axis (see Fig 2.(a),(b)). This is in agreement with theory, which predicts a predominantly perpendicular transition for both N_2 and CO_2. Using an IR pump/EUV probe, we also measure the angle-dependent photoionization for the dissociative channels $N^+ + N$ and $O^+ + CO$. We find that both for N_2 and CO_2, the molecular dissociative channels exhibit parallel transitions.

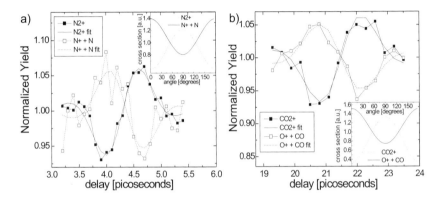

Fig. 2. EUV ionization yields from transiently aligned (a) N_2, (b) CO_2. Solid (hollow) symbols: nondissociative (dissociative) ionization channels; Lines: fits. Inset: angular dependent cross-sections extracted from data.

Conclusions

We have demonstrated a novel method for measuring the angular dependence of molecular photoionization[5]. This method does not require fragmentation of the molecule to determine the orientation at the time of ionization, and thus allows for the first time to measure the angular dependence of nondissociative ionization for N_2^+ and CO_2^+. In addition, this method allows determination of transition dipoles for dissociative ionizations in cases where the dissociation timescale is comparable to or longer than the molecular rotation period, as for the $O^+ + CO$ channel.

Acknowledgements. We thank C. La-O-Vorakiat and A. Sandhu for their assistance on the experiment. We gratefully acknowledge NSF and DOE for financial support.

1   E. Gagnon et al, Science **317**, 1374 (2007).
2   J. Ortigoso et al, J. Chem. Phys. **110**, 3870 (1999).
3   D. Pavicic et al, Phys. Rev. Lett. **98**, 243001 (2007).
4   F. LéPine et al, J Mod Opt. **54**, 953 (2007).
5   I. Thomann et al, submitted to JPC (2008).

Field-free unidirectional molecular rotation

Sharly Fleischer, I. Sh. Averbukh and Yehiam Prior
Department of Chemical Physics, Weizmann Institute of Science, Rehovot, Israel 76100
yehiam.prior@weizmann.ac.il, phone +972-8-934-4008, fax +972-8-934-4126

Abstract: By varying the polarization and delay between two ultrashort laser pulses, we control the plane, speed, and sense of molecular rotation. This control may be implemented to individual components within a molecular mixture.

The interest in laser induced field-free alignment and control of molecular rotation led to several applications which are based on the time-dependent alignment[1,2,3], where the ensemble averaged alignment factor $\langle \cos^2 \theta \rangle$ was the main observable. Preparation of molecules in a state of fast, unidirectional spinning is a much more complicated task that has been so far achieved via a sophisticated control over time-dependent field polarization[4]. Here we show that field-free, unidirectional molecular rotations can be induced by two properly polarized and time delayed ultrashort laser pulses.

An ultrashort, linearly polarized strong laser pulse, induces in linear molecules coherent molecular rotations which, due to quantum revivals, are manifested in a series of aligned ("cigar") and antialigned ("disk") angular distributions. (figure 1).

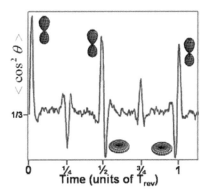

Figure 1: Alignment factor as a function of time measured in N_2 gas at room temperature. The two angular distributions ("Cigar" and "Disk" discussed above) are depicted near their corresponding peaks.

In order to induce unidirectional rotation we apply the first pulse polarized in z-direction, and let the molecular angular distribution evolve under field-free conditions towards an aligned "cigar" (or an anti-aligned "disk") state. At the time of alignment, the molecules are confined in a narrow cone around the polarization direction of the first pulse. At this moment, a second pulse, linearly polarized at a different angle, is applied, and thereby induces molecular rotation in the plane defined by the two polarization vectors.

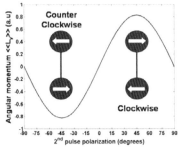

Figure 2: Mean angular momentum along the y-axis as a function of the polarization angle of the second pulse applied to the pre-aligned molecules ("Cigar" distribution) around the half rotational revival time (see fig. 1). The induced sense of rotation is depicted in the figure.

For a relative angle of +45 degrees between the two polarizations in the xz plane, the molecular ensemble rotates mostly clockwise (positive angular momentum), and the opposite is seen for the -45 degrees case. For a second pulse polarized parallel to the first one, no angular momentum along the y-axis is induced. In all cases, the mean angular momentum along the x-axis remains zero.

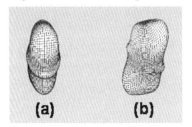

Figure 3: Calculation of $\left|\psi_{rot}\right|^2$ averaged of 1 T_{rev} of propagation. The anisotropic distribution is clearly seen from the comparison between the two view angles (90 degrees apart) (a) and (b).

As was mentioned above, molecules in this state are constrained to rotate in a plane, leading to highly anisotropic angular distribution. In figure 3 we observe the calculated $\left|\psi_{rot}\right|^2$ subjected to 2 pulses polarized 45 degrees at the right delay to induce unidirectional clockwise rotation, averaged over 1 revival time of propagation. This anisotropy in angular distribution is expected to yield anisotropic collisional cross section and corresponding anisotropic diffusion in space.

Up to this point we have considered the case where the second pulse, polarized at 45 degrees to the first pulse, is applied at the point of maximal alignment. If the second pulse is applied at the point of antialignment ("disk") its effect on the angular momentum gained by the molecular ensemble is opposite. This can be seen in figure 4, where the second pulse, polarized at $+45^0$ to the first one is applied at different delays. Since around half revival time (and full revival time) the molecules alternate between two dramatically different angular distributions, the effect of the second pulse is expected to depend on the exact delay.

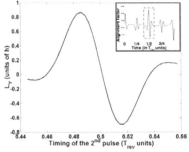

Figure 4: Calculated ensemble averaged angular momentum $<< L_y >>$ for different time delays around the half T_{rev} (shown in the inset). The second pulse is polarized 45^0 to the first one.

As can be seen in figure 4, by varying the time delay between the two pulses around the half revival time, one can induce clockwise or counterclockwise (positive or negative $<< L_y >>$ respectively) unidirectional rotation, depending on the angular distribution attained by the molecules at the time of application of the second cross polarized pulse. Based on this observation and on our previous works[3] we can induce selective unidirectional rotation in mixtures of molecular isotopes and nuclear spin isomers. Due to the rational ratio of the revival times of molecular isotopes, there is a specific time where one isotope completes an integer number of revival periods while the other completes and integer and a half of its own periods. At this time we find the two isotopic species at opposite angular distributions ("disk versus "cigar"), making them amenable for opposite induction of the sense of rotation (clockwise versus counterclockwise). The same can be done for nuclear spin isomers, based on their different behavior around $\frac{1}{4}$ and $\frac{3}{4}$ T_{rev}. The case of nuclear spin isomers is shown in figure 5.

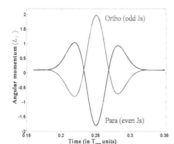

Figure 5: The mean angular momentum along the y-axis of two spin isomers of $^{15}N_2$ is plotted against the delay between the pulses for a second pulse polarized at +45 degrees to the first and applied at $\frac{1}{4}T_{rev}$.

When the second pulse is applied at specific times around $\frac{1}{4}$ and $\frac{3}{4}T_{rev}$ but is polarized at an angle to the first one, i.e. at +45 degrees, different senses of unidirectional rotation are clearly seen (Figure 5) for the individual nuclear spin species (Ortho, Para).

In conclusion, we show that two, time delayed, properly polarized ultrashort laser pulses can induce unidirectional rotational molecular motion in a gas of linear molecules. Based on the different rotational constants of molecular isotopes and the different symmetry properties of nuclear spin isomers and corresponding transient angular distributions, we can induce selective senses of rotation to the different species. Such opposite sense of rotation can be used for physical separation of the different species in the sample through scattering from a surface. Moreover, unidirectional rotation is expected to yield interesting optical features such as induced optical activity.

This work was supported in part by the Nancy and Steve Grand Center for Sensors and Security, and by a grant from the Israel Science Foundation. This research is made possible in part by the historic generosity of the Harold Perlman Family.

[1] I.V. Litvinyuk, et Al., PRL, **90**, 23 (2003).
[2] R.A.Bartels et al.PRL, **88**, 019303 (2002).
[3] Sharly Fleischer et al. PRL **99**, 093002 (2007), PRA **74**, 041403 (2006).
[4] J. Karczmarek et al., PRL, **82**, 3420 (1999).

Attosecond coincidence spectroscopy of diatomic molecules

M. Lezius[1], Z. Ansari[2], M. Böttcher[2], B. Manschwetus[2], W. Sandner[2], A. Verhoef[1], G. G. Paulus[3], A. Saenz[4], D. B. Milosevic[5], and H. Rottke[2]

[1] Max-Planck-Institut for Quantum Optics, D-85748 Garching, Germany
[2] Max-Born-Institute, Max-Born-Str. 2A, D-12489 Berlin, Germany
[3] Institute of Optics and Quantum Optics, Friedrich-Schiller-University, 07783 Jena, Germany
[4] Institut für Physik, Humboldt-University, D-10117 Berlin, Germany
[5] Faculty of Science, University of Sarajevo, 71000 Sarajevo, Bosnia and Herzegovina
corresponding author e-mail: matthias.lezius@mpq.mpg.de

Abstract. : Sub-cycle ionization of Ar_2 by few-cycle laser fields is investigated with COLTRIMS. Low energy photoelectrons show clear deviations from double slit interference. We suggest that breakdown of the single-active electron approximation could be responsible for such effect.

Introduction

Recent attosecond experiments have been based on precise GV/cm electric field synthesis (see e.g.[1]). Few cycle fields have been produced via control over the carrier-envelope offset phase in combination with enhanced spectral broadening and mirror compression techniques. With such techniques single events of strong field tunneling have been timed with sub-cycle accuracy [2]. Sub-femtosecond ionization has become an attractive tool for time-resolved spectroscopy, in particular via rescattering of the liberated sub-femtosecond electron wavepacket onto the parent molecule, and via attosecond XUV pulse generation from HHG, which is a well-known daughter process of rescattering [3]. Attosecond XUV pulses have been used for attosecond time domain spectroscopy, but presently only for a limited number of text-book cases (see e.g. [4, 5]). Successful application of attosecond spectroscopy to molecular systems is pending for many cases. In our recent experiment we have applied sub-cycle ionization to systems consisting of two identical atoms (Argon dimers), thereby asking for geometrical information about the system via the interference pattern of the outgoing electrons. In principle this information could be retrieved with attosecond timing accuracy. We have expected that the photoelectron energy spectra would be fully determined by two-center interference. We found that this is only partially true.

Experimental Methods

To investigate the interference effect in strong-field ionization of molecules we have used a reaction microscope. We have applied it here to measure photoelectron momentum distributions in coincidence with Ar^+ and Ar_2^+ ions. Ar_2 is usually present to an amount of a few percent in supersonic expansions of Argon. The internuclear separation of the Van-der-Waals bound dimers is distributed from 6.8 to 9.4 Å. After collimation the supersonic beam crosses a Ti:sapphire laser beam in its focal spot where an intensity of 2×10^{14} W/cm^2 is reached. The laser pulses have a FWHM of 5.5 fs and a central wavelength of 770 nm. A weak electric field (7.5 V/cm) is applied across the interaction region to extract photoelectrons and ions. After acceleration and subsequent field-free drift ions and electrons are detected by two position sensitive multichannel plate detectors. A homogeneous magnetic field of 8.5 Gauss parallel to the electric field additionally guides photoelectrons to the detector. For each individual particle the time-of-flight is recorded along with the position where it hits the detector. From such data the full momentum vector for each particle can be reconstructed.

Results and Discussion

Fig.1 shows a comparison between the photoelectron momentum distributions (in cylindrical coordinates) for Ar and Ar$_2$, with our theoretical simulations. In our model we derive the molecular momentum distribution $|M_{A-A}(\mathbf{p},\mathbf{R})|^2$ based on the atomic momentum distribution $|M_A(\mathbf{p})|^2$ multiplied with appropriate Franck-Condon factors and additional R-dependent double slit interference terms. The latter depend on the respective atomic orbitals from which electrons can be removed ($3p\sigma_u$, $3p\pi_u$, $3p\pi_g$). The calculation is based on strong field approximation (SFA) to the transition state matrix to the continuum. In particular, three final states can be accessed and lead to stable Ar$_2^+$. Least square agreement with the experimental data can be reached when using a weighted superposition of these three transitions at an internuclear distance of about 7.4 Å. For non-aligned molecules we calculate momentum distributions according to

$$|M_{A-A}(\mathbf{p},\mathbf{R})|^2 = \begin{aligned} & 0,11 \cdot |M_A(\mathbf{p})|^2 \left|S_{fi}^{3p\sigma_u}\right|^2 \left[\tfrac{1}{6}+\tfrac{1}{2}\int \tfrac{d\Omega_{\hat{R}}}{4\pi}\cos(\mathbf{pR})\cos^2\theta_R\right] \\ & +0,57 \cdot |M_A(\mathbf{p})|^2 \left|S_{fi}^{3p\pi_u}\right|^2 \left[\tfrac{1}{3}+\tfrac{\sin(pR)}{2pR}-\tfrac{1}{2}\int \tfrac{d\Omega_{\hat{R}}}{4\pi}\cos(\mathbf{pR})\cos^2\theta_R\right] \\ & +0,32 \cdot |M_A(\mathbf{p})|^2 \left|S_{fi}^{3p\pi_g}\right|^2 \left[\tfrac{1}{3}-\tfrac{\sin(pR)}{2pR}+\tfrac{1}{2}\int \tfrac{d\Omega_{\hat{R}}}{4\pi}\cos(\mathbf{pR})\cos^2\theta_R\right] \end{aligned}$$

(1)

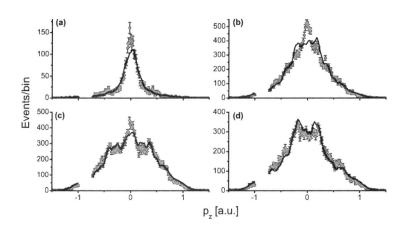

Fig. 1. : Least square fit (full black line) of a sum of simulated Ar$_2$ momentum distributions to the measured Ar$_2$ photoelectron momentum distribution (gray dots with error bars). Cuts shown are made at the radial momenta $p_r = 0.01 \pm 0.01$a.u. (a), $p_r = 0.05 \pm 0.01$ a.u. (b), $p_r = 0.09 \pm 0.01$ a.u. (c), and $p_r = 0.15 \pm 0.01$ a.u. (d). The least square fit was made for graph (c). The simulated spectra for all other cuts use the same fit parameters as in (c).

It can be seen that the experimental data deviate from theory significantly only for small electron momenta. We interpret this as a manifestation of an energy exchange between the outgoing tunnelling electron and the remaining Ar_2^+. If timescales between departure and typical charge oscillations in the remaining system are comparable, deviations from a double slit interference could become likely. Moreover, intramolecular recollisions during ionization are not taken into account within our SFA model. Such recollisions could be 10-20 times more relevant for dimers than for isolated atoms, and could lead to internal excitation, thereby obscuring interference.

Conclusions

Our experiment suggests that directly emitted electrons can be used to retrieve information about emitter geometries on attosecond timescales. However, we find that particle momenta should be high enough to avoid energy exchange during the ionization. The observed low momentum deviations appear to be a clear manifestation of a failure of the single-active electron approximation for tunnelling, and it is quite interesting that they can be observed even in physically quite well separated systems like Van-der-Waals clusters. Because sub-cycle tunnelling is a basic ingredient for most attosecond experiments, breakdown of SFA close to threshold ATI may affect the interpretation of various future molecular attosecond experiments. Our result also adds support to our previous observations that non-adiabatic multielectron effects in molecular strong field ionization should not be neglected [6].

Acknowledgements. We acknowledge support by the Deutsche Forschungsgemeinschaft, in particular M. L. by the DFG Cluster of Excellence: Munich-Centre for Advanced Photonics, G. P. acknowledges support by the Welch Foundation, D. B. M. by the Volkswagenstiftung and by the Federal Ministry of Education and Science, A. S. by the Stifterverband für die Deutsche Wissenschaft. The work has been submitted for publication [7].

1　Goulielmakis, E., et al., Attosecond control and measurement: Lightwave electronics. Science, 2007. 317(5839): p. 769-775.
2　Baltuska, A., et al., Attosecond control of electronic processes by intense light fields. Nature, 2003. 421(6923): p. 611-5.
3　Corkum, P.B. and F. Krausz, Attosecond science. Nature Physics, 2007. 3(6): p. 381-387.
4　Niikura, H., et al., Probing molecular dynamics with attosecond resolution using correlated wave packet pairs. Nature, 2003. 421(6925): p. 826-829.
5　Uiberacker, M., et al., Attosecond real-time observation of electron tunnelling in atoms. Nature, 2007. 446(7136): p. 627-632.
6　Lezius, M., et al., Nonadiabatic multielectron dynamics in strong field molecular ionization. Phys.Rev.Letters, 2001. 86(1): p.51
7　Ansari, Z., et al., Interference in strong field ionization of a two-center atomic system. 2008, submitted.

Real-time Evolution of the Valence Orbitals in a Dissociating Molecule as Revealed by Femtosecond Photoelectron Spectroscopy

Philippe Wernet[1], Michael Odelius[2], Kai Godehusen[1], Jérôme Gaudin[1], Olaf Schwarzkopf[1], and Wolfgang Eberhardt[1]

[1] BESSY, Albert-Einstein-Str. 15, D-12489 Berlin, Germany
[2] FYSIKUM, Stockholm University, AlbaNova, S-106 91 Stockholm, Sweden
 E-mail: wernet@bessy.de

Abstract. We follow in real time the evolution of the valence orbitals of Br_2 molecules as the bonds break during dissociation with femtosecond vacuum-ultraviolet (VUV) photoelectron spectroscopy and with simulations of the nuclear and electron dynamics.

Introduction

How does a chemical bond break? The obvious answer to this question is given by laser femtochemistry [1], a combination of femtosecond (fs) laser technology and laser spectroscopy that allows for tracking in real time the nuclear dynamics during chemical reactions such as molecular dissociations. However, it is the valence electrons that form the bonds and their evolution during dissociation remains elusive when probing the nuclear dynamics. So how does a chemical bond actually break?

Photoelectron spectroscopy is one of the most direct ways of characterizing chemical bonding in molecules: With sufficiently high photon energy all valence states can be probed and their orbital energies and populations are reflected in the measured photoelectron binding energies and intensities.

We used time-resolved photoelectron spectroscopy based on a pump-probe scheme to watch in real time a chemical bond break. We followed the evolution of all valence orbitals during the ultrafast (<100 fs) dissociation of Br_2 molecules in the gas phase all the way from the photo-excited molecules to the free atoms. This allowed for unprecedented insight into how and on what time scale valence electrons rearrange when a covalent bond breaks [2].

Experimental Methods

We used a fs laser-pump and VUV-probe scheme with VUV pulses from high-order harmonic generation (HHG). Br_2 molecules are excited by absorption of one laser-pump photon (wavelength 400 nm) to the $^1\Pi_u$ dissociative state (population of the antibonding $8\sigma_u$ orbital at the equilibrium inter-nuclear distance of 2.3 Å). The molecules are ionized after a well-defined fs time delay by a VUV-probe pulse (15th harmonic of the fundamental, photon energy 23.5 eV, pulse duration 120 fs, generation medium Xe, approximate flux 10^9-10^{10} photons/s at 1kHz repetition rate in a focus smaller than 100x100 μm^2). Photoelectron energies are analyzed with a time of flight spectrometer. Dissociation from the $^1\Pi_u$ state results in two ground-state Br atoms.

Nugent-Glandorf and Strasser et al. have demonstrated the feasibility of this experiment in pioneering investigations [3, 4] but the greatly improved temporal resolution (135±5 fs) and sensitivity in our experiment combined with a theoretical description allows for unprecedented insight into the evolution of the electronic structure during Br_2 dissociation. For in-depth analysis we also calculated the electronic structure of an ensemble of dissociating Br_2 molecules. Classical molecular dynamics simulations provided distributions of nuclear distances versus time delay and binding energies and photoelectron intensities were calculated in fs steps from the excited molecule to the free atom.

Results and Discussion

Measured and calculated valence-electron binding energies are depicted in Figure 1 (left).

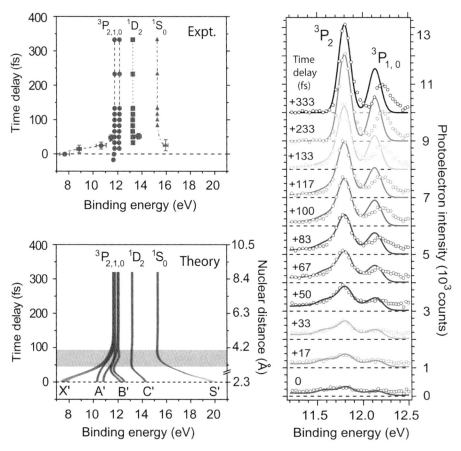

Fig. 1. Measured (left, top) and calculated (left, bottom) valence orbital energies as a function of pump-probe delay time and inter-nuclear distance. Right: Measured (circles) and calculated (solid lines) evolution of part of the valence band photoelectron spectrum.

Clearly, states start shifting immediately after excitation and by up to 4 eV while some essentially remain at constant energy. States are mixing and merging after approximately 50 fs when the atoms already travel at constant speed (~2.1 Å/ 100 fs) and until less than 100 fs (nuclear distance close to 4 Å) when the electronic structure of the free atoms is essentially established. Those states corresponding to σ orbitals show the largest shifts because they arise from atomic orbitals oriented along the bond whose overlap changes most with changing bond length. Horizontal cuts in Figure 1 (left) correspond to photoelectron spectra at the respective delay time (Figure 1 right) and vertical cuts reflect the population dynamics of states with the respective binding energy (not shown here).

The evolution of the photoelectron spectrum close to 12 eV (Figure 1 right) from a broad distribution to the well-defined peaks of the free atoms demonstrates how we can spectroscopically distinguish the valence states of the excited molecule (0 fs delay) and of the free atoms (+333 fs delay).

These observations motivate an electronic-structure based definition of a transition period lasting from 50 fs (~3 Å) to less than 100 fs after excitation (~4 Å) during which valence electrons rearrange to form the orbitals of the free atoms. Our methodology, the analysis of the data and the calculations will be presented in detail in a forthcoming publication [2]. Our approach can be contrasted with laser femtochemistry with a particular emphasis on the sensitivity of the different approaches to different effects (electronic versus nuclear dynamics) [2].

Conclusions

We mapped in real time the evolution of the valence electronic structure of a diatomic molecule as the covalent bond breaks. Orbital energies and population dynamics of the valence states of Br_2 molecules during dissociation were followed with femtosecond photoelectron spectroscopy and simulations of the nuclear and electron dynamics. Our results allow for an unprecedented insight into how a chemical bond breaks.

Acknowledgements. J.G. acknowledges support by the European Union Marie Curie program. M.O. acknowledges support from the Swedish research council.

1   A. H. Zewail, Science **242**, 1645, 1988.
2   Ph. Wernet, M. Odelius, K. Godehusen, J. Gaudin, O. Schwarzkopf, W. Eberhardt, submitted, 2008.
3   L. Nugent-Glandorf, M. Scheer, D. A. Samuels, A. M. Mulhisen, E. R. Grant, X. Yang, V. M. Bierbaum, S. R. Leone, Phys. Rev. Lett. **87**, 193002, 2001.
4   L. D. Strasser, F. Goulay, S. R. Leone, J. Chem. Phys. **127**, 184305, 2007.

Transient Waveguiding in a Rotationally Excited Molecular Gas

Francesca Calegari[1], Caterina Vozzi[1], Sergei Gasilov[1], Enrico Benedetti[1], Giuseppe Sansone[1], Mauro Nisoli[1], Sandro De Silvestri[1], and Salvatore Stagira[1]

[1] National Laboratory for Ultrafast and Ultraintense Optical Science - CNR-INFM
   Dipartimento di Fisica, Politecnico di Milano, Milano, I-20133, Italy
   E-mail: francesca.calegari@polimi.it

Abstract. Transient waveguiding and spectral broadening of a delayed probe pulse were observed in the wake of a laser filament propagating in Nitrogen and Oxygen. The observed effects are ascribed to the excitation of rotational wavepackets.

Introduction

Propagation of light filaments in atmosphere has been a subject of intense studies [1, 2] for many applications including light discharge control, atmospheric remote sensing and THz waves generation. For these reasons, several theoretical works have been developed on filamentation in molecular gases. However, the majority of these studies neglects the retarded part of the rotational response excited in the wake of the filament. This response shows periodic revivals [3], that can be thought of as field-free recurrences of molecular alignment.

In this work we report on the experimental investigation of the spatio-temporal effects occurring in the wake of a filamenting pulse propagating in a molecular gas. In particular, we demonstrate that the evolution of the rotational wave packet excited by the filamenting pump affects substantially a time-delayed probe pulse propagating in the wake of the filament. Besides a strong spectral modulation, the probe experiences a tight spatial confinement in the core of the wake, which can be compared to the propagation inside a transient waveguide [4].

Experimental Methods

The experiment has been performed using an amplified Ti:sapphire laser system, which generates 60-fs pulses at 800 nm with a repetition rate of 10 Hz. The laser pulses were split in two beams: the first pulse (pump, maximum energy 0.9 mJ) was focused in a gas cell, generating a filament; the second pulse was frequency doubled in 1-mm thick BBO crystal followed by an IR-absorbing filter (energy 10 J) and was used to probe the effects occurring in the filament wake. The two beams were collinearly recombined by a dichroic beam splitter and their foci were overlapped into the gas cell. The probe was extracted after the cell by a second dichroic beam splitter; its spatial shape and spectrum were recorded as a function of pump-probe delay. We performed the experiments both in N_2, at a pressure of 2.5 bar, and in O_2 at 2 bar.

Results and Discussion

Figure 1(a) shows the simulated molecular alignment factor induced by the pump pulse in the core of the filament generated in Nitrogen; only the evolution around the first half

84

revival is displayed. A maximum (point A) and a minimum (point B) in the alignment factor are present. Figure 1(b) shows the spectrum of the probe pulse propagating in the wake of the filament at a time delay corresponding to point A (solid line); as can be seen, the probe presents a remarkable broadening with respect to its original spectrum (dashed line) in correspondence of the maximum alignment; this effect is ascribed to modulation of the refractive index due to excitation of a rotational wavepacket in the gaseous medium [5].

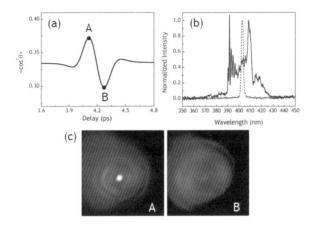

Fig. 1. (a) Simulated alignment factor around the first half revival in N_2 after excitation with the filamenting pump. (b) Probe spectrum with (solid line) and without (dashed line) rotational excitation at time delay A in panel (a). (c) Spatial profiles of the probe beam at delays A and B of panel (a).

The noticeable amount of spectral broadening (about 700%) is not compatible with a free propagation of the probe, since its natural divergence would limit the interaction with the excited region to a small distance. Nevertheless, as shown on the left of Figure 1(c), the probe beam undergoes a strong spatial confinement in correspondence of the maximum alignment (delay A) since it experiences a convex refractive index profile and, as a result, it is guided inside the filament wake. This spatial confinement extends the interaction of the probe with the rotationally excited gas, thus accounting for the observed spectral broadening. On the other hand, at the minimum of the half revival (delay B), an antiguiding refractive index profile is present. As a result, in this case the probe beam is deviated away from the center of the filament wake and its spatial pattern assumes an annular shape, as displayed on the right of Figure 1(c).

We further explored these effects in Oxygen, where quarter revivals are stronger owing to a different nuclear spin statistic with respect to Nitrogen. Figure 2(a) shows the simulated alignment factor induced by the filamenting pump pulse in O_2 around the first quarter revival; in this case a maximum (delay B) and two minima (delays A and C) are observed. A huge spectral broadening of the probe pulse was observed for a pump-probe delay corresponding to point B, as shown in Figure 2(b). The probe spatial profiles recorded in Oxygen are reported in Figure 2(c); as can be seen, the probe beam is guided in correspondence of the maximum alignment (delay B) and antiguided in correspondence of the minimum alignment (delays A and C).

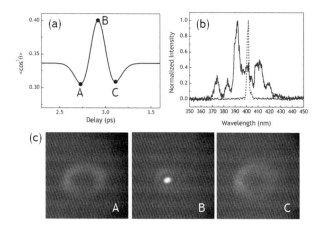

Fig. 2. (a) Simulated alignment factor around the first quarter revival in O_2 after excitation with the filamenting pump. (b) Probe spectrum with (solid line) and without (dashed line) rotational excitation at time delay B in panel (a). (c) Spatial profiles of the probe beam at delays A, B and C of panel (a).

In order to demonstrate that the experimental finding can be assigned to the rotational Raman response excited in the filament wake, we developed a numerical model in cylindrical symmetry for the propagation of pump and probe pulses. Calculated probe beam evolution, performed for different pump-probe delays, is in very good agreement with experimental findings and shows that a guided propagation in the filament wake is obtained at maxima of rotational revivals.

Conclusions

In conclusion, we demonstrated that the retarded Raman response of a molecular gas in the wake of a light filament can dramatically affect the spectral bandwidth and spatial pattern of a collinear time-delayed probe pulse. In particular we showed that light confinement and a simultaneous significant spectral broadening can be achieved in the filament wake at maxima of rotational revivals for Nitrogen and Oxygen.

1 J. Kasparian, M. Rodriguez, G. Méjean, J. Yu, E. Salmon, H. Wille, R. Bourayou, S. Frey, Y. B. André, A. Mysyrowicz, R. Sauerbrey, J. P. Wolf, and L. Woste, Science 301, 61, 2003.
2 A. Couairon and L. Bergé, Phys. Rev. Lett. 88, 135003, 2002.
3 A. M. Zheltikov, Opt. Lett. 32 , 2052, 2007.
4 F. Calegari, C. Vozzi, S. Gasilov, E. Benedetti, G. Sansone, M. Nisoli, S. De Silvestri, and S. Stagira, Phys. Rev. Lett. 100, 123006, 2008.
5 R. A. Bartels, T. C. Weinacht, N. Wagner, M. Baertschy, C. H. Greene, M. M. Murnane, and H. C. Kapteyn , Phys. Rev. Lett. 88, 013903, 2001.

Molecular Recollision Interferometry in High Harmonic Generation

Xibin Zhou, Robynne Lock, Nick Wagner, Wen Li, Henry C. Kapteyn, and Margaret M. Murnane

JILA and Department of Physics, University of Colorado
Boulder, CO 80309-0440
E-mail:xibin.zhou@colorado.edu

Abstract. Using extreme-ultraviolet interferometry, we measure π phase shifts in high harmonics generated from transiently aligned molecules. This data directly reflects the quantum interferences in the electron wave packet due to the two-center molecular structure.

There has been considerable recent interest in using high-harmonic generation (HHG) to observe molecular structure and dynamics. Lein et. al. introduced the two-center model which explains HHG from molecules as quantum interference of harmonic emission from two distinct regions in the molecule [1,2]. The two-center model has been used to explain the changes in HHG intensity as a molecule rotates [3]. Measuring the HHG phase would further test this model, which predicts π phase shifts in the harmonic emission above certain harmonic orders. In addition, tomographic imaging of molecules must currently assume information about the HHG phase [4]. However, only the HHG intensity has been measured in published measurements of HHG from molecules to date. In this work, we directly measure the phase of HHG from aligned molecules, by observing the quantum interference between HHG from regions of aligned and isotropic molecules. We unambiguously observe a π phase shift in the harmonic emission.

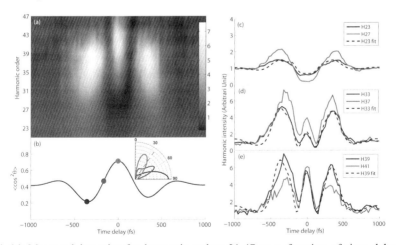

Fig. 1 (a) Measured intensity for harmonic orders 21-47 as a function of time delay. (b) Theoretical alignment parameter $<\cos^2\theta>$. See text for the inset. (c) –(e) Lineouts of harmonic orders 23, 27, 33, 37, 39, and 41. The dashed lines show fits.

For our experiment, a pump beam transiently aligns the molecules in part of the sample, and a probe beam generates harmonics. We use two glass plates tilted at slight angles, and partially inserted into the focusing laser beam (2~3 cm after a lens of 30 cm focal length), to split the focus into two elliptical focal spots with diameters (full width at half maximum) of ~80 μm. The harmonics generated in the two regions then interfere in the far field. The pump beam is focused non-collinearly into the molecular gas jet at the position of one of the probe foci. Blocking the harmonic generating beam for the isotropic molecules allows for measurement of the intensity.

The net harmonic intensity for orders 21-47 as a function of time delay between pump and probe pulses for the ¾ revival in CO_2 is shown in Fig 1(a). To aid in understanding the relation between molecular alignment and harmonic intensity, Fig. 1(b) shows the alignment parameter $<\cos^2\theta>$, while Figs. 1(c)-(e) show lineouts for each harmonic order. For harmonic orders 29 and below, the harmonic intensity follows an inverse of the alignment parameter. However, for harmonic orders 31 and above, the intensity goes through a minimum, increases when the molecules are best aligned, and goes through another minimum before increasing again. This previously unreported anomalous peak is strongly suggestive of a phase shift due to quantum interference.

The interference patterns observed when harmonics from aligned and isotropically distributed CO_2 molecules interfere in the far field confirm this phase shift. Figs. 2(a) and 2(c) show the interference patterns for harmonic orders 27 and 33 respectively. When strongly aligned, the fringes clearly shift for order 33; however, there is no such shift for order 27. The duration of the phase shift exactly corresponds to the duration of the anomalous peak for order 33 (Fig. 1(d)). The magnitude of the phase shift is determined by integrating the fringes from -100fs to 100fs and comparing with the integration of the fringes from outside this time window. Figs. 2(b) and 2(d) plot the results. The sine function fit yields HHG phase shifts of 0.035 ± 0.5 radians and 3.4 ± 0.3 radians for orders 27 and 33 respectively. We also investigated HHG intensity and phase at half revival of N_2O and N_2 moleucles. For N_2O, a phase shift and anomalous peak occurred for order 31 and above during the ½ rotational revival. However, in N_2 we observed no phase shift in the harmonics range from 19 to 31.

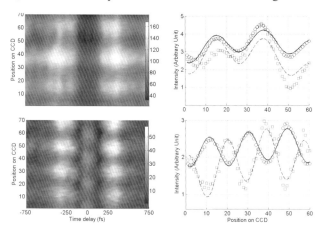

Fig. 2 (a), (c) Interference patterns as functions of time for harmonic orders 27 (a) and 33 (c). (b), (d) Intensity-scaled integrated fringes in the -100fs to 100fs interval (squares) and fit (dash line). Integrated fringes outside this time window are also shown (circles) with fit (solid line) for orders 27 (b) and 33 (d). A clear π phase shift is seen for order 33 when the molecules are aligned.

The inset of Fig. 1(b) aids in explaining the origin of the anomalous peak and the phase shift, by plotting the angular distributions for three different time delays. The dashed line corresponds to the critical alignment angle – where, according to the two-center molecular HHG model, the sign (phase) of the harmonic emission changes for order 33 due to quantum interferences in the radiating electron wave packet. When well aligned, most of the CO_2 angular distribution falls on one side of the critical angle. At the other two time delays, most of the CO_2 angular distribution falls on the other side of the critical angle. Consequently, the 33rd harmonic emission will undergo a phase reversal during a rotational alignment. For harmonic orders 29 and below, the critical angle is smaller. Under our alignment conditions, the entire molecular distribution never crosses the 29th order critical angle, and therefore no phase shift is observed. Finally, when the angular distribution is centered on the critical angle, HHG is minimized due to destructive interference. This explains the anomalous peaks in Fig. 1.

To further test the two-center interference model, we fit the CO_2 data to

$$HHG(t) = \left| \int \rho(\theta,t) A \sin(\pi R \cos\theta / \lambda) d\theta \right|^2 + C \tag{1}$$

where $\rho(\theta,t)$ is the time-dependent angular distribution calculated by numerically solving time dependent schrödinger equation, A is a normalization constant, R is the distance between the two centers, λ is the wavelength of the recolliding electron, and C is an offset constant. Fits for harmonic orders 23, 33 and 39 are shown in Figs. 1(c)-(e). In the least-square fitting process we use A, $B=R/\lambda$ and C as the fitting parameter. The data agrees well with the two-center model, strongly suggesting that quantum interferences during recombination determine the primary features of the angular modulation of HHG, and both the phase and amplitude of the returning electron wave packet are not sharply depends on the molecular orientation. From the fits, we extract the value of R/λ for each harmonic order, allowing us to determine the kinetic energy of the electron i.e. $E_k = nh\omega - \delta I_p$, where n is the harmonic order and I_p is the ionization potential. A value for δ of 0 corresponds to a physical picture of the returning electron emitting harmonics as it returns to the edge of the molecular potential, whereas a value for δ of 1 corresponds to the emission occurring when the electron returns to the "bottom" of the molecular potential. We found that δ increases with increasing harmonic order, indicating a decreaseing influence (dispersion) of the molecular potential on the recolliding electron for high order harmonics.

In summary, we have directly observed π phase shifts in harmonic emission from molecules for the first time. We also use the measured HHG phase and intensity to extract information about the effect of the molecular potential on the recombining electron in CO_2 [5].

Acknowledgements We gratefully acknowledge the financial support of DOE.

1   M. Lein et al, "Interference effects in high-order harmonic generation with molecules," Phys. Rev. A **66**, 023805 (2002).

2   M. Lein et al, "Role of the intramolecular phase in high-harmonic generation," Phys. Rev. Lett. **88**, 183903 (2002).

3   T. Kanai, et al., "Quantum interference during high-order harmonic generation from aligned molecules," Nature **435**, 470 (2005).

4   J. Itatani, et al., "Tomographic imaging of molecular orbitals," Nature **432**, 867 (2004).

5   X. Zhou et al., "Molecular recollision interferometry in high harmonic generation," Phys. Rev. Lett. **100**, 073902 (2008)

Multi-Electron Dynamics in Molecular High Harmonic Generation

Gerald Jordan[1], and Armin Scrinzi[2]

[1,2] Photonics Institute, Vienna University of Technology, 1040 Vienna, Austria
[1] E-mail: gerald.jordan@tuwien.ac.at
[2] E-mail: scrinzi@tuwien.ac.at

Abstract. We demonstrate the significance of multi-electron dynamics, in particular core polarization by the laser, in molecular HHG, as various simplifying models of increasing complexity fail in reproducing the multi-electron spectra obtained using the MCTDHF method.

Introduction

Tomographic orbital imaging (demonstrated for N_2 [1] and CO_2 [2]) uses electron recollision-induced high harmonic radiation to retrieve the electronic molecular orbital including the phase information. This technique would allow to watch the breaking and formation of molecular bonds on a femtosecond time scale. The reconstruction scheme relies on the 3-step-model [3] for high-harmonic generation and the single active electron approximation, which are both known to work well for multi-electron atoms.

In the case of molecules, however, multi-electron effects like correlation and polarization are expected to be more important due to the closer-lying energy levels. It is therefore essential to cross-check single-electron models against presently available multi-electron calculations. Further details of our investigation are described in [4].

MCTDHF

Using the multi-configuration time-dependent Hartree-Fock method [5,6], we are capable of solving the time-dependent few-electron Schrödinger equation in 3 dimensions. The method consists in making an ansatz for the f-electron wave function as a sum of all Slater determinants formed from n single-electron orbitals

$$\Psi(q_1, \ldots, q_f; t) = \frac{1}{\sqrt{f!}} \sum_{j_1} \cdots \sum_{j_f} C_{j_1 \cdots j_f} \phi_{j_1}(q_1; t) \cdots \phi_{j_f}(q_f; t). \tag{1}$$

Both the coefficients and orbitals are time-dependent and are dynamically adapted to provide an approximation to the wave function which is optimal in a variational sense. This allows to keep the number of orbitals relatively small and makes multi-electron calculations feasible.

System and Models

We study high harmonics from diatomic homonuclear molecules with 2 and 4 active electrons, respectively. The molecules have an internuclear distance of $R = 2.8$ a.u. and the molecular axis is aligned parallel to the laser polarization direction. The ionization potential of $I_p = 0.58$ a.u. is chosen to agree with N_2.

Using $n = f + 4$ orbitals, we obtain harmonic spectra from MCTDHF wave functions. While full numerical convergence could not be reached, the spectral shape is already well reproduced by the Hartree-Fock ($n = f$) calculations.

The multi-electron solution is compared with several simplifying models of increasing complexity:

- Lewenstein model [7]. The Lewenstein model is a single-electron model and neglects the distortion of the continuum by the molecular binding potential. For the multi-electron molecules, we tried different definitions of a single-electron orbital to use as a ground state in the Lewenstein formula.

- Multi-electron Lewenstein model [8,9]. This model includes multi-electron effects on the SFA level which show up in the recombination matrix element as cross-terms between different orbitals.

- A single-electron TDSE with an effective Hamiltonian modelling a realistic molecular binding potential.

Results and Discussion

The resulting spectra are shown in Figure 1 for the 4-electron molecule. We find that neither of these simplifying descriptions can quantitatively or qualitatively reproduce the harmonic spectra of the multi-electron calculation.

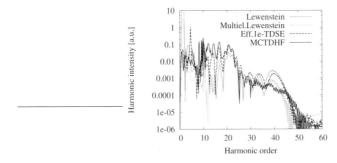

Fig. 2. Comparison of harmonic spectra for a molecule with 4 active electrons using MCTDHF and simplifying models.

If the basic physical picture of high harmonic generation is still valid for molecules, the harmonic spectra are determined by 2 ingredients: the recollision electron current and the recombination matrix element of the molecular ion. We find that accounting for the influence of the molecular binding potential in the effective single-electron TDSE gives a sufficiently accurate description of the time structure of the electron recollision current.

Hence we are left with the second part, i.e. recombination, as the one affected most by multi-electron effects. This is consistent with the large corrections from the cross-terms already observed when comparing the single- and multi-electron Lewenstein model.

Going beyond SFA, we show that these cross-terms are highly sensitive to the polarization of the molecular ion target by the laser field, which in turn results in severe changes of the harmonic spectrum. To demonstrate this effect, we try to fit the MCTDHF multi-particle wave function with a product wave function consisting of an ionic core and an active-electron orbital

$$\Psi'(q_1,\ldots,q_f;t) = \frac{1}{N(t)}\mathscr{A}[\Phi(q_1,\ldots,q_{f-1};t)\phi(q_f;t)] \tag{2}$$

For the core we use either the static ionic ground state Φ_{stat} or the dynamic ion wave function Φ_{dyn} time-evolving in the same laser pulse. Choosing ϕ as the corresponding Dyson orbital maximizes the overlap between Ψ' and the true, multi-electron solution Ψ. We find that only the dynamic approach can adequately reproduce the exact spectra both in shape and magnitude (see Figure 2). In this case, the cross-terms between active electron and core become small, and the Dyson orbital alone provides already a good approximate spectrum. In this sense only, molecular high harmonic generation can still be considered a single-electron process. However, to determine this single-electron orbital, the full multi-electron wave function is required, which incorporates the full multi-electron dynamics.

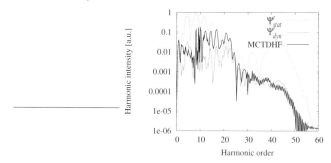

Fig. 1. Comparison of the spectra of the 4e molecule from the product wave functions (2) (with static Ψ'_{stat} and dynamic ion core Ψ'_{dyn}) with the MCTDHF solution.

Conclusions

In conclusion, we have found strong indications that *dynamic* multi-electron effects play a decisive role in the high harmonic response of molecules. Before tomographic imaging can be meaningfully applied to a specific system, the presence and consequences of such effects have to be clarified.

1   J. Itatani *et al.*, Nature 432, 867, 2004.
2   R. Torres *et al.*, Phys. Rev. Lett. 98, 203007, 2007.
3   P. B. Corkum, Phys. Rev. Lett. 71, 1994, 1993.
4   G. Jordan, and A. Scrinzi, New J. of Physics 10(2), 025035, 2008.
5   J. Caillat *et al.*, Phys. Rev. A 71, 012712, 2005.
6   G. Jordan, J. Caillat, C. Ede, and A. Scrinzi, J. Phys. B 39, 341, 2006.
7   M. Lewenstein, P. Balcou, M. Y. Ivanov, A. L'Huillier, and P. B. Corkum, Phys. Rev. A 49, 2117, 1994.
8   R. Santra and A. Gordon, Phys. Rev. Lett. 96, 073906, 2006.
9   S. Patchkovskii, Z. Zhao, T. Brabec, and D. M. Villeneuve, Phys. Rev. Lett. 97, 123003, 2006.

Probing the dynamics of plasma mirrors on the attosecond time scale

C. Thaury[1], F. Quéré[1], H. George[1], R. A. Loch[2], J-P. Geindre[3], P. Monot[1], and Ph. Martin[1]

[1] CEA IRAMIS, Service des Photons Atomes et Molécules, F-91191 Gif-sur-Yvette, France
[2] Laser Physics and Nonlinear Optics Group, Faculty of Science and Technology, MESA$^+$ Institute for Nanotechnology, University of Twente, The Netherlands
[3] Laboratoire pour l'Utilisation des Lasers Intenses, CNRS, Ecole Polytechnique, 91128 Palaiseau, France
E-mail: fabien.quere@cea.fr

Abstract. We demonstrate that the generation of high-order harmonics (HHG) of a laser on plasma mirrors preserves the coherence of the laser. We then exploit this coherence to study the dynamics of the plasma electrons.

Introduction

When an intense ultrashort laser pulse hits an optically-polished solid target, it generates a dense plasma that acts as a mirror, known as a plasma mirror. At high enough intensities, high-order harmonics of the incident frequency, associated in the time-domain to attosecond pulses, can be generated upon reflection on this mirror [1]. The phase properties of these harmonics are largely determined by the dynamics of the plasma, which depends itself on the laser intensity [2]. The study of this phase as a function of the intensity, can thus provide information on the laser-driven plasma dynamics. In addition, knowing this intensity-dependence phase is essential to understand the properties of any laser-driven harmonic source. Indeed, because high laser intensities are required, HHG occurs around the focus of ultrashort laser pulses, and the laser intensity thus has strong spatial and temporal variations. These lead to non-trivial spatial and temporal phases of the harmonics, which affect the divergence, spectral width, as well as the exact coherence degree of the source [2].

Mutual coherence of harmonic sources generated on plasma mirrors

Our experiment uses 60 fs pulses with a high temporal contrast, to produce high-order harmonics on plasma mirrors through Coherent Wake Emission (CWE) [3], at intensities going from a few $10^{16} W/cm^2$ to a few $10^{17} W/cm^2$. Groups of harmonics in the beam diverging from the plasma mirror are selected with different kinds of thin metallic filters, and the resulting spatial profile in the far field is then measured at a distance $D = 38$ cm from the source. To study the mutual coherence of CWE, we have implemented a new and remarkably simple technique. A transmission grating is placed into the beam before focusing (Fig. 1(a)). It is designed in such a way that at the focus of the beam, diffraction produces a central focal spot surrounded by two slightly weaker satellites spots at a distance $a = 40 \mu m$ (Fig. 1(b)). Temporally synchronized and phase-locked laser pulses reach these three spatially-separated foci of identical shapes. High-order harmonics are generated on each focal spots upon reflection onto a plasma mirror.

When measuring the spatial profile of the harmonic beam in the far-field, fringes with an almost perfect contrast can be observed, which result from the interferences between the three spatially-separated XUV sources (Fig. 1(c-d)). This simple result demonstrates the mutual coherence of several harmonic sources generated on a plasma mirror.

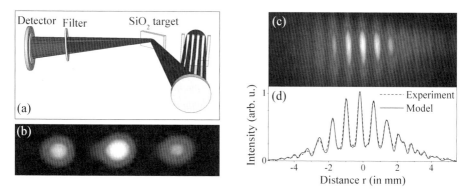

Fig. 1. Coherent harmonic beams from plasma mirrors (a) Experimental set-up. (b) Intensity distribution in the focal plane. In this case, the intensity ratio between the central and lateral spots is 0.57. (c) Single shot far-field interference pattern of harmonics 8 to 10. (d) Lineout of the fringes and theoretical fit with phase shifts $\Delta\Phi_\omega = 0.23\omega/\omega_L$ rad.

Probing the dynamics of the plasma electrons

In the case of three sources of different intensities shown in Fig. 1, the interference pattern in the far field is given by the following elementary equation [4]:

$$I(\mathbf{r}) = \int d\omega F_\omega(\mathbf{r})\left[1 + 2\alpha_\omega + 4\sqrt{\alpha_\omega}\cos(\Delta\phi_\omega)\cos(\mathbf{k}_\omega\mathbf{r}) + 2\alpha_\omega\cos(2\mathbf{k}_\omega\mathbf{r})\right] \quad (1)$$

where $\mathbf{k}_\omega = (\omega/cD)\mathbf{a}$ (with c the speed of light and $\mathbf{a}$ the vector joining the central focal spot to a lateral one), and $F_\omega(\mathbf{r})$ is the spatial intensity profile of frequency ω in the detection plane for a single focal spot. α_ω is the intensity ratio between the central harmonic source and the two lateral ones, and $\Delta\phi_\omega = \phi_\omega(I) - \phi_\omega(\alpha_{\omega_L}I)$ the phase shift between these sources for frequency ω (where ω_L is the driving laser frequency).

Equation (1) shows that the interference pattern contains two sets of fringes for each ω, with spatial frequencies $|\mathbf{k}_\omega|$ and $2|\mathbf{k}_\omega|$. Fringes of frequency $|\mathbf{k}_\omega|$ correspond to the interference of the central harmonic source with either of the two lateral ones, while those of frequency $2|\mathbf{k}_\omega|$ correspond to the interference between the two outer sources. In usual two-source interferometry, phase terms are encoded in shifts of the fringes. Because of the symmetry of the present configuration with three sources, the phase shift $\Delta\phi_\omega$ affects the contrast of the $|\mathbf{k}_\omega|$ fringes, but not their positions.

This phase shift can be extracted from the interfrogramm by performing a Fourier analysis [4]. The results of this measurement are shown in Fig. 2, for harmonics 11, 12 and 13, and for different intensity ratios α_{ω_L}. $\Delta\phi_\omega$ decreases from $\approx \pi$ to slightly less than $\pi/2$ as the laser intensity ratio α_{ω_L} goes from about 0.57 to 0.74. For each α_{ω_L}, the phase shift $\Delta\phi_\omega$ between the central harmonic source and the two lateral ones is observed to vary linearly with harmonic frequency ω (see Fig. 2(a)). A linear phase shift $\Delta\phi_\omega$ corresponds to a simple translation by $\tau = \Delta\phi_\omega/\omega$ in the time domain. In other words, these results show that the light fields $E^0(t)$ and $E^1(t)$, corresponding to the superposition of harmonics 11 to 13, emitted respectively by the central and lateral sources, are identical and simply shifted in time by a delay τ, *i.e.* $E^0(t) = E^1(t + \tau)$. Fig. 2(b) shows that τ decreases as intensity ratio α_{ω_L} increases, as expected.

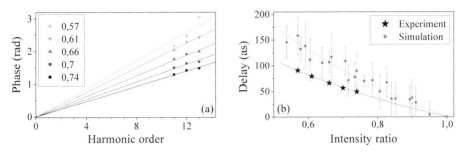

Fig. 2. Relative phases $\Delta\phi_\omega$ between the central and lateral sources. In (a), this phase shift is plotted as a function of the harmonic order (for orders 11 to 13), for different laser intensity ratio α_{ω_L} which values are given in the legend. These values are an average over 5 laser shots, and have an uncertainty of $\approx 10\%$. The lines are linear fits, $\Delta\phi_\omega = \omega\tau$. In (b), the delay τ deduced from these data is shown as a function of α_{ω_L}. These measurements are compared to the results of PIC simulations performed at different laser intensities. In this case, the error bars correspond to the fluctuations in delay between pairs of attosecond pulses in the two trains.

Numerical simulations with Particles-in-Cells (PIC) codes provide a physical interpretation of this delay τ [2]. In Coherent Wake Emission, trains of attosecond pulses are emitted by collective electron oscillations at the plasma-vacuum interface, which are triggered once every optical cycle in the wake of electron bunches [3]. These electron bunches are formed as the electric field pulls some electrons into vacuum, and then, when it changes sign, pushes them back into the overdense plasma [5]. When the peak intensity of the laser pulse is changed, the return time of these so-called "Brunel electrons" to the plasma changes by a fraction of the laser optical period. The same time shift is transferred to the attosecond pulses. According to simulations, this is the main effect of an intensity variation of the driving laser on the attosecond pulse train.

The delay associated to this effect, obtained from 1D PIC simulations, is shown in Fig. 2(b), as a function of the intensity ratio α_{ω_L}. It differs from the experimental values by a factor 1.6, but the evolutions with α_{ω_L} are similar. This qualitative agreement is satisfactory, since a PIC code is only a model of the actual plasma, which does not take all physical processes into account. This comparison with simulations shows that our measurements give access to the change in return time of Brunel electrons with laser intensity, with sub-laser cycle temporal resolution. More generally, these results prove that HHG from plasma mirrors can be used to probe the coherent dynamics of plasmas driven by high-intensity lasers.

1 C. Thaury *et al.*, in Nature Physics 3, 424-429, 2007.
2 F. Quéré *et al.*,in Physical Review Letter 100, 095004, 2008.
3 F. Quéré *et al.*, in Physical Review Letter 96, 125004, 2006.
4 C. Thaury *et al.*, in Nature Physics, doi:10.1038/nphys986, 2008.
5 F. Brunel, Physical Review Letter. 59, 52-55, 1987.

Shaping Entangled Photon Pairs

Florian Zäh[1], and Thomas Feurer[1]

[1] University of Bern, Institute of Applied Physics
Sidlerstr. 5, 3012 Bern, Switzerland
E-mail: florian.zaeh@iap.unibe.ch

Abstract. We demonstrate automated amplitude and phase modulation of entangled photon pairs with attosecond precision, different autocorrelation measurements, and the observation of nonlocal effects, such as an increase of the coherence time due to spectral filtering.

Introduction

Nonlocal effects in quantum mechanics have been investigated ever since Einstein, Podolsky, Rosen, and Bohr have started the discussion. Entangled photon pairs produced by spontaneous parametric down-conversion (SPDC) have become one of the main workhorses to observe nonlocal effects, and nonlocal correlations were investigated with respect to polarization, energy, or wave vector. In SPDC a pump laser photon of frequency ω_p and wave vector k_p produces a pair of correlated photons (signal and idler) which have frequencies ω_s and ω_i and wave vectors k_s and k_i. The three photons must obey energy conservation and conservation of wave vector (phase matching). While the pair as a whole must have well defined properties, the individual photons have largely undetermined properties. In the meantime, many experiments have shown clear evidence of nonlocal effects such as violating Bell's inequalities. If time-energy entanglement is used, nonlocal effects have for example been demonstrated through the appearance of fourth-order interferences among distant interferometers. Recently, the Silberberg group has demonstrated in a beautiful set of experiments that techniques very similar to those used in femtosecond pulse shaping can be employed to manipulate the wave function of a time-energy entangled photon pair [1]. Based on these experiments, we demonstrate simultaneous phase- and amplitude shaping of entangled photon pairs with an attosecond precision. This allows for a pulse shaper-assisted realization of unbalanced interferometers and the shaper-assisted detection of nonlocal modulations of the two-photon wave function.

Experimental Methods

The down-converted entangled photon pairs are produced by focusing a 5 W continuous wave pump laser in a PPKTP crystal. The spectrally broadband light is send through a standard prism compressor consisting of 4 prisms, with the prism separation such that all material dispersion is compensated for. A spatial light modulator is positioned in the symmetry plane of the prism compressor. Time coincidences are detected by sum-frequency generation of the shaped down-converted light in a second PPKTP crystal and the sum-frequency photons are recorded by a single-photon counter. This experimental setup is similar to that reported by the Silberberg group [1]. The measured quantity, the second order coherence function, is simplified in the case of a delta-like pump to the form

$$G^{(2)}(t_s - t_i \approx 0) = \left| \int d\omega_s M_s(\omega_s) M_i(\omega_p - \omega_s) \xi(\omega_s) \right|^2 . \tag{1}$$

$M_j(\omega_j), j = s, i$ denote some transfer function experienced by the photons, and $\xi(\omega)$ is a function determined by the phase matching condition of the crystal.

Results and Discussion

First, we demonstrate coherent phase and amplitude modulation of time-energy entangled two-photon wave functions based on computer-controlled pulse shaping techniques [1, 2]. The combination of the two allows generating transfer functions which imitate most linear optical elements and more complex linear optical arrangements, such as interferometers. Typically, unbalanced interferometers are used to measure correlations and an example for a shaper-assisted (without any moving parts) autocorrelation of the two-photon wave function is shown in figure 1. The delay was scanned from -200 to +200 fs for both interferometric (left side) and intensity-like (right side) autocorrelation.

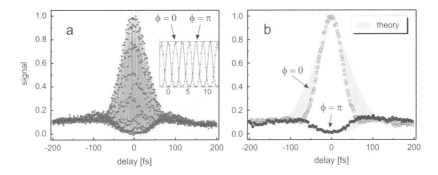

Fig. 1. Interferometric and intensity like autocorrelations of an entangled two-photon wave packet. Both- interferometer exit ports are measured.

By dynamically adjusting the transfer function the autocorrelation trace can be recorded with a minimum delay increment of 40 as, which is determined mostly by the spectral resolution of the setup. Contrary to the mechanical version of a Michelson interferometer the pulse shaper allows to delay the slowly varying envelope without changing the phase delay of the carrier wave. On the right side in figure 1 the result of such a measurement is presented. The rapid oscillations have disappeared and only an intensity-like autocorrelation trace remains. Moreover, the pulse shaper allows switching between the two exit ports of the interferometer by applying an additional phase shift of π to the delayed replica in the transfer function. The resulting curve is the complementary signal to the previous measurement in the sense, that the sum of their energies is equal to the amount of energy entering the interferometer. From the two intensity-like autocorrelation traces, we can directly derive the visibility of the two-photon wave function. All experiments agree well with the appropriate simulations (grey background).

Finally, we demonstrate the possibility to perform shaper-based quantum optical experiments. To that end two different transfer functions were applied to the signal and the idler part of the spectrum. Through spectral amplitude filtering in one beam path it is demonstrated that the other beam path is modified accordingly when the coincidences

97

are detected. Recently, Bellini et. al performed similar experiments [3] based on filters and standard mechanical interferometers. Here, we replaced all optical elements by applying the corresponding transfer functions to the pulse shaping apparatus.

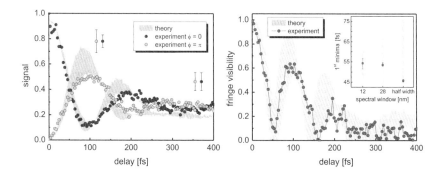

Fig. 2. Due to the entanglement the spectral narrowing of the idler photon influences the intensity-like autocorrelation of the signal photon and the extracted visibility function. The inset shows the position of the first minimum of the visibility as a function of the width of the spectral slit.

An example is shown in figure 2. Amplitude and phase modulation are required, first, to spectrally filter the idler side of the spectrum and, second, to realize a shaper-assisted interferometer on the signal part of the spectrum. For a fixed width of the spectral slit the shaper-based interferometer performs an intensity-like measurement of both exit ports of the simulated interferometer. The result of such a measurement is shown on the left side of figure 2. From these two traces the fringe visibility is constructed and the position of the first minimum is a reasonable measure for the coherence time of the wave function. The inset shows the position of the first minimum as a function of the width of the spectral slit, clearly demonstrating that the temporal coherence decreases as the width of the spectral slit increases.

Conclusions

We demonstrated that time-energy entangled photon pairs can be modulated in phase and amplitude similar to broadband laser pulses. The wave function was analyzed through autocorrelation methods and shaper-based quantum optics experiments were demonstrated.

Acknowledgements. This work was supported by NCCR Quantum Photonics (NCCR QP), research instrument of the Swiss National Science Foundation (SNSF).

1   B. Dayan, A. Pe'er, A.A. Friesem, and Y. Silberberg, "Temporal Shaping of Entangled Photons," Phys. Rev. Lett. 94, 073601 (2005).
2   A.M. Weiner, "Femtosecond pulse shaping using spatial light modulators," Rev. Sci. Instrum. 71, 1929-1960 (2000).
3   M. Bellini, F. Marin, S. Viciani, A. Zavatta, and F. T. Arecchi, "Nonlocal Pulse Shaping with Entangled Photon Pairs," Phys. Rev. Lett. 90, 043602 (2003).

Part II

Ultrafast X-ray and Electron Science

Ultrafast Structural Dynamics of Polar Solids Studied by Femtosecond X-Ray Diffraction

T. Elsaesser[1], C. von Korff Schmising[1], N. Zhavoronkov[1], M. Bargheer[1,2], M. Woerner[1], M. Braun[3], P. Gilch[3], W. Zinth[3], I. Vrejoiu[4], D. Hesse[4], M. Alexe[4]

[1] Max-Born-Institut für Nichtlineare Optik und Kurzzeitspektroskopie, D-12489 Berlin, Germany, E-mail: elsasser@mbi-berlin.de
[2] Institut für Physik, Universität Potsdam, D-14469 Potsdam, Germany
[3] BioMolekulare Optik und MAP, Department für Physik, Ludwig-Maximilians-Universität, D-80538 München, Germany
[4] Max-Planck-Institut für Mikrostrukturphysik, D-06120 Halle, Germany

Abstract. We discuss recent progress in ultrafast x-ray diffraction, addressing photoinduced structural dynamics in ferroelectric superlattices and polar molecular crystals. Elongations of coupled phonon modes affecting ferroelectric polarizations and structural changes connected with the solvation of molecular dipoles are determined quantitatively.

Introduction

Function of physical, chemical, and biological systems is frequently connected with ultrafast changes of electronic and/or nuclear structure. X-ray diffraction with a femtosecond time resolution has developed into an important tool to probe such changes most directly by determining transient atomic and/or molecular positions [1,2]. Presently, ultrafast x-ray diffraction undergoes a rapid development based on novel approaches for generating ultrashort hard x-ray pulses in laser-driven [3] or accelerator based sources [4,5]. In this paper, we discuss the results of recent prototype experiments addressing the photoinduced structural dynamics of polar solids. In a first series of measurements, we study lattice dynamics in superlattices consisting of metal/ferroelectric nanolayers. We demonstrate that ultrafast intensity changes on two Bragg reflections provide direct information on the coupled dynamics of the two phonon modes relevant for ferroelectricity in perovskites. In a second experiment, we unravel the structural changes connected with the solvation of a molecular dipole in a crystalline environment.

Experimental Techniques

In our laser-driven x-ray plasma source, pulses of 45 fs duration, up to 5 mJ energy, and a 1 kHz repetition rate are focused onto a copper tape (thickness 20 μm) to generate Cu K_α emission (E=8.05 keV) in the forward direction [3]. A polymer tape protects the focussing lens against debris emanating from the target. The x-ray spectrum consists of strong K_α and K_β components and a much weaker background emission of the hot plasma which extends up to photon energies of 100 keV. The total K_α photon flux is up to 6×10^{10} photons/s. The pulse duration is estimated to be less than 200 fs. The x-ray emission originates from a point-like area of 10 μm diameter only, facilitating the imaging of the x-ray output by x-ray optics [6]. Advanced multilayer mirrors collect a solid angle of $\sim 10^{-3}$ with reflectivities of up to 0.8 for K_α radiation. This results in an effective x-ray flux in the small spot on the sample (spot size 30 μm) of up to several 10^6 photons/s. We apply a pump-probe scheme where a sub-50 fs optical pump pulse induces structural dynamics that are probed by diffracting a femtosecond hard x-ray pulse from the excited sample. Changes in intensity and position of a single or several Bragg diffraction peaks are measured as a function of pump-probe delay using an x-ray CCD camera as a position sensitive detector. The relative timing between optical pump and x-ray probe pulses is known with an accuracy of less than 100 fs from all-optical measurements in the same setup.

Lattice dynamics in ferroelectric nanolayers

Displacive ferroelectricity of crystals with a perovskite structure is essentially determined by two lattice coordinates: the tetragonal distortion η, the ratio of the out-of- and in-plane lattice constant, and the soft mode ξ, the displacement of positive and negative ions within the unit cell, causing the macroscopic electric polarization P. The two phonon modes are anharmonically coupled, i.e., an elongation along η is connected with a change of ξ. Such elongations result in changes of P and should thus allow for polarization switching on a time scale set by the periods of lattice vibrations. Applying ultrafast x-ray diffraction, we directly map real-time lattice and polarization dynamics of nanolayered perovskites. We studied a $PbZr_{0.2}Ti_{0.8}O_3/SrRuO_3$ (PZT/ SRO) superlattice sample containing n=15 pairs of PZT/SRO layers of 5nm/6nm thickness. Optical excitation of the metallic SRO layers launches a coherent superlattice phonon mode that modulates the SRO and PZT layer thicknesses periodically by up to 2 percent with a period of 2 ps, thereby changing the tetragonal distortion η. The time-dependent x-ray intensities (Fig. 1) measured on two different superlattice Bragg reflections [7] provide complementary information on the coupled dynamics of the lattice coordinates η and ξ [8]. The tetragonal distortion η

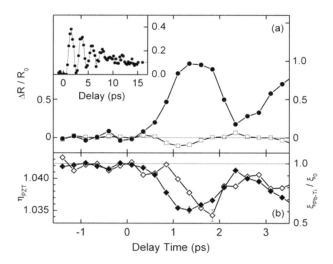

Fig. 1. (a) Transient change of the x-ray reflectivity of a PZT/SRO superlattice on the (0 0 56) (solid circles) and (0 0 55) (open squares) superlattice peaks of the diffraction pattern. Inset: Time evolution of the (0 0 56) reflectivity change of delay times up to 15 ps. (b) Transient elongations along two different lattice coordinates, the tetragonality η(t) (solid diamonds, left ordinate scale) and the soft mode coordinate ξ(t) (open diamonds, right ordinate scale).

displays a maximal change after ~1.5 ps. The concomitant motion along the (coupled) soft mode results in a reduction of the elongation by up to 100 percent with an additional 0.5 ps delay. As a result, the ferroelectric polarization P is reduced, i.e., switched off on a time scale of a few picoseconds. The ultrafast character of the polarization changes induced is highly relevant for ferroelectric switching devices and can be optimized by tailoring the superlattice structure.

Structural dynamics of polar dipole solvation

Solvation of molecular dipoles, i.e., the lowering of the dipole energy by a rearrangement of the surrounding polar environment represents a basic process occurring in any polar medium. In our experiments, we directly map for the first time rearrangement processes that occur during a charge-transfer induced dipole solvation process [9]. Single crystals of 4-(diisopropylamino)benzonitrile (DIABN) [10] are studied as a model system undergoing photoinduced intramolecular charge transfer. Charge transfer in the excited state leads to a doubling of the ground state dipole moment of approximately 8.5 Debye and occurs with a rate depending on the polarity of the environment. A 50 fs pulse at a wavelength of 400 nm excites a small fraction of 10^{-4} of the chromophores in the crystal which represents a highly polar ordered environment. The resulting change of the crystal structure is monitored by femtosecond x-ray pulses diffracted from or transmitted through the excited crystal. The (004) and (006) Bragg peaks of the monoclinic crystal structure display a strong 10 percent change of intensity that builds up with a time constant of 10 ps. This time constant is identical to the charge transfer time of DIABN measured independently in femtosecond pump-probe studies of transient electronic and vibrational absorption. The strong changes of the diffracted and transmitted x-ray intensity are due to the collective response of unexcited molecules to the local photoinduced dipole change. In contrast, geometry changes of the diluted excited chromophores themselves play a negligible role. The local dipole change results in a torque exerted on the unexcited surrounding dipoles, causing the angular reorientation of molecules towards a spherical symmetry. In this process, the separation of lattice planes in the crystal remains practically unchanged. A theoretical analysis of the x-ray data gives rotation angles up to 10 degrees, depending on the distance from the excited dipole.

1   A. Rousse, C. Rischel, and J. C. Gauthier, Rev. Mod. Phys. **73**, 17, 2001.
2   C. von Korff Schmising, M. Bargheer, M. Woerner, and T. Elsaesser, Z. Kristallogr. **223**, 283, 2008.
3   N. Zhavoronkov, Y. Gritsai, M. Bargheer, M. Woerner, T. Elsaesser, F. Zamponi, I. Uschmann, and E. Förster, Opt. Lett. **30**, 1737, 2005.
4   B. Beaud et al., Phys. Rev. Lett. **99**, 174801, 2007.
5   A. M. Lindenberg et al., Science **308**, 392, 2005.
6   M. Bargheer, N. Zhavoronkov, R. Bruch, H. Legall, H. Stiel, M. Woerner, and T. Elsaesser, Appl. Phys. B **80**, 715, 2005.
7   M. Bargheer, N. Zhavoronkov, Y. Gritsai, J. C. Woo, D. S. Kim, M. Woerner, and T. Elsaesser, Science **306**, 1771, 2004.
8   C. von Korff Schmising, M. Bargheer, M. Kiel, N. Zhavoronkov, M. Woerner, T. Elsaesser, I. Vrejoiu, D. Hesse, and M. Alexe, Phys. Rev. Lett. **98**, 257601, 2007.
9   M. Braun, C. von Korff-Schmising, M. Kiel, N. Zhavoronkov, J. Dreyer, M. Bargheer, T. Elsaesser, C. Root, T. E. Schrader, P. Gilch, W. Zinth, and M. Woerner, Phys. Rev. Lett. **98**, 248301, 2007.
10 W. Frey, C. Root, P. Gilch, and M. Braun, Z. Kristallogr. NCS **219**, 291, 2004.

Atomic Motion in Laser Excited Bismuth Studied with Femtosecond X-Ray Diffraction

Paul Beaud[1], Steven L. Johnson[1], Christopher J. Milne[2], Faton S. Krasniqi[1], Ekaterina Vorobeva[1], and Gerhard Ingold[1]

[1] Swiss Light Source, Paul Scherrer Institut, CH-5232 Villigen, Switzerland
[2] Laboratoire de Spectroscopie Ultrarapide, Ecole Polytechnique Fédérale de Lausanne, CH-1015 Lausanne, Switzerland

Abstract. Asymmetric grazing incidence femtosecond x-ray diffraction is applied to investigate carrier transport, carrier relaxation and phonon coupling in laser excited bismuth crystals.

Introduction

The femtosecond lattice dynamics of bismuth is an ongoing area of interest. The time dependent response has been measured either by optical methods in the near infrared [1–3] or by probing directly the structural dynamics using x-ray pulses [4,5]. Progress has also been made in developing computational models [5-7] that can be directly compared to the experiments. Strong optical excitation of the crystal results in large amplitude coherent A_{1g} optical phonons, corresponding to motion of atoms along the body diagonal of the rhombohedral unit cell. Recent experiments performed at 7 K have shown evidence of impulsive excitation of the E_g phonon mode and coupling between the A_{1g} and E_g optical phonon modes [3]. These observations may be explained by 3-D density functional theory (DFT) calculations showing that at strong laser excitations the phonon modes become strongly anharmonic and coupling between them significant [6].

Experimental Methods

Very recently we have used highly asymmetric x-ray diffraction from single crystals of bismuth to measure high-amplitude A_{1g} phonon dynamics. These results allow us to estimate the pulse length and timing stability of the femtosecond hard x-ray pulses generated at the Swiss Light Source (SLS) by implementing the laser electron beam slicing technique first demonstrated at the ALS [8]. It delivers at 2 kHz temporally and spatially stable x-ray pulses tunable in energy from $4 - 12$ keV with a duration of 140 ± 30 fs and low timing drifts of 30 fs rms measured over several days [9].

The use of bismuth crystals in conjunction with highly asymmetric x-ray diffraction allows us to vary the probe depths over a nanoscale range (7 - 150 nm) simply by tuning the x-ray incidence angle (0.4 - 2°). This is a significant advantage over previous studies, which used either symmetric x-ray diffraction from thin films to obtain dynamics averaged over the 50 nm film thickness, or optical probes that are limited to the near surface region. The depth-resolved data can be compared to models and yield an estimate of the effective diffusion rate of 2.3 ± 0.3 cm^2/s for the highly excited carriers and an electron-hole thermalization time of 260 ± 20 fs [10].

In addition, X-ray diffraction also allows investigation of atomic motion perpendicular to the [111] direction associated with A_{1g} optical phonon mode. This is of particular interest in view of the recently reported evidence of impulsive excitation of the E_g phonon mode and A_{1g} - E_g coupling [3].

Results and Discussion

As shown in Figure 1, diffraction from the (1-10) and (1-21) Bragg planes reveal an initial drop over the first 400 fs and continue at later times to decrease at a slower rate. No coherent time structure is observed that could be attributed to the E_g phonon (~1.62 THz [6]) or to coherent coupling from the A_{1g} phonon. However, at room temperature these effects may be concealed due to the large thermal population of the E_g mode [3]. To definitely exclude any significant coupling from the large amplitude coherent A_{1g} phonon to the atomic motion in the (111) plane we applied a sequence of two pulses to control the population of the A_{1g} phonon mode. In this double-pump control experiment, the amplitude of the coherent atomic motion can be manipulated through the delay of the second pulse by keeping the electronic excitation constant. Any significant coupling from the A_{1g} mode to the motion in the perpendicular plane should lead to a change of the corresponding Bragg intensity when the pulse sequence is adjusted to either maximize or to cancel the coherent A_{1g} amplitude.

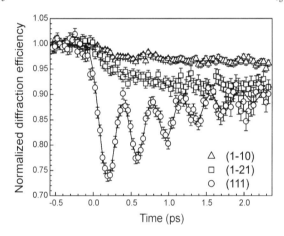

Fig. 1. Normalized integrated diffracted intensity from the bulk (○) Bi(111), (□) Bi(1-21) and (△) Bi(1-10) Bragg reflection at 7.15 keV as a function of pump-probe delay for an absorbed excitation fluence of 1.4 mJ/cm². The curve is a fit to the (111) data using a simple model of displacive excitation yielding a phonon frequency of 2.58 ± 0.04 THz and a relaxation of the electronic excitation within 3.2 ± 0.4 ps.

The control of the A_{1g} amplitude for two different delay settings of the exciting pulse pair is shown in Figure 2 (lower panel). The corresponding (1-21)-transients are shown in the upper panel of the figure. Within the accuracy of our experiment we do not observe any change of the (1-21)-diffraction signal and we conclude that at room temperature coupling from the A_{1g} to the perpendicular E_g phonon mode is not measurable.

The observed decrease in diffraction efficiency appears instead to be an increase of the direction projected incoherent rms atomic motion, driven by the electronic excitation of the crystal. At least two possible mechanisms for this increase exist. First, both experiments and model calculations predict a strong softening of phonon modes which involve atomic motion perpendicular to the [111] direction upon electronic excitation [7]. This drop in frequency of these modes results in a larger rms displacement of atoms, that causes the observed drop in x-ray intensity via the Debye-Waller factor. Another possible mechanism is electron-phonon scattering that actually transfers energy from the carriers to the lattice at a rate that depends on the statistical distribution of electrons and holes. Work to quantify the relative contributions of these two mechanisms is underway [11].

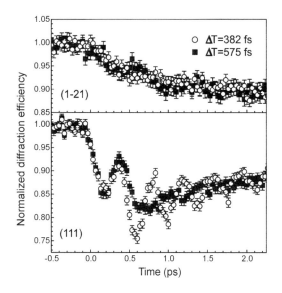

Fig. 2. Measured response of the Bi(111) (lower) and Bi(1–21) diffraction signals (upper panel) controlled by a 2-pulse excitation scheme for the two delay settings of the pump pulse pair. The absorbed energy density was 1.09 mJ/cm^2 for the first and 0.77 mJ/cm^2 for the second pulse, respectively.

Acknowledgements. We would like to thank Daniel Grolimund and Camelia Borca for assistance, and Rafael Abela for helpful discussions and continuous support.

1   H. J. Zeiger, *et al.*, Phys. Rev. B **45**, 768 (1992).

2   M. Hase, M. Kitajima, S. Nakashima, K. Mizoguchi, Phys. Rev. Lett. **88**, 067401 (2002).

3   O. V. Misochko, K. Ishioka, M. Hase, and M. Kitajima, J. Phys.: Condens. Matter **18**, 10571 (2006).

4   K. Sokolowski-Tinten, *et al.*, Nature (London) **422**, 287 (2003).

5   D.M. Fritz, *et al.*, Science **315**, 633 (2007).

6   E. S. Zijlstra, L. L. Tatarinova, and M. E. Garcia, Phys. Rev. B **74**, 220301 (2006).

7   É. D. Murray, S. Fahy, D. Prendergast, T. Ogitsu, D. M. Fritz, and D. A. Reis, PRB **75**, 184301 (2007).

8   R. W. Schoenlein, *et al.* , Science **287**, 2237 (2000).

9   P. Beaud, S. L. Johnson, A. Streun, R. Abela, D. Abramsohn, D. Grolimund, F. Krasniqi, T. Schmidt, V. Schlott, and G. Ingold, Phys. Rev. Lett. **99**, 174801 (2007).

10   S. L. Johnson, P. Beaud, C. J. Milne, F. S. Krasniqi, E. S. Zijlstra, M. E. Garcia, M. Kaiser, R. Abela, and G. Ingold, Phys. Rev. Lett. **100**, 155501 (2008).

11   S. L. Johnson, *et al.*, to be published.

Femtosecond X-ray Diffraction Study of the Ultrafast Coupling between Magnetization and Structure in the Ferromagnet SrRuO$_3$

C. v. Korff Schmising[1], M. Bargheer[1,2], A. Harpoeth[1], N. Zhavoronkov[1], Z. Ansari[1], M. Woerner[1], T. Elsaesser[1], I. Vrejoiu[3], D. Hesse[3] and M. Alexe[3]

[1] Max-Born-Institut für Nichtlineare Optik und Kurzzeitspektroskopie, 12489 Berlin, Germany
   E-mail: woerner@mbi-berlin.de
[2] Universität Potsdam, Institut für Physik,14469 Potsdam, Germany
[3] Max-Planck-Institut für Mikrostrukturphysik, Weinberg 2, 06120 Halle, Germany

Abstract. Femtosecond optical excitation of magnetically ordered SrRuO$_3$ nanolayers leads to an ultrafast demagnetization and a concomitant magnetoelastic contractive stress. The resulting ultrafast structural response of the sample is imaged by femtosecond X-ray diffraction.

Introduction

The correlation of electrons leading to phenomena like ferroelectricity, superconductivity and ferromagnetism and its ultrafast interaction with structural degrees of freedom play a fundamental role in solid state physics, e.g., the metal-insulator transition in VO$_2$ [1]. In particular, materials with a perovskite crystal structure display both strongly correlated electrons leading to electronic phase transitions and structural changes of the crystal lattice in response to electronic correlations. A prominent example is the itinerant ferromagnet SrRuO$_3$ (SRO) [2] which becomes ferromagnetic below the Curie temperature of $T_C \approx 165$ K. SRO exhibits the "Invar effect" [3]: the itinerant electron magnetic substance has a negative contribution to thermal expansion and compensates the normal lattice expansion due to lattice vibrations. Nanolayers of SRO show an extraordinarily strong dependence of the saturation magnetization as a function of the tetragonal distortion of the crystal, i.e., at $T = 0$ the crystal exhibits a huge magneto-elastic effect [4,5]. Recent femtosecond magneto-optical Kerr measurements [6] show that the macroscopic magnetization of SRO can be reduced almost instantaneously (≈ 200 fs) by optical excitation with pump pulses of moderate fluence (30 μJ cm^{-2}). This raises the question if and how fast an ultrafast manipulation of the magnetic system causes mechanical stress in the crystal.

Experiment

Here, we present an ultrafast time resolved x-ray structure analysis to directly measure the femtosecond buildup of optically induced uniaxial stress in the ferromagnetic nanolayers of a SRO/SrTiO$_3$ superlattice (SL). The sample studied here was fabricated by pulsed laser deposition and consists of 10 periods of 15.24 nm STO and 7.48 nm SRO. The equilibrium structure of the sample is derived from high resolution static x-ray reflectivity measurements in combination with dynamic x-ray diffraction theory described in Refs. [7,8]. In the femtosecond experiments, the sample is excited by a 50 fs pump pulse which interacts exclusively with the SRO layers. The ultrashort hard X-ray probe pulses (Cu K$_\alpha$, photon energy 8.05 keV, $\lambda = 0.154$ nm) are derived from a laser-driven plasma source and are diffracted of the excited sample to image the resulting structural changes. Changes of the diffracted intensity are measured as a function of pump-probe delay. For the experiments presented here, we placed the sample in a cryostat to control the sample temperatures between $T = 20$ and 300 K.

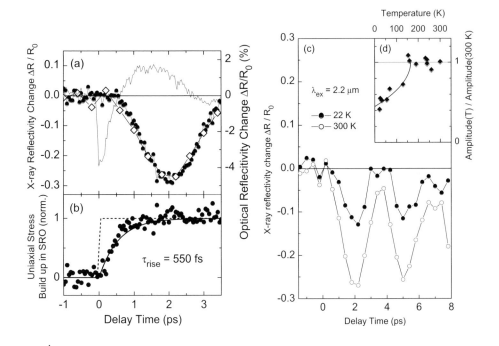

Fig. 1. (a) Transient change of the x-ray reflectivity $\Delta R/R_0$ after ultrafast optical excitation with $\lambda_{ex} = 800$ nm (circles) and $\lambda_{ex} = 2.2\mu$m (diamonds) of the SRO layers in a STO/SRO superlattice ($T = 300$ K). Thin line: (800 nm)-pump / (800 nm)-probe experiments serve for an exact determination of time delay zero. (b) Time evolution of the uniaxial stress buildup in the SRO layers derived from the curves in (a) according to the differential equation of a forced oscillator. (c) Measured transients of the x-ray reflectivity with $\lambda_{ex} = 2.2\ \mu$m for two different lattice temperatures. (d) Lattice temperature dependence of the amplitude of the SL phonon oscillation normalized to its value at room temperature.

Results and Discussion

Pump-probe measurements were performed in a broad range of excitation wavelengths between 800 nm and 2.2 μm. In Fig. 1(a), we present the transient change of the x-ray reflectivity $\Delta R/R_0$ at room temperature after ultrafast optical excitation at $\lambda_{ex} = 800$ nm (circles) and $\lambda_{ex} = 2.2\ \mu$m (diamonds). The particular SL Bragg reflection was chosen to be most sensitive to the particular SL phonon which modulates the layer thicknesses d_{SRO} and d_{STO} while keeping the SL period $d_{SL} = d_{SRO} + d_{STO}$ constant [8]. The transients show a delayed rise and oscillations with a period determined by d_{SL} and the velocity of sound. The zero time delay is measured precisely, using an all-optical reflectivity measurement with the attenuated x-ray generating 800 nm beam as a probe [thin line in Fig. 1(a)]. Data taken with other pump wavelengths show the same time behavior.

The amplitude of the signal is determined by the absorbed pump fluence, independent of the pump wavelengths. We derive the driving force $F(t)$ (uniaxial stress) acting on the SL phonon oscillator from the differential equation $d^2\Delta d_{SRO}/dt^2 - \omega_0^2 \Delta d_{SRO} = F(t)$ with the frequency ω_0 of the experimentally observed SL oscillation. The force $F(t)$ [Fig. 1(b)] shows a rise time of $\tau_{rise} = 550$ fs. This time constant represents the electron-lattice equilibration time, reflecting the build-up of an (incoherent) phonon population, which is the predominant source of the uniaxial stress at room temperature. This stress generation mechanism is described by the (lattice) Grüneisen constant, similar to femtosecond electron diffraction experiments on aluminum [9].

The situation changes drastically if we cool the sample down to temperatures $T < T_C$ at which the electronic system is in the ferromagnetic phase. Results from a series of measurements with a constant pump fluence at an excitation wavelength $\lambda_{ex} = 2.2\mu m$ are presented in Fig. 1(c,d). For this pump wavelength, the amplitude of the SL phonon depends on the lattice temperature T with a strong reduction below T_C. In contrast, the rise time of the x-ray reflectivity change as well as the phase is temperature independent, pointing to a negligible change of the electron-lattice equilibration time. For a constant amount of absorbed pump energy and a temperature independent Grüneisen constant, one expects the SL phonon driving stress amplitude to be independent of the lattice temperature. Thus, the observed behavior gives evidence of additional contributions to the photogenerated stress. It has been shown that a significant magneto-elastic effect exists in the ferromagnetic phase of SRO [4,5]. As a result, a reduction of the magnetization leads to a contraction of SRO. Recent femtosecond studies have revealed a quasi-instantaneous reduction of the magnetization upon optical excitation at 800 nm [6]. The data points in Fig. 1(d) follow the temperature dependent magnetization (solid line), revealing a contractive magneto-elastic contribution to the stress driving the SL phonon.

Conclusion

We have identified different mechanisms of ultrafast stress generation in the itinerant ferromagnet SRO. Measuring photoinduced structural dynamics in real-time by ultrafast x-ray diffraction, we provide the first direct evidence for a subpicosecond magnetostriction in nanolayered SRO. The amplitude of the magnetostrictive component of transient stress decreases with increasing temperature, mimicking the temperature dependent magnetization.

1 A. Cavalleri et al., in Physical Review Letters, Vol. 87, 237401, 2001.
2 P. B. Allen et al., in Physical Review B, Vol. 53, 4393, 1996.
3 T. Kiyama et al., in Physical Review B, Vol. 54, R756, 1996.
4 Q. Gan et al., in Applied Physics Letters, Vol. 72, 978, 1998.
5 K. Terai et al., in Japanese Journal of Applied Physics, Vol. 43, L227, 2004.
6 T. Ogasawara et al., in Physical Review Letters, Vol. 94, 087202, 2005.
7 C. von Korff Schmising et al., in Physical Review Letters, Vol. 98, 257601, 2007.
8 C. von Korff Schmising et al., in Applied Physics B, Vol. 88, 1, 2007.
9 S. Nie et al., in Physical Review Letters, Vol. 96, 025901, 2006.

Electron-Phonon Energy Transfer in Bismuth Observed by Ultrafast Electron Diffraction

Ivan Rajkovic[1], Manuel Ligges[1], Ping Zhou[1], Thomas Payer[1,2], Frank Meyer zu Heringdorf[1,2], Michael Horn-von Hoegen[1,2], and Dietrich von der Linde[1]

[1] Institut für Experimentelle Physik, Universität Duisburg-Essen, Lotharstrasse 1, 47048 Duisburg, Germany
E-mail: ivan.rajkovic@uni-duisburg-essen.de

[2] Center for Nanointegration, Duisburg-Essen (CeNIDE), Universität Duisburg-Essen, Lotharstrasse 1, 47048 Duisburg, Germany

Abstract. We describe time resolved electron diffraction on bismuth films. Lattice heating following femtosecond laser excitation is observed via the transient Debye-Waller effect. Our results indicate different heating processes with different time constants.

Introduction

The development of ultrashort electron and X-ray pulses has made it possible to observe the atomic motion in crystals with a temporal resolution less than one picosecond. Several types of ultrafast processes, e.g. phase transitions [1], carrier relaxations [2] and coherent acoustic phonons [3], have previously been studied using time resolved electron diffraction. Here we present the experimental results of time resolved electron diffraction on bismuth films with a temporal resolution less than one picosecond.

Experimental Methods

In our experiments we used a Ti:Sapphire chirped pulse amplification laser system working at 800 nm with a pulse duration of 40 fs and a repetition rate of 1 kHz. The electron source consists of a photocathode coated with a 20 nm silver film, illuminated by the third harmonic of the laser pulses. The photoelectrons released from the cathode are accelerated to a kinetic energy of 30 keV over a distance of 3 mm. The electron beam emerges from the anode through a 100 μm pinhole. Using a magnetic lens the electron beam is focused to a spot of 450 μm diameter on the MCP plate, which serves as a detector for the diffraction patterns. The use of an electron source with high pulse repetition frequency relaxes space charge restrictions and enables efficient signal averaging techniques to be used for the recording of the electron diffraction signals. Thus subtle reversible structural changes can be measured.

To study the electron-phonon energy transfer in bismuth we used free standing Bi films with a thickness of 15 and 22 nm positioned directly after the magnetic lens. The bismuth sample was epitaxially grown by evaporating bismuth onto polished NaCl [4]. The thin film sample was than floated off in the water and picked up by a support mesh. The inset in Fig. 1(a) shows a recording of the transmitted electron diffraction pattern of this film (electron probe pulses only, no laser excitation). The diffraction pattern with 12-fold symmetry is characteristic of a Laue pattern from a crystalline Bi film growing in different crystal domains [5].

Ultrafast Heating of Bismuth

We are interested in the relaxation processes following photo-excitation of the Bi-film by a femtosecond laser pulse at 800 nm. The electronic excess energy is transferred to the lattice by electron-phonon scattering processes leading to an increase of the lattice temperature. The temporal evolution of the lattice temperature can be monitored by electron diffraction. The thermal motion of the lattice leads to a decrease of the diffraction intensity in the various diffraction spots *(hkl)* according to the Debye-Waller formula:

$$\frac{I_{hkl}}{I_0} = exp\left(-\frac{4\pi^2}{3}\frac{<u^2>}{d_{hkl}^2} \right) \tag{1}$$

Here, $<u^2>$ is the mean square atomic displacement which is proportional to the lattice temperature T when the lattice is in thermal equilibrium, and d_{hkl} is the distance of the lattice planes corresponding to the Miller indices *(hkl)*.

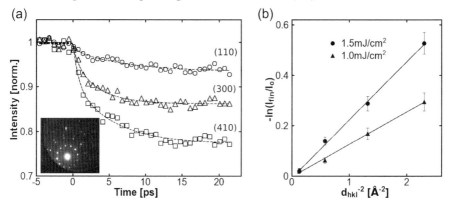

Fig. 1. (a) Temporal evolution of the normalized diffraction intensity for three diffraction orders (given in hexagonal notation). Data points are fitted with a sum of two exponentials. The insert in the lower left corner shows the diffraction pattern. (b) Dependence of the measured Debye-Waller factor on the lattice plane distance for two different pump fluences

In Fig. 1(a) the measured integrated diffraction intensity of different spots is plotted as a function of the delay time between the laser pump pulse and the electron probe pulse. A fast initial drop in about 1 ps followed by slower decrease is observed. The diffraction intensity reaches its final value after about 20 ps. No further changes could be observed over a delay time range of 200 ps. The final values for different diffraction orders are in excellent agreement with the Debye-Waller law given above (Fig. 1(b)), supporting the interpretation of the observed decrease in the diffraction intensity as being due to lattice heating. The observed temperature increase is approximately 130 K and 220 K for the pump fluence of 1 mJ/cm^2 and 1.5 mJ/cm^2, respectively. The calculated values for the temperature increase from the absorbed energy (112 K and 168 K) are in agreement with the measured data. The film used in this experiment had a thickness of 22 nm.

Because in our experiment the laser pump pulse is incident from the backside of the sample at an angle of 45° the excitation process is stretched to approximately 500 fs across the excited area of the sample which is probed by an electron probe pulse at normal incidence on the front. The pulse duration of the electron pulse is estimated to be about 600 fs, being due to the velocity spread of the photo-emitted electrons at the cathode and some residual space charge broadening (we used approximately 2000 electrons per pulse).

The two different time constants obtained from the fit curves of our measured data (Fig. 1(a)) are 0.7 ps and 6 ps. This result indicates that two distinct heating processes are operative. We suggest the following tentative interpretation.

The initial fast heating could be due to the lattice excitation resulting directly from the emission of phonons during the relaxation of the hot electrons. The slow component could be due to the so called delayed Auger heating as observed, for example, in Si [6]. When electron-hole pairs recombine by an Auger process the recombination energy is transferred to a third carrier giving rise to a source of hot carriers on the time scale of the Auger recombination. It is conceivable that such a process can also occur in a semimetal such as Bi, given the fact that thin bismuth films can develop a band-gap and become a semiconductor [7].

Another possibility is the following. It is well known that in Bi the fully symmetric A_{1g} optical phonon mode can be coherently excited by laser excitation [8,9]. The relaxation of this coherent mode could lead to a distinct contribution to the lattice heating on a time scale different from the electron-phonon scattering time.

Acknowledgements. We would like to gratefully acknowledge financial support by the Deutsche Forschungsgemeinschaft through SFB 616 "Energy dissipation at surfaces". We would also like to acknowledge Prof. R.J. Dwayne Miller and his group from the University of Toronto for their assistance with this project.

1   B. J. Siwick, J. R. Dwyer, R. E. Jordan, R. J. Dwayne Miller, Science **302**, 1382, 2003.

2   M. Harb, R. Ernstorfer, T. Dartigalongue, C. T. Hebeisen, R. E. Jordan, and R. J. Dwayne Miller, J. Phys. Chem. B **110**, 25308, 2006.

3   H. Park, X. Wang, S. Nie, R. Clinite, J. Cao, Solid State Communications **136**, 559, 2005.

4   T. Payer, I. Rajkovic, M. Ligges, D. von der Linde, M. Horn-von Hoegen, and F.-J. Meyer zu Heringdorf, to be published

5   G. Jnawali, H. Hattab, B. Krenzer, and M. Horn-von Hoegen, Phys. Rev. B **74**, 195340, 2006.

6   M. C. Downer and C. V. Shank, Phys. Rev. Lett. **56**, 761, 1986.

7   C. A. Hoffman, J. R. Meyer, F. J. Bartoli, X. J. Yi, C. L. Hou, H. C. Wang, J. B. Ketterson, and G. K. Wong, Phys. Rev. B **48**, 11431, 1993.

8   M. Hase, K. Mizoguchi, H. Harima, S. Nakashima, and K. Sakai, Phys. Rev. B **58**, 5448, 1998.

9   K. Sokolowski-Tinten, C. Blome, J. Blums, A. Cavalleri, C. Dietrich, A. Tarasevitch, I. Uschmann, E. Förster, M. Horn-von-Hoegen, and D. von der Linde, Nature **422**, 287, 2003

Atomic View of the Photoinduced Collapse of Gold and Bismuth

R. Ernstorfer[1], M. Harb[1], C.T. Hebeisen[1], G. Sciaini[1], T. Dartigalongue[1], I. Rajkovic[2], M. Ligges[2], D. von der Linde[2], Th. Payer[3], M. Horn-von-Hoegen[3], F.-J. Meyer zu Heringdorf[3], S. Kruglik[1], R.J.D. Miller[1]

[1]Institute for Optical Sciences and Departments of Chemistry and Physics, University of Toronto, 80 St. George St., Toronto, Ontario M5S 3H6, Canada
E-mail: dmiller@lphys.chem.utoronto.ca
[2]Fachbereich Physik, Universität Duisburg-Essen, 47057 Duisburg, Germany
[3]Fachbereich Physik and Center for Nanointegration, Universität Duisburg-Essen (CeNIDE), 47057 Duisburg, Germany

Abstract. Two different mechanisms of photoinduced melting were studied by femtosecond electron diffraction. The structural response of gold indicates an electronically-induced increase of the melting temperature. Bismuth was found to disorder within one vibrational period.

Introduction

In recent years, time-resolved diffraction techniques (electrons and X-rays) have become available to study the dynamics of photoinduced phase transitions. Intense optical excitation initially creates a strong non-equilibrium between the electrons and the lattice. The structural response of crystalline systems under this condition strongly depends on the effect of electronic excitation on the interatomic potential energy landscape. In free-electron metals like aluminum, the lattice stability appears to be mostly unaffected by electronic excitation. Subsequent to the optical excitation, electron-phonon scattering heats the lattice and results in thermal melting on the picosecond timescale [1]. In contrast, the excitation of semiconductors weakens the covalent bonding, softens the lattice and, at an excitation level of approximately 10% of the valence electrons, leads to the collapse of the transverse acoustic phonon bands, resulting in electronically-driven non-thermal melting [2]. We used femtosecond electron diffraction (FED) to study the photoinduced melting of two other classes of systems, namely gold, a noble metal, and bismuth, a semimetal with a Peierls-distorted crystal structure. In the case of gold, it has been theoretically predicted that strong electronic excitation induces an increase in the interatomic forces resulting in an increase of its melting point [3]. On the other hand, impulsive electronic excitation of bismuth shifts the minimum of its potential energy surface and launches coherent, large-amplitude optical phonons, equivalent to excited-state wave packet motion in a molecular system. For excitation below the melting threshold, a softening of the interatomic potential with increasing carrier density has been observed [4].

Experimental Methods

An essential requirement for any FED experiment is the ability to generate high density electron pulses with sub-picosecond durations. As the photoinduced melting of the samples is irreversible and the available sample area is limited, the number of electrons per pulse must be high enough to obtain data with a sufficient signal-to-noise ratio

113

in few shots. In order to limit space-charge broadening of the electron pulse during propagation, we employ a compact design comprising of a 55 kV DC electron gun and a magnetic lens. The photocathode-to-sample distance is < 30 mm. Approximately 10 individual acquisitions are averaged per given time delay. The electron pulse durations have been characterized using ponderomotive scattering of the electron pulse by an optical pulse [5]. The samples, free-standing, 111-oriented gold and bismuth films of 20 and 30 nm thickness, respectively, are excited with the second harmonic of a Ti:sapphire laser system. The instrument response function (given by electron and pump pulse durations as well as the shift of the sample position during the raster scan) is ≤ 450 fs, sufficient to resolve the dynamics of interest. The fastest dynamics of Bi were further verified with 200 fs electron pulses and an instrument response function of < 300 fs.

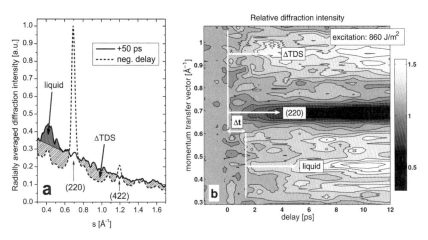

Fig. 1. (a) Diffraction pattern of 111-oriented (neg. delay) and hot liquid (+50 ps) gold. Excitation conditions: 860 J/m^2. (b) Temporal evolution of the diffraction pattern shown as relative diffraction intensity I(s,t)/I(s,t<0), where I(s,t<0) denotes the diffraction signal averaged over all negative delay points.

Results and discussion

Fig. 1a shows the radially averaged diffraction pattern of the gold film before and 50 ps after optical excitation with an incident fluence of 860 J/m^2. The temporal evolution of the diffraction signal (Fig. 1b) shows three distinct features in the scattering range 0.3 to 1.1 Å^{-1}: an overall rise of the scattering background, the rise of a broad peak (liquid structure factor) characteristic for the liquid product state, and the decay of the (220) Bragg peak. The temporal evolutions of these three features differ significantly. Quantifying the dynamics by fitting with monoexponential functions reveals that the increase of the diffuse scattering shows the fastest dynamics ($\tau = 3.5\pm0.6$ ps). In comparison, the decay of the Bragg peak, being a probe for lattice heating as well as disordering, is slightly slower ($\tau = 4.1\pm0.3$ ps). The rise of the liquid signature is delayed by 1.4 ± 0.3 ps compared to onset of the diffuse scattering and (220) dynamics and shows the slowest dynamics ($\tau = 7.3\pm0.5$ ps). The retardation of the formation of the liquid signature reflects the time required to heat the lattice above the melting temperature, which is an indication of a thermal melting process. A study of the excitation intensity dependence was conducted up to 2000 J/m^2, at which point the absorbed energy corresponds to fourteen times the heat required to melt the sample; the overall dynamics accelerate while the different time constants remain in their relative order. For the highest excitation conditions employed, we estimate that the lattice temperature reaches the melting point about 250 fs after the optical excitation, while we observe a time constant of 2.2±0.4 ps for the rise of the liquid signature. This indicates that the lattice remains in its fcc structure for an extended period of time. This observation is consistent with the theoretically predicted effect of electronic bond hardening in gold [3]. The highest excitation conditions employed in this work correspond to an initial electron temperature of approximately 4 eV, which instantaneously increases the melting temperature from 1340 K to 2400 K according to the ab initio calculations.

In contrast to gold, the structural response of bismuth indicates an electronically-driven, non-thermal melting mechanism. Fig. 2 shows the diffraction intensity of the Bragg spots of a bismuth film, which decays with a time constant of 470 ± 70 fs. Optical excitation of Bi launches optical phonons with a frequency of 2.9 THz (in the low-excitation limit). Increasing electronic excitation flattens the interatomic potential resulting in a red-shift of the phonon frequency [4]. The lowest phonon frequency that has been observed under reversible excitation conditions corresponds to a vibrational period of 470 fs. This indicates that the disordering of the Bi lattice occurs roughly on the time scale of one vibrational period.

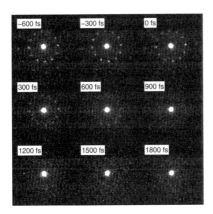 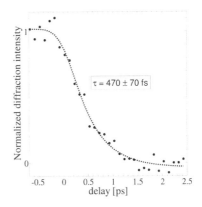

Fig. 2. Left panel: Raw diffraction data of the Bragg spots of a 30 nm thick 111-oriented Bi film at selected time points. Right panel: Normalized diffraction intensity above background as a function of time delay (dots). The data is fitted with a monoexponential function convoluted with the instrument response function (dashed line). Excitation conditions: incident fluence 220 J/m^2.

1   B. Siwick, J.R. Dwyer, R.E. Jordan and R.J.D. Miller, Science 302, 1382 (2003).
2   M. Harb, R. Ernstorfer, C.T. Hebeisen, G. Sciani, W. Peng, T. Dartigalongue, M.A. Eriksson, M.G. Lagally, S.G. Kruglik, R.J.D. Miller, Phys. Rev. Lett. 100, 155504 (2008).
3   V. Recoules, J. Clerouin, G. Zerah, P.M. Anglade, and S. Mazevet, Phys. Rev. Lett. 96, 055503 (2006).
4   D.M. Fritz et al., Science 315, 633 (2007).
5   C.T. Hebeisen, G. Sciaini, M. Harb, R. Ernstorfer, T. Dartigalongue, S.G. Kruglik, and R.J.D. Miller, Opt. Express 16, 3334 (2008).

Four-dimensional Visualization of Transitional Structures in Phase Transformations by Electron Diffraction

Peter Baum[1,2], Ding-Shyue Yang[1], and Ahmed H. Zewail[1]

[1] California Institute of Technology, 1200 E. California Bld, Pasadena CA 91125, USA
[2] Ludwig-Maximilians-Universität München, Oettingenstr. 67, 80538 München, Germany
E-mail: peter.baum@lmu.de

Abstract. Imaging with ultrashort electron pulses allows visualizing atomic-scale motions in all four dimensions of space and time. We report the transitional structures and mechanism of the ultrafast insulator-to-metal phase transformation in crystalline vanadium dioxide.

Introduction

Dynamical changes in molecules and condensed matter involve motion of atoms and electrons from initial to final conformations on multidimensional energy landscapes. Understanding such complex behavior requires knowledge of transitional structures and separation of time scales for atomic movements. Ultrafast electron pulses, because of their short De Broglie wavelength, allow for visualizing atomic-scale motions in all four dimensions of space and time. Here we report, from femtoseconds to nanoseconds, the transitional structures during the ultrafast insulator-to-metal phase transformation in vanadium dioxide. By excitation with tilted optical pulses and observation of 16 dynamical Bragg diffractions, the femtosecond primary vanadium-vanadium bond dilation, the displacements of atoms in picoseconds, and the sound wave shear motion on hundreds of picoseconds were resolved in four dimensions, elucidating the nonconcerted mechanism of the structural pathway.

Ultrafast Electron Crystallography on VO$_2$ Single Crystals

Vanadium dioxide is a crystalline material with a symmetry-raising phase transformation from a monoclinic to a tetragonal structure at 68°C, which goes along with a transition from metallic to insulator behavior. Static structures show that the vanadium atoms arrange in pairs in the low-temperature monoclinic phase. In contrast, all V–V distances become equal and the symmetry is raised in the high-temperature tetragonal phase. The present study is performed on single VO$_2$ crystals with ultrafast electron diffraction [1]. Figure 1 shows some of the observed diffraction patterns at room temperature for different surface normal directions and electron directions (zone axes). All spots were identified as monoclinic VO$_2$ and the single-crystal order is evident in the well-indexed diffraction patterns. In order to observe all possible directions of atomic motions, a total of 16 Bragg diffractions of different Miller indices were examined on the femtosecond to nanosecond time scale. The phase transformation was initiated by femtosecond near-infrared excitation (800 nm), and the structural dynamics were followed by diffracting femtosecond electron packets (~500 electrons/pulse) after a variable delay time. To overcome the group velocity mismatch between excitation and electron pulses, we tilted the optical

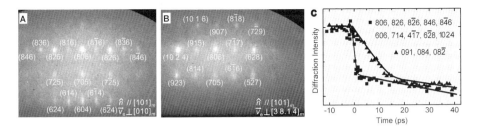

Fig. 1. Structural dynamics of VO_2. A-B: Experimentally obtained electron diffraction patterns. C: Intensity changes of Bragg spots after femtosecond laser excitation. Some spots decay within ~300 fs (squares), but others show slower picosecond dynamics (triangles). Shown here are the (606) and (091) Bragg reflections. The two distinct classes of Miller indices reveal sequential atomic motion along different crystal directions.

pulses to achieve temporal synchrony along the entire probed crystal surface, in order to achieve femtosecond resolution [2].

Immediately after laser excitation, all investigated Bragg spots show ultrafast intensity changes (Fig. 1C). Notably, two different types of dynamics were observed: an extremely fast decay within 307 fs was measured for (806), (826), ($8\overline{2}6$), (846), ($8\overline{4}6$), (606), (714), ($4\overline{1}7$), (10 2 4), and ($6\overline{2}8$), but other Bragg reflections, such as (091), (084), and ($08\overline{2}$), lack femtosecond dynamics and show decay with a time constant of ~9 ps. No shift in position or change in width was measurable at early times, demonstrating that lattice expansion or disorder is insignificant in such time range. In consequence, the observed intensity changes are associated with atomic motions within the unit cell, apparent by constructive or destructive interference in the structure factors.

The observed two types of dynamics reveal stepwise atomic motions along different directions. All Bragg spots that show the femtosecond behavior involve nonzero values of *(hkl)*, whereas those displaying the slower picosecond behavior have a zero component of *h*. Atomic movement along a certain direction can only affect such Bragg spots that have nonzero contributions in the corresponding Miller indices. The initial femtosecond motion is therefore along the **a** axis, which is the direction of the V–V bond in the monoclinic structure. In contrast, the picosecond structural transformation projects along the **b** and **c** axes.

Visualization of Atomic Motion in Four Dimensions

The measurements provide the following dynamical picture (Fig. 2). The initiating excitation at 1.55 eV primarily involves the d$\|$ band, which arises from bonding of the vanadium pairs. From a chemical perspective, the excitation is to an antibonding state, which instantly results in a repulsive force on the atoms, and they separate along the bond direction. In sequence and on a distinct time scale, the unit cell transforms towards the configuration of the tetragonal phase. The phase transformation thus proceeds by a non-direct pathway on the multidimensional potential energy landscape and not by direct structural conversion.

The transient behavior on the longer, up to nanosecond time scale reveals another dimension of movement (see Fig. 2). In addition to atomic rearrangements within the unit cell, the formation of the final tetragonal structure requires large-scale shear motion, which involves tilt of the principle crystal axes. Experimentally, on the 100-

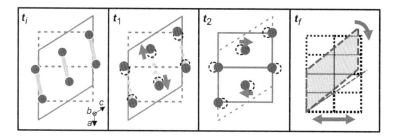

Fig. 2. Transitional structures during the ultrafast phase transformation. The elementary steps of the metal-to-insulator phase transformation in vanadium dioxide follow a nonconcerted mechanism with a sequence of transitional structures, first (t_1) involving local separation of the vanadium atoms within ~300 fs, followed by atomic rearrangements within the unit cell (t_2) and long-range shear motion with sound wave velocity (t_f). The V–V bond dilation is the initial step of the insulator-to-metal transformation.

ps time scale, some Bragg spots show an intensity increase or decrease when probed along different directions. Such behavior reveals shear motion with the associated enhancement or suppression of Bragg diffractions.

Both the initial femtosecond atomic rearrangements and the slower shear dynamics show a marked threshold of excitation fluence (6 ± 1 mJ/cm^2). For all investigated Bragg spots, the threshold for the sub-ns component is the same or higher than that of the femtosecond component, suggesting that the shear is a consequence of the initial atomic motions. The absorbed optical excitation energy at threshold is remarkably similar to the total heat required to initiate the phase transformation thermally. This observation suggests that the phase transition is critically dependent on the total number of carriers, thermal and optical, and that the described transitional structures have general significance.

Perspectives

These studies demonstrate the ability to decipher atomic motions in all dimensions during structural phase transitions in condensed matter. For vanadium dioxide, the elementary steps follow a nonconcerted mechanism with a sequence of transitional structures, first involving local displacements on the femtosecond and picosecond time scale followed by long-range shear rearrangements on the sub-ns time scale at the speed of sound. The V–V bond dilation is the initial step of the insulator-to-metal transformation. With atomic-scale spatial and temporal resolutions, and possible extension to the attosecond domain [3], we expect many future studies of complex condensed matter systems with electron diffraction [4].

1 P. Baum, D.-S. Yang, A. H. Zewail, Science **318**, 788, 2007.
2 P. Baum and A. H. Zewail, Proc. Natl. Acad. Sci. USA **103**, 16105, 2006.
3 P. Baum and A. H. Zewail, Proc. Natl. Acad. Sci. USA **104**, 18409, 2007.
4 A. H. Zewail, Annu. Rev. Phys. Chem. 57, **65**, 2006.

Ultrashort soft x-ray pulses from a femtosecond slicing source for time-resolved laser pump- x-ray probe experiments

N. Pontius[1], C. Stamm[1], T. Kachel[1], R. Mitzner[2], T. Quast[1], K. Holldack[1], S. Khan[1,3], H.A. Dürr[1], W. Eberhardt[1]

[1] BESSY GmbH, Albert-Einstein-Straße 15, 12489 Berlin, Germany
E-mail: pontius@bessy.de
[2] Physikalisches Institut der Universität Münster, 48149 Münster, Germany
[3] now at: Institut für Experimentalphysik, Universität Hamburg, 22761 Hamburg, Germany.

Abstract. The new femtosecond-slicing source generates energy-tuneable femtosecond x-ray pulses which are used for time-resolved soft x-ray spectroscopy. We report on the experimental setup and show first results using the laser pump and x-ray probe technique.

Introduction

Combining ultrashort pulse duration with continuously tuneable photon energies in the soft x-ray range makes a powerful probe technique for studying ultrafast processes in solids, liquids or gases. Along with element specificity, soft x-ray spectroscopy provides insight into geometric and electronic properties as well as the magnetic structure. Thus it delivers complementary information compared to existing time-resolved techniques probing in the IR to VUV range.

X-ray pulses from state of the art synchrotrons, however, have a minimum time duration of at least 10 ps, too long to study ultrafast processes in a time-resolved experiment. To overcome this obstacle the special technique of electron beam slicing can be employed [1]. With this technique it is possible to produce x-ray pulses with duration as short as 100 fs from the synchrotron. In 2004 this technique was implemented at BESSY [2,3]. The unique feature of the BESSY 'femtoslicing' scheme is the use of a helical undulator to produce the fs-x-ray pulses from the sliced electron beam. This allows full control over the polarization state of the x-ray pulses (linear or circular) and opens up the field of magnetization dynamics for ultrafast studies employing x-ray magnetic circular dichroism (XMCD).

The "Femtoslicing" Source

Figure 1 shows the femtoslicing scheme used at BESSY. Femtosecond pulses from a Ti:sapphire laser system (wavelength $\lambda = 780$ nm, pulse duration $\tau = 50$ fs FWHM, pulse energy E = 2 mJ, repetition rate = 3 kHz) co-propagate with electron bunches in a planar U139 undulator (the 'modulator'). The electric field of a 50 fs laser pulse induces an energy modulation of up to $\pm 1\%$ to the 1.7 GeV electrons within an ultrashort 'slice' of the 50-70 ps wide bunch. Within our scheme the energy modulated electrons are separated from the major (99.9%) nonsliced part of the bunch by a dipole bending magnet. It serves for angular separation within the helical UE56 soft x-ray source (the 'radiator'). Picosecond radiation from the main bunch is blocked by front end apertures. The improved time resolution is accompanied by a sizable reduction in X-ray intensity since only 10^{-3} of the electrons of the initial bunch are involved in the fs-pulse generation.

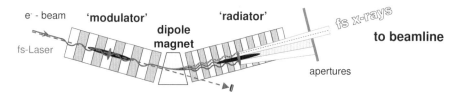

Fig. 1. Femtoslicing scheme used at BESSY

The Laser Pump- X-Ray Probe Experimental Setup

An overview over the whole laser pump- x-ray probe experimental setup is provided in figure 2. Pulses from the fs-laser system are split into two parts. The 'slicing'-part (~ 90%) is guided into the electron storage ring to produce the fs x-ray probe pulse. The 'pump'-part (~ 10%) is transferred to the experiment to excite the sample. Using

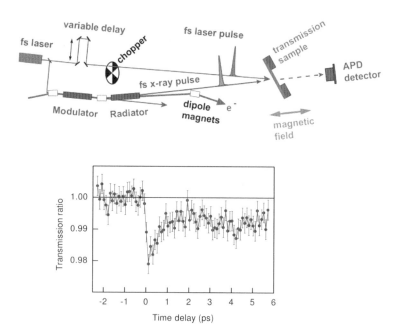

Fig. 2. Top: overview over the whole experimental setup. Bottom: X-ray transmission signal of the pumped sample normalized on that of the unpumped sample measured at the L_3 - edge of a 30nm thick nickel sample (852eV) recorded with linearly polarized x-ray pulses. It indicates a modification of the Ni electronic structure within the first 120 fs [4].

identical laser pulses for electron beam slicing and for sample excitation ensures intrinsic synchronisation of x-ray and laser pulses. Delay between both is controlled by changing path length of the laser pulse to the experiment. The synchronized chopper serves to alternate the measurement between the pumped and unpumped state on a shot-to-shot bases for proper normalization.

Summary

The BESSY 'femtoslicing' source provides ultrashort x-ray pulses of ~100fs for user experiments. Energy tunability (400–1200 eV) and full polarization control makes it an ideal tool for ultrafast soft x-ray spectroscopy.

1   R. W. Schoenlein et al., in Science **287**, 2237, 2000.
2   S. Khan et al., in Phys. Rev. Lett. **97**, 074801, 2006.
3   K. Holldack et al., in Phys. Rev. ST Accel. Beams **8**, 040704, 2005
4   C. Stamm et al., in Nature Mater. **6**, 740, 2007.

Femtosecond X-Ray Absorption Spectroscopy of a Photoinduced Spin-Crossover Process

C. Milne[1], V.-T. Pham[1], W. Gawelda[1,3] A. El Nahhas[1], R. M. van der Veen[1,2], S. L. Johnson[2], P. Beaud[2], G. Ingold[2], C. Borca[2], D. Grolimund[2], R. Abela[2], M. Chergui[1], and Ch. Bressler[1]

[1] Laboratoire de Spectroscopie Ultrarapide (LSU), Ecole Polytechnique Fédérale de Lausanne (EPFL), BSP, CH-1015 Lausanne, Switzerland
[2] Swiss Light Source, Paul Scherrer Institut, CH-5232 Villigen-PSI, Switzerland
E-mail: christian.bressler@epfl.ch

Abstract. We present ultrafast x-ray absorption studies of photoexcited aqueous iron tris-bipyridine with 160 fs and with 70 ps temporal resolution to monitor the structural evolution in this spin-crossover complex.

Introduction

Time-resolved x-ray absorption fine structure (XAFS) spectroscopy with picosecond temporal resolution has recently been established as a new method to observe electronic and geometric structures of short-lived reaction intermediates in solution [1,2]. It combines an intense femtosecond laser source synchronized to the 50-100 ps x-ray pulses delivered at the microXAS beamline of the Swiss Light Source (SLS). To go to the sub-picosecond time resolution, femtosecond x-ray pulses can be extracted via the time-slicing scheme [3,4]. We used this scheme in the hard x-ray domain at the microXAS beamline (SLS), to investigate the light-induced magnetization of aqueous Iron(II)-tris(2,2')-bipyridine ($[Fe^{II}(bpy)_3]^{2+}$), which is completed within 300 fs. This is the first example of a femtosecond hard x-ray absorption experiment of a molecule in liquids.

Light-driven spin crossover at room temperature in solution

$[Fe^{II}(bpy)_3]^{2+}$ represents the simplest molecule of iron-based light-induced spin-cross over complexes, itself being a typical example of a low spin (LS) compound, which can undergo a spin change to a quintet state upon irradiation [5]. The optical absorption spectrum of aqueous $[Fe^{II}(bpy)_3]^{2+}$ is characterized by an intense broad band centred at 520 nm due to the singlet Metal-to-Ligand-Charge-Transfer (^{1}MLCT) state. Photoexcitation into this band (or to higher energies) is followed by a cascade of intersystem crossing (ISC) steps through singlet, triplet and quintet MLCT and ligand-field (LF) states, which brings the system to the lowest-lying (HS) quintet state, $^5T_2$, with almost unit quantum yield in < 1 ps [6]. This state relaxes non-radiatively to the LS ground state within 0.6 ns in aqueous solutions at room temperature. Using picosecond XAS, we recently determined that in the HS state, an elongation of ~0.2 Å of the Fe-N bond distances occurs [2].

However, the pathway and time scale of the cascade from the ^{1}MLCT to the $^5T_2$ state are still not known, as ultrafast optical spectroscopy can neither resolve the intermediate steps nor determine their structures. In order to address these issues and to probe the relaxation processes, we have implemented femtosecond XANES spectroscopy.

Femtosecond X-Ray Absorption Spectroscopy

Tunable hard x-ray pulses with 140 (30) fs pulse width in the 5-20 keV range are extracted via the time-slicing scheme at the microXAS beamline, as described in Ref. 4. We performed the experiments at the iron K edge around 7 keV with approximately 10 photons/pulse onto the sample at 2 kHz repetition rate within a bandwidth of approximately 2 eV. The instrument response width, given by the cross correlation between the x-ray and the laser pulse widths, is 160-200 fs in agreement with jitter and long term (days) drift measurements [4]. Time delay scans at a fixed energy were recorded, next to energy scans at a fixed time delay. The precise time zero (and thus the selected time delay) was quantified from the data after fitting the data to a rate model (see below).

Results

Fig. 1a shows the transient absorption spectrum after 50 ps, which has been established as being due to the HS structure [2]. It shows that the strongest absorption change at a multiple scattering edge feature near 7126 eV reflects the increased Fe-N bond distance determined to be 0.2 Å [2]. We therefore used this feature to investigate the temporal evolution of the relaxation process from the ^{1}MLCT to the ^{5}T$_2$ by scanning the laser-x-ray time delay, which also confirms that this process terminates below 300 fs (Fig. 1b). In order to quantify the time required for this spin crossover process, we have calculated the rate equations for this process. Hereby the following reaction cycle using the input from our optical studies [6] was applied (using the indicated lifetimes and an overall cross correlation time of 200 fs [4]):

$$^1GS + h\nu \text{ (400 nm)} \rightarrow {}^1MLCT \text{ (20 fs)} \rightarrow {}^3MLCT \text{ (120 fs)} \rightarrow {}^5T \text{ (665} \qquad (1)$$
$$\text{ps)} \rightarrow {}^1GS$$

With this we calculated the population dynamics of all intermediate states given above, and determined the final HS signal, which is shown in Fig. 1b together with the data. In addition, we fit the final arrival time to the ^{5}T state (previously fixed to the ^{3}MLCT departure time of 120 fs) yielding 130 (60) fs. This result implies that the electron back transfer from the ligand system simultaneously triggers the excitation of a second electron from the bonding t_{2g} orbital, so that both electron spins are parallel in the antibonding e_g orbital, together with two unpaired electrons from the bonding t_{2g} orbitals. Any possible intermediate steps (in the metal-centered states) are considerably faster than 60 fs.

Conclusions

Via femtosecond x-ray absorption spectroscopy we determined the spin-crossover process to occur within 130 fs, which is identical to the electron back transfer time from the ligand system to the metal-centered orbitals. This represents the first direct measurement of the ^{5}T arrival time. Optical methods so far failed to observe this rise time, as this state remains optically silent throughout the visible/IR range. While the XAFS experiment determines the Fe-N *distance* to increase by 0.2 Å with the 130 fs population time, it does not measure vibrational relaxation within the HS state due to the poor cross correlation time, but only the ensemble and time averaged distance change. Nevertheless, the quantitative analysis reveals that the system has undergone a prompt ISC step ^{3}MLCT $\rightarrow$ ^{5}T involving two electrons simultaneously, thus directly coupling both states and bypassing intermediate (single electron) steps. We are currently investigating the anticipated vibrational cooling (of about 1.5 eV) within the promptly populated hot ^{5}T state.

This experiment exploiting femtosecond hard x-radiation is a promising approach to add a new observable for our understanding of spin-dynamics in molecular systems. With this and new intense femtosecond x-ray sources becoming available (e.g. XFELs) it is possible to apply this method to more complex systems, e.g., to study ISC processes in biological systems, including heme-proteins.

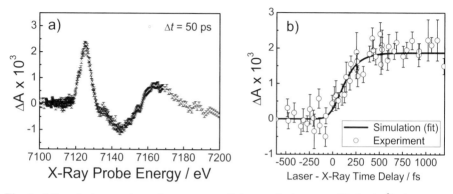

Fig. 1. a) Transient x-ray absorption spectrum of photoexcited aqueous [Fe(bpy)$_3$]$^{2+}$ after 50 ps. b) Femtosecond time trace of the evolution of the strongest transient x-ray absorption feature in a) (near 7125 eV) together with a fit according to reaction sequence via eq. (1).

Acknowledgements. This work was funded by the Swiss National Science Foundation (FNRS), via contracts 116023, 110464, 107956, 620-066145, and 105239, and by the SBF *via* contract COST D35 C06.0016..

1   C. Bressler, M. Chergui, "Ultrafast X-Ray Absorption Spectroscopy", Chem. Rev. **104**, 1781 (2004)
2   W. Gawelda, V.-T. Pham, M. Benfatto, Y. Zaushitsyn, M. Kaiser, D. Grolimund, S. L. Johnson, R. Abela, A. Hauser, C. Bressler, M. Chergui, M. "Structural Determination of a Short-Lived Excited Iron(II) Complex by Picosecond X-Ray Absorption Spectroscopy", Phys. Rev. Lett. **98**, 057401 (2007)
3   R. W. Schoenlein, S. Chattopadhyay, H. H. W. Chong, T. E. Glover, P. A. Heimann, C. V. Shank, A. A. Zholents, M. S. Zolotorev, "Generation of femtosecond pulses of synchrotron radiation", Science **287**, 5461 (2000)
4   P. Beaud, S. L. Johnson, A. Streun, R. Abela, D. Abramsohn, D. Grolimund, F. Krasniqi, T. Schmidt, V. Schlott, G. Ingold, G. *Spatiotemporal Stability of a Hard-X-Ray Undulator Source Studies by Control of Coherent Optical Phonons*, Phys. Rev. Lett. **99**, 174801 (2007)
5   A. Hauser, C. Enachescu, M. L. Daku, A. Vargas, N. Amstutz, "Low-temperature lifetimes of metastable high-spin states in spin-crossover compounds: The rule and exceptions to the rule", Coord. Chem. Rev. **250**, 1642-1652 (2006)
6   W. Gawelda, A. Cannizzo, V.-T. Pham, F. Van Mourik, C. Bressler, M. Chergui, "Ultrafast Nonadiabatic Dynamics of [FeII(bpy)3]2+ in Solution". J. Am. Chem. Soc. **129**, 8199-8206 (2007)

Probing Reaction Dynamics of Transition-Metal Complexes *in Solution* via Time-Resolved Soft X-ray Spectroscopy

Nils Huse[1], Tae Kyu Kim[2], Munira Khalil[3], Lindsey Jamula[4],
James K. McCusker[4], and Robert W. Schoenlein[1]

[1] Chemical Sciences Division, Lawrence Berkeley National Laboratory, 1Cyclotron Road,
  Berkeley, CA 94720, USA
  E-mail: nhuse@lbl.gov
[2] Department of Chemistry, Pusan National University, Geumjeong-gu, Busan 609-735, Korea
[3] Department of Chemistry, University of Washington, Seattle, WA 98195, USA
[4] Department of Chemistry, Michigan State University, East Lansing, MI 48824, USA

Abstract. We report the first time-resolved soft x-ray measurements of *solvated* transition-metal complexes. L-edge spectroscopy directly probes dynamic changes in ligand-field splitting of $3d$ orbitals associated with the spin transition, and mediated by changes in ligand-bonding.

Introduction

A major goal in chemistry is to quantitatively probe chemical bonding dynamics at the molecular level. That is: to follow valence-charge distributions in real time, to observe the formation and dissolution of chemical bonds, and to understand their cooperative relationship with atomic rearrangement and the formation of new molecular structures. The emerging technique of ultrafast soft x-ray spectroscopy provides important new insight to molecular dynamics because x-ray core-level transitions (e.g. $2p_{1/2} \rightarrow$ LUMO) are element specific, symmetry and spin state selective, and effective for probing lighter elements (C, N, O etc.) that are biologically relevant. Thus, the x-ray near-edge spectral region (XANES) provides a quantitative fingerprint of the local chemical bonding geometry (orbital hybridization etc.) in the vicinity of the atom(s) of interest. This is particularly powerful for understanding molecular dynamics in solution, where much important chemistry occurs, and where the solvent environment substantially influences reaction dynamics (often precluding quantitative interpretation of transient absorption spectra in the visible regime). A significant experimental challenge for soft x-ray spectroscopy in liquid-phase has been the requirement for sub-micron liquid films (due to large absorption cross-sections) that remain stable in a vacuum environment under typical photo-excitation conditions.

Here we report the first time-resolved Fe L-edge spectroscopy of the ultrafast photo-induced intersystem crossing occurring in the solvated transition-metal complex $[Fe(tren(py)_3)]^{2+}$. Interest in the Fe^{II} complex stems from the rapid spin transition ($\Delta S=2$) that is thought to result from changes in the ligand-field splitting of the Fe-$3d$ orbitals, and associated changes in Fe-$3d$/N-$2p$ hybridization. These changes are mediated by a dilation of the octahedrally-coordinated Fe-N bonds, and optically triggered by the initial metal-to-ligand charge transfer excitation [1–3]. Time-resolved measurements at the Fe L-edges directly probe the evolution of the ligand-field split $3d$ orbitals via selective transitions from the spin-orbit-split Fe-$2p_{1/2}$/$2p_{3/2}$ levels.

The $[Fe(tren(py)_3)]^{2+}$ complex in Fig. 1A has an octahedral ligand geometry around a central Fe atom [1]. Hybridization with the nearest ligand N-$2p$ orbitals lifts the degeneracy of the Fe-$3d$ orbitals to form t_{2g} and e_g levels as pictured schematically in

Fig. 1C. The associated Fe-N charge-transfer transition has a maximum absorption at 560nm. Optical spectroscopy has extracted a 300fs time constant, and this has been interpreted as a signature of intersystem crossing to a high-spin state with a change in spin of $\Delta S=2$, accompanied by an ~0.2Å elongation in the Fe-N bond distance and an octahedral distortion as revealed by hard x-ray (EXAFS) spectroscopy [1–3]. The ultrafast intersystem crossing and the structural changes point to an intricate interplay of electronic, atomic, and magnetic structure that is yet to be deciphered.

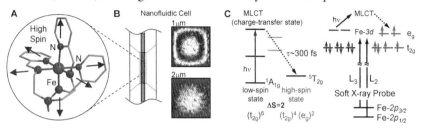

Fig. 1. Schematic of [Fe(tren(py)$_3$)]$^{2+}$, arrows indicate the metal-ligand bond change associated with the spin transition (**A**). Nanofluidic cell layout and transmission interferograms that indicate liquid film thickness (**B**). Simplified energy levels of intersystem crossing after 560nm excitation and L-edge spectroscopy of the Fe^{II} complex with soft x-ray probe pulses (**C**).

Experimental Methods

Soft x-ray spectroscopy experiments are made possible by a novel nanofluidic cell design (Fig. 1B) in which a liquid film is held between two 100nm silicon nitride membranes. The interior cell pressure is balanced against the pressure of the sample chamber, thereby controlling the film thickness with sub-200nm accuracy [4]. A third silicon-nitride membrane isolates the low-pressure sample chamber from the ultrahigh vacuum of the femtosecond soft x-ray beamline at the Advanced Light Source. The Fe^{II} complex is dissolved in acetonitrile at 100mM concentration, and excited with ultrashort 560nm pulses at 1kHz. Transient absorption changes at the Fe L$_2$ and L$_3$ edges are probed with tunable 70ps synchrotron pulses at 2kHz as a function of x-ray probe energy and time delay between excitation and probe pulses. From the spectral measurements of both the ground-state and photo-excited sample, transient x-ray absorption spectra of the high-spin state at the Fe L$_2$ and L$_3$ edges are reconstructed.

Results and Discussion

Figure 2A shows the absorption of the high and low-spin state after the photo-induced metal-to-ligand charge transfer. These are to our knowledge the first time-resolved solution-phase transmission spectra ever recorded in the soft x-ray region. The L$_3$ edge of the high-spin state shows a pronounced red-shift by 1.6eV with no discernable change in linewidth relative to the low-spin state. However, more complex absorption features on the blue side of the maximum are present. The L$_2$ edge exhibits only a small red-shift by about 0.4eV upon laser excitation but the absorption cross-section is diminished by nearly a factor of 2. The time delay scans in Fig. 2B show changes in sample absorption at specific probe photon energies, consistent with the spectral scans. While the absorption decreases at the spectral absorption maxima of the low-spin state after excitation, it increases around 707eV where the absorption of the high-spin state is maximal. The delay scans reflect the 70ps width of the x-ray probe pulses and demonstrate that the Fe^{II} compound reaches its meta-stable high-spin state within the time resolution of the experiment. Differential spectra at 5ns delay demonstrate that the evolution of the high-spin electronic structure is complete within 70ps.

Our time-resolved liquid-phase measurements are consistent with static x-ray measurements of a similar compound in crystalline phase [5] (where the high-spin state is trapped at low temperature), and offer an important opportunity to investigate the solvent influence on the ultrafast dynamics of the electronic and spin states. Our experimental results are in good agreement with theoretical calculations of the high- and low-spin electronic structure [5]. In particular, the size of the red-shift of both L_3 and L_2-edges is reproduced. Crystal field multiplet calculations attribute our observations to changes in $3d$ occupancy. The red-shift of the L_3 edge is caused by a change in ligand-field splitting of 0.5eV and a core-hole shift of about 1eV. The red-shift of the L_2 edge seems to directly reflect the change in ligand-field splitting.

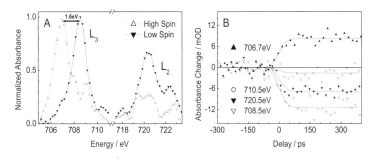

Fig. 2. L_2 and L_3 absorption spectra of the initial low spin complex, and the transient photo-induced high-spin complex at 150ps delay (**A**). Sample absorption change as a function of delay between the 560nm excitation pulse and the soft x-ray probe pulse at specific x-ray energies. Solid lines are fits to the data based on an instantaneous absorption change convolved with a 70ps pulse, with negligible relaxation within the measurement window (**B**).

Conclusions

We report the first time-resolved soft x-ray spectroscopy of solution-phase molecular dynamics. Changes in ligand-field splitting and spin-state populations in $3d$ orbitals of the Fe^{II} complex are directly probed via transient absorption changes of the Fe L_2 and L_3 edges following photo-induced metal-to-ligand charge transfer. With the emergence of high-flux ultrafast soft x-ray sources, details on interplay between atomic structure, electronic states, and spin contributions will be revealed. Our experimental approach opens the door to femtosecond soft x-ray investigations of liquid phase chemistry that have previously been inaccessible.

Acknowledgements. This work was supported by the Department of Energy under Contract No. DE-AC02-05CH11231.

1   J. E. Monat and J. M. McCusker, J. Am. Chem. Soc. **122**, 4092, 2000.

2   M. Khalil, M. A. Marcus, A. L. Smeigh, J. K. McCusker, H. H. W. Chong, and R. W. Schoenlein, J. Phys. Chem. A **110**, 38, 2006.

3   W. Gawelda, V.-T. Pham, M. Benfatto, et al., Phys. Rev. Lett. **98**, 057401, 2007.

4   D. Kraemer, M. L. Cowan, A. Paarmann, N. Huse, E. T. J. Nibbering, T. Elsaesser, and R. J. D. Miller, Proc. Nat. Acad. Sci. **105**, 437, 2008.

5   V. Briois, Ch. Cartier dit Moulin, J Ph. Sainctavit, Ch. Brouder, and A.-M. Flank, J. Am. Chem. Soc. **117**, 1019 1995.

Sub-20-fs Optical Pump-X-ray Probe Spectroscopy beyond the Si K Edge

Enikoe Seres[1], Christian Spielmann[1,2]

[1] Physikalisches Institut EP1, Universität Würzburg, D-97074 Würzburg, Germany
E-Mail: enikoe.seres@physik.uni-wuerzburg.de
[2] Institute for Optics and Quantum Electronics, Friedrich-Schiller-University Jena, Max-Wien-Platz 1, 07743 Jena, Germany
E-Mail: spielmann@ioq.uni-jena.de

Abstract. Time resolved x-ray absorption spectroscopy is a powerful tool to follow the fast atomic, molecular changes on the atomic time and spatial scale. Here we demonstrate time-resolved XANES and EXAFS relying on a laser driven ultrafast XUV source. We have examined the electronic and structural changes in amorphous Si with a temporal resolution of 20 fs.

Introduction

The development of femtosecond lasers brought new possibilities into time-resolved spectroscopy. Structural techniques such as X-ray diffraction (XRD) or X-ray absorption spectroscopy (XAS) deliver direct structural information and temporal resolution of XRD and XAS has been extended to a few 100 femtoseconds [1-6]. Here we investigate the structural and electronic dynamics of a polycrystalline silicon sample with time-resolved x-ray absorption spectroscopy having a temporal resolution of less than 20 femtoseconds in the spectral range up to 4 keV.

The static structure of amorphous silicon has been well characterized with x-ray spectroscopy [7] in the last few years. Information about the electronic structure can gained from the x-ray absorption spectrum in a range up to 20 – 50 eV above the absorption edge (x-ray absorption near edge structure XANES). Structural information can be easily extracted from the absorption spectrum in an extended range (extended x-ray absorption fine structure spectroscopy, EXAFS). For a good spatial resolution it is necessary to record the absorption spectrum up to several 100eV's above the edge. Using an ultrafast broad-band, high harmonic generation based x-ray source opened the way to monitor the nuclear motion of atoms at their natural time and spatial scale with time resolved EXAFS [6, 8] for the first time.

High harmonic generation (HHG) is capable generating XUV pulses which can be a short as a few hundreds of attoseconds. The major drawback of the first realized HHG sources was their limited photon energy range and flux [9]. Adequate optimization of the setup allowed the generation coherent femtosecond x-ray pulses in an energy range up to several keV [10-12]. To reach these parameters it was necessary to employ different ways for phase-matched HHG such as nonadiabatic self phase matching (NSPM) [10, 13] or quasi phase matching (QPM) [14, 15]. For the measurement described in this paper we only applied NSPM, allowing the generation of sub 20-fs x-ray pulses in a wide range, with a simple and reliable setup.

Experimental Methods

A 1 kHz, 3 mJ, 12 fs Ti:sapphire amplifier system served as a source for this experiment. The major fraction of the energy is used for driving the x-ray source. The intense laser beam was tightly focused into a He gas jet, resulting in an estimated peak intensity of 2×10^{16} W/cm^2. This intensity was high enough supporting the generation of x-ray photons in a range up 3.5 keV [12]. The spectral characterization was performed with a scanning grazing incidence monochromator equipped with a channeltron detector. A small fraction (approx. 100μJ) of the laser output beam is split off for pumping the sample a 100-nm-thick a-Si foil. The sub-20 fs laser pump pulses have been loosely focused onto the silicon sample obtaining fluencies in the range of 0.1 to 5 mJ/cm^2, which are two orders of magnitude below the damage threshold.

The x-ray absorption spectrum of a 100 nm thick Si foil above the Si K-edge (1.8 keV) is shown in Figure 1a. Our x-ray source delivered enough photons for an x-ray absorption measurement in a range up to 3 keV [10, 12]. The excitation of the carriers in silicon causes a modification of density of states in the valence and conduction band and induces structural distortions such as coherent phonons [8, 16] and non-thermal melting for higher intensities. To reveal the dynamic of the motion we recorded the difference in absorption spectra for the pumped and un-pumped case; the spectra in a range of 80 to about 3500 eV, i.e. covering the L- and K-edge of silicon very well. The delay has been varied in 20 fs steps in a range of -30 to 400 fs for spectra above the K-edge and from -400 to 800 fs for the L-edge measurement. From these spectra we also evaluated XANES range [17].

Results and Discussion

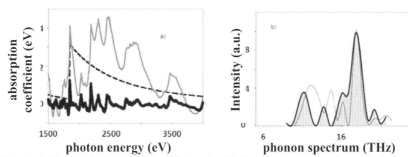

Fig. 1. a) Measured absorption coefficient for Si above the K edges (grey line), the atomic absorption coefficient for Si (dashed line) and the calculated EXAFS signal (black line). b) Dynamic of the atomic motion recorded from time-resolved EXAFS spectra above the Si L- [8] and K-edge (filled area).

In our EXAFS evaluations we applied a two dimensional Fourier-transform on the whole set of data. For the data obtained in the longer delay and reduced energy range, i.e. EXAFS above the L-edge we found a sharp maximum at 3.6 THz and at 17 THz [8]. For the evaluation of the data above the K-edge we observed again a distinct maximum at about 16 THz. The slow component could not be resolved for this measurement due to the limited delay range. Nevertheless the two data sets are in reasonable agreement, as depicted in Figure 1 b. The measured frequencies agree also very well with the measured phonon frequencies in Si. This coincidence makes us confident, that we have directly observed coherent phonons in the time-domain.

The pump photons are absorbed via single and two photon absorption, creating a non-thermal distribution in the conduction band, which relaxes over different channels and time scales. The excitation of the electrons creates occupied states in the conduction band and unoccupied states in the valence band. The modified electronic structure results in a shift of the x-ray absorption edge and/or modify its spectral shape. Here we studied the shift of the edge. Evaluating the shift of the K- and L-edge we observed a similar temporal evaluation as depicted in Fig. 2. Further measurements are necessary to study the evolution of the structural dynamics in more detail.

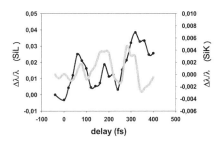

Fig. 2. Time-resolved XANES near the Si L-edge (black) and the Si K-edge (grey).

Conclusions

Taking the advantages of the high harmonics generation technique, our 3keV XUV source, with its sub 20 fs pulses made possible the direct measurement of the atomic motion and electronic relaxations after an excitation of short IR pulse in a Si beyond the Si K-edge.

Acknowledgements. This work has been sponsored by the Austrian Science Fund under Grant No. F016 ADLIS and DFG under Grant No. SP 687/1-3.

1   A. Rousse, et al., Rev. Mod. Phys. 73, 17, 2001.
2   C. W. Siders et al., Science 286 1340-1342, 1999.
3   M. Bargheer, et al., Science 306, 1771, 2004.
4   F. Raksi , et al., J. Chem. Phys. 104, 6066-6069, 1996.
5   A. M. Lindenberg, et al., Science 308, 392, 2005.
6   C. Bressler, M. Chergui, Chem. Rev. 104, 1781, 2004.
7   C. J. Glover, G et al., Nucl. Instrum. Methods B 199, 195, 2003.
8   E. Seres, C. Spielmann, Appl. Phys. Lett. 91, 121919, 2007.
9   T. Brabec, F. Krausz, Rev. Mod. Phys. 72, 545, 2000.
10   J. Seres, et al., Nature 433, 596, 2005.
11   J. Seres, et al., New J. Phys. 8 251, 2006.
12   E. Seres, et al., Appl. Phys. Lett. 89, 1819, 2006.
13   G. Tempea, et al., Phys. Rev. Lett. 84 4329, 2000.
14   X. Zhang, et al., Nature Phys. 3, 270, 2007.
15   J. Seres et al., Nature Phys. 3, 878, 2007.
16   P. Stampfli, K. H. Bennemann, Phys. Rev. B 46, 10686, 1992.
17   E. Seres, C. Spielmann J. Mod. Opt. to be published

Capturing Transient Solute Structures in Solution by Pulsed X-ray Diffraction

Jae Hyuk Lee[1], Tae Kyu Kim[2], Joonghan Kim[1], Qingyu Kong[3], Marco Cammarata[3], Maciej Lorenc[3], Michael Wulff[3], and Hyotcherl Ihee[1]

[1] Center for Time-resolved Diffraction, Department of Chemistry, KAIST, Daejeon 305-701, Korea
E-mail: hyotcherl.ihee@kaist.ac.kr
[2] Department of Chemistry, Pusan National University, Busan 609-735, Korea
[3] European Synchrotron Radiation Facility, BP220, Grenoble Cedex 38040, France

Abstract. The structural kinetics of photo-induced elimination reaction of 1,2-diiodotetrafluoroethane in solution is examined by transient X-ray diffraction. The structure of $\cdot CF_2CF_2I$ is determined to be a classical mixture of anti and gauche forms and their decay mechanism is also tracked in real time.

Introduction

The elucidation of temporally varying molecular structures during ultrafast processes is essential for understanding the mechanism and function of molecular reactions. Although ultrafast optical spectroscopy has been provided a wealth of information about various photo-induced reactions with a time resolution down to tens of femtoseconds [1,2], its observables are not generally linked with detailed structural information in terms of bond angles and lengths except for a few favorable cases in time-resolved vibrational spectroscopy and multidimensional spectroscopy measurements. By contrast structural dynamics and kinetics information over wide time ranges up to microseconds are often conveniently obtained by replacing the optical probe pulse with the X-ray pulses as in transient X-ray diffraction (TXD) [3–5] and time-resolved X-ray absorption spectroscopy [6,7].

Here we report about the structural dynamics for the photo-induced elimination reaction of $C_2F_4I_2$ dissolved in methanol by TXD. In previous studies in gas [8] and solution [9] phases, the reaction is initiated by forming a short-lived C_2F_4I radical which subsequently dissociated to C_2F_4 and I. The structure of the C_2F_4I radical (either bridged structure or classical mixture of anti and gauche conformers) is a crucial issue for the stereochemical control of the reaction. The structure of the intermediate in the solution study could not be determined by time-resolved spectroscopy [9] and only gas-phase electron diffraction revealed the classical mixture of the intermediate structure [8]. However, the structure of intermediate in solution remains to be determined. In addition, the solvent may influence the reaction kinetics such as secondary C–I bond dissociation. A comparative study of the iodine elimination reaction of $C_2F_4I_2$ by TXD can give this information directly as shown here.

Experimental Methods

TXD data were collected on the beamline ID09B at the European Synchrotron Radiation Facility (ESRF) using a pump–probe scheme where a laser pulse (2 ps, 267 nm, 50 μJ/pulse) was used to initiate the iodine elimination reaction of $C_2F_4I_2$ and X-

131

ray pulses (5×10^8 photons per pulse, 100 ps (fwhm), 3% bandwidth around the wavelength of 0.68 Å, repeated at 986.3 Hz) from the synchrotron were used to interrogate the evolving structures in the sample. The sample was a kept in a recirculating liquid jet with 60 mM of $C_2F_4I_2$ dissolved in methanol. Diffraction data were collected for several time delays from – 100 ps up to 1 µs, and each delay was interleaved by a measurement at –3 ns, which served as a reference for the unperturbed sample.

Results and Discussion

Figure 1(a). shows the difference diffraction intensities ($q\Delta S(q,t)$, $q=4\pi/\lambda\sin(\theta)$ where λ is the X-ray wavelength, 2θ the scattering angle and t the time-delay). Structure determination in solution requires quantitative analysis of three contributions to the difference diffraction signal that are constraint to energy conservation, on the short time scales presented here, in the X-ray illuminated volume. We fitted the experimental difference intensities with theoretical difference intensities including the changes from three components: (i) the solute-only term, (ii) the solute-solvent cage term, and (iii) the solvent-only term. More details about the global-fitting can be found in our previous publication [3,4]. The first term is due to the changes within the solute that can be described by Debye scattering from isolated solute molecules. The second term, the cage term, can be simulated by Molecular Dynamics simulation by considering the interactions between solute-solvent atomic pairs. The last term, the hydrodynamics term, is from the temperature and pressure change in the solvent which can be obtained from a separated experiment in which pure methanol solvent is vibrationally excited by near-infrared light [5].

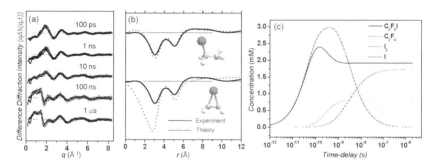

Fig. 1. (a) Time-resolved X-ray diffraction data at selected time-delays for $C_2F_4I_2$ in methanol. Difference-diffraction intensities, $q\Delta S(q,t)$ excited minus nonexcited (circle) and theoretical fits from global-analysis (line). (b) Determination of transient structure of $C_2F_4I_2$ elimination reaction in solution. The solute only terms for two candidate models (classical and bridged forms) are compared. (c) Reaction kinetics for iodine elimination from $C_2F_4I_2$.

To determine the structure of the early intermediate(s), we performed two separate global-fittings using two candidate models of the transient structure: the classical mixture (anti- and gauche-conformers) and the bridged structure. As a result, the χ^2 value for the fit with the classical model was smaller than that of the bridged structure at all investigated time delays. Moreover, the χ^2 value ($\chi^2=1.72$) for the classical model was nearly a factor of two smaller than for the bridged ($\chi^2=3.24$) at 100 ps where C_2F_4I is the major chemical species. When solute only contribution is extracted from the total difference diffraction intensity, the difference between classical and bridged forms is enhanced. This feature can be seen in Figure. 1(b) in which sine-Fourier transform of the solute only intensity is shown. It is intuitively helpful to understand the measured changes in real-space as arising from changes in atom-atom correlations inside the solute. In Figure.1(b), the negative peak near 5 Å corresponding to the I···I bond distance in parent $C_2F_4I_2$ which is therefore common in both models, but the broad negative peak between 2.0 and 3.0 Å can only match the classical model. This is because the features in this peak are very sensitive to the iodine atom position relative to the two carbon and four fluorine atoms. All these findings imply that the classical mixture structure of intermediate reproduces the experimental curve with higher fidelity than the bridged structure does. It should be noted that in the solution phase we could not refine the structure of the solute molecule as in crystallography due to the complex interplay of factors contributing to the total difference diffraction intensity [3].

From global-fitting of the time-resolved diffraction signal, we can also unravel the spatio-temporal photoreaction of $C_2F_4I_2$ as shown in Figure. 1(c). Upon 267 nm photoexcitation, the classical mixture of C_2F_4I and I species are dominant at 100 ps. 20 % of transient C_2F_4I radical dissociated in a second step to C_2F_4 and I in nanoseconds with a time constant of 153 ps which is six times greater than that in the gas phase (26 ps) [3]. Subsequently, two I atoms form molecular iodine (I_2) in about 100 ns by nongeminated recombination [10].

Conclusions

In this report, we have shown the simultaneous determination of accurate structural information and reaction kinetics for the iodine elimination reaction for $C_2F_4I_2$ in methanol by TXD. The results clearly demonstrate that TXD offers a complementary tool for structural dynamics in solution which is inaccessible by time-resolved optical spectroscopy.

Acknowledgements. This work was supported by the Creative Research Initiatives (Center for Time-Resolved Diffraction) of MOST/KOSEF, the EU grant FLASH (FP6-503641) and a grant from the center of Molecular Movie, the Niels Bohr Institute, Copenhagen.

1   H. Lee, Y. C. Chang, and G. R. Fleming, Science **316**, 1462, 2007.

2   P. Kukura, D. W. McCamant, S. Yoon, D. B. Wandschneider, and R. A. Mathies, Science **310**, 1006, 2005.

3   H. Ihee, M. Lorenc, T. K. Kim, Q. Y. Kong, M. Cammarata, J. H. Lee, S. Bratos, and M. Wulff, Science **309**, 1223, 2005.

4   T. K. Kim, M. Lorenc, J. H. Lee, M. Lo Russo, J. Kim, M. Cammarata, Q. Kong, S. Noel, A. Plech, M. Wulff, and H. Ihee, Proc. Natl. Sci. USA **103**, 9410, 2006.

5   M. Cammarata, M. Lorenc, T. K. Kim, J. H. Lee, Q. Y. Kong, E. Pontecorvo, M. Lo Russo, G. Schiro, A. Cupane, M. Wulff, and H. Ihee, J. Chem. Phys. **124**, 124504, 2006.

6   C. Bressler, and M. Chergui, Chem. Rev. **104**, 1781, 2004.

7   A. Cavalleri, M. Rini, H. H. W. Chong, S. Fourmaux, T. E. Glover, P. A. Heimann, J. C. Kieffer, and R. W. Schoenlein, Phys. Rev. Lett. **95**, 067454, 2005.

8   H. Ihee, V. A. Lobastov, U. M. Gomez, B. M. Goodson, R. Srinvasan, C. Y. Ruan, and A. H. Zewail, Science **291**, 458, 2001.

9   M. Rasmusson, A. N. Tarnovsky, T. Pascher, V. Sundström, and E. Åkesson, J. Phys. Chem. A **106**, 7090, 2002.

10 S. Aditya, and J. E. Willard, J. Am. Chem. Soc. **79**, 2680, 1957.

Structural kinetics in protein-coated gold nanoparticles probed by time-resolved x-ray scattering

A. Plech[1], H. Ihee[2], M. Cammarata[3], A. Siems[1], V. Kotaidis[1], F. Ciesa[1], J. Kim[2], K. H. Kim[2], J. H. Lee[2]

[1]Center for Applied Photonics and Dept. of Physics, University of Konstanz, Universitätsstr. 10, D-78457 Konstanz
[2]Center for Time-Resolved Diffraction, Dept. of Chemistry (BK21), Korea Advanced Institute of Science and Technology (KAIST), Daejeon, 305-701, Republic of Korea
[3]ESRF, BP. 220, 6, rue J. Horovitz, F-38043 Grenoble

Abstract. Laser-excited gold nanoparticles in aqueous suspension act as nanoscale heat sources on a short time to the surrounding. By employing pulsed small-angle x-ray scattering the structural kinetics of an adsorbed protein layer is temporally resolved. The scattering data reveals a layer expulsion from the surface.

Introduction

The usage of labeled metal nanoparticles for photothermal treatment of targeted tissue has been proposed as novel nanoscale applications as local damage can be caused by local scale heating, or the pressure wave created by explosive boiling of the surrounding aqueous environment [1]. Strong non-equilibrium excitation of gold nanoparticles can be achieved by pulsed laser heating [2]. The nanoscale provides cooling rates of 10^{13} K/sec in aqueous suspension. Consequently the phase transition in the adjacent water layer is of explosive nature with strong supersaturation. The heat pulse acts with a duration of several hundred picoseconds on the surrounding, for instance on proteins in biological tissue. We present a study on the structural relaxations of protein-coated gold nanoparticles by using pulsed x-ray scattering. X-ray scattering allows determining directly structural parameters of the protein coating and the spatiotemporal distribution of energy, i.e. the temperature history of the system.

Experimental Methods

We have employed a pump-probe technique using femtosecond laser pulses as heat stimulus and pulsed x-rays as probe for the nanoscale structure relaxations. By a combination of x-ray techniques it was possible to measure the particle temperature, pressure transients in the water phase, vapor bubble morphology and particle shape [3-5]. Briefly, a femtosecond amplifier system (Dragon, KMLabs) has been synchronized to the pulses of x-rays emitted by the European Synchrotron Radiation facility, beamline ID09B, and used for a stroboscopic pump-probe experiment. The bandwidth of the x-ray beam is relaxed to 2.6 % in order to increase the flux and maximize the signal-to-noise ratio. The particle suspension is excited at 390 nm, within the interband (IB) absorption of gold. The IB absorption is hardly sensitive on sizes and shapes. Therefore the average absorbed energy per unit cell is well defined. A suspension of protein conjugated gold nanoparticles is pumped through the interaction area of the focused x-ray and laser pulses. While the particles have been produced in advance of the experiment by an adaptation of the Turkevich method [6], they have been conjugated just prior to the experiment with bovine serum albumin (BSA; Roth, protease free). Mercaptosuccinic acid has been used as linker. It adsorbs covalently to gold and the BSA is attached to the carboxylic groups. By this linker the protein denaturation due to strong adsorption is reduced. The adsorption isotherms have been derived by absorption spectroscopy and dynamic light scattering. For a suspension of gold particles (from a 2 mM monomer solution) a maximum coverage is achieved at a 3 μM BSA concentration, which is close to the calculated full coverage in dense packing at 2-2.5 μM.

Results and Discussion

The mechanism of thermal laser excitation of gold nanoparticles is well understood with a rapid heating of the lattice and a subsequent cooling step in the order of 250 ps for the 16 nm particles. The laser fluence is precisely tuned and can serve as temperature scale. The scattered x-rays are analyzed in small angle scattering (SAXS) [7] and wide angle scattering geometry (WAXS) [3]. The measurement of the gold (111) peak position allows to determine the particle temperature and the threshold fluence for particle melting. In SAXS one can determine the mesoscale changes of density around the particles. By the simultaneous solution of the heat transfer equations the temperature profile within a region in the aqueous phase close to the particle surface is derived. One point to mention is the importance of the thermal boundary resistance (Kapitza resistance), which acts as a barrier for heat flow to the surrounding. A consequence is that for very short heat pulses the system always shows a temperature discontinuity across the interface, which reduces the water temperature strongly relative to the particle temperature. This effect is size dependent and for 16 nm particles the temperature outside the particles is at maximum 18 % of the initial temperature inside the particles.

In the present case the laser pulses have been stretched to 4 ps. In this case irreversible structural modifications of the particles can be ignored [8] and the particle dynamics are completely reversible. A comparison of the small angle scattering signal between bare particles and the protein coated particles shows, that strong modifications of the scattering patterns are induced for the coated particles, while no changes are observed at the same time for bare particles. The laser fluence in fig. 1 has been set to just below the threshold for vapor bubble formation around the particles [5]. Consequently only a faint structural change may be expected, while the particles have a high temperature. This is indeed the case for the bare particles, as seen for the Porod invariant in fig. 1. The Porod invariant P

$$P = \int_0^\infty I(q)q^2 dq = 2\pi^2 \Delta\rho^2 r_e^2 \Phi(1 - \Phi) \qquad (1)$$

is a measure for the global scattering length density contrast $r_e \Delta\rho$ change after laser excitation, Φ the filling fraction. The scattering contrast can arise from the protein shell around the gold particles with an average protein density exceeding the water density. Reported values for the (hydrated) density of BSA are in the range of 1.35 g/cm^3 [9]. A transient scattering contrast can arise from the appearance of a vapour bubble around the heated particle. The coated particles show a large signal, with a sub-nanosecond transient and a constant signal towards long time delays. The transient is caused by the onset of bubble formation.

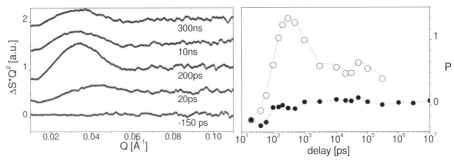

Fig. 1. Difference SAXS of a laser excited gold nanoparticle suspension (left) with albumin coating. The curves have been scaled for clarity. The Porod invariant at a laser fluence of 93 J/m^2 is shown for the bare (filled symbols) and the bioconjugated particles (open symbols).

At quasi equilibrium after 10-100 nanoseconds the remaining scattering distribution is a clear indication of the complete expulsion of the protein layer from the surface. The laser excitation shows that the protein is detached due to the fast heating. A similar observation has been described for DNA conjugated with gold particles and a disruption of the thiol bond [10]. More interestingly this phenomenon was initiated at a local temperature at the outside of the particle far above 100 °C, just below the spinodal decomposition point of the water phase. We find that for small laser fluence these irreversible structural modifications are completely absent. This observation points towards an unexpectedly large tolerance of the protein layer to very short heating stimuli, not in agreement with the temperature of static denaturation.

Conclusions

Time-resolved x-ray scattering provides valuable information on heat induced structural relaxations in bio-conjugated nanoparticles.

Acknowledgements. The research is funded by the Ministry of Science, Education and the Arts Baden-Württemberg. We acknowledge ESRF for provision of beamtime and funding, and M. Wulff and F. Ewald for their assistance. This work was also supported by the Creative Research Initiatives (Center for Time-Resolved Diffraction) of MOST/KOSEF.

1 C. M. Pitsillides, E. K. Joe, X. Wei, R. R. Anderson, and C. P. Lin, Biophys. J. **84**, 4023, 2003; J. Neumann and R. Brinkmann, J. Biomed. Opt. **10**, 24001, 2005.

2 G. V. Hartland, M. Hu, and J. E. Sader, J. Phys. Chem. B **107**, 7472, 2003.

3 A. Plech, V. Kotaidis, S. Grésillon, C. Dahmen, and G. von Plessen, Phys. Rev. B **70**, 195423, 2004.

4 V. Kotaidis, and A. Plech, Appl. Phys. Lett. **84**, 213102, 2005.

5 V. Kotaidis, C. Dahmen, G. von Plessen, F. Springer, and A. Plech, J. Chem. Phys. **124**, 184702, 2006.

6 J. Kimling et al., J. Phys. Chem. B **110**, 15700, 2006.

7 A. Plech, V. Kotaidis, K. Istomin, and M. Wulff, J. Synchr. Rad. **14**, 288, 2007.

8 A. Plech, V. Kotaidis, M. Lorenc, and J. Boneberg, Nature Phys. **2**, 44, 2006.

9 B. Jachimska, M. Wasilewska, and Z. Adamczyk, Langmuir, ASAP May 2008.

10 P. K. Jain, W. Qian, and M. A. El-Sayed, J. Am. Chem.Soc. **128**, 2426, 2006.

X-ray induced transient optical reflectivity for fs-X-ray/optical cross-correlation at Free-Electron Lasers

C. Gahl,[1,3] A. Azima,[5] M. Beye,[2] M. Deppe,[2] K. Döbrich,[1] U. Hasslinger,[2] F. Hennies,[2,4] A. Melnikov,[1] M Nagasono,[2] A. Pietzsch,[2] M. Wolf,[1] W. Wurth,[2] and A. Föhlisch[2]

[1] Fachbereich Physik, Freie Universität Berlin, Berlin, Arnimallee 14, 14195 Berlin, Germany
[2] Institut für Experimentalphysik, Universität Hamburg, Luruper Chaussee 149, 22761 Hamburg, Germany
[3] Max-Born-Institut für nichtlineare Optik und Kurzzeitphysik, Max-Born-Str. 2A, 12489 Berlin, Germany
[4] MAX-lab, Lund Universitet, Ole Römers väg 1, Box 118, 221 00 Lund, Sweden
[5] HASYLAB/DESY, Notkestr. 85, 22607 Hamburg, Germany
E-mail: gahl@mbi-berlin.de; alexander.foehlisch@desy.de

Abstract. Using the high peak brilliance of the X-ray Free-Electron Laser at Hamburg, we have studied the X-ray pulse induced transient optical reflectivity on GaAs and establish a novel tool for fs X-ray/optical cross-correlation.

Introduction

Due to their high peak brilliance and ultrashort pulse length over a wide range of photon energies Free-Electron lasers (FEL) currently are developing to versatile tools for time-resolved studies of ultrafast dynamics on atomic length scales. Since a FEL based on the principle of self-amplified spontaneous emission (SASE) is not intrinsically synchronized to a conventional femtosecond laser, it becomes especially important to determine timing and pulse characteristics for pump-probe experiments. We developed a method for fs Xray/optical cross-correlation based on the X-ray pulse induced transient optical reflectivity ($\Delta R/R$) on GaAs [1].

Experimental Methods

As depicted schematically in Fig. 1, extreme ultraviolet radiation pulses of 39.5 ± 0.5 eV with a duration of < 50 fs impinge on the GaAs(100) crystal at 41.5° incidence angle with the electric field vector in the surface plane. Reaching fluences up to 16 mJ/cm^2 we stayed below the optical damage threshold of GaAs [3]. The X-ray pulse induced change of optical reflectivity $\Delta R/R$ was probed at an angle of incidence of 53° by delayed optical pulses at 800 nm or optionally 400 nm with a duration of 120 – 150 fs (fwhm). The optical pulse energies were detected in a reference path and after reflection with two fast photodiodes enabling pulse-per-pulse measurements even in a multibunch operation mode. Additionally, operating the optical laser at twice the FEL repetition rate within each pulse train of 30 radiation bursts results in alternating measurements of the reflectivity with and without the X-ray pump pulse.

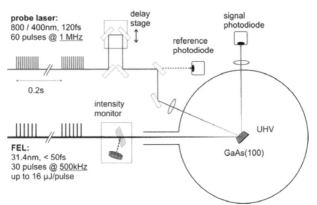

Fig. 1. Transient X-ray induced optical reflectivity (ΔR/R) measurement - schematic overview: Extreme ultraviolet FEL pulses (39.5 eV, < 50 fs, < 16 μJ) impinge onto a crystalline GaAs(100) surface and generate photoexcited carriers. The transient changes of the dielectric function are probed by visible laser pulses (800 nm or 400 nm, 120 fs, < 10 nJ) reflected from the GaAs surface at 53° as a function of their temporal delay relative to the FEL radiation pulse. The visible laser operates at twice the repetition rate (1 MHz) of the FEL (500 kHz) to measure the pumped and unpumped surface as a reference.

In Fig. 2 the optical reflectivity change after intense X-ray excitation is summarized. First an ultra fast drop in optical reflectivity occurs, which recovers within a few picoseconds. Depending on the X-ray fluence and probe wavelength, ΔR/R can even overshoot to positive values before the system approaches equilibrium on the time scale of more than 100 ps as shown for 800 nm probe. In particular the fast drop of reflectivity at τ_{drop}=160±44 fs allows us to apply X-ray pulse induced transient optical reflectivity as a tool for fs cross-correlation between the X-ray and optical laser pulses at the interaction point of the experimental set-up. The physical origin of the ultrafast change of ΔR/R is the absorption of the FEL radiation leading preferentially to Ga 3d vacancies (atomic photoionization cross section for Ga 3d: 3.6 Mbarn [4]) which decay on the timescale of a few femtoseconds via Auger processes and autoionization into valence excitations further thermalizing by electron-electron scattering. An excitation fluence of 10 mJ/cm^2 creates within the 10 nm thick surface layer produces an electron-hole pair density in the order of $1 \cdot 10^{21}$ cm^{-3} corresponding to ~1 % of the valence electrons. In comparison to the mechanism of X-ray induced optical transient reflectivity, optical laser excitation at and above the optical damage-threshold photo generated free carriers exceeding 10^{22} cm^{-3} , which results in similar ΔR/R transients as we find for X-ray excitation below the damage threshold. In the case of optical excitation the free carrier absorption changes the dielectric function resembling the Drude model. However, beyond that screening of ionic potentials and electron many-body effects are important as they modify the band structure [7]. For X-ray excitation the distortion of the valence electronic structure is created through photoionization and ultra fast Auger decay of inner shell vacancies, electronic screening and electron–electron scattering as well as the structural changes to the crystal lattice need to be considered.

A first application of our finding was the single shot X-ray optical cross-correlation measurement in a time-to-space mapping geometry to determine the arrival time of individual X-ray and optical femtosecond pulses [8].

138

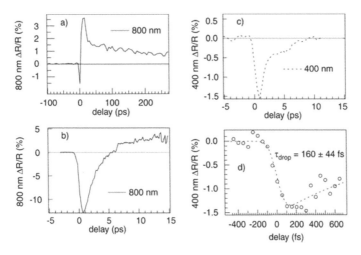

Fig. 2. Timescales of transient X-ray induced optical reflectivity (ΔR/R):
a) Ultra fast femtosecond excitation equilibrates on the picosecond timescale due to electron-phonon coupling. b) & c) show the ultrafast X-ray induced modifications probed at two different optical wavelength. Electronic relaxation into the conduction band minimum lead for 800 nm probe to increased reflectivity after 10 ps. d) Ultra fast drop in optical reflectivity allows for cross-correlation between optical and X-ray pulses from FLASH.

Conclusions

In conclusion, we have monitored the X-ray pulse induced ultrafast dynamics in GaAs through transient optical reflectivity, which provides us with a valuable tool to perform X-ray/optical cross-correlation at present and futures X-ray Free-Electron Laser sources. The physical origin is the ultra fast X-ray induced electron dynamics created through the high peak brilliance of the Free-Electron LASer at Hamburg FLASH.

Acknowledgements.

1 C. Gahl, A. Azima, M. Beye, M. Deppe, K. Döbrich, U. Hasslinger, F. Hennies, A. Melnikov, M. Nagasono, A. Pietzsch, M. Wolf, W. Wurth, and A. Föhlisch, "A femtosecond X-ray/optical cross-correlator" Nature Photonics **2**, 165 (2008).

2 S. Düsterer et al. Spectroscopic characterization of vacuum ultraviolet free electron laser pulses, Opt. Lett. **31**, 1750 (2006).

3 A. Cavalleri, et al., Ultra fast x-ray measurement of laser heating in semiconductors: Parameters determining the melting threshold, Phys. Rev. B **63**, 193306 (2001).

4 J.-J. Yeh and I. Lindau, Atomic Subshell Photo ionization Cross Sections and Asymmetry Parameters: $1 < Z < 103$, At. Data Nucl. Data Tables **32**, 1 (1985).

5 M. O. Krause and J. H. Oliver, Natural widths of atomic K and L levels, K alpha X-ray lines and several KLL Auger lines, J. Phys. Chem. Ref. Data **8**, 329 (1979).

6 M. Krumrey, E. Tegeler, J. Barth, M. Krisch, F. Schäfers, and R. Wolf, Schottky type photodiodes as detectors in the VUV and soft Xray range, Applied Optics **27**, 4336 (1988).

7 L. Huang, J. P. Callan, E. N. Glezer, and E. Mazur, GaAs under intense ultra fast excitation: Response of the dielectric function, Phys. Rev. Lett. **80**, 185 (1998).

8 T. Maltezopoulos et al., Single-shot timing measurement of extreme-ultraviolet free-electron laser pulses, New. J. Physics **10**, 033026 (2008).

Autocorrelation Experiments with Soft X-ray FEL Pulses

R. Mitzner[2], W. Eberhardt[1], M. Neeb[1], T. Noll[1], M. Richter[3], S. Roling[2], M. Rutkowski[2], B. Siemer[2], A.A. Sorokin[3], K.Tiedtke[4], H. Zacharias[2]

[1] BESSY GmbH, Albert-Einstein-Str. 15,12489 Berlin, Germany
[2] Physikalisches Institut, Universität Münster, D-48149 Münster, Germany
[3] PTB, Abbe-Str. 2-12,D-10587 Berlin, Germany
[4] DESY, Nottkestr. 85, 22603 Hamburg, Germany
E-mail: mitzner@bessy.de

Abstract. We report first direct measurements of the average temporal coherence and pulse length of soft X-ray fs pulses from the free-electron laser at DESY (FLASH) by means of linear and nonlinear autocorrelation.

Introduction

Among the new ultrashort X-ray sources the free-electron laser (FEL) at DESY (FLASH) has outstanding features. Providing pulse energies up to 100 µJ and pulse lengths down to 20 fs [1] FLASH is ready for new class of soft X-ray experiments e.g. nonlinear X-ray optics [2] , X-ray pump-probe experiments, two beam interferometry and holography. In order to provide two correlated pulses with a precise and variable delay as well as to temporally characterize (autocorrelation) the X-ray pulses an optomechanical beam splitter and delay unit (autocorrelator) has been constructed and installed at FLASH. Based on geometrical wave front beam splitting and grazing incident angles it covers the energy range of FLASH (20 –200 eV) with an efficiency of better than 50 % [3].

The objective of the experiments presented here is to characterize the temporal properties of the delivered FEL pulses at 24 nm. Furthermore, this investigation serves to prove the usefulness of the new X-ray autocorrelator for time resolved pump-probe experiments. A fundamental property of a FEL pulse is its spatial and temporal coherence being essential for interferometric and holographic experiments. The spatial coherence has been previously investigated as a function of various FEL parameters by applying a double slit geometry [4]. The good spatial coherence has been used in our experiments to record the interference pattern of the overlapped half beams as a function of their path length difference (delay). We are thus able to extract the average coherence time of the FEL pulse. Furthermore, the knowledge of the pulse length is crucial for many experiments e.g. any kind of time resolved measurements and multiphoton processes [2, 5]. While the nonlinear autocorrelation is a well established method to determine the pulse length in the UV and visible spectral regions there is a lack of efficient nonlinear (nonresonant) detection processes in the soft X-ray regime. Here we used direct two-photon double ionization (2PDI) of He [6] to measure the nonlinear autocorrelation function at 24 nm (51.6 eV).

Experimental

The autocorrelator and the detection chambers have been installed at beamline BL3 approx. 70 m behind the undulator (for details of the autocorrelator and the FEL see [1,3]). The spatio–temporal coherence of the FEL–pulses is determined from two-beam interference pattern generated by overlapping the split beams directly on a soft X-ray sensitive CCD (fig. 1.a). In the 2PDI experiment (fig. 1.b) a spherical multilayer mirror has been used to refocus the X-ray pulse into the helium gas to achieve an intensity of $1 \cdot 10^{14}$ Wcm^{-2}. The details of the detection of the ions are described in [2].

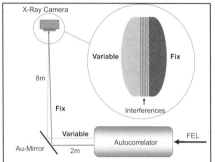 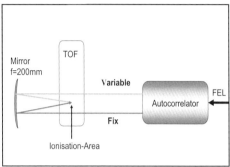

Fig. 1.a Two beam interference – linear autocorrelation, **b** 2PDI of helium - nonlinear autocorrelation

Results and Discussion

Figure 2.a displays the interference pattern observed for spatially fully overlapped partial beams (crossing angle 0.51 mrad) at zero path length difference (delay). The spacing (s) of the fringes is governed by the wavelength λ and the angle α between the two wave fronts ($s = \lambda / \sin\alpha$) resulting in a spacing of about 45 μm. The visbility v of the fringes versus the delay i.e. the mutual temporal coherence (Fig. 1.b) is extracted from a delay scan with a crossing angle of 0.18 mrad. Here a fringe spacing

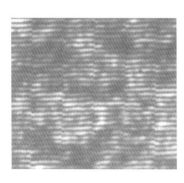 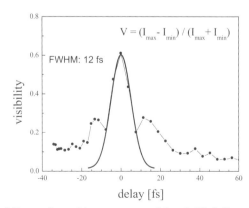

$$V = (I_{max} - I_{min}) / (I_{max} + I_{min})$$

FWHM: 12 fs

Fig. 2.a Interference fringes of spatially fully overlapped beams at zero delay, **b** Visibility of the fringes against the delay between both partial beams (separation of the overlapped points in the beam 0.8 mm)

of 130 μm is measured. The averaged visibility at zero delay, i.e. the transversal coherence for a separation of 0.8 mm, is determined to v = 0.63. With increasing delay between the partial beams the visiblity decreases rapidly. Nevertheless even at 60 to 80 fs delay interference fringes are clearly observable at individual shots. The central peak of the temporal coherence function can be well described by a Gaussian function with a FWHM of 12 fs corresponding to a coherence time of 6 fs. This is in good agreement both with a recently deduced value for 13.7 nm [1] and bandwidth considerations [7].Obviously, the visibility is not a monotonous function of the delay between the partial beams but shows distinct maxima and minima. This is a clear indicate for an multiple pulse structure. A detailed discussion is given in [7].

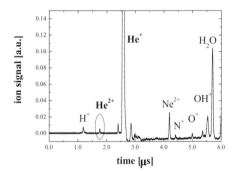 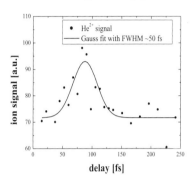

Fig. 3.a Ion time-of-flight (TOF) spectrum of helium and b nonlinear autocorrelation curve of the FEL-pulse both taken at a photon energy of 51.6 eV and an intensity of about 10^{14} Wcm^{-2}.

The result of a first nonlinear autocorrelation is displayed in Fig. 3. From the Gaussian fit an overall pulse length (FWHM) of about 35 fs is deduced. Nevertheless further optimization is needed as the first data set is affected by the poor statistics of the nonlinear process and the difficulty to overlap the tiny foci (~5 micrometer).

Conclusions

The new beam splitter and delay line for the soft X–ray has been successfully tested as a useful tool for time resolved two pulse soft X-ray experiments. In first autocorrelation experiments both the spatio-temporal coherence as well as the overall pulse length of FEL pulses at 24 nm have been measured for the first time.

Acknowledgements. We would like to thank our colleagues from BESSY, the University of Münster and the FLASH team for their support. The project has been founded by the Bundesministerium für Bildung und Forschung.

1   W. Ackermann et.al, Nature Phot. **1**, 336 (2007).
2   A. A. Sorokin et.al, Phys. Rev. Lett. **99**,213002 (2007).
3   R. Mitzner , M. Neeb, T. Noll, N. Pontius, W. Eberhardt, Proc. Of SPIE 59200D.
4   R. Ischeberg et.al, NIM A 507, 175 (2003).
5   Wabnitz et.al, Nature **420**, 482(2002.
6   Y. Nabekawa et.al, Phys. Rev. Lett. **94**, 0043001 (2005).
7   R. Mitzner et.al, submitted.

Ultrafast coherent X-ray diffractive imaging with the FLASH Free-Electron Laser

H.N. Chapman[1,3], S. Bajt[2], A. Barty[3], W.H. Benner[3], M.J. Bogan[3], S. Boutet[4], A. Cavalleri[5], S. Düsterer[2], M. Frank[3], J. Hajdu[4,6], S.P. Hau-Riege[3], B. Iwan[6], S. Marchesini[7], K. Sokolowski-Tinten[8], M.M. Siebert[6], R. Treusch[2], & B.W. Woods[3]

[1]Centre for Free-Electron Laser Science, University of Hamburg and DESY, Hamburg, Germany.
[2]HASYLAB, DESY, Hamburg, Germany
[3]Lawrence Livermore National Laboratory, Livermore CA, USA.
[4]SLAC, Menlo Park CA, USA.
[5]Department of Physics, Clarendon Laboratory, University of Oxford, Oxford, UK
[6]Uppsala University, Uppsala, Sweden.
[7]Lawrence Berkeley National Laboratory, Berkeley CA, USA.
[8]Institut für Experimentelle Physik, Universität Duisburg-Essen, Germany
E-mail: henry.chapman@desy.de

Abstract. High-resolution ultrafast coherent diffractive imaging has been carried out at the FLASH FEL. Reconstructed images show no effect of sample destruction. Time resolved imaging was achieved by time-delay holography and with a synchronized optical laser.

Single pulse diffractive imaging

The ultrafast pulses from future X-ray free-electron lasers (XFELs) will enable extraordinary new capabilities in the imaging of non-periodic objects at extremely high spatial and temporal resolution. Single-particle diffractive imaging with XFELs may provide near-atomic resolution [1]. In this method radiation damage limits are overcome by obtaining coherent diffraction patterns from objects with pulses that are shorter than the timescales for radiation-induced changes to occur in the object at the resolution length scale. The specimen will be completely destroyed by the pulse, but that destruction will only happen after the pulse has passed through the object. Dynamic processes may also be imaged at high temporal resolution due to short X-ray pulse durations of these sources of $10-100$ fs, using X-ray optical systems to split the beam into a pump and a probe, or by synchronizing the FEL pulses with optical laser pulses.

Diffractive imaging at FLASH

We are utilizing the FLASH FEL at DESY, Hamburg, to develop the experimental methods for ultrafast FEL diffractive imaging. The pulses are almost totally transversely coherent, giving a forward-directed far-field scattering pattern that is proportional to the square of the absolute value of the Fourier transform of the object. This coherent diffraction pattern is recorded with a bare CCD detector. If these Fourier magnitudes are sampled at a sufficient density it is possible to apply phase-retrieval techniques to reconstruct a real-space image. For single-pulse imaging this method has several advantages over other forms of x-ray microscopy: focusing is carried out as part of the reconstruction process, so injected clusters or particles can easily be imaged; the resolution is dependent only on the largest scattering angle

143

recorded and not limited by the quality or design of an x-ray lens; there is indeed no optical element near the sample that could be destroyed by the intense pulse or radiate or scatter light onto the detector. We do filter the scattered radiation using a combination of a coating on the CCD (or a free-standing foil in front of the CCD) and a multilayer mirror [2]. The mirror has a hole for the strong undiffracted beam to pass through without any interaction.

We have demonstrated that low-noise coherent diffraction patterns can be recorded with single pulses as short as 10 fs duration and that these patterns can be phased to give high resolution images [3]. The images show no indication of any sample destruction, although the sample is completely vaporized by the pulse. Images have been reconstructed from samples fixed to membranes and placed in the beam path and from particles that are injected across the beam. Particles from solution are first aerosolized by electrospray ionization and then drawn into a focused beam using an aerodynamic lens [4]. In this configuration we use 100 kHz pulse trains from FLASH to increase the rate at which the particles, for which the arrival time is unsynchronized with the FEL, are hit by a pulse. A time-of-flight mass spectrometer, which records an ion spectrum every pulse in the train, is used to determine if only one pulse hits a particle over the exposure time of the CCD. Exposure times are usually less than 700 pulses. It has been possible to reconstruct images of cells and particles that were injected across the beam. This method of sample introduction rapidly cools the cell by flowing with expanding background gas which may preserve its structure of the initially wet cell much like plunge freezing in liquid nitrogen. The sample travels at about 100 m/s and is in vacuum for less than 1 ms before interaction with the FEL pulse.

Time-resolved Imaging

We are particularly interested in studying the dynamics of particles and matter irradiated by intense FEL pulses and understanding the resolution limits imposed by the sample destruction. We invented a new method, time-delay holography [5], to obtain precise measurements of the rate of the FEL-induced explosion of particles. In this method, the FEL pulse impinges on the sample and diffracts from it. A multilayer mirror is placed behind the sample and reflects the diffracted and strong undiffracted beams back towards the sample. The latter beam diffracts again from the sample, at a delay simply given by the travel time from the sample to the mirror and back. The prompt and delayed diffracted beams continue to propagate back towards the source. We use the mirror with the hole to reflect these beams, which interfere at the CCD. Although there is a delay between the prompt and delayed diffraction, both beams have similar path lengths as they both are reflected from the mirror. The interference pattern can be thought of as a hologram consisting of the coherent addition of a known reference beam (the prompt diffraction from the unexploded particle) and the unknown beam (the delayed diffraction from the exploded particle). Additionally, it is straightforward to measure the difference in the phase shift of the diffracted beam as it travels through the exploding particle compared with the initial state. We have used these techniques to measure the rate of explosion of spherical latex particles and find reasonably good agreement with a model of the hydrodynamic explosion [6]. At the shortest delay measured of 350 fs we find that 140-nm spheres expand no more than 6 ± 3 nm. We expect for cellular imaging that the rate of explosion of resolution elements (or, for example, organelles) will depend on the surrounding matrix. We are currently investigating the effect of a tamper around test particles decreases the explosion rate according to predictions [7].

We have also studied the evolution ablation of materials by optical laser pulses in a pump-probe configuration. A linearly-polarized laser (Nd:YLF operating at 527 nm, 10 ps pulse duration) impinged on a silicon membrane. The transmitting coherent FEL diffraction pattern was measured at a wavelength of 13.5 nm after a given delay. We observed light-induced periodic structures that strongly diffracted the FEL pulse. These structures were observable after about a 10 ps delay and became most prominent after 200 ps, as shown in Fig. 1. Additionally we observed strong speckles most prominent after about 300 ps and structure scales of 200 nm, persisting for 1 ns. To better study thes formation of structures we have performed correlation spectroscopy and phased the coherent diffraction patterns [8].

In the future we will continue to apply these methods to the study of laser-matter interactions and photoinduced processes at resolutions lengths below 10 nm, using shorter wavelengths at FLASH and higher peak intensities with improved condensing optics.

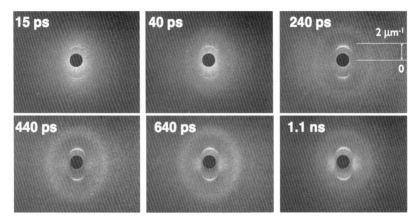

Fig. 1. . Coherent diffraction patterns of laser-induced ablation of Si foils measured at a wavelength of 13.5 nm. The delay between the optical pump and the FEL probe is indicated. The polarization of the pump is vertical in these images.

1   R. Neutze, R. Wouts, D. van der Spoel, E. Weckert, and Ja. Hajdu, "Potential for biomolecular imaging with femtosecond x-ray pulses," Nature **406** 753–757 (2000).

2   S. Bajt *et al*. "A camera for coherent diffractive imaging and holography with a soft-X-ray free electron laser," Appl. Opt. **47**, 1673–1683, (2008).

3   H.N. Chapman *et al*. "Femtosecond diffractive imaging with a soft-x-ray free-electron laser," Nat. Phys **2** 839–843 (2006).

4   M.J. Bogan *et al*. "Single particle x-ray diffractive imaging," Nano Lett. **8** 310– 316 (2008).

5   H.N. Chapman *et al*. "Femtosecond time-delay x-ray holography," Nature **448** 676–679 (2007).

6   S.P. Hau-Riege, R.A. London, A. Szoke, "Dynamics of biological molecules irradiated by short x-ray pulses," Phys. Rev. E **69** 051906 (2004).

7   S.P. Hau-Riege, R.A. London, H.N. Chapman, A. Szoke, and N. Timneanu, "Encapsulation and diffraction-pattern-correction methods to reduce the effect of damage in x-ray diffraction imaging of single biological molecules," Phys. Rev. Lett. **98** 198302 (2007).

8   A. Barty *et al*. "Ultrafast single-shot diffraction imaging of nanoscale dynamics," Nature Photonics **2**, 415-419 (2008).

Lensless Microscopy and Holography with 60 nm Resolution using Tabletop Coherent Soft X-Rays

Daisy A. Raymondson[1], Richard L. Sandberg[1], William F. Schlotter[2], Kevin Raines[3], Chan La-O-Vorakiat[1], Ariel Paul[1], Anne E. Sakdinawat[4], Margaret M. Murnane[1], Henry C. Kapteyn[1], and Jianwei Miao[3]

[1] Department of Physics and JILA, University of Colorado and NIST, Boulder, Colorado, USA
E-mail: daisy.raymondson@colorado.edu
[2] Stanford Synchrotron Radiation Laboratory, SLAC, Menlo Park, California, USA
[3] Department of Physics and Astronomy and California NanoSystems Institute, University of California, Los Angeles, California, USA
[4] Center for X-ray Optics at Lawrence Berkeley National Laboratory, Berkeley, California, USA

Abstract. We demonstrate high numerical aperture lensless diffractive imaging and Fourier transform holography using high harmonic generation from an ultrafast laser system. The HHG source operates at 29 nm or 13 nm wavelength and achieves 60 nm resolution.

Introduction

Lensless imaging (also known as x-ray diffractive microscopy) is a relatively new technique that is particularly suited to x-ray imaging, enabling high-resolution imaging of noncrystalline, aperiodic samples such as biological materials [1]. This technique provides a large depth of field, insensitivity to alignment, and (as we demonstrate here) near-wavelength-limited resolution without the need of complex optical systems. In past work, we demonstrated lensless diffraction imaging using a tabletop high harmonic (HHG) soft x-ray source at 29 nm, achieving resolutions ~200 nm [2]. In this work, we significantly enhance our image resolution by implementing a new high numerical aperture (up to NA=0.5) scheme and field curvature correction. We achieve a resolution of 60 nm using a 29 nm HHG source[3]. We also report the first demonstration of Fourier transform holography [4] with a tabletop soft x-ray source, acquiring images with a resolution $\approx$ 90 nm [5]. As higher brightness, shorter wavelength, tabletop sources become available, this high numerical aperture scheme – a first for diffractive imaging with x-rays – will make possible near-wavelength resolution below 10 nm. Important applications of this work exist in nanotechnology, lithography, materials science, and biological imaging. Moreover, the ultrafast (fs) duration of the HHG pulses will enable imaging of dynamic systems with high temporal resolution.

Experimental Setup

The experimental setup for lensless diffractive microscopy is compact and straightforward. The sample is illuminated by coherent, narrowband, soft x-ray light. A set of narrowband multilayer mirrors acts as both a monochromator and a condenser, to gently focus the beam onto a sample. Samples are positioned using a closed loop x-y piezo stage, and are illuminated with either a HHG beam at 29 nm or

13 nm wavelength. The diffraction of the soft x-ray light from the sample is then recorded on a large-area CCD camera (13.5 μm pixel size).

Although the CCD camera records only the intensity of the light scattered by the sample, images can retrieved through one of two separate techniques reported here: lensless diffractive microscopy [1,2] or Fourier transform holography [4]. Both of these image retrieval techniques use the fact that the diffraction pattern is a two-dimensional Fourier transform of the transmitted soft x-ray wavefront immediately after the sample. In lensless diffractive microscopy, the image of the sample is retrieved by "oversampling" the diffraction pattern [1,2]. If the diffraction pattern is oversampled – meaning that the sampling rate on the CCD is greater than twice the Nyquist frequency – the image can be reconstructed using an iterative phase retrieval algorithm [1]. For Fourier transform holography, the sample is surrounded by one or more small (~ 100 nm in diameter) references holes. Each reference hole produces a strongly diffracting spherical reference wave that interferes with the diffracted light from the sample on the CCD camera. The phase information from the sample is then recorded in this interference, and an image of the sample can be retrieved by performing a simple 2-D Fourier transform.

Results and Discussion

Figure 1 shows an image obtained using lensless diffractive microscopy with the 29 nm HHG source [3], indicating a resolution of 94 nm. The sample is a "waving stick figure" transmission pattern, which was fabricated using e-beam lithography on a composite sample consisting of 200 nm of gold on a 100 nm Si_3N_4 substrate.

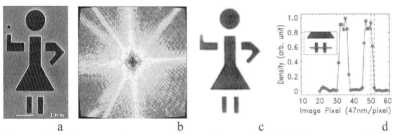

| a | b | c | d |

Figure 1. a) Electron microscope image of a waving stick figure transmission sample; b) curvature corrected diffraction pattern using a 29 nm HHG source c) reconstructed image, showing a 94 nm resolution, as shown in (d)

Figure 2. a) SEM image of FTH sample with five reference holes, b) hologram after curvature correction (log scale) obtained with the 29 nm HHG source, c) autocorrelation reconstruction of the sample, d) line cut through one sub-image showing 90 nm resolution [5].

Figure 2 shows the results from a 5-hole Fourier transform holography sample using a 29 nm HHG source. The test pattern and five reference holes were fabricated using a focused gallium-ion mill in a Au thin film (400 nm) on a Si_3N_4 substrate (100 nm). The image retrieved with a simple Fourier transform has ~90 nm resolution [5]. The iterative phase retrieval algorithm was applied to refine the resolution to 60 nm, shown in Figure 3. This near-diffraction-limited resolution of 2λ was made possible by the high-NA (0.52) configuration, a first for soft x-ray diffractive imaging.

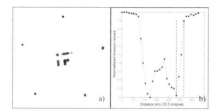

Figure 3. a) image of the test sample after refinement with the phase retrieval algorithm, b) line cut showing 60 nm resolution from the 29 nm HHG source [6].

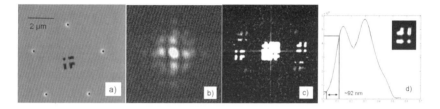

Figure 4. a) SEM image of FTH sample with five reference holes b) hologram after curvature correction (log scale), taken with the 13 nm HHG source, c) autocorrelation reconstruction of the sample, d) line cut through one sub-image showing 92 nm resolution [6].

Figure 4 shows the first demonstration of Fourier transform holography with a 13 nm tabletop source [6]. 92 nm resolution is also obtained from a simple Fourier transform at this wavelength, and straightforward extensions will allow near-wavelength resolution as in the 29 nm results.

In the future, shorter wavelength tabletop x-ray sources will enable versatile soft x-ray lensless imaging with sub-10nm resolution for dynamics imaging in biology, materials science, and nanotechnology.

Acknowledgements The authors thank Y. Liu and F. Salmassi at the CXRO for providing multilayer optics. This work was supported by the NSF Engineering Research Center for Extreme Ultraviolet Science and Technology, the DOE NNSA, and the JILA Instrument Shop and the Lehnert laboratories.

1   J. Miao et al., Nature **400**, 342-344, 1999.
2   R. Sandberg et al., Proc. Natl. Acad. Sci. USA **105**, 24, 2008.
3   R. Sandberg et al., Phys. Rec. Lett. **99**, 098103, 2007.
4   W. Schlotter et al., Ap. Physics Lett. **89**, 163112, 2006.
5   R. Sandberg et al., in preparation.
6   D. Raymondson et al., in preparation.

Nanoscale Heat Transport Probed with Ultrafast Soft X-Rays

Mark Siemens[1], Qing Li[1], Margaret Murnane[1], Henry Kapteyn[1], Ronggui Yang[1], and Keith Nelson[2]

[1]University of Colorado at Boulder, 440 UCB, Boulder, CO 80309, USA
[2]Massachusetts Institute of Technology, Cambridge, MA 02139, USA
E-mail: siemens@colorado.edu

Abstract. We characterize heat transport in nanostructures using coherent soft x-rays to probe thermally induced surface deformation. By varying the substrate temperature, we observe the transition from diffusive to quasi-ballistic heat transport.

On macroscopic (micron and longer) length scales, thermal transport can be related to integrated or averaged phonon properties, and is well described by the diffusive Fourier heat conduction theory. However, on length scales that are short compared with the phonon mean free path, thermal transport is not expected to be diffusive, but rather ballistic. For intermediate length scales comparable to the phonon mean free path, a quasi-ballistic transition region is predicted. Although various methods for predicting heat flow in this regime have been proposed [1, 2], no experiments have yet validated the predictions for phonon heat transport surrounding a nanoscale heat source. This issue is technologically relevant as it relates directly to the question of thermal dissipation in a nanoscale transistor.

High harmonic generation (HHG) is an ideal coherent light source for studying nanoscale thermal transport effects for many reasons. First, the spatial coherence of HHG in a waveguide enables highly sensitive interferometric measurements of the temperature dependent surface profile. Moreover, the short wavelength of HHG (tunable from 100-1 nm) allows for tighter focusing and better phase sensitivity in interferometric experiments. Finally, the femtosecond and sub-femtosecond duration of HHG pulses holds promise for achieving unprecedented temporal resolution in dynamics experiments. We previously demonstrated the use of HHG beams as a sensitive probe of the changing diffraction in micro-patterned surfaces due to surface acoustic waves [3], and in a Gabor holography geometry to interferometrically observe laser heating-induced surface displacements and subsequent acoustics [4].

To facilitate access to the ballistic regime of heat transport, we use sapphire as the substrate material because the mean free path of the dominant heat-carrying phonons is long ($\Lambda = 150$ nm at room temperature, and is much longer at lower temperatures) and it is transparent to the 800nm optical pumping wavelength. An array of thin nickel lines (20 nm high x 1 μm wide, with 3 μm between lines) was then fabricated on the surface by electron beam lithography and liftoff. This sample was heated by an 800 nm pulse from a Ti:Sapphire laser amplifier. The sapphire substrate transmits the 800 nm pump light, while the nickel lines are strongly absorbing and serve as a heat source. The laser induced thermal excitation and subsequent relaxation of the nickel lines can be probed by measuring the diffraction of a 30 nm HHG beam from the sample (see Fig. 1). The diffraction efficiency of the nickel grating depends strongly on small thermally-induced changes in height of the nickel lines. By assuming a linear coefficient of thermal expansion, we can determine the temperature change corresponding to a given change in diffraction.

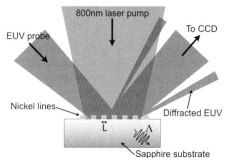

Fig. 1. Experimental geometry: a sample consisting of nickel lines of width "L" of 1μm on sapphire is illuminated with an 800 nm pump pulse, and then probed using EUV light. The phonon mean free path "Λ" in sapphire is 150 nm at room temperature.

By monitoring the change in the probe diffraction from the grating as a function of pump-probe delay, we obtain a transient signal shown in Figure 2. The oscillation is due to surface acoustic wave (SAW) propagation in the nanostructure. The SAW is generated by the patterned surface strain caused by the local heating; we confirmed that this is the source of the oscillation by comparing the measured frequency of oscillation with that predicted by the nickel line geometry and the effective transverse speed of sound in the nanostructure (1.3 GHz). We measure the same frequency regardless of the sample temperature or the pump intensity. Along with the oscillatory component, we also see an exponential decay as the heat in the nickel lines dissipates into the unheated sapphire substrate. This thermal decay is an indicator of phonon energy transport efficiency between the nickel lines and the sapphire substrate.

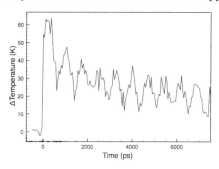

Fig. 2. Dynamic change of the 1μm nickel lines at room temperature. The signal consists of a non-thermal oscillation at 1.34 GHz due to SAW propagation, as well as a thermal decay.

At room temperature, heat transport in this sample is diffusive since the characteristic length (nickel wire width L=1 μm) is much greater than the phonon mean free path (Λ~150 nm) in sapphire. In order to observe ballistic heat transport in this sample, the phonon mean free Λ must be increased. Since Λ~1/T at temperatures between 76 and 300 K [5,6], this can be accomplished by cooling the sample. L/$\Lambda\approx$1 at 110 K, so quasi-ballistic heat transport is expected at such low temperatures.

To measure the transition in heat transport regimes, we fit thermal response curves (shown in Fig. 2) at intermediate temperatures to a decaying exponential (Fit = $I_0\exp[-t/\tau]$, where I_0 is a fitting constant). The decay time τ as a function of temperature is plotted in Fig. 3 as the points with error bars. From a Newton's law of cooling analysis, we expect this decay time to be inversely proportional to the diffusivity K=k/C, where "C" is the volumetric specific heat and "k" is the thermal conductivity of sapphire. This prediction obeys diffusive heat behavior, where k = CvΛ/3 [5], and is depicted on Fig. 2 as the line. At temperatures above 130 K, the decay time we measure matches the bulk, diffusive transport prediction. Below 130 K, we observe significant deviation from bulk behavior, consistent with a transition to the quasi-ballistic regime of heat transport. Λ~500 nm in sapphire at 130 K, so the onset of quasi-ballistic transport occurs when L/Λ~2.

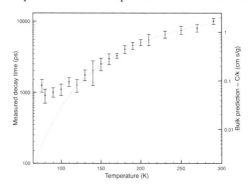

Fig. 3. Thermal decay time from the exponential decay fit as a function of temperature (points-left axis), and predicted bulk behavior from 1/K=C/k for sapphire (line- right axis).

In summary, coherent HHG beams are a sensitive probe of nanoscale dynamics, including thermal transport and surface acoustic wave propagation. We have used the change in diffraction from a microstructured grating to observe the transition between diffusive and ballistic heat transport. It is straightforward to extend this technique to study heat transport in nanoscale wire arrays, because the technique is limited only by the probe wavelength of 30 nm. These experiments will further understanding of heat transport fundamentals and thermal management at nanoscale dimensions.

Acknowledgements. This work was funded by the DOE Division of Chemical Sciences, Geosciences, and Biosciences and the National Science Foundation Engineering Research Center for Extreme Ultraviolet Science and Technology.

1   R. G. Yang, G. Chen, M. Laroche and Y. Taur, "Simulation of nanoscale multidimensional transient heat conduction problems using ballistic-diffusive equations and phonon Boltzmann equation", Journal of Heat Transfer-Transactions of the ASME 127, 298 (2005)

2   G. Chen, "Nonlocal and nonequilibrium heat conduction in the vicinity of nanoparticles", Journal of Heat Transfer 118, 539 (1996).

3   R. Tobey, E. Gershgoren, M. Siemens, M. M. Murnane, H. C. Kapteyn, T. Feurer and K. A. Nelson, "Photothermal and Photoacoustic Transients probed with Extreme Ultraviolet Radiation", Applied Physics Letters 85, 564-566 (2004).

4   R. Tobey, M. Siemens, O Cohen, M. M. Murnane, H. C. Kapteyn, and K. A. Nelson "Ultrafast Extreme Ultraviolet Holography: Dynamic Monitoring of Surface Deformation", Optics Letters 32, 286 (2007).

5   G. Chen, *Nanoscale Energy Transport and Conversion* (Oxford University Press, 2005)

6   G. Chen, D. Borca-Tasuica and R.G. Yang, "Nanoscale Heat Transfer," in "Encyclopedia of Nanoscience and Nanotechnology", Vol. 7, pp. 429-459 (American Scientific Publishers, 2004).

Relativistic attosecond electron pulses from cascaded acceleration using ultra-intense radially polarized laser beams

Charles Varin1, Pierre-Louis Fortin2, and Michel Piché2

1 University of Ottawa, Ottawa, Ontario K1N 6N5, Canada
  E-mail: cvarin@uottawa.ca
2 Centre d'optique, photonique et laser, Université Laval, Québec, Qc G1V 0A6, Canada
  E-mail: mpiche@phy.ulaval.ca

Abstract. Attosecond electron pulses with peak energy above 200 MeV could be produced with ultrafast 100-TW radially polarized laser beams in a two-stage configuration. Such electron beams would be collimated and potentially quasi-monoenergetic.

Introduction

In the past years, electron pulses have been used to probe ultrafast electron dynamics in molecules and atoms on the picosecond [1] and the femtosecond [2] timescales. Avenues to reach the attosecond timescale have been proposed [3]. Typically, those techniques employ multi-keV-energy electron beams corresponding to de Broglie wavelengths of the order of 10^{-12} m. For ultrafast electron diffraction from inner atomic shells or subatomic structures, higher kinetic energies are required. Multi-MeV 560-fs electrons pulses have been successfully used for electron diffraction experiments [4]. Innovative laser acceleration schemes could help in reaching the multi-hundred-MeV energy range with attosecond duration.

In a previous work, it has been demonstrated that collimated multi-MeV attosecond electron pulses could be produced using intense ultrafast radially polarized laser beams [5]. The proposed scheme takes advantage of the strong longitudinal electric field component present at the center of the lowest-order mode of this family of laser beams. While energies in the tens of MeV range are expected from electrons initially at rest, the peak energy could be increased above 200 MeV in a two-stage configuration of the scheme [6], as described herein.

Radially Polarized Laser Beams

Radially polarized laser beams have received much attention in the past years. The lowest-order radially polarized mode, also known as the TM_{01} beam, can be generated from two cross-polarized TEM_{01} and TEM_{10} Gauss-Hermite modes of identical waist and phase (see Figure 1). In the paraxial approximation, this particular type of beam is characterized by a doughnut-shaped intensity profile. Its specific field configuration allows it to be focused to a tighter spot size than the fundamental TEM_{00} Gaussian beam mode [7]. Because of its unique properties, it can effectively find applications in microscopy and laser processing and also be used for particle acceleration.

Electron Acceleration With Radially Polarized Laser Beams

Tightly focused radially polarized laser beams exhibit a strong longitudinal electric field component at center [7]. When this component is of relativistic strength, it can push electrons initially at rest at the beam waist out of the focal region and accelerate them along the beam propagation axis [5]. The use of a few-cycle laser beam and a compact initial electron cloud forces the particles to effectively interact with a single half-cycle of the laser field and form a pulse of attosecond duration. Transverse electric and magnetic field components provide for the lateral confinement of the accelerated particles.

$$\mathrm{TEM}_{01}(\vec{y}) + \mathrm{TEM}_{10}(\vec{x}) = \mathrm{TM}_{01}(\vec{r})$$

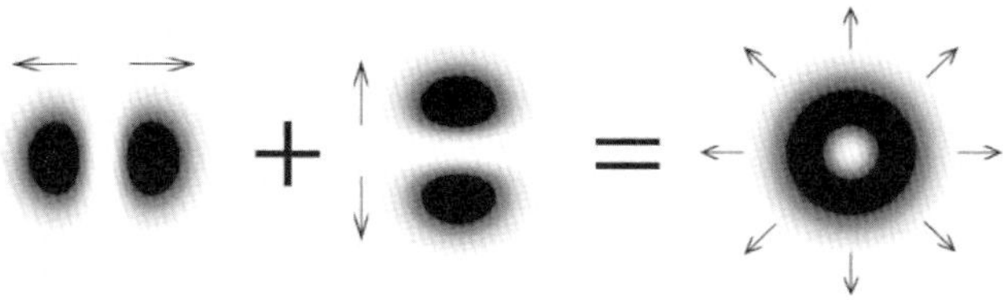

Fig. 1. Schematic construction of the lowest-order radially polarized, or TM_{01}, laser beam.

Electron acceleration with radially polarized laser beams effectively takes place when beam power exceeds

$$P_0 = (\pi^5/2\,\eta_0)(w_0/\lambda_0)^4 (mc^2/e)^2 \tag{1}$$

where $\eta_0 = 120\pi\,\Omega$, w_0 is the beam spot size at the waist, λ_0 is the laser wavelength, c is the speed of light *in vacuo*, and m and e are the electron rest mass and charge, respectively. The maximum energy gain (in MeV) available during the acceleration along the TM_{01} beam propagation axis is given by

$$\Delta W_{max}[MeV] = (2\,\eta_0/\pi)^{1/2} P[TW]^{1/2}, \tag{2}$$

where $P[TW]$ is the laser power in terawatts.

The validity of equations (1) and (2) has been confirmed by the numerical integration of the three-dimensional time-dependent Maxwell-Lorentz force equations [5]. It has been observed that best performance can be reached by optimizing the beam spot size at the waist for a given power and laser pulse duration. For a 100-TW 12-fs (FWHM) laser pulse, the optimal energy gain is effectively obtained for beam spot sizes between 2 and 3 μm ($\lambda_0 = 0.8$ μm). Calculations predict 20-MeV 25-as electron pulses with an energy spread below 1% and mrad divergence.

Cascaded Electron Acceleration with Radially Polarized Beams

The longitudinal electric field component of radially polarized laser beams can also be used to accelerate relativistic electron pulses. This feature allows for a cascaded implementation of the acceleration scheme. In Figure 2 the electron energy gain is shown as a function of the initial electron energy. The results indicate that the energy gain rapidly saturates with increasing values of initial electron energy.

Numerical simulations for a 100-TW 12-fs (FWHM) laser pulse suggest that a two-stage implementation of the acceleration scheme could produce electron pulses with duration in the 100-as timescale, peak energy above 200 MeV, 20% energy spread, and beam divergence less than 1 mrad.

153

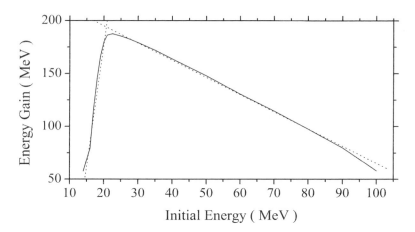

Fig. 2. Energy gained by an electron accelerated along the propagation axis of a 100-TW 12-fs (FWHM) radially polarized laser beam ($w_0 = 3$ μm and $\lambda_0 = 0.8$ μm) as a function of its initial energy. The pulse carrier-phase has been optimized for highest energy gain.

Conclusions

It has been shown that attosecond electron pulses with peak energy in the multi-hundred-MeV range can be obtained by two intense ultrafast radially polarized laser beams. It appeared that in such a case transverse confinement of electrons should be improved to keep both beam divergence and energy spread as low as possible.

For a given radially polarized beam power, there exists an optimal beam spot size that gives the best performance in terms of electron pulse duration and energy. Following such guidelines, sub-attosecond electron pulses with peak energy in the GeV range are predicted for petawatt laser power.

Acknowledgements. This work has been supported by the Fonds québécois de recherche sur la nature et les technologies, the National Sciences and Engineering Research Council of Canada, and the Canadian Institute for Photonic Innovations.

1 H. Ihee *et al.*, "Direct imaging of transient molecular structures with ultrafast diffraction," Science **291**, 458 (2001).

2 B. J. Siwick, J. R. Dwyer, R. E. Jordan, and R. J. D. Miller, "An atomic-level view of melting using femtosecond electron diffraction," Science **302**, 1382 (2003).

3 P. Baum and A. H. Zewail, "Attosecond electron pulses for 4D diffraction and microscopy," Proceedings of the National Academy of Sciences **104**, 18409 (2007).

4 J. B. Hastings *et al.*, "Ultrafast time-resolved electron diffraction with megavolt electron beams," Appl. Phys. Lett. **89**, 184109 (2006).

5 C. Varin and M. Piché, "Relativistic attosecond electron pulses from a free-space laser-acceleration scheme," Phys. Rev. E **74**, 045602 (2006).

6 C. Varin and M. Piché,"Electron Acceleration staging with radially polarized laser beams," To be published.

7 R. Dorn, S. Quabis, and G. Leuchs, "Sharper focus for a radially polarized light beam," Phys. Rev. Lett. **91**, 233901 (2003).

Attosecond Free Electron Pulses for Diffraction and Microscopy

Peter Baum[1,2] and Ahmed H. Zewail[1]

[1] California Institute of Technology, 1200 E. California Bld, Pasadena CA 91125, USA
[2] Ludwig-Maximilians-Universität München, Oettingenstr. 67, 80538 München, Germany
E-mail: peter.baum@lmu.de

Abstract. In synthesized gratings of optical fields, free non-relativistic electrons compress to pulses of ~15 attosecond duration. Such pulses have potential to advance ultrafast electron diffraction and microscopy to the domain of attosecond electron dynamics.

Introduction

The structural dynamics of complex molecular or condensed-matter systems proceed through intermediates and transition states on a multidimensional energy landscape [1-2]. For the electron dynamics in atoms, clusters, or special materials, a similar behavior is expected when multiple electrons interact through Coulomb forces, which may induce sequential or concerted interactions. In contrast to the femtosecond time scale of ultrafast nuclear motions, electron dynamics are expected to involve the attosecond time domain. For the investigation of such processes, attosecond optical pulses at 30-100 eV have been generated and applied in spectroscopy. The achievable pulse duration is currently limited to ~100 attoseconds by the central energy required to support one optical cycle. Another approach is the use of transient recollision electrons that are generated from atoms or diatoms in high-intensity optical fields. Both methods, however, are not well suited to directly visualize complex electron systems in space and time, because the De Broglie wavelength of the attosecond optical fields (10-40 nm) or transient electrons (~0.1 nm) are too long to provide sufficient resolution. Here we consider the generation of 15-attosecond free electron pulses at keV energy, and their utilization for advancing electron microscopy and diffraction to the attosecond domain. A compression methodology for electron packets in synthesized optical intensity gratings is discussed, and we report simulations showing the feasibility of such approach under realistic experimental conditions.

Attosecond Free Electron Pulses from Synthesized Optical Gratings

For electron diffraction and microscopy applications on attosecond dynamics, the applied electron pulses must match three criteria. First, they should have temporal durations of below 100 attoseconds in order to freeze the relevant transient configurations in time. Second, nonrelativistic electron pulses are needed to obtain a good scattering cross section and detectable diffraction angles. Third, the attosecond electron pulses must be well synchronized with an optical field or secondary electron/x-ray pulses for excitation and imaging without jitter. These three requirements are met in the approach presented here [3].

The concept is centered on the use of synthesized optical fields to compress a long electron packet to pulses of sub-wavelength duration (see Fig. 1). A femtosecond electron packet of 31 keV energy (light grey), such as generated by photoemission

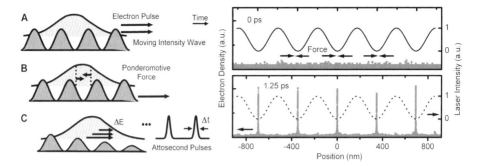

Fig. 1. Compression of electron pulses in a moving intensity grating. (A) A femtosecond electron packet (light gray) copropagates with the maxima and minima of a moving intensity grating (dark gray), such as synthesized from two counter-propagating laser pulses of different frequencies [3]. (B) Within every optical cycle, parts of the electron packet are accelerated or retarded, depending on position. (C) The resultant momentum distribution induces compression to attosecond duration (Δt) after some propagation. (D) Trajectory simulations show the expected compression at the minima of the intensity grating.

and static acceleration, is superimposed with an optical intensity grating (dark gray). The direct interaction between electrons and an electromagnetic wave averages to zero, because of the high velocity mismatch (c as opposed to 0.33 c), but an intensity distribution can be synthesized to form an intensity grating that propagates at matched speed with the electrons. The necessary field is a superposition of two counter-propagating laser pulses of different frequencies. In such a moving intensity grating, the ponderomotive force accelerates charged particles out of the regions of high intensity. Those parts of the electron packet located at a falling slope feel an accelerating ponderomotive force, and those parts located at a rising slope are retarded (Fig. 1B, arrows). These forces accumulate over time to a macroscopic momentum distribution that is directly correlated with the original electron position. After the laser intensity fades away, the initially extended electron packet contracts and self-compresses to attosecond pulses (see Fig. 1 C), with an inter-pulse spacing given by the wavelength of the intensity wave.

Numerical Simulations and Discussion

The quality of compression depends on the linearity of the induced velocity dispersion, which is the derivative of the sinusoidal intensity distribution and approximately linear around the minima, where the pulses compress. In principle, this would lead to a perfect compression, but in practice the attainable pulse duration is limited by the energy distribution within the initial electron packet (realistically ~0.4 eV). To account for such effects, simulations of electron trajectories in the moving intensity wave were obtained. The electric and magnetic fields of two colliding laser pulses of 5 μJ energy and 300 fs duration, focused to 50 μm diameter, and a Gaussian distribution of electrons at 31 keV energy (spread 0.4 eV) with a duration of 300 fs and a transversal diameter of about 10 μm were assumed initially. Figure 2 depicts the results. Just before compression, trailing electron have a slightly higher energy than leading ones and start catching up with them (A). At the best compression (B), the electron density peaks up (B) and disperses (C) again. Figure 2D shows the simulation results for the electron density; at the moment of best

156

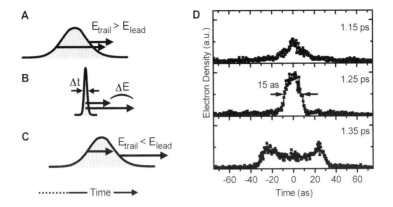

Fig. 2. Concept of electron packet self-compression to attosecond duration. (A-C) Before compression, trailing electrons have a higher speed than leading ones. At compression, electrons of all energy coincide in time. After that, the packet disperses again. (D) Results of numerical simulations, demonstrating the feasibility of 15-attosecond pulses under realistic conditions.

compression, an electron pulse duration of 15 attoseconds is achieved, evaluated as full width at half maximum. Higher grating intensities can yield sub-attosecond (zeptosecond) pulses.

Consideration of Coulomb repulsion reveals that the total electron density is limited to about 1-100 electrons per attosecond pulse, depending on the geometry. Because μJ-level optical pulse energies are well feasible to obtain at up to MHz repetition rates, the low single-pulse electron flux can be compensated for, as it was demonstrated in ultrafast electron microscopy for even single electrons [4]. The attosecond electron packets are precisely located at the minima of the moving intensity wave and are therefore synchronized to the optical phase of the involved laser fields. This allows for utilizing them for pumping and/or probing in combination with secondary electron/x-ray pulses or an optical field.

Perspectives

Attosecond electron pulses, generated in free space, have potential to extend the methodology of electron microscopy and diffraction to the domain of attosecond electron dynamics. In contrast to attosecond optical pulses generated from high laser harmonics, electrons offer a much shorter de Broglie wavelength (for example 7 pm at 31 keV), readily attainable energy scaling, and a large cross-section for elastic and inelastic interaction with matter. Because the electron wavelength is short compared to atomic dimensions, four-dimensional imaging is possible with sub-atomic resolutions of space and time.

1   P. Baum, D.-S. Yang, and A. H. Zewail, Science **318**, 788, 2007.
2   J. C. Polanyi and A. H. Zewail, Acc. Chem. Res. **28**, 119, 1995.
3   P. Baum and A. H. Zewail, Proc. Natl. Acad. Sci. USA **104**, 18409, 2007.
4   A. H. Zewail, Annu. Rev. Phys. Chem. **57**, 65, 2006.

Electronically Driven Structural Dynamics of Si Resolved by Femtosecond Electron Diffraction

Maher Harb[1], Weina Peng[2], Germán Sciaini[1], Christoph T. Hebeisen[1], Ralph Ernstorfer[1], Thibault Dartigalongue[1], Mark A. Eriksson[2], Max G. Lagally[2], Sergei G. Kruglik[1], and R. J. Dwayne Miller[1]

[1]Institute for Optical Sciences and Departments of Physics and Chemistry, University of Toronto, 80 St. George Street, Toronto, Ontario M5S 3H6, Canada
[2]University of Wisconsin-Madison, Madison, Wisconsin 53706, USA
 E-mail: dmiller@Lphys.chem.utoronto.ca

Abstract: Femtosecond electron diffraction studies on (001)-oriented single crystalline Si found that at low excitation, longitudinal and transverse [001] acoustic phonon modes were generated. At ~11% valence excitation, the lattice collapsed non-thermally in <500 fs.

Introduction

There has been a considerable amount of studies suggesting that Si undergoes an order-to-disorder phase transition at excitation levels corresponding to the promotion of ~10% of the valence electrons [1]. This electronically-driven phase transition, referred to as non-thermal melting, occurs on a time scale much faster than the ps dynamics involved in converting electronic energy into lattice vibrations through electron-phonon interactions [2]. Experimentally, non-thermal melting of Si has been indirectly observed using all-optical pump-probe methods. Here, we use femtosecond electron diffraction [3] to resolve a variety of structural dynamics associated with photoexcited Si including the non-thermal disordering process.

Experimental methods

The experimental setup is based on a fs Ti:Sapphire laser where the 775 nm output is split into pump and probe arms. The frequency doubled pump arm excites the sample. The probe arm drives a Non-collinear Optical Parametric Amplifier, whose output at 500 nm generates electrons via two-photon photoemission from an Au photocathode. Photoemitted electrons are accelerated to 55 keV by a DC field and collimated with a magnetic lens before diffracting off the sample in transmission mode. Diffraction images are detected using a micro-channel plate/phosphor screen and recorded with a CCD camera. The electron pulse duration is characterized using the recently developed electron-laser cross-correlation technique based on ponderomotive scattering [4]. In this study, the system was configured to deliver 200 fs pulses containing ~6000 electrons in a 150 μm spot.

Excitation of the [001] acoustic phonon modes

The samples are 30±3 nm thick free standing films of (001)-oriented single crystalline Si fabricated using the approach described in ref. 5. The sample was excited at an absorbed fluence of 3 mJ/cm^2, which corresponds to a temperature rise of ~400 K.

1The normalized intensities of the (220) diffraction spots are shown in Fig. 2 as function of time delay. The dynamics display several interesting features: 1) A fast change in intensity which amounts to ~20% difference for one of the spots. 2) Changes in intensity are accompanied by oscillations. 3) Opposite spots exhibit dissimilar trends. 4) No further changes in intensity were observed when extending the experiment to 1.5 ns. The latter feature indicates that after the shift in intensity has developed, the deposited energy lies entirely in the vibrational modes of the lattice. Cooling of the sample back to room temperature by heat diffusion was estimated to occur at the ~100 μs time scale.

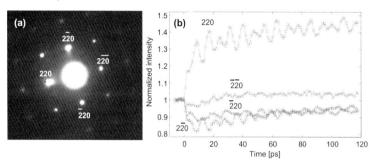

Fig. 1. (a) Diffraction pattern of (001)-oriented single cystalline Si. **(b)** Kinetics of the normalized intensity of the different (220) spots.

The observed intensity changes reflect minute lattice deformations which are amplified at the detection side due to the geometry of the diffraction setup. The well known Bragg law for constructive interference from lattice planes with interplanar distance d is $2d\sin\theta = n\lambda$, where θ is the Bragg angle and λ is the de Broglie wavelength of the incident electrons. The ideal condition for diffraction from the (220) planes, at $\theta = 0.76°$, cannot be simultaneously satisfied for the 4 spots. Deviation from the Bragg angle causes changes in the diffracted intensity according to

$$I_g = \frac{\sin^2\left(\pi L\left(s^2 + \xi_g^{-2}\right)^{1/2}\right)}{\xi_g^2\left(s^2 + \xi_g^{-2}\right)}, \tag{1}$$

where s is the deviation vector, L is the sample thickness, g is the reciprocal lattice vector, and ξ is the extinction distance [6]. According to Eq. 1, a shear deformation of ~0.1° could result in ~50% change in intensity due to the effective tilting of the (220) lattice planes.

From the Fourier transform of the kinetics, we extracted two frequency components corresponding to periods of 7.6 ps and 11.4 ps. The calculated periods of oscillation of the [001] longitudinal (L) and transverse (T) acoustic modes from $T=2L(C_{11}/\rho)^{-1/2}$, and $T=2L(C_{44}/\rho)^{-1/2}$, where C_{11} and C_{44} are elastic constants of the Si lattice and ρ is the density of Si, are 7.8±0.7 ps and 11±1 ps respectively. It is well known that thermoelastic stress launches acoustic waves that propagate along the surface normal [7]. We believe that the shear component arises from a slight mis-cut in the film surface relative to the (001) planes. No attempt was made to solve the thermoelastic equations of the excited membrane. The results reveal that electron diffraction is naturally sensitive to the detection of shear deformations in single crystalline structures, and can therefore be used to characterize the elastic properties of nanostructures.

Non-thermal collapse of the lattice

In this part of the study, the Si sample was excited at an absorbed fluence of 68 mJ/cm^2, which corresponds to the promotion of ~11% of the valence electrons to the conduction band. The diffraction intensity profile of the (220) order is shown in Fig. 2a for selected time points. The photoinduced dynamics appear as loss in intensity of the (220) order and an increase in scattering intensity in a broad detection range relative to the reference profiles. This emerging structure is consistent with the structure factor of liquid Si based on X-ray diffraction measurements [8]. The decay and rise traces associated with the decrease in order and the increase in disorder are 1:1 correlated as shown in Fig 2b. The diffraction signal from the initial lattice configuration was found to decay in 320 ± 50 fs, accompanied by a rise in scattering intensity in the 0.22-0.60 Å^{-1} detection range with a time constant of 400 ± 110 fs. These sub-500 fs dynamics cannot be explained by the thermal relaxation mechanisms which transfer heat from the hot carriers to the lattice on a time scale of a few ps [2], indicating that the observed phase transition is electronic in nature.

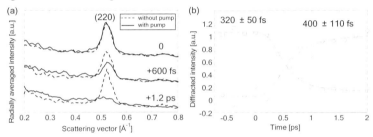

Fig. 2. (a) (220) diffraction profiles taken at selected time points (solid) and reference profiles taken without laser excitation (dashed). **(b)** Decay in of (220) order (squares) and rise in scattering intensity in the 0.22-0.60 Å^{-1} range (diamonds). Solid lines are best fits to the data using convolution of exponential decay (rise) function and Gaussian instrumental response of 300 fs.

Our recent work on polycrystalline Si revealed that the non-thermal disordering process occurs above 6% valence excitation [9]. In that study, the different diffraction orders were found to decay concertedly, suggesting a modified (as opposed to a flattened) interatomic potential and a relaxation process which is not inertial but rather involves multiple scattering towards the disordered state.

References

1. P. Stampfli and K. H. Bennemann, Appl. Phys. A, Vol. 60, 191, 1995.
2. M. Harb et al., J. Phys. Chem. B, Vol. 110, 25308, 2006.
3. J. R. Dwyer et al., Phil. Trans. R Soc. A, Vol. 364, 741, 2006.
4. C. T. Hebeisen et al., Optics Express, Vol. 16, 3334, 2008.
5. M. Roberts et al., Nature of Materials, Vol. 5, 388, 2006.
6. P.E. Champness, Electron Diffraction in the Transmission Electron Microscope, Garland Science, 2001.
7. C. Thomson et al., Phys. Rev. B, Vol. 34, 4129, 1986.
8. Y. Kita, J. B. Van Zydveld, Z. Morita, and T. Iida, J. Phys., Vol. 6, 811, 1994.
9. M. Harb et al., Phys. Rev. Lett., Vol. 100, 155504, 2007.

Picosecond electron deflectometry of optical-field ionized plasmas

Martin Centurion[1], Peter Reckenthaeler[1], Sergei A. Trushin[1], Ferenc Krausz[1,2] and Ernst Fill[1]

[1]1Max-Planck-Institut für Quantenoptik, , Hans-Kopfermann-Str. 1, D-85748 Garching,
 E-mail: martin.centurion@mpq.mpg.de
[2] 2Ludwig-Maximilians-Universität München, Am Coulombwall 1, D-85748 Garching,
 Germany

Abstract. We demonstrate a new method to image optical-field ionized (OFI) plasmas. Ultrashort electron pulses are directed onto a laser plasma. The deflection allows determination of features such as the field distribution or plasma charge.

Introduction

OFI plasmas continue to be of high interest in various fields. A number of methods have been applied to investigate the parameters of OFI plasmas such as Thomson scattering [1], ion spectrometry [2] and recording the emission of X-rays from the plasma [3]. The time-resolved plasma density profile has been measured by optical interferometry and holography [4,5]. Radiography using energetic (MeV) protons has been used to diagnose density perturbations and transient fields in high-density plasmas with a temporal resolution of 100 ps [6-8].

Experimental Methods

We demonstrate a new technique of diagnostics of OFI plasmas with a temporal resolution of 2.7 ps and very high sensitivity, viz. deflectometry using monoenergetic 20 keV electron pulses. These Pulses are directed onto an OFI nitrogen plasma generated by a 50 fs titanium-sapphire laser pulse. The electrons are deflected by the fields resulting from charge separation, and the resulting distortion of the electron beam yields time-resolved images of the plasma. Pump-probe experiments of the plasma evolution capture changes within a few picoseconds with high spatial resolution. This direct time-resolved imaging of plasma fields reveals features not accessible to other methods. Such knowledge can help improving parameters for optimizing particular applications. As an example, X-ray lasers and soft X-ray sources may greatly benefit from a better understanding of the dynamics of laser-generated plasmas. The new technique has the potential to lead to better control of plasma electron- and ion accelerators.

Results and Discussion

The high sensitivity of electron deflectometry is based on the fact that even small charge imbalances within the plasma are observable as distortions in the spatial profile of the electron beam. Application of the method is illustrated in Fig. 1, which shows the evolution of a nitrogen OFI plasma for about 200 ps. On the left side in the panels the shadow of the gas nozzle is seen. The laser traverses the electron beam from below and is focused into the nitrogen gas jet which expands from the nozzle.

161

Initially, a small depleted region appears in the electron beam in the area of the laser focus. This "hole" expands for approximately 50 ps, after which time a spot develops in the center. The spot becomes brighter than the initial electron beam for T > 100 ps, and its intensity increases up to 200 ps. For later times (not shown in the figure) the brightness of the spot slowly decreases.

The observations can be explained by a cloud of electrons expanding away from a central positively charged core. The fields resulting from this charge separation focus the electron beam strongly in front of the detector. As the plasma electron cloud expands farther, the focusing gets weaker and finally the focus reaches the detector which results in the bright spot seen for time delays larger than about 100 ps. The expanding lobes can be explained by ionization of the background gas along the line of the laser propagation.

T = 0.0 ps   T = 33.3 ps   T = 66.7 ps   T = 100.0 ps   T = 133.3 ps   T = 166.7 ps

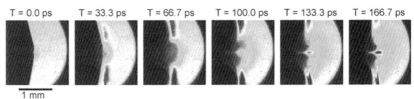

1 mm

Fig. 1. Pump-probe radiographies of plasma evolution. The figure shows the changes in the electron beam due to the charge distribution in the plasma as a function of the time-delay between the electron and laser pulses. The dark area in the left part of the pictures is the shadow of the gas nozzle. The laser is incident from below with a polarization perpendicular to the image plane, focused in front of the nozzle. The area of each image is 4.4 mm x 2.8 mm. The scale bar shows the calibration of the normalized intensity of the images.

For the analysis of the patterns, the equation of motion for the expanding electrons and their self-generated fields was solved numerically in a Lagrangian coordinate system. The pattern on the detector for a given time-delay was then simulated by propagating a collimated electron beam through the calculated fields using the GPT (General Particle Tracer) code [9], a well-established simulation tool for particle propagation. The calculations have successfully reproduced the main features of the experiment: the hole in the electron beam, the side-lobes and the focused spot. The ionizing laser pulse traverses the gas jet and creates a thin plasma cylinder with a diameter of approximately 20 μm, as the Rayleigh length of the laser focus (600 μm) is larger than the radius of the gas jet (150 μm). The side lobes result from ionization of the surrounding background gas. The initial plasma density and electron temperature are proportional to the gas density and laser intensity, respectively, and thus depend on the position along the axis. In the calculations a background gas density of 5% was assumed. The best fit between experiment and simulation was achieved for 5×10^7 charges (5×10^{-4} of the total number of electrons in the plasma), and hot electrons with an exponential energy distribution and a temperature of 250 eV. From the simulation we can extract the field distribution. The field reaches a value of 10^8 V/m at the center of the gas jet where the density is highest and decreases with radial distance from the positive charges. The results of the simulation, along with additional experimental results are shown in ref. 10.

Conclusions

We demonstrate a new method of imaging optical-field ionized plasmas with picosecond temporal resolution. The sensitivity of subrelativistic electrons to relatively small fields makes this method applicable to investigate the dynamics of low-density plasmas. New features of OFI plasma were observed using this method, such as a cloud of electrons separating from the plasma core far beyond the Debye length. Additionally, we have observed the effect of both electric and magnetic fields on the probe electron beam. The relevant physical parameters, such as the fields, number of charges and electron temperature, were calculated numerically.

Acknowledgements. This work was funded in part by DFG under contract SFB Transregio 6039 and by the DFG Cluster of Excellence "Munich Centre for Advanced Photonics" - MAP (www.munich-photonics.de). M. C. is supported by a research fellowship from the Alexander von Humboldt Foundation. P. R. is supported by a scholarship in the frame of the International Max Planck Research School on Advanced Photon Science - IMPRS-APS (www.mpq.mpg.de/APS). S.A.T. thanks the Deutsche Forschungsgemeinschaft for a research fellowship (project FU 363/1).

1 T.E. Glover, T.D. Donnelly, E.A. Lipman, A. Sullivan, and R.W. Falcone, "Subpicosecond Thomson scattering measurements of optically ionized helium plasmas," Phys. Rev. Lett. 73, 78-81 (1994).

2 S. Augst, D.D. Meyerhofer, D. Strickland, and S.L. Chin, "Laser ionization of noble gases by Coulomb-barrier suppression, " J. Opt. Soc. Am. B 8, 858-867 (1991).

3 E. Fill, S. Borgström, J. Larsson, T. Starczewski, C.-G. Wahlström, and S. Svanberg,"XUV spectra of optical-field-ionized plasmas," Phys. Rev. E 51, 6016-6026 (1995).

4] S. Tzortzakis, B. Prade, M. Franco and A. Mysyrowicz, "Time evolution of the plasma channel at the trail of a self-guided IR femtosecond laser pulse in air, " Opt. Commun. 181, 123-127 (2000).

5 M. Centurion, Y. Pu, Z. Liu, D. Psaltis, and T. W. Hänsch, "Holographic recording of laser-induced plasma," Opt. Lett. 29, 772-774 (2004).

6 A. J. Mackinnon, P. K. Patel, R. P. Town, M. J. Edwards, T. Phillips, S. C. Lerner, D. W. Price, D. Hicks, M. H. Key, S. Hatchett, S. C. Wilks, M. Borghesi, L. Romagnani, S. Kar, T. Toncian, G. Pretzler, O. Willi, M. Koenig, E. Martinolli, S. Lepape, A. Benuzzi-Mounaix, P. Audebert, J. C. Gauthier, J. King, R. Snavely, R. R. Freeman, and T. Boehlly,"Proton radiography as an electromagnetic field and density perturbation diagnostic (invited)," Rev. Sci. Instrum. 75, 3531 (2004).

7 C. K. Li, F. H. Seguin, J. A. Frenje, J. R. Rygg, R. D. Petrasso, R. P. J. Town, P. A. Amendt, S. P. Hatchett, O. L. Landen, A. J. Mackinnon, P. K. Patel, V. A. Smalyuk, T. C. Sangster, and J. P. Knauer,"Measuring E and B Fields in Laser-Produced Plasmas with Monoenergetic Proton Radiography," Phys. Rev. Lett. 97, 135003 (2006).

8 A. J. Mackinnon, P. K. Patel, M. Borghesi, R. C. Clarke, R. R. Freeman, H. Habara, S. P. Hatchett, D. Hey, D. G. Hicks, S. Kar, M. H. Key, J. A. King, K. Lancaster, D. Neely, A. Nikkro, P. A. Norreys, M. M. Notley, T. W. Phillips, L. Romagnani, R. A. Snavely, R. B. Stephens, and R. P. J. Town,"Proton Radiography of a Laser-Driven Implosion," Phys. Rev. Lett. 97, 045001 (2006).

9 http://www.pulsar.nl/gpt

10 M. Centurion, P. Reckenthaeler, S. A. Trushin, F. Krausz, E. E. Fill, "Picosecond electron deflectometry of optical field ionized plasmas," Nature Photon. 2, 315 (2008).

Part III

Correlated Electron Systems, Magnetization and Spin Dynamics

Clocking the Collapse of a Mott Gap

S. Wall[1], D. Brida[2], H. P. Ehrke[1], A. Ardavan[1], S. Bonora[2], H. Matsusaki[3], H. Uemura[3], Y. Takahashi[4], T. Hasegawa[4], H. Okamoto[3,4], G. Cerullo[2], and A. Cavalleri[1]

[1] Department of Physics, Clarendon Laboratory, University of Oxford OX1 3PU, UK.
    E-mail: a.cavalleri1@physics.ox.ac.uk
[2] Dipartimento di Fisica, Politecnico di Milano, P.za L. da Vinci 32, 20133 Milano, Italy.
[3] Department of Physics, University of Tokyo, Japan.
[4] Correlated Electron Research Center, Tsukuba, Japan.

Abstract. We probe the dynamics of band gap collapse in the Mott (electronic) insulator ET-F_2TCNQ and Peierls (structural) insulator K-TCNQ. The collapse of the Mott gap is a purely electronic process occurring within 20 fs. In the Peierls insulator, the gap collapse takes 300 fs, only after a lattice distortion has relaxed.

Mott insulators are fractionally filled electronic materials in which Coulomb repulsion between carriers does not allow conductivity. Chemically doping holes into the material can turn it metallic, allowing holes to delocalize across several sites. However, a pure filling controlled transitions of this type is difficult to identify in isolation from other degrees of freedom, which also rearrange after doping (e.g. lattice distortions). With photo-doping, one can transiently change the filling of the material while preserving other properties. Here we demonstrate that the photo-induced formation of the metallic state in a prototypical Mott insulator can occur within 20 fs, significantly faster than in Peierls (structural) insulators [1,2,3].

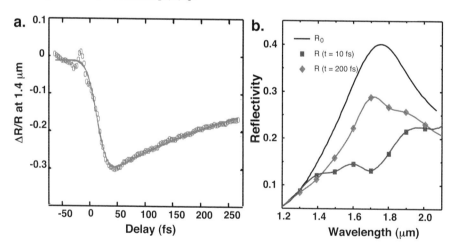

Fig. 1. a Temporal response after photo-doping with resonant 10 fs pulses. **b.** Spectrally resolved dynamics of the Mott gap (R_0 taken from [4]).

ET-F_2TCNQ is a half-filled 1D Mott insulator with low electron lattice coupling. Conduction electrons are localized on individual ET sites due to the onsite Coulomb repulsion, $U = 1.5$ eV, being significantly bigger than the electrons kinetic energy, $t = 200$ meV. Recently, it has been shown that photo-excitation of ET-F_2TCNQ melts the insulating state and creates a transient metallic state which decays after a few picoseconds [4]. This is evidenced from a Drude-like response at long wavelengths and a collapse of a charge-transfer resonance at 1.7 μm. The charge transfer resonance corresponds to an inter site transition $(ET^+ - ET^+) \rightarrow (ET^{2+} - ET^0)$ i.e. the creation of a hole and doubly occupied site and is the optical manifestation of the Mott gap.

In this work, we use a newly developed optical parametric amplifier (OPA) [5], which generates 10 fs pulses in the spectral region between 1 and 2 μm. This allows us to probe the dynamics of the charge transfer resonance with unprecedented temporal resolution. Figure 1a shows the temporal evolution of the Mott gap at 1.4 μm during resonant photo-doping. The fitted line gives a rise time of 20 fs and a 300 fs decay. This rise time is consistent with the electron hopping time, $h/t \approx 20$ fs, where h is Planck's constant. Figure 1b shows the spectrally resolved response of the charge transfer resonance, which occurs at 0.7 eV due to the existence of a Mott gap. The resonance almost entirely vanishes after photo-doping, and recovers after a few hundred femtoseconds. We also find that the formation of the metallic begins to saturate when pumped above 3.5 mJ cm^{-2}, or approximately 1 photon for every 10 ET site.

To emphasize the difference between this material and other photo- induced transient states, we also probed the response of the spin- Peierls insulator K-TCNQ. Below 395 K K-TCNQ undergoes a structural dimerization, which doubles the unit cell size and creates a bandgap. A broad peak in the absorption spectra at 1.5 μm corresponds to the charge transfer between neighbouring TCNQ molecules [6]. Therefore, probing in the same spectral region still gives information on the electron delocalisation dynamics.

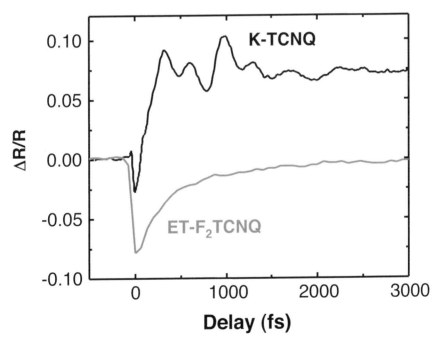

Fig. 2 Temporal dynamics of K-TCNQ at 1.3 μm after short pulse visible excitation, and ET-F$_2$TCNQ at 1.4 μm.

Figure 2 shows the dynamics of photo-induced response of both K-TCNQ and ET-F_2TCNQ. K-TCNQ is pumped above gap with a short pulse visible OPA. At 1.3 μm we see an increase in the signal due to a red shift of the gap. The rise time of the product state formation is approximately 300 fs, substantially longer than the 20 fs found in ET-F_2TCNQ. In addition to this, we see that the product state is much longer lived, with a lifetime of many hundreds of picoseconds.

Residual vibrational coherence is also observed in the metastable product phase, corresponding to lattice modes which are also seen in static Raman data. These measurements show the importance of electron phonon coupling. In K-TCNQ, despite prompt photo-doping, the material's properties are dictated by the slower dynamics of the structural relaxation and not the faster electronic time scales.

From our experiments, a new scenario emerges for photo-induced phase transitions in electronic insulators, in which the elementary timescale for the formation of a metallic electronic structure is dictated by the electron transfer integral t. Furthermore, because the structural degree of freedom does not stabilize the photo- initiated metallic phase, a rapid recovery into the parent phase takes place.

In Summary, by using 10 fs pulses in the spectral region between 1 and 2 microns we have measured the the collapse of the Mott gap in ET- F_2TCNQ occurs within 20 fs, comparable to the electron hopping time. The electronic nature of the phase transition is contrasted to the spin-Peierls insulator K-TCNQ, where lower energy and, therefore, slower time scale structural processes dictate the charge delocalisation dynamics.

1   A. Cavalleri, Cs Toth, C.W. Siders, J.A. Squier, P. Forget, J.C. Kieffer Phys. Rev. Lett. 87, 237401 (2001).

2   A. Cavalleri, T. Dekorsy, H.H.W. Chong, J.C. Kieffer, R.W. Schoenlein Phys. Rev. B 70, 161102 (2004).

3   P.Baum, J. Yang, A.H. Zewail Science 318, 788 (2007).

4   H Okamoto, H Matsuzaki, T Wakabayashi, Y Takahashi, T Hasegawa Phys. Rev. Lett. 98, 037401 (2007).

5   D. Brida, G. Cirmi, C. Manzoni, S. Bonora, P. Villoresi, S. De Silvestri, G. Cerullo Opt. Lett. 33 741 (2008).

6   K Ikegami, K Ono, J Togo, T Wakabayashi, Y Ishige, H Matsuzaki, H Kishida, H Okamoto Phys. Rev. B 76 085106 (2007).

Coherent Orbital Waves in Manganites

S. Wall[1], D. Polli[2], M. Rini[3], P. Dharmalingam[1], A. T. Boothroyd[1], Y. Tomioka[4], Y. Tokura[4], R. W. Schoenlein[3], G. Cerullo[2], and A. Cavalleri[1]

[1] Department of Physics, Clarendon Laboratory, University of Oxford OX1 3PU, UK
  E-mail: s.wall1@physics.ox.ac.uk
[2] Dipartimento di Fisica, Politecnico di Milano, P.za L. da Vinci 32, 20133 Milano, Italy
[3] Material Science Division, Lawrence Berkeley National Laboratory, Berkeley, 94720, USA
[4] Correlated Electron Research Center, Tsukuba, Japan

Abstract. High temporal resolution pump-probe experiments are used to excite and observe high-frequency orbital oscillations in the room temperature phase of 2D and 3D Colossal Magneto-Resistive manganites.

The choice of ground state in correlated-electron materials is determined by a balance between many degrees of freedom including spin, electron, lattice, orbital. Slight changes in one parameter can trigger collective changes in the material, often leading to exotic states of matter and excitations.

The perovskite manganite $Pr_{1-x}Ca_xMnO_3$ is a classic example. As a function of doping, x, and temperature the material passes through many insulating and magnetic states. Most manganites exhibit charge ordering at some critical temperature, T_{CO}. Below this temperature, electrons are localized on manganese sites in such a way that their position is ordered throughout the material. In addition to their physical ordering, the electron's occupied orbital is also ordered. At $x = 0.3$, the system is at its most delicate, and can be turned metallic under the application of a magnetic field, a phase which is not accessible under temperature change or doping. This effect, known Colossal Negative Magneto Resistance (CMR), 'melts' the charge ordering in the material and can change the resistivity by 9-order-of-magnitude at low temperatures [1]. Beyond the CMR effect, several other perturbations can 'melt' charge ordering and switch to 'hidden' metallic state, such as laser irradiation [2,3] and resonant phonon pumping[4].

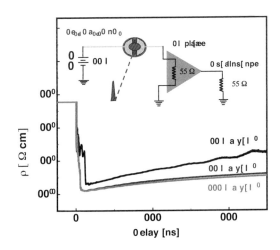

Fig. 1. Nanosecond transient resistivity measurement after photo-excitation.

Here we report on high fluence, high temporal pump-probe measurements on the cubic manganite $Pr_{0.7}Ca_{0.3}MnO_3$. Figure 1 shows the time resolved change in resistivity as measured by a current amplifier after photodoping with 800 nm pulses at 77 K on a nanosecond timescale. We see a prompt 5 orders of magnitude drop in resistivity as the system moves from the anti-ferromagnetic charge and orbitally ordered insulating state to a transient metallic state. This measurement demonstrated the transient formation of a metallic state, the conductivity of which begins to saturate at pump fluences above 30 mJ cm^{-2}.

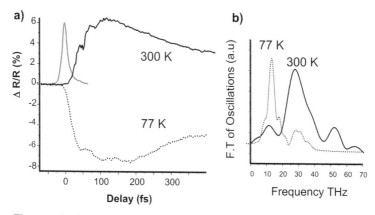

Fig. 2. a. Time resolved change in reflectivity at 77 K (lower) and room temperature (upper). **b.** Fourier transform of background subtracted oscillations.

To study the ultrafast dynamics during the photo-induced phase transition we use two independent non-collinear optical parametric amplifiers which produce a 7 fs pump pulse at 580 nm and a 12 fs probe pulse at 660 nm, which allow us to directly access the broad charge-transfer resonance. At 77 K we observe coherent oscillations accompanying the phase transition. A Fourier transform of these oscillations (fig. 2b) gives a frequency of 14 THz. This mode can be assigned to an A_g bending mode [5]. This mode directly modifies the bond angle between oxygen and manganese ions and it has been shown that direct modulation of this bond angle by exciting the appropriate IR mode can also induce the phase transition [4].

Figure 2 also shows the experiment at room temperature. In this case, the structural oscillations are no longer apparent but, instead, high frequency oscillations at 33 THz are observed. The frequency of this mode is well above the highest frequency optical phonon in this material (approximately 17 THz). Excitations around this frequency has been observed in a similar manganite using static Raman measurements and has been assigned to a manganese inter d-band excitation called a d-d exciton or orbiton [6]. However, this assignment has triggered much debate and an alternative explanation of peaks based on a non-linear two phonon process has also been used to explain the data [7]. Figure 3 shows the photo-induced response of a 2D layered manganite $La_{0.5}Sr_{1.5}MnO_4$ at room temperature. Again we see high frequency (25 THz) oscillations above the phonon modes of the material suggesting that high-frequency excitations are fundamental in manganites.

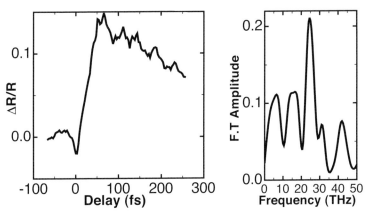

Fig. 3. Time resolved change in reflectivity at room temperature in $La_{0.5}Sr_{1.5}MnO_4$ together with the Fourier transform of the oscillations.

To summarize our results, we have used femtosecond spectroscopy to highlight the ultrafast time-dynamics of the charge de-localization process in manganites. We find that impulsive doping in the low temperature phase of $Pr_{0.7}Ca_{0.3}MnO_3$ triggers coherent waves at phonon frequencies (14 THz). At room temperature, for two different manganites, we observe a high frequency, non-structrual, oscillations. We believe these to be a fundamental excitation of the system and assign them to orbiton excitons.

1   Y. Tomioka et. al. Physical Review B 53 R1689 (1996).
2   K. Miyano et. al. Physical Review Letters 78 4257 (1997).
3   D. Polli et. al. Nature Materials 6 643 (2007).
4   M. Rini et. al. Nature 446 72 (2007).
5   M. Iliev et. al. Physical Review B 57 2872 (1998).
6   E. Saitoh et. al. Nature 410 180 (2001).
7   M. Grüninger et al. Nature 418 39 (2002)

Ultrafast terahertz response driven by photo-induced insulator to metal transition in layered organic salt

H. Nakaya[1], Y. Takahashi[1], S. Iwai[1]*, K. Yamamoto[2], K. Yakushi[2], and S. Saito[3]

[1] Department Physics, Tohoku University, Sendai 980-8578, Japan
*E-mail: s.iwai@sspp.phys.tohoku.ac.jp
[2] Institute for Molecular Science, Okazaki, 444-8585, Japan[2]
[3] Institute of Informations and Communications Technology, Kobe, 651-2492, Japan

Abstract. Photo-induced insulator to metal transition in two-dimensional organic salt α-(ET)$_2$I$_3$ (ET: [bis(ethylenedithio)]tetrathiafulvalene) was investigated by near-IR-pump and THz-probe spectroscopy. Photo-induced microscopic and semi-macroscopic metallic domains were characterized by a transient THz spectrum.

Introduction

Ultrafast photo-induced insulator to metal (I-M) transitions (PIMTs) have attracted considerable interests for application to switching devices and basic research on non-equilibrium phase dynamics in solids [1]. PIMTs have been demonstrated in organic and inorganic highly correlated systems using various ultrafast detection methods. Among them, pump-probe spectroscopy in terahertz (THz) region is a powerful technique for revealing electronic nature of the photo-induced metallic state, because I-M transition is characterized by low energy (<100 meV) spectra. To now, however, there are few THz spectroscopic studies on the photo-induced I-M transition. In this study, ultrafast spectral responses reflecting the I-M transition in THz region was investigated using near-IR (NIR) pump and THz-probe spectroscopy in a typical two-dimensional organic salt (ET)$_2$X (ET: [bis(ethylenedithio)]tetrathiafulvalene, X: monoanion) shown in Fig. 1. Very recently, this type of 3/4 filling organic salt shows potential for application to organic device [2]. α-(ET)$_2$I$_3$ transforms to charge ordered (CO) insulator at temperatures below T_{CO} (transition temperature to CO state=135 K). The I-M transition in α-(ET)$_2$I$_3$ is a first-order transition, but the lattice effect is not crucial, i.e., the CO in this compound can be described by the Wigner crystallization. The photo-induced melting of the CO state and generation of the metallic state have been studied by near-mid IR pump-probe spectroscopy in this compound [3-5]. Ultrafast (< 30 fs) response and slower (~500 fs) time

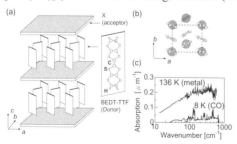

Fig.1 (a) Schematic illustration of a structure of layered ET salt (ET)$_2$X. (b) Molecular

evolution are attributable to an electronic response and a lattice stabilization, respectively [3-5].

Experimental Methods

For NIR-pump (1400 nm) and THz-probe spectroscopy, a 1 kHz Ti: Al_2O_3 regenerative amplifier system (Spectra Physics Hurricane) was employed as the light source. The NIR pump was generated using handmade OPA. The THz probing pulse was generated by differential frequency generation in a ZnTe crystal with the thickness of 1 mm, and focused in a pumped area of 1.5 mm diameter on the sample ($2x2x0.025$ mm^3). The transmitted THz light was detected by an EO sampling method. Time resolution is ~ 1ps.

Results and Discussion

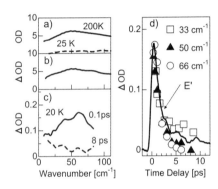

Fig.2 (a) Steady state and (c) transient absorption spectra of α-(ET)$_2$I$_3$ in THz region. (b) A difference between the absorption of 20 K and 124 K. (d) Time evolution of ΔOD measured at 33, 50, 66 cm^{-1}. A solid curve shows the time evolution of the normalized THz amplitude $E'=A(1-E_{TH}/E_0)$ THz light after photo excitation (see text).

Fig. 2(a) shows polarized absorption spectra of α-(ET)$_2$I$_3$ for //b axis in a THz region (15-100 cm^{-1}, 0.45-3 THz). The absorption spectra for T=150 K and 20 K are represented as solid and dashed curves, respectively. Marked change in absorbance (metallic: OD~4 at 150 K, CO: OD~0.1 at 25 K) reflects the I-M transition. A differential spectrum $(R_M-R_I)/R_I$, in which R_M and R_I represent the reflectivity of the metallic and the CO insulator phases, respectively, are displayed in Fig. 2(b). Transient absorption (ΔOD) spectra measured at 20 K are shown in Fig. 2(c). The absorption increase at t_d (time delay after excitation)=0.1 ps in a whole spectral region suggests photo-induced melting of the CO state and formation of the metallic state. The photo-induced absorption change (ΔOD~0.15) is apparently smaller than that for the thermal I-M transition.

However, considering that the penetration depth of the pump light (~ 1 µm), magnitude of the photo-induced absorption change on the sample surface can be evaluated as large as that for the thermal I-M transition.

Time evolutions of the photo-induced absorption change measured at 33, 50, and 66 cm^{-1} are shown in Fig. 2(d). The decay curves are reproduced by two-component exponential function whose time constants (fractions) are 1 ps (0.83) and 13 ps (0.17) at 20 K. The time constants, reflecting a relaxation time of the photo-induced metallic state, are approximately equal to those of the photo-induced reflectivity change measured at near and mid IR regions [3-5]. The solid curve shows a time evolution of the photo-induced change of the transmitted THz amplitude E_{TH_z} which is normalized as $E'(t)=A(1-E_{TH}(t)/E_0)$, where E_0 and A represent amplitude before photo-excitation and normalized coefficient, respectively. The $E'(t)$ is approximately equal to the $\Delta OD(t)$.

It is noteworthy that relaxation time of the photo-induced metallic state increases with temperature as shown in Fig. 3(a). The increase of the relaxation time near T_c has been ascribed to a critical slowing down (CSD) [3, 5]. The short-lived photo-induced metallic state observed at low temperature is a microscopic metallic domain with a scale of ~10 nm, whereas a long-lived metallic state detected near T_c is a macroscopic metallic domain.

Figs. 3(b) and 3(c) illustrate ΔOD spectra measure at 124 K and 20 K for $t_d = 0.1$ ps. The ΔOD spectrum for 124

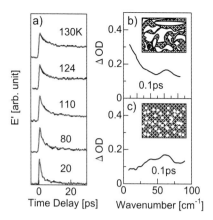

Fig. 3 (a) Time evolution of normalized THz amplitude E' at various temperatures. (b) (c) ΔOD spectra at (b) 124 K and (c) 20 K.

K has a large spectral weight in the low energy region <30 cm^{-1}, reflecting that the photo-induced macroscopic metallic state involves the large Drude component, while the electronic nature of the microscopic metallic is different from the Drude metal.

Conclusions

We have measured the transient THz absorption spectra of the photo-induced metallic states in a single crystal of charge ordered organic salts α-(ET)$_2$I$_3$. Spectral and the dynamic aspects of the transient absorption for the macroscopic metallic domain (124 K) are different from those for the microscopic metallic domain (20 K), i.e., the spectrum for the macroscopic domain shows a large spectral weight in the low energy (<30 cm^{-1}) region and slow decay, although the spectrum for microscopic domain is characterized by the small spectral weight in the low energy and the fast decay.

1 Eds. M. Gonokami and S. Koshihara, "Special Topics on Photoinduced Phase Transition and Their Dynamics", J. Phys. Soc. Jpn. **75**, 011001 (2006).
2 F. Sawano, I. Terasaki, H. Mori, M. Watanabe, N. Ikeda, Y. Nogami, and Y. Noda, "An organic thyristor "Nature **437**, 522 (2005)
3 S. Iwai, K. Yamamoto, A. Kashiwazaki, F. Hiramatsu, H. Nakaya, Y. Kawakami, K. Yakushi, H. Okamoto, H. Mori, Y. Nishio, "Photoinduced melting of stripe-type charge order and metallic domain formation in layered BEDT-TTF-based salts ", Phys. Rev. Lett. **98**, 097402 (2007).
4 H. Nakaya, F. Hiramatsu, Y. Kawakami, S. Iwai, K. Yamamoto, K. Yakushi, "30 fs infrared spectroscopy of photo-induced phase transition in 1/4 filling organic salt" , J. Lumin., **128**, 1065(2008).
5 S. Iwai, K. Yamamoto, F. Hiramatsu, H. Nakaya, Y. Kawakami, K. Yakushi, "Hydrostatic pressure effect on photoinduced insulator-to-metal transition in layered organic salt" Phys. Rev. B **77**, 125131 (2008).

Photo-induced macroscopic oscillation between insulator and metal in layered organic Mott insulator

Y. Kawakami[1], S. Iwai[1]*, N. Yoneyama[2], T. Sasaki[2], and N. Kobayashi[2]

[1] Department Physics, Tohoku University, Sendai 980-8578, Japan
  *E-mail: s.iwai@sspp.phys.tohoku.ac.jp
[2] Institute for Materials Research, Sendai, 980-8577, Japan

Abstract. Photo-induced insulator to metal transition in two-dimensional organic Mott insulator κ-(d-ET)$_2$Cu[N(CN)$_2$]Br was investigated by mid-IR pump-probe spectroscopy. Photo-induced macroscopic GHz oscillation between the Mott insulator and the metal, reflecting the competitive phase diagram, was observed.

Introduction

Control of coherent phonons in solids enables us to expect optical modulation of material phases in solids [1]. However, most observations of coherent phonons are limited to those in the potential minimum of a ground state and a photo-induced state. To now, a macroscopic coherent oscillation between the material phases has not been detected. Observation of such macroscopic oscillation between the phases is important in realizing coherent control of a macroscopic phase transition.

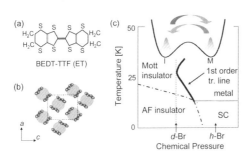

Fig. 1 (a) Molecular structure of ET. (b) Molecular arrangement of κ-(d- ET)$_2$Cu[N(CN)$_2$]Br. (c) Phase diagram of κ-ET salts for the chemical pressure or equivalently the effective band width. Schematic illustration of the macroscopic coherent oscillation between a Mott insulator and a metallic state is also displayed.

[bis(ethylenedithio)]-tetrathiafulvalene (BEDT-TTF) (hereafter ET, Fig. 1(a)) salt κ-(ET)$_2$X (X denotes anion) is a typical two-dimensional organic compound [2, 3]. In this compound, each molecule with 3/4 electron filling is dimerized to form 1/2 filling band as shown in Fig. 1(b). As a result of on-site Coulomb interaction on the dimer site (U_{dimer}), π electron is localized on each dimer site, forming a Mott insulator. The phase diagram described by changing a chemical pressure or equivalently the effective band width t/U_{dimer} (t: transfer energy between the dimer) consists of a Mott insulator, a metal, and a superconductor(SC) as indicated by Fig. 1(c). In the phase diagram, κ-(d-ET)$_2$Cu[N(CN)$_2$]Br (d-Br) is a Mott insulator which is near the insulator-metal (or SC) boundary, whereas κ-(h-ET)$_2$Cu[N(CN)$_2$]Br (h-Br) is a metal. Here, d- and h-ET represent deuterated and normal BEDT-TTF molecules, respectively.

Characteristic feature of this competitive phase diagram is the curved 1'st order transition line (bold line in Fig. 1(c)). Thereby instability of the Mott insulator phase is thermally tuneable in d-Br, although the insulator to metal (I-M) transition does not occur thermally in this compound. The Mott insulator - metal boundary in d-Br enables us to expect the realization of macroscopic oscillation because of the small barrier between the insulator and the photo-induced metal at ~25 K.

Experimental Methods

In near IR (0.89 eV) pump - mid-IR (0.12- 0.9 eV) probe spectroscopy, a 1 kHz Ti: Al_2O_3 regenerative amplifier system (Spectra Physics Hurricane) with optical parametric amplifiers was used as a light source. Averaged sample size of d-Br and h-Br is (2x2x1 mm). The time resolution is ~0.15 ps.

Results and Discussion

Fig. 2 (a) shows polarized reflectivity (R) spectra for $//c$ axis of d-Br(solid curve) and h-Br(dashed curve), respectively [3]. Open and closed circles in Fig. 2(c) shows the transient reflectivity ($\Delta R/R$) spectrum of d-Br at t_d(pump-probe delay time)=0.1 ps and 2 ps, respectively at 10K. The $\Delta R/R$ spectrum at t_d=0.1 ps shows the decrease of R immediately after the excitation, reflecting the bleaching of the steady state absorption. Subsequently, the R below 0.4 eV increase until t_d=2 ps. The $\Delta R/R$ spectrum measured at t_d=2 ps is analogous to the $[R(h$-Br$)-R(d$-Br$)]/R(d$-Br$)$ (Fig. 2(b)), where the $R(h$-Br$)$ represents the R spectrum of h-Br, showing the metallic or SC phase even at low temperature. Such spectral coincidence reminds us the photo generation of the metallic state.

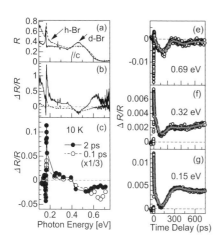

Fig. 2 (a) Steady state R spectra of d-Br and h-Br at 10 K. (b) $[R(h$-Br$)-R(d$-Br$)]/R(d$-Br$)$ (solid curve) , $[R(d$-Br, 80 K$)-R(d$-Br, 10 K$)]/R(d$-Br, 10K$)$ (dashed curve), (c), (d), (f) Time evolution of the reflectivity change at 0.69 (e), 0.32 (f), and 0.15 eV (g), respectively.

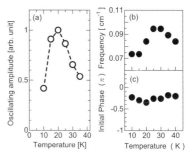

Fig. 3 Temperature dependence of (a) oscillating amplitude (excitation intensity I_{ex}=0.01 mJ/cm2) and (b) frequency (I_{ex}=0.1 mJ/cm^2), and (c) initial phase.

177

It is noteworthy that rise of a temperature does not cause the I-M transition in this compound, and the increasing temperature shows different spectral change as shown in the dashed curve in Fig. 2(b) ([$R(d$-Br, 80 K)-$R(d$-Br, 10 K)]/$R(d$-Br, 10K)]). In the spectral region 0.2-0.4 eV and <0.6 eV, the increase of R indicates the generation of the metallic state, whereas the decrease of R is attributable to the temperature rise. Therefore, we confirmed that the observed spectral change can not be ascribed to the increasing temperature.

Figs. 2(c), 2(d), and 2(f) shows the time evolutions of $\Delta R/R$ measured at 0.69 (2(e)), 0.32 (2(f)), and 0.15 eV (2(g)) Oscillating structure in the time scale of 100 ps is detected. Such long-period oscillation in the time evolutions have been observed in various compounds and interpreted as a coherent acoustic phonon (propagated shock wave) induced by the impulsive structural change, whose frequency depends on a probe energy reflecting the dispersion relation of the acoustic phonon [4]. However, in our case, the oscillating frequency is independent of probing energy, indicating that the oscillation can not be attributed to the shock wave. The oscillation is suggested to be a reciprocating motion between the Mott insulator and the metal, as shown in Fig. 1(c), because the spectrum of the oscillating amplitude approximately coincides with the [$R(h$-Br)-$R(d$-Br)]/$R(d$-Br)] (Fig. 2(b)).

To reveal the I-M reciprocating motion, temperature dependence of the oscillation was investigated. Here, oscillation amplitude (A), frequency (ω), initial phase (ϕ) were obtained by fitting the oscillating component using an equation $A\sin(\omega t - \phi)\exp(-\tau/t)$. The oscillating amplitude is markedly enhanced at 25 K. Furthermore, the oscillating frequency changes approximately 20% with temperature, showing a maximum at 25 K. Considering that the d-Br is close to the phase boundary at 25 K as shown in Fig.1(c), the increase of the oscillation amplitude and the frequency at 25 K are attributable to the reduction of the barrier height between I and M states near the phase boundary.

Conclusions

Photo-induced insulator to metal transition was observed in dimer Mott insulator κ-(d-ET)$_2$Cu[N(CN)$_2$]Br. The oscillation with a period of ~200 ps is attributable to the macroscopic coherent oscillation between the Mott insulator and the metallic state. The temperature dependence of the oscillating amplitude and the frequency reflect s the structure of the competitive phase diagram.

1 S. Iwai, Y. Ishige, S. Tanaka, Y. Okimoto, Y. Tokura, and H. Okamoto, "Coherent Control and Lattice Dynamics in Photoinduced Neutral to Ionic Transition in Charge-Transfre Compound", Phys. Rev. Lett. **96**, 057403 (2006).

2 K. Kanoda, "Metal-Insulator Transition in κ-(ET)$_2$X and (DCNQI)$_2$M : Two Contrasting Manifestation of Electron Correlation", J. Phys. Soc. Jpn. **75**, 051007 (2006).

3 T. Sasaki, I. Ito, N. Yoneyama, N. Kobayashi, N. Hanasaki, H. Tajima, Y. Iwasa, "Electronic correlation in the infrared optical properties of the quasi-two-dimensional κ-type BEDT-TTF dimer system", Phys. Rev. B**69**,064508(2004).

4 A. M. Lindenberg, I. Kang, S. L. Johoson, T. Missalla, P. A. Heimann, Z. Chang, J. Larsson, P. H. Bucksbaum, H. C. Kapteyn, H. A. Padmore, R. W. Lee, J. S. Wark, and R. W. Falcone, "Time-Resolved X-Ray Diffraction from Coherent Phonons during a Laser-Induced Phase Transition ", Phys. Rev. Lett. **84**, 111(2000).

THz Slow Motion of an Ultrafast Insulator-Metal Transition in VO$_2$: Coherent Structural Dynamics and Electronic Correlations

R. Huber[1], C. Kübler[1], H. Ehrke[1], R. Lopez[2], A. Halabica[3], R. F. Haglund, Jr.[3], and A. Leitenstorfer[1]

[1] Department of Physics and Center for Applied Photonics, University of Konstanz, Universitätsstraße 10, 78464 Konstanz, Germany
E-mail: rupert.huber@uni-konstanz.de
[2] Department of Physics and Astronomy and Institute of Advanced Materials, Nanoscience and Technology, University of North Carolina, Chapel Hill, North Carolina 27599, USA
[3] Department of Physics and Astronomy and Institute for Nanoscale Science and Technology, Vanderbilt University, Nashville, Tennessee 37235, USA

Abstract. The multi-THz conductivity of VO$_2$ recorded during a photoinduced insulator-metal transition directly reveals the femtosecond dynamics of V-V stretching modes and electronic correlations. We suggest a novel qualitative model for the nonthermal phase transition.

Introduction

The insulator-metal transition in VO$_2$ represents one of the most intriguing pheno-mena in strongly correlated electron systems [1-9]. The material undergoes a first-order transition from a high-temperature metallic to a low-temperature insulating phase at T_L = 340 K [1], while dimerization of V atoms reduces the crystal symmetry from rutile (R) to monoclinic (M1). The underlying microscopic driving force has been a subject of controversy. A structural Peierls instability [2] as well as Coulomb repulsion and charge localization typical of a Mott insulator have been proposed [3]. Ultrafast optical triggering of the insulator-metal transition combined with structural rearrangement has promised new insight into the time ordering of the microscopic mechanisms. Yet sufficient resolution to reveal the inherent time scales has been achieved solely in femtosecond optical reflectivity data [4]. The elementary processes, however, remain elusive for visible light pulses.

Ultrabroadband THz pulses, in contrast, have been exploited to directly trace lattice polarizability and electronic conductivity – an order parameter of the insulator-metal transition – on the femtosecond scale [6]. Here, we report multi-THz measurements of VO$_2$ monitoring a femtosecond insulator-metal transition initiated by a 12-fs light pulse [8]. The infrared (IR) conductivity simultaneously resolves the spectral signatures of electronic and ionic degrees of freedom, precisely unraveling their interplay.

2D multi-THz study of the ultrafast insulator-metal transition

The sample is a 120-nm-thick film of polycrystalline VO$_2$ on a diamond window. Light pulses of 12-fs duration (λ_{center} = 800 nm) from a low-noise Ti:sapphire amplifier photoexcite the dielectric phase [10]. Multi-THz transients generated via optical rectification [11] are subsequently transmitted through the sample. Phase-matched electro-optic sampling yields amplitude and phase of the probe electric field in the time domain for variable delay times τ_D between excitation of the film and electro-optic detection of its response [11]. This two-dimensional (2D) scheme affords direct access to the complex conductivity spectrum $\sigma(\omega)$ and its pump induced changes $\Delta\sigma(\omega,\tau_D)$ with femtosecond resolution [6]. Fig. 1(a) depicts the real part $\sigma_1(\omega)$ of the THz conductivity of the unexcited VO$_2$ film below T_c. The three prominent maxima at $\hbar\omega = 50$, 62, and 74 meV are due to the polarizability of the highest-frequency transverse optical phonons associated with vibrations of the oxygen cages enclosing the V atoms. Photoinjection of electron-hole pairs into the M1 phase ($T_L = 250$ K) induces ultrafast changes of the complex mid infrared conductivity. Fig. 1(b) displays the spectrally resolved real part $\Delta\sigma_1(\omega,\tau_D)$ of this quantity. We consider two domains, P and E, reflecting different physical processes: Features in P relate to IR-active phonon resonances and expose the dynamics of the lattice. Since spectral region E is free of phonon absorption, the pump-induced THz conductivity in this region derives solely from electronic contributions.

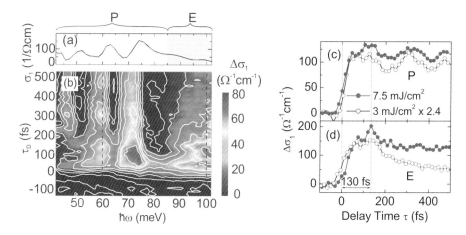

Fig. 1. 2D optical pump – THz probe data: (a) equilibrium conductivity of insulating VO$_2$. (b) gray scale plot of the pump-induced changes of the conductivity $\Delta\sigma_1(\omega,\tau_D)$ for a pump fluence of $\Phi = 3$ mJ/cm^2 ($T_L = 250$ K). The spectral domains P and E comprise predominantly changes of the phonon resonances and the electronic conductivity, respectively. (c),(d) cross sections through 2D data along the time axis τ_D for a photon energy of (c) $\hbar\omega = 60$ meV and (d) $\hbar\omega = 100$ meV. Data for two pump fluences are compared: $\Phi = 3$ mJ/cm^2 [see Fig. 1(b)] (open circles) and $\Phi = 7.5$ mJ/cm^2 (filled circles). The traces taken at $\Phi = 3$ mJ/cm^2 are upscaled in amplitude by a factor of 2.4.

The distinctive origins of the signals in the two spectral windows are underscored by their qualitatively different dynamics: Ultrafast photodoping induces a quasi-instantaneous onset of conductivity in region E due to directly injected mobile carriers, followed by a prompt decay within 0.4 ps. A cross section through the 2D data at a THz photon energy of $\hbar\omega = 100$ meV is given by the open circles in Fig. 1(d). In contrast, the phonon contribution (domain P) is more long lived. Photoexcitation induces an increase of polarizability on the low-frequency side of each phonon resonance. This effective change in frequency is superimposed on a remarkable coherent modulation of $\Delta\sigma_1(\omega,\tau_D)$, along the pump-probe axis τ_D, most notably at $\hbar\omega = 60$ meV [Fig. 1(c)]. The Fourier transform of the oscillations along τ_D

180

is centered at 6 THz, characteristic of A_{1g} lattice modes associated with stretching and tilting of V-V dimers [5]. These vibrations map the M1 structure onto the R lattice.

Lattice and electronic dynamics also exhibit differing dependences on the pump fluence Φ [see Fig. 1(c) and (d)]. For fluences in excess of a temperature-dependent threshold ($\Phi_c = 5.3$ mJ/cm^2 at $T_L = 250$ K), the electronic conductivity shows one cycle of modulation in phase with the coherent lattice motion [filled circles in Fig. 1(d)]. Subsequently, the THz conductivity settles at a constant value for at least 10 ps, indicating the transition of the electronic system into a metallic phase. In sharp contrast, the lattice dynamics is qualitatively unchanged with Φ.

Qualitative model of the ultrafast phase transition

Our femtosecond measurements of the mid-IR dielectric function of VO$_2$ suggest a novel qualitative picture for the photoinduced insulator-metal transition [8]. The model is inspired by recent cluster dynamical mean-field theory approximating the dielectric phase as a molecular crystal of V-V dimers [7]. In the earliest stage, the dynamics initiated by the femtosecond pulse resembles the local excitation of the V-V dimers into an antibonding state, triggering coherent wave packet motion of the stretching vibration. At moderate excitation fluence, the strong correlations between the two binding electrons of each dimer are re-established on a sub-picosecond time scale and the mid-IR electronic conductivity vanishes rapidly. However, if the density of excited lattice sites exceeds a threshold value, wave packet oscillations and thermal fluctuations drive the system to a point where electronic correlations can no longer be restored. The mid-IR electronic conductivity then settles close to the steady-state value characteristic of the metallic phase after approximately one V-V oscillation cycle, even though the lattice is still far from equilibrium.

We expect that this novel picture may enable the development of a quantitative theory of the phase transition. In an analogous way, femtosecond field-resolved 2D studies may substantially improve our understanding of phase transitions in other strongly correlated electron systems including non-conventional superconductors.

Acknowledgements. This work was supported by the Alexander von Humboldt Foundation and the German Research Foundation (DFG) via the Emmy Noether Program.

1 F. J. Morin, Phys. Rev. Lett. **3**, 34 (1959).

2 J. B. Goodenough, Phys. Rev. **117**, 1442 (1960).

3 A. Zylberstejn et al., Phys. Rev. B **11**, 4383 (1975); D. Paquet et al., Phys. Rev. B **22**, 5284 (1980).

4 J. Y. Suh et al., J. Appl. Phys. **96**, 1209 (2004).

5 A. Cavalleri et al., Phys. Rev. B **70**, 161102(R) (2005).

6 R. Huber et al., Nature **414**, 286 (2001); Phys. Rev. Lett. **94**, 0274011 (2005).

7 S. Biermann et al., Phys. Rev. Lett. **94**, 026404 (2005).

8 C. Kübler et al., Phys. Rev. Lett. **99**, 116401 (2007).

9 P. Baum et al., Science **318**, 788 (2007).

10 R. Huber et al., Opt. Lett. **28**, 2118 (2003).

11 R. Huber et al., Appl. Phys. Lett. **76**, 3191 (2000); C. Kübler et al., Appl. Phys. Lett. **85**, 3360 (2004).

Nonthermal Melting of Orbital Order in La$_{1/2}$Sr$_{3/2}$MnO$_4$ by Coherent Excitation of a Mn-O Stretching Mode

Raanan I. Tobey, Dharmalingam Prabhakaran, Andrew T. Boothroyd, Andrea Cavalleri

Department of Physics, Oxford University, Oxford UK
E-mail: r.tobey1@physics.ox.ac.uk

Abstract. An ultrafast electronic phase transition, associated with melting of orbital order, is driven in La$_{1/2}$Sr$_{3/2}$MnO$_4$ by selectively exciting the Mn-O stretching mode with femtosecond pulses at 16 μm wavelength. The energy coupled into this vibration is less than 1% of that necessary to induce the transition thermally. Nonthermal melting of this electronic phase originates from coherent lattice displacements comparable to the static Jahn-Teller distortion.

We directly drive large amplitude coherent oscillations of infrared active vibrational modes in La$_{1/2}$Sr$_{3/2}$MnO$_4$ and monitor its effects on the electronic properties. This single-layer manganite exhibits long-range homogeneous electronic and orbital ordering which is directly connected to optical response[1]. Upon vibrational excitation, we measure transient reflectivity over a broad spectral range and detect the formation of a long-lived electronic phase with optical properties similar to those observed in the high-temperature, orbitally-melted insulator. The occurrence of ultrafast melting of orbital order is further substantiated by the prompt loss of birefringence, which we measure with femtosecond pulses at 650 nm. Importantly, the energy deposited into the solid accounts for only minimal heating. We thus conclude that selective vibrational excitation drives the transition nonthermally.

Figure 1(a) shows the three dimensional lattice structure of single-layer La$_{1/2}$Sr$_{3/2}$MnO$_4$. The manganese cations occupy the centre of oxygen octahedra forming planes separated by lanthanum and strontium dopants. Within the planes, lattice-commensurate ordering of the high-lying Mn$3d$ e$_g$-like orbitals develops below T$_{oo}$=220 K, yielding a long-range orbital pattern as shown in figure 1(b). The electronic properties are dominated by the crystal-field-split Mn$3d$ orbitals, which are strongly hybridized with $2p$ orbitals from neighbouring oxygen atoms, and are thus very sensitive to distortions in the Mn-O bond.

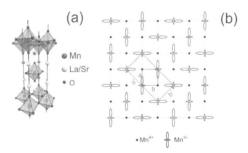

Fig 1: (a) Three dimensional structure of single layer La$_{1/2}$Sr$_{3/2}$MnO$_4$. (b) Orbital ordering below T$_{oo}$ = 220K breaks the symmetry of the ab plane resulting in optical birefringence.

In our experiments, large amplitude, coherent distortions of the Mn-O bonds were driven with femtosecond pulses at 16 μm (625 cm^{-1}, 77 meV) wavelength, resonant with the IR-active, 625 cm^{-1} stretching vibration. High-intensity pulses at this wavelength were generated with an optical parametric amplifier (OPA), which was pumped by an Ti:Sapph laser generating 2.5mJ pulses at 1 KHz repetition rate. The maximum fluence in the mid-IR was 2 mJ/cm^2. A second Optical Parametric Amplifier, pumped by the same Ti:Sa laser, was used to generate broadband probe pulses, which in our experiments were tuned between 10 μm (0. 12 eV photon energy) and 600 nm (2.2 eV). The single crystal samples were grown by the floating zone technique, cleaved and polished along [001], and mounted in a closed loop cryostat.

Excitation resonant with the Mn – O vibration results in a prompt transition to a long-lived phase that survives for hundreds of picoseconds. In Figure 2 we show the reflectivity of La$_{1/2}$Sr$_{3/2}$MnO$_4$ measured at 100-ps time delay for photon energies between 120 meV and 2.2eV. We also show a comparison between reflectivity changes obtained by vibrational excitation and by static temperature tuning. We note that the spectral response of the transient phase is the same as the change in static reflectivity when La$_{1/2}$Sr$_{3/2}$MnO$_4$ is heated between 90 K and room temperature (solid line), suggesting a similarity between the two phases. Additionally, the reflectivity cannot be reconciled with changes observed for smaller temperature increases. This is shown by the 110 K – 90 K differential reflectivity (dashed), where the sample remains in the orbitally ordered phase.

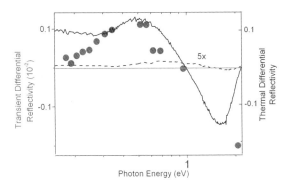

Fig 2: Reflectivity of the phono – induced phase closely matches the spectral weight distribution of the thermally tuned transition (solid line).

The spectral response suggests that vibrational excitation fully relaxes the Jahn-Teller-distorted low-temperature state, and leads to a metastable phase where orbital order is melted. Loss of orbital ordering can be directly confirmed by measuring time-dependent birefringence, which at 650 nm is proportional to the orbital-order parameter, as measured with resonant x-ray diffraction. Figure 3 shows time-dependent birefringence measured at 650 nm. A prompt drop of birefringence is observed, directly indicating melting of orbital order. This state persists for hundreds of picoseconds. In figure 3, we also include a cross correlation of mid-IR pump and 650 nm probe pulses, as measured with the Kerr effect in ZnTe. The data show that birefringence is lost on a timescale identical to the risetime of the excitation pulse, indicative of the ultrafast nature of this process. This is the first indication that the electronic phase transition is non-thermal, in that it occurs on a timescale that is significantly shorter than the known thermalization time for hot optical phonons in solids[2].

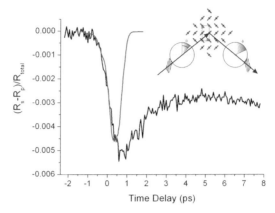

Fig 3: Ultrafast melting of orbital order is directly seen in the loss of birefringence.

Additionally, the observed effects cannot be explained by heating of $La_{1/2}Sr_{3/2}MnO_4$ above T_{oo} ~ 220 K. At 16 μm, we use fluences less than 2-mJ/cm^2, corresponding to energy densities below 6 KJ/cm^3 over the 300 nm absorption depth. This results in heating that never exceeds 2 K, assuming that the energy is distributed among all degrees of freedom of the system.

However, the peak electric field amplitude is nearly 10 MV/cm. Assuming ionic bonding between $Mn^{3.5+}$ and O^{2-} ions, and deriving a value for the polarizability from the dielectric constant at this frequency, we estimate a coherently driven lattice displacement between 2 pm and 10 pm, about 1% - 5% of the equilibrium Mn – O distance. This field-induced displacement is comparable to the static Jahn-Teller distortion, which for manganites is about 10pm^3. We conclude that selective vibrational excitation destabilizes the charge and orbitally ordered state of low-temperature $La_{1/2}Sr_{3/2}MnO_4$ by electric-field induced coherent displacements rather then by heating of the solid.

In summary, we have shown mode selective excitation of an orbital melting transition in single layer $La_{1/2}Sr_{3/2}MnO_4$. The electronic phase transition is detected by measuring a prompt change in the mid-IR optical properties and in the ultrafast loss of birefringence. The transition is spectrally selective and efficiently driven when the pump frequency matches that of the 625 cm^{-1} stretching vibration. The transition is non-thermally driven by coherent atomic displacements comparable to the Jahn-Teller distortion.

[1] T. Ishikawa, K. Ookura, and Y. Tokura, Physical Review B **59**, 8367 (1999).

[2] D. v. d. Linde, J. Kuhl, and H. Klingenberg, Physical Review Letters **44**, 1505 (1980).

[3] Satpathy S, Popovic Z, and Vukajlovic F, Physical Review Letters **76**, 960 (1996).

Ultrafast Gigantic Photo-Response in Charge-Ordered Organic Salt (EDO-TTF)$_2$PF$_6$ on 10-fs time scales

J. Itatani[1,2,*], M. Rini[1], A. Cavalleri[3], K. Onda[2,4], T. Ishikawa[4], S. Ogihara[4], S. Koshihara[2,4], X. F. Shao[2,5], Y. Nakano[2,5], H. Yamochi[2,5], G. Saito[6], and R. W. Schoenlein[1]

[1] Lawrence Berkeley National Laboratory, 1 Cyclotron Road, MS2R300, Berkeley, CA 94720 USA
[2] ERATO, Japan Science and Technology Agency, 3-5 Sanbanchou, Chiyoda-ku, Tokyo 102-0075, Japan
[3] Department of Physics, Clarendon Laboratory, University of Oxford, Parks Road, Oxford OX1 3PU, United Kingdom
[4] Department of Materials Science, Tokyo Institute of Technology, 2-12-1-H61, Oh-okayama, Meguro-ku, Tokyo 152-8551, Japan
[5] Research Center for Low Temperature and Materials Sciences, Kyoto University, Saikyo-ku, Kyoto 606-8502, Japan
[6] Division of Chemistry, Graduate School of Science, Kyoto University, Sakyo-ku, Kyoto 606-8502, Japan
* Current address: Institute for Solid State Physics, University of Tokyo, 5-1-5 Kashiwanoha, Kashiwa, Chiba 277-8581, Japan
Email: jitatani@issp.u-tokyo.ac.jp

Abstract. The initial dynamics of photo-induced phase transition in charge-ordered organic salt (EDO-TTF)$_2$PF$_6$ was investigated using 10-fs near-infrared laser pulses. We observed sub-20-fs gigantic photo-responses ($|\Delta R/R|>100\%$) due to intra-molecular vibration and a clear signature of a structural bottleneck (~50 fs) for the first time.

Introduction

The recent discovery of the gigantic photo-responses in a 1/4-filled organic salt (EDO-TTF)$_2$PF$_6$ has uncovered a new class of molecular systems whose physical properties can be drastically altered under optical excitation on femtosecond time scales [1]. Underlying physics of these gigantic photo-responses are considered to be the light-induced melting of its unusual [0110]-type charge order in an EDO-TTF tetramer at low temperature (T<280 K). The charge-order melting induces an imbalance in the electron-lattice coupling, which initiates the photo-induced phase transition (PIPT) process. Previous studies have investigated this insulator-to-metal PIPT on the 100-fs time scales, focusing on the sub-picosecond dynamics and the photo-induced quasi-stable state [2]. They have revealed that (i) PIPT is accompanied by strong coherent intermolecular vibrations, and (ii) the reflectivity spectrum of the photo-induced quasi-stable phase differs from the thermally-induced phase especially in the infrared. Recent theoretical studies also support these observations, and suggest that the photo-induced metallic-like state is indeed the [1010]-type charge ordered state [3]. While there has been a significant progress in understanding the photo-induced phase in (EDO-TTF)$_2$PF$_6$, little is known about the initial dynamics of the PIPT, which is of fundamental interest for understanding the interplay between various molecular/lattice/electronic degrees of freedom in strongly correlated systems. It is also crucial to examine if the PIPT is driven purely by electronic processes or by the photo-induced change of molecular conformation in order to understand the nature of the phase transition [4].

We have investigated the ultrafast dynamics of the reflectivity changes in $(EDO\text{-}TTF)_2PF_6$ during the insulator-to-metal PIPT initiated with 10-fs laser pulses. We observed (i) a large reflectivity modulation ($\Delta R/R \sim 100\%$) on 10-fs time scales, and (ii) a clear signature that the PIPT is driven by the change of molecular conformation.

Experimental Methods

We investigated the early dynamics of PIPT with 10 fs pump-probe spectroscopy. We produced near-infrared 10 fs pulses using a Ti:sapphire chirped-pulse amplifier with hollow-fiber pulse compression. We used a hollow-core fiber (core diameter: 150 μm) filled with Argon (~2 bar) and dispersion compensation mirrors to obtain nearly transform limited 10-fs pulses at the sample position in a cryostat. Figure 1 shows the spectrum of the compressed pulses and their autocorrelation trace (inset above) compared to the reflectivity of $(EDO\text{-}TTF)_2PF_6$ in the metallic (T=290 K, dotted line) and insulator (T=180 K, solid line) phases. The laser spectrum covers the highest charge transfer band ($D^0D^+D^+D^0 \Rightarrow D^0D^{2+}D^0D^0$) peaked around 1.4 eV and the isosbestic point of the reflectivity curves at 1.6 eV where the thermally-induced insulator-to-metal transition results in $\Delta R=0$. The sample was kept at 180 K during the experiment to photo-excite the insulating ground state. The repetition rate of the laser was kept at 1 kHz since the relaxation from the photo-induced quasi-stable state to the insulating ground state occurs less than 1 ms.

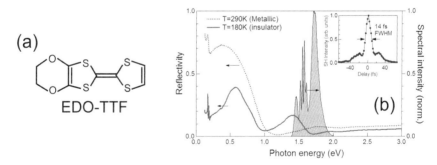

Fig. 1. (a) The structure of a single EDO-TTF molecule, (b, left axis) reflectivity at 290K and 180K, and (b, right axis and inset) spectrum of 10-fs pulses and their autocorrelation.

Results and Discussion

Figure 2(a) shows the normalized reflectivity change ($\Delta R/R$) in the spectral domain at the delay of 150, 610, and 1500 fs. They are similar to the thermally-induced reflectivity change (thick gray curve) on their signs and the location of the isosbestic point, except for the relative magnitudes. This result is consistent with the case of 150-fs pump pulses [1]. Figure 2(b) shows the time-resolved values of $\Delta R/R$ at the early stage of PIPT. Around 1.65 eV where the absorption is dominated by intramolecular bands, we observed huge ($\Delta R/R \sim 100\%$) and extremely fast (T$\sim$20 fs) modulation. This modulation was rapidly dumped as the $\Delta R/R$ values reach the quasi-static value around τ_d=80 fs. Based on this time scale, this modulation is likely due to the coherent vibrational motion of carbon double bonding in EDO-TTF molecules induced via the impulsive Raman process. At the photon energies below 1.6 eV where the absorption is dominated by inter-molecular charge transfer, the value of $\Delta R/R$ goes negative. Since there is no periodic modulation on 10-fs time scales in this spectral region, we can clearly resolve the response of PIPT that is slower than the laser pulse duration. The existence of this delayed response, or the structural bottleneck time [4], suggests that PIPT is mediated by molecular conformational changes in contrast with prompt electronic transfer. The time scale of the structural bottleneck time was estimated to be $\sim$ 50 fs, defined by the delay where the $\Delta R/R$ value reached 90% of the asymptotic value.

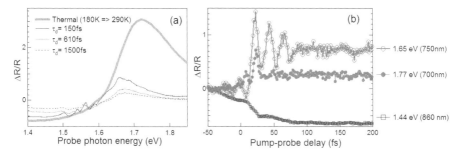

Fig. 2. (a) Reflectivity change in the spectral domain at different delays. (b) Reflectivity change in the time domain.

Conclusions

We have investigated the initial dynamics of PIPT in charge-ordered organic salt (EDO-TTF)$_2$PF$_6$ using 10-fs laser pulses. We have observed gigantic photo-responses that are modulated by the impulsive Raman excitation of intra-molecular vibrational modes of carbon double-bondings. This oscillation is rapidly dumped as the reflectivity spectrum of the charge transfer band reaches to asymptotic values in 50 fs. The results suggest that the molecular conformation after charge-order melting initiates the PIPT.

Acknowledgements. This work was supported by the U.S. Department of Energy under Contract No. DE-AC02-05CH11231.

1   M. Chollet, L. Guerin, N. Uchida, S. Fukaya, H. Shimoda, T. Ishikawa, K. Matsuda, T. Hasegawa, A. Ota, H. Yamochi, G. Saito, R. Tazaki, S. Adachi, and S. Koshihara, Science **307**, 86 (2005).

2   K. Onda, T. Ishikawa, M. Chollet, X. Shao, H. Yamochi, G. Saito, and S. Koshihara, J. Phys. Conf. Ser. **21**, 216 (2005).

3   K. Onda et al., in preparation.

4   A. Cavalleri, Th. Dekorsy, H. H. W. Chong, J. C. Kieffer, and R. W. Schoenlein, Phys. Rev. **B 70**, 161102(R), (2004).

Teasing a Quasiparticle: Ultrafast Nonlinear Response of the Fröhlich Polaron in GaAs

P. Gaal[1], W. Kuehn[1], K. Reimann[1], M. Woerner[1], T. Elsaesser[1], R. Hey[2]

[1] Max-Born-Institut für Nichtlineare Optik und Kurzzeitspektroskopie, 12489 Berlin, Germany
E-mail: woerner@mbi-berlin.de
[2] Paul-Drude-Institut für Festkörperelektronik, 10117 Berlin, Germany

Abstract. Ultrafast acceleration of polarons in a strong THz field results in an oscillatory occurrence of midinfrared gain/absorption with the LO phonon frequency. THz-pump–midinfrared-probe measurements give a first insight into the internal motion of a quasiparticle.

Introduction

A charged particle modifies the structure of the surrounding medium. In turn, the medium acts back on the particle. In a polar solid a free electron distorts the crystal lattice, displacing the atoms from their equilibrium positions [Fig. 2(a)]. One considers the electron together with its surrounding lattice distortion a quasiparticle [1], the Fröhlich polaron [2]. The basic properties of polarons and their drift motion in a weak electric field are well known. However, their nonlinear high-field properties relevant for transport on nanometer length and ultrashort time scales are not understood.

Experiment

Here, we show for the first time that a high electric field in the terahertz range drives the polaron in a GaAs crystal into a highly nonlinear regime where—in addition to the drift motion—the electron is impulsively moved away from the center of the surrounding lattice distortion [3]. The experimental setup is sketched in Fig. 1. Both a high-field THz transient generated by four-wave mixing in a dry nitrogen plasma [4] and a synchronized midinfrared (MIR) transient generated by difference frequency mixing in GaSe [5] are focused collinearly onto the sample, a 500 nm thick layer of Si-doped GaAs with a doping concentration of 10^{17} cm^{-3}. Subsequently, the time-dependent electric field of the THz and MIR pulses is measured with electro-optic sampling in a thin ZnTe crystal [5]. Both the THz-pump and the MIR-probe beam are chopped with different frequencies allowing for independent measurements of $E_{\text{THz}}(t)$, $E_{\text{MIR}}(t, \tau)$, and $E_{\text{Both}}(t, \tau)$. The latter transient is measured when both pulses are applied. τ is the delay between the THz and the MIR pulse and t is real time. The time dependent electric field $E_{\text{NL}}(t, \tau)$ radiated from the nonlinear intraband polarization $E_{\text{NL}}(t, \tau) = E_{\text{Both}}(t, \tau) - E_{\text{THz}}(t) - E_{\text{MIR}}(t, \tau)$ is shown as the solid line in Fig. 1(a). The sample shows a coherent, nonlinear emission, which is for this particular τ in phase with the MIR pulse, demonstrating a THz-field induced MIR gain of the sample. From $E_{\text{NL}}(t, \tau)$ we calculate the time-integrated midinfrared transmission change $\Delta T / T_0(\tau)$ [circles in Fig. 1(b)]. The nonlinear signal shows an oscillatory behavior changing periodically with the frequency of the GaAs LO phonon between gain and absorption.

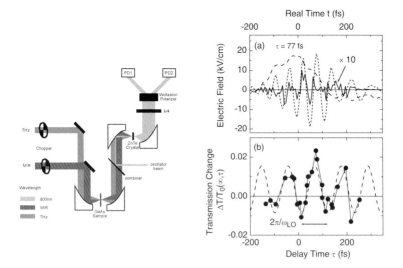

Fig. 1. Left: THz-pump–MIR-probe setup. Both terahertz $E_{THz}(t)$ and midinfrared transients $E_{MIR}(t, \tau)$ propagate collinearly through a 500-nm thick n-type GaAs sample and are measured subsequently by electro-optic sampling in a thin ZnTe crystal. τ is the delay between the THz and the MIR field. Right: Experimental results. (a) $E_{THz}(t)$ (dashed line), $E_{MIR}(t, \tau)$ (dotted line), and $E_{NL}(t, \tau)$ (solid line) for $\tau = 77$ fs. (b) Transmission change $\Delta T/T_0(\infty, \tau)$ of the midinfrared pulse as a function of τ (dots). Dashed line: sine wave with the LO phonon frequency for comparison.

Discussion

In thermal equilibrium, a Fröhlich polaron is characterized by a self-consistent attractive potential [Figs. 2(b) and (c)] for the electron caused by a surrounding cloud of longitudinal optical (LO) phonons. The electron-phonon coupling strength $\alpha = 0.067$ determines the polaron binding energy of 5 meV and its radius of 2.7 nm (at room temperature) [6]. The oscillatory behavior of transmission observed here for the first time is a manifestation of the highly nonlinear response of polarons to a strong external field. In our experiments, the polaron potential is strongly distorted by the femtosecond THz field [see Fig. 1(a)]. First, it accelerates the electron, leading to a finite distance of the electron from the center of the polaron [along the coordinate r shown in Fig. 2(g)]. This distance is generated impulsively, i.e., on a time scale short compared to the LO phonon oscillation period. As soon as the kinetic energy of the electron reaches $\hbar\omega_{LO}$, the electron velocity saturates by transferring energy to the lattice. Due to the impulsive character of this transfer, coherent LO phonon oscillations appear as a stern wave of the moving electron [Fig. 2(f)]. With increasing strength of such oscillations, the related electric field (polarization) alters the motion of the electron so that electron oscillations occur along the coordinate r with the frequency ω_{LO} [Fig. 2(g)]. Thus, on top of the drift motion of the entire quasiparticle the electron oscillations along the internal coordinate $r(t)$ are connected to a periodic modulation of the momentary electron velocity $v_e(t) = dr(t)/dt + v_{polaron}(t)$. This modulates the transmission of the midinfrared probe pulses [circles in Fig. 2(b)] in an oscillatory manner as the differential midinfrared mobility depends on $v_e(t)$ and changes its sign around $v_e = \sqrt{2\hbar\omega_{LO}/m}$.

189

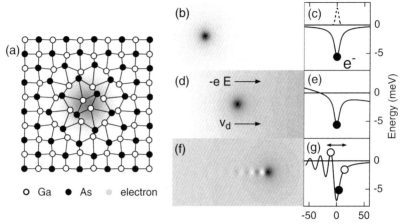

Fig. 2. (a) Lattice distortion of the Fröhlich polaron in GaAs. (b) and (c) Self-induced polaron potential [(b) contour plot and solid line in (c)] and electron wave function (dashed line) of a polaron at rest. (d) and (e) Linear transport: for low applied electric fields the total potential is the sum of the applied potential and the zero field polaron potential. (f) and (g) Nonlinear transport: in a strong DC field [which has been subtracted from the potentials shown in (f) and (g)] the drifting electron (dot) is quasi-stationarily displaced from the minimum of the LO phonon cloud and generates coherent phonon oscillations in its stern wave. As the amplitude of coherent LO phonons exceeds a certain threshold, the polaron potential eventually causes electron oscillations (shown as open circles) along the relative coordinate r on top of the drift motion of the entire quasiparticle.

Conclusions

Our results highlight the quantum-kinetic character of the nonlinear polaron response: the crystal lattice responds with coherent vibrations to the impulsive motion of electric charge. The time scale of such non-instantaneous process is inherently set by the LO phonon oscillation period and the picosecond decoherence of the LO phonon excitation. It is important to note that this nonlinear phenomenon occurs at comparably low electric field amplitudes of the order of $|E| = 10$ kV/cm = 0.1 V/(100 nm). Thus, the quantum kinetic response plays a key role for high-frequency transport on nanometer length scales, in particular for highly polar materials such as GaN and II-VI semiconductors. Furthermore, the impulsive generation of coherent excitations with tailored external fields, e.g., phase-shaped terahertz pulses, may allow for optical modulation and switching in the femtosecond time domain.

1   R. Huber *et al.*, Nature **414**, 286, 2001.
2   H. Fröhlich, Advances in Physics **3**, 325, 1954.
3   P. Gaal *et al.*, Nature **450**, 1210, 2007.
4   T. Bartel *et al.*, Optics Letters **30**, 2805, 2005.
5   K. Reimann *et al.*, Optics Letters **28**, 471, 2003.
6   F. M. Peeters and J. T. Devreese, Physical Review B **31**, 4890, 1985.

Time-resolved X-ray Absorption Spectroscopy of Photoinduced Insulator-Metal Transition in a Colossal Magnetoresistive Manganite

M. Rini[1], R. Tobey[2], S. Wall[2], Y. Zhu[1], Y. Tomioka[3], Y. Tokura[3], A. Cavalleri[2] and R.W. Schoenlein[1]

[1] Lawrence Berkeley National Laboratory, Berkeley, CA 94720, USA
[2] Department of Physics, Clarendon Laboratory, University of Oxford, Oxford OX1 3PU, United Kingdom
[3] Correlated Electron Research Center, AIST, Tsukuba, Ibaraki, 305-8562 Japan
E-mail: mrini@lbl.gov

Abstract. We studied the ultrafast insulator-metal transition in a manganite by means of picosecond X-ray absorption at the O K- and Mn L-edges, probing photoinduced changes in O-$2p$ and Mn-$3d$ electronic states near the Fermi level.

Manganites pose an important testing ground for studying electron correlation phenomena and exhibit close analogies with a wide variety of materials in the correlated-electron family, including high-T_c superconducting cuprates. In these systems, the strong interplay between charge, spin, orbital and lattice degrees of freedom results in rich phase diagrams. Phase competition at the boundaries between these phases leads to a number of remarkable phenomena including colossal magnetoresistance (CMR) [1]. Arguably, the most striking aspect of the physics of manganites is the occurrence of a number of metal-insulator transitions, initiated for instance via perturbations of temperature, magnetic field, pressure, and irradiation with light [1].

The importance of ultrafast time-resolved measurements in understanding complex materials lies in the capability to excite them (perturbatively) on time-scales shorter than the underlying correlations and then disentangle the interactions by probing their time response as the correlation develops. Ultrafast x-ray techniques in particular present an exceptional opportunity for studying the correlated interplay between atomic and electronic structures.

Here we applied time-resolved x-ray absorption spectroscopy to study the photoinduced phase-transitions in the CMR $Pr_{1-x}Ca_xMnO_3$ (Fig. 1). In $Pr_{1-x}Ca_xMnO_3$, at the optimal doping level x=0.3, the insulating phase adjoins a "hidden" metallic state of the system characterized by enormous changes in resistivity [2]. The metallic phase can be reached by application of external perturbations such as a magnetic field, as in the CMR effect shown in Fig. 1(b). An important development has been the demonstration that a first-order insulator-metal phase transition can be induced optically, by the photoinjection of carriers into the insulating state [3]. Even more remarkably, our recent time-resolved reflectivity and transport studies (Fig. 1 (c)) show that the transition to the metallic state occurs on a sub-ps time scale and can be initiated by selective vibrational excitation of the Mn-O stretching mode [4].

191

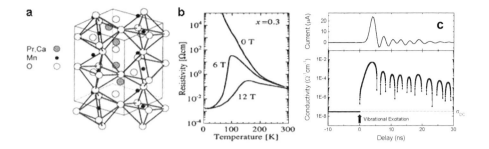

Fig. 1. (a) Perovskite crystal structure of $Pr_{0.7}Ca_{0.3}MnO_3$. (b) CMR effect in $Pr_{0.7}Ca_{0.3}MnO_3$ shown by temperature dependent resistivity measurements under 0, 6, 12 T magnetic field [2]. (c) 5-order-of-magnitude changes in sample resistivity are observed when the Mn-O stretching phonon mode is resonantly excited by 17-μm laser pulses. X-ray measurements will give better insights into the dynamics of relevant electronic states and of changes in local structure distorsion following vibrational excitation.

We studied the photoinduced phase transition in $Pr_{0.7}Ca_{0.3}MnO_3$ by means of infrared-pump/X-ray absorption spectroscopy (time-resolved XANES). X-ray absorption spectroscopy has proven to be a powerful tool for addressing important questions about the physics of manganites, providing important insights into the electronic structure and the lattice distortions of these oxides [5]. Since the valence and conduction bands in CMR manganites are comprised of hybridized Mn-3d and O-2p states, complementary information can be obtained from time-resolved XANES measurements at the O K-edge and Mn L-edges. O K-edge spectra probe unoccupied states of mixed O-2p and Mn-3d character near the Fermi level, and the low-energy pre-edge structure directly reflects changes in the hybridization of these states [5], which influences charge localization/conductivity and d-electron hopping probability. Recent static measurements in manganites have shown that the O K-edge XANES spectrum exhibits substantial changes across the insulator-metal transition and is particularly sensitive to the presence of local Jahn-Teller distorsions [5]. On the other hand, Mn L-edge XANES probes the unoccupied states of predominantly metal-3d character and the chemical edge shift provides information about changes in the Mn oxidation state [6]. Time-resolved XANES measurements are thus particularly sensitive to local structural distortion of the Mn-O complex resulting from the photo-excitation, e.g. polaron effects or changes in the Mn-O-Mn bond angle which strongly influence the electron hopping probability [1]. Local distortions will be manifest in changes to the *3d-2p* hybridization which is at the heart of the insulator-metal transition in these compounds.

$Pr_{0.7}Ca_{0.3}MnO_3$ samples were excited with 100 fs pulses at 1.55 eV. We followed the phase transition dynamics by probing with 70 ps X-ray pulses tuned through the O K-edge at 530 eV and the Mn L-edges at 640 eV. Measurements were taken in Total Electron Yield (TEY) mode at 80 K. Following 800-nm excitation, we observed photoinduced absorption changes at the O K-edge and at the Mn L-edge (Fig 2(a,b)). Fig. 2(c) shows time-dependent pump-probe signals at probe energy of 529 eV, measured as a function of the delay time between laser and X-ray pulses. The solid line is a fit of the data using a cross correlation width of 70 ps, corresponding to the temporal resolution dictated by the X-ray pulse width. The intensity dependence of the measured signals exhibits the same features as in previous femtosecond reflectivity studies [4], with a threshold and saturation behavior which is characteristic of a phase transformation to the metallic state. The extension of these experiments to the femtosecond regime techniques is in progress and will monitor the phase-transition dynamics with appropriate time-resolution and access time scales faster than electron-lattice relaxation processes, disentangling effects due to the phase transition from laser-induced heating effects.

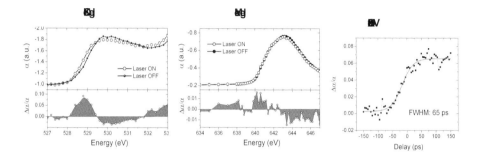

Fig. 2. Picosecond time-resolved XANES in $Pr_{0.7}Ca_{0.3}MnO_3$. (a,b) Upper panels: Static absorption spectra (solid circles) at the O K-edge (a) and Mn L-edge (b) and spectra measured 300 ps after 800-nm excitation (open circles) at a fluence of $\sim$30mJ/cm^2. Lower panels: corresponding relative change of absorption ($\Delta\alpha/\alpha$). (c) Relative change of absorption ($\Delta\alpha/\alpha$) as a function of delay between the 800-nm pump and the X-ray probe pulses, measured at the representative wavelength of 529 eV. Measurements were taken at 80 K in Total Electron Yield (TEY) mode.

Acknowledgements. This work was supported by the U.S. Department of Energy under Contract No. DE-AC02-05CH11231

1 Y. Tokura, Colossal magnetoresistive oxides (Gordon and Breach Science Publishers, 2000).
2 Y. Tomioka et al., Phys. Rev. B **53**, 1689-1692 (1996).
3 M. Fiebig et al., Science **280**, 1925-1928 (1998).
4 M. Rini et al., Nature, **449**, 72-72 (2007).
5 N. Mannella et al., Phys. Rev. B **71**, 125117 (2005).
6 Y. S. Lee et al., Phys. Stat. Sol. A, **196**, 70-73 (2003).

X-ray Absorption Spectroscopy on the fs Time Scale: Ultrafast Electron and Spin Dynamics in Nickel

C. Stamm[1], N. Pontius[1], T. Kachel[1], K. Holldack[1], T. Quast[1], R. Mitzner[1,2], S. Khan[1,3], M. Wietstruk[1], H. A. Dürr[1], W. Eberhardt[1]

[1] BESSY, Albert-Einstein-Str. 15, 12489 Berlin, Germany
[2] Physikalisches Institut der Universität Münster, Wilhelm-Klemm-Str. 10, 48149 Münster, Germany
[3] now at: Institut für Experimentalphysik, Universität Hamburg, Luruper Chaussee 149, 22761 Hamburg, Germany
E-mail: christian.stamm@bessy.de

Abstract. We present femtosecond x-ray absorption experiments investigating the electron and spin dynamics in a thin nickel film after excitation by an optical fs laser pulse. By changing between linearly and circularly polarized x-rays, the electron and spin dynamics can be investigated individually. A temporal response as fast as 120 fs is found.

Introduction

Ultrafast pulses of a few 100 fs duration from a synchrotron can be produced by the beam slicing technique [1]. A first attempt of fs time-resolved absorption demonstrated the potential of this new type of source [2]. In 2004, we have implemented "femtoslicing" with an elliptical undulator as photon source, a device that allows full polarization control of the generated x-rays [3]. The key parameters of our facility are tunable photon energy in the soft x-ray range, linear and circular polarization, a pulse duration around 100 fs, and intrinsic synchronization with a fs laser pump pulse. This unique combination of features allows us for the first time to investigate a fundamental problem in the field of magnetization dynamics, the ultrafast demagnetization of a thin ferromagnetic nickel film [4], with fs x-rays [5].

Experimental Setup

The measurements were performed at the BESSY femtoslicing source, where a third generation synchrotron is combined with an amplified fs laser system to produce x-ray pulses of $\approx$ 100 fs duration [3]. About 10-15% of the laser pulse intensity is used to pump the sample, a 15 nm Ni film evaporated onto a 500 nm Al foil. The fs x-ray pulse then probes the electronic and spin states by means of near-edge x-ray absorption spectroscopy (NEXAFS) and x-ray magnetic circular dichroism (XMCD). X-ray anbsorption is measured by transmitting the beam through the sample and measuring the remaining intensity with an avalanche photodiode.

Results

Two main effects are observed when the fs laser excites the 15 nm nickel film. The NEXAFS at the Ni L_3 edge exhibits a transient absorption line change as shown in figure 1. We find an apparent shift of the absorption line to lower photon energies

194

following the laser excitation. The dynamic evolution of this novel effect can now be investigated by tuning the x-ray energy to the maximum of the transient signal, indicated by the arrow in figure 1(b).

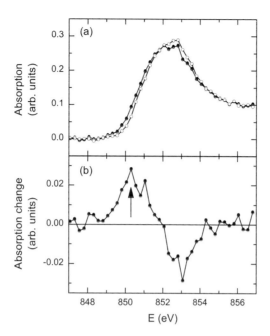

Fig. 1. X-ray absorption on nickel. Linearly polarized femtosecond x-rays are transmitted through a 15 nm Ni film under normal incidence. In (a), the full circles denotes the absorption acquired with a delay time of 200 fs after sample excitation, and the open circles the spectrum of the unperturbed sample. The difference spectrum is plotted in (b). The maximum of the transient signal is found in the leading edge of the absorption line (see arrow).

In order to measure the response of the spin system, XMCD was measured by magnetizing the sample along the x-ray propagation direction and probing with circularly polarized x-rays tuned to the peak of the L_3 absorption edge. Switching the applied field to align the Ni spins parallel and antiparallel to the x-ray helicity and measuring the difference in x-ray absorption for the two orientations, results in a dichroic signal indicating the magnetic state of the sample. The XMCD signal at the L_3 edge is given by a linear combination of spin and orbital magnetic moments [5], and its decrease thus indicates the loss of ferromagnetic order.

Varying the pump-probe delay time, one finds the dynamic evolution of the NEXAFS and XMCD signals, corresponding to the electronic and spin response of the sample. Figure 2 shows the measured data as well as fit curves obtained by using a three-temperature model describing the electron, spin, and lattice sub-systems [4,5]. The time constants of the initial response are 120 ± 50 fs for the electronic and 120 ± 70 fs for the spin component. While the initial response happens on the same time scale, the NEXAFS signal almost immediately starts to relax, whereas the XMCD saturates at 30% of its original value. Relaxation to the initial state usually takes several 100 ps for both components.

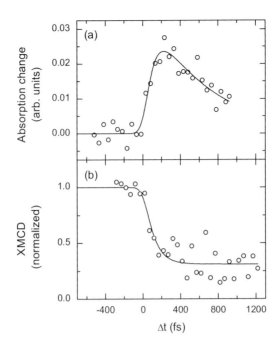

Fig. 2. Temporal evolution of the absorption change (a) and XMCD signal (b) after excitation with a fs laser pulse with a fluence of 8 mJ/cm^2. The experimental data are shown as dots, and fits according to a three-temperature model as lines. Linearly (circularly) polarized x-rays were used in (a) and (b), respectively. In (b), a magnetic field of 0.24 Tesla was applied parallel to the x-ray propagation direction, which enclosed an angle of 30° with the sample surface normal. The XMCD signal was normalized by the signal from the unpumped sample.

Discussion

The observed transient changes at the L_3 absorption line can be reproduced in a cluster model calculation by increasing the valence electron localization, i.e. the nickel metal becomes more atom-like on a fs time scale. This in turn could result in an increased spin-orbit coupling. As the LS coupling is also thought to be responsible for the transfer of spin moment to the lattice, our results could point towards a new mechanism of ultrafast demagnetization, which we find to proceed on the same time scale of 120 fs.

Acknowledgements. We thank the BESSY staff for their continuing support.

1   R. W. Schoenlein et al., Science **287**, 2237 (2000).
2   A. Cavalleri et al., Phys. Rev. Lett. **95**, 067405 (2005).
3   S. Khan et al., Phys. Rev. Lett. **97**, 074801 (2006).
4   E. Beaurepaire et al., Phys. Rev. Lett. **76**, 4250 (1996).
5   C. Stamm et al., Nature Mater. **6**, 740 (2007).

Ultrafast Photoinduced Ferromagnetic Order in a Magnetic Semiconductor Heterostructure

Ingrid Cotoros[1], Jigang Wang[1], Xinyu Liu[2], Jacek K. Furdyna[2], and Daniel S. Chemla[1]

[1] Department of Physics, University of California at Berkeley and Materials Science Division, Lawrence Berkeley National Laboratory, Berkeley, California 94720, USA
E-mail: iacotoros@lbl.gov
[2] Department of Physics, University of Notre Dame, Notre Dame, Indiana 46556, USA

Abstract. We report ultrafast enhancement of ferromagnetism in GaMnAs via photo-excited holes. The ultrafast magnetization increase close to the critical Curie temperature constitutes the first transient evidence of photoinduced phase transition from para- to ferromagnetic state.

Introduction

A long, outstanding issue in correlated spin dynamics is to explore the possibilities for ultrafast photo-initiation of a collective phase from an uncorrelated ground state, as well as to measure the ultrafast time scales relevant to the establishment of such quantum ordering. Carrier-mediated ferromagnetism in manganese doped III-V semiconductors [1] offers unique opportunities for ultrafast manipulation of magnetism. The ferromagnetic interaction between localized Mn moments is mediated by hole spins through the strong exchange interaction between Mn and hole spins, making the magnetic properties sensitive functions of hole density and polarization. Recently, ultrafast photo-excitations have shown to transiently enhance ferromagnetism via photoexcited carriers [2], opening many fascinating avenues for the study of ultrafast magnetic phase transitions and critical phenomena. However, no experiments so far have shown signatures of ultrafast photoinduced paramagnetic to ferromagnetic phase transition. Here, we report on the observation of ultrafast enhancement of ferromagnetism in a III-Mn-V ferromagnetic semiconductor heterostructure, and on the transient signatures of ultrafast ferromagnetic order formation from an uncorrelated ground state.

Experimental Methods

We perform two-color, ultraviolet pump/ near-infrared probe magneto-optical Kerr effect (MOKE) spectroscopy to study the magnetization dynamics in GaMnAs. Our sample was grown by molecular beam epitaxy and has a Curie temperature (T_C) of 77K. The experimental setup consisted of a femtosecond oscillator with 120 fs pulse duration and a BBO crystal in the pump path to double the photon energy. The pump beam at 3.1eV was linearly polarized, and had peak fluence of about 10 μJ/cm^2. A small fraction of the fundamental beam at 1.55eV was used as a probe, detecting magnetization via the polar MOKE angle θ_k. The magnetic origin of the transient MOKE response was confirmed by separate measurements showing the overlap of the pump-induced rotation and ellipticity through the entire scan range. At temperatures bellow T_C, the spontaneous magnetization lies in the sample plane, along an easy axis direction. An external magnetic field can be applied perpendicular to the sample plane to either align the sample magnetization along it or tilt it out of the sample plane.

Results and Discussion

Figure 1(a) shows the temporal profile of transient MOKE angle dynamics at 60K, with an external field of 1T aligning the magnetization perpendicular to the sample. In this configuration, $\Delta\theta_k$ measures magnetization amplitude variations versus pump-probe delay. A transient magnetization increase is clearly present on the 100 ps timescale, after initial fast demagnetization.

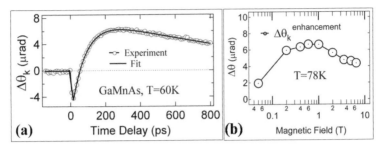

Fig. 1. (a) Time-resolved MOKE dynamics showing transient enhancement of magnetization, on the 100 ps timescale, after initial sub-ps demagnetization. Thick line is the fit allowing for extraction of magnitudes of magnetization enhancement and demagnetization components. (b) Magnetic field dependence at 78K of the photoinduced magnetization enhancement amplitude.

Fitting the temporal profile with a demagnetization and a magnetization enhancement component - as shown by the thick line in Fig. 1(a), one can extract the amplitude of the magnetization increase. MOKE signals $\Delta\theta_k$ measured by the probe arise from the macroscopic magnetization amplitude changes, through the coupling of the electronic states near the band edge, spin split by the average localized Mn spins field. Therefore, the photoinduced positive $\Delta\theta_k$ seen in Fig. 1(a) directly reflects ultrafast enhancement of ferromagnetism. We attribute the observed ultrafast enhancement of ferromagnetic phase to the transient strengthening of the hole-Mn interaction by photo-excited holes, leading to an increase of T_C [2]. Simulations based on a modified Weiss mean field model indicate that the transient ΔT_C increase in current experimental condition is about 1K.

It is interesting to consider the magnetic field dependence of the ferromagnetic enhancement at temperatures slightly above T_C, to explore possible transient signatures of a paramagnetic to ferromagnetic phase transition via pump-induced ΔT_C. Figure 1(b) shows the field dependence of the photoinduced magnetization at 78K ($T_C = 77K$). The photoinduced transient magnetization increase is relatively constant across a magnetic field of 0.2T to 1T, but drops off at low and high fields. At the high fields end, the drop off can be attributed to the larger static magnetization achieved, resulting in smaller photoinduced changes. However, at low fields the situation is very different. It is critical to note that in the absence of a magnetic field, even though spontaneous symmetry breaking (i.e. ferromagnetic correlation) is present, the transient pump-induced macroscopic magnetization detected will always be zero, because of the lack of ordering between individual magnetic domains. Therefore, the results in Fig. 1(b) clearly indicate that, unlike under static equilibrium conditions, at ultrafast time scales a small but finite magnetic field is needed to ensure single domain growth, in order to observe the magnetic phase transition.

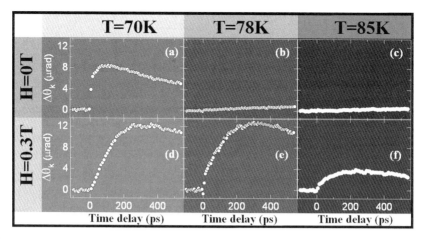

Fig. 2. Transient magnetization enhancement at 70K, 78K, and 85K, under 0T (a-c) and 0.3T (d-f) external magnetic field. The response at 78K resembles the bellow-T_C response at 70K under a small magnetic field, both in terms of amplitude and build-up time.

Figure 2(a-f) showcases the transient signature of ultrafast photo-creation of ferro-magnetic order above T_C. We plot the transient magnetization dynamics, emphasizing magnetization enhancement, under 0T and 0.3T magnetic field, at three different temperatures: bellow T_C (70K), slightly above T_C (78K), and above T_C (85K). Most intriguingly, the photoinduced magnetization enhancement at 78K is almost identical to the bellow-T_C response at 70K under a small magnetic field - both in terms of amplitude and of magnetization build-up time [Fig. 2(d) and (e)]. In the absence of the magnetic field, the 78K response is negligible, and similar to the above-T_C, 85K response [Fig. 2(b) and (c)]. These results clearly reveal that the photo-excited magnetic state at 78K is similar to the 70K ferromagnetic state, even though the unexcited, ground state at 78K is paramagnetic, just like the state at 85K. The data also corroborates our interpretation that in order to detect macroscopic magnetization changes, an above-threshold, external magnetic field is needed to activate the growth of magnetic domains. The single phase in these domains is nucleated via long range correlation enabled by photoexcited holes.

Conclusions

We have observed ultrafast photo-enhanced ferromagnetism and ultrafast photoinduced paramagnetic to ferromagnetic phase transition in GaMnAs. We explain these by a transient strengthening of the Mn-hole correlation via photoexcited holes, resulting in a transient increase ΔT_C of about 1K. Exploring the magnetic field and temperature dependence of the transient magnetization enhancement, we reveal transient signatures of the photoinduced magnetic phase transition. Our measurements reveal new fundamental collective magnetic processes on ultrafast time scales and provide exciting prospects for studying competing phases in strongly correlated magnetic materials.

1   H. Ohno et al., Phys. Rev. Lett. 68, 2664 (1992); Appl. Phys. Lett. 69, 363 (1996)
2   J. Wang et al., Phys. Rev. Lett. 98, 217401 (2007)

Non-equilibrium spin-dynamics of Gd(0001) studied by time-resolved SHG and magnetic linear dichroism in 4f core-level photoemission

A. Melnikov[1], H. Prima-Garcia[2], M. Lisowski[1], T. Gießel[2], R. Weber[2], R. Schmidt[2], C. Gahl[2], N. M. Bulgakova[3], U. Bovensiepen[1], and M. Weinelt[1,2]

[1] Freie Universität Berlin, Fachbereich Physik, Arnimallee 14, 14195 Berlin, Germany
[2] Max-Born-Institut, Max-Born-Straße 2 A, 12489 Berlin, Germany
[3] Institute of Thermophysics SB RAS, 1 Lavrentyev Ave., 630090 Novosibirsk, Russia
 E-mail: alexey.melnikov@physik.fu-berlin.de

Abstract. Spin dynamics following short pulse laser excitation was studied for Gd(0001). The dynamics of Gd valence-spins differs fundamentally from that in $3d$ ferromagnets. This is attributed to interaction between Gd $5d$ and $4f$ magnetic moments and the large $4f$ spin-lattice interaction time of 80 ps deduced from laser-pump synchrotron-probe $4f$ photoemission.

Further development of magnetic devices stimulates the study of spin dynamics on timescales below the nanoseconds of conventional switching by external magnetic field pulses. This goal requires ultrafast excitation and detection of magnetic and spin dynamics which are achieved with femtosecond laser pulses. In the present work we investigate the spin dynamics in Gd(0001). Gd is a prototype system for a Heisenberg ferromagnet with its spins localized in the ion core. The magnetic moment per atom of $7.55\,\mu_B$ arises mainly from the half-filled $4f$ shell contributing $7.0\,\mu_B$. Ferromagnetic order below the Curie temperature T_C=293 K is generated by indirect exchange interaction which aligns the $4f$ magnetic moments through a spin polarization of $5d$ electrons (adding $0.55\,\mu_B$ to μ) via strong intra-atomic $4f$-$5d$ exchange interaction (Fig. 2a, inset). To understand the magnetism and spin dynamics in Gd, we utilize complementary time-resolved (TR) techniques that probe either $4f$ or $5d$ magnetic moments and respective partial magnetizations M_d and M_f.

At first we focus on the magnetic linear dichroism (MLD) in $4f$ core-level photoemission (PE) [1]. This surface sensitive probe is based on the difference of PE spectra for opposite orientations of M, $S^{\uparrow}(E)$ and $S^{\downarrow}(E)$, that scales with M. Here, E is the electron kinetic energy. The experiment was performed at the undulator beamline U125 PGM-1 of the Berlin synchrotron facility (BESSY) operated in single bunch mode with a repetition rate of 1.25 MHz. To study the spin dynamics, a laser excitation ($h\nu$=1.55eV, pulse duration 100 fs, absorbed fluence F=3.5 mJ/cm^2) is synchronized with the X-ray pulses of synchrotron radiation ($h\nu$=60 eV, 50 ps) as illustrated in Fig. 1a. Typical PE spectra and the corresponding MLD are shown in Fig. 1b and c, respectively. It is clearly seen from Fig. 1c that optical excitation reduces the MLD. In addition the spectrum shifts to higher *energy* due to laser-induced space charge effects. The latter problem is bypassed when analyzing an integral quantity, the *dichroic contrast* defined as

$$DC = \sqrt{\int \left(S^{\uparrow}(E) - S^{\downarrow}(E)\right)^2 dE \Big/ \int \left(S^{\uparrow}(E) + S^{\downarrow}(E)\right)^2 dE} \sim M_f. \qquad (1)$$

To confirm that DC monitors the magnetic ordering of the $4f$ shell ($DC{\sim}M_f$), we have measured its dependence on the sample temperature T under steady state conditions

(Fig. 2b). With increasing T, DC reduces down to zero at $T=T_C$ following the temperature dependence of the spontaneous magnetization.

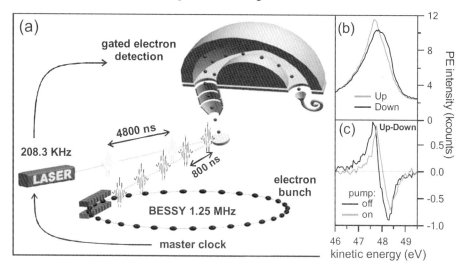

Fig. 1. (a) Scheme of the experiment. (b) 4f core level PE spectra measured at $h\nu=60$eV in the absence of laser excitation and for opposite directions of M. (c) Black, the difference of spectra shown in (b); gray, the same at 50 ps after the laser excitation.

After femtosecond laser excitation, M_f drops by 20% within the 50 ps time resolution defined by the synchrotron bunch length and recovers on a 300 ps timescale (Fig. 2a). Recent experiments suggest that this drop consists of a fast disordering of 4f spins by optically excited 5d-electrons with partial transfer of M_f to M_d due to strong 4f-5d coupling and a continuing thermal disordering on a 80 ps timescale.

The contribution M_d of 5d electrons to the surface magnetization is probed by TR magneto-induced second harmonic (SH) generation (SHG) [2]. After IR ($h\nu=1.55$ eV) laser excitation at $F=1$ mJ/cm^2, M_d exhibits an ultrafast 60% breakdown to a nearly constant level persisting for several tens of ps (Fig. 2a). This breakdown might be attributed to the fast transfer of M_d from the surface to the bulk by hot electrons. In addition exchange-scattering of non-thermalized electrons and interaction of hot electrons with the lattice could provide the transfer of 5d spin polarization to the lattice angular momentum on a 1 ps time scale [3]. The recovery proceeds on a 100 ps timescale [3]. This differs fundamentally from the behavior observed in itinerant ferromagnets such as Ni where M_d exhibits a slower break-down and much faster recovery [4].

Combining 4f and 5d spin dynamics we now focus on the recovery of the magnetization within 500 ps. In Fig. 2a we compare the measured transients with $\Delta M/M$ evaluated from the equilibrium $M(T)$ dependence (Fig. 2b) and the transient lattice temperature T_1 (Fig. 2c). The latter is calculated in the two-temperature model including light absorption and heat diffusion and convoluted with a Gaussian to account for the 50 ps resolution of the MLD experiment (Fig. 2a). This proves that the recovery of M_d is delayed when compared to the laser spot cooling and moreover that the observed reduction of M_f is only half the amount expected from the thermal model. We conclude that equilibration of the 4f spin system and thus M_f with the lattice is achieved only after ~80ps, where the experimental and model curves start to match. Such "inertia" of M_f could arise owing to a weak coupling of the localized 4f spins to the lattice, which is well known for Gd: its half-filled 4f shell results in the orbital moment $L=0$ and consequently small spin-lattice coupling. We therefore conclude that the 4f spin-lattice interaction time is on the order of 80 ps. The 5d magnetic moments show in accordance with the 4f spin dynamics a delayed recovery. M_d matches M_f first after 200 ps. The even slower recovery of M_d may reflect differences between dynamics of surface and bulk magnetic system.

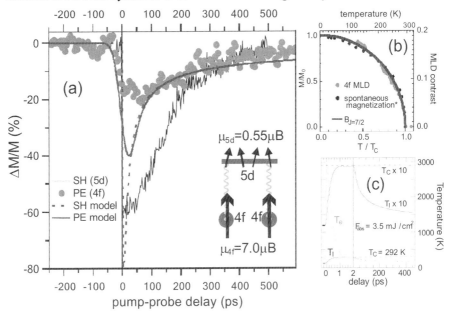

Fig. 2. (a) Transient variation of $\Delta M/M$ of the 5d electrons (thin solid line) measured by TR SHG and of the 4f electrons (gray dots) measured by TR MLD The calculated transient (dashed gray curve) uses was convolved with a Gaussian of 50 ps FWHM (solid gray curve). (b) Temperature dependence of the MLD contrast (gray circles) compared to the spontaneous magnetization M/M_0 (black circles) and Brillouin function $B_{J=7/2}$ (solid line) after Ref. [5]. (c) Calculated transient temperature of Gd valence electrons T_e and lattice T_l.

We arrive at the following conclusions. (i) Time-resolved magneto-induced SHG and MLD in 4f core-level PE are complementary surface sensitive tools for studies of spin dynamics of localized ferromagnets. (ii) Spin dynamics of the valence electrons in lanthanides fundamentally differs from that in itinerant ferromagnets which we attribute to strong interaction of μ_{5d} with the much larger μ_{4f} magnetic moments. (iii) The spin dynamics in lanthanides essentially proceeds in a non-equilibrium regime lasting 80 ps after laser excitation, attributed to the slow 4f spin-lattice interaction.

1   A. Melnikov et al., Phys. Rev. Lett. **100**, 107202 (2008).
2   A. Melnikov et al., Phys. Rev. Lett. **91**, 227403 (2003).
3   M. Lisowski et al., Phys. Rev. Lett. **95**, 137402 (2005).
4   U. Conrad, J. Güdde, V. Jähnke, E. Matthias, Appl. Phys. B **68**, 511 (1999).
5   H.E. Nigh, S. Legvold, and F.H. Spedding, Phys. Rev. **132**, 1092 (1963).

Ultrafast Spin Control by Charge-separated States in Colloidal ZnO Quantum Dots

Nils Janßen[1], Tobias Hanke[1], Florian Sotier[1], Tim Thomay[1], Kelly M. Whitaker[2], Daniel R. Gamelin[2], and Rudolf Bratschitsch[1]

[1] Department of Physics and Center for Applied Photonics, University of Constance, D-78457 Konstanz, Germany
E-mail: Rudolf.Bratschitsch@uni-konstanz.de
[2] Department of Chemistry, University of Washington, Seattle, WA 98195, USA
E-mail: Gamelin@chem.washington.edu

Abstract. Time-resolved Faraday rotation measurements are performed to reveal the ultrafast spin dynamics in colloidal ZnO quantum dots, which are shown to be strongly influenced by hole trapping on the dot surface.

Introduction

Single spins confined within semiconductor quantum dots (QDs) are promising candidates for qubits [1]. ZnO possesses a wide bandgap of 3.4 eV and an exciton binding energy of ~ 60 meV, and holds promise for possible room-temperature operation of future spintronic devices

For the fabrication of spintronic devices based on ZnO quantum dots it is important to know about the mechanisms involved in spin dephasing. Time-resolved Faraday rotation (TRFR) is a leading technique to directly observe the ultrafast spin dynamics of carriers. Here, we present experimental results describing spin dephasing in colloidal ZnO QDs measured by TRFR spectroscopy.

Experimental Methods

In this work freestanding ZnO QDs with a diameter of 7.0 ± 1.7 nm capped with dodecylamine (DDA) were prepared as colloidal suspensions in toluene and characterized as described previously [2]. The sample for TRFR was prepared by drop coating the toluene solution of QDs on a c-plane sapphire substrate, thus forming a frozen matrix of DDA with embedded ZnO colloids.

In TRFR spin-polarized electrons and holes are optically excited by a spectrally narrow (~0.2 nm) circularly polarized pump pulse resonant with the fundamental bandgap in the ultraviolet. Driven by an applied magnetic field in the Voigt geometry the carriers precess coherently (Larmor spin precession). By measuring the polarization rotation of a linearly polarized probe pulse transmitted through the sample one is able to monitor the carrier spin precession over time. An analysis of the damped oscillating signal provides the spin splitting (g-factor) and the spin dephasing time T_2^* of the carriers.

Results and Discussion

Fig. 1 shows the time-resolved Faraday rotation signal of an ensemble of ZnO quantum dots. The oscillation frequency exhibits an electronic g-factor of 1.96, corresponding well to previous electron paramagnetic resonance measurements [3,4]. The spin dephasing time T_2^* shows a biexponential decay of two different timescales. A fast decay on the order of 100 ps is followed by long lasting coherence of the electrons exceeding one nanosecond.

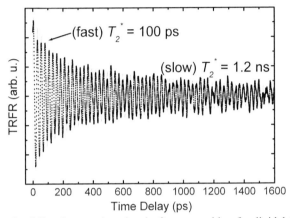

Fig. 1: Time-resolved Faraday rotation signal of an ensemble of colloidal ZnO quantum dots recorded at T = 10 K and B = 1.4 T.

To reveal the involved dephasing processes transient transmission measurements have been performed under essentially identical conditions (Fig. 2). Band edge bleaching is observed, also exhibiting a biexponential decay with timescales of $\tau_1 = 250$ ps and $\tau_2 \geq 7$ ns. The fast decay is attributed to recombination of the fundamental band edge excitons.

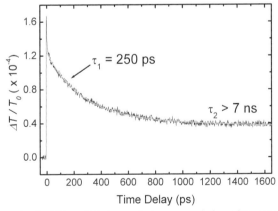

Fig. 2: Time-resolved differential transmission recorded under essentially identical conditions at T = 10 K and B = 1.4 T.

Hence, the fast component in the TRFR signal can be attributed predominantly to exciton recombination. In a competing process the optically excited hole is trapped by states at the surface of the QD leaving the electron behind. This configuration leads to slow radiative recombination with photon emission in the visible [5]. Time-resolved photoluminescence measurements reveal a lifetime of this visible emission of several hundred nanoseconds [5]. In this case, the electron spin-dephasing time is not limited by the carrier lifetime, and long coherence times are observed. The measured T_2^* thus reflects the intrinsic spin-dephasing dynamics of the conduction-band electrons in the hole-trapped metastable state. Polydispersity of the dot sizes as a source for spin dephasing due to inhomogeneous broadening of the g-factor can be neglected since we apply spectrally narrow pump and probe pulses (~ 0.2 nm). Only a small ensemble of dot sizes is excited and probed in that way. Another important dephasing mechanism in ZnO quantum dots is electron-nuclear hyperfine coupling involving ^{67}Zn [3]. Further studies are necessary to reveal the precise origin of the spin dephasing.

Conclusions

Ultrafast spin dynamics in ZnO QDs have been measured for the first time using time-resolved Faraday rotation in the ultraviolet. Our results reveal that metastable charge-separated states offer interesting possibilities for the control and study of carrier spin dynamics in semiconductor nanostructures.

Acknowledgements. Financial support from the Deutsche Forschungsgemeinschaft (DFG) through SFB 513 and the priority program SPP1285, the Kompetenznetz Funktionelle Nanostrukturen Baden-Württemberg, and a grant from the Ministry of Science, Research, and Arts Baden-Württemberg, as well as from the US NSF (CRC-0628252 to DRG and DGE-0504573 (IGERT fellowship) to K.M.W.), the Research Corporation, the Dreyfus Foundation, and the Sloan Foundation is gratefully acknowledged. R. B acknowledges the support of the Centre for Junior Research Fellows of the University of Constance and the continuous support of Alfred Leitenstorfer and Ulrich Rüdiger.

1   V. Cerletti, W. A. Coish, O. Gywat, D., Nanotchnology **16**, R27-R49, 2005

2   N. S. Norberg, D. R. Gamelin, J. Phys. Chem. B **109**, 20810-20816, 2005

3   W. K. Liu, K. M. Whitaker, A. L. Smith, K. R. Kittilstved, B. H. Robinson, D. R. Gamelin, Phys. Rev. Lett. **98**, 186804, 2007

4   W. K. Liu, K. M. Whitaker, K. R. Kittilstved, D. R. Gamelin, J. Am. Chem. Soc. **128**, 3910-3911, 2006

5   A. Van Dijken, E. A. Meulenkamp, D. Vanmaekelbergh, A. Meijerink, J. Phys. Chem. B **90**, 123, 2000

Ultrafast electronic and spin dynamics in thin iron films: electron-magnon and electron-phonon interactions

E. Carpene, E. Mancini, C. Dallera, M. Brenna, E. Puppin and S. De Silvestri

ULTRAS, CNR-INFM, Dipartimento di Fisica, Politecnico di Milano, p.zza Leonardo da Vinci 32, 20133 Milano, Italy
E-mail: ettore.carpene@fisi.polimi.it

Abstract. The electronic and spin dynamics in thin iron films have been investigated by means of time-resolved reflectivity and magneto-optical Kerr effect. The electron-magnon and the electron-phonon coupling times are extrapolated and their influence is discussed.

Introduction

We used an all optical pump-probe approach to investigate the electronic and spin dynamics in thin iron films by means of time-resolved reflectivity and time-resolved magneto-optical Kerr effect (TR-MOKE). From the transient reflectivity experiment we deduced the electron-phonon coupling time of 240±10 fs, a value remarkably similar to the theoretical estimate. The magneto-optic measurements revealed a rapid demagnetization that has been interpreted in terms of electron-magnon scattering with a characteristic time constant shorter than 100 fs, while the subsequent spin order is re-established on a time scale of about 900 fs and has been attributed to Elliot-Yafet spin-flip events [1,2]. ...is not re-established within the investigated time window of 60 ps (this is attributed to thermal effects).

Experimental Methods

The experiments have been conducted on thin Fe(100) films (about 7 nm thick) epitaxially grown on MgO(100) substrates. The optical analysis has been performed *ex situ* with an amplified Ti:Sapphire laser, generating 60 fs pulses centered at 800 nm (1.55 eV). TR-MOKE has been performed applying the external magnetic field along the Fe(100) easy axis and parallel to the film surface. Both pump and probe wavelengths were 800 nm with crossed polarizations. Bleaching effect has been ruled out comparing the results obtained with different pump wavelengths. The temporal evolution of the magnetization has been deduced from the hysteresis loops measured at different pump-probe delays. The variation of the remanence M(t) as a function of the delay represents the magnetization dynamics, while the center of the loops shift following the transient reflectivity signal. This technique permits to simultaneously extract the time-resolved reflectivity and the magnetization under the same experimental conditions. Time-resolved reflectivity has also been measured with a separate experiment, using the 800 nm pump beam, and varying the probe wavelength from 500 nm (2.48 eV) to 710 nm (1.75 eV).

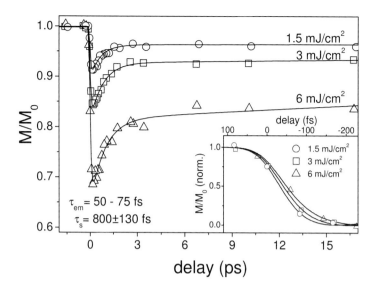

Fig. 1. Normalized magnetization of the Fe(100) film vs pump-probe delay at three different pump fluences. The inset reports the comparison between the hysteresis loops measured with no pump (solid line) and 160 fs after the 6 mJ/cm² pump pulse (dashed line).

The temporal evolution of the magnetization is shown in Fig. 1 for different pump fluences. We point out that $M(t)/M_0$ scales proportionally to the pump fluence, excluding the presence of non-linear (i.e. multi-photon) effects. Regardless of the pump intensity, three distinct regimes can be identified: (i) the demagnetization time is approximately 100 fs, (ii) a subsequent partial recovery of the spin order is observed within about 3 ps, (iii) the original value of $M(t)/M_0$

Results and Discussion

From the time-resolved reflectivity measurements at various probe wavelengths we have deduced an electron-phonon relaxation time of 230 fs. Fig. 2 reports the transient reflectivity curves for three different probe photon energies. Using the theoretical approach proposed by P. B. Allen [3] the electron-phonon scattering time has been estimated to range from 150 fs to 290 fs, in agreement with our experimental result. Adapting the theory of electron-phonon scattering to the electron-magnon interaction, we also extrapolated an electron-magnon interaction time of 60-80 fs, nicely matching the observed demagnetization time. These results suggest that laser-induced ultrafast demagnetization of epitaxial Fe thin films can be understood in terms of electron-magnon interaction, taking place on a time scale <100 fs. Hot electrons can efficiently excite magnons, leading to a rapid reduction of the magnetization. The subsequent recovery of the spin order takes place with a characteristic time constant of ~900 fs, that is longer than the electron-phonon relaxation time (230 fs), and supports the picture of Elliott-Yafet spin-flip scattering.

Conclusions

From the time-resolved reflectivity measurements at various probe wavelengths we have deduced an electron-phonon relaxation time of 230 fs. Fig. 2 reports the transient reflectivity curves for three different probe photon energies. Using the theoretical approach proposed by P. B. Allen [3] the electron-phonon scattering time has been estimated to range from 150 fs to 290 fs, in agreement with our experimental result. Adapting the theory of electron-phonon scattering to the electron-magnon interaction, we also extrapolated an electron-magnon interaction time of 60-80 fs, nicely matching the observed demagnetization time. These results suggest that laser-induced ultrafast demagnetization of epitaxial Fe thin films can be understood in terms of electron-magnon interaction, taking place on a time scale <100 fs. Hot electrons can efficiently excite magnons, leading to a rapid reduction of the magnetization. The subsequent recovery of the spin order takes place with a characteristic time constant of ~900 fs, that is longer than the electron-phonon relaxation time (230 fs), and supports the picture of Elliott-Yafet spin-flip scattering.

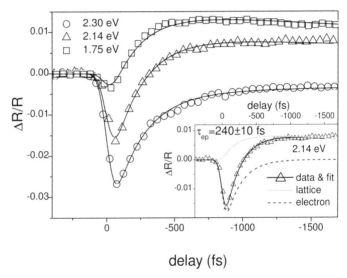

delay (fs)

Fig. 2. Time-resolved reflectivity curves (and corresponding fits) obtained with 1.55 eV pump photons (800 nm) and various probe photon energies. The extrapolated electron-phonon relaxation time is 230±20 fs. The inset shows the electronic (dash line) and lattice (gray solid line) contribution to the reflectivity curve measured with 2.14 eV probe photons.

1   R. Elliott, "Theory of the Effect of Spin-Orbit Coupling on Magnetic Resonance in Some Semiconductors", Phys. Rev. 96, 266, 1954.

2   Y. Yafet, "g Factors and Spin-Lattice Relaxation of Conduction Electrons" in Solid State Physics, Edited by F. Seitz and D. Turnbull, Vol. 14, 1, Academic, New York, 1963.

3   P. B. Allen, "Theory of Thermal Relaxation of Electrons in Metals", Phys. Rev. Lett. 59, 1460, 1987.

Laser Induced Alignment of Water Spin Isomers

E. Gershnabel[1], and I. Sh. Averbukh[1]

[1] Dept. of Chemical Physics, The Weizmann Institute of Science, Rehovot 76100, ISRAEL
  E-mail: erez.gershnabel@weizmann.ac.il, ilya.averbukh@weizmann.ac.il

Abstract. We consider laser alignment of ortho and para spin isomers of water molecules by using strong and short off-resonance laser pulses. A single pulse is found to create distinct transient alignment and antialignment of the isomeric species. We demonstrate selective alignment of one isomeric species (leaving the other species randomly oriented) by a pair of two laser pulses.

Introduction

According to quantum mechanics, a molecule that contains two identical atoms whose nuclei have nonzero spins can exist in the form of spin isomers. In particular, a water molecule exists in one of the two spin isomers, ortho or para with parallel or antiparallel proton spins, respectively, where in the gas phase the ortho-para conversion is highly improbable.

Recently Andreev *et at* [1], demonstrated an expected difference in response of the ortho and para forms of water to an inhomogeneous dc electric field. We study the effect of off-resonance femtosecond laser pulses interacting with the ortho and para water molecules. In particular, we are interested in the laser induced alignment of the molecules, where we define alignment as the angular localization of one of the principal axes of the molecule. By antialignment we refer to the confinement of the molecular axis in a plane perpendicular to the laser pulse polarization. The water molecule wavefunction is given by a product

$$\left|\Psi\right\rangle = \left|\Psi_{rot}\right\rangle \left|\Psi_{spin}\right\rangle \left|\Psi_{vib}\right\rangle \left|\Psi_{el}\right\rangle \tag{1}$$

of the rotational (*rot*), nuclear spin (*spin*), vibrational (*vib*) and electronic (*el*) wavefunctions.

The rotational part of the water Hamiltonian is given by the asymmetric rigid rotor Hamiltonian:

$$\hat{H} = \frac{\hat{J}_a^{\,2}}{2I_a} + \frac{\hat{J}_b^{\,2}}{2I_b} + \frac{\hat{J}_c^{\,2}}{2I_c} \tag{2}$$

The water molecule has C_{2V} symmetry, and the corresponding rotational wavefunctions are classified according to the irreducible representations of this group, i.e. A_1, A_2, B_1 and B_2. According to the Pauli Exclusion Principle, the total wavefunction should change sign after permutation of the hydrogen nuclei (equivalent to rotation by π about the symmetry axis). Rotational states of A_1 and A_2 do not change sign as the result of the permutation, and they correspond to the para states with antiparallel spins, while the B_1 and B_2 states correspond to the ortho states (with the total spin 1). Here we consider molecules in the ground electronic and vibrational states.

Molecular alignment method is based upon off-resonance laser pulses, which induce a molecular dipole moment and interact with it. The Hamiltonian for interaction of the rigid rotor with a laser pulse is given by [2],

$$H_{int} = -\frac{1}{4}\sum_{\rho,\rho'} \varepsilon_{\rho} \alpha_{\rho\rho'} \varepsilon_{\rho'} \tag{3}$$

where the indexes refer to the space fixed Cartesian coordinates, ε is the laser pulse envelope and α is the polarizability tensor. Such interaction is symmetric with respect to the group operations and does not mix the para and ortho rotational states. The water molecule geometry is shown in Figure 1.

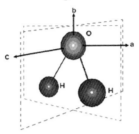

Fig. 1. A schematic illustration of the water molecule. The highest polarizability component is along the a axis.

Results and Discussion

Applying a linear polarized $10^{13}W/cm^2$, $20fs$ laser pulse, we expect the molecular highest polarizability axis (i.e. the a axis in Figure 1) to be aligned along the laser field polarization axis. Since this pulse is short with respect to the water rotational dynamics, it is considered as a delta pulse. We denote the angle between the molecular a axis and the laser field axis as θ, and calculate the alignment factor $<\cos^2\theta>$ as a function of time ($<>$ denotes quantum and thermal averaging over the molecular ensemble). The alignment factor is plotted in Figure 2 for temperature of $20K$.

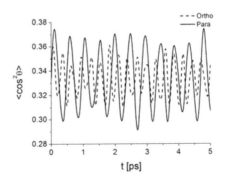

Fig. 2. The alignment factor after the laser pulse at temperature of 20K. Different dynamics of the ortho and para isomers is revealed, while free evolution of excited rotational wavepackets. Simultaneous alignment and anti-alignment for the two species can be observed.

It is seen that after the application of the pulse, the ortho and para rotational wavepackets reveal different dynamics, and the moments of simultaneous transient alignment and antialignment for the two species can be found. For instance, at $t=2ps$, the para molecules are aligned while the ortho molecules become antialigned. This dynamics can be further exploited by applying additional pulse (of the same intensity and duration) at $t=1.9ps$, where the ortho and para molecules are on the way of becoming antialigned and aligned, respectively (at this time the alignment factor of them both is similar and a bit larger than 1/3). The results are given in Figure 3, again for the temperature of 20K. It can be seen that the second pulse enhanced a bit the alignment of the para molecules, but suppressed the alignment of the ortho molecules, leaving them randomly oriented.

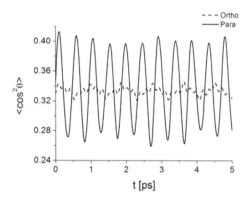

Fig. 3. Alignment factor after the second pulse versus time after the second pulse at 20K. While the transient alignment of the para molecules was increased, the ortho alignment was reduced, and the ortho molecules became almost uniformly oriented as before the first pulse.

Conclusions

In conclusion, we demonstrated the possibility for selective control of the alignment of ortho and para water molecules. Transient alignment and antialignment were found to happen simultaneously for different isomer species at certain time moments. Enhancement and suppression of the alignment effect was observed for different isomers using a second delayed laser pulse. Ortho and para alignment control allows further manipulation of these species, including spatial separation using external fields.

1   S. N. Andreev, V. P. Makarov, V. I. Tikhonov, and A. A. Volkov, arXiv:physics/0703038v1, 2007.
2   T. Seideman and E. Hamilton, Adv. At. Mol. Opt. Phys., **52**, 2005.

Memory Effects in Photo-induced Femtosecond Magnetization Rotation in a Ferromagnetic Semiconductor

Ingrid Cotoros[1], Jigang Wang[1], Xinyu Liu[2], Jacek K. Furdyna[2],
Jaroslav Chovan[3], Ilias E. Perakis[3] and Daniel S. Chemla[1]

[1] Department of Physics, University of California at Berkeley and Materials Science Division,
Lawrence Berkeley National Laboratory, Berkeley, California 94720, USA
E-mail: iacotoros@lbl.gov
[2] Department of Physics, University of Notre Dame, Notre Dame, Indiana 46556, USA
[2] Department of Physics, University of Crete, Heraklion, Greece

Abstract. We report on the first observation of photo-induced femtosecond cooperative magnetization rotation at the non-equilibrium, charge distribution timescale, in ferromagnetic semiconductor GaMnAs, enabling the ultrafast detection of four-state magnetic memory.

Introduction

Mn-doped III-V semiconductors with carrier-mediated ferromagnetic order [1], such as GaMnAs, offer fascinating opportunities for non-thermal, potentially femtosecond manipulation of magnetism. The magnetic properties of these materials show large responses to external excitations, such as light or electrical gate and current, via carrier density tuning [2-4]. At the same time, the strong coupling between carriers (holes) and Mn ions (e.g., on the order of 1eV in GaMnAs) could support a femtosecond cooperative magnetic response to photoexcited carriers. Indeed, the possibility of existence of an early non-equilibrium, non-thermal femtosecond regime for collective magnetization rotation in (III,Mn)V semiconductors has been predicted theoretically [5]. However, all the prior studies of photoexcited magnetization rotation in ferromagnetic (III,Mn)V systems only show dynamics on the few picosecond timescale, which mainly accesses the quasi-thermal equilibrium, lattice-heating regime [6].

Custom-designed (III,Mn)V hetero- and nanostructures exhibit rich magnetic memory effects (i.e. GaMnAs-based four-state magnetic memory). Thin film GaMnAs grown on GaAs substrate under compressive strain has four equivalent in-plane easy axis directions, along which the magnetization naturally aligns. "Giant" magneto-transport and magneto-optical effects in this system allow for ultra-sensitive detection of the magnetization direction, i.e. magnetic memory readout [7, 8]. However, all demonstrated memory detection schemes so far have been static measurements. Accessing and understanding the memory-dependent collective magnetization dynamics at the femtosecond timescale is critical for terahertz detection of magnetic memory, and for developing realistic "spintronic" devices and large-scale functional systems.

Here we report on the first observation of photoinduced femtosecond collective magnetization rotation which distinctly depends on the magnetization memory state the system is prepared in. The temporal profiles of magnetization dynamics show an initial, distinct temporal regime of the magnetization rotation within the first 200 fs, during the highly non-equilibrium photoexcitation and non-thermal carrier redistribution times, which enables the femtosecond detection of magnetic memory in GaMnAs.

Experimental Methods

We perform UV pump/ NIR probe, vectorially resolved Magneto-Optical Kerr Effect (MOKE) spectroscopy in ferromagnetic semiconductor GaMnAs. The main sample studied was grown by low-temperature molecular beam epitaxy, and consisted of a 73 nm $Ga_{0.925}Mn_{0.075}As$ layer on a 10 nm GaAs buffer layer. The magnetic origin of the transient MOKE response was confirmed by separate measurements showing the overlap of the pump-induced rotation and ellipticity through the entire scan range. Our structure exhibits four equivalent in-plane easy axis directions [Fig. 1(a)]. An external magnetic field can be applied perpendicular to the sample plane, with a very small in-plane component, to align the in-plane magnetization component (M_{IP}) along any of the four easy axis directions. Varying the external magnetic field, one can detect the magnetic state that M_{IP} occupies by static measurement such as Hall resistivity [Fig. 1(b)]. The abrupt jumps in Hall resistivity value correspond to abrupt consecutive 90° M_{IP} switchings between the four available states.

Results and Discussion

Figure 1(c) presents the static magnetization curve (i.e. MOKE signal θ_k, proportional to the out of plane magnetization component M_z) when the field is varied along the hard axis, perpendicular to the sample plane. Unlike in the static Hall resistivity measurements, there are no indications of the memory state that M_{IP} is in. However, in Figure 1(d), the pump-induced MOKE signal $\Delta\theta_k$, proportional to ΔM_z, shows clearly distinct dynamics for the four magnetic states accessed with |B|=0.2T. Moreover, two distinct temporal regimes are observed: a substantial, fast magnetization rotation taking place during the first 200 fs, followed by a much slower magnetization change afterwards.

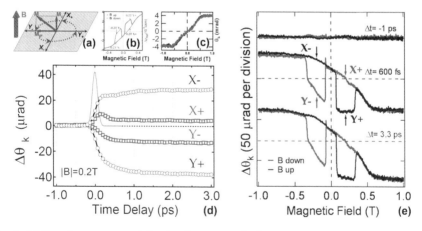

Fig. 1. (a) Sample geometry and four in-plane easy axis directions. (b) Hall magneto-resistivity measurement showing hysteresis loops as consequence of abrupt consecutive 90° M_{IP} switching between the four available states. (c) Static magnetization curve along the hard axis measured by MOKE. (d) Temporal profiles of photoinduced $\Delta\theta_k$ for the four distinct magnetic states accessed with |B|=0.2T. Shaded area is the pump-probe cross correlation. (e) Photoinduced femtosecond four-state magnetic hysteresis: magnetic field scans of $\Delta\theta_k$ at three time delays. Traces are offset for clarity.

Figure 1(e) presents magnetic field scan traces taken at three time delays between the pump and the probe: -1 ps, 600 fs, and 3.3 ps, representing the magnetic memory dependent photo-induced MOKE response (i.e. time-resolved). ΔM_z is negligible at -1 ps. However, a mere 600 fs after photoexcitation, a clear photo-induced four-state magnetic hysteresis is observed, with four abrupt jumps, arising as a direct consequence of magnetic memory effects (i.e. M_{IP} switching). It is critical to emphasize again that the steady-state MOKE measurement, without the pump excitation [Fig. 1(c)], does not show any sign of magnetic switching or memory behavior. In addition, the photo-induced hysteresis loops at Δt=3.3 ps sustain similar shapes, with only slightly larger amplitudes than those at Δt=600 fs, indicating that the dynamic magnetic processes sensitive to the abrupt switchings of M_{IP} occur at the femtosecond timescale.

Next we discuss the origin of the femtosecond magnetization rotation effect observed. It has been shown theoretically [5] that the Mn spin in GaMnAs can respond quasi-instantaneously to an effective femtosecond magnetic field pulse generated by the photoexcited hole spin and by second-order nonlinear optical excitation. Such effective magnetic field pulse may be thought of as a femtosecond modification of the magnetic anisotropy fields during the photoexcitation. Due to the hole-mediated effective exchange interaction between Mn spins, the anisotropy fields in GaMnAs are a direct consequence of the coupling of several valence bands by the spin-orbit interaction. Consequently, the anisotropy fields are sensitive functions of the transient hole distribution and coherences. One therefore expects an effective magnetic field pulse ΔB to be turned on by the optical excitation of holes in high momentum states, which then exerts a spin torque on the magnetization vector. Since this mechanism is carrier-mediated, the appearance of ΔB is quasi-instantaneous, limited only by the pulse duration of ~100 fs. We present elsewhere theoretical simulations for this effect, as well as a discussion on the origin of the discontinuity that reveals the two temporal regimes in collective magnetization rotation [9].

Conclusions

We have observed photo-induced magnetization rotation in ferromagnetic semiconductor GaMnAs at the femtosecond time scale. We show that a substantial, fast magnetization rotation takes place during the first 200 fs, distinctly depending on the magnetic state the system is prepared in. We explain this rotation by the quasi-instantaneous apparition of a transient magnetic field pulse turned on by the optical excitation. Our measurements reveal a new regime of fundamental collective magnetic processes at the femtosecond timescale, and provide exciting prospects for terahertz magnetic memory detection.

1   H. Ohno, Science 281, 951 (1998)
2   H. Ohno et al., Nature 408, 944 (2000)
3   S. Koshihara et al., Phys. Rev. Lett. 78, 4617 (1997)
4   J. Wang et al., Phys. Rev. Lett. 98, 217401 (2007)
5   J. Chovan et al., Phys. Rev. Lett. 96, 057402 (2006)
6   e.g. J. Qi et al., Appl. Phys. Lett. 91, 112506 (2007)
7   H. X. Tang et al., Phys. Rev. Lett. 90, 107201 (2003)
8   A. V. Kimel et al., Phys. Rev. Lett. 94, 227203 (2005)
9   J. Wang et al., arXiv:0804.3456v1

Part IV

Physics - Condensed Phase and Low Dimensional Systems

Transient Dielectric Function of Fs-Laser Excited Bismuth

Andrei V. Rode[1], Davide Boschetto[2], Thomas Garl[2], Antoine Rousse[2]

[1]Laser Physics Centre, The Australian National University, Canberra, ACT 0200, Australia
[2]Laboratoire d'Optique Appliquée, ENSTA/Ecole Polytechnique, Chemin de la huniere, Palaiseau, France
E-mail: *avr111@rsphysse.anu.edu.au*

Abstract: We present time-resolved dual-angle single-wavelength pump-probe study of the reflectivity of femtosecond laser excited bismuth crystal. The recovered real and imaginary parts of the dielectric function show complex a transition to a new quasi-stable excited state at about 20 ps. This state lasts for about 4 ns after the excitation and is not an intermediate state between the solid and the liquid phase. The results suggest that the photo-excited state does not correspond to neither a warm nor a liquid phase, and open general questions on photo-excited phase transition.

Introduction

Studies of nonequilibrium, nonthermal states hold the promise of discovering new transient phases, metastable states, and chemical reaction pathways. The absence of thermal equilibrium provides a way to observe the resulting dynamics of the electronic and the atomic subsystems, which determines the basic electronic, magnetic, and optical properties of the materials. Observation of coherent displacement and oscillation of phonons, which appear as oscillation of reflected light after excitation by an ultrashort laser pulse, gives access to study lattice dynamics in solids through the electron-phonon energy coupling.[1,2] Semimetallic bismuth with a small overlap in the energy of the conduction band and valence bands is one of the most studied elements for unique electrical and thermal properties and their applications. Ultrafast laser-induced coherent lattice vibrations in solids, changes in long-range order,[2,3] as well as direct measurements of the atomic positions within the unit cell,[4,5] were established using time-resolved x-ray diffraction. However, the role of excitation electrons and oscillation of optical phonons in the solid-melt phase transition, and in particularly non-thermal melting such as found in tetrahedrally bonded semiconductors,[6,7] is still obscure. Density functional theory (DFT) calculations suggested that bismuth undergoes a structural phase transition to a higher symmetry state, and it was not clear if the level of excitation could be achieved without thermal melting of the material.[4]

   A direct way to test if the transient state in Bi is an intermediate state between the solid and the liquid or it is a transient solid-state phase is to measure the real and imaginary parts of the dielectric function. In this paper we present the results of time-resolved measurements of the dielectric constant of femtosecond laser excited Bi crystal.

Experimental

We performed dual-probe reflectivity measurements at 800 nm to measure the changes in the reflectivity of bismuth after the excitation with femtosecond pulses.

The dielectric constant at 800 nm was recovered from the time-resolved reflectivities of two optical probes measured with an accuracy of $\Delta R = 10^{-5}$ employing dual-angle reflectometry, with 40 fs time resolution. We have chosen p-polarisation for the pump for effective absorption and thus effective excitation of coherent oscillation of the A_{1g} phonon mode along the c-axis in Bi; subsequently the probes had to be s-polarised on the target surface.

The pump was directed to a crystal surface at a close to normal angle of incidence and focused over a 125 μm spot size (FWHM), while the two probes were focused down to 40 μm spots and the angles of incidence were 19.5° and 34.5°; at these angles the geometrical time resolution was kept ~40 fs. The pump beam spot size was much larger than that of the probe beam in order to probe a central excited part of the sample. Spatial overlapping of two probes was double-checked by time synchronisation using an initial negative drop in reflectivity of the excited crystal. This negative dip in reflectivity related to a coherent displacement of atoms by the polarisation force during the pulse,[8] so that negative drop was shorter than the excitation pulse. The laser fluence was much lower than the damage threshold, which was confirmed by reflectivity recovery to the initial value, and by observations of the crystal surface in the visualisation system.

The pump pulse was chopped at 500 Hz and the probe reflectivity measured using two digital lock-in amplifiers, which allowed us to improve the signal-to-noise ratio and to observe reflectivity changes ΔR with the accuracy $\Delta R/R_0 < 10^{-5}$. Fig.1 shows the recorded reflectivity behaviour at the excitation fluence 6.9 mJ/cm^2, and the dynamics of real and imaginary parts of the dielectric function are shown in Fig. 2.

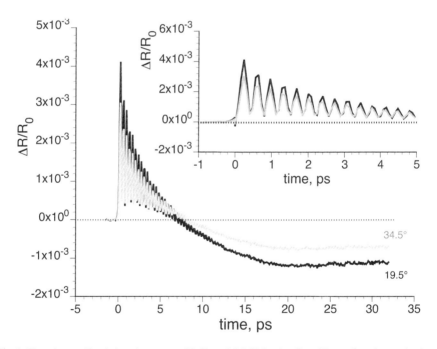

Fig.1. Transient reflectivity changes at 19.5° and 34.5° in the first 32 ps after the excitation. The inset shows the first negative dip during the pulse and the following coherent oscillations in the first 5 ps.

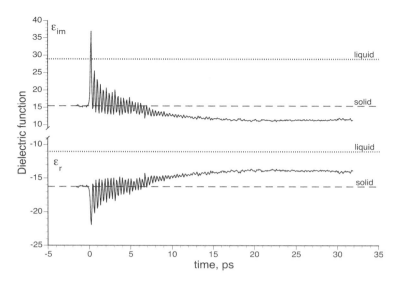

Fig. 2. Dynamics of real (left) and imaginary (right) parts of the dielectric function. Values for solid (solid line) and liquid (dotted line) are also shown.

Conclusions

Our findings show that the laser-induced oscillations of coherent phonons followed by a transition to a new quasi-equilibrium state of Bi lasting up to 4 ns. While the real part of the dielectric function does move toward the liquid value, the imaginary part moves to lower values, away from the range defined by the solid and the liquid dielectric constants. There was no single moment in time when both, the real and the imaginary parts had values within the solid-liquid gap. The value of the complex dielectric constant of the transient state $\varepsilon^{tr} = 17.9 \pm 0.2$ at 800 nm was out of the solid-liquid gap between the of $\varepsilon^{sol} = 22.4$ and $\varepsilon^{liq} = 30.9$. This suggests that the excitation of coherent oscillations in Bi followed by a solid-plasma transition into a quasi-steady state. The transient state does not lead to melting even at the absorbed excitation laser fluence as high as 4 mJ/cm², supplying the energy into the 28-nm skin layer far above the energy density required for an equilibrium melting.

Acknowledgements. The support of Programme International De Cooperation Scientifique (PICS, France) is gratefully acknowledged. One of the authors (T. G.) acknowledges support of the European contract FLASH (No. MRTN-CT-2003-503641).

[1] T. K. Cheng, et. al., , Appl. Phys. Lett. **57**, 1004 (1990).

[2] A. Rousse, et. al., Nature **410**, 65 (2001).

[3] K. Sokolowski-Tinten, et. al., Nature **422**, 287 (2003).

[4] D. M. Fritz, et.al., Science **315**, 633 (2007).

[5] P. Beaud, et.al., Phys. Rev. Lett. **99**, 174801 (2007).

[6] A. M. Linderberg, et. al., Science **308**, 392 (2005).

[7] V. Recoules, et. al., Phys. Rev. Lett. **96**, 055503 (2006).

[8] D. Boschetto, E. G. Gamaly, A. V. Rode, B. Luther-Davies, D. Glijer, T. Garl, O. Albert, A. Rousse, J. Etchepare, Phys. Rev. Lett. **100**, 027404 (2008).

Coherent A_{1g} and E_g Phonons of Antimony

Kunie Ishioka[1], Masahiro Kitajima[1], and Oleg V. Misochko[2]

[1] Advanced Nano-characterizatrion Center, National Institute for Materials Science,
305-0047 Tsukuba, Japan
E-mail: ishioka.kunie@nims.go.jp
[2] Institute of Solid State Physics, Russian Academy of Sciences, 142432 Chernogolovka, Russia

Abstract. Femtosecond dynamics of coherent phonons of antimony is investigated as a function of temperature and optical polarization. Results indicate that coherent A_{1g} phonons couple to photoexcited electrons much more strongly than E_g phonons.

Introduction

Semimetals Bi and Sb have been model systems for the coherent phonon studies. Both crystals have an A7 crystalline structure and sustain two Raman active optical phonon modes: the totally symmetric A_{1g} mode in which two atoms beat out-of-phase along the trigonal axis and the doubly degenerate E_g mode with the atom movements perpendicular to the trigonal axis. In the present study, we investigate the femtosecond dynamics of coherent optical phonons in Sb systematically, as a function of pulse duration, optical polarization, temperature and pump power, to reveal the generation mechanism.

Experimental Methods

Pump-probe reflectivity measurements are performed on the (0001) and ($0\bar{1}11$) surfaces of single crystal Sb in a cryostat. An output of Ti:sapphire laser (800 nm, 60 fs, 86 MHz) is used as the light source. Pump beam is modulated at 1.98 kHz with an optical chopper for lock-in detection. Time delay t between the pump and probe pulses is scanned by a translational stage (slow scan).

Results and Discussion

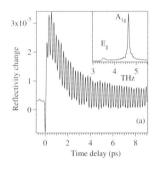 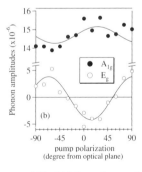 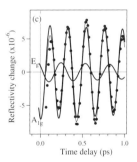

Fig. 1. (a) Isotropic reflectivity change of Sb (0001) surface at 8 K, and its FT spectrum (inset). (b) Pump-polarization dependence of the initial amplitudes of the reflectivity oscillations obtained from Sb ($0\bar{1}11$) surface at 8 K. (c) Decomposition of the oscillatory component of the reflectivity change (filled circles) into A_{1g} and E_g components.

Figure 1(a) show a typical reflectivity change of Sb, which features the oscillations of the coherent A_{1g} and E_g phonons at 4.6 and 3.5 THz, as well as the non-oscillatory component due to photoexcited carriers. Experimental distinction of the generation mechanism of the coherent phonons can be given by the optical polarization dependence of the phonon amplitudes. Displacive excitation of coherent phonons (DECP) [1] would be dominated by the polarization dependence of the optical absorption, while impulsive stimulated Raman scattering (ISRS) [2] by the symmetry of the Raman tensor. The amplitude of the E_g phonon follows $\cos 2\theta$ with θ the pump polarization angle and has no isotropic component, as shown in Fig. 1(b), indicating that the E_g phonon is generated via ISRS. The amplitude of the A_{1g} phonon from the $(0\bar{1}11)$ surface also shows a slight but clear polarization dependence, which is an indication of the ISRS contribution to the coherent A_{1g} phonon.

Initial phases of coherent phonons have been often used for experimental distinction between the generation mechanisms. The initial phases of the A_{1g} and E_g phonons are very close to $-\cos\omega_a t$ and $\sin\omega_e t$, respectively, as shown in Fig. 1(c). Similar phase difference between the A_{1g} and E_g modes have also been observed in our previous study on Bi [3]. The results appear as if A_{1g} and E_g phonons are induced on the excited and the ground electronic states via DECP and ISRS, respectively [1,4]. In terms of the resonant ISRS model [4,5], the lifetime of the photoexcited electrons to which the A_{1g} phonon is coupled is considerably longer than the phonon period, while that to which the E_g phonon is coupled is significantly shorter.

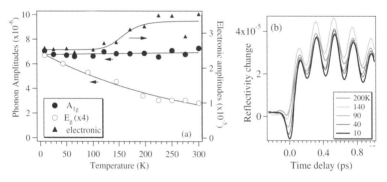

Fig. 2. (a) Temperature dependences of the initial amplitudes of the reflectivity change due to A_{1g} and E_g phonons (left-hand axis), and of the maximum height of the non-oscillatory reflectivity change due to the photoexcited electrons (right-hand axis). (b) Reflectivity change of Sb measured at different temperature.

We observe different temperature dependences between the A_{1g} and E_g amplitudes, as shown in Fig. 2(b). Interestingly, the amplitude of the electronic transient also shows a quite different temperature dependence. Because the amplitude of DECP-driven coherent phonons should be proportional to the carrier density, different temperature dependence between the phononic and electronic amplitudes is another indication of non-DECP contribution to the generation of the coherent A_{1g} and E_g phonons. We note that the negative drop in the reflectivity at $t=0$ depends sensitively on temperature, as shown in Fig. 2(b). Since the coherent phonons show different temperature dependence as discussed above, the drop is attributed to electronic, not phononic [6], excitation.

221

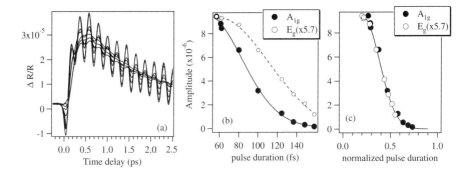

Fig. 3. (a) Reflectivity change of Sb (0001) surface at 8 K with different pulse durations from 60 fs to 160 fs. (b) The initial amplitude of coherent phonons as a function of pulse duration. (c) The initial amplitude as a function of pulse duration normalized with the phonon period.

Fig. 3 clearly shows that coherent phonons can be effectively generated only with optical pulses shorter than the half phonon period, while the dynamics of photoexcited electrons, observed as the non-oscillatory reflectivity change, is independent of the pulse duration for 60 - 160 fs. The result implies that the electron-phonon coupling contributing to the coherent phonon generation is active only in the time scale of 10 fs or shorter, and the slower electron dynamics contribute neither to DECP nor ISRS.

To summarize the experimental results, coherent E_g phonon is induced via ISRS, while the generation of coherent A_{1g} phonon is dominated by DECP. Such a drastic difference between different phonon modes is indeed expected in a highly anisotropic crystal like Sb; A7 crystalline structure is a rhombohedrally distorted variant of a cubic primitive structure, and the distortion is a Peierls distortion. The totally symmetric A_{1g} mode is the Peierls distortion mode parallel to the trigonal axis. A recent theoretical study showed that the potential energy surface along the trigonal axis is significantly affected by an electronic excitation for that reason [7].

Conclusions

Coherent phonons of different symmetries of Sb are dominated by different generation mechanisms because of the anisotropic electron-phonon coupling. Coherent A_{1g} phonon is generated primarily via DECP (but partly contributed by ISRS) because the phonon mode is associated with Peierls distortion. The generation of the coherent E_g phonon, which is perpendicular to the Peierls distortion, can be understood as purely ISRS.

1   H. J. Zeiger, J. Vidal, T. K. Cheng, E. P. Ippen, G. Dresselhaus and M S. Dresselhaus, Phys. Rev. B45, 768, 1992.
2   L. Dhar, J. A. Rogers, and K. A. Nelson, Chem. Rev. 94, 167, 1994.
3   K. Ishioka, M. Kitajima, and O. Misochko, J. Appl. Phys. 100, 093501, 2006.
4   T.E. Stevens, J. Kuhl, and R. Merlin, Phys. Rev. B 65, 144304, 2002.
5   D.M. Riffe and A.J. Sabbah, Phys. Rev. B 76, 085207, 2007.
6   D. Boschetto, E. G. Gamaly, A. V. Rode, B. Luther-Davies, D. Glijer, T. Garl, O. Albert, A. Rousse and J. Etchepare, Phys. Rev. Lett. 100, 027404, 2008.
7   E. S. Zijlstra, L. L. Tatarinova and M. E. Garcia, Phys. Rev. B 74, 220301, 2006.

Mode selective Excitation of Coherent Phonons in Bismuth by Femotosecond Pulse Pair

K. G. Nakamura[1], H. Takahashi[1], K. Ishioka[2], M. Kitajima[2], J. C. Delagnes[3], H. Katsuki[3], K. Hosaka[3], H. Chiba[3], K. Ohmori[3], K.Watanabe[4], and Y. Matsumoto[4]

[1] Materials and Structures Laboratory, Tokyo Institute of Technology, Yokohama 226- 8053, JAPAN
E-mail: nakamura@msl.titech.ac.jp
[2] Advanced Nano Characterization Center, National Institute for Materials Science, Tsukua 305- 0047, JAPAN
[3] Institute of Molecular Science, Okazaki 444-8585, JAPAN
[4] Department of Chemistry, Kyoto University, Kyoto 606-8502, JAPAN

Abstract. Coherent phonons (A_{1g} and E_g modes) of bismuth are excited by irradiation of two femtosecond pulses. Amplitude and phase are controlled by change of intervals in a time range of vibrational period and/or optical cycle.

Introduction

Impulsive excitation of materials with a femtosecond laser pulse generates coherent phonons, which all excited phonons have same phase [1,2]. Interfererene of coherent phonons of a specific mode generated at different timing by irradiation of a pair of femtosecond pulses with some delays or a pulse shaping technique has been demonstrated.

Bismuth is one of the most extensively studied materials on coherent phonons, and has two Raman-active optical phonon modes (A_{1g} and E_g modes). Both modes can be excited by femotosecond laser irradiation, but the detection of the E_g mode is difficult in ordinary (isotropic) reflectivity measurements at room temperature, because the amplitude of the A_{1g} mode is by far larger. Anisotropic reflectivity measurements using an electro-optic (EO) sampling cancels the A_{1g} contribution and enables to detect the E_g mode efficiently. Amplitude control of A_{1g}-mode coherent optical phonon has bee demonstrated by Hase *et al.* [3] However, simultaneous control of amplitude of both the modes has not been accomplished yet.

In this paper we demonstrate the mode selective excitation of optical coherent phonons in a Bi(111) single crystal by using a pair of femtosecond pulses.

Experimental Methods

We measured transient reflection change by using used a femtosecond timeresolved pump-probe-technique. The laser used was a mode-locked Ti:sapphire laser with a wavelength centered at 800 nm, providing 80 fs pulses at a repetition rate of 80 MHz. The sample used was a Bi(111) single crystal. Measurements were carried out at room temperature.

The pump beam was split into two beams by a Michelson interferometer to produce double-pulse pump-probe experiments with a separation time of Δt. In the double-pulse pump-probe experiments, the first pump pulse generates coherent phonons at $t = 0$ (phonons excited first), and the second pump pulse generates phonons at $t = \Delta t$ (phonons excited second). The separation time Δt was varied by moving a mirror of an arm in the Michelson interferometer. EO sampling was used. Power of each the pump

223

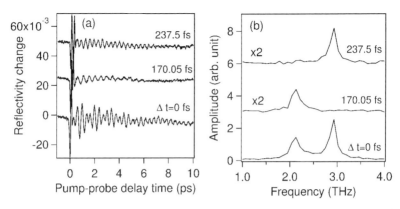

Fig. 1. Transient reflectivity change of Bi(111) by controlled emtosecond double pulse excitation (a) and its Fourier transform amplitude spectra (b).

pulse was about 15 mW. In this technique, the probe beam polarization is set perpendicular to the pump beam polarization, and the difference between orthogonal components of the reflected probe beam was measured via a polarizing beam splitter.

Results and Discussion

The reflectivity change was measured as a function of the delay time (pumpprobe delay time) by changing the optical length of the probe beam. Figure 1(a) shows the reflectivity change induced by the double pulse with $\Delta t = 0$, which corresponds to that obtained by a single pulse experiment. The reflectivity change consists of two damping oscillations. Figure 1 (b) shows a Fourier Transform (FT) amplitude spectrum of the reflectivity change (Fig. 1(a)). There are two peaks at 2.13 and 2.93 THz, which are assigned to E_g mode (TE=476 fs) and A_{1g} mode (TA=343 fs), respectively. TE and TA are vibrational periods of phonons with E_g ad A_{1g} modes, respectively. The phonon decay times, which are obtained by curve fitting of the time domain data (Fig. 1(a)) with a combination of two damping oscillations, are 2.9 and 4.6 ps for E_g and A_{1g} modes, respectively. The coherent phonons of Bi(111) at room temperature obtained in the present study are comparable to those of Bi film [3], although no coherent E_g phonons were reported in EO sampling measurement in Bi(111) at room temperature [4].

Figure 1 (a) also show the reflectivity change induced by the double pulse with $\Delta t = 170.5$ fs (TA/2) and 237.5 fs (TE/2). At double-pulse delay time of 170.5 fs, the second phonons of the A_{1g} mode destructively interfere the first phonons and the oscillation components with the A_{1g} mode disappeared. Figure 1 (b) clearly shows there is only one oscillation of the E_g mode. On the other hand, at with $\Delta t = 237.5$ fs (TE/2), the E_g-mode coherent phonon oscillation is significantly depressed. Figure 2 shows amplitude of coherent phonons (A_{1g} and E_g modes) as a function of time interval between two excited pulses. Mode selective excitation of different modes was performed. Isotropic measurements of reflectivity change were also performed at liquid helium temperature. Coherent phonons of the both modes were observed at low temperatures. Similar mode selective excitation was also observed by using a pulse-shaping technique with control of the pulse interval in time scale of optical cycle.

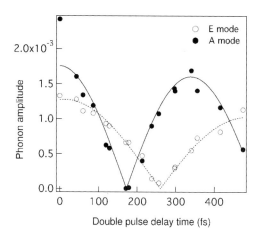

Fig. 2. Control of amplitude of coherent phonons (A_{1g} and E_g modes) of Bi

Conclusions

Femtosecond time-resolved reflectivity measurement has been performed on Bi(111) by using a pair of femtosecond laser pulses. Mode selective excitation of coherent phonons (A_{1g} and E_g modes) has been demonstrated in bismuth by controlling timing of the pulses for excitation.

Acknowledgements. KGN and HT thank Dr. Y. Hironaka, J. Saito, M. Hasegawa, and S. Ishii for their help at the early stage of this experiment. This work was financially supported by KAKENHI #18340083 and Collaborative Research Project of MSL, Tokyo Tech.

1   T. K. Cheng, S. D. Broson, A. S. Kazeroonian, J. S. Moodera, G. Dresselhaus, M. S. Dresselhaus, and E. P. Ippen, Appl. Phys. Lett., 57, 1004, 1990.
2   H. J. Zeiger, J. Vidal, T. K. Cheng, E. P. Ippen, G. Dresselhaus, and M. S. Dresselhaus, Phys. Rev. B 45, 768, 1992.
3   M. Hase, K. Mizoguchi, H. Harima, S. Nakashima, M. Tani, K. Sakai, and M. Hango, Appl. Phys. Lett., 69, 2474, 1996.
4   K. Ishioka, M. Kitajima, and O. V. Misochko, J. Appl. Phys., 100, 09350, 2006.
5   A. Q. Wu and X. Xu, Appl. Surf. Sci., 253, 6301 (2006).

Ultrafast Dynamics of Electron-Hole Plasma Coupled to Optical Phonons in a ZnO Thin Film

Hideki Ichida[1], Shuji Wakaiki[2], Kohji Mizoguchi[3], DaeGwi Kim[2], Yasuo Kanematsu[1], and Masaaki Nakayama[2]

[1] Venture Business Laboratory, Center for Advanced Science and Innovation, Osaka University, 2-1 Yamada-oka, Suita, Osaka, 565-0871, Japan
[2] Department of Applied Physics, Graduate School of Engineering, Osaka City University, 3-3-138 Sugimoto, Sumiyoshi-ku, Osaka, 556-8585, Japan
[3] Department of Physical Science, Graduate School of Science, Osaka Prefecture University, 1-1 Gakuen, Naka-ku, Osaka, 599-8531, Japan

Abstract. We have investigated ultrafast photoluminescence dynamics of electron-hole plasma coupled to longitudinal optical phonons in a ZnO thin film. The dynamical change of the electron-hole-pair density is characterized by time-resolved photoluminescence spectra measured with an optical-Kerr-gating method.

Dynamical property of interactions between photogenerated carriers and phonons is one of attractive subjects in ultrafast phenomena [1]. In semiconductors, the electron-hole plasma (EHP) which corresponds to a collective phase of ionized dense electrons and holes is formed under a high-density-excitation condition above a Mott transition density [2]. The photoluminescence (PL) dynamics of EHP has been intensively investigated from the aspect of many-body effects such as the band-gap renormalization, phase-space filling, and relaxation processes of hot carriers. Since the EHP generated by an ultrashort pulse laser is instantaneously coupled to longitudinal optical (LO) phonons, it is expected that the phonon-side band of the EHP-PL band is affected by the EHP–LO-phonon coupled mode. For example, the energy of the phonon-side band will depend on the frequency of the coupled mode that temporally changes during the PL-decay process. Although PL properties of the phonon-side band due to the EHP were reported in II-VI semiconductors such as CdS and CdSe [3,4], the dynamics has not been revealed until now. In the present work, we have focused on the time-resolved PL spectra of the phonon-side band coupled to the EHP in a ZnO thin film.

The samples used is a crystalline ZnO thin film with the thickness of 200 nm grown on a (0001) Al_2O_3 substrate at 650 °C prepared by an rf-magnetron sputtering method. The ZnO thin film was just oriented along the [0001] axis, which was confirmed from x-ray diffraction patterns [5]. Time-resolved PL spectra were measured by using the following optical-Kerr-gating method. The excitation pulse was provided by the fourth harmonic pulse (350 nm) from an optical parametric amplifier operated by a Ti:sapphire regenerative amplifier laser system with the center wavelength of 775 nm, the repetition rate of 1 kHz and the pulse width of ~200 fs. The maximum excitation density was 460 $\mu J/cm^2$. The gating pulse with the wavelength of 775 nm also delivered from a regenerative amplifier laser system was delayed by a variable optical delay line. We used an yttria-stablized zirconia as a Kerr-gating material. The time-gate width was ~0.7 ps. The time-resolved PL spectra were measured by using an intensified CCD camera system attached to a single polychromator with a spectral resolution of 0.8 nm. The sample temperature was kept at 10 K by a constant He-flow cryostat.

Figure 1 (a) shows the excitation-density dependence of the time-integrated PL spectra of the ZnO thin film at 10 K. At the relatively low excitation densities of 0.25 I_0 and 0.3 I_0, the EHP-PL band marked with the open circle was observed as a main PL band. The EHP-PL band is located at the lower energy side than the free exciton band by about 150 meV. The peak energy of the EHP–PL band shifts to the low energy side and its spectral shape broadens with an increase in excitation density. This behavior indicates that the origin of the EHP-PL band is attributed to the formation of the EHP. At the excitation density of 0.45 I_0 (180 $\mu J/cm^2$), the EHP-L_+-PL band marked with the solid circle appears at the low energy side of the EHP-PL band with the energy difference of about 80 meV, which is almost consistent with the LO-phonon energy of ZnO that is 72 meV. Thus, the EHP-L_+-PL band seems to be the phonon-side band of the EHP-PL band. However, the energy difference between the two PL bands depends on the excitation density, which is a key result in this work. Figure 1(b) shows the energy difference between the EHP-PL and EHP-L_+-PL bands as a function of photogenerated electron-hole-pair density estimated from the excitation density. It is found that the energy difference between the two PL bands becomes larger with increasing excitation density. The dashed curves indicate the calculated dispersion curve of the LO-phonon-plasmon coupled mode in ZnO as a function of electron-hole-pair density. It is obvious that the experimental results are good in agreement with the dispersion relation of the upper branch of the coupled mode. This agreement reveals that the EHP-L_+-PL band is assigned to the side band of the EHP band associated with the LO-phonon-plasmon coupled mode.

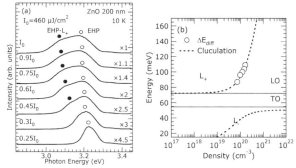

Fig.1. (a) Time-integrated PL spectra of the ZnO thin film at 10 K under various excitation densities. (b) Energy difference between EHP-PL and EHP-L_+-PL bands as a function of electron-hole-pair density estimated from the excitation density. The dashed curves indicate the calculated dispersion curve of the LO-phonon-plasmon coupled mode of ZnO, and the solid lines denote the LO- and TO-phonon energies.

Figure 2(a) shows the image plots of the time-resolved PL spectra of the ZnO thin film under the excitation density of 350 $\mu J/cm^2$ at 10 K. We clearly observed the dynamical changes of the EHP-PL and EHP-L_+-PL bands. In Fig. 2(a), the dashed and dotted lines indicate the temporal changes of the peak energies of the EHP-PL and EHP-L_+-PL bands, respectively. The difference of the peak energies between two PL bands changes with time delay. In Fig. 2(b), the closed circles indicate the temporal trace of the energy difference between the two PL bands obtained from the time-resolved PL spectra. The energy difference between the two PL bands becomes smaller with increasing time delay, and then eventually reaches to the LO-phonon energy of ZnO. This behavior reflects a decrease of the photogenerated electron-hole-pair density of the EHP with an increase in time delay. From the energy difference between the EHP-PL and EHP-L$_+$-PL bands, we can estimate the temporal change of the electron-hole-pair density from the dispersion relation of the LO-phonon-plasmon coupled mode shown in Fig. 1(b). The open circles shown in Fig. 2(b) denote the estimated electron-hole-pair density as a function of time delay. It is found that the electron-hole-pair density quickly decreases with the decay time of 1.2 ps, which is estimated from the single-exponential fitting shown by the dashed line in Fig. 2(b). The decay time of the photogenerated electron-hole-pair density is almost equal to that measured by an optical-pump-THz probe method in a ZnO thin film [6]. This result demonstrates that the dynamical change of the photogenerated electron-hole-pair density of the EHP and its decay time can be probed by the time-resolved PL spectra of the PL side band of the EHP due to the LO-phonon-plasmon coupled mode.

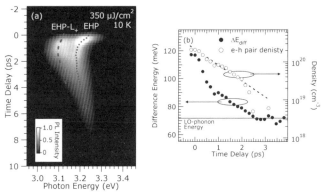

Fig. 2. (a) Image plots of the time-resolved PL spectra of the ZnO thin film under the excitation density of 350 μJ/cm^2 at 10 K. The dotted and dashed lines indicate the peak energies of the EHP-PL and EHP-L$_+$-PL bands, respectively. (b) Energy difference between the EHP-PL and EHP-L$_+$-PL bands as a function of time delay (solid circles). The solid line denotes the LO-phonon energy of ZnO. The open circles indicate the electron-hole-pair density estimated from the dispersion relation of the LO-phonon-plasmon coupled mode of ZnO shown in Fig. 1(b). The dashed curve indicates a single exponential function.

Acknowledgements. This research was supported by the Grant-in-Aid for Creative Scientific Research (No. 17GS1204) from Japan Society for the Promotion of Science and Grant-in-Aid for Young Scientists (B) (No. 20760015) from the Ministry of Education, Culture, Sports, Science and Technology.

1. J. Shah, *Ultrafast Spectroscopy and Semiconductors and Semiconductor Nanostructures*, 2nd. ed., (Springer, Berlin, 1999).
2. C. F. Klingshirn, *Semiconductor Optics*, 3rd. ed., (Springer, Berlin, 2007).
3. H. Saito and E. O. Göbel, Phys. Rev. B **31**, 2360 (1985).
4. M. S. Brodin, N. V. Volovik, and M. I. Strashinkova, JETP Lett. 23, 227 (1976).
5. D. Kim, T. Shimomura, S. Wakaiki, T. Terashita and M. Nakayama, Physica B **376**, 741 (2006).
6. E. Hendry, M. Koeberg, and M. Bonn, Phys. Rev. B **76**, 045214 (2007).

Large-amplitude coherent phonons in semimetals

Oleg V. Misochko[1], Michael V. Lebedev[1], Kunie Ishioka[2], Masahiro Kitajima[2], Sergey V. Chekalin[3], Thomas Dekorsy[4]

[1]Institute of Solid State Physics, Russian Academy of Sciences, 142432 Chernogolovka, Russia
 E-mail: misochko@issp.ac.ru
[2]National Institute for Materials Science, 1-2-1 Sengen, Tsukuba, 305-0047 Ibaraki, Japan
[3]Institute of Spectroscopy, Russian Academy of Science, 142190 Troitsk, Russia
[4]Physics Department, Konstanz University, 78457 Konstanz, Germany

Abstract. We report on the time-resolved dynamics of two, different in symmetry, large-amplitude coherent phonons in Bi and Sb. A systematic study was made of the variation of the nonlinear ultrafast lattice dynamics with pulse duration, excitation strength, temperature, and probe wavelength.

Introduction

There have been extensive experimental studies of the ultrafast dynamics of crystal lattice by making use of coherence of phonons [1,2]. At low excitation strength, when the oscillation amplitude is small, the vibrational properties are considered as intrinsic properties of the crystal [1,3]. Such simple and relatively well understood picture changes radically as one enters the non-linear regime, where the intense laser pulses produce the high density of phonons and free carriers thereby modifying the phonon-phonon and electron-phonon coupling. At this high excitation level, the interatomic forces that bind solids and determine many of their properties are substantially altered due to large-amplitude lattice excitation resulting in a variety of novel effects [4-8].

Semimetals are the prototype materials in which coherent optical phonons are generated upon femtosecond laser excitation. Both Bi and Sb have a rhombohedral crystal structure with the unit cell containing two atoms. The group theory predicts for Γ-point a fully symmetric A_{1g} and a doubly degenerate E_g mode. The atoms are displaced along the trigonal axis in the A_{1g} mode and perpendicular to that axis in the E_g mode. In these crystals, femtosecond excitation drives both the fully symmetric A_{1g} and doubly degenerate E_g coherent phonons. The vibrational excitation in the case of A_{1g} phonons is generally believed to be displacive [1,3]: the population redistribution of electrons alters the potential energy surface of the lattice and gives rise to a restoring force that drives coherent atomic motion. The dynamics of this mode are determined by the curvature and minima location of the altered lattice potential. In contrast, the excitation of E_g phonons is thought to be impulsive [1,3]: owing to the large spectral bandwidth available in a short laser pulse, it can generate the phonons with the two, different in frequency, electromagnetic fields acting as a driving force.

Experimental Methods

In our study we utilized conventional pump-probe technique to perform time-resolved measurements of transient reflectivity changes. The detected signal was the change in intensity of the reflected probe beam as a function time delay t after the arrival of the pump pulse. The light sources were Ti-sapphire regenerative amplifiers producing 50 and 130 fs transform-limited pulses at a repetition rate of 1, 100 and 250 kHz. The laser wavelength was 800 nm and the pump pulse was tailored with a pulse shaper comprising a spatial light modulator. Time delay between the pump and probe pulses was scanned by either a translational stage (slow scan), or a retroreflector mounted on the shaker (fast scan). Employing isotropic detection scheme we measured the phonons having non-zero diagonal matrix elements, whereas anisotropic detection allowed discriminating between the fully symmetric phonons and the non-fully symmetric phonons with zero trace.

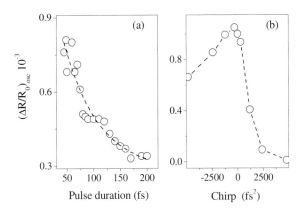

Fig. 1. Pulse duration (a) and chirp (b) dependence for A_{1g} coherent phonons in Bi at T=5K.

Results and Discussion

We measured how the large-amplitude phonons in Bi and Sb depend on pumping pulse duration, pulse chirp, pulse energy and temperature. In addition, we disperse the probe beam to observe the large amplitude vibrations at different wavelengths. First, we stretched the pulse duration to see how coherent amplitude depends on the duration of the pump pulse. Independent of the phonon symmetry, the coherent amplitude decreases exponentially with a characteristic time of T/4 where T is the vibrational mode period as shown in Fig.1 (a) for fully symmetric phonon in Bi. Next, we introduced chirp into the pump pulse. We observed that at the high excitation there is a dependence on the chirp sign in Bi: the coherent oscillations disappear for a larger chirp with a rate that depends on the chirp sign: a single intense pulse with a negative chirp leads to considerably more coherent fully symmetric motion than observed for positively chirped pulse as shown in Fig.1 (b). What is even more interesting at small chirps the coherent amplitude is appeared to be larger for negatively chirped pulse than for transform limited one.

We observed that in Bi at higher excitation strength, the oscillation amplitude grows and then saturates (there is no such saturation in Sb), and in both crystals the oscillation frequency softens becoming more chirped and the oscillation lifetime significantly decreases for each symmetry mode. Also in both crystals, the coherent amplitude for doubly degenerate E_g mode grows faster than for A_{1g} mode when temperature is lowered, while the lifetime of coherent phonons becomes larger. At the excitation level slightly lower the Lindemann stability limit, the oscillation pattern for each symmetry mode (A_{1g} and E_g) in both crystals changes dramatically. For short time delays, the oscillations die out, however, at some time later the oscillations revive. We studied how the collapse and revival [7,8] parameters depend on the exciting pulse characteristics and observed that the collapse time is independent on the pulse chirp and the pulse energy, at least for Bi. On the other hand, we found that the collapse time in Bi is strongly dependent on the size of the exciting laser spot. We detected that the ooscillations at the centre of the pumped area are more chirped than those at the edges. This variation of the phonon chirp results in a π-shift for a certain delay, and therefore in the appearance of collapse-revival pattern. Thus, the collapse and revival pattern observed near the Lindemann stability limit is most likely caused by a polarization beating (not by a quantum beating) and it is a measure of the spatial coherence.

Finally, by analyzing the probe wavelength dependence of the oscillations we showed that the amplitude of the oscillations depends on, whereas the frequency and lifetime are independent of the probe wavelength. The spectral dependence of coherent amplitude was observed to be essentially similar to that of the electronic part at a small time delay.

Conclusions

The large-amplitude phonons in semimetals (as well as small-amplitude ones) are excited primarily impulsively for low symmetry mode and displacively for high-symmetry mode. Their amplitudes decrease exponentially with increasing pulse duration and increase with higher excitation strength. For the fully symmetric phonons in Bi there is saturation in the pump dependence, which is absent for their counterpart in Sb demonstrating a superlinear behavior. In bismuth, the negatively chirped pulses are more effective in creating the large-amplitude coherent phonons than the positively chirped ones. The collapse and revival pattern observed near the Lindemann stability limit is most likely caused by a polarization beating, and not by a quantum beating as it thought before [7,8]. The observed spectral dependence of the coherent amplitude can be attributed to the different electron-phonon coupling for various groups of probed charge carriers.

Acknowledgements. This work was supported by the Russian Foundation for Basic Research (06-02-16186 and 07-02-00148) and the Deutsche Forschungsgemeinschaft (DE567/9).

1 R. Merlin, Solid State Commun. **102**, 207, 1997.

2 T. Dekorsy, G.C.Cho, and H Kurz in Light Scattering in Solids VIII, eds. M. Cardona and G. Güntherodt, Springer, Berlin, 2000.

3 K. Ishioka, M. Kitajima, and O.V. Misochko, J. Appl. Phys., **100**, 093501, 2006.

4 M. F. DeCamp, D. A. Reis, P. H. Bucksbaum, and R. Merlin, Phys. Rev. **B 64**, 092301 (2001).

5 M. Hase, M. Kitajima, S. Nakashima, and K. Mizoguchi,, Phys. Rev. Lett. **88**, 067401, 2002; ibid. **93**, 109702, 2004.

6 S. Fahy and D. A. Reis, Phys. Rev. Lett. **93**, 109701, 2004.

7 O.V. Misochko, M. Hase, K. Ishioka, and M. Kitajima, Phys. Rev. Lett. **92**, 197401, 2004.

8 O.V. Misochko, K. Ishioka, M.Hase, and M. Kitajima, J. Phys.: Condens. Matter **18**, 10571, 2006.

Laser-Induced Undoing of a Peierls Distortion

Eeuwe S. Zijlstra, Nils Huntemann, and Martin E. Garcia

Theoretische Physik, Universität Kassel, Heinrich-Plett-Str. 40, 34132 Kassel, Germany
E-mail: zijlstra@physik.uni-kassel.de

Abstract. On the basis of *ab initio* calculations we predict that in Arsenic under pressure a solid-solid phase transition from the A7 structure, which is stabilized by the Peierls mechanism, into the simple cubic structure, which has no Peierls distortion, can be induced by an ultrashort laser pulse. A comparison between the local density approximation and the generalized gradient approximation shows that our prediction does not depend on the approximation used.

Introduction

In Arsenic a series of pressure-induced phase transitions exists. At ambient conditions As crystallizes in the A7 structure. At 25 GPa there is a transition to the simple cubic (sc) phase [1], followed by transitions [2] to the so-called As(III) structure at 48 GPa and to body-centered cubic As at 97 GPa. The fact that the A7 structure is electronically stabilized by the Peierls mechanism makes it a candidate for an ultrafast laser-induced solid-solid phase transition. This is of fundamental interest, because the majority of the experimentally studied laser-induced phase transitions shows ultrafast melting. Examples of laser-induced solid-solid phase transitions include a metallic-to-semiconductor phase transition in SmS [3] and a monoclinic-to-rutile transition in VO_2 [4]. The main goals of this paper are (i) to analyze the possibility to induce a solid-solid phase transition by an ultrafast laser pulse in As under pressure using all-electron density functional theory calculations in the local density approximation and (ii) to compare our results with previously published data [5], which we have obtained using the generalized gradient approximation.

Structure of Arsenic

First we describe the relationship between the A7 and sc structures, the two phases of As between which we will predict a laser-induced transition. Starting from an sc atomic packing, the A7 structure can be derived in two steps:

1. The sc lattice is deformed by elongating it along one of the body diagonals. The magnitude of this deformation is usually expressed by c/a, where c is twice the length of the above-mentioned body diagonal (In the A7 structure the unit cell is doubled in this direction) and a is the length of the shortest lattice vector connecting atoms in planes perpendicular to this body diagonal. In the sc lattice $c/a = \sqrt{6} \approx 2.45$. In As at ambient conditions $c/a = 2.86$.

2. A Peierls instability causes the atomic planes perpendicular to the elongated diagonal to be displaced along this diagonal, in alternating directions. The magnitude of these displacements is expressed by z: A value of 0.25 indicates no Peierls distortion. Deviations from $z = 0.25$ give the magnitude of the displacement of the atoms in units of c. In the sc lattice $z = 0.25$ and in As at ambient conditions $z = 0.228$.

Method

We computed total energies of As in the sc and A7 structures as a function of the atomic volume using the all-electron density functional theory program WIEN2k [6]. For the sc structure this is a straightforward task. For the A7 structure this involved the optimization of c/a and z for each volume. All computational details are the same as in [5]. The only difference is that in the present work the exchange and correlation potential was approximated by the local density approximation [7], whereas in [5] we have used the generalized gradient approximation.

The effect of the excitation by an ultrashort laser pulse was simulated by heating the electrons to 3000 K. In this case, the phase stability is governed by the free energy $F = E - T_e S_e$, where T_e is the electronic temperature and S_e is the electronic entropy.

Effect of Pressure

We fitted the computed energies to analytical functions. Because the fitting parameters summarize the new data presented in this paper, they are given in Table 1. For an explanation of our fitting procedure we refer the interested reader to [5].

Table 1. Fitting parameters for the electronic ground state ($T_e = 158$ K) and the laser-excited state ($T_e = 3000$ K). For a description of these parameters we refer the interested reader to [5].

T_e (K)	V_0 (a_0^3)	B_0 (GPa)	B_0'	V_c (a_0^3)	β_c	A_c (Ry)
158	129.068	92.284	4.285	109.58	1.410	−0.04658
3000	129.690	90.449	4.316	113.63	1.601	−0.05732

From the analytical dependence $E(V)$, the pressure $P = -\left(\frac{\partial E}{\partial V}\right)$ and the enthalpy $H = E + PV$ were calculated. The difference between the enthalpies of the A7 and sc structures is plotted in Figure 1. It can be seen that a pressure-induced phase transition takes place at 21.7 GPa, which is lower than the experimental transition pressure of 25 ± 1 GPa [1] and the generalized gradient approximation value of 26.3 GPa [5].

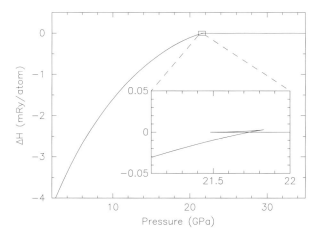

Fig. 1. Enthalpy difference $\Delta H = H_{A7} - H_{sc}$ as a function of pressure. The inset shows the indicated region enlarged. The transition from the A7 to the sc structure takes place at 21.7 GPa.

Laser-Induced Phase Transition

For the subsequent analysis we assumed that the laser pulse width is much shorter than the timescale of ionic motion (~ 130 fs). Therefore the system does not have time to expand, so that structural changes take place at constant volume. In Figure 2 we plotted the free energy differences between the A7 and sc structure before and after laser excitation. An arrow indicates the possibility to induce the A7 to sc transition in As by a laser pulse. At this specific volume, the applied pressure before laser excitation is 3.6 GPa lower than the ordinary transition pressure (21.7 GPa) and the energy absorbed from the laser pulse is 3.4 mRy/atom, which corresponds to a laser fluence of 2.3 mJ/cm^2 when the laser light has a wavelength of 800 nm [5].

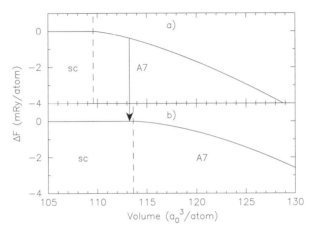

Fig. 2. Free energy difference $\Delta F = F_{A7} - F_{sc}$ as a function of volume a) before and b) after laser excitation. The arrow indicates a laser-induced phase transition from the A7 into the sc structure.

Conclusion

Using the local density approximation [7] we demonstrated that it is possible to induce a solid-solid phase transition in As under pressure with an ultrashort laser pulse. Our results are in qualitative agreement with earlier density functional theory results [5], which we have obtained using the generalized gradient approximation.

Acknowledgements. This work has been supported by the Deutsche Forschungsgemeinschaft (DFG) through the priority program SPP 1134 and by the European Community Research Training Network FLASH (MRTN-CT-2003- 503641).

1 H. J. Beister, K. Strössner, and K. Syassen, Phys. Rev. B 41, 5535, 1990.

2 R. G. Greene, H. Luo, and A. L. Ruoff, Phys. Rev. B 51, 597, 1995.

3 R. Kitagawa, H. Takebe, and K. Morinaga, Appl. Phys. Lett. 82, 3641, 2003.

4 A. Cavalleri et al., Phys. Rev. Lett. 87, 237401, 2001.

5 E. S. Zijlstra, N. Huntemann, and M. E. Garcia, New J. Phys. 10, 033010 (2008).

6 P. Blaha, K. Schwarz, G. K. H. Madsen, D. Kvasnicka, and J. Luitz, WIEN2k, An Augmented Plane Wave + Local Orbitals Program for Calculating Crystal Properties, Karlheinz Schwarz, Techn. Universität Wien, Austria, 2001.

7 J. P. Perdew and Y. Wang, Phys. Rev. B 45, 13244, 1992.

Ultrafast dynamics of coherent optical phonons in α-quartz

K. von Volkmann, T. Kampfrath, M. Krenz, M. Wolf, and C. Frischkorn

Freie Universität Berlin, Fachbereich Physik, Arnimallee 14, 14195 Berlin, Germany
E-mail: christian.frischkorn@physik.fu-berlin.de

Abstract. Femtosecond laser excitation of α-quartz launches coherent optical phonons modulating the refractive index of the sample. The observed oscillations in the transmission and ellipticity of probe light decays due to phonon-phonon scattering. With decreasing temperature, the vibrations shift towards higher energies and are accompanied by a rise of the phonon lifetime caused by lattice stiffening and freezing of phonon modes, respectively.

Introduction

Coherent lattice vibrations induced by intense ultrashort laser pulses have been observed in various insulators, semiconductors and metals [1]. The generation mechanism of these phonons has been frequently described in terms of impulsive stimulated Raman scattering (ISRS). Here, we use transient transmission and ellipticity measurements to find rather strong optical phonon modes at THz frequencies with transmission modulations $\Delta T/T$ of up to 20 %. The observed signal amplitude crucially depends on the symmetry of the respective phonon mode. α-quartz may serve as an ideal candidate for excitation with pulse sequences and arbitrarily shaped pump pulses exploring optical control scenarios of coherent lattice motion [2].

Experiment

Our experimental setup is based on a Ti:sapphire amplifier system which generates 20-fs, 1-mJ, 800-nm pump pulses at a repetition rate of 1 kHz. The probe pulses (12 fs, 1 nJ, 800 nm, 80 MHz) are taken directly from the seed oscillator [3,4]. In transient transmission experiments, the transmitted intensity of the time-delayed probe pulse is monitored, while in the ellipticity setup, the transmitted probe light is sent through a quarter-wave plate, then split (by a Wollaston prism) into two paths with orthogonal polarizations and detected by a pair of balanced photodiodes. Unlike in the case of a conventional probe delay line based on a computer-controlled linear stage, the pump-probe delay is varied here using a shaker setup, which significantly increases the duty cycle, but also imposes to account for non-equidistant time steps. Details on the data acquisition and evaluation in the latter setup will be presented elsewhere [5]. The quartz sample [$SiO_2(0001)$, 1 mm thickness] is mounted in a He-cooled cryostat for temperature dependent measurements from a few Kelvin to room temperature.

Results and Discussion

Figure 1 shows the pronounced oscillatory behavior in the transmitted intensity and ellipticity of the probe light with obvious differences between both probe methods. Altogether and in accordance with Raman spectra [6], resonance frequencies are

235

obtained at 3.8, 6.2, 10.6, 12.1, and 13.9 THz whereby the phonon amplitude varies strongly with the probe method. In particular, the phonon modes at 6.2 and 10.6 THz are only observed in transient transmission, while only ellipticity measurements show the resonance at 12.1 THz. All modes have in common that their amplitudes increase linearly with pump fluence. Together with sine-like oscillations rather than cosine-like, this is consistent with the ISRS excitation mechanism of the coherent lattice motion. The alternative scenario based on displacive excitation of coherent phonons (DECP) would involve a resonant electronic excitation, which is rather unlikely for α-quartz with an indirect band gap of 8.9 eV [7].

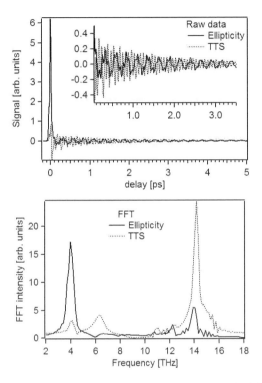

Fig. 1. (top) Oscillations in the transmitted intensity (TTS) and ellipticity of the probe pulse after fs-laser excitation of the quartz sample. The first oscillation after excitation around 0 ps originates from a coherent nonlinear effect of third order (stimulated Rayleigh-wing scattering [8].) (bottom) Fourier transforms of the time traces for both probe methods.

Furthermore, it is noteworthy that each of the observed phonon frequencies remains constant over the entire range of pump-probe delays and independent of the pump fluence. This indicates that potential anharmonic effects of the coherent lattice motion upon ISRS in the electronic ground state do not contribute to a measurable extent.

The decay mechanism of the excited phonon modes has been addressed with pump-fluence and temperature dependent measurements. The decay rate $\Gamma = 1/\tau$ is found to be fluence independent ruling out electron-hole (el-h) pair excitations to dominate the decay channels. Alternative mechanisms and consistent with this finding are phonon-phonon (ph-ph) interaction and scattering with sample inhomogenities. Yet ph-ph scattering should exhibit a clear temperature dependence of the phonon lifetime τ with a potential freezing-out of modes below a certain (and mode-specific) threshold. Figure 2 displays τ for the most dominant phonon modes at 6.2 and 13.9 THz as a function of temperature illustrating a pronounced increase of the lifetime with decreasing temperature. Below 50 K, the lifetime levels off which suggests the freezing of lattice motion at these low temperatures. In addition, the phonon frequencies are found to blue-shift by up to 10 % with respect to the room temperature value as the sample temperature decreases to the few Kelvin range [5]. A lattice stiffening with decreasing temperature is held responsible for this observation.

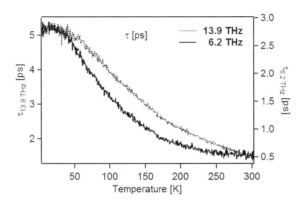

Fig. 2. Temperature dependence of the phonon lifetime τ exemplarily shown for the two modes at 6.2 and 13.9 THz. Note the leveling-off below 50 K.

Conclusions

In summary, we have investigated the ultrafast dynamics of coherent lattice vibration in α-quartz using transient transmission and ellipticity measurements. The excitation mechanism via impulsive stimulated Raman scattering consistently describes the experimental findings. Temperature dependent measurements yield valuable information on the decay mechanism of the coherent phonon modes.

Acknowledgements. Financial support from the Deutsche Forschungsgemeinschaft through Sfb450 is gratefully acknowledged.

1   R. Merlin, Solid State Comm. **102**, 207 (1997).

2   M. M. Wefers, H. Kawashima, and K. A. Nelson, J. Chem. Phys. **108**, 10248 (1998).

3   K. Reimann, R. P. Smith, A. M. Weiner, T. Elsaesser, and M. Woerner, Opt. Lett. **28**, 471 (2003).

4   T. Kampfrath, D. O. Gericke, L. Perfetti, P. Tegeder, M. Wolf, and C. Frischkorn, Phys. Rev. E **76**, 066401 (2007).

5   K. von Volkmann, T. Kampfrath, M. Krenz, A. Grujic, C. Frischkorn, and M. Wolf, (in preparation).

6   J. F. Scott and S. P. S. Porto, Phys. Rev. **161**, 903 (1967).

7   J. R. Chelikowsky and M. Schlüter, Phys. Rev. B **15**, 4020 (1977).

8   A. Dogariu. T. Xia, D. J. Hagan, A. A. Said, E. W. Van Stryland, and N. Bloembergen, J. Opt. Soc. Am. B **14**, 796 (1997).

Influence of Lattice Heating Time on Strain Wave Dynamics in InSb

Faton S. Krasniqi, Steven L. Johnson, Paul Beaud, Maik Kaiser, Daniel Grolimund, and Gerhard Ingold

Swiss Light Source, Paul Scherrer Institute, CH-5232 Villigen PSI, Switzerland
E-mail: faton.krasniqi@psi.ch

Abstract. Time resolved x-ray diffraction with sub-picosecond time resolution is used to investigate the fluence dependence of the lattice heating in InSb.

Introduction

The heating of a crystal with a sub-picosecond laser pulse with pulse energies just below the damage threshold can induce strain waves that propagate with the speed of sound [1,2]. These transient coherent lattice dynamics have been studied for many years in a variety of materials by observing the resultant changes in the optical properties (reflectivity) [1,3,4] and by x-ray diffraction [2,5-7]. Generally, these kind of studies are useful in deducing the acoustic properties of the excited crystal and testing theories of electron-phonon coupling [3,5].

A simple model that describes the generation of a strain wave in a laser excited solid was proposed by Thomsen *et al.* [1]. They solved the elastic equations assuming that thermal stress is instantaneously generated in an absorbing solid. The general validity of the model has been verified by both optical scattering and x-ray diffraction techniques [1-7]. In principle one may expect that the model predicts reasonably the structure of the strain wave after the carrier-lattice thermalization. During thermalization, on the other hand, one expects that the instantaneous heating assumption may overestimate the strain. In polar semiconductors such as InSb, the excess energy of photo-excited carriers (out of equilibrium with the lattice) is first transferred to small momentum longitudinal optical (LO) phonons that subsequently decay into acoustic phonons due to the anharmonicity of the crystal potential. The intrinsic lifetime of LO phonons is expected to govern the carrier-lattice thermalization dynamics and thus the strain evolution [8].

The goal of this study is the investigation of the strain wave dynamics in InSb during the lattice heating time, the time needed for the excitation energy to be transferred to the lattice.

Experimental Methods

To investigate the fluence dependence of the lattice heating time in laser excited InSb we have employed time resolved x-ray diffraction using the femtosecond 'slicing' source at the SLS. The FEMTO slicing source delivered temporally and spatially stable x-ray pulses at 1 kHz, tunable in energy from 4 – 12 keV with a duration of 140 ± 30 fs [9]. The operating energy of 5.9 keV was selected by a double multilayer monochromator with a bandwidth $\Delta E/E$=1.2%. The x-ray diffraction measurements were performed on an asymmetrically cut InSb single crystal (500 μm thick) in order to match the penetration depths of the laser and the x-rays. The x-rays were diffracted from (111) planes and the crystal was cut such that the asymmetry angle was 15.5°. In the experiment, we have measured the evolution of x-ray diffraction intensity as a function of laser fluence, keeping the incident angle fixed to about +0.06° from the Bragg peak.

Results and Discussion

The measured time resolved x-ray diffracted intensities indicate that at low excitation fluences, during the first 15 ps following laser excitation, the rate of decrease of the diffracted signal is slower than at higher fluences. To understand this observation, a model for laser excited strain waves has been developed assuming that the energy transfer from photo-excited carriers to the lattice is mediated mainly by LO phonons [10]. This model is an extension of the Thomsen model producing Thomsen-like strain profiles in the limit of instantaneous heating. The lattice heating time is largely dependent on the decay time of LO phonons to other lattice modes. Figure 1a compares the strain waves using heating times of 50 fs and 8 ps with that predicted by the Thomsen model [1]. It is obvious that the inclusion of the heating time in the strain wave smoothes the boundary between the expansion and compression regions.

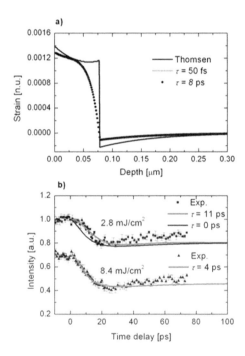

Fig. 1 a) Comparison of the laser-generated strain wave 20 ps after laser excitation using a heating time of $\tau = 50$ fs and 8 ps with a Thomsen strain profile, for InSb at laser fluence of 3mJ/cm^2. b) Comparison between the measured time dependent diffracted intensities and simulations which assume a strain history that depends on the lattice heating time.

X-ray diffraction in the presence of laser induced strain is calculated using the Takagi-Taupin dynamical theory for depth dependent strain gradients [11]. To take into account the bandwidth of the multilayer monochromator, the simulated rocking curves are convolved with a Gaussian function corresponding to $\Delta E/E \approx 1.1\%$.

In Figure 1b we show the comparison of the time evolution of the measured diffracted intensities with those simulated for two laser fluences, 2.8 mJ/cm^2 and 8.4 mJ/cm^2. The simulations assume a strain history that depends on the lattice heating time. A good agreement is found at the lowest fluence by assuming a heating time of about 11±4 ps, whereas for the fluence of 8.4 mJ/cm^2 good agreement is found for a heating time of about 4±1.5 ps. The discrepancy between the measurements and simulations (<4%) for time delays larger than 30 ps is due to the neglect of heat diffusion in the strain wave profiles. In the framework of our model these observations can be attributed to the phonon dynamics that occurs following the carrier energy relaxation. Since the lattice heating time is largely dependent on the optical phonon decay time, these results suggest that the lattice heating time is decreased with increasing excitation energy. This implies that the phonon decay time decreases with increasing excitation energy. This conclusion is supported by time resolved Raman studies of phonon lifetimes in polar semiconductors such as GaN [12], where it was found that the LO phonon decay time decreases with increasing laser fluence.

Conclusions

In conclusion, we have studied femtosecond laser-induced strain wave dynamics by using time resolved x-ray diffraction. In the framework of a model for laser-induced strain waves which assumes an energy transfer from photo-excited carriers to the lattice mediated mainly by LO phonons, our results indicate that the lattice heating time decreases with increasing excitation energy. This implies that the lifetime of LO phonons decreases with increasing excitation energy, similar to previous observations in other polar semiconductors [12].

1   C. Thomsen, H. T. Grahn, H. J. Maris and J. Tauc, Phys. Rev. B **34**, 4129, 1986.

2   Ch. Rose-Petruck, R. Jimenez, T. Guo, A. Cavalleri, C. W. Siders, F. Ráksi, J. A. Squier, B. C. Walker, K. R. Wilson, C. P. J. Barty, Nature 398, **310**, 1999.

3   G. Tas and H. J. Maris, Phys. Rev. B **49**, 15046, 1994.

4   O. B. Wright, Phys. Rev. B **49**, 9985, 1994.

5   A. M. Lindenberg, I. Kang, S. L. Johnson, T. Missalla, P. A. Heinmann, Z. Chang, J. Larsson, P. H. Bucksbaum, H. C. Kapteyn, H. A. Padmore, R. W. Lee, J. S. Wark and R. W. Falcone, Phys. Rev. Lett. **84**, 111, 2000.

6   D. A. Reis, M. F. DeCamp, P. H. Bucksbaum, R. Clarke, E. Dufresne, M. Hertlein, R. Merlin, R. Falcone, H. Kapteyn, M. M. Murnane, J. Larsson, Th. Missalla and J. S. Wark, Phys. Rev. Lett. **86**, 3072, 2001.

7   M. F. DeCamp, D. A. Reis, D. M. Fritz, P. H. Bucksbaum, E. M. Dufresne and R. Clarke, J. Synchrotron Rad. **12**, 177, 2005.

8   J. Shah, Ultrafast Spectroscopyof Semiconductors and Semiconductor Nanostructures, Springer, Berlin Heidelberg, 1996.

9   P. Beaud, S.L. Johnson, A. Streun, R. Abela, D. Abramsohn, D. Grolimund, F. Krasniqi, T. Schmidt, V. Schlott, and G. Ingold, Phys. Rev. Lett. **99**, 174801, 2007.

10 F. S. Krasniqi, S. L. Johnson, P. Beaud, M. Kaiser, D. Grolimund, and G. Ingold., submitted to Phys. Rev. B.

11 J. Gronkowski, Phys. Rep. 206, **1**, 1991.

12 K. T. Tsen, J. G. Kiang, D. K. Ferry and H. Morkoc, Appl. Phys. Lett. **89**, 112111, 2006.

Soft X-Ray Thomson Scattering in Warm Dense Matter at FLASH

R.R. Fäustlin[1], S. Toleikis[1], Th. Bornath[2], L. Cao[3], T. Döppner[4], S. Düsterer[1], E. Förster[3], C. Fortmann[2], S.H. Glenzer[4], S. Göde[2], G. Gregori[5], A. Höll[2], R. Irsig[2], T. Laarmann[1], H.J. Lee[6], K.-H. Meiwes-Broer[2], A. Przystawik[2], P. Radcliffe[1], R. Redmer[2], H. Reinholz[2], G. Röpke[2], R. Thiele[2], J. Tiggesbäumker[2], N.X. Truong[2], I. Uschmann[3], U. Zastrau[3], Th. Tschentscher[1]

[1] DESY, Notkestr. 85, 22607 Hamburg, Germany
  E-mail: roland.faeustlin@desy.de
[2] Universität Rostock, Universitätsplatz 3, 18051 Rostock, Germany
[3] Friedrich-Schiller-Universität Jena, Max-Wien-Platz 1, 07743 Jena, Germany
[4] LLNL, 7000 East Av., Livermore, CA 94550, USA
[5] University of Oxford, Parks Road, Oxford OX1 3PU, United Kingdom
[6] University of California, Berkeley, CA 94720, USA

Abstract. We present the attempt to diagnose electron temperature and density of a plasma via Thomson Scattering in the Warm Dense Matter Regime using soft x-ray Free Electron Laser radiation. A preliminary Self Thomson Scattering experiment has already been conducted. In a current pump-probe experiment, together with an optical heating laser, we will record the temporal evolution of the plasma achieving a resolution of approximately 250fs.

The Free Electron Laser in Hamburg (FLASH)

The advent of ultrafast accelerator based radiation sources opens new possibilities for research in the fields of physics, chemistry and biology. One most prominent of such sources is the Free-Electron-Laser (FEL) facility FLASH (Free Electron Laser Hamburg) [1-2]. FLASH is currently the world's only self-amplified spontaneous emission (SASE) FEL [3] providing users with photons ranging from 6.5 to 47 nm wavelength. These photons are packed into pulses with 10 to 50 fs duration and up to 100 µJ energy. Multiple pulses can be combined with 1 µs spacing to form pulse trains repeating every 0.2 s. FLASH reaches a peak brilliance of 10^{29}-10^{30} [photons/(s mrad^2 mm^2 0.1% BW)].

Thomson Scattering in Warm Dense Matter

We utilize FLASH to implement Thomson scattering in near-solid density plasmas as a diagnostic which allows the determination of temperature and density in warm dense matter (WDM) plasmas (free electron density of $n_e=10^{21}$-10^{26} cm^{-3} with temperatures of several eV). The WDM regime [4] at near-solid density ($n_e=10^{21}$-10^{22} cm^{-3}) is of special interest, because here the transition from an ideal plasma to a degenerate, strongly coupled plasma occurs. A systematic understanding of this largely unknown WDM domain is crucial for the modeling and understanding of contemporary plasma experiments, such as laser shock-wave or Z-pinch as well as for inertial confinement fusion (ICF) where the plasma evolution crosses this regime. In order to study basic plasma properties from scattered spectra, it is necessary that the radiation source penetrates dense plasmas, i.e. the frequency of the probing light has to be larger than the density dependent electronic plasma frequency. Available

241

sources to probe such plasmas are backlighter systems [5] in the x-ray regime as well as short wavelength FEL radiation. Experiments with x-rays in solid density plasmas have been successfully carried out investigating non-collective Thomson scattering [6] from the thermal electron distribution as well as collective Thomson scattering [7] from the collective electron mode (plasmon). Novel FEL sources, such as FLASH [8], are suited perfectly for Thomson scattering experiments, particularly due to their high photon energy, narrow bandwidth, high repetition rate and high intensity.

This experiment's goal is to resolve the collective plasmon resonances in the scattered spectrum and consequently determine electron density and temperature of a near solid density plasma. The scattered spectrum is described by the structure factor $S(k, \omega)$ (figure 1b). From the measured spectrum the electron temperature T_e (detailed balance in the intensity asymmetry of the up and down shifted plasmons) and the electron density n_e (frequency shift ω_{res} of the plasmons) can be determined [8]. In order to obtain a strong signal from plasmon scattering but not from bound electrons a dense, low Z target material is used: a hydrogen pellet source [9].

Experimental Setup

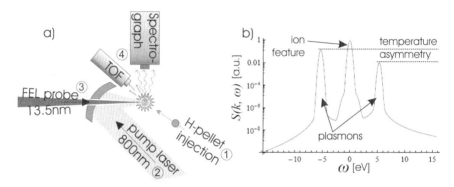

Fig. 1. a) Cartoon illustrating the Thomson Scattering pump probe experiment; b) Schematic of a Thomson Scattering Spectrum.

The experimental setup at FLASH is depicted in figure 1a. The FEL beam with a wavelength of 13.5 nm is focused over 2 m down to a focal spot size of 20-30 µm FWHM. At this point the beam hits the horizontally injected hydrogen pellet which has been transformed into a WDM state by the pump laser (800nm, 100 fs, 3×10^{15} W/cm²). A transmission [10] or reflection grating spectrograph, optimized for efficiency, detects the light scattered perpendicularly to the FEL propagation axis. The time delay between optical laser and FEL can be adjusted to record the temporal evolution of the hydrogen plasma's density and temperature in a multi-shot pump-probe scheme. That way, a temporal resolution of 250 fs is possible, limited by the synchronization jitter of the laser pulses. In addition, a field free electron time-of-flight spectrometer records the photoelectron sidebands which can be obtained when a soft x-ray and an optical photon pulse are temporally and spatially superposed. This indicates the relative timing between the pulses.

Furthermore, numerical simulations with the HELIOS radiation-hydrodynamics code [11] have been conducted to provide more detailed information on the properties of the WDM generated (figure 2a). On top of that, a code calculating Thomson

Scattered spectra from plasma profiles is being produced now. It will facilitate the prediction of experimental spectra as well as their interpretation.

Results

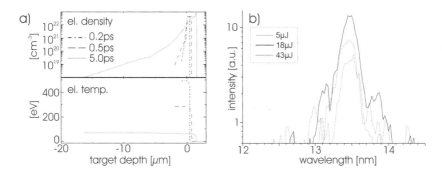

Fig. 2. a) Electron density and temperature profiles of the Hydrogen target after pump laser (800nm, 100 fs, 3×10^{15} W/cm^2) irradiation as simulated with HELIOS and indicating the generation of Warm Dense Matter. b) Normalized spectra of the scattered photons for different FEL intensities. The side peaks originate from the support grid of the transmission grating.

Until now basic components of the experiment have been tested and the interaction of the FEL beam itself with the hydrogen target has been investigated, i.e. self Thomson Scattering. With pulse energies of up to 100 µJ, energy densities of 10^{13}-10^{14} W/cm^2 can be reached, leading to a partially ionized target. The measured spectra (figure 2b) show artifacts originating from the transmission spectrograph's support grid which inhibit the investigation of plasmon features. An improved spectrograph has been built by now.

However, the intensity from elastic scattering should rise linearly with higher FEL energy, but at an average energy of 38 µJ the scattered intensity does not increase any more but starts to decrease instead. This non-linear behavior is not understood so far and will be investigated further.

1   V. Ayvazyan, et al., in The European Physical Journal D, Vol. 37, 297, 2006.
2   W. Ackermann, et al., in Nature Photonics, Vol. 1, 336, 2007.
3   E. L. Saldin, E. A. Schneidmiller, M. V. Yurkow, The Physics of Free Electron Lasers, Springer, Berlin, 2000.
4   R.W. Lee, et al., in Journal of the Optical Society of America B, Vol. 20, 770, 2003.
5   O.L. Landen, et al., in Review of Scientific Instruments, Vol. 72, 627, 2001.
6   S.H. Glenzer, et al., in Physical Review Letters, Vol. 90, 175002, 2003.
7   S.H. Glenzer, et al., in Physical Review Letters, Vol. 98, 065002, 2007.
8   A. Höll, et al., in High Energy Density Physcis, Vol. 3, 120, 2007.
9   A. Przystawik, et al., in HASYLAB Annual Report 2007
10 J. Jasny, et al., in Review of Scientific Instruments, Vol. 65, 1631, 1994.
11 J.J. MacFarlane, et al., in Journal of Quantitative Spectroscopy and Radiative Transfer, Vol. 99, 381, 2006.

Magnon-Enhanced Phonon Damping at Gd(0001) and Tb(0001) surfaces

Alexey Melnikov[1], Alexey Povolotskiy[2], and Uwe Bovensiepen[1]

[1] Freie Universität Berlin, Fachbereich Physik, Arnimallee 14, 14195 Berlin, Germany
  E-mail: uwe.bovensiepen@physik.fu-berlin.de
[2] St. Petersburg State University, Laser Research Institute, St. Petersburg, 198505, Russia

Abstract. Damping of coherent phonons at Gd(0001) and Tb(0001) surfaces is investigated by optical femtosecond time-resolved second harmonic generation. At low temperatures the damping rate increases monotonous with temperature. Close to the Curie temperature the phonon damping rate features an anomalous decrease, which is more pronounced for Tb than for Gd. This systematic variation is based on the difference in orbital momentum of the elements and identifies phonon-magnon scattering as the origin of the anomaly.

Introduction

Scattering of elementary excitations like electron-hole pairs, phonons, and magnons occurs on the femtosecond time scale. Interaction of these excitations with, e.g., phonons brings up the question of the efficiency of the individual interaction processes like phonon-phonon, phonon-magnon, and phonon-electron scattering. An investigation of this problem by pump-probe experiments benefits from two distinguishable phonon subsets. One set is excited specifically and monitored in the experiment. A second set represents the thermal bath which mediates damping of the excited one. Such a situation has been realized in case of Bi and similar systems for coherent (non-thermal) phonons generated with femtosecond laser pulses. It has been shown, that for such simple systems the damping rate of optical phonons increases continuously with the equilibrium temperature T, which has been explained quantitatively by anharmonic decay of the coherent optical phonon into two acoustic phonons at half the frequency and opposite momenta [1]. For materials where vibrations couple efficiently to other degrees of freedom, like electron-hole-pairs in metals or magnons in case of ferromagnets, additional damping channels might become relevant.

Experimental Methods and Data Analysis

Lanthanide surfaces are prepared under ultrahigh vacuum conditions by growing Tb(0001) and Gd(0001) films of 20 nm thickness epitaxially on a W(110) single crystal. Pump-probe experiments are carried out using a cavity-dumped Ti-Sapphire oscillator operating at 1.52 MHz repetition rate and delivers pulses of 35 fs duration at 42 nJ. Coherent optical phonons at frequencies of 3 THz can be generated on these surfaces [2], which are metallic and ordered ferromagnetically. Thus, these surfaces facilitate an investigation of phonon damping through magnetic and electronic excitations. We determine the damping time of coherent phonons in pump-probe experiments by analysis of the optical second harmonic (SH) yield of 800 nm laser pulses generated resonantly through $5d_{z^2}$ surface states [3].

Typical results for the transient behavior of the SH electric field $\Delta(t) = \sqrt{I(t)/I_0} - 1$ is shown in Fig. 1a below for the case of Gd(0001). Here, $I(t)$ is the measured time-dependent SH intensity and I_0 is the intensity before excitation. After optical excitation the signal can be described by the sum of an oscillatory contribution stemming from the coherent phonon and a part that varies continuously within 3 ps which represents equilibration of hot electrons and phonons (dashed line). The oscillatory part that remains after subtraction of the continuous variation is fitted by an exponentially damped oscillation (Fig. 1b) which results in the coherent phonon damping time τ. Measurements have been carried out as a function of T on Gd(0001) and Tb(0001). Albeit not shown in Fig. 1, the $\Delta(t)$ data on Tb(0001) also feature the oscillatory component with a center frequency of 3 THz.

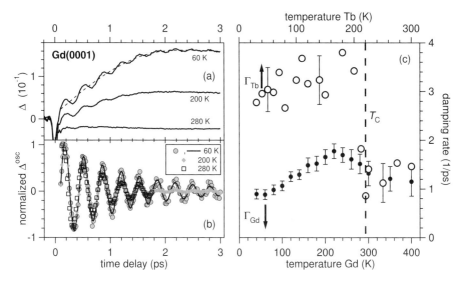

Fig. 1. Left: (a) Transient second harmonic field $\Delta(t)$ of Gd(0001) for different equilibrium temperatures T. (b) Oscillatory part with the amplitude normalized at the first maximum. The weaker damping at higher T can be recognized from a larger normalized amplitude at 1.5 ps if compared to the 200 K data. (c) Temperature dependent damping rates Γ that are determined by fitting the oscillatory data in (b) by a damped oscillation. Both datasets show an anomaly near the respective T_C which is more pronounced for Tb compared to Gd.

Results and Discussion

Here we report on temperature-dependent τ ranging from 1 to 0.3 ps which have been studied between 50 and 400 K. For both surfaces we observe an anomaly of the damping rate $\Gamma = 1/\tau$ in the vicinity of the Curie temperature T_C, which is 293 K for Gd and 225 K for Tb (Fig. 1c). The damping of the coherent phonon due to (*i*) interaction with thermal phonons [1] and by (*ii*) electron-hole pair excitations becomes more efficient for larger T and the damping rate increases monotonous with T [4]. Defect mediated damping is generally assumed to be temperature independent. Since magnons, i.e. spinwaves, represent excitations of the ferromagnetically ordered system at $T<T_C$, the variation in the damping rate near T_C quantifies the magnon mediated damping. The responsible interaction is phonon-magnon coupling. This interaction is based on dynamical magnetoelastic effects which scale with the magnetization M and are mediated by the spin-orbit coupling $\lambda \mathbf{S} \cdot \mathbf{L}$. Since the orbital momentum from 4f electrons is in Gd L=0 (half filled 4f shell, $4f^7$) and in Tb L=3 ($4f^8$), the magnon-mediated damping is expected to be more efficient for Tb than for Gd. Our finding of a several times larger decrease in damping rate for Tb than for Gd at the respective Curie temperature is a direct evidence that this damping contribution originates from phonon-magnon coupling. Above the Tb and Gd Curie temperatures the damping rates are identical within the experimental accuracy. This finding corroborates the above conclusion because (i) the magnetic subsystem is not present at $T>T_C$ and (ii) the electronic and vibrational properties of both surfaces are very similar [5] and electron and phonon mediated damping is expected to be similar for both systems.

Conclusions

For Gd and Tb surfaces phonon-magnon scattering at a scattering rate Γ_m is the dominating contribution to optical phonon damping almost up to T_C. While for Gd Γ_m is comparable with the total scattering rate below 80 K, the strong spin-orbit interaction in Tb due to L=3 for the 4f electrons leads to a much more efficient phonon damping at all temperatures below T_C. In combination with identical scattering rates for both surfaces above the respective T_C, this evidences that magnon-mediated damping is the largest contribution over a wide temperature range, which highlights the relevance of magnon-mediated processes in general.

Acknowledgements. We thank I. Radu for his contributions in the early stage of the experiment. Continuous support by M. Wolf and G. Kaindl and funding by the DFG through SPP 1133 are gratefully acknowledged.

1   M. Hase, K. Mizoguchi, H. Harima, S.I. Nakashima, K. Sakai,
    Phys. Rev. B **58**, 5448 (1998).

2   U. Bovensiepen, J. Phys.: Cond. Matt. **19**, 083201 (2007).

3   A. Melnikov, O. Krupin, U. Bovensiepen, K. Starke, M. Wolf, E. Matthias,
    Appl. Phys. B **74**, 723 (2002).

4   M. Hase, K. Ishioka, J. Demsar, K. Ushida, M. Kitajima, Phys. Rev. B **71**, 184301 (2005).

5   I. D. Hughes, M. Däne, A. Ernst, W. Hergert, M. Lüders, J. Poulter, J. B. Staunton, A. Svane, Z. Szotek, W. M. Temmerman, Nature **446**, 650 (2007).

Ultrafast Coherent Interactions in Quantum Wells Studied by Two-Dimensional Fourier Transform Spectroscopy

Tianhao Zhang,[1] Irina Kuznetsova,[2] Lijun Yang,[3] Alan D. Bristow,[1] Xingcan Dai,[1] Xiaoqin Li,[4] Torsten Meier,[5] Peter Thomas,[2] Shaul Mukamel,[3] Richard P. Mirin[6] and Steven T. Cundiff[1]

[1]JILA, University of Colorado and National Institute of Standard and Technology, Boulder, CO 80309-0440 U.S.A.
[2]Department of Physics and Material Sciences Center, Philipps University, Renthof 5, D-35032 Marburg, Germany
[3]Department of Chemistry, University of California, Irvine, CA 92697-2025 U.S.A.
[4]Department of Physics, University of Texas, Austin, TX 78712-0264 U.S.A.
[5]Department Physik, Fakultät für Naturwissenschaften, Universität Paderborn, Warburger Strasse 100, D-33098 Paderborn, Germany
[6]National Institute of Standards and Technology, Boulder, CO 80305 U.S.A.
E-mail: cundiffs@jila.colorado.edu

Abstract. Many-body effects dominate the polarization studies of heavy- and light-hole excitons. Accurate simulations require Coulomb correlations beyond Hartree-Fock approximation. Raman coherences are isolated with a new two-dimensional projection.

Coherent excitation of carriers has been studied extensively by transient four-wave mixing (TWFM) spectroscopy. While versatile, this technique has several limitations. Developed from TFWM, two-dimensional Fourier-transform (2DFT) spectroscopy[1,2] is the optical analogue of multi-dimensional nuclear magnetic resonance spectroscopy. Multi-dimensional techniques unfold coherent peaks associated with populations (diagonal features) from coherent coupling between states (off-diagonal features), typically by plotting the signal versus both the absorption and emission photon energies. The advantage of this technique is that it can clearly separate a) self-coherent peaks from cross-coherent peaks, b) quantum beats from polarization beats, c) different many-body interactions,[3,4,5] and d) homogeneous from inhomogeneous broadening.[6]

2DFT spectroscopy has recently become a rigorous experimental test for many-body interactions of semiconductor excitons, in particular the heavy- and light-hole coherences. (See Fig 1 (a) for a typical absorption spectrum.) 2DFT spectroscopy has pushed the theoretical modelling to the level of many-body correlations that include all excitonic and biexcitonic contributions. These solutions go beyond phenomenological approaches of many-body effects,[6] such as excitation-induced dephasing and shift or local-field corrections.

The similarity of 2DFT spectroscopy to other techniques like multi-dimensional NMR provides a wealth of expertise applicable to the spectroscopy of semiconductors. For example, the use of different two-dimensional (2D) projections allows for the separation of otherwise co-contributing Feynman pathways within the heavy- and light-hole system.[7] Separation of these pathways can also been modelled with a microscopic many-body theory based on the semiconductor Bloch equations.

Our experimental setup (described in detail in ref. [2]) uses heterodyne detection of the spectrally resolved TFWM signal over a range of time delays. A Fourier transform produces a 2D map that represents the section of the density matrix allowed by the specified geometry and time ordering of the pulses. A three-pulse box geometry is used, where the signal is detected in the $-k_A+k_B+k_C$ direction and scanning pulse A provides rephasing (photon-echo) information.

The non-rephasing information is similarly obtained by scanning pulse *B*. In this standard $S_l(\omega_\tau,T,\omega_t)$ experiment the 2D map is equivalent to a plot of the absorption versus emission photon energies. Alternatively, the time delay between pulses *B* and *C* can be scanned, yielding a 2D map of the mixing versus emission photon energies, denoted $S_l(\tau,\omega_T,\omega_t)$.

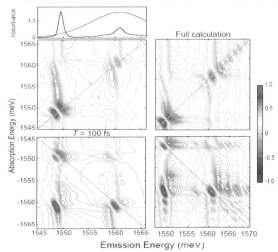

Fig.1 (a) Absorbance spectrum of the HH and LH excitons. Real-part of the $S_l(\omega_\tau,T,\omega_t)$ 2DFT spectra for co-circular excitation in the (b) non-rephasing and (c) rephasing configurations. Theoretical comparison: (d) to (b) and (e) to (c).

The samples are grown by molecular beam epitaxy, consisting of 10 period GaAs/AlGaAs multiple quantum wells, with 10 nm wells and barriers. The samples are attached to wedged sapphire disks and the substrate is removed by lapping and etching. Measurements are performed at temperatures below 10 K.

Numerical simulations are based on a one-dimensional tight-binding model. The microscopic semiconductor Bloch equations include exciton and biexciton formation and Coulomb correlations due to exciton-exciton and exciton-biexciton interaction; further details are given in ref. 6 and 7.

The interplay of heavy- and light-hole excitons is examined for a variety of polarizations, using the real-part of the $S_l(\omega_\tau,T,\omega_t)$ projections. Experimental rephasing and non-rephasing spectra are shown in Fig. 1 (b) and (c), with the associated theoretical plots in (d) and (e). The diagonal features have dispersive line shapes, that appear as derivatives of the TFWM pulse in the cross-diagonal direction for the rephasing and the diagonal direction for the non-rephasing. Simple models give absorptive line shapes, but the introduction many-body interactions (even at the Hartree-Fock level) give dispersive shapes too. Indeed from previous work this effect appears to be dominated by excitation-induced shifts.[3] Compared to co-linear polarized excitation (not shown), co-circular excitation is expected to decouple the HH and LH exciton resonances due to selection rules, if many-body correlations and valence band-mixing effects are neglected. The full simulation exhibits the non-diagonal peaks seen in the experiment for the co-circular case. Thus they can be explained by many-particle correlations within the $\chi^{(3)}$ limit. Neglecting these higher-order terms leads to poor comparison to the experimental data; see ref. 4 for further comparison.

Amplitude $S_l(\tau,\omega_T,\omega_t)$ spectra are shown in Fig. 2 (a). This projection can isolate the so-called Raman coherences[7] that contribute to the cross peaks in $S_l(\omega_\tau,T,\omega_t)$ spectra. Raman coherences correspond to Feynman pathways that oscillate with the mixing time T, and are normally obscured by stronger non-oscillating pathways. In the $S_l(\tau,\omega_T,\omega_t)$ spectra, the latter collapse into the zero-energy peaks along the center of the plot, thus exposing the Raman coherences. The overwhelming amplitude of the non-oscillating coherences is one reason why the zero-energy peaks saturate this figure. An asymmetry is observed in the amplitude of the HH-LH and LH-HH Raman coherences [peaks indicated in Fig.2 (b)]. This is most likely due to their respective dephasing rates and the overlap of the heavy-hole continuum with the light-hole exciton. The experimental data is well reproduced by the microscopic many-body theory, as shown in Fig. 2 (b). Calculations are performed in the coherent limit and exciton dephasing rates are estimated from $S_I(\omega_\tau, T, \omega_t)$ experiments.

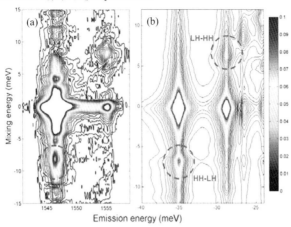

Fig.2 (a) Experimental and (b) theoretical amplitude of the $S_I(\tau, \omega_T, \omega_t)$ 2DFT spectra for co-linear excitation. The dashed circles show the excitonic Raman coherences.

It would also be interesting to investigate the real-part of the $S_I(\tau, \omega_T, \omega_t)$ spectra. This requires all three pump pulses in the box geometry be phase-locked. Apparatus is being presently being built to perform these types of measurements. Additionally, the new apparatus would allow for examination of two-quantum transitions in biexcitons.[8]

Two-dimensional Fourier-transform spectroscopy has been applied to exciton dynamics in semiconductor quantum wells. It clearly reveals the dominance of many-body interactions that are accurately modeled by the semiconductor Bloch equations that include all exciton and biexciton Coulomb correlations within the $\chi^{(3)}$ limit. The experiments have been performed for various polarizations, time ordering and 2D projections. Both $S_I(\omega_\tau, T, \omega_t)$ and $S_I(\tau, \omega_T, \omega_t)$ projections demonstrate that two-dimensional Fourier-transform spectroscopy is presently the most stringent tool for many-body interactions in semiconductors. Also, the future application of the $S_I(\tau, \omega_T, \omega_t)$ projection may yield useful information about other systems, such as photosynthetic complexes,[9] which have been studied with the $S_I(\omega_\tau, T, \omega_t)$ projection.

1   C. N. Borca, T. Zhang, X. Li and S. T. Cundiff, Chem. Phys. Lett. **416**, 311 (2005).
2   T. Zhang, C. N. Borca, X. Li and S. T. Cundiff, Opt. Exp. **13**, 7432-7441 (2005).
3   X. Li, T. Zhang, C. N. Borca, and S. T. Cundiff, Phys. Rev. Lett. 96, 057406 (2006).
4   T. Zhang, I. Kuznetsova, T. Meier, X. Li, R. P. Mirin, P. Thomas and S. T. Cundiff, Proc. Natl. Sci. USA **104**, 14227 (2007).
5   I. Kuznetsova, P. Thomas, T. Meier, T. Zhang, X. Li, R. P. Mirin, and S. T. Cundiff, Solid State Commun. **142**, 154 (2007).
6   I. Kuznetsova, T. Meier, S. T. Cundiff, and P. Thomas, Phys. Rev. B **76**, 153301 (2007).
7   L. Yang, I. V. Schweigert, S. T. Cundiff and S. Mukamel, Phys. Rev. B **75**, 125302 (2007).
8   K. W. Stone, K. Gundogdu1, D. B. Turner, K. A. Nelson, X. Li, and S. T. Cundiff, "Two-quantum 2D optical FT spectroscopy of biexcitons in GaAs quantum wells," submitted for publication.
9   G. S. Engel, T. R. Calhoun, E. L. Read, T.-K. Ahn, T. Mančal, Y.-C. Cheng, R. E. Blankenship and G. R. Fleming, Nature **446**, 782 (2007).

Two-quantum Two-dimensional Fourier Transform Electronic Spectroscopy of Biexcitons in GaAs Quantum Wells

Katherine W. Stone[1], Kenan Gundogdu[1], Daniel B. Turner[1], Xiaoqin Li[2], Steven T. Cundiff[3], and Keith A. Nelson[1]

[1] Department of Chemistry, MIT, Cambridge MA 02139, USA
  E-mail: kanelson@mit.edu
[2] Department of Physics, University of Texas at Austin, Austin TX 78712, USA
  E-mail: elaineli@physics.utexas.edu
[3] JILA, University of Colorado, and National Institute of Standards and Technology, Boulder CO 80309, USA
  E-mail: cundiffs@jila.colorado.edu

Abstract. Using two-dimensional Fourier transform electronic spectroscopy, we directly observe two-quantum biexciton-ground state coherences which are typically temporally and spectrally convolved with one-quantum excitonic coherences.

Introduction

Two-dimensional Fourier Transform spectroscopy (2D FTS) is an ultrafast four-wave mixing measurement that has proven itself a very powerful tool for probing electronic and vibrational dynamics in large biomolecules and inhomogeneously broadened condensed systems[1, 2]. In semiconductors, complex dynamics arise due to long-range many-body interactions among excitons. These interactions can be understood phenomenologically in terms of modifications of exciton properties through local field effects [3], renormalization effects [4, 5], and in terms of the formation of entirely new quasiparticles, biexcitons [6]. Ultrafast electronic 2D FTS has enabled rigorous comparison of distinct many-body effects by isolating them in distinct experimental features and by providing exciton phase information [7]. However, direct time-resolved observations of biexcitons have been elusive because there has not been any optical analog of multiple-quantum 2D NMR.

Here we present 2D FTS measurements in which we directly observe biexciton coherences in GaAs quantum wells using a two-quantum technique. Extending ultrafast electronic 2D FTS to include two-quantum coherences, which are studied routinely in NMR, will greatly enhance its versatility by allowing access to biexcitons and other multiply excited states that are "dark" to one-photon transitions and by providing an additional conjugate frequency axis along which complex spectra can be spread out. Furthermore, multiple-quantum coherences are often used in quantum information processing and thus their decoherence rates are important for practical as well as fundamental reasons.

Experimental Methods

These experiments were performed using a spatiotemporal pulse shaper [8] in which a single beam of light with a single ultrashort pulse is transformed into multiple beams with single or multiple pulses. The pulse shaper generates and controls the timing and optical phases of all the pulses in each beam so coherent evolution may be examined during any of the specified time periods between pulses in any beam.

The output of a femtosecond Ti:Sapphire oscillator was directed through a passive spatial shaping element forming the four beams needed for the fields E_A, E_B, E_C, and E_{LO} which were directed into the spatiotemporal pulse shaper. The shaped fields were directed into the sample in a non-collinear phase-matching geometry so that the signal emerged from the sample as a coherent beam in a direction (collinear with E_{LO}) given by the wavevector matching condition $k_{sig} = k_A + k_B - k_C$. One-quantum coherences excited by the first field, E_A, are converted to two-quantum coherences by the second field, E_B, which evolve during time delay τ_2, the time between E_B and the third field, E_C. The spectrally-resolved, interferometrically detected signal emitted after E_C during time t was collected while sweeping τ_2.

The sample consisted of ten layers of 10 nm thick GaAs, separated by 10 nm thick barriers of $Al_{0.3}Ga_{0.7}As$, and was held at a temperature of 10 K. Layering of the two materials with slightly different bandgap confines the electronic potential in one dimension and causes the degenerate valence bands to split into heavy-hole (HH) and light-hole (LH) bands. The HH and LH exciton transition frequencies are 372.2 and 373.7 THz (1.539 and 1.546 eV), respectively. The polarizations of the input beams, controlled via individual quarter-wave plates, selected whether homogeneous (HH-HH) biexciton coherences or mixed (HH-LH) biexciton coherences were observed.

Results and Discussion

The biexciton-ground state coherence frequency, ω_b, is given by the biexciton energy $E_b = \hbar\omega_b = 2\varepsilon_e - \varepsilon_B$ where ε_e is the exciton energy and ε_B the biexciton binding energy. Note that these coherences are detected in the rotating wave frame since the pulse shaper can vary the pulse envelopes without varying the optical phases. Therefore, we subtract a user-defined carrier frequency ($\omega_0 = 368.0 \pm 0.048$ THz or 814.7 nm) from the emission frequency axis in the 2D spectra in Fig. 1A and B. The spectra are the magnitudes of the complex signal field displayed across two frequency dimensions (ω_2 and ω) which are the Fourier-transform conjugates of the two-quantum coherence time and signal emission time (τ_2 and t). After carrier frequency subtraction, the HH and LH emission frequencies are 4.3 and 5.8 THz, respectively.

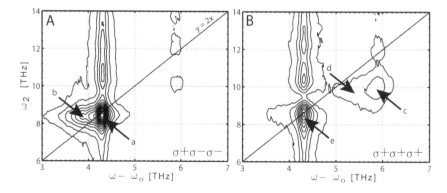

Fig. 1. 2D spectral magnitudes for cross-circular (A) and co-circular (B) excitation pulse sequences. The contour lines are at 6.25% intervals.

When the fields are cross-circularly polarized, homogeneous biexciton coherences are selected as shown in Fig. 1A where the peak is centered at 8.3 THz (*a*) in ω_2. Using the position of *a* and the HH exciton emission we obtain a biexciton binding energy of 1.2 meV. The third field can also excite additional exciton coherences, resulting in an exciton-biexciton coherence during the emission time, which appears as a shoulder (*b*) red-shifted from the main feature due to the small biexciton binding energy in GaAs quantum well structures. By subtracting the peak position of *a* and *b* along ω we obtain a biexciton binding energy of 1.5 meV. The difference in the measured binding energy reflects the many-body phenomena relevant to four-particle and six-particle correlations.

Fig. 1B shows the signal when the fields are co-circularly polarized. Mixed biexciton coherences, *c* and *d*, are located at 9.7 THz in ω_2 and result from Feynman pathways analogous to those producing features *a* and *b*. We calculated a mixed biexciton binding energy of 1.4 meV. Integration over the emission frequency yields the two-quantum absorption lineshape which was fit to a single Lorentzian in order to extract the biexciton dephasing rate. The homogeneous and mixed biexciton dephasing rates are 2.08 ± 0.11 ps and 1.03 ± 0.05 ps, respectively.

The interaction-induced coherence, resulting from the excitonic interactions described by local-field and renormalization effects, is observed at twice the exciton emission frequency (*e*) along ω_2. These interaction-induced coherences, long used to explain the signal at negative delay in two-pulse four-wave mixing experiments [9], oscillate at a frequency given by twice the exciton energy.

Conclusions

The two-quantum 2D FTS measurements presented here permit direct observation and characterization of multi-exciton coherences. Comprehensive analysis of the real and imaginary parts of the complex 2D FT spectra will provide further insight into the many-body effects in this prototype system. More complex nanostructures also will be of interest, for instance semiconductor double quantum wells in which coherence and energy transfer among excitons and biexcitons in different wells should be observable directly.

Acknowledgements. This work was partially supported by NSF grant no. CHE-0616969. D. Turner wishes to thank the NDSEG Fellowship Program for financial support.

1   T. Brixner, J. Stenger, H. M. Vaswani, M. Cho, R. E. Blankenship, and G. R. Fleming, Nature **434**, 625, 2005.

2   H. Chung, M. Khalil, and A. Tokmakoff, J Phys. Chem. B **108**, 15332, 2004.

3   M. Wegener, et. al., Phys. Rev. A **42**, 5675, 1990.

4   J. M. Shacklette and S. T. Cundiff, J. Opt. Soc. Am. B **20**, 764, 2003.

5   H. Wang, et. al., Phys. Rev. Lett. **71**, 1261, 1993.

6   K. Bott, O. Heller, D. Bennhardt, S. T. Cundiff, P. Thomas, E. J. Mayer, G. O. Smith, R. Eccleston, J. Kuhl, and K. Ploog, Phys. Rev. B **48**, 17418, 1993.

7   T. Zhang, I. Kuznetsova, T. Meier, X. Li, R. P. Mirin, P. Thomas, and S. T. Cundiff, Proc. Nat. Acad. Sci. **104**, 14190, 2007.

8   J.C. Vaughan, T. Hornung, K.W. Stone, and K.A. Nelson, J. Phys. Chem. A **111**, 4873, 2007.

9   K. Leo, M. Wegener, J. Shah, D. S. Chemla, E. O. Gobel, T. Damen, S. Schmitt-Rink, and W. Schafer, Phys. Rev. Lett. **65**, 1340, 1990.

Three-Pulse Echo Peak Shift Spectroscopy of Disordered Semiconductor Quantum Wells and Dense Atomic Vapors

S.T. Cundiff[1], V.O. Lorenz[1], S.G. Carter[1], Z. Chen[1], S. Mukamel[2] and W. Zhuang[2]

[1] JILA, National Institute of Standards and Technology and University of Colorado, Boulder, Colorado, 80309-0440 USA
E-mail: cundiffs@jila.colorado.edu
[2] Department of Chemistry, University of California, Irvine, Calfornia, 92697-2025 USA
E-mail: smukamel@uci.edu

Abstract. Three-pulse echo peak shift spectroscopy yields the correlation function of the frequency fluctuations due to acoustic phonons for excitons in disordered semiconductor quantum wells and fluctuations due to atomic motion in a potassium vapor.

The dephasing of optical transitions is often treated in the Markovian approximation, where the dephasing events are assumed to be infinitely fast and there is no memory, i.e., the phase after the event does not depend on what it was prior to the event. Physically, dephasing occurs because the transition frequency fluctuates during a dephasing event. The Markovian approximation means that the two-time correlation function of the frequency fluctuations is a delta function. Given sufficiently high time resolution, the Markovian approximation is not valid, resulting in non-exponential temporal dynamics. In the non-Markovian regime, the dephasing cannot be described by a simple rate, but rather the correlation function must be determined [1]. Three-pulse echo peak shift (3PEPS) spectroscopy is capable of measuring the correlation function [2,3]. We apply 3PEPS to two systems that display non-Markovian dynamics, a disordered semiconductor quantum well [4] and a dense atomic vapor [5].

In 3PEPS spectroscopy, a 3-pulse transient four-wave-mixing (FWM), also known as photon echo, experiment is performed. The incident pulses are approximately 100 fs in duration and have nJ energy. The signal is recorded as the delay between the first two pulses, τ, is scanned. Typically the signal is recorded in two directions, $\mathbf{k}_+ = -\mathbf{k}_a + \mathbf{k}_b + \mathbf{k}_c$ and $\mathbf{k}_+ = \mathbf{k}_a - \mathbf{k}_b + \mathbf{k}_c$ where τ is defined to be positive for ka arriving first and negative for kb arriving first. As these scans are symmetric about $\tau = 0$, recording both reduces error in determination of $\tau = 0$. These scans are performed as the delay between the second and third pulses, T, is varied. The echo peak shift is the τ for which the signal is the maximum. The echo peak shift as a function of T maps out the correlation function for the frequency fluctuations.

Figure 1 shows typical FWM signals from the heavy-hole exciton resonance in a disordered semiconductor quantum well and the EPS for a series of temperatures and excitation densities. The disorder localizes excitons and produces inhomogeneous broadening. Two main processes lead to decoherence, migration of excitons between localization sites and the interaction of excitons with acoustic phonons. The EPS display a sharp decrease followed by a slow increase. After ~ 4 ps, there is an exponential decay to an offset of ~ 100 fs. We attribute the exponential decay to migration of the excitons between localization sites, which results in spectral diffusion (SD). Perhaps the most intriguing feature of the EPS is the increase during the first few picoseconds. This increase cannot be explained by a SD model. To model the non-Markovian dynamics at short times, we use an underdamped oscillator model [1] in which acoustic phonons modulate the exciton energy. This model does give an increasing EPS.

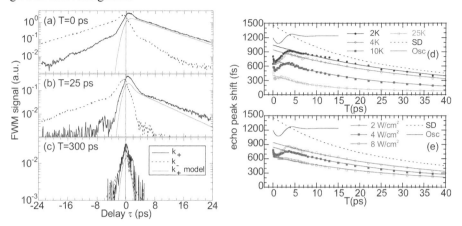

Fig. 1. Time-integrated FWM signals from a disordered GaAs quantum well as a function of τ for T = (a) 0, (b) 25, and (c) 300 ps (logarithmic vertical scale). The solid (dashed) lines represent the $\mathbf{k}_+$ ($\mathbf{k}_-$) direction. The solid gray lines represent the calculated $\mathbf{k}_+$ signal. The vertical lines at the center are at $\tau = 0$. The sample temperature is 2 K. (d) Echo peak shift as a function of T for a series of temperatures (2, 4, 10, and 25 K) at an excitation intensity of 2 W/cm². (e) EPS as a function of T for a series of excitation intensities (2, 4, and 8 W/cm²) at a temperature of 4 K. The markers represent experimental points and the lines represent exponential fits to the data for $T > \sim 5$ ps. The dashed black and solid gray lines represent the calculated EPS from the spectral diffusion (SD) model and the underdamped oscillator (Osc) model, respectively.

Figure 2 shows the transient FWM signal from a dense potassium vapor and the EPS for a series of temperatures. At these temperatures and densities, Doppler broadening is negligible and decoherence is due to resonance broadening, i.e., "collisions" between like atoms. Previous studies have shown that the signatures of non-Markovian dynamics are evident in the FWM signal as non-exponential behavior at short times and that the experimental results are well described by assuming an exponentially decaying correlation function [6]. 3PEPS removes the need to assume a correlation function. An atomic vapor is an ideal system for studying non-Markovian dynamics because there is a good separation of timescales, with the pulse being much shorter than the "collision duration" which in turn is much shorter than the decoherence rate.

These experimental results are in qualitative agreement with an exciton model combined with stochastic frequency fluctuations obtained from molecular dynamics simulations. The measured EPS from the vapor shows a single exponential decay at low temperatures (and densities) while at higher temperatures it become bi-exponential (Fig. 2(e)). The molecular dynamics simulations provide snapshots of the configuration of 20 potassium atoms. For each snapshot, the resonant couplings are determined from the potential energy surfaces for the potassium dimer. The couplings, and hence the eigenenergies, fluctuate as the atoms move. From the energy fluctuations, a correlation function and EPS can be determined (Fig. 2(f)). The calculations also show a biexponential behavior. By artificially turning off the long range part of the potential, we can show that the slower component is due to long range dipole-dipole interactions [5].

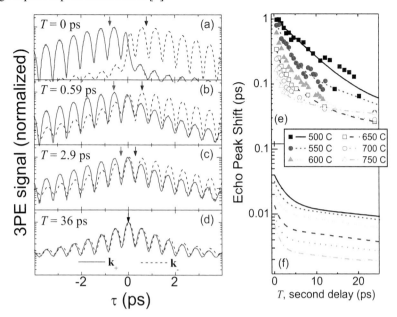

Fig. 2. (a-d) Experimental 3PE signals from a dense potassium vapor at 600 C versus the delay between the first two pulses, τ, for 4 values of the delay between the second and third pulse, T. The beats are due to simultaneous excitation of the D_1 and D_2 lines. The arrows indicate the peak of the signal without beats. (e) Experimental peak shifts for a series of temperatures and densities. (f) Calculated peak shifts from molecular dynamics simulations in an exciton picture.

3PEPS spectroscopy is a powerful tool to extract information about decoherence processes in a variety of systems. Here we have used it to study the decoherence of excitons in disordered quantum wells and in a dense atomic vapor.

1   S. Mukamel, *Principles of Nonlinear Optical Spectroscopy* (Oxford University Press, New York, 1995).

2   A. M. Weiner, S. D. Silvestri, and E. P. Ippen, "Three-pulse scattering for femtosecond dephasing studies: theory and experiment," J. Opt. Soc. Am. B **2**, 654 (1985).

3   M. Cho, J.-Y. Yu, T. Joo, Y. Nagasawa, S. A. Passino, and G. R.Fleming, "The Integrated Photon Echo and Solvation Dynamics," J. Phys. Chem. **100**, 11944 (1996).

4   S.G. Carter, Z. Chen and S.T. Cundiff, "Echo peak-shift spectroscopy of non-Markovian exciton dynamics in quantum wells," Phys. Rev. B **76**, 121303(R) (2007).

5   V.O. Lorenz, S. Mukamel, W. Zhuang and S.T. Cundiff, "Ultrafast Optical Spectroscopy of Spectral Fluctuations in a Dense Atomic Vapor," Phys. Rev. Lett. **100**, 013603 (2008).

6   V.O. Lorenz and S.T. Cundiff, "Non-Markovian Dynamics in a Dense Potassium Vapor," Phys. Rev. Lett. **95**, 163601 (2005)..

Coherently controlled ballistic charge currents in unbiased bulk silicon and single-walled carbon nanotubes

M. Betz,[1] L. Costa,[1] M. Spasenović,[1] R. W. Newson,[1] J.-M. Ménard,[1] C. Sames,[1] A. D. Bristow,[1] and H. M. van Driel [1]

[1] Department of Physics and Institute for Optical Sciences, University of Toronto, Toronto, ON M5S 1A7, Canada
E-mail: mbetz@ph.tum.de

Abstract. Phase-related fundamental and second harmonic femtosecond pulses induce directional charge motion in group IV materials at 300 K. THz emission reveals peak current densities of 0.5 kA/cm^2 in silicon and currents of 1 nA per carbon nanotube.

Introduction

Charge transport in group IV-semiconductors plays a central role in solid-state physics. In particular, silicon integrated devices are the cornerstone of electronics while carbon nanotubes might represent important building blocks for future devices. Control over charge motion in these materials has only been achieved electronically. In contrast, in III-V semiconductors efficient ways to all-optically generate directional currents have been realized utilizing harmonically related femtosecond pulses [1,2]. These schemes rely on quantum interference of one- and two-photon absorption.

Here, we show that the concept of all-optical coherent control of electrical currents is rather general in nature and can be extended even to indirect bandgap materials such as silicon [3,4] and molecular systems such as carbon nanotubes [5]. The efficiency of this current injection is surprisingly comparable to that in established III-V optoelectronic materials. As a result, we demonstrate an attractive non-contact method to generate electrical currents in important classes of semiconductors.

Experimental Methods

The optical source for current injection is a 250 kHz Ti:sapphire amplifier combined with an OPA which produces 150 fs pulses with an average power of 50 mW and a tuning range between 1380 nm and 1800 nm (ω). A BBO crystal generates several mW of second harmonic light (2ω). A phase-stable superposition of the co-polarized ω and 2ω pulse trains is achieved in a two-color Michelson interferometer and focused to a 100 μm spot on the semiconductor specimens at room temperature (see the scheme in Fig.1(a)). The far field THz emission related to ultrafast charge displacement is analyzed by electro-optic sampling in a 500 μm ZnTe crystal using a small fraction of the 800 nm, 150 fs amplifier pulse as a probe.

Results and Discussion

First, we analyze current injection in carbon nanotubes. CVD-grown single-walled nanotubes are standing vertically and tightly-packed on a silicon substrate to between 150 and 200 μm in length. The distribution of the tube diameters is 2.5 +/- 1.5 nm.

The wavelength of the OPA is tuned to 1400 nm. The beam produces peak focused intensities for the ω and 2ω pulses of 10 and 0.15 GW/cm^2, respectively.

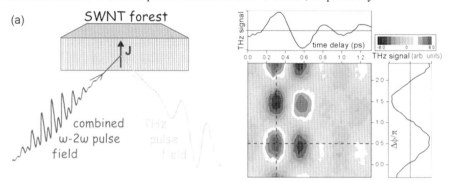

Fig. 1. Illustration of the excitation of the nanotubes and THz emission due to ultrafast charge displacement. (b) Time dependent THz radiation field as a function of time delay between 1400/700 nm pump and 800 nm probe pulse and the phase parameter, $\Delta\phi$. Top panel: Time-resolved THz trace for constant $\Delta\phi$ represented by the horizontal dashed line. Right panel: $\Delta\phi$ dependence of the THz field for constant time delay represented by the vertical dashed line.

Fig. 1(b) displays the THz radiation field from the nanotubes as a function of $\Delta\phi = \phi_{2\omega} - 2\phi_\omega$ and the time delay between 1400/700 nm pump and 800 nm probe beam. Both pump beams impinge on the side facet of the nanotube forest and are co-polarized along the tube axes. A typical THz pulse trace as a function of probe pulse time delay with constant $\Delta\phi$ is shown in the top panel, corresponding to the horizontal dashed line on the main panel. While the THz signature lasts for > 1 ps, this reflects the limited bandwidth of the THz detection rather than dynamics of the current which is expected to rise with the 150 fs pulse and decay with the carrier momentum relaxation time. The right panel of Fig 1(b) shows the dependence of the THz field with $\Delta\phi$ for constant pump/probe delay. The current reverses direction as the phase varies, and, more generally, follows a $\sin(\Delta\phi)$ dependence. This finding is consistent with a coherently controlled photocurrent which, in GaAs, is known to scale as $\dfrac{dJ_i^{inj}}{dt} = 2\eta_{ijkl}E_j^\omega E_k^\omega E_l^{2\omega}\sin\Delta\phi$. Here, η_{ijkl} is the injection current tensor related to the third order susceptibility. The strength of the emission points to peak currents as large as 1 nA per nanotube. Sample rotation indicates that current injection is strongly redu-ced for light polarization perpendicular to the tubes, i.e. η_{ijkl} is highly anisotropic.

For current injection in silicon and the photon energies used here, the fundamental beam is well below both the indirect ($E_{G,i} = 1.1$ eV) and the direct ($E_{G,d} = 3.5$ eV) gap, whereas the second harmonic beam generates electron-hole pairs via phonon-assisted indirect transitions. As a result of the momentum redistribution due to phonon scattering, coherent control of directional currents might be considered unlikely. Nonetheless, we have recently demonstrated that coherent control of electrical currents also extends to excitation across an indirect bandgap [3]. Under similar excitation conditions as used for the data of Fig. 1(b), the THz emission of a 125 mm thick silicon wafer is a factor of ~2 smaller than for carbon nanotubes, while it shows very similar temporal behaviour and phase-dependence. The field strength is indicative of a peak current density of 0.5 kA/cm^2.

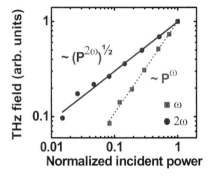 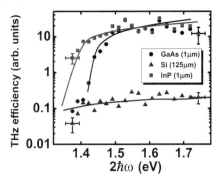

Fig. 2. Dependence of THz emission on pump power and photon energy. Left panel: THz emission from a 125 μm thick silicon wafer as a function of average power of one of the pump beams when the other pump beam's average power is fixed. The lines indicate linear and square root power laws. Right panel: THz field strengths obtained from (001) GaAs, InP and Si layers as a function of photon energy $2\hbar\omega$. The field strengths are normalized by $P_\omega (P_{2\omega})^{0.5}$ to account for variation in pump power with tuning.

Fig. 2 shows how the amplitude of the THz field varies with average pump power, $P^{\omega,2\omega}$, and $2\hbar\omega$. The THz field emitted from a 125 μm thick [100] oriented silicon wafer varies as $P^\omega \sqrt{P^{2\omega}}$, consistent with a third-order optical nonlinearity. For comparison, data are also shown for the THz signals emitted by 1 μm thick GaAs and InP samples under identical conditions. For small $2\hbar\omega$, the strong dispersion is consistent with the onset of current injection at the band-gap (1.42 eV for GaAs, 1.34 eV for InP). In silicon, current injection is possible across the entire accessible spectral range. As $2\hbar\omega$ changes from 1.75 eV to to 1.35 eV, the injection efficiency decreases by a factor of ~ 3. This is likely related to the decreasing velocity of carriers generated in silicon when $2\hbar\omega$ approaches the indirect bandgap. The peak THz field emitted by the GaAs or InP is ~50 times larger than in silicon. However, given the stronger absorption (for GaAs the single and two-photon absorption are an order of magnitude larger), the current density generated per electron-hole pair is only ~7 times higher in GaAs despite the indirect nature of optical transitions in silicon.

Conclusions

In conclusion, we have extended all-optical coherent control of ballistic electrical currents to the technologically important materials silicon and carbon nanotubes. These results offer new insight into optoelectronic functionalizations of silicon and bring new understanding to the field of electronic transport in carbon nanomaterials.

1 R. Atanasov, et al., Phys. Rev. Lett. **76**, 1703 (1996).

2 A. Haché, et al., Phys. Rev. Lett. **78**, 306 (1997).

3 L. Costa, et al., Nature Physics **3**, 632 (2007).

4 M. Spasenović, et al., Phys. Rev. B **77**, 085201 (2008).

5 R. W. Newson, et al., Nano Lett. **8**, 1586 (2008).

Ultrafast dynamics of coherent phonons in the aligned single-walled carbon nanotubes

Keiko Kato[1], Kunie Ishioka[1], Masahiro Kitajima[1,2], Jie Tang[3], and Hrvoje Petek[4,5]

[1] Advanced Nano Characterization Center, National Institute for Materials Science, 305-0047, Tsukuba, Japan
E-mail:KATO.Keiko@nims.go.jp
[2] Department of Applied Physics, School of Applied Science, National Defense Academy of Japan, 239-8686, Yokosuka, Japan
E-mail: kitaji@nda.ac.jp
[3] Innovative Materials Engineering Laboratory, National Institute for Materials Science, 305-0047, Tsukuba, Japan
[4] Department of Physics and Astronomy, University of Pittsburgh, Pittsburgh, PA 15260, USA
[5] Donostia International Physics Center DIPC, P. Manuel de Lardizabal 4, 20080 San Sebastian, Spain

Abstract. Sub-10-fs pulses allow real time observation of coherent phonons in aligned bundles of single-walled carbon nanotubes. A complex polarization dependence of the G band phonon amplitude in the transient reflectivity is explained by the superposition of G-band phonons with different symmetries.

Introduction

Quasi-one-dimensional carbon nanotubes (CNTs), rolled-up graphene sheets, have attracted much attention due to their unique electronic and structural properties. Because of their reduced dimensionality, confinement effects play an important role in their physical properties, such as ballistic transport, excitons, Peierls distortions, and Kohn anomalies. These physical properties are closely related to the carrier-scattering processes such as electron-electron (e-e), electron-phonon (e-ph) or impurity scattering, which are dominated by the one-dimensional structure.

Especially, e-ph interaction plays an important role in the electronic transport in CNTs. In order to understand e-ph interactions, the real-time observation of electron and phonon dynamics is the most direct way. Recently, the observations of coherent phonons in single-walled carbon nanotubes (SWCNTs) have been reported in randomly oriented CNTs [1,2]. Gambetta *et al.* reported the anharmonic coupling between radial breathing modes (RBM) and G mode through the modulation of the G mode frequency with the RBM oscillation [1].

Here, we present ultrafast dynamics of coherent phonons in aligned bundles of SWCNTs measured by time-resolved reflectivity measurement, in order to understand phonon dynamics which are dominated by the one-dimensional structures. We have studied polarization dependence of the coherent phonons in SWCNTs. Electronic excitation is strongly dependent on the axis of carbon nanotubes. The amplitude of G mode shows an anomalous polarization dependence on the relative angle between the pump and probe polarization which cannot be explained with any Raman tensor. From the comparison with the polarization dependence in graphite, the observed polarization dependence of SWCNTs can be explained by superposition between G modes with different symmetries.

Experimental Methods

To detect coherent phonons in SWCNTs, pump-probe time-resolved reflectivity measurements were carried out under ambient conditions with sub-10 fs UV laser pulses at 395 nm. The excitation pulses were produced by frequency doubling of light from a 65 MHz Ti:sapphire laser oscillator. The linearly polarized pump and probe pulses were focused by a 50 mm focal-length mirror to a 10 μm spot on the sample, with respectively, angles of 20° and 5° from the surface normal. The pump power could be varied up to a maximum of 50 mW, while the probe power was kept below 3 mW. For the detection of all phonon modes in SWCNTs irrespective of phonon symmetries, the transient reflectivity was chosen. In the transient reflectivity, the difference (ΔR) between the signal reflected from sample and the reference (R) without sample was measured as a function of the pump-probe delay.

Results and Discussion

The reflectivity of the probe pulse from aligned SWCNTs shows a polarization dependence that can be used to define the direction of the alignment. The reflectivity of the probe pulse, without pump laser, is plotted in Fig. 1 as a function of the angle (θ) between the laser polarization and the axis of SWCNTs. As expected from the previous work [3], the polarization dependence of reflection shows a $\cos(2\theta)$ dependence with a minimum (maximum) when the polarization direction of the laser is parallel (perpendicular) to the long axis of SWCNTs. This result is consistent with the antenna effect [3].

After excitation of SWCNTs by the pump light, we can observe the ensuing carrier and coherent phonon dynamics through the delay-dependent changes in the amplitude of the reflected probe light. Fig. 2(a) shows the experimental results for the time-resolved reflectivity of SWCNTs taken in the transient reflectivity measurement. The transient photo-induced reflectivity consists of two components. The first component is the initial, non-oscillatory response due to the excitation and the subsequent relaxation of excited carriers, while the second component is the oscillatory signal due to the coherent lattice vibrations.

To clear oscillation frequency in ΔR/R amplitude, the Fourier transform (FT) of the oscillatory component is shown in Fig. 2(b). Three different oscillatory components with different frequencies were observed. The multiple peaks in the low-frequency region correspond to RBMs of SWCNTs with different diameters. The small peak at 40 THz is assigned as the defect-induced (D) mode. The strongest peak at 47 THz is assigned as G mode.

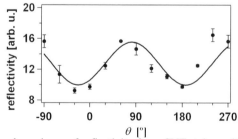

Fig. 1. Polarization dependence of reflectivity from CNTs taken without the pump laser. Solid circles are experimental results plotted as a function of the angle between the laser polarization and the axis of CNTs. The solid line is the fitting result with $\cos(2\theta)$.

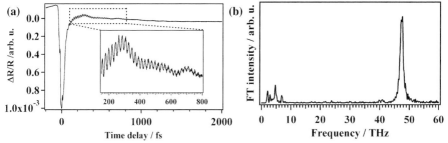

Fig. 2. (a) Differential reflectivity ($\Delta R/R$) dynamics of SWCNTs taken with the transient reflectivity. Inset shows the enlarged image which clearly shows the signal consists of high- and low-frequencies components. FT spectrum of the trace is shown in Fig. 2(b).

When the polarization dependence is measured, RBM and G modes show the different polarization dependence on the angle between pump and probe (ρ).

Phonon amplitude of G modes strongly depends on ρ, it takes the maximum value at $\rho = 0°$ and 90° and completely disappears at 60° while those of RBMs appear at all angles. While the polarization dependence in graphite of G mode obeys E_{2g} Raman tensors [5], the observed polarization dependence of CNTs cannot be explained with the Raman tensor. While G mode in graphite has only one phonon mode, that in SWCNTs has several modes with different symmetries [6]. The observed anomalous polarization dependence suggests the superposition between G-modes with different symmetries.

Conclusions

We have observed coherent phonons in aligned bundles of SWCNTs with time-resolved reflectivity measurement. With the use of aligned bundles of SWCNTs, we observed strong polarization dependence of both the reflectivity and the amplitude of coherent phonons. The polarization dependence of the G mode taken by the transient reflectivity shows anomalous behavior, which cannot be explained with the Raman tensor of either A or E symmetry, but requires the superposition of both responses.

Acknowledgements. K.K. would like to thank Dr. R. Saito, Dr. M. Hase, Dr. O. V. Misochko, Dr. S. Nakashima, and Dr. T. Kitamura. This research was supported by NIMS project research fund and Grants-in-Aid for Scientific Research (No. 18340093) from MEXT and JSPS, and NSF grant CHE-0650756.

1   A. Gambetta, C. Manzoni, E. Mennna, M. Meneghetti, G. Cerullo, G. Lanzani, S. Tretiak, A. Piryatinski, A. Saxena, R. L. Martin, and A. R. Bishop, Nature Phys. **2**, 512, 2006.

2   Y. S. Lim, K. J. Yee, J. H. Kim, E. H. Haroz, J .Sahver, J. Kono, S. K. Doorn, R. H. Hauge, and R. E. Smalley, Nano Lett. **6**, 2696, 2006

3   H. Ajiki and T. Ando, Physica B **201**, 349, 1994.

4   S. M. Bachilo, M. S. Strano, C. Kitterell, R. H. Hauge, R. E. Smalley, and R. B. Weismann, Science, **298**, 2361, 2002.

5   K. Ishioka, M. Hase, M. Kitajima, L. Wirtz, A. Rubio, and H. Petek, Phys. Rev. B **77**, 121402(R), 2008.

6   R. Saito, G. Dresselhaus, and M. S. Dresselhaus, *Physical Properties of Carbon Nanotubes*, (Imperial College Press, 1998).

Evidence for electron correlation in (6,5) carbon nanotubes from pump-probe spectroscopy with broadband pulses

L. Lüer[1], J. Crochet[2], T. Hertel[2], D. Polli[3], G. Lanzani[3]

[1]National Laboratory for Ultrafast and Ultraintense Optical Science, INFM-CNR, Dipartimento di Fisica, Politecnico di Milano, Italy
Email: larry@bimore.eu
[2]Department of Physics and Astronomy & Vanderbilt Institute of Nanoscale Science and Engineering (VINSE), Vanderbilt University, 6301 Stevenson Center Lane, Nashville, TN 37235, USA
E-mail: tobias.hertel@vanderbilt.edu
[3]CNISM and Dipartimento di Fisica, Politecnico di Milano, P.za L. da Vinci 32, 20133 Milano(Italy)
E-mail: guglielmo.lanzani@fisi.polimi.it

Abstract. Pump-probe spectroscopy with 10 fs time resolution is performed on (6,5) carbon nanotubes. We decompose the spectra into contributions from the first and second exciton, demonstrating their electronic correlation.

Introduction

In semiconducting carbon nanotubes (CNT), the low-energetic optical excitations have been shown to be excitons [1]. Therefore, CNT present the opportunity to study the photophysics of excitons in a highly symmetric, quasi monodimensional model system. However, spectroscopic studies up to now were complicated by spectral congestion (due to the presence of many tube types) and by limitations in combining spectral and temporal resolution in a single measurement. This is especially true for excited state spectroscopy, where for each tube type, several spectral features due to ground state bleach (GB), stimulated emission (SE) and photoinduced absorption (PA) occur. Here, we present pump-probe spectra of CNT of a single chirality (6,5), after pumping the second excitonic transition.

Global fitting is applied to reproduce the time-resolved pump-probe spectra $\Delta A(\lambda, t)$ using the Beer-Lambert law with two absorbing species, characterized by cross-sections $\sigma_1(\lambda)$ and $\sigma_2(\lambda)$, and populations $c_1(t)$ and $c_2(t)$, respectively. The latter are obtained by numerically fitting a kinetic model to the data described below. The fit is considered successful if the residuals are free from population kinetics, but show only coherences and noise.

Experimental Methods

Time-resolved pump-probe spectra were obtained using broad band, few-cycle optical pulses and a single-shot detection technique, thereby allowing for very low irradiation dose and high sensitivity [2]. We study colloidal CoMoCAT/Na-Cholate suspensions obtained by density gradient ultracentrifugation, enriched with tubes of the (6,5) type [3].

Results and Discussion

In Fig.1a, we show a two-dimensional pump-probe spectrum, given as differential absorption ΔA as a function of probing wavelength λ and the pump-probe time delay t. At early times after the pump pulse, the spectrum is dominated by a reduced absorption band peaking at 572 nm, which forms immediately upon pumping. Since this reduced absorption coincides with the ground state absorption of the second excitonic transition, E_{22}, the band is straight-forwardly assigned to a bleach (GB) of the fundamental transition. As long as the primary excitation has not yet undergone the $E_{22} \rightarrow E_{11}$ transition, an additional contribution from SE is expected. Note that due to a very low Stokes shift in CNT, no significant spectral shift will occur between GB and SE. After about 100 fs, two increased absorption bands form at 530 and 620 nm, respectively. They reach a maximum PA at around 150 fs, see Fig. 2a. Both PA and GB decay to zero on a 4 ps time scale. No long-lived photoexcitations can be observed.

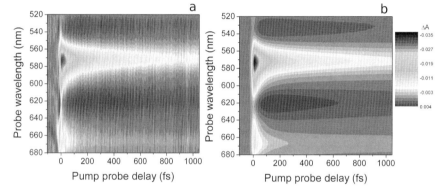

Fig.1. a) Time-resolved pump-probe spectra of (6,5) CNT after pumping the E_{22} state. Note the presence of strong coherent oscillations (G-mode). b) Global fitting according to the model described in the text.

We performed a global fitting procedure, which assumes that the single pump-probe spectra are composed of contributions from the E_{11} and E_{22} excitonic states. In order to model the time-dependent populations $c(E_{11})$ and $c(E_{22})$, we use the kinetic model shown in Fig. 2a: after pumping the E_{22} state, population is transferred to the E_{11} state with rate constant k_{21}. E_{11} decays to the ground state with dispersive rate constant $k_1 = k_1^{'} \cdot t^{-0.45}$ [3] but also takes part in exciton-exciton annihilation with rate k_a. By this bimolecular mechanism, known in CNT [4], one exciton decays to the ground state (G.S.), while the other regenerates the E_{22} state. Since we know k_1 from the literature, we have only k_{21} and k_a as parameters for the fitting of the kinetic model. In Fig.1b, we show that this model yields time-resolved pump-probe spectra that are indistinguishable from the measured ones. We obtain $k_{21} = 2.6 \cdot 10^{13} s^{-1}$, in close agreement with the value obtained in HiPCO nanotubes [5].

As results of the global fitting, we obtain the individual spectral contribution of the E_{22} and E_{11} states to the pump-probe spectra, given as relative cross-sections (Fig. 2b). It is evident that the negative contribution that resembles ground state absorption, is present in both E_{22} and E_{11} states, in a ratio 2:1, respectively. This shows for the first time that in CNT, both E_{11} and E_{22} excitons share a common ground state, comparable to conjugated polymers: transferring population from the E_{22} to the E_{11} state, leaves the GB contribution unaltered but only removes the SE contribution from the E_{22} state. Since the Stokes shift in CNT is negligible, the Einstein coefficients for stimulated absorption and stimulated emission are the same, which explains the 2:1 ratio.

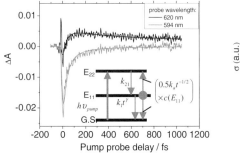 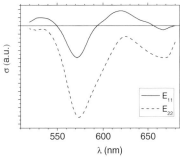

Fig.2. a): Single time traces, obtained as horizontal cuts through Fig. 1a, together with the respective fits (cuts through Fig. 1b). Grey curves: probe wavelength 594 nm, pure E_{22}-related signal, black curves: probe wavelength 620 nm, predominantly E_{11}-related signal. The inset shows the kinetic model. b): Relative excited state absorption cross-sections of first and second exciton (solid and dashed curve, respectively), as obtained from the global fit.

Conclusions

Using time-resolved pump-probe spectroscopy with 10 fs resolution, we demonstrate that both the second and the first excitonic transition in semiconducting (6,5) carbon nanotubes share a common ground state. Our results show that electron-electron interaction between different manifolds (first and second valence-conduction bands) needs to be taken into account when describing excitons in CNT.

1   F. Wang, G. Dukovic, L. E. Brus, T. F. Heinz, Science 308, 838 (2005).
2   D. Polli, L. Lüer, G. Cerullo, Rev. Sci. Inst. 78(10), Art.No. 103108 (2007)
3   Z. Zhu, J. Crochet, M. S. Arnold, M. C. Hersam, H. Ulbricht, D. Resasco, and T. Hertel, J. Phys. Chem. C, 111, 3831-2825 (2007)
4   ] F. Wang, G. Dukovic, E. Knoesel, L. E. Brus, T. F. Heinz, Phys. Rev. B 70, 241403(R) (2004)
5   C. Manzoni, A. Gambetta, E. Menna, M. Meneghetti, G. Lanzani, and G. Cerullo, Phys. Rev. Lett. 207401 (2005)

Ultrafast Relaxation of Excited Dirac Fermions in Epitaxial Graphene

Dong Sun[1], Zong-Kwei Wu[1], Charles J. Divin[1], Xuebin Li[2], Claire Berger[2], Walt A. de Heer[2], Phillip N. First[2], and Theodore B. Norris[1]

[1] Center for Ultrafast Optical Science, University of Michigan, Ann Arbor, MI 48109-2099
 E-mail: sundong@umich.edu
[2] School of Physics, Georgia Institute of Technology, Altanta, GA 30332

Abstract. Nondegenerate ultrafast pump-probe spectroscopy of epitaxial graphene is used to study hot electron relaxation and interlayer thermal coupling. The DT spectra are understood in terms of hot thermal carrier distributions with no electron-hole interaction.

Introduction

Two-dimensional graphene layers are the subject of considerable research at present, due to their unusual band structure and the potential for future electronic-device applications [1-3]. Graphene grown expitaxially on SiC substrates and patterned via standard lithographic procedures is being investigated as a platform for carbon-based nanoelectronics and molecular electronics [1-3]. In high-speed devices, hot electrons are generated due to the presence of time-dependent high electric fields. It is critical therefore to understand both the cooling of the electrons due to coupling to lattice phonons, and the coupling of different layers in epitaxial graphene samples.

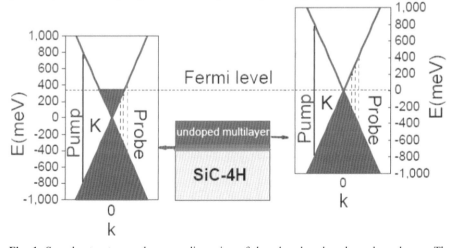

Fig. 1. Sample structure and energy dispersion of doped and undoped graphene layers. The sample has a buffer layer on the SiC substrate followed by 1 heavily doped layer and multiple undoped layers on top. The Fermi level is indicated by a dashed line lying 348 meV above the Dirac point of the doped graphene layer and passing through the Dirac point of the undoped graphene layers. The blue solid line shows the transitions induced by the 800-nm optical pump pulse; the three dashed lines correspond to probe transitions at different energies with respect to the Fermi level (discussed in the text).

Experimental Methods

The sample is an ultrathin epitaxial graphene film produced on the C-terminated face of single-crystal 4H-SiC by thermal desorption of Si. Details of the growth process and sample characterization may be found in ref. [1]. The structure of the sample is shown in Figure 1; a 100-fs near-infrared (800-nm) optical pulse (from a 250-kHz regenerative amplifier) excites quasiparticles from the valence to the conduction band across the Dirac point; the optical response is measured via the differential transmission (DT) of a mid-infrared probe pulse from a tunable femtosecond OPA as a function of pump-probe delay. The elevated temperature of the quasiparticles is manifested primarily through the modification of the probe beam absorption by Pauli blocking of interband transitions.

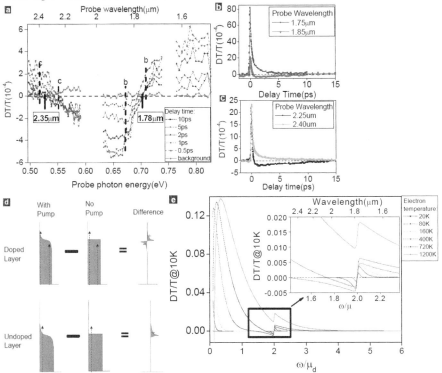

Fig. 2. a, DT spectrum on epitaxial graphene at 10K, with 500- μ W 800-nm pump with less than 100-fs pulse width at probe delays of 10 ps, 5 ps, 2 ps, 1 ps , 0.5 ps, and background (50 ps before the pump arrives). The arrows at 1.78 μ m and 2.35 μ m indicate where the DT signal flips sign. b, DT time scan of the two probe wavelengths marked in part a at the red (1.85 μ m) and blue side (1.75 μ m) of the 1.78 μ m DT zero crossing. c, Time scan of the two probe wavelengths marked in part a at the red (2.40 μ m) and blue side (2.25 μ m) of the 2.35 μ m DT zero crossing. In all figures, the dashed line (brown) marks where the DT signal is zero. The DT tails in b and c are simply fitted by a sigmoidal curve. d. Distribution of carriers after (left), before (middle) the excitations of the pump beam and the carrier distribution difference (right) for doped (upper) and undoped (lower) layer respectively. The + and - give the sign of the DT/T signal corresponding to the probe transitions showed by the arrows on the left. Simulated DT/T curves at different electron temperatures with lattice temperature at 10K. In the inset, the DT/T curves for low electron temperatures are expanded in the vicinity of the two DT zero crossings. Both figures share the same legend.

Results and Discussion

The DT spectra (shown in Figure 2a) exhibit two zero crossings, at 1.78 µm and 2.35 µm respectively. Figures 2b and 2c show DT time scans for selected probe wavelengths on both sides of the two zero crossings. From the simplest point of view, the differential probe transmission spectrum arises from the change in carrier occupation functions as shown in Figure 2 (d). Following the excitation of quasiparticles into the conduction band by the pump pulse, electron-electron scattering on a time scale short compared to 150 fs establishes a hot thermal distribution characterized by an electron temperature T_e. Since the carrier occupation probability above the Fermi energy is increased, the DT signal is positive due to reduced probe absorption. The probe DT is negative below the Fermi level; however, since the heating of the electron plasma reduces the occupation probability for low energies. Thus the upper zero crossing at 1.78 µm probe wavelength is interpreted as arising from the smearing of the Fermi level in the doped layers. Assuming no bandgap, we find the Fermi level to be 348 meV above the Dirac point for the doped layers. We note additionally that there is no peak in the DT spectrum near the Fermi level; this indicates that there is no Fermi edge singularity due to electron-hole interactions in the interband absorption spectrum of graphene, consistent with the expected massless nature of the quasiparticles. At very long probe wavelengths, one may expect the DT spectra to be determined primarily by the carrier occupations in the undoped layers; however, for probe wavelengths below the Fermi level of the doped layer, the contribution of the doped layer to the DT is negative. Thus one expects that for some probe energy the net DT signal should flip sign; this is the origin of the lower zero crossing at 2.35 µm. Figure 2e shows calculated DT/T curves at different electron temperatures for a lattice temperature of 10 K, based on the theoretical electron temperature dependent dynamic conductivity [4], using the transfer matrix method. The simulated DT spectra show the upper and lower zero crossings at energies (2 μ_d and 1.5 μ_d respectively) close to those observed in the experiment. Additional simulations performed by excluding various contributions to the total conductivity reveal that our DT spectra are well described by interband transitions and the single-particle density of states for linear dispersion, and no electron-hole interaction.

Conclusions

In summary, we have observed the ultrafast relaxation dynamics of hot Dirac fermionic quasiparticles in multilayer epitaxial graphene. The DT spectra are well described by interband transitions with no electron-hole interaction. Following the initial thermalization and emission of high-energy phonons, the cooling is determined by electron-acoustic phonon scattering.

1 M. L. Sadowski, G. Martinez, M. Potemski, C. Berger and Walt A. de Heer, in Solid State Communication, Vol. 143, 123, 2007.

2 C. Berger, Zhimin Song, Xuebin Li, Xiaosong Wu, Nate Brown, Cecile Naud, Didier Mayou, Tianbo Li, Joanna Hass, Alexei N. Marchenkov, Edward H. Conrad, Phillip N. First, and Walt A. de Heer, in Science, Vol. 312, 1191, 2006.

3 K. S. Novoselov, A. K. Geim, S. V. Morozov, D. Jiang, Y. Zhang, S. V. Dubonos, I. V. Grigorieva, and A. A. Firsov, in Science, Vol. 306, 666, 2007.

4 S. A. Mikhailov and K. Ziegler, in Physical Review Letters, Vol. 99, 016803, 2007.

Radiationless Transitions and Angular Momentum Transfer in Semiconductor Nanocrystals

Cathy Y. Wong, Michelle C. Nagy, Jeongho Kim and Gregory D. Scholes

Department of Chemistry, 80 St. George Street, Institute for Optical Sciences, and Centre for Quantum Information and Quantum Control, University of Toronto, Toronto, Ontario M5S 3H6 Canada
E-mail: gscholes@chem.utoronto.ca

Abstract. Measurements of ultrafast relaxation processes for population in the exciton fine structure states of CdSe nanocrystals are reported and discussed. Relationships between the mechanism of these dynamics and size and shape of nanocrystals are described.

Introduction

Nanoscale systems are forecast to be a means of integrating desirable attributes of molecular and bulk electronic structure regimes into easily processed, yet sophisticated, materials [1]. The optical properties of nanoscale systems, such as semiconductor nanocrystals (NCs), are dictated by rules that combine aspects of molecular chromophores with those of bulk systems. Thus the physical size and shape of such materials significantly influence the properties of electronic levels and excited states, but those electronic states are founded on the large-scale periodicity of the three-dimensional crystal lattice. Therefore spectroscopic properties of NCs relate to those of the corresponding bulk materials, but details such as the fine structure are accentuated. This new size regime also offers opportunities for exploring the control of optical properties of inorganic nanoparticles by synthesizing more complex systems. In recent work we have employed synthesis, theory, and spectroscopy in concert to study ultrafast radiationless transitions which cause relaxation among the fine structure states of CdSe semiconductor NCs.

A differentiating property of nanocrystals is that the exciton is coupled to only a few nuclear degrees of freedom, and this coupling is weak. Moreover, the frequencies characteristic of these motions are small, and they correspond largely to two kinds of vibrational modes, the longitudinal optical phonon (at ~200 cm^{-1}) and size-dependent quantized acoustic phonons (torsional modes at ~10 cm^{-1}) [2]. These frequencies are small compared to the dominant absorption features seen in NC spectra, so it was postulated that radiationless relaxation rates measured after exciting high into the NC absorption spectrum would be slow (the phonon bottleneck). However, contrary to that prediction, relaxation rates are fast in colloidal NCs and have been explained to be promoted by a distinctly different mechanism than the one operating in molecules: the Auger mechanism [3]. In contrast to that body of work, we have been examining the relaxation processes among the band of "fine structure" states associated with the lowest energy exciton transition, Figure 1a. Although these states are completely obscured by inhomogeneous line broadening in the frequency domain, they can be likened to a "Jablonski diagram" for NCs. Here we elucidate the relative rates for relaxation pathways among these states and develop a comparison and contrast with analogous, but much simpler, processes in molecules (internal conversion and intersystem crossing). The present contribution will report on these relaxation processes, how they depend on NC size and shape, and their origin, with emphasis on processes that change the sign of the total angular momentum projection. Such phenomena are associated with exciton spin relaxation.

Results and Discussion

A cross polarized, heterodyned, transient grating method has been developed to measure relaxation within the fine structure of NCs [4]. The inset of Figure 1b shows signals measured using this technique with a core-shell-shell (CSS) dot sample, shown schematically in Figure 1c. The cross polarized data can be fit to biexponential decays, corresponding to two relaxation processes: a fast process, and a slower secondary process. These processes in NCs have been largely overlooked until recently [5, 6].

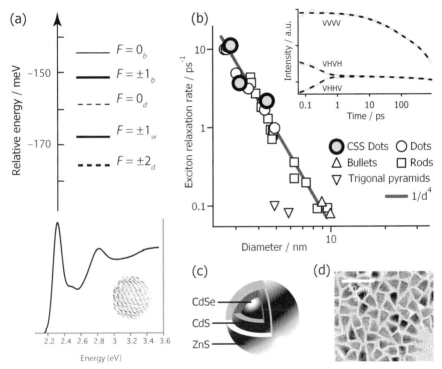

Fig. 1. (a) Exciton fine structure states, categorized by total angular momentum, F, and absorption spectrum for Wurtzite CdSe quantum dots. The fine structure is hidden under the inhomogeneously broadened first absorption band. (b) Dominant fine structure relaxation rate versus sample diameter for various NC shapes and CSS quantum dots. Inset: Experimental data collected with CSS dots (solid grey) fit to biexponentials (black dashed). (c) Schematic of a CSS dot. (d) Transmission electron micrograph of a more complex CdSe nanocrystal.

Figure 1b shows a plot of the dominant radiationless relaxation rate, measured in samples of various shape and size. We have found that the time scales for the dominant relaxation process are strongly pathway dependent and time constants range from 100s of femtoseconds to tens of picoseconds. The rate constant plotted in Figure

1b turns out to be strongly dependent only on the diameter of nanocrystals (and not on length at all) [6], which has enabled us to employ this trend as a 'calibration curve' to estimate how exciton size depends on shape. In NCs with more complex shape, like those shown in Figure 1d, the exciton size cannot be approximated as a sphere, indicating that the exciton expands in a shape dependent manner [7, 8].

In the studied CSS dots, the shells are arranged such that each successive layer is of increasing bandgap. This confines the exciton to the inner CdSe core, the size of which is determined by measuring the absorption maximum. New measurements of CSS dot samples of different sizes yield relaxation rates which agree with the size dependence of our previous measurements of organically passivated NCs, as emphasized in Figure 1b. These results indicate that the measured relaxation process is independent of surface states; CSS dots should ideally not have any surface states, yet they show dynamics which agree with those of (imperfectly) passivated samples. As well, we note that the relaxation rate scales with the size of the exciton confined in the core, not with the size of the NC as we would expect were the size-dependence of the relaxation caused predominantly by coupling to phonons.

The processes we are observing include the various relaxation pathways within the fine structure states of the lowest exciton. For example, such relaxation processes decide the pathways taken from optical excitation to photoluminescence: say, $F = +1 \rightarrow +2$ versus $+1 \rightarrow -2$, etc. in the "resonant Stokes shift". Note that the sign of the total angular momentum flips, and it is this phenomenon that enables us to record these dynamics, even though the states involved are completely obscured by inhomogeneous line broadening [4]. The spacing between these levels is small, so there is no reason to expect a phonon bottleneck; the relaxation mechanism is therefore more closely related to that operative in molecules.

What is interesting about NCs as compared to organic molecules is that we postulate that there are a few different kinds of relaxation mechanisms, somewhat analogous in spirit to internal conversion and intersystem crossing in molecules. The corresponding relaxation rates are strongly dependent on electronic matrix elements (involving a combination of exchange, strain, and spin-orbit coupling contributions) as well as coupling to phonons.

Although the methodology is challenging, studying these ultrafast relaxation processes provides deeper insights into the electronic structure and properties of NC excitons because it is those details that decide the relaxation rates among the fine structure states. Since we are examining fine structure states specifically, we have a precise window into properties specific to excitons, such as electron correlation effects and how the wavefunctions can be weakly coupled by perturbations that are too small to detect by other means.

1   G. D. Scholes and G. Rumbles, Nature Mater. **5**, 683, 2006.

2   M. R. Salvador, M. W. Graham, and G. D. Scholes, J. Chem. Phys. **125**, 184709, 2006.

3   V. I. Klimov, Annu. Rev. Phys. Chem. **58**, 635, 2007.

4   G. D. Scholes, J. Kim, and C. Y. Wong, Phys. Rev. B **73**, 195325, 2006.

5   V. M. Huxter, V. Kovalevskij, and G. D. Scholes, J. Phys. Chem. B **109**, 20060, 2005.

6   J. Kim, C. Y. Wong, S. Nair, K. P. Fritz, S. Kumar, and G. D. Scholes, J. Phys. Chem. B **110**, 25371, 2006.

7   G. D. Scholes, J. Kim, C. Y. Wong, V. M. Huxter, P. S. Nair, K. P. Fritz, and S. Kumar, Nano Lett. **6**, 1765, 2006.

8   J. Kim, P. S. Nair, C. Y. Wong, and G. D. Scholes, Nano Lett. **7**, 3884, 2007.

Ultrafast Carrier Dynamics in Semiconductor Nanowires

R. P. Prasankumar[1], S. G. Choi[1], G. T. Wang[2], P. C. Upadhya[1], S. A. Trugman[1], S. T. Picraux[1], and A. J. Taylor[1]

[1] Center for Integrated Nanotechnologies, Los Alamos National Laboratory, Los Alamos, New Mexico 87545
 E-mail: rpprasan@lanl.gov
[2] Sandia National Laboratories, P. O. Box 5800, MS-1086, Albuquerque, New Mexico 87185

Abstract. Ultrafast wavelength-tunable optical measurements on semiconductor nanowires allow us to independently probe the dynamics of electrons, holes, and defect states. These investigations reveal the influence of two-dimensional confinement on carrier dynamics in these nanosystems.

Introduction

Semiconductor nanowires (NWs) have attracted much recent attention as growth techniques have matured, enabling the fabrication of high quality NWs and nanowire-based devices. This has made fundamental studies possible [1-4] and led to a number of potential applications for these one-dimensional systems. Arguably the most promising application of semiconductor NWs is in the area of nanophotonics [1]. However, further development of NW-based nanophotonic devices will depend critically on an understanding of carrier relaxation in these systems. For example, GaN-based photonic devices typically exhibit "yellow luminescence" (YL), in which a broad luminescence band centered at 550 nm that is believed due to deep acceptor states reduces device efficiency [5, 6]. Ultrafast optical spectroscopy can shed light on this problem by measuring carrier transfer to and from these states, which will be important in understanding this phenomenon and optimizing device performance.

In this work, we use ultrafast optical spectroscopy to examine carrier dynamics in semiconductor nanowires while varying different growth parameters. Recent advances in NW synthesis have enabled the fabrication of high quality vertically aligned NWs with well controlled diameters. This allows us to examine the size dependence of carrier dynamics in Ge NWs. In addition, we can separately probe electron and hole dynamics parallel and perpendicular to the NW axis by varying the pump and probe wavelength, polarization, and incidence angle. Finally, time-resolved ultraviolet (UV)-pump, visible-probe measurements on GaN NWs enable us to examine the effects of growth temperature on carrier dynamics in the defect states responsible for YL.

Experimental Methods

Ultrafast optical experiments were performed at room temperature on single crystal germanium and gallium nitride nanowires grown by chemical vapor deposition using the vapor-liquid-solid (VLS) mechanism. Our two GaN NW samples were grown at 800 and 900 °C, respectively, on r-plane sapphire substrates using seed Ni nanoparticles [5]. The GaN NWs have average diameters of ~50-150 nm and lengths of ~5 μm. Gold nanoparticles were used to seed the growth of vertically aligned Ge

NWs on Si (111) substrates [7]. Two samples were fabricated, with average NW diameters of 18 and 30 nm, respectively, and lengths of ~1.6 μm. The optical pump-probe system is based on a 100 kHz regeneratively amplified Ti:sapphire laser system producing 50 fs, 10 μJ pulses at 800 nm that concurrently pumps two optical parametric amplifiers, enabling measurements with independently tunable pump and probe wavelengths from 400 nm to 3 μm. Third harmonic generation of the fundamental 800 nm beam was also implemented to enable UV pumping at 267 nm.

Results and Discussion

Previous photoluminescence (PL) measurements on GaN NWs grown at different temperatures demonstrated that NWs grown at 800 °C exhibited YL comparable in strength to the band edge luminescence (BEL), while growth temperatures of 900 °C increased the BEL by over two orders of magnitude while the YL emission remained essentially unchanged [6]. Therefore, we performed 267 nm pump, 550 nm probe experiments in transmission on GaN NWs aimed at measuring the effects of growth temperature on carrier transfer rates to and from the YL band after above-band gap excitation. It is clear from Figure 1 that the dynamics differ greatly between NWs grown at 800 °C and 900 °C. The long-lived negative signal in GaN NWs grown at 800 °C (Fig. 1(a)) may indicate that carriers relax into defect states of unknown origin, from which they can be promoted to higher energy states through induced absorption of the probe photons. These long-lived defect states are most likely not YL states, which would give a positive bleaching signal, but may instead be surface-related or due to carbon incorporation at lower growth temperatures [6]. In contrast, the positive signal observed from NWs grown at 900 °C (Fig. 1(b)) indicates that photoexcited carriers saturate YL defect states that recover within 43 ps. This supports the PL data by demonstrating that 900 °C growth eliminates the relaxation pathway observed in 800 °C grown NWs, likely by creating higher quality NWs with fewer defect states. Our results thus suggest that the growth temperature may provide a degree of control over defect states in GaN NWs.

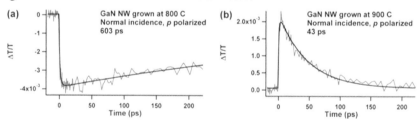

Fig. 1. Comparison of dynamics in GaN NWs grown at (a) 800 °C and (b) 900 °C.

Previous theoretical work on bulk Ge [8] demonstrated that after 800 nm excitation, electrons scatter to the conduction band minimum at the L point and holes scatter to the valence band maximum at the Γ point within 4 ps due to the indirect nature of the band gap. A visible probe will thus only be sensitive to electron dynamics at the L point, while a near-IR probe is only sensitive to hole dynamics at the Γ point [5]. Figure 2(a) depicts 800 nm pump, 550 nm probe measurements performed in reflection on 18 and 30 nm diameter Ge NWs and an undoped Ge substrate. It is clear that the transient signal decays most rapidly for 18 nm NWs, followed by 30 nm NWs and finally bulk Ge (this was also observed for a 1200 nm probe). The size dependence of the relaxation times agrees well with a model for surface recombination in NWs, strongly indicating that after the first few picoseconds, surface-related phenomena dominate electron and hole dynamics in Ge NWs [5].

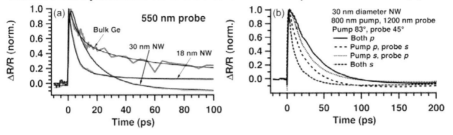

Fig. 2. Ultrafast optical pump-probe measurements on Ge NWs. (a) Dependence of electron dynamics on NW diameter. (b) Dependence of hole dynamics at a large incidence angle on pump and probe polarizations.

Carrier dynamics perpendicular and parallel to the axis of the vertically aligned Ge NWs can be probed by varying the incidence angle and polarization of the pump and probe beams (Figure 2(b)). In this configuration, s polarization refers to a beam polarized perpendicular to the NW axis, while p polarization refers to a beam that has components polarized both perpendicular and parallel to the NW axis. From Fig. 2(b), it can be seen that the $\Delta R/R$ signal decays most rapidly when both pump and probe are s polarized, while the decay is slowest when both pump and probe are p polarized (also observed at 550 nm). This suggests that carriers propagating orthogonal to the NW axis rapidly recombine at the surface, while carriers propagating parallel to the NW axis can travel farther before recombining. However, further investigation is necessary for a complete understanding of this phenomenon.

Conclusions

In conclusion, we have performed ultrafast optical spectroscopy on Ge and GaN NWs while varying growth and experimental conditions. Time-resolved measurements on GaN NWs indicate that the growth temperature alters the role of defect states in carrier relaxation. In Ge NWs, we are able to isolate electron and hole dynamics perpendicular and parallel to the NW axis as a function of NW diameter, revealing the influence of surface-related phenomena and two-dimensional confinement on carrier relaxation. Future experiments will focus on varying material parameters such as the nanowire composition and heterostructuring to further elucidate carrier dynamics in these novel nanosystems. Sandia is a multiprogram laboratory operated by Sandia Corporation, a Lockheed Martin Company, for the United States Department of Energy under contract DE-AC04-94AL85000.

1. D. J. Sirbuly et al., J. Phys. Chem. B 109, 15190 (2005).
2. J. C. Johnson et al., Nano Lett. 4, 197 (2004).
3. P. Parkinson et al., Nano Lett. 7, 2162 (2007).
4. R. P. Prasankumar et al., Nano Lett. 8, 1619 (2008).
5. G. T. Wang et al., Nanotechnology 17, 5773 (2006).
6. A. A. Talin et al., Appl. Phys. Lett. 92, 093105 (2008).
7. J. W. Dailey et al., J. Appl. Phys. 96, 7556 (2004).
8. D. W. Bailey and C. J. Stanton, J. Appl. Phys. 77, 2107 (1995).

Time-resolved photoemission spectroscopy in graphite

Tadashi Togashi[1,2], Kazuya Yamamoto[4,2], Yukiaki Ishida[2] Masashi Tnanaka[3], Toshiyuki Taniuchi[3], Atsushi Shimoyamada[3], Takayuki Kiss[3], Kyoko Ishizaka[3], Ashish Chainani[2], Yasitaka Takata[2], Makoto Nakajima[3], Tohru Suemoto[3], Tetsuya Ishikawa[1,2], and Shik Shin[3,2]

[1] XFEL Project Head Office RIKEN, 1-1-1 Sayo-cho, Sayo-gun, Hyogo 679-5148, Japan
   E-mail: tadashit@spring8.or.jp
[2] SPring8 Centre RIKEN, 1-1-1 Sayo-cho, Sayo-gun, Hyogo 679-5148, Japan
[3] Institute for Solid State Physics University of Tokyo, 5-1-5 Kashiwanoha Kashiwa, Japan
[4] Graduate School of Engineering, Osaka Prefecture University

Abstract. We study electron dynamics in graphite with time resolved photoemission spectroscopy (TRPES). Using the fourth harmonic (5.95 eV) and fundamental of a Ti:sapphire laser as a pump-probe, respectively, we observe a biexponential decay of excited electrons.

Introduction

Graphite has captivated continuous attention in scientific research over the past several decades. More recently, graphene, a one-atom-thick crystalline layer of graphite has ignited enormous fundamental and applied interest in graphite and related materials [1]. A breakdown of the Born-Oppenheimer approximation (BOA, which assumes that the lighter electrons adjust adiabatically to motion of the heavier nuclei) was found in graphene [2]. It was shown to originate in the excited electrons, which deviate from an adiabatic ground state, and a strongly coupled electron-phonon interaction. Several other studies have also reported anomalous quasi-particle lifetime behaviour observed with angle resolved photoemission spectroscopy [3, 4]. Time resolved measurements of electron states have been recently studied with photoemission spectroscopy [5, 6] and tera-hertz spectroscopy [7]. In this work, we study the electron dynamics in graphite in order to elucidate the electron-phonon/collective mode coupling, using time resolved pump-probe photoemission spectroscopy (TRPES). The fourth harmonic (4ω: 5.90 eV) and fundamental (ω: 1.48 eV) of a Ti:sapphire laser have been employed as pump and probe, respectively. We have adapted a high-resolution hemispherical electron analyzer.

Experiment

Figure 1 shows the experimental setup of the TRPES system. We employed a Ti:sapphire laser system (RegA9000, Coherent Inc.) producing $\sim$3.2-μJ pulse energy with 250-kHz repetition rate and 170-fs pulse duration. The deep ultraviolet pulses for probing photoemission spectra were generated by focusing second harmonics (2ω) of ω on a β-BaB$_2$O$_4$ (BBO) type I crystal. The wavelength of second harmonic generation is limited to 204.5 nm because of phase matching condition in BBO [8]. The conversion efficiency is reduced on shortening the wavelength. Therefore we tuned the laser system as 840-nm wavelength and generated 210-nm probe pulses as 4ω. The temporal profile of 4ω was estimated to be a Gaussian pulse of 112-fs duration by cross correlation trace

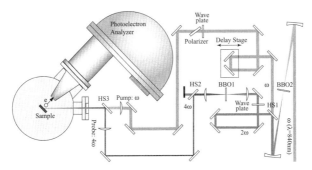

Fig. 1. Schematic layout of the TRPES system. BBO1 and BBO2: β-BaB$_2$O$_4$ for 2ω and 4ω generations respectively, HS1 and HS2: dielectric mirrors to separate 2ω from ω and 4ω from 2ω respectively, HS3: dielectric mirror for combining 4ω with ω

of 3ω generated as difference frequency mixing between ω and 4ω. The fundamental pulses (ω) remaining after the second harmonic generation, which was divided by harmonic separator from 2ω, were used for pump. The probe and pump pulses were combined by a dielectric mirror and focused onto a sample in the measurement chamber through CaF$_2$ window by 1000-mm and 700-mm focal length lenses respectively. For spatial and temporal adjustment of the overlap between two pulses ω and 4ω at the sample position, transition IR reflectivity of gallium arsenide (GaAs) was measured at same position of sample in the measurement chamber. The measurement chamber including the hemispherical electron analyzer (SES2002, VG Scienta) has a vacuum better than 1×10^{10} Torr. The energy resolution of the TRPES system was estimated to be 15 meV by fitting the photoemittion spectrum of gold, which was measured by 4ω, with Fermi-Dirac function. The spectral width of 4ω amounts to the energy resolution of the system because of high resolution of the electron analyzer [9]. The HOPG sample was attached to a copper plate and the measurements were performed at 200 K. The clean surface was obtained by cleaving and/or annealing in the preparation chamber. We set the pump power to 200 mW with the attenuator (half wave plate and polarizer in Fig. 1). The spot size of the pump was estimated to be 1.4-mm diameter at the sample position with a knife-edge method, which gives a 52-μJ/cm^2 photon flux.

Results and Discussion

Figure 2 (a) shows the TRPES spectra of HOPG measured at room temperature. The spectrum before 3.7 ps from pumping (-3.7-ps delay) is that measured by the probe in the absence of the pump. The spectral intensity of the unoccupied states grows up rapidly, upon shortening the interval between the pump and the probe, and becomes maximum at the time of pumping (0 delay). The spectra after the pumping shows gradual decay processes in the unoccupied states. Finally, the spectrum after $\sim$ 20-ps delay returns to the same shape as in the absence of the pump. Figure 2 (b), which shows time resolved photoelectron intensity map of HOPG as functions of delay time and binding energy, indicates decay processes at each energy position. More than a single time-scale decay can be recognized in the trace at each energy position. For instance at 8 meV above Fermi level (E_F) (Fig 2 (c)), Fast and slow decay timescales

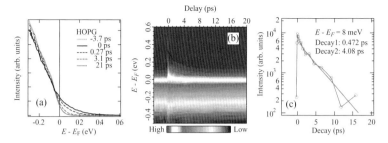

Fig. 2. (a): TRPES spectra of HOPG pumped by the 1.48 eV pulse at room temperature. The 0 eV of the horizontal axis stands for Fermi level (E_F). (b): Time resolved photoelectron intensity map of HOPG as functions of delay time and energy from E_F. (c): Decay trace at 8 meV above E_F.

are estimated to be 0.472 ps and 4.08 ps respectively by fitting with a dual exponential function. The fast decay corresponds to initial thermalization of electron and hole. The electrons excited immediately by ultrafast pulses are in non-equilibrium state and then their energy is redistributed possibly by electron-electron scattering. After the thermalization, the hot electrons constitute a Fermi-Dirac distribution on cooling and diffuse due to an additional interaction such as electro-phonon, electron-electron and/or electron-collective mode with $1 \sim 10$ ps decay scale. Because the decay process of the hot electrons directly connects to the transport property of materials, it is important to investigate the timescale and energy dependence of the decay not only of the occupied states, but also for the unoccupied states populated upon pumping.

Conclusions

We observed the electron state dynamics excited the ultrafast pulse in graphite using time resolved photoemission spectroscopy system. The density of state on the unoccupied state rose up within 0.1 ps. Then the life times of decay presence could consist of at lease two different time scale, $0.1 \sim 1$ ps fast and $1 \sim 10$ ps slow decays.

1   A. K. Geim and K. S. Novoselov, Nature Materials 6, 183, 2007
2   S. Pisana, M. Lasseri, C. Casiraghi, K. Novoselov, A. K. Geim, A. C. Ferrari, and F. Muri, Nature Materials 6, 198, 2007
3   A. Bostwick, T. Ohota, T. Seyller, K. Horn, and E. Rotenberg, Nature Physics 3, 36, 2007
4   Sugawara, T. Sato, S. Souma, T. Takahashi, ans H. Suematsu, Phys. Rev. Lett. 98, 036801, 2007.
5   S. Xu, J. Cao, C. C. Miller, D. A. Mantell, R. J. D. Miller, ans Y. Gao, Phy. Rev. Lett. 76, 483, 1996.
6   G. Moos, C. Gahl, R. Fesel, M. Volf, and T. Hertel, Phys. Rev. Lett. 87, 267402, 2001.
7   T. Kampfrath, L. Perfetti, F. Schapper, C. Frischkorn, and M. Wolf, Phys. Rev. Lett. 95, 187403, 2005.
8   V. G. Dmitriev, G. G. Gurzadyan, and D. N. Nikogosyan, Handbook of Nonlinear Optical Crystals, Springer Verlag, 1999.
9   A. Chainani, T. Yokoya, T. Kiss and S. Shin, Phys. Rev. Lett. 85, 1966, 2000.

Exciton Dephasing in Semiconducting Single-Walled Carbon Nanotubes

Y.-Z. Ma[1], M. W. Graham[1], A. A. Green[2], S. I. Stupp[2], M. C. Hersam[2], G. R. Fleming[1]

[1] Department of Chemistry, University of California, Berkeley, and Physical Biosciences Division, Lawrence Berkeley National Laboratory, Berkeley, California 94720-1460, USA
   E-mail: grfleming@lbl.gov
[2] Department of Materials Science and Engineering, Northwestern University, Evanston, Illinois 60208-3108, USA

Abstract. Two-pulse four-wave mixing experiments at various excitation intensities and temperatures enable the contributions of exciton-exciton and exciton-phonon scattering to exciton dephasing to be separated. We identify the dominant phonon mode, and estimate the homogeneous linewidth.

Introduction

The diameter-tunable optical properties of semiconducting single-walled carbon nanotubes (SWNTs) are governed by excitons with an anomalously large binding energy [1]. Application of a variety of time-resolved optical spectroscopic techniques have enabled elucidation of the different dynamical processes involved in exciton population relaxation in structurally distinct nanotube species [2]. Investigations of the exciton dynamics in the coherent regime have not been reported. As demonstrated previously for numerous systems, such studies can provide detailed information about the interactions between various quasi-particles such as excitons, charged carriers and phonons, and their roles in exciton dephasing process [3]. In this contribution, we present an experimental study of ultrafast exciton dephasing in semiconducting SWNTs using a femtosecond two-pulse degenerate four-wave mixing (FWM) technique.

Experimental Methods

The sample used in this study is highly enriched in a single semiconducting tube type, the (6, 5) nanotube, which was isolated through density-gradient ultracentrifugation [4]. In order to suppress laser light scattering, an isolated aqueous solution was used to fabricate a thin SWNT-polymer composite film using water soluble polyvinyl-pyrrolidone polymer. The two-pulse degenerate FWM technique employed here has been described extensively elsewhere [3]. In short, the light source was an optical parametric amplifier pumped by a 250 kHz Ti:Sapphire regenerative amplifier [5], which was tuned to a central wavelength of 990 nm in order to resonantly excite the lowest excitonic transition of the (6, 5) tube. Two nearly equal intensity laser pulses of 45 fs duration with wavevectors k_1 and k_2 are focused to the sample, and the intensity of the diffracted signal is detected at the phase-matching direction $2k_2 - k_1$ as a function of time delay between the two pulses, τ_{12}. The detection involves an InGaAs photodiode and a lock-in amplifier.

277

Results and Discussion

The experiment was performed at several lattice temperatures ranging from 77 to 292 K, and at each temperature five different excitation intensities between 1.70 and 12.21 $\mu J/cm^2$ were employed. Representative time-integrated FWM signal measured at 77, 200 and 292 K with the lowest excitation intensity ($\sim 1.70 \mu J/cm^2$) is shown in Figure 1. It is evident that the decay of the FWM signal depends strongly on temperature, manifested by a significantly faster decay with increasing temperature. Quantitative analysis of the data shows that the decay time τ_d decreases from 107 fs at 77 K to 54 fs at 200 K, and further to 35 fs at 292 K. Such a trend of the decay time change with temperature is also observed at other excitation intensities employed.

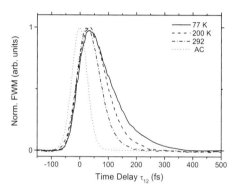

Fig. 1. FWM signals measured at 77, 200 and 292 K for the lowest excitation intensity employed, 1.70 $\mu J/cm^2$. The dotted line is the auto-correlation function of the two laser pulses obtained with a thin nonlinear crystal. All the data have been normalized at the signal maxima

At a given lattice temperature, the decay of the FWM signal is further found to accelerate remarkably with excitation intensity. For a same seven-fold intensity change, the variation of the decay time τ_d obtained at 292 K is about three times smaller than the corresponding value at 77 K. This strong temperature effect on the decay time can also be clearly seen from a plot of the decay rate $1/\tau_d$ versus excitation intensity (see Figure 2a). At 77, 100 and 130 K, the rate increases linearly with intensity, whereas at higher temperatures obvious deviation from linearity is observed.

By extrapolating the decay time determined at different excitation intensities to the limit of zero-intensity for each lattice temperature, we can examine the contribution of phonons to the exciton dephasing without other effects. The obtained decay rate is plotted as a function of temperature in Figure 2b. As shown by the solid line, this temperature dependence can be best described by $1/\tau_d = a + b\exp(-\Delta E/k_B T)$ with $a = 12.7 \pm 0.2$ ps^{-1}, $b = 809.3 \pm 170.0$ ps^{-1}, and $\Delta E = 855.5 \pm 43.4$ cm^{-1}. As the second term in this fitting function is equivalent to the thermal occupation function of optical phonons [3, 6], the obtained ΔE value is simply the frequency of the phonon mode involved in the dephasing process. According to the Raman spectra of both SWNT bundles and single nanotube [7], it is straightforward to identify it as the out-of-plane, transverse optical (TO) phonon mode. We therefore conclude that this TO mode, rather than the much stronger ones such as the radial breathing mode and the G-band mode, plays a dominant role in the exciton dephasing process within the temperature range of our study. The data shown in Figure 2b further suggests that the contribution from acoustic phonons to the dephasing process, which should give rise to a linear dependence of the rate on temperature, is insignificant.

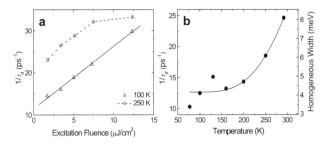

Fig. 2. (a) Dependence of the decay rates at 100 K (triangles) and 250 K (circles) on excitation intensity. The solid line is a linear fit to the data obtained at 100 K, and the dashed line is drawn to guide the eye. (b) The decay rates of the FWM signals in the zero-intensity limit versus temperature. The solid line is a least-squares fit. The corresponding homogeneous widths are depicted by the scales on the right side

From the decay rates shown in Figure 2b, we also estimate the intrinsic homogeneous linewidths Γ_h (full width at half maximum) by assuming strongly inhomogeneous broadening. In this case, Γ_h was calculated through $\Gamma_h = 2\hbar/T_2 = 2\hbar/4\tau_d$ [3]. Despite the fact that this simple relation is strictly valid only for independent two-level systems, our calculated linewidths at different temperatures (see Figure 2b, the label on the right side) agree very well with the results of single tube PL measurements on pillar-suspended tubes [8].

Conclusions

In summary, we have demonstrated for the first time that both exciton-exciton and exciton-phonon scattering profoundly affect the exciton dephasing in semiconducting SWNTs. The out-of-plane TO phonon with a frequency of 855 cm^{-1} is identified to be the dominant mode. The estimated homogeneous linewidths at different temperatures are in excellent agreement with the experimental data of a single tube PL experiment.

Acknowledgements. This research was supported by the NSF.

1   M. S. Dresselhaus, G. Dresselhaus, R. Saito, and A. Jorio, Annu. Rev. Phys. Chem. **58**, 719, 2007.

2   Y.-Z. Ma, T. Hertel, Z. V. Vardeny, G. R. Fleming, and L. Valkunas, in Carbon Nanotubes: New Topics in the Synthesis, Structure, Properties and Applications, Edited by A. Jorio, G. Dresselhaus, and M. S. Dresselhaus, Vol. 111, 321, 2008.

3   J. Shah, Ultrafast spectroscopy of semiconductors and semiconductor nanostructures, 2nd ed., Springer, Berlin, 1999.

4   M. S. Arnold, A. A. Green, J. F. Hulvat, S. I. Stupp, and M. C. Hersam, Nature Nanotech. **1**, 60, 2006.

5   Y.-Z. Ma, L. Valkunas, S. L. Dexheimer, and G. R. Fleming, Mol. Phys. **104**, 1179, 2006.

6   D.-S. Kim, J. Shah, J. E. Cunningham, T. C. Damen, W. Schäfer, M. Hartmann, and S. Schmitt-Rink, Phys. Rev. Lett. **68**, 1006, 1992.

7   M. S. Dresselhaus, G. Dresselhaus, and A. Jorio, Annu. Rev. Mater. Res. **34**, 247, 2004.

8   J. Lefebvre, P. Finnie, and Y. Homma, Phys. Rev. B **70**, 045419, 2004.

On the Absence of Carrier Multiplication in InAs Core/Shell/Shell Nanocrystals

Meirav Ben-Lulu[1], David Mocatta[2], Mischa Bonn[3] Uri Banin[2], and Sandy Ruhman[1]

[1] Institute of Chemistry and the Farkas Center for Light Induced Processes, The Hebrew University, Jerusalem 91904, Israel
E-mail: sandy@fh.huji.ac.il
[2] Institute of Chemistry and the Center for Nanoscience and Nanotechnology The Hebrew University of Jerusalem, Jerusalem 91904, Israel
[3] FOM Institute for Atomic and Molecular Physics (AMOLF), Kruislaan 407, 1098 SJ Amsterdam, The Netherlands

Abstract. An ultrafast pump-probe methodology for detecting spontaneous carrier multiplication is applied to InAs/CdSe/ZnSe Core/Shell1/Shell2. Contrary to previous reports no carrier multiplication following above-band gap photoexcitation is observed, questioning the ubiquity of this phenomenon.

Introduction

Carrier multiplication (CM) is a process in which absorption of an energetic photon leads to generation of more than 1 exciton (fig. 1, left panel). In semiconductor (SC) quantum dots (QDs) this process was found to be highly efficient. Here we report a protocol for probing CM in QDs using femtosecond transient absorption (TA) spectroscopy. Demonstration of CM has relied on disparity of single and multiple exciton state lifetimes in SC QD's [1]. While single excitons live for nanoseconds, multiple excitons decay to long-lived monoexciton states via Auger recombination (AR) which is effectively the reverse process of CM, within tens of picoseconds (fig.1). Multiple exciton states are however generated not only through CM but also by consecutive multiphoton absorption, posing a significant obstacle in the path of quantifying CM from TA data. Above band gap (BG) irradiation necessarily leads to single as well as multiple exciton states with probabilities which follow Poisson statistics.

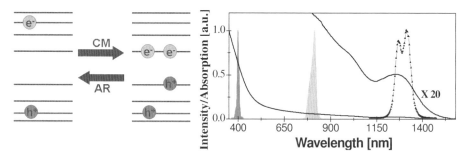

Fig. 1. left panel - A schematic diagram of the CM process, creating 2 band edge excitons from an energetic exciton. The reverse process is an auger recombination. right panel - Absorption spectra of the CSS sample (seen in black. there is a magnified section of the band gap area), along with the pump (grey and light grey pulses) and probe (triangles) pulses.

In InAs the degeneracy of the lowest transition dictates that multiexcitons contribute twice as much as single excitons to the band edge bleach. Accordingly, the ratio between single and multiple excitons generated by ultrafast photoexcitation has been derived from the ratios of fast and slow decay components of that feature in these samples. This can be derived similarly from photoluminescence kinetics. Finally, CM has been quantified from excess in the measured ratio relative to that expected from the Poissonian distribution of exciton states directly excited by the pulse. That requires knowledge of the NC absorption cross sections (σ) at all excitation wavelengths, and from them a calculation of the photoexcited number distribution which is also a function of sample optical density due to variations in the pump fluence. But the absolute σ, and perfect kinetic separation of single and multiple exciton decays are inherently difficult, and intertwined. To overcome these difficulties we have adopted an approach which requires only knowledge of the relative σ for absorption at the excitation energies - a ratio which can be derived directly from the absorption spectrum. Our approach consists of comparing pump-probe experiments at different excitation wavelengths above and below 2BG, in which the direct distribution of absorbed photon numbers are identical! That requires no more than the conservation of the product of the σ times the pump beam photon flux, and the sample OD, at each excitation wavelength. Notice that neither of these requires precise quantification of the σ itself, but only the ratio of its change from one wavelength to another.

Experimental Methods

The sample studied consisted of InAs/CdSe/ZnSe core-shell-shell (CSS) QDs [2], with a radius of 6.7nm (core radius was 5.9nm, see fig. 1, right panel). Excitation pulses at 800 and 400 nm were obtained from an amplified Titanium Sapphire laser producing 0.5 mJ, 30 fsec pulses centered at 790 nm, directly or by frequency doubling respectively. Those at 350nm were generated by quadrupling the signal output from a parametric amplifier (TOPAS). Probe pulses were generated by interference filtering from a supercontinuum generated in 1mm of N-SF6. A ratio above 2 was maintained between the diameters of pump and probe beams in the sample to ensure we are probing a region of nearly constant photon flux. Probe and reference were recorded on amplified Germanium photodiodes and lock-in amplified before digitization.

Results and Discussion

Pump-probe data is depicted in fig. 2. Panel A exhibits 800nm pump - NIR probe results for an optically thin sample (OD800nm=0.23). At low fluences the bleach amplitude increases linearly with fluence and exhibits a slow decay extending far beyond our range of delay variation. As the fluence is increased, a stage of rapid bleach decay assigned to AR is observed. And for higher fluences the bleach rises until it reaches saturation, due to the band edge degeneracy explained above. A decay time of 53psec was extracted for the AR by fitting a bi-exponential decay. Panel B exhibits a comparison of 800 and 400 nm excitation results. Sample OD was 1.2 at the pump wavelength. Optically thick samples were chosen to maximize the signal amplitudes. A perfect agreement at all delays is obtained for both excitation wavelengths. In particular no excess in bleach is observed at early times for 400 nm excitation relative to that obtained with 800 nm pump pulses. Also no rapid phase of bleach decay is apparent for the lowest 400 nm excitation fluence. All of these findings are consistent with total absence of CM up to photon energies of 3.25EBG. Finally, even when exciting the crystals with photons containing 3.54 eV of energy - 3.7E_{BG}, no difference other than the scaling of signal amplitude is observed. This is demonstrated in fig. 2 panel B along with matching 800 and 400 nm scans. It is important to point out that in all previous reports on InAs the threshold for appearance of CM was close to twice the band gap, which we have exceeded here considerably.

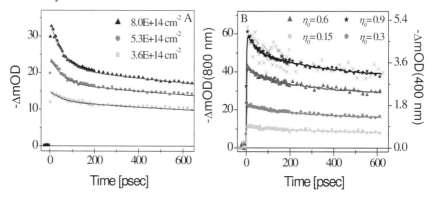

Fig. 2. A exhibits TA signals obtained by 800nm pumping of a diluted sample (OD800nm=0.23) in varying pump fluences. B exhibits 3 TA sets of signals taken each in 2 pumping wavelengths, but at the same number of photons absorbed per QD, on an optically thick sample (ODλpump=1.2). The grey crosses represent a signal obtained by a 350nm pumping. The left Y axis is scaled to -ΔmOD units of 800nm measurements, and, the right Y axis is scaled to −ΔmO.D. units of 400nm measurements. (350nm signal was scaled to - ΔmOD units of 800nm measurements).

Conclusions

We have presented a systematic protocol for detecting CM in SC QDs. It's application to InAs CSS systems clearly show no sign of CM, in contrary to a previous report on InAs QDs [3]. This study joins another recently report on the absence of CM in another SC QDs [4]. In light of these findings it is imperative to clarify the existence and efficiency of CM in SC QDs, particularly in view of its practical importance. Accordingly, we will apply the described protocol to reveal whether our findings carry over to other SC systems, or for that matter to other InAs core/shells.

1   R. D. Schaller, V. I. Klimov, Phys. Rev. Lett. **92**, 186601, 2004.
2   A. Aharoni, T. Mokari, I. Popov, U. Banin, J. Am. Chem. Soc. **128**, 257, 2006.
3   R. D. Schaller, J. M. Pietryga, V. I. Klimov, Nano Lett. **7(11)**, 3469, 2007.
4   G. Nair, M. G. Bawendi, Phys. Rev. B **76**, 081304, 2007.

Temporal dynamics of polaritons in a strongly-coupled organic-semiconductor microcavity

T. Virgili[1], S. Ceccarelli[2], D. Polli[1], G. Lanzani[1], G. Cerullo[1], D.G. Lidzey[2]

[1] IFN, CNR Dipartimento di Fisica, Politecnico di Milano, P.zza Leonardo Da Vinci 32, 20132 Milano, Italy
  Email:tvirgili@polimi.it
[2] Department of Physics and Astronomy, University of Sheffield, Hicks Building, Hounsfield Road, Sheffield S37RH United Kingdom

Abstract. Using pump-probe spectroscopy, we investigate exciton-polariton dynamics in a strongly-coupled organic microcavity. We observe Rabi oscillations, decay of polaritons and the signature of the upper-branch cavity polaritons scattering to the exciton reservoir with phonon emission.

Introduction

A microcavity is a structure in which two mirrors are placed in close proximity and are separated by a thin-film of an active semiconductor material. Within the so-called strong-coupling regime, the confined optical modes of the cavity undergo a coupling with the excitons within the semiconductor, resulting in the formation of cavity polariton states. Such cavities are characterized by two polariton branches (lower and upper), which are split at the point of exciton-photon resonance by the Rabi-splitting energy [1]. In this work we used pump-probe spectroscopy with 5fs time resolution to investigate exciton polariton dynamics in an organic strongly coupled microcavity. We observe Rabi oscillations in the time domain and the scattering of the upper-branch cavity polaritons back to the exciton reservoir.

Experimental Methods

The microcavities we have studied are based on a double dielectric-mirror structure containing a thin-film of a J-aggregate cyanine dye suspended in a gelatin matrix. This material (called F1) has been selected for study due to its narrow linewidth (thus facilitating strong-coupling), its high oscillator strength and ability to form optical-quality thin-films. The microcavity used in the experiment is shown schematically in the inset of Fig. 1. Here the bottom DBR is composed of 11 $\lambda/4$ pairs of silicon oxide and silicon nitride. The J-aggregate containing thin organic film was coated onto this mirror by spin-coating from solution. Onto this, a second DBR mirror was deposited by thermal evaporation and was composed of 12 $\lambda/4$ layers of TeO_2 and LiF. The cavity studied here had a relatively modest Q-factor of 155.

Figure 1 plots the energy of the two peaks observed in reflectivity measurements increasing the angle of incidence (defined away from the cavity normal). The data are presented in the form of a dispersion plot.

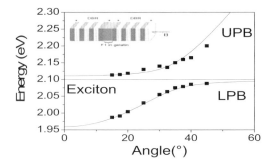

Fig.1: Dispersion curves constructed from the reflectivity spectra. The solid lines are a fit to the dispersion using a standard two-level model

Here, anticrossing behaviour between the exciton and the photon modes is clearly seen, with the upper (UPB) and lower (LPB) polaritonic branches marked. At exciton-photon resonance (at $\theta = 30°$) a doublet of cavity polaritons states is observed, which have equal linewidth and intensity. The two polaritons are separated equidistantly in energy around the peak of the J-aggregate absorption by the Rabi splitting energy, which in this cavity is equal to 80 meV.

Discussion and Conclusion

Using pump-probe spectroscopy with 5 fs time resolution [2], we investigate exciton- polariton dynamics in this strongly-coupled organic microcavity. In fig 2 we plot $\Delta T/T$ measured at 2.08 eV when the cavity is placed at 30° respect to the direction normal to the cavity. The slow-rise of the $\Delta T/T$ transients recorded from the cavity is associated with the time required for the optical field to build up within the cavity. The pump-probe signal of the polariton-like state (probed at 30°) has a fast initial decay component that disappears with ~200 fs time constant, which is superimposed on a slower decay with few ps lifetime (similar to that recorded from the photon-like mode). We associate the fast transient with the decay of the lower-branch cavity-polaritons and the longer transient to the decay of excitons in reservoir states that are populated by scattering from upper-branch cavity polaritons which lie at higher energies.

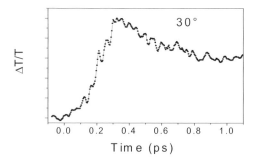

Fig.2: $\Delta T/T$ at 2.08 eV on the cavity placed at 30°.

284

The energy of the various coherent vibrational modes that contribute to the measured $\Delta T/T$ signal can then be extracted by performing a series of Fourier transforms as a function of pump-probe delay after subtracting from the $\Delta T/T$ signal the long decay. Figure 3 shows the Fourier transform for two temporal windows :0-500 fs and 500 fs-1000 fs. It can be seen that in the first 500 fs, two oscillatory modes are observed positioned around 40 and 88 meV.

We can straightforwardly associate the mode at 88 meV with the quantum beating of the laser-light transmitted through the two optical modes of the cavity which have an energy separation of $\hbar\Omega_{Rabi} = 80$ meV. Even more interesting is the vibrational mode observed at 40 meV. The energy of this mode is equal to approximately half of the Rabi-splitting energy, and thus coincides with the energy separation between the upper polariton branch and the top of the uncoupled exciton reservoir. We attribute this signal to the signature of the upper-branch cavity polaritons scattering back to the exciton reservoir following the emission of an energy quanta which excites a discreet molecular vibrational mode.

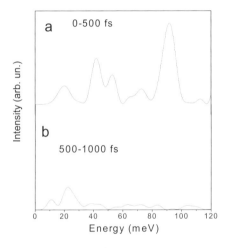

Fig 3: Fourier transforms of the oscillatory component of the signal for two different temporal windows: 0-500 fs (a) and 500fs-1000fs (b).

1. D.G. Lidzey, D.D.C. Bradley, M.S. Skolnick, T. Virgili, S. Walker and D.M. Whittaker, "Strong exciton-photon coupling in an organic semiconductor microcavity" Nature **395**, 53 (1998)
2. C. Manzoni ,D. Polli and G. Cerullo "Two-color pump-probe system broadly tunable over the visible and the near infrared with sub-30 fs temporal resolution" Rev. Sci. Instrum. **77**, 023103 (2006)

Three-Dimensional Electronic Four Wave-Mixing Spectroscopy in GaAs Quantum Wells

Daniel B. Turner, Katherine W. Stone, Kenan Gundogdu and Keith A. Nelson

Department of Chemistry, Massachusetts Institute of Technology, 77 Massachusetts Ave, Cambridge MA 02139, USA
E-mail: kanelson@mit.edu

Abstract. Three-dimensional electronic Fourier transform four wave-mixing spectroscopy of GaAs quantum wells is demonstrated for the first time using wave vector beam shaping and femtosecond spatiotemporal pulse shaping to create four fully phase-coherent, non-collinear optical fields. This technique is an optical analogue of multidimensional and multi-quantum NMR. From a study of GaAs quantum wells, an internally calibrated, yet previously unmeasured two-dimensional projection correlating events between the first two time periods is presented.

Introduction

Two-dimensional electronic Fourier transform spectroscopy (2D FTS), subject to a recent review [1], has begun to make clear the fuliginous many-body effects present in quantum wells and other highly correlated systems. Spreading the complex spectrum into two dimensions using a demanding experimental procedure has shown couplings among electronic states [2] and excitation-induced shifts in exciton energy and dephasing rates due to interactions with other excitons [3]. Since the emission dimension (the Fourier transform of the third time period during which the coherent signal field is radiated in response to three input fields that are separated by the first two time periods) is necessarily included in any coherent 2D measurement, correlations among events in the first and second time periods are absent. To reveal any possible correlations and many-body effects between these time periods, a 3D measurement is required.

Here we demonstrate for the first time 3D electronic spectroscopy of excitons in GaAs quantum wells. In these third-order measurements, each peak corresponds to exactly one Feynman pathway. Since peaks in the first two frequency dimensions are Fourier transforms of coherent oscillations during their respective time periods, they are independent of any uncertainty in the carrier frequency determination, unlike peaks in the emission frequency dimension.

Multidimensional measurements beyond two dimensions have been previously demonstrated in other fields such as NMR [4] and IR spectroscopy [5]. In the IR case, the high-energy states involved were simple overtones of the fundamental vibrational modes of the systems; in the present system under the proper excitation conditions, the high-energy two-quantum states involved are biexcitons, which only exist due to a many-body effect – biexciton correlations. 2D electronic rephasing spectra have been taken of photosynthetic systems for different waiting times but neither the fully coherent 3D spectrum nor the coherent 2D spectral surface correlating events in the first two time periods was produced [6]. Furthermore, this experiment only required phase stability between the members of two pairs of beams but not between the pairs. In the case of two-quantum SIII measurements—in which the phases of the biexciton coherences evolve during the second time period—phase stability during all time periods is necessary. Fortunately, this is an inherent trait of the spatiotemporal pulse shaper because all of the pulses traverse common optical elements.

Experimental Methods

The apparatus combines crossed phase masks—to create the BOXCARS geometry—and a 2D femtosecond pulse shaper detailed previously [7] used in diffraction mode to control the field in each beam, including all temporal delays and phase shifts. Using the pulse shaper to control the delays has the advantage that only the pulse envelope (and not the phase) is delayed; this is the equivalent of the rotating frame in NMR. The pulse shaper can change the order of arrival of the fields at the sample; thereby it has the ability to reconfigure the experiment (SI, SII or SIII) using only software control. Phase cycling [8] is used to reduce undesirable scattered light by both the sample and the SLM surface. Scanning the first two time periods and detecting the emitted field through spectral interferometry with a CCD and spectrometer, with a typical data acquisition time of about ten hours, creates a three-dimensional spectral solid. Measurements have been performed using various polarization and timing parameters. With the user-defined carrier frequency set to 368.00 ± 0.05 THz, coherent evolution is slowed so that heavy-hole (H) and light-hole (L) excitons oscillate at 4.3 and 5.9 THz, respectively. The optical carrier frequency error is due to the spectral distribution across the width of a single 24-micron wide SLM pixel. The exciton density was about 10^{11} excitons/cm^2/well, the sample consisted of ten layers of 10 nm thick GaAs, separated by 10 nm thick $Al_{0.3}Ga_{0.7}As$ barriers, and the sample was cooled to 10 K in a cold-finger cyrostat.

Results and Discussion

A rephasing scan was performed where all fields were co-circularly polarized. The 3D spectral solid is shown below with a 2D projection onto the (ω_1, ω_2) plane. The shoulder along the ω_2 axis up to approximately 2 THz corresponds to the |L><H| exciton quantum beat (or Raman) frequency during the second time period in the Liouville pathway as depicted by the double-sided Feynman diagram shown in the insert. The shoulder appears at the difference frequency of about 1.6 THz. This pathway is one of only two that contribute to the Raman frequency; the other excites the excitons in the opposite order and so it contains the opposite coherence, |H><L|, during the second time period. All other pathways are in population states during this time period and thus contribute to the large signal at zero frequency.

These measurements contain more information than shown because both the real and imaginary spectral surfaces are measured using this technique. Previous studies have illustrated the sensitivity of the full complex response to the many-body interactions present in these strongly correlated systems [9].

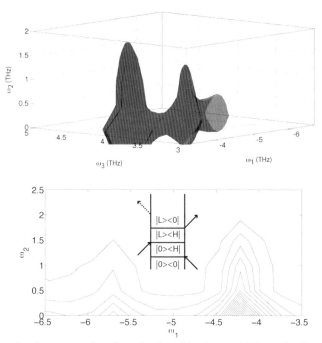

Fig. 1. 3D rephasing spectral surface (top) and its (ω_1, ω_2) 2D projection (bottom)

Conclusions

The passive phase stability of four fully phase-coherent fields controlled by a femtosecond pulse shaper is exploited to create a 3D spectral solid via a four wave-mixing process in GaAs quantum wells. Using the nonstandard 2D SIII projection to measure the biexciton binding energy may be valuable. Similar 3D measurements are potentially useful for other systems, such as light-harvesting complexes or nonlinear wave packet interferometry measurements [10] where correlations among events in all time periods could be important.

Acknowledgements. This work was supported in part by the National Science Foundation grant CHE-0616939. D. Turner acknowledges support from the NDSEG Fellowship Program. We thank X. Li and S. T. Cundiff for related discussions.

1   S. T. Cundiff, Optics Express **16**, 439, 2008

2   T. Zhang, C. N. Borca, X. Li and S. T. Cundiff, Optics Express **13**, 7432, 2005

3   J. M. Shacklette and S. T. Cundiff, Phys. Rev. B **66**, 045309, 2002

4   R. R. Ernst, G. Bodenhausen and A. Wokaun, *Principles of Nuclear Magnetic Resonance in One and Two Dimensions*, (Oxford U. Press, Oxford, UK, 1987)

5   F. Ding and M. T. Zanni, Chem. Phys. **341**, 95, 2007

6   G. S. Engel *et al.*, Nature **446**, 782, 2007

7   K. Gundogdu, K. W. Stone, D. B. Turner and K. A. Nelson, Chem. Phys. **341**, 89, 2007

8   J. C. Vaughan, T. Hornung, K. Stone and K. A. Nelson, J. Phys. Chem. A **111**, 4873, 2007

9   X. Li, T. Zhang, C. Borca and S. T. Cundiff, Phys. Rev. Lett. **96**, 057406, 2006

10 T. S. Humble and J. A. Cina, J. Phys. Chem. B **110**, 18879, 2006

Ultrafast carrier dynamics in spherical CdSe core/ elongated CdS shell nanocrystals

Maria Grazia Lupo[1,2], Luigi Carbone[2], Liberato Manna[2], Roberto Cingolani[2]
Margherita Zavelani-Rossi[1], Guglielmo Lanzani[1]

[1]NNL-National Nanotecnology Laboratories- CNR-INFM, Università degli Studi di Lecce, Via per Arnesano 23, Lecce, Italy
 E-mail: mariagrazia.lupol@polimi.it
[2]Dipartimento di Fisica, Politecnico, Piazza L. da Vinci 32, 20133 Milano, Italy

Abstract. The ultrafast carrier dynamics in CdSe/CdS dot/rod nanocrystals have been investigated by sub-ps-pump-probe experiments in the visible spectral range. Photobleaching dynamics due to state filling is explored and a picture for the electron-hole dynamics at the interface, within the nanostructure, is obtained.

Introduction

Progress in colloidal synthesis makes it possible to prepare highly crystalline and fairly monodisperse nanocrystals, with controllable size and shape, interesting for scientific and technological applications.
A novel class of quantum rods (QRs) consisting of two different materials grown with a strongly asymmetric shape have been recently introduced: CdSe/CdS asymmetric core/shell nanorods have been synthesized with a new, seeded-type growth approach. Starting from preformed spherical CdSe nanocrystal seed (core) and a mixture of hot surfactants, a CdS rod (shell) grows asymmetrically along the unique c-axis of the CdSe wurtzite structure (Fig. 1b). The aspect ratio (A.R.) can be changed from 2.5 to 35.

These rods have a narrow distribution of lengths and diameters, high crystallinity and present the appealing characteristics of strong and tunable light emission from green to red varying CdSe seeds diameter. They spontaneously self assemble on substrates (Fig. 1a) and can be easily organized in close-packed ordered arrays over large areas. The photoluminescence quantum yield is very high, up to 70% for rods with A.R. lower than 10 [1].

In these rods electrons can penetrate the elongated CdS shell whereas holes are confined inside the CdSe core: the charge delocalization allows the control of wavefunction confinement and an extremely large tunable quantum confined Stark effect with varying AR. This behaviour makes these nanocrystals particularly interesting both for technological applications and for a better understanding of physics in semiconductor nanoparticles.

In particular, elongated QR are particularly suited as switchable single photon source and ultrasensitive charge detectors [2].

Experimental Methods

We use femtosecond pump-probe transient spectroscopy in the visible spectral range, to analyze ultrafast carrier dynamics of these new heterostructures and to obtain informations about different mechanisms responsible of radiative and non radiative

289

recombination and to investigate the dynamical processes leading to build-up and decay of the optical gain.

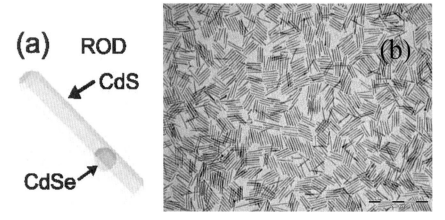

Fig. 1. (a) Sketch of a CdSe/CdS nanorod. (b) Transmission electron micrograph of self-assembled Cdse/CdS nanorods.

Results and Discussion

The analysis of differential transmission dynamics in the emission spectral range give evidence of stimulated emission (SE) and optical gain. We found a gain lifetime of about 200ps at fluence giving rise to an average number of hole-electron pair of about 10. Compared to other nanostructures previously reported in literature[3] these rods present a longer gain lifetime probably due to a reduction of Auger recombination, one of the main non radiative processes in nanocrystals. Both aged and fresh samples show an analogous behaviour. These characteristics are a straight outcome of charge separation in core and shell and make these heterostructure very promising for lasing.

Differential transmission measurements at different pump fluence and at probe wavelengths corresponding to the band edge transitions in core and shell have outlined different mechanisms of charge generation and relaxation: at low fluence excitation takes place into the shell and the population of the Cdse band edge transition is due to hole transfer (HT) from CdS to CdSe (Fig. 2a). At pump fluence generating few excitons for each rod (Fig. 2b)., the HT is still evident but superimposed to some other path of relaxation: with increasing the fluence, the radiation may generate excitons in the CdSe core. Bleaching at 590 nm is due to intraband relaxation of charge from the higher energy states to the band edge. At high pump fluence (Fig. 2c) the main mechanism responsible of bleaching at 590 nm is intraband relaxation of charge generated in the CdSe core. When excitation takes place in the CdS shell, holes relax to the band edge of the CdSe core and we measured a time constant of about 0.7 ps associated with hole transfer at the interface.

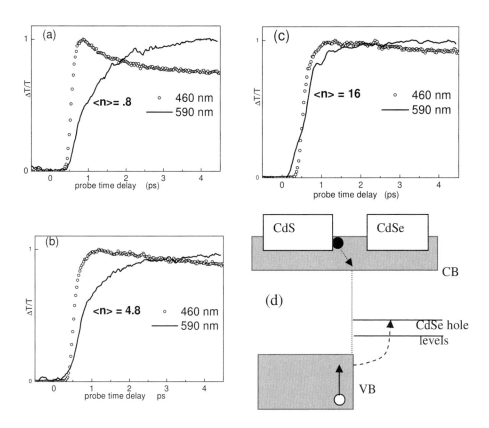

Fig. 2. (a-c) Bleaching ΔT/T kinetics recorded at the band edge transition in CdSe (590 nm) and CdSe (460 nm) for different pump fluence: (a) at low pump fluence (< n > is the average number of excitons generated at a determined pump fluence value) the decay at 460 nm and the time rise at 590 nm strictly correspond; (b) at pump fluence generating few excitons for each rod; (c) at high pump fluence. (d) Sketch of levels involved in hole transfer.

Conclusions

Our analysis of ultrafast dynamics in CdSe/CdS dot-rods suggests these nanocrystals are good candidates for lasing and enlighten some important mechanisms responsible of their optical properties.

1   L. Carbone, C.Nobile, M. De Giorgi, F. Della Sala, G. Morello, P. Pompa, M. Hytch, E. Snoeck, A. Fiore, I. R. Franchini, M. Nadasan, A. F. Silvestre, L. Chiodo, S. Kudera, R. Cingolani, R. Krane, L. Manna, Nano Lett. 7, 2942-2950, (2007).

2   J. Muller, M. Lupton, P. G. Lagoudakis, F. Schindler, R. Koeppe, A.L. Rogach, J. and Feldmann, D.V. Talapin, and H. Weller, Nano Lett. 5, 2044-2049, (2005).

3   H.Htoon, J.A.Hollingworth, A.V. Malko, R. Dickerson, and V.I. Klimov, Appl. Phys. Lett. 82, 4776-4778, (2003).

Time-Resolved Optical Studies of InGaAs/GaAs Quantum Wells in High Magnetic Fields

J. Lee[1], D. H. Reitze[1], S. McGill[2], J. Kono[3], A. A. Belyanin[4], G. S. Solomon[5]

[1] Department of Physics, University of Florida, Gainesville, Florida 32611
  E-mail: jinho_lee@magnet.fsu.edu
[2] National High Magnetic Field Laboratory, Tallahassee, Florida, 32310
  E-mail: mcgill@magnet.fsu.edu
[3] Department of Electrical and Computer Engineering, Rice University, Houston, Texas 77005
  E-mail: kono@rice.edu
[4] Department of Physics, Texas A&M University, College Station, Texas 77843
  E-mail: belyanin@physics.tamu.edu
[5] National Institute of Standards & Technology, Gaithersburg, MD 20899-8423 USA
  E-mail: glenn.solomon@nist.gov

Abstract. The ultrafast dynamics of high density electron-hole plasmas in $In_{0.2}Ga_{0.8}As$/GaAs multiple quantum wells are probed in high magnetic fields by time-resolved methods. Complementary transient absorption (TA) and time-resolved photoluminescence (TRPL) experiments reveal complicated, Landau-level-dependent recombination dynamics. Evidence for delayed stimulated emission is observed. Correlations between the TA and TRPL indicate an intra-Landau level relaxation bottleneck.

Introduction

A strong magnetic field B applied perpendicular to the plane of a semiconductor quantum well (QW) generates a quantized series of Landau levels (LL), evolving from a step-like 2D density-of-states at low field into a highly degenerate δ-function-like 0D density-of-states at high fields. Under intense ultrafast excitation, large filling factors ($\nu = N_e h / eB$) can be achieved even at large magnetic fields, resulting in a high density electron-hole (e-h) plasma in higher lying LLs which can subsequently undergo recombination through spontaneous, stimulated, or cooperative processes [1].

  We recently reported on the first observation of cooperative emission (superfluorescence) from a dense photo-excited electron-hole plasma in InGaAs/GaAs multiple quantum wells in strong perpendicular magnetic fields [2]. Here, we probe the dynamics of emission processes using transient absorption (TA) and time-resolved photoluminescence (TRPL).

Experimental Methods

Our experiments were performed in a 17.5 T superconductor magnet using a 150 fs, 800 nm Ti:Sapphire chirped pulse amplifier (CPA) to pump the sample and a broadly tunable optical parametric amplifier (OPA) to probe the temporal dynamics as a function of laser fluence F and magnetic field B. The pump wavelength was 800 nm, generating carriers high in the bands in the well and barrier layers; carrier densities in excess of $10^{12}/cm^2$ are achieved even at the lowest F. A 15-period quantum well sample consisting of 80 Å $In_{0.2}Ga_{0.8}As$ wells separated by 150 Å GaAs barriers was held at 15 K during the experiments. For the TA experiments, the OPA was tuned to span a specific energy and interrogated the sample as a function of delay time after the arrival of the excitation pulse. The probe spectrum was collected via an optical fiber on the opposite side of the sample and analyzed as a function of time delay. For the TRPL measurements, the pump was used to excite the sample and a Hamamatsu streak camera temporally resolved both the in-plane and perpendicular plane emission from the QW, collected by using a micro-prism mounted at the side of the sample and directed through a low dispersion graded-index 600 μm core diameter multimode fiber to an imaging spectrometer. The time resolution of the streak camera was 50 ps.

Results and Discussion

Figure 1(a) plots the energies of the LLs as measured in absorption (with low intensity whitelight) and PL (with high intensity fs pulses). A large red shift is seen in the PL from higher LLs due to band gap renormalization. Figures 1(b) displays the transient absorption dynamics probing at the lowest heavy hole level in the absence of a

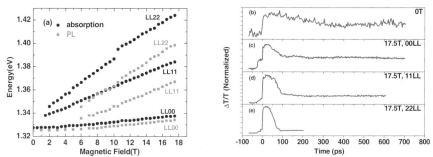

Fig. 1. (a) Landau fan diagram showing absorption peaks (circles) and PL peaks (squares) for the 00, 11, and 22 LLs. (b-e) Normalized ΔT/T probing (b) the lowest heavy hole absorption peak for $B = 0$ T and (c)-(e) the 00, 11, and 22 LL absorption peaks for $B = 17.5$ T.

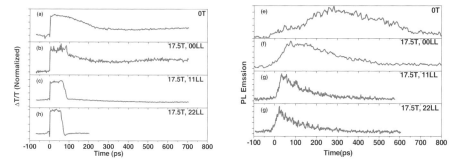

Fig. 2. (a)-(d) Normalized ΔT/T probing the energy where the maximum PL emission is observed. (a) the lowest heavy hole state $B=0$T and (b)-(d) the 00, 11, and 22 LL PL energies for $B=17.5$T. (e – h) Spectrally- and time-resolved PL from (a) the lowest heavy hole state at $B=0$T and the (b)-(d) 00, 11, and 22 LLs for $B=17.5$T.

magnetic field for $F = 4.1$ mJ/cm^2. Figs. 1(c)-(e) display the dynamics for $B = 17.5$ T and $F = 2.2$ mJ/cm^2 with the probe tuned to the low excitation density (white light) absorption peak for the 00, 11, and 22 LLs. As noted above, at high excitation densities, band gap renormalization red shifts the energy levels with respect to the low density case, thus the probe does not directly interrogate the carrier occupancy. With no field, an increase in transmission occurs at $t = 0$, with a ~270 ps relaxation time. When the field is applied, larger transmission increases are observed, and increasingly fast relaxation times occur at progressively higher LLs. In addition, the residual long-lived transmission increase (>600 ps) is diminished relative to the peak transmission at small time delays. Similar data is shown in Fig. 2 (a)-(d), but with the probe tuned to the energy where the maximum PL emission is observed. In this case, the probe is directly interrogating the occupancy of the electron-hole plasma. The data is qualitatively similar, with one striking difference, however. At high magnetic fields, the transmission signal exhibits an abrupt reduction which occurs at earlier times for higher LLs (90 ps, 00LL; 75 ps, 11LL; 50 ps, 22LL). In addition, the transition time from high to low transmission becomes increasingly more rapid as the LL index increases.

Figure 2(e-h) displays time-resolved PL for $F = 0.5$ mJ/cm^2 for (e) $B = 0$ T, heavy hole emission and (f)-(h) for the 00, 11, and 22 LLs at $B = 17.5$ T. For zero field, a long lived emission is observed, with a build-up time of 300 ps. At 17.5 T, the 00LL exhibits a faster build-up time (100 ps) and somewhat shorter relaxation time (250 ps) than the zero field case. This trend continues at higher LLs, with the 11 and 22LLs displaying faster build-up and relaxation times.

The rapid recovery of the transmission signal, particularly evident in the 11 and 22 LLs, indicates that a sudden and rapid recombination occurs approximately 60-90 ps after excitation. This picture is qualitatively corroborated by the time-resolved PL emission. Keeping in mind that the observed PL response has an instrumental resolution of 50 ps, the correlation between the carrier dynamics and the light emission (with faster emission and relaxation times occurring for higher lying LLs) suggests that the inter-LL recombination rate is much larger than the intra-LL relaxation rate. Once the initial relaxation into LLs occurs, carriers remain there until they undergo radiative recombination. A linewidth analysis of the PL emission peaks and earlier work [2,3] reveals decreased linewidths relative to PL observed at much lower pumping fluence, indicating that the emission mechanism is stimulated amplification of spontaneous emission.

Conclusions

We have performed transient absorption and time-resolved PL measurements on InGaAs/GaAs. Our measurements reveal correlations between the PL emission characteristics and the lifetimes of the high density magneto-plasmas.

Acknowledgments. The authors acknowledge support from the National Science Foundation and the National High Magnetic Field Laboratory.

1   A. A. Belyanin, V. V. Kocharovsky, and Vl. V. Kocharovsky, Quantum Semiclass. Opt. **9**, 1 (1997); Quantum Semiclass. Opt. **10**, L13 (1998); Laser Phys. **13**, 161 (2003).

2   Y. D. Jho, X. Wang, J. Kono, D. H. Reitze, X. Wei, A. A. Belyanin, V. V. Kocharovsky, Vl. V. Kocharovsky, and G. S. Solomon, Phys. Rev. Lett. **96**, 237401 (2006).

3   Y. D. Jho, X. Wang, J. Kono, D. H. Reitze, X. Wei, A. A. Belyanin, V. V. Kocharovsky, Vl. V. Kocharovsky, and G. S. Solomon, J. Mod. Optics **53**, 2325 (2006).

Femtosecond Formation of Ultrastrong Light-Matter Interaction

G. Günter[1], A. A. Anappara[2], J. Hees[1], S. Leinß[1], L. Sorba[2], G. Biasiol[2,3], A. Tredicucci[2], A. Leitenstorfer[1], and R. Huber[1]

[1] Department of Physics and Center for Applied Photonics, University of Konstanz, Universitätsstraße 10, 78464 Konstanz, Germany; Email: rupert.huber@uni-konstanz.de
[2] NEST CNR-INFM and Scuola Normale Superiore, Piazza dei Cavalieri 7, I-56126 Pisa, Italy
[3] Lab. Nazionale TASC CNR-INFM, Area Science Park, I-34012 Trieste, Italy

Abstract. Intersubband cavity polaritons in a GaAs/AlGaAs quantum well waveguide structure are photogenerated by a 12-fs near-infrared pulse. Multi-THz field transients directly trace the non-adiabatic switch-on of light-matter coupling with a vacuum polariton splitting of up to 44% of the bare photon frequency. We study the decay of a coherent photon state during quasi-instantaneous switching with sub-cycle resolution.

Introduction: Cavity QED in the ultrastrong coupling regime

Microcavities provide an elegant way to tailor the photonic density of states and control light-matter interaction. If the dipole coupling is strong enough to exceed damping, the interaction of an elementary excitation with single photon fluctuations of the quantum vacuum gives rise to new eigenstates: so-called cavity polaritons. These mixed light-matter waves display anticrossing with a mode separation known as vacuum-field Rabi splitting Ω_R [1]. Recently mid-infrared intersubband resonances of semiconductor quantum wells (QW) coupled to the photon mode of a planar waveguide have entered a unique regime of ultrastrong interaction, where Ω_R represents a significant fraction of the bare eigenfrequencies ω_{12} themselves [2].

The resulting squeezed two-mode quantum vacuum is expected to give rise to a variety of novel quantum electrodynamical (QED) effects [3]. In particular, ultrafast non-adiabatic switching of the Rabi frequency Ω_R has been predicted to release correlated photon pairs out of the vacuum, reminiscent of the intriguing, yet unobserved dynamic Casimir effect [3]. While electronic means offer a versatile approach to control Ω_R, e.g. by an external gate voltage [4], such techniques are too slow to study the predicted QED phenomena. Here, we report an all-optical pump – multi-THz probe scheme for the first implementation of non-adiabatic control of ultrastrongly coupled cavity polaritons.

Multi-THz waveguide with semiconductor quantum wells

The sample contains 50 identical, undoped GaAs QWs separated by $Al_{0.33}Ga_{0.67}As$ barriers. Electronic wave functions are quantized along the growth direction [Fig. 1(a)] forming subbands. They are unpopulated in thermal equilibrium. Yet radiative transitions between subbands of quantum number n = 1 and n = 2 become possible when optical excitation promotes electrons from the valence into the conduction band. The intersubband resonance features a narrow absorption line centered about a photon energy of $\hbar\omega_{12}$ = 120 meV (wavelength λ = 10 μm) and a strong dipole moment oriented along the growth direction. The multi-QW structure is designed as a planar step index waveguide for mid-infrared light [2]. Radiation is confined between

295

a top-cladding ($Al_{0.33}Ga_{0.67}As$)-air interface ($n_{air} = 1$, $n_{QW} = 3.1$) on one side and a low refractive index AlAs layer ($n_{AlAs} = 2.9$) on the other. The effective thickness of the entire waveguide is chosen to be $\lambda/2$ at an internal angle of $\theta = 65°$. Hence, photon modes with electric field components in growth direction (TM polarization) may resonantly couple to intersubband transitions provided the subbands are populated.

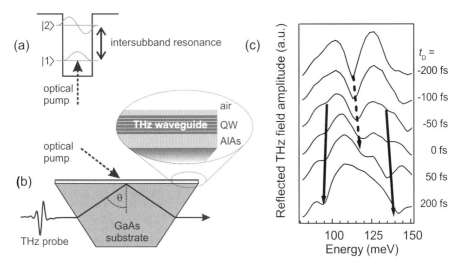

Fig. 1. (a) Schematic band diagram of a GaAs/AlGaAs QW; broken arrow: interband excitation populates subband $|1\rangle$; double-headed arrow: intersubband transition. (b) Sample geometry: The THz waveguide is resonant with the intersubband transition for light propagation at $\theta = 65°$. THz probe pulses coupled through the substrate are reflected off the waveguide. (c) 2D optical pump – THz probe data: amplitude reflectivity spectra at $T_L = 300$ K for various delay times t_D. The arrows serve as guides to the eye for the position of the cavity and polariton resonances.

Non-adiabatic formation of intersubband cavity polaritons

We employ a low-noise Ti:sapphire system generating amplified 12-fs light pulses centered at a photon energy of 1.55 eV [5] to selectively photoinject 4×10^{12} cm^{-2} electrons from the valence band into the lowest conduction subband of the QWs [Fig. 1(a)], activating the intersubband boson field. The subsequent ultrafast dynamics of the non-equilibrium cavity is traced by multi-THz spectroscopy [6]: A second part of the laser output generates phase-locked THz pulses covering the spectral window from 80 meV to 150 meV by optical rectification in a 50-μm-thin GaSe emitter [7]. TM polarized field transients are coupled through the prism-shaped substrate and internally reflected off the photoexcited area of the waveguide under incidence angles around $\theta = 65°$ [Fig. 1(b)]. The pulse front of the near-infrared pump is tilted to match the geometry of the THz phase surfaces. The oscillating electric field of the reflected THz transient is resolved in the time domain via phase-matched electro-optic sampling [7].

Fourier transformation provides amplitude [Fig. 1(c)] and phase spectra in the mid infrared. The eigenmodes of the cavity are identified via their characteristic minima in the amplitude reflectivity [2]. We repeat the experiments for various delay times t_D between near-infrared pump and multi-THz probe pulses. For $t_D \leq -50$ fs the spectra are dominated by a single dip at 113 meV. On resonance the amplitude reflectivity reaches values below 5%. Since there is no material excitation in the equilibrium sample throughout the frequency window shown, the observed feature is unambiguously assigned to the photonic waveguide mode.

Photoinjection of electrons into the lower subband induces dramatic changes of the spectra of order unity. The initial bare photon eigenstate is replaced on a ten-femtosecond-scale by two coupled cavity polariton modes appearing simultaneously at asymmetric energy positions of 94 meV and 136 meV, respectively. Most remarkably, these new resonances do not gradually develop out of the bare cavity mode. Rather a discontinuous switching to a cavity polariton system is seen with our sub-cycle time resolution. A slight additional shift of both new resonances occurs within the subsequent window of 200 fs. The maximum polariton splitting amounts to as much as 50 meV, corresponding to a fraction of 44% of the bare photon frequency. This value is comparable to the record achieved in delta-doped structures to date and clearly fulfills the criteria of ultrastrong light-matter coupling. Antiresonant terms in the light-matter Hamiltonian going beyond the rotating-wave description of the quantum vacuum are indeed expected to manifest themselves in an asymmetric splitting [3]. The vacuum Rabi frequency is continuously tuned via the near-infrared pump power (not shown). We also confirmed the characteristic anticrossing angular dispersion of the cavity polaritons by systematic tilting of the sample.

Non-adiabatic switching of the light-matter Hamiltonian is predicted to give rise to spectacular, yet unexplored effects on pre-existing photon states as well as the quantum vacuum itself. For example, a release of squeezed vacuum radiation is expected under such conditions [3]. In a preliminary model experiment, we prepare a coherent photon state in the bare cavity and perturb the radiative relaxation by non-adiabatic switching. On a sub-cycle scale, the photon state converts into ultrastrongly coupled coherent polaritons. Electro-optic sampling directly monitors this phenomenon for the first time (not shown). A rigorous description will have to consistently take into consideration cavity QED effects as well as carrier dynamics. The latter may be responsible for the slight additional increase of Ω_R on the 100 fs scale [Fig.1(c)]. Quantitative modeling with state-of-the-art theory [3] is currently under way.

In conclusion, we have demonstrated the first non-adiabatic switching scheme of cavity-polaritons, reaching the regime of ultrastrong light-matter coupling on a sub-cycle scale. The experiments provide a benchmark for latest cavity-QED theories in the ultrastrong coupling regime and point out a viable route towards novel quantum optical phenomena such as the observation of Casimir-type vacuum radiation.

Acknowledgements. We thank S. De Liberato and C. Ciuti for helpful discussions and an ongoing collaboration. This work is supported by the German Research Foundation (DFG) via the Emmy Noether Program and SFB767.

1   C. Weisbuch et al., Phys. Rev. Lett. 69, 3314 (1992).

2   D. Dini et al., Phys. Rev. Lett. 90, 116401 (2003).

3   C. Ciuti, G. Bastard, and I. Carusotto, Phys. Rev. B 72, 115303 (2005); S. De Liberato et al., Phys. Rev. Lett. 98, 103602 (2007).

4   A. A. Anappara et al., Appl. Phys. Lett. 87, 051105 (2005); Appl. Phys. Lett. 89, 171109 (2006).

5   R. Huber et al., Opt. Lett. 28, 2118 (2003).

6   R. Huber et al., Nature 414, 286 (2001); Phys. Rev. Lett. 94, 0274011 (2005); C. Kübler et al., Phys. Rev. Lett. 99, 116401 (2007).

7   C. Kübler, et al., Semicond. Sci. Technol., 20, 128 (2005) and references therein.

Ultrafast Bleaching and Gain
in a Single Semiconductor Quantum Dot

Rudolf Bratschitsch[1], Florian Sotier[1], Tim Thomay[1], Tobias Hanke[1], Jan Korger[1],
Suddhasatta Mahapatra[2], Alexander Frey[2], Karl Brunner[2], and Alfred Leitenstorfer[1]

[1] Department of Physics and Center for Applied Photonics, University of Konstanz, D-78464
Konstanz, Germany
E-mail: rudolf.bratschitsch@uni-konstanz.de
[2] Experimental Physics III, University of Würzburg, D-97074 Würzburg, Germany
E-mail: brunner@physik.uni-wuerzburg.de

Abstract. The transient quantum dynamics in a single CdSe/ZnSe quantum dot is investigated
via femtosecond spectroscopy. Ultrafast Coulomb renormalization and single exciton gain are
observed with these first resonant pump-probe measurements on a single-electron system.

Introduction

Semiconductor quantum dots are promising systems for robust and scalable quantum
information processing [1, 2]. Ultrafast sequences of coherent quantum operations
may be envisioned with *femtosecond* light pulses, if the involved quantum states are
separated by at least tens of meV. Therefore, small quantum dots with high
confinement potentials are favourable. Due to their large Coulomb correlation
energies, CdSe quantum dots are ideal candidates.

Experimental Methods

We have performed two-color femtosecond pump-probe spectroscopy on a *single*
self-assembled CdSe/ZnSe quantum dot. The transient quantum dynamics was probed
with *resonant* excitation and detection. Figure 1 shows the microphotoluminescence

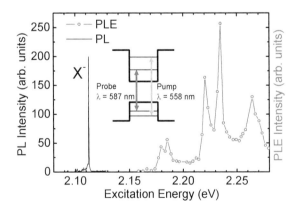

Fig.1. Photoluminescence (PL) and photoluminescence excitation spectrum (PLE) of a single
CdSe/ZnSe quantum dot, Inset: Schematic energy level structure of the quantum dot with
indicated pump and probe wavelength.

(PL) spectrum of the quantum dot. The fundamental exciton line emits at 2.112 eV. This system was resonantly excited via a transition in the p-shell at 2.22 eV (see Fig. 1 for PLE spectrum) with femtosecond laser pulses of a duration of 750 fs and a narrow spectral bandwidth of 2.6 meV. The transient transmission around the exciton transition was probed by a time-delayed 190 fs light pulse with a spectral width of 10.4 meV (inset of Fig. 1).

Results and Discussion

Figure 2 shows differential probe transmission spectra, recorded for pump-probe

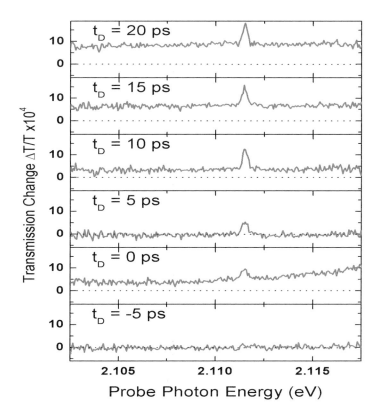

Fig.2 Differential transmission spectra of the CdSe/ZnSe quantum dot for different pump-probe time delays from -5 ps to +20 ps.

delay times from -5 to +20 ps. At negative delays of a few ps, small oscillatory signatures around the exciton resonance are discernible. These features are related to the perturbed free induction decay. From zero time delay to a few ps, a prominent bleaching of the exciton transition is observed. After 10 ps the positive transmission signal has increased by approximately a factor of two. It reaches a constant plateau that slowly decays with the radiative lifetime of the exciton of 500 ps.

Our interpretation of the ultrafast dynamics observed is as follows: The resonant generation of a highly energetic exciton ("hot exciton") by the pump pulse results in an *instantaneous bleaching* of the exciton resonance at 2.112 eV via Coulomb renormalization of the transition. As this hot exciton relaxes to its ground state within a few ps, the original transition is restored under inversion conditions. Therefore, a delayed probe pulse causes stimulated emission at the exciton transition, which leads to *gain* at 2.112 eV. This interpretation is supported strongly by the observation of a retarded increase in differential transmission between 5 ps and 15 ps.

Conclusions

This experiment represents a significant step towards *ultrafast quantum optical devices based on single electrons* in semiconductor quantum dots, which are initialized, controlled and read out by *femtosecond* light pulses.

1   X. Li, Y. Wu, D. G. Steel, D. Gammon, T. H. Stievater, D. S. Katzer, D. Park, C. Piermarocchi, and L. J. Sham, Science 301, 809 (2003).
2   J. R. Petta, A. C. Johnson, J. M. Taylor, E. A. Laird, A. Yacoby, M. D. Lukin, C. M. Marcus, M. P. Hanson, A. C. Gossard, Science 309, 2180 (2005).

Probing coherent optical phonons by Extreme Ultraviolet radiation based on high-order harmonic generation

E. Papalazarou[1], D. Boschetto[1], J. Gautier[1], C. Valentin[1], M. Marsi[2], Ph. Zeitoun[1], and Ph. Balcou[3]

[1] Laboratoire d'Optique Appliquée, Ecole Nationale Supérieure de Techniques Avancées, Ecole Polytechnique, CNRS, Chemin de la Hunière, F-91761 Palaiseau Cedex, France
E-mail: evangelos.papalazarou@ensta.fr

[2] Laboratoire de Physique des Solides, Université Paris-Sud, Bâtiment 510, F-91405 Orsay Cedex, France

[3] Centre Lasers Intenses et Applications, Université de Bordeaux, CNRS, CEA, 351 Cours de la Libération, F-33405 Talence Cedex, France

Abstract. We report an experimental approach we used in order to time-resolve coherent optical phonon oscillations in a bismuth (111) oriented bulk crystal by extreme ultraviolet (XUV) and soft x-ray femtosecond pulses based on high-order harmonic generation in rare gases.

Introduction

Since almost three decades, following the development of femtosecond lasers, many groups have shown an intensive interest in studying the response of materials in the time domain, thanks to time-resolved pump-probe techniques based on ultrashort optical pulses. They have demonstrated that the propagation of light pulses through solids or reflection by the solid surface is accompanied by intense lattice vibrations (i.e. phonons) in THz frequencies showing a high degree of spatial and temporal coherence [1]. Consequently, femtosecond optical pulses in visible and NIR (1.5 – 2 eV) were used to excite and time-resolve coherent electronic and lattice dynamics in a number of materials [2-5] revealing the relaxation mechanisms after photoexcitation.

Recently, femtosecond table-top laser-produced plasma x-ray pulses [6], extended phonon observations in the keV spectral region and up to wavevectors comparable to the reciprocal lattice constant. Moreover, the x-ray short wavelengths permitted the direct stroboscopic-like observations of the atomic motion in crystals. Consequently, ultrafast atomic vibrations and phase transitions can be monitored with femtosecond temporal resolution, while knowledge of the structural changes can be directly derived by means of time-resolved x-ray diffraction [7].

On the other hand, the tremendous progress during the past few years of ultrashort light sources [8] in the XUV and soft x-rays, based on high-order harmonic generation (HHG) in gases, opened up a whole range of possibilities for investigations in material sciences. Characteristics such as the ultrashort pulse duration and the high spatial and temporal coherence represent distinctive advantages of HHG over other sources of XUV/soft x-ray radiation (e.g. synchrotron radiation).

We present an experimental approach to probe coherent optical phonons by using a tabletop XUV/soft x-ray radiation source based on HHG. For this first approach optical phonons are coherently excited by NIR (1.55 eV) pulses and probed by XUV/soft x-ray femtosecond pulses with an energy centered at 39 eV and a bandwidth

~15 eV (FWHM). This was possible thanks also to technical improvements aiming to reduce the effects of the harmonic photon flux instability like for instance the normalization of the detected signal by the photocurrent measured on the mirror focusing the XUV/soft x-ray radiation on the specimen.

Experimental Methods

The pump-probe experiments were performed using a 1 kHz Ti:Sapphire high-power laser system at 810 nm central wavelength, 35 fs pulse duration and compressed output energy of 10 mJ/pulse. The total energy used for the experiment was 4 mJ/pulse. Coherent optical phonons were excited in bismuth single crystal oriented along the (111) direction by focusing 30 μJ pulses on a target area of ~0.5 mm. High-order harmonics are generated by focusing the major fraction of the laser energy (80%) into an argon filled gas-cell. The NIR beam was filtered out by a 250-nm-thick aluminum filter, whereas the transmitted XUV radiation was further focused by a broadband off-axis parabolic mirror (f = 65 mm, θ = 5°) multilayer-coated with negligible pulse broadening in the femtosecond and attosecond regimes [9]. Pump and probe beams were focused on the target having an incidence angle of 75° and 79° respectively from the surface normal. Coherent lattice dynamics were probed through changes in the optical reflectivity of the XUV/soft x-ray beam induced by the variations of the refractive index at the surface of the crystal. The time-resolved transient reflectivity change $\Delta R/R$ of the XUV/soft x-ray pulses was measured by using a calibrated XUV sensitive photodiode.

Results and Discussion

Figure 1 shows the normalized reflectivity change as a function of the pump-probe temporal delay. High-amplitude optical phonon oscillations can clearly been seen (filled dots). The inset shows its Fourier-transformed spectrum. Femtosecond x-ray diffraction experiments have shown that after femtosecond pulse excitation at 800 nm the bismuth atoms start oscillating around a new equilibrium position, due to modification in the inter-atomic potential. Therefore, we aspect the change in reflectivity to be in the following way

$$\frac{\Delta R}{R} = A\Theta(t) + B\exp(-t/\tau_q)\cos(\omega t + \varphi) \qquad (1)$$

where $\Theta(t)$ is the step-function and A the step-amplitude, whereas B and τ_q are the amplitude and decay time of the optical phonons, respectively, ω is the phonon frequency and φ is the initial phase of coherent oscillations with respect to time zero. The first term in equation (3) takes into account the modification in inter-atomic equilibrium position, whereas the second term takes into account the coherent phonon mode.

The fit gives a phonon relaxation time of (0.6±0.2) ps. The frequency of (1.73±0.71) THz is in agreement with the earlier measured red-shifted A_{1g}-mode of oscillations for similar high excitation fluencies [7, 10, 11]. Finally, the change in reflectivity is measured to be as large as $\Delta R/R \approx 0.3$. It could possibly be attributed to near-resonance of the high harmonics with the $5d_{3/2}$ states around 27 eV, enhancing the time-dependent change in refractive index and thus the change in reflectivity.

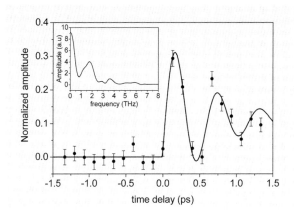

Fig. 1. The measured optical phonon oscillations in bismuth (111) using HHG pulses. The pump fluence is ~6 mJ/cm². The solid line corresponds to a fit to the measured data curve. Inset: Fourier-transform spectrum.

Conclusions

In conclusion, we reported an approach to time-resolve coherent lattice dynamics probed by ultrashort XUV pulses based on HHG. This technique discloses the possibility of studying optical phonons with femtosecond resolution in a previously unexplored range of wavevectors, away from the center of the Brillouin zone.

Acknowledgements. The authors gratefully acknowledge support from the Marie Curie Research Training Networks XTRA (MRTN-CT-2003 505138) and FLASH (MRTN-CT-2003503641) as well as the TUIXS European project (FP6 NEST-Adventure n. 012843).

1 S. De Silvestri, J.G. Fujimoto, E.P. Ippen, E.B. Jr. Gamble, L.R. Williams, and N.A. Nelson, Chem. Phys. Lett., **116**, 1985, 1985.

2 G.C. Cho, W. Kutt, and H. Kurz, Phys. Rev. Lett., **65**, 764, 1990.

3 T.K. Cheng, S.D. Brorson, A.S. Kazeroonian, J.S. Moodera, G. Dresselhaus, M.S. Dresselhaus and E.P. lppen, Appl. Phys. Lett., **57**, 1004, 1990.

4 H. J. Zeiger, J. Vidal, T. K. Cheng, E. P. Ippen, G. Dresselhaus, and M. S. Dresselhaus, Phys. Rev. B **45**, 768, 1992.

5 R. Merlin, Solid State Commun. **102**, 207, 1997.

6 A. Rousse, C. Rischel, S. Fourmaux, I. Uschmann, S. Sebban, G. Grillon, Ph. Balcou, E. Foerster, J.P. Geindre, P. Audebert, J.C. Gauthier, and D. Hulin, Nature, **410**, 65, 2001.

7 K. Sokolowski-Tinten, C. Blome, J. Blums, A. Cavalleri, C. Dietrich, A. Tarasevitch, I. Uschmann, E. Förster, M. Kammler, M. Horn-von-Hoegen, and D. von der Linde, Nature (London) **422**, 287, 2003.

8 T. Brabec and F. Krausz, Rev. Mod. Phys. **72**, 545, 2000.

9 A.-S. Morlens, R. López-Martens, O. Boyko, P. Zeitoun, P. Balcou, K. Varjú, E. Gustafsson, T. Remetter, A. L'Huillier, S. Kazamias, J. Gautier, F. Delmotte, and M.-F. Ravet, Opt. Lett., **31**, 1558, 2006.

10 S. Fahy and D. A. Reis, Phys. Rev. Lett, **93**,109701, 2004.

11 D. Boschetto, E.G. Gamaly, A.V. Rode, B. Luther-Davies, D. Glijer, T. Garl, O. Albert, A. Rousse, and J. Etchepare, Phys. Rev. Lett. **100**, 027404, 2008.

Part V

Chemistry - Condensed Phase

Real-Time Monitoring of Structural Evolution in *Cis*-Stilbene Photoisomerization by Ultrafast Time-Domain Raman Spectroscopy

Satoshi Takeuchi[1], Sanford Ruhman[2], Takao Tsuneda[3], Mahito Chiba[4], Tetsuya Taketsugu[5], and Tahei Tahara[1]

[1]Molecular Spectroscopy Laboratory, RIKEN, 2-1, Hirosawa, Wako 351-0198, Japan
 E-mail: stake@riken.jp and tahei@riken.jp
[2]Department of Physical Chemistry, Hebrew University, Jerusalem 91904, Israel
[3]School of Engineering, The University of Tokyo, Tokyo 113-8656, Japan
[4]National Institute of Advanced Industrial Science and Technology, Tsukuba 305-8568, Japan
[5]Graduate School of Science, Hokkaido University, Sapporo 060-0810, Japan

Abstract. We studied the vibrational structure of reactive S_1 *cis*-stilbene through wavepacket motions generated impulsively at various delay-times. They showed gradual frequency downshift, demonstrating highly anharmonic nature of the excited-state potential and structural evolution with photoisomerization.

Introduction

Cis-stilbene is a prototypical molecule showing ultrafast olefinic photoisomerization (Figure 1). It is known that the photoisomerization of *cis*-stilbene proceeds in a nearly barrierless way with a time constant as short as 1 ps, which is significantly shorter than the photoisomerization time of *trans*-stilbene (ca. 100 ps). Accordingly, there have been growing interests in the way how the cis isomer reacts so fast. Obviously, the photoisomerization of *cis*-stilbene involves not only a simple twisting around the central C=C bond but also skeletal deformations of the whole molecule. Because of this complicated nature of the "reaction coordinate of polyatomic molecules", our understanding of the structural change and relevant potential energy surface (PES) has been so far limited to a qualitative or even conceptual level. To obtain deeper understanding of the dynamics and mechanism of the *cis*-stilbene photoisomerization, it is crucial to track the structural change of the molecule in the course of the reaction. In other words, it is highly desirable to probe the structures of non-stationary excited states that continuously evolve during the reaction. Motivated by this challenging idea, we studied the ultrafast structural dynamics and shape of the reactive PES of *cis*-stilbene by femtosecond time-domain Raman spectroscopy using 11-fs pulses, and succeeded in newly disclosing a structural evolution of the "isomerizing" excited-state molecule.

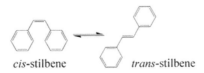

cis-stilbene   *trans*-stilbene

Fig. 1. Photoisomerization of stilbene between the cis and trans isomers.

Experimental

Femtosecond time-domain Raman measurements were carried out by using a setup based on a non-collinear optical parametric amplifier (NOPA). Briefly, the fundamental pulse at 800 nm was generated by a Ti:sapphire regenerative amplifier system. A portion of this fundamental output was converted to the third harmonic at 267 nm, and the uv pulse was used for $S_1 \leftarrow S_0$ photoexcitation of *cis*-stilbene. The rest of the fundamental output was used to drive the NOPA. The NOPA output was tuned to 620 nm, and it was compressed down to 11 fs by using prism and grating pairs. This ultrashort pulse was divided into two, and they were used for generation of the nuclear wavepacket in S_1 *cis*-stilbene and for probing the wavepacket motion, respectively.

Results and Discussion

Previously, we measured the $S_n \leftarrow S_1$ absorption of *cis*-stilbene with 40-fs time resolution, and observed a beating feature due to the 240-cm^{-1} wavepacket motion in the reactive S_1 state [1]. Since the wavepacket motion damped much faster than the isomerization time, we concluded that it is not directly correlated with the isomerization. Because of this rapid damping, the wavepacket motion only reflected the nuclear motion in the early time region after photoexcitation, and hence, it monitored only the vicinity of the Franck-Condon region of the PES where the wavepacket was initially prepared by photoexcitation.

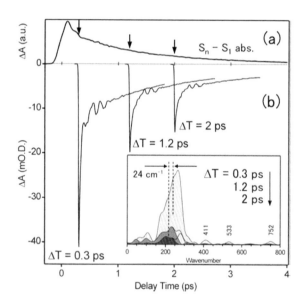

Fig. 2. Time-domain vibrational data of *cis*-stilbene in hexadecane. (a) Temporal behavior of the S_1 absorption, and (b) the change of the S_1 absorption that is induced by the second pulse irradiated at $\Delta T = 0.3$, 1.2, and 2 ps after uv photoexcitation. (inset) Fourier-transform power spectra for the three delay times. uv pulse: 267 nm, 150 fs; vis pulse: 620 nm, 11 fs.

To overcome this limitation and get structural information of the reactive S_1 state that isomerizes in the picosecond time scale, we carried out ultrafast time-domain Raman measurements using three laser pulses [2]. In this experiment, we first generated S_1 cis-stilbene by uv irradiation, and then introduced an ultrashort visible pulse to generate an S_1 nuclear wavepacket at various delay times (ΔT) by the impulsive Raman process. The wavepacket motion was monitored by the third pulse as beating features in the $S_n \leftarrow S_1$ absorption change.

Figure 2 depicts time-domain vibrational traces of cis-stilbene in hexadecane for three delay times, $\Delta T = 0.3$, 1.2, and 2 ps. Each trace clearly shows a beating feature due to the same predominant mode around 240 cm^{-1}. Remarkably, Fourier transform analysis indicated that the (center-of-mass) frequency of this mode exhibits a significant downshift as large as 24 cm^{-1} in a few picoseconds: 239 cm^{-1} ($\Delta T = 0.3$ ps) $\rightarrow$ 224 cm^{-1} ($\Delta T = 1.2$ ps) $\rightarrow$ 215 cm^{-1} ($\Delta T = 2$ ps). Further experiments revealed that the solvent dependence of the rate of the frequency downshift coincides with the solvent dependence of the isomerization rate (S_1 lifetime). Therefore, the observation of the frequency downshift indicates that the force constant of the predominant mode substantially decreases through anharmonic coupling with another coordinate that corresponds to a slow structural change of the molecule due to the isomerization. In other words, the structural evolution along the isomerization coordinate was successfully tracked through the frequency change of the observed wavepacket motion, which can be regarded as a spectator. We also examined the PES and vibrational structure of the reactive S_1 state by a time-dependent DFT method. The vibrational analysis at geometries along the reaction coordinate has reproduced the significant frequency downshift of the corresponding mode and revealed the complicated nature of the structural change, which is fully consistent with our experimental observation. These experimental data combined with high-level quantum chemical calculations demonstrate highly anharmonic nature of the multi-dimensional S_1 PES of cis-stilbene, and reveal a structural evolution occurring with isomerization in the picosecond time scale.

Acknowledgements. This work is supported by Grant-in-Aids for Scientific Research on Priority Area "Molecular Science for Supra Functional System" (No.19056009) from MEXT of Japan. S.T. acknowledges financial support by a Grant-in-Aid for Scientific Research (B) (No. 19350017) from the Japan Society for the Promotion of Science.

1   K. Ishii, S. Takeuchi, and T. Tahara, "A 40-fs time-resolved absorption study on cis-stilbene in solution: observation of wavepacket motion on the reactive excited state", Chem. Phys. Lett. **398**, 400, 2004.
2   S. Fujiyoshi, S. Takeuchi, and T. Tahara, "Time-resolved impulsive stimulated Raman scattering from excited-state polyatomic molecules in solution", J. Phys. Chem. A **107**, 494, 2003.

Origin of Negative and Dispersive Features in Resonance Femtosecond Stimulated Raman Spectroscopy

Renee R. Frontiera, Sangdeok Shim and Richard A. Mathies

Department of Chemistry, University of California, Berkeley, CA 94720, USA
E-mail: rich@zinc.cchem.berkeley.edu

Abstract. Negative anti-Stokes femtosecond stimulated Raman features seen off-resonance and dynamic dispersive lineshapes seen on resonance are experimentally characterized and explained by multiple four-wave mixing processes that contribute to the total signal. Hot luminescence terms which cause dispersive lineshapes are separated from positive resonance Raman features on the Stokes side by varying the time delay between pulses.

Introduction

Femtosecond stimulated Raman spectroscopy (FSRS) is a valuable new time-resolved technique for the structural investigation of a wide variety of reactive and dynamic systems [1]. FSRS enables the collection of vibrational spectra with excellent time (50 fs) and spectral (10 cm^{-1}) resolution. Raman spectra arise from the interaction of a sample in either its ground or excited state with a picosecond Raman pump pulse and a broadband femtosecond probe pulse. Highly resolved spectra result from the creation of a vibrational coherence that is initiated with high temporal precision. Previous experiments have followed the structural evolution of rhodopsins, carotenoids, and proton transfer reactions by examining vibrational features on the Stokes side. The novel time and energy resolution capabilities of FSRS would facilitate an even better understanding of reactive internal conversion processes if the vibrational energy distribution could be analyzed on the anti-Stokes side as well.

A recent theoretical analysis of the FSRS process proposed that eight four-wave mixing terms contribute to the total FSRS signal [2]. These terms include two resonance Raman gain terms, one of which involves a long-lived vibrational coherence resulting in well-resolved peaks; two inverse Raman terms, which result in the annihilation of a probe photon at Raman resonances, one of which gives narrow peaks; and four hot luminescence terms involving population of an excited electronic state, two of which involve the creation of a vibrational coherence on the excited state. Here we explore the experimental contribution of these terms in resonance FSRS.

Results and Discussion

Figure 1 presents resonance FSRS spectra of rhodamine 6G (R6G) in methanol with pump wavelengths across the red edge of the absorption band, taken using the setup described elsewhere [3]. The methanol feature at 1033 cm^{-1} is negative [4] at all pump wavelengths, resulting from an annihilation of probe photons due to an inverse Raman process involving a long-lived vibrational coherence on the ground state. The R6G solute band lineshapes are complicated by the addition of a resonant excited state. At 580 nm, where R6G absorbs minimally, the low frequency modes at 604 and 761 cm^{-1} are negative as expected, while the higher frequency modes (eg. 1504 and

1647 cm^{-1}), where the probe wavelength overlaps with the absorption band, are positive or slightly dispersive. As the Raman pump wavelength is tuned into the absorption band, and the probe wavelengths are more strongly resonant, the features at 604 and 761 cm^{-1} become dispersive, progress to purely positive peaks, and display oppositely phased dispersion. Similar trends are seen for the higher energy vibrations, such as the features at 1647 and 1504 cm^{-1}, although they begin as positive features with 580 nm excitation and evolve through dispersive shapes as the Raman pump becomes more resonant.

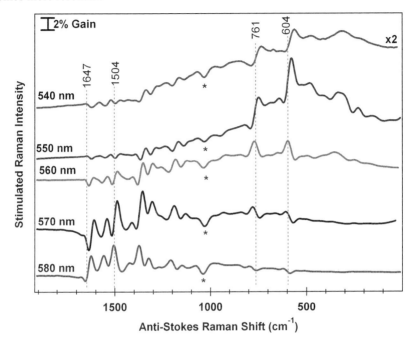

Fig. 1. Resonance anti-Stokes FSRS spectra of rhodamine 6G in methanol at the indicated pump wavelengths. Solvent features are marked with an asterisk.

From these spectra we conclude that the hot luminescence terms proposed by Sun *et al.* [2] make a significant contribution to the resonance FSRS spectra. When the probe window overlaps with the ground state absorption, the hot luminescence terms which involve a long-lived vibrational coherence on the excited state become strong contributors to the resonance FSRS lineshape and intensity, causing the dispersive features.

Figure 2 presents the resonance FSRS spectra of R6G on the anti-Stokes and Stokes sides as the time delay between the Raman pump and probe is varied. The lineshapes are wider at positive time delays due to truncation of vibrational dephasing by the Raman pump duration. The lineshapes on the anti-Stokes side do not change aside from the widening. However, the Stokes lineshapes are dispersive at positive time delays and purely positive at negative time delays. The third field interaction in the hot luminescence Stokes terms must occur with the higher vibrational level in the excited state coherence. As the vibrational coherence on the excited state dephases and the molecules relax, these hot luminescence pathways are no longer possible as there is no excited vibrational state for the Raman pump to interact with. Thus these terms contribute less when the pump maximum occurs after the probe. Therefore, due to the nature of the four-wave mixing processes involved, we can separate the hot luminescence and resonance Raman processes on the Stokes side, which should prove useful when exploiting resonance enhancement to probe reaction dynamics.

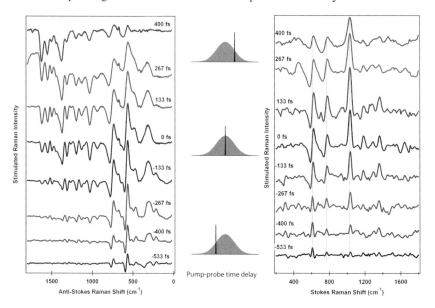

Fig. 2. Baseline subtracted anti-Stokes (left) and Stokes (right) FSRS spectra of rhodamine 6G in methanol at selected time delays between the 540 nm picosecond Raman pump and femtosecond probe pulses. A positive time delay indicates that the Raman pump maximum arrives before the probe pulse.

Conclusions

We have demonstrated that the terms proposed by Lee and coworkers do contribute significantly to the FSRS signal on resonance, and can be used to explain the off resonant negative anti-Stokes peaks and the on resonant dispersive peaks on the Stokes and anti-Stokes sides. On the Stokes side, the resonance Raman terms may be separated from the hot luminescence terms by simply changing the time delay between the Raman pump and probe. These results should prove useful in performing and interpreting resonance FSRS studies, and in further exploring the excited state vibrational coherence on reactive potential energy surfaces.

1 P. Kukura, D. W. McCamant, R. A. Mathies, Annu. Rev. Phys. Chem. **58**, 461, 2007.
2 Z. Sun, J. Lu, D. H. Zhang, and S. Y. Lee, J. Chem. Phys. **128**, 144114, 2008.
3 S. Shim and R. A. Mathies, Appl. Phys. **89**, 121124, 2006.
4 E. Ploetz, S. Laimgruber, S. Berner, W. Zinth, and P. Gilch, Appl. Phys. B: Lasers Opt. **87**, 389, 2007.

Reactive Dynamics in Nanoscale Water droplets Confined in Inverse Micelles

Minako Kondo, Ismael A. Heisler and Stephen R. Meech

School of Chemical Sciences and Pharmacy, University of East Anglia, Norwich NR4 7TJ, UK
E-mail: s.meech@uea.ac.uk

Abstract. Ultrafast excited state reactions of Auramine are studied in inverse micelles with water droplets between 1 and 10nm. Dynamics, inhomogeneous and a function of droplet size, are discussed in terms of interfacial and confinement effects.

Introduction

The molecular dynamics of constrained liquids (e.g. in micelles or nanoporous media) have been extensively studied by ultrafast optical Kerr effect,[1] solvation dynamics,[2-4] transient infra-red spectroscopy[5] and molecular dynamics simulation.[6,7] Constrained liquids have been shown to have dramatically different, and often much slower, dynamics than those of the liquid in the bulk phase. Two effects may contribute, first an intrinsic size effect due to the constraining curved interface [8] and second strong specific interactions with the interface,[9] which may cause the liquid dynamics in the first one or two molecular layers be dramatically slowed. Here we address the question of the influence of the observed inhomogeneity and slowing down of liquid dynamics on elementary chemical reactions in constrained media. It is established that liquid dynamics can have an important, even controlling, influence of the rate of reactions in liquids. An understanding of how such effects are manifested in constrained media will be critical in understanding the microscopic dynamics of reactions in living cells, which often occur in highly restricted microenvironments.

In our experiments the constrained liquid environment is provided by nanometre sized water droplets contained in inverted AOT micelles dispersed in heptane. In this system the droplet size can readily be controlled with the droplet radius being selectable between 1 and 10 nm. The liquid dynamics in this system have been extensively studied through ultrafast solvation dynamics and infra-red spectroscopy. The reaction we choose to study is the excited state conformational change in the dye auramine O (AO). This excited state process has been fully characterized in bulk liquids.[10] The initially excited state relaxes via a barrierless internal reorganisation to a non-emissive state, which subsequently decays back to the ground state. The initial rapid excited state decay is extremely sensitive to both the medium viscosity and to solvent polarity making it a useful probe of complex media. In this report we describe a 60 fs resolution time resolved fluorescence up-conversion study of AO dynamics in AOT micelles between 1 and 10 nm in radius.

Experimental Methods

AOT was purified and carefully dried as previously described, and the water content measured with calibrated IR spectroscopy. Samples were prepared with a known ratio of the concentration water and surfactant, w_0, to yield dispersed water droplets of radius r_w where $r_w = 0.18w_0$ in nanometres. The AO was initially dispersed in the water at a concentration which ensured less than 1 AO per micelle, even at the largest r_w.

Excited state dynamics were measured by the fluorescence up-conversion method.[11] The time resolution of our experiment has been optimised through careful control of the dispersion in both the 400 nm excitation and 800 nm up-conversion beams, the use of all reflective optics and thin crystals. The ultimate time resolution of our apparatus is 40 fs, but with additional optical filters to eliminate scatter from the AOT sample we here obtain 60 fs resolution (as judged from up-converted Raman scattering from the solvent). Examples of the experimental data are shown in Fig. 1a.

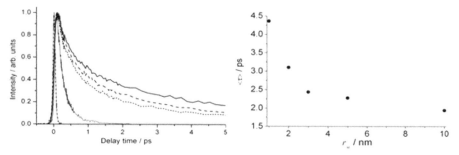

Fig. 1. (a) Time resolved fluorescence of AO in bulk water and AOT micelles with $r_w = 5$, 2 and 1 nm (short dash, long dash and solid line respectively). The up-converted Raman scatter is also shown and was used for deconvolution. (b) the mean relaxation time as a function of r_w.

Results and Discussion

From inspection of Fig. 1(a) it is immediately apparent that the rate of excited state reorganisation in AO has been dramatically slowed compared to bulk water. In addition the dynamics are a strong function of the size of the dispersed water droplet. The mean reaction time increases from 400fs to 2 ps between bulk water and the largest droplet with $r_w = 10$ nm. As the droplet radius is decreased further the reaction time increases in a nonlinear fashion, reaching 4.5 ps at $r_w = 1$ nm (Fig 1(b)).

Also evident from Fig. 1(a) is the inhomogeneity in the measured decay profile. This is also reflected in the strong wavelength dependence. By recording decay profiles across the spectrum the spectral evolution can be determined. The time dependent emission spectra are shown in Fig. 2, where the data have been fit to a log normal function. The mean frequency and width of the resulting function are shown in Fig 2(b) and (c) respectively. These results suggest the existence of two distinct timescales for AO dynamics in AOT micelles. The spectral shift occurs on a sub picosecond time scale while the excited state lifetime is on the order of a few picoseconds. Both are a function of r_w, with the rate of spectral shift being faster in the larger micelles. We propose that the rate of spectral evolution reflects intramolecular reorganisation on the upper surface, while the decay time reveals changes in the coupling of dark and directly excited states.

The effect on the excited state dynamics of incorporation of AO into the micelle is very large (Fig. 1) even for the $r_w = 10$ nm micelle. Such a large effect immediately suggests the incorporation of AO into the headgroup/water interface, with the much slower decay reflecting the changed environment. The increasing suppression of the reaction with decreasing r_w is also very significant. This size effect may be due to either the fundamental size effect on liquid dynamics, or an effect of size on the structure of the interface. We are currently unravelling these possibilities by modifying the micelle headgroup charge, the properties (pH, counterion...) of the water droplet and the nature of the probe molecule

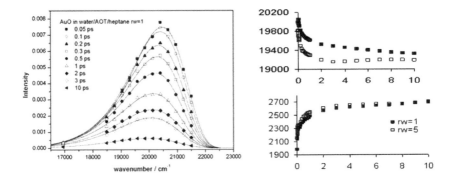

Fig. 2 (a) Time resolved fluorescence spectra of AO in an $r_w = 1$ AOT micelle. Date for 0.1 (■), 0.5 (○), 2 (▲), 10 (□), 30 (♦) ps after excitation, all fit to the log-normal function. (b) log normal width parameter and (c) mean frequency of the log-normal plot as a function of time after excitation. Both (b) and (c) are in wavenumber and picoseconds. Filled symbols $r_w = 1$, open $r_w = 5$ nm

Acknowledgements. We are grateful to EPSRC for financial support and MK thanks JSPS for the award of a studentship.

(1) RA Farrer, JT Fourkas:. Acc of Chem. Research **36**, 605, 2003.

(2) D Pant, RE Riter, NE Levinger: J. Chem. Phys., **109**, 9995, 1998.

(3) RE Riter, EP Undiks, NE Levinger, JACS **120**, 6062, 1998.

(4) S Sen, P Dutta, D Sukul, K Bhattacharyya, J. Phys. Chem. A, **106**, 6017, 2002.

(5) IR Piletic, DE Moilanen, DB Spry, NE Levinger, MD Fayer, J. Phys. Chem. A **110**, 4985, 2006.

(6) MR Harpham, BM Ladanyi, NE Levinger: , J. Phys. Chem., B **109**, 16891, 2005.

(7) MR Harpham, BM Ladanyi, NE Levinger, KW Herwig: J. Chen. Phys., **121**, 7855, 2004.

(8) DE Moilanen, NE Levinger, DB Spry, MD Fayer, JACS **129**, 14311, 2007

(9) N Nandi, K Bhattacharyya, B Bagchi, Chem. Rev. **100**, 2013, 2000.

(10) MJ van der Meer, H Zhang, M Glasbeek: J. Chem. Phys **112**, 2878, 2000.

(11) H Rhee, T Joo: Opt. Lett., **30**, 96, 2005.

Symmetry Dependent Solvation of Donor-Substituted Triarylboranes

U. Megerle[1], C. Lambert[2], E. Riedle[1], S. Lochbrunner[1,3]

[1] LS für BioMolekulare Optik, LMU München, Oettingenstr. 67, D-80538 Munich, Germany
E-mail: Uwe.Megerle@physik.uni-muenchen.de
[2] Institut für Organische Chemie, Universität Würzburg, Am Hubland, D-97074 Würzburg, Germany
[3] present address: Institut für Physik, Universität Rostock, Universitätsplatz 3, D-18055 Rostock, Germany

Abstract. Femtosecond transient absorption reveals an accelerated solvation for a highly symmetric donor-substituted triarylborane compared to its less symmetric counterpart. We explain this by ultrafast intramolecular charge delocalization over the subchromophores of the symmetric compound.

1. Contribution of internal charge mobility to solvation

In modern multi component materials, which are applied e.g. in organic electronics, the interaction between the various components on the nanoscale is a key aspect regarding the material properties. Solvation processes resulting from such interactions affect the electronic excitations and have a strong impact on the dynamic material response. A fascinating possibility to alter solvation processes is to make the charges or excitations within the solvated molecule mobile. This can elegantly be achieved in highly symmetric molecules with several equivalent active sites. Solvation is well investigated in terms of solvent properties and electronic dipole changes in the solute. However, almost no data is available about the influence of molecular symmetry on the dynamics.

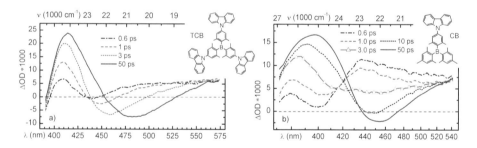

Fig. 1. Transient absorption spectra of TCB (a) and CB (b) in the polar solvent benzonitrile after 360 nm excitation.

We investigate this aspect by comparing the ultrafast solvation dynamics of the highly symmetric triple carbazole substituted triarylborane (TCB) with a single carbazole substituted triarylborane (CB) (see Fig. 1). Donor-substituted triarylboranes (TABs) are promising candidates for both electron and hole transporting as well as light emitting materials in organic light emitting diodes [1]. Their HOMO-LUMO gap can be tuned by the polarity of the surrounding medium to vary the emission from violet up to green. This allows for a fine tuning of the level alignment in such compounds and makes them particularly interesting for our femtosecond pump-probe experiments.

CB is highly solvatochromic, i.e. the optical transitions shift with the solvent polarity, since the static dipole moment reverses its orientation and increases strongly upon photoexcitation. For the highly symmetric TCB no static dipole is expected. Surprisingly, TCB does show similarly strong solvatochromic shifts, both for the charge transfer absorption and emission band [1]. This indicates that the interaction with the environment induces a spontaneous symmetry breaking already in the ground state leading to a sizeable static dipole moment. As a consequence, only one of the subchromophores of TCB is excited and thereby its dipole moment is also reversed and increased in analogy to CB. From our femtosecond experiments we conclude that this localized excitation rapidly reacts to the solvent arrangement and relocates to one of the other available sites on a ps time scale [2]. This means that indeed the internal charge mobility is decisive for the overall solvation.

2. Solvatochromic shifts in the femtosecond transient absorption spectra

The lowest energy absorption band of both compounds is found in the near UV and can be assigned to a charge transfer from the nitrogen to the boron center. In the nonpolar cyclohexane, the maximum of this band is found at 362 nm for CB and 385 nm for TCB. The 1600 cm^{-1} red shift results from the electronic coupling between the subchromophores of TCB, which causes a splitting of the excited state [1]. Such a coupling is a prerequisite for an exchange of excitation energy from one chromophore to another.

In our pump-probe experiment we excite CB and TCB dissolved in various selected polar and nonpolar solvents with sub-100 fs pulses at 360 nm. The transient absorption changes between 350 and 700 nm are monitored by means of a supercontinuum generated in a calcium fluoride disk. In the nonpolar cyclohexane, the transient spectra observed after photo excitation remain unchanged up to delay times of 100 ps (data not shown). The spectra can be decomposed into the signatures of the ground state bleach, the stimulated emission (SE) and the excited state absorption (ESA). Only the latter is associated with positive absorption changes and dominates at wavelengths longer than 410 nm for CB and 440 nm for TCB. Below these values, a negative transient absorption is found with a band structure that reproduces the steady state absorption and emission.

In polar solvents like benzonitrile, we find a pronounced evolution of the spectra between 0 and ~50 ps (see Fig. 1). The red shift of the local minimum observed in the visible region of the transient spectra is an unambiguous signature for solvation. For example, TCB in benzonitrile shows initially a local minimum at 440 nm (compare 600 fs delay) that subsequently shifts to 480 nm. However, this minimum cannot be directly identified with the position of the SE maximum, since the ESA strongly distorts the spectral shape and in addition experiences a solvation dependent shift. For CB, the superposition of the SE with the ESA even causes the local minimum to disappear for certain delay times (e.g. 3 ps). Therefore, the time dependent Stokes shift cannot directly be inferred from kinetic traces at a fixed probe wavelength or from the shift of the transient absorption minimum.

We use a global fit procedure that models the transient spectra assuming that the spectral shift of the two contributing signals in the visible, namely the SE and the ESA, depends exponentially on the time (Fig. 2a). For delay times longer than 50 ps, the steady state fluorescence spectrum multiplied by λ^4 is used to simulate the SE signature. By subtracting it from the final transient curve we obtain the broad ESA band (solid lines in Fig. 2a). By optimizing the common time constant and the total shifts of the SE and the ESA a very good agreement with the experimental data is achieved.

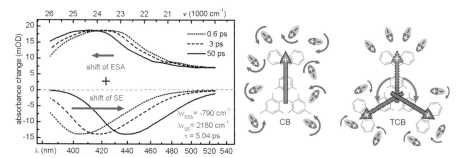

Fig. 2. (a) Simulation (global fit) of the transient spectra of CB in benzonitrile (compare Fig. 1b). The sum of the two contributions reproduces well the experimental data.

(b) Schematic representation of the different solvation processes of CB (left) and TCB (right).

For the solvents dichloromethane, tetrahydrofuran, benzonitrile and chloroform we thereby find the solvation times 1.3 ps, 1.5 ps, 5 ps, and 8 ps in the case of CB and 1.1 ps, 1.1 ps, 3 ps, and 4 ps in the case of the symmetric TCB. These times are in the same regime as the average solvation timescales obtained with a coumarin solute ranging from ~600 fs for dichloromethane to ~5 ps for benzonitrile [3]. However, the solvation of TCB is always faster than that of the less symmetric CB. The difference is more pronounced if the solvation is slow.

We suggest that the acceleration of the solvation of TCB is due to a charge transfer between the equivalent subchromophores of the symmetric molecule. Thus, the electric dipole moment of the excited state can change its orientation and adapt to the arrangement of the solvent dipoles, which might still reflect the ground state charge distribution of the solute (Fig. 2b). This step is fast compared to solvation since it corresponds predominantly to a rearrangement of electron density instead of the reorientation of molecules. The results show that the electronic delocalization in a symmetric molecule and the solvation dynamics are intimately connected. In particular, the symmetry of a molecule can be exploited to accelerate the solvation process if the response time of the environment is too slow.

1   R. Stahl, C. Lambert, C. Kaiser, R. Wortmann, and R. Jakober, "Electrochemistry and Photophysics of Donor-Substituted Triarylboranes: Symmetry Breaking in Ground and Excited State", Chem. Eur. J. **12**, 2358 (2006).

2   U. Megerle, F. Selmaier, C. Lambert, E. Riedle, and S. Lochbrunner, "Symmetry Dependent Solvation of Donor-Substituted Triarylboranes", *submitted to* Phys. Chem. Chem. Phys. (2008).

3   M.L. Horng, J.A. Gardecki, A. Papazyan and M. Maroncelli, "Subpicosecond Measurements of Polar Solvation Dynamics: Coumarin 153 Revisited", J. Phys. Chem. **99**, 17311 (1995).

Substitution- and Temperature-Effects on Hemithioindigo Photoisomerization – The Relevance of Energy Barriers

T. Cordes[1*], T. Schadendorf[2], M. Lipp[1], K. Rück-Braun[2], and W. Zinth[1]

[1] LS für BioMolekulare Optik, LMU München, Oettingenstr. 67, D-80538 Munich, Germany
[2] TU Berlin, Straße des 17. Juni 135, D-10623 Berlin, Germany
 E-mail: thorben.cordes@physik.uni-muenchen.de

Abstract. The kinetics of the *Z/E*-photoisomerization of Hemithioindigo-compounds (HTI) with variations of substitution and temperature are investigated using transient absorption spectroscopy in the visible spectral range. The existence of potential energy barriers in the excited state is demonstrated. These barriers mainly determine the photoisomerization time. Effective tuning of the barriers can be achieved by adequate chemical substitution.

Introduction

Photoinduced isomerizations play an important role in many biological systems (e.g. photosynthesis, vision) or in chemical processes [1]. The detailed investigation of reaction mechanisms is hence essential for the understanding of all underlying principles. Prototype molecules with conjugated double bonds are stilbene [2] or hexatriene, which have both been used as model systems for biologically relevant systems [1-2]. Interesting properties arise from the structural motive of a C=C double bond: the light-switchable structures often show a photochromic behaviour, meaning that the two isomeric forms *Z/E* exhibit different absorption properties. These features allow new applications: optical data storage [3] or non diffraction-limited microscopy [4]. The light-induced alteration of structure and function of photochromic compounds has also been used to modulate and to switch chemical and biological reactions (On/Off-switching) by incorporation into biomolecules, which allows investigations with very high temporal resolution [5].

We recently studied photochromic HTI-compounds that undergo a lightinduced *Z/E*-isomerization on a picosecond timescale [5-8]. The use of transient absorption spectroscopy could reveal that HTI ω-amino-acids are suitable to act as trigger molecule in chromopeptide-structures [5]. The present study reveals mechanistic details about the photochemical reaction course of the sequential *Z→E* photoprocess [6]: the variation of substitution [7,8] of HTI-molecules with polar substituents and the influence of the temperature on the reaction rate are studied. The results clearly show the relevance of energy barriers for the HTI-photoisomerization.

Experimental Methods

The experiment were conducted using a home-built Ti:Sapphire laser system and transient-absorption spectrometer (TA) as described in refs. [5-8]. To study the *Z→E*-process sample solutions of the pure *Z*-Isomer (> 95%) were excited at 400 nm. A white light continuum in combination with multichannel detection allowed recording the time dependent absorbance changes in a wavelength range between 350 – 650 nm. All experiments concerning the substitution-dependence were conducted in dichloromethane at 295 K. The temperature-dependent experiments were performed in dichloromethane in a temperature range from 284 – 303 K.

Results and Discussion

Fig. 1a illustrates the photochromic behavior of HTI and the absorbance changes due to the $Z{\rightarrow}E$-isomerization process initiated by 400 nm light.

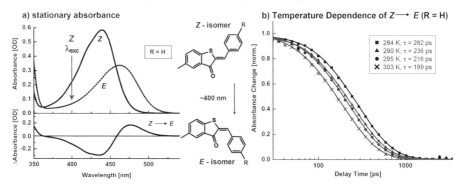

Fig. 1. Stationary (a) and transient absorption data (b) of a reference HTI-molecule shown in the middle of the figure. b) Temperature dependent kinetic traces (λ_{Probe} = 526 nm) of the same compound. The normalized absorbance changes can be taken as a qualitative value for the isomerization speed. Reaction times from a global fitting routine are displayed in the figure.

The transient kinetics with varying temperature, shown in Fig. 1b, reveal that the $Z{\rightarrow}E$-isomerization reaction is an activated process: the reaction time decreases with increasing temperature (Fig. 1b). The reaction times are changed by a factor of 1.4 within the studied temperature range. This behaviour clearly proofs that the reaction rate of the photoisomerization is controlled via a potential barrier. The activation energy of the process is found to be 14 kJ/mol (Arrhenius-analysis, correlation coefficient of R = 0.96) which is comparable with values for stilbene (gas phase, 14 kJ/mol [2]). Please note, that this value does not consider the temperature-dependence of the solvent-viscosity. A viscosity corrected value is expected to be somewhat smaller in the range of 10 kJ/mol.

The height of the reaction barrier can be varied applying different polar substituents to the stilbene-part of the HTI-molecule (Fig. 2a). It can be seen that electron donating groups (p-OMe) strongly increase the reaction speed while electron withdrawing groups (p-Cl, p-Br, p-CN) decelerate the reaction kinetics (Fig. 2a). The time constants obtained from a global fitting routine vary by more than two orders of magnitude. It could also be shown that the substituent effects in the stilbene-part obey the empirical Hammett-relation [7]. A different behaviour has been found for thioindigo-substitution: while bromium slows the reaction kinetics in the stilbene-part as shown in Fig. 2a, it accelerates the reaction in the thioindigo-moiety [8]. Electron donating groups show an inverse effect on the isomerization rate when being attached to the different parts of the asymmetric HTI-molecule [8].

All results are consistent with the existence of potential barriers in the excited electronic state. In combination with former studies [6-8], it can be concluded that the potential barrier is found between the fluorescing state CTC and the twisted dark phantom state P* [2,6-8] as shown in Fig. 2b.

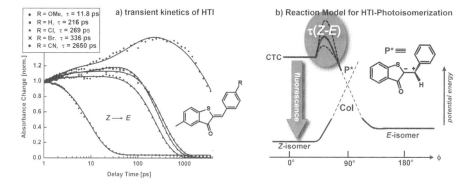

Fig. 2. a) TA-data of HTI-molecules with a differing substitution as indicated in the figure. The temporal evolution of the absorbance changes (λ_{Probe} = 526 nm) can be taken as a qualitative value of the isomerization time (time constants from a global fitting routine). b) Schematic reaction model for the photoisomerization processes together with a postulated structure of the P*-state.

The transition from CTC to P* is hence the slowest and rate-limiting step of the photoreaction and is connected to strong geometrical changes: the central double bond and most likely additional single bonds have to be twisted. The P*-state can not be observed in our time-resolved experiments due to its slow formation but short lifetime < 1 ps. It decays presumably via a conical intersection (CoI) into the ground-state of both isomers Z/E. The postulated zwitterionic character explains the observed behavior upon substitution (Fig. 2a). In the case of a stabilization of the zwitterionic P*-structure due to adequate substitution, the energy of P* drops and reduces the barrier between CTC and P*. This also accelerates the reaction rate of the photoisomerization reaction. For a destabilization of P* the inverse behavior is found.

Conclusions

We could demonstrate that the kinetics of HTI photoisomerization are highly sensitive to substitution and to the specific environment. Our results could show that the reaction rate of the rate-limiting reaction step is controlled by an energy barrier on the excited state potential surface separating a fluorescing species and the dark phantom state P*. Substituent-effects strongly influence this barrier height, which was found to be in the order of ~10 kJ/mol for an unsubstituted compound.

1   C. Dugave, L. Demange, Chem. Rev. **103,** 2475, 2003.
2   H. Meier, Angew. Chem. Int. Ed. Engl. **31,** 1399, 1992.
3   S. Kawata Y. Kawate, Chem. Rev. **100,** 1777, 2000.
4   S. W. Hell, Science **316,** 1153, 2007.
5   T. Cordes, D. Weinrich, S. Kempa, K. Riesselmann, S. Herre, C. Hoppmann, K. Rück-Braun, W. Zinth, Chem. Phys. Lett. **428,** 167, 2006.
6   T. Cordes, B. Heinz, N. Regner, C. Hoppmann, T. E. Schrader, W. Summerer, K. Rück-Braun, W. Zinth, ChemPhysChem **8,** 1713, 2007.
7   T. Cordes, T. Schadendorf, B. Priewisch, K. Rück-Braun, W. Zinth, J. Phys. Chem. A **112,** 581, 2008.
8   T. Cordes, T. Schadendorf, K. Rück-Braun, W. Zinth, Chem. Phys. Lett. **455,** 197, 2008.

Vibrational Coherence Decay in Metal Carbonyls: Solvent Dependence of Coherence Lifetimes Studied with MDIR

Matthew J. Nee, Carlos R. Baiz, Jessica M. Anna, Robert McCanne, and Kevin J. Kubarych

Department of Chemistry, University of Michigan, Ann Arbor, Michigan 48109, USA
E-mail: mattnee@umich.edu

Abstract. Multidimensional infrared spectra of a metal carbonyl in different solvents are presented as a function of the waiting time, t_2. The evolution of each peak is related to excited-state coherences and the relative correlation between excited state vibrational energies.

Introduction

Recent studies of electronic coherence transfer in photosynthetic systems [1-3] have generated interest in characterizing different types of energy transfer between chromophores. Although vibrational coherence and coherence transfer have been studied previously by multidimensional infrared (MDIR) spectroscopy [4], recent technological advances in our lab enable us to view fully the evolution of the two-dimensional infrared (2DIR) spectrum as a function of the waiting time, t_2, [Fig. 1(a)]. The spectra presented here show the first fully resolved oscillations which are characteristic of excited-state coherences with lifetimes equal to or shorter than the timescale of intramolecular vibrational redistribution. Solvents which create extensive broadening in the diagonal features show a markedly shorter coherence lifetime.

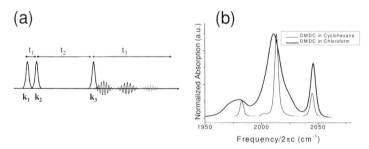

Fig. 1. MDIR uses a three pulse sequence (a). The three fields with wavevectors ($\mathbf{k}_1$, $\mathbf{k}_2$, $\mathbf{k}_3$) are separated by the times t_1 and t_2 as indicated; FTIR spectra of solutions of DMDC (b) in cyclohexane (thick line) and in chloroform (thin line) show that not all transitions are broadened to the same extent by changes in solvation environment.

Experiment

For each set of experiments, a 2DIR spectrum was collected at each of many t_2 delays. The procedures involved in the acquisition of each 2DIR spectrum have been detailed elsewhere [5]. The signal and idler from each of two OPAs are difference-frequency mixed produce 100-fs pulses of 5000-nm radiation. Three such IR pulses [Fig. 1(a)] generate a four-wave-mixing signal in a background-free phase matching direction that is upconverted to the visible using a chirped 800-nm pulse (350-ps pulsewidth) temporally synchronized with the third IR pulse. 2DIR spectra [FTIR spectra, Fig. 1(b)] of samples containing 100-μm pathlength, 6-mM solutions of $Mn_2(CO)_{10}$ (dimanganese decacarbonyl, DMDC) in cyclohexane were taken for the range $t_2 = -1$ to 18 ps. 3-mM solutions of DMDC in chloroform ($CHCl_3$) were also analyzed.

Results

Fig. 2(a) shows a series of four absolute-value rephasing ($\mathbf{k}_s = -\mathbf{k}_1 + \mathbf{k}_2 + \mathbf{k}_3$) 2DIR spectra of DMDC in cyclohexane at different values of t_2. The intensities of the off-diagonal peaks oscillate as a function of t_2 as plotted in Fig. 2(b). Each peak oscillates at the difference frequency between the two energy levels involved in the coherence [Fig. 2(c)] and decays with increasing delay owing to population relaxation and orientational dephasing.

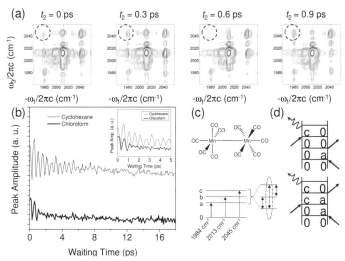

Fig. 2. A series of four 2DIR spectra of DMDC in cyclohexane (a), taken at different waiting times, t_2. Integrating the volume of the circled peak at each time delay shows oscillatory behavior (b) in cyclohexane and in chloroform caused by the waiting time excited-state coherence circled in (c). Two Liouville paths (d) contribute to that 2DIR peak. One involves a ground-state population during t_2, the other, a coherence between states a and c. In chloroform, the oscillations arising from the coherence decay much more rapidly than in cyclohexane.

Changing the solvent to chloroform changes the FTIR spectrum as seen in Fig. 1(b). In DMDC, two of the three peaks are broadened significantly, but the highest frequency transition retains approximately the same lineshape. This disparity in linewidth indicates a fundamental difference in the way that the solvent interacts with each C≡O vibrational mode. Fitting the diagonal peak volumes shown in Fig. 3 to biexponential decays give fast decay rates ranging from 2.3-3.5 ps in cyclohexane to 1.2-2.4 ps in chloroform. The peak at $\omega_1 = \omega_3 = 2044$ cm^{-1} is least effected by the solvent (decay time confidence intervals in cyclohexane and chloroform overlap), consistent with the mode-dependent broadening seen in the FTIR spectra in Fig. 1(b). All coherence lifetimes are shortened considerably in chloroform owing to the increased broadening as previously observed in 2DIR spectra [6]. The decay of these oscillations, which likely derives from both dephasing and coherence transfer to bath or dark modes, may reveal information about excited-state energy level fluctuations relative to one another, rather than referenced to the ground state as in a frequency-frequency autocorrelation function. Figs. 3(c) and (d) show Fourier transforms of the time-domain peak amplitudes shown in Fig. 2(b). In chloroform, the off-diagonal peak followed in Fig. 2 still maintains relatively long-lived coherence (decay time, τ_c = 3.5 ps in cyclohexane; 0.8 ps in chloroform) between the two modes *a* and *c*, which have parallel dipoles. In contrast, the coherence between modes *a* and *b* (perpendicular dipoles) has a considerably shorter decay time in chloroform (~0.3 ps) relative to cyclohexane solution (2.4 ps). Both solvent effects on the individual modes and the relative orientations of the dipoles involved in a vibrational coherence appear to contribute to the resulting decay rates. Fleming and coworkers indicate that different *excitons* in photosynthetic complexes may fluctuate in phase with one another [2,3]. Here, we are able to cleanly resolve our *vibrational* transitions, while tuning the system-bath coupling by choice of solvent. To understand the role that quantum coherences can play in energy transfer, we must be able to model the coherence lifetimes and their dependence on solvent environment (dipole moment, polarizability, etc.); these experiments are a step in that direction. A full analysis must include the possibility of coherence transfer between pulses as well as the influence of the six Raman active (IR dark) modes in the carbonyl stretch region of DMDC.

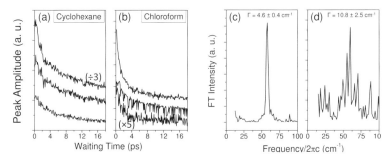

Fig. 3. Peak amplitude volumes as a function of t_2 for the diagonal peaks in the 2DIR spectra of DMDC in cyclohexane (a) and chloroform (b). The diagonal peaks at ω_1 and ω_3 equal to 1983 cm^{-1} (bottom lines), 2015 cm^{-1} (top lines), and 2044 cm^{-1} (middle lines) are scaled as indicated. Fourier transforms of the oscillations of the peak circled in Fig. 1(a) of DMDC in cyclohexane (c) and chloroform (d) show considerable broadening in chloroform as well.

1   K. Hyeon-Deuk, Y. Tanimura, and M. Cho, "Ultrafast exciton-exciton coherent transfer in molecular aggregates and its application to light-harvesting systems," J. Chem. Phys. **127**, 075101 (2007).

2   H. Lee, Y.-C. Cheng, and G. R. Fleming, "Coherence Dynamics in Photosynthesis: Protein Protection of Excitonic Coherence," Science **316**, 1462 (2007).

3   G. S. Engel, T. R. Calhoun, E. L. Read, T.-K. Ahn, T. Mančal, Y.-C. Cheng, R. E. Blankenship, and G. R. Fleming, "Evidence for wavelike energy transfer through quantum coherence in photosynthetic systems," Nature **446**, 782 (2007).

4   M. Khalil, N. Demirdröven, and A. Tokmakoff, "Vibrational coherence transfer characterized with Fourier-transform 2D IR spectroscopy," J. Chem. Phys. **121**, 362 (2004).

5   M. J. Nee, R. McCanne, and K. J. Kubarych, "Two-dimensional infrared spectroscopy detected by chirped pulse upconversion," Opt. Lett. **32**, 713 (2007).

6   M. Khalil, N. Demirdröven, and A. Tokmakoff, "Coherent 2D IR Spectroscopy: Molecular Structure and Dynamics in Solution," J. Phys. Chem. A **107**, 5258 (2003).

Generation of Narrowband Ultrashort Pulses Tunable in the mid-IR and the Application to Vibrational Energy Transfer in a Modified Amino Acid

Karin Haiser[1], Florian O. Koller[1], Markus Huber[1], Tobias E. Schrader[1], Nadja Regner[1], Wolfgang J. Schreier[1] and Wolfgang Zinth[1]

[1] Lehrstuhl für BioMolekulare Optik, Department für Physik, Ludwig-Maximilians-Universität München, Oettingenstr. 67, D-80538 München, Germany
E-mail: karin.haiser@physik.uni-muenchen.de

Abstract. Difference frequency mixing of pulses with adjustable chirp is applied to produce narrowband and widely tunable pulses in the mid-IR. They are used for selective excitation of vibrational modes in IR-pump-IR-probe experiments in a modified amino acid.

Introduction

Two-dimensional vibrational spectroscopy [1] as well as pump-probe experiments in the infrared require synchronised pulses with diverse properties. For a selective excitation of specific vibrational modes in polyatomic molecules in the condensed phase the spectral width of the pump pulses should be in the order of $10\ \mathrm{cm}^{-1}$ to $20\ \mathrm{cm}^{-1}$ matching the width of typical molecular bands. Efficient multi-channel probing of multiple vibrational modes requires however synchronized broadband infrared probing pulses with a spectral width in the $200\ \mathrm{cm}^{-1}$ range to allow the simultaneous detection of various vibrational modes.

In most experiments of 2D-IR spectroscopy narrow mid-IR pulses are generated by spectrally filtering broadband infrared pulses with a suitable Fabry-Perot interferometer [2], a process which implies a strong reduction of pulse intensity. In this paper we present an alternative approach for the generation of narrowband mid-IR pulses and show an application of this new light source for the study of ultrafast energy transfer processes in the modified amino acid Fmoc-p-Nitro-Phenylalanine (Fmoc-Phe(NO$_2$)).

Experimental Methods

The mid-IR pulse generation system is based on a Ti:sapphire laser system with a regenerative amplifier (Spitfire Pro, Spectra Physics; repetition rate of 1 kHz, central wavelength of ~ 800 nm and durations of 65 fs), where pulse energies of about 360 μJ are available for the infrared generation process. In the first stage of frequency conversion pairs of near infrared pulses (Signal and Idler) are generated in a double-stage optical parametric amplifier. In a standard infrared generation process these pulses are directly converted to the mid-infrared by a difference frequency mixing process [3]. For the generation of narrowband pulses we make use of the idea that frequency mixing of suitably chirped pulses may lead to spectral narrowing. This approach was already successfully applied to the generation of narrowband Terahertz and UV pulses [4,5]. In the employed set-up Signal and Idler pulses are propagated through dispersive media (Si-prisms, see inset in Fig.1), which stretches the pulse duration (by group velocity dispersion) and imposes a nearly linear chirp. If the chirp of both pulses is matched the subsequent difference frequency mixing leads to the generation of narrowband mid-IR pulses. Their bandwidth and duration can be adjusted by the amount of dispersive material [6]. Bandwidths of about 20 cm^{-1} tunable over the whole mid-infrared range from 3 μm to 11 μm can be generated (see Fig.1).

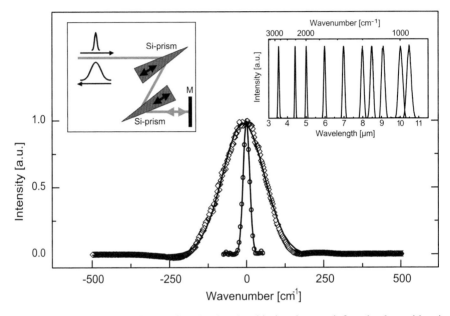

Fig. 1. Spectral narrowing of mid-infrared pulses by chirping the near-infrared pulses with pairs of prisms. The tunable wavelength of the narrowed mid-IR pulses in the range of 3 μm to 11 μm is demonstrated in the right inset. The spectra are normalized to equal intensity.

Results and Discussion

The new mid-IR pulse source has been applied to the study of excitation transfer in the modified amino acid Fmoc-p-Nitro-Phenylalanine. The mid-IR of Fmoc-Phe(NO$_2$) exhibits three dominating absorption bands (see inset of Fig.2), which can be assigned to NO$_2$-stretching (1346 cm^{-1} and 1518 cm^{-1}) as well as to CO-stretching modes (1716 cm^{-1}). Selective excitation of the NO$_2$-group or the CO-group is achieved by the tunable narrowband infrared pulses. Subsequent relaxation and transfer processes lead to the redistribution of vibrational excitation within the molecule.

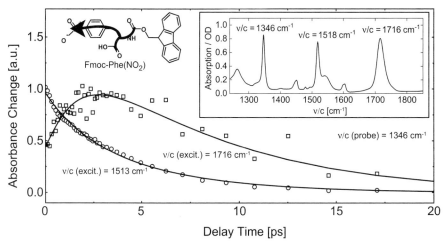

Fig. 2. Results of a IR-pump-IR-probe experiment on Fmoc-p-Nitro-Phenylalanine (Fmoc-Phe(NO$_2$)): The absorption change at 1346 cm^{-1} is shown as a function of delay time for two excitation wavenumbers that correspond to different vibrational modes of the molecule. The relevant region of the infrared absorption spectrum is shown in the inset. Excitation of the antisymmetric NO-stretching mode at 1513 cm^{-1} leads to instantaneous absorption in the probed symmetric NO-stretching mode at 1346 cm^{-1}, while exciting the CO-mode at 1716 cm^{-1} shows a delayed response in the NO$_2$-group.

When the symmetric NO$_2$-stretching band at 1346 cm^{-1} is detected (see Fig.2) we obtain different absorption dynamics depending on whether the antisymmetric NO$_2$-stretching band or the CO-stretching band was originally excited. (i) After the excitation of the antisymmetric NO$_2$-stretching band at 1513 cm^{-1} instantaneous absorption features are observed which point to a strong coupling of the symmetric and antisymmetric NO$_2$-stretching vibrations. (ii) If we excite the CO-stretching mode at 1716 cm^{-1}, a delayed response of the absorption changes in the range of the symmetric NO$_2$-band is found. The observed coupling between the NO$_2$- and the CO-stretching mode is weak. Absorption changes in the NO$_2$-range are seen only during the decay of the CO-mode. Apparently the decay of the CO-stretching vibration leads to the population of low-lying modes which are coupled to the NO$_2$-group.

1 R. M. Hochstrasser, Proc. Natl. Acad. Sci. U.S.A. 104, 14190 (2007), and references therein.
2 P. Hamm, M. H. Lim and R. M. Hochstrasser, J. Phys. Chem. B, 102, 6123-6138 (1998).
3 T. Elsaesser and M. C. Nuss, Optics Letters, 16, 6 (1991).
4 K. J. Kubarych, M. Joffre, A. Moore, N. Belabas, and D. M. Jonas, Optics Letters, 30, 10, 1228-1230 (2005).
5 S. Laimgruber, H. Schachenmayr, B. Schmidt, W. Zinth, and P. Gilch, Applied Physics B, 85, 557-564 (2006).
6 F. O. Koller, K. Haiser, M. Huber, T. E. Schrader, N. Regner, W. J. Schreier and W. Zinth, Optics Letters, 32, 22 (2007).

Ultrafast Exciton Dynamics of J- and H-Aggregates of Porphyrin Catechol in Aqueous Solution

Sandeep Verma, and Hirendra Nath Ghosh*

Radiation & Photochemistry Division, Bhabha Atomic Research Centre
Trombay, Mumbai – 400085, INDIA
E-mail: hnghosh@barc.gov.in

Abstract. Porphyrin catechol found to form J- and H-aggregates in different pH at certain concentration. Ultrafast exciton dynamics of J- and H-aggregates found to be 200 fs and 100 fs respectively as monitored by femtosecond visible spectroscopy.

Introduction

The formation of porphyrin aggregates in solution through noncovalent interaction is a well known phenomena that has a deep impact on many physicochemical properties which include spectroscopic, photophysical as well as their specific interactions [1]. Among the aggregates, highly ordered J- or H-aggregates are found to have unique optical and electronic properties in connection with performing energy- and electron-transfer functions in recent years. Porphyrin dyes are found to be suitable molecule to sensitize TiO_2 nanoparticles in dye-sensitized solar cell due to their broad absorption in the visible region. Earlier we have demonstrated ultrafast interfacial electron transfer (IET) dynamics of porphyrin catechol (TPP-Cat) (Figure 1) sensitizing TiO_2 nanoparticles [2]. However it is widely known as that porphyrin molecules are prone to form aggregation at certain concentration and pH range. The aggregation behavior of TPP-Cat might affect the IET dynamics on TiO_2 nanoparticle surface. So it is very important to study the aggregation behavior of TPP-Cat molecule in solution phase and study the excited state dynamics of the aggregated molecules. Earlier Misawa and Kobayashi [3] has reported S_2 exciton and S_1 exciton lifetimes of J-aggregated of tetraphenyl porphyrin tetrasulfonic acid (TPPS) and reported to be 1.3 ps and 40 ps respectively. However till date not many reports are available on both S_2 exciton and S_1 exciton relaxation dynamics of both J- and H-aggregated porphyrins.

In the present investigation we have carried out pH dependence study of optical absorption of TPP-cat to find out optical absorption of non-aggregated, J- and H-aggregated porphyrin molecules. We have also carried out femtosecond transient absorption studies in the visible and near-IR region to find out the excited state behavior and relaxation dynamics of non-aggregated, J- and H-aggregated porphyrin molecules in ultrafast time scale.

Results and Discussion

Figure 1 shows the optical absorption spectra of TPP-Cat (50 μM) in different pH. It is clearly seen at pH 2.5 the optical absorption spectra of TPP-cat molecules resembles to the optical spectra of non-aggregated protonated molecules. However at pH 5.8 we can observe an additional new sharp blue shifted absorption peak appeared at 395 nm in optical absorption spectra, which can be attributed to H-aggregates of TPP-Cat. On the other hand at pH 1.8 a new red shifted peak appears at 490 nm with an additional peak at 730 nm in the optical absorption spectra, which can be attributed to J-aggregate of TPP-Cat.

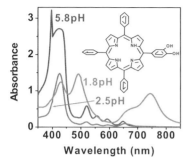

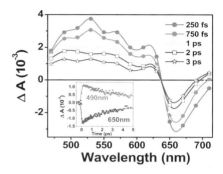

Fig. 1. Figure 1: Optical absorption spectra of porphyrin catechol (TPP-cat) at different pH in water. The concentration of TPP-cat has been kept 50 μM

Fig 2. Transient absorption spectra of TPP-cat in water at pH 2.5 (non-aggregated) at different time delay after excitation at 400 nm. Inset: Transient decay kinetics at 490 nm and bleach recovery kinetics at 650 nm)

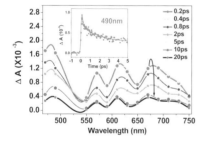

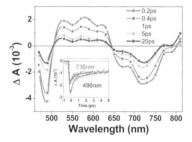

Figure 3. Transient absorption spectra of TPP-cat in water at pH 5.8 (H-aggregated) at different time delay after excitation at 400 nm. Inset: Transient decay kinetics at 490 nm.

Figure 4: Transient absorption spectra of TPP-cat in water at pH 1.8 (J-aggregated) at different time delay after excitation at 400 nm. Inset: Bleach recovery kinetics at 490 nm and 730 nm.

We have also carried out ultrafast transient absorption studies of TPP-cat of all the three different of TPP-cat molecules exciting 400 nm laser pulse. Figure 2 shows the transient absorption spectra of protonated non-aggregated TPP molecules at different time delay after 400 nm laser excitation. The positive transient absorption has been attributed to the excited singlet state absorption (ESA) and negative absorption band has been attributed to the bleach of the protonated Q-band. Figure 2 inset shows the transient decay kinetics at 490 nm and bleach recovery kinetics at 650 nm. Both the kinetics can be fitted muti-exponentially with time constants 1.5 ps (56%), 10 ps (32%) and > 1ns (12%). The shorter components are attributed to vibration relaxation and non-radiative transition due to H-bonding network with the bulk water molecules and the longer component is due to excited state lifetime of non-aggregated TPP molecules [2].

Figure 3 shows transient absorption spectra of H-aggregated TPP-cat molecules at pH 5.8. We can observe excited absorption band at 490 nm and another broad transient band due to excited Q band from 550 nm to 750 nm. Figure 3 inset shows the transient decay kinetics at 490 nm. The band at 490 nm can be attributed to excited absorption due to H-aggregates. The transient decay kinetics at 490 can be fitted to 100 fs (70%), 1.5 ps (10%), 10 ps (10%), 50 ps (8 %) and > 1ns (2%). In the present kinetic decay trace we can observe two extra components in addition to other three components, which we have observed in non-aggregated porphyrin (Figure 2). However the 100 fs component can be attributed to exciton relaxation dynamics due to the soret band (S_2 band) and 50 ps component can be attributed to exciton relaxation dynamics due to the Q-band (S_1 band) of H-aggregates. Figure 4 shows transient absorption spectra of J-aggregated TPP-cat molecules at pH 1.8. The transient spectra shows bleach at 490 nm and 730 nm and a positive absorption band at 520 nm -620 nm. Figure 4 inset shows the bleach recovery kinetics at 490 nm and 730 nm. The bleach kinetics at 490 nm can be fitted muti-exponentially with time constants 200 fs (76 %), 1.5 ps (9%), 10 ps (6%) 180 ps (5%), >1 ns (4%). Again here the 200 fs component can be attributed to exciton relaxation dynamics due to the soret band (S_2 band) and 180 ps component can be attributed to exciton relaxation dynamics due to the Q-band (S_1 band) of J-aggregates.

Conclusion:

In conclusion, we have shown that porphyrin catechol (Tpp-cat) molecules form both J- and H-aggregates in different pH at certain concentration. We have carried out femtosecond transient absorption spectroscopy for non-aggregated, J- and H-aggregated TPP-cat molecules. Transient measurements revealed that the exciton relaxation dynamics due Soret band (S_2 band) is 100 fs for H-aggregates and 200 fs for J-aggregate. On the other hand the exciton relaxation dynamics due Q band (S_1 band) is 50 ps for H-aggregates and 180 ps for J-aggregates. In turn this observation will help to study interfacial electron transfer dynamics of aggregated porphyrin on TiO_2 nanoparticle surface.

References:

1   W. I. White, in The Porphyrins; Dolphin, D., Ed.; Academic Press: New York, Vol. **5,** 303.1978.

2   G. Ramakrishna, S. Verma, D. A. Jose, D. KrishnaKumar, A. Das, D. K. Palit, and H. N. Ghosh, J. Phys. Chem. B, **110,** 9012, 2006.

3   K. Misawa, and T. Kobayashi, J. Chem Phys **110,** 5844. 1999.

4   M. Chachisvilis, O. Kuhn, T. Pullerits, and V. Sundstrom, J. Phys. Chem. B **1997,** 101, 7275.

Chirp Effect on Vibrational Wave Packets in Large Molecules: a Multimode Perspective

Amir Wand[1], Ofir Shoshanim[1], Oshrat Bismuth[1], Shimshon Kallush[2], Ronnie Kosloff[2] and Sanford Ruhman[1]

Institute of Chemistry and [1]Farkas Center for Light Induced Processes, [2] Fritz-Haber Research Center for Molecular Dynamics, The Hebrew University, Jerusalem 91904, Israel.
E-mail: sandy@fh.huji.ac.il

Abstract. Chirp effects on generation of ground state vibrational wave packets in polyatomic molecules are experimentally and theoretically investigated. Theory indicates optimal chirp is quadratic, depends on all displaced modes, and follows the evolving multidimensional potential energy gap. Experiments on 2 organic dyes reveal that 5 fsec pulses are unable to track this evolution, resulting in optimal GVD which maximizes the pulses chirp rate.

Introduction

Predictions that negative chirp (NC) can boost an ultrashort pulse's propensity for generating ground state wave packets through resonant impulsive Raman scattering (RISRS) [1], have been demonstrated in molecular systems [2]. Most experiments involve polyatomics in solution, but interpretations have focused on single dominant vibrational modes. This includes a recent study on Oxazine-1 [3] where the optimal chirp (OC) for RISRS was systematically mapped.

Current study, combining theory and experiment, seeks to answer the following: What determines the OC for a polyatomic chromophore? What molecular characteristic is the chirp following? Would the same chirp be optimal for all displaced vibrations, and how does electronic T_2 effect the questions above?

To address these questions, experiments on two dyes (Betaine-30 and Oxazine-1 in EtOH) have been conducted, guided by theoretical analysis and simulations. They were chosen since they differ both in the frequency range of principle displaced modes and in dephasing rates, while exhibiting prominent RISRS activity [3,4].

Theory

Analysis based on the TD Schrödinger equation (TD-SE) for N-displaced harmonic vibrations coupled to an electronic transition shows that the momentum jump in ground state mode j is dependent on all other modes in the molecule, and that the physical quantity that must be followed by the chirp is the multi dimensional vertical potential difference: $\frac{d}{dt}\langle\psi_j^g|\hat{P}|\psi_j^g\rangle = -\frac{2}{\hbar}\Im\langle\psi_j^g|\hat{P}|\psi_j^e\rangle\varepsilon(t)\prod_{m\neq j}\langle\psi_m^g|\psi_m^e\rangle$.

To test the predicted dependence, the momentum kick imparted by an 8 fsec pulse tuned on resonance with a dual mode chromophore was simulated with and without electronic dephasing. Two extreme cases of displacements in the 600 and 1,500 cm^{-1} modes were tested. Results are shown in figure 1 in terms of the amplitude of $<P>$ in the ground state for each mode as a function of the GVD.

As shown, a dominant high frequency mode dictates an OC for all others, while a low frequency one allows for different OCs for each. Dephasing reduces the amplitudes of vibronic coherences in both modes (particularly in second case) and also shifts OC toward the transform-limit (TL). This must be due to rapid erasure of transition dipole coherence, which is the vehicle of all chirp control schemes, leading to a narrower "window" of coherent dynamics. These results are trivially obtained from the analytical expression and are in accordance with our intuitive understanding of the wave packet dynamics.

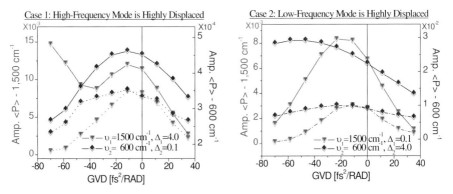

Fig. 1. Simulations results for two extreme cases of displacements (displacements are shown inside panels) - with (dashed lines) and without (solid lines) electronic dephasing.

Experiment

A single stage NOPA centered at ~550 nm and pumped by a 30 fsec amplified Ti:Sapphire system was used to derive pump, probe and reference pulses. Precompression in BK7 prisms was followed by a deformable mirror 4-f shaper, allowing nearly full compression to the TL of ~6.5 fs. Chirp was varied by insertion or removal of differing amounts of fused silica in the pump beam path, covering a GVD range of (-140)-(+140) fs^2/rad, and characterized using X-PG-FROG. Transmitted probe and reference pulses were co-dispersed in an imaging spectrograph to obtain delay dependent ΔOD spectra from 500-700 nm. Analysis of the results involves subtraction of a slowly varying background and Fourier transformation of the isolated residuals.

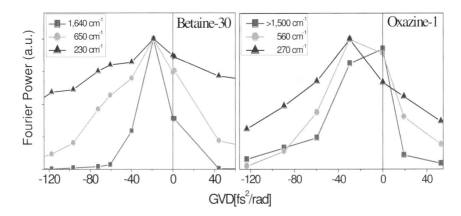

Fig. 2. Modulation depth vs. GVD of pulse for a few representative modes in two systems.

The results, depicted in figure 2, demonstrate a substantial enhancement of the amplitude of ground state vibronic coherences using NC, particularly for high frequency modes of both molecules. Both systems demonstrate a nearly uniform OC, of only ~20 fs^2/rad, regardless of the vibrational frequencies (250-1,600 cm^{-1}). This OC broadens the pulse by a factor of only ~1.5. While in Betaine this might be indicative of rapid dephasing, [5,6], this is not the case for Oxazine-1 which exhibits a narrow absorption band with pronounced vibronic structure.

Conclusions

To account for similarity of OC in both dyes, the instantaneous shift in the vertical difference potential for Betaine was calculated based on RR data [5] (Fig 3). In addition, we present a graph of the chirp rate (CR) of Gaussian pulses as a function of the GVD.

The first suggest that a ~5-7 fs pulse isn't broad enough to compensate for the changes in the difference potential as required from the OC. It also demonstrates that for short times, the OC for following ΔP should be quadratic and not linear. The latter shows that for a given TL pulse width, CR reaches extrema at distinct values of GVD which broaden the pulse by $\sqrt{2}$ in agreement with the data of both dyes. Accordingly we conclude that the measured OC reflects the limited bandwidth of our source, not the polyatomic molecular systems under study. Nonetheless NC clearly enhances coherences of ground state wave packets in all active modes, and can still serve to separate excited and ground state modes in a polyatomic to some extent.

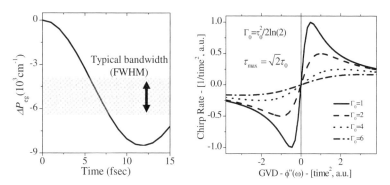

Fig. 3. Demonstration of the limitations of excitation pulse to dynamically follow the instantaneous potential gap in Betaine (see text for details).

References:

1   Ruhman S. and Kosloff R., J. Opt. Soc. Am. B **7**, 1748 (1990).
2   Bardeen C.J., Wang Q. and Shank C.V., J. Phys. Chem. A **102**, 2759 (1998).
3   Malkmus S., Durr R., Sobotta C. and Pulvermacher H., Zinth W. and Braun M., J. Phys. Chem. A **109**, 10488 (2005).
4   Kovalenko S.A., Eilers-Konig N., Senyushkina T.A. and Ernsting N.P., J. Phys. Chem. A **105**, 4834 (2001).
5   Zhao X., Burt J.A. and McHale J.L., J. Chem. Phys. **121**(22), 11195 (2004).
6   Hwang H. and Rossky, J. Phys. Chem. B **108**, 6723 (2004).

Determining Vibrational Huang-Rhys Factors by Photon Echo Spectroscopy

N. Christensson, A. Yartsev and T.Pullerits

Department of Chemical Physics, Lund University, P.O. Box 124, SE-22100, Lund, Sweden
E-mail: Niklas.Christensson@chemphys.lu.se

Abstract. Electronic and vibrational dephasing dynamics of Rhodamine 800 has been studied with 3PEPS. With careful analysis, the S-factors of the vibrational modes can be accurately determined. The vibrational dephasing rate displays abnormal frequency dependence.

Introduction

The interaction of electronic states with the surrounding bath of nuclear motions gives rise to dephasing and population relaxation in the condensed phase. The time dependence of the interaction of the nuclear bath with the chromophore gives rise to time dependent fluctuations of the transition frequency. The auto correlation of these fluctuations, C(t), contains the information on the timescale and amplitude of the nuclear motions affecting the electronic states. The normalised correlation function, M(t), can be determined in the time domain by the three-pulse photon echo peak shift (3PEPS) experiment. In the impulsive limit it has been shown that this measurement can directly follow the slow decay of correlation function on timescales longer than the bath correlation time [1, 2]. However, intermolecular modes generally have a frequency of a few hundred wave numbers and are thus comparable to the width of the pulses used in experiments. To what extent it is possible to correctly determine the absolute coupling strength of these modes from time domain experiments with finite pulses is the main topic of this work.

To get and independent measure of the spectra of the vibrational modes of the molecule in this study we employ Fluorescence Line Narrowing (FLN) [3].

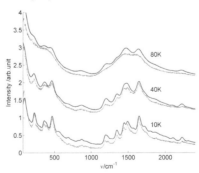

Fig. 1. Temperature dependence of the FLN signal of Rhodamine 800 in ethanol matrix. Excitation laser line is at 713.8 nm. Blue dotted lines are experiment points and the dashed lines are simulation based on the theory in ref [3]. The simulations have been displaced vertically by 0.15.

We extract the frequencies and Huang-Rhys factors of all modes that couple to the electronic transition in Rhodamine 800. Figure 1 shows the temperature dependence from 10 to 80 K of the FLN signal of Rhodamine 800 excited at 713.8 nm using a narrow band CW Ti:Sapphire laser. The low temperature trace reveals a large number of vibrational (under-damped) modes above 100 cm^{-1} that couple to the electronic transition. We resort to simulations to accurately determine the correct shape of the spectral density[3].

Figure 2 shows the 3PEPS signal of Rhodamine 800 in ethanol at room temperature recorded with ~20 fs pulses centred at 700 nm (red edge of the absorption spectrum) with a FWHM of 34 nm. The experimental set-up has been describe previously [4]. The peak shift shows a multi-exponential decay originating from solvation and a complex beating pattern of multiple vibrations. A least–square curve fitting routine was employed to extract the frequencies and the dephasing times of all vibrations in the signal. We directly identify the 5 vibrational modes below 500 cm^{-1} seen in the FLN experiment. The apparent dephasing rate of each of the modes is shown as a function of the vibrational frequency in figure 3.

We start with a direct simulation of the 3PEPS signal based on the non-linear response function formalism in the impulsive limit [1, 5]. To phenomenologically account for the finite duration of the pulses we introduce an effective spectral density via a multiplication by the power spectra of the pulses used in the experiments. The results are shown together with the experimental trace in figure 2 and we find a nice agreement between the experiment and the simulations.

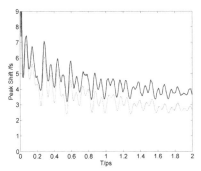

Fig. 2. 3PEPS signal of Rhodamine 800 in ethanol at 290 K. Dotted line show the experimental data points and the solid line shows the simulated trace. The simulations have been displaced vertically by 1 fs for clarity.

However, it's desirable to approach the inverse problem, i.e to obtain the S-factors directly from the time domain data. To do so we assume that the peak shift signal is equal to the transition frequency correlation function. Any mode that enters the correlation function does so with relative amplitude that corresponds to its reorganisation energy. To obtain the reorganisation energy of the modes probed by the experiment we determine the total reorganisation energy by simulations of the peak shift in the impulsive limit. Since the finite bandwidth of the pulses discriminate against the high frequency vibrational modes we divide the amplitude of each mode by the amplitude of the power spectra of the pulses at the vibrational frequency. We thus arrive at the following equation:

$$S_j = \frac{A_j^{vib}}{\sum A_k} \lambda_{tot} \frac{1}{v_j E(\Omega - v_j)} \qquad (1)$$

where S is the Huang-Rhys factor, v_j is the mode frequency, A_j is the amplitude obtained from the fit of the peak shift,, λ_{tot} is the total reorganisation energy, E is the power spectra of the pulses and Ω is the position of the pure electronic transition.

The simulations allow for a determination of the reorganisation energy of the overdamped modes of the peak shift. When the reorganisation energy of these modes are known, we can directly calculate the relative value of the reorganisation energy of the vibrational modes as the ratio of the amplitudes obtained from the curve fitting of the 3PEPS trace. Dividing the reorganisation energy of each mode by the laser pulse envelope we obtain the S-factors of the modes impulsively excited by the laser pulses. Table 1 show the S-factors obtained, in an independent fashion, from the 3PEPS and the FLN measurements. If we use the method for obtaining the coupling strength without simulations proposed by Christensson [6], we can directly obtain the S-factors from curve fitting of the peak shift and from the knowledge of the duration of the pulses. This would amount to a considerable simplification compared to other methods like FLN or resonance Raman spectra where simulations of the experimental signals are needed to obtain the S-factors.

Table 1. Comparison of the Huang-Rhys factor obtained from FLN and Photon echo experiments for the modes that can be impulsively excited with the fs pulses

v /cm^{-1}	91	219	345	373	455
S(Echo)	0.155	0.120	0.093	0.051	0.104
S(FLN)	0.143	0.137	0.091	0.059	0.124
"diff" %	+8	-12	+2	-13	-16

Conclusions

We have shown by a direct comparison of Fluorescence Line Narrowing spectra and three pulse photon echo peak shift that the amplitude of the high frequency vibrational modes are scaled by the power spectra of the laser pulses. Having obtained the total reorganisation energy of the overdamped solvent mode via impulsive limit simulations we use a simple expression to obtain the S-factors of the impulsively excited vibrational modes.

1   Joo, T.H., Y.W. Jia, J.Y. Yu, M.J. Lang, and G.R. Fleming, "Third-order nonlinear time domain probes of solvation dynamics". J. Chem. Phys., 104(16): p. 6089 (1996).

2   deBoeij, W.P., M.S. Pshenichnikov, and D.A. Wiersma, "On the relation between the echo-peak shift and Brownian-oscillator correlation function". Chem. Phys. Lett., 253(1-2): p. 53 (1996).

3   Personov, R.I., ed. *Site selection spectroscopy of complex molecules in solutions and its applications*. Spectroscopy and excitation dynamics of condensed molecular systems, ed. V.M. Agranovich and R.M. Hochstrasser. (North-Holland: Amsterdam, 1983).

4   Dietzek, B., N. Christensson, P. Kjellberg, T. Pascher, T. Pullerits, and A. Yartsev, "Appearance of intramolecular high-frequency vibrations in two-dimensional, time-integrated three-pulse photon echo data". Phys. Chem. Chem. Phys., 9(6): p. 701 (2007).

5   Mukamel, S., *Principles of Non-linear Optical Spectroscopy* (New York: Oxford University Press, 1995).

6   Christensson N., Dietzek B., Yartsev A. and Pullerits T. submitted (2008)

7   May, V. and O.Kuhn, *Charge and Energy Transfer Dynamics in Molecular Systems* (Berlin: Wiley/VHC, 1999).

Observation of High-Frequency Coherent Vibrational Motion with Strongly Chirped Probe Pulses

D. Polli[1], D. Brida[1], G. Lanzani[1], and G. Cerullo[1]

[1] Dipartimento di Fisica, Politecnico di Milano, Piazza L. da Vinci 32, 20133 Milano, Italy
  E-mail: dario.polli@polimi.it

Abstract. We observe time-domain coherent vibrational wavepackets at 1585-cm^{-1} frequency (21-fs period) using broadband probe pulses strongly chirped up to 150-fs duration. The results are explained using the chronocyclic (Wigner) representation of the chirped pulse.

During the last decade, the availability of broadly tunable few-optical-cycle light pulses has dramatically improved the resolution of time-domain vibrational spectroscopy, allowing the direct detection of vibrational motions at frequencies as high as 2100 cm^{-1} (16 fs period) and providing new insights into chemical and structural rearrangements in all phases of matter [1]. The pump pulse excites a molecule on a timescale shorter than that of nuclear vibrational motion, generating coherent vibrational wavepackets in both the excited and the ground potential energy surfaces. The wavepacket motion is tracked by following the time-dependent transmission modulations of a delayed probe pulse. Traditionally, both pump and probe pulses with nearly transform-limited (TL) duration are used and the time resolution of the experiment is taken as the instrumental response function, which is the cross-correlation function between pump and probe pulses. While the role of pump pulse chirp in changing the relative weight of ground and excited state coherences has been studied [2], the effects of probe pulse chirp have not been investigated.

In this work, we combine sub-10-fs visible pulses with a high-sensitivity broadband detection setup using a fast spectrometer to study coherent vibrational dynamics in organic molecular systems. We demonstrate that it is possible to observe high-frequency coherent wavepacket dynamics employing strongly chirped broadband probe pulses with duration almost 10 times longer than the period of the detected vibrational mode. This surprising result is explained by analysing the chirped probe pulse in terms of its chronocyclic (Wigner) representation.

The experimental setup starts with a commercial regeneratively-amplified Ti:Sapphire laser delivering 150-fs pulses at 790 nm with 500-μJ energy and 1-kHz repetition rate. A home-built non-collinear optical parametric amplifier then generates ultrabroadband pulses with μJ-level energy and spectrum spanning the 500-700 nm wavelength range. The pulses are compressed down to nearly TL ≈6-fs duration, measured by Frequency Resolved Optical Gating (FROG), by chirped mirrors and sent to a degenerate pump-probe setup. After the sample the probe beam is sent to an optical multichannel analyser with fast electronics, allowing single-shot recording of the probe spectrum at the full 1-kHz repetition rate [3]. Differential transmission (ΔT/T) maps as function of probe wavelength and delay are acquired. The probe pulse is chirped by inserting different blocks of dispersive material along its path before the sample.

We studied thin films with blends of poly-phenylene vinylene (PPV) and C_{60}, with 1:3 molecular weight ratio. The 6-fs pump pulse resonantly excites PPV, creating a singlet exciton; rapid electron transfer (ET) to the fullerene then occurs with ≈50 fs time constant, leaving the polymer chain in a metastable polaron excited state [4]. The vibrational coherence created by the pump pulse in the excited state is rapidly quenched by the ET process, while the ground state coherence, induced via impulsive stimulated Raman scattering, is preserved. The periodic modulations in the ΔT/T signal observed at time delays longer than ≈200 fs are thus solely attributed to ground state vibrational coherence.

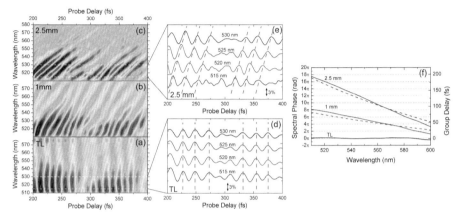

Fig. 1. (a, b, c) ΔT/T plots as a function of probe wavelength and delay in PPV-C_{60} for different values of probe pulse chirp; (d, e) pump-probe dynamics at selected probe wavelengths (as indicated) for transform-limited (d) and strongly-chirped (e) probe pulses; (f) solid lines: spectral phase of the Fourier transform at 1585 cm^{-1} frequency (C=C stretching) and corresponding group delay for the transform-limited probe pulse (TL) and the chirped ones; dashed lines: simulated group delays introduced by the sapphire plates.

Figure 1(a) shows the experimental results for the oscillatory component of the ΔT/T signal in the 510-580 nm wavelength region and in the 200-400 fs temporal window using TL probe pulses. We observe a complex vibrational pattern due to the beating of two modes at $v_1 \approx 1320$ cm^{-1} ($T_1 \approx 25$ fs) and $v_2 \approx 1585$ cm^{-1} ($T_2 \approx 21$ fs), corresponding to the single and double carbon bond stretching respectively. The phase of the oscillations is almost flat across the whole probe spectrum (see also Fig. 1(d)), in accordance with expectations for a ground-state mode to the red of the steady-state absorption peak.

Figures 1(b-c) show the results for strongly-chirped probe pulses obtained by dispersing them in sapphire plates with thickness 1 mm (Fig. 1(b)) and 2.5 mm (Fig. 1(c)), corresponding to group delay dispersions of 100 fs^2 and 250 fs^2, respectively; the corresponding probe pulsewidths, measured by FROG, are 55 fs and 150 fs. Astonishingly, even with probe pulse that are ≈7 times longer than the period of oscillation, the oscillatory pattern is still visible with high contrast, although it is now strongly bent (note the 2π phase drift occurring in the 15-nm spectral region plotted in Fig. 1(e)). This observation is possible only thanks to the combination of high temporal and spectral resolution of our instrument, because the fringe pattern would be cancelled in open-band or pass-band (≈10 nm bandwidth) detection schemes. Fig. 1(f) shows the probe wavelength dependence of the spectral phase φ for the 1585 cm^{-1} mode (solid line) and the corresponding group delay (GD) calculated as

GD$=\varphi^*T_2/2\pi$. The results are in very good agreement with the calculated GD introduced by the sapphire windows (dashed lines in Fig. 1(f)).

To understand our results, we analysed the broadband probe pulse using the chronocyclic representation [5], also known as Wigner Distribution (WD), which provides information on the temporal distribution of the different spectral components of the pulse. Fig. 2(b) shows the WDs for the TL pulse and the strongly chirped one. While for the TL pulse all the spectral components occur in a short time window (see dashed line in Fig. 1(a)), the WD for the chirped pulse is elongated but maintains a narrow shape. Therefore, while the wavelength integrated pulse temporal profile is long ($\approx$150 fs, see solid line in Fig. 2(a)), cuts at selected wavelengths are still very short, almost comparable to the TL pulse duration (see Fig. 2(c)). It is thus possible to preserve a very high temporal resolution in the pump-probe experiments even with a strongly chirped probe pulse.

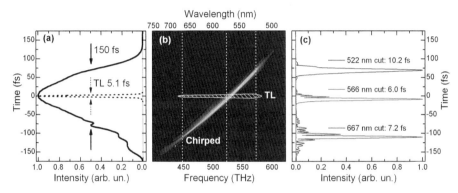

Fig. 2. Chronocyclic representation of the TL and chirped probe pulses (panel (b)) and corresponding wavelength integrated pulse profiles (panel (a)) and cuts at selected wavelengths of the chirped pulse (panel (c)).

In conclusion, we have shown that it is possible to measure very fast dynamics and/or high frequency coherent vibrational oscillations using strongly chirped broadband pulses, provided that a suitable detection setup with high spectral resolution is used. This result becomes important in those spectral ranges where it is easy to generate broadband spectra (e.g. by self-phase modulation) but temporal compression is not easy to implement.

References

1   S. De Silvestri, G. Cerullo, G. Lanzani, eds., Coherent Vibrational Dynamics, CRC Press, Boca Raton, 2008.
2   C. J. Bardeen, Q. Wang, and C. V. Shank, Phys. Rev. Lett. **75**, 3410 (1995).
3   D. Polli, L. Lüer and G. Cerullo, Rev. Sci. Instrum. **78**, 103108 (2007).
4   Ch. Brabec, G. Zerza, G. Cerullo, S. De Silvestri, S. Luzzati, J.C. Hummelen, S. Sariciftci, Chem. Phys. Lett. **340**, 232 (2001).
5   J. Paye, IEEE J. Quantum Electron. **28**, 2262 (1992).

Coherent Transfer of Molecular Vibrations in the Electronic Excited States

Chul Hoon Kim, Sohyun Park, Intae Eom, and Taiha Joo

Department of Chemistry, Pohang University of Science and Technology (POSTECH), Pohang 790-784, South Korea
E-mail: thjoo@postech.ac.kr

Abstract. Coherent wave packet motions in the electronic excited states prepared by impulsive nuclear rearrangements such as electronic transition, internal conversion, and chemical reaction are observed exclusively by ultrafast 35 fs time-resolved spontaneous fluorescence. Direct information on the excited state dynamics, reaction coordinates, and coupling between excited states can be obtained.

Introduction

When a molecule is excited impulsively by a short pulse of light, coherent vibrational wave packets are created, which are manifested in the oscillations of the time trace in various time-resolved spectroscopies, pump/probe transient absorption (TA) being a typical example. The nature of the vibrational modes excited and their decay provide a wealth of information on the dynamics and molecular structures of the states involved. In general, the coherent nuclear wave packet can be launched by any nuclear rearrangements such as chemical reaction and internal conversion that occur faster than roughly half of the vibrational period. Excited state intramolecular proton transfer (ESIPT) is one such example. Lochbrunner et al. have observed characteristic coherent oscillation components in the excited state absorption and stimulated emission from the ESIPT reaction product keto form of the 2-(2'-hydroxyphenyl)benzothiazole by TA experiments [1,2]. They showed that these oscillations are nearly the same as the low frequency skeletal motions which modulate the distance between the proton donor and acceptor groups.

In this work, we report the observation of the coherent wave packet motions of the molecules in the excited states created by various processes such as chemical reaction and internal conversion from S_n ($n \geq 2$) to S_1. These observations give direct information on the structures of the excited states and the reaction coordinates by examining the modes excited. We used femtosecond time-resolved spontaneous fluorescence (TRF), because interpretation of a TA signal is not always straightforward due to the fact that a TA signal consists of several contributions originating from the ground state, excited state, and all product states. On the other hand, TRF provides information on the dynamics of the excited state exclusively, although typical time resolution of the TRF measurement is limited to around 200 fs, which is not high enough to observe coherent vibrational motions.

Experimental Methods

A home-built Kerr lens mode-locked cavity-dumped Ti:sapphire laser and a home-built cavity-dumped optical parametric oscillator (OPO) employing a periodically-poled lithium niobate were used as light sources. Details of the OPO have been described elsewhere [3].

TRF was measured by fluorescence up-conversion technique. Detailed description of the up-conversion apparatus employing noncollinear sum frequency generation has been described elsewhere [4]. To achieve ultrafast time resolution of ~35 fs comparable to that of TA experiment, we carefully minimized the group velocity dispersion, group velocity mismatch, and phase front mismatch in a nonlinear crystal. Instrument response of the apparatus was estimated to be as short as 35 fs (full width at half maximum) from the cross-correlation of the scattered pump and gate pulses.

Results and Discussion

First, we prepared rhodamine B in S_1 state by the internal conversion from S_n ($n \geq 2$) state. It is well known as the Kasha's rule [5] that the internal conversion from the higher excited state S_n ($n \geq 2$) to S_1 is ultrafast to give fluorescence from the S_1 state exclusively. Figure 1. shows the TRF of rhodamine B in methanol detected at different wavelengths followed by the excitation to S_2 state. At early times, TRF signals show a detection wavelength independent slow rise of around 130 fs due to the vibronic relaxation in S_1. Stokes shift of rhodamine B is rather small to give minor detection wavelength dependence due to the intermolecular solvation process. At longer times, the TRF signals show the typical behavior in picosecond time scales due to the dielectric relaxation of the solvent; rise at longer detection wavelengths and a decay at shorter wavelengths. More interestingly, oscillations of the TRF signals at 155, 203, and 277 cm^{-1} are observed at all detection wavelengths, which are due to the coherent wave packet motions. Thus, the internal conversion from S_2 to S_1 is much faster than the 100 fs period of the highest frequency vibrational modes observed. Once the structure of the S_1 state is calculated quantum mechanically, assignments of the observed vibrational modes lead to the qualitative description of the structure of the S_2 state, since the vibrational excitation in the product state (S_1) is directly proportional to the projection of the displacement between the reactant (S_2) and the product onto the vibrational normal modes of the product state [6], in analogy to the Franck-Condon principle in the electronic transition of the molecules.

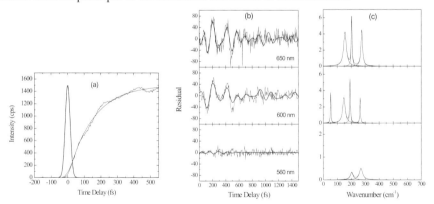

Fig. 1. (a) TRF of rhodamine B in methanol. The exponential fit and the instrument response are also indicated. (b) Residuals from the exponential fits of the TRF signals at different wavelengths. (c) Frequency spectra of the oscillation components obtained by the linear prediction singular value decomposition method.

We also prepared a molecule in the excited state impulsively by an ultrafast chemical reaction. Figure 2 shows the intramolecular charge transfer (ICT) of laurdan. The ICT of laurdan is essentially the same as the ICT of much studied dimethylaminobenzonitrile (DMABN). The ICT state of DMABN is usually called a twisted intramolecular charge transfer (TICT) state. The TICT hypothesis, however, is highly controversial, and recent work found strong evidence that the perpendicular twist of the amino group is not necessary to reach the ICT state [7].

To study the molecular dynamics of the ICT reaction, we have obtained the TRF spectra over the whole emission range of the reactant and the product directly at 50 fs resolution without the spectral reconstruction method, which are required to separate the ICT reaction and the solvation dynamics. The ICT reaction in methanol occurs by 50 fs, 300 fs, and 10 ps, where the 10 ps time constant correlates well with the dielectric relaxation time of the solvent molecules. Moreover, when the fluorescence is detected at the product ICT state with 35 fs resolution, we have observed an oscillation at 510 cm^{-1} due to the wave packet motion of the ICT state. *Ab initio* calculations identified that the vibration corresponds to the twisting motion of the amino group, which unambiguously identifies the twisting motion as the ICT reaction coordinate.

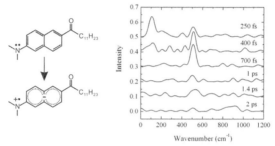

Fig. 2. Frequency spectra of the oscillation components in the TRF of laurdan detected at the emission of the intramolecular charge transfer state obtained by the sliding window (500 fs width) Fourier transformation of the residual from the exponential fits

Conclusions

Coherent nuclear motions are observed for molecules in the excited state in a variety of systems prepared by direct electronic excitation, non-adiabatic electronic transition, and chemical reactions. This was possible by the ultimate time-resolution in the TRF measurement. The wave packet dynamics provides a wealth of information on the dynamics of the system and the structures of the states involved.

1   S. Lochbrunner, A. J. Wurzer, and E. Riedle, J. Chem. Phys. **112**, 10699, 2000.
2   S. Lochbrunner, K. Stock, and E. Riedle, J. Mol. Struct. **700**, 13, 2004.
3   C. K. Min and T. Joo, Opt. Lett. **30**, 1855, 2005.
4   H. Rhee and T. Joo, Opt. Lett. **30**, 96, 2005.
5   M. Kasha, Discuss. Faraday. Soc. **9**, 14, 1950.
6   J. R. Reimers, J. Chem. Phys. **115**, 9103, 2001.
7   K. A. Zachariasse, S. I. Druzhinin, W. Bosch, and R. Machinek, J. Am. Chem. Soc. **126**, 1705, 2004.

342

Ultrafast Isomerization Dynamics of Biomimetic Photoswitches

J. Briand[1], D. Sharma[1], J. Léonard[1], J. Helbing[2], A. Cannizzo[3], M. Chergui[3], V. Zanirato[4], S. Haacke[1] and M. Olivucci[5]

[1] Institut de Physique et Chimie des Matériaux de Strasbourg, UMR 7504 ULP – CNRS, F-67034 Strasbourg, France
[2] Physikalisch-Chemisches Institut, Universität Zürich Witerthurerstr. 190, CH-8057 Zürich, Switzerland
[3] Laboratoire de Spectroscopie Ultrarapide, ISIC – EPFL, CH-1015 Lausanne, Switzerland
[4] Dipartimento di Scienze Farmaceutiche, Università di Ferrara, 44100 Ferrara, Italy
[5] Dipartimento di Chimica, Università degli Studi di Siena, 53100 Siena, Italy & Chemistry Dept., Bowling Green State University, Bowling Green, OH 43403, USA
Email: stefan.haacke@ipcms.u-strasbg.fr

Abstract. Femtosecond UV-VIS and mid-IR experiments show that a new class of biomimetic photoswitches photo-isomerizes in less than 300 fs. In close analogy to rhodopsin, the isomerization is driven by ultrafast motion along the stretch and the torsional coordinates.

Introduction

The ultrafast reaction dynamics of the methylated methoxy-IP Schiff base (MeO-IP) compound (Inset fig. 1) is presented. This molecule is part of a family of indanylidene-pyrroline (IP) compounds recently synthesized with the aim of mimicking the photochemistry of retinal Schiff bases in rhodopsins [1]. Combined *ab initio* quantum mechanics/molecular mechanics (CASPT2/CASSCF/AMBER) calculations in methanol indicate that the relaxation in the excited state is barrierless for torsion around the central double bond, leading to a conical intersection (CI) at almost 90° torsion. This motivated the expectation for the photo-initiated twist to occur on a sub-picosecond time scale. Absorption spectra of the Z and E forms are dominated by intense $\pi–\pi^*$ transitions at 392 and 385 nm, respectively. The isomerization quantum yield is 0.20-0.22 [1].

The present work shows that the Z→E photo-isomerization occurs indeed within less than 300 fs; similar to the record values held by retinal proteins [2]. As for the latter, the excited state dynamics leading to the CI seems to be well described by a two-mode scenario involving motion along stretch and torsional coordinates.

Results and discussion

We used polychromatic fluorescence up-conversion [3] to obtain the time-resolved spectra and kinetic traces of Z-MeO-IP after 400 nm excitation (fig. 1A). The fluorescence, centred at 530 nm, covers most of the visible spectral region, rises instantaneously and shows a bi-phasic decay. The decay constants, obtained from bi-exponential fits convoluted with the 120 fs instrument response function, are $\tau_1 < 40$ fs (limited by IRF) and $\tau_2 = 300 \pm 30$ fs. At wavelengths > 600 nm, a mono-exponential fit with the longer time suffices to reproduce the data.

We attribute the observed fluorescence to the directly excited singlet state S_1, as it lies to the red of the absorption band and appears promptly. The fast fluorescence decay at short wavelengths could reflect an initial ultrafast relaxation along the stretching coordinate, or may be due to molecules with high torsional kinetic energy affording fast isomerisation (see below). On the other hand, the slower component represents an excited state population decaying non-radiatively within 0.3 ps.

Transient absorption data show prompt excited state absorption (ESA, <440 nm) and stimulated emission (170 fs spectrum, fig. 1B). Fits to the kinetic traces yield sub-100 fs decay times for both ESA and SE, in agreement with τ_1 found in the fluorescence decay at short wavelengths. However, the slower 0.3 ps component is not observed since a broad spectrum of an induced absorption (IA) dominates the signal (400 fs spectrum). Fits yield a decay time of $\tau_{IA} = 160\pm40$ fs for IA, with IA's onset delayed by 180 ± 20 fs to that of SE (560 nm trace, fig. 1C). Note that fitting the IA without such a delay yields unsatisfactory results. As no ESA is found in the electronic structure calculations at these wavelengths and angles, we suggest that the IA is associated with a transient S_0 absorption in Z or E conformations. IA's wavelength-dependent onset time (dashed line in fig.1C) and the associated time scale are consistent with a wave-packet like evolution of molecules arriving in S_0, and evolving along the torsional coordinate. The earliest arrival at 180 ± 20 fs observed in the red part of the spectrum gives a value for the isomerisation time. This is consistent with the above sub-100 fs fluorescence component. The wavepacket then causes the abrupt bleach reduction at ~ 450 fs (389 nm trace). The IA merges into hot ground state absorption (GSA), which relaxes within ~5 ps. For delay times > 25 ps, we observe the stationary E-Z difference spectrum with positive and negative lobes at 418 and 370 nm, respectively.

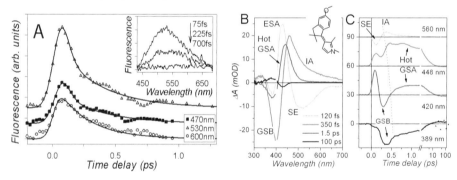

Fig. 1. A) Fluorescence kinetic traces of Z-MeO-IP at selected wavelengths (vertically shifted for sake of clarity). Solid lines are best fits to the data. Inset: Time-resolved fluorescence spectra. B&C: Magic angle transient absorption – spectra and kinetic traces. S_1-S_n ESA (425 nm), S_1-S_0 SE (460-600 nm), and IA (440-620 nm) evolving into hot GSA (420-450 nm). Inset: Structure of the Z-MeO-IP compound.

UV-pump-midIR-probe spectra were recorded in the 1450-1650 cm^{-1} spectral region, where the C=C stretch band of Z-MeO-IP at 1575 cm^{-1} is spectrally isolated and strongly red-shifted in E-MeO-IP (FTIR-spectra in Fig. 2). Immediately after excitation a bleach spectrum is observed which is a close replica of the Z-absorption spectrum (Fig. 2b), in line with the expectation that the C=C double bond character is lost in S_1. Then, a broad product absorption band grows at lower wavenumbers, which progressively narrows and blue-shifts (Fig. 2c). After 30 ps the pump-probe signal closely resembles the steady-state E-Z difference spectrum, (Fig. 2d). An analysis of

344

the photoproduct absorption band reveals that its maximum integrated intensity is reached on a timescale of a few hundred femtoseconds. The induced absorption signal must be due to molecules in the electronic ground state, because of the absence of a C=C double bond in the excited state, and the agreement between the timescales for its appearance with the decay of fluorescence (<0.3 ps). The steady state difference spectrum is only observed after dissipation of the excess energy to the solvent and the full solvation of the molecule. A moment analysis of the pump-probe data shows that the main shift of the photoproduct band takes place on a timescale of 6-9 ps, again in good agreement with the time scales of vibrational cooling and solvation inferred from the hot GSA band in the UV-VIS.

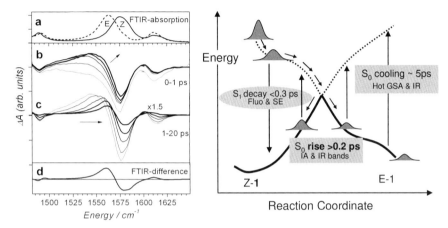

Fig.2. Left- UV-pump/mid-IR probe absorption changes monitor the bleach of the C=C stretch vibration (Z-form at 1575 cm^{-1}) and formation of the E form (b&c). The >20 ps spectrum closely resembles the FTIR difference spectrum (d). Right- Schematic view of the observed transitions and of the related reaction times. The vibrationally hot IR bands and GSA are represented on the product side only, but can arise from both Z and E conformers.

The different stages of the photoreaction are depicted in fig. 2-right. The sub-100 fs decay time is most probably due to molecules with a torsion angle <-30°, and reflects relaxation along the C-C stretching coordinate. Progression towards the CI gives rise to the red-shifted fluorescence decaying within 0.3 ps. At these times, the IA from molecules that have crossed the CI sets in at longer wavelengths. The hot GSA and hot vibrational bands correspond to molecules in the Z- and E-conformations, close to their equilibrium structure exhibiting 6-9 ps relaxation time.

Conclusions

A combination of UV-VIS-midIR femtosecond absorption and fluorescence studies reveal that methylated MeO-IP exhibits photo-isomerization in less than 0.3 ps. This is consistent with preliminary simulations of wave packet dynamics propagating on barrierless S_1 potential energy surface. Additional data shows that these photoswitches exhibit ultrafast E→Z photoconversion too. Future work will be directed towards the effect of chemical substitutions within the IP framework.

1   F. Lumento, V. Zanirato, et al. , Ang. Chemie Int. Ed. **46**, 414, 2007.
2   R. W. Schoenlein, L. A. Peteanu, et al., Science **254**, 412, 1991.
3   A. Cannizzo, O. Bräm, et al., Optics Lett. **32**, 3555, 2007.

Broadband femtosecond fluorescence up-conversion and photon echo experiments in the UV

O. Bräm, A. Cannizzo, A. Ajdarzadeh Oskouei, A. Tortschanoff, F. van Mourik, and M. Chergui

Ecole Polytechnique Fédérale de Lausanne (EPFL), Laboratoire de Spectroscopie Ultrarapide, ISIC, FSB, BSP; CH-1015 Lausanne, Switzerland
E-mail : olivier.braem@epfl.ch

Abstract. The fs-time-resolved study of a small UV dye in different solvents with fluorescence up-conversion and photon-echo techniques in the UV range provides new insight in cooling relaxation and solvation dynamics of non-polar molecules in polar solvents.

Introduction

The development of ultrafast spectroscopic techniques has provided a powerful tool for direct investigations of intra-molecular and solvation dynamics. In the Vis-IR range, one of the most established techniques is fluorescence up-conversion (FlUC) [1, 2], which allows time-resolving the fluorescence Stokes shift and band narrowing. Complementary information can be obtained through Photon Echo Peak Shift (PEPS), which measures the fluctuation correlations of the chromophore environment [3-5]. Only few experiments have been carried out in the UV range and, in the case of FlUC, nearly always with single wavelength detection. The UV range is interesting since UV dyes are typically small and simple molecules, and thus easier to characterize from a theoretical and experimental point of view. In addition, the extension towards the UV allows the investigation of systems previously inaccessible spectroscopically, and opens up the possibility to study proteins by probing aromatic amino acids.

We recently implemented a broadband femtosecond FlUC and a PE set-up in the UV-range [6, 7], both with sub-50 fs temporal resolution, to carry out the study of solvation and cooling dynamics of several UV dyes in different solvents.

Results and discussion

PE and FlUC experiments were performed on a photo-stable non-polar dye molecule, p-terphenyl (pTP), which absorbs around 280 nm and emits at 340 nm with a quantum yield of 0.93 (see figure 1A). It consists of a chain with three phenyl rings with a non planar, twisted, configuration in the ground state. After excitation to the first excited state it becomes planar [8].

Figure 1A presents a selection of FlUC emission spectra at different time delays after excitation at 290 nm, along with the steady state spectrum, for pTP in ethanol (EtOH). Figure 1A shows that almost the whole Stokes shift (5000 cm^{-1}) between excitation wavelength and steady state fluorescence maximum occurs "instantaneously", meaning that the excess of energy put into the system is mostly dissipated to different intramolecular modes within our temporal resolution. The spectral resolution of 6 nm of the detection system permits to observe the vibronic structure present in the steady

346

state spectrum, clearly showing up in less than 1 ps. Figure 1B shows the shift of the first spectral moment M_1 as a function of time (normalized according to the formula $S(t) = [M_1(t)-M_1(\infty)]/[M_1(0)-M_1(\infty)]$), which reflects the solvation correlation function. It reveals a red-shift of the center of mass of the spectra with a decay time of 500 fs, accompanied by a spectral profile evolution as pointed out by the change in relative peak height at the blue side of the spectra.

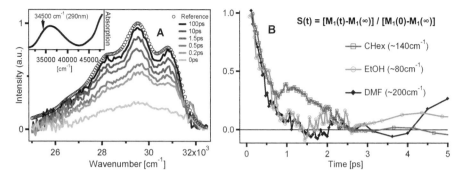

Fig. 1. A) transient fluorescence spectra of pTP in ethanol after 290 nm excitation, at different time-delays as indicated in the legend. The open circles indicate the steady state (c.w.) spectrum. The inset shows the absorption spectrum and the excitation wavelength. B) Solvation correlation function S(t) (see text for more details), as a function of delay time for pTP in cyclohexane, ethanol and dimethylformamide.

Time-resolved FlUC with broadband detection permits spectral evolution studies of ultrafast emission and our data demonstrates its capability to follow the dynamics of the excited state in the UV region. Similar measurements in cyclohexane (fig. 1B) reveal a slightly slower cooling in agreement with the non-polar nature of this solvent which is expected to interact very weakly with pTP.

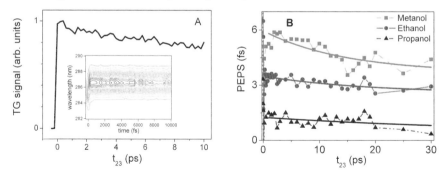

Fig. 2. A) Transient grating signal (TG) as a function of t_{23} for pTP in ethanol in the direction of $-k_1+k_2+k_3$. Inset shows the spectrally resolved TG signal (excitation pulse at 286.5 nm and 4 nm bandwidth). B) Photon-echo peak shift (PEPS) against t_{23} for pTP in three different solvents: methanol, ethanol and 2-propanol. Curves are vertically displaced for clarity.

In general, in PEPS experiments, signals both from excited- and ground-state frequency gratings are expected but, in the case of pTP, we know from the FlUC results, that only contributions from the ground state hole have to be taken into account, since in less than 100 fs the excited state population grating gets out of the spectral detection window, which is determined by the spectral pulsewidth. To verify this result we measured the spectrally resolved Transient Grating (TG) signals (i.e. with $t_{12}=0$) as a function of t_{23} on pTP in EtOH (figure 2A). Indeed no spectral evolution is observed after ~100 fs, apart from a picosecond intensity decay which we assign to a spectral diffusion of the ground state hole.

Figure 2B presents PEPS results obtained on pTP in different solvents. Discarding analysis of the component at very short times, which comprises also solvent contributions, we focus on the ps timescales, which are characteristic of solvent reorganization. A clear decay takes place on the timescale of 10-20 ps and qualitatively, it becomes faster with decreasing viscosity of the solvents (methanol < ethanol < 2-propanol), as would be expected. We attribute this decay to the dissipation of the ground state grating due to solvent induced spectral diffusion processes. In contrast to FlUC experiments, the PEPS succeeds to capture the characteristic solvent dynamics.

Conclusions

We reported a comparative study of a non-polar UV dye in different solvents with recently developed UV-broadband FlUC and UV-PEPS set-ups both with high temporal resolution (~50 fs). This approach allows us to describe the whole sequence of relaxation process: after excitation, the system undergoes an extremely fast departure of the excited state from the Franck Condon region, toward the new equilibrium configuration. The system is, however, vibrationally hot from the point of view of the lower frequency modes, which relax in ~500 fs. On the 10-20 ps time scale, spectral diffusion driven by solvation fluctuations occurs, bringing both the excited and the unexcited molecules to a fully equilibrated configuration. The present demonstration of these UV techniques is a nice example of non-polar solvation in polar solvents and opens up the possibility to investigate the dynamics of proteins using tryptophan as a probe.

1   J. Shah, in Ieee Journal of Quantum Electronics, Vol. 24, 276, 1988.

2   S. K. Pal, J. Peon, B. Bagchi, and A. H. Zewail, in Journal of Physical Chemistry B, Vol. 106, 12376, 2002.

3   W. P. de Boeij, M. S. Pshenichnikov, and D. A. Wiersma, in Chemical Physics Letters, Vol 253, 53, 1996.

4   M. H. Cho, J. Y. Yu, T. H. Joo, Y. Nagasawa, S. A. Passino, and G. R. Fleming, in Journal of Physical Chemistry, Vol. 100, 11944, 1996.

5   W. P. de Boeij, M. S. Pshenichnikov, and D. A. Wiersma, in Annual Review of Physical Chemistry, Vol. 49, 99, 1998.

6   A. Ajdarzadeh Oskouei, O. Bräm, A. Cannizzo, F. van Mourik, A. Tortschanoff and M. Chergui, in Chemical Physics, Vol. 350, 104, 2008

7   A. Cannizzo, O. Bräm, G. Zgrablic, A. Tortschanoff, A. Ajdarzadeh Oskouei, F. van Mourik, and M. Chergui, in Optics Letters, Vol. 32, 3555, 2007.

8   G. Heimel, M. Daghofer, J. Gierschner, E. J. W. List, A. C. Grimsdale, K. Mullen, D. Beljonne, J. L. Bredas, and E. Zojer, in Journal of Chemical Physics, Vol. 122, 54501, 2005.

Intramolecular Vibrational Energy Redistribution Measured by Femtosecond Pump-Probe Experiments in a Hollow Waveguide

Alexander Kushnarenko, Vitaly Krylov, Eduard Miloglyadov, Martin Quack and Georg Seyfang

Laboratory of Physical Chemistry, ETH-Zurich, Wolfgang-Pauli-Strasse 10, CH-8093 Zurich, Switzerland
E-mail: martin@quack.ch and seyfang@ir.phys.chem.ethz.ch

Abstract. We report femtosecond pump-probe experiments to investigate the intramolecular vibrational energy redistribution in the gas phase for CF_3CHFI, CHBrFI, CHBrClF and benzene (C_6H_6). To increase the measured probe signal the experiments have been performed in a hollow waveguide.

In reaction rate theory it was already realized early on that intramolecular vibrational energy redistribution (IVR) is essential for the understanding of unimolecular reactions by statistical and dynamical theories [1-3]. Mode selective chemistry may be expected if the time scale for IVR is comparable to or slower than the time scale for the unimolecular reaction. IVR after near-infrared (IR) overtone excitation has been investigated successfully by two significantly different methods: I. If a molecular Hamiltonian H_{mol} for the intramolecular couplings is derived from high resolution IR-spectra the time dependent dynamics of the initially excited vibrational levels can be obtained [3-5]. II. Intramolecular relaxation times can also be determined in femtosecond pump-probe experiments, where the population of the initial level, excited by the strong near-IR pump pulse, is followed by a time delayed weak probe pulse [6-11].

We have applied sensitive UV-absorption to investigate IVR after excitation of the first overtone of the CH-stretching vibration in femtosecond pump-probe experiments. Because of Franck-Condon factors with the electronically excited S_1-state, the relaxation of the vibrational energy out of the initially excited CH-stretching state to the low lying background states with excitation of CI-stretching mode in iodides or related modes in other molecules is accompanied by a change of the UV-absorption spectrum, usually a spectral broadening and a shift of the absorption to longer wavelengths. In the experiments, presented here, we have investigated the collisionless IVR process for CF_3CHFI, CHBrFI, CHBrClF, C_6H_6 and some of its deuterated isotopomers.

Apart from applying the UV-absorption technique, the IVR process can also be followed by time resolved IR-absorption [11]. In general the absorption cross section of vibrational transitions is reduced as compared to the one of electronic transitions. But a higher selectivity can be obtained for the IR-detection if the spectrally broad probe pulse is dispersed behind the probe cell by a monochromator. Choosing a detection band width of approximately 10 cm^{-1} the time dependent population of individual vibrational levels can be measured.

In pump-probe experiments the measured signal is determined by the density of the probe molecules, the absorption cross section at the pump wavelength σ_{pump} and

349

the probe wavelength σ_{probe} and the spatial overlap integral for the fluence distribution of the pump and the probe beam. Therefore, the experiments are usually performed with strongly focused laser beams in a confocal arrangement. But in the gas phase the molecular density is very small as compared to the liquid phase and only a small probe signal can be expected. To increase the effective probe volume we have coupled the two laser beams to a hollow waveguide of inner radius a = 125 μm and a length L_{wg} = 500 mm. For well chosen focusing conditions 98% of the intensity of a Gaussian laser beam can be coupled to the zero order Bessel mode J_0 inside the waveguide and an optimum overlap between pump- and probe beam is maintained for the whole length of the waveguide. For the waveguide diameter used in our experiment an enhancement factor $\gamma_{enhance}$ = 10 for the measured probe signal as compared to the confocal arrangement is calculated. In Figure 1a different decay signal functions measured by time resolved UV-absorption for C_6H_6 are compared. One function is obtained with conventional confocal geometry and the two other decay functions are measured in a hollow waveguide with L_{wg} = 500 mm and two different inner radii. An enhancement factor of $\gamma_{enhance}$ = 10.5 is obtained for a = 125 μm and of $\gamma_{enhance}$ = 5.5 for a = 160 μm, which is in good agreement with our theoretical calculations. For the confocal geometry a relaxation time is barely determined whereas the experiments with the two different waveguides give an identical single exponential relaxation time of approximately 26 ps for the slower part of the relaxation kinetics. Clearly visible in the measured decay functions is the strong signal in the temporal overlap region and two signals arising from rotational coherences around 44 and 88 ps.

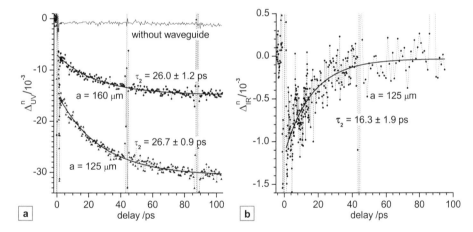

Fig. 1. Measured relaxation in C_6H_6 (p = 10kPa) after excitation of the first overtone of the CH-stretching vibrations at 6006cm^{-1}. a) UV-detection (37000cm^{-1}) with confocal geometry (upper trace) and with a hollow waveguide of L_{wg} = 500 mm and an inner radius of a = 160 μm (middle trace) or a = 125 μm (lower trace). Clearly visible is the strong signal in the temporal overlap region and the two rotational coherences around 44 and 88 ps. b) IR-detection (2800cm^{-1}, $v_{CH}=2 \rightarrow v_{CH}=3$) with a waveguide of L_{wg} = 500 mm and a = 125 μm. The full line represents a fit of an exponential decay function to the experimental data resulting in a relaxation time of τ_2 = 16±2 ps

Figure 1b shows the measured decay function for the initially excited first overtone of the CH-stretching vibration in C_6H_6. Due to the lower absorption cross section the signal-to-noise ratio is strongly reduced. The measured decay and the fitted decay time $\tau_2 = 16.3$ ps are different from those obtained with the UV-detection for the same pump conditions. This different behaviour is not unexpected as in general the dynamics of different vibrational levels is probed by the IR- or UV-absorption technique. Of course, processes faster than our time resolution, defined here by the overlap of pump and probe pulses, about 150 fs, cannot be observed.

Table 1. Measured relaxation times (n/o: not observable), preliminary results

Molecule	Pump $\tilde{v}$ /cm^{-1}	Probe $\tilde{v}$ /cm^{-1}	Relaxation time		
			τ_1 / ps	τ_2 / ps	τ_3 / ps
CHBrClF	5930	38 500	n/o	8.5 ± 0.5	n/o
CHBrFI	5930	28 570	n/o	31.2 ± 0.9	n/o
CF$_3$CHFI	5886	32 300	n/o	4.5 ± 0.5	n/o
C$_6$H$_6$	5900	37 000	< 1	17 ± 2	n/o
	6006	2 800	-*	16.3 ± 1.9	n/o
		37 000	< 0.5	9 ± 1	91 ± 9
C$_6$H$_5$D	5990	37 000	< 0.3	4.7 ± 0.3	n/o
C$_6$HD$_5$	5790	37 000	2.1 ± 0.8	n/o	n/o
	5980	37 000	1.2 ± 0.3	8.8 ± 1.2	n/o
	6100	37 000	< 3	21 ± 10	n/o

*can not be obtained in the experiment with IR-probing due to time resolution and the presence of a strong coherent signal

The relaxation times measured for the different molecules are summarized together with the pump and probe conditions in Table 1. For the halogenated methanes and for CF$_3$CHFI a simple decay process is measured which can be represented by a single decay time τ_2. For the benzene molecules a more complicated relaxation kinetics is obtained which has to be presented by up to three different relaxation times and the IVR processes are strongly dependent on the wavelength of the pump laser, indicating a dependence of the measured relaxation kinetics upon the initially excited vibrational level.

Acknowledgements. Our work is supported financially by the ETH Zurich and the Schweizerischer Nationalfonds.

1   M. Quack and J.Troe, in Theoret. Chem.: Advances and Perspectives, **6B**, 199 (1981).

2   M.Quack, Il Nuovo Cimento, **63B**, 358, (1981).

3   M.Quack, "Molecular Femtosecond Quantum Dynamics between Less than Yoctoseconds and More than Days: Experiment and Theory" Chapter 27 in Femtosecond Chemistry, J.Manz, and L. Woeste, editors, Verlag Chemie (Weinheim), 781 (1995).

4   M.Quack, Annu. Rev. Phys. Chem., **41**, 839 (1990).

5   J.Pochert, M.Quack, J.Stohner, and M.Willeke, J. Chem. Phys., **113**, 2719 (2000).

6   D.Bingemann, M.P.Gorman, A.M.King, and F.F.Crim, J. Chem. Phys., **107**, 661 (1997).

7   R. von Benten, A.Charvat, O. Link, B.Abel and D.Schwarzer, Chem.Phys.Lett., **386**, 325 (2004).

8   T.Ebata, M.Kayano, S.Sato and N.Mikami, J. Phys. Chem **A 105**, 8623 (2001).

9   V.Krylov, M.Nikitchenko, M.Quack, and G.Seyfang, Proc. SPIE **5337**, 178 (2004).

10 V.Krylov, A.Kushnarenko, E.Miloglyadov, M.Quack, and G.Seyfang, Proc. SPIE **6460**, 64601D (2007).

11 H.S.Yoo, M.J.De Witt, and B.H.Pate, J. Phys. Chem. **A 108**, 1348 (2004).

Femtosecond Fluorescence Spectroscopy of N^6,N^6-Dimethyladenine: New Explanation of the "Dual Fluorescence" Dynamics from Decay and Rise Time Measurements at Threshold

Nina K. Schwalb[1] and Friedrich Temps[1]

[1] Institut für Physikalische Chemie, Christian-Albrechts-Universität zu Kiel, Olshausenstr. 40, D-24098 Kiel, Germany
E-mail: schwalb@phc.uni-kiel.de

Abstract. Femtosecond time-resolved fluorescence measurements on N^6,N^6-dimethyladenine in a wide wavelength range following excitation at threshold and much higher show identical dynamics, requiring a new explanation for the so-called "dual fluorescence" of the molecule.

Introduction

N^6,N^6-dimethyladenine (DMA) differs from its parent molecule adenine (A) only by the double methylation of the exocyclic amino group, yet this minor modification alters its spectroscopic properties truly dramatically. In particular, DMA has been found to show so-called "dual fluorescence" [1-3]. While its UV absorption is quite similar to that of A, its fluorescence spectrum exhibits – apart from the expected A-like emission band peaking at $\approx$330 nm – a strongly red-shifted second band with maximum between 450 and 500 nm (depending on the solvent) and a Stokes shift of $\approx$15 000 cm^{-1} (Fig. 1). The "normal" fluorescence was attributed to the

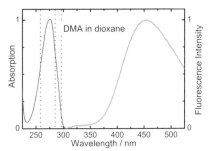

Fig. 1. Normalized absorption (left) and static fluorescence spectrum (right) of DMA in dioxane. Dotted lines indicate the experimental excitation wavelengths.

local excited (LE) state, whereas the red-shifted band was proposed to arise from an intramolecular charge transfer (ICT) state [1-3] populated from the LE state through an excited state reaction. Previous work assumed this to be accompanied by a 90° out-of-plane twist of the exocyclic dimethylamino group.

We recently reported first femtosecond fluorescence measurements on DMA in H_2O [4], which showed that the emission in the UV is very short-lived, with a lifetime ≤ 1 ps and only minor contributions from two other decay components with time constants up to 60 ps. In contrast, the time profiles of the red-shifted emission could basically be described solely with the 60 ps decay component. The results could be interpreted with a modified kinetic scheme built on the literature proposal. However, as the DMA had been excited at a wavelength of $\lambda_p = 258$ nm, well above the electronic origin, it was somewhat difficult to draw firm conclusions on the mechanism.

Here, we report on new excited state lifetime measurements on DMA at several wavelengths in the range 294 nm $\geq \lambda_p \geq$ 258 nm. At 294 nm, the molecules are excited just barely above the electronic origin of the first excited state, allowing for a critical test of the ensuing dynamics.

Experimental Method

Measurements were carried out by femtosecond UV/VIS fluorescence up-conversion spectroscopy using a Ti:Sa laser pumped frequency-doubled non-collinear optical parametric amplifier for the excitation and the Ti:Sa laser for the gate. Solutions of DMA in dry dioxane as an aprotic nonpolar solvent or water (bidest.) at 10^{-3} M concentration were continuously pumped through the 1 mm long sample cell. The excitation power was kept ≤ 0.2 mW. Fluorescence-time profiles were monitored for each λ_p and solvent at five to seventeen emission wavelengths in the range 290 nm $\leq \lambda_f \leq$ 600 nm. The time resolution after deconvolution was $\approx$150 fs. The results reported in the following are for dioxane. Similar data have also been obtained with H_2O.

Results

Figure 2 shows the measured fluorescence decays of DMA at six resp. five selected fluorescence wavelengths, $\lambda_f = 290$ nm (258 nm pump only), 350 nm (close to peak of the UV fluorescence), $\approx$410 nm (intermediate region), $\approx$450 nm (close to the red-shifted emission peak), and 510 nm (long wavelength wing of red emission band) for high and low excitation energy ($\lambda_p = 258$ nm, Fig. 2a, and $\lambda_p = 294$ nm, Fig. 2b). The differences between the decay curves at the selected emission wavelengths are indeed striking: The shorter the fluorescence wavelength, the faster becomes the decay. The UV emission decays to virtually zero in ≤ 5 ps. The red-shifted emission reaches a plateau on the time scale of Fig. 2, followed by a very slow decay ($\tau \approx 1.3$ ns). The amplitude of the long-lived component rises with the emission wavelength.

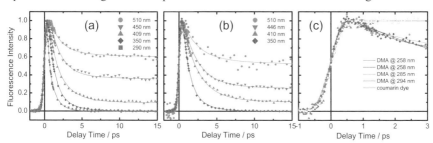

Fig. 2. Measured fluorescence decays of DMA in dioxane with $\lambda_p = 258$ nm (a) and $\lambda_p = 294$ nm (b) at six resp. five selected fluorescence wavelengths λ_f (colored lines and symbols). (c): Fluorescence rise profiles of DMA ($\lambda_f = 510$ nm) in dioxane using $\lambda_p = 258$ nm (1 and 10 mM solution), 285 nm, and 294 nm (1 mM solutions) compared with the fluorescence rise profile of a coumarin dye added after the DMA measurements.

At $\lambda_p = 258$ nm, the DMA molecules are prepared with almost 6000 cm^{-1} of excess energy in the excited state, whereas $\lambda_p = 294$ nm reaches the excited state just above its origin (see Fig. 1). Nevertheless, as can be seen, the fluorescence decay curves at the two pump wavelengths are virtually identical. Global fits to the decay curves at three different excitation wavelengths show that distinctive dynamical processes take place in the molecules on five time scales. The obtained lifetimes are $\tau_1 \approx 0.19$ ps ($\lambda_p = 258$ nm only), $\tau_2 = 0.63$ ps ($\lambda_p = 285$ and 258 nm only), $\tau_3 = 1.3 - 2.1$ ps, $\tau_4 = 7 - 10$ ps, and $\tau_5 \approx 1300$ ps. The amplitudes of the respective exponential decay components vary strongly with the fluorescence wavelength, but only slightly with the excitation wavelength.

In addition, we examined the rise times of the red fluorescence at the different pump wavelengths by comparison with the rise time from a strongly fluorescent coumarin dye as time standard. Results at $\lambda_f = 510$ nm are displayed in Fig. 3. All DMA profiles appear identical, and show no time lag compared to the coumarin dye. This demonstrates that the red fluorescence of DMA appears promptly, without sizable time delay, and indicates the absence of a significant energy barrier, even when the molecules are excited just above their electronic origin.

Discussion

The present results close to the electronic origin of the optically bright state suggest that the so-called "dual fluorescence" of DMA needs a new explanation. Towards these ends, we propose a model built on the recent findings for A [5]. We assume that the optically bright state in the case of DMA is identical to that of A, where the excitation at energies well above the origin leads to the $^1\pi\pi^*$ (L_a) state. It is reasonable to assume that the $^1\pi\pi^*$ (L_a) state of DMA exhibits a similar conical intersection (CI) with the S_0 ground state as in A, where out-of-plane motion of the H atom at C^2 is involved. This gives rise to the sub-picosecond ($\tau_2 \approx 0.63$ ps) deactivation of the $^1\pi\pi^*$ (L_a) state and corresponding decay of the UV fluorescence.

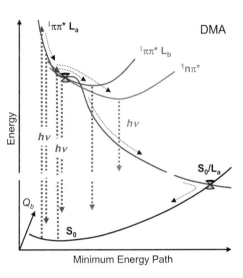

Fig. 3. Proposed scheme for the electronic deactivation and "dual fluorescence" of DMA.

The ultrafast τ_1 component (≤ 0.2 ps) observed only at high excitation energy is attributed to the rapid departure of the initial wavepacket from the Franck-Condon region. En route to the CI, the wavepacket crosses the $^1\pi\pi^*$ (L_b) and the $^1n\pi^*$ state. We assume therefore that a fraction of the excited wavepacket crosses from the $^1\pi\pi^*$ (L_a) to the $^1\pi\pi^*$ (L_b) state and to the $^1n\pi^*$ state. The crossing is facilitated by a local minimum of the $^1\pi\pi^*$ (L_a) state in the vicinity of the CI with the $^1\pi\pi^*$ (L_b) and $^1n\pi^*$ states, similar to what has been found for the case of 7H-adenine (7A) [5]. The lifetime of $\tau_3 = 1.3 - 2.1$ ps is tentatively attributed to that local minimum, which is close to the electronic origin of DMA and is the optically excited state at $\lambda_p = 294$ nm. The time constant $\tau_4 = 7 - 10$ ps likely belongs to the minimum of the $^1\pi\pi^*$ (L_b) state. The strongly red-shifted long-lived fluorescence has to be assigned to the $^1n\pi^*$ state. We note that the $^1n\pi^*$ state resembles the previously proposed ICT state. In the case of 7A, the $^1n\pi^*$ state has a minimum with the exocyclic NH_2 group perpendicular to the molecular plane.

Acknowledgements. We thank the Deutsche Forschungsgemeinschaft for support.

1   B. Albinsson, J. Am. Chem. Soc. **119**, 6369, 1997.

2   J. Andréasson, A. Holmén, and B. Albinsson, J. Phys. Chem. B **103**, 9782, 1999.

3   A. B. J. Parusel, W. Rettig, and K. Rotkiewicz, J. Phys. Chem. A **106**, 2293, 2002.

4   N. K. Schwalb and F. Temps, Phys. Chem. Chem. Phys. **8**, 5229, 2006.

5   L. Serrano-Andrés, M. Merchán, and A. C. Borin, Chem. Eur. J. **12**, 6559 2006.

Assignment of the Excited-State Infrared-Spectra in the Course of the Ring Opening Reaction of a Photochromic Dihydroazulene

Tobias E. Schrader[1], Uli Schmidhammer[2], Wolfgang J. Schreier[1], Florian O. Koller[1], Igor Pugliesi[1]

1 LS für BioMolekulare Optik, LMU München, Oettingenstr. 67, D-80538 Munich, Germany
2 Laboratoire de Chimie Physique, UMR8000 CNRS-Université Paris Sud, Bât 349, F-91405 Orsay, France
E-mail: Igor.pugliesi@physik.uni-muenchen.de

Abstract: With femtosecond infrared spectroscopy and ab initio calculations we could assign the transient spectrum at 1 ps to the ring opened product of the dihydroazulene photo induced reaction. Thus, ring-opening proceeds within 1 ps.

Introduction

Photochromic compounds have attracted much attention in the past due to their possible application in data storage [1]. Among them the photochromic reaction of the spiropyran-merocyanine chemical ring opening has been studied by time resolved techniques in the mid-infrared spectral range [2]. The ring opening reaction of fulgimides provides another intensely investigated example [3]. For the ring-opening reaction of 1,1-dicyano-2-(4-cyanophenyl)-1,8a-dihydroazulene (CN-DHA) to its vinylheptafulvene conformer (CN-VHF) a detailed mechanistic picture based on a structure sensitive technique is still missing. We report here on transient infrared (IR) spectra after photo-excitation of CN-DHA on a time scale up to 2.7 ns after initiation of the ring opening. Since the calculated spectrum of the excited *s-cis*-CN-VHF agrees well with the transient spectrum recorded after 1 ps, we conclude that the ring opening is mostly completed at that time after photo-excitation of CN-DHA.

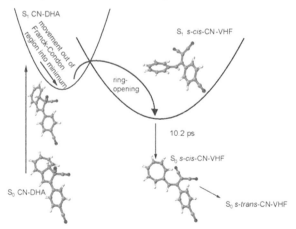

Fig. 1. A schematic overview over the ring opening reaction of CN-DHA. The time constant of 10.2 ps measured in the solvent acetonitrile is inferred from ref. [6]. The structures shown are the calculated minimum energy structures of the TD-DFT calculations.

355

Materials and Methods

We performed pump probe experiments with a time resolution of 200 fs. For details see [4]. The pump pulses were centred at 398 nm and the mid-IR probe pulses in the fingerprint region were centred at 1355 cm^{-1}, 1485 cm^{-1} and 1600 cm^{-1}, respectively. The solvent in this study was perdeuterated acetonitrile-d3. Additionally, ab initio calculations for the spectra in the excited state have been carried out using the time dependent density functional theory (TD-DFT) routines implemented in Turbomole [5]. The Karlsruhe SVP-basis set was employed in all calculations.

Results and Discussion

In Fig. 1 a reaction model of the photo-conversion from CN-DHA to CN-VHF is shown as inferred from pump probe spectroscopy in the UV-visible spectral range [6,7,8]. The structures shown are the equilibrium geometries as now calculated by DFT/ TD-DFT. It was up to now not clear whether these structures agree with the ones found in the experiment. Recent advances in computational chemistry have made it possible to calculate excited state normal modes and associated IR intensities in the DFT framework. This offers a means for direct comparison of the excited state transient IR spectra with theory.

Fig. 2 provides this comparison between experimental data and theoretically calculated spectra. After the decay of the cross phase modulation effects of the solvent reliable transient spectra could be recorded for delay times ≥ 1 ps. A spectrum recorded at 1 ps is shown in Fig. 2B. Comparing it with the calculated S$_1$-IR spectra of the closed CN-DHA and the opened compound s-cis-CN-VHF clearly shows that after 1 ps the ring opening reaction has already taken place. This is in agreement with the visible pump probe data in acetonitrile [6].

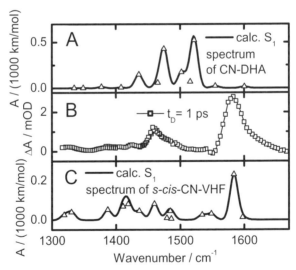

Fig. 2.A) Calculated excited state IR-spectrum of CN-DHA. The open triangles indicate the spectral positions and intensities of each calculated band. The calculated frequencies were scaled with a factor of 0.965. The black line represents a gaussian convolution of the calculated spectrum with a line width of 25 cm^{-1} (FWHM). B) Transient spectrum recorded at 1 ps delay time after photo-excitation. C) Calculated excited state IR-spectrum of the s-cis-isomer of CN-VHF.

We can furthermore assign some of the vibrational absorption bands in the excited state of cis CN-VHF. For example the band with the strongest intensity seen in the experiment around 1583 cm^{-1} can be assigned to the theoretical band of strongest oscillator strength in the fingerprint region with an unscaled frequency of 1641 cm^{-1}. The corresponding mode is mainly a C-C stretch and C-H bending vibration located on the cyano-benzene ring.

The theoretical calculations furthermore offer an insight into the possible reaction mechanism. Excitation of the CN-DHA leads to a flattening of the seven membered ring and the sp^3-hybridized carbon centre becomes nearly sp^2-hybridized. The ring opening reaction happens by charge transfer of an electron from the seven membered ring onto the C-(CN)$_2$-group. The excess charge is stabilized by the adjacent cyano-phenyl group that has rotated into the C-(CN)$_2$ plane. The internal conversion to the ground state of the opened CN-VHF form is accompanied by the rotation of the cyano-phenyl group out of the C-(CN)$_2$ plane. These findings agree with the interpretation of the UV-Vis time resolved data that provides a dissociative electron transfer in the excited state supported by out of plane vibrations and solvation [6,7].

Conclusions

In summary, by comparing calculated S$_1$ state spectra with transient infrared spectra we were able to conclude that the ring-opening of CN-DHA occurs within only 1 ps. Furthermore, we are able to shed light onto the structural changes accompanied with the ring-opening in the excited state. With the presented minimum energy structures shown in Fig. 1 the paving stone is laid for the theoretical investigation of the reaction path. This knowledge could be important for the application of these systems in data storage as it could help to improve the design of these molecules in terms of read-out speed and reliability.

Acknowledgements. We would like to thank Prof. J. Daub for providing the sample.

1   J. Ern, M. Petermann, T. Mrozek, J. Daub, K. Kuldova, and C. Kryschi, in Chem. Phys. **259**, 331, 2000.

2   M. Rini, A. K. Holm, E. T. J. Nibbering, and H. Fidder, in J. Am. Chem. Soc. **125**, 3028, 2003.

3   F. O. Koller, W. J. Schreier, T. E. Schrader, A. Sieg, S. Malkmus, C. Schulz, S. Dietrich, K. Ruck-Braun, W. Zinth, and M. Braun, in J. Phys. Chem. A **110**, 12769, 2006.

4   T. Schrader, A. Sieg, F. Koller, W. Schreier, Q. An, W. Zinth, and P. Gilch, in Chem. Phys. Lett. **392**, 358, 2004.

5   Turbomole 5.9.1, R. Ahlrichs, M.Bär, M. Häser, H. Horn, and C. Kölmel, in Chem. Phys. Lett. **162**, 165, 1989.

6   V. De Waele, M. Beutter, U. Schmidhammer, E. Riedle, and J. Daub, in Chem. Phys. Lett. **390**, 328, 2004.

7   U. Schmidhammer, V. de Waele, O. Poizat, G. Buntinx, J. Daub, E. Riedle, in J. Phys. Chem. A, to be submitted

8   V. de Waele, U. Schmidhammer, T. Mrozek, J. Daub, E. Riedle, in J. Am. Chem. Soc. **124**, 2438, 2002.

Time-resolved coincidence imaging of ultrafast molecular dynamics

Arno Vredenborg, Wim G. Roeterdink and Maurice H.M. Janssen

Laser Centre and Department of Chemistry, Vrije Universiteit, De Boelelaan 1083, 1081 HV
  Amsterdam, The Netherlands
  E-mail: mhmj@chem.vu.nl

Abstract. We report on femtosecond molecular dynamics experiments in NO_2 using a novel photoelectron-photoion coincidence imaging apparatus.

Introduction

Ultrafast photon induced molecular dynamics can be probed in great detail with time-resolved coincidence imaging of ions and electrons [1-3]. The three-dimensional energy- and angle-resolved detection of ions and the correlated electrons gives full information about the energy and angular distribution of the photofragments over the various accessible reaction channels. By varying the pump-probe delay time, the transition state properties of the reaction can be elucidated. Moreover, the coincident angular detection of electrons and ionic fragments enables the determination of the molecular frame angular distribution of the electron. These distributions directly reflect the contributions of the molecular orbitals involved in the photoionization process. Although the time resolved coincidence imaging technique is experimentally very involved, the results can be very intuitive.

In this contribution we present the dissociation pathways in NO_2 observed after excitation with 400 nm femtosecond laser pulses.

Experimental Methods

In our laboratory in Amsterdam a new time-resolved photoelectron-photoion coincidence imaging machine has become operational for the study of femtosecond molecular photodynamics. The novel apparatus has been recently described in full detail [2]. In the detection chamber, two position- and time-sensitive particle detectors are mounted perpendicular to the molecular beam. The electrons and ions are detected in a velocity map imaging configuration with additional lenses to provide low extraction fields. The electron detector uses small pore micro-channel-plates (5 μm) to obtain high resolution electron images. The new apparatus has an electron time-of-flight (TOF) resolution of 18 ps on a typical TOF of 15 ns. The electron energy resolution is nearly laser bandwidth limited with $\Delta E/E \approx 3.5$ % for electron energies near 2 eV.

The commercial laser system (Spectra Physics) consists of a Titanium-Sapphire oscillator (Mai-Tai) which seeds the chirped regenerative amplifier (Spitfire Pro) running at 5 kHz. The pulse duration at 800 nm of the amplified regen pulses is approximately 130 fs with a pulse energy of 500 μJ. In the experiment reported here we used the second harmonic at 399.8 nm with a typical pulse energy of 15 μJ. Furthermore, two multi-stage noncollinear opto-parametric amplifiers (NOPA) are available to generate tunable short (30 fs) laser pulses in the visible wavelength region.

Results

In this contribution we report on the single color multiphoton dynamics in NO_2. A full paper reporting two color pump-probe time-resolved dynamics has been reported recently [3]. The dominant product in the single color 400 nm multiphoton excitation of NO_2 is NO^+, which accounts for 95% of the events originating from NO_2. The coincident electron and ion image of NO are shown in Fig. 1 together with the corresponding kinetic energy distributions. The minimum photon energy to create the NO^+ fragment with a zero kinetic energy electron is 12.38 eV [3]. A four photon excitation excites the NO_2 molecule with 12.4 eV in total, which is sufficient to give rise to slow photoelectrons and ions. The excess energy for this photon process is only 20 meV and a minor contribution of this photon process is observed.

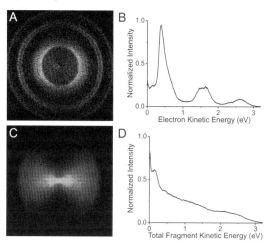

Fig. 1. The photoelectron slice image of electrons in coincidence with NO^+ is given in panel A with the kinetic energy distribution shown in panel B. The ion image of NO^+ and the corresponding total fragment kinetic energy distribution is shown in panels C and D, respectively. The images result from single color 400 nm excitation of NO_2.

The images in Fig. 1 clearly show high kinetic energy electrons and ions and therefore the dominant process is a five photon excitation at 15.5 eV. The excess energy is equal to the energy of a photon or 3.1 eV. The electron image shows three distinct kinetic energy bands, while the ion distribution is very broad.

The energy correlation plot for the NO^+ fragment is given in Fig. 2 A. The vertical axis of the correlation plot shows the kinetic energy of the electron and the horizontal axis the total kinetic energy of $NO^+ + O$. The diagonal line indicates the maximum excess energy of a five photon process. The correlation plot shows that the electrons obtain well defined kinetic energies according to the three photoelectron bands shown in Fig. 1B. On the other hand, the kinetic energy distribution of the fragments ranges from zero to the maximum available energy. In a single color multiphoton experiment the electron is ejected instantaneously, i.e. within the temporal width of the laser pulse. Subsequently, the NO_2^+ is produce in three (pre)dissociative states giving rise to (fast) $NO^+ + O$ fragments. The Recoil Frame Photoelectron Angular Distribution (RF-PAD) of the 0.3-0.5 eV photoelectron band is given in Fig. 2 B. The RFPAD does not change with the kinetic energy of the fragments and is approximately the same for all three photoelectron bands. In the case of fast dissociation, the direction of the ion is approximately equal to the molecular frame orientation. Therefore, the angular distribution of the electron in the molecular frame is only determined by the molecular orbital from which the electron originates.

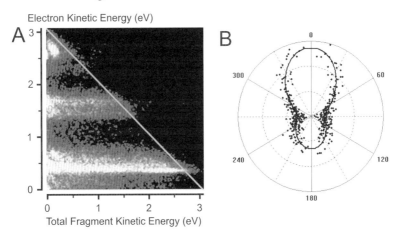

Fig. 2. A) The energy correlation plot of (e,NO$^+$) events. B) The RF-PAD of the events with 0.3-0.5 eV electron kinetic energy.

Discussion

The origin of the photoelectron band at 1.3-1.8 eV corresponds to a state located at 14.2-13.7 eV, which is in agreement with previous experiments for the photoionization of NO$_2^+$ into the b^{3}A$_2$ excited state [4]. This b^{3}A$_2$ state was found to have a very short lifetime of about 100 fs and lead to fast NO$^+$ + O fragments. The photoelectron band at 0.3-0.75 eV corresponds to a state located at 15.2-14.75 eV, which correlates with the fourth exited state, the B^{1}B$_2$ state. Furthermore, the photoelectron band at 2.3-2.8 eV corresponds to a state located at 13.2-12.7 eV, which correlates with the first excited state of NO$_2^+$, the a^{3}B$_2$, state. The coincidence correlation plot in Fig. 2A shows that the photoelectron bands appear to be split which is not observable in the total photoelectron spectrum shown in Fig. 1B. A more detailed analysis of the energy correlation will be discussed elsewhere.

The photoelectron-photoion coincidence imaging technique provides a very powerful and detailed insight in the complex femtosecond dynamics of multichannel multiphoton processes in molecular systems.

1 J.A. Davies, J.E. LeClaire, R.E. Continetti and C.C. Hayden in Journal of Chemical Physics, **111**, 1, 1999
2 A. Vredenborg, W.G. Roeterdink and M.H.M. Janssen in Review of Scientific Instruments, **79**, 063108, 2008.
3 A. Vredenborg, W.G. Roeterdink and M.H.M. Janssen in Journal of Chemical Physics, **128**, 204311, 2008.
4 J. H. D. Eland and L. Karlsson in Chemical Physics, **237**, 139, 1998.

Ultrafast time and frequency domain vibrational dynamics of the CaF$_2$/H$_2$O interface

Ali Eftekhari-Bafrooei, Satoshi Nihonyanagi and Eric Borguet

Department of Chemistry, Temple University, 1901 N 13th street, Philadelphia, PA 19122, USA
Email: eborguet@temple.edu

Abstract: The structure of water at the CaF$_2$/KOH interface was studied by vibrational sum-frequency-generation (SFG) spectroscopy and ultrafast SFG-Free Induction Decay, suggesting the presence of weakly hydrogen bonded OH at high pH.

1. Introduction:

The interaction of water with solid surfaces is of great importance to areas ranging from environmental chemistry to electrochemistry [1,2]. However, it is challenging to investigate the first layer of interfacial water molecules. Vibrational Sum-Frequency Generation (VSFG) is one the few techniques that addresses this challenge.

VSFG, whereby an IR pulse (ω_{IR}) excites a vibrational coherence which is upconverted by a visible pulse (ω_{VIS}) to the sum frequency ($\omega_{SFG} = \omega_{IR} + \omega_{VIS}$), is a versatile technique for probing interfacial phenomena. VSFG has high surface specificity because, as a second order nonlinear optical phenomenon, it is forbidden in centrosymmetric media, e.g. bulk water. VSFG also provides chemical information with submonolayer sensitivity [3]. A number of frequency domain VSFG measurements, i.e., VSFG spectroscopy, have been reported for various interfaces since the first application of vibrational SFG spectroscopy was described [2-4]. VSFG spectroscopy with ultrashort laser pulses can also be used for probing the ultrafast surface vibrational dynamics. For example, the SFG-Free Induction Decay (FID) of a vibrational coherence can be measured by introducing a delay between the IR pulse that creates the decay and the visible pulse that probes it. In the absence of rephasing, the macroscopic vibrational polarization decays with time due to dephasing. While time and frequency domain measurements contain identical information, if they probe the entire time and spectral dependence of the system, we have recently found that time domain measurements can offer information that can not be easily extracted from frequency domain measurements.

Interestingly, the mineral fluorite (CaF$_2$) is observed to be hydrophobic at both the macroscopic and microscopic level. As such it offers an interesting system to investigate the structure and dynamics of water at a hydrophobic interface. Becraft et. al. observed a narrow peak at ~3657 cm^{-1} which dominates the SFG spectrum of CaF$_2$/H$_2$O at high pH (>13)[5]. They assigned it to the Ca-OH species resulting from ion exchange of surface fluorite with hydroxide ions[5]. However, the assignment of the peak is not straightforward. It is not clear, for example, why a surface hydroxyl group would not be solvated, i.e. remain free of hydrogen bonds, in the presence of an aqueous environment.

In this study, by tuning the IR excitation to the region of free (dangling) OH we measured the SFG spectrum and the SFG-Free Induction Decay (FID) of CaF$_2$/KOH at high pH. Our SFG and macroscopic contact angle results suggest that the peak at around 3680 cm^{-1} observed in this study arises from weakly hydrogen bonded OH

361

oscillators similar to free (dangling) OH observed at air/water and water/OTS interfaces[3.]

2. Experimental:

The CaF_2 prism was cleaned in concentrated sulfuric acid for 30 min and rinsed with de-ionized (DI) water. The prism was then soaked in KOH (pH=13) solution for 10 min and rinsed with DI water again before taking the contact angle and SFG measurements.

A femtosecond regenerative amplifier (Quantronix, Integra-E) seeded by a Ti:S oscillator (Coherent, Miraseed) produces 3.5 mJ at 815 nm with a duration of 100 fs at a repition rate of 1 khz. 90 % of the output of a 1 kHz femtosecond regenerative amplifier energy is used to pump an optical parametric amplifier (Light Conversion, TOPAS) which generates tunable IR radiation (1180 – 2640 nm). Mid-infrared pulses, ~200 cm^{-1} FWHM, were obtained by difference frequency generation in an $AgGaS_2$ crystal with a pulse energy of 10 μJ at 2.7 μm. The remaining 10 % of the regenerative amplifier output is used as the visible light for SFG measurements. For frequency domain SFG measurement, the visible pulse was spectrally narrowed to a FWMH of ~1.7 nm (25 cm^{-1}) by a pair of band-pass filters (CVI). For time domain measurements (SFG-FID), the ~100 femtosecond regenerative amplifier pulses were used. The visible and infrared beams, with energies of ~2 μJ/pulse and ~10 μJ/pulse, respectively, were overlapped in time and space at the sample surface. In the experiments reported here, the SFG, visible and IR beams were polarized *s, s, p* respectively.

3. Results and Discussion

After treating the CaF_2 in KOH (pH=13), the surface of CaF_2 became hydrophobic as was evidenced from the high contact angle (~60-70°) of a water droplet spread on the prism surface. The SFG spectrum and SFG-FID from the CaF_2/KOH interface at different pH were measured with the IR pulse centered at 3700 cm^{-1}. A typical SFG spectrum and SFG-FID at pH=13 are shown in Figure 1.

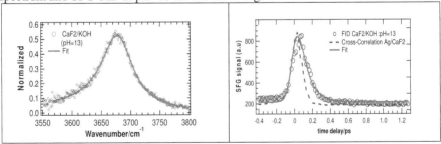

Figure 1. SFG spectrum (left) and SFG-FID (right) of the CaF_2/KOH interface at pH=13

The SFG spectrum shows a relatively narrow (~60 cm^{-1}) peak at 3679 cm^{-1}. The peak position and the linewidth are very close to that of free OH feature observed at air/water and water/OTS interfaces[3]. The peak position is ~20 cm^{-1} blueshifted with respect to the peak observed peak by Becraft et. al[5]. Our observation of hydrophobic CaF_2 at high pH, together with peak position suggest that the peak at 3679 cm^{-1} is due to free or weakly hydrogen bonded OH of water, rather than surface bound hydroxyls. The decrease in the intensity of the peak, the increase in the linewidth and the redshift in the peak position as the pH is decreased from 13 to 11 (data not shown) suggest that only at high enough pH are a fraction of the interfacial water molecules weakly hydrogen bonded (i.e., almost free OH).

To complement the SFG spectral results, and to investigate the dynamics of the system, the SFG-FID of the same system was also measured. The total dephasing time (T_2) extracted from the decay of SFG-FID was ~190 fs for pH=13. The reported dephasing of the free OH at water/OTS interface is ~300 fs.[6]. A shorter dephasing in the case of CaF_2/KOH compared to water/OTS suggests that the observed peak at CaF_2/KOH is not characteristic of a totally free OH but has some degree of interaction with crystalline CaF_2 and/or hydrogen bonding with neighboring water molecules. The dephasing time was observed to decreasing as the pH was reduced (data are not shown). These observations, accompanied by redshifts and broadening of the SFG spectra, support the hypothesis that the observed peak at CaF_2/KOH arises from weakly hydrogen bonded species and suggests that the degree of hydrogen bonding at this interface decreases as pH increases.

4. Conclusion:

We have observed that CaF_2 surfaces become hydrophobic upon treating with basic solutions. The existence of the hydrophobic surface suggests that the hydrogen bonding network of interfacial water molecules is disrupted, resulting in some water molecules being incompletely hydrogen bonded. Our macroscopic observation is supported by SFG measurements. SFG spectra of the CaF_2/KOH interface at high pH showed a narrow peak which we assigned to weakly hydrogen bond OH oscillators. The dephasing of the mode measured by SFG-FID further supports our hypothesis for the assignment. The degree of hydrogen bonding increases upon decreasing the solution pH, as evidenced by both an increase in linewidth of the SFG spectra and a shortening of the dephasing time observed by SFG-FID.

5. References:

1. Wu, L. M.; Forsling, W., *Surface Complexation of Calcium Minerals In Aqueous-Solution.3. Ion-Exchange And Acid-Base Properties of Hydrous Fluorite Surfaces.* Journal of Colloid And Interface Science **1995,** 174, (1), 178-184.

2. Bain, C. D., *Sum-Frequency Vibrational Spectroscopy of The Solid-Liquid Interface.* Journal of The Chemical Society-Faraday Transactions **1995,** 91, (9), 1281-1296.

3. Shen, Y. R.; Ostroverkhov, V., *Sum-frequency vibrational spectroscopy on water interfaces: Polar orientation of water molecules at interfaces.* Chemical Reviews **2006,** 106, (4), 1140-1154.

4. Gragson, D. E.; McCarty, B. M.; Richmond, G. L., *Ordering of interfacial water molecules at the charged air/water interface observed by vibrational sum frequency generation.* Journal of The American Chemical Society **1997,** 119, (26), 6144-6152.

5. Becraft, K. A.; Richmond, G. L., *In situ vibrational spectroscopic studies of the CaF_2/H_2O interface.* Langmuir **2001,** 17, (25), 7721-7724.

6. McGuire, J. A.; Shen, Y. R., *Ultrafast vibrational dynamics at water interfaces.* Science **2006,** 313, (5795), 1945-1948.

Non-Condon vibronic coupling of coherent molecular vibration in MEH-PPV induced by a visible few-cycle pulse laser

Takayoshi Kobayashi [1,2], Jun Zhang [1], and Zhuan Wang [1,2]

[1] Department of Applied Physics and Chemistry and Institute for Laser Science,The University of Electro-Communications, 1-5-1 Chofugaoka, Chofu, Tokyo, 182-8585, Japan
[2] JST，ICORP，Ultrashort Pulse Laser Project，4-1-8 Honcho，Kawaguchi，Saitama，Japan
E-mail: kobayashi@ils.uec.ac.jp

Abstract. The dependence of coherent vibrational amplitudes at 128 wavelengths in MEH-PPV (EL polymer) with 1-fs resolution gave the evidence of non-Condon effect due to ^1B_u-exciton strongly coupled with m^1A_g state essential in the third-order nonlinearity.

Introduction

One of the most important derivatives of electroluminescent polymer is poly-[2-methoxy, (5-2' -ethyl-hexyloxy)-p-phenylenevinylene] (MEH-PPV) because of its high solubility in common solvents and high electroluminescence efficiency. The mechanism of the efficient luminescence in MEH-PPV is described in terms of excitons [1-3]. Several research works on MEH-PPV have been made until now using ultrashort pulse lasers with several tens to hundred fs pulse duration [4,5]. It is of vital importance to study the mechanism of vibronic coupling, which determines the rate of radiationless processes, to obtain the strategy of obtaining even higher efficiency. Wer studied the vibronic mechanism using a 5-fs pulse laser with multi-wavelength information to clarify the mechanism.

Experimental Methods

A 5.7 fs noncollinear optical parametric amplifier (NOPA) was used as a light source of the pump-probe experiment as described in our previous papers [7-9]. The pulse energies of the pump and probe were about 35 and 5 nJ, respectively. A 128-channel lock-in amplifier was used as a phase-sensitive broad band detector. Chloroform solutions of MEH-PPV were spin-coated onto quartz plates to form 0.5-1.0 μm thick films. All the experiments were performed at room temperature (293 ± 1 K).

Results and Discussion

Figures 1(a) and 1(b) show the real-time traces of the pump-probe experiment and their Fourier power spectra, respectively, at 128 wavelengths which were discussed using a model with two harmonic potential curves belonging to the ground and the excited states. Since in the two-potential curve case the derivative of the susceptibility $\chi^{(1)}$ with respect to the normal coordinate is proportional to the derivative of $\chi^{(1)}$ with respect to optical frequency, hence $\partial \chi^{(1)}/\partial Q \propto \partial \chi^{(1)}/\partial \omega_v$ is satisfied. Therefore the probe photon energy dependence of the vibrational amplitude, which is proportional to the Raman susceptibility, can be discussed in terms of the derivative of the susceptibility $\chi^{(1)}$. Then spectral change by modulation of Q (vibration) by the amount of ΔQ is given by $(\partial \chi / \partial Q)\Delta Q$ which is proportional to $\partial \chi / \partial \omega$ and to the frequency derivative of absorption spectrum. In case (3), where $\partial \chi / \partial Q$ is zero, then the dominant contribution of spectral change associated with the modulation of Q by ΔQ is given by a term proportional to $(1/2)(\partial^2 Q / \partial Q^2)(\Delta Q)^2$, which is then proportional to $\partial^2 \chi / \partial \omega^2$.

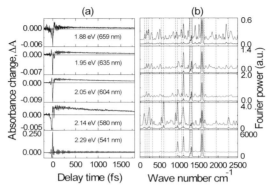

Fig. 1. (a) The change in the absorbance as a function of pump-probe delay time at five different probe wavelengths, and (b) Fast Fourier transform (FFT) power spectra (solid thick lines) of traces as shown in (a). The intensity is magnified by 10 to show the details (solid thin lines). The dashed lines mark the reproducible peaks.

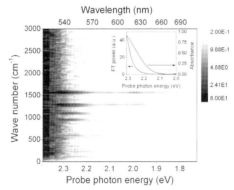

Fig. 2 Contour map of the Fourier amplitude of vibration in the absorbance change. The Fourier power is indicated on the logarithmic scale with a step of 1.44. The inset shows the probe-photon energy dependence of ground-state absorption intensity (line 1) and amplitude of 1587 cm^{-1} (line 2).

In the above discussion for all of the cases it was assumed that the Condon approximation is satisfied because the integrated intensity of the absorbance change covering the relevant electronic state does not change during the wave-packet motion. The deviation from the Condon approximation can then be introduced by taking another electronic state (a third state), which is radiatively coupled, to the two states between which transition intensity is being monitored. In the present case the two corresponding states are the ground state and the lowest excited state. Even though the model is simple but essential point of the vibronic coupling can be studied using the harmonic oscillator model, because the system studied has relatively low Huang-

Rhys factors of all the modes studied and excitations only taking place at the lowest few vibrational levels determined by the laser spectrum and the fundamental spectrum of the polymer.

The Fourier transform (FT) was performed in the probe delay time range 50~1800 fs, results being shown in Fig. 2. The probe photon energy dependences of all the modes studied (961, 1278, 1315, and 1587 cm^{-1}) are much steeper than the absorption spectrum with respect to the increasing photon energy as shown in the inset of Fig. 2 for 1587 cm^{-1}. All of the probe photon energy dependence of the four most intense modes have similar features, and can be well reproduced by the sum of the zero-th, first, and second derivatives of the absorption spectrum, one example of which is shown in Fig. 4(a) for 1587 cm^{-1}. The contributions of zeroth, first, and the second derivatives are separated as a function of wavelength. As can be seen in the inset of the amplitude of the vibrational mode with the frequency of 1587 cm^{-1} has a very large positive value in the spectral range. It indicates that the transition probability is increased by the deformation induced by the molecular vibration. Therefore, this intense signal observed in the ground-state absorption spectral region can be explained by the non-Condon mechanism with both potential minimum displacement and curvature change in the potential curve upon photo-excitation as discussed earlier. The non-Condon effect is due to the strongly allowed electronic state(s) existing in the higher energy. Several theoretical and experimental papers have verified that such Ag symmetry states are located above ^{1}Bu state [1,6]. They are called m^1A_g and n^1A_g states, which are contributing with essential importance to the third-order optical nonlinearity [6].

Conclusions

In conclusion, coherent molecular vibration in MEH-PPV was studied by using the extremely short pulse laser and the 128-wavelength simultaneous detection system. From the spectral dependence of the molecular vibrational amplitudes of these modes the contributions of the time-dependent Franck-Condon and non-Condon (Hertzberg-Teller) mechanisms were obtained. From these results it was concluded that so-called essential state in the third-order non-linearity is important even in the linear spectroscopic vibronic coupling.

Acknowledgements. This work was partly supported by the 21st Century COE program on "Coherent Optical Science" and partly supported by the grant from the Ministry of Education (MOE) in Taiwan under the ATU Program at National Chiao Tung University. A part of this work was performed under the joint research project of the Laser Engineering, Osaka University under contract subject B1-27.

1   M. Liess, et al., Phys. Rev. B **56**, 15712 (1997).
2   S. Abe, J. Phys. Soc. Jpn. **58**, 62 (1989).
3   T. Ogawa and T. Takagahara, Phys. Rev. B **44**, 8138 (1991).
4   X. Yang, T. E. Dykstra, and G. D. Scholes, Phys. Rev. B **71**, 045203 (2005).
5   A. Ruseckas, et al., Phys. Rev. B **72**, 115214 (2005).
6   S. Mazumdar and F. Guo, J. Chem.Phys. **100**, 1665 (1994)

Specific Channel of Energy Dissipation in Carotenoids: Coherent Spectroscopic Study

Masazumi Fujiwara[1], Kensei Yamauchi[1], Mitsuru Sugisaki[1], Andrew Gall[2], Bruno Robert[2], Richard J. Cogdell[3], and Hideki Hashimoto[1,*]

[1] CREST-JST and Department of Physics, Osaka City University, Sumiyoshi, Osaka 558-8585, Japan E-mail: *hassy@sci.osaka-cu.ac.jp
[2] CEA, Institut de Biologie et Technologies de Saclay and CNRS, Gif-sur-Yvette, F-91191, France
[3] IBLS, Glasgow Biomedical Research Centre, University of Glasgow, Glasgow G12 8QQ, Scotland, UK

Abstract. We investigate transient grating signals in β-carotene homologues by using sub-20-fs optical pulses. The results clearly show that shorter-chain carotenoids have a specific major channel of energy dissipation to the environment (the central C=C stretching mode) in the ground-state vibrational manifold, whereas the longer-chain carotenoids do not.

Introduction

Carotenoids are important pigments involved in the primary reactions of photosynthesis. Very efficient and ultrafast energy transfer between carotenoid's excited state to bacteriochlorophyll's excited state is realized in light-harvesting pigment-protein complexes [1]. It has been proposed that a specific vibrational mode plays a central role in this energy transfer and also in the energy relaxation simultaneously occurring with this energy transfer. An understanding how the energy given by the excitation light is transfered or dissipated requires not only information on the population dynamics but also a detailed description of their coherence behavior.

We have recently developed an ultrashort-pulse-generation system that can produce sub-20 fs pulses across the whole visible wavelength region [2]. Using this system coherent nuclear motions of β-carotene of up to 1500 cm^{-1} and that are coupled to its optical transitions have been observed. We have extended this study using a series of β-carotene homologues (Fig. 1) in order to investigate how the coherence dynamics depend on the extent of the π-conjugation length.

Experiments

β-carotene homologues were synthesized and purified by the protocol reported in ref. [3]. TG signals of the homologues were measured using the experimental setup described in ref. [2]. The sample was dissolved in THF and flowed through an optical flow cell. The sample concentrations were adjusted so that the optical densities were 1.0–1.4 at their absorption maxima. The time durations of the pulses were determined by an SHG-intensity autocorrelation technique. The excitation and the detection wavelength were degenerated and set to 491 nm for C36, 520 nm for C40, 551 nm for C44, and 564 nm for C50. All the time durations of the pulses used in the measurements were under 18 fs.

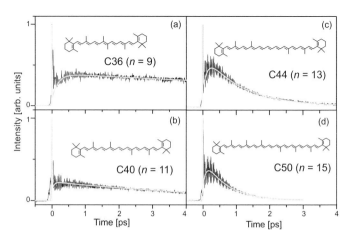

Fig. 1. The TG signals (black line) observed with (a) C36, (b) C40, (c) C44, and (d) C50. Chemical structures of the homologues with the numbers of their conjugation lengths (n) are shown as the insets.

Results and Discussion

Figure 1 shows the experimental TG traces. With all the homologues, a sharp peak was observed at $t = 0$ (coherent spike), and the subsequent coherent oscillations were superimposed on a slowly varying component. This slowly varying component (gray line) originates from the population relaxation dynamics of the excited electrons. The decay time of the slowly varying component decreases with increasing the conjugation length. This decay time of each homologue observed here is consistent with the S_1–S_0 lifetime reported previously [1].

FFT power spectra of the coherent oscillations basically corresponded to the resonance Raman spectra (data not shown). In the coherent oscillations four main vibrational modes are included, and they are, respectively, solvent THF mode (ν_{THF} at ~ 912 cm^{-1}), the in-plane methyl-rocking (ν_3 at ~ 1000 cm^{-1}), central C-C stretching (ν_2 at ~ 1150 cm^{-1}), and central C=C stretching (ν_1 at ~ 1500 cm^{-1}) modes. This coincidence means that the coherent oscillations come from the ground-state vibrational modes.

The coherent oscillations contain information on how vibrational modes behave in the time domain. These temporal behaviors can be determined using wavelet transformation (WT) [4]. A single exponential decay function has been fitted to the temporal profile of each vibrational mode in the WT spectrograms. The dephasing times of the four main vibrational modes of each homologue are shown in Fig. 2. One clearly sees that the dephasing times of the vibrational modes of the homologues decrease as the conjugation length increases.

The dephasing time determined here is a total dephasing time (T_2) which consists of a population-induced dephasing time (T_1) and a pure dephasing time (T_2^*), where it is given by $T_2^{-1} = (2T_1)^{-1} + T_2^{*-1}$ [5]. The population lifetime of the vibrationally hot ground state has been reported to be ~ 10 ps ($T_2^* \ll 2T_1$). Therefore the decoherence of the ground-state vibrational modes are mainly caused by system-bath interactions because the pure dephasing directly reflects the environmental contribution.

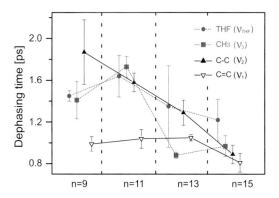

Fig. 2. The total dephasing times of the four main vibrational modes of the homologues as a function of the conjugation length.

The conjugation-length dependence of the dephasing times of the two polyene-backbone central stretching modes (the v_1 and v_2 modes) is of our particular interest because they deeply relate to the relaxation dynamics in carotenoids [1]. Figure 2 show remarkable difference between these two modes. The dephasing time of the v_2 mode strongly depends on the conjugation length whereas that of the v_1 mode does not. Assuming that system-bath interaction increases dephasing rate, we can conclude that the energy dissipation occurs mainly through v_1 mode in the shorter-chain carotenoids. In contrast the longer-chain carotenoids do not appear to have such a specific dissipation channel.

Conclusions

Transient grating signals in β-carotene homologues have been measured by using sub-20-fs excitation pulses. The total dephasing times of the homologues have been determined by applying wavelet transformation analyses to the coherent oscillations, which were observed in the experimental TG traces. The dephasing time of the v_2 mode depends strongly on the conjugation length whereas the v_1 mode does not. This trend clearly shows that shorter-chain carotenoids have a specific major channel (the v_1 mode) to the environment, whereas the longer-chain carotenoids do not.

Acknowledgements. Financial support is gratefully acknowledged below. HH and MS thank the JSPS and MEXT (Grants No. 17204026, 17654083, 18340091, 18654074). HH, MS, AG, RJC thank a Strategic International Cooperative Research grant from JST. RJC thanks the BBSRC. AG and BR thank the ANR (Caroprotect project).

1   T. Polívka and V. Sundström, in Chemical Reviews Vol. 104, 2021, 2004.
2   M. Sugisaki *et al.* in Physical Review B Vol. 75, 155110, 2007.
3   K. Yanagi *et al.* in Physical Review B Vol. 71, 195118, 2005.
4   P. S. Addison, The Illustrated Wavelet Transform Handbook, Taylor & Francis, New York, 2002.
5   S. Mukamel, Principles of Nonlinear Optical Spectroscopy, Oxford University Press, New York, 1995.

Coherent phonons in cyanine dye monomers and J-aggregates

T. Virgili[1], S. Ceccarelli[2], L. Lüer[1], G. Lanzani[1], G. Cerullo[1], D.G. Lidzey[2]

[1] IFN, INFM, CNR Dipartimento di Fisica, Politecnico di Milano, P.zza Leonardo Da Vinci
32, 20132 Milano, Italy
Email: tvirgili@polimi.it
[2] Department of Physics and Astronomy, University of Sheffield, Hicks Building, Hounsfield
Road, Sheffield S37RH United Kingdom

Abstract. Using pump-probe spectroscopy, we investigate coherent oscillations in cyanine dye, in monomeric form and in J-aggregate. We identify a low energetic intramolecular mode amplified in the J-aggregate film producing a modulation of the excitonic coupling.

Introduction

J-aggregates of cyanine dyes are 1-D supra-molecular structures formed by self-organization of the dye molecules dissolved in polar solvents. They are preferentially one-dimensional aggregates where the transition dipoles of the lowest energetic singlet transition are arranged in a head-to tail fashion. They have been extensively studied both experimentally and theoretically as a model system for one-dimensional Frenkel excitons . In the aggregate, new collective excited states are generated, due to the coherent delocalization of the electronic wavefunction, which leads to linear and nonlinear optical properties much different from the original monomer dye.

Experimental Methods

The dye in the form of monomer is studied in methanol solution at low concentration to avoid any aggregation. J-aggregates are obtained by mixing in polar solvent the dye dispersed in gelatine matrix. The thin films containing J-aggregates are achieved by spin coating this solution onto glass substrates.

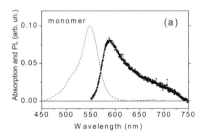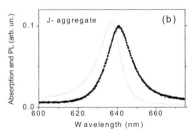

Fig.1: (a) Absorption spectrum (dash line) and photoluminescence spectrum (solid line) are shown for the monomer. (b) Absorption spectrum (dash line) and photoluminescence spectrum (solid line), are shown for the J-aggregate.

Fig 1 shows the absorption spectrum and the emission spectrum for the monomer (Fig1a) and for the J-aggregate (Fig 1b). The linear absorption of the J-aggregate

370

presents a peak at 636 nm (1.95 eV), with a linewidth of 12 nm (40 meV), much narrower than the molecular absorption linewidth (193 meV). The red-shift (100 nm) with respect to the absorption of the monomer and the narrowing of the band is expected from J-aggregation. The emission spectrum consists of one narrow band at 640 nm (linewidth of 40 meV), this band is assigned to one-exciton emission.

Results and Discussion

Using pump-probe spectroscopy with sub 10 fs time resolution, we have studied coherent oscillations in this cyanine dye, in monomeric form as well as in J-aggregates. Fig 2(a) shows the transient transmission spectrum for the monomer (dashed line) and for the J-aggregate (solid line). The spectrum for the monomer is the superposition of the bleaching signal due to the depopulation of the ground state after the pulse excitation, and stimulated emission.

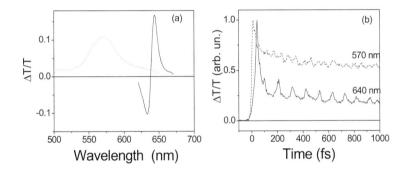

Fig.2: (a) Transient transmission spectrum at 100 fs probe delay respect to the pump for the monomer (dash line) and for the J-aggregate (solid line). (b) Transient transmission decay at 570 nm for the monomer (dash line) and at 640nm for J-aggregate (solid line).

For the J-aggregate the spectrum consists of a photoinduced absorption band and the bleaching signal. The origin of the induced absorption is the optical transition from the lowest state of the one-exciton band to the two excitons band. From the energetic difference between those two bands [1], we can calculate the exciton size of the aggregate. We estimate that about 10 molecules are coherently coupled in the exciton state.

After subtracting the temporal decay, Fourier transform is performed on the pump-probe data (Fig 2b). Fig. 3 shows the results. For the monomer the plot presents different features, which corresponds to vibrational intramolecular modes of the molecule [2]. For the J-aggregate there is instead just one dominant peak, at 320cm^{-1} and a weaker feature at 260 cm^{-1}. Note that our temporal resolution is the same as that used for the monomer. The result suggests that a substantially different electron-phonon coupling regime is taking place in the exciton.

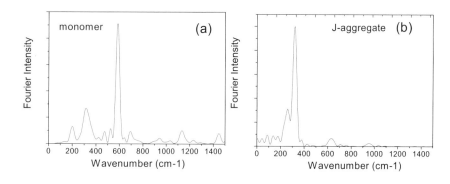

Fig.3: (a) Fourier spectrum for the monomer. (b) Fourier spectrum for the J-aggregate.

We conjecture that in the aggregate only those intra-molecular phonons that modulates the excitonic coupling do appear in the spectrum. Their action happens through the modulation of the $\pi-\pi$stacking, because of the change in the molecular configuration,
but they cannot strictly be classified as "lattice" (inter-molecular) phonons, which would modulate the inter-molecular separation (lattice constant). Their effect is to introduce a periodic modulation of the excitonic coupling $V=V_0 cos(\omega t)$, and thus of the electronic transition energy. We thus observe an "*electronic*" phenomenon, which is fundamentally different from the *vibronic* mechanism observed in single molecules, where a modulation of the Franck-Condon envelope is usually observed.

Conclusions

In conclusion, we have used pump-probe spectroscopy with sub 10 fs time resolution to study the photophysics of a cyanine dye, both in monomeric form and in the form of a J-aggregate. By comparing their time-domain vibrational dynamics, we find that the aggregate is dominated by strong *intra-molecular* modes. Their effect is to introduce a periodic modulation of the excitonic coupling potential that in turn modulates the electronic transition energy.

Acknowledgements. Ing. D. Polli for helping in the experimental set-up and G. Gerelli for interesting discussions.

References

1. M.V. Burgel, D.A.Wiersma and K. Duppen *"The dynamics of one-dimensional excitons in liquids"*, J. Chem.Phys. **102**, 20, (1995)
2. C.Guo, M.Aydin, H.Zhu and D.L.Akins *"Density Functional Theory Used in Structure Determinations and Raman Band Assignments for Pseudoisocyanine and its aggregate"* J.Phys. Chem. B **106**, 5447 (2002)

Ultrafast Dynamics in Na-doped water Clusters

H.T. Liu[1], J.P. Müller[1], C.P. Schulz[1], C. Schröter[1], N. Zhavoronkov[1], and I.V. Hertel[1,2]

[1] Max-Born-Institut, Max-Born-Strasse 2a, D-12489 Berlin, Germany
[2] Fachbereich Physik, Freie Universität Berlin, Arnimallee 14, D-14195 Berlin, Germany
E-mail: htliu@mbi-berlin.de

Abstract. The lifetimes of the first electronically excited state of $(H_2O)_n \cdots Na$ clusters (n up to 40) are measured using two colour pump-probe spectroscopy. The lifetimes obtained are compared to those of water cluster anions [1].

Introduction

Gas phase clusters of polar solvent molecules doped with an alkali metal atom are model systems for studying the behaviour of loosely bound electrons in a polar environment. In the past years, our group has investigated spectroscopic properties of size-selected $(H_2O)_n \cdots Na$ and $(NH_3)_n \cdots Na$ clusters, such as the ionization potentials [2, 3] and the energy of the first electronic excited state [4–6]. It has been shown that pump-probe experiments provide a suitable tool to study the dynamics of the electronically excited state [7]. Previous work on $(NH_3)_n \cdots Na$ clusters has shown that the lifetimes of first electronically excited states strongly decreases for larger n. For $n \geq 4$ they are on the order of picoseconds and lower. Similar results have been obtained for the anions of pure water clusters [1, 8]. These short lifetimes are presumably provoked by a fast internal conversion, which is strongly correlated to the DOS of the intra molecular vibrations [7]. In the present work we focus on the water clusters. The lifetimes of the first electronically excited state of $(H_2O)_n \cdots Na$ clusters (corresponding to the 3p ← 3s transition in free sodium atoms) with n up to 40 are measured by two colour pump-probe spectroscopy (800/400 nm) with 30 fs pulses. The observed lifetimes are compared to those of water cluster anions, which are expected to have a similar electronic structure. It turns out that the energetics of $(H_2O)_n^-$ compares very well with the sodium doped water clusters. However, lifetimes of $(H_2O)_n^-$ reported in the literature [1, 8] are significantly larger.

Experimental Methods

A detailed description of the experimental setup can be found in an earlier publication [9]. A brief overview will be given here. Water clusters are created by expanding water vapour through a conical nozzle with 50 µm diameter into vacuum. The water oven can be heated up to a temperature of 200°C, which corresponds to a saturation pressure of 10 bar. To avoid clogging the nozzle is heated separately and kept 10 to 20°C warmer than the water oven temperature. The supersonic jet is collimated by a skimmer and the water cluster beam traverses to a pick-up oven containing sodium. The number of sodium atoms picked up by the water clusters is controlled by the sodium vapour pressure [9]. The sodium-water clusters formed reach the detector chamber, where they are excited with a femtosecond pump pulse (800 nm) and ionized with a probe pulse (400 nm). Finally, the cluster ions are mass-separated in a

373

time-of-flight mass spectrometer, recorded by channel plates, and registered by a fast acquisition card.

The femtosecond laser pulses are generated by a commercial Ti:Sapphire laser system, which provides sub 30 fs pulses at a wavelength of 800 nm and pulse energy of 0.7 mJ at a repetition rate of 1 kHz. This laser beam is split into two arms. One is used directly as the pump pulse for exciting the clusters. The other beam is focused into a BBO crystal producing 400 nm second harmonic to probe the system. The pump and probe pulses are nearly collinear and intersect the cluster beam perpendicularly. The spatial overlap of the beams is optimized directly by maximising pump-probe ion intensities. The time delay is controlled by a stepper motor driven delay table with a time resolution of 0.6 fs.

Results and Discussion

Previous spectroscopic studies have revealed a broad absorption band of the lowest electronically excited state of $(H_2O)_n \cdots Na$ clusters in the near infrared spectral region, with centre wavelengths is around 800 nm [6]. Thus, pulses at the fundamental wavelength (800 nm) of a Ti:Sapphire laser can conveniently be used to pump the clusters from the ground state to the lowest electronically excited state, while the SHG at 400 nm probe pulse readily ionises the excited $(H_2O)_n \cdots Na$ clusters. The lifetimes τ of the lowest electronically excited state $(H_2O)_n \cdots Na$ are then determined by evaluating the decay of the transient $(H_2O)_n \cdots Na^+$ ion signal, as measured by tuning the time delay Δt between pump and probe pulse. The ion signals observed exhibit a single exponential decay for all cluster sizes as a function of Δt. Size-dependent decay times of the ion signal are summarized in Figure 1 for $(H_2O)_n \cdots Na$ clusters with up to 40 water molecules. The error bars are within the size of the plotted dots. For comparison, Figure 1 also shows the lifetimes of water cluster anions taken from reference 8.

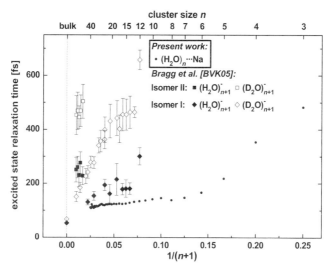

Fig. 1. Size dependent lifetime trends of the lowest electronically excited state of $(H_2O)_n \cdots Na$ clusters, which are plots against $1/(n+1)$. The lifetimes of water cluster anions are taken from reference 8. Trends and the comparison with water cluster anions are discussed in section 3.

From the curve of Fig. 1 we learn that the lifetime trends of $(H_2O)_n \cdots Na$ clusters are somewhat different compared to water cluster anions. Bragg et al. [1] have concluded that their isomer I of $(H_2O)_n^-$ clusters (considered to be internal states) extrapolate linearly with $1/n$ toward to the known conversion lifetime of 50 fs in bulk water. The isomer II of $(H_2O)_n^-$ clusters (considered surface-bound) have larger sizes ($n = 60-100$), and their decay time is more or less independent of the cluster size. For our measurements of $(H_2O)_n\cdots Na$ clusters, the first excited states of $(H_2O)_n\cdots Na$ are shorter living than water cluster anions. The decay time of $(H_2O)_n\cdots Na$ clusters decreases massively with $n \leq 8$ (to about 140 fs), however, the lifetime of $(H_2O)_9\cdots Na$ is distinctly longer than $(H_2O)_8\cdots Na$. For $n \geq 9$ the lifetimes change much slower but still constantly drop as the cluster size increases. For clusters with $n \geq 24$ the experimental results may indicate a somewhat more rapid decrease with size. Whether this trend continues and can be rationalized by an influence of the solvation shells of water around the Na^+ ion will have to be discussed in the light of further measurements.

The lifetime trends of $(H_2O)_n\cdots Na$ clusters discussed here show strong similarities to those of sodium ammonia clusters reported earlier [7]. In the latter case, there is strong evidence that the energy of the excited state is converted into vibrations of the solvent molecules in the electronic ground state, as born out by simple DOS calculations. For water clusters such a model is still missing but a similar result is expected. The measurement of deuterated sodium water clusters $(D_2O)_n\cdots Na$ is planned in the near future.

Conclusions

The lifetimes of the first electronically excited state of $(H_2O)_n\cdots Na$ clusters with n up to 40 are measured by two colour pump-probe spectroscopy. The lifetimes of those clusters show a similar trend as water cluster anions as a function of cluster size but differ in magnitude. The decay time is decreasing massively for n up to 8 (about 140 fs), the lifetime of (H2O)9$\cdots$Na exhibit a longer value than (H2O)8$\cdots$Na. For n $\geq$ 9, the lifetimes change much slower. The similarities of the lifetime behaviour of sodium doped water and ammonia clusters give evidence that the short lifetimes of the lowest excited state are caused by internal conversion. Extension of the experiments are in progress and in depth theoretical simulations, possibly *ab initio* based, would be highly desirable.

Acknowledgements. This work is financially supported by the Deutsche Forschungsgemeinschaft through SFB 450. H.T. Liu is grateful for support by Alexander von Humboldt foundation.

1   J.R.R. Verlet, A.E. Bragg, A. Kammrath, O. Cheshnovski, and D.M. Neumark, Science **307**, 93, 2005.

2   C.P. Schulz, R. Haugstätter, H.U. Tittes, and I.V. Hertel, Phys. Rev. Lett. **57**, 1703, 1986.

3   I.V. Hertel, C. Hüglin, C. Nitsch, and C.P. Schulz, Phys. Rev. Lett. **67**, 1767, 1991.

4   C.P. Schulz and C. Nitsch, J. Chem. Phys. **107**, 9794, 1997.

5   P. Brockhaus, I.V. Hertel, and C.P. Schulz, J. Chem. Phys. **110**, 393, 1999.

6   C.P. Schulz, C. Bobbert, T. Shimosato, K. Daigoku, N. Miura, and K. Hashimoto, J. Chem. Phys. **119**,11620, 2003.

7   C.P. Schulz, A. Scholz, and I.V. Hertel, Isr. J. Chem. **44**, 19, 2004.

8   A.E. Bragg, J.R.R. Verlet, A. Kammrath, O. Cheshnovski, and D.M. Neumark, J. Am. Chem. Soc. **127**, 15283, 2005.

9   C. Bobbert and C. P. Schulz, Eur. Phys. J. D **16**, 95, 2001.

Electronic Excitations in Pentacene Films: Singlet versus Triplet Dynamics

Henning Marciniak[1], Bert Nickel[2], and Stefan Lochbrunner[1]

[1] Institut für Physik, Universität Rostock, Universitätsplatz 3, 18055 Rostock, Germany
E-mail: stefan.lochbrunner@uni-rostock.de
[2] Fakultät für Physik und CeNS, Ludwig-Maximilians-Universität, Geschwister-Scholl-Platz 1,
80539 München, Germany
E-mail: nickel@lmu.de

Abstract. Polarization dependent femtosecond spectroscopy shows that photoexcited excitons in microcrystalline pentacene films decay within 70 fs to a non fluorescing singlet species. On the picosecond time scale, a small fraction of triplet excitons forms.

Introduction

Pentacene is due to its extraordinary high hole mobility in the crystalline phase of more than 1 cm^2/Vs one of the most promising candidates for applications in organic electronics like transistors[1]. Therefore pentacene serves as model compound to understand the electronic properties of organic crystals. In optoelectronic devices the behavior of electronic excitations is crucial for the performance. While the pentacene monomer fluoresces with a high quantum yield, pentacene films show only an extremely low photoluminescence pointing to the relevance of ultrafast relaxation processes. To investigate this dynamics, we apply pump-probe absorption spectroscopy with a time resolution of 30 fs to microcrystalline pentacene films.

Ultrafast Electronic Dynamics

50 nm thick microcrystalline films are prepared by vapor deposition of pentacene on transparent polymer substrates. The transient absorption is measured after photoexcitation at the lowest absorption band with an excitation wavelength of 669 nm. First the results obtained with normal incidence and parallel polarizations for pump and probe are discussed (Fig. 1, black traces). The transient spectrum taken at a delay of 8 ps shows a strong ground state bleach and excited state absorption (ESA) around 630 nm but no indication for stimulated emission [2]. Kinetic traces measured at 620 nm and 680 nm reveal an ultrafast change in the ESA and a decay of the transmission with a time constant of 70 fs (Fig. 1, rhs.). This 70 fs transmission decay appears dominantly in the spectral region of the low energy side of the first steady state absorption band. In this region emission from the originally excited excitons is expected. We interpret the 70 fs dynamics therefore as an ultrafast relaxation of the photoexcited excitons to a non fluorescing species.

Indications for such an ultrafast process have been found before and were interpreted as fission of the primarily generated Frenkel excitons into triplet excitons [3]. Since per each singlet exciton two triplet excitons are generated the total spin is conserved. In tetracene crystals this process occurs on the sub-nanosecond time scale and is associated with a thermal activation energy of 0.21 eV [4]. In the pentacene monomer the triplet energy is slightly less than half of the first electronically excited singlet state. Extrapolating the fission rate of tetracene to the energetics of pentacene

376

predicts a sub-100 fs decay for the singlet excitons. Consequently the ESA occurring in the transient spectra around 630 nm was attributed to triplet excitons [3]. However, this assignment does not match to the triplet spectrum of pentacene monomers in solution [5]. There the energy difference between triplet and singlet absorption is about 10 times larger than the energy difference between the ground state bleach and the ESA in the transient spectra of the microcrystalline films.

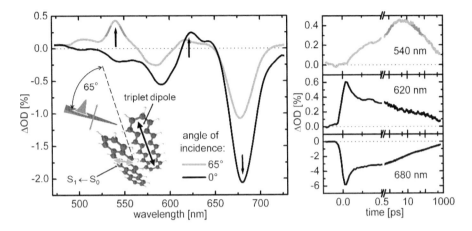

Fig. 1. Left: Transient spectra of a thin pentacene film 8 ps after excitation at 669 nm with parallel polarized pump and probe beam for different angles of incidence. Right: Time scans at different probe wavelengths taken under the same conditions. Grey curves denote measurements with an angle of incidence of 65° and black curves with normal incidence.

It is also problematic to understand the dynamics on longer time scales in terms of triplet excitons. The decay of the ESA as it is observed at 620 nm and the bleach recovery at 680 nm (see Fig. 1) depend on the excitation fluence. We have shown previously, that the dynamics after the first picosecond is due to diffusion controlled exciton exciton annihilation and a limited lifetime of the excitons of 850 ps [2]. Since the energy of two triplet excitons should be below the energy of one singlet exciton two encountering triplets cannot create a highly excited state and it is unlikely that they annihilate efficiently. In summary, a more specific triplet signature is needed to get a clear picture of the electronic dynamics in pentacene crystals.

Delayed Triplet Formation

A delicate balance determines the strength of the triplet absorption in pentacene films. The transition dipole of the lowest triplet-triplet absorption band is quite big and the absorption cross section is more than 10 times larger than that of the lowest singlet absorption [5]. The dipole is aligned along the long molecular axis and the pentacene molecules in crystalline films with their typical herringbone-structure are standing almost upright on the substrate [6]. At normal incidence of pump and probe pulses the electric fields are almost perpendicular to the triplet transition dipole and absorption signatures from triplet excitons are strongly suppressed.

To obtain triplet specific signatures, measurements were performed with an angle of incidence of 65° for the laser beams. The results are shown in Fig. 1 as grey traces. The transient spectra exhibit similar features as the spectra measured at normal incidence but an additional ESA band is clearly observed around 540 nm which does not appear at normal incidence. The spectral position is slightly red shifted by 0.16 eV with respect to the triplet absorption of the monomer. Therefore we assign this signature to triplet excitons while the ESA around 630 nm results from a species with singlet character. Taking the large triplet absorption cross section into account, the moderate strength of the band indicates that the triplet population is a minority.

The kinetic trace measured at 540 nm (see Fig. 1) shows an absorption rise on the picosecond timescale and not within 70 fs. This indicates that the triplet excitons are formed in a subsequent process after the first ultrafast relaxation step. The corresponding time constant does not significantly contribute to the dynamics in other spectral regions. This is a further indication that the fission into triplets is a minor relaxation channel. At longer delay times a signal decay is observed at 540 nm. However, we observe a corresponding signal rise at 650 nm (not shown) indicating that the triplet excitons are trapped or experience another relaxation process but do not decay within the first nanosecond.

Conclusions

The presented measurements show that in microcrystalline pentacene films the primarily excited Frenkel excitons decay within 70 fs to a non fluorescing singlet species. Contrary to expectations fission into triplets is only a minor channel. We propose that the ultrafast relaxation of the photoexcited excitons leads to a species very similar to excimers. Neighboring pentacene molecules form dimers which are bound stronger in the electronically excited state than in the ground state. The relaxation time probably reflects geometric rearrangements like a rotation around the long molecular axis and a reduction of the intermolecular distance to increase the overlap of the π-orbitales and the interaction strength. The resulting gain in energy should be in the order of a few 100 meV. In this case the excimer energy is smaller than two times the triplet energy and fission into triplets should be a thermally activated process. This results in a picosecond rate for the triplet formation directly after the excimers are generated and as long as the released energy has not yet dissipated. After this energy is distributed over a larger sample region the local temperature has dropped so far that the fission rate becomes irrelevant. Accordingly, only a minor fraction of the excitons is transformed into triplets. Like in H-aggregates the radiative transition of the excimer to the ground state is electric dipole forbidden due to symmetry reasons. The excimer formation therefore provides an efficient mechanism to turn off the emission within 70 fs as it is observed in the experiment.

Acknowledgements. We thank Florian Selmeier for experimental support.

1   C. D. Dimitrakopoulos, S. Purushothaman, J. Kymissis, A. Callegari, and J. M. Shaw, Science **283**, 822, 1999.

2   H. Marciniak, M. Fiebig, M. Huth, S. Schiefer, B. Nickel, F. Selmaier, and S. Lochbrunner, Phys. Rev. Lett. **99**, 176402, 2007.

3   C. Jundt, G. Klein, B. Sipp, J. Le Moigne, M. Joucla, and A. A. Villaeys, Chem. Phys. Lett. **241**, 84, 1995.

4   R. P. Groff, P. Avakian, and R. E. Merrifield, Phys. Rev. B **1**, 815, 1970.

5   C. Hellner, L. Lindqvist, and P. C. Rodberge, J. Chem. Soc., Faraday Trans. 1 **68**, 1928, 1972.

6   S. Schiefer, M. Huth, A. Dobrinevski, and B. Nickel, J. Am. Chem. Soc. **129**, 10316, 2007.

Photoreaction from a light generated non-equilibrium state

Simone Draxler[1], Stephan Malkmus[1], Thomas Brust[1], Jessica A. DiGirolamo[2], Watson J. Lees[2], Markus Braun[1], and Wolfgang Zinth[1]

[1] BioMolekulare Optik, Fakultät für Physik, Ludwig-Maximilians-Universität München, Oettingenstr. 67, D-80538 München, Germany, E-mail: Simone.Draxler@physik.uni-muenchen.de or Markus.Braun@physik.uni-muenchen.de
[2] Department of Chemistry and Biochemistry, Florida International University, 11200 SW 8th St., Miami, FL, 33199, USA

Abstract. We report on the acceleration of the S_1 photoreaction combined with the dramatic increase of the photochemical quantum efficiency, when the reaction is directly preceded by another ultrafast photoreaction.

Introduction

Most ultrafast photochemical reactions occurring in the femtosecond time regime involve reaction dynamics, which proceed directly via a conical intersection connecting the excited electronic state with the product state. Afterwards a certain vibrational excess population is found, which dissipates to the surroundings on the 10 ps time scale. Photochemical reactions in the picosecond time domain often occur via barriers on the potential energy surface of the excited electronic state [1]. Here the vibrational population prior to the optical excitation may influence the reaction dynamics and will lead to an Arrhenius type behaviour.

In this study we will investigate a photoreaction which is known to be controlled by a potential barrier in the excited electronic state [2]. We will show that this reaction can strongly be modified by a pre-excitation process, which speeds up the reaction rate and increases the photochemical quantum efficiency by more than a factor of 3.

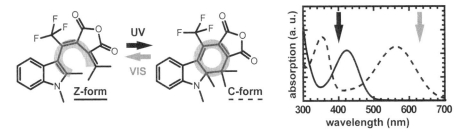

Fig. 1. Left: Chemical structures of Z- and C-form molecules of tri-fluorinated indolylfulgide; the active centre, a cyclohexadiene motif is highlighted. Right: Absorption spectra of Z-form molecules (solid line) and C-form molecules (dashed line) of tri-fluorinated indolylfulgide dissolved in 1,4-dioxane. Wavelengths of the pre-excitation pulse for ring-closure (black) and pump pulse for ring-opening (grey) are marked by arrows.

Materials and Methods

The time-resolved experiments are performed on the tri-fluorinated indolylfulgide [3] where the two investigated forms are shown in Fig. 1 (left side). This molecule is an ideal candidate for the proposed investigation since (i) the two molecular isomers show clearly separated absorption bands (Fig. 1 right), which allow selective excitation. (ii) The Z to C ring-closure reaction is ultrafast and is finished within 1 ps, while it generates a ground state vibrational excess population, which relaxes with a time constant of about 15 ps. This reaction dynamics is characteristic for similar fulgides and has been studied in detail for another indolylfulgimide [4]. (iii) The absorption spectra allow selective triggering of the ring-opening reaction (which proceeds via a barrier on the excited state potential surface [2]) by light in the wavelength range between 550 and 700 nm.

In the experiment Z-form molecules of indolylfulgide are dissolved in 1,4-dioxane. For the time-resolved absorption experiment we use a homebuilt Ti:Sa laser-system, operated at 1 kHz. In order to perform experiments with pre-excitation the standard pump-probe setup [4] (delay between the 630 nm pump and the white light probing pulse Δt_2) is supplemented by a pre-excitation pulse (400 nm), which occurs at Δt_1 before the 630 nm pump pulse [5]. The pre-excitation pulse is generated as the second harmonic of the fundamental of the Ti:Sa system. In the experiment we recorded the absorption changes as a function of Δt_2 for certain settings of the pre-excitation delay Δt_1. In Fig. 2 we plotted (symbols) the pre-excitation induced absorption changes, which are obtained by evaluating the difference between absorption traces recorded with and without 630 nm pump pulse.

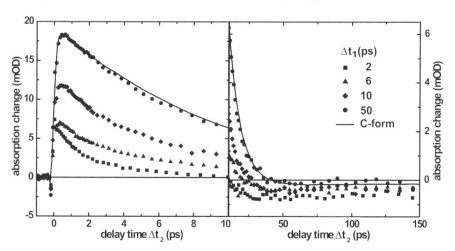

Fig. 2. Transient absorption signals for the ring-opening reaction of pre-excited, vibrationally hot C-form molecules; left part for small delay times Δt_2, right part, long Δt_2 (enlarged scale). Different symbols represent the delay times Δt_1 (see legend) between the pre-excitation and the pump pulse. Solid line: scaled transient absorption signal for the ring-opening reaction of C-form molecules.

Results and Discussion

Without pre-excitation the ring-opening reaction (solid curve in Fig. 2) occurs with a time constant of 9 ps. The weak offset shown at late delay times is proportional to the product formation and represents a photochemical quantum efficiency in the range of 4%. The curve recorded with pre-excitation and long delay ($\Delta t_1 = 50$ ps) is nearly identical to the solid curve. For smaller pre-excitation delays ($\Delta t_1 < 10$ ps), the ring-opening reaction is accelerated considerably (see Fig. 2). In addition to this acceleration we find a strong change of the final signal offset. A detailed analysis shows that the photochemical quantum efficiency is increased to 14% ($\Delta t_1 = 2$ ps), which corresponds to a more than threefold increase as compared to the case without pre-excitation.

For simulation of the observed photochemistry, we assume that the pre-excitation process leads to a vibrational excess population, which we describe by an elevated temperature decaying exponentially on the picosecond time scale. The simulation shows that it is not possible to model the data by using a single temperature for all vibrational modes. Motivated by the observation of different barriers for reactive and non-reactive processes [2] we assume that the excess energy is not evenly distributed over all modes but that vibrations with different effective temperatures exist: With these assumptions we found a good agreement between simulation and measured data for a starting temperature for the photochemically active modes of about 2080 K, whereas the starting temperature of modes dominating internal conversion is around 410 K. So shortly after ring-closure the vibrational distribution in the ground state is far away from being thermal. Most of the energy is stored in promoting, photochemically active modes. An obvious explanation for the fact that these modes preferentially carry vibrational excess population is, that the same modes have also been important for the preceding ring-closure reaction.

Conclusion

The pre-excitation has a strong impact on the ring-opening reaction. When the delay time Δt_1 decreases, the ring-opening reaction becomes faster and the quantum efficiency is increased by more than a factor of 3. In addition we observe mode specific photochemistry: The promoting, photochemically active modes of the photoreaction are efficiently heated (excited) by the preceding ring-closure reaction. They start at a temperature of 2080 K versus 410 K for non-active modes.

Acknowledgements. This work was supported by Deutsche Forschungsgemeinschaft through the DFG-Cluster of Excellence Munich-Centre for Advanced Photonics and the SFB 749.

1   Z. Lan, A. Dupays, V. Vallet, S. Mahapatra, W. Domcke, J. Photochem. Photobiol. A-Chem. **190**, 177-189, 2007.

2   T. Brust, S. Draxler, S. Malkmus, C. Schulz, S. Dietrich, K. Rück-Braun, M. Braun, W. Zinth, J. Mol. Liq. doi: 10.1016/j.molliq.2008.02.011, 2008.

3   M.A. Wolak, N.B. Gillespie, C.J. Thomas, R.R. Birge, W.J. Lees, J. Photochem. Photobiol. A-Chem. **144**, 83-91, 2001.

4   S. Draxler, T. Brust, S. Malkmus, F.O. Koller, B. Heinz, S. Laimgruber, K. Rück-Braun, W. Zinth, M. Braun, et al., J. Mol. Liq. doi:10.1016/j.molliq.2008.02.001, 2008.

5   S. Malkmus, F.O. Koller, S. Draxler, T.E. Schrader, W.J. Schreier, T. Brust, J.A. DiGirolamo, W.J. Lees, W. Zinth, M. Braun, Adv. Funct. Mater. **17**, 3657-3662, 2007.

Excited-State Nuclear Wavepacket Motion of an Ultrafast Inorganic Molecular Switch

Munetaka Iwamura, Hidekazu Watanabe, Kunihiko Ishii, Satoshi Takeuchi, and Tahei Tahara

Molecular Spectroscopy Laboratory, RIKEN, 2-1 Hirosawa, Wako, Saitama 351-0198, Japan
E-mail: tahei@riken.jp

Abstract. Ultrafast photo-induced structural change of $[Cu(dmphen)_2]^+$ was studied by pump-probe spectroscopy with 25-fs time-resolution. The observed nuclear wavepacket motion unveiled a new mechanism of photo-induced Jahn-Teller distortion that is a key of inorganic molecular switches.

Introduction

Transition metal complexes show faster and more complicated excited-state dynamics than organic compounds, and the elucidation of ultrafast dynamics is essential to understand the property and function of metal complexes. One of the most intriguing ultrafast dynamics of metal complexes is the photo-induced structural change of copper complexes, which is the change of the dihedral angle (θ) between the two ligand planes. In general, Cu(I) complexes have tetrahedral-like structures ($\theta = 90°$) while Cu(II) complexes have square planar-like structures ($\theta = 0°$). Thus, oxidization (or reduction) of copper complexes induces a significant structural change, which is now considered one of the most promising components of molecular machines [1,2]. Because the photo-induced structural change of Cu(I) complexes is driven by metal-to-ligand charge transfer (MLCT) excitation which formally oxidizes the central copper, the excited-state structural change of Cu(I) complexes provides a system where we can investigate the essential dynamics of the molecular machine that utilizes the copper complexes.

$[Cu(dmphen)_2]^+$ (dmphen = 2,9-dimethyl-1,10-phenanthroline) is a prototypical Cu(I) complex that exhibits the ultrafast photoinduced structural change. Femtosecond and picosecond time-resolved spectroscopy have been applied to elucidate the mechanism of the ultrafast structural change of this metal complex [3-5]. Especially, we recently reported the emission dynamics of $[Cu(dmphen)_2]^+$ in the femtosecond and picosecond time region with $S_2 \leftarrow S_0$ excitation, and clarified that the 'flattening' structural change occurs in the S_1 state with 660 fs before the intersystem crossing [3]. Importantly, the fluorescence spectral change that corresponds to the structural change clearly showed an isoemissive point, which indicated that a short-lived precursor state appears in the S_1 potential before the structural change. This observation suggested that there exists a shallow potential minimum at the perpendicular geometry on the S_1 potential energy surface which traps the S_1 state for a finite waiting time.

In this study, we investigated the nuclear dynamics of $[Cu(dmphen)_2]^+$ immediately after photoexcitaion by ultrafast pump-probe spectroscopy with 25-fs time-resolution. With direct $S_1 \leftarrow S_0$ photoexcitation, we successfully observed the excited-state wavepacket motion that precedes the ultrafast structural change. This result clearly demonstrates the existence of a hollow on the S_1 potential at the perpendicular configuration, which is not expected in the ordinary picture based on the (pseudo-) Jahn-Teller effect.

Experimental Methods

Ultrafast pump-probe measurements of $[Cu(dmphen)_2]^+$ was performed by the apparatus that has already been reported [6]. Briefly, two NOPAs were excited by the second harmonic of a Ti:sapphire regenerative amplifier (Legend, Coherent). The output of the first NOPA was tuned at 550 nm and was used to photoexcite $[Cu(dmphen)_2]^+$ through $S_1 \leftarrow S_0$ transition. The output of the second NOPA was set at 650 nm and was used to probe transient absorption. The time resolution of the measurements was 25 fs (FWHM of the cross correlation between the pump and probe). All the measurements were performed in solution at room temperature. $[Cu(dmphen)_2]^+$ was synthesized as $[Cu(dmphen)_2]PF_6$ according to the literature [7]. It was purified several times by recrystallization before use.

Results and Discussion

$[Cu(dmphen)_2]^+$ exhibits a weak absorption around 550 nm, which has been assigned to the optical forbidden transition to the S_1 ($^1A_2$) state at the perpendicular D_{2d} structure. This transition gains finite transition intensity due to either vibronic coupling with the S_2 ($^1B_2$) state or the large amplitude flattening vibration. The transient absorption measurements in the range of 450 – 750 nm revealed that the absorption spectral change reflected three dynamics having time constants of 900 fs, 9 ps, and 41 ns. These dynamics were attributed to the flattening structural change, $S_1 \rightarrow T_1$ intersystem crossing, and $T_1 \rightarrow S_0$ relaxation process of $[Cu(dmphen)_2]^+$ (inset of Fig. 1). We carried out ultrafast pump-probe measurements with 25-fs time-resolution with direct $S_1 \leftarrow S_0$ photoexcitation to examine the nuclear dynamics that precedes ultrafast structural change in the S_1 state.

Fig. 1 shows the pump-probe signal measured at 650 nm, which clearly exhibits an oscillatory feature due to the nuclear wavepacket motion of the S_1 state. The damping time of the nuclear wavepacket motion is in the sub-picosecond range, which implies that this is the wavepacket motion of the perpendicular S_1 state before the structural change. Observation of the nuclear wavepacket motion before the structural change directly manifests that the perpendicular S_1 state is a bound state. In other words, the observed oscillation is ascribed to the vibration of the S_1 state that is trapped in the shallow potential minimum at the perpendicular structure immediately after photoexcitaiton. The Fourier transform of the oscillatory feature of the pump-probe signal showed that two vibrations at 292 and 242 cm^{-1} predominantly contributed to the signal, with relatively weak contributions from the vibrations at 417, 183, 125 and 83 cm^{-1}. A quantum chemical calculation for the S_1 state nicely provided a potential curve that is very consistent with the results of ultrafast spectroscopy: A shallow local minimum exists at the perpendicular configuration and the global energy minimum is located at a flattened geometry. The prominent components in the observed wavepacket motions were attributed to b_1 vibrations in the S_1 state. Because the symmetry of the two vibrations is the same as that of the flattening structural change, it is considered that the initial wavepacket motion effectively leads to the structural distortion of $[Cu(dmphen)_2]^+$ through the anharmonic coupling.

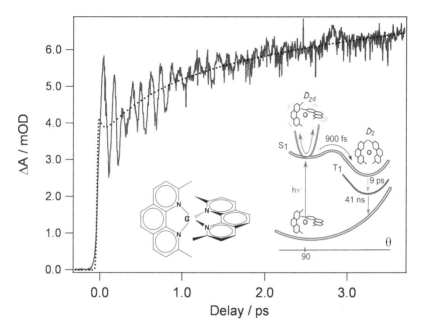

Fig. 1. 1. Ultrafast pump-probe signal of [Cu(dmphen)$_2$]$^+$ (pump 550 nm; probe 650 nm; in dichloromethane). Inset: schematic potential curve of [Cu(dmphen)$_2$]$^+$.

Conclusions

The present work unambiguously showed the existence of a "hollow" on the S$_1$ potential at the perpendicular structure. This finding significantly changes the ordinary picture of photoinduced structural change of Cu(I) complexes based on the (psedo-) 'Jahn-Teller' effect, because the Jahn-Teller distortion is induced instantaneously by the instability of the molecule at the unperturbed perpendicular structure.

Acknowledgements. This work is supported by Grant-in-Aid for Scientific Research on Priority Area "Molecular Science for Supra Functional Systems" (No. 19205005) from MEXT of Japan.

1   V. Balzani, A. Credi, F. M. Raymo, and J. F. Stoddard, Angew. Chem. Int. Ed., **39**, 3348, 2000.
2   O. Sato, Acc. Chem. Res., **36**, 692, 2003.
3   M. Iwamura, S. Takeuchi, and T. Tahara, J. Am. Chem. Soc. **129**, 5248, 2007.
4   G. B. Shaw, C. D. Grant, H. Shirota, E. W. Castner Jr., G. J. Meyer and L. X. Chen, J. Am. Chem. Soc., **129**, 2147, 2007.
5   Z. A. Siddique, Y. Yamamoto, T. Ohno, and K. Nozaki, Inorg. Chem.**42**, 6366, 2003.
6   S. Takeuchi and T. Tahara, Chem. Phys. Lett., **326**, 430, 2000.
7   D. R. McMillin, M. T. Buckner, and B. Tae Ahn, Inorg. Chem. **16**, 943, 1976.

Femtosecond Electronic Dynamics via a Conical Funnel

Eric Ryan Smith, William K. Peters, and David M. Jonas[1]

[1] Department of Chemistry and Biochemistry, University of Colorado, Boulder, Colorado 80309-0215, USA
E-mail: david.jonas@colorado.edu

Abstract. Femtosecond polarization spectroscopy measures electronic wavepacket motion after vibrational wavepackets are excited near an energetically inaccessible conical intersection in a free-base naphthalocyanine. Partial equilibration via the conical funnel takes place within ~100 fs.

Introduction

The breakdown of the Born-Oppenheimer adiabatic separation between fast electronic and slow nuclear motions is important throughout chemistry. In the adiabatic picture, slow nuclei move on a potential energy surface given by the nuclear coordinate dependent energy of the electronic quantum state. This adiabatic picture breaks down when the electronic motions are slowed by near electronic degeneracy. With several vibrational degrees of freedom, crossing between potential curves of the same symmetry cannot be avoided: the geometry of the crossing region resembles the vertex of a right circular cone, where two conical surfaces smoothly connect through a point.[1] As a path between electronic states, these "conical intersections" have been implicated in many photochemical reactions.[2,3] We have recently measured the dynamics after electronic-vibrational wavepackets are excited at a conical intersection in a four-fold symmetric silicon naphthalocyanine [SiNc], and found a surprisingly fast (~100 fs) complete equilibration for the stabilization energy (~1 meV) measured from the asymmetric vibrational quantum beat amplitudes.[4]

It has long been suggested that near misses between potential surfaces can also act as "conical funnels" between electronic states.[5] The new experiments described here impulsively excite electronic-vibrational wavepackets near an energetically inaccessible conical intersection in a two-fold symmetric free-base naphthalocyanine [H_2Nc]. Calculations indicate the conical intersection is inaccessible because it requires a double hydrogen atom transfer, but the width of the linear absorption spectrum indicates an energetic separation of less than 150 cm 1. Because excited state absorption is insignificant for the free-base naphthalocyanine, interpretation of the experiments is simplified. For impulsive excitation, the anisotropy, $r(T)$, is given by

$$r(T) = [1 + c(T) + d(T)]/10 , \qquad (1)$$

where $c(T)$ and $d(T)$ quantify the decay of electronic coherence and the electronic population difference between the two electronic states, respectively. The initial anisotropy of 3/10 reflects coherent excitation of two electronic states with equal magnitude perpendicular transition dipole moments [$c(0) = d(0) = 1$]. An anisotropy of 1/10 is characteristic of electronic delocalization in a plane and indicates complete electronic equilibration [$c = d = 0$].

385

Experiment and Results

We carried out pump-probe anisotropy measurements using near transform-limited pulses that coherently excited both quasi-degenerate Q_x and Q_y electronic transitions of a free-base 2,3 naphthalocyanine in benzonitrile solution. The concentration was low to avoid aggregation (maximum optical density of 0.03) and the pulse energy was limited to avoid saturation (~300 pJ in a 40 µm spot). Free-base naphthalocyanine has the same point group symmetry as a rectangle, and the Q transitions are analogous to the transitions from the ground state (n_x=1, n_y=1) to the excited states (n_x=2, n_y=1) and (n_x=1, n_y=2) for a particle in a rectangular box. Signals with the probe polarization parallel, perpendicular, and at the magic angle to the pump polarization were measured for pulses centered on the absorption maximum (blue tuned - 29 fs pulses) and at lower photon energy (red tuned - 38 fs pulses). The magic angle signal (which is largely time independent) measures isotropic dynamics, while the pump-probe polarization anisotropy, $r = (S_{\parallel} - S_{\perp})/(S_{\parallel} + 2S_{\perp})$, reflects rotational motion. The pump-probe polarization anisotropy in Fig. 1 indicates electronic dynamics on a ~100 fs timescale, but shows that equilibration via the conical funnel is incomplete, and slightly slower, for low energy excitation. The prominent 176 cm-1 vibration seen in Fig. 1 has a quantum beat anisotropy of r_{vib} = 8±2. This anisotropy is indicative of an asymmetric (diamond) vibration in the D_{2h} point group of the ground electronic state (rectangular symmetry) and provides further information on the electronic splitting between Q_x and Q_y. The amplitude of the quantum beat indicates a Jahn-Teller stabilization energy of 9 cm^{-1} (~1.1 meV). In contrast to SiNc, where three asymmetric vibrations are observed, only one low frequency asymmetric vibration is seen for H$_2$Nc. This suggests that attachment to the central silicon atom couples the macrocycle distortion to internal distortions of the smaller rings in SiNc, thus exciting the higher frequency vibrations. We thus assign the one asymmetric vibration observed in H$_2$NC to a diamond vibration of the naphthalocyanine macrocycle.

Fig. 1. Comparison of pump-probe polarization anisotropy for red-tuned (top trace) and blue-tuned (bottom trace) experiments. Both traces are characterized by fast decay from an initial value of ~0.3 and by modulation with a 190 fs period (corresponding to a vibrational frequency of 176 cm^{-1}) due to an asymmetric vibrational motion. The red-tuned anisotropy does not drop to 1/10 on the timescale of the experiment, indicating incomplete electronic equilibration for excitation energies below the absorption maximum.

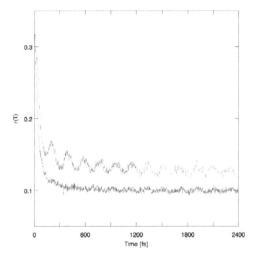

Modeling

The simplest quantitative explanation of both anisotropy decays treats two coordinates in a diabatic electronic basis (basis states polarized along x and y independent of coordinates). The diabatic states are energetically split by a "static" rectangular distortion with a 100 cm^{-1} splitting [δ] caused by the two central hydrogen atoms and vibronically coupled by a Jahn-Teller active asymmetric vibration (diamond type deformation) with a 9 cm^{-1} stabilization energy [$(D\omega)=(1/2)\omega d^2$]. The Hamiltonian is:

$$\hat{\mathcal{H}} = \begin{bmatrix} \hat{\mathcal{H}}_0 + (\delta/2) & \omega d\hat{q} \\ \omega d\hat{q} & \hat{\mathcal{H}}_0 - (\delta/2) \end{bmatrix} \tag{2}$$

where $\hat{q}$ is the operator (in dimensionless normal coordinates) for the diamond vibration and $\hat{\mathcal{H}}_0 = (1/2)\omega(\hat{q}^2 + \hat{p}^2)$ is the harmonic vibrational Hamiltonian of the ground electronic state. Diagonalization in the diabatic basis automatically includes the non-adiabatic derivative coupling arising from the coordinate dependence of the electron configuration coefficients. With a thermal distribution of initial vibrational states, this Hamiltonian predicts the fast initial electronic realignment for both excitation energies, the slight slowing of this initial realignment for red-tuned excitation, complete (within experimental error) electronic equilibration for blue-tuned excitation, and the incomplete electronic equilibration seen in the anisotropy with red-tuned excitation. The vibrational anisotropies agree with the red-tuned experiment only if a splitting between 50 and 100 cm^{-1} is assumed, while the red-tuned anisotropy at 2.4 ps pump-probe delay requires a splitting of 100 to 150 cm^{-1}. The intersection of these two error bounds indicates a splitting of 100 cm^{-1}.

Discussion and Conclusion

The adiabatic surfaces that result from the above static splitting and vibrational coordinate dependent coupling have a one-dimensional coordinate dependent electronic character and do not exhibit a conical intersection. Their closest energetic approach is 100 cm^{-1}. As the anisotropy measures realignment of the initially excited electronic coherent superposition state in the molecular frame, dephasing and population transfer between diabatic states dominate this observable. There is little vibrational wavepacket motion in this system (less than 1/3 the zero point width). As in the SiNc conical intersection,[4] the width of the vibrational wavepacket, not the velocity, dominates the initial electronic dynamics at the conical funnel because it controls the distribution of adiabatic electronic energy gaps and the distribution of adiabatic electronic character. This suggests that only small stabilization energies are needed to drive rapid equilibration of electronic character via conical funnels.

Acknowledgement. This work was supported by the National Science Foundation.

1   D. R. Yarkony, Rev. Mod. Phys. **68**, 985 (1996).

2   F. Bernardi, M. Olivucci, M.A. Robb, Chem. Soc. Rev. **25**, 321 (1996).

3   A.J. Wurzer, T. Wilhelm, J. Piel, and E. Riedle, Chem. Phys. Lett. **299**, 296 (1999).

4   D.A. Farrow, W. Qian, E.R. Smith, A.A. Ferro, and D.M. Jonas, J. Chem. Phys. **128**, 144510 (2008).

5   M. Klessinger and J. Michl, Excited States and Photochemistry of Organic Molecules VCH, New York, 1995.

A new technique to measure time-resolved circular dichroism : ultrafast conformational dynamics of 1,1'-bi-2-naphthol

Claire Niezborala and François Hache

Laboratoire d'Optique et Biosciences, Ecole Polytechnique, CNRS, INSERM
91128 Palaiseau, France
E-mail: francois.hache@polytechnique.edu

Abstract. Using a new technique to measure time-resolved circular dichroism in a pump-probe experiment, we study the conformational relaxation in photoexcited (R)-(+)-1,1'-Bi-2-naphthol. The dihedral angle decreases of *ca* 20-30° in 100 psec in Ethanol.

Introduction

Circular dichroism (CD) is known to be a sensitive probe of molecular conformation. The idea of using CD in a pump-probe configuration in order to study the dynamics of structural changes in molecules or in biomolecules is therefore quite natural [1]. The most common technique used to otain time-resolved CD measurement consists in modulating the probe polarization and measuring the modulated part of the signal transmitted by a chiral sample as a function of the pump-probe delay. Unfortunately, such an experiment is prone to many artefacts [2,3] and therefore very difficult to implement. In this context, we have recently developed a new technique which allows measurement of time-resolved CD in a much more user-friendly, artefact-free way. It can be easily applied to all spectral ranges and used to study many chemical or biophysical samples. As a first application, we have studied the conformational dynamics of the excited state of (R)-(+)-1,1'-Bi-2-naphthol and measured the relaxation time of the dihedral angle in various solvents.

Time-Resolved Circular Dichroism

CD can be obtained through modulated circular polarizations but this methos induces many artefacts. To get rid of them, we have developed a new technique which only involves non-modulated, linearly-polarized beams. The experimental set-up we are using is depicted in figure 1.

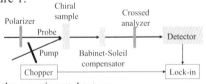

Fig. 1. Scheme of the experimental set-up.

The sample is placed between crossed polarizer and analyzer and the probe ellipticity is measured through a Babinet-Soleil compensator (BS) by recording the transmitted intensity (hereafter called the "PM" signal) as a function of the BS retardation φ. A mechanical chopper is inserted on the pump path and a lock-in amplifier allows us to measure the modulated part of the PM signal (hereafter called the "LI" signal). We briefly recall the principle of the measurement which is detailed in [4]. When measuring simultaneously the PM and the LI signals for small BS retardations, we obtain two parabolas:

$$PM = \varphi^2 + C_1 \text{ and } LI = -\delta\alpha L\,\varphi^2 - \frac{\delta CD}{2}\varphi + C_2.$$

In these equations, C_1 and C_2 are constant, $\delta\alpha L$ is the pump-induced absorption change and δCD the pump-induced CD change. Examination of the formulas shows that comparing the quadratic coefficients of the PM and LI parabolas directly yields $\delta\alpha L$ and that the LI parabola is shifted compared to the PM one by $\delta CD\,/4\delta\alpha L$. This shift is easily measurable (especially when $\delta\alpha L$ is smaller than 0.1) and this technique allows $\delta\alpha L$ and δCD to be detected in the 10^{-4} range. Examples of the measured parabolas are given in the inset of figure 2.

This technique has many advantages compared to the probe polarization modulation ones. First, as it only involves linearly-polarized beams, it much less prone to artefacts. Second, it is much less sensitive to laser fluctuations because information is obtained through a parabolic fit of the raw data. Furthermore, this technique can be employed under various experimental conditions and particularly in all spectral ranges. In this context, starting from a commercial Titanium-Sapphire laser (1kHz, 0.6 mJ, 150 fs), we have extended the probe spectral range of our set-upto the far UV by implementing a two-stage optical parametric amplifier followed by a sum-frequency generation stage which provides sub-picosecond pulses tunable between 220 and 350 nm. This range gives access to relevant domains in biophysics (absorption bands of the aromatic amino acids in the near UV and peptide backbone bands in the far-UV).

Conformational dynamics in excited state Binaphthol

We have taken advantage of this new time-resolved CD set-up tuned in the UV to investigate the dynamics of the change in the dihedral angle in photoexcited (R)-(+)-1,1'-Bi-2-naphthol. Pump pulses are obtained after frequency tripling the output of the Titanium-Sapphire laser. Pulse energy is about 200 nJ. Two probe wavelengths were investigated: 237 nm, which corresponds to a bleaching of the ground state absorption and 245 nm for which we observed excited-state absorption. Several solvents were also investigated ($c = 2.2 \times 10^{-4}$ M). Experimental results are summarized in figure 2. In this figure, the pump-induced CD change is plotted as a function of the pump-probe delay. It is important to note that the absorption change is the same for all the solvents and that there is no noticeable dynamics on the 400 psec timescale (curves not shown). On the contrary, we observe a quite different behaviour for the CD curves: considering first the data at 237 nm, a large difference is noticeable for the two solvents and a 100 ± 20 psec dynamics is measured for Ethanol. Another measurement in an Ethanol/Ethylene glycol mixture (32:1) yields an intermediate time of 130 ± 20 psec. These differences indicate that what we observe is not a mere electronic relaxation but involves changes in the molecule conformation. This conformational change is seen to be solvent dependent: the higher the viscosity, the longer the relaxation time.

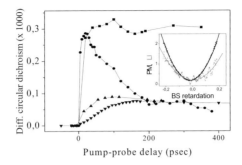

Fig. 2. Time-resolved CD (in optical density) of (R)-(+)-1,1'Bi-2-naphthol (red dots: λ = 237 nm, ethanol, black squares: λ = 237 nm, ethylene-glycol, blue triangles down: λ = 245 nm, ethanol, green triangles up: λ = 245 nm, cyclohexane). The inset shows raw data: the shift between the two parabolas allows the measurement of the differential CD.

These conformational changes are likely to be due to a change in the dihedral angle. In order to analyze this feature more precisely, we have developed a model calculation based on the coupling of the ^1B_b transitions of the naphthyl moieties. Adjustment of the steady-state absorption and CD spectra yields a dihedral angle of 104°, close from the measured value. Let us now consider the case of Binaphthol in Ethanol probed at 237 nm. The change in CD can be separated in two contributions: bleaching of the ground state CD and excited-state CD. Only the latter is expected to change with time. Considering the sign of the CD, the observed relaxation actually denotes an increase (in absolute values) of the excited-state CD. According to our model calculation this increase is consistent with a decrease of the dihedral angle around 20-30°. These values are in agreement with previous estimations in Binaphthyl [5,6].

Finally, we note that the relaxation time measured in Ethanol is much longer than the one measured for Binaphthyl [5]. This is a consequence of the hydrogen-bonds which are formed between the hydroxy groups of Binaphthol and the solvent molecules. The two time-resolved CD curves displayed in figure 2 for λ = 245 nm confirm this affirmation. In this excited-state absorption region we observe the onset of the excited-state CD which is much more rapid in an apolar solvent (cyclohexane) than in a polar solvent (ethanol).

1   J. W. Lewis, R. A. Golbeck, D. S. Kliger, X. Xie, R. C. Dunn and J. D. Simon, J. Phys. Chem. **96**, 5243, 1992.
2   X. Xie and J. D. Simon, J. Opt. Soc. Am. B **7**, 1673, 1990.
3   T. Dartigalongue and F. Hache, J. Opt. Soc. Am. B **20**, 1780, 2003.
4   C. Niezborala and F. Hache, J. Opt. Soc. Am. B **23**, 2418, 2006.
5   D. P. Millar and K. B. Eisenthal, J. Chem. Phys. **86**, 5076, 1985.
6   S. Fujiyoshi, S. Takeuchi and T. Tahara, J. Phys. Chem. A **108**, 5938, 2004.

Picosecond Time-Resolved Vibrational Circular Dichroism Spectroscopy

Mathias Bonmarin and Jan Helbing

Physikalisch-Chemisches Institut, Winterthurerstrasse 190, CH-8057 Zürich, Switzerland
E-mail: j.helbing@pci.uzh.ch

Abstract. An experimental setup for the recording of time-resolved vibrational circular dichroism (VCD) spectra is presented. Both static and transient VCD signals can be measured with high sensitivity. First results for a transition metal complex show chiral transients that are clearly distinct from the conventional pump-probe signals.

Introduction

Chiral molecules and supermolecular structures play an important role in living matter, and chiral-sensitive spectroscopic methods are particularly useful for their detection. For peptides and proteins, for example, UV-circular dichroism (CD) spectra are routinely used to analyse secondary structure content[1]. First measurements of transient CD signals with picosecond time resolution were reported by Xie and Simon in the visible domain [2] and the technique has been more recently extended to the UV with improved time-resolution and sensitivity[3]. Vibrational circular dichroism (VCD) is the difference of absorption of a vibrational transition between left and right circularly polarized infrared light. Similar to 2D-IR spectroscopy, VCD is intrinsically sensitive to molecular structure via the coupling of vibrational modes yielding precious information on the conformation of biomolecules. Since only the electronic ground state is involved, comparison of VCD spectra with theoretical calculations is often much more reliable than in the case of electronic CD spectroscopy. On the other hand, the typical VCD signal (10^{-3}-10^{-5} of the absorbance) is 10-100 times smaller than in the UV, increasing the technical difficulties in implementing transient measurements. We report here the development of a set-up which is able to record photo-induced changes in VCD spectra and show the first results obtained for a test molecule. Our technique follows conceptually the original implementation of transient electronic CD spectroscopy [4] and consists in measuring the difference in transmission of left and right circularly polarized mid IR pulses after photo triggering a reaction with a femtosecond UV or visible laser pulse.

Experimental Methods

The setup is depicted in the left part of Fig. 1. At present, we disperse the broad band mid-IR pulses from an optical parametric amplifier (OPA) before the sample with a home-built monochromator to avoid artifacts, which may arise from the reflection of a circular polarized beam by polarisation sensitive optics. The bandwidth and time duration of the probe light can be varied with the slit size, and wavelength scans are carried out by rotating the grating. The data presented here was recorded with pulses of 3 cm^{-1} spectral width and a time-resolution of approximately 8 ps.

Behind the monochromator, a portion of the beam is split off and recorded on a single shot basis by a nitrogen-cooled MCT detector as a reference to correct for intensity fluctuations. The probe beam is focused onto the sample flow cell to a spot size of 100μm, and imaged onto a second MCT element. In order to produce a train of circular polarized pulses of alternating handedness, the amplifier of the femtosecond laser system (1.01 kHz) is synchronized to the 50 kHz intrinsic frequency of a photo-elastic modulator (PEM). The trigger sent to the Pockels cells of the amplifier cavity is electronically delayed such that the IR pulses cross the modulator exactly when it acts as a quarter waveplate of opposite sign for consecutive pulses. The remaining time jitter between laser beam and modulator of 12 ns due to the free-running 80 MHz laser oscillator is too small to affect the polarisation of the probe pulses. For transient measurements, a visible pump beam is focused onto the sample in spatial overlap with the IR beam.

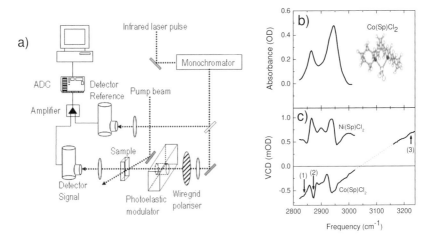

Fig. 1. .a) Schematic view of the setup. b) Absorption spectrum of the $Co(Sp)Cl_2$ molecule in the C-H stretch region. c) VCD spectrum of $Co(Sp)Cl_2$ and $Ni(Sp)Cl_2$ recorded with the pump-probe setup. (1), (2) and (3) indicate the probe wavelengths used in the pump-probe scans. The absorption spectra of the $Ni(Sp)Cl_2$ (not shown) is almost identical to that of $Co(Sp)Cl_2$.

Results and Discussion

To demonstrate the experimental feasibility of transient VCD, we chose a molecule which is known to exhibit a relatively large static VCD signal, the cobalt complex with the chiral ligand (-)-sparteine ($Co(Sp)Cl_2$). The VCD signal in the ground state of this open shell transition metal complex ($\Delta\varepsilon \approx 0.5$ lmol^{-1}cm^{-1} in the C-H stretch region) has been attributed to the closeness of vibrational transitions and magnetic dipole-allowed electronic transitions [5]. The absorption spectrum in the C-H stretch region and the corresponding VCD are shown in Fig. 2 b) and c). For comparison, the VCD spectrum of $Ni(Sp)Cl_2$ is plotted as well. Despite an almost identical absorption spectrum, the VCD spectrum of the nickel complex is of opposite sign and of very different shape, indicating a strong influence of the d-electron configuration. Therefore a change in the VCD signal of the cobalt complex is expected after pumping a d-d transition in the visible.

Excitation of the cobalt complex at 600 nm causes an immediate red-shift of the C-H stretch band, leading to a negative transient absorption signal at 2870 cm^{-1} and a positive one at 2842 cm^{-1} (Fig. 2, top, (1) and (2)). This red shift subsequently decreases on a 20 ps timescale due to electronic relaxation and energy dissipation to the solvent (chloroform). The transient VCD signal, on the other hand, diminishes (becomes less negative) at both probe frequencies as a result of excitation. The different character of transient absorption and transient CD is even more apparent at 3222 cm^{-1}(3), where there are no vibrational transitions (and therefore no pump-probe signal) but a large positive electronic CD-band (see Fig. 1c). At this probe-wavelength the CD signal becomes less positive as a result of visible excitation, confirming the overall pump-induced decrease in the chiral signal. We can therefore exclude transient linear dichroism and birefringence effects, which arise mainly from imperfect modulation of the circularly polarised probe beam, and may give rise to dominating artifacts in time-resolved circular dichroism measurements [5]. Steady state control experiments confirm our observation and indicate that the change in VCD-signal of the sparteine complex persisting after tens of picoseconds is due to a laser-induced increase in sample temperature.

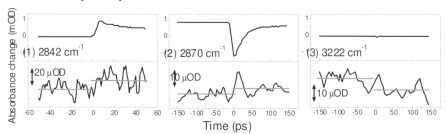

Fig. 2 Transient absorption signals (top) and transient VCD signals (bottom) for three different probe frequencies (1): 2842 cm^{-1}, (2): 2870 cm^{-1} and (3): 3222 cm^{-1}, also indicated by arrows in Fig. 1c. Less than 2% of the molecules are excited at 600 nm.

Conclusions

We have shown for the first time that time-resolved VCD spectroscopy is experimentally feasible. Equilibrium and time-resolved circular dichroism spectra can be recorded simultaneously with the same set-up. Benefiting form unusually large VCD in the electronic ground state, we could record VCD transients for a chiral transition metal complex with sufficient signal to noise for an excitation density in the probed sample volume of less than 2%. Ongoing improvements of our set-up and in particular broad-band detection should make this technique a promising new tool for detecting conformational changes in biomolecules and chiral intermediates in chemical reactions with unprecedented specificity.

Acknowledgements. This work was supported by the Swiss National Science Foundation, grant number 200021-111902/1.

1   K. Nakanishi, N. Berova, and R. W. Woody, *Circular Dichroism Principles and Applications* (VCH Publishers, Inc., New York, 1994).

2   X. Xie and J. D. Simon, J. Am. Chem. Soc. **112**, 7802-7803 (1990).

3   T. Dartigalongue, C. Niezborala, and F. Hache, Phys. Chem. Chem. Phys. **9**(13), 1611-1615 (2007).

4   X. Xie and J. D. Simon, Review of Scientific Instruments **60**(8), 2614-2627 (1989).

5   Y. He, X. Cao, L. A. Nafie, and T. B. Freedman, " J. Am. Chem. Soc. **123**, 11320-11321 (2001).

Chemistry - Advanced Spectroscopy, Molecular Control, Hydrogen Bonding, Liquids and Interfaces

Automated 2D infrared and electronic spectroscopies using pulse shaping

Martin T. Zanni

[1] Department of Chemistry, University of Wisconsin-Madison, 1101 University Ave. Madison,
Wisconsin 53706-1396
E-mail: zanni@chem.wisc.edu

Abstract. We present a method for collecting 2D infrared and visible spectroscopies that uses a pulse shaper and a pump-probe beam geometry. This approach reduces the technical hurdles for implementing these techniques and makes many new experiments possible.

1. Introduction

An exciting new class of ultrafast laser techniques in physics, chemistry and biological research is multidimensional spectroscopy (nD). The most common versions are 2D infrared and electronic (visible) spectroscopies, called 2D-IR or 2D-E, respectively. While interpretation of the 2D spectra depends on the application, the two techniques use virtually identical sequences of femtosecond pulses and both transform the data in the same way. While having many applications, their wide implementation has been hindered by the extensive infrastructure and effort required to implement these methods. In this article and corresponding presentation, we present a methodology for measuring 2D optical spectroscopies using commonly available pulse shapers and a partially collinear pump-probe beam geometry (Fig. 1)[1,2]. This simple combination of technology and beam geometry eliminates many of the technical hurdles to implementing 2D-E spectroscopy, making it possible for many research groups to easily and quickly exploit this powerful spectroscopy in their research. In the accompanying talk, we demonstrate one application, which is using 2D-IR spectroscopy to follow the folding kinetics of amyloid fiber formation.

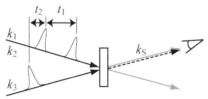

Fig. 1. Pump–probe beam geometry used to collect 2D-IR spectra. The "pump" pulses are created from a pulse shaper.

2. Experimental Methods

We demonstrate how collecting both 2D-IR and 2D-E spectroscopies can be implemented using pulse shaping and a collinear beam geometry. The collinear beam geometry consists of a pump and probe pulse, of which the pump pulse is controlled by a pulse shaper. For both experiments, we used standard Ti:Sapphire laser systems. For 2D-IR, mid-IR pulses are generated by difference frequency mixing (DFM) the signal and idler outputs of a BBO based OPA, while 800 nm pump and probe pulses are used in the 2D-E experiments. The pump pulses traverse through a standard 4-f geometry pulse shaper, which utilizes 130 pixels of a liquid crystal modulator for 2D-E and a 500 "pixel" Ge acoustooptic modulator for the 2D-IR experiments.[3,4] The pump and probe pulses then overlap in the sample, which is a Rb gas cell for 2D-E and a dilute solution of $W(CO)_6$ for demonstrating 2D-IR spectroscopy via pulse shaping. The probe is dispersed in a spectrometer with either a multichannel array detector or line camera.. We emphasize that our approach is applicable to many types of electronic and infrared absorbing samples besides these model systems.

3. Results and Discussion

There are many pulse shapes that could be used to collect 2D-IR and 2D-E spectra.[1] Shown in Fig. 2 is the 2D-E spectra generated by creating a pair of femtosecond pump pulses using the pulse shaper (Fig. 1). The time-domain data along t_1 is then Fourier transformed to give ω_1 (referred to as ω_{pump}). The spectrometer optically Fourier transforms the signal along t_3 to give ω_3 (ω_{probe}). The resulting 2D-E spectrum displays diagonal and cross peaks as expected.[2]

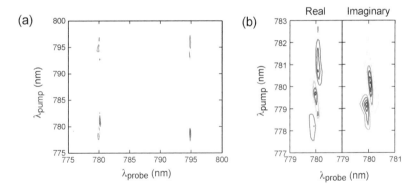

Fig. 2. (a) 2D-E spectrum of Rb vapor using a 130 pixel LCM pulse shaper. (b) Detail of the peak at $\lambda_{pump} = \lambda_{probe} = 780$ nm.

There are several nice features about using pulse shaping to collect 2D-IR and 2D-E spectroscopies. First, the non-rephasing and rephasing pathways are measured simultaneously, so that the resulting spectra are absorptive with the maximum possible frequency resolution. Second, since the time-zero of the pump pulses can be perfectly set and the probe pulse also serves to heterodyne the signal, the spectra are automatically phased, which can be a tedious process in other beam geometries. Third, high optical density samples can be studied for the same reason.[5] Fourth, phase cycling can be implemented to shift into the rotating frame, subtract background, and remove scatter. Fifth, in combination with polarization control, signal intensity can be improved.[6] Sixth, it is easy to implement new pulse sequences, which only requires programming not optical realignment. Seventh, data collection is very rapid since there are no moving parts. Finally, and perhaps most importantly, alignment is much more straightforward and robust than 2D spectroscopies using boxcar and other beam geometries, so that more time can be spent studying science.

4. Conclusions

In this paper, we introduce a simple way of implementing 2D-IR and 2D-E spectroscopies. By using a pulse shaper to generate collinear pulse pairs, spectra can be collected using a simple pump-probe beam geometry rather than with four individual optical paths in a four-wave mixing geometry. There are probably hundreds of research groups across the world that currently own a pulse shaper suitable to perform 2D-E spectroscopy.

Acknowledgements. We appreciate the collaboration of Niels Damrauer and his group in the demonstration of 2D-E spectroscopy via pulse shaping. This research is supported by the National Science Foundation and the Packard Foundation.

References

1 S.-H. Shim, D. B. Strasfeld, Yun L. Ling and M. T. Zanni, Proceedings of the National Academy of Sciences, 104, 14197 (2007).
2 E. M. Grumstrup, S.-H. Shim, M. A. Montgomery, N. H. Damrauer, and M. T. Zanni, Optics Express, 15, 16681 (2007).
3 Shim, D. B. Strasfeld, E. C. Fulmer, and M. T. Zanni, Opt. Lett. **31**, 838 (2006).
4 S.-H. Shim, D. B. Strasfeld, M. T. Zanni, Optics Express, 14, 13120 (2006).
5 W.V. Xiong, D. B. Strasfeld, S.-H. Shim and M. T. Zanni, Vibrational Spec., *Submitted.*
6 W. V. Xiong and M. T. Zanni, Optics Letters, In press.

Relaxation-Assisted Dual-Frequency Two-Dimensional Infrared Spectroscopy: Measuring Distances and Bond Connectivity

Igor V. Rubtsov[1], Sri Ram G. Naraharisetty[1], Christopher Keating[1], Beth A. McClure[2], Jeffrey J. Rack[2], Valeriy M. Kasyanenko[1]

[1] Department of Chemistry, Tulane University, New Orleans, LA 70118, USA
  E-mail: irubtsov@tulane.edu
[2] Department of Chemistry and Biochemistry, Ohio University, Athens, OH 45701, USA
  E-mail: rack@helios.phy.ohiou.edu

Abstract. Potential of a novel relaxation-assisted 2DIR spectroscopy method is demonstrated on several molecular systems, including model compounds, peptides, and transition metal complexes. The cross-peaks for modes separated by distances greater than 11Å can be easily detected using RA 2DIR. A correlation of the energy transport time (arrival time) with the intermode distance is shown in several molecular systems for a number of mode pairs.

Introduction

Implementation of weak IR modes as structural reporters for 2DIR is very attractive as it has a potential of measuring a large number of structural constraints for a molecule, but practical use of such modes is difficult as the direct mode coupling decays steeply with the intermode distance.[1,2] We have recently demonstrated a new approach, relaxation-assisted 2DIR method (RA 2DIR), that permits significant amplification of the observed cross-peaks.[3] The method utilizes vibrational energy (population) transport in molecules for generating stronger cross peaks. Fig. 1a shows the basic ideas behind the amplification. In short, vibrational relaxation and IVR processes cause energy to relax form the initially excited mode ω_1 and travel across the molecule. When the excess energy arrives to the site where the ω_2, the probed mode, is located the cross peak grows due to a strong coupling between the modes excited at the ω_2 site and the ω_2 mode itself. We have recently reported a 6-fold amplification for the CN/CO cross-peak amplitude in p-acetylbenzonitrile where the C≡N and C=O modes were separated by ca. 6.5Å.[3] It is not clear however that there will be always a monotonic correlation between the arrival time and the intermode distance. Nevertheless, if such correlation is found for a variety of fragments bridging the two modes in question it will be possible to use RA 2DIR for rapid assessment of the bond connectivity patterns, for assigning modes in molecules, and/or for measuring delocalization sizes for modes.

Results and Discussion

We report here on the correlation between the arrival time, the T-delay time (Fig. 1b) at which the cross peak reaches maximum, and intermode distance in several molecular systems. Fig. 1c shows dual-frequency RA 2DIR spectrum of PBN (see the structure in Fig. 1d) which contains over 25 cross peaks. Examples of the dependences of the cross-peak amplitude on the time delay, T, is shown in Fig. 1d for the C≡N/C=O and C≡N/amide-I (Am-I) modes. The C≡N/Am-I cross-peak demonstrates a dramatic enhancement (ca. 18 times) with the maximum reached at T = 10.6 ps. Similar dependences were measured for most of the cross-peaks seen in Fig. 1c. A correlation between the arrival time and the distance is observed for many mode pairs, which is schematically shown in Fig. 1d. The numbers in bold (Fig. 1d) represent the time needed for energy to cross the molecule, calculated from the given cross peaks arrival times. The additivity for the total time observed experimentally for many mode pairs reports on a monotonic correlation of the arrival time with the distance. However, for some modes probed, the correlation with the distance was not observed or obscured as, for example, for the mode at 1402 cm^{-1} (Fig. 1d). We attribute this effect to strong delocalization of such modes. The DFT calculations

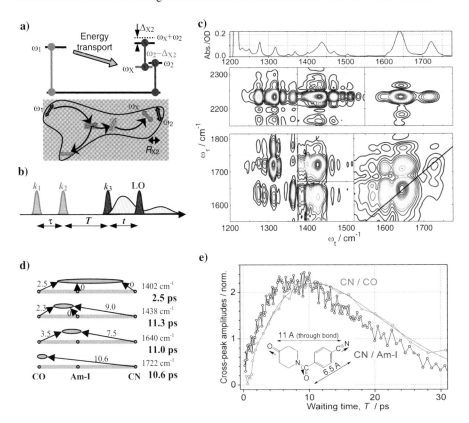

Fig. 1. a) Energy diagram and cartoon describing the RA cross-peak formation at (ω_2, ω_1) frequencies. Thin, curved arrows show schematically the energy relaxation and transport pathways. **b)** Experimental pulse sequence for the dual-frequency 2DIR measurements. **c)** Dual-frequency RA 2DIR absolute-value spectrum measured for the PBN compound (the structure in shown in Fig.1e inset). FTIR linear spectrum is shown in the upper panel. **d)** Schematic representation of the experimental results: thick grey horizontal lines symbolize the PBN molecular skeleton with CO and CN groups located at the left and at the right sides, respectively. The numbers by the arrows represent the arrival times in picoseconds for the modes shown at the right. The ovals show qualitatively the sizes of mode delocalization for the modes denoted at the right. The numbers in bold represent the time needed for energy to cross the molecule, calculated from the given cross peak arrival times. **e)** Normalized absolute-value amplitudes of the CN/Am-I and CN/CO cross-peaks for PBN compound as a function of time delay T. The structure of the PBN compound is shown in the inset.

with B3LYP hybrid functional performed with Gaussian 03 software (Gaussian Inc.) support this conclusion. We found that the modes in the fingerprint region, while difficult to assign, are very useful for dual-frequency measurements (Fig. 1c). The RA 2DIR method can either use a mode for connectivity measurements if the mode properties are determined or can help characterizing those properties.

Monotonic dependence of the energy arrival time with the distance is observed in Leu-d_{10}-methyl ester for the C-D (five transitions) and C=O mode pairs. It is shown that RA 2DIR data correlate well with the distances determined using the assignment of the CD transitions based on the anharmonic DFT calculations.

The RA 2DIR spectrum of Ru(bpy)$_2$OSO transition metal complex (Fig. 2) shows multiple cross peaks, including peaks between C=O, C-O, and phenyl stretches and the C-H bending modes located on the methylsufinylbenzoate ligand. These cross peaks show no enhancement *via* RA 2DIR, which reports on delocalization of these modes over the ligand caused by a small size of the ligand and its structure.

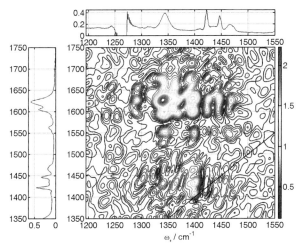

Fig. 2. RA 2DIR spectrum of ruthenium bisbipyridyl ortho-methylsufinylbenzoate (Ru(bpy)$_2$OSO) in chloroform measured at waiting time, T, of 15 ps. The linear IR spectrum is shown in the attached panels.

In summary, the large amplification of the cross-peak amplitude (ca. 18-fold) shown in this work enhances the applicability of 2DIR spectroscopy by permitting the long–range measurements and measurements using weak IR modes. With a proper calibration, the RA 2DIR method is expected to permit speedy assessments of the bond connectivity patterns and distances, which indicates an analytical potential of the method. A correlation of the arrival time and the intermode distance has been experimentally detected on multiple mode pairs in several compounds.

Acknowledgements. I.V.R. gratefully acknowledges support from the NSF (award number 0750415) and the Louisiana Board of Regents (RCS grant).

1   D.V. Kurochkin, S.G. Naraharisetty, and I.V. Rubtsov, J. Phys. Chem. A **109**, 10799, 2005.

2   S. G. Naraharisetty, D.V. Kurochkin, and I. V. Rubtsov, Chem. Phys. Lett. **437**, 262, 2007.

3   D. V. Kurochkin, S. G. Naraharisetty, and I. V. Rubtsov, Proc. Natl. Acad. Sci. U.S.A. **104**, 14209, 2007.

Triggered-exchange Two-dimensional Infrared Spectroscopy of Metal Carbonyl Photodissociation Dynamics

Carlos R. Baiz, Matthew J. Nee, Robert McCanne and Kevin J. Kubarych

^{1}Department of Chemistry, 930 N. University Ave., University of Michigan, Ann Arbor, MI
40109 USA
E-mail: kubarych@umich.edu

Abstract. We present an ultrafast study of the dissociation dynamics of a metal carbonyl complex using transient Fourier transform 2DIR spectroscopy.

Introduction

Multidimensional IR spectroscopy offers a structurally-specific dynamical probe of ultrafast condensed-phase processes[1]. Triggered-exchange 2DIR spectroscopy (TE-2DIR) provides a direct correlation between reactant and product vibrational modes in an optically-triggered ultrafast chemical process, while avoiding ambiguities of one-dimensional transient absorption techniques[2].Similarly, transient 2DIR (t-2DIR) offers dynamical information on transient species beyond one-dimensional pump-probe methods[3]. Photodissociation dynamics of $Mn_2(CO)_{10}$ have been previously studied with one-dimensional pump-probe methods where UV excitation is followed by either IR or visible probes[4,5]. Interpretation of these spectra is complicated by several processes that occur upon excitation and contribute to changes in transient absorption. Excitation at 400 nm primarily breaks the Mn–Mn bond yielding $2Mn(CO)_5$. A full understanding of the photochemistry of $Mn_2(CO)_{10}$ requires characterization of the excited electronic state lifetime, photodissociation quantum yield, geminate recombination rate and yield, and carbonyl vibrational relaxation pathways and rates. The results presented here are the first to embed non-equilibrium electronic excitation within the three-pulse Fourier transform approach to 2DIR spectroscopy, offering the advantages of spectral resolution, as well as temporal dynamic range and background free detection.

Experimental Methods

Triggered-exchange and transient 2DIR spectra of 6-mM $Mn_2(CO)_{10}$ in cyclohexane [FTIR shown in Fig. 1(a).] were obtained at different t_2 delays [Fig. 1(b)]. Briefly, a dual OPA/DFG generates tuneable mid-IR pulses (1 µJ, ~100 fs pulse width, 100 cm^{-1} FWHM), where the first two pulses are obtained from one OPA/DFG, and the third and reference pulses from the second OPA/DFG. The output from a regeneratively-amplified Ti:Sapphire laser was frequency-doubled in a 0.4-mm BBO crystal to obtain 400-nm UV pulses (>30 µJ, ~60 fs pulse width, 1 kHz repetition rate) and chopped to 500 Hz to obtain difference spectra. A wire-guided liquid jet[6] provides adequate linear flow velocity and excellent film stability (<60-nm RMS fluctuations) was used in order to have a continuous flow of sample. The heterodyned signal electric field was measured by chirped-pulse upconversion[7] employing a 1340x100-pixel silicon CCD detector. The data are plotted as the difference of absolute value

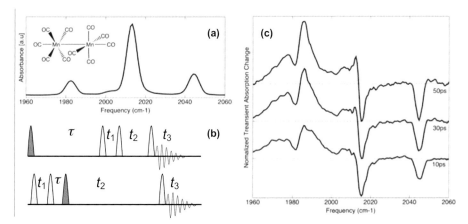

Fig. 1.(a) Linear FTIR spectrum of Mn2(CO)$_{10}$ in cyclohexane. (b) Triggered-exchange 2DIR pulse sequence. The first two pulses label the reactant vibrations, the third pulse, ET (400 nm), triggers the photodissociation and the fourth pulse causes IR emission from the products. All pulses are vertically polarized. (c) Normalized transient IR absorption spectra of Mn2(CO)$_{10}$ at different pump-probe delay times following 400-nm excitation.

rephasing ($k_s = -k_1 + k_2 + k_3$) spectra with and without UV pump pulse excitation. This technique enables acquisition of a full triggered-exchange 2DIR correlation spectrum in less than 15 seconds. One-dimensional transient absorption spectra [Fig. 1(c)] are obtained by recording the transmission of an IR probe following UV excitation.

Results and Discussion

The transient absorption spectrum shows three transient bleaches at 2045, 2015, and 1982 cm^{-1}, corresponding to the depletion of ground-state Mn$_2$(CO)$_{10}$, and a broad transient absorption band centred at 1985 cm^{-1} attributed to the formation of Mn(CO)$_5$. Spectral narrowing of this transient absorption feature at longer delays is caused by vibrational energy relaxation of the product[3]. The transient absorption peak at 2010 cm^{-1} exhibits a large growth between t_2=10 ps and 50 ps. We attribute this peak to anharmonically-shifted hot-band transitions of the 2015 cm^{-1} mode in Mn$_2$(CO)$_{10}$. Although this feature may be due either to vibrationally excited photoproduct or to hot reactant following geminate rebinding, the observed growth of the signal (as opposed to its decay) suggests a rebinding origin.

A transient 2DIR spectrum [Fig. 2(a)] with $\tau = 25$ ps and $t_2 = 1$ ps shows a spectrally broad induced 2D feature centred on the 1985-cm^{-1} photoproduct absorption and overlaps somewhat the bleach of the ground state. A t_2-dependent experiment reveals the vibrational relaxation of the photoproduct, and this rate is found to decrease with increased delay τ. Since the photoproduct cools with increasing τ, our results are in agreement with the expected trend of faster relaxation at higher temperature. Triggered-exchange 2DIR for $\tau = 1$ ps, and $t_2 = 25$ ps is shown in Fig. 2(b). The most prominent feature of the data is a pronounced induced peak corresponding to excitation at the 2015-cm^{-1} reactant band and detection at the 1985-cm^{-1} product band. A weak positive cross-peak is also observed for excitation at 1982 cm^{-1} and emission at 1985 cm^{-1}. This cross peak suggests that reactant excitation remains in the carbonyl

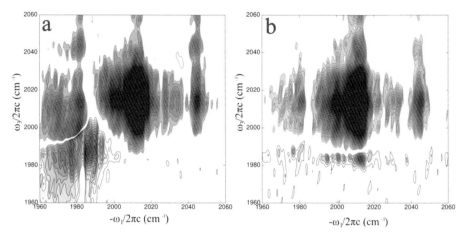

Fig. 2.(a) Transient 2DIR spectrum of $Mn_2(CO)_{10}$ at $\tau = 25$ ps, $t_2 = 1$ ps.(b) Triggered-exchange 2DIR spectrum at $t_2 = 25$ ps. Positive peaks are indicated in solid contours whereas negative peaks (bleaches) are dashed.

stretching band due to weak coupling to low-frequency modes or to the solvent. Due to the spectral overlap of reactant and product spectra, a substantial bleach signal obscures a full resolution of the triggered-exchange peak. Nevertheless, this data indicate the feasibility of Fourier transform TE-2DIR to track the fastest chemical reaction dynamics with spectral resolution limited by the molecule and temporal resolution limited by the cross-correlation of the IR and trigger pulses.

Conclusions

Triggered-exchange Fourier transform 2DIR spectroscopy offers valuable insight into $Mn_2(CO)_{10}$ dissociation dynamics beyond 1D pump-probe by directly correlating reactant and product vibrational modes. Geminate rebinding appears to significantly contribute to hot-band transitions of the central peak in $Mn_2(CO)_{10}$ due to energy redistribution into normal modes. Although several spectrally and temporally overlapping contributions to the photodissociation of $Mn_2(CO)_{10}$ make transient spectra difficult to interpret, TE-2DIR offers a general approach to disentangle competing dynamical processes. Transient 2DIR enables the determination of vibrational relaxation rates in non-equilibrium systems. Future experiments will deconvolve these processes to obtain a better understanding of the reaction dynamics of this and other systems, with particular attention to solvent effects and the role of vibrational coherence.

1. M. Cho, Chem. Rev. 108, 1331 (2008).
2. J. Bredenbeck, J. Helbing, K. Nienhaus, G. U. Nienhaus, P. Hamm, Proc. Natl. Acad. Sci. USA, 104, 14243 (2007).
3. C. Kolano, J. Helbing, M. Kozinski, W. Sander, P. Hamm, Nature, 444, 269 (2006).
4. D. A. Steinhurst, A. P. Baronavski, J. C. Owrutsky, Chem. Phys. Lett. 361, 513 (2002).
5. A. Waldman, S. Ruhman, S. Shaik, G. N. Sastry, Chem. Phys. Lett. 230, 110 (1994).
6. M .J. Tauber, R. A. Mathies, X. Chen, S. E. Bradforth, Rev. Sci. Instrum. 74, 4958 (2003).
7. M.J. Nee, R. McCanne, K. J.Kubarych, M. Joffre, Opt. Lett. 32, 713 (2007).

Observation of Quantum Coherence in Light-Harvesting Complex II by Two-Dimensional Electronic Spectroscopy

Tessa R. Calhoun[1], Naomi S. Ginsberg[1], Gabriela S. Schlau-Cohen[1], Yuan-Chung Cheng[1], Matteo Ballottari[2], Roberto Bassi[2], and Graham R. Fleming[1]

[1] Department of Chemistry, University of California, Berkeley and Physical Biosciences Division, Lawrence Berkeley National Laboratory, Berkeley CA 94720, USA
  E-mail: GRFleming@lbl.gov
[2] Department of Science and Technology, University of Verona, 37134 Verona, Italy

Abstract. Two-dimensional Fourier transform electronic spectroscopy is employed to investigate quantum beating in the major light-harvesting complex II. Long-lived excitonic coherence is observed for the first time in a higher plant system between two different types of chlorophyll molecules.

Introduction

Photosynthesis is highly evolved in its ability to transfer energy with near-unity quantum efficiency through a matrix of pigment-protein complexes allowing almost every photon absorbed by the light-harvesting antennae to be transferred to the reaction center [1]. Although the mechanisms by which nature achieves this feat are not fully understood, recent studies on the Fenna-Matthews-Olson complex from the *Chlorobium tepidium* green sulfur bacterium and the bacterial reaction center of *Rhodobacter Sphaeroides* indicate that quantum excitonic coherence may be a fundamental contributor [2,3]. Furthermore, the coherence examined in these experiments is surprisingly long-lived, suggesting that the protein support-structure enveloping the chromophores plays an active role in preserving superpositions of the excitons within it [2,3]. This phenomenon, however, has not been observed in any complex from higher plants, and neither of these bacterial systems plays a dominant energy transfer role in photosynthesis. If excitonic coherence is indeed a vital component to the high quantum efficiency of photosynthesis, it must be present in antennae complexes.

Over half of all chlorophylls, the main light harvesting molecule in green plants and algae, reside in the light-harvesting complex II (LHCII) of Photosystem II. LHCII is far more complex than the previous systems in which excitonic coherence has been observed; it contains a total of 42 excitonic states from the lowest energetic, Q_y, transitions of two different chlorophyll (Chl) variants, Chl a and b. We explore excitonic coherence in LHCII with 2D electronic spectroscopy because it has proven valuable in studying similar complex systems [4] as well as in investigating excitonic coherence [2].

Experimental Methods

The theory and apparatus used for the 2D FT electronic spectroscopy have been described in detail elsewhere [5]. Briefly, three successive pulses, generated from a non-collinear optical parametric amplifier pumped by 3.4 kHz 800 nm pulses produced from a home-built Ti:sapphire oscillator/regenerative amplifier system, are incident on the sample, the signal from which is then spectrally resolved and heterodyne-detected on a CCD camera. The time delay between pulses 1 and 2, the coherence time, is varied for each population time, the delay between pulses 2 and 3. The resulting spectrum for each population time is then Fourier-transformed along the coherence time axis to produce the final 2D spectrum. The 18 fs pulses used in this experiment were centered at 640 nm with a FWHM of ~90 nm in order to excite the Q_y transitions of all of the Chl a and Chl b molecules (645 – 680 nm) as well as their higher energy vibronic tails. The total pulse energy incident on the sample was ~30 nJ. The population times collected ranged between 0 and 500 fs in 10 fs increments, and all experiments were performed at 77 K. Recombinant LHCII from *Arabidopsis thaliana* in a solution containing pH=7.6 Hepes buffer and n-dodecyl alpha-D-maltoside detergent was combined 30/70 (v/v) with glycerol in a 200 μm quartz cell.

Results and Discussion

Electronic coherence manifests itself in 2D experiments as an amplitude oscillation of peaks on and off of the diagonal. Due to the complexity of the electronic structure of LHCII, it is imperative for analysis that beating frequencies due to peaks on the diagonal be isolated from interfering with those of nearby cross peaks. Recent theoretical simulations have illustrated that this can be achieved through analysis of the nonrephasing contribution to the 2D spectrum alone [6]. Slices of the nonrephasing spectra along the diagonal for each population time are shown in Figure 1. Oscillations can clearly be seen in both Chl a features at ~14,750 cm^{-1} and ~15,000 cm^{-1} and, though less intense, in the peak at ~15,500 cm^{-1} due to Chl b. The coherence is long-lived with the beating persisting beyond the extent of the experiment at 500 fs. Furthermore, the peak amplitude oscillates only as a function of time, not exciton energy, indicating that the beating is due to electronic, not phonon, coupling. These oscillations can allow the excited energy populations to transfer more rapidly among all of the excitons than they would through a purely dissipative mechanism.

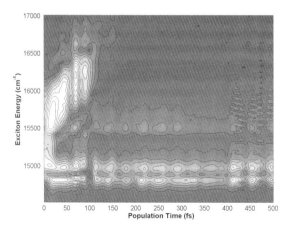

Fig. 1. Diagonal slices from the nonrephasing contribution to the 2D signal for population times from 0 to 500 fs. A spline interpolation was applied along the population time axis which is shown as a function of exciton energy, corresponding to the diagonal frequency in the original 2D spectra.

Conclusions

Our 2D electronic spectroscopy studies on LHCII reveal excitonic coherence in both Chl *a* and Chl *b*, exemplifying the breadth of this phenomenon in photosynthetic complexes. The coherence observed is also long-lived, living beyond the 500 fs of this experiment. Further analysis of the emergent beat frequencies, obtained by performing a Fourier-transform along the population time axis, is underway. A display of the data in this format may, in principle, also resolve individual exciton energy levels.

Acknowledgements.   The authors thank Rienk van Grondelle for valuable discussions. This work was supported by the Office of Basic Energy Sciences, Chemical Sciences Division, U.S. Department of Energy (contract DE-AC03-76SF000098). N. S. G. also thanks the Lawrence Berkeley National Laboratory Glenn T. Seaborg postdoctoral fellowship for support.

1   R. E. Blankenship, *Molecular Mechanisms of Photosynthesis*, (Blackwell Science, Oxford/Malden, 2002)

2   G. S. Engel, T. R. Calhoun, E. L. Read, T.-K. Ahn, T. Mančal, Y.-C. Cheng, R. E. Blankenship, and G. R. Fleming, "Evidence for wavelike energy transfer through quantum coherence in photosynthetic systems," Nature **446**, 782 (2007).

3   H. Lee, Y.-C. Cheng, and G. R. Fleming, "Coherence Dynamics in Photosynthesis: Protein Protection of Excitonic Coherence," Science **316**, 1462 (2007).

4   D. Zigmantas, E. L. Read, T. Mančal, T. Brixner, A. T. Gardiner, R. J. Cogdell, and G. R. Fleming, "Two-dimensional electronic spectroscopy of the B800-B820 light-harvesting complex," PNAS **103**, 12672 (2006).

5   T. Brixner, T. Mančal, I. V. Stiopkin, and G. R. Fleming, "Phase-stabilized two-dimensional electronic spectroscopy," J. Chem. Phys. **121**, 4221 (2004).

6   Y.-C. Cheng and G. R. Fleming, "Coherence Quantum Beats in Two-Dimensional Electronic Spectroscopy," J. Phys. Chem. A **112**, 4254 (2008).

Vibrational Beating in Two-Dimensional Electronic Spectra

Alexandra Nemeth[1], Franz Milota[1], Tomáš Mančal[2], Vladimír Lukeš[3],
Harald F. Kauffmann[1,4], and Jaroslav Sperling[1]

[1] Department of Physical Chemistry, University of Vienna, Währingerstraße 42, 1090 Vienna, Austria; E-mail: alexandra.nemeth@univie.ac.at
[2] Institute of Physics, Faculty of Mathematics and Physics, Charles University, Ke Karlovu 5, 121 16 Prague, Czech Republic
[3] Department of Chemical Physics, Slovak Technical University, Radlinského 9, 91237 Bratislava, Slovakia
[4] Ultrafast Dynamics Group, Faculty of Physics, Vienna University of Technology, Wiedner Hauptstraße 8-10, 1040 Vienna, Austria

Abstract. We trace vibrational wavepacket motion in two-dimensional electronic spectra of a two-level electronic system. The vibronic evolution induces a periodic beating pattern of the diagonal-to-antidiagonal peak width ratio in the absorptive signal part and a periodic tilt of the nodal line in the dispersive signal part. These modulations can be assigned to periodic modulations of the relative amplitudes of rephasing and non-rephasing contributions to the signal.

Heterodyne detected two-dimensional (2D) photon-echo (PE) spectroscopy is a four-wave-mixing (FWM) technique tracing the correlation between electronic transition frequencies and the effect of population and coherence evolution on this correlation. Three pulses are incident on the sample and the 2D-PE spectra are obtained by Fourier transform with respect to t_1 (time-separation between the first two pulses) and the detection time t_3, leading to a 2D plot in the conjugated frequencies ω_1 and ω_3. An additional time-delay (t_2) between the two observation windows allows to follow the dynamics of populations and/or coherences created by the first two interactions. Recently, effects of electronic coherence in electronic multi-level systems have received considerable experimental and theoretical attention [1-3]. It is however not established

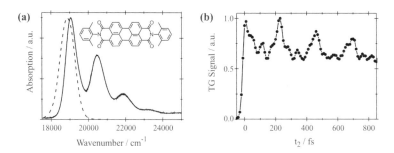

Figure 1: (a) Linear absorption spectrum of PERY (structure shown in the inset) in toluene, along with the laser pulse spectrum (dashed line). (b) Normalized transient grating signal of PERY in toluene.

which kind of spectral signatures are characteristic for the coherent evolution of vibrational states as opposed to electronic coherences, and whether these two types can be clearly distinguished. In this study we investigate coherent vibrational contributions to 2D-PE spectra on a simple model system with a two-level electronic structure [4].

PERY [5] (cf. Fig. 1a) is in many respects a well-suited candidate for our purpose. Quantum chemical calculations (on the TDDFT level) predict only one excited state with significant oscillator strength. The small apparent Stokes shift indicates solute-solvent interactions to be weak, suggesting that vibrational wavepackets are likely to contribute to time-domain signals [6]. As illustrated by the transient grating signal shown in Fig. 1b, the one-dimensional FWM signals of PERY are strongly modulated by a low-frequency mode of roughly 140 cm^{-1}, corresponding to a 240 fs period [6]. In our experiments, we record electronic 2D spectra along the vibrational beating pattern employing sub-20 fs pulses centered at 18800 cm^{-1} [7]. The PE experiments are realized with a diffractive optics based set-up [8,9]. The signal field is reconstructed for a particular setting of time delays by spectral interferometry with a local oscillator pulse. The complex 2D-PE spectra are dissected into absorptive and dispersive parts by projecting (phasing) onto frequency resolved pump-probe signals [10].

Fig. 2 shows the experimental and simulated absorptive and dispersive 2D signal parts at two selected t_2 time-delays. The absorptive part features only positive contributions, according to the (ground-state bleach and stimulated emission) Liouville-space pathways that contribute to the echo signal of a two-level absorber. The ellipticity of the absorptive signal peak shows a pronounced modulation, arising mainly from an antidiagonal breathing. In the dispersive signal part the zero-crossing between positive and negative contributions experiences a periodic change in its orientation. The diagonal-to-antidiagonal peak width ratio extracted from a t_2-sequence recorded in 50 fs steps is shown in Fig. 3a. The angle of the nodal line with respect to the ω_1-axis in the dispersive signal part is shown in Fig. 3b. Both transients follow the modulation period observed in the one-dimensional FWM signals (cf. Fig. 1b).

The oscillations in the absorptive and the dispersive signal part can be attributed to periodic enhancement and decrement in the rephasing and non-rephasing signal parts,

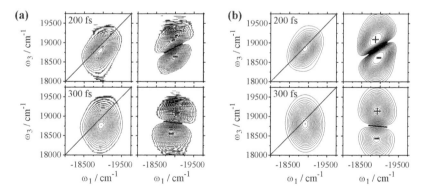

Figure 2: Experimental (a) and simulated (b) 2D electronic spectra of PERY in toluene, dissected into absorptive (right) and dispersive (left) part at t_2-delays of 200 and 300 fs.

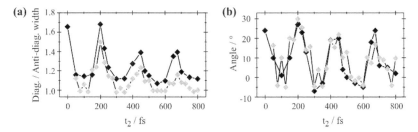

Figure 3: (a) t_2-dependency of the diagonal-to-antidiagonal peak width ratio of the absorptive 2D peak. (b) Angle of the nodal line in the dispersive 2D signal with respect to the ω_1-axis. Black symbols refer to experimental data, gray symbols to calculations.

with the rephasing part being strongest in the maxima of the one-dimensional FWM signals and the non-rephasing part being strongest in the minima of the one-dimensional FWM signals. Rephasing and non-rephasing contributions refer to pulse orderings 1-2-3 and 2-1-3, respectively. In order to obtain purely absorptive lineshapes, rephasing and non-rephasing contributions have to be equally weighted [11]. The observed oscillation of the 2D lineshape is qualitatively similar to the one observed for the coherent evolution of electronic states [2], but follows the low-frequency vibrational beats in our case. To support our interpretation of the data, we perform simulations of linear and non-linear spectra in the impulsive limit, applying standard semi-classical second order cumulant expansion treatment of the electron-phonon interaction [12]. Both the solvent and the intra-molecular oscillatory modes of PERY are treated within the Brownian oscillator model. This simple treatment reproduces the experimental trends with very good agreement (cf. Fig. 2 and 3). To the best of our knowledge, these are the first experimental signatures of vibrational coherence in 2D electronic spectra.

Acknowledgements. Support by the Austrian Science Foundation (FWF) within the projects No. P18233 and F016/18 ADLIS is gratefully acknowledged. T. M. was supported by the Czech Science Foundation (GACR) grant 202/07/P278 and by the research plan MSM0021620835 of the Ministry of Education of Czech Republic.

1 G. S. Engel *et al.*, Nature, Vol. 446, 782, 2007.
2 A. V. Pisliakov, T. Mančal, and G. R. Fleming, J. Chem. Phys., Vol. 124, 234505, 2006.
3 D. Egorova, M. F. Gelin, and W. Domcke, J. Chem. Phys., Vol. 126, 74314, 2007.
4 A. Nemeth *et al.*, Chem. Phys. Lett. (2008), doi:10.1016/j.cplett.2008.05.057
5 N,N'-bis(2,6-dimethylphenyl)perylene-3,4,9,10-tetracarboxylicdiimide
6 D. S. Larsen, K. Ohta, and G. R. Fleming, J. Chem. Phys., Vol. 111, 8970, 1999.
7 F. Milota *et al.*, Ultrafast Phenomena XV, Springer Series in Chemical Physics 88, 2007.
8 M.L. Cowan, J. P. Ogilvie, and R. J. D. Miller, Chem. Phys. Lett., Vol. 386, 184, 2004.
9 T. Brixner, T. Mančal, I. V. Stiopkin, and G. R. Fleming, J. Chem. Phys., 121, 4221, 2004.
10 D. M. Jonas, Annu. Rev. Phys. Chem., Vol. 54, 425, 2003
11 M. Khalil, N. Demirdöven, and A. Tokmakoff, Phys. Rev. Lett., Vol. 90, 047401, 2003.
12 S. Mukamel, Principles of nonlinear optical spectroscopy, Oxford University Press, Oxford 1995.

Double-Quantum Coherence Spectroscopy of Chromophore Aggregates

Darius Abramavicius and Shaul Mukamel

Chemistry Department, University of California Irvine, California, USA
E-mail: smukamel@uci.edu

Abstract. Double-quantum four wave mixing signals contain signatures of exciton correlations and can distinguish between couplings arising from transition- and excited-state charge densities. The former (resonant couplings) determine single-exciton properties, while the latter affect the two-exciton manifold.

Photosynthetic antennas consist of coupled pigments embedded in proteins [1]. Their optical properties are strongly affected by the couplings between transition-charge densities of different pigments. These couplings control single-exciton absorption, exciton delocalization and energy transport.

The Frenkel exciton model is widely used to simulate the optical signals of coupled chromophores [1]. The model is based on the Heitler-London approximation, where the wavefunction of the aggregate is constructed out of direct products of wavefunctions of isolated molecules. The Hamiltonian of a complex in this basis reads [1-3]

$$\hat{H}^{(e)} = \sum_m \varepsilon_m \hat{B}_m^\dagger \hat{B}_m + \sum_{m,n}^{m \neq n} J_{mn} \hat{B}_m^\dagger \hat{B}_n + \frac{1}{2} \sum_{m,n}^{m \neq n} K_{mn} \hat{B}_m^\dagger \hat{B}_n^\dagger \hat{B}_m \hat{B}_n, \tag{1}$$

where $\hat{B}_m^\dagger$ is the exciton creation ($\hat{B}_m$ - annihilation) Pauli operators of two-level systems, ε_m is the excitation energy of chromophore m, J_{mn} is the resonant excitonic coupling between transition charge densities of two chromophores and K_{mn} coupling is given by Coulomb interaction between excited state charge densities of two chromophores. In this basis the Hamiltonian is block-diagonal with single-, double-, and higher-exciton manifolds. In the single-exciton manifold, the diagonal elements of the Hamiltonian are the transition energies of individual molecules, ε_m and the off diagonal elements reflect inter-molecular interactions J. Both types of couplings affect the double-exciton manifold. A state with two chromophores m and n excited, (mn), is characterized by the diagonal element, its energy, $\varepsilon_m + \varepsilon_n + K_{mn}$. The off-diagonal elements are induced by J coupling. The J and K couplings thus have different physical significance: J couplings are all off-diagonal elements of the Hamiltonian, mixing exciton states; K couplings represent the shifts of the double-exciton states (also known as bi-exciton binding energies).

In a dipole approximation for the charge densities the couplings are given by:

$$J_{mn} = \frac{1}{4\pi\varepsilon\varepsilon_0} \left(\frac{\boldsymbol{\mu}_m \cdot \boldsymbol{\mu}_n}{|\boldsymbol{R}_{mn}|^3} - 3 \frac{(\boldsymbol{\mu}_m \cdot \boldsymbol{R}_{mn})(\boldsymbol{\mu}_n \cdot \boldsymbol{R}_{mn})}{|\boldsymbol{R}_{mn}|^5} \right), \tag{2}$$

$$K_{mn} = \frac{1}{4\pi\varepsilon\varepsilon_0} \left(\frac{\boldsymbol{d}_m \cdot \boldsymbol{d}_n}{|\boldsymbol{R}_{mn}|^3} - 3 \frac{(\boldsymbol{d}_m \cdot \boldsymbol{R}_{mn})(\boldsymbol{d}_n \cdot \boldsymbol{R}_{mn})}{|\boldsymbol{R}_{mn}|^5} \right), \tag{3}$$

where $\boldsymbol{\mu}_n$ is a transition dipole moment of chromophore n and $\boldsymbol{d}_n$ a corresponding dipole moment in the excited state. Most simulations only include the J couplings: the permanent dipole moments are neglected. However, electronic structure calculations of Bacteriochlorophyll molecules (BChls), which are the main pigments in photosynthetic complexes, show that their excited state dipole is comparable with the transition dipole [4]. Additional K couplings then affect double-exciton energies: they dominate for pigments with very different energies (like in heteronuclear NMR), where the J couplings are not effective.

To probe the K couplings we propose to use the double-quantum coherence technique $\mathbf{k}_{III} = \mathbf{k}_1 + \mathbf{k}_2 - \mathbf{k}_3$ (Fig. 1C) [5,8]. The signal is described by two Feynman diagrams (Fig. 1D). Both involve excited state absorption and monitor the double-exciton resonances during delay t_2; during t_3 single- and double-exciton resonances are mixed. The two diagrams have opposite signs and cancel at $t_3 = 0$. The signal is thus generated by exciton correlations when the double-exciton energies are not additive, i. e. $\varepsilon_f \neq \varepsilon_e + \varepsilon'_e$. Fourier transform with respect to $t_2 \to \Omega_2$ and $t_3 \to \Omega_3$ generates the two-dimensional signal, directly displaying double-exciton resonances along Ω_2. Since during t_1 the evolution of single-exciton coherences is equivalent to phase-rotation, we set $t_1 = 0$. Simulations were based on the response function formalism (Spectron code) as described in ref [6].

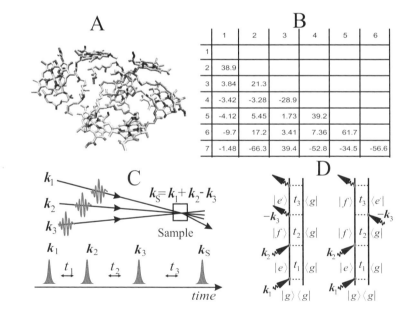

	1	2	3	4	5	6
1						
2	38.9					
3	3.84	21.3				
4	-3.42	-3.28	-28.9			
5	-4.12	5.45	1.73	39.2		
6	-9.7	17.2	3.41	7.36	61.7	
7	-1.48	-66.3	39.4	-52.8	-34.5	-56.6

Fig. 1. A: configuration of BChls in photosynthetic Fenna-Matthews-Olson complex; B: estimated $K = K_0$ couplings of the complex in [cm^{-1}]; C: scheme of the coherent double-quantum technique; D: its Feinman diagrams of the contributing Liouville space pathways.

We apply this technique to the Fenna-Matthews-Olson (FMO) complex of a green sulfur bacteria [1]. This is a trimer of small noninteracting identical subunits, each consisting of seven BChls (see Fig. 1A). A single unit is usually sufficient for simulations of optical signals. The broad absorption band of the complex extends from 11800 cm^{-1} to 12800 cm^{-1}. We use the single-exciton Hamiltonian of ref. [7]. K couplings were calculated by Eq. (3) using dipole moments obtained from charge distributions of ref. [4]; the permanent dipole of a BChl molecule is perpendicular to its N atoms plane, pointing towards Mg; the dipole amplitude was set to 4.9 D. The calculated K couplings (a set K_0) are given in Fig. 1B.

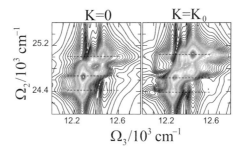

Fig. 2. Simulation results of double-quantum signal of Fenna-Matthews-Olson complex using $K = 0$ and $K = K_0$ as of Fig. 1B.

Fig. 2 compares the two-dimensional signal for $K = K_0$ (Fig. 1B *right*) and for K switched off (*left*). Three main double-exciton resonances along Ω_2 can be observed at $K = 0$ as marked by dashed lines. The peak pattern along both Ω_2 and Ω_3 reveals and elaborate exciton resonances. The $K = K_0$ simulations show a similar peak pattern, while some double-exciton signatures are now blue-shifted as can be seen from the dashed lines. Peak intensities change significantly for $\Omega_3 > 12400 cm^{-1}$.

Recently performed two dimensional photon echo experiments revealed single-exciton dynamics and relaxation pathways [7] This dynamics is governed by the J couplings and is not affected by K. Our simulations address a different class of phenomena, by using a different technique which shows a high sensitivity to K couplings, and thus provides a direct probe of excited state permanent dipole moments. Excited state dipole moments can be extracted by comparing the clear and well-resolved peak pattern with experiment. Both intermolecular J and K couplings can be obtained using a combination of different two-dimensional techniques: $k_I = -k_1 + k_2 + k_3$ is most sensitive to J couplings [7], whereas the presented double-quantum technique is a direct probe of K [5]. By combining these techniques the entire exciton Hamiltonian can be derived from experiment.

Acknowledgements. The research was supported by the National Science Foundation (CHE-0745892) and the National Institute of Health (GM59230).

1   H. van Amerogen, L. Valkunas, R. van Grondelle, Photosynthetic Excitons, World Scientific, Singapore, 2000.
2   V. Chernyak, W. M. Zhang, S. Mukamel, J. Chem. Phys. v 109, 9587, 1998.
3   S. Mukamel, Principles of Nonlinear Optical Spectrscopy, Oxford University Press, New York, 1995.
4   M. E. Madjet, A. Abdurahman, T. Renger, J. Phys. Chem. B, v 110, 17268, 2006.
5   Z. Li, D. Abramavicius, S. Mukamel, J. Am. Chem . Soc. v 130, 3509, 2008.
6   D. Abramavicius, L. Valkunas, S. Mukamel, European. Phys. Lett., v 80, 17005, 2007.
7   T. Brixner, J. Stenger, H. M. Vaswani, M. Cho, R. E. Blankenship, G. R. Fleming, Nature, v 434, 625, 2005.
8   D. Abramavicius, D. V. Voronine, S. Mukamel, Proc. Nat. Acad. Sci. USA, (in press).

414

Chain Length Dependence of Two-Dimensional Infrared Spectral Pattern Characteristic to 3_{10}-Helix Peptides

Hiroaki Maekawa[1], Fernando Formaggio[2], Claudio Toniolo[2], and Nien-Hui Ge[1]

[1] Department of Chemistry, University of California, Irvine, CA 92967-2025, USA
E-mail: nhge@uci.edu
[2] Institute of Biomolecular Chemistry, CNR, Padova Unit, Department of Chemistry,
University of Padova 35131 Padova, Italy
E-mail: claudio.toniolo@unipd.it

Abstract. Two-dimensional infrared spectra of Z-(Aib)$_n$-OtBu (n = 3, 5, 8 and 10) were measured to investigate how they depend on the peptide chain length. The onset of the 3_{10}-helical spectral signature appears to occur at the pentapeptide.

Introduction

Two-dimensional infrared (2D IR) spectroscopy, with its high sensitivity to molecular structure, has made considerable progress in elucidating peptide and protein conformations. Recent studies established the relationship between polypeptide secondary structures and 2D IR spectral patterns of the amide-I modes: a Z-shaped pattern appears in the 2D absorptive spectra of the antiparallel β-sheet [1], and a doublet cross-peak pattern distinctively shows up for the 3_{10}-helix, which can serve to differentiate it from the α-helix [2, 3]. To apply these structure-spectrum relationships as standard and practical criterions to assign peptide secondary structures, it is important to experimentally investigate how many amino acid residues are required to exhibit the characteristic 2D spectral patterns. The chain length dependence would also provide a foundation for interpreting time-resolved 2D IR spectra and understanding secondary structure formation in protein folding.

In this study, we report 2D IR spectroscopy of Aib homopeptides, Z-(Aib)$_n$-OtBu (Z, benzyloxycarbonyl; Aib, α-aminoisobutyric acid; OtBu, $tert$-butoxy; n = 3, 5, 8 and 10), in CDCl$_3$ and examine how the spectral patterns evolve with the chain length. The structural rigidity of Aib and its high helical propensity make Aib-containing oligopeptides attractive models for developing and refining experimental and theoretical approaches to peptide conformational analysis. The origin of the chain length dependence of the 2D IR spectra is discussed in relation with the size of coupled vibrational networks, the strengths of couplings, and the site dependence of local mode frequencies.

Experimental Methods and Simulations

All third-order 2D IR experiments were performed using 100-fs, 6 μm IR pulses from our home-built mid-IR source. Rephasing (R) and nonrephasing (NR) pulse sequences were utilized. The signal field was measured by spectral interferometry and Fourier transformed along the scanning delay time to give 2D spectra in (ω_t, ω_τ). The polarization configuration of the excitation pulses and the signal was set to parallel.

Measurements also were taken under the double-crossed polarization to suppress the diagonal peaks [4]. The Aib homopeptides were synthesized and characterized as described previously [5]. The peptide concentration was below 10 mM and the path length of the sample cell was 250 μm.

We employed a vibrational exciton model to simulate linear and 2D IR spectra. The fundamental frequencies of the amide-I local modes were estimated by incorporating the effects of intramolecular hydrogen bonding and intermolecular solvation [3]. The vibrational coupling between the nearest oscillators was obtained from *ab initio* calculations [6]. The coupling between the non-nearest oscillators was approximated by the transition dipole couplings. To include inhomogeneity of the local mode frequencies and couplings, 25 000 different structures were generated by fluctuating the peptide backbone dihedral angles (ϕ, ψ) according to Gaussian distributions with the standard deviations σ. The capping urethane and ester C=O groups were also incorporated into the exciton model.

Results and Discussion

Figure 1 presents the measured and simulated linear and 2D IR spectra. In the linear IR spectra (A–D), the protecting urethane and ester C=O groups are not resolved and one apparent peak appears at 1718–1720 cm^{-1}. For $n = 3$, the amide-I band exhibits two partially resolved peaks at 1671 and 1685 cm^{-1}. One broad band with a weak shoulder is seen for $n = 5$. For $n = 8$ and 10, the amide-I band shapes are almost the same, with a single peak at 1666 and 1662 cm^{-1}, respectively.

The real parts of R spectra (a–d) exhibit positive peaks involving 0–1 vibrational coherences. The negative peaks involve 1–2 coherences and are anharmonically-shifted along the ω_t axis. Two closely spaced amide-I peaks were observed for $n = 3$. As the peptide chain lengthens, a single pair of positive and negative amide-I peaks appears in the $n = 8$ and 10 spectra. The real parts of NR spectra (e–h) are separately plotted to reveal the many fine spectral features that are not discernable in the absorptive spectrum (not shown). For an inhomogeneously broadened system, the absorptive spectrum is dictated by the more intense R spectrum. Examining the R and NR contributions separately, and comparing them with the simulation, can provide us with more spectral constraints than the absorptive spectrum alone.

The chain length dependence of the cross-peak pattern is observed in the absolute magnitude of the R spectra (i–l). The cross-peak pattern for $n = 3$ exhibits three peaks. The weakest peak, close to the diagonal, merges with the two stronger peaks and evolves into a doublet peak pattern as the chain length increases. The pentapeptide exhibits spectral features that are similar to those of the longer peptides, but are quite different from the tripeptide. Overall, the simulations reasonably reproduce the observed features in the linear and 2D IR spectra. The amide-I local mode frequencies obtained from our simulations show strong site dependence for short peptides. The general trend agrees well with the results from scaled semi-empirical AM1 calculations after taking into account the solvent effects heuristically [7]. The results suggest that the Aib homopeptides adopt the 3_{10}-helical conformation in CDCl$_3$, with average dihedral angles (ϕ, ψ) $\sim$ ($-57°$, $-31°$) and $\sigma \sim 8.5°$. The average angles and conformational disorder exhibit slight chain length dependence. There is more noticeable discrepancy between the simulated and experimental cross-peak patterns of $n = 3$. The tripeptide can be floppier than the longer chain peptides with some population not falling within the single β-turn conformation, because it is fully exposed to solvent and can form at most one intramolecular hydrogen bond.

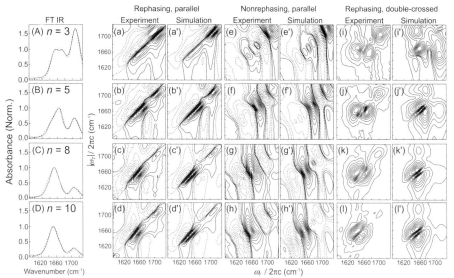

Fig. 1. Chain length dependent linear and 2D IR spectra of Z-(Aib)$_n$-OtBu (n = 3, 5, 8, and 10, from top to bottom) in CDCl$_3$: (A–D) measured (solid) and simulated (dashed) linear IR spectra; (a–d) real parts of R spectra measured with the parallel polarization; (e–h) real parts of NR spectra; (i–l) absolute magnitude cross-peak patterns measured with the R sequence and double-crossed polarization. Positive and negative contour lines are drawn in solid and dashed lines, respectively. Simulated 2D IR spectra (a′–l′) are shown next to the measured spectra.

The onset of the 3$_{10}$-helix spectral signatures appears to occur upon the formation of a complete 3$_{10}$-helical turn, when the number and strength of intermode vibrational couplings and conformational stability in a pentapeptide begin to resemble those in a longer peptide. Our results show that 2D IR spectroscopy is highly sensitive to peptide conformation, disorder, and size, and thus, is ideally suited for probing the earliest steps in the coil-to-helix transition.

Acknowledgements. This research was supported by grants from the American Chemical Society Petroleum Research Fund (39148-G6) and the National Science Foundation (CHE-0450045) to N.-H. G.

1   N. Demirdöven, C. M. Cheatum, H. S. Chung, M. Khalil, J. Knoester and A. Tokmakoff, J. Am. Chem. Soc. **126**, 7981, 2004.

2   H. Maekawa, C. Toniolo, A. Moretto, Q. B. Broxterman and N.-H. Ge, J. Phys. Chem. B **110**, 5834, 2006.

3   H. Maekawa, C. Toniolo, Q. B. Broxterman and N.-H. Ge, J. Phys. Chem. B **111**, 3222, 2007.

4   S. Sul, D. Karaiskaj, Y. Jiang and N.-H. Ge, J. Phys. Chem. B **110**, 19891, 2006.

5   C. Toniolo, G. M. Bonora, V. Barone, A. Bavoso, E. Benedetti, B. Di Blasio, P. Grimaldi, F. Lelj, V. Pavone and C. Pedone, Macromolecules **18**, 895, 1985.

6   S. Ham and M. Cho, J. Chem. Phys. **118**, 6915, 2003.

7   H. Maekawa, F. Formaggio, C. Toniolo and N.-H. Ge, J. Am. Chem. Soc. **130**, 6556, 2008.

Two-dimensional infrared spectroscopy of Glycine-L-Alanine-Methylamide.

M. Candelaresi[1], P. Foggi[1,2] , M. Lima[1], D.J. Palmer[1], B. Mennucci[3], C. Cappelli[3], S. Monti[4]

[1]LENS, Polo Scientifico Universitario, Via Nello Carrara 1, 50019, Sesto F.no (Firenze) Italy.
[2]Chemistry Department, University of Perugia, Via Elce di Sotto 8, 06100 Perugia, Italy.
[3]Chemistry Department, University of Pisa, Via Roma 67, 56126 Pisa, Italy.
[4]CNR, Chemical-Physics Process Institute, Via G. Moruzzi 15624 Pisa, Italy.

Abstract. Two-dimensional pump-probe infrared spectroscopy is utilized to study the structural proprieties of Glycine-L-Alanine-Methylamide in D_2O and DMSO solutions. Experimental results are compared to computational predictions. Preliminary calculations confirm the presence of a CO-CO coupling stronger in D_2O than in DMSO as experimentally observed.

Introduction

A multidisciplinary approach to the study of the structural and dynamical proprieties of biomolecules makes possible to understand their biological activity in details even at the atomic level. On short time scales (ps), the molecular properties can be now determined by two dimensional infrared (2D-IR) spectroscopy[1]. This recent technique demonstrated already to be a powerful tool in the investigation of many systems of different complexity like liquids[2,3,4], polypeptides[5,6], DNA[7], proteins[8], and membranes[9]. It is well known that *Amide I* mode, in the 6 μm region, is very sensitive to the secondary structure of a polypeptide. Through the analysis of this mode, it is possible to obtain details concerning the conformations and the nature of the interactions when varying the properties of the surrounding solvent molecules[10]. Experimental data can be then compared to calculations to evaluate the dominant contributions responsible for a specific conformation. The present study reports on the solvent dependent C=O-C=O coupling of Gly-L-Ala-Methylamide (GAM) in D_2O and DMSO.

Experimental

The laser setup consists of a Ti:Sapphire oscillator and amplifier system. The output is a 1kHz train of 600 μJ pulses centered at 800 nm with temporal width of ≈ 60 fs. The output is split into equal parts to pump two different optical parametric amplifiers (OPA). Mid-infrared pulses tunable from 3 to 6 μm are obtained by difference-frequency generation of signal and idler pulses in a $AgGaS_2$ crystal. The infrared output from both OPA systems has a spectral width of 200 cm^{-1} and energy of the order of 1 μJ/pulse. The pump beam passes through a half-wave plate controlling its polarization relative to that of the probe and then a variable delay line. In order to accomplish narrow band pump experiments, the pump pulse passes through an infrared Fabry-Perot interferometer. In this configuration the pump has a bandwidth of ≈ 15 cm^{-1}(FWHM) and a duration of ~800 fs (FWHM). The dipeptide with ^{13}C labeled C=O on the L-alanine moiety (GA*M) was purchased by Biosyntan. The sample was repeatedly dissolved in DCl-D₂O solutions and lyophilized to remove the trifluoroacetic acid.

Results and Discussion

In Figure 1 (left) linear and 2D-IR spectra of GA*M in DMSO and D_2O acquired in parallel and perpendicular polarization at 0 delay time are shown.

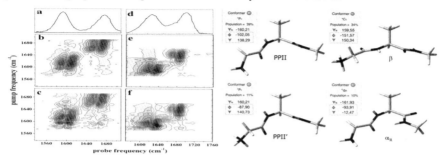

Fig. 1 Left :(a) Linear spectra of GA*M in D_2O. (b,c) 2D-IR spectra recorded at 0 time with parallel and perpendicular polarization respectively. (d) linear spectra of GA*M in DMSO. (e,f) 2D-IR spectra recorded at 0 time in parallel and in perpendicular polarization respectively. Right: The most stable conformations obtained by *ab initio* calculations.

The linear spectra show two well separated bands at 1604 cm⁻¹ and 1680 cm⁻¹ in D_2O and 1631 cm⁻¹ and 1687 cm⁻¹ in DMSO. These frequency values, compared to those measured on GAM (1645 cm⁻¹ and 1677 cm⁻¹ in D_2O and 1670 cm⁻¹ and 1697 cm⁻¹ in DMSO), show a large isotopical shift of ≈ 40 cm⁻¹ in both solvents. In addition our spectra confirm that the *Amide I* frequency of the residue closer to the N terminal (Gly) has a higher frequency than the C terminal one (Ala)[11].

Table 1. Structural parameters of GAM obtained by PCM-QM calculations.

Conformer	D₂O				DMSO			
	PPII	PPII˙	β	α_R	PPII	PPII˙	β	α_R
%	40	10	32	18	32	25	39	4
θ	49	53	63	70	58	36	65	109
Anis. cal.	0.058	0.017	-0.076	0.130	-0.032	0.192	0.093	-0.136
Anis. avg.	0.02				0.04			
Anis. exp.	0.00				0.08			

The larger splitting in D_2O can be attributed to a strong coupling between the two *Amide I* modes. Coupling between the two modes can be better determined by non-linear infrared spectroscopy. 2D-IR spectra show well defined cross peaks even at zero delay time. From their intensity measured in parallel and perpendicular polarization it is possible to evaluate the off-diagonal anisotropy which results to be 0.00 and 0.08 in D_2O and in DMSO respectively. In principle, from the anisotropy it is possible to calculate the angle between the two transition dipole moments. However, such value can be either the result of a single conformation or of the average of different conformations. At this purpose we performed *ab initio* calculations in order to evaluate the most stable conformers in the two solvents.

From these calculations, GAM has four stable conformations in D_2O and DMSO (see right part of Fig.1, for their definition, and Table 1 for their relative abundance). There is a good agreement between the calculated and the experimentally observed anisotropy. A preliminary analysis of the calculated vibrational frequencies shows that the splitting of two Amide I modes is larger in D_2O than in DMSO and its value is closer to the experimental one.

MD simulations allow to give an interpretation of the different frequency shift between the two *Amide I* modes of GAM in terms of the different arrangement of the solvent surrounding the dipeptide. D_2O molecules establish specific interactions with C=O bonds. The C=O···H radial distribution has a sharp peak around 1.8 Å, while at longer distances this value drops almost to zero indicating that there are few water molecules surrounding the C=O. In DMSO the C=O···H radial distribution shows only a broader peak around 2.5 Å, indicating the presence of solvent molecules at medium distances. The C=O oscillators seems therefore to couple more efficiently in D_2O for a reduced screening effect, this causing a larger splitting between two *Amide I* modes.

DMSO is a hydrogen bond acceptor, as a consequence DMSO molecules are arranged around NH group of GAM. The absence of specific interactions with C=O bonds causes the ordering of DMSO preferentially in the second solvation sphere. MD simulations indicate that both GAM oxygens are engaged in weak interactions with DMSO which orients methyl hydrogens towards the peptide oxygens. The calculations provide a good explanation of what experimentally observed.

Acknowledgements. Authors wish to thank Prof. R.Righini and Dr. V.V.Volkov for helpful discussions. This work has been partially supported by the European Community under the contract RII3-CT-2003-506350.

1   M. Cho, Chem. Rev. **108,** 1331-1418 (2008).

2   J. Park, J.-H., Ha, R. M. Hochstrasser, J. Chem. Phys. **112,** 7281-7292 (2004).

3   M. L. Cowan, B.D. Bruner, N. Huse, J.R. Dwyer, B. Chugh, E. T. J. Nibbering, T. Elsaesser, R. J. D. Miller, Nature **434,** 199-202 (2005).

4   J. B. Asbury, T. Steinel, C. Stromberg, S. A. Corcelli, C. P. Lawrence, J. L. Skinner, M. D. Fayer, J. Phys. Chem. A **108,** 1107-1119 (2004).

5   P. Hamm, M. Lim, F. DeGrado, M.R. Hochstrasser, J. Chem. Phys. **112,** 1907-1916 (2000).

6   W. A. Smith, A. Tokmakoff, Angew. Chem. Int. **46,** 7984-7987 (2007).

7   A. Krummel, M. T. Zanni, J. Phys. Chem. B **110,** 13991-14000 (2006).

8   P. Hamm, M. Lim, F. M. R. Hochstrasser, J. Phys. Chem. B **102,** 6123-6138 (1998).

9   V. V. Volkov, D. J. Palmer, R. Righini, Phys. Rew. Lett. **99,** 078302-1-078302-4 (2007).

10   A. Barth, C. Zscherp, Rev. Biophys. **35,** 369-430 (2002).

11   S. Woutersen, P. Hamm, J. Chem. Phys. **114,** 2727-2737 (2001).

How do vibrations change their composition upon electronic excitation? – EXSY-T2D-IR measurements challenge DFT calculations.

Andreas Messmer[1,2], Ana-Maria Blanco Rodríguez[3], Jakub Šebera[4], Peter Hamm[1], Antonín Vlček Jr[3,4], Stanislav Záliš[4], and Jens Bredenbeck[2]

[1] Institute for Physical Chemistry, University of Zurich, Winterthurerstr. 190, CH-8057 Zurich, Switzerland

[2] Institute for Biophysics, Johann Wolfgang Goethe-University Frankfurt, Max von Laue-Str. 1, D-60438 Frankfurt (Main), Germany
E-mail: messmer@biophysik.org

[3] School of Biological and Chemical Sciences, Queen Mary, University of London, Mile End Road, London E1 4NS, United Kingdom

[4] J.Heyrovský Institue of Physical Chemistry, Academy of Sciences of the Czech Republic, Dolejškova 3, CZ-18223 Prague, Czech Republic

Abstract. The composition of excited state vibrations can be disentangled by projecting ground state vibrations on them using exchange transient two-dimensional IR spectroscopy. The results challenge excited state DFT calculations.

Introduction

Vibrations are sensitive probes of molecular structure and dynamics. Changes in conformation, the surrounding or the electronic state of a molecule lead to a change of the internuclear forces and thus a different vibrational Hamiltonian. In order to achieve a deeper understanding of the electronically excited state not only the vibrational eigenenergies but also the compositions of the transient vibrations are of great interest. Exchange transient two-dimensional IR (EXSY-T2D-IR) spectroscopy is able to reveal information on both of them [1, 2].

Experimentally, the vibrations of the sample in the ground state are first selectively labelled by IR excitation, followed by an additional pump pulse in the UV/VIS range, which creates an electronically excited state e.g. by a charge transfer process. The result is read out by a broadband IR probe pulse. In the case of uncoupled oscillators, just a frequency shift will occur and a one-to-one correlation between ground state and electronically excited state vibration is obtained [3]. In the case of coupled vibrations remixing of the modes might occur [4,5]. Mathematically speaking, the ground state vibrations ψ_i get projected onto the basis set of excited state vibrations φ_j. From the EXSY-T2D-IR spectra the remixing coefficients $c_{ij}=|\langle\psi_i|\varphi_j\rangle|^2$ can be extracted by evaluating the cross peaks between the labelled ground state vibration (pump axis) and the bands of the electronically excited state (probe axis).

In the presented study we investigate the vibrational remixing of *fac*-[Re(NCS)(CO)$_3$(2,2'-bipyridine)] caused by a charge transfer process. Vibrational states of metal carbonyl complexes are used to distinguish electronic states [5] and furthermore report on the changed electron density in different parts of the molecules. Knowledge about the composition and therefore localization of vibrational eigenstates is of course a prerequisite.

Results and Discussion

The time resolved IR spectrum of *fac*-[Re(NCS)(CO)₃(2,2'-bipyridine)] is shown in fig. 1a. Since it is a difference spectrum, the ground state bands have negative sign whereas the excited state signals are positive. The ground state spectrum consists of four bands: the two bands overlapping at 1924 cm⁻¹ as well as the one at 2027 cm⁻¹ are carbonyl vibrations whereas the band at 2099 cm⁻¹ belongs to the NC stretching mode. The excited state bands are denoted by φ_1 to φ_4 ordered by increasing energy (fig. 1a).

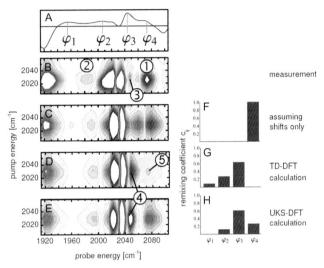

Fig. 1. A. Time resolved IR spectrum of *fac*-[Re(NCS)(CO)₃(2,2'-bipyridine)] for parallel polarization between UV pump and IR probe pulse (delay 3 ps). B. Measured EXSY-T2D-IR spectrum for all polarizations beeing parallel to each other (delay between IR pump and IR probe 5 ps; delay between UV pump and IR probe 3 ps). Red color: positive sign; Blue color: negative sign. C-F. Simulated EXSY-T2D-IR spectrum based on the assumption that the bands only shift in energy (C), the TD-DFT calculation (D), the UKS-DFT calculation (E). F-H. Corresponding bar charts representing the remixing coefficients for the A'(1) ν(CO) mode. F. Assuming shifts only; G. TD-DFT calculation; H. UKS-DFT calculation. Description of the numbers are given in the text.

TD-DFT as well as UKS-DFT specified in [5] match well the experimental excited-state vibrational frequencies and predict strong remixing of the vibrational eigenmodes upon excitation. However, the composition of the excited state vibrational modes deviates strongly in the two calculations. It therefore seems that the composition of the vibrations is a much more sensitive parameter, and a bigger challenge to theory than the energies of the vibrations. Exemplarily, the remixing coefficients for labelling the vibration A'(1) ν(CO) at 2028 cm⁻¹ in the electronical ground state are shown in fig. 1g and 1h for TD-DFT and UKS-DFT, respectively. Both calculations predict a major contribution of A'(1) ν(CO) to the excited state vibration φ_3 and a minor one to φ_1 and φ_2 (TD-DFT) or φ_2 and φ_4 (UKS-DFT).

In following section the measured EXSY-T2D-IR spectra (fig. 1b) are compared to different simulations (fig. 1c-e) in order to extract how the vibrations change upon excitation. Based on the eigenenergies and the composition of the vibrational normal modes obtained by DFT, the time resolved IR spectra were fitted by adapting the transition dipole moments (data not shown) and the EXSY-T2D-IR spectra were simulated (fig. 1d and 1e); an additional simulation (fig. 1c) was performed assuming that the composition of the vibrations remains and only the vibrational energies change upon excitation. For the vibration A'(1) ν(CO) shown here it is assumed that the energy shifts to φ_4 (fig. 1f).

The measurement (fig. 1b) shows a very strong excited state cross peak with φ_4 (pos. 1) indicating that the A'(1) ν(CO) mode mainly shifts to φ_4. This interpretation is supported by the simulation based on the assumption that the bands shift only (fig. 1c), which is also able to explain the weak cross peak with φ_2 (pos. 2) since coupling in the excited state leads to additional cross peaks. However, the weak cross peak with φ_3 cannot be explained by this simple simulation (pos. 3). Studying the dependence of the spectra on the time delay between the UV pump and IR probe pulse a rise of this signal within the first 4ps after the UV pump pulse is observed (data not shown). This behaviour is expected for a cross peak caused by population transfer and is consistent with measured population transfer in the ground state.

In contrast to the simple assumption of shift only, both DFT calculations lead to spectra (fig. 1d and 1e) that differ considerably from the measured ones. They predict a dominant contribution of A'(1) ν(CO) to φ_3 leading to intense exchange cross peaks between the labelled band and φ_3 (fig. 1 pos. 4) that are not observed in the experiment (fig. 1 pos. 3). Additionally, the cross peak with φ_4 is much too weak in the simulation based on the TD-DFT calculation (fig. 1 pos. 5). Furthermore, the polarization dependence of the spectra based on the ab initio calculations is not in agreement with the measured ones (data not shown).

Conclusions

We are able to show by means of EXSY-T2D-IR spectroscopy and the comparison to simulated spectra that the vibrational normal modes do not change their composition upon electronic excitation to the extent predicted by DFT calculations and the experimental data are better explained by the assumption that no remixing occurs. The assignment of the excited state bands (i.e. φ_3 being localized mostly on the NCS ligand and φ_4 being a CO vibration) based on our experiments are supported by the influence of the solvent on the energies of φ_{1-4} [5]. Present excited-state DFT calculations thus fail to predict correctly the character of the vibrations even if they reproduce well the frequencies. It is demonstrated that EXSY-T2D-IR spectroscopy is able to provide an important benchmark for quantum chemical calculations since both energies and remixing coefficients are experimentally accessible.

1   J. Bredenbeck, J. Helbing, R. Behrendt, C. Renner, L. Moroder, J. Wachtveitl, and P. Hamm, J. Phys. Chem. B **107**, 8654 (2003).

2   J. Bredenbeck, J.Helbing, and P. Hamm, J. Am. Chem. Soc. **126**, 990 (2004).

3   J. Bredenbeck, J. Helbing, K. Nienhaus, G.U. Nienhaus, and P. Hamm, Proc. Natl. Acad. Sci. **104**, 14243 (2007).

4   D. M. Dattelbaum, K. M. Omberg, J. R. Schoonover, R.L. Martin, and T. J. Meyer in Inorg. Chem. **41**, 6071 (2002).

5   A. M. Blanco Rodríguez, A. Gabrielsson, M. Motevalli, P. Matousek, M. Towrie, J. Šebera, S. Záliš, and A. Vlček, J. Phys. Chem. A **109**, 5016 (2005).

Propagation and beam geometry effects on 2D Fourier transform spectra of multi-level systems

Byungmoon Cho, Michael K. Yetzbacher, Katherine A. Kitney, Eric R. Smith, and David M. Jonas[1]

[1]Department of Chemistry and Biochemistry, University of Colorado, Boulder, Colorado, 80308, USA
E-mail: david.jonas@colorado.edu

Abstract. We calculate 4-level two-dimensional (2D) Fourier transform relaxation spectra with propagation and beam geometry distortions, which are 14% for an optical density of 0.2 and 25% for a crossing angle of 10°.

Introduction

Multidimensional spectra can be distorted by pulse propagation and noncollinear detection[1, 2]. It is desirable to perform measurements on samples with optical density (OD) ~1, where the signal is maximized. At such optical densities, propagation effects have long been known to distort the true microscopic response, complicating the measurement[1]. For example, such effects can become particularly significant for weakly absorbing samples in non-transparent solvents[3], strongly absorbing semiconductor,[4] and neat liquids[5]. It may be desirable to include signal distortions in these cases in order to relate the spectroscopic measurement to the nonlinear susceptibility. Previous calculations of 2D spectra with propagation and detection effects treated peak shape distortions[2]. Here we study a model system loosely based on the coupled CO stretches of acetyl-acetonato dicarbonyl rhodium(I)[6]. We find that the detection beam geometry can significantly affect cross-peaks and present a new 2D transformation to (mostly) remove this distortion.

Methods

The energy level scheme for the model system from which the spectra are calculated is shown in Figure 1. Coupling between transitions is only due to coherence pathways; population and coherence transfer between the singly excited states, α and β, is neglected. We calculate the signal due to three excitation fields, E_a, E_b and E_c as a function of three time variables and treat propagation and detection using 3DFT algorithms[2, 7]. The signal is

$$S(-\omega_t, -\omega_a, -\omega_b, -\omega_c) = (i\omega_t / 2\varepsilon_0 c)[\Pi^{(3)}(\chi^{(3)}\hat{E}_a\hat{E}_b\hat{E}_c)]\hat{E}_d\Phi^{(3)} \tag{1}$$

where $\Pi^{(3)}$ is the path length dependent propagation function, $\chi^{(3)}$ is the nonlinear susceptibility, $\Phi^{(3)}$ is the directional filter which accounts for the beam geometry, and E_d is the interference detection field. From the 3D frequency domain signal, we transform to the 3D time domain, extract a 2D time domain signal, and transform to obtain the 2D spectrum as a function of the excitation and detection frequencies.

The undistorted spectrum is S2Dideal, the spectrum usually measured is S2D0. This spectrum is transformed with experimentally accessible 2D functions yielding S2D++ and S2DF for the OD and directional filtering corrections, respectively. The OD correction has been described previously[2]. The correction for the beam geometry distortions is $\exp[(\omega\det - \omega\exc)^2 w^2 \sin^2(\beta)/2c^2]$, where w is the beam waist, 2β is the crossing angle, and c is the speed of light. Like the previously proposed correction for absorptive distortions, the correction for beam geometry becomes exact at large mixing times, T.

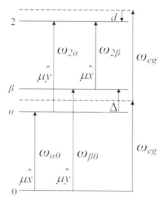

Fig. 1. Two ground transitions are denoted by $\omega_{\alpha 0}$ and $\omega_{\beta 0}$ which differ in energy by Δ. The average of two ground state transitions is ω_{eg}. $\omega_{2\alpha}$ and $\omega_{2\beta}$ denote the transitions from the singly excited states, α and β, to the doubly excited state which differ from ground transitions by energy d. $\mu\hat{x}$ and $\mu\hat{y}$ indicate the equal magnitude, perpendicular transition dipole moments. The demodulation frequency is, $\omega_{eg}/2\pi c = 2049.5$ cm^{-1}; frequency splitting, $\Delta = 69.8$ cm^{-1}; and red-shift of the doubly excited state, $d = 26.8$ cm^{-1}. The total dephasing rates $\Gamma_{\alpha 0} = \Gamma_{\beta 0} = \Gamma_{2\alpha} = \Gamma_{2\beta} = 1/(4.08\text{ps})$ are the same for all optical transitions. $\Gamma_{\alpha\beta} = (1/2.04\text{ps})$ is the dephasing rate for the quantum beats between singly excited levels. The population lifetimes are 1 ns.

Results and discussion

We first calculated the effects of OD with the collinear beam geometry. Absorptive attenuation is greatest at the peak center. An OD of 0.5 leads to changes in peak amplitude of 30% and peak width of 56%. The tranformation from S2D0 to S2D++ uses only the complex valued refractive index and sample length at mixing time T=472fs. S2D++ recovers the ideal amplitude within 8% and width within 11%.

Fig. 2 illustrates the directional filter distortion due to noncollinear beam geometry which only changes the peak amplitude (not shape) in this limit. The largest change in relative peak height occurs for the cross-peak at excitation frequency 2090 cm^{-1} and detection frequency 1990 cm^{-1}, which drops from 0.10 to 0.07 relative to the highest peak. This would lead to underestimation of the coupling. 2D correction for the 3D directional filter uses only the excitation and detection frequencies to approximate the distortion. Transformation to S2DF recovers the ideal relative amplitude within 5%.

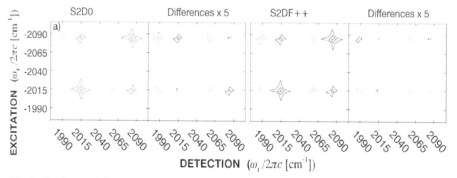

Fig 2. Real part of distorted (left panel) and transformed (right panel) spectra are shown with the differences to undistorted spectra multiplied by 5 on the right half of each panel. Crossing angle 10°, beam waist 150μm. Mixing time T=472fs. Contour levels are -60, -7, -1, 0, 7, 60 % with dotted contours indicating negative amplitude.

Fig. 3 shows the result of combined distortion (S2D0, OD of 0.2, crossing angle 10°, beam waist 150μm) and the successive application of corrections described above. S2D++ reduces the maximum error from 14% to 4%. However, the relative amplitude errors for three of the four cross-peaks are larger in S2D++ than in S2D0. In other words, the transformed cross-peaks have errors which are larger relative to their heights and could lead to an overestimate of the coupling. S2DF++ corrects this relative error in cross-peak amplitude while retaining the 4% maximum deviation in S2D++.

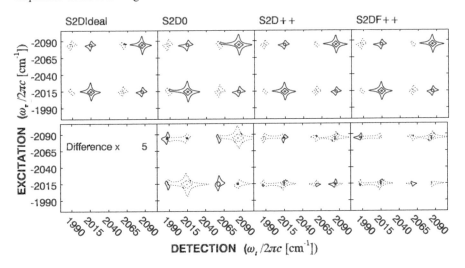

Fig. 3. Left to right, real parts of S2Dideal, S2D0, S2D++, and S2DF++ (top row) at mixing time T=472 fs. The differences to S2Dideal are multiplied by 5 (bottom row). Contour levels are -60, -7, -1, 0, 7, 60 % with dotted contours indicating negative amplitude.

Conclusion

OD distortion affects the ground state transitions strongly, leading to peak broadening, but affects excited state absorption transitions only weakly, leading to changes in off-diagonal amplitudes relative to the diagonal peaks. Collinear spectra improve when previously described OD corrections are applied. Directional filtering also significantly suppresses off-diagonal peak amplitudes. In the case of well-separated peaks, directional filtering distortions can be corrected to within 5% of the cross-peak amplitude.

References

1   Keusters, D. and W.S. Warren, Chemical Physics Letters, 2004. **383**(1-2): p. 21-24.
2   Yetzbacher, M.K., Belabas, N., Kitney, K.A.,and Jonas, D.M., Journal of Chemical Physics,2007. **126**(4) p.04451
3   Kozinski, M., S. Garrett-Roe, and P. Hamm, Chemical Physics, 2007. **341**(1-3): p. 5-10.
4   Borca, C.N., Zhang, T.H.,Li, X.Q.,and Cundiff,S.T., Chemical Physics Letters, 2005. **416**(4-6): p. 311-315.
5   Cowan, M.L.,Bruner, B.D.,Huse, N.,Dwyer, J.R.,Chugh, B.,Nibbering,E.T.J.,Elsaesser,T., and Miller,R.J.D., Nature, 2005. **434**(7030): p. 199-202.
6   Khalil, M., N. Demirdoven, and A. Tokmakoff, Journal of Chemical Physics, 2004. **121**(1): p. 362-373.
7   Belabas, N. and D.M. Jonas, Journal of the Optical Society of America B-Optical Physics, 2005. **22**(3): p. 655-674.

Difference 2D-IR spectroscopy on the chromophore in bacteriorhodopsin

Esben Ravn Andresen[1], Jan Helbing[1], and Peter Hamm[1]

[1] Physikalisch-Chemisches Institut, Universität Zürich,
Winterthurerstrasse 190, 8057 Zürich, Switzerland
E-mail: phamm@pci.uzh.ch

Abstract. Difference two-dimensional-infrared (2D-IR) spectroscopy has been employed to measure the 2D-IR spectrum of the C=C and C=ND chromophore bands in bacteriorhodopsin, demonstrating that couplings between individual bands in a mid-size protein can be measured.

Introduction

Two-dimensional infrared (2D-IR) spectroscopy [1] is the optical analogue to two-dimensional NMR spectroscopy (2D-NMR); whereas 2D-NMR spectroscopy measures the couplings between nuclear spins, 2D-IR spectroscopy measures the coupling between molecular vibrations and can be employed in a faster time regime, down to $\sim$ 1 ps, due to the short duration of the laser pulses being used. To date, an important application of 2D-IR spectroscopy has been the study of fast conformational dynamics of peptides. The next logical step would be to study larger systems *i.e.* proteins. Because of the large number of bonds in the protein, it is difficult to obtain any specific information from static spectra; peaks from individual bonds are usually buried in the background absorption from the rest of the protein. A way of adressing single bonds in proteins is to incorporate amino acids with non-natural side groups with absorptions in non-congested spectral regions, or isotope labelling. In photoactive molecules like bacteriorhodopsin (bR), there is the additional possibility to form a difference spectrum, *i.e.* the spectrum after photoactivation minus the spectrum before photoactivation, which singles out the peaks that change [3]. This concept has previously been used extensively in various difference FTIR studies of bR. The same concept should allow for difference 2D-IR spectroscopy, whereby one can measure the 2D-IR spectrum of bands that change upon photoactivation. bR [2] is a transmembrane protein (26 kDa, 248 amino acid residues) found in the bacterium *halobacterium salinarium* where it functions as a light-activated proton pump that establishes a proton gradient across the cell membrane for ATP production. The initial step of the photocycle involves the chromophore, retinal, which within few hundred femtoseconds switches from the all-*trans* state to the energetically higher-lying 13-*cis* state, which stores energy for the rest of the thermally activated proton pumping cycle.

Results

Figures 1a and 1b present the FTIR spectrum and the static 2D-IR spectrum of light-adapted bR, respectively. The main features in the 2D-IR spectrum are the strong diagonal signals around 1550 cm^{-1} and 1650 cm^{-1}, stemming from amide I and II bands. These broad features arise from bonds from the entire protein and thus bury any site-specific information in the spectra.

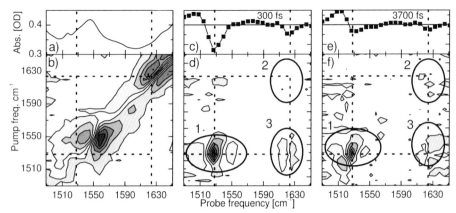

Fig. 1. (a) FTIR spectrum of bR; (b) static 2D-IR spectrum of light-adapted bR; (c) visible-pump-IR-probe spectrum of bR for 300 fs delay; (d) difference 2D-IR spectrum of bR with visible pump-IR-pump delay of 300 fs and IR-pump-IR-probe delay of 1200 fs; (e) visible-pump-IR-probe spectrum of bR for 3700 fs delay; (f) difference 2D-IR spectrum of bR with visible-pump-IR-pump delay of 3800 fs and IR-pump-IR-probe delay of 1200 fs. All measurements were performed on $\sim$ 1 mM bR in D_2O and all polarizations are parallel. The dashed lines in all figures mark the frequencies of the C=C and C=ND chromophore bands at 1528 cm^{-1} and 1624 cm^{-1}.

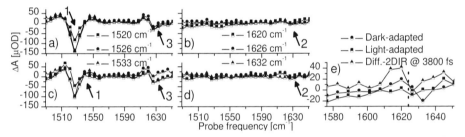

Fig. 2. Slices along the probe frequency axis of the (a and b) difference 2D-IR spectrum in Fig. 1d; (c and d) slices of the difference 2D-IR spectrum in Fig. 1f. The legends denote the pump frequencies; (e) comparison of the cross-peak at position 3 in various 2D-IR spectra for pump frequencies of 1526 cm^{-1}. The vertical line marks the frequency of the C=ND band.

To measure difference 2D-IR spectra, we initiated the photocycle with 590 nm-pulses of 100 fs duration from an noncolinear optical parametric amplifier. We then used IR pump and probe pulse to record the 2D-IR spectrum at a specified delay time before and after the visible pump. By taking the difference of 2D-IR spectra recorded after and before photoactivation, the difference 2D-IR spectrum could be measured. The results are presented in Fig. 1c-f. Slices of the difference 2D-IR spectra along the probe frequency axis are presented in Fig. 2 to provide a clearer picture of the essential features.

Figure 1c presents the visible-pump-IR-probe spectrum at 300 fs delay, and Fig. 1d presents the corresponding difference 2D-IR spectrum. The significant signals in Fig. 1c are coming from the two chromophore bands at 1528 cm^{-1} and 1624 cm^{-1} (marked by the dotted lines in the figure). It is known, that the 1528 cm^{-1}-band has contribution from the stretching vibration of C_{13}=C_{14} bond, around which the photoisomerization occurs, and that the 1624 cm^{-1}-band is mostly the C_{15}=ND vibration. As such, both bands are sensitive to the conformation of the chromophore and undergo changes when the chromophore isomerizes within the first few hundred fs of the bR photocycle, in Fig. 1c mostly visible as the bleach of these bands. Fig. 1d should then reveal the 2D-IR spectrum of these bands. Indeed, the peaks that show up in Fig. 1d coincide with the expected positions of the chromophore bands (marked by the dotted lines), demonstrating that these bands are singled out by forming the difference spectrum. A diagonal peak at position 1 is clearly visible. A further diagonal peak at position 2 is observed, albeit comparable to the noise level (it is more easily seen in Fig. 2b and 2d), and a cross-peak is observed below the diagonal at position 3, but not at the corresponding position above the diagonal. We would have expected Fig. 1d to look essentially like the negative 2D-IR spectrum of the all-*trans* chromophore. The spectrum in Fig. 1d, however does contain signal from the 13-*cis* product, as apparent from the triple-stucture at position 1. This is likely due to the fact that during the IR-pump, IR-probe pulse sequence, which exceeds 1 ps, the visible-pump-IR-probe spectrum changes towards the spectrum in Fig. 1e, for which the delay was 3700 fs, where the product bands have become sharper due to partial cooling of the 13-*cis* product. The corresponding difference 2D-IR spectrum is shown in Fig. 1f.

The appearance of a cross-peak at a higher probe frequency after pumping at a lower frequency, but not the opposite seems to indicate that other effects than only direct through-space or through-bond coupling are going on during the duration of the measurement. One possible explanation for the assymetry of the cross-peaks would be if the C=ND vibration relaxes faster than the employed IR pulse sequence, $\sim$ 1 ps. Direct population transfer between the two bands are unlikely, as that would produce the stronger cross-peak at the lower probe frequency.

The observed cross-peak is so strong, that we would expect to see it in the static 2D-IR spectra as well. Indeed, we have observed a similar cross-peak, albeit slightly shifted, in the static 2D-IR spectrum of dark-adapted bR in which half of the bR molecules contain retinal in its 13-*cis* configuration. However, the cross-peak is absent in the static 2D-IR spectra of light-adapted bR in which all molecules contain all-*trans* retinal. In Fig. 2e, the cross-peaks in the different cases are compared. This would point to the chromophore conformation having influence on the molecular couplings in the system.

Conclusion and outlook

The results presented here yield promise that we will be able to single out individual bands in mid-size proteins without artificial labels and measure their couplings with difference 2D-IR spectroscopy.

Acknowledgements. Funding by the Danish National Science Research Foundation. We are grateful to Martin Engelhard for providing the sample.

1 S. Woutersen and P. Hamm, in J. Phys. Cond. Matt., Vol. 14, R1035, 2002.
2 J. K. Lanyi, in Annu. Rev. Physiol., Vol. 66, 665, 2004.
3 J. Herbst, K. Heyne, and R. Diller, Science, Vol. 297, 822, 2002.

Coherent Control of Retinal Isomerization in Bacteriorhodopsin in the High Intensity Regime

A. C. Florean[1], D. Cardoza[2], J. L. White[2], R. J. Sension[1] and P. H. Bucksbaum[2,3]

[1] Department of Physics, University of Michigan, Ann Arbor, MI 48109, USA
  E-mail: rsension@umich.edu
[2] Department of Physics, Stanford University, Stanford, CA 94305, USA
[3] PULSE Center, SLAC, Menlo Park, CA 94025, USA

Abstract. We use a learning algorithm to optimize retinal isomerization in bacteriorhodopsin. At low energies the learning algorithm fails to converge. At higher energies it converges to a transform-limited pulse with the photoproduct yield increasing linearly with excitation energy beyond the saturation of the first excited state. The results are interpreted including the influence of one-photon and multiphoton transitions.

Introduction

The initial steps of the bacteriorhodopsin (bR) photocycle (see Figure 1) have been studied extensively over the last several decades [1]. Recently these studies have been extended to move beyond observation and include optical control and optimization of the excited state dynamics [2, 3]. In one recent study a learning algorithm was used to optimize the amplitude- and phase of an ultrafast pulse to manipulate the isomerization efficiency of bR. This study concentrated on the "biologically relevant" low intensity regime (2.7 x 10^{14} photons/cm^2) where excitation is limited to linear one-photon absorption. However, the potential application of bR based photoswitches suggests the utility of a more extensive exploration of pulse parameters to manipulate the initiation of the photocycle. In the work reported here we describe phase-only control experiments at pump fluence levels up to 1.5 x 10^{17} photons/cm^2 (upper estimate), to explore the possibility of additional pathways toward the control of the isomerization process and production of the 13-cis (K_{590}) conformation.

Experimental Methods

A non-collinear optical parametric amplifier was used to produce 600 nJ, 25 fs pulses centered at 570 nm. These pulses were shaped using a programmable acousto-optic modulator (AOM). A white light probe pulse spanning wavelengths from 460 nm to 870 nm was generated in a sapphire plate and overlapped with the pump in a 1 mm flow cell containing bR membranes in a buffer solution (1 OD at 570 nm). The bR was illuminated before the experiment and maintained at room temperature. The measured probe and pump diameters at the focus in the flow cell were approximately 25 and 20 μm respectively. After the cell the probe was sent into a spectrometer or monochromator. The prism monochromator allowed detection of any wavelength between 460 nm and 870 nm with high sensitivity, while the spectrometer monitored the entire difference spectrum on each shot.

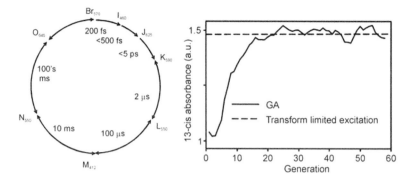

Fig. 1. Left: bR photocycle Right: a typical genetic algorithm (GA) run. The dashed-line shows the *cis* state absorbance resulting from excitation by a transform limited pulse.

A genetic algorithm (GA) was used to optimize the production of the K_{590} cis-retinal product state measured at 650 nm 40 ps after excitation. To assure broad sampling of the parameter space the GA was used with three different parameterizations for the excitation pulse. These basis sets used polynomial functions up to the 7th order, harmonic functions, or the phase difference between adjacent pixels respectively. These different parameterizations are used to avoid spurious results in the search for a global maximum.

Results and Discussion

GA searches to locate pulses capable of optimizing the production of the cis-retinal photoproduct were carried out over a broad intensity range. Searches carried out at low excitation energies, below ca. 50 nJ, failed to converge. GA searches at higher pulse energies consistently converged to a transform-limited pulse. The convergence of the GA is illustrated in Figure 1 for one typical run. For a given photon flux we also found that application of linear chirp consistently decreases the photoproduct production. These results suggest that peak energy is the key control parameter optimizing the photoproduct production. This observation was tested by monitoring the signal intensity as a function of pulse energy for several probe wavelengths and time delays as shown in Figure 2.

The initial bleach of the ground state all-trans conformation probed at 570 nm 200 fs after excitation exhibits a clear nonlinear saturation with increasing pump energy. The population of the I_{460} excited state intermediate was probed via excited state absorption at 487 nm and stimulated emission at 850 nm. Both of these signals exhibit identical power dependence within the accuracy of the measurements and saturate with increasing pump pulse energy. The saturation behavior is more pronounced than that observed for the ground state bleach at 570 nm. In contrast to the saturation behavior observed for the I_{460} intermediate, the magnitude of the absorption due to the 13-cis photoproduct is linear with increasing pump energy over the entire energy range shown. Signals corresponding to the J_{625} intermediate state and the ground state bleach 30 ps after excitation are also linear in pump energy over this range (data not shown). From this data we conclude that high intensity opens a multiphoton pathway that accesses the 13-cis conformation with near unit efficiency, while bypassing the I_{460} emissive state. The combined contributions of the multiple pathways keep the 13-cis yield approximately linear in pump energy despite the saturation of the initial excitation processes. The involvement of higher excited states has been reported previously in a two-color pump experiment following the excited state dynamics [4].

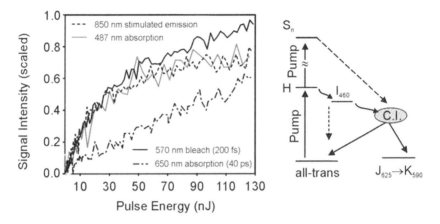

Fig. 2. Left: Intensity scans for the initial bR bleach at 570 nm, the I_{460} population monitored at 487 and 850 nm, and the production of K_{590} probed at 650 nm. Right: Schematic diagram illustrating the multiphoton pathway identified by the GA bypassing the I_{460} intermediate state.

Conclusions

Figure 2 (right) summarizes the isomerization paths suggested by our data. A one-photon transition excites population to the H state. At low pump energies no sensitivity of the K state 13-cis population yield with respect to the phase of the excitation pulse is observed. Above this level the GA consistently finds that a transform-limited (TL) pulse is optimal for maximizing the isomerization yield. For this optimal pulse the yield is linear with respect to the pulse intensity, well beyond the point where the initial one-photon excitation (all-trans → H) starts to saturate. These results suggest that optimal access to the multiphoton pathway requires absorption before excited state wave packet moves out of a region of favorable overlap. The promotion of population to a higher excited state, S_n is followed by rapid internal conversion and evolution through a conical intersection to the ground state photoproduct. The results reported here may have relevance for bio-photonic molecular applications requiring high yield, linear optical signals.

Acknowledgements. This research was supported by the NSF through grants CHE 0718219, PHY 0649578 and the FOCUS Center at the University of Michigan. We also thank J. K. Lanyi for providing the bR membrane samples.

1. H. Abramczyka, J. Chem. Phys. 20, 11120 (2004).
2. G. Vogt, P. Nuernberger, T. Brixner, and H. Gerber, Chem. Phys. Lett. 433, 211 (2006).
3. V. I. Prokhorenko, A. M. Nagy, S. A. Waschuk, L. S. Brown, R. R. Birge, and R. J. D. Miller, Science 313, 1257 (2006).
4. S. L. Logunov, V. V. Volkov, M. Braun, and M. A. El-Sayed PNAS 98, 8475 (2001).

Quantum Control of the Photoinduced Wolff Rearrangement of Diazonaphthoquinone in the Condensed Phase Using Mid-Infrared Spectroscopy

Daniel Wolpert[1], Marco Schade[2], Gustav Gerber[1], and Tobias Brixner[1,2]

[1] Physikalisches Institut, Universität Würzburg, Am Hubland, 97074 Würzburg, Germany
[2] Institut für Physikalische Chemie, Universität Würzburg, Am Hubland, 97074 Würzburg, Germany
E-mail: brixner@phys-chemie.uni-wuerzburg.de

Abstract. A shaped ultraviolet pump – mid-infrared probe setup is employed for spectroscopy and quantum control of the photoinduced Wolff rearrangement of diazonaphthoquinone in the condensed phase.

Diazonaphthoquinone (DNQ) derivatives are the photoactive compounds of photoresist materials used in 80 % of the world market's production of integrated circuits [1]. The photolithography process is based on a photoreaction known as the Wolff rearrangement [2] in which a diazo group (N_2) is separated first and a ketene is finally formed (Fig. 1a). The existence of a carbene intermediate is still disputed. Despite DNQ's tremendous commercial significance no femtosecond spectroscopy of its photochemistry has been performed to date.

Here we report mid-infrared transient absorption experiments on DNQ dissolved in methanol in order to monitor the ultrafast kinetics of the reactant molecule and products and establish the reaction time constants and quantum yield. In addition we perform quantum control of this photoreaction using shaped femtosecond laser pulses. Whereas control concepts have proven very successful for gas-phase photo-dissociation reactions and liquid-phase selective excitation [3], controlling bond breakage in liquids is not an established technique. The motivation for control of liquid-phase femtochemistry is that here the particle density is high enough for processing macroscopic amounts of chemical substances.

The photoreaction is monitored by ultrafast vibrational spectroscopy in the mid-infrared. All presented experiments were performed with a Ti:Sa amplifier system (1 kHz, 800 nm, 80 fs). A fraction of the laser pulse energy was frequency-doubled to 400 nm and employed as pump pulses for the photoexcitation of 2-Diazo-1-naphthoquinone dissolved in methanol. Mid-infrared probe pulses were generated in a home-built two-stage optical parametric amplifier (OPA) which was tuned to the 2100 cm^{-1} region. The time resolution of the experiment was 300 fs. Due to the clear distinction of different vibrational bands, in contrast to the strong overlap of the electronic bands in the UV-VIS spectral region, different species can be identified unambiguously.

Besides bleach of the DNQ diazo vibration a vibrational band appears in the low wavenumber region and shifts to 2128 cm^{-1} (Fig. 1b) which can be assigned to the C=C=O stretching mode of the ketene. The data is fitted very well by a model that attributes the strong shift of this band towards higher wavenumbers to vibrational cooling (Fig. 1c). The time constant for relaxation of the vibrationally hot ketene was

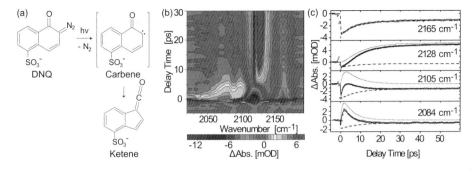

Fig. 1. Ultrafast mid-infrared spectroscopy. (a) Scheme of the photoreaction of 2-diazo-1-naphthoquinones (DNQ). (b) Spectrally and time-resolved transient absorption signal of DNQ dissolved in methanol. The data shows bleaching (negative changes in optical density) of the ground-state vibrational modes of DNQ after photoexcitation with 400 nm fs pulses. A new band appears in the lower wavenumber region and shifts to 2128 cm^{-1} within about 10 ps (c) Temporal evolution of the absorption changes at fixed spectral positions. Bleach contributions (negative) in all transients are related to the DNQ reactant. The new absorption band corresponds to the ketene (positive) and shifts to higher wavenumbers, which can be observed in the transients from 2084 cm^{-1} to 2128 cm^{-1}.

found to be 9.1 ps. From the partial bleach recovery at 2165 cm^{-1} the quantum yield for ketene formation is determined to be 0.32, and the time for vibrational relaxation in the DNQ ground state is 8.7 ps. Our observations indicate that there exists a direct pathway to the ketene without the intermediacy of the carbene.

The effect of shaped excitation pulses on the reaction of DNQ is investigated by variation of the linear chirp and in closed-loop optimization experiments employing a learning algorithm. In all our control experiments the delay position was fixed to 50 ps after excitation so that the yield of the vibrationally relaxed ketene photoproduct at 2128 cm^{-1} could be probed. Since the phase-shaped 800 nm pulses were frequency-doubled, normalization was necessary due to the dependence of the SHG efficiency on the pulse shape.

In Fig. 2a the dependence of the normalized ketene yield on the quadratic spectral phase is shown for different excitation intensities. For small pulse intensities starting with 0.6×10^{10} W/cm^2 almost no dependence on the linear chirp is found. When the pulse intensity is increased the chirp dependence becomes more pronounced, until at the highest intensity employed a distinct minimum can be observed at a negative quadratic spectral phase around -700 fs^2. Hence, a slightly down-chirped pulse leads to less effective photoproduct formation than the transform-limited pulse at zero quadratic phase. For mechanistic insight, we conducted pump-probe experiments with chirped pump pulses and observed the same temporal evolution of transient spectra as for unshaped pump pulses (apart from the overall increased/decreased signal). This allows for the conclusion that the reduction/enhancement of the product yield is due to an intrapulse dumping mechanism [4]. For negative chirp ($b_2 < 0$) the momentary frequency within the pulse decreases, so that the shrinking energy gap between the electronic potential energy surfaces can be matched and thus, population is efficiently brought back to the S_0 ground state. As a consequence, less photoproduct can be formed in consecutive reaction steps. As intrapulse dumping needs two interactions within the same laser pulse, it strongly depends on the laser intensity, which is exactly what we observe in the experiment. In contrast, for positive chirp much more photoproduct is formed, and hence the bandwidth-limited pulse is not optimal for driving the photoreaction.

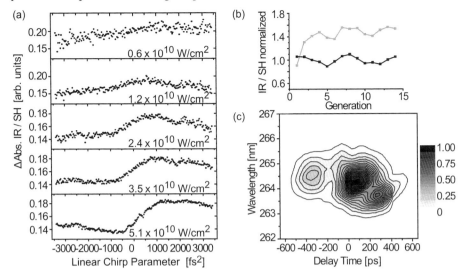

Fig. 2. Quantum control experiments. (a) Variation of the linear chirp parameter. At higher excitation intensities the chirp dependence is more pronounced. (b) Adaptive optimization of the normalized ketene yield (grey) and result for unshaped pulses (black) for comparison. (c) XFROG characterization of the optimal pulse found by the evolutionary algorithm.

In a closed-loop optimization the normalized ketene yield could be increased by about 50 % (Fig. 2b) with respect to the transform-limited pulse. In Fig. 2c an XFROG measurement of the optimal pulse reveals an up-chirped multipulse structure (wavelength decreasing with time). This frequency ordering is consistent with the previously found intrapulse dumping mechanism as it reduces population transfer back into the ground state.

Mid-infrared spectroscopy is a versatile tool for the observation of structural changes of molecules and the unambiguous identification of emerging photoproducts. Thus, it is an ideally suited method for delivering a feedback signal for open-loop and closed-loop quantum control experiments involving bond breakage, in extension of previous work on pure selective excitation control. This approach was successfully demonstrated for the photoinduced Wolff rearrangement reaction of DNQ yielding a ketene product. The kinetic evolution was determined via spectroscopy, and the yield of this reaction (that is also employed in commercial photolithography) could be increased with shaped pulses.

Acknowledgments. We would like to thank Florian Langhojer for his contributions to the pulse-shaping experiments.

1   A. Reiser, J. P. Huang, X. He, T. F. Yeh, S. Jha, J. Y. Shih, M. S. Kim, Y. K. Han, and K. Yan, Euro. Pol. J. **38**, 619, 2002.

2   W. Kirmse, Eur. J. Org. Chem. **2002**, 2193, 2002.

3   P. Nuernberger, G. Vogt, T. Brixner, and G. Gerber, Phys. Chem. Chem. Phys. **9**, 2470 2007.

4   G. Cerullo, C. J. Bardeen, Q. Wang, and C. V. Shank, Chem. Phys. Lett. **262**, 362, 1996.

Coherent control of matter waves passing through a conical intersection in β-carotene

Jürgen Hauer[1], Tiago Buckup[1], Judith Voll[2], Regina de Vivie-Riedle[2], and Marcus Motzkus[1]

[1] Department of Physical Chemistry, Philipps Universität Marburg, Hans Meerwein Str., D-35043, Germany
E-mail: motzkus@staff.uni-marburg.de
[2] Theoretical Femtoscience, Ludwig-Maximilians-Universität, D-81377 München, Germany
E-mail: imeier@princeton.edu

Abstract. The interplay between structural and electronic dynamics near a conical intersection in β-carotene is disclosed by coherent control. A low-frequency coupling mode is found to determine the ultrafast relaxation rate between the involved electronic states.

Introduction

Coherent control as a field of current research has expanded significantly in recent years. The core competence remains the steering of photo-induced processes into a desired channel while suppressing unwanted pathways. In the field of ultrafast spectroscopy, coherent control has also already proven to be an indispensable tool. The behavior of a molecule after Fourier-limited or unmodulated excitation can reveal insightful differences to the trajectory after a shaped excitation pulse. This idea forms the conceptual basis of Quantum Control Spectroscopy (QCS) and has already yielded important results on molecular vibrational dynamics and biological function unattainable by conventional spectroscopic techniques [1,2].

In this work, we employ QCS to unravel the ultrafast dynamics near a conical intersection between two excited electronic states. The system under investigation is β-carotene as depicted in figure 1a.

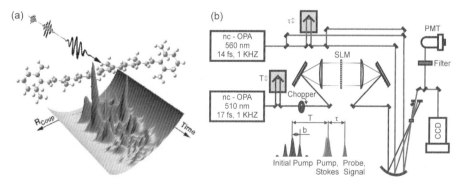

Fig. 1. . Wavepacket dynamics near β-carotene's S_2-S_1 conical intersection in (a). (b) shows the layout for a Pump-DFWM experiment.

In photosynthetic light harvesting, the carotenoid's energy after photo-excitation is passed to chlorophyll via electron transfer. In a competing channel, the system may also undergo rapid (180 fs) internal conversion to the lower lying S_1-state. It has been experimentally demonstrated that the ratio between these reaction pathways can be influenced by selectively addressing a low frequency torsional mode [2]. Building up on these findings, we pursue the following questions: *What is the precise role of the low-frequency torsional mode? How does this mode relate to the properties of the conical intersection between S_2 and S_1?*

Experimental Methods

To address these problems, one must employ a technique which allows for the discrimination of excited state modes against their ground state counterparts. Transient absorption does not fulfill this criterion since molecular oscillations can not be unambiguously assigned to the electronic state of their origin in this method. The problem is resolved by exploiting excited state resonances in a Pump-Degenerate Four-Wave-Mixing (Pump-DFWM) experiment [3]. In this technique, an initial pump pulse (see figure 1b) promotes population to an electronically excited state. The subsequent DFWM-sequence interacts with the excited state population after an initial pump delay T and consists of a DFWM-pump and Stokes pulse arriving simultaneously and a probe pulse delayed by τ. If the DFWM-sequence is resonant with an excited state transition missing in the ground state spectrum, the signal will be enhanced upon prior interaction with the initial pump. Thus, the observed phenomena like wavepacket motion can be unambiguously attributed to the excited state. The aim of the control experiment depicted below is to determine the effect of phase-modulating the initial pump pulse. The pulses in the DFWM-sequence are all kept Fourier-limited and serve detection purposes only, while the initial pump pulse exerts coherent control. The latter is achieved by turning the initial pump into a multipulse by a phase function $\phi(\omega)=sin(b\omega+c)$. b describes the sub-pulse spacing and c their relative phase (see figure 1b).

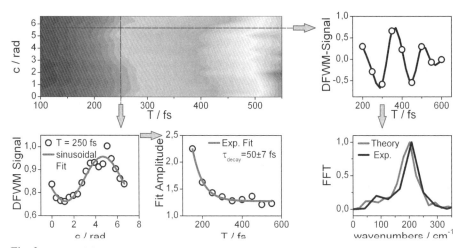

Fig. 2. A scan of the relative sub-pulse phase c against the initial pump delay time T is shown in the 2D-plot above. Slices along the c- and the T-axis reveal the loss of controllability due to the conical intersection (vertical dashed line) or an overdamped low-frequency mode coupling S_2 and S_1 respectively (horizontal dashed line).

The modulation of the initial pump pulse creates a tailored wavepacket on S_2, which seems to dephase due to dissipative propagation on the S_1 and/or the non-adiabatic passage between S_2-S_1. Only by QCS it is possible to determine the rate of this coherence loss $\tau_{decay}=50\pm7$ fs. Slices taken at specific values of c (see horizontal dashed line in figure 1) reveal another intriguing aspect of the ultrafast dynamics near the mentioned conical intersection. After subtraction of the population-related exponential rise, the signal consists of a damped low frequency oscillation in the range of 200 cm^{-1}, the origin of which becomes clear after a theoretical investigation of the $S_2 \rightarrow S_1$ non-adiabatic passage.

Based on quantum chemical and experimental data, a quantum dynamical model was constructed describing the light induced excitation to the S_2-state, as well as the coupling to the S_1-state by internal conversion. In quantum dynamical simulations we observe S_2-lifetime control depending on the multipulse-parameters b and c. By plotting the S_2 excited state wave packets of the propagations in the Wigner representation, the induced mechanism could be analyzed: In an anharmonic potential an initial state with minimum spatial and momentum uncertainty is spread with time in both spaces. We observe that by appropriate choice of the pulse train parameters, the wave packet can be refocused by the subpulses (see figure 1a). These theoretical results are in good agreement with the experimental observations.

Conclusions

Quantum Control Spectroscopy (QCS) is employed to reveal two aspects of the $S_2 \rightarrow S_1$ non-adiabatic passage in β-carotene. First, the time scale of the loss of phase memory imprinted on the wavepacket on S_2 is determined to be in the range of 50 fs. Second, the low-frequency coupling mode between S_2 and S_1 is identified at 200 cm^{-1}, indicating a slow and over-damped back and forth motion in the vicinity of the conical intersection. It has to be noted that the demonstration of these two points is only feasible by QCS and hidden from conventional spectroscopic techniques.

1   T. Buckup, T. Lebold, A. Weigel, W. Wohlleben, and M. Motzkus in Journal of Photochemistry and Photobiology A-Chemistry **180**, 314, 2006.

2   J. L. Herek, W. Wohlleben, R. J. Cogdell, D. Zeidler, and M. Motzkus in Nature **417**, 533, 2002.

3   J. Hauer, T. Buckup, and M. Motzkus in Journal of Physical Chemistry A **111**, 10517, 2007

Mode selective single-beam coherent anti-Stokes Raman scattering

Paul J. Wrzesinski[1], Haowen Li[2], D. Ahmasi Harris[1], Bingwei Xu[1], Vadim V. Lozovoy[1] and Marcos Dantus[1*]

1 Department of Chemistry, Michigan State University, East Lansing MI 48824
2 BioPhotonic Solutions Inc. Okemos MI
* Corresponding email: dantus@msu.edu

Abstract: We report the detection of chemicals using a single-beam coherent anti-Stokes Raman scattering (CARS) technique. Characteristic Raman lines for several chemicals were successfully obtained from a 12 m standoff distance.

1. Introduction

Standoff detection of chemicals and explosives remains a significant challenge in science and technology. Currently, several spectroscopic techniques are being explored for standoff detection. These techniques take advantage of the high intensity and directionality of lasers. Laser-induced breakdown spectroscopy (LIBS) [1] and spontaneous Raman spectroscopy (SRS) [2] are presently among the most developed. Many difficulties arise with standoff detection because the signal needs to be collected from large distances in the presence of other sources of light, and the desired molecules are in a highly complex environment.

A newer approach to standoff explosive detection is coherent anti-Stokes Raman scattering (CARS). CARS is a third order nonlinear process, which typically involves the interactions of the materials with three laser beams (pump, Stokes, and probe). The anti-Stokes Raman signal carries the fingerprint vibrational responses of the material providing molecular information. This is a distinct advantage over LIBS, which generally provides only atomic information. This can be used to distinguish different molecules, including geometric isomers, based on their unique spectroscopic information. Furthermore, due to the coherent stimulation of the Raman process, CARS requires much less power than LIBS and SRS, thereby making it much safer. Additionally, it is possible to control the coherent excitation of the sample to prevent false positive/negative results.

Recently, Scully and coworkers reported the first femtosecond adaptive spectroscopic technique (FAST) CARS spectrum from anthrax markers obtained at 0.2 m [3]. However, the FAST CARS approach requires three different femtosecond lasers with different wavelengths and a complex experimental set-up. Fortunately, it is possible to achieve single-beam CARS (SB-CARS) following the approach developed by Silberberg [4]. Single-beam CARS removes the complexity of multi-beam CARS. The use of one single laser beam makes the CARS technique capable of standoff detection.

Here we present remote molecular identification based on an SB-CARS implementation, using the sub-5 fs pulses emerging from a hollow waveguide, and describe the key experimental elements that make this approach possible. Data obtained at a standoff distance of 12 m (presently limited by laboratory space) from a number of compounds in solid, liquid, and gaseous states are shown, included among them are isomeric pairs.

2. Experimental setup

Femtosecond laser pulses from a Micra oscillator (Coherent) are shaped by a 4f reflective pulse shaper with a 128-pixel phase only programmable liquid crystal spatial light modulator (SLM) (CRi). The shaped pulses are amplified (Legend USP, Coherent) and focused into an Argon-filled hollow waveguide. The first shaper uses multiphoton intrapulse interference phase scan (MIIPS)[6, 7] to compensate for all orders of phase distortions of the laser pulse in the amplifier cavity, making the amplifier laser output transformed limited. In our experiments, a hollow waveguide generates an output pulse from 500 to 1000 nm [5]. Optimum continuum generation in the hollow waveguide was obtained when the output pulses were chirped by -500 fs^2. The output laser beam is then collimated and shaped by a second 4f pulse shaper, which has a 640-pixel amplitude and phase dual mask SLM. After the second shaper, the laser beam is reflected from a beam splitter and loosely focused on the sample at a distance of 12 m by a home-built Newtonian telescope. The second shaper also uses MIIPS to measure and compensate the phase distortion of the hollow waveguide output and those introduced by propagation in air to the target. Pulses as short as 4.8 fs have been measured from this setup. The second shaper is also used to control the phase and polarization [8] of the ultra-broadband laser pulse. Wavelengths shorter than 765 nm are blocked at the shaper Fourier plane as they overlap the CARS spectrum. Further pulse shaping can be done in the form of binary phase functions [9] tuned to selective excite individual vibrational modes, while eliminating the non-resonant background. The laser power is kept below the damage threshold of all the samples; no detectable damage or continuum light generation is observed during the CARS measurements. A back mirror is placed behind the sample to retro-reflect the CARS signal back to the beam splitter for standoff detection. After the signal passes a horizontal polarizer and a low pass filter, a compact QE65000 spectrometer (Ocean Optics) is used to record the SB-CARS spectrum.

3. Results and Discussion

Figure 1A shows the unprocessed SB-CARS spectra of liquid chloroform (CHCl$_3$) with and without correction of chromatic dispersion. The SB CARS signal increases by more than one order of magnitude upon MIIPS compensation.

Figure 1B, shows the SB-CARS measurements of several liquid, samples at a distance of 12 m. Liquid toluene, o- and m-nitrotoluene are measured using 1 mm path length quartz cuvettes. CARS spectra from solids and vapors were also measured (data not shown). In Figure 1B, the non-resonant background is removed by subtraction of a CARS spectrum obtained without the applied polarization and phase gates. Figure 1C shows unprocessed spectra in with selective excitation of the v_1 breathing or v_{12} in-plane bending modes of m-xylene, along with the complete elimination of non-resonant background, as the result of binary phase shaping [10].

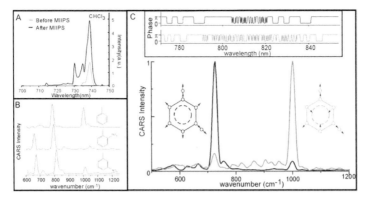

Figure 1 A) SB-CARS unprocessed spectra of $CHCl_3$ with and without compensation of chromatic dispersion using MIIPS B) SB-spectra from liquid toluene, o-nitrotoluene and m-nitrotoluene; C) Mode selective excitation of m-xylene.

Our approach is being developed for the detection of films and residues on solid targets. We have determined that the signal from SB-CARS is generated from the first 10-100 micron of material. Provided the solid target has a partial reflection the CARS signal is not affected by the distance. Based on our observations from 0.05 m to 10 m, we conclude that a 100 m distance detection would also be feasible provided we use a much larger telescope and detection system. Remote detection using single-beam CARS at 100 m is currently under development in our lab [11].

4. References

[1] F. C. DeLucia, A. C. Samuels, R. S. Harmon, R. A. Walters, K. L. McNesby, A. LaPointe, R. J. Winkel, and A. W. Miziolek, "Laser-induced breakdown spectroscopy (LIBS): A promising versatile chemical sensor technology for hazardous material detection," IEEE Sensors Journal **5**, 681-689 (2005).

[2] J. C. Carter, S. M. Angel, M. Lawrence-Snyder, J. Scaffidi, R. E. Whipple, and J. G. Reynolds, "Standoff detection of high explosive materials at 50 meters in ambient light conditions using a small Raman instrument," Applied Spectroscopy **59**, 769-775 (2005).

[3] D. Pestov, R. K. Murawski, G. O. Ariunbold, X. Wang, M. C. Zhi, A. V. Sokolov, V. A. Sautenkov, Y. V. Rostovtsev, A. Dogariu, Y. Huang, and M. O. Scully, "Optimizing the laser-pulse configuration for coherent Raman spectroscopy," Science **316**, 265-268 (2007).

[4] N. Dudovich, D. Oron, and Y. Silberberg, "Single-pulse coherently controlled nonlinear Raman spectroscopy and microscopy," Nature **418**, 512-514 (2002).

[5] M. Nisoli, S. DeSilvestri, and O. Svelto, "Generation of high energy 10 fs pulses by a new pulse compression technique," Applied Physics Letters **68**, 2793-2795 (1996).

[6] V. V. Lozovoy, I. Pastirk, and M. Dantus, "Multiphoton intrapulse interference. 4. Characterization of the phase of ultrashort laser pulses," Optics Letters **29**, 775-777 (2004).

[7] B. Xu, J. M. Gunn, J. M. Dela Cruz, V. V. Lozovoy, and M. Dantus, "Quantitative investigation of the multiphoton intrapulse interference phase scan method for simultaneous phase measurement and compensation of femtosecond laser pulses," Journal of the Optical Society of America B-Optical Physics **23**, 750-759 (2006).

[8] D. Oron, N. Dudovich, and Y. Silberberg, "Femtosecond phase-and-polarization control for background-free coherent anti-Stokes Raman spectroscopy," Physical Review Letters **90**, 213902 (2003).

[9] V. V. Lozovoy, B. W. Xu, J. C. Shane, and M. Dantus, "Selective nonlinear optical excitation with pulses shaped by pseudorandom Galois fields," Physical Review A **74** (2006).

[10] H. Li, D.A. Harris, B. Xu, P. Wrzesinski, V. Lozovoy, and M. Dantus, " Coherent mode-selective Raman excitation toward standoff detection," Optics Express (Submitted January 2008)

[11] This work is funded by a grant from the Army Research Office. The content of the information does not necessarily reflect the position or the policy of the Government, and no official endorsement should be inferred.

Early Time Vibrationally Hot Ground-State Dynamics in β-Carotene Investigated with Pump-Degenerate Four-Wave Mixing (Pump-DFWM)

Tiago Buckup, Jürgen Hauer, Jens Möhring and Marcus Motzkus

Physikalische Chemie, Philipps Universität Marburg, D-35043 Marburg, Germany
E-mail: motzkus@staff.uni-marburg.de

Abstract. Pump-DFWM is used to study the early events in structural and electronic population dynamics of the S_2, S_1 and hot-S_0 states of β-carotene. New evidence to the existence of a long-lived hot-S_0 is discussed.

Introduction

Carotenoids perform a variety of critical functions in nature including acting as photoprotection chromophores, structural elements and accessory pigments in light harvesting. Their photochemistry depends strongly on the deactivation dynamics of the first two excited low-lying singlet states [1]. In the current standard energy model for carotenoids, the excitation by a blue-green photon in the visible region is to the S_2 state ($1B_u^+$), the first allowed one-photon transition from the ground state S_0 ($1A_g^-$). The population then decays through two channels: fast internal conversion to the lower-lying singlet excited state S_1 ($2A_g^-$) or directly back to the ground state emitting fluorescence. This simple three level model was challenged by theoretical calculations of *Tavan and Schulten* who predicted the presence of 2 or 3 additional dark states lying in the vicinity of the S_2 state, which could also be active in the deactivation network [2]. Several experimental studies have addressed the spectral characteristics of these dark states. The red-wing of the S_0-S_2 absorption band (usually called S*) and its possible linkage to the proposed dark states were hotly debated [3-6]. In this work, we investigate further this hypothesis using a time-resolved method, called pump-DFWM [7,8].

Experimental Methods

In Pump-DFWM, the initial pump promotes the system to its first optically allowed state S_2 (Fig.1(a)). Since the Degenerate-Four-Wave-Mixing sequence (DFWM) is set resonant with the $S_1 \rightarrow S_n$ excited state transition, the signal will be dramatically enhanced only if the initial pump is on. In this set up, two different time scales are considered. The first is the initial pump delay T between the initial pump pulse and the DFWM-sequence's pump and Stokes pulses. Along this axis, the evolution of the deactivation network can be monitored. The other time axis along the probe delay τ is spanned between the simultaneously arriving pump and Stokes pulses and the probe pulse. If τ is varied, the molecular vibrations are detected as fast oscillations superimposed on the slowly decaying DFWM-signal. After a fast Fourier transformation of these oscillations at different values of T, the structural dynamics of the system can be visualized as seen in figure 1(c).

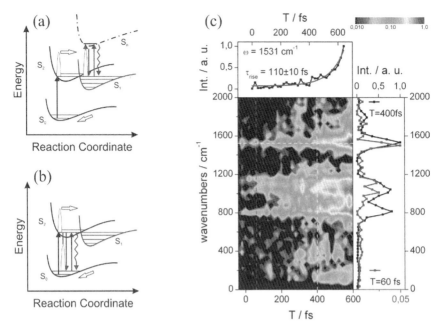

Fig. 1. The employed excitation scheme is shown in (a). For small values of the initial Pump delay T (~60fs) (b), modes in the ground-state S_0 are excited via stimulated emission pumping DFWM (SEP-DFWM). For larger initial Pump delays T, excited state dynamics become observable. When the probe delay time τ between the probe and the other DFWM pulses is scanned for every pump delay T, Fourier spectra as seen in (c) are obtained. The temporal evolution of the excited state vibrational modes can be investigated. The observed excited state modes exhibit individual growth times.

Results and Discussion

Figures 2(a-e) show mono-exponential fits to the Pump-DFWM signal at different detection wavelengths. The axis of the probe delay τ within the DFWM sequence is plotted against the initial pump delay T. This perspective allows one to observe the change in electronic structure of the investigated molecule. A common feature of figures 2(a-e) is the slow increase of the signal along the T-axis after T > 160 fs. This is due to the population build-up on S_1. Hence, the increase of the signal along the T-axis is explained by the electronic population sliding into the Franck-Condon window of detection defined by the spectral properties of the DFWM sequence. For early initial pump delays T when S_2 is still populated however, a signal with a long life time along the τ-axis occurs. Considering that the relative intensity of this signal increases when detected closer to the $S_0 \rightarrow S_2$ resonance (see figure 2(f)), we propose that for T < 180 fs the DFWM sequence dumps the excited state population back to a vibrationally hot ground state. A similar mechanism has been exploited in the frequency domain for stimulated emission pumping DFWM (SEP DFWM) (Fig.1 (b)) [9].

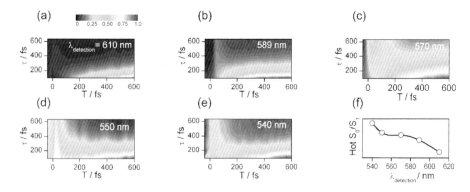

Fig. 2. The slowly decaying part of the Pump-DFWM signal is plotted against the probe delay τ for different values of the initial pump delay T. The electronic population dynamics is thus visualized. In figures (a) to (e), the dependence on the detection wavelength is shown. The increase of the signal with T is due to population of the resonantly excited S_1 state. The color code is set relative to the maximum in each plot. The long lived component for initial pump delay $T < 100$ fs is attributed to a long lived vibrationally hot ground state as argued by the contribution of this component at different detection wavelengths shown in (f).

Conclusions

Pump-DFWM combines high-spectral (down to $10\ cm^{-1}$) and temporal resolution (20 fs) to yield information on electronic as well as structural dynamics. In comparison to, for example, time-resolved stimulated Raman techniques, the results obtained with Pump-DFWM are more comprehensive due to the possibility to spectrally disperse the signal and hence to obtain simultaneously information from different Franck-Condon regions of the excited state surface. This was exploited here to investigate the early population dynamics in β-carotene, providing new evidence to the existence of a long lived vibrationally hot ground state.

1 T. Polivka, and V. Sundstrom, in Chemical Review, Vol. 104, 2021, 2004.

2 P. Tavan, and K. Schulten, in Journal of Chemical Physics, Vol. 85, 6602, 1986.

3 G. Cerullo, D. Polli, G. Lanzani, S. De Silvestri, H. Hashimoto, and R. J. Cogdell, in Science, Vol. 298, 2395, 2002.

4 T. Buckup, J. Savolainen, W. Wohlleben, J. L. Herek, H. Hashimoto, R. R. B. Correia, and M. Motzkus, in Journal of Chemical Physics, Vol. 125, Art. No. 194505, 2006.

5 K. Furuichi, T. Sashima, and Y. Koyama, in Chemical Physics Letters, Vol. 356, 547, 2002.

6 D. Niedzwiedzki, J. F. Koscielecki, H. Cong, J. O. Sullivan, G. N. Gibson, R. R. Birge, and H. A. Frank, in Journal of Physical Chemistry B Vol. 111, 5984, 2007.

7 J. Hauer, T. Buckup, and M. Motzkus, in Journal of Physical Chemistry A, Vol. 111, 10517, 2007.

8 T. Hornung, H. Skenderovic, and M. Motzkus, in Chemical Physics Letters, Vol. 402, 283, 2005.

9 P. H. Vaccaro, Molecular Dynamics and Spectroscopy by Stimulated Emission Pumping, World Scientific Publisher, New York, 1994.

Surface Femtochemistry: Investigation and Optimization of Bond-Forming Chemical Reactions

Patrick Nuernberger[1,2], Daniel Wolpert[1], Horst Weiss[3], and Gustav Gerber[1]

[1] Physikalisches Institut, Universität Würzburg, Am Hubland, 97074 Würzburg, Germany
[2] Institut für Physikalische Chemie, Universität Würzburg, Am Hubland, 97074 Würzburg, Germany
[3] BASF AG, Polymer Research Division, 67056 Ludwigshafen, Germany
E-mail: gerber@physik.uni-wuerzburg.de

Abstract. We investigate femtosecond laser-induced surface reactions by varying the properties of the surface, the reactant gases, and the laser. In optimal control experiments, we selectively manipulate the bond-forming catalytic reactions.

Femtosecond spectroscopy techniques have made it possible to manipulate several chemical processes on metal surfaces [1,2], and a also few chemical reactions could be induced, e.g. the oxidation of carbon monoxide in the presence of oxygen [3]. We have recently reported the successful initiation and quantum control of the reaction of hydrogen and carbon monoxide on a Pd(100) single crystal with femtosecond lasers [4]. This reaction is further investigated by variations of the surface material, the adsorbed gas species, and the properties of the laser, in order to infer information about the role of these reactants involved in the reaction mechanism.

In our experiments, a femtosecond laser (80 fs, 800 nm) is focused onto a single crystal in a time-of-flight mass spectrometer (TOF-MS) used to detect ions produced by the laser. Dosed with mass flow controllers, different gases are effusively streamed via a nozzle and a skimmer onto the surface. Experiments are performed at a temperature of 290 K and at a vacuum chamber pressure of up to 10^{-4} mbar with the highest amounts of gas. Laser intensities are chosen so that the formation of metal ions is marginal. The laser pulses can be additionally phase-shaped in a liquid-crystal display spatial light modulator.

When molecular hydrogen is streamed onto the laser-irradiated surface, three immense peaks appear in the ion spectrum, which correspond to H^+, H_2^+, and H_3^+ [Fig. 1(a)]. If only CO enters the chamber, C^+, O^+, and CO^+ can be observed. Additional ion peaks appear when mixtures of CO and H_2 are employed, which can be attributed to (multiply) hydrogenated species of carbon, oxygen, and carbon monoxide. The existence of these ions substantiates the bond-forming reaction mechanism involved in these experiments.

The assignment of the observed product ions is verified by replacing H_2 with D_2. As expected, the peaks in the ion spectrum [Fig1(b)] are separated by 2 atomic mass units (amu), so that e.g. the heavy water ion D_2O^+ appears at 20 amu. One can also infer that the bond-forming reactions are less efficient compared to the experiment with H_2, as can e.g. be seen by the very weak D_3O^+ signal at 22 amu.

To further investigate the role of the surface in these reactions, additional experiments have been performed, which provide multiple evidence for the surface's importance: Firstly, a variation of the linear polarization of the laser, performed since signals in laser surface spectroscopy are often polarization-sensitive [5], leads to a

445

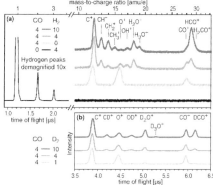

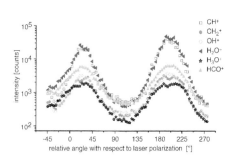

Fig. 1. Ion spectra (vertically offset for clarity) with different amounts of CO and H_2 or D_2, respectively. Gas amounts streaming through the nozzle are given in standard cubic centimeters per minute (sccm). (a) Spectra with H_2 only (lowest curve), CO only, an equal mixture of CO and H_2 and an H_2 excess (from bottom to top). The signal at early flight times is just shown for the H_2-only measurement, and is demagnified by a factor of 10 to give an impression of the hydrogen ion peaks; (b) Spectra for CO and D_2.

Fig. 2. Integrated signal of the product ion peaks as a function of linear polarization. $0°$ corresponds to vertical laser polarization at the mirror in front of the vacuum chamber. Due to the relative geometry of the laser beam and the Pd surface, the maximum of the signal is not at $0°$, but rather at polarization angles where the polarization component parallel to the normal of the palladium surface is maximal. Note the logarithmic ordinate.

strong dependence of the overall signal, which is highest when the field component parallel to the surface normal is maximal (Fig. 2). This observation may be interpreted as an effect of CO molecules adsorbing with the carbon binding to the metal, and the molecular axis pointing along the direction of the surface normal.

Secondly, when CO is replaced by carbon dioxide (data not shown), no CO_2-related ions could be observed for all provided amounts of CO_2 and all applied laser intensities, indicating that CO_2 does not adsorb on the Pd surface or at least not in such a way that it could be detected by laser ionization in an analogous manner as CO.

Thirdly, replacing the Pd(100) with a Pt(100) single crystal, CH^+, OH^+, and H_2O^+ ions are formed, but very inefficiently (Fig. 3). An excess of H_2 leads to a shrinking of all CO-related peaks, in contrast to experiments with the Pd surface, which confirms that the surface properties are essential for the initiated catalytic reactions.

To get an estimate for the time scale underlying the observed reactions, pump-probe experiments with two identical pulses are performed (data not shown). The relative ratio of the peaks does not change drastically for different pump-probe delays, and the transients resemble the cross-correlation of the two pulses. The reaction mechanism hence is not a phonon-mediated heating process (lasting typically tens of picoseconds), but possibly involves electronic excitations of the adsorbate, analogous to the laser-induced oxidation of CO to CO_2 which is due to laser-generated hot substrate electrons, giving rise to sub-picosecond reaction dynamics along a new reaction pathway by coupling to the adsorbate [3]. However, our experiments failed with 400 nm pulses, an indication that the dominant mechanism may not be hot substrate electrons, but rather an excitation of the adsorbed molecules.

The relative proportion of the two gases can be easily changed, allowing an analysis of the reaction mechanism by employing optimal control as a tool to

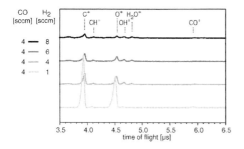

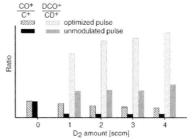

Fig. 3. Ion spectra (vertically offset for clarity) with different amounts of carbon monoxide and hydrogen streamed onto a Pt(100) crystal. For all curves, the amount of CO is held constant at 4 sccm. Spectra with little H₂ (lowest curve), an equal mixture of CO and H₂, and excesses of H₂ are shown (from bottom to top).

Fig. 4. Variation of adsorbate composition after an adaptive optimization. After maximization of the ratio DCO^+/CD^+ with 4 sccm CO and 4 sccm D_2, the amount of D_2 is reduced. The graph shows both the fitness goal DCO^+/CD^+ (gray) and the ratio CO^+/C^+ (black). With decreasing D_2 amounts, the optimization effect diminishes, and the optimal pulse has no effect on the CO^+/C^+ ratio in the absence of D_2.

determine whether the dissociation of CO is the sole step dominating the reaction. In an adaptive quantum control experiment, an optimized pulse shape is determined which increases an ion ratio relative to an unmodulated pulse [4]. After that, we test the optimization effect with respect to variations in the adsorbate composition. With this procedure, we want to explore whether the optimization is insensitive to hydrogen or deuterium and is achieved via control of CO dissociation only.

In the associated experiment, a mixture of CO and D_2 is used and the ion ratio DCO^+/CD^+ is maximized. Subsequent reduction of the D_2 amount changes the conditions on the surface and the optimization effect diminishes (gray bars in Fig. 4). A simultaneous analysis of the ratio CO^+/C^+ (black bars) reveals that the optimal pulse obtained with D_2 also influences this ratio, but has no effect on it anymore if D_2 is absent. Therefore, the optimal pulse does not simply control CO dissociation independent of the other experimental conditions, but is adapted to the situation during the optimization, i.e. with both CO and D_2 being present at the surface.

In summary, the experiments have demonstrated the feasibility of laser-induced catalytic reactions of carbon monoxide and hydrogen (or deuterium) under high vacuum conditions, and have disclosed the importance of the surface material, its adsorbates and the laser polarization. Application of an optimal control scheme has further shown that the reaction mechanism is nontrivial and sensitive to the adsorbate composition.

1   H. Petek and S. Ogawa, Annu. Rev. Phys. Chem. **53**, 507, 2002.

2   Y. Matsumoto and K. Watanabe, Chem. Rev. **106**, 4234, 2006.

3   M. Bonn, S. Funk, C. Hess, D. N. Denzler, C. Stampfl, M. Scheffler, M. Wolf, and G. Ertl, Science **285**, 1042, 1999.

4   P. Nuernberger, D. Wolpert, H. Weiss, and G. Gerber, in Ultrafast Phenomena XV, Edited by. P. Corkum, D. Jonas, R. J. D. Miller, and A. M. Weiner, Springer Series in Chemical Physics **88**, 237, 2007.

5   H.-L. Dai and W. Ho (Eds.), Laser spectroscopy and photochemistry on metal surfaces, World Scientific, Singapore, 1995.

Coherent Control of the Exciton Dynamics in the FMO Protein

M.T.W. Milder[1], B. Brüggemann[2], M. Miller[3], and J.L. Herek[1,4]

[1] FOM-Institute for Atomic and Molecular Physics (AMOLF) , Kruislaan 407, 1098SJ, Amsterdam, The Netherlands
[2] Humboldt Universität, Institut für Physik AG Halbleitertheorie, Newtonstr. 15, D-12489, Berlin, Germany
[3] University of Southern Denmark, Department of Biochemistry and Molecular Biology, Campusvej 55, DK-5230, Odense, Denmark
[4] Optical Sciences Group, MESA+ Institute for NanoTechnology, University of Twente, 7500 AE Enschede, The Netherlands
E-mail: milder@amolf.nl

Abstract. We have achieved first steps toward coherent control of excitonic energy migration in the FMO pigment-protein complex, by combining femtosecond pulse shaping with a feedback loop using an evolutionary algorithm. The experimental conditions achieved, with a rotating sample, a cryostat, and a pulse shaper, are sufficient for closed loop optimizations.

Introduction

In 1975 Roger Fenna and Brian Matthews solved the X-ray structure of the water soluble bacteriochlorophyll a (BChla) protein [1]. It was isolated a decade before from green sulfur bacteria by John Olson, hence the name the Fenna-Matthews-Olson (FMO) protein [2]. The protein complex is a trimer with three identical subunits, each consisting of seven BChla pigments surrounded by a protein shell. Within green sulfur bacteria the complex is part of the photosynthetic pathway and takes care of the transfer of energy from the light-harvesting antenna to the reaction center. Since the crystal structure of the FMO complex was the first of the photosynthetic proteins to be resolved, it has been explored by a wide range of spectroscopic studies. One of which, 2D electronic spectroscopy performed by Brixner et al., revealed directly the coupling between the exciton levels [3]. Combination of the results of experiments with simulations linked the spectroscopic properties with the structure of the system.

Experimental Methods

The seven BChla pigments in the FMO protein give rise to seven excitonically coupled states. The resulting low temperature transient absorption spectrum (77K) shows three distinct peaks at 805, 815 and 825 nm respectively. These "exciton peaks" show up in the transient spectra obtained by pump-probe spectroscopy and change in time representing the energy decay along the seven exciton levels. Our aim is to influence the pathway of energy decay in this system by using coherent control. This technique uses shaped broadband laser pulses to steer a quantum system into a desired direction. The "optimal" pulse for this preset output state is often obtained in a closed loop experiment (fig.1.). In these experiments a signal representing the desired target state is used as a feedback signal, which is optimized by an evolutionary algorithm [4].

Coherent control of excitonic wavepackets has been demonstrated before in biological systems both by experiments [5] and by computational studies [6]. We try to apply this to the FMO complex by monitoring the exciton decay by pump-probe spectroscopy and shaping the pump pulse.

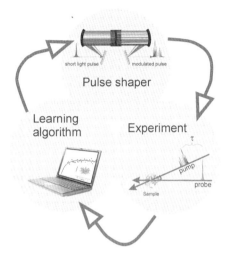

Fig. 1. Closed loop "blind" optimization of the highest exciton energy band in the FMO complex at 805 nm using phase only shaping. Feedback signals are derived from the 77K transient absorption spectrum.

Results and Discussion

The excitation energy in the FMO complex decays downwards starting from the state at 805 nm within ~10 ps. As a target we chose to optimize the intensity of the band at 805 nm at 3 ps delay. The pulse shaper in our experiment (CRi SLM-640) is capable of shifting the pulse in time by ± 6 ps simply by adding a linear chirp. To prevent the trivial solution of the arrival of the pump pulse 3 ps later we therefore restricted the linear chirp to have zero slope, in order to fix the center of mass in the time domain around t_0. The resulting optimal pulse shape of the optimization still partly shows the trivial solution of a peak at 3 ps (fig.2.). However, there is significant additional structure to the pulse that suggests the involvement of additional processes. We are currently trying to see if the optimized pulse induces any coherent processes by comparing data and simulations of the time-dependent pump-probe spectra with the transform limited pulses and with the optimized pulses.

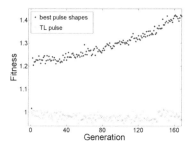 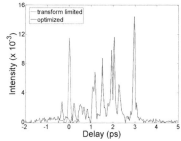

Fig. 2. [Left] Learning curve of the optimization of the 805 nm exciton band at 3 ps pump-probe delay. The monitor the laser drifting the fitness value of the TL pulse was measured before each new generation. [Right] XFROG signal showing the TL pulse (dashed line) and the optimized pulse shape (solid line) at 400 nm.

Conclusions

We have performed femtosecond pump-probe studies using transform limited and phase-shaped pulses in an attempt to manipulate the pathway of excitonic energy migration within the FMO complex. While the learning curve is complicated due to an underlying trivial control effect, the resulting complex pulse shapes suggest a mechanism involving low-frequency vibrational modes. Further experiments combined with simulations are needed to fully understand this effect.

1   R. Fenna, B. Matthews, Nature **258**, 573, 1975

2   J.Olson, C. Romano, Biochimica Biophysica Acta **59**, 726, 1962

3   T. Brixner, J. Stenger, H. M. Vaswani, M. Cho, R.E. Blankenship and G. R. Fleming, Nature, **434**, 625, 2005

4   R.S. Judson, H. Rabitz, Phys. Rev. Lett. **68**, 1500, 1992

5   J. L. Herek, W. Wohlleben, R. J. Cogdell, D. Zeidler, M. Motzkus, Nature **417**, 533, 2002

6   B. Brüggemann and V. May, J. Phys. Chem. B **108**, 10529, 2004

Coherent Control of Population Transfer in an Ionic Multilevel System using Phase- and Amplitude-Shaped Femtosecond Pulses

Andreas Galler[1] and Thomas Feurer[1]

[1] Institute of Applied Physics,
   University of Bern, CH-3012 Bern, Switzerland
   E-mail: andreas.galler@iap.unibe.ch

Abstract. We demonstrate selective control of population transfer in a multilevel system through phase- and amplitude-modulated femtosecond pulses. A combination of adiabatic rapid passage and amplitude modulation allows controlling the final population of individual states.

Introduction

Coherent control is of great interest for spectroscopic applications and more so for controlling dynamical processes in chemistry, physics, and biology [1,2]. One major goal of coherent control is to selectively populate a single excited state in a manifold of excited states and, moreover, to achieve a 100% population transfer to that particular state. A very efficient tool to realize high population transfer rates is the adiabatic rapid passage (ARP) [3]. Here, the quantum system is irradiated with a frequency swept laser pulse with the sweep rate slow enough to ensure adiabatic conditions. While the final population transfer for transform-limited pulses depends on the pulse area and, thus, is very sensitive to intensity fluctuations, it is always close to 100% for chirped pulses fulfilling the ARP condition irrespective of their pulse area and, therefore, robust against fluctuations. It has been shown the ARP conditions can be used to selectively populate one out of two excited state levels in a V-type level system [4]. Simply a change of the sign of the frequency sweep, i.e. the chirp, determines whether the lower or the upper excited state level is populated. If the instantaneous frequency crosses first the lower resonance and then the upper resonance all population will end up in the lower excited state and vice versa. In a more recent publication [5] it was shown that a similar degree of selectivity can be obtained by combining ARP with optical interference. Here, two identical but time-delayed replica of a chirped pulse are overlapped such that one of the two resonance frequencies is missing in the two-pulse spectrum. Contrary to the previous scheme there is no need to reverse the sign of the chirp; in other words, the chirp remains constant in the ARP regime and amplitude shaping through optical interference is used to remove the frequency component which corresponds to the undesired resonant transition. In principle, this scheme has the potential to be extended to more than two levels. Especially, when the interferometer in reference [5] is replaced by a pulse shaper and amplitude modulation rather than optical interference is used to remove the undesired frequency components. Here, we apply this scheme to coherently control the population dynamics in the complex multilevel system of the Er3+ ion in a crystalline host.

Experiment and Simulation

We use an Erbium doped Yttrium Aluminum Oxide crystal and the quantum system of interest here is the Er^{3+} ion. The strong electric crystal field of AlO_3 removes some of the degeneracy in the Er^{3+} levels through Stark splitting. While the ground state level $^4I_{15/2}$ splits into 8 Stark-levels, the excited state of interest, i.e. the $^4I_{9/2}$ state, splits into 5 Stark-levels. Coupling between the two level manifolds is provided by a spectrally broadband femtosecond light field. A single excited Stark-level leads to a maximum of 8 possible emission lines yielding a total number of 40 possible emission lines originating from all excited Stark-levels. Another interesting feature of the Er^{3+} ion is a possible two photon absorption reaching the $^2H_{9/2}$ state. Through a subsequent radiationless decay either the $^2H_{11/2}$ or the $^4S_{3/2}$ state is populated. Spontaneous emission from those states leads to two closely spaced green fluorescence lines. All experiments were performed with a Ti:Sapphire oscillator (KML) delivering pulses with an energy of approximately 1nJ, a bandwidth of 80 nm at a centre wavelength of 820 nm. The pulses were phase- and amplitude modulated in a standard pulse shaping apparatus consisting of a double display spatial light modulator (Jenoptik SLM640-d) in the symmetry plane of a 4f zero-dispersion compressor. They were then focused to the sample by an additional lens. The experiments are performed at room temperature. The fluorescence was detected perpendicular to the pump beam direction with a 2048 pixel Avantes spectrometer covering a spectral range of approximately 900 nm from 250 nm to 1150 nm and having an average resolution of about 0.5 nm per pixel. The simulations solve the space-time evolution of a multi-level quantum system and its driving fields. The multi-level system can be described mathematically by a set of differential equations in the rotating wave approximation. That is, Maxwell's wave equation together with Bloch's quantum-mechanical description of the field-matter interaction is solved simultaneously.

Results and Discussion

In a first series of measurements the pulses were phase modulated with a quadratic phase of different magnitude ranging from positive to negative values. The fluorescence spectra showed a significant dependence on the chirp parameter. While a transform limited pulse led only to small population transfer, a chirped pulse was able to increase the excitation rate quite substantially. The sign of the chirp played only a minor role. Further experiments were designed to investigate whether the combination of frequency sweep and amplitude modulation would be sufficient to switch selected resonant transitions on and off at will. A result is shown in figure 1a). The vertical axis refers to the position of the spectral blocking and the horizontal axis to the wavelength of the fluorescence detected.

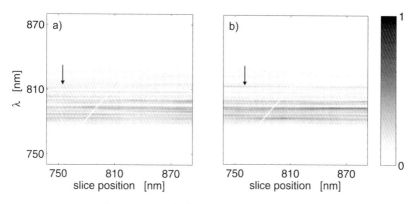

Fig. 1.) slice scan of Er^{3+} using a -10^4fs^2 chirped pulse when no extra frequency is blocked, and b) when the 810nm line is blocked in addition to the scanned slice.

A quadratic chirp, i.e. a linear frequency sweep, of -10^4fs^2 was imposed on an otherwise transform-limited femtosecond laser pulse. Its spectral width of 80 nm was sufficiently broad to cover all resonant transitions between the $^4I_{15/2}$ and the $^4I_{9/2}$ Stark levels. Generally, the fluorescence intensity of the low wavelength (high frequency) contributions is stronger than that of the longer wavelength part which is related to the negative quadratic chirp. As the spectral block is scanned through the excitation spectrum the fluorescence from the corresponding transitions disappears. In other words, blocking a specific resonance frequency in the chirped pulse has the effect that this specific level is not populated at all. In a subsequent experiment an additional frequency component at 810 nm was blocked throughout the whole scan. The result is shown in figure 1b). The fluorescence around 810 nm has disappeared and the fluorescence at 813 nm is substantially enhanced. This effect is corroborated by the simulations and confirms that the above mentioned scheme is indeed capable of selectively populating a desired state in a manifold of excited levels. More systematic studies also varying the temperature in order to simplify the observed fluorescence spectrum and to alleviate the spectroscopic assignment were done. They prove that selective excitation of a specific target state, which under the pure ARP conditions is not populated, can be reached.

Acknowledgements. This work was funded by the SNF.

1   H. Rabitz, R. de Vivie-Riedle, M. Motzkus, and K.L. Kompa, "Whither the future of controlling quantum phenomena?" Science 288, 824 (2000).
2   J.L. Herek, W.Wohlleben, R.J. Cogdell, D. Zeidler, and M.Motzkus, "Quantum control of energy flow in light harvesting," Nature 417, 533 (2002).
3   N. Dudovich, B. Dayan, S.M. Gallagher Faeder, and Y. Silberberg, "Transform-Limited Pulses Are Not Optimal for Resonant Multiphoton Transitions," Phys. Rev. Lett. 86, 47 (2001).
4   R. Netz, T. Feurer, G. Roberts, and R. Sauerbrey, "Coherent population dynamics of a three-level atom in spacetime," Phys. Rev. A 65, 043406 (2002).
5   R. Netz, A. Nazarkin, and R. Sauerbrey, "Observation of Selectivity of Coherent Population Transfer Induced by Optical Interference," Phys. Rev. Lett. 90, 063001 (2003).

Coherent control of the efficiency of an artificial light-harvesting complex

Janne Savolainen[1,2], Riccardo Fanciulli[2], Niels Dijkhuizen[2], Ana L. Moore[3], Jürgen Hauer[4], Tiago Buckup[4], Marcus Motzkus[4], and Jennifer L. Herek[1]

[1] Optical Sciences, Department of Science and Technology, MESA+ Institute for Nanotechnology, University of Twente, Postbus 217, 7500 AE Enschede, The Netherlands
E-mail: janne.savolainen71@gmail.com
[2] FOM Institute for Atomic and Molecular Physics (AMOLF), 1098 SJ Amsterdam, The Netherlands
[3] Department of Chemistry and Biochemistry, Arizona State University, Tempe, Arizona 85287, USA
[4] Physichalische Chemie, Fachbereich Chemie, Philipps-Universität, 35032 Marburg, Germany

Abstract. Coherent control over the branching ratio between competing pathways for energy flow is realised for artificial light-harvesting complex. Direct insights to the mechanism featuring quantum interference of a low-frequency mode are presented.

Introduction

Conversion of light energy into chemical potentials using artificial photosynthesis is an important challenge of science and technology today [1]. Nature has inspired systems based upon complicated natural light-harvesting complexes (LHCs) reduced to their basic elements [2]. In this contribution we perform Quantum Control Spectroscopy (QCS) on a dyad molecule that closely mimics the early-time photophysics of the LH2 photosynthetic antenna complex [3]. The system we study is inspired by the LH2 complex from the purple bacterium *Rhodopseudomonas acidophila* and consists of a single donor (carotenoid) and single acceptor (purpurin) moiety thus reducing the structural complexity significantly compared to the LH2. In order to control the pathways of energy flow in the dyad molecule we use femtosecond pulse shaping in a adaptive learning loop [4]. We extract recognisable features from the resulting pulse shapes, simplify the parameter space accordingly. This strategy provides a powerful spectroscopic tool that is sensitive to the function of the artificial LHC, thereby revealing important characteristics affecting the efficiency of the lightharvesting process. Furthermore, we show that it is possible to enhance or suppress the functional channel by pulse shapes exploiting different control mechanisms [5]. Ultimately, this approach may lead to the discovery of new design principles to aid the development of more efficient artificial light-harvesting systems.

Experimental Methods

The control measurements were made with a transient absorption setup using tailored pump pulse and an unmodulated probe pulse. An amplified fs laser system (Clark CPA2001), non-collinear optical parametric amplifier (NOPA), and a pulse shaper introduced into the beam path provided the pump pulses. In the pulse shaper a 640-pixel liquid-crystal spatial light modulator (SLM; Cambridge Research Instruments) is placed in the Fourier plane of a 4-f zero-dispersion compressor. Only phase shaping is used.

A robust calibration method, where each SLM pixel is calibrated and correction for any phase distortion is made by optimising second-harmonic generation in a non-linear crystal, ensured that the shaping introduced no effect on the amplitude of the pump pulse.

Results and Discussion

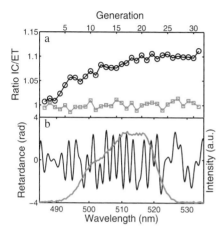 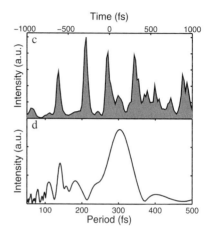

Fig. 1. Optimisation of the ratio IC/ET. (a) The learning curve shows an improvement of 10% in the fitness value of the best pulses of each generation (black circles), and the fitness of the TL pulse (grey squares), measured prior to each new generation. The initial increase of the fitness value (from TL to the first generation) due to the stretching of the pulse is subtracted from the data. (b) Pump spectrum (grey line); best found phase shape (black line). (c) Best found pulse shape. (d) FFT of the cross-correlation of the best pulse.

Fig. 1(a) shows a learning curve of an optimisation in which the target was to maximise the ratio between the two competing energy flow pathways of the dyad, internal conversion and energy transfer (IC/ET). A total of 10% increase of the ratio is obtained in the fitness values of the best pulse shapes (circles) after 31 generations. The fitness value of the transform limited (TL) pulse (squares) was determined prior to each new generation, providing an excellent indicator that the experimental conditions remained constant during the optimisation. The best found phase shape and the used pump pulse spectrum are plotted in Fig. 1(b) and the cross-correlation with an unshaped pulse in Fig. 1(c). Fig. 1(d) shows the power spectrum of the measured cross-correlation signal. Similarly, we explored a target objective aimed to improve the relative yield of the energy transfer, using fitness function ET/IC [5]. As in the case of the IC/ET optimisation, a multi-pulse structure with varying time separation between the pulses is resolved. The four-pulse structure has a total duration that is significantly shorter than the best pulse from the IC/ET optimisation, and the most pronounced sub-pulse spacing is approximately 200 fs.

The experiments show that both product channels (ET and IC) in the artificial LHC

455

are susceptible to coherent control. Using the strategy of sequentially moving from blind optimisations to a restricted parameter space and analysing the optimisations using Fourier analysis, we find that for both product channels a pulsetrain structure with varying subpulse spacings (~300 fs for IC and ~200 fs for ET) is responsible for the control. We propose a mechanism that incorporates impulsive stimulated Raman scattering (ISRS) of low-frequency skeletal modes in the ground state. Wavepacket generation on specific vibrational modes by shaped pulses, which turns into enhancement of vibrational coherence under near-electronic resonant condition, has been demonstrated in various molecules including carotenoids. By periodically modulating the phase of the laser pulse over its spectrum, it is possible to prepare wave packets selectively and, under near-resonant conditions, to enhance wave-packet excitation of Raman-active modes. The leading pulses prepare a wave packet in the vibrationally hot ground state of the carotenoid. By matching the frequency of the pulse train to a ground-state vibrational mode (e.g., a low-frequency twisting of the backbone), we introduce momentum along a trajectory that may take the wave packet toward Franck-Condon regions not accessed by a Fourier-limited pulse. Subsequently, this push leads to an altered evolution on the excited state, either toward or away from the conical intersection between S_2 and S_1. In the former case, an excitation of vibrationally hot ground state modes could lead to a more efficient IC process, averting the ET pathway.

Conclusions

The experiments show that both product channels (ET and IC) in the artificial LHC are susceptible to coherent control. Using the strategy of sequentially moving from blind optimisations to a restricted parameter space and analysing the optimisations using Fourier analysis we find that for both product channels, a pulse train structure with varying sup-pulse spacings (around 300 fs for IC and 200 fs for ET) is responsible for the control. Thus, we have found important directions on the fitness landscape describing a smaller search space still containing the optimal solution. Many repeated runs of the recorded phase shapes from the optimisations were performed indicating that the results are repeatable and robust. The efficiency of ET in the dyad depends strongly on the photophysics of the carotenoid moiety. By reducing the parameter space in combination with the Fourier analysis of obtained pulse shapes we were able to track down the functionally important features of this molecular system. A mechanism based on the periodic excitation pulse enhancing the vibrational coherence of low-frequency wavepackets via ISRS process is most likely responsible for the control, analogous to that proposed earlier for LH2 [6].

1   P.V. Kamat, J Phys Chem C, 111, 2834-2860 (2007).
2   D. Gust, T. A. Moore, and A. L. Moore, Accounts Chem Res, 34, 40-48 (2001).
3   J. Savolainen, et al. J Phys Chem B, 112, 26782685 (2008).
4   R. S. Judson, and H. Rabitz, Phys Rev Lett, 68, 1500-1503 (1992).
5   J. Savolainen, et al., PNAS, 105, 7641-7646 (2008).
6   J. L. Herek, et al., Nature 417, 533-535 (2002).

Strong Field Coherent Control Using 2D Spatio-Temporal Mapping

B. D. Bruner, H. Suchowski, A. Natan and Y. Silberberg

Dept. of Physics of Complex Systems, Weizmann Institute of Science, Rehovot, 76100 Israel
e-mail: Barry.Bruner@weizmann.ac.il

Abstract: Multiphoton excitation in Rubidium can be effectively controlled using simple pulse shaping parameters. Interplay between ionization and dynamic Stark shifts is revealed by mapping onto 2D landscapes using a recently developed spatio-temporal coherent control technique.

Introduction

Nonlinear interactions, even in relatively simple atomic systems, may be acutely sensitive to a small number of analytically derived pulse shaping parameters. Spatio-temporal coherent control is well-suited to address these problems. By simultaneously scanning both quadratic phase and a second, arbitrary phase parameter using a pulse shaper, we can map the parameter space onto a two-dimensional landscape in a very straightforward manner [1,2]. This type of multiparameter visualization is extremely valuable under strong field conditions, where effects such as Stark shifts and power broadenings contribute to the deterioration of the spectral selectivity that is essential in weak field control schemes [3,4].

Experimental Methods

The experimental setup has been described in detail in our earlier work [1,2]. Pulses from a Ti:Sapphire laser amplifier (30 fs, 1mJ pulses at 1 kHz) are shaped using a programmable liquid crystal Spatial Light Modulator (SLM), aligned in a standard 4-f configuration. Additional spectral envelope filtering is accomplished using slits in the Fourier plane at the SLM. These shaped pulses enter the spatio-temporal setup, where the beam is temporally focused using a 600 l/mm grating aligned perpendicular to the optical axis of an X10 telescope. The system of interest is placed within the temporal Rayleigh range, where the spatially-dependent spectral phase of the pulse can be written as the sum of the contributions from the SLM and from temporal focusing:

$$\varphi_{total}(\omega, z) = \varphi_{SLM}(\omega) - 2\beta z \omega^2 \qquad (1)$$

Here, φ is the spectral phase, ω is the optical frequency, $\beta*z$ is the GDD coefficient (where β is dependent on the geometry of the spatio-temporal setup), and z is the displacement along the propagation axis.

Results and Discussion

Strong field excitation in Rb_{85} gas was investigated by recording spatio-temporal CCD images for various intensities, as shown in Figure 1. The spectrum was centred at 780 nm with a 20 nm FHWM bandwidth, so that both the $5S_{1/2} \rightarrow 5P_{3/2}$ (780 nm) and the $5P_{3/2} \rightarrow 5D_{1/2}$ (776 nm) transitions fall well within the laser spectrum.

Population in the excited $5D_{1/2}$ state is measured indirectly by imaging the spatial fluorescence pattern (at 420 nm from the $6P_{1/2}$ level) using a camera mounted above the Rb_{85} cell. In the weak and intermediate field regimes, excitation occurs only for positive chirp, corresponding to a sequential population transfer via resonant one-photon interactions. As the intensity is increased, transfer via an adiabatic two-photon crossing becomes efficient for negative chirp. This is a chirped-pulse variant of the Stimulated Raman Adiabatic Passage (STIRAP) technique of population transfer [5], which can be performed in one shot using our spatio-temporal control method.

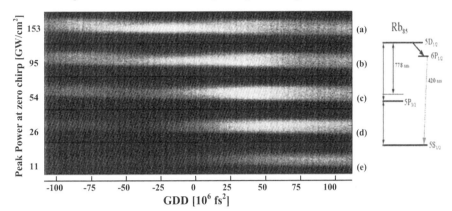

Fig. 1. Weak and strong field population transfer in three-level Rb_{85}. The unperturbed energy level structure of the atom is shown at right in (f). Spatio-temporal CCD images were acquired for the following zero-chirp field intensities, (a) 153 (b) 95 (c) 54 (d) 26 (e) 11 GW/cm^2. For weak to intermediate fields, excitation is possible only for positive chirp values, whereas for stronger intensities, population transfer with negatively chirped pulses also becomes efficient.

A more complete picture is obtained by scanning a π-step phase across the spectral envelope using the SLM. The envelope shape was symmetric, centred near 780 nm, with a FWHM of 10 nm. For each position of the π-step, the spatial fluorescence pattern is imaged as described previously and integrated along the transverse axis. The data is plotted in a 2D grid as a function of π-step position and chirp. In the weak field, resonant enhancement is induced by the π-step near the two transitions at 776 nm and 780 nm [6]. When the π-step is placed outside both of the resonant lineshapes, no enhancement via the intermediate state is observed, and we recover the low intensity result from Figure 1e, where only positively chirped pulses excite the atom.

As the intensity is increased, changes in the energy level structure emerge in the 2D mappings. With the π-step phase located far from the resonances, we again recover the high intensity features of Figure 1, where both positively and negatively chirped pulses lead to population transfer. Near the zero-chirp axis, the peak pulse intensity is high enough to ionize the atom, and no excited state fluorescence was observed. However, the three-photon ionization process is suppressed when the π-step phase is located within one of the resonant transitions. This suggests that ionization can be coherently controlled using simple pulse shapes that maintain the two-photon component while suppressing the three-photon component of the interaction.

The shape of the resonant enhancement features are altered dramatically in strong fields. The π-step enhances the signal at the wavelengths where resonant transitions occur most favourably. At zero-chirp, both resonances are significantly broadened and are spectrally blue-shifted by about 1 nm. As expected for dressed atoms [3], maximum enhancement does not occur at zero chirp. For example, we observe maximum enhancement of the overall $5S_{1/2} \rightarrow 5D_{1/2}$ transfer at positive chirp with the π-step located on the $5P_{3/2} \rightarrow 5D_{1/2}$ resonant transition. A similar enhancement occurs at negative chirp, with the π-step located on the $5S_{1/2} \rightarrow 5P_{3/2}$ resonance.

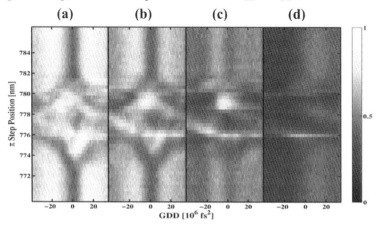

Fig. 2. 2D spatio-temporal maps of Rb_{85}. The spectral position of the π-step and the chirp are plotted along the vertical and horizontal axes, respectively. The peak intensities at zero chirp are (a) 130, (b) 60, (c) 24, (d) 10 GW/cm^2. The curvature and broadening of the resonant enhancement features shows the effect of the Stark shift on the dressed atomic energy levels.

The overall curvatures of the resonances are strongly dependent on the fluence, and the magnitude and sign of the chirp. For high fluence and large negative chirp, the red-detuned intermediate level plays less of a factor in the dynamics, and the three-level system is effectively reduced to a two-level system. Enhancement is then observed at the two-photon non-resonant frequency. In this case, as with STIRAP, population is adiabatically transferred via the so-called counterintuitive path. For large positive chirps, the adiabatic intuitive path applies, with sequential absorption via the blue-shifted resonant transitions. For lower intensities (Figures 2c and 2d), the peak curvatures are less pronounced as the resonant transition frequencies approach their weak field, intensity independent values.

In conclusion, multiphoton processes in resonant media were examined under strong field excitations. It is expected that 2D spatio-temporal mappings will continue to reveal new information about the dynamics of atoms and provide the ability to coherently control these systems using simple control parameters.

[1] H. Suchowski, A. Natan, B. D. Bruner, Y. Silberberg, J. Phys. B. **41** (7), 074008/1-9 (2008).

[2] H. Suchowski, D. Oron and Y. Silberberg, Opt. Comm. **264** (2), 482-487 (2006).

[3] M. Wollenhaupt, A. Praekelt, C. Sarpe-Tudoran, D. Liese, T. Baumert, Appl. Phys. B **82** (2), 183-188 (2006).

[4] C. Trallero-Herrero, D. Cardoza,, T. C. Weinacht, J. L. Cohen, Phys. Rev. A **71** (1), 013423/1-6 (2005).

[5] D. J. Maas, C. W. Rella, P. Antoine, E. S. Toma, and L. D. Noordam, Phys. Rev. A **59** (2), 1374-1381 (1999).

[6] N. Dudovich, B. Dayan, S. M. Gallagher Faeder and Y. Silberberg, Phys. Rev. Lett. **86** (1), 47-50 (2001).

Control of Excited-State Population and Vibrational Coherence with Shaped-Resonant and Near-Resonant Excitation

Tiago Buckup[1], Jürgen Hauer[1], Carles Serrat[2] and Marcus Motzkus[1]

[1] Physikalische Chemie, Philipps Universität Marburg, D-35043 Marburg, Germany.
  E-mail: motzkus@staff.uni-marburg.de
[2] ICFO-Institut de Ciències Fotòniques, 08860 Castelldefels, Barcelona, Spain, and
  Tecnologies Digitals i de la Informació, Universitat de Vic, 08500 Vic, Spain.
  E-mail: carles.serrat@icfo.es

Abstract. The enhancement of vibrational coherence and population transfer in solution using tailored pulses has been investigated numerically and experimentally. The general control mechanism is based on the control of the absorption coefficient after excitation with multipulses.

Introduction

Wave packet phenomena in molecules are a clear manifestation of the wave nature of matter. The manipulation of molecular vibrations has been a long term goal of coherent control, especially because it may represent motion along a reaction coordinate. In this context, pulse shapes for manipulating vibrations can be predicted to be trains of pulses with temporal spacing between the sub-pulses equal to an integer of the vibrational phase. Recently, as a result of coherent control experiments performed in closed loop approach, optimal solutions have been found which can be also interpreted as pulse trains [1,2]. If the manipulation of molecular vibrations with pulse trains is expected to be a general excitation approach on the long-standing aim of mode selective chemistry, it is necessary to understand its fundamental limits. In this work, the interaction of pulse trains with matter is discussed under the light of time-resolved nonlinear experiments (Transient Absorption) and density matrix simulations. Emphasis is put into the role of electronic coherence between excited- and ground-state and to the influence of the electronic resonance.

Experimental Methods

The pump and the probe pulses used in the transient absorption experiment were generated in two home-built, single-stage noncollinear optical parametric amplifiers (nc-OPA). The phase of the pump pulse was modulated with a liquid-crystal shaper with 128 pixels (CRI LCM). The phase applied to the pump pulse was a sinusoidal function, leading to well defined sequence of pulses. The sub-pulse spacing (parameter b in Fig.1(b)) corresponded to the vibrational period (56 fs) of the ring-breathing mode of Nile Blue (LD 6900). The probe pulse remained transform-limited in all experiments. The effect of the multipulse excitation on the amplitude of the excited state absorption and of the oscillations was investigated with this fixed sub-pulse spacing for several excitation wavelengths (see Fig.1).

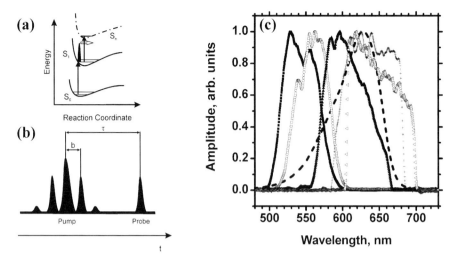

Fig. 1. (a) A tailored pump was used to excite the S_0-S_1 transition and an unshaped probe to test the population and vibrational coherence in the excited state. (b) Pulse sequence used in the transient absorption. (c) Linear absorption spectrum of Nile Blue in methanol (traced line) and the five different pump excitation spectra (symbols) used in this work.

Results and Discussion

Recently we showed that it is possible to enhance the amplitude of vibrations as well as the excited state population in transient signals using pulse trains with the same total incident energy as transform limited ones [3,4]. Here we report the excitation wavelength dependence of this effect and compare the experimental findings with a density matrix system based on 2 electronic levels, each with two vibrational levels ($|1\rangle$ and $|2\rangle$ in ground-state and $|3\rangle$ and $|4\rangle$ in the excited-state). The enhancement of the ESA signal level and the FFT amplitude is shown in Fig. 2 for several excitation spectra. The ESA signal (Fig.2(a)) is enhanced when the pump is blue-detuned from the absorption center of Nile Blue. For a resonant excitation and slightly red-detuned spectrum at about 640 and 655 nm, respectively, the multipulse can not enhance the ESA signal. When the oscillatory amplitude (vibrational coherence) is analyzed (Fig. 2(c)), a similar wavelength dependence is obtained. For blue-detuned multipulse excitation, the amplitude of the oscillations is enhanced up to 80%.

The density matrix simulation describes qualitatively the wavelength dependence of the population transfer enhancement (Fig. 2(b)) as well as the vibronic coherence (Fig. 2(d)). The small deviation for blue-detuned excitation (<570 nm) is due to the reduced number of vibrational levels of the simulation. Furthermore, the density matrix approach shows that the enhancement is due to the oscillatory behaviour of the total electronic coherence during the near-resonant excitation, since in this case the consecutive sub-pulses in the pulse train can benefit from the variations induced in the absorptive properties of the system. As a general trend, we observed that the enhancement increases for higher values of the electronic coherence time ($T_{2, elec}$). This behaviour seems reasonable, since $T_{2, elec}$ describes the coupling efficiency between the electronic levels and therefore has influence on the population transfer. A maximum value for the enhancement is obtained when the total pulse train duration is of the order of the electronic coherence time.

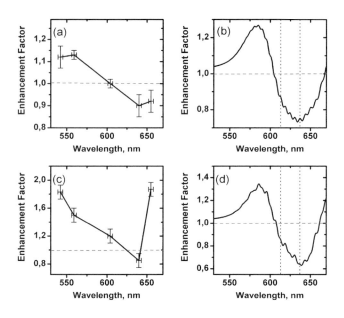

Fig. 2. Enhancement of the ESA signal at 550 nm measured with multipulse excitation (b = 56 fs) with different spectra. The enhancement of the population: (a) experiment and (b) simulation. The enhancement of the coherence: (c) experiment and (d) simulation. All values are compared to the signal obtained after TL-excitation with the respective spectrum (traced red line). The dotted black lines in (b) and (d) at 637 nm and 613 nm represent the position of the $|1\rangle{\rightarrow}|3\rangle$ and $|1\rangle{\rightarrow}|4\rangle$ transitions, respectively.

Conclusions

We showed that by periodical excitation with ultrashort pulses we are able to manipulate the ultrafast vibrational dynamics, generating this way an ultrafast modulation of the S_0-S_1 absorption. This dynamic absorption leads to enhancement of population transfer to the excited state as well as vibrational coherence in the excited state. These experimental observations are corroborated in a density matrix approach. Our findings suggest that multipulses are a versatile tool for the enhancement of population as well as vibrational coherence in the excited state under near-resonant excitation conditions [5].

1   J. L. Herek, W. Wohlleben, R. J. Cogdell, D. Zeidler, and M. Motzkus, in Nature Vol. 417, 533 2002.

2   V. I. Prokhorenko, A. M. Nagy, S. A. Waschuk, L. S. Brown, R. R. Birge, and R. J. D. Miller, in Science, Vol. 313, 1257, 2006.

3   J. Hauer, T. Buckup, and M. Motzkus, in Journal of Chemical Physics, Vol. 125, Art.No.061101 2006.

4   J. Hauer, H. Skenderovic, K. L. Kompa, and M. Motzkus, in Chemical Physics Letters, Vol. 412, 523, 2006.

5   T. Buckup, J. Hauer, , C. Serrat and M. Motzkus, in Journal of Physics B: Atomic, Molecular and Optical Physics, Vol. 41, Art.No.071024, 2008.

Pump-push-probe transient spectroscopy of isolated conjugated oligomers

Jenny Clark[1], Juan Cabanillas-Gonzalez[1], Tersilla Virgili[1], Luca Bazzana[2,] Guglielmo Lanzani[1]

[1] Dipartimento di Fisica IFN, CNR, Politecnico di Milano, Piazza Leonardo da Vinci 32, Milano 20133, Italy
E-mail: jenny.clark@polimi.it
[2] LUCEAT S.p.A., Viale G. Marconi, 31, Dello (BS) Italy

Abstract. We use a transient pump-push-probe technique to study intrinsic charge photogeneration and subsequent recombination in isolated conjugated molecules. Furthermore, we demonstrate stimulated emission switching with large on/off ratio in doped polymer optical fibers.

Introduction

The field of semiconducting conjugated polymers (CP) is now well established. The materials have been successfully used in a variety of applications including light emitting diodes, photovoltaic cells, thin film field effect transistors, light emitting transistors and lasers. Despite much work, however, the processes of charge transport, charge generation and recombination and the role of intermolecular interactions are not fully understood. Within that context, here we study charge generation and recombination in molecules well dispersed in liquid or solid matrices using a time-resolved pump-push-probe technique, with the push arriving a few picoseconds after the pump. The reason for dispersing the molecules is to study the intrinsic charge generation, without intermolecular effects, which are dominant in films. Furthermore, using the three-pulse technique, we demonstrate ultrafast stimulated emission switching with large on/off ratios in polymer optical fibers (POF) doped with conjugated materials. These last results are exciting as they pave the way towards all-optical amplification and switching in POF networks.

Here we use the commonly studied poly(9,9-dioctylfluorene) (PFO) and its oligomers (Fig. 1(a)). In such materials, excitation of an electron from the ground state (S0) to the first exited state (S1) forms an exciton due to the low dielectric constant of the materials. Fig. 1(b) (red markers) shows the transient absorption spectra of PFO dispersed in a matrix. The bands are attributed to excitons only, with stimulated emission (positive) in the visible and photo-induced absorption from S1 to the next dipole-coupled state (Sn) in the near infra-red [1]. On the other-hand, pure films of the material (black markers) demonstrate a second photo-induced absorption band, which overlaps with the stimulated emission band and is due to the formation of long-lived charges.

The formation of charges in conjugated materials such as PFO, is thought to arise due to excitation to a high energy state (Sn) followed by a process similar to 'autoionization' [1]. The Sn state can be attained through either two-photon absorption or via a process of 'sequential excitation'[2]. In well-dispersed molecules, charge generation also occurs, however the recombination of the charges is fast, occurring within one picosecond.

463

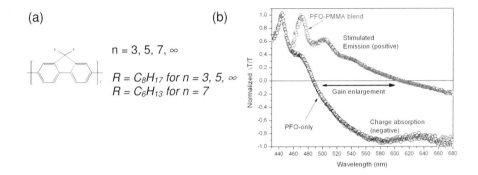

Fig. 1. (a) Structure of the materials used in this study. Trimer (n=3), pentamer (n=5), heptamer (n=7) and polymer (n=∞) and (b) Pump-probe spectra at 2ps after excitation in a blend film of poly(9,9-dioctylfluorene) (PFO) in polymethylmethacrylate (PMMA) and a pure film of PFO.

To verify these findings, and to study the processes in more detail, we use a pump-push-probe technique to excite the molecules. The pump excites to S1 and creates excitons. The push excites from S1 to Sn and we study the transient absorption spectra and dynamics to look at the interplay between three processes: internal conversion from Sn to S1, charge generation from Sn and recovery of S1 from charge recombination.

Experimental Methods

The 780nm, 140fs output from a 1kHz CLARK-MXR regenerative Titanium:Sapphire system is split in three to produce (1) a first pump at 390nm using second harmonic generation in a BBO crystal (2) a push at 780nm and (3) a white light supercontinuum probe produced by focussing the 780nm onto a sapphire plate. The three beams are overlapped spatially on the sample. The delay between the pump and push is fixed at 2ps and the delay between the pumps and probe is varied using a computer-controlled delay state. Changes in transmission when one or two pumps are incident on the sample are measured. Samples included dilute solution, pure film (spin-cast), blend film (spin-cast 10% weight active material in poly(methylmethacrylate), PMMA) and doped polymer optical fibers produced using a standard preform drawing technique. The active materials were purchased from American Dye Source.

Results and Discussion

Figure 2(a) and (b) show pump-push-probe dynamics in the stimulated emission band of a pure film of the trimer (part a) and in a film made of trimer dispersed in poly(methylmethacrylate) (part b). In both cases, the effect of the push at 780nm is to reduce the stimulated emission signal to a negative value. The fact that the signal goes negative means that there is an underlying absorption band, which we attribute to the formation of charges. In the pure film, the stimulated emission does not recover after the second pump on these time-scales, while in the blend film the recovery is complete. The results indicate that the charges are stabilised in the pure film due to interchain interactions, while in the blend film, the charges recombine rapidly.

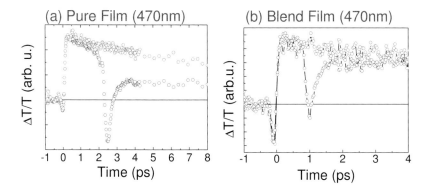

Fig. 2. Pump-probe dynamics and pump-push-probe dynamics of a (a) pure film of trimer and (b) a blend film of trimer in PMMA (10% by weight).

Similar meausurements were performed on the pentamer and heptamer of PFO (not shown). There does not appear to be a change in recovery time of the stimulated emission when increasing the oligomer length.

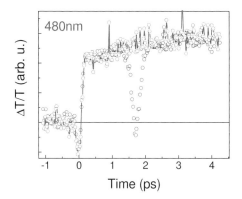

Fig. 3. Pump-probe dynamics (markers and line) and pump-push-probe dynamics (markers only) on a 2mm section of polymer optical fiber doped with heptamer.

Finally, Figure 3 shows pump-push-probe dynamics in the stimulated emission band of the heptamer doped into a polymer optical fiber. We see again the reduction of the signal upon arrival of the second pump. The on/off ratio is high and the switch recovers within a picosecond. We have therefore demonstrated an ultrafast stimulated emission switch within a POF, which has great potential for ultrafast optical signal processing applications.

1   T. Virgili, D. Marinotto, C. Manzoni, G. Cerullo and G. Lanzani, Phys. Rev. Lett. 94 (11), 117402 (2005).
2   C. Silva, A. S. Dhoot, D. M. Russell, M. A. Stevens, A. C. Arias, J. D. MacKenzie, N. C. Greenham, R. H. Friend, S. Setayesh and K. Müllen, Phys. Rev. B. 64, 125211, (2001)

Vibrational Energy Relaxation in Liquid-to-Supercritical Ammonia Studied by Femtosecond Mid-Infrared Spectroscopy

Jörg Lindner[1], Tim Schäfer[2], Dirk Schwarzer[2], and Peter Vöhringer[1]

[1] Abteilung für Molekulare Physikalische Chemie, Institut für Physikalische und Theoretische Chemie der Rheinischen Friedrich-Wilhelms-Universität, Wegelerstraße 12, 53115 Bonn, Germany
 E-mail: p.voehringer@uni-bonn.de
[2] Arbeitsgruppe Reaktionsdynamik, Max-Planck-Institut für biophysikalische Chemie, Am Fassberg 11, 37077 Göttingen, Germany

Abstract. Femtosecond mid-infrared spectroscopy was used to explore the vibrational energy relaxation dynamics of NH_2D in fluid NH_3. The density and temperature dependence of the ND-stretching lifetime suggests that hydrogen bonding is unimportant for vibrational energy transfer.

Introduction

An ammonia (NH_3) molecule exhibits three hydrogen atoms and a lone electron pair. Just like in the case of water (H_2O), these features enable the NH_3 molecules to act simultaneously as a hydrogen-bond donor and a hydrogen-bond acceptor. Therefore, liquid NH_3 is often referred to in chemistry textbooks as an associated liquid forming extended networks of hydrogen bonds (H-bond) similar to H_2O. Indeed, x-ray diffraction reveals a crystal structure of solid NH_3, in which each molecule is H-bonded to six nearest neighbors thereby acting as a triple H-donor and a triple H-acceptor at the same time [1]. Yet, experimental verification for the existence of H-bonds in the liquid is much more difficult to obtain. Femtosecond (fs) mid-infrared (MIR) spectroscopy has proven to be valuable tool to explore the structure and the dynamics of H-bonds in fluids; in particular, when implemented in a density (ρ) and temperature (T) dependent approach [2]. Recently, we have been able to carry out fs MIR studies in the OH-stretching region of the water isotopomer, HOD, dissolved in liquid-to-supercritical heavy water [2]. The combined T and ρ-dependence of the OH-stretching lifetime enabled us to unequivocally disentangle the dynamical interrelation between vibrational energy relaxation (VER) and spectral diffusion dynamics without performing highly sophisticated and instrumentally demanding multi-dimensional techniques of fs-MIR-spectroscopy.

To our knowledge, equivalent experiments have not yet been performed on NH_3. To test the significance of H-bonding for VER in this fluid and to provide a direct comparison to our previous studies on fluid H_2O, we decided to carry out fs-MIR-spectroscopy on the stretching vibrations in liquid-to-supercritical NH_3. Since the NH-stretching region of NH_3 is heavily perturbed by Fermi coupling between the stretching fundamental, v_1, and the first overtone of the anti-symmetrical bending mode, $2v_4$, we focused on the ND-stretch of NH_2D (the solute) in liquid-to-supercritical NH_3 (the solvent).

Results and Discussion

Linear mid-IR absorption spectra of NH_2D in fluid NH_3 are shown in Fig. 1A for different thermodynamic conditions of the mixture. The spectral bandwidth of the ND-stretching resonance is about 60 cm^{-1}, which is about twice as narrow as the OD-stretching resonance of HOD in H_2O at similar densities. In addition, a magnitude of ~50 cm^{-1} for the bathochromic shift of the ND-band in the fluid relative to the gas-phase (vertical arrow in Fig. 1A) is about a factor 2.5 smaller as compared to water under similar conditions [2]. Furthermore, although slightly weaker in the supercritical phase, the overall ρ-dependence of the ND-resonance frequency is similar to water. This is shown in the inset of Fig. 1B, where the ND resonance frequency is plotted versus the average interparticle distance obtained from the bulk density. Taken together, these features may or may not characterize a fluid of weakly interacting particles. Obviously, linear spectroscopy alone is insufficient to establish the importance of H-bonding to the vibrational dynamics in fluid ammonia.

Transient differential absorbance spectra of NH_2D in fluid NH_3 at 233 K and 1 kbar following an ND-stretching excitation with fs pulses centered at 2450 cm^{-1} are shown in Fig. 1C. Since the excitation bandwidth covers the entire ND-resonance, the ground state bleach peaks at the frequency of maximal absorbance of the linear spectrum recorded under identical thermodynamic conditions. The maximum of the transient absorption (negative absorbance) due to the excited ND-stretching state occurs at a probe frequency of 2348 cm^{-1}. From these data a reliable estimate of -97 cm^{-1} for the diagonal anharmonicity of the ND-stretch of the solute can be obtained, which agrees very well with predictions from coupled cluster theory [3]. From Fig. 1C, it can further be concluded that under the given thermodynamic conditions, depopulation of the ND-stretch excited state is likely to be rate-determining for VER back to the ground state. This is because the excited-state absorption decays on exactly the same time scale as the ground-state bleach recovers, thereby giving rise to a pronounced isosbestic point at a frequency of 2398 cm^{-1}. Finally, under the same thermodynamic conditions, spectral diffusion within the ND-stretching resonance of NH_2D in NH_3 must be much faster than VER. If these two distinct dynamical processes would occur on similar time scales, the spectral shape of the pump-induced absorbance should vary with the pump-probe time-delay, which is clearly not observed here.

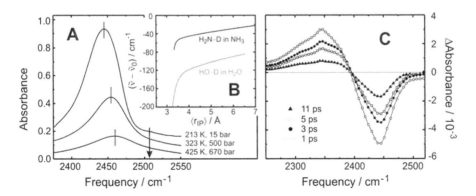

Fig. 1. A: Linear absorption spectra of NH_2D in NH_3. B: Spectral position of the ND-resonance as a function of the average interparticle distance obtained from the bulk density. C: fs-MIR pump-probe spectra for various time delays.

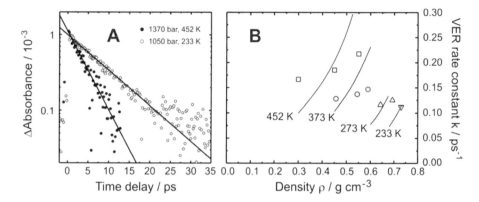

Fig. 2. A: Semi-logarithmic representation of the excited-state absorption decay. B: Dependence of the VER rate coefficient on the bulk density for various temperatures.

A semi-logarithmic plot of the transient absorption decay (see Fig. 2A) clearly demonstrates the mono-exponential behavior of the excited state depopulation and emphasizes once more the time scale separation between spectral diffusion and VER. The same figure also highlights the temperature dependence of the relaxation kinetics. Obviously, since the two transients were recorded at roughly the same solvent densities, the VER dynamics become faster with increasing T. This finding is in stark contrast to the isobaric and isochoric T-dependence we recently reported for VER of HOD in liquid-to-supercritical D_2O. In the case of water, VER slows down upon heating [2]. This is because for a given bulk density, ρ, the average number of H-bonds per H_2O molecule decreases with increasing T. The surprising temperature dependence observed here for the relaxation kinetics of NH_2D in NH_3 is much more reminiscent of a solute that vibrationally relaxes in a non-associating solvent.

To further substantiate this interpretation, the ρ and T-dependence of the VER rate was interpreted in terms of a simple "breathing-sphere" model consisting of a spherically symmetric solute oscillator embedded in a Lennard-Jones (LJ) bath. The spectral density of the fluctuating solvent forces exerted onto the solute vibration was calculated from classical molecular dynamics (MD) simulations that employed binary LJ-interaction parameters satisfying the critical data of NH_3. The VER rate constant is compared in Fig. 2B with the simulations. Obviously, the model is able to reproduce the T-dependence of the rate constant quite nicely. This finding fully confirms the initial interpretation that the influence of H-bonding on the dynamics of VER in fluid NH_3 is minor. However, the simulations are unable to reproduce the density dependence for a given temperature in a satisfactory manner. We believe that the rather weak experimental ρ-dependence as compared to the simulations is brought about by detuning effects, i.e. energetic shifting of solute and solvent vibrational states. By increasing the density, the resonance between donor and acceptor vibrational energy gaps weakens thereby effectively decelerating the apparent rate for VER. We will follow up on this explanation in the near future through MD-simulations involving more realistic and flexible models for fluid NH_3.

1   R. Boese, N. Niederprum, D. Blaser, A. Maulitz, M. Y. Antipin, and P. R. Mallinson, J. Phys. Chem. B **101**, 5794, 1997.

2   D. Schwarzer, J. Lindner, and P. Vöhringer, J. Phys. Chem. A **110**, 2858, 2006.

3   J. M. L. Martin, T. J. Lee, and P. R. Taylor, J. Chem. Phys. **97** , 8361, 1992.

Probing Intermolecular Couplings in Simulations of the Two-Dimensional Infrared Photon Echo Spectrum of Liquid Water

Alexander Paarmann[1], Tomoyuki Hayashi[2], Shaul Mukamel[2], and R. J. Dwayne Miller[1]

[1] 1Institute for Optical Sciences, Departments of Chemistry and Physics, University of Toronto, 80 St George Street, Toronto, Ontario, M5S3H6 Canada.
Email: dmiller@1phys.chem.utoronto.ca
[2] Department of Chemistry, University of California, Irvine, California 92697-2025, USA.

Abstract. Simulations of the 2D-IR photon echo spectrum of the OH stretching vibration in liquid water were presented, explicitly including intermolecular coupling and nonadiabatic effects. Close agreement with the experimental polarization anisotropy was found for surprisingly small intermolecular couplings. Increased sensitivity of the anharmonic two-dimensional OH stretching potential causes the fast spectral diffusion dynamics.

Introduction

The unique properties of liquid water ultimately originate from the correlations and intermolecular couplings in the extended hydrogen bond network. Two-dimensional photon echo spectroscopy of the OH stretch vibration has proven the most direct probe of fluctuations and correlations in the local structures. The recent experimental observation of extremely fast memory loss in pure water [1,2] is still pending a full understanding of the underlying mechanisms and theoretical description due to difficulties in treating the resonant energy transfer in the liquid.

Here, we present our most recent simulation results [3] using numerical integration of the Schrödinger equation to calculate two-dimensional photon echo and polarization anisotropy response of the OH stretch vibration in liquid water. This is the first work explicitly treating intermolecular vibrational coupling and anharmonic effects, as well as fluctuations of the transition frequencies and dipole moments and their anharmonicities. Relatively small intermolecular couplings are found sufficient to reproduce experimental energy transfer time scales due to rapid fluctuations of the two-dimensional OH stretch potential, distinctly different from one-dimensional stretch potentials as in HOD. These fluctuations also lead to the fast loss of correlations observed in the two-dimensional spectra.

Theory

We use molecular dynamics simulation of 64 H_2O molecules at room temperature in combination with an *ab initio* electrostatic map to obtain local OH stretch frequencies, dipole moments, and anharmonicities of both. The distribution of fundamental frequencies is shown in Fig. 1 (a). We then construct the time-dependent anharmonic vibrational Hamiltonian using the 128 local anharmonic modes (symmetric and asymmetric stretch for 64 molecules) as basis. Intermolecular coupling is included using dipole dipole coupling. To study different coupling regimes, we adjusted the dielectric constant in the dipole dipole coupling term. Statistical analysis of the couplings allows extraction of the average next neighbour coupling κ.

Calculation of the 3rd order vibrational signal requires summation of 6 different Liouville diagrams. The propagation of the time dependent multimode system is performed using numerical integration of the Schrödinger equation [4]. For propagation of the two-particle excitations we use the split operator method, splitting the two-particle Hamiltonian into harmonic and anharmonic parts. Fluctuations and anharmonicities of the transition dipole moments are explicitly treated within the calculated nonlinear response.

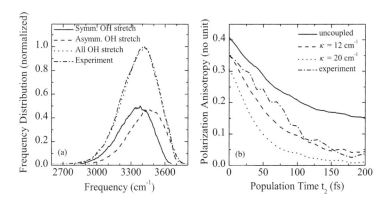

Fig. 1. (a) OH stretch frequency distributions in liquid water generated from all molecular configurations using our electrostatic map; in blue, experimental linear spectrum [5]. (b) Simulated polarization anisotropy for different coupling regimes; in blue, experimental result [2].

Results and Discussion

In Figure 1 (b), we show the simulated polarization anisotropy, calculated from spectrally integrated pump probe signal. The uncoupled system displays a fast initial signal decay due to strong fluctuations of the transition dipole moments accounting for of the signal decay. The remainder of the signal decay is due to fast librational and slow reorientation dynamics. Upon intermolecular coupling, the signal decays speeds up with increasing coupling strength and long lived components vanish entirely. This effect can clearly be assigned to resonant intermolecular energy transfer. Additionally, the initial polarization anisotropy is reduced with increasing coupling strength κ. We find excellent agreement with the experiment for κ as small as 12 cm^{-1}.

The two-dimensional spectra ($\kappa = 12$ cm^{-1}) for population times t_2=0, 100, 200, 500 fs are displayed in Fig. 2. At t_2=0, two peaks corresponding to the fundamental transition and excited state absorption are observed. Both are stretched along the diagonal, indicating initial inhomogeneity of the OH stretch ensemble. As a function of t_2, both peaks become vertical on different time scales across the spectrum; no single time scale of loss of correlations can be extracted. The red side of the spectrum shows faster dynamics where correlations decay on a 100 fs time scale, whereas on the blue side of the spectrum correlations persist up to   200 fs. These findings are in close agreement with our recent experimental results [2].

470

The fast loss of correlations is not dominated by energy transfer but rather due to rapid fluctuations of the two-dimensional OH stretching potential in H_2O, leading to mixing between states for fundamental and overtone states. This effect is distinctly different from one-dimensional OH stretching potentials as in HOD. It results in increased sensitivity of the fundamental transitions leading to fast loss of correlations in the two-dimensional spectra. Similarly, strong fluctuations of the transition dipole moments lead to modulations of the intermolecular couplings, opening up many energy transfer pathways, ultimately allowing rapid energy transfer and, consequently, some contribution to memory loss from spatial averaging.

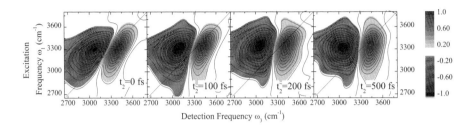

Fig. 2. Simulated 2D-IR spectra of the OH stretch vibration in liquid water for $\kappa = 12$ cm^{-1}.

Conclusions

We presented simulations of the two-dimensional photon echo spectra and polarization anisotropy of the OH stretch vibration in liquid water. Our new simulation procedure using numerical integration of the Schrödinger equation allows explicit treatment of intermolecular vibrational coupling for the first time. Average next neighbour couplings of only 12 cm^{-1} were found fully sufficient to reproduce experimental energy transfer time scales. The simulated two-dimensional spectra are in close agreement with recent experimental results.

Acknowledgements. This research was supported by the Canadian Institute of Photonics Innovation, Photonics Research Ontario, and the Natural Sciences and Engineering Research Council of Canada. S.M. gratefully acknowledges the support of NIH Grant No. GM59230 and NSF Grant No. CHE-0745892. A.P. thanks Thomas la Cour Jansen for helpful discussions.

1   M. L. Cowan, B. D. Bruner, N. Huse, J. R. Dwyer, B. Chugh, E. T. J. Nibbering, T. Elsaesser, and R. J. D. Miller, Nature (London) 434, 199 (2005).
2   D. Kraemer, M. L. Cowan, A. Paarmann, N. Huse, E. T. J. Nibbering, T. Elsaesser, and R. J. Dwyane Miller, Proc. Natl. Acad. Sci. U.S.A. 105, 437 (2008).
3   A. Paarmann, T. Hayashi, S. Mukamel, R. J. D. Miller, J. Chem. Phys. 128, 191103 (2008).
4   T. l. C. Jansen and J. Knoester, J. Phys. Chem. B 110 , 22910 (2006).
5   S. Ashihara, N. Huse, A. Espagne, E. T. J. Nibbering, and T. Elsaesser, J. Phys. Chem. A 111, 743 (2007).

Heterogeneous Dynamics of Coupled Vibrations

Dan Cringus, Thomas l. C. Jansen, and Maxim S. Pshenichnikov

Zernike Institute for Advanced Materials, University of Groningen, Nijenborgh 4, 9747 AG Groningen, The Netherlands
E-mail: M.S.Pchenitchnikov@RuG.nl

Abstract. Frequency-dependent dynamics of coupled stretch vibrations of a water molecule are revealed by 2D IR correlation spectroscopy. These are caused by non-Gaussian fluctuations of the environment around the individual OH stretch vibrations.

Introduction

Two-dimensional (2D) IR spectroscopy is capable of revealing the finest details of transient molecular dynamics that are otherwise hidden beneath broad featureless absorption bands [1, 2]. A few well-documented examples of additional information contained in the 2D spectra are; the coupling between vibrations resulting in cross peaks, correlation between fluctuations of the environment near different vibrations affecting the tilt of the cross peaks, and the environmental dynamics obtained from changes in the shapes of diagonal peaks. As any optical spectroscopy is sensitive to the eigenfrequencies but not to local vibrations, for such a coupled system it is of paramount importance to understand which part of the dynamics is caused by the environment and which originates from the coupling. Conventionally, the coupling between chromophores is taken to be constant in time while solvent dynamics are assumed to be Gaussian [3]. Under these circumstances, there will be no difference in dynamics between local and eigenmodes. However, the central question remains whether these assumptions are valid for real systems.

In this contribution we demonstrate both experimentally and theoretically that coupling and non-Gaussian character of bath fluctuations do have clear signatures in 2D IR spectra. The detailed line shape analysis of the diagonal contributions clearly revealed dissimilarity between correlation functions corresponding to different eigenmodes. A comprehensive theoretical analysis based on MD simulations and numerical integration of the Schrödinger equation (NISE) allowed us to identify underlying physical processes accountable for such behavior. In a broader sense, the heterogeneous dynamics of the coupled vibrations provide additional valuable information on the environment that is hardly available from an isolated vibration. Our findings are highly relevant for a better understanding of the dynamics of complex systems as proteins, polymers, and light-harvesting complexes.

Experimental Methods

As a model system, we chose a water molecule diluted in acetonitrile at room temperature. In this system, two OH vibrations are coupled via the oxygen atom while intermolecular water-water interactions are reduced to naught by use of extremely low water concentration. Environmental fluctuations are provided by acetonitrile molecules that are weakly bonded to water. The 2D spectra were obtained following the standard protocol [2] using 70-fs, 3 μm IR pulses from a home-build optical parametric oscillator.

Results and Discussion

The correlation spectra obtained at different evolution times t_{23} (i.e. delays between the pulse pairs) are shown in the top panel of Fig.1. Two peaks at ~3540 and ~3630 cm^{-1} correspond to the symmetric and asymmetric stretching OH vibrations, respectively. The cross-peak at ω_1~3540 and ω_3~3630 cm^{-1} is already visible at an early waiting time which indicates that these two modes share the same ground state. The second cross-peak at ω_1~3630 and ω_3~3540 cm^{-1} is suppressed at t_{23}<500 fs because of its coincidental overlap with the excited state absorption at ~3560 cm^{-1} by the asymmetric stretch (blue contours). With the increase of the evolution time, amplitudes of the cross-peaks increase because of the intramolecular energy exchange between the two modes [4].

At zero delay, both diagonal peaks are diagonally-elongated which is indicative of inhomogeneous broadening. However, the homogeneous contribution to the symmetric mode seems to exceed that for the asymmetric mode. This becomes more apparent at longer evolution times where the asymmetric mode still carriers some inhomogeneity while the symmetric one is almost round and, therefore, any initial inhomogeneity is washed away. To quantify these observations, we extended the line-shape analysis presented in [5] to account for multiple transitions in a 2D spectrum. Namely, each cut at frequency ω_1 through a 2D spectrum was fitted to a combination of Gaussians from which the positions of extrema were found (Fig.1). The slope of the curve connecting the extrema can be regarded as an indicator of phase memory at the given frequency and is exactly equal to the instantaneous value of the frequency correlation function (CF) in case of Gaussian dynamics.

The results of such analysis (Fig.2a) convincingly confirm that the symmetric mode dephases much quicker than the asymmetric one, at all evolution times. This could not be predicted by a standard theory [3] as both modes exert similar environmental dynamics. Therefore, we performed combined MD and quantum mechanical simulations. First, the GROMACS 3.1.4 program was run with standard force fields for water and acetonitrile. Second, the vibrational frequencies, coupling, and transition dipole moments were extracted from the MD trajectory using a recently published electrostatic *ab initio* map [6]. Finally, a fluctuating vibrational Hamiltonian was constructed with these parameters, adding a constant anharmonicity (210 cm^{-1}) [4] to include the doubly excited states. The linear

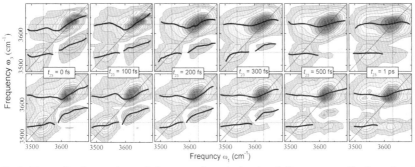

Fig.1. Normalized experimental (top panel) and simulated (bottom panel) absorptive 2D correlation spectra at different waiting times t_{23}. Solid curves connect the maxima of the ω_1=const cuts through the data. Polarizations of all pulses are linear and identical.

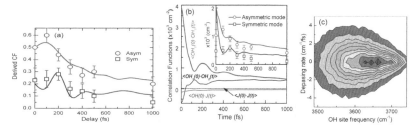

Fig.2. (a): CFs as directly derived from experimental (symbols) and simulated (curves) 2D spectra. (b): Calculated CFs for the OH site frequency (magenta) and coupling (dotted curve), and cross-CFs between both OH site frequencies (navy) and OH site frequency and coupling (green). Inset: CFs of the asymmetric (red) and symmetric (blue) modes with experimental data (symbols). (c): Joint distribution of dephasing rate and the OH site frequency.

absorption and 2D IR spectra were simulated with the NISE scheme [7].

The simulated 2D spectra (Fig.1, bottom panel) agree very well with the experimental observations given the fact that no free parameters were used. This also holds for the CFs derived from the calculated 2D patterns (Fig.1, black curves). The calculated CF of the asymmetric mode decays indeed much slower than the symmetric one (inset in Fig.2b) which is fully consistent with the experimental observations. The experimental values of the CFs (Fig.2a) rescaled with a single identical factor fall excellently on the simulated curves (inset in Fig.2b, symbols).

The librational motion of water molecules dominates at short timescales (<100 fs), while translational motion takes over at longer timescales (Fig.2b, magenta curve). The recurrence at ~200 fs observed experimentally, is due to underdamped low-frequency O..N stretching mode. The cross-correlation between the OH site frequencies is dominated by anticorrelated librational motion at short times and correlated translational motion at long times (navy curve). Quite unexpectedly, the cross-correlation between the fluctuating coupling and site frequency (green curve) alone is *not sufficient* to account for differences between the symmetric and asymmetric CFs. More careful theoretical analysis revealed that the joint distribution function of the dephasing rate and the OH site frequency (Fig.2c) is much broader at low than at high frequencies (or, in other words, environmental fluctuations are not Gaussian). The fast dephasing at low frequencies is predominantly affecting the symmetric mode because out of the two OH site frequencies it always inherits the lowest one due to the mode coupling. Similarly, the slow dephasing at high frequencies determines the slower dephasing of the asymmetric mode.

1   D.M. Jonas, An.Rev.Phys.Chem. 54, 425 (2003)

2   M. Khalil, N. Demirdoven, A. Tokmakoff, J.Phys.Chem. A107, 5258 (2003).

3   S. Mukamel, "Principles of Nonlinear Optical Spectroscopy", Oxford University Press, USA, New-York,1995.

4   D. Cringus, T. l. C. Jansen, M.S. Pshenichnikov, D.A. Wiersma, J.Chem.Phys.127, 084507 (2007).

5   K. Lazonder, M.S. Pshenichnikov, D.A. Wiersma, Opt.Lett. 31, 3354 (2006).

6   B.M. Auer, J.L. Skinner, J. Chem. Phys.127,104105 (2007).

7   T. l. C. Jansen and J. Knoester, J. Phys. Chem. B 110, 22910 (2006).

Immobilized water in hydrophobic hydration

Yves L.A. Rezus and Huib J. Bakker

FOM Institute for Atomic and Molecular Physics, Kruislaan 407,
1098 SJ Amsterdam, The Netherlands
E-mail: bakker@amolf.nl

Abstract. Using femtosecond mid-infrared spectroscopy, we find that water molecules in the hydration shells of hydrophobes show much slower orientational dynamics than pure liquid water. Each methyl group is observed to immobilize four water OH groups.

Introduction

The dissolution of hydrophobic compounds in water is accompanied by an anomalously large increase in the heat capacity of the solution. In the 1940s Frank and Evans introduced a model to account for this observation: they proposed that the water molecules around hydrophobic groups form rigid, icelike structures, which they denoted as icebergs [1]. Later molecular-scale studies were however not conclusive: neutron diffraction studies did not find evidence for the presence of ice layers around hydrophobic solutes, while NMR studies did observe a clear decrease in the *average* mobility of the water molecules. Here we use polarization-resolved mid-infrared pump-probe spectroscopy to study the rotational motion of water molecules in the solvation shells of hydrophobic solutes. An advantage of this method over NMR is that the dynamics of the molecules is probed on a time scale that is shorter than the molecular exchange time between the bulk liquid and the solvation shell. As a result, the dynamics of the hydration shells can be distinguished from the rest of the liquid.

Experiment

We study solutions of hydrophobic solutes of varying concentrations dissolved in 4% $HDO:H_2O$[2]. The O-D stretch vibration of the HDO molecules has a strong absorption around 2500 cm^{-1}, and the orientational dynamics of these molecules is monitored using pump-probe spectroscopy. An intense femtosecond pump pulse, tuned in resonance with the O-D vibration, is used to excite a significant fraction of the HDO molecules. The pump-induced absorption changes are monitored by delayed probe pulses that are polarized parallel and perpendicular to the pump. These two probe signals are used to construct the anisotropy parameter that reflects the orientation of the excited O-D groups.

Results and Discussion

In Fig. 1a anisotropy decays are shown for TMAO solutions at four different concentrations. We observe a biexponential decay composed of a fast component with a time constant of ~2 ps and a slow component with a time constant >10 ps.

The presence of this long time constant shows that a fraction of the water molecules are strongly immobilized by TMAO. The fast component has also been observed for pure water, showing that the orientation of a large part of the molecules is unaffected.

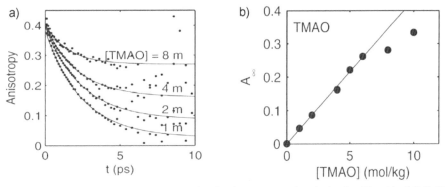

Fig. 1. a) Anisotropy decay of HDO molecules in aqueous trimethylamine-N-oxide (TMAO) solution of varying concentrations; b) Amplitude of immobilized water molecules as a function of TMAO concentration.

For each concentration we have fit the anisotropy to a monoexponential decay with an offset ($R=Ae^{-t/\tau} + A_\infty$). The offset represents the slow component, as its time constant falls outside our experimentally accessible time range. In Fig. 1b the amplitude of the slow component is plotted as a function of the solute concentration. Up to high concentrations we observe the dependence to be linear. From the linear relation it follows that the slow component is due to water molecules that are part of the solvation shell of the TMAO molecule. The slope of the linear part of Fig. 1b shows that the solvation shell of a TMAO molecule contains approximately 12 strongly immobilized OH groups.

For each of the solutes we have determined the number of OH groups immobilized per solute molecule. The results are summarized in Fig. 2, where the number of immobilized water molecules is plotted versus the equivalent number of CH_3 groups in the solute molecule. The observed linear relation shows that the immobilized water molecules are part of the hydration shell around the *hydrophobic* methyl groups of the solutes. Apparently the *hydrophilic* groups of the solutes do not lead to the immobilization of water molecules. The slope of the graph in Fig. 2 has a value of 3.9, indicating that every methyl group is responsible for the immobilization of approximately 4 water OH groups.

Slowing down of the orientational dynamics of water in aqueous solutions of hydrophobic species was also observed in a previous NMR study [5]. For aqueous solutions of tetramethylurea, the NMR measurements showed a dramatic slowing down of the water reorientation as the tetramethylurea concentration is increased: at a TMU concentration of 2.5 m the average reorientation is slowed down by a factor of 2. In contrast, in the same study the average reorientation time of the water molecules in aqueous urea was observed to increase only negligibly with increasing urea concentration. This agrees with a previous femtosecond study that showed that urea does not affect the orientational mobility of the majority of water molecules [6].

The immobilization of water near hydrophobic molecular groups cannot be explained from a strengthening of the hydrogen bonds: the hydrogen bonds are very similar to those in pure liquid water. Instead, the immobilization most likely results from a steric effect. Sciortino *et al.* have shown that the relatively high orientational mobility of pure water is related to the presence of defects (i.e., five-coordinated water molecules) in the tetrahedral hydrogen-bond network of liquid water [3].

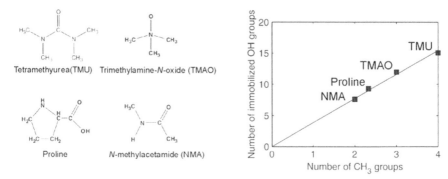

Fig. 2. a) Molecular structure of the solutes studied; b) Number of immobilized water OH groups as a function of the equivalent number of CH3 groups in the solute molecule.

Recently, Laage and Hynes proposed a detailed mechanism for water reorientation involving five-fold coordinated water molecules [4]. In this mechanism the pathway for reorientation involves a rotating water molecule that concertedly breaks a hydrogen bond with an over-coordinated first-shell neighbor and reforms one with an under-coordinated second-shell neighbor. These studies together with our results indicate that the immobilization of water molecules around a hydrophobic solute likely arises from a steric effect: the hydrophobic group prevents a fifth water molecule from approaching a tetrahedrally coordinated water molecule, and as such prevents the molecule to reorient.

Conclusions

The present results provide a molecular picture of the icebergs proposed by Frank and Evans [1]. The "icebergs" consist of four strongly immobilized water OH groups for every methyl group in solution. This immobilization is the consequence of a decrease in the configurational space available to water molecules around hydrophobic solutes. This notion may also explain the observation of Frank and Evans of a decrease in entropy upon dissolution of hydrophobic compounds in water. The structure of the iceberg, however, is not the ordered structure observed in ice, but likely resembles the disordered hydrogen-bond network of bulk water: the icebergs are ice-like from a dynamical perspective, but water-like as far as their structure is concerned.

1 H. S. Frank and M.W. Evans, J. Chem. Phys. **13**, 507, 1945.
[2] Y.L.A. Rezus and H.J. Bakker, Phys. Rev. Lett. **99**, 148301, 2007.
[3] F. Sciortino, A. Geiger, and H. E. Stanley, Nature (London) **354**, 218, 1991.
[4] D. Laage and J. T. Hynes, Science **311**, 832, 2006.
[5] A. Shimizu, K. Fumino, K. Yukiyasu, and Y. Taniguchi, J. Mol. Liq., **85**,269, 2000.
[6] Y.L.A. Rezus and H.J. Bakker, Proc. Natl. Acad. Sci. U.S.A. **103**, 18417, 2006.

Collective Breakdown of H-Bonding in Ice

Hristo Iglev and Marcus Schmeisser

Physik-Department E11, Technische Universität München, D-85748 Garching, Germany
E-mail: higlev@ph.tum.de

Abstract. We report on bulk melting of ice after ultrafast temperature jump. Our experiments show that homogeneous melting occurs only for an energy deposition beyond the superheating limit of 330 ± 10 K. The process includes two steps: (i) initial formation of water clusters with more than 10^3 water molecules, and (ii) secondary melting at the already generated phase boundaries.

Introduction

Melting of ice is the most common structural transition in nature. The process usually starts at the surface at temperatures above the melting point. We demonstrated that ultrafast heating of bulk ice can avoid the surface melting, leading to substantial superheating of the solid phase [1]. For higher excitation levels strong evidence for homogeneous bulk melting was observed [2]. It is of fundamental interest to understand the different melting processes [3,4].

Here, we present an experimental study on melting of bulk ice. The OH-stretching vibration is applied for rapid heating and for fast and sensitive probing. We have shown that a comparison of time-resolved and steady state spectra of ice allows the determination of temperature and pressure changes in the sample [1,2]. It is important to recall the isochoric character of an ultrafast temperature jump. In our data analysis the steady-state differential spectra are calculated by subtraction of two absorption spectra for certain temperature differences (first fitting parameter) taking into account a down-shift of the spectrum at higher temperature according to an assumed pressure increase (second parameter). The fraction x of molten ice is introduced as a third fitting parameter. The contribution to the calculated difference spectrum is generated by the difference of the steady-state absorption spectrum of the ice sample at the initial temperature and that of the produced water, both scaled with the relative amount x of melting. According to energy arguments the temperature of the molten component is assumed to be at the common melting point. The values of the fitting parameters are obtained by a global fitting procedure to the time-resolved differential spectra.

Results and Discussion

Data on the melting dynamics measured after excitation at 3290 cm^{-1} of HDO:D$_2$O ice (15 M) at 255 K are presented in Fig. 1a. The time-resolved differential spectra measured at various delay times are shown. The solid lines represent the calculated differential spectra obtained by a superposition of the absorption of superheated ice (dashed lines) and liquid water (dotted curves). The good agreement with the experimental data gives evidence that the water absorption already appears within the first few picoseconds. The finding supports the assumption that liquid water is formed without evidence for pre-molten species for 15 ps or later.

The extracted temporal evolution of the molten fraction is shown in Fig. 1b. The time dependence can be fitted with a simple relaxation model (solid curve) with two assumed exponential time constants. The first kinetic step (i) involves melting of approximately 12±2 % of the probed ice volume with a time constant of 5±2 ps. Obviously, breaking of straight H-bonds in the ice lattice already occurs during the initial thermalization process when energy is transferred from the excited OH-stretching mode to intermolecular bridge bond vibrations (measured thermalization time of 5.4±0.5 ps [1]). The similarity of both time constants supports the assumption that the thermalization of the H-bonded network determines the kinetics of the initial melting. Theoretical calculations predict that melting within a superheated crystal is initiated by nucleation seeds [4].

The initial melting is followed by a second, slower process (ii) that increases the molten component by another 19±2 % with an effective time constant of 33±5 ps. A similar number of 37±5 ps was reported for surface melting of ice after ultrafast heating [5]. The accordance suggests that this time constant is connected to secondary interfacial melting at the phase boundaries obviously generated during the initial melting.

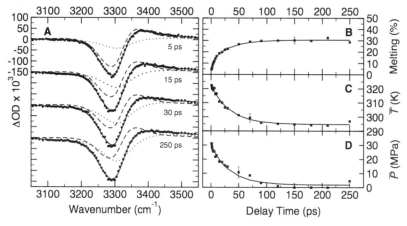

Fig. 1 Melting of HDO:D$_2$O ice measured after ultrafast laser heating. **a**, Transient absorption changes at various delay times; experimental points, calculated lines. The solid curve is a superposition of the absorption of superheated ice (dashed line) and molten liquid (dotted curve). **b**, Temporal evolution of the amount of molten liquid. **c, d**, Time-evolution of the residual ice, average temperature (c) and average pressure (d) in the probing volume.

The time evolution of the average temperature of the residual ice component in the probing volume is presented in Fig. 1c. The data show that even at short delay times the signal amplitude does not exceed the value corresponding to the superheating limit $T_{LS}= 330\pm10$ K [2]. The finding provides evidence that the melting starts already during the initial thermalization process and consumes parts of the deposited energy. The secondary melting (ii) is accompanied by a decrease of the average ice temperature in the probing volume with the same time constant of 33 ± 5 ps (see Fig. 1c). The average temperature drops from 330 K to approximately 295 ± 2 K and remains almost constant during later times. The initial melting step (i) is obviously continued by a secondary process (ii) that consumes thermal energy of the superheated ice component and seems to be governed by heat conduction.

The extracted average pressure of the residual ice component in the probing volume is presented in Fig. 1d. The data show a rapid pressure decay with a time constant of 35 ± 10 ps. The feature is much faster than the pressure decrease for the superheating case without melting and presents striking thermodynamic evidence for structural changes in the sample.

As mentioned above, the initial homogeneous melting is believed to form water clusters in the ice sample. Estimates based on the nucleation theory and on the observed temperature relaxation suggest a cluster size in the order of a few nm. The result supports the argument that the formation of melting nuclei requires breaking of the ice lattice in regions including more than 10^3 water molecules. Obviously, melting is a collective process and breaking of isolated hydrogen bonds is not sufficient to initiate homogeneous nucleation. This conclusion may explain the unexpectedly large thermal stability of the linear H-bonds of a water molecule to its neighbours in the ice lattice. The observed catastrophic breaking of H-bonds opens the way for experimental manipulations of H-bonded materials aiming at a more comprehensive understanding of hydrogen bonding in nature.

Conclusions

We report on bulk melting of ice after ultrafast temperature jump. The process includes two steps: (i) homogeneous melting with time constant of 5 ps and (ii) secondary melting at the already generated phase boundaries with time constant of 33 ps. Our data give evidence that the formation of melting nuclei requires breaking of the ice lattice in regions including more than 10^3 water molecules.

References

1 H. Iglev et al. , Nature **439**, 183, 2006.
2 M. Schmeisser, H. Iglev and A. Laubereau, J. Phys. Chem. B **111**, 11271, 2007.
3 C. Caleman and D. van der Spoel, Angew. Chem. Int. Edition **47**, 1417 , 2008.
4 S. N. Luo, A. Strachan, and D.C. Swift, Model. Simul. Sci. Eng. **13**, 321, 2005.
5 Ch. Y. Ruan et al., Science, **304**, 80, 2004.

The Dynamics of Aqueous Hydroxide Ion Transport Probed via Ultrafast Vibrational Echo Experiments

Sean T. Roberts[1], Poul B. Petersen[1], Krupa Ramashesha[1], and Andrei Tokmakoff[1]

[1] Department of Chemistry and George Harrison Spectroscopy Laboratory, Massachusetts Institute of Technology, Cambridge MA 02139
E-mail: tokmakof@mit.edu

Abstract. We use peakshift, transient grating, and 2D IR measurements to probe the dynamics of NaOD solutions. Our experiments suggest that OD⁻ possesses a stable solvation shell and signatures of fast intermolecular proton transfer are observed.

Introduction

Compared to ions of similar size and charge density, the aqueous hydroxide ion possesses an anomalously fast diffusion constant due to its ability to accept a proton from a neighboring water molecule, leading to the translocation of the ion. Despite this being a fundamental reaction of acid-base chemistry, the mechanism by which hydroxide ions are conducted through water is still highly contested and poorly understood. Lewis dot diagrams predict that the hydroxide oxygen has three lone pairs and hence can accept three hydrogen bonds. To date only three coordinate structures have been observed in gas phase clusters [1]. In contrast, neutron scattering [2] experiments suggest that OH^- is hypercoordinated in the liquid, accepting four hydrogen bonds from its neighbors. *Ab initio* molecular dynamics simulations have observed both three and four coordinate structures and suggest that proton transfer proceeds only when OH^- forms thee hydrogen bonds [3]. Moreover, proton transfer in the three coordinate state proceeds very rapidly, on the order of 180 fs, but is gated by the exchange between three and four coordinate structures which occurs over a few ps. As of yet, no direct evidence of this exchange exists due to the difficulty of developing a probe that is both structurally sensitive and has adequate time resolution.

Femtosecond infrared spectroscopy is a powerful tool to study proton transfer since short pulses on the order of ~50 fs can be generated and the OH stretching frequency of water molecules, $\Box_{OH}$, is sensitive to its hydrogen bonding partner. Short, linear hydrogen bonds of the type formed between water and an OH^- ion appear red shifted from the main OH stretching band whereas weak hydrogen bonds of the type formed by the proton of the OH^- ion appear at high frequency (see Fig 1A). Time dependent changes in $\Box_{OH}$ result from the dynamics of the solvent that drive molecules into and out of the solvation shells of OH^- ions.

We present the results of multiple of third order IR spectroscopy measurements of the OH stretch of dilute HOD (~1%) dissolved in NaOD:D_2O solution. By using an isotopically dilute system, we break the symmetry of the HOD molecule, creating a localized stretching coordinate and eliminate resonant energy transfer between molecules. With increasing NaOD concentration, photon echo peakshift measurements indicate the environment surrounding HOD molecules is static up to

481

~1.5 ps, suggesting that OD⁻ ions possess a stable solvation shell over this timescale. Pump probe and transient grating measurements show the appearance of fast vibrational relaxation attributable to HOD molecules hydrogen bonded to OD⁻ ions. 2D IR measurements display a large off diagonal intensity that relaxes on a ~100 fs timescale, and may be an indicator of rapid proton exchange.

Nonlinear Infrared Spectroscopy of HOD in NaOD Solution

Shown below are the results of both linear and nonlinear IR spectroscopy measurements of dilute HOD dissolved in various concentrations of NaOD. With increasing concentration, the FT-IR spectrum shown in Fig. 1A undergoes a number of changes, including a large increase of intensity on the red side of the spectrum from the formation of strong hydrogen bonds between HOD molecules and the oxygen of OD⁻ ions [4]. Also, a small shoulder appears near 3600 cm⁻¹ which has been attributed to the OH⁻ ion since its hydrogen atom can only form weak hydrogen bonds due to the ion's overall negative charge.

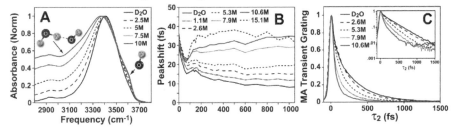

Fig. 1. Linear and nonlinear infrared measurements of the OH stretch of HOD:NaOD solution as a function of NaOD concentration. (A) FT-IR. (B) Three Pulse Photon Echo Peakshift. (C) Magic Angle Transient Grating (inset: log scale).

The three pulse photon echo peakshift (PS) taken using 45 fs pulses centered at 3350 cm⁻¹ is seen in Fig. 1B. Although the initial decay of the PS remains fairly independent of NaOD concentration, the offset at long waiting times increases linearly with concentration. The increasing offset is consistent with a slowing of the dynamics of the hydrogen bonding network, which agrees with the fact that solution viscosity increases with NaOD concentration. Given the local nature of the probe in our experiments, this result suggests that OD⁻ ions possess a stable solvation shell up to ~1.5 ps.

Fig. 1C shows the concentration dependence of the magic angle transient grating (TG) decay. The TG for HOD:D_2O is fit well by a single squared exponential with a time constant of 600 fs. This is somewhat shorter than the previous estimated value of 700 fs for the HOD lifetime. However, it is well known that the observed timescale for the TG decay will vary depending on where the pulse spectrum is centered relative to the sample absorption maximum due to absorption induced heating effects [5]. As the concentration increases the decay becomes bi-exponential and can be fit well with a squared sum of exponentials with time constants of 120 and 600 fs for all concentrations. This result is similar to that of Nienhuys et. al. [6] who reported transient hole burning measurements on dilute HOD in 10M NaOD solution showed the existence of two decay components with timescales similar to those reported here.

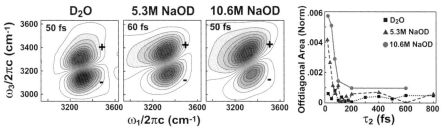

Fig. 2. 2D IR spectra at short waiting times as a function of NaOD concentration. Peaks corresponding to photobleaching and excited state absorption are labeled "+" and "-" respectively. To the right is the integrated area between ω_1 = 3000 & 3100 cm^{-1} and ω_3 = 3500 & 3600 cm^{-1} normalized by the integrated absolute value spectrum.

The origin of the fast decay seen in the TG measurement is displayed in the 2D IR spectra shown in Fig. 2. The antidiagonal linewidth at a particular value along the diagonal of the 2D spectrum is related to the homogeneous linewidth for molecules at that frequency. With increasing concentration the initial antidiagonal linewidth of the red side of the 2D spectrum, which corresponds to molecules hydrogen bonded to OD$^-$, increases dramatically. As the waiting time increases the offdiagonal intensity relaxes over ~100 fs, consistent with the fast decay seen in the TG measurement. This differs greatly from the 2D lineshape for HOD:D$_2$O which instead broadens on the blue side of the spectrum with waiting time due to the transient nature of broken hydrogen bonds [7]. The ability to sequentially drive transitions at 3600 cm^{-1} and 3100 cm^{-1} within a 50 fs window is suggestive of large frequency shifts that occur as OH bonds are broken and reformed during a proton transfer process.

1   W. H. Robertson, E. G. Diken, E. A. Price, J.-W. Shin and M. A. Johnson, Science **299**, 1367, 2003.
2   A. Botti, F. Bruni, S. Imberti, M. A. Ricci and A. K. Soper, J. Chem. Phys. **120**, 10154, 2004.
3   A. Chandra, M. E. Tuckerman and D. Marx, Phys. Rev. Lett. **99**, 145901, 2007.
4   D. Schloberg and G. Zundel, J. Chem. Soc., Faraday Trans. 2 **69**, 771, 1973.
5   S. Yeremenko, M. S. Pshenichnikov and D. A. Wiersma, Phys. Rev. A **73**, 021804, 2006.
6   H.-K. Nienhuys, A. J. Lock, R. A. van Santen and H. J. Bakker, J. Chem. Phys. **117**, 8021 2002.
7   J. D. Eaves, J. J. Loparo, C. J. Fecko, S. T. Roberts, A. Tokmakoff and P. L. Geissler, Proc. Natl. Acad. Sci. U. S. A. **102**, 13019, 2005.

Glasslike Behaviour in Aqueous Electrolyte Solutions

David A. Turton,[1] J. Hunger,[2] G. Hefter,[3] Richard Buchner,[2] Klaas Wynne[1]

[1] Department of Physics, SUPA, University of Strathclyde, Glasgow G4 0NG, UK
[2] Institut für Physikalische und Theoretische Chemie, Universität Regensburg, D-93040 Regensburg, Germany
[3] Chemistry Department, Murdoch University, Murdoch, WA 6150, Australia

Abstract. Ultrafast optical Kerr effect studies and dielectric relaxation spectroscopy applied to the relaxation dynamics of aqueous solutions, resolves the apparent conflicts between viscosity and rotational relaxation, and implies a jamming transition at high concentration.

When salts are added to water, the viscosity typically increases, suggesting that the ions alter the hydrogen-bond network. However, ultrafast infrared pump-probe measurements on electrolyte solutions have found that ions do not influence the rotational dynamics of water molecules suggesting that there is no enhancement or breakdown of the hydrogen-bond network in liquid water. [1] Here we report ultrafast optical Kerr effect (OKE) and dielectric relaxation (DR) spectroscopy measurements, which show that salt solutions behave like a supercooled liquid approaching a glass transition, where rotational and translational molecular motions become decoupled. [2] The rotational motions of bulk water molecules – observed as an α-relaxation in DR – are essentially independent of concentration. The translational motions seen in OKE spectroscopy can be understood as a β-relaxation. [3] This insight reconciles previously conflicting viscosity data, nuclear magnetic resonance relaxation, [4] and ultrafast infrared spectroscopy [1] data in a single unifying picture.

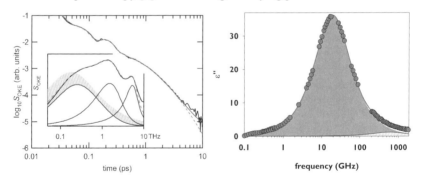

Figure 1. (left) OKE signal of water and the fit with the sum of a Cole-Cole function and five Brownian oscillators. Inset is the frequency-domain representation including the fit, which includes α termination. The fit has been decomposed to show the β-relaxation mode and the first two Brownian oscillators. The shaded part shows the modification to the Cole-Cole function by the inertial rise and α termination. [3] (right) Dielectric loss spectrum, $\varepsilon''(\nu)$, of water fitted to the sum of two Debye equations with relaxation times of 8.38 ps (dark grey) and 0.3 ps (light grey). [2]

OKE spectroscopy measures the two-point time-correlation function of the anisotropic part of the polarisability tensor in the time domain, while DR spectroscopy measures the two-point correlation function of the dipole-moment vector in the frequency domain. In principle, both techniques measure the same dynamics – molecular reorientations – but with different amplitudes. [5] The DR spectrum of room-temperature water (see Figure 1) can be fit by a single Debye function $(1-i\omega t_1)^{-1}$ with $t_1 = 8.38$ ps up to ~100 GHz where a weak secondary relaxation with $t_1' = 0.3$ ps appears. [2] The ultrafast OKE decay is most satisfactorily fit by a Cole-Cole function $[1-(i\omega t_2)^\beta]^{-1}$ with $t_2 = 0.61$ ps and $\beta = 0.86$. [3]

The measured relaxation times t_1 and t_2 clearly differ. As we have shown elsewhere, [6] DR and OKE (or Raman) spectroscopy do not measure the same dynamics. The water molecule has a large dipole moment but an almost isotropic polarisability tensor. [7] Therefore, DR spectroscopy is sensitive to diffusive **orientational** relaxation of water molecules – often referred to as α relaxation – giving a timescale of $t_1/3 = 2.8$ ps. Because of the near-isotropic molecular polarisability tensor, OKE is insensitive to pure rotational (single molecule) motions and instead measures only interaction-induced effects due to **translational** motions of pairs and larger groups of water molecules. [3] The good fit of the Cole-Cole function to the OKE data for water with $t_2 = 0.61$ ps shows that OKE spectroscopy measures a β relaxation related to the formation of transient cages in the liquid. The Cole-Cole exponent β indicates the degree of heterogeneity in the liquid.

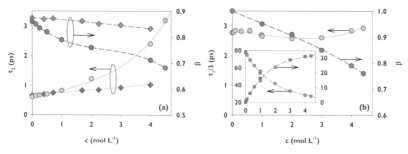

Figure 2. Fit parameters (Cole-Cole function) for OKE and DR data as a function of salt concentration. **(a)** Fit parameters (t_2 and β) for OKE data on aqueous NaCl (◆) and MgCl$_2$ (●) solutions. The values of the relaxation-time t_2 have been fit to Eq (1) with c_0 fixed at 12.7 M (NaCl, dotted curve) or 12 M (MgCl$_2$, solid curve). **(b)** Fit parameters ($t_1/3$ and β) for DR data in aqueous MgCl$_2$ solutions. The lines are in all cases added only as a guide to the eye. **(Inset)** The static dielectric constant measured with DR spectroscopy (●) and the calculated concentration of water bound to the Mg^{2+} cation (■) as functions of concentration. [2]

OKE and DR data were collected on a series of aqueous NaCl and MgCl$_2$ solutions at concentrations up to 4.5 M. The Cole-Cole function could be fit (in the time domain) to the OKE datasets at all concentrations. The concentration-dependent fit parameters are shown in Figure 2(a). The t_2 relaxation time is seen to rise monotonically with concentration, becoming five times larger in 4.5 M MgCl$_2$ solution than in pure water. The Cole-Cole β exponent decreases with concentration showing that the environment of the water molecules becomes increasingly inhomogeneous. The DR data for aqueous solutions of MgCl$_2$ are well fit with the Cole-Cole function and also show increasing inhomogeneity with concentration (see Figure 2(b)). Compared with t_2, the relaxation time $t_1/3$ shows little variation with concentration. Thus, $t_1/3$ is essentially decoupled from the solution viscosity. As can be seen in Figure 2(a), the rate of change of t_2 increases with concentration with a rapid acceleration at high concentration. This behaviour cannot be explained simply by the increase in bound water because – as the inset of Figure 2(b) shows – the rate of change of the concentration of bound *vs*. bulk water decreases with concentration.

Pure water has a glass transition temperature of ~135 K (obscured by a change of structure at 225 K), which increases with the addition of salts and other solutes. [8] Measurements of electrical conductivity and viscosity as a function of temperature in aqueous electrolyte solutions have shown temperature-dependent behaviour consistent with the empirical Vogel-Fulcher-Tammann (VFT) equation, which predicts a critical glass-transition temperature. For viscosities under isothermal conditions, this can be expressed as a function of electrolyte concentration as

$$\eta / \eta_0 = 1 + P\left\{ e^{Qc_0/(c_0-c)} - e^Q \right\} \tag{1}$$

where c_0 is a glass-transition (jamming) concentration and P and Q are parameters.

The fits of viscosity data to Eq (1) produce glass-transition concentrations c_0 of 12.7, 12.0, and 11.9 M for NaCl, $MgCl_2$, and $FeCl_3$ respectively, indicating the increased tendency for glass formation. The fits also give Q-parameters of 0.6, 3.4, and 3.9 respectively, consistent with an increasingly rapid approach to the glassy state. The NMR quadrupolar relaxation rates of $^{25}Mg^{2+}$ and $^{35}Cl^-$ ions in aqueous solution were also fit with Eq (1), with an identical glass-transition concentration c_0 of 12.0 M. The Q-parameter was found to be somewhat larger at 4.5 (Mg^{2+}) and 4.1 (Cl^-), which could be due to ion-ion interactions at the highest concentrations. [4] Thus, it is reasonable and consistent to describe concentrated electrolyte solutions in terms of liquids close to a glass transition.

The t_2 relaxation-time parameters measured with OKE spectroscopy are also consistent with a c_0 of 12.0 M, although the Q-parameter was found to be smaller at 2.0. Thus, we find that the β relaxation measured using OKE follows the same trend as the macroscopic shear viscosity and NMR quadrupolar relaxation. Crucially, the three measurements are consistent with an identical glass-transition concentration. These results are consistent with a slowing down of the translations of all water molecules in the electrolyte solution, and a complete arrest of translational motion at the glass-transition concentration. In contrast, with DR the expected slowing down with increasing salt concentration is not seen. These observations are therefore consistent with the picture of a "jamming" transition as observed for granular materials and colloidal suspensions. This picture is consistent with the colloidal-suspension picture proposed previously. [1]

Acknowledgment. We gratefully acknowledge funding for this project from the Engineering and Physical Sciences Research Council (EPSRC).

[1] A. W. Omta, M. F. Kropman, S. Woutersen, and H. J. Bakker, Science 301 (2003) 347-349.

[2] D. A. Turton, J. Hunger, G. Hefter, R. Buchner, and K. Wynne, J. Chem. Phys. 128 (2008) 161102.

[3] D. A. Turton and K. Wynne, J. Chem. Phys. 128 (2008) 154516.

[4] R. Struis, J. Debleijser, and J. C. Leyte, J. Phys. Chem. 93 (1989) 7943-7952.

[5] G. Giraud and K. Wynne, J. Chem. Phys. 119 (2003) 11753-11764.

[6] T. Fukasawa, T. Sato, J. Watanabe, Y. Hama, W. Kunz, and R. Buchner, Phys. Rev. Lett. 95 (2005) 197802.

[7] W. F. Murphy, J. Chem. Phys. 67 (1977) 5877-5882.

[8] C. A. Angell, Chem. Rev. 102 (2002) 2627-2649.

Mid-IR-Induced Nuclear Wavepacket Motion of a Hydrogen Bonding System: Effects of Mechanical and Electrical Anharmonic Couplings

Kunihiko Ishii, Satoshi Takeuchi, and Tahei Tahara

Molecular Spectroscopy Laboratory, RIKEN (The Institute of Physical and Chemical Research), 2-1 Hirosawa, Wako, Saitama 351-0198, Japan
E-mail: kishii@riken.jp

Abstract. Coherent hydrogen-bond stretching vibration was observed by probing ultrafast visible absorption change after impulsive mid-infrared excitation of the OH-stretching mode of an intramolecularly hydrogen-bonded molecule. The underlying mechanism of the wavepacket formation was discussed on a theoretical basis considering mechanical and electrical anharmonicities.

Introduction

Anharmonic coupling between vibrational modes is an important property for understanding vibrational energy transfer or nuclear wavepacket dynamics in polyatomic molecules and hydrogen-bonding systems. Anharmonic coupling is reflected in the bandshape of IR absorption bands, e.g. the OH-stretching band of hydrogen-bonded molecules. However, they show broad and structureless feature in the condensed phase and are often difficult to analyze.

We investigated anharmonic interactions using a time-domain vibrational spectroscopy. In our method, the OH-stretching mode was excited by an ultrashort IR pulse, and vibrational coherence of low-frequency modes was detected by monitoring absorption change of an ultrashort visible probe pulse. We carried out an anharmonic vibrational analysis based on the density functional theory (DFT) to characterize the anharmonic interaction giving rise to the low-frequency vibrational coherence. Previously, consideration about anharmonic interactions was limited to the mechanical anharmonicity in analysis of ultrafast IR pump-probe data of a hydrogen-bonding system [1]. In this study, we evaluated the contribution of the *electrical* anharmonicity as well. It will help complete understanding of the optical process behind the wavepacket formation upon IR-excitation.

Methods

The IR-pump visible-probe measurement was performed using outputs of two home-built non-collinear optical parametric amplifiers (NOPAs) pumped by a Ti:sapphire regenerative amplifier (Legend, Coherent) [2]. The output of the first NOPA was tuned to near IR region (~1.05 μm) and it was used to generate a mid-IR pump pulse (~3.4 μm) by difference frequency mixing with the fundamental output of Ti:sapphire laser (800 nm) using a 0.5 mm-thick $LiIO_3$ crystal. A visible probe pulse (~550 nm) was generated with the second NOPA. The time resolution of the measurement was about 60 fs. The sample was quinizarin (1,4-dihydroxy-9,10-anthraquinone) dissolved in deuterated chloroform ($CDCl_3$).

Anharmonic Vibrational Analysis. A DFT vibrational analysis was carried out with Gaussian 03 program [3] using the B3LYP functional and the 6-311+G(d,p) basis set. Anharmonic coupling strength between the OH-stretching mode (v_i) and low-frequency modes (v_j) were evaluated through the transition moments of the combination transitions. An analytical expression for the transition moment of the combination transition, $\langle n_i=1, n_j=1 | \mu | n_i=0, n_j=0 \rangle = \langle 1,1 | \mu | 0,0 \rangle$, can be obtained using the perturbation theory as

$$\langle 1,1 | \mu | 0,0 \rangle \cong \frac{1}{4hv_j} \frac{\partial \mu}{\partial Q_i} \frac{\partial^3 V}{\partial Q_i^2 \partial Q_j} + \frac{1}{4hv_i} \frac{\partial \mu}{\partial Q_j} \frac{\partial^3 V}{\partial Q_i \partial Q_j^2} + \frac{1}{2} \frac{\partial^2 \mu}{\partial Q_i \partial Q_j}. \qquad (1)$$

In the right-hand side of this equation, the first two terms correspond to the mechanical anharmonicity and the last term represents the electrical anharmonicity. Numerical evaluation of each term was done by the finite difference procedure [1] to calculate the coordinate derivatives of V and μ from the Gaussian output.

Results and Discussion

Quinizarin has two intramolecular hydrogen-bonding sites and shows a broad IR absorption band at ~3000 cm^{-1}, which is an indication of the hydrogen-bond formation. The red-edge of its lowest energy electronic transition peaked at 484 nm was probed in the transient absorption measurement after pumping the hydrogen-bonded OH-stretching band. Fig. 1(a) shows the pump-probe signal measured at 546 nm, which mainly consists of initial spike at the time origin and subsequent decaying components. These responses are highly likely due to the intramolecular dissipation of the vibrational energy that was initially deposited in the OH-stretching mode. A triple-exponential function with three decay constants, 0.07 ps, 1.4 ps, 6.3 ps, was used to reproduce this decay and it was subtracted from the observed data to obtain oscillatory components shown in Fig.1(b). The Fourier-transformed amplitude spectrum of the oscillatory component (inset of Fig. 1(b)) shows several peaks. Apart from the solvent Raman bands, a clear peak was found at 314 cm^{-1}. According to the DFT vibrational analysis, this band was assigned to an intramolecular hydrogen-bond stretching mode at 319 cm^{-1}.

The transition moment of the combination transition with the symmetric OH-stretching mode calculated using Eq.(1) is tabulated in Table 1 for totally-symmetric low-frequency vibrations. For all calculated five low-frequency modes, the first mechanical term in Eq. (1), μ_{iij}, has a dominant contribution. On the other hand, the electrical term (μ_{elec}) has relatively small, but not negligible, contribution (8-57 % of μ_{iij}). In total, the anharmonic coupling strength of the 319 cm^{-1} mode is the largest, which is in harmony with the experimental observation.

Table 1. Transition moments of combination transitions derived from the DFT result (B3LYP/6-311+G(d,p)).

v_j (cm^{-1})	μ_{iij} (10^{-3} Debye)	μ_{ijj} (10^{-3} Debye)	μ_{elec} (10^{-3} Debye)
319	-35	0.026	5.8
426	-24	0.27	8.5
438	-11	-0.065	-0.85
467	-7	-0.021	-4.0
523	9	0.015	-4.0

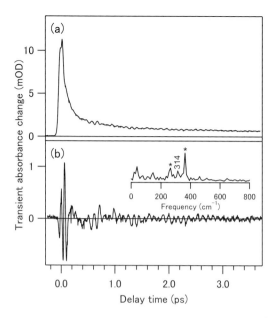

Fig. 1. (a) IR-pump visible-probe signal of quinizarin in $CDCl_3$ (7×10^{-2} mol dm^{-3}) probed at 546 nm. (b) Oscillatory component of the transient signal. Inset: Fourier transform of the oscillatory component. Asterisks indicate solvent bands.

In conclusion, the hydrogen-bond stretching vibration at 314 cm^{-1} is strongly coupled to the OH-stretching mode mainly via the mechanical anharmonic interaction. This can be understood with a simple idea of dynamic modulation of the hydrogen-bond strength by the hydrogen-bond stretching mode. As hydrogen-bond donor and acceptor sites approach to each other, the hydrogen bond gets stronger and correspondingly the force constant of the OH-stretching mode decreases, because of the mechanical anharmonic coupling. On the other hand, under the influence of the electrical anharmonicity, an increase in the hydrogen-bond strength results in enhanced OH-stretching transition intensity. These considerations correspond to a dynamic representation of the well-known spectral changes of OH-stretching IR absorption bands upon hydrogen-bond formation, i.e. a red-shift and an intensity enhancement.

Acknowledgements. This work was supported by Grant-in-Aid for Science Research on Priority Area "Molecular Science for Supra Functional Systems" (No.19056009) from MEXT of Japan.

1   K. Heyne, N. Huse, J. Dreyer, E. T. J. Nibbering, T. Elsässer, and S. Mukamel, J. Chem. Phys. **121**, 902, 2004.
2   S. Takeuchi and T. Tahara, Chem. Phys. Lett. **326**, 430, 2000.
3   M. J. Frisch, G. W. Trucks, H. B. Schlegel et al., Gaussian 03, Gaussian, Inc., Wallingford, 2004.

Ultrafast Photodecomposition of Dibenzoyl Peroxide studied by Time-Resolved Infrared Spectroscopy

C. Reichardt[1], T. Schäfer[1], J. Schroeder[2], P. Vöhringer[3], D. Schwarzer[1]

[1] AG Reaktionsdynamik, Max-Planck-Institut für biophysikalische Chemie, Am Fassberg 11, 37077 Göttingen, Germany, E-mail: dschwar@gwdg.de
[2] Institut für Physikalische Chemie, Georg-August-Universität Göttingen, Tammannstraße 6, 37077 Göttingen, Germany
[3] LS Molekulare Physikalische Chemie, Institut für Physikal.& Theoret. Chemie, Universität Bonn, Wegelerstr. 12, 53115 Bonn, Germany

Abstract. The photodissociation of dibenzoyl peroxide is controlled by its S_1-lifetime and in 0.4 ps leads to a benzoyloxy/phenyl radical pair plus CO_2 via concerted bond breakage of the O-O and the phenyl-C(carbonyl) bond.

Introduction

Whereas the thermal decomposition of dibenzoyl peroxide (DBPO) is well understood and proceeds via bond breakage of the O-O bond yielding benzoyloxy radicals which in inert environment subsequently decarboxylate on a microsecond time scale to form phenyl radicals (pathway 1 in Figure 1), the primary steps of fragmentation after UV excitation of DBPO are not completely understood [1]. Early chemical trapping experiments showed that the photodissociation not only produces benzoyloxy radicals, but also in a two-bond cleavage phenyl radicals and CO_2 as expressed by pathway 2 in Figure 1. Recent femtosecond pump-probe experiments [2] in the visible to near-IR spectral range suggest an alternative mechanism in which the DBPO photofragmentation initially follows pathway 1 with O-O-bond scission being faster than 200 fs, yielding highly vibrationally excited benzoyloxy radicals. A fraction f of these radicals rapidly decomposes to phenyl plus CO_2 (vertical arrow in Figure 1), while the remaining population by vibrational energy relaxation cools down to thermal equilibrium and subsequently decarboxylates on a microsecond timescale. This analysis, however, rests upon the correct assignment of electronic spectra to intermediates which due to spectral overlap with other species is difficult.

In the present work we use time-resolved IR spectroscopy to answer the question whether in the fragmentation of DBPO all bonds break sequentially as suggested by femtosecond UV/VIS pump-probe experiments [2] or whether possibly concerted bond cleavage is involved.

Fig. 1. Reaction pathways of DBPO decomposition

Results and Discussion

In Figure 2 the linear absorption spectra of DBPO and ^{13}C-DBPO are compared with the corresponding transient difference spectra recorded 1.3 ns after UV excitation, respectively. The transient spectra show negative peaks caused by bleaching of the peroxide and positive absorptions due to the formation of photoproducts. Most prominent is the bleaching of the CO stretching bands but also phenyl ring vibrations at 1453 cm^{-1} clearly show the disappearance of DBPO and ^{13}C-DBPO, respectively. The strongest increase of absorption arises from the asymmetric stretching vibration of newly formed CO$_2$ at 2337 cm^{-1} and ^{13}CO$_2$ at 2273 cm^{-1}, respectively. In the fingerprint region strong and sharp bands at 1476 (DBPO) and 1461 cm^{-1} (^{13}C-DBPO) and, additionally, weaker but broader absorptions at 1535 (DBPO) and 1508 cm^{-1} (^{13}C-DBPO) appear as marked by asterisks. Because of the pronounced isotope shift of 15 cm^{-1} for the sharp and 27 cm^{-1} for the broad peaks we attribute these bands to the benzoyloxy radical. DFT calculations [3] support this assignment as demonstrated in the insert of Figure 2b. Weak absorptions at 1737 (DBPO) and 1694 cm^{-1} (^{13}C-DBPO) in Figure 2b indicate the formation of phenyl benzoate which within this time range can only be formed by geminate recombination of the benzoyloxy-phenyl radical pair. Further evidence supporting these assignments is provided from the time evolution of the transient absorption bands. Immediately upon excitation by the UV pump pulse, CO stretching bands in the electronic ground and first excited state of DBPO appear as a superposition of bleach and absorption, respectively (Figure 3a). With increasing pump-probe delay the excited state population decays and only the ground state bleach remains. An exponential fit to the integrated band intensity (insert in Figure 3a) gives a S$_1$ lifetime of 0.4 ± 0.2 ps. Then, on a longer timescale the new CO stretching band of phenyl benzoate emerges at 1745 cm^{-1} (Figure 3b). The time constant of 75 ± 10 ps derived from exponentially fitting the temporal evolution of the band integral (insert in Figure 3b) is typical for geminate recombination.

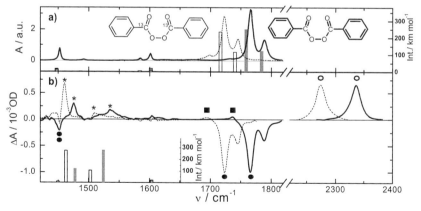

Fig. 2. (a) Linear absorption spectra of DBPO (full line) and ^{13}C-DBPO (dashed) in CD$_3$CN (bars indicate frequencies and intensities of DFT calculations); (b) transient difference spectra 1.3 ns after 267 nm excitation of DBPO (full line) and ^{13}C-DBPO (dashed line) in CD$_3$CN (at 2220-2380 cm^{-1} in n-heptane); symbols indicate bleaching of DBPO (filled circles) and formation of the products CO$_2$ (open circles), benzoyloxy radical (stars), and phenyl benzoate (squares); the insert shows calculated frequencies and intensities of the benzoyloxy radical.

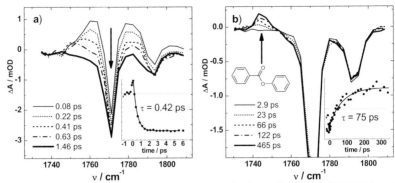

Fig. 3. Transient difference spectra of the CO stretch absorption band of DBPO in n-heptane recorded at selected pump-probe delay times (arrows indicate chronology); (a) at early times the S_1 decay of DBPO is monitored (the insert shows the temporal evolution of the integrated band intensity; (b) at later times formation of the recombination product phenyl benzoate is observed (insert: band intensity around 1745 cm^{-1}).

After UV excitation of ^{13}C-DBPO also the benzoyloxy radical is formed on a subpicosecond timescale as shown in the contour diagram of Figure 4. The absorption of the radical at 1461 cm^{-1} is superimposed by bleaching of the sharp CH bending vibration of the parent molecule at 1453 cm^{-1}. Nevertheless, Figure 4 clearly evidences that the initially broad radical band narrows within several tens of ps: a clear signature of vibrational energy relaxation of a hot species. The absorption reaches a maximum at 50 ps and, subsequently, decreases by about 15% with a time constant of 80 ± 20 ps. which coincides with the time constant observed for the geminate recombination of the benzoyloxy-phenyl radical pair to phenyl benzoate.

Our findings clearly show that the primary steps of the DBPO photofragmentation involve a concerted two-bond cleavage into phenyl, benzoyloxy, and CO_2 with a rate constant limited by the S_1-lifetime of 0.4 ± 0.2 ps. 15-20% of the radical pairs geminately recombine to phenyl benzoate on a timescale of 75 ps.

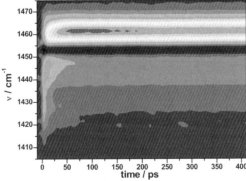

Fig. 4. Spectral evolution of the 1461 cm^{-1} absorption band of the benzoyloxy radical formed after photodissociation of ^{13}C-DBPO in CD_3CN.

1   K. Fujimori, in Organic Peroxides, Edited by W. Ando, Wiley, New York, 1992.

2   B. Abel, J. Assmann, M. Buback, M. Kling, R. Oswald, S. Schmatz, J. Schroeder, and T. Witte, J. Phys. Chem. A **107**, 5157, 2003.

3   C. Reichardt, J. Schroeder, P. Vöhringer, and D. Schwarzer, Phys. Chem. Chem. Phys. **10**, 1662, 2008.

Electron detachment of OH⁻ (aq)

Christian Petersen, Jan Thøgersen, Svend Knak Jensen, and Søren R. Keiding

University of Aarhus, Langelandsgade 140, DK-8000, Århus C, Denmark
E-mail: thogersen@chem.au.dk

Abstract. Transient absorption spectroscopy is used to study the photo-induced electron detachment of OH⁻ (aq). The electron is detached by exciting OH⁻ (aq) to its charge-transfer-to-solvent state. The primary quantum yield of OH(aq) is 62 %, while recombination with e⁻(aq) reduces the yield to 37 % after 5 ps and 13 % after 200 ps. The yield of hot (OH⁻)* ions is 38%. Rotational anisotropy measurements of OH⁻ (aq) and OH(aq) yield a 1.9 ps reorientation time for OH⁻ (aq), whereas no rotational anisotropy is resolved for the OH(aq) radicals.

Introduction

The detachment of the electron from OH⁻ (aq) at 6.2 eV is believed to proceed through a short-lived charge-transfer-to-solvent (CTTS) state. The electron in the CTTS state is quasi-bound by the potential from lingering water molecules reminiscent of the hydration of ground state OH⁻(aq). Solvent fluctuations rapidly cause the electron to be adiabatically released from the CTTS state and a complex, or contact pair, (OH:e⁻)$_{aq}$ is formed[1]. However, the exact nature of the CTTS state and the contact pair is unknown.

Here we use femtosecond transient absorption spectroscopy to study the primary reaction dynamics of the photo-induced electron detachment of OH⁻ following closely the work of Crowell *et al.*[2]. A 200 nm pump pulse initiates the detachment and the resulting photoproducts are monitored by probing their time dependent absorption in the spectral range from 193 to 800 nm. The complete coverage of this spectral region enables direct observation of transient species and possibly any spectral dynamics reflecting the relaxation or hydration of the photoproducts. In addition, it allows the determination of the time dependent quantum yields of OH⁻, OH and e⁻. We have also measured the rotational anisotropy of the fragments. The rotational anisotropy of OH⁻ and OH may show if the orientation of the excited OH⁻ molecules is transferred to the OH fragment and thereby elucidate the processes between the excitation of OH⁻ and the actual electron detachment.

Experimental Methods

The transient absorption spectrometer utilizes a 1 kHz Titanium-Sapphire laser system emitting 100 fs pulses with pulse energy of 0.75 mJ. The 800 nm output pulses are frequency quadrupled to generate the 200 nm excitation pump pulses. The pump pulses are focused through the sample by an f = 50 cm concave mirror. An optical parametric amplifier (OPA) generates the probe pulses across the spectral range from 460 – 800 nm. Probe pulses ranging from 230 nm to 460 nm are produced by frequency doubling, while the spectral region between 193 nm and 300 nm is covered by frequency mixing the OPA pulses with either 400 nm or 266 nm pulses. The probe beam is focused onto the sample and subsequently detected by a photodiode. The polarization of the pump pulses is at the magical angle (54.7°)

relative to the polarization of the probe pulse. In the measurements of the rotational anisotropy the pump and probe pulses are either polarized parallel or perpendicular to one another. The sample consists of a ~0.15 mm thick jet of 40 mM KOH(aq).

Results and Discussion

Figure 1 shows selected wavelengths of the transient absorption spectrum from 193 – 800 nm induced by the 200 nm excitation pulse. Separate measurements have been extended to 200 ps (not shown). The interval from 193 to 220 nm involves the absorption of OH^-, OH and e^-. The interval from 220 – 400 nm relates to the absorption of OH and e , and above 400 nm the absorption is entirely due to e^-.

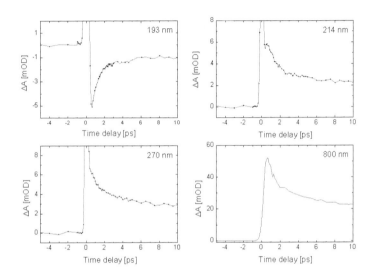

Fig. 1. Selected transient absorption data taken at 193, 214, 270 and 800 nm.

The data at 193 nm display an initial negative absorption of –5.5 mOD caused by the excitation of ground state OH^- molecules. The absorption recovers to a nearly constant level of – 0.5 mOD on two time scales of 2 ps and 20 ps. At 214 nm the absorption shows a positive transient peaking after 0.5 ps. The absorption is predominantly, but not entirely, due to OH and e^-. Comparing the absorption from 193 to 225 nm to the total absorption of equilibrated OH^-, OH and e^- confirms that these species are the only contributors to the absorption after 5 ps. However, at delays shorter than ~4 ps the transient spectra are poorly matched by any combination of equilibrated OH^-, OH and e^- spectra. Nor can the spectra be assigned to any of the products from the photolysis of the water solvent [3]. Accordingly, we infer that other species are present during the first ~4 ps. Based on a detailed analysis of the transient absorption spectrum we tentatively assign the short-lived transient to unequilibrated OH^- molecules. The hot OH^- molecules result from the conversion of OH^- molecules in the CTTS state to hot $(OH^-)^*$ molecules that aided by the strong solute-solvent coupling rapidly relax to ground state OH^-. The prompt (<0.5 ps) appearance of the short-lived transient suggests that the relaxing OH ions are formed by rapid geminate recombination of closely spaced OH:e⁻ pairs, or due to OH⁻ molecules undergoing direct solvent assisted conversion from the CTTS state. The transient absorption pertaining to OH and e⁻ is illustrated by the measurement at 270 nm. The 0.5 ps width of the t = 0 peak is slightly larger than the time resolution of 0.3 ps indicating the immediate (<0.3 ps) appearance of the OH and e⁻ detachment products. The coherence peak completely dominates the first 0.5 ps thereby obscuring the onset of the absorption of OH and e⁻. Hence, apart from setting an upper limit for the CTTS state life time of 0.3 ps, the absorption traces in this spectral region provide no additional information concerning the OH⁻ CTTS state. The prompt onset of absorption of 5.5 mOD at 270 nm is followed by a double exponential drop with time constants of 2.2 ps (60%) and 30 ps (40%). The absorption at 800 nm is entirely due to hydrated electrons. It peaks after 0.5 ps setting an upper limit for the lifetime of the CTTS state of 0.3 ps in agreement with the measurement at 270 nm. Subsequently, the absorption declines to a nearly constant level of 10 mOD following a curve with time constants identical to those of the 270 nm transient. The course of the absorption also matches the recovery of the absorption below 200 nm associated with the ~20 ps component of the OH⁻ absorption. Accordingly, the decline in the hydrated electron concentration is entirely dominated by recombination with OH.

Information about the rotational motion of OH⁻ and OH is obtained from rotational anisotropy measurements at 193 and 260 nm, respectively. The rotational anisotropy of OH⁻ increases from r(0 ps) = 0 to a maximum of r(2 ps) = 0.2 followed by an exponential decay with a 1.9 ps time constant. In contrast, OH exhibits no rotational anisotropy after the time resolution of 0.3 ps. Consequently, the rotational anisotropy of OH⁻ is either not transferred to the OH radical during electron detachment, or alternatively, rapid rotational diffusion of OH removes the anisotropy within 0.3 ps after the excitation of OH⁻.

Conclusions

The photo-detachment of OH⁻ at 200 nm has two primary reaction channels. The hydroxide ion, OH⁻ is excited to the CTTS state from which 62% of the molecules detach to form an OH:e⁻ contact pair and 38% return to the ground state via solvent assisted conversion, giving rise to hot OH⁻ molecules. The combined rate of solvent assisted conversion and contact pair formation depopulates the CTTS state in less than 0.4 ps. The contact pairs, or rather two separated OH and e⁻ fragments move around under the influence of a weakly attractive potential before they recombine geminately leaving a quantum yield of OH radicals of 13 % after 200 ps Rotational anisotropy measurements of OH⁻ and OH give an OH⁻ reorientation time of τ = 1.9 ps and further show that the rotational anisotropy of OH⁻ is not transferred to the OH radical.

1. R. A. Crowell, R. Lian, M. C. Sauer, D. A. Oulianow, I. A. Shkrob, Chemical Physics Letters, Vol. 383, 481, 2003.

2. R. A. Crowell, R. Lian, I. A. Shkrob, D. M. Bartels, X. Y. Chen, S. E. Bradforth, Journal of Chemical Physics, Vol. 120, 11712, 2004.

3. C. L. Thomsen, D. Madsen, S. R. Keiding, J. Thogersen, O. Christiansen, Journal of Chemical Physics, Vol. 110, 3453, 1999.

Pathways of Vibrational Relaxation after N-H Stretching Excitation in Intermolecular Hydrogen Bonds

Valeri Kozich, Jens Dreyer, and Wolfgang Werncke

Max-Born-Institut, Max-Born-Strasse 2A, D-12489 Berlin, Germany
E-mail: werncke@mbi-berlin.de

Abstract: Pathways of vibrational relaxation of azaindole dimers after NH stretching excitation have been studied by picosecond infrared-pump/anti-Stokes resonance Raman-probe spectroscopy. Our measurements indicate mode-relaxation via several vibrations with N-H bending character.

Introduction

N-H hydrogen bonds play an important role in the biological function of DNA base pairs [1]. Because of their comparatively simple and well defined structure 7-azaindole dimers have been used as model systems for the much more complex base pairs [2,3]. An extreme shortening of the lifetime of the N-H stretching vibration from about 10 ps to about 100 fs has been observed for dimeric in comparison to monomeric azaindole [4]. This has been explained by frequency down-shifting the N-H stretching vibration due to hydrogen bonding to the frequency range of combination tones of fingerprint modes. Because of efficient vibrational anharmonic coupling accompanied by a small energy mismatch with the $v = 1$ state of the N-H stretching vibration efficient relaxation pathways are created.

However, experimental observation of the modes accepting the energy from the initially excited N-H stretching mode is still lacking. Here we report on the first observation of vibrational population kinetics in a wide frequency range of 600 – 1700 cm^{-1} applying picosecond infrared-pump/anti-Stokes resonance Raman-probe spectroscopy.

We show that modes in the frequency region 1400 –1600 cm^{-1} containing significant contributions of N-H bending motions primarily accept the energy, which relaxes further via secondary processes of intramolecular vibrational energy redistribution (IVR).

Methods

Azaindole was dissolved in C_2Cl_4 at a concentration of 67 mmol/l. In this case the concentrations of the dimers and monomers were about 26 and 15 mmol/l, respectively. For infrared-pump/anti-Stokes resonance Raman-probe spectroscopy [5] we use a twin optical parametric generator/amplifier (OPG/OPA) system. Pulses at 1 kHz repetition rate were generated at 3.2 μm (3150 cm^{-1}) for selective excitation of the N-H stretching dimer band, which is well separated from the corresponding monomer band. Pulses at 340 nm were used for resonance Raman probing. The temporal resolution is characterized by the width of the cross correlation function of pump and probe pulses of about 1 ps (fwhm). Vibrations of the azaindole dimer have been assigned by DFT normal mode calculations together with stationary infrared and resonance Raman measurements [6]. Experimental work was accompanied by calculations of cubic vibrational anharmonic couplings with the N-H stretching vibration.

Results and discussions

Immediately after infrared pumping at 3150 cm^{-1} we observe three anti-Stokes Raman bands at 1590, 1500 and 1420 cm^{-1}. Frequencies and assignments of the corresponding modes are given in Table 1. The bands at 1590 and 1500 cm^{-1} are doublets and spectrally not resolved in the time-resolved experiment. The temporal evolution of vibrational excess populations which have been obtained from the corresponding transient anti-Stokes Raman intensities of the bands is presented in Fig. 1. The data were analyzed by a three-level model with rise (τ_1) and decay times (τ_2) of the excited vibrational levels. According to their fast rise times τ_1 close to the known decay time of the N-H stretching vibration of the dimers of about 100 fs, the two bands are due to the modes primarily accepting the energy from the excited N-H stretching vibration, in particular the modes at 1590 cm^{-1}. With the estimated excitation density of the N-H stretching vibration of 20 % we calculate that the doublet of modes at 1590 cm^{-1} accept about 30 % of the initial N-H stretching mode population. According to Table 1 these are modes with pronounced N-H bending contribution.

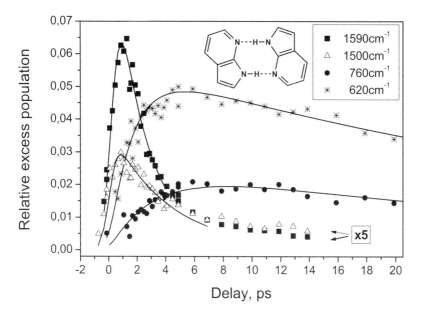

Fig. 1. Temporal evolution of the excess populations of the vibrations at 620, 780, 1500 and 1590 cm^{-1}. Solid lines: Fits with rise (τ_1) and decay times (τ_2) of the excited vibrational levels using a three-level model. The excess populations at 1590 and 1500 cm^{-1} are multiplied by a factor of 5. Inset: molecular structure

After about 1 ps population of the modes at 620 and 760 cm^{-1} is observed (cf. Fig. 1). From their rise times which are slow compared to the decay time of the N-H stretching vibration (cf. Table 1) it follows that these modes are not populated directly from the N-H stretching vibration. In contrast, rise times close to the decay times of the primary accepting modes indicate that they are populated from the primary accepting modes via a secondary step of relaxation. Their slow decay time is probably mainly due to energy transfer to the solvent.

Table 1. Frequencies of the modes, (the complete assignment of the vibrations is given in Ref. 4), rise (τ_1) and decay times (τ_1) of the anti-Stokes Raman bands, assignments of the modes

Wavenumber [cm^{-1}]	τ_1 [ps]	τ_2 [ps]	Assignment (nomenclature according to Ref. 7)
1589 /1600	0.3±0.2	2.5±0.5	8a pyridine + δNH / 8b pyridine + δNH
1499 /1508	0.3±0.2	4±1	19a+ vC=C pyrrole / δNH + δCH pyridine
1420	0.3±0.2	4±1	19b + vpyridine-NH+δNH
764	3.0±0.5	35±5	δ-1 azaI
621	3.0±0.5	35±5	δ-6b pyridine

The function of the modes at 1508, 1589 and 1600 cm^{-1} as primary accepting modes is in accordance with our calculations of vibrational anharmonic coupling constants with the N-H stretching vibration predicting strong coupling with combination tones of modes with N-H bending character in the fingerprint region. Due to symmetry reasons each combination tone interacting with the primarily excited N-H stretching vibration consists of pairs of Raman- and infrared-active vibrations. Thus, excitation of a Raman-active mode is accompanied by excitation of an infrared-active counterpart, which can not be observed in our Raman experiment. Consequently, the primary step of energy transfer from the N-H stretching vibration is mode-selective with excitations distributed among other accepting modes which are infrared-active and – according to our calculations – with pronounced N-H bending character too.

In conclusion, our combined experimental and theoretical approach indicates mode-selective relaxation of the N-H stretching vibration in intermolecular hydrogen bonds. It highlights the important role of energy transfer from stretching to a manifold of in-plane bending motions in dimers with N-H intermolecular hydrogen bonds.

1   G.A. Jeffrey and W. Saenger, Hydrogen Bonding in Biological Structures (Springer, 1991).

2   T. Elsaesser, N. Huse, J. Dreyer, J.R. Dwyer, K. Heyne and E.T.H. Nibbering, Chem. Phys. **341**, 175, 2007.

3   Oh-Hoon Kwon, and A.H. Zewail, Proc. Nat. Acad. Sci. USA **104**, 8703, 2007.

4   J.R. Dwyer, J. Dreyer, ET.J. Nibbering, and T. Elsaesser, Chem. Phys. Lett, **432**, 146, 2006.

5   V. Kozich, J. Dreyer, S. Ashihara, W. Werncke, and T. Elsaesser, J. Chem. Phys. **125**, 074504, 2006.

6   J. Dreyer, J. Chem. Phys. **127**, 054309, 2007.

7   G. Varsanyi, Vibrational spectra of benzene derivatives (Academic, New York 1969).

GHz Longitudinal and Transverse Acoustic Waves and Structural Relaxation Dynamics in Liquid Glycerol

Christoph Klieber[1], Thomas Pezeril[1], Stéphane Andrieu[2], and Keith A. Nelson[1]

[1] Department of Chemistry, Massachusetts Institute of Technology, Cambridge, MA 02139, USA
[2] Laboratoire de Physique des Matériaux, Université H. Poincaré, 54506 Vandoeuvre, France
  E-mail: cklieber@mit.edu

Abstract. Novel picosecond ultrasonic techniques for longitudinal and transverse acoustic pulse generation have been employed to probe structural relaxation dynamics in liquid glycerol at gigahertz frequencies.

Introduction

Collective dynamics of glass-forming liquids remain a major scientific challenge in condensed matter [1]. Ultimately, those dynamics are described in terms of density and shear relaxation, so longitudinal and transverse acoustic modes carry the key information. However, significant dynamical gaps remain unexplored due to the experimental inaccessibility of fast time scales (high acoustic frequencies) on which much of the correlation among motions of particles is lost. To close this gap, we have used a unique pulse shaping and spectroscopic approach where we adapt different laser ultrasonics techniques [2-4] to directly probe the GHz frequency longitudinal and transverse responses of glass forming liquids.

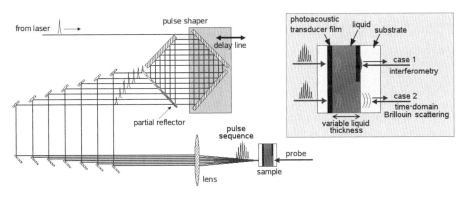

Fig. 1: Schematic illustration of optical pulse shaper and GHz coherent acoustic wave spectrometer. Box: Sample design with front-back excitation-probe geometry and two different probing techniques for longitudinal (case 1) and transverse (case 2) acoustic waves.

Experimental Method and Results

In both front-back pump-probe type measurements, longitudinal and transverse acoustic waves are optically generated in a thin metal transducer film upon ultrafast laser irradiation. A novel pulse shaping method is used to generate multiple-cycle acoustic waves at GHz frequencies [4]. These acoustic waves are transmitted into and through an adjacent liquid layer and either into an opaque metallic film, at the backside of which they are detected interferometrically (longitudinal, case 1 in Figure 1), or into a transparent substrate in which they are detected through time-domain coherent depolarized Brillouin scattering (transverse, case 2 in Figure 1). For transverse wave generation, a film with canted out-of-plane crystallographic orientation [3] is used. The density and shear relaxation dynamics of the liquid are revealed through the measured acoustic

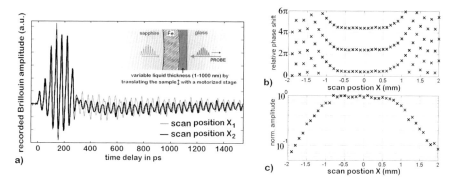

Fig. 2: a) Time derivative of the reflected depolarized probe laser intensity for two distinct liquid thicknesses. The sample was translated to bring regions of different liquid layer thickness into the beam paths. The acoustic generation signature is visible as the strong modulation starting at $t = 0$ and is followed by oscillations due to coherent shear Brillouin light scattering of the probe light off of the strain pulse as it propagates through the liquid and into the glass substrate. The phase shift between the two acoustic signals is due to the difference in liquid layer thickness. The inset illustrates the sample design as described in the text. Plots b) and c) show relative phases and attenuation information for a large number of liquid layer thicknesses. Comparison of two data directly yields the speed of sound and acoustic attenuation at a given frequency and temperature.

frequency-dependent damping and dispersion.

Fourier spectral analysis of acoustic waves transmitted through different liquid thicknesses allows the extraction of a relative phase and attenuation for each set of recorded data. Liquid layers of different thicknesses are accessed by means of confining the liquid between two substrates (with one of them slightly curved) and moving the sample in the plane of the liquid layer by a motorized stage as shown in the inset in Figure 2 a). From the comparison of any two individual sets of data from different sample regions with different liquid layer thickness, it is possible to calculate the phase difference and attenuation due to the thickness difference. The phase differ-

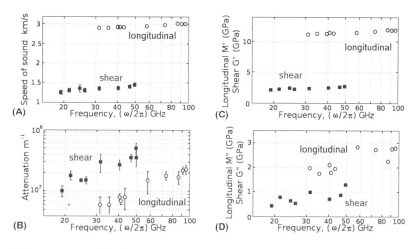

Fig. 3: The rates measured longitudinal and shear sound velocities (**A**) and acoustic attenuation (**B**) for glycerol at room temperature as a function of frequency. The complex acoustic moduli spectra, the longitudinal modulus $\hat{M}(\omega) = M' + iM''$ (**C**) and the shear modulus $\hat{G}(\omega) = G' + iG''$ (**D**), can be calculated as described in the text.

ence and attenuation directly yield the acoustic speed and attenuation rate at the selected acoustic frequency and sample temperature without the necessity of knowing any further material parameters for the specific sample [5]. An example for such a pair of raw data sets is depicted in Figure 2 a). The phase and amplitude results from many data sets recorded at different sample positions X (i.e. different liquid layer thicknesses) are shown in Figures 2 b) and 2 c).

The liquid response can be expressed in terms of a complex modulus, $\hat{M}(\omega) = \rho \hat{v}^2$ (with given sample density ρ), being the longitudinal modulus $\hat{M}(\omega)$ for longitudinal and the shear modulus $\hat{G}(\omega)$ for transverse waves. The complex velocity is defined by $\hat{v}(\omega) = \omega/\hat{q}(\omega)$ through the complex wave vector $\hat{q}(\omega) = q' + iq'' = \omega/v(\omega) + i\alpha(\omega)$, where $v(\omega)$ is the frequency dependent speed of sound and $\alpha(\omega)$ the frequency dependent acoustic attenuation. The obtained results are shown in Figure 3.

Discussion

Glass-forming liquids show two prominent features in their relaxation dynamics, a fast (picosecond time scale) "beta" regime revealed through the present GHz-frequency acoustic measurements and a slower, highly non-exponential, temperature-dependent "alpha" regime revealed through complementary MHz-frequency photoacoustic methods [6]. Phenomenological models and, in recent years, a first-principle theoretical framework (mode-coupling theory) have been advanced to describe the temperature-dependent dynamics as reflected in the frequency-dependent moduli [7]. Our experiments yield direct access to structural relaxation dynamics throughout both regimes over a wide temperature range, and permit direct testing of theoretical predictions of a quantitative relationship between the fast and slow dynamics (power-law exponent relations reminiscent of critical phenomena) that has remained untested to date due to the absence of the GHz-frequency acoustic data that we can now record. Work toward a comprehensive assessment of the theory is currently under way.

Acknowledgements. This work was supported in part by DOE Grant No DE-FG02-00ER15087 and NSF grants CHE-0616939 and DMR-0414895.

References

1   P. Debenedetti, and F. Stillinger, "Supercooled liquids and the glass transition," Nature **410**, 259-267 (2001).

2   H. Maris, "Picosecond Ultrasonics," Sci. Am. **278**, 86-89 (1998).

3   T. Pezeril, P. Ruello, et al., "Generation and detection of plane coherent shear picosecond acoustic pulses by lasers: Experiment and theory," Phys. Rev. B **75**, 174307 (2007).

4   J. D. Choi, T. Feurer, et al. "Gerneration of ultrahigh-frequency tunable acoustic wave," Appl. Phys. Lett. **87**, 081907 (2005).

5   R. M. Slayton, and K.A. Nelson, "Picosecond acoustic transmission measurements. II. Probing high frequency structural relaxation in supercooled glycerol," J. Chem. Phys., **120**, 3919 (2004).

6   Y. Yang and K.A. Nelson, "Impulsive stimulated light scattering from glass-forming liquids: I. Generalized hydrodynamics approach," J. Chem. Phys. **103**, 7722-7731 (1995) and Y. Yang and K.A. Nelson, "Impulsive stimulated light scattered from glass-forming liquids: II. Salol relaxation dynamics, nonergodicity parameter, and testing of mode coupling theory," J. Chem. Phys. **103**, 7732-7739 (1995).

7   W. Gotze, "Recent tests of the mode-coupling theory for glassy dynamics," J. Phys.: Condens. Matter **11**, A1-A45 (1999).

Frequency dependence of the molecular reorientation of liquid water

Huib J. Bakker and Rutger L.A. Timmer

FOM Institute for Atomic and Molecular Physics, Kruislaan 407,
1098 SJ Amsterdam, The Netherlands
E-mail: bakker@amolf.nl

Abstract. Using multi-color femtosecond mid-infrared spectroscopy, we observe that the anisotropy dynamics of the O-D stretch vibration of HDO:H_2O shows a strong dependence on the frequency of the excitation. The results indicate that a sub-ensemble of the molecules shows a fast reorientation that is accompanied by a large change of the vibrational frequency. In contrast to recent theoretical predictions [1,2], we find the probability for reorientation to be strongly frequency dependent.

Introduction

Femtosecond mid-infrared spectroscopy allows the study of the correlation between the orientational mobility of water molecules and the strength of their hydrogen-bond interactions. Interestingly, previous studies of this correlation have led to contradictory results, seemingly dependent on the water isotope that was studied. For the O-H stretch vibration of HDO:D_2O, a frequency dependence of the reorientation was observed [3], while for the O-D vibration of HDO:H_2O no frequency dependence was found [4,5]. In the latter studies the O-D absorption band was probed over a broad frequency range, but excited only at its center frequency [4,5]. Here we investigate the frequency dependence of the anisotropy by also varying the excitation frequency.

Experiment

The pump and probe pulses have independently tunable wavelengths near 4 μm (2500 cm^{-1}). The pump pulses have a duration of ~200 fs, a pulse energy of 10 μJ, and a bandwidth of 70 cm^{-1}. The probe pulses have a duration of ~100 fs, a pulse energy per 0.1 μJ and a bandwidth of 200 cm^{-1}. The pump-induced bleaching of the v=0->1 transition and induced absorption of the 1->2 transition are monitored by delayed probe pulses that are polarized parallel and perpendicular to the pump. The broadband probe pulses are dispersed and detected with a 2x32 mercury-cadmium-telluride detector array. The probe signals are used to construct the anisotropy parameter.

Results and Discussion

In Fig. 1 the measured spectral dependence of the anisotropy is shown for three different pump frequencies and five different delays. When the O-D stretch vibration is pumped in the red wing of the absorption band or close to its central frequency (top panel), the anisotropy is the same at all probe frequencies, except in the frequency region where the bleaching changes into an induced absorption, where the anisotropy

502

is erratic due to the vanishing of the isotropic signal. When the pump is tuned to the blue wing of the absorption spectrum (middle and lower panels), the anisotropy is strongly frequency dependent. At early delay times the anisotropy in the center and the red wing is significantly lower than 0.4 meaning that quickly after the excitation the center and the red wing contain a large fraction of reoriented molecules.

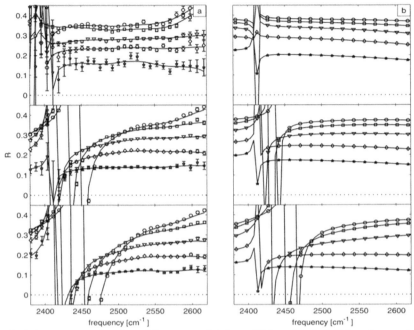

Fig. 1. Anisotropy as a function of frequency at delays of 0.1 ps (circles), 0.2 ps (squares), 0.5 ps (triangles), 1 ps (diamonds), and 2 ps (stars). Shown are results obtained with a pump frequency of 2450 cm^{-1} (upper panel), 2550 cm^{-1} (middle panel), and 2650 cm^{-1} (lower panel). Figure (a) shows the experimental results, figure (b) shows the results obtained with a model that includes frequency jumping from the blue wing to all other frequencies in the absorption band, and a Gaussian spectral diffusion process with a time constant of 700 fs.

When probing in the blue wing, the number of directly excited molecules is large and the *relative* contribution of molecules that have reoriented and have jumped to all possible frequencies in the absorption band will be small. Hence, in the blue wing the initial anisotropy is relatively high. At later delays (>1 ps), the anisotropy becomes the same at all frequencies, due to the slow components (~1 ps) of the spectral diffusion of liquid water [6]. If the absorption band is pumped in the center or in the red wing, the initial anisotropy will be close to 0.4 and has the same dynamics at all probe frequencies (Fig. 2a). This frequency independence follows from the fact that the excited molecules do not directly reorient, but first have to diffuse spectrally to the blue wing before they can undergo the reorientation and frequency jumping process. After the jump the O-D oscillator can return at all frequencies in the absorption band, leading to the same decay of the anisotropy at all frequencies, with an effective time constant of 2.5 ps. This finding agrees with previous femtosecond studies of the O-D stretch vibration of HDO:H$_2$O in which the anisotropy dynamics were also observed to be the same at all probe frequencies following excitation in the center [4,5].

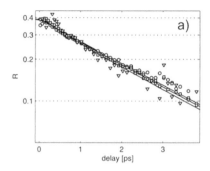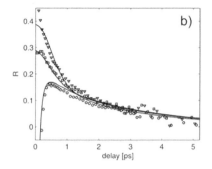

Fig. 2. Anisotropy as a function of delay for pump frequencies of 2500 cm^{-1} (a) and 2650 cm^{-1} (b), and 2650 cm^{-1} and probe frequencies of and probe frequencies of 2500 cm^{-1} (circles), 2550 cm^{-1} (squares), and 2600 cm^{-1} (triangles). The solid curves in the figures are calculated with the same model that was used to calculate Fig. 1b.

In Fig. 2b the anisotropy dynamics are shown at different probe frequencies following excitation in the blue wing. When probing in the red wing, the anisotropy acquires a low initial value, because the signal is dominated by molecules that have jumped from the excited blue wing of the spectrum to lower frequencies. Then the anisotropy increases because of the spectral equilibration with the molecules in the blue wing that have not reoriented and that still have a high value of the anisotropy. After reaching a maximum at ~500 fs, the anisotropy shows the same decay of 2.5 ps that is observed at all other pump and probe frequencies for delays >1 ps. This time constant is determined by the fraction of the molecules that show the reorientation and frequency jump, and the rate at which molecules diffuse to this state. The experimental results are all well reproduced with a model that includes the spectral diffusion and a frequency-dependent probability for frequency jumping and reorientation (Fig. 1b and solid curves of Fig. 2).

Discussion and Conclusion

Our results are consistent with the recent observation that the blue wing of the absorption band shows a very rapid spectral diffusion effect that is absent in the red wing [6]. Our findings also agree with the recently proposed molecular jump model for reorientation [1,2]. In this model the reorientation of a water molecule proceeds via a bifurcated hydrogen bond with two other water molecules. The bifurcated transition state decays by breaking one of the hydrogen bonds while strengthening the other bond, which leads to a large change in the frequency of the O-D vibration. According to the theoretical work, the probability to evolve to the bifurcated transition state does not depend on frequency, which disagrees with the present findings. We hope that the present results will stimulate further theoretical investigations of the relation between the hydroxyl frequencies of the water molecule and the probability for reorientation.

1 D. Laage and J. T. Hynes, Science **311**, 832, 2006.
2 D. Laage and J. T. Hynes, Chem. Phys. Lett. **433**, 80, 2006.
3 H.-K. Nienhuys, R.A. van Santen, and H.J. Bakker, J. Chem. Phys. **112**, 8487, 2000.
4 T. Steinel, J. B. Asbury, J. Zheng, and M. D. Fayer, J. Phys. Chem. A **108**, 10957, 2004.
5 Y. L. A. Rezus and H. J. Bakker, J. Chem. Phys. **123**, 114502, 2006.
6 J. J. Loparo, S. T. Roberts, and A.Tokmakoff, J. Chem. Phys. **125**, 194522, 2006.

Ultrafast Temperature Jumps in Liquid Water Studied by Infrared-Pump and X-ray Absorption-Probe Spectroscopy

G. Gavrila[1], Ph. Wernet[1], K. Godehusen[1], C. Weniger[1], E. T. J. Nibbering[2], Th. Elsaesser[2], and W. Eberhardt[1]

[1]BESSY, Albert-Einstein-Str. 15, D-12489 Berlin, Germany
[2]Max-Born-Institut für Nichtlineare Optik und Kurzzeitspektroskopie, Max-Born-Str. 2 A, D-12489 Berlin, Germany

Abstract. We report the first time-resolved x-ray absorption study of liquid water. Structural changes in the hydrogen-bond network as induced by resonant femtosecond-infrared excitation are monitored via transient x-ray absorption at the oxygen K-edge.

Liquid water consists of an extended molecular network with highly polar water molecules that are coupled via intermolecular hydrogen bonds. Structural dynamics of this network including the breaking and making of hydrogen bonds occur on femto- to picosecond time scales set by the strength of intermolecular interactions and the different vibrational and translational motions. Ultrafast vibrational spectroscopy has provided detailed information on such couplings and the resulting ultrafast processes [1]. Transient vibrational spectra give, however, only indirect spectroscopic insight into time-dependent molecular arrangements. In contrast, x-ray diffraction and x-ray absorption have recently been applied to water [2,3], providing direct access to the time-averaged equilibrium structure through radial distribution functions of the oxygen atoms [2] and local configurations of hydrogen bonds [3]. In view of the extremely fast structural fluctuations, an extension of x-ray methods into the ultrafast time domain holds great promise for unraveling structural dynamics in liquid water on their intrinsic time scales.

Here we investigate the structural changes in the hydrogen bond network of liquid water upon an ultrafast temperature jump of a few degrees Kelvin. We apply femtosecond infrared (IR) pulses to excite the intramolecular O-H stretching vibration, and monitor the transient response in the oxygen K-edge x-ray absorption spectrum (Fig. 1 (a)) with picosecond x-ray pulses. Picosecond x-ray absorption spectroscopy has successfully been used before in studies of transient structure of electronically excited molecules [4]. Here, we used it for the first time to unravel ultrafast structural fluctuations in the electronic ground state of a molecular liquid with an unprecedented combination of ultrafast IR and x-ray spectroscopies.

The experimental set-up consists of an amplified femtosecond Ti:sapphire laser system driving an optical parametric amplifier to generate intense mid-infrared pulses. For our experiment the infrared pulse is tuned to the O-H stretching band ($3400\ cm^{-1}$) of liquid water. The infrared pulse energies are about 2.2 μJ at the sample and the spot size is in the order of 100 μm (diameter for the full width at half maximum intensity). The x-ray probe pulses are generated in the soft x-ray undulator beamline UE56-1 PGM-B at the synchrotron radiation source BESSY. We used the single bunch mode of the electron storage ring, with one electron bunch in the ring and with a corresponding repetition rate of 1.25 MHz (800 ns between two x-ray pulses). The x-ray pulse width was 70 ps. X-ray spot sizes larger than the infrared spot and single-bunch mode were used to avoid x-ray irradiation induced changes of the samples. Samples consisted of thin liquid water films with typical thicknesses of 100-500 nm held in the vacuum chamber between two lithographically made x-ray transparent silicon nitride (Si_3N_4) membranes.

The oxygen K-edge absorption spectrum (Fig. 1 (a)) arises from transitions of oxygen 1s electrons into empty molecular orbitals of the probed water molecules. Transitions to these empty states are particularly sensitive to hydrogen bonding with the nearest neighbours and the spectrum reflects the ensemble average of spectral contributions from water molecules in various configurations with frozen-in geometries. The three prominent features, labelled as the pre-, main- and post-edge and, are related to different hydrogen bonding configurations [3]. In particular, the pre-edge peak at 535 eV (see Fig. 1 (b)) can be assigned to locally asymmetric configurations with one strong and one weak/broken hydrogen bond on the hydrogen side (donor hydrogen bond).

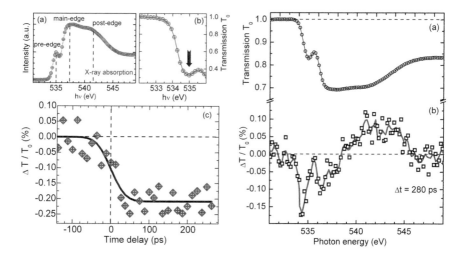

Fig. 1. (a) Oxygen K-edge x-ray absorption spectrum of liquid water [7]. (b) Steady-state transmission close to the pre edge (535 eV). (c) Change of x-ray transmission $\Delta T/T_0$ at the pre-edge as a function of pump-probe delay. Measured data are shown as markers and the fit using a Gaussian (FWHM=70 ps) broadened step function is shown as solid line.

Fig. 2. (a) X-ray transmission of liquid water at the oxygen K edge. (b) Transient changes of the x-ray transmission at a pump-probe delay of 280 ps. (Transmission change $\Delta T/T_0$, $\Delta T = (T-T_0)$ with T_0 and T transmission before and after excitation). Measured raw data are shown as markers (squares) and smoothed data are shown as solid line [7].

The change of x-ray transmission $\Delta T/T_0$ with $\Delta T=(T-T_0)$ at the pre-edge photon energy of 535 eV (see Fig 1 (b)) is plotted as a function of pump-probe delay (symbols) in Fig. 1 (c). T_0 and T are the transmission before and after excitation. The sample transmission decreases in a step-like fashion by approximately 0.2 % and stays constant up to the longest measured delay time of 280 ps. The time evolution of the transmission decrease follows the time-integrated cross-correlation function of the femtosecond pump and the 70 ps x-ray probe pulse (solid line). The width of the measured transmission step is hence determined by the duration of the probe pulses.

Upon femtosecond excitation, the O-H stretching vibration displays a population relaxation with a 200 fs time constant by vibrational redistribution through the intramolecular bending and intermolecular librational and hydrogen bond vibrational modes of the hydrogen bond network. Full equilibration into a heated water sample is reached within a few picoseconds [5]. The related ultrafast temperature jump is monitored here for the first time with the changes in x-ray transmission at the oxygen K edge. The associated structural changes are characterized in more detail with the transient x-ray absorption spectrum measured at a fixed pump-probe delay of 280 ps (Fig. 2 (b)). The pre-edge transmission decrease and the concurrent post-edge increase directly indicate that the number of molecules in locally asymmetric configurations with one weak/broken donor hydrogen bond increases [3], fully consistent with a temperature jump in the probed water volume. With the heat capacity of water and with our experimental parameters we estimate a temperature jump of 1–2 K, averaged over the entire probed volume. A comparison with steady-state x-ray absorption spectra measured at different temperatures [6] corroborates this estimate [7]. This demonstrates that our method serves as a very sensitive probe of transient structural changes in liquid water.

In conclusion, we report on the first infrared pump – x-ray probe study of structural dynamics in liquid water on a picosecond time scale. We demonstrate the feasibility of such combined laser-synchrotron experiments with very high structural sensitivity. Our results pave the way for future experiments with substantially shorter x-ray pulses such as generated with a femtosecond slicing scheme. In this way, the ultrafast structural changes of the hydrogen bond network during and immediately after the disposal of excess energy will become accessible. This establishes a direct link between femtosecond vibrational spectroscopy and structural information.

Acknowledgements. We thank Nils Huse for valuable support during the early stages of this experiment and Karsten Holldack, Christian Stamm, Ulrich Schade and Torsten Quast for their assistance. We also gratefully acknowledge financial support from the Deutsche Forschungsgemeinschaft (No. SPP1134).

1   E. T. J. Nibbering, T. Elsaesser, Chem. Rev. **104**, 1887, 2004.

2   T. Head-Gordon, G. Hura, Chem. Rev. **102**, 2651, 2002.

3   Ph. Wernet, D. Nordlund, U. Bergmann, M. Cavalleri, M. Odelius, H. Ogasawara, L. Å. Näslund, T. K. Hirsch, L. Ojamäe, P. Glatzel, L. G. M. Pettersson, A. Nilsson, Science **304**, 995, 2004.

4   M. Khalil, M.A. Matcus, A. L. Smeigh, J. K. McCusker, H.H. W. Chong, R. W. Schoenlein, J. Phys. Chem. A **110**, 38, 2006.

5   S. Ashihara, N. Huse, A. Espagne, E. T. J. Nibbering, T. Elsaesser, J. Phys. Chem. A **111**, 743, 2007.

6   U. Bergmann, D. Nordlund, Ph. Wernet, M. Odelius, L. G. M. Pettersson, A. Nilsson, Phys. Rev. B **76**, 024202, 2007.

7   Ph. Wernet, G. Gavrila, K. Godehusen, C. Weniger, E.T.J. Nibbering, T. Elsaesser, W. Eberhardt, Appl. Phys. A, DOI 10.1007/s00339-008-4726-5

Influence of the Environment on Reaction Dynamics: Excited State Intramolecular Proton Transfer in the Gas Phase and in Solution

C. Schriever, S. Lochbrunner[1], and E. Riedle
Lehrstuhl für BioMolekulare Optik, Ludwig-Maximilians-Universität München,
Oettingenstr. 67, D-80538 Munich, Germany
[1] present address: Institut für Physik, Universität Rostock, Universitätsplatz 3,
   D-18055 Rostock, Germany
e-mail: christian.schriever@physik.uni-muenchen.de

Abstract. Femtosecond transient absorption reveals very similar excited state intramolecular proton transfer and associated wavepacket dynamics in the gas phase and in solution. There are striking differences for the kinetics associated with the subsequent internal conversion.

Unified probe process in the gas phase and in solution

Ultrafast molecular processes are governed by intramolecular motions at the speed of skeletal vibrations as well as the interaction with the surrounding medium. In particular only little is known about the influence of the environment on the coherent wavepacket motion and how environment induced variations of the wavepacket motion change the outcome of a process. To understand this interplay we investigate the ultrafast intramolecular excited state proton transfer (ESIPT) of 2-(2'-hydroxy-phenyl)benzothiazole (HBT; cf. Fig. 1a) in the gas phase and compare the dynamics to the one found in solution [1]. ESIPT is a prototypical process for very fast chemical reactions since it leads to the breaking of the bond between the reactive hydrogen atom and the donating oxygen atom (enol tautomer, UV absorption) and the simultaneous formation of a new bond between the reactive hydrogen atom and the accepting nitrogen atom (keto tautomer, fluorescence in the visible).

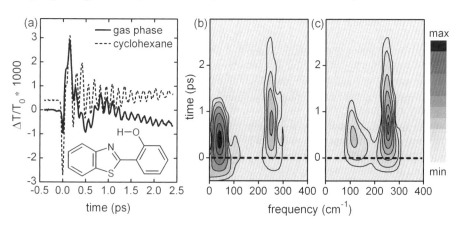

Fig. 1. (a) Transient absorption probed at 510 nm after exciting HBT at 350 nm in the gas phase and in a cyclohexane solution. The depicted sliding window Fourier transforms [(b) gas phase, (c) solution] are obtained after subtracting the exponential contributions from the transients.

It is crucial to use the same probe process for both experiments; Otherwise the probe process projects the wavefunction onto different manifolds of final states, resulting in different signatures even if there are no changes in the molecular dynamics. We choose transient absorption as probe signal since it provides a rich amount of spectroscopic information revealing detailed insight into the dynamics and it allows for a time resolution superior to most other techniques. A 20 fs time resolution and a sensitivity of 10^{-6} ΔOD allow us for the first time to compare directly the coherent wavepacket motion of low vapor pressure, medium sized molecules in the gas phase to the one in solution [2].

Mechanism of the excited state intramolecular proton transfer: Transfer time and coherent wavepacket motion

We measure the evolution of the transient HBT absorption at various probe wavelengths after photoexcitation at 325 and 350 nm (first absorption band) in the gas phase as well as in a cyclohexane solution. Figure 1 shows the transient transmission change at 510 nm for 350 nm excitation and the sliding window Fourier transforms of the oscillatory components. The transients show a transmission decrease at time zero which is caused by the excited state absorption (ESA) and occurs immediately after the pump pulse has promoted the molecule to the S1 state. The transmission increase follows with a delay of 35 fs and originates from the emission that occurs when the electronically excited keto form is populated by the ESIPT. The delay is therefore identified with the transfer time [1]. A comparison with a precision of better than 5 fs shows that it occurs in both environments with the identical delay of 35 ± 5 fs [2]. From the similarity of the proton transfer time and the subsequent signal signatures which result from the coherent wavepacket motion during and after the proton transfer we conclude that the ESIPT proceeds in the same way in the gas phase and in solution and the environment has only a negligible influence on the ultrafast reaction.

For the transients recorded in cyclohexane solution we find dominant modes at 113 cm^{-1} and 255 cm^{-1}. A comparison to ab-initio calculations reveals that the modes are in-plane deformations of the molecular skeleton which reduce the donor acceptor distance. This contraction allows for a barrierless reaction path and an efficient mixing between the electronic configurations of the enol and the keto form. In the gas phase we observe the 255 cm^{-1} oscillation with the same frequency and relative phase as in solution (Fig. 1a). A significant contribution of the 113 cm^{-1} mode cannot be established. In the gas phase a very strong modulation with an even lower frequency of 41 cm^{-1} is observed which is absent in solution (see Fig.1). However, this mode is not related to the ESIPT but to the IC as it is discussed below. It most likely obscures contributions from the 113 cm^{-1} mode in the gas phase. The dephasing of the oscillatory contributions occurs with a time constant of ~1 ps for the 255 cm^{-1} mode both in the gas phase and in solution. The dephasing times in both environments slightly decrease with the excess energy.

These findings clearly show that the vibrational dephasing is in this case intrinsic to the molecule and purely of intramolecular origin. High vibrational levels are populated which experience strong coupling to other modes and anharmonicities at the S$_1$ keto minimum of the potential energy surface (PES). In line with this minor influence of the solvent, the vibrational frequencies also do not change. This is in striking contrast to the electronic dephasing which is accelerated by many orders of magnitude via the solvent interaction.

Internal conversion through a conical intersection

In cyclohexane solution the decay of the electronically excited state, as measured by the stimulated emission decrease, occurs with a time constant of ~100 ps. It reflects the internal conversion (IC) back to the electronic ground state. For the isolated molecule, we observe that the emission decays much faster with a time constant of 2.6 ps for excitation at 325 nm and somewhat slower for 350 nm excitation. Both high level calculations of the PES and classical mechanics trajectories support that the 41 cm^{-1} mode found in the transients is associated with a torsion of the molecule around the phenyl-thiazole bond leading to a conical intersection at a torsional angle of 90° [3]. Upon IC the molecule can return to the S_0 enol configuration or form the metastable S_0 trans-keto tautomer. Clear spectral signatures for the latter process are found in our data. The time constant of the trans-keto formation corresponds to the decay of the stimulated emission, supporting this model.

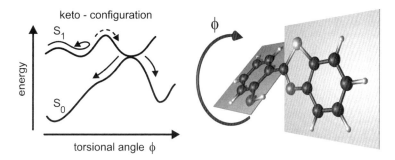

Fig. 2. Model for the potential energy surface along the torsion around the phenyl-thiazole bond leading to a conical intersection of the S_1 and the S_0 state.

The 41 cm^{-1} contribution to the transients shows that at least the first step in the IC of the isolated molecule is associated with wavepacket dynamics. Contrary to the ESIPT, the reaction path seems to involve a significant energy barrier which leads to the excess energy dependence (see Fig. 2). The pronounced initial decay of the gas phase transients indicates that a compact wavepacket is partially leaving the detection window of the probe process before it spreads and dephases. For later times, the emission decay can be described as a rate like process. The 41 cm^{-1} motion is associated with a large amplitude twisting of the entire molecular skeleton and should be subject to frictional forces in solution which cause a significant dissipation of kinetic energy. The wavepacket motion along this coordinate is overdamped in solution and no oscillatory behavior is observed. The excess energy of the torsional motion is much higher in the gas phase and the IC much faster than in solution. Friction changes the character of the process from a more or less ballistic wavepacket motion to a rate governed process in solution.

1   S. Lochbrunner, A. J. Wurzer, and E. Riedle, J. Phys. Chem. A **107**, 10580 (2003).

2   C. Schriever, S. Lochbrunner, E. Riedle, D. J. Nesbitt, Rev. Sci. Instrum. **79**, 013107 (2008).

3   M. Barbatti, A. J. A. Aquino, H. Lischka, C. Schriever, S. Lochbrunner, and E. Riedle, in preparation.

Ultrafast 2D-IR spectroscopy of a molecular monolayer

Jens Bredenbeck[1,2], Avishek Ghosh[1], Marc Smits[1], Mischa Bonn[1].

[1]FOM Institute for Atomic and Molecular Physics, Kruislaan 407, 1098 SJ, Amsterdam, the Netherlands
[2]Institut für Biophysik, Universität Frankfurt, Max von Laue-Str. 1, 60438 Frankfurt, Germany
E-mail: bredenbeck@biophysik.uni-frankfurt.org

Abstract. We report on ultrafast 2-dimensional vibrational surface spectroscopy, providing information on coupling and energy transfer between vibrations of surface molecules. As a 4th order technique, it is bulk-forbidden in centrosymmetric materials and hence surface specific.

Introduction

Coupling and energy flow through vibrational modes at surfaces and interfaces are important in areas as diverse as heterogeneous catalysis, electrochemistry, and membrane biophysics and –chemistry [1]. Furthermore, vibrational coupling patterns contain information on molecular structure, a feature already explored in bulk experiments to measure structure parameters with femtosecond time resolution [2]. However, measuring vibrational mode coupling at surfaces is challenging, because it requires both distinguishing the signal of a small number of surface molecules from a much larger bulk response and recording this signal within typical vibrational lifetimes (i.e. on sub-picosecond timescales).

For bulk studies, femtosecond two-dimensional infrared (2D-IR) spectroscopy is ideally suited to reveal vibrational mode coupling. In 2D-IR, a vibrational mode A is excited, and the effect of this excitation on a different mode B is probed. If the modes are uncoupled, mode B remains unaffected by excitation of mode A, and a spectral response is only observed for mode A at identical pump and probe frequencies, i.e. on the diagonal of the 2D-IR spectrum. Inversely, the off-diagonal peaks between modes A and B are determined by the strength of their coupling and depend on their relative orientation and distance. As such, 2D-IR spectroscopy is increasingly useful in determining (sub-)molecular structures and dynamics; 2D-IR analogues of NMR experiments like NOESY, COSY and EXSY have been demonstrated [2-6]. Here we introduce femtosecond sumfrequency generation 2D-IR spectroscopy (SFG-2D-IR) with submonolayer sensitivity and surface specificity. Closely related surface 2D vibrational techniques have been recently proposed theoretically [7,8].

Experimental Method

In bulk 2D-IR spectroscopy, a sequence of coherent interactions between the sample and the IR laser fields is designed such that an odd (typically third) order coherence is detected. To apply 2D-IR spectroscopy to surfaces, we gain monolayer sensitivity and interface specificity through an additional interaction with a nonresonant near-IR laser pulse. This additional interaction upconverts the third-order coherence to a fourth order coherence, which radiates a field in the visible, at the sum frequency of near-IR and IR. This upconversion process is beneficial in two ways. Firstly, it ensures surface specificity for materials whose optical response is dominated by dipole contributions.

Most materials are centrosymmetric, and the upconverted, even-order response can only originate from the surface molecules [9]. Secondly, the upconverted signal is background free and lies in the visible spectral range, where CCD cameras with high quantum efficiencies are available, readily providing sub-monolayer sensitivity. Specifically, we used a commercial Ti:Sapphire femtosecond amplified laser system to generate mid-IR pump and probe pulses. The pump pulse was shaped using a Fabry-Perot filter, resulting in a tunable narrowband ~15 µJ, 20 cm^{-1} pulse to excite specific vibrations. After a variable delay, the surface was probed by simultaneous mid-IR and near-IR pulses. The ~10 µJ IR probe pulse sustained 200 cm^{-1} bandwidth. To maintain spectral resolution in the upconversion step, a narrow-band (~12 cm^{-1}) 800 nm pulse was used. The resulting sum frequency generation (SFG) spectrum reveals the effect of the pump pulse on all resonances within the probe bandwidth.

Results and Discussion

We investigated a dodecanol monolayer on water – a model system for biological membranes – in the region of the C–H stretching modes [10]. The self-assembled monolayer was prepared by putting a small crystal of 1-dodecanol in contact with water. Fig. 1a and b show the IR spectrum of bulk dodecanol and the static 1D SFG spectrum of the monolayer. The C–H stretching region (Fig. 1a) features vibrational modes assigned to symmetric and antisymmetric CH_2 and CH_3 stretching (ss and as) perturbed by Fermi resonances (fr) with bending modes. For symmetry reasons, only some of these modes are SFG active and appear in the static SFG spectrum in Fig. 1b. Figs. 1c and d show SFG-2D-IR spectra of the self-assembled monolayer. In addition to the diagonal peaks, several off-diagonal peaks appear. As SFG selection rules apply for the probe process, off-diagonal peaks reporting on vibrational coupling appear at the two frequencies corresponding to the two peaks in the static SFG spectrum. The collective molecular alignment present at the interface (as opposed to the bulk) allows us to enhance the sensitivity to specific modes, by controlling the polarization of the pump laser pulse, as evident from a comparison of Figs. 1c and d.

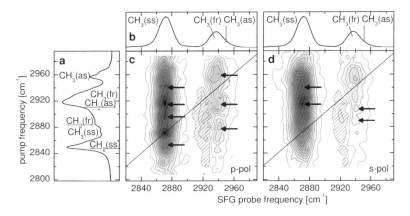

Fig. 1. Fig. 1. SFG-2D-IR spectra of a dodecanol monolayer on D_2O. Plain (shaded): pump-induced decrease (increase) in SFG intensity. (a) IR spectrum of crystalline dodecanol at 150 K. (b) SFG spectrum of the monolayer, polarizations: SFG/VIS/IR: s/s/p. (c) SFG-2D-IR spectrum, p-pol. pump, t = 0.7 ps. (d) s-pol. pump, in-plane CH_2 modes are efficiently excited, leading to the dominance in the 2D spectrum of the cross-peak between the CH_2(as) and CH_3(ss) modes. Lines: diagonal; arrows: off-diagonal peaks indicative of vibrational coupling.

When comparing the new surface 2D-IR technique to bulk 2D-IR spectroscopy, a few striking characteristics are apparent from the data: Firstly, the different selection rules for pump interactions (IR activity) and probe interactions (SFG activity requiring IR and Raman activity as well as broken centrosymmetry at the surface) make off-diagonal peaks appear that report on vibrational coupling between SFG inactive modes that are visible only in the IR absorption spectrum (shown along the pump axis) and SFG active modes (shown along the probe axis). Secondly, there is remarkably little increase in SFG-2D-IR intensity (positive red features in Figs. 1c and d). In third order bulk 2D-IR spectroscopy, an individual 2D-IR spectral response consists of a positive and a negative feature due to bleach, stimulated emission and excited state absorption. These paired features are largely suppressed in SFG-2D-IR spectra owing to the relative insensitivity of SFG to excited state transitions (the homodyne SFG signal is proportional to the square of the population difference between vibrational levels, heterodyne detection will allow to detect excited state dynamics as well). Thirdly, coherent interferences between the different modes play an important role in the SFG-2D-IR spectrum. These coherent interferences result from collective molecular alignment at the interface in combination with homodyne detection. Interestingly, the sign of off-diagonal peaks resulting from coherent interference depends on relative orientation of interfering oscillators, directly revealing structural information. Also interference terms could be suppressed by heterodyne detection.

Conclusions

In summary, we have demonstrated the implementation of ultrafast surface 2D-IR spectroscopy. We expect this technique to be useful for a variety of applications, including the study of: the structure and reactivity of (mixed) molecular adsorbate layers in catalytic systems, the structures and interactions of membranes and membrane proteins as well as the structure and dynamics of interfacial water in various systems.

1   T. Komeda, Y. Kim, M. Kawai, B. N. J. Persson, and H. Ueba, Science **295**, 2055, 2002.

2   R. M. Hochstrasser, Proc. Nat. Acad. Sci. USA **104**, 14190, 2007.

3   M. T. Zanni and R. M. Hochstrasser, Curr. Opin. Struct. Biol. **11**, 516, 2001.

4   M. L. Cowan, B. D. Bruner, N. Huse, J. R. Dwyer, B. Chugh, E. T. J. Nibbering, T. Elsaesser, and R. J. D. Miller, Nature **434**, 199, 2005.

5   J. D. Eaves, J. J. Loparo, C. J. Fecko, S. T. Roberts, A. Tokmakoff, and P. L. Geissler, Proc. Nat. Acad. Sci. USA **102**, 13019, 2005.

6   C. Kolano, J. Helbing, M. Kozinski, W. Sander, and P. Hamm, Nature **444**, 469, 2006.

7   Y. Nagata, Y. Tanimura, and S. Mukamel, J. Chem. Phys. **126**, 204703, 2007.

8   M. Cho, J. Chem. Phys. **112**, 9978, 2000.

9   Y. R. Shen, Nature **337**, 519, 1989.

10 J. Bredenbeck, A. Ghosh, M. Smits, and M. Bonn, J. Am. Chem. Soc. **130**, 2152, 2008.

Frozen Dynamics and Insulation of Water at the Lipid Interface

Artem A. Bakulin, Dan Cringus, Maxim S. Pshenichnikov, and Douwe A. Wiersma

Zernike Institute for Advanced Materials, University of Groningen, Nijenborgh 4, 9747 AG Groningen, The Netherlands
E-mail: a.a.bakulin@rug.nl

Abstract. 2D IR correlation spectroscopy reveals extremely slow dynamics and splitting of the OH-stretching mode of water in anionic micelles. Water at the lipid interface behaves as if the molecules were isolated in a "frozen" environment.

Introduction

The properties of the dense and flexible 3D hydrogen bond network among water molecules are substantially altered in proximity of chemical or biological interfaces. At the same time, these properties play an important role in a wide range of processes like chemical reactions, protein secondary structure stabilization, proton and energy transfer, etc. The recent sum-frequency generation studies on water-silica [1] and water-air [2] interfaces have shown that despite noticeable differences in spectral responses of bulk and surface water, the OH stretch vibrational dynamics of the surface water molecules are hardly distinguishable from those in the bulk. Yet, similar experiments on water-lipid interfaces exposed a noticeable decrease of population relaxation rates at high frequencies [3]. This was attributed to the decelerated spectral diffusion that otherwise equalizes frequency-dependent relaxation rates. However, conclusions about dephasing (that is, the T_2 time) were based on the OH-stretch vibration lifetimes (i.e., T_1) while these two values are not necessarily interconnected.

In this contribution, we present a 2D IR correlation spectroscopy study of the effect of the lipid interface on dynamics of the neighbouring water hydrogen-bond network. 2D correlation spectroscopy is capable of disclosing features otherwise hidden either beneath the broad absorption or in the broadband nonlinear response. As a model system for a membrane surface we used reverse micelles (fig. 1a) - the nanosize water droplets covered with a monolayer of lipid-like surfactant and floating in a nonpolar solvent. By varying the relative concentrations of H_2O and surfactant (AOT), the amount of interface water can be accurately controlled. To probe to the dynamics of hydrogen bond network, the environmentally sensitive OH vibration of an H2O molecule was excited by a pair of 70-fs IR pulses, allowed to evolve, and finally probed by another pulse pair.

Results and Discussion

The Figure 1b presents FTIR spectra of the OH stretch mode for 1 nm and 10 nm diameter H_2O micelles, and for bulk H_2O. The absorption spectrum for large micelles (d=10 nm diameters) follows the absorption spectrum of neat H_2O because 90% of water molecules are displaced from the micelle interface and create a "bulk-like" core. In contrast, in the small micelles (d=1 nm) 90% of water molecules border the lipid membrane which leads to a clear blue shift of the OH stretch frequency indicating weaker hydrogen bonding.

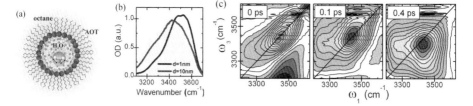

Fig. 1. a) Schematics of the reverse micelle. b) Absorption spectra of bulk (grey shape), 10 nm and 1 nm reversed micelles in the OH stretch region. c) Normalized absorptive components of 2D correlation spectra (the x,x,x,x polarization geometry) of H_2O confined in 10 nm diameter micelles at waiting times 0 ps, 0.1 ps, and 0.4 ps. The equilateral contours are drawn with the 8% step of the maximal amplitude.

Figure 1c shows the absorptive part of the 2D correlation spectra of H_2O confined in d=10 nm micelles at different evolution times. A strong positive signal (red) along the diagonal represents ground state bleaching and stimulated emission at the $|0>-|1>$ transition. The excited state $|1>-|2>$ absorption is also visible as a negative (blue) signal shifted along the vertical axis for the anharmonicity value of ~200 cm^{-1}. At waiting times longer than 0.1 ps the excited state absorption gradually disappears because of the ~0.2 ps population relaxation time of the stretch mode [4] and overlap with an additional bleaching induced by the overall heating of the sample. The diagonally-elongated shapes of the peaks are indicative of the inhomogeneous broadening of the absorption line. However, the antidiagonal width is also substantial showing significant dephasing during the coherence interval. At 0.1 ps inhomogeneity decreases dramatically while after 0.4 ps the 2D spectrum entirely loses any signs of correlation. Therefore, the water phase memory in the large micelles decays at ~0.1 ps timescale which is in a good agreement with the previous study on bulk H_2O [4].

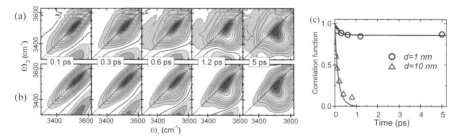

Fig. 2. Normalized experimental (a) and calculated (b) absorptive components of 2D correlation spectra (x,x,x,x polarization) of H_2O confined in d=1 nm micelles at waiting times 0.1, 0.3, 0.6, 1.2 and 5 ps. (c) Correlation functions derived from the 2D IR spectra for d=1 nm (circles) and d=10 nm (triangles) micelles. Solid curves are exponential fits to data.

Figure 2a depicts absorptive 2D correlation spectra of H2O confined in the d=1 nm micelles. Similarly to Fig.1c, the spectra display the on-diagonal $|0>-|1>$ bleaching peak and $|1>-|2>$ induced absorption shifted by the anharmonicity value. However, in a striking contrast with the d=10 nm case, the correlation spectra are considerably narrower at short waiting times and retain the elliptical shape for much longer evolution times. Fortunately, even at waiting times longer than 1 ps the thermal response remains spectrally separated from the population peak which allows tracking the latter up to 5 and even 10 ps (not shown). The diagonal peak shapes observed in the 2D spectra directly demonstrate the still-existing correlation between the excitation and probing. This implies that at 10 ps time scale there are no dynamical processes which wash out the structural variations in the hydrogen bond network and thereby scramble the OH bond stretching frequency.

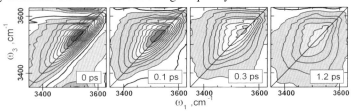

Fig. 3. Absorptive components of the 2D spectra for the d=1 nm micelles obtained in the cross-polarization geometry (x,x,y,y). Waiting times are 0 ps, 0.1 ps, 0.3 ps and 1.2 ps.

The results of the eccentricity analysis [5] are presented in Fig.2c by symbols. Consistently with previous quantitative estimates, the correlation function for the large water droplets decays with the ~0.15 ps. In contrast, the correlation function for the small micelles, after a small decrease in the first 0.3 ps, levels off and stays invariable on a few ps timescale. With this correlation function, the 2D spectra are successfully modelled as a simple combination of the population (correlated) and thermal (uncorrelated) responses (Fig.2b). The long phase memory in the small micelles signifies frozen dynamics of the OH stretch vibrations in the interface-bounded water. This also shows that there are no intermolecular communications amongst the interface H_2O molecules nor between the interface and bulk water because otherwise the correlation would be wiped out by intermolecular energy transfer [4].

The correlation spectra measured in the cross-polarization geometry (Fig.3) show a noticeable off-diagonal peak at ω_1=3450, ω_3=3540 cm^{-1} (the second symmetric peak is shadowed by the excited state absorption). The amplitude of this peak is non-zero at early evolution times which signifies a common ground state of both transitions. Furthermore, its amplitude rises with a ~200 fs time constant until it equalizes with the amplitude of the diagonal 3540 cm^{-1} peak. Such behavior is typical for a pair of coupled dipoles with orthogonal directions of the dipole moments such as the H_2O symmetric and asymmetric stretch modes. The splitting originates from isolation of the water molecules near the interface and their double bonding to the membrane.

Conclusions

We directly demonstrate that the phase memory of water near the lipid interface is retained for over 10's of ps which is a factor of 100 slower than in bulk [4]. The observed splitting of the interfacial H_2O stretch onto the symmetric and asymmetric modes signifies the weakening of the hydrogen bond network and dominant bonding of the water molecules to the lipid membrane.

1   J.A. McGuire, Y.R. Shen, Science **313**, 1945, 2006.
2   M. Smits, A. Ghosh, M. Sterrer, M. Muller, M. Bonn, Phys. Rev. Lett. **98** 98302, 2007.
3   A. Ghosh, M. Smits, J. Bredenbeck, M. Bonn, J. Am. Chem. Soc. **129**, 9608, 2007.
4   M.L. Cowan, B.D. Bruner, N. Huse, J.R. Dwyer, B. Chugh, E.T.J. Nibbering, T. Elsaesser, R. J. D. Miller, Nature **434**, 199, 2005.
5   K. Lazonder, M.S. Pshenichnikov, D.A. Wiersma, Optics Letters **31**, 3354, 2006.

Ultrafast vibrational dynamics of interfacial water

Avishek Ghosh[1,2], Richard K. Campen[1], Maria Sovago[1], Jens Bredenbeck[1] and
Mischa Bonn[1,2]

[1] FOM-Institute for Atomic and Molecular Physics AMOLF, Kruislaan 407, 1098 SJ
Amsterdam, The Netherlands;
  E-mail: ghosh@amolf.nl
[2] Chemistry Dept., Leiden University, P.O. Box 9502, 2300 RA Leiden, The Netherlands

Abstract. We report investigations on the ultrafast vibrational dynamics of water molecules at
model biological interfaces and the neat water/air interface, using a newly developed surface-
specific 4th-order femtosecond infrared pump-probe spectroscopic technique. The vibrational
relaxation rates and mechanisms depend strongly on the nature of the interface. Whereas water
at the neat water/air interface exchanges vibrational energy rapidly with the bulk, the water
molecules at model biological interfaces are energetically decoupled from the bulk.

Introduction

The interaction between water and lipid headgroups is very important in biology [1-3].
It is challenging, however, to investigate and characterize the structure and dynamics
of the ~1 molecular layer of water interacting with the lipid headgroups. Vibrational
sum frequency generation (SFG) has enabled the investigation of the vibrational
spectrum of ~1 monolayer of water molecules directly interacting with the lipids,
owing to its unique selection rules. In this spectroscopic technique, infrared light
pulses resonant with the O-H stretch vibration are overlapped with visible pulses at
the interface, which results in the generation of light with a frequency equal to the
sum of the two incident frequencies. The nonlinear optical process of SFG is
forbidden in centrosymmetric media but allowed wherever inversion symmetry is
broken, for instance at interfaces, making this technique highly surface-sensitive. The
static SFG spectrum of interfacial water in the O-H stretch frequency range has been
intensely investigated the past decades. Similar to the water-air interface [4], the static
SFG spectrum of the water-lipid interface [5] is characterized by two broad peaks in
the hydrogen-bonded region between 3100 and 3500 cm^{-1}, thus masking the true
dynamics of interfacial, lipid-bound water molecules. Questions arise pertaining to
the heterogeneity of water molecules and the timescales of exchange between possible
subensembles. To address these issues, we have recently developed the technique of
femtosecond time-resolved sum frequency generation spectroscopy (TR-SFG) that
allows us to study interfacial water dynamics directly [6].

Experimental Methods

In the TR-SFG experiments, the O-H stretch vibration of water molecules is excited
by an intense infrared pump pulse. The vibrational relaxation of these excited O-H
vibrations of interfacial water molecules is then monitored in real-time using an SFG
probing scheme (see figure 1). The fact that real membranes are chemically
heterogeneous with respect to lipid composition, has motivated us to perform time-

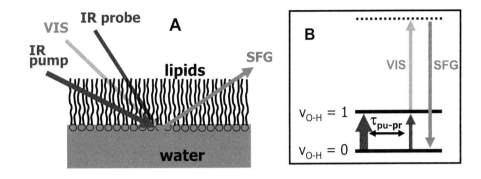

Figure 1 (A) Schematic for the time-resolved pump-probe SFG technique applied to a water-lipid interface. (B) Energy levels involved in the IR pump-SFG probe technique. The SFG response is monitored as a function of the delay between the IR pump pulse and the SFG probe pulse pair (IR and visible).

resolved IR-pump-SFG-probe studies on water molecules at a variety of lipid interfaces in which the lipid head-group structure differs. TR-SFG experiments were performed on Langmuir monolayer films of 1,2-Dimyristoyl-Glycero-3-Phospho-L-Serine (DMPS, Sodium salt, net negatively-charged headgroup), 1,2-Dipalmitoyl-3-Trimethylammonium-Propane (DPTAP, Chloride salt, net positively-charged headgroup), 1,2-Dipalmitoyl-sn-Glycero-3-Phosphocholine (DPPC) and 1,2-Dipalmitoyl-sn-Glycero-3-Phosphoethanolamine (DPPE, both zwitterionic headgroups), prepared on ultrapure Millipore water subphases.

Results and Discussion

Using tr-SFG, it has been shown that water molecules at the neat water surface [6] and water-silica interfaces [7], exchange vibrational energy with the underlying bulk water molecules on sub-100 fs timescales. However, in the tr-SFG transients for water-DMPS no such ultrafast energy exchange with the bulk was observed. In fact, we have shown that the water molecules at this water/lipid interface are energetically decoupled from the bulk and that vibrational relaxation proceeds by coupling of the O-H oscillator to the hydrogen-bond mode [5]. In figure 2, a comparison of the vibrational dynamics at the air-water and lipid-water interfaces is shown. Water at the DPTAP interface exhibits similar dynamics to water at DMPS, indicating that the water behavior for charged lipids is independent of the type and the total number of charges on the headgroup. For both charged lipids, the vibrational relaxation process can be explained by efficient energy flow from the excited O-H oscillators into the hydrogen-bond mode: we find no evidence for significant intramolecular energy flow. Interestingly, the dynamics of water at DPPC and DPPE show several spectral features that suggest additional modes participate in vibrational relaxation. These observations are consistent with water molecules strongly interacting with lipid functional groups in the DMPS/DPTAP case and enjoying more structural flexibility in the case of DPPC/DPPE.

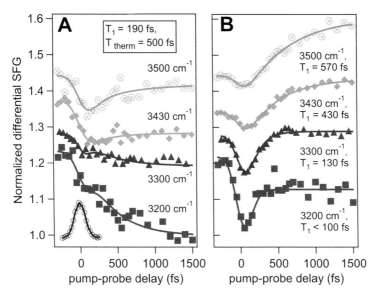

Figure 2. Pump-probe SFG transients for (A) the neat water-air interface and (B) the water-DMPS interface. The infrared-infrared-visible SFG crosscorrelation trace (lower left) determines the time-zero and illustrates the time resolution of the experiment. All traces are offset from 1.0 for clarity.

Conclusions

In summary, we have investigated the local hydrogen-bonding environment of water surrounding a variety of lipid headgroups, using a novel time-resolved spectroscopic technique to probe water interfacial vibrational dynamics. Future work is aimed at answering some long-standing questions concerning the molecular description of lipid-water interaction at biological membranes.

Acknowledgements. This work is part of the research program of the Stichting FOM with financial support from NWO.

1 Mulkidjanian, A. Y.; Heberle, J.; Cherepanov, D. A. Biochim. Biophys. Acta-Bioenergetics **2006**, 1757, 913.

2 Poolman, B.; Spitzer, J. J.; Wood, J. A. Biochim. Biophys. Acta-Biomembranes **2004**, 1666, 88.

3 Freites, J. A.; Tobias, D. J.; von Heijne, G.; White, S. H. Proc. Natl. Acad. Sci.USA **2005**, 102, 15059.

4 Du, Q.; Superfine, R.; Freysz, E.; Shen, Y. R. Phys. Rev. Lett. **1993**, 70, 2313.

5 Ghosh, A.; Smits, M.; Bredenbeck, J.; Bonn, M. J. Am. Chem. Soc. **2007**, 129, 9608.

6 Smits, M.; Ghosh, A.; Sterrer, M.; Muller, M.; Bonn, M. Phys. Rev. Lett. **2007**, 98, 098302.

7 McGuire, J. A.; Shen, Y. R. Science **2006**, 313, 1945.

Ultrafast Dynamics at Liquid Interfaces Investigated with Femtosecond Time-Resolved Multiplex Electronic Sum-Frequency Generation (TR-ESFG) Spectroscopy

Kentaro Sekiguchi, Shoichi Yamaguchi, and Tahei Tahara

Molecular Spectroscopy Laboratory RIKEN (The Institute of Physical and Chemical Research), 2-1 Hirosawa, Wako, Saitama, 351-0198, Japan
E-mail: tahei@riken.jp

Abstract. We developed a new nonlinear spectroscopy, femtosecond time-resolved electronic sum-frequency generation (TR-ESFG) spectroscopy, to investigate ultrafast dynamics at liquid interfaces. Transient electronic spectra at the air/water interface were obtained for the first time.

Introduction

Investigation of molecular properties at liquid interfaces, especially at aqueous solution interfaces, has been an important issue that has fundamental significance in chemical kinetics, biophysics, and atmospheric chemistry. Even-order nonlinear optical spectroscopy allows us to selectively obtain signals from molecules at interfaces. Especially for ultrafast dynamics of excited molecules, time-resolved second harmonic generation (TR-SHG) spectroscopy has been utilized to monitor interesting phenomena that happen only at the interfaces.

In this study, we developed femtosecond time-resolved electronic sum-frequency generation (TR-ESFG) spectroscopy [1] based on the steady-state ESFG method that we have developed earlier [2,3]. The advantage of TR-ESFG spectroscopy over conventional TR-SHG is that it provides *spectral* information. We examined the ultrafast dynamics of a surface active dye, Rhodamine 800 (R800), at the air/water interface and successfully measured time-resolved interface-selective electronic spectra for the first time.

Experimental Methods

The energy diagram and experimental configuration are schematically shown in Fig. 1. A femtosecond Ti:sapphire regenerative amplifier was used as the light source, and its output (800 nm, 1 kHz, 1 mJ) was divided into three. The first part was used for the excitation of an optical parametric amplifier to generate a signal output at 1380 nm. It was frequency-doubled to 690 nm, and used as the pump pulse, ω_p. The second part of the regenerative amplifier output was used as the narrow-band ω_1 probe pulse. The third part was focused into water to generate a white light continuum, which was used as the broad-band ω_2 probe pulse. After the excitation of molecules with the pump pulse, sum-frequency signals generated by the two probe pulses were introduced to a single polychromator and detected by a CCD. Obtained spectra were normalized by the quartz standard spectrum to correct the spectral distortion due to the intensity distribution of the ω_2 pulse. The spectra after this correction are called the "ESFG spectra" that correspond to $|\chi^{(2)}|^2$ spectra.

520

(a)

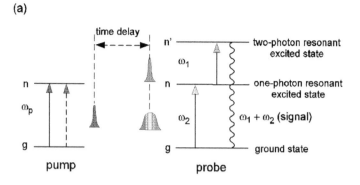

(b)

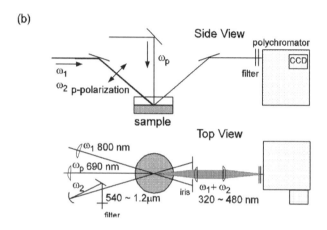

Fig. 1. (a) Energy diagram and (b) experimental configurations for the TR-ESFG spectroscopy.

The ESFG spectrum was measured with and without the pump pulse alternately, and the difference spectrum ($\Delta|\chi^{(2)}|^2$ spectrum) was obtained by subtracting the steady-state ESFG spectrum (which was observed without the pump irradiation) from the spectrum measured with the pump pulse. The time-resolution of TR-ESFG measurements was about 400 fs (FWHM). The ω_p pulse was circularly polarized and does not induce in-plane anisotropy. The ω_1 and ω_2 pulses were linearly polarized and set at p-polarization. The pulse energies of ω_p, ω_1, ω_2 pulses were typically 7.5 μJ, 10 μJ, and 7.5 μJ, respectively.

Results and Discussion

The obtained TR-ESFG spectra of R800 at the air/water interface are shown in Fig. 2. In addition to the negative signals due to the ground state bleaching, positive signals with a finite rise time were clearly observed. We carried out a global fitting analysis using exponential functions and found that all the temporal change can be represented with three time constants 0.32 ps, 6.4 ps, and 0.85 ns.

By comparing the result with that obtained by transient absorption measurements for R800 in bulk water [4], we ascribed the 0.32-ps component to the lifetime of R800 dimers in the lowest excited singlet (S_1) state. They dissociate and generate monomers

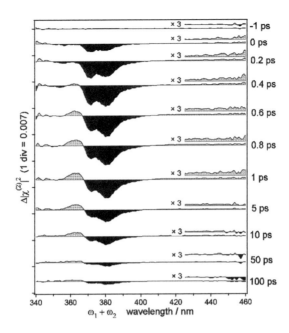

Fig. 2. TR-ESFG signals of R800 at the air/water interface. The time delays between the pump (ω_p) and probe (ω_1 and ω_2) pulses are shown at the side of each spectrum. The positive and negative signals are represented with the area painted grey and black, respectively. Expanded signals are also shown in the wavelength region of 425 – 460 nm. (ω_p, ω_1, and ω_2 were 690 nm, 800 nm, and 540 – 1200 nm, respectively.)

in the S_1 state, and 0.85 ns dynamics is ascribed to the lifetime of the S_1 monomer. The 6.4-ps component is attributed to an interface-specific deactivation process of the S_1 monomers, because the corresponding dynamics is missing in the bulk.

The present work demonstrated that TR-ESFG spectroscopy allows us to investigate ultrafast dynamics at liquid interfaces as thoroughly as we do for the dynamics in bulk solutions with conventional transient absorption spectroscopy.

Acknowledgements. K.S. acknowledges the Special Postdoctoral Researchers Program of RIKEN. S.Y. acknowledges a Grant-in-Aid for Young Scientists (B) (No. 15750023) from the Ministry of Education, Culture, Sports, Science and Technology (MEXT) of Japan. This work is supported by a Grant-in-Aid for Scientific Research on Priority Areas "Molecular Science for Supra Functional Systems" (No. 19056009) from MEXT of Japan and a Grant-in-Aid for Scientific Research (A) (No. 19205005) from Japan Society for the Promotion of Science (JSPS).

1   K. Sekiguchi, S. Yamaguchi, and T. Tahara, J. Chem. Phys. **128**, 114715 (2008).
2   S. Yamaguchi, and T. Tahara, J. Phys. Chem. B **108**, 19079 (2004).
3   Yamaguchi, and T. Tahara, J. Chem. Phys. **125**, 194711 (2006).
4   Sekiguchi, and S. Yamaguchi, and T. Tahara, J. Phys. Chem. A **110**, 2601 (2006).

Femtosecond spectral phase shaping for CARS spectroscopy and imaging

Sytse Postma, Alexander C. W. van Rhijn, Jeroen P. Korterik, Jennifer L. Herek, and Herman L. Offerhaus

Optical Sciences Group, Department of Science and Technology, MESA+ Institute for Nanotechnology, University of Twente,
P.O. Box 217, 7500 AE Enschede, The Netherlands
E-mail: s.postma@utwente.nl

Abstract. Coherent Anti-Stokes Raman Scattering (CARS) is a third-order non-linear optical process that provides label-free, chemically selective microscopy by probing the internal vibrational structure of molecules. Due to the resonant enhancement of the CARS process, faster imaging is possible compared to Raman microscopy. CARS is unaffected by background fluorescence, but the inherent non-resonant background signal can overwhelm the resonant signal. We demonstrate how simple phase shapes on the pump (and probe) beam reduce the background signal and enhance the resonant signal. We demonstrate chemically selective microscopy using these shaped pulses on plastic beads.

Introduction

Coherent anti-Stokes Raman scattering (CARS) has been used successfully in spectroscopy and microscopy since the development of (tunable) pulsed laser sources. In resonant CARS, molecular vibrations are coherently excited by a pump (ω_p) and Stokes (ω_s) pulse. Subsequently a probe (ω_{pr}) pulse, which is often similar to the pump pulse, generates the anti-Stokes signal ($\omega_c = \omega_p - \omega_s + \omega_{pr}$). The resonant CARS signal is accompanied by an inherent non-resonant background. Here we demonstrate how spectral phase shaping strategies can amplify the resonant features in the spectrum to such an extent that spectroscopy and microscopy can be done at high spectral resolution, even on the integrated spectral response [1]. We use this technique for chemical selective imaging of polystyrene (PS) and polymethyl-methacrylate (PMMA)

Setup

We use a tunable Ti:Sapphire oscillator with a FWHM of 20 nm (80 MHz repetition rate). The liquid crystal device (LCD) of the reflective spectral phase shaper has 4096 pixels with a pixel size of 1 μm by 6 mm and a pitch of 1.8 μm. Effectively the spectral phase shaper has ~600 degrees of freedom for pattering. For the absolute positioning of phase profiles the complete number of pixels can be used, which implies a positioning precision of 14 GHz (0.5 cm^{-1}). Further details of the spectral shaper setup can be found in an earlier publication [2]. The shaped Ti:Sapphire pulses are used as the pump and probe pulses in the CARS process. The Stokes pulse is generated by a 15 ps (1 cm^{-1}) Nd:YVO laser. A reflective objective of 0.65 NA is used to focus the light on the sample. The collection objective is a 0.65 NA regular glass objective. The collected light is detected by a spectrometer or a photomultiplier tube.

Spectroscopy

Our spectroscopic method is based on sweeping a π-phase step through the spectrum of the broadband (pump and probe) pulse and recording the CARS spectrum for each position of the step [1]. In this 2D plot the signal from vibrational resonances and the signal from the non-resonant background can be easily identified as distinct features. The difference between the positive and negative π-phase step rejects purely non-resonant features. Figure 1 shows the spectra for (a) a positive step sweep, (b) a negative step sweep, (c) the difference between (a) and (b) and (d) the integrated CARS signals. The horizontal axes represent the frequency of the phase step and the vertical axes represent the CARS spectra or the integrated signal. The intensities in the figures have been normalized to the unshaped CARS intensity. The strong resonance is clearly identified, the weaker resonances are lost in the noise.

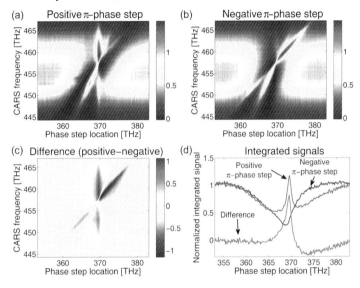

Fig. 1. CARS spectra for a π-phase step sweep on acetone. a) Positive π-phase step sweep. b) Negative π-phase step sweep. c) The difference between the positive and negative sweeps. d) The integrated signal for the three cases.

The imaging of the plastic beads is based on the integrated difference signal. Figure 1(d) shows that, for a slightly red shifted frequency for the phase step in comparison with the location of the vibrational resonance, the difference signal is positive. For blue shifted frequencies the difference signal is negative. In a sample of mixed 4 μm PS and PMMA beads dried on a glass substrate, an ordinary transmission image can not resolve the different types of beads. Figure 2 shows integrated difference between the CARS signal for a positive and a negative phase profile. Figure 2(a) shows the CARS image for transform limited pulses (flat phase profile). Figure 2(b-d) show difference CARS images, with (b) the difference of two π-phase steps at 372.7 THz, (c) the difference of two π-phase steps at 369.0 THz, and (d) the difference of two phase structures with two phase steps. The images have been normalized by the maximum CARS signal for the transform limited pulse.

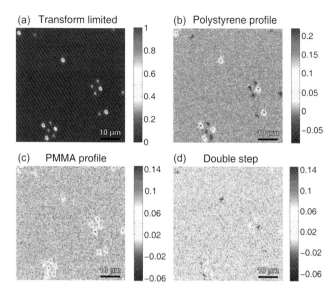

Fig. 2. Spectral phase profiles applied to the pump and probe pulses, which result in chemical contrast as a result of the applied phase profile. a) flat phase profile. b) PS enhanced phase profile. c) PMMA enhanced phase profile. d) PMMA enhanced and PS decreased phase profile.

In this particular case the transform limited CARS image also shows a difference between the PS and PMMA beads, because PS has a larger cross-section than PMMA for the chosen spectra of the pump and probe and Stokes. Figure 2(b) shows a CARS image for an applied phase profile that enhances the resonant signal of the PS beads and it results in a net negative result for the PMMA beads. Figure 2(c) shows a CARS image for an applied phase profile that enhances the resonant signal of the PMMA beads, which results in a similar net result for the PS beads. Figure 2(d) shows the same image for an applied phase profile that enhances the PMMA beads and at the same time results in a net negative result for the PS beads.

The subtraction of opposite phase profiles results in a removal of all pure non-resonant contributions. The concurrent phase step in the probe pulse results in less (resonant) CARS signal, which reduces the signal to noise ratio. The low signal to noise ratio is especially visible in the case for the double step image.

Conclusions

We demonstrate a method for chemically selective imaging based on detection of the integrated CARS signal, by applying simple phase profiles to the pump and probe pulses of the CARS process. For the future we are planning to expand these techniques for use in the fingerprint region for biological purposes.

[1] S. Postma, A. C. W. van Rhijn, J. P. Korterik, P. Gross, J. L. Herek, and H. L. Offerhaus, Opt. Express **16**, 7985 (2008).
[2] S. Postma, P. van der Walle, H. L. Offerhaus, and N.F. van Hulst, Rev. Sci. Instrum. **76**, 123105 (2005).

Part VII

Biological Systems, Molecular Light Harvesting and Charge-Transfer Complexes

Energy transfer along a poly(Pro) - peptide

Wolfgang Zinth[1], Wolfgang J. Schreier[1], Tobias E. Schrader[1], Florian O. Koller[1], Markus Löweneck[3], Hans-Jürgen Musiol[2] and Luis Moroder[2]

[1] LS für BioMolekulare Optik, LMU München, Oettingenstr. 67, D-80538 Munich, Germany
[2] Max-Planck Institut für Biochemie, Am Klopferplatz 18 A, D-82152 Martinsried, Germany
[3] Senn Chemicals AG, Guido Senn Strasse 1,CH-8157 Dielsdorf, Switzerland
E-mail: wolfgang.zinth@physik.uni-muenchen.de

Abstract. Using a novel molecular thermometer, p-nitro-phenylalanine, we investigate the transport of vibrational excess energy along a poly(Pro) sequence. Time resolved IR-spectroscopy reveals that heat transfer proceeds at a speed of several Å per picosecond.

Introduction

Many biological processes rely on the directed supply of energy to the molecular reaction sites. In this respect the speed of the energy transport, its efficiency, pathways of preferred energy flow and even solitonic energy transport are widely discussed. A number of experiments has shown, that vibrational excess energy released in a chromophore in solution is dissipated within about 10 ps to the surrounding solvent. Similar times for vibrational cooling have been found for chromopeptides or chromoproteins. Very little is known about heat diffusion over longer distances in proteins. Recently the transport of vibrational energy (heat flow) released after photoexcitation/photoisomerization of an azobenzene was studied in a 3_{10}-helix, covalently attached to the chromophore [1]. Isotopic labelling was used to monitor the heat flow. In the present contribution we apply an alternative approach to study the transport of excess energy in a peptide. We introduce a modified amino acid containing a NO_2-group as a local sensor (molecular thermometer) for vibrational excess population or local temperature. We study the heat transport along a rigid poly-Pro sequence with time resolved IR-spectroscopy on the picosecond time range and discuss the results using a model for directed and isotropic heat transport.

Materials and Methods

The azobenzene peptides, (Ac-AMPB-$(Pro)_n$-Phe(NO_2)-$(Pro)_{7-n}$-Asp-NH_2, with n = 1, 5 or 7) have been synthesized following analogous procedures as reported for other azobenzene peptides [3] and are dissolved in DMSO-d6. The amino acid containing the sensor - a p-nitro-phenylalanine, Phe(NO_2) - is placed at different positions (see Fig. 1) along the poly(Pro) sequence. The strong oscillatory strength of the NO_2-stretching vibrations and the clear separation from other bands in the peptide allow sensitive detection of the local temperature change. It has been shown recently that the NO_2-group is well coupled to the peptide backbone: experiments on Fmoc-Phe(NO_2) indicate that vibrational excitation of the backbone CO vibration is coupled to the reporting NO_2-group within 2 ps.

Transient absorption experiment are performed by the pump-probe technique using a Ti-sapphire laser system (operated at 1 kHz) with excitation pulses at 405 nm in the $n\pi^*$-band of the AMPB-chromophore. Probing in the mid-IR is performed as described recently [2]. After photo-excitation and ultrafast internal conversion the AMPB-moiety acts as a heat source, that is covalently linked with the peptide chain.

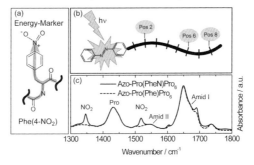

Fig. 1. Schematic representation of the energy-marker molecule (a) and the AMPB-peptide (b). Three sequences have been synthesized with the Phe(NO₂)-amino acid incorporated at the indicated positions along the peptide chain. (c) Infrared absorption spectra of a peptide with the marker at Position 2 and a corresponding peptide missing the NO_2-group.

Results and Discussion

In Fig. 2 we show the transient absorption data recorded in the range of the symmetric NO_2-mode for three positions of Phe(NO_2). When the marker molecule Phe(NO_2) or its directly coupled surrounding undergoes a temperature rise, the coupling of the NO_2-stretching modes to thermally excited low frequency vibrations changes the NO_2 absorption band. From steady state experiments performed at different temperatures we know that the band is red-shifted resulting in an absorption decrease at the original band position and an increase at lower frequencies. This explains the signature (absorption decrease) observed in Fig. 2 taken nearby the peak of the original NO_2 absorption band. When the Phe(NO_2) is closer to the AMPB-chromophore (Pos. 2), the signal rise occurs earlier than for more distant locations. As illustrated in Fig. 2 the maximum excess heat arrives at position 2 after about 3.5 ps. At positions 6 and 8 the excess heat is further delayed by about 2.5 and 3.5 ps respectively. At later times the signal decays due to energy transfer along the residual peptide and to the surrounding solvent on a time scale of 10 – 20 ps. In addition it can be seen, that the measured absorption change corresponding to the maximum temperature increase at the Phe(NO_2) location is smaller for Pos. 6 and 8. At very late times the samples show similar absorption changes as expected for a thermal equilibrium where most of the excess energy resides in the surrounding solvent.

A qualitative model of the heat transport would suggest, that the excess energy is released from the chromophore by isomerization and internal conversion to the attached peptide and to the surrounding solvent. This process happens on the 1 ps time scale. We assume that the energy is essentially thermalized within the vibrational system and that the observed absorption changes originate from the population of low lying frequency modes coupling anharmonically to the monitor bands. With increasing distances from the heat source (e.g. along the peptide) the heat wave arrives more and more delayed. Simultaneously the peak temperature at the sensor will decrease for larger distances since the heating energy is distributed over a larger volume. From the results shown in figure 2 one may estimate the speed of the heat transfer from Pos. 2 to Pos. 6 to be in the order of a few Å per picosecond.

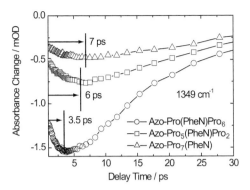

Fig. 2. Comparison of the transient absorption data recorded in the range of the symmetric NO$_2$-mode for three positions of the Phe(NO$_2$)-molecule in the peptide chain. The data were corrected for the induced absorption caused by neighboring hot proline.

To model the heat transport process we assume that azobenzene acts as an ultrafast local heat source and consider two situations: (i) In a 3-d heat diffusion model the azobenzene heat source feeds the excess energy isotropically into its surroundings (solvent and peptide). There is a fast distribution of the excess energy over a considerable volume and the peak temperature drops rapidly with distance from the heat source. For large distances from the heat source, the simulation does not fit to the observations. (ii) In a 1d-model heat diffuses predominantly along the peptide chain (we assume a rod-like structure and a heat conductivity as found in water), before it is conducted to the surrounding solvent. This model qualitatively describes the experimental observations. Along the peptide the decrease of peak temperature is slower as compared to the 3d-case. Within this model it is not surprising that the excess heat has been observed as far away from the source as 25 Å.

Conclusions

Heat transfer in molecular systems of biological relevance can occur within few picoseconds over distances of several nanometers. The experimental observations together with simulations within a model of 1d-heat conduction along the poly(Pro) peptides suggest that heat transfer may occur preferentially along defined molecular structures.

Acknowledgements. We would like to thank the German Science Foundation DFG for supporting the studies via SFB 533 and SFB 749.

1   V. Botan, E. H. G. Backus, R. Pfister, A. Moretto, M. Crisma, C. Toniolo, P. H. Nguyen, G. Stock, and P. Hamm in Proceedings of the National Academy of Sciences U. S. A. 104, 12749, 2007.

2   T. Schrader, A. Sieg, F. Koller, W. Schreier, Q. An, W. Zinth, and P. Gilch in Chemical Physics Letters 392, 358, 2004.

3   C. Renner, J. Cramer, R. Behrendt, and L. Moroder in Biopolymers 54, 501, 2000.

Energy transport in peptide helices around the glass transition

Ellen H.G. Backus[1], Phuong H. Nguyen[2], Virgiliu Botan[1], Rolf Pfister[1], Alessandro Moretto[3], Marco Crisma[3], Claudio Toniolo[3], Gerhard Stock[2], and Peter Hamm[1]

[1] Physikalisch-Chemisches Institut, Universität Zürich, Winterthurerstrasse 190, CH-8057 Zürich, Switzerland
E-mail: e.backus@pci.uzh.ch
[2] Institut für Physikalische und Theoretische Chemie, J.W. Goethe Universität, Max-von-Laue-Strasse 7, D-60438 Frankfurt, Germany
[3] Institute of Biomolecular Chemistry, Padova Unit, CNR, Department of Chemistry, University of Padova, via Marzola 1, I-35131 Padova, Italy

Abstract. The energy transport through a small helical peptide has been studied as function of temperature. Diffusive transport dominates at high temperature, while ballistic transport seems to be important at low temperature.

Introduction

Proteins are molecular machines which need energy to function, but also only function in narrow temperature ranges. Apparently, Nature has invented a way to transport efficiently energy to and from the active site of a protein. To study part of this process, we have investigated energy transport in a small helix, a dominant structural element of a protein. Our model system consists of 8 amino acids (one of them labelled with $^{13}C=O$) attached to an azobenzene photoswitch, which is used to deposit energy in the molecule (Fig. 1 top). Heat in the molecule is detected with infrared spectroscopy, making use of the shift of vibrational bands upon local heating [1]. The isotope labelled amino acid is either the second or the fourth residue counted from the N-terminal azobenzene, resulting in the molecules Aib16 and Aib34. Isotope labelling shifts the frequency of the Amide I band to lower frequency, well-separated from the other Amide I bands, resulting in a locally specific thermometer (Fig. 1 middle). Combining this isotope labelling with femtosecond pump-probe spectroscopy, spatial and temporal resolution can be obtained.

Results and Discussion

The bottom panel of Fig. 1 depicts pump-probe data at four different times after excitation of the azobenzene group at two different temperatures. At time zero the azobenzene moiety isomerizes from *cis* to *trans* depositing a huge amount of heat (estimated temperature close to 1000 K) in the molecule. Clearly, band 1 and band 3 respond immediately to the heat in the azobenzene, because already a signal is observed at time zero. For band 5 a signal is observed only after 3 ps at high temperature.

In Fig. 2 (top) the bleach intensity for bands 1, 3, and 5 is plotted as a function of time for 220 and 303 K. At 303 K the maxima of band 1, 3, and 5 are delayed with respect to each other, showing the propagation of heat through the molecule. We can describe the curves very well with a diffusive model depicted in the inset. In this model [2] we need two time constants, one for the heat transport between neighbouring C=O groups (1.5 ps) and one for cooling to the solvent (7 ps). In contrast, at 220 K the diffusive model does no longer work. The intensity of band 3 can be modelled correctly by reducing the propagation rate, but then the maximum comes too late (inset of Fig. 2). Apparently, the energy transport is no longer diffusive.

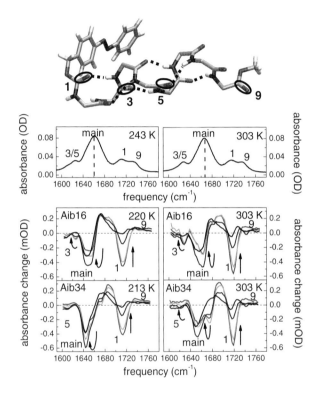

Fig. 1. Top) X-ray diffraction structure of the backbone of the molecule showing the azobenzene, the ester connection group, the helix with the 8 residues and the OMe endgroup. Middle) FTIR spectra of Aib16 at 243 and 303 K ('main' means all C=O groups except number 1, 9, and the labelled one). Bottom) Transient pump-probe spectra of Aib16 and Aib34 at two different temperatures at 1, 3, 11 and 40 ps after excitation of the azobenzene group with 420 nm. The numbers 1 to 9 refer to C=O groups in the molecule counted from the azobenzene.

To get a model independent measurement for the energy transport, we plot in the bottom right panel of Fig. 2 the ratio of the increase of band 3 (from t=0 to t=3 ps) and the intensity of band 1 (t=0) for six different temperatures. This plot suggests that the heat transport is more or less constant from 220 to 270 K after which it suddenly increases with temperature. Also the vibrational frequency of the main band shows a discontinuity around this temperature as shown in the bottom left panel. Below and above 280 K the vibrational frequency shift linear as a function of temperature, but the slope is different in these two temperature regimes. With NMR spectroscopy we see a similar effect on the chemical shift of the NH protons. In combination with MD simulations we conclude that the molecule is less flexible at low temperature and becomes more flexible around 270 K. Similar sudden changes in the behaviour of molecules have been observed with for example x-ray crystallography and infrared spectroscopy for proteins in aqueous solution around ~200 K, the so called glass transition [3].

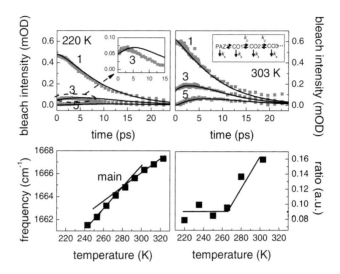

Fig. 2. Top) Time dependence of the bleach intensity (opposite sign as Fig. 1) for C=O 1, 3, and 5 as a function of time at 220 and 303 K. Bottom) Frequency of the main band (left) and ratio of the increase of band 3 and the intensity of band 1 (right) as a function of temperature.

Conclusions

Low-frequency vibrational modes are delocalized over large parts of (bio)polymers, while high-frequency modes are not [4]. Therefore, vibrational energy in these delocalized low-frequency modes dominates the energy transport. After excitation, part of the energy is in this type of modes and will be transported through the chain in a ballistic-like manner. The part in higher frequency modes can participate in the transport after it relaxes to low-frequency modes. At low temperatures, the molecule is rigid, and the vibrational energy re-distribution is inefficient. As the temperature increases above the glass transition, the molecule is flexible enough to get energy out of higher-frequency modes into the low-frequency modes. This re-feeding makes the transport more efficient and diffusive-like, in the sense that it fits a rate equation model. Apparently, the energy transport properties of molecular devices can be regulated by engineering the flexibility of a molecule.

1   P. Hamm, S.M. Ohline, and W. Zinth, J. Chem. Phys. **106**, 519 (1997).
2   V. Botan, E.H.G. Backus, R. Pfister, A. Moretto, M. Crisma, C. Toniolo, P.H. Nguyen, G. Stock, and P. Hamm, Proc. Natl. Acad. Sci. U.S.A. **104**, 12749 (2007).
3   D. Ringe and G.A. Petsko, Biophys. Chem. **105**, 667 (2003).
4   X. Yu and D.M. Leitner, J. Phys. Chem. B **107**, 1698 (2003).

Ultrafast Vibrational Dynamics of Adenine-Thymine Base Pairs in Hydrated DNA

J. R. Dwyer[1,2], Ł. Szyc[1]. E. T. J. Nibbering[1], and T. Elsaesser[1]

[1] Max-Born-Institut für Nichtlineare Optik und Kurzzeitspektroskopie, D-12489 Berlin,
Germany, E-mail: elsasser@mbi-berlin.de
[2] Department of Physics and Astronomy, University of British Columbia, Vancouver, B.C.,
Canada, V6T 1Z1

Abstract. We report femtosecond two-color pump-probe studies of the congested N-H/O-H stretching absorption of high-quality thin films of DNA oligomers in a broad hydration range. Different vibrational excitations are separated and their characteristic relaxation times identified.

Introduction

Hydrogen bonds play a key role for the structure of DNA and its interaction with an aqueous environment. Intermolecular hydrogen bonds define the planar Watson-Crick geometry of adenine-thymine (A-T) and guanine-cytosine base pairs in the double helix [1]. Moreover, the interaction of water molecules with different parts of the DNA structure is mediated through hydrogen bonding [2]. Linear vibrational spectroscopy has been applied to identify particular functional groups and characterize different DNA structures. Such work has led to conflicting conclusions as the vibrational spectra of DNA are highly congested, both in the fingerprint range and in the range between 3000 and 3600 cm^{-1} where different N-H and O-H stretching bands display a strong overlap. Ultrafast vibrational spectroscopy allows for a much more specific insight as different types of excitations are separated via their intrinsic dynamics and vibrational couplings are determined in a quantitative way [3,4].

So far, ultrafast vibrational spectroscopy has mainly concentrated on fingerprint vibrations [4]. Here, we present femtosecond two-color pump-probe studies of high-frequency N-H and O-H stretching excitations in artificial DNA oligomers containing 23 A-T pairs. Transient vibrational spectra, vibrational relaxation and anisotropy decay times are measured for different levels of DNA hydration. We discern N-H stretching excitations of the A-T base pairs from O-H stretching excitations of water molecules even at a high hydration level and determine the ultrafast dynamic properties of these N-H stretching excitations.

Experimental Methods

In our experiments, we study artificial DNA oligomers containing 23 alternating A-T pairs. Thin supramolecular DNA films of approximately 10 micron thickness and high structural quality were prepared on 500 nm thick Si_3N_4 substrates. The sample preparation preserves the double helix structure and allows changes in water content to drive conformational changes well-known from native DNA. The substrate is fully transparent in the frequency range studied here and makes a negligible contribution to the nonlinear response measured in the ultrafast experiments. The DNA sample was part of a closed sample cell in which a well-defined moisture level was maintained.

Independently tunable pump and probe pulses were generated in two parametric frequency converters driven by amplified pulses from a Ti:sapphire laser (repetition rate 1 kHz). The energy of the pump pulses of 200 cm^{-1} bandwidth was ~2 μJ, the temporal width of the cross-correlation with the probe pulses was 150 fs. After interaction with the sample, the probe pulses were dispersed in a monochromator and detected with a 16-element HgCdTe detector array.

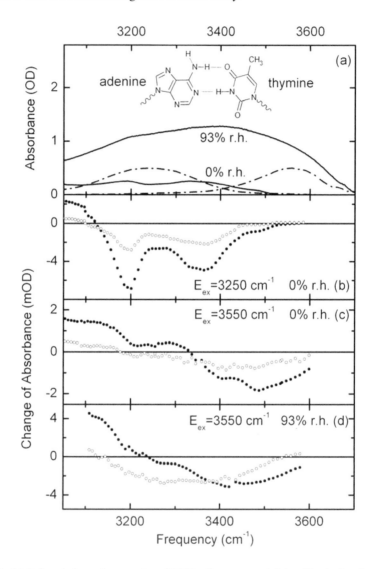

Fig. 1. (a) Infrared absorption spectra of DNA oligomers containing 23 adenine-thymine base pairs for 0% and 93% relative humidity (r.h., solid lines). Dash-dotted lines: spectra of the femtosecond pump pulses. (b,c) Transient infrared spectra at 0% r.h. for pump-probe delays of 200 fs (solid circles) and 1 ps (open circles) after excitation centered at $E_{ex}=3250$ cm^{-1} and $E_{ex}=3550$ cm^{-1}. The change of absorbance is plotted as a function of probe frequency. (d) Transient infrared spectra at 93% r.h. for the same pump-probe delays after excitation centered at $E_{ex}=3550$ cm^{-1}.

Results and Discussion

The (linear) vibrational absorption spectra of the DNA films shown in Fig. 1 (a) (solid lines) display a broad absorption band between 3050 and 3700 cm^{-1}, i.e., in the range of N-H and O-H stretching absorption. With increasing humidity, both a reshaping of the spectral envelope with enhanced high-frequency components and an increase of the overall absorption strength occur.

The (nonlinear) transient spectra in Figs. 1 (b,c) reveal different vibrational transitions contributing to the overall absorption. For 0% relative humidity (r.h.), there is a pronounced bleaching on vibrational fundamentals (v=0 to 1 transitions) with maxima at 3200 cm^{-1}, 3350 cm^{-1} (Fig. 1 b), and 3500 cm^{-1} (Fig. 1 c). The enhanced absorption at low frequencies is due to the v=1 to 2 transitions of the different oscillators and decays with the vibrational lifetimes of ~0.5 ps (not shown). The fundamental transitions display a negligible spectral diffusion up to pump-probe delays of 10 ps. This behavior is attributed to both the absence of structural fluctuations of the hydrogen-bonded dimers and a limited Coulomb interaction with the spatially separated ionic phosphate groups of the DNA backbone that undergoes low-frequency (sub-20 cm^{-1}) fluctuating motions.

Taking into account the vibrational spectra of isolated A-T pairs in the gas phase [5], we assign the band around 3200 cm^{-1} to a superposition of the symmetric NH$_2$ stretching vibration of adenine and the stretching vibration of the hydrogen bonded NH group of thymine. The peaks at 3350 cm^{-1} and 3500 cm^{-1} are attributed to the asymmetric NH$_2$ stretching vibration of adenine and the O-H stretching absorption of the residual H$_2$O molecules (~2 water molecules per base pair), respectively. The O-H stretching absorption occurs at higher frequencies than in bulk water [6] due to the strongly modified hydrogen bonding with the phosphate groups of DNA. This behavior is similar to water in micelles where water molecules interact with the micelle's polar head groups [7,8]. Time-resolved pump-probe transients at fixed probe frequencies of 3200 cm^{-1} and 3335 cm^{-1} (not shown) demonstrate a femtosecond decay of polarization anisotropy to a constant residual value of ~0.18, pointing to a pronounced coupling of the different N-H stretching excitations. In contrast, the anisotropy of the O-H streching component has a constant value of 0.4.

For 93% r.h., the broad O-H stretching component is well-pronounced, in particular at a 200 fs delay (Fig. 1 d). In contrast to 0% r.h., this band exhibits a distinct spectral diffusion towards smaller frequencies. The subpicosecond time scale of such spectral evolution is again much slower than in bulk water [6], and in line with the comparably slow spectral diffusion of water in small micelles [8].

1   J. D. Watson and F. H. C. Crick, Nature **171**, 737, 1953.

2   M. Ouali, H. Gousset, F. Geinguenaud, J. Liquier, J. Gabarro-Arpa, M. Le Bret, and E. Tallandier, Nucleic Acids Res. **25**, 4816, 1997.

3   E. T. J. Nibbering and T. Elsaesser, Chem. Rev. **104**, 1887, 2004.

4   A. T. Krummel, P. Mukherjee, and M. T. Zanni, J. Phys. Chem. B **107**, 9165, 2003.

5   C. Pluetzer, I. Huenig, K. Kleinermanns, E. Nir, and M. S. de Vries, ChemPhysChem **4**, 838, 2003.

6   M. L. Cowan, B. D. Bruner, N. Huse, J. R. Dwyer, B. Chugh, E. T. J. Nibbering, T. Elsaesser, R. J. D. Miller, Nature **434**, 199, 2005.

7   D. Cringus, A. Bakulin, J. Lindner, P. Vöhringer, M. S. Pshenichnikov, and D. A. Wiersma, J. Phys. Chem B **111**, 14193, 2007.

8   H. S. Tan, I. R. Piletic, R. E. Riter, N. E. Levinger, and M. D. Fayer, Phys. Rev. Lett. **94**, 057405, 2005.

Ultrafast Vibrational Dynamics in the AppA Blue Light Sensing Protein

Allison Stelling,[2] Minako Kondo,[1] Kate L. Ronayne,[3] Peter J. Tonge[2] and Stephen R. Meech[1]

[1] School of Chemical Sciences and Pharmacy, University of East Anglia, Norwich NR4 7TK, UK
E-mail: s.meech@uea.ac.uk
[2] Department of Chemistry, Stony Brook University, Stony Brook, New York 11794-3400, USA
E-mail: ptonge@notes.cc.sunysb.edu
[3] Central Laser Facility, Harwell Science and Innovation Campus, Didcot, Oxon OX11 0QX, UK

Abstract. The mechanism of blue light sensing in the photoactive protein AppA is investigated by transient infra-red spectroscopy. Modes associated with the flavin excited state and perturbation of the protein are detected.

Introduction

AppA is one of a number of recently characterised blue light sensing proteins. These have important roles in, for example, regulating phototropism, circadian rhythms and photosystem biosynthesis. Specifically AppA is a transcriptional anti-repressor found in photosynthetic bacteria.[1,2] Under low light and oxygen conditions it is bound to the transcriptional repressor PspR, preventing PspR from binding to DNA. However, under strong light (or high oxygen) the AppA-PspR complex dissociates, allowing PpsR to repress the synthesis of photosynthetic proteins. AppA is bifunctional, being sensitive to light and oxygen levels, but here we are concerned with the chromophore containing BLUF (blue light sensing using flavin adenine dinucleotide (FAD)) domain, the portion of the protein responsible for sensing light levels.

  In many photoactive proteins, such as the rhodopsins and photoactive yellow protein the driving force for formation of the signaling state is a large scale structural change in the excited electronic state of the chromophore. Intriguingly in the planar FAD chromophore no such possibility exists. Moreover, in AppA, unlike the flavin based LOV domain and cryptochrome light sensing proteins, no excited state photochemistry is observed. Instead the only spectroscopic signature of formation of the signaling state is a small red shift in the electronic absorption spectrum between the dark(d) and light(l) adapted states. Recent ultrafast time resolved studies in the visible region showed that the red shift occurs on a 1 ns time scale.[3] It was proposed that the primary step was an electron and proton transfer between the FAD chromophore and a nearby highly conserved tyrosine residue. Importantly, structural information for both l and dAppA exist. Most agree that an important structural change is the reorientation of the Q63 residue adjacent to the flavin ring.[4] Such a reorientation (illustrated in Figure 1) might be a result of changes in the extensive network of protein – chromophore H-bonds in AppA. Ultrafast vibrational spectroscopy is a powerful tool for studying changes in the proton transfer network following photoexcitation of chromoproteins [5,6] and this technique has now been applied to FAD and AppA.[7,8]

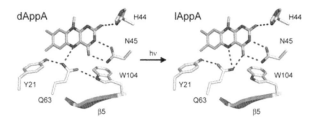

Fig. 1. An illustration of the possible transformation in Q63 orientation between dark (dAppA) and light (lAppA) forms of AppA. Possible H-bond interactions are shown as dash lines. Data for this figure were based on the structure presented in [4].

Experimental Methods

Solutions of AppA and its mutants were prepared in D_2O solution at $0.1 - 1$ mM concentrations, and therefore probably exist as dimers or higher aggregates. dAppA photodynamics were measured in a flow cell under low irradiation conditions such that negligible photoconversion occurred during the experiment. Mutants were studied in a raster scanned cell, and the photoactive mutant was exposed to radiation for less than 1 minute before being replaced and allowed to recover in the dark. Excitation was at 400 nm and transient IR difference spectra were measured between 1600 and 1750 cm^{-1}.

Results and Discussion

The transient IR difference spectra have been measured between 1 ps and 1 ns after excitation for dAppA, lAppA, the photoactive mutant W104F and the photoinactive mutant Q63L.(7) Both dAppA and W104F are observed to exhibit the characteristic red shift under continued irradiation at 400 nm, while the absorption spectra for lAppA and Q63L are independent of irradiation The transient IR data for the light and dark adapted forms are shown in Figure 2

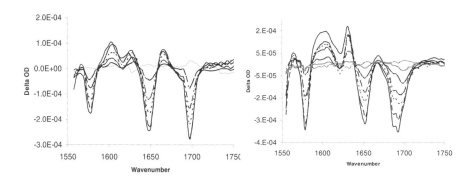

Fig 2 (a) Transient IR spectra for dAppA recorded between 1 ps and 2 ns after excitation, characterized by an instantaneous bleach and slow recovery. (b) The same for lAppA between 1ps and 200 ps after excitation.

The four bleach bands observed in both cases are readily assigned to modes of the isoalloxazine ring of the flavin chromophore on the basis of our previous study [8]: the two highest frequency bands are mainly localised on the carbonyl groups; the two bleach modes below 1600 cm^{-1} are largely due to ring CN stretches; the transient absorption around 1600 cm^{-1} is ascribed to carbonyl modes in the excited state.

There are three significant differences between the spectra of the light and dark adapted states. First the highest frequency C=O mode evolves into a doublet under irradiation. Second the ground state recovery is more rapid in the light adapted form, suggesting quenching of the excited state. Finally the dark adapted form has a transient absorption at 1666 cm^{-1} which is absent in lAppA. In a study of the photoactive W104F mutant the same three changes were observed on irradiation.[7] Significantly in neither the dark nor the light adapted transient spectra is there any evolution in the shape of the spectrum as a function of time during at least 1 ns.

The doubling of the carbonyl mode and the quenching are provisionally ascribed to structural disorder induced by irradiation. The doubling of the carbonyl mode suggests that FAD in lAppA occupies at least two distinct sites. The observation of quenching of the flavin excited state in the light adapted form may suggest an enhanced electron transfer rate. One possibility is that the separation between FAD and a tyrosine or tryptophan residue is reduced as a result of disorder. Alternatively disorder may allow mobile water molecules into the FAD binding site; rapid reorientation of water molecules could contribute to enhanced electron transfer rates.

The absence of any observable intermediates being formed in dAppA during the one nanosecond time window of our experiments is surprising. It was previously proposed that the primary step involved electron and proton transfer from the nearby tyrosine residue (Y21) to the flavin ring. Such an electron transfer reaction is feasible on both structural and energetic grounds. The absence of a spectrum assignable to a radical in the IR may simply reflect the kinetics or a low oscillator strength in the spectral region investigated, or suggest an alternative mechanism; model studies are needed

The most interesting observation concerns the instantly formed (<200fs) 1666 cm^{-1} transient in dAppA which is absent in lAppA. This has no equivalent for FAD in solution. The transient is observed in the photoactive W104F mutant but not in the inactive Q63L. Evidently this band is a marker mode for AppA in its photoactive form. One plausible explanation is that this mode is associated with ultrafast reorganization of the FAD – protein H-bonding network, which may be involved in the primary events in the AppA photocycle. A search for other intermediates is in progress using mutagensis, isotope editing and model flavins.

Acknowledgements. We are grateful to EPSRC, NIH and JSPS for financial support.

(1) MA Van der Horst, KJ Hellingwerf. Acc. Chem. Res. **37**, 13, 2004.

(2) S Braatsch, G Klug, Photosynthesis Research **79**, 45, 2004.

(3) M Gauden, S Yeremenko, W Laan, et al., Biochemistry **44**, 3653, 2005.

(4) S Anderson, V Dragnea, S Masuda et al., **44**, 7998, 2005.

(5) D Stoner-Ma, AA Jaye, P Matousek, et al., J. Amer. Chem. Soc. **127**, 2864, 2005.

(6) D Stoner-Ma, EH Melief, J Nappaet al., J. Phys Chem. B **110**, 22009, 2006.

(7) J. Nappa, A. Stelling, K. L. Ronayne et al., J. Amer. Chem. Soc. **129**, 15556, 2007.

(8) M Kondo, J Nappa, KL Ronayne etal., J. Phys Chem. B **110**, 20107, 2006.

Direct observation of ligand transfer and bond formation in cytochrome *c* oxidase using mid-infrared chirped-pulse upconversion

Johanne Treuffet, Kevin J. Kubarych,* Jean-Christophe Lambry, Eric Pilet,[†] Jean-Baptiste Masson, Jean-Louis Martin, Marten H. Vos, Manuel Joffre, Antigoni Alexandrou

Laboratoire d'Optique et Biosciences, Ecole Polytechnique, CNRS, INSERM
91128 Palaiseau, France
E-mail: antigoni.alexandrou@polytechnique.edu

Abstract. We time resolved the CO ligand transfer process in the bimetallic active site of cytochrome *c* oxidase, using mid-infrared chirped-pulse upconversion to observe the full vibrational signature of Fe-CO bond breaking and Cu_B-CO bond formation.

Introduction

Cytochrome *c* oxidase (CcO) is the key enzyme of the respiratory chain that reduces oxygen to water and couples the energy of this reaction to the pumping of four protons across the membrane to set up a transmembrane proton gradient. The active site includes the Fe atom of a heme molecule and a close-lying (~5 Å) Cu atom. A doorstep role for ligands on their way into and out of the active site has also been attributed to Cu. Indeed, it has been shown that CO binds to Cu after photodissociation from the Fe on its way out of the protein. The ultrafast sub-ps dynamics of this process, however, has been observed either indirectly using experiments in the visible sensitive to the heme electronic transitions [1] or with picosecond time resolution [2]. Apart from implications for the functioning of CcO, this transfer process of a diatomic molecule between two well-identified binding sites can be seen as a model system for bond breaking and formation processes.

Experimental approach: mid-infrared chirped-pulse upconversion

Chirped-pulse upconversion of mid-IR pulses into the visible and detection based on highly mature silicon CCD technology was demonstrated for mid-IR electric field characterization [3]. In the present work, we used chirped-pulse upconversion in a pump-probe experiment to measure the spectral amplitude of the transmitted mid-IR probe pulses [see Fig. 1(left)]. The use of a highly chirped, 150-ps pulse ensures that the mid-IR pulse overlaps only with a well-defined frequency inside the chirped pulse. This technique benefits from high quantum yield, low-noise CCD technology for detection in the visible and has already been successfully applied to two-dimensional IR spectroscopy [4]. It allows for high-resolution, high-sensitivity measurements over a broad spectral range currently unattainable with IR-detector technology. In our case, 125 pixels are available over the full 200-cm^{-1} width of the probe spectrum, the detection range being limited solely by the probe spectral width (total CCD pixel number: 1340x100).

Beef heart CcO is a ~150 kD membrane protein that is not easily obtained at very high concentrations. To increase the absorption and hence the differential signal, we resorted to relatively large sample thicknesses (200 μm) which translate into very weak transmission levels (about 2% at 1960 cm^{-1} and 0.8% at 2064 cm^{-1}) even after H_2O substitution by D_2O. Consequently, the power of the chirped pulse used for the upconversion was increased to 500 μJ in order to maximize the signal arriving at our 16-bit CCD and minimize the photon shot noise. The noise level we achieved in the differential transmission spectra is $5x10^{-5}$ ($2.2x10^{-5}$ OD) for a total acquisition time of 1h, i.e. 3.3 min per pump-probe delay value excluding the time needed for the delay variation, and is limited by the photon shot noise [5].

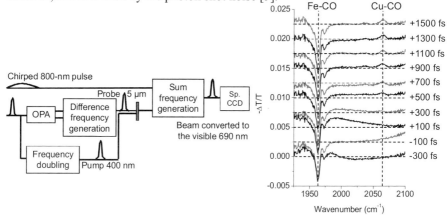

Fig. 1. Left: Scheme of the experimental setup (Sp.: spectrometer). Right: Differential transmission spectra for pump-probe delays between -300 and +1500 fs. The spectra have been displaced vertically for clarity.

Results and Discussion

Differential transmission spectra are shown in Fig. 1 (right). The probe spectrum was chosen to encompass both the Fe-CO and the Cu_B-CO vibrational frequency (probe centered at 2030 cm^{-1} with a FWHM of 80 cm^{-1}). The most prominent feature of the spectra is the gradual appearance, after the pump-probe delay of +300 fs, of an induced absorption peak at 2064 cm^{-1} representing the arrival of CO at Cu_B. The departure of CO from its initial Fe binding site is observed as an induced transmission peak at 1963 cm^{-1}. Both the Fe-CO and Cu_B-CO vibrational frequencies are in agreement with previous observations [2].

The dynamics of the signal at 2064 cm^{-1} due to the arrival of CO at the Cu_B binding site is shown in Fig. 2A. We fitted the signal with Gaussians and then plotted the area under the Gaussian as a function of delay time. If the transfer process were purely ballistic, we would expect a Heaviside-like form convoluted with the instrument response function (dashed line). The measured dynamics is clearly slower. If, on the other hand, the dynamics were purely diffusive, an exponential signal rise would be expected (dotted line). The difference between the experimental data and this fit is first negative and then positive (Fig. 2B). Indeed, a pure delay is observed before the signal starts rising after 300 fs. This feature cannot be reproduced by an exponential fit and is indicative of a ballistic contribution to the transfer process. We therefore resorted to a transfer model including both a ballistic and a diffusive component described by a delayed exponential (solid line in Fig. 2A) which gave the best agreement with the experiment (see residuals in Fig. 2B) [5].

In addition, molecular-dynamics calculations including a Morse potential for the binding of CO to Cu_B support this interpretation [5]. During the first hundreds of fs, the various trajectories show a similar behavior indicating a ballistic process with almost no CO molecules binding to Cu_B before 500 fs. The various CO molecules then reach the final Cu_B-C distance of 1.9 Å after a time ranging from 500 to 2000 fs (Fig. 2C) suggestive of the exponential rise observed in the experiment. It is interesting to note that the apparent diffusive component in the transfer process may be due, at least partly, to the distribution of initial conditions and of the positions of the atoms surrounding Cu_B due to thermal fluctuations at room temperature.

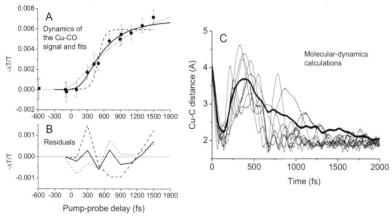

Fig. 2. A) Area under the Cu_B-CO vibrational peak at 2064 cm^{-1} as a function of pump-probe delay time. The lines are fits with i) the instrument response function displaced by 500 fs (dashed), ii) a 1200-fs exponential convoluted with the instrument response function (dotted), and iii) a 450-fs exponential displaced by 200 fs and convoluted with the instrument response function (solid). B) Residuals corresponding to the different fits. C) Molecular-dynamics calculations: time evolution, after CO dissociation at t=0, of the distance between the Cu_B and the CO carbon atom. 8 individual trajectories (thin lines) as well as the average (thick line) of 30 trajectories calculated for different initial conditions are shown.

Thus, both experimental and calculated data strongly suggest a ballistic contribution to the transfer process. The two metallic atoms in the active site of this enzyme constitute a unique arrangement for the observation of a bond breaking and formation process where the reactants are held in place by the protein structure. These results open the way for the application of the chirped-pulse upconversion technique to a large variety of proteins and ligands and mid-IR experiments in general.

* Present address: Department of Chemistry, University of Michigan, Ann Arbor, MI 48109-1055
† Present address: LCBM, Commissariat à l'Energie Atomique, 38054 Grenoble Cedex 09, France
[1] U. Liebl, G. Lipowski, M. Négrerie, J.-C. Lambry, J.-L. Martin, M. H. Vos, Nature **401**, 181-184 (1999).
[2] R. B. Dyer, K. A. Peterson, P. O. Stoutland, W. H. Woodruff, Biochemistry **33**, 500-507 (1994).
[3] K. J. Kubarych, M. Joffre, A. Moore, N. Belabas, D. M. Jonas, Opt. Lett. **30**, 1228-1230 (2005).
[4] M. J. Nee, R. McCanne, K. J. Kubarych, M. Joffre, Opt. Lett. **32**, 1228–1230 (2007).
[5] J. Treuffet, K. J. Kubarych, J.-C. Lambry, E. Pilet, J.-B. Masson, J.-L. Martin, M. H. Vos, M. Joffre, A. Alexandrou, Proc. Natl. Acad. Sci. USA **104**, 15705 (2007).

Tryptophan Residues as Natural Ultrafast Voltmeters in Retinal Proteins

J. Léonard[1], E. Portuondo-Campa[2], A. Cannizzo[2], F. Van Mourik[2], J. Tittor[3], S. Haacke[1], M. Chergui[2]

[1] Institut de Physique et Chimie des Matériaux de Strasbourg, UMR 7504 ULP – CNRS, F-67034 Strasbourg, France
E-mail: leonard@ipcms.u-strasbg.fr
[2] Laboratoire de Spectroscopie Ultrarapide, ISIC – EPFL, BSP, CH-1015 Lausanne, Switzerland
[3] Max-Planck-Institut für Biochemie, 82152 Martinsried, Germany

Abstract. The comparison between UV transient absorption spectra of wild type bacteriorhodopsin and two tryptophan-mutant proteins gives evidence for the possibility to use tryptophans as ultrafast probes for the photo-induced dipole moment change in retinal proteins.

Retinal in bacteriorhodopsin (bR) is a model system for studying the enzymatic role of the protein environment in the ultrafast, selective isomerisation of the protonated Schiff base retinal (PSBR). The ~0.5 ps photo-isomerization upon excitation at ~550 nm is preceded by a large dipole moment change [1] of the chromophore. In bR, due to their proximity, PSBR and the two tryptophans Trp86 and Trp182 are expected to exhibit significant dipolar interactions, including the formation of an excitonic complex for the near-UV transitions. Experiments of ultrafast transient UV spectroscopy upon VIS excitation of PSBR have revealed new ultrafast dynamics, and an exciton coupling model involving three interacting dipole moments (PSBR, Trp182 and Trp86) was introduced to rationalize the differential UV absorption [2,3].

Our present study includes the two mutants W86F and W182F, in which a spectroscopically silent phenylalanine residue is inserted in place of Trp86 and Trp182 respectively. By direct subtraction of the pump-probe signals of wt bR and mutants obtained under very controlled conditions we are now able to isolate the spectral contribution of specific aminoacids close to the retinal. In particular, Trp86 shows an induced absorption band (300-320 nm), which was predicted to be related to the dipole moment change $\Delta\mu$ of PSBR[2]. It appears as an efficient probe for $\Delta\mu$ during the isomerisation process. This work exemplifies the possibility to use Trp residues as natural ultrafast voltmeters in proteins by the means of ultrafast transient UV absorption spectroscopy.

An amplified Ti:Sa laser system (1-kHz, 1-mJ) is used to operate two identical NOPA's. The first one delivers 20-fs pump pulses at a central wavelength of 550 nm. The second one is frequency doubled yielding 3- to 4-nm broad probe pulses, at a central wavelength adjustable from 260 nm to 320 nm. At each of the 20 selected probe wavelengths, the transient absorption signals are recorded spectrally integrated (photodiode) as a function of pump-probe delay for wt bR, W86F, and W182F successively in three experimental runs in a row. The time resolution is measured to be 90-100 fs for all probe wavelengths.

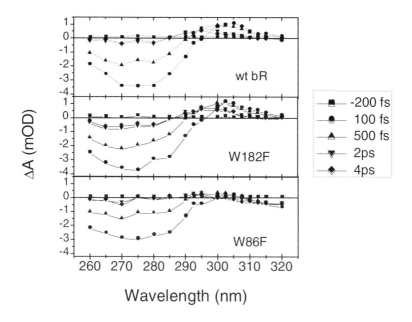

Fig. 1. UV (260-320nm) absorption spectra measured at 5 selected delay times after excitation at 550nm. The spectra for wild type (wt) bR (top), and for the W182F (middle) and W86F (bottom) mutants are reconstructed from a set of 20 single-wavelength pump-probe experiments. At each wavelength, the transient absorptions of the 3 samples are acquired in a row.

For each sample, Fig. 1 displays the transient absorption spectra reconstructed from the single-wavelength measurements. The main difference between the three samples is a positive signal observed around 300-305 nm in wt bR and in W182F, but not in W86F, where in addition negative transient absorption is measured around 310-320 nm. Apart from that, all three samples show a dominant negative signal of comparable amplitude and time behaviour between 260 nm and 285 nm. This bleach signal previously observed in wt bR and W182F in the range of 260 - 285 nm decays with two time constants of 0.4 ps and 3.5 ps corresponding to the formation of the J and K isomerized states [3]. But the occurrence of a similar (somewhat weaker) bleach signal in the W86F mutant now clarifies that it is mostly due to retinal.

The contribution of each of the two Trp's to the overall signal can be visualized by linear superpositions of the raw spectra (results in Fig. 2-left). The 3-interacting-dipole model introduced previously [2] indicates that Trp86 mostly interacts with retinal through state dipole interaction (Stark shift), whereas excitonic coupling dominates the interaction between Trp182 and retinal. Hence, in the case of Trp86, the difference between the transient spectra obtained for wt bR and W86F can safely be considered as the isolated transient response of Trp86. A key feature delivered by this analysis is the positive absorption band observed in the Trp86 contribution at λ>295 nm, with wavelength-independent kinetic traces in the range 305-320 nm (see Fig. 2-left-top). Figure 2-right displays the kinetic trace of the Trp86 contribution spectrally averaged from 305 to 320 nm. A biexponential fit yields the well-known retinal isomerization time 450 +/- 50 fs and a second time constant much longer than our observation time window (up to 20 ps, not shown here). The amplitude associated wih the fast decay component is 2.2 +/- 0.3 times larger than that of the slow one.

From the above interacting-dipole model, it appears that the positive induced absorption shown in Fig. 2-right is a measurement of the transient Stark shift of the Trp86 absorption band induced by the dipole moment change of PSBR. The latter decreases while isomerization proceeds as observed by the 450 fs decay time. The remaining positive absorption at later times suggests that the S_0 dipole moment of J and K states remains larger than that of all-trans bR, but other scenarios can be envisaged as well. In addition, the signal rises within the instrument response function indicating that a possible twist-induced charge translocation [4] has to occur on a sub-80 fs time scale. This conclusion now based upon a sound assignment of the Trp86 contribution rectifies results from our previous work.

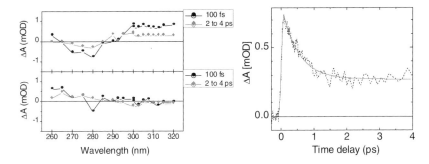

Fig. 2. Left: Individual spectral contributions of Trp86, and Trp182 obtained by linear superposition of the raw data displayed in Figure 1: Trp86 = wt bR - W86F, Trp182 = wt bR - W182F. Right: transient contribution of the Trp86 residue in the range of 305-320 nm. Overlaid is a fit to a two-component exponential decay convolved with a Gaussian response function.

This new systematic measurement and comparison of the transient UV absorption spectroscopy of wt bR and the two mutants W86F and W182F allows us to isolate the spectral contribution of the Trp's interacting with the PSBR. The results demonstrate that transient absorption spectroscopy of Trp's can reveal ultrafast changes of dipole moments on nearby moieties. The strong distance dependence of excitonic interactions introduces a natural selectivity among the many Trp's that the protein may contain (8 in bR). A further refined analysis of the excitonic interaction between Trp's and PSBR is in progress to explore the possibility to turn this approach into a quantitative measurement of the time-dependent $\Delta\mu$.

Acknowledgements. J.L. acknowledges support from the DYNA programme of the European Science Foundation.

1   A. Colonna, G. I. Groma, J.-L. Martin, M. Joffre, M. H. Vos, J. Phys. Chem. B, **111**, 2707 -2710 (2007).

2   S. Schenkl, F. van Mourik, G. van der Zwan, S. Haacke, M. Chergui, Science, 309, 917 (2005).

3   S. Schenkl, F. van Mourik, N. Friedman, M. Sheves, R. Schlesinger, S. Haacke, and M. Chergui, PNAS **103**, 4101 (2006).

4   R. Gonzalez-Luque, M. Garavelli, F. Bernardi, M. Merchán, M. Robb, M. Olivucci, PNAS, **97**, 9379 (2000).

Interrogating Fiber Formation Kinetics with Automated 2D-IR Spectroscopy

David B. Strasfeld[1], Yun L. Ling[1], Sang-Hee Shim[1] and Martin T. Zanni[1]

[1] Department of Chemistryt of Chemistry, University of Wisconsin-Madison, 1101 University Ave. Madison, Wisconsin 53706-1322
E-mail: zanni@chem.wisc.edu

Abstract. A new method for collecting 2D-IR spectra that utilizes both a pump-probe beam geometry and a mid-IR pulse shaper is used to gain a fuller understanding of fiber formation in the human islet amyloid polypeptide (hIAPP). We extract structural kinetics in order to better understand aggregation in hIAPP, the protein component of the amyloid fibers found to inhibit insulin production in type II diabetes patients.

1. Introduction

The formation of amyloid fibers by the human islet amyloid polypeptide (hIAPP) has been indicated as a primary cause of β-cell death in type II diabetes patients. The structural kinetics that dictate this transformation remain obscure due to interrogation primarily by circular dichroism (CD) and a fluorescence shift in protein bound dyes [1,2]. These two spectroscopies cannot monitor different secondary structural confirmations independently and simultaneously. Two dimensional infrared (2D-IR) spectroscopy offers an improved capacity to resolve protein secondary structures and added structural insight from coupling indicative cross peaks. The monitoring of amyloid formation with 2D-IR spectroscopy is hindered by two inherent difficulties: the fact that the kinetics of amyloid fiber formation are not perfectly reproducible and the enormous background noise attributable to light scattering by the amyloid fibers. We have learned to overcome these impediments by automating data collection [3] with a mid-IR pulse shaper. The issue of a scattering background is resolved by rotating the phases of subsequent pulses in the pump-pulse train, which, in turn, shifts the scatter frequency away from desired signal elements. Replacing mechanical stages with a pulse shaper greatly reduces the time necessary to generate a 2D spectrum, allowing us to take a single scan in <1 second and collect 2D-IR spectra rapidly enough to monitor aggregation.

2. Experimental Methods

The 37 residue hIAPP was purchased from BaChem and dissolved in deuterated hexafluoroisopropanol. An aliquot of the stock solution was separately dialysized with DCl and D_2O, respectively, to remove residual TFA. Each sample was then dried before being redissolved in D_2O to initiate aggregation. Aggregation, which took place between two CaF_2 plates separated by a 56 mm Teflon spacer, was observed by continuously collecting 2D-IR spectra. The 2D-IR spectra were generated using mid-IR light from a difference frequency OPA. A portion of the mid-IR light was split off to serve as the probe beam, while the rest was sent into a Ge-AOM based pulse shaper[4] to generate the double pulse that serves as the pump. The signal, self-heterodyned by the probe beam, was then frequency dispersed in a monochromator and detected with a 32 element MCT array. Each 2D spectrum was generated by scanning over 1.92 ps of the first coherence time in 9 fs steps. The phase of the pulse sequence was cycled between 0 and π, necessitating the generation of 426 unique pulses and allowing for scans to be taken in 0.43 second intervals. A single Fourier transform of the time/frequency data results in a 2D IR spectrum.

3. Results and Discussion

Figure 1a shows three 2D-IR spectra collected during the course of amyloid formation. The 2D-IR spectra cover the amide I region of the infrared spectrum. This region of the spectrum has proven to be uniquely suited to identifying secondary structural components and has, consequently, been extensively employed in elucidating protein structure and dynamics. A feature centered at $\omega_{pump} = \omega_{probe} = 1644$ cm^{-1} with an on-diagonal bandwidth of 33 cm^{-1} dominates the earliest spectrum shown in Figure 1. The breadth of the on-diagonal linewidth of

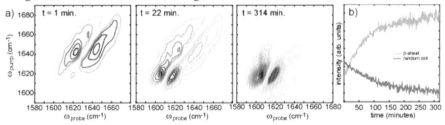

Figure 1. (a) Representative 2D-IR spectra taken at t = 1, 22, and 314 minutes during aggregation. (b) The intensity change of the β-sheet and random coil overtone features plotted as a function of aggregation time.

this feature indicates a structural disorder and distribution of environments most often associated with random coil secondary structure. The random coil feature eventually gives way to a split feature with fundamental centered at $\omega_{pump} = \omega_{probe} = 1618$ cm^{-1} and bearing an on diagonal bandwidth of 24 cm^{-1}. The split diagonal pair and its attendant cross peaks are indicative of β-sheet structure. Utilizing our unique capacity to take a 2D-IR spectrum every ~500 ms, we observed the transition from random coil dominated spectra to β-sheet dominance over the course of fiber formation. By monitoring intensity changes in unique portions of the 2D spectra, we can extract the kinetics of those conformations that yield features in the region of the spectrum that we are observing. Such kinetics are included in Fig 1b for the β-sheet anti-symmetric mode, random coil and β-sheet symmetric/anti-symmetric cross peak.

Augmenting our studies with the local structural information obtained from $^{13}C=^{18}O$ isotope labeling has further informed our understanding of how hIAPP folds and associates to form amyloid fibers. Figure 2a shows the structure of a ten monomer hIAPP aggregate derived from solid state NMR [5]. The alanine residues at positions 13 and 25 are highlighted within the structure in Fig 2a. Spectra of ala-25 and ala-13 labelled hIAPP at low pH are featured in Fig. 2b. These spectra display a splitting of the ala-25 labelled peak that is not present in the ala-13 labelled spectrum.

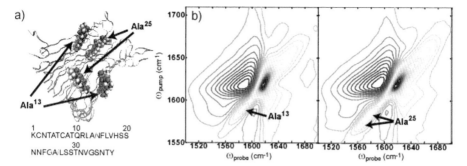

Figure 2. (a) hIAPP aggregate structure from solid state NMR. (b) A 2D-IR spectrum of 13C=18O labeled alanine-25 and alanine-13 with arrows indicatin shifted, labelled peaks.

4. Conclusions

We have found that automated 2D-IR utilizing a pump-probe beam geometry and pulse shaping is uniquely suited to monitoring structural kinetics not previously available to multi-dimensional IR techniques. Applying our method to hIAPP aggregation has clarified the competing kinetics of β-sheet evolution and random coil diminution in this system. Furthermore, by isotope labeling individual residues in hIAPP, we can acquire localized structural information, such as was obtained for ala-13 and ala-25 labelled hIAPP.

Acknowledgements. We acknowledge the contributions of Dan Raleigh and Peter Marek from SUNY – Stony Brook for synthesizing the isotope labeled hIAPP. Funding comes from the NSF (CHE-0350518), the Packard Foundation, and the NIH (R21AI064797, R01DK079895).

References

1 S. A. Jayasinghe and R. Langen, Biochem. 44, 12113 (2005).
2 J. D. Knight, J. A. Hebda, and A. D. Miranker, Biochem. 45, 9496 (2006).
3 S. H. Shim, D. B. Strasfeld, Y. L. Ling, and M. T. Zanni, Proc. Natl. Acad. Sci. USA 104, 14197 (2007).
4 S. H. Shim, D. B. Strasfeld, E. C. Fulmer, and M. T. Zanni, Opt. Lett. 31, 838 (2006).
5 S. Luca, W. M. Yau, R. Leapman, and R. Tycko, Biochem. 46, 13505 (2007).

Coherent Control of Chirality-Induced 2D Electronic Spectroscopy Signals

Dmitri V. Voronine[1], Darius Abramavicius[2], and Shaul Mukamel[2]

[1] Institut für Physikalische Chemie, Universität Würzburg, Am Hubland, Würzburg, 97074, Germany
  E-mail: dmitri.voronine@gmail.com
[2] Department of Chemistry, University of California Irvine, Irvine CA 92697-2025, USA
  E-mail: dariusa@uci.edu, smukamel@uci.edu

Abstract. Chirality-induced 2D electronic spectra, calculated to first order in the optical wavevector, reveal new features, not available from non-chiral signals. Coherent control is used to optimize the resolution of various energy transfer pathways in the Fenna-Matthews-Olson complex from photosynthetic green sulfur bacteria.

Ultrafast 2D electronic spectroscopy recently performed on several photosynthetic systems provides valuable insights into the mechanisms of light harvesting [1-4]. Energy transfer pathways can be directly observed through the temporal evolution of the cross peaks [1, 5], and oscillations of these provide information on electronic quantum coherences [2]. The spectra of photosynthetic molecular aggregates consist of bands of closely-lying excited states. Spectral congestion due to the large number of states and the solvent and protein environments complicates the analysis. Weak cross-peaks overlap with strong diagonal peaks preventing the unambiguous assignment of exciton features. 2D four-wave mixing (FWM) spectroscopy provides rich information through snapshots of vibrational and electronic dynamics. These can be considerably simplified by using coherent control techniques which manipulate the interference of various Liouville space pathways [6].

We have recently demonstrated the simplification of 2D spectra at zero population delay time (t_2) in a chiral porphyrin dimer by designing linear combinations of two-dimensional electronic chirality-induced (2D ECI) tensor components, which lead to the control of spectral features through interferences [7]. Linear combinations of 9 2D ECI tensor components were optimized using genetic algorithm. The ratios of the amplitudes of weak cross peaks were chosen as control targets. Here we apply this method to simulate t_2-dependent 2D electronic spectra of the light-harvesting Fenna-Matthews-Olson (FMO) complex from the photosynthetic green sulphur bacteria *Chlorobium tepidum*. We use the Hamiltonian obtained by fitting the linear and 2D xxxx spectra by Brixner et. al [1] and an overdamped Brownian oscillator bath spectral density. The 7 one-exciton states contributing to the linear absorption and circular dichroism spectra are shown in Figure 1. At 77 K the linear absorption spectra form 2 overlapping peaks and 2 shoulders. The peaks are labelled 1 to 7 increasing in energy. The largest contributions of the 7 constituent bacteriochlorophyll (BChl) molecules (numbered 1 to 7 according to Fenna et. al. [8]) to the exciton states are indicated on each peak (in decreasing order of strength). The simulated linear absorption and CD fit the experiments well. CD provides a better resolution of the exciton states. We extend the CD technique to the nonlinear regime, and apply coherent control algorithms to simplify the chiral 2D spectra.

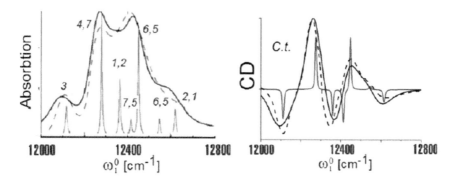

Fig. 1. Linear absorption and circular dichroism (CD) spectra of the FMO complexes from the *Chlorobium tepidum* green sulfur bacteria: experimental (dashed) and simulated (solid) spectra. An overdamped Brownian oscillator model at 77 K ($\Lambda^{-1} = 100$ fs, $\lambda = 55$ cm^{-1}) was used for the line broadening. Stick spectra are shown in red. BChls with the largest contributions to the corresponding exciton states are indicated on each peak (in decreasing order of the contribution)

The couplings between the various BChls of the complex, observed as cross-peaks in the 2D spectra, determine the energy flow timescales and the efficiency of photosynthesis. Two main energy transfer pathways are revealed in time-resolved electronic 2D spectra of FMO, which are spatially separated and occur on different time scales. Analysis of exciton delocalization patterns in 2D ECI spectra allows to visualize the energy transfer through space, and reveals the slower energy transfer pathway which was not well resolved in the non-chiral 2D xxxx spectra shown in Figure 2.

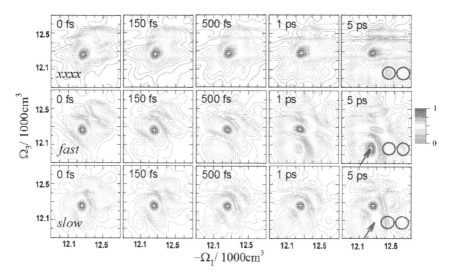

Fig. 2. Coherent control of energy transfer in FMO. Absolute magnitude of 2D signals: non-chiral xxxx and optimized 2D ECI (fast) and (slow) pathways

551

The xxxx spectra are dominated by strong diagonal peaks, and clearly indicate only the fast energy transfer pathways (left circle). Using coherent control with a genetic learning algorithm, we have separated different energy transfer channels by optimizing a linear superposition of tensor components. The control target aims at manipulating various peaks in these 2D spectra. We chose it to be the ratios of cross peak absolute values of the 2D signals at delay time $t_2 = 5$ ps.

The two cross peaks contributing to the fast and slow energy transfer pathways (1,5) and (1,7) are marked in Figure 2. by the left and right circles, respectively. These cross peaks are due to energy transfer from higher to lower energy states and they grow as a function of t_2. The coefficients obtained from the optimizations at $t_2 = 5$ ps were then used to obtain the 2D spectra at other t_2-delay times. In the optimization of the fast energy transfer pathway, the ratio of (1,5) to (1,7) is maximized. The left circle and arrow indicate cross peaks (1,5) and (1,2), respectively. The states 1, 2, and 5 are involved in the fast pathway and the corresponding cross peaks are enhanced. For the slow pathway the ratio of (1,7) to (1,5) is maximized. The right circle and arrow indicate cross peaks (1,7) and (1,3), respectively. These participate in the slow energy transfer pathway. As a result both pathways are separately optimized and well-resolved. In addition, there is a new spectral feature in the fast pathway: a node between the arrow and the left circle, corresponding to the elimination of the (1,3) cross peak of the slow pathway. Similarly, in the optimization of the slow pathway we see nodes corresponding to the elimination of the (1,2) and (1,5) cross peaks of the fast pathway.

Using a coherent control algorithm we demonstrated optimal laser polarization configurations which enhance chirality-sensitive spectral features, revealing a slow energy transfer pathway which was not resolved in the xxxx spectra.

Acknowledgements. The support of the National Institutes of Health (GM59230) and the National Science Foundation (CHE-0745891) is gratefully acknowledged.

1   T. Brixner, J. Stenger, H. M. Vaswani, M. Cho, R. E. Blankenship, and G. R. Fleming, in Nature **434**, 625, 2005.

2   G. S. Engel, T. R. Calhoun, E. L. Read, T.-K. Ahn, T. Mancal, Y.-C. Cheng, R. E. Blankenship, and G. R. Fleming, in Nature **446**, 782, 2007.

3   E. L. Read, G. S. Engel, T. R. Calhoun, T. Mancal, T. K. Ahn, R. E. Blankenship, and G. R. Fleming, in PNAS **104,** 14203, 2007.

4   D. Zigmantas, E. L. Read, T. Mancal, T. Brixner, A. T. Gardiner, R. J. Cogdell, and G. R. Fleming, in PNAS **103**, 12672, 2006.

5   M. Cho, H. M. Vaswani, T. Brixner, J. Stenger, and G. R. Fleming, in J. Phys. Chem. B **109**, 10542, 2005.

6   S. Mukamel, Principles of Nonlinear Optical Spectroscopy, Oxford University Press, New York, 1995.

7   D. V. Voronine, D. Abramavicius, and S. Mukamel, in J. Chem. Phys. **125**, 224504, 2006.

8   R. E. Fenna, and B. W. Matthews, in Nature **258**, 573, 1975.

Two-Photon Two-Color Generation of Zeaxanthin Radical Cation in CP29 Light Harvesting Complex

Sergiu Amarie[1], Laura Wilk[2], Tiago Barros[2], Werner Kühlbrandt[2], Andreas Dreuw[1] and Josef Wachtveitl[1]

[1] Institute for Physical and Theoretical Chemistry, Johann Wolfgang Goethe-University Frankfurt, Max von Laue-Str. 7, 60438 Frankfurt am Main, Germany
E-mail: wveitl@theochem.uni-frankfurt.deu
[2] Max Plank Institute of Biophysics, Department of Structural Biology, Max von Laue-Str. 3, 60438 Frankfurt am Main, Germany
E-mail: Werner.Kuehlbrandt@mpibp-frankfurt.mpg.de

Abstract. Recent theoretical and experimental studies reveal the central role of zeaxanthin radical cations (Zea^{+}) in regulation of photosynthetic light harvesting. Two-color two-photon spectroscopy on LHC-II protein CP29 reveals the in-situ photodynamics of Zea^{+}.

Introduction

Feedback deexcitation (qE), a component of nonphotochemical quenching (NPQ) is an important process in the photoprotection of plants during which the light harvesting antennae switch to specific states to dissipate excitation energy of chlorophylls (Chls) as heat [1]. Along with other factors, activation of qE is correlated with zeaxanthin formation. It is thought that zeaxanthin plays a direct role in quenching the excitation energy [2]: a quenching complex forms between Chl *a* and zeaxanthin during the induction time of qE, and zeaxanthin acts as terminal quencher either via excitation energy or electron transfer, the latter leading to carotenoid radical cation formation. This hypothesis was first predicted by theory [3], and has been confirmed experimentally [2]. The corresponding experiment was performed on intact spinach thylakoid membranes and the near-IR region was probed after excitation at 664 nm under quenched and unquenched conditions. The spectral differences observed between the quenched and unquenched states were ascribed to the formation of zeaxanthin radical cations [2]. Recent studies in which zeaxanthin radical cation signature has been used to infer qE quenching suggest that minor LHCs (CP24, CP26 and CP29) provide sites for charge transfer quenching [4]. Nevertheless, it remains to be shown where exactly and in which pigment-protein complex of the photosynthetic apparatus qE actually occurs. In the presented work, we use multiple light pulses to investigate the ultrafast dynamics of carotenoid radical cation formation in CP29 revealing the spectroscopic signature of zeaxanthin radical cations in zeaxanthin enriched complexes. This is a further crucial step for the understanding of the qE protection mechanism in higher plants.

Experimental Methods

Recombinant CP29 from *Arabidopsis thaliana* was refolded in the presence of isolated pigments with adjusted chlorophyll content in order to resemble native CP29. Exposure of CP29 to zeaxanthin led to zeaxanthin-containing CP29 (CP29-Zea) without introducing further influences like light or pH gradient. Pigment composition was determined by HPLC analysis. The typical sample OD was 0.8/mm at 665 nm. Sample stability was confirmed by measuring the absorption spectra before and after the time resolved measurements. The time resolved measurements were performed using a CLARK CPA 2001 (Dexter, MI) laser/amplifier system operating at a repetition rate of ~1 kHz at a central wavelength of 775 nm. The amplified pulses were divided into three parts: two pump pulses and one probe pulse. The first excitation pulses were generated using a noncollinear optical parametric amplifier with excitation energy of 60 nJ at 490 nm. The second pump pulse, with energy of 200 nJ at a central wavelength of 775 nm, is delayed 40 fs with respect to the first pulse. For the probe pulses a white light continuum was generated by focusing amplified 775 nm light into a 5 mm sapphire window and a RG 830 filter was used to select the useful spectral region.

Results and Discussion

Excitation of carotenoids with a pump pulse centered at 490 nm promotes the carotenoids into the S_2 excited state. Applying a second pulse (775 nm), while the S_2 state is occupied results in a further excitation of the carotenoid molecules into the S_N state. The relaxation of this higher excited state can occur either back to S_2 or an electron can be released leading to the formation of a carotenoid radical cation (Fig. 1). The second pathway is supported by previous studies [5], and leads to a characteristic spectral signature in the near-IR region. The photoinduced absorption originating from the carotenoid radical cation is clearly separated from the strong excited state absorption of the $S_2 \rightarrow S_N$ transition by their sequential occurrence, the first has a lifetime on the order of µs while the latter is very short lived, with $\tau \approx 100$ fs (Fig. 1).

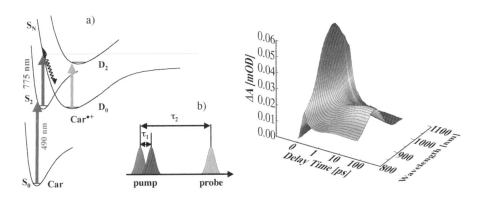

Fig. 1. Two-photon two-color schematic diagram (left) and temporal evolution of β-Car$^{\cdot+}$ in dichloromethane (right).

Application of this pulse sequence to the CP29 sample leads to the spectroscopic generation of a transient excited state absorption which corresponds to sums of all different carotenoid cations present in CP29 (lutein, neoxanthin and violaxanthin). Figure 2 shows the HPLC analysis (left) and transient absorption spectra (right) recorded at 40 ps time delay following two-photon two-color excitation of CP29 (blue) and CP29Zea (red). The positive component of the difference transient absorption spectrum CP29Zea-CP29 (black triangles) corresponds to the zeaxanthin radical cation. A Gaussian fit of the data points shows a maximum at 980 nm.

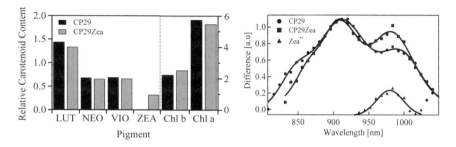

Fig. 2. HPLC analysis (left) and difference transient absorption spectra (right) for CP29 (circle) and CP29Zea (square).

Conclusions

The results presented in this ultrafast two-photon two-color study clearly show the spectral position of zeaxanthin radical cation in isolated CP29 at 980 nm. Combining with previous results [4], which show the transient zeaxanthin radical cation also at 980 nm, we can conclude that zeaxanthin bound to CP29 may act as a terminal quencher. We will perform further two-photon two-color experiments in combination with Chl lifetime measurements on individual complexes (CP26 and CP24), allowing for additional critical evaluation of the role of zeaxanthin radical cations in qE component of NPQ.

Acknowledgements. This work has been supported by the SFB 472 ("Molecular Bioenergetics"), as well as by the CEF "Macromolecular Complexes" of the University of Frankfurt.

1   P. Müller, X. P. Li and K. K. Niyogi, Plant Physiol. **125,** 1558, 2001.
2   N. E. Holt, D. Zigmantas, L. Valkunas, X. P. Li, K. K. Niyogi and G. R. Fleming, Science **307,** 433, 2005.
3   A. Dreuw, G. R. Fleming, M. Head-Gordon, J. Phys. Chem. B **107,** 6500, 2003.
4   T.K. Ahn, T.J. Avenson, M. Ballottari, Y.C. Cheng, K.K. Niyogi, R. Bassi, G.R. Fleming, Science **320,** 794, 2008.
5   S. Amarie, J. Standfuss, T. Barros, W. Kuhlbrandt, A. Dreuw and J. Wachtveitl, J. Phys. Chem. B **111,** 3481, 2007.

Rebinding of Proximal Histidine in the Cytochrome *c'* from *Alcaligenes xylosoxidans* Acts as a Molecular Trap for Nitric Oxide

Byung-Kuk Yoo[1], Jean-Louis Martin[1], Colin R. Andrew[2], Michel Negrerie[1]

[1] Laboratory for Optics and Biosciences, INSERM U696, CNRS UMR 7645, Ecole Polytechnique, 91128 Palaiseau Cedex, France
E-mail: michel.negrerie@polytechnique.fr
[2] Department of Chemistry and Biochemistry, Eastern Oregon University, La Grande, Oregon 97850, USA

Abstract: Transient absorption spectra on cytochrome *c'* and their kinetics were recorded to identify the formation of 5-coordinate (5c)-NO and 5c-His hemes from 4c-heme (99% and 1% amplitudes; 7-ps and 100-ps time constants, respectively). We demonstrate that proximal histidine precludes NO rebinding at the proximal site.

Introduction

Nitric oxide (NO) acts as a second messenger in several physiological systems [1] by binding to NO-receptors, NO-sensors and some c-type cytochromes. Unexpectedly, the crystal structure of *Alcaligenes xylosoxidans* cytochrome *c'* (AXCP) has revealed that NO is bound to the proximal side of the ferrous heme replacing the endogenous His ligand [2]. Here we explored the NO binding and release mechanisms of these bacterial NO sensors with time-resolved absorption studies in a wide time range with femtosecond resolution. We have probed NO dynamics and the proximal His motion and have found that after NO dissociation from the heme iron, the structure of the proximal heme pocket of AXCP confines NO close to the iron so that an ultrafast (7 ps) and complete (~99%) geminate rebinding occurs [3]. The very minor 1% population escaping could be due to the proximal His reattachment according to the previously proposed "kinetic trap" hypothesis [4], this small fraction of NO that escapes the heme pocket is prevented from direct bimolecular recombination to the proximal heme face by the motion and binding of the His ligand once NO has left the heme pocket. NO can only rebind from solution through the 6c-NO distal intermediate formation. Probing the formation of the 5c-His proximal species from the photodissociated 4c-heme necessitates the identification of the 1% population corresponding to NO escaping the heme pocket. Here we present a spectral demonstration that NO cannot rebind if it goes out of the heme pocket because of His reattachment. This demonstration is supported by the identification of transient species in the picoseconds time-range after photodissociation of NO from the 5c-NO complex, and the associated kinetics.

Experimental Methods

Purified AXCP was put in a 1-mm optical path length quartz cell, degassed and reduced by the sodium ascorbate (5 mM final concentration). The photodissociation of NO was achieved with an excitation pulse at 564 nm whose duration was ~40 fs

556

with a repetition rate of 30 Hz. Transient spectra were recorded simultaneously to kinetics with a CCD detector as a time-wavelength matrix data. Transient spectral analysis of the data was performed by singular value decomposition (SVD) of the time-wavelength matrix such that all spectral components were identified in the time-range 1 ps-1 ns. Only two SVD spectral components represented the coordination changes of the heme.

Results and Discussion

The absorbance decay associated with NO geminate rebinding can be assigned to 99% SVD 1 component (solid line; Fig 1-A) with a time constant of 7 ps. The spectrum of the 1% population is represented as SVD 2 component (dot; Fig. 1-A). The bleaching at 415 nm is assigned to the 4c-heme and the absorbance at 435 nm

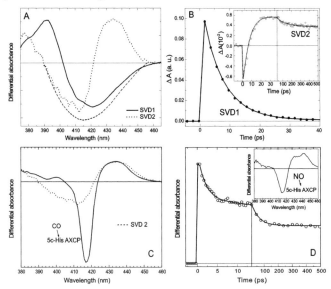

Fig. 1. A Normalized transient spectra obtained from SVD analysis, B Kinetics of SVD 1 and SVD 2 (inset) components. C Normalized transient spectra in case of CO compared with SVD 2, D Kinetics of NO rebinding to the 5c-His AXCP with its transient spectrum (inset) after photodissociation from His-6c-NO obtained at low NO concentration (0.1 %).

is due to the 5c-His heme. The dashed line represents the calculated 4c-heme spectrum (long wavelength part of SVD 1 and short part of SVD 2 spectra in Fig. 1-A). SVD 1 represents the rebinding of NO to the heme [τ_{gem} = 7 ps], whereas SVD 2 represents the rebinding of His to the heme [τ_1 = ~100 ps] (Fig. 1-B). In order to assign the transient SVD 2 spectra we compared it to that of CO rebinding to AXCP. The bleaching part of SVD 2 is different from the spectrum of CO rebinding to the 5c-His AXCP and two induced absorptions at 435 nm are well matched (Fig. 1-C). Kinetics of NO rebinding to the 5c-His AXCP after photodissociation of 6c-His-NO is obtained (τ_2 = 52 ps) as shown in 1-D with the corresponding transient spectral component (inset). The rebinding of NO to the 5c-His heme appears slower than to the 4c-heme of AXCP with an associated spectrum different from SVD 2. This shows that SVD 2 is not due to 6c-His-NO. If NO escapes the heme pocket, His rebinds to the heme and precludes direct NO rebinding from the solution. NO must then rebind to the distal side, confirming the mechanism of "ligand trap". This property of AXCP is more likely that of a NO transporter rather than of a NO reductase. The present results shows that in AXCP the distal side controls the initial NO binding, whilst the proximal heme pocket controls the release of NO, with the ability of trapping and gating NO, with a virtually unidirectional release of NO. The complete kinetic behavior of AXCP is depicted in Fig. 2.

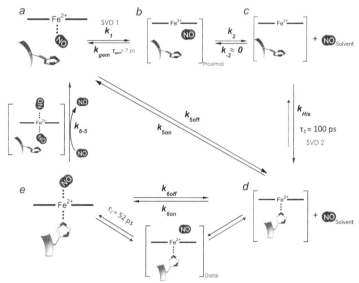

Fig. 2. Model for the binding and release of NO to and from the different transient forms and coordination states of AXCP. The various species are: (**a**) 5c-NO AXCP with NO at the proximal side which is the starting species of our experiments; (**b**) transient 4c-AXCP with NO dissociated and located within the heme pocket; (**c**) transient 4c-AXCP with NO dissociated and located in the solvent; (**d**) 5c-His resting state AXCP; (**e**) 6c-His-NO AXCP. The rate constants: k_1, thermal dissociation of NO; k_{-1}, rebinding rate; k_2 and k_{-2}, rates for exiting and accessing the heme pocket of the 4c-AXCP; k_{His}, rate for His rebinding after NO release (the shorter arrow indicates that the back process, albeit formally possible, has a low probability to occur and is not observed); k_{6on}, NO binding rate from the solvent to 6c-His-NO AXCP; k_{6off}, rate of NO release from 6c-His-NO AXCP; k_{6-5}, rate of conversion of 6c-His-NO to 5c-NO AXCP; k_{5off}, overall rate of NO release from 5c-NO AXCP to the solvent; k_{5on}, overall rate of NO binding from the solvent leading to 5c-NO AXCP.

Conclusions

We demonstrated that the proximal His rebinds in about 100 ps after the exit of NO from the heme pocket, and thus precludes the rebinding of NO from the solution, so that the rebinding of proximal His functions as a molecular trap for NO-dissociated heme. This behavior represents a unique kinetic trap, and has not been observed among other heme proteins.

1 J. W. Denninger and M. A. Marletta, in Biochim. Biophys. Acta **1411**, 334, 1999.

2 D. M. Lawson, C. E. Stevenson, C. R. Andrew, and R. Eady, in EMBO J. **19**, 5661, 2000.

3 S. G. Kruglik, J.-C. Lambry, S. Cianetti, J.-L. Martin, R. R. Eady, C. R. Andrew, and M. Negrerie, in J Biol. Chem., **282**, 5053 2007.

4 C. R. Andrew, K. R. Rodgers, and R. R. Eady, in J. Am. Chem. Soc. **125**, 9548, 2003.

Two-Dimensional Electronic Spectroscopy of the Low-Light Adapted Light Harvesting Complex 4

Elizabeth L. Read[1,2], Gabriela S. Schlau-Cohen[1,2], Gregory S. Engel[3], Toni Georgiou[4], Miroslav Z. Papiz[4], Graham R. Fleming[1,2]

[1] Department of Chemistry, University of California, Berkeley, CA 94720, USA
  E-mail: grfleming@lbl.gov
[2] Physical Biosciences Division, Lawrence Berkeley National Laboratory, Berkeley, CA 94720
[3] Current Address: Department of Chemistry, The University of Chicago, Chicago, IL, 60637
[4] Department of Synchrotron Radiation, STFC Daresbury Laboratory, Warrington, Cheshire WA4 4AD, United Kingdom

Abstract. Two-dimensional electronic spectroscopy of Light Harvesting Complex 4 from photosynthetic bacteria reveals excited state dynamics on two timescales and resolves exciton states with little to no oscillator strength. The results suggest a molecular structure in which the pigment dipole organization within the circular complex has more tangential than radial character.

Introduction

In purple photosynthetic bacteria, excitonically coupled bacteriochlorophyll (BChl) pigments are arranged in cyclic pigment-protein light-harvesting complexes. The so-called peripheral complexes, most commonly LH2, absorb sunlight and funnel energy to photosynthetic reaction centers, where charge separation occurs. Several purple bacteria have been shown to express different types of peripheral light-harvesting complexes when grown under low light intensities. *Rhodopseudomonas palustris* produces a low-light complex known as LH4, which exhibits only a single absorption band in the near-IR region. The origins of LH4's unique spectrum are not decisively known, however biochemical analysis and electron microscopy point to a novel pigment architecture in the complex, with four BChls per repeating subunit rather than three [1].

In this work, we employ two-dimensional (2D) Fourier transform electronic spectroscopy to investigate the energy level structure and energy transfer dynamics of LH4. 2D spectroscopy yields frequency maps of electronic states, molecular coupling, and dynamical processes with femtosecond time resolution. This information, in combination with knowledge of protein structures, is crucial to understanding the design principles governing natural light-harvesting.

Experimental Methods

LH4 from *Rp. palustris* was isolated as described previously [1]. The sample was dissolved in a pH 8 buffer solution of 20mM Tris-HCl and mixed with glycerol (30:65 by volume) to form a glass in a 200 μm-thick quartz cell for 2D measurements at 77 K. The sample OD was 0.4 at 800 nm.

Two-dimensional spectra were collected in a passively phase-stabilized manner as described in Brixner, et al [2]. Three 43 fs laser pulses centered at 806 nm with FWHM 29 nm were focused on the sample in a box geometry to generate the third order signal. A fourth pulse (passing through the sample) acted as the local oscillator for heterodyne detection. The spectrally resolved signal in amplitude and phase was measured while scanning the coherence time, τ (between pulses 1 and 2), for 17 population time points, T (between pulses 2 and 3), from 0 to 50 ps. The coherence time was scanned from 0 to +/-600 fs (where the minus sign denotes arrival of pulse 2 before pulse 1). The pulse ordering used selected for rephasing ($+\tau$) or nonrephasing ($-\tau$) signals.

Results and Discussion

Experimental correlation (sum of rephasing and nonrephasing) 2D spectra of LH4 from *Rp. palustris* at 77K are presented in Figure 1 for three representative population times. The peak-shapes show rapid coherent evolution within the first 100 fs marked by initial decay of the negative excited state absorption (ESA) feature, followed by picosecond-scale intra-band energy transfer evidenced by widening of the positive band below the diagonal and resurgence of the ESA.

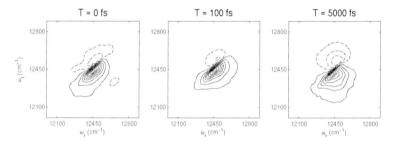

Fig. 1. Experimental 2D electronic correlation spectra (real part) of LH4 at 77 K for T=0, 100, and 5000 fs. Solid and dashed contours denote positive and negative signals, respectively.

  Separating correlation spectra into rephasing and nonrephasing components has been shown to improve resolution of individual exciton states in inhomogeneously broadened systems [3]. The nonrephasing experimental 2D spectrum of LH4 at T=0 (Figure 2, right panel) reveals low-amplitude diagonal peaks on the red edge of the band, signaling weakly absorbing states in this region, as were previously predicted [1,4]. In addition, the pronounced ESA exhibited at higher energy, on the diagonal in the nonrephasing spectrum and above the diagonal in the rephasing spectrum, arises from strong coupling through dark states on the blue edge of the exciton manifold. Theoretical simulations performed based on Zigmantas, et al. [5], show that the positioning of ESA in 2D spectra is strongly sensitive to pigment organization. Reproducing the high-energy ESA requires a pigment organization in the complex with pigment dipoles aligned more "in-line" (head-to-tail or head-to-head) than "sandwiched" (side-by-side), reminiscent of the tangential orientation of dipoles around the protein ring in LH2.

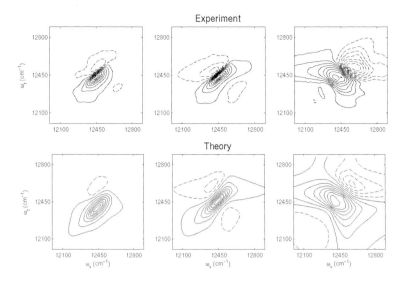

Fig. 2. Experimental (top) and theoretical (bottom) 2D electronic spectra of LH4 at 77 K and *T*=0 fs. Shown are correlation, rephasing, and nonrephasing signals, from left to right. Theoretical spectra are calculated for a 4-BChl subunit, based on the structure of LH2.

Conclusions

The 2D electronic spectra of LH4 from *Rp. palustris* show unique sensitivity to the exciton level structure and exhibit both rapid coherent dynamics and ps-timescale population transfer. The observed weakly-absorbing red-edge states and dark blue-edge states may play a role in light-harvesting by accepting and transferring energy initially absorbed by bright states, as was observed previously [5]. The level structure suggests a pigment dipole organization with mixed in-line and sandwich character, but with the in-line coupling dominating. The results demonstrate that 2D electronic spectroscopy can be used to help characterize supramolecular structures of chromophore aggregates, when high-resolution crystallographic information is not available.

Acknowledgements. US Department of Energy grant No. DE-AC03-76SF000098, National Science Foundation.

1   N. Hartigan, H. A. Tharia, F. Sweeney, A. M. Lawless, and M. Z. Papiz, Biophys. J. **82**, 963 (2002).

2   T. Brixner, I. V. Stiopkin, G. R. Fleming, J. Chem. Phys. **121**, 4221 (2004).

3   E. L. Read, G. S. Schlau-Cohen, G. S. Engel, J. Wen, R. E. Blankenship, and G. R. Fleming, Biophys. J. in press (2008).

4   W. P. F. de Ruijter, S. Oellerich, J.-M. Segura, A. M. Lawless, M. Z. Papiz, and T. J. Aartsma, Biophys. J. **87**, 3413 (2004).

5   D. Zigmantas, E. L. Read, T. Mancal, T. Brixner, A. T. Gardiner, R. J. Cogdell, G. R. Fleming, Proc. Natl. Acad. Sci. **103**, 12672 (2006).

Three-Pulse Photon Echo Spectroscopy as a Probe of Flexibility and Conformational Heterogeneity in Protein Folding

Emily A. Gibson and Ralph Jimenez

JILA and Department of Chemistry and Biochemistry, University of Colorado and National Institute of Standards and Technology, University of Colorado, Boulder Colorado, 80309, USA
E-mail: rjimenez@jila.colorado.edu

Abstract. We investigate the equilibrium unfolding of Zn-cytochrome c by three-pulse photon echo peak shift (3PEPS) spectroscopy, revealing denaturant-dependent timescales of protein motion and inhomogenous broadening. The results are consistent with a two-state model.

Introduction

The current understanding of protein folding is dominated by the concept of a "folding landscape" for each system [1]. One challenge is to obtain knowledge of the topology for this landscape under various denaturant and temperature conditions. Experimental studies attempt to glean information on the landscape either from kinetics measurements, or from the structural characterization of intermediate species under equilibrium conditions. However, with these methods it is difficult to characterize two critical aspects of folding: the conformational diversity and the evolution of protein fluctuations and flexibility. These factors distinguish folding from ordinary chemical reaction kinetics [2]. Here, we present photon echo measurements on the Zn-heme cofactor to obtain insight into the folding of cytochrome c.

3PEPS is an electronic four-wave mixing technique that has been used to characterize dynamics in solvents, glasses, and proteins. It can provide valuable new information on unfolded states and the folding process because it provides information on the spectrum of fs-ns motions of the system (a measure of flexibility) and the inhomogeneous broadening of the electronic transition (a measure of conformational diversity/entropy) [3].

3PEPS measurements typically reveal 2-3 time scales of collective coordinates corresponding to protein motions coupled to the electronic transition. Comparisons of timescales observed for various proteins binding the same cofactors (e.g. antibodies) have revealed differences in their flexibilities. 3PEPS has a number of advantages compared to transient absorption or emission measurements: 1) the peak shift decay is typically insensitive to the decay of excited state population, so the complex fluorescence decay kinetics usually observed in proteins are not convoluted with time scales of protein motions 2) fluorescent cofactors are not necessary 3) 3PEPS can be used to quantify inhomogeneous broadening, which characterizes the static disorder of the system. From a model based on impulsive excitation of a two-level system coupled to a classical bath of harmonic oscillators [4], it is found that the initial peak shift is related to the homogeneous linewidth, Γ, *i.e.* $\tau^*(0) \approx (\Gamma\pi)^{-1}$, at long delay times, the asymptotic peak shift is related to the inhomogeneous linewidth, Δ_{in}, *i.e.* $\tau^*(T\rightarrow\infty) \propto \Delta_{in}^2(\Gamma + \Delta_{in}^2)^{-1}$, and the peak shift decay time scales reflect those in the time-correlation function of the electronic energy gap. A more complete analysis using realistic excitation pulses and including the high frequency vibrations of the chromophore is available.

Methods

Zn cyt c was prepared from horse-heart cyt c, according to published methods [5]. Denatured and intermediate unfolded Zn cyt c samples were prepared by adding GuHCl to pH 7 100 mM phosphate buffer solution to obtain concentrations of 4.5 M and 2.5 M GuHCl respectively.

Experiments were performed with the second harmonic of a cavity-dumped, mode-locked Ti:sapphire oscillator. Pulses were tunable from 400-425 nm with a bandwidth of 6-7 nm. Material dispersion in the set-up was compensated with a fused silica prism pair to provide near-transform limited pulses of 45 fs at the sample. 3PEPS measurements were performed with an arrangement described previously [6]. For each sample, the center wavelength of the excitation light was tuned slightly to the red side of the Soret band.

Sample concentrations of 150 µM in a spinning cell were used for all measurements. A laser repetition rate of 40 kHz was used to avoid artifacts associated with Zn-porphyrin triplet states.

Results and Discussion

The sensitivity of the Soret UV-vis absorption band of heme proteins to conformation is well known. In this case (Fig. 1), it blue-shifts (from 422 nm to 416 nm) and broadens (from 907 cm^{-1} to 1050 cm^{-1} FWHM) upon unfolding. Interestingly, the intermediate state spectrum, corresponding to that of the mid-point of the denaturation process, is slightly broader (1162 cm^{-1} FWHM) than that of the unfolded form. The spectral line broadening results from changes in homogeneous and inhomogeneous linewidths, resolved by the 3PEPS data (Fig. 1). The data are fit by nonlinear least squares to a functional form with multiple exponential decay components, reported in Table 1. In addition, all traces have a damped sinusoidal component with a frequency of 152 cm^{-1} and a damping time constant of 50-58 fs. This oscillation results from a vibrational mode of the Zn-heme.

The magnitude of the initial peak shift is similar in all three cases, showing that the homogeneous linewidth, which is likely dominated by vibrational modes of the Zn-porphyrin cofactor, only changes slightly upon unfolding. The timescale of the fastest (30-50 fs) decay component is coupled to the fit of the oscillatory mode, but in similar

Table 1. Fit parameters of peak shift data to function: $\tau^*(T) = \tau(\infty) + \sum_i A_i \exp(T/\tau_i)$

Sample	τ_i(fs)	A_i(fs)	$\tau(\infty)$ (fs)
Folded	53 (±12) 3027 (±546)	11.7 (±3.7) 1.8 (±0.1)	4.2 (±0.06)
Unfolded	32 (±9) 5153 (±1940)	9.6 (±1.6) 1.7 (±0.2)	5.7 (±0.15)
Intermediate	34 (±24)	12.8 (±4.7)	7.75 (±0.05)

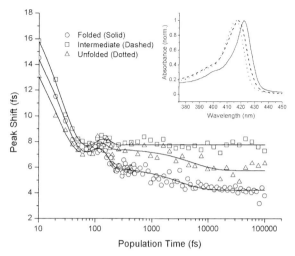

Fig 1. Peak shift traces for folded, intermediate (2.5 M GuHCl) and unfolded (4.5 M GuHCl) equilibrium states of Zn cyt c, and nonlinear least squares fits (solid lines). Inset: UV-vis spectra of Soret-band region.

protein/cofactor systems, this timescale results from high frequency chromophore and protein motions. The most interesting changes are observed at longer timescales. In particular, the increase in the intermediate time constant of the decay upon unfolding corresponds to an increase in flexibility (lower frequency motions) of the unfolded state. The increase in the asymptotic peak shift (larger inhomogeneous broadening) is consistent with the larger conformational diversity expected of an unfolded protein relative to the folded state. The largest asymptotic peak shift, observed for the intermediate state, is consistent with the larger absorption bandwidth of this sample relative to the folded or unfolded states. The intermediate state shows a larger inhomogeneous broadening relative to the folded or unfolded states. This effect probably arises because cyt c unfolding in GuHCl does not show an intermediate equilibrium state. At the midpoint of the denaturation process, the sample consists of a nearly 1:1 mixture of folded and unfolded forms, both of which are probed by the laser. The folded and unfolded forms are more spectrally distinct than conformations within either the folded or unfolded ensembles. Thus, the larger asymptotic peak shift of the intermediate state reflects the heterogeneity due to a mixture rather than conformational diversity of a folding intermediate.

In conclusion, photon echo spectroscopy reveals differences in the sub-picosecond dynamics of the folded and unfolded forms, and changes in conformational diversity leading to large changes in inhomogeneous broadening upon unfolding.

Acknowledgements. This work was supported by NSF grant DBI 0454763.

1 J. N. Onuchic, Z. Luthey-Schulten, and P. G. Wolynes, Ann. Rev. Phys. Chem. **48,** 545, 1997.

2 B. A. Shoemaker and P. G. Wolynes, J. Mol. Biol. **287,** 657, 1999.

3 R. Jimenez and F. E. Romesberg, J. Phys. Chem. B **106,** 9172, 2002.

4 M. H. Cho, J. Yu, T. Joo, Y. Nagasawa, S. A. Passino, and G.R. Fleming, J. Phys. Chem., **100,** 11944, 1996.

5 J. M. Vanderkooi, F. Adar, and M. Erecinska, Eur. J. Biochem., **64,** 381, 1976.

6 B. M. Cho, C. F. Carlsson, and R. Jimenez, J. Chem. Phys. **124,** 144905, 2006.

Ultrafast Rebinding of CO to Carboxymethyl Cytochrome c Probed by Femtosecond Vibrational Spectroscopy

Jooyoung Kim, Jaeheung Park, Taegon Lee, and Manho Lim*

Department of Chemistry and Chemistry Institute for Functional Materials, Pusan National University, Busan 609-735, Korea
E-mail: mhlim@pusan.ac.kr

Abstract. The direct observation of the dissociated CO from CO-cm cyt. c shows that the fraction of CO escapes into the bulk solution on a ps time scale and the remainder geminately rebinds (GR) within 1 ns. Absence of primary docking site-like structure in cyt. c and inefficient GR of CO in Mb suggests that the primary docking site in Mb indeed plays a physiologically important role in efficient expulsion of toxic CO in Mb.

Introduction

Rebinding dynamics of ligand photolyzed from myoglobin (Mb) has been widely used to investigate the means by which Mb modulates the rate of binding of small ligands to the heme. Heme pocket, a small vacant site near the active binding site of Mb, has been suggested to serve as a station mediating ligand binding and escape [1]. Only 4% of CO photolyzed from MbCO geminately rebinds to the heme even the ligand resides in the vicinity of the heme for hundreds of ns after photolysis [2, 3]. While heme pocket may have well-reserved structure for ligand accommodation, it likely preserves structural flexibility for efficient ligand pathways to the solvent. On the contrary, heme environment of cytochrome c (cyt. c), a ubiquitous small soluble electron transfer protein carrying one heme cofactor, is known to have relatively rigid structure for efficient low driving force electron transfer [4]. Therefore, comparative study on the dynamics of ligand rebinding in these proteins offers a unique chance to understand the relation between protein dynamics and function [4]. In particular, the functional role of heme pocket in Mb and Hb can be assessed by probing deligated CO that resides in hydrophobic interior of the protein immediately after photodissociation before escapes to the solvent.

Experimental Methods

Two identical home-built optical parametric amplifiers, pumped by a Ti:S amplifier, are used to generate a visible pump pulse and a mid-IR probe pulse. The broadband transmitted probe pulse is detected with a 64-elements $N_2(l)$-cooled HgCdTe array detector. The instrument response function is typically 150 fs. In native cyt. c, the heme iron is coordinated by two internal axial ligands, His-18 and Met-80, and can not bind CO. Carboxymethyl cyt. c (cm cyt. c), a chemically modified form of cyt. c, possesses a vacant sixth site to the heme iron that is available to bind external ligands. Cm cyt. c was prepared as described in the literature [5]. About 5 mM of cm cyt. c was prepared by adding 0.5 M KCN and then 0.5 M bromoacetic acid to 10 mM cyt. c dissolved in a pD 7.4, 0.1 M phosphate buffer. The mixed solution was incubated overnight to ensure complete carboxymethylation of the sulfur atom of Met-80 after detachment of Met-80-iron bond. The 5-coordinate protein was reduced to the ferrous form by ten-fold equivalent sodium dithionate and combined with CO that was bubbled into the solution for about half an hour. The ferrous CO adduct (CO-cm cyt. c) was confirmed by Soret peak at 414 nm. UV-Vis and FT-IR spectroscopy show that the integrity and concentration of the sample are maintained throughout the experiments. During data collection the sample cell was rotated sufficiently fast so that each photolyzing laser pulse illuminated a fresh volume of the sample. The temperature of the sample cell was kept at 283 ± 1K.

Results and Discussion

Figure 1 shows transient absorption spectra of stretching mode of ^{13}CO bound to (right side) and photolyzed from (left side) cm cyt. c in D_2O at 283 K following excitation with 575 nm pulse. The negative-going features arise from the loss of bound CO and are denoted to A state. The bleach appears faster than the time resolution of our instrument suggesting that ultrafast photodissociation of CO in cyt. c as observed in Mb and Hb. While the broad bleach spectra have a single feature, there is a small but significant spectral evolution (see the peak shift of the bleach in the figure), suggesting that the single feature consists of several spectral components. At least four Gaussians were required to fully describe the spectral evolution of the bleach, consistent with four distinct bands observed in NO bound cm cyt. c (data not shown), indicating that the protein has at least four conformations differing near the bound site. Rebinding kinetics of CO to cm cyt. c, monitored by probing the intensity change of the CO stretching mode, exhibits slight conformation dependence and is much faster than that to MbCO (see the inset in the figure 1). It is even faster than that to a model heme, microperoxidase-8-CO (MpCO), a fragment of cytochrome c consisting of a heme with an 8 amino acid peptide that covalently linked to a heme through a histidine. Mp mimics the heme protein structure in the absence of a specifically organized distal pocket and is an excellent model system for obtaining the reaction characteristics of the heme free from conformational change and protein structure. The geminate rebinding (GR) yield and rate of CO to Mp increases with the viscosity of the solvent (data not shown), suggesting that GR of CO is enhanced by retardation of its diffusion. Similarly, the organized protein matrix near the active site in cm cyt. c serves as an efficient trap for the dissociated CO, thereby accelerating GR of CO to the heme. More than 82% of the dissociated CO geminately rebinds to cm cyt. c, consistent with 89% GR yield measured at 4 ns [4].

The positive going features, representing vibrational spectra of CO photolyzed from CO-cm cyt. c (left side of te figure 1), appear in less than 0.3 ps after photolysis indicating that they arise from CO located very near the Fe atom of the heme. They are similar to but different in detail from CO residing in the primary docking site after photolysis of MbCO [1]. The spectral features can be decomposed into two bands (denoted to C states) plus their hot bands. One band appearing later delay (denoted to C_s) is very similar to that of CO dissolved in water and is ascribed to the ligand escaped to bulk solution. Evidently a fraction of the dissociated CO reaches bulk solution in ps time scale and heme environments of cm cyt. c has direct ligand channels to the solvent. According to a solution NMR structure of ligand-bound cyt. c [6], portion of the distal site is exposed to the solvent. This exposed channel is likely used as a direct pathway into the solvent for the dissociated CO. The other band (denoted to C_1) is asymmetric in shape and represented by a sum of two Gaussians. It decays in less than 1 ns suggesting that almost all the dissociated CO in C_1 band geminate rebinds in ps time scale. The asymmetric shape of the C_1 band indicates that it likely arises from CO in various conformations of the protein and/or several cavities in a protein. It appears that, while the dissociated CO toward the open channel escapes, the entire trapped CO in cavities of cm cyt. c geminately rebind within 1 ns. Efficient GR of trapped CO in cm cyt. c is a stark contrast to the behavior observed in Mb, in which the dissociated CO resides in the primary docking site located about 3 Å from the binding site [3]. Clearly the heme pocket docking site of Mb is structured such a way that suppresses GR of CO while holding CO close proximity of the Fe atom for 180 ns. Heme pocket docking site indeed plays a functionally important role in efficient expulsion of toxic CO in Mb.

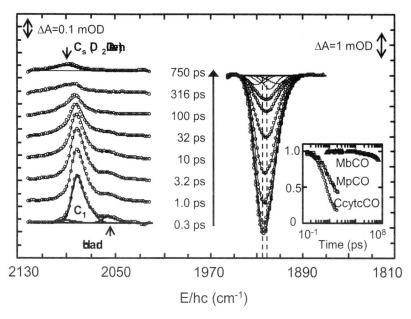

Fig. 1. Representative time-resolved spectra of CO after photolysis of CO-cm cyt. c in D_2O at 283 K. (Inset) The survival probability of deligated heme after the photolysis of CO.

Acknowledgements. This work was supported by the Korea Science and Engineering Foundation (KOSEF) grant funded by the Korean government (MOST) (R01-2007-000-20853-0) and Korea Research Foundation Grant funded by the Korean Government (MOEHRD, Basic Research Promotion Fund) (KRF-2007-C00146-000-I00694).

1   M. Lim, T. A. Jackson, P. A, Anfinrud, Nature Struc. Biol., Vol 4, 209 (1997).

2   E. Henry, J. H. Sommer, J. Hofrichter, W. A. Eaton, J. Mol. Biol., Vol 166, 443 (1983).

3   F. Schotte, J. Soman, J. S. Olson, M. Wulff, P. A. Anfinrud, J. Struc. Biol., Vol 147, 235 (2004).

4   G. Silkstone, A. Jasaitis, M. H. Vos, M. T. Wilson, Dalton Trans., Vol 3489, 3489 (2005).

5   A. Schejter, P. George, Nature Vol 206, 1150 (1965).

6   Y. Yao, C. Qian, K. Ye, J. Wang, Z. Bai, W. Tang, J. Biol. Chem., Vol 7, 539 (2002).

Real-time observation of the bond length modulation of carbon double bond during the photoisomerization of bacteriorhodopsin

Takayoshi Kobayashi[1,2,3,4], Atsushi Yabushita[3]

[1] JST, ICORP, Ultrashort Pulse Laser Project, 3 Bancho-Builing, 5 Banchi, 3 Bancho, Chiyoda-ku, Tokyo, 102-0075, Japan
[2] Department of Applied Physics and Chemistry and Institute of Laser Research, The University of Electro-Communications, 1-5-1, Chofugaoka, Chofu-shi, Tokyo 182-8585, Japan
[3] Department of Electrophysics, National Chiao-Tung University, 1001 Ta-Hsueh Road, Hsinchu, Taiwan 30050, R.O.C.
[4] Institute of Laser Engineering, Osaka University, 2-6 Yamada-oka, Suita, Osaka 565-0871 Japan
E-mail : kobayashi@ils.uec.ac.jp

Abstract. The observation of the real time frequency of C=C stretching mode shows that the bond length is modulated in the order of 10mÅ by torsion of the $C_{13}=C_{14}$ double bond with a period of 200 fs.

Introduction

Bacteriorhodopsin, bR568, undergoes the following photochemical cycle:

$$bR_{568} \xrightarrow{h\nu} H\,(\tau < 50\,fs) \rightarrow I_{460}\,(50\,fs < \tau < 100\,fs) \rightarrow J_{625}\,(100\,fs < \tau \lessapprox 1\,ps) \rightarrow K_{610}$$
$$\rightarrow L_{543} \rightarrow M_{412} \rightarrow N_{560} \rightarrow O_{640} \rightarrow bR_{568}$$

Here, H denotes the Franck-Condon excited state of bR_{568} with an all-trans retinal Schiff base. The subscripts represent the absorption maxima of the species. Several vibrational spectroscopy studies using ultrashort pulses [1-5] show that H and I_{460} are excited-state species with a chromophore of an all-*trans* conformation and that K_{610} is a ground-state species with the chromophore already converted to the 13-*cis* conformation. However, primary molecular processes between the H and K_{610} states have been controversial.

Two models have been proposed for the photoisomerization process, the 2-state [6-8] model and the 3-state model [9-12]. However, recent data from the pump-probe experiments on bR_{568} shows that there is no spectral change in the stimulated emission spectrum during the period of the exited state lifetime (30 fs - 1 ps delay time region) [9-11]. Also spontaneous fluorescence lifetime measurements [12] show that there is almost no dynamic Stokes shift during the period of the exited state lifetime. If bR follows the 2-state model, K state should appear even during the period of the exited state lifetime, which makes the stimulated emission spectrum modification and dynamic Stokes shift. Therefore, these experimental results cannot be explained by the formerly accepted 2-state model but instead support the 3-state model [13]. Pump-probe experiments for bacteriorhodopsin analogues containing synthetic $C_{13}=C_{14}$ locked chromophores with *cis* and *trans* configurations [14,15] showed the appearance of an I_{460}–like species within 30 fs. The close resemblance in the initial transient spectral evolution of both native and the artificial pigments indicates that the ultrafast process from H to I_{460} does not involve $C_{13}=C_{14}$ torsion, so I_{460} is still on all-*trans* conformation. Furthermore, recent papers claimed that the first event is not the *trans-cis* isomerization but a skeletal stretching from the H state [14-16]. Recent *ab-initio* calculations of photoisomerization dynamics have suggested that skeletal deformation takes place about 50 fs before $C_{13}=C_{14}$ torsion takes place [17,18].

Results and Discussion

Transient absorption spectra of bacteriorhodopsin were measured in broadband using sub-5-fs visible laser pulse generated by non-collinear optical parametric amplifier. Fourier transformed spectra of the transient absorption spectra shows that the transient absorption signal is modulated by the several molecular vibrations modes including in-plane C=C-H bending coupled with C-C stretching mode ($1,150$-$1,250cm^{-1}$), hydrogen-out-of-plane (HOOP) mode (800-$1,050cm^{-1}$), and C=C stretching mode ($1,500$-$1,550cm^{-1}$). Spectrogram analysis of the transient absorption spectra shows real-time modulations of the molecular vibration frequencies (see Fig. 1).

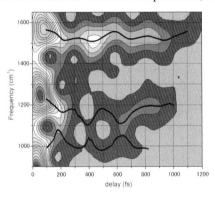

Fig. 1. Spectrogram calculated from the measured transient absorption probed at 585 nm.

The band around 1250 cm^{-1} splits into two peaks after 100 fs. The frequency of one peak gradually increases, while the frequency of the other peak gradually decreases and it becomes as low as 1130 cm^{-1} at 200 fs after excitation. The temporal shift of the frequency of the latter mode (C=C-H in-plane bending) is correlated with that of the HOOP band. These frequencies are modulated with 200-fs period of the C-C=C torsion. Noticeably, the peak positions of the HOOP mode and the in-plane C=C-H bending modes are well separated in the <100 fs region, but not in the 150-200 fs region. These mode frequencies approach each other and merge into a single peak, since there is neither an in-plane nor an out-of-plane mode in a distorted configuration. The two modes reappear in the slightly longer than 200 fs region, planarity is recovered. This suggests that the *trans*→13-*cis* isomerization occurred within <200 fs according to the 2-state model. However the product K_{610} has not yet been formed. Therefore, the coincident increase and decrease of the frequencies of the HOOP and in-plane C=C-H bending modes strongly indicate that the torsion modifies both the out-of-plane and in-plane bending-mode frequencies with the period of 200 fs.

Molecular vibration frequency of the C=C stretching mode also modulated associated with the torsion mode during the photoisomerization. When we utilize the empirical equation representing the relation between the bond length and frequency [19] the change of frequency is estimated to be related to change in the bond length of

$C_{13}=C_{14}$ double bond. The equation is given by $v = C_i\left[a(P+1)(\chi/d)^{3/2}+b\right]^{1/2}$, where $a = 1.67\times10^5, b = 0.30\times10^5, \chi = 2.5, C_i = 1.66, d = 1.489 - 0.151P$. When the carbon bond is single bond and double bond, P is 0 and 1, respectively. The observed change of frequency of the C=C stretching mode revealed that the bond length of $C_{13}=C_{14}$ double bond was changed about 10mÅ. From these data the relations among the real-time torsion angle, bond order, change in the equilibrium distance between two carbon atoms composing $C_{13}=C_{14}$ double bonds can be discussed as shown in Fig. 2 Thus we could determine the C=C bond length during the torsion motion around the double bond with 1-fs resolution with about 0.01Å accuracy and precision.

Acknowledgements. This research is partly supported by a Grant-in-Aid for Specially Promoted Research (#14002003), the program for the Promotion of Leading Researches in Special Coordination Funds for Promoting Science and Technology from the Ministry of Education, Culture, Sports, Science and Technology. This work is also supported partly by the ICORP program of Japan Science and Technology Agency (JST) and the Grant MOE ATU Program in NCTU.

1   G. H. Atkinson, T. L. Brack, D. Blanchard and G. Rumbles, Chem. Phys. **131**, 1-15 (1989).

2   R. van den Berg, D. J. Jang, H. C. Bitting and M. A. El-Sayed, Biophys. J. **58**, 135-141 (1990).

3   S. J. Doig, P. J. Reid and R. A. Mathies, J. Phys. Chem. **95**, 6372-6379 (1991).

4   R. Diller, S. Maiti, G. C. Walker, B. R. Cowen, R. Pippenger, R. A. Bogomolni and R. M. Hochstrasser, Chem. Phys. Lett. **241**, 109-115 (1995).

5   G. H. Atkinson, L. Ujj and Y. Zhou, J. Phys. Chem. A **104**, 4130-4139 (2000).

6   L. A. Peteanu, R. W. Shoenlein, Q. Wang, R. A. Mathies and C.V. Shank, Proc. Natl. Acad. Sci. USA **90**, 11762-11766 (1993).

7   R. W. Shoenlein, L.A. Peteanu, Q. Wang, R. A. Mathies and C. V. Shank, J. Phys. Chem. **97**, 12087-12092 (1993).

8   R. A. Mathies, C. H. Brito Cruz, W. T. Pollard and C. V. Shank, Science **240**, 777-779 (1988).

9   K. C. Hasson, F. C. Gai and P. A. Anfinrud, Proc. Natl. Acad. Sci. USA **93**, 15124-15129 (1996).

10 G. Haran, K. Wynne, A. H. Xie, Q. He, M. Chance and R. M. Hochstrasser, Chem. Phys. Lett. **261**, 389-395 (1996).

11 F. Gai, K. C. Hasson, J. C. McDonald and P. A. Anfinrud, Science **279**, 1886-1891 (1998).

12 M. Du, and G. R. Fleming, Biophys. Chem. **48**, 101-111 (1993).

13 W. Humphrey, H. Lu, I. Logunov, H. J. Werner and K. Schulten, Biophys. J. **75**, 1689-1699 (1998).

14 Q. Zhong, S. Ruhman and M. Ottolenghi, J. Am. Chem. Soc. **118**, 12828-12829 (1996).

15 T. Ye, N. Friedman, Y. Gat, G. H. Atkinson, M. Sheves, M. Ottolenghi and S. Ruhman, J. Phys. Chem. B **103**, 5122-5130 (1999).

16 L. Song, and M. A. El-Sayed, J. Am. Chem. Soc. **120**, 8889-8890 (1998).

17 T. Vreven, F. Bernardi, M. Garavelli, M. Olivucci, M. A. Robb and H. B. Schlegel, J. Am. Chem. Soc. **119**, 12687-12688 (1997).

18 M. Garavelli, F. Negri and M. Olivucci, Am. Chem. Soc. **121**, 1023-1029 (1999).

19 R. H. Baughman, J. D. Witt and K. C. Yee, J. Chem. Phys. **60**, 4755-4759 (1974).

Electron Transfer in Photosynthetic Reaction Centers: Optimization in Model and Nature

Benjamin P. Fingerhut[1], Wolfgang Zinth[2], and Regina deVivie-Riedle[1]

[1] Department of Chemistry, Ludwig-Maximilians-Universität München,
D-81377 München, Germany
E-mail: Benjamin.Fingerhut@cup.uni-muenchen.de
[2] Institut für BioMolekulare Optik, Ludwig-Maximilians-Universität München,
D-80538 München, Germany
E-mail: Zinth@physik.uni-muenchen.de

Abstract. We discuss the principles of optimal charge separation processes in bacterial reaction centers. Non-adiabatic electron transfer (ET) theory is combined with a Darwinian optimization. Our results reveal that the ET cascade is stable with respect to severe perturbations in the last ET step to maintain efficient charge separation.

Introduction

The development and the application of photochemical solar energy conversion processes is an important alternative approach to the exploitation of fossil energy carriers. Thereby the basic operational principles in the natural photo solar energy converting system may serve as a guideline. Despite the diversity of photosynthetic organisms functional principles are maintained as building blocks [1] for the light reaction and especially for energy conversion. They comprise the biophysical processes of light absorption, transfer of electronic excitation energy, initial charge separation and directed charge transfer. We focus on the initial charge separation to generate and maintain a transmembrane charge gradient.

The systems considered are bacterial reaction centers (RC) which can be regarded as prototypes of light converting nano- machines, optimized during evolution. We present a flexible model to simulate energy converting photosynthetic reaction centers. For the kinetics of the primary energy conversion of light into a potential gradient, we assume Marcus-type electron transfer processes. To optimize the conversion process in the theoretical framework, we set up, in analogy to the Darwinian evolution, a genetic algorithm (GA).

Setup of Rate Equations for the Genetic Algorithm

In non-adiabatic electron transfer (ET) theory the rate constants for the electron transfer from an excited donor i to an acceptor j can be calculated from the Fermi Golden Rule

$$\gamma_{ij} = \frac{1}{\tau_{ij}} = \frac{2\pi}{\hbar} FC_{HT,ij} V_{ij}^2 \qquad (1)$$

The coupling strength V_{ij} as well as the Franck-Condon factors $FC_{HT,ij}$ of the vibrational wave functions of the donor i and acceptor j include a set of free variables, the free energy ΔG_{ij} and the tunnelling distance d_{ij}. For the forward ET we consider zeroth-order overlap between neighboring chromophores. For the loss channel of charge recombination additionally first-order terms, which can be identified as Super-Exchange coupling, and second-order terms are considered (a detailed description of the model for ET in bacterial RC will be given in a forthcoming publication [2]). We calculate the time evolution of the probability by solving the rate equations for the number of states m.

$$\frac{dN_i}{dt} = \sum_{j=0}^{m} \gamma_{ij} N_j \quad \text{with} \quad \gamma_{ii} = -\sum_{j \neq i} \gamma_{ji} \tag{2}$$

The variables ΔG_{ij} and d_{ij} are optimized with a single-objective GA for maximum quantum yield Φ and by use of a multi-objective GA (NSGAII [3]) for Φ and maximal lifetime of the charge separated (CS) state or maximal lifetime τ_{34}, respectively. The search space is limited according to the physical boundary conditions ($\Delta G_{ij} \geq$ 1 kJ/mol, $d_{ij} \geq 2$ Å, membrane size $L \leq 40$ Å).

Results and Discussion

The optimal ET rates in the parameter regime of different bacterial RC's were calculated ($m = 6$). We concentrate on the two organisms *Rhodobacter sphaeroides* (*Rb. sph.*) and *Blastochloris viridis* (*B. viridis*) as prototypes in the family of bacterial RC's. They are characterized by the deactivation rate γ_{chem}, the available chemical potential ΔG_{chem} and the excited state lifetime τ_{P*} (Table 1). The solution of nature is also found as optimum by the GA. The lifetime of the photoexcited special pair in *B. viridis* of 3.2 ps is in very good agreement with the experimental value of 3 ps [4]. Like in nature the second ET step is the fastest (0.9 ps).

The investigation shows that this reaction sequence is the optimal solution for the biological system consisting of four chromophores. The results confirm the basic assumptions on the structure and the reaction principles of the model system and document that photosynthetic electron transfer is guided by stepwise Marcus theory.

Table 1. Optimized quantum yield Φ and optimal ET rates in bacterial reaction centers calculated by single-objective GA: algorithm parameters are individuals - 120, reproducing individuals - 100, populations - 500, mutation rate - 0.001, crossover rate - 0.93

	B. viridis	*Rb. sph.*	*artificial B. viridis*
organism specific parameters			
γ_{chem} (cytochrtome c)	190 ns	1 μs	1 μs
excitation wavelength λ	957 nm	861 nm	957 nm
ΔG_{chem}	0.88 eV	0.70 eV	0.88 eV
$\tau_{P*} [ps]$	297.80	500.00	297.80
optimized parameters			
Φ	0.9761	0.9853	0.9696
τ_{12} [s]	$3.23 * 10^{-12}$	$3.92 * 10^{-12}$	$3.75 * 10^{-12}$
τ_{23} [s]	$9.08 * 10^{-13}$	$1.45 * 10^{-12}$	$2.12 * 10^{-12}$
τ_{34} [s]	$5.06 * 10^{-11}$	$7.66 * 10^{-11}$	$9.38 * 10^{-11}$

Charge separated (CS) state lifetime. Beyond the central ET motif, which is reproduced well by the GA, mayor deviations occur in the ET lifetime from Bacteriopheophytine H_A to the Quinone Q_A (exp. 200 ps [1] compared to τ_{34}=51-94 ps) and hence in the lifetime of the CS state (τ_{40}). In order to inspect the sensitivity of the ET cascade on the ET rates we performed optimizations of Φ by use of the NSGAII-algorithm in the parameter regime of *B. viridis*. Our calculations show that $\Phi > 97\%$ can be maintained under the experimentally observed conditions (Fig. 1, top). The Pareto front exhibits only slight dependence on τ_{34} and Φ of 95 % can be achieved with τ_{34}-lifetimes of up to 1 ns.

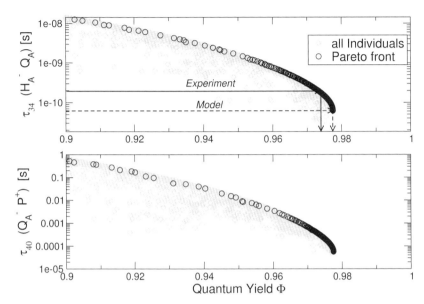

Fig. 1. Pareto front of quantum yield Φ and H_A^--Q_A lifetime (τ_{34})(top) resp. charge separated state lifetime Q_A^--P^+ (τ_{40})(bottom), optimized by multi-objective GA (NSGAII): the algorithm parameters are individuals - 600, generations - 500, mutation rate - 0.01, crossover rate - 0.93

The situation is comparable when τ_{40} is introduced as a second objective (Fig. 1, bottom). The short CS lifetime of 35.3 μs corresponds to the optimal solution, found by the GA. This lifetime can be exceeded up to tens of milliseconds on cost of $\Delta\Phi < 2\%$.

Conclusions

Within our flexible model we are able to describe the ET in bacterial RC. The optimal solution, found by evolutionary algorithms reflects the solution for ET realized in nature. By the optimization of Pareto fronts, we are able to show that the four chromophore ET cascade is stable to large distortions in the last ET step.

Acknowledgements. The authors would like to acknowledge the DFG for financial support through the SFB749 and the Munich Center of Advanced Photonics (MAP).

1 W. Zinth and J. Wachtveitl, ChemPhysChem 6, 871, 2005.
2 B. P. Fingerhut, W. Zinth and R. deVivie-Riedle, manuscript in preparation.
3 K. Deb, S. Agrawal, A. Pratap and T. Meyarivan, IEEE Trans. Evol. Comput. 6, 182, 2002.
4 P. Huppmann *et al.*, J. Phys. Chem. A 107, 8302, 2003.

Coherently Controlled Release of Drugs in Ophthalmology

Tiago Buckup, Jens Möhring, Volker Settels, Jens Träger, Hee-Cheo Kim, Norbert Hampp and Marcus Motzkus

Physikalische Chemie, Philipps Universität Marburg, D-35043 Marburg, Germany.
E-mail: motzkus@staff.uni-marburg.de

Abstract. The photocleavage of a coumarin derivative dimer is a promising mechanism for laser controlled drug release in medical applications. We investigate the efficiency of the two-photon induced cleavage in open- and closed-loop control schemes.

Introduction

Coherent control as a field of modern research has diversified greatly in recent years. In this contribution we show a novel application of coherent control in the field of ophthalmology, where tailored pulses are used for efficient drug release in the treatment of cataract. Cataract is an opacification of the human eye lens and is the most important reason for acquired blindness in the world. Although the treatment with the replacement of the opaque lens by polymeric intraocular lens (IOL) is well established, the major complication of cataract surgery is the regrowth of epithelial cells in the visual axis. Recently, it has been proposed this secondary cataract can be treated if a photosensible drug delivery device is linked to the polymeric IOL (Fig. 1).[1-3] The dimer of the coumarin molecule is an ideal release system because it can be cleaved via two-photon excitation using intense visible light. With a two-photon excitation step resonant in the UV, the cleavage of coumarin dimer and, consequently, the drug release can be precisely controlled, while, at the same time, the strong UV absorption of the human eye is circumvented. In order to turn this promising technique into a viable tool for medical technology, it is important to understand how the reaction efficiency can be enhanced. In this work we investigate in open- and closed-loop control schemes the optimal pulse solution to achieve an efficient two-photon induced cleavage.

Experimental Methods

The optimization of the dicoumarin cleavage was performed using the shaped emission of a noncollinear optical parametric amplifier at 530 nm in an accumulative absorption setup based on a capillary as probe holder. In this setup, a defined sample volume is irradiated for few seconds with a tailored pulse and the reaction conversion efficiency is probed at 310 nm, the absorption maximum of the coumarin monomer. This accumulative exposure to thousands of laser pulses increases the sensitivity of the turnover measurement in quantum control experiments dramatically [4]. After one laser pulse is evaluated, the sample volume is exchanged and a new pulse form can be tested. In order to take in account the pulse duration modification due to phase shaping, a two-photon dependent signal was measured in a SiC photodiode. The coumarin dimer was obtained as described in Ref.[5] and dissolved in HPLC-graded isopropanol to yield a 3 mM solution.

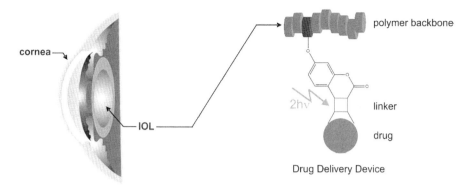

Fig. 1. Principle of two-photon absorption controlled drug delivery from IOL. The polymer backbone serves as material for the IOL, while the linker system can be cleaved with tailored femtosecond pulses via a two-photon transition. The drug can be this way more efficiently delivered.

Results and Discussion

Pulse optimizations using an evolutionary algorithm (EvoAlg) were performed employing different phase parameterizations. In order to explore the parameter space, several fitness functions were tested in the EvoAlg using relations between the relative turnover of the reaction and the two-photon dependent signal (TPA) from a SiC diode. We obtained that closed-loop optimizations targeting, for example, the maximization of reaction turnover and minimization of the two-photon photodiode signal shows that Fourier-limited pulse (~18 fs) is not the best suitable pulse to steer the reaction. Figure 2a shows the density of individuals from multiple EvoAlg runs with different target functions. The density plot shows a robust optimization of the turnover rate against the TPA signal for low TPA values, indicating an improved quantum yield for longer pulses. This impression is further supported by principle component analysis of the evolutionary phase functions. It should be noted that there is no turnover offset present in the energy dependency also shown in the figure. The two main branches in the figure can be associated with two target definitions respectively. Simple turnover optimization leads finally to a approximately transform limited case at high TPA values, whereas |Turnover-TPA| x (TPA+Turnover) yields a clear phase effect compared to the amplitude effect illustrated by the energy dependence.

The closed-loop results are corroborated by open-loop scans using defined parameterizations. Figure 2 shows the enhancement of the ratio between the reaction yield and the two-photon photodiode signal for several values of the quadratic phase. For example, at $\phi = 3000$ fs^2, which expands our initial sub-20 fs to approximately 600 fs, the ratio reaches about a factor 8. This result shows that though fewer molecules are excited via two-photon absorption with the stretched pulse, the reaction yield becomes about 8 times more efficient compared to a transform limited pulse. This is in agreement with our previous results where the reaction yield was compared under femtosecond and nanosecond excitation. [3] The optimization can be rationalized as a repump mechanism (Figure 2 b), where a longer pulse is able to reinduce the dissociation process for excited dimers that did not react with the initial two-photon excitation. Open-loop scans suggest that the optimal pulse duration for maximization of the reaction quantum yield (relative turnover normalized by the two-photon signal from a SiC diode) lies between 1-2 ps.

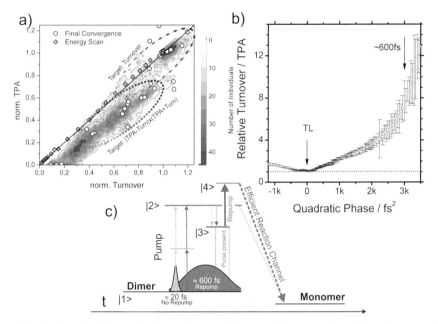

Fig. 2. a) Density of individuals during multiple Evolutionary optimizations. The target "Turnover" resembles approximately the TL case, whereas the more complex target function leads to optimization of the turnover to TPA ratio. **b)**Ratio between the total reaction yield and the two-photon absorption signal for different values of the quadratic phase. Longer pulses cleaves more efficiently the dimer of Coumarin. **c)** Schematic energy level diagram showing the repump model. For pulses stretched in time the excited dimers which were not initially cleaved are re-excited and eventually react to form monomers.

Conclusions

The potential of the two-photon induced cleavage of coumarin dimer was already shown in the treatment of vision diseases. The efficiency of the treatment is strongly dependent on the cleavage efficiency. Here we have shown that by applying coherent control, the reaction yield can be enhanced when the excitation pulse is stretched in time by almost an order of magnitude, making this way laser treatment in ophthalmology much safer.

1   J. Trager, H. C. Kim, and N. Hampp, in Nature Photonics, Vol. 1, 509, 2007.

2   H.-C. Kim, S. Kreiling, A. Greiner, and N. Hampp, in Chemical Physics Letters, Vol. 372, 899, 2003.

3   T. Buckup, J. Dorn, J. Hauer, S. Hartner, N. Hampp, and M. Motzkus, in Chemical Physics Letters, Vol. 439, 308, 2007.

4   F. Langhojer, F. Dimler, G. Jung, and T. Brixner, in Optics Letters, Vol. 32, 3346, 2007.

5   S. Haertner, H.-C. Kim, and N. Hampp, in Journal of Photochemistry and PhotoBiology A:Chemistry, Vol. 187, 242, 2007.

Light Harvesting, Energy Transfer and Photoprotection in the Fucoxanthin-Chlorophyll Proteins of *Cyclotella meneghiniana*

Nina Gildenhoff[1], Sergiu Amarie[1], Anja Beer[2], Kathi Gundermann[2], Claudia Büchel[2], Josef Wachtveitl[1]

[1] Institut für Physikalische und Theoretische Chemie, Max von Laue-Strasse 7, Goethe Universität Frankfurt, 60438 Frankfurt am Main, Germany
[2] Institut für Molekulare Biowissenschaften, Siesmayerstraße 70, Goethe-Universität Frankfurt, 60438 Frankfurt am Main, Germany
E-mail: wveitl@theochem.uni-Frankfurt.de

Abstract. The excitation energy transfer and the protective role of diadinoxanthin and diatoxanthin in two different fucoxanthin-chlorophyll-proteins have been investigated using femtosecond transient absorption spectroscopy.

Introduction

Fucoxanthin-chlorophyll proteins (FCPs) are found in diatoms and brown algae and their function is light harvesting as well as protection against photodamage. Like light harvesting complexes (LHCs) in higher plants the FCPs are membrane intrinsic antenna proteins but their pigmentation differs from the LHCs. Chlorophyll *b* is replaced by Chlorophyll *c* and the major carotenoid is fucoxanthin. The chlorophyll (Chl) to carotenoid (Car) ratio is ~1 whereas LHCs contain far more Chls than Cars. FCPs also contain diadinoxanthin and diatoxanthin in substoichiometric amounts. Under high light conditions the amount of diatoxanthin increases by de-epoxidation of diadinoxanthin, it might thus act as a quencher for chlorophyll (Chl) fluorescence. Hence the photoprotection in diatoms might be based on a mechanism similar to the xanthophyll cycle in higher plants [1].

The centric diatom *Cyclotella meneghiniana* contains two different FCPs, which are discriminable by their oligomeric state [2]. FCPa consists of trimers whereas FCPb contains higher oligomers. In the present work FCPa and FCPb are studied by time resolved absorption and fluorescence spectroscopy. Additionally, selective excitation of fucoxanthin and diadinoxanthin at 500 nm and primarily fucoxanthin at 550 nm reveals different deexcitation pathways involving an intramolecular charge transfer (ICT) state.

Experimental Methods

FCPa and FCPb were purified from the diatom *Cyclotella meneghiniana* both from high light (HL) and low light (LL) cultures according to Beer et al. [1].
The time resolved measurements using femtosecond pump/probe technique were performed with a setup described before [3]. All FCP samples were excited with pulses centered around 500 nm and 550 nm respectively using a noncollinear optical parametric amplifier (NOPA) with pulse energies around 20 nJ. Single filament white light supercontinuum generated in a sapphire crystal served as probe light covering a spectral range of 450-1100 nm. To prevent multiple excitation of molecules and

degradation of the sample a continuous exchange of the sample between the laser pulses was achieved by continuously moving the cuvette. All FCPs were dissolved in a buffer (25mM Tris, 2mM KCl, 0,03% ß-DDM, pH 7.4) with an optical density of ~0.9/mm at 671 nm.

Results and Discussion

Stationary UV-vis-spectroscopy shows slight differences between FCPa and FCPb (Figure 1a), but both spectra contain the characteristic contributions of the pigments contained in the proteins: The Q_y band of Chl a appears around 670 nm and the Q_y-absorption of Chl c is visible as a small peak at 633 nm. The Soret bands of the chlorophylls appear below 475 nm. The absorption between 475 nm and 565 nm belongs to the absorption of the S_2 state of the Cars. In this region a shoulder at 486 nm is only visible in the case of FCPa which might reflect a higher amount of diadinoxanthin in FCPa [1].

Figure 1b) depicts the transient absorbance changes after excitation at 500 nm. The negative contribution (A) is assigned to the ground state bleach of the S_2 absorption band of fucoxanthin. At later times the excited state absorption (ESA) of Chl a (B) is dominating in this spectral region. The positive absorbance changes belonging to region C are due to the ESA of Cars and of Chl a at later times. This positive signal is superimposed by the evolution of a negative signal (D), which originates from the ground state bleach and stimulated emission signal of Chl a.

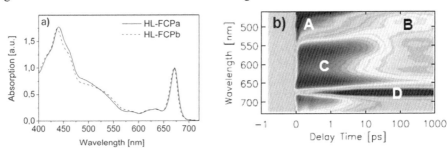

Fig. 1. Stationary absorption spectra of FCPa (—) and FCPb (---) (a) and transient absorbance changes of HL-FCPa excited at 500 nm (b). A: bleach of fucoxanthin S_2; B: ESA of Chl a; C: ESA of carotenoids and Chl a (at later times); D: bleach and SE of Chl a. The time axis is linear up to 1 ps and logarithmic for longer delay times.

The temporal response of the absorbance changes of HL-FCPa and HL-FCPb upon excitation of fucoxanthin and diadinoxanthin at 500 nm are in figure 2a) and 2b). The onset of the Chl a bleach signal at 678 nm appears exclusively due to the energy transfer from fucoxanthin/diadinoxanthin to Chl a, since the 500 nm excitation pulse does not directly excite the Chls. The bleach signal appears with a slight delay within the first few ps, i.e. the time needed for the excitation energy to reach the Chl a. The comparison of FCPa und FCPb at 678 nm for long delay times shows pronounced differences in the Chl a dynamics.

For FCPa and FCPb samples we obtained pronounced differences between the two excitation wavelengths. The most prominent differences appeared in the Chl a ESA region and in the Car ESA region around 550 nm, where we assume the ESA of the S_1 state of diadinoxanthin and diatoxanthin. Furthermore, we needed five time constants ($\tau_1 < 150$ fs, $\tau_2 = 0.6$ ps, $\tau_3 = 3.1$ ps, $\tau_4 = 25$ ps and $\tau_5 = \infty$) to achieve a good fit for the data obtaind from the 500 nm excitation of FCPa. The data obtained from the 550 nm excitation of FCPa could be fitted with very similar time constants and associated fit amplitudes except for $\tau_4 = 25$ ps, which was not required for a good fit.

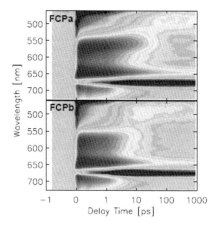 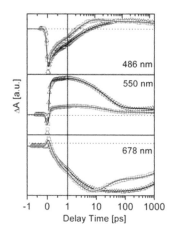

Fig. 2. Transient absorbance changes of HL-FCPa (Δ) and HL-FCPb ($\square$) excited at 500 nm (left) and transients at selected probing wavelengths. The solid lines represent the results from the global fit analysis.

Conclusions

The differences in the Chl a lifetime together with small differences observed in the diadinoxanthin spectral region may be indicative for a different stoichiometry and association of the Cars in the trimeric FCPa than in the oligomeric FCPb. This might refer to different functions of the two protein complexes within the photosynthetic apparatus.

We assume a hot EET from the vibrationally excited S_2 state of fucoxanthin to diadinoxanthin and diatoxanthin upon 500 nm excitation since there should be very little direct excitation of diadinoxanthin molecules due to absorption properties and stoichiometry. In the NIR data (data not shown) we could not confirm a slow S_1/ICT state of fucoxanthin [5] so we assign the $\tau_4 = 25$ ps time constant to the intrinsic lifetime of the S_1 state of diadinoxanthin.

For an improved band assignment of the different carotenoids contained in FCPs and a better understanding of the molecular reaction mechanisms, we intend to characterize the isolated pigments of the FCPs by time resolved spectroscopy.

1   A. Beer, K. Gundermann, J. Beckmann and C. Büchel, Biochemistry **45**, 13046-13053 (2006)
2   C. Büchel, Biochemistry **42**, 13027-13034 (2003)
3   S. Amarie, J. Standfuss, T. Barros, W. Kühlbrandt, A. Dreuw and J. Wachtveitl, J. Phys. Chem. B **111**, 3481-3487 (2007)
4   T. Polívka, I.H.M. van Stokkum, D. Zigmantas, R. van Grondelle, V. Sundström and R.G. Hiller, Biochemistry **45**, 8516-8526 (2006)
5   E. Papagiannakis, I.H.M. van Stokkum, H. Fey, C. Büchel, R. van Grondelle, Photosynth. Res. **86**, 241-250 (2005)

Primary Reaction Dynamics of Green Absorbing Proteorhodopsin WT and D97N Mutant Observed by fs Infrared and Visible Spectroscopy

Karsten Neumann[1], Mirka-Kristin Verhoefen[1], Ingrid Weber[2], Clemens Glaubitz[2], and Josef Wachtveitl[1]

[1] Institut für Physikalische und Theoretische Chemie, Johann Wolfgang Goethe-Universität, Max-von-Laue-Str. 7, 60438 Frankfurt am Main, Germany
[2] Institut für Biophysikalische Chemie, Johann Wolfgang Goethe-Universität, Max-von-Laue-Str. 9, 60438 Frankfurt am Main, Germany
E-mail: wveitl@theochem.uni-frankfurt.de

Abstract. We study the D97N mutant of the light driven proton pump proteorhodopsin and compare the results to experiments on the wild type at two pH values. The application of transient absorption spectroscopy in the visible and infrared spectral range provides detailed information on the first steps in the photocycle.

Introduction

Proteorhodopsin (PR) is a new member of the type I retinal binding protein family and functions as a light-driven proton pump [1]. Upon light excitation the chromophore isomerizes (from all-*trans* to 13-*cis*) leading to the first ground state intermediate K (PR$_K$) followed by a series of distinguishable intermediates similar to that of bacteriorhodopsin. During this so called photocylce a proton is transported across the membrane.

In PR, the primary proton acceptor Asp-97, which plays a crucial role in this process, has an unusually high pK$_a$ of about 7.6. Measurements of the pumping activity via BLM techniques showed that the pumping direction can be inverted by lowering the pH [1]. However, the issue of variable vectoriality is still under discussion, since the inversion of H$^+$ pumping could not be confirmed by photocurrent measurements of oriented membranes [2]. Studies of the primary dynamics of PR pointed out that already the first step of the photocycle is highly affected by the pH value [3]. A branched reaction model was derived, which explains the different dynamics for pH 9 and pH 6. In the present study we investigate the primary dynamics of the green absorbing variant of wild type (PR wt) as well as the D97N mutant in the first 2 ns combining fs time resolved spectroscopy in the IR and visible spectral range to gain a more detailed picture of the primary reaction dynamics.

Experimental Methods

Expression and purification of green absorbing PR and the D97N mutant was performed as described in [4,5]. Both the experimental setups and the data analysis have been described elsewhere [5]. All samples were excited at a central wavelength of 525 nm with pulse energies around 100 nJ. The measurements were carried out in D$_2$O buffer solution (500 mM NaCl and 0.1 % n-dodecyl-ß-D-maltoside, 20 mM TRIS).

Results and Discussion

The spectral evolution of PR wt at pD 9.2 (left), pD 6.4 (middle) and PR D97N at pD 7.4 (right) in the spectral range of 440 nm to 750 nm is shown in Figure 1 (top). The general spectral features are similar to those obtained for the samples in H_2O [3,4]. The global analysis reveals a biexponential decay of the excited state and the formation of the PR_K intermediate at long delay times. The derived time constants are significantly greater for the experiments under D_2O conditions than under H_2O conditions.

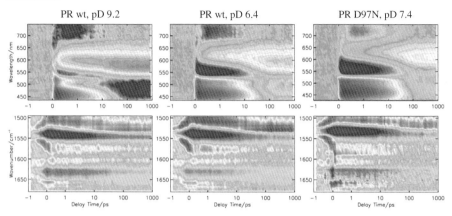

Fig. 1. Transient absorbance changes of solubilized wild type PR in D_2O at pD 9.2 (left) and pD 6.4 (middle) and D97N mutant (right) after photoexcitation at 525 nm. The time axis is linear in the range of -1 ps to +1 ps and logarithmic for longer delay times.

Vis pump / IR probe spectroscopy (Fig. 1 bottom) was carried out in two spectral regions. In the range of 1500 cm^{-1} to 1580 cm^{-1} the C=C stretching vibration of the retinal is observed, the region of 1590 cm^{-1} to 1650 cm^{-1} shows contributions of the C=N stretching vibration of the chromophore and the response of the amide I mode of the protein backbone. Apart from the shift of the C=C absorption band the results for PR wt at pD 9.2 and pD 6.4 as well as PR D97N at pD 7.4 show no significant deviations. This is further substantiated by the global fit analysis: the decay associated spectra (DAS) represent the same processes and the associated decay time constants are in the same range, thus the interpretation is analogous [5]. The DAS of the infinite time constant τ_∞ reflects the $PR_K - PR$ difference spectrum and the shortest time constant τ_1 is assigned to the formation of this photoproduct, which is in agreement with a recent study by Amsden et al. [7]. The two subsequent time constants describe the loss of the initial bleach signal of the C=C and the C=N stretching vibration and are thus attributed to the recovery of the PR ground state vibrations due to vibrational cooling. A negative band around 1660 cm^{-1} appears on the picosecond time scale and can be ascribed to the response of the protein backbone to the retinal isomerization. Whereas for both pD values of PR wt all time constants are the same within the experimental error, the shortest time constant τ_1 is about a factor of 2 greater for PR D97N.

Conclusion

The $PR_K - PR$ difference spectra of PR wt in the IR are found to be virtually independent of the pD value. Consequently, the quantum yield of K formation can readily be compared in the infrared data, whereas the different superposition of pD dependent bands in the visible makes this evaluation difficult. The comparison reveals that the quantum yield of photoproduct formation is essentially the same for both pD values. This leads to the conclusion that shifted ground state absorption bands cause the differences in the $PR_K - PR$ difference spectrum for changed pD values in the visible spectral range. The quantum efficiency of the primary reaction, i.e. the formation of the K intermediate, is therefore not affected by the protonation state of the primary proton acceptor Asp-97 [5].

Due to the fact that the DAS and decay time constants for the IR data of both pD 6.4 and pD 9.2 are substantially identical, it must be concluded that the processes observed in the IR are not affected by the pD value. The comparison of PR D97N to PR wt reveals that the time constants τ_2 and τ_3 are conserved, whereas τ_1 is a factor of 2 greater. This is in perfect agreement with the assumption that τ_2 and τ_3 are connected to the recovery of the ground state vibrations due to vibrational cooling within the retinal and the surrounding protein. The process related to τ_1 is mainly correlated with the formation of the K-intermediate and shows a significant difference between PR D97N and PR wt. This might indicate that the photo induced isomerization of the retinal is slowed down in the PR D97N sample, whereas it is independent of the protonation state of Asp 97 in PR wt. A definite statement for this behaviour cannot be given at this point.

The transient absorption data in the visible show a pronounced pD-dependence. In addition, the comparison of the spectra recorded in D_2O and H_2O reveal a kinetic isotope effect for PR wt and D97N. The time constants associated to the biphasic decay of the excited state in D_2O are increased by a factor of about 2.3 for PR wt at pD 9.2, a factor of about 1.6 for PR wt at pD 6.4 and a factor of about 1.3 for PR D97N at pD 7.4. Obviously, the isotope effect is less pronounced under acidic conditions or if the primary proton acceptor Asp-97 is mutated. This is explained by a H-bonding network in the binding pocket of PR which is strongly influenced by Asp-97 and catalyzes the primary dynamics. In contrast to the IR, where only minor differences between the PR samples are found, the data obtained in the visible significantly differ between the various samples. This supports our conclusion that both experiments yield complementary information and are therefore not directly comparable to each other.

In summary, it can be concluded that the control of the reaction kinetics is more complex and depends on electrostatic interactions as well as the assembly of hydrogen bonds in the retinal binding pocket. Since in PR both the residues Asp-97 and Asp-227 are assumed to interact with the proposed water cluster, subsequent investigations on the PR D227N mutant will give further insights into the mechanism of the retinal isomerization.

1   T. Friedrich *et al.,* Journal of Molecular Biology, **321**, 821, 2002.

2   A.K. Dioumaev *et al.*, Biochemistry, **42**, 6582, 2003.

3   M.O. Lenz *et al.*, Biophysical Journal, **91**, 255, 2006.

4   M.O. Lenz *et al.*, Photochemistry and Photobiology, **83**, 226, 2007.

5   K. Neumann *et al.*, Biophysical Journal, **94**, 4796, 2008.

6   J. J. Amsden *et al.*, Journal of Physical Chemistry B, **111**, 11824, 2007.

Photodynamics of a Collagen Model Peptide

Lisa Lorenz[1], Karsten Neumann[1], Ulrike Kusebauch[2], Luis Moroder[2] and
Josef Wachtveitl[1]

[1] Institute of Physical and Theoretical Chemistry, Goethe University Frankfurt
  Max von Laue Str. 7, 60438 Frankfurt, Germany
  E-mail: wveitl@theochem.uni-frankfurt.de
[2] Max-Planck-Institute of Biochemistry
  Am Klopferspitz 18, 82152 Martinsried, Germany
  E-mail: moroder@biochem.mpg.de

Abstract. *Trans/cis* isomerization of a specially designed collagen sample with an azobenzene clamp is examined by time-resolved spectroscopy. The bistable functionality of the azobenzene switch is conserved upon covalent attachment to the peptide, making this model collagen molecule suitable for investigating tertiary structure formation.

Introduction

Azobenzene derivatives have been widely used for triggering folding and unfolding of model peptides with well defined secondary structures such as β-hairpins or α-helices [1,2]. Due to the ultrafast isomerization of this photoswitch, conformational changes can be monitored with high temporal resolution. The chromophore is either inserted directly into the peptide backbone or attached as a clamp on the side chains of amino acid residues. It can be optimized in geometry, flexibility and polarity for triggering the reaction of interest.

A new and challenging aim would be to use this system for studying the folding process of tertiary structures. To explore these complex events in macromolecules such as natural collagen, small model peptides have been designed and synthesized that allow monitoring of elementary steps of folding and/ or unfolding of a triple helix by time-resolved spectroscopy [3]. For this purpose in the present study a specially designed thiol-reactive azobenzene-based crosslinker (AB-switch) is used, that clamps two mercapto-prolines of a collagen single strand (Figure 1). Three of these azobenzene-clamped strands combine into a triple helix that is stabilized by steric restrictions and hydrogen bonds. The structure of the peptide and the position of the AB-switch have been optimized for the formation of the triple helix in solution and the possibility to cause changes in the tertiary structure upon photoinduced isomerization [4,5].

First experiments on the dynamics of the photoswitch as chloride derivative and upon its incorporation into the collagen molecule have been performed by pump probe spectroscopy as proof of concept.

Experimental Methods

Time-resolved measurements in the visible and in the mid infrared regions have been performed with setups described in [6-8]. For the *trans/cis* isomerization the pump wavelength (λ_{exc}) has been adjusted by sum frequency mixing to selectively excite the $\pi\pi^*$-transition ($\lambda_{exc} = 345$ nm) or the $n\pi^*$-transition ($\lambda_{exc} = 430$ nm), respectively. Two dimensional movement of the sample and cw-illumination with a spectrally filtered mercury-xenon lamp (400-480 nm and 320-380 nm, respectively) guaranteed a con-

583

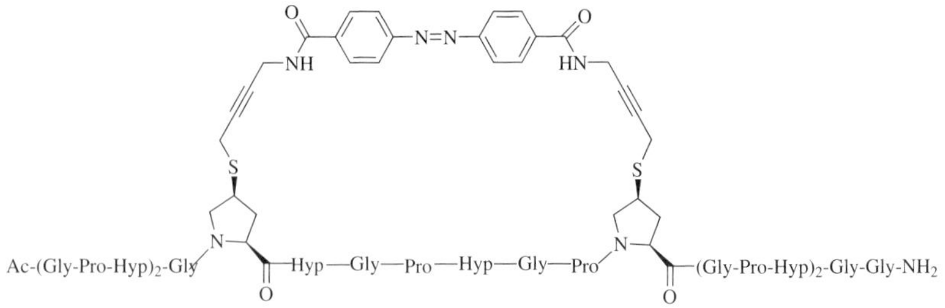

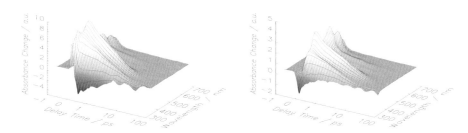

Figure 1: Molecular structure of the azocollagen single strand

stant concentration of the *trans/cis* isomer before every excitation pulse. Synthesis of the samples is described in [4]. The azobenzene chloride derivative has been dissolved in deuterated dimethyl sulfoxide (d_6-DMSO), whereas a 1:1 mixture of deuterated ethylene glycol (d_6-EG) and deuterated water (D_2O) has been used for the preparation of the collagen sample.

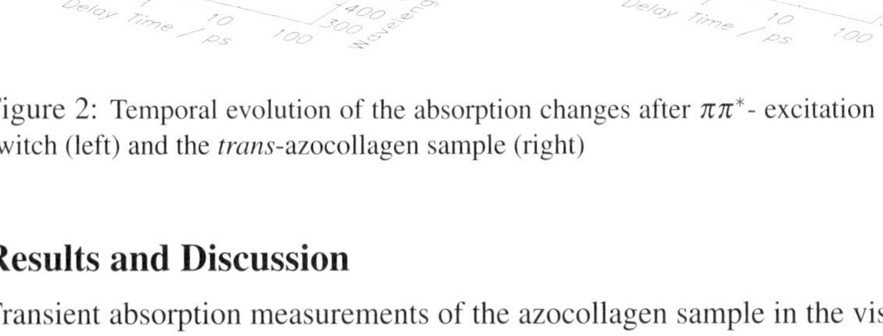

Figure 2: Temporal evolution of the absorption changes after $\pi\pi^*$- excitation of the *trans*-AB-switch (left) and the *trans*-azocollagen sample (right)

Results and Discussion

Transient absorption measurements of the azocollagen sample in the visible after $\pi\pi^*$-excitation show that the properties of the AB-switch are conserved upon binding to the peptide strand (Figure 2). Only minor differences between the dynamics of the chromophore itself and the azobenzene-peptide have been found between 340 and 680 nm. After photoexcitation at 345 nm the ground state bleach becomes visible on the blue side and the broad excited state absorption dominates the plot in the first ten picoseconds. Since this spectral region is neither sensitive to the peptide nor to the tertiary structure of the sample, measurements in the mid-infrared region are essential. At frequencies between 1750 and 1250 cm^{-1}, the peptide as well as the AB-switch show characteristic bands, which allow to monitor simultaneously the dynamics of the switch and the peptide after photoexcitation. Transient data of the *cis-to-trans* isomerization of the *cis*-AB-switch in the mid-IR from 1350 to 1750 cm^{-1} after excitation at 430 nm are shown in Figure 3. The experimental data can be fitted with a set of four exponential decay times (τ_D). Isomerization proceeds with $\tau_1 < 1$ ps and $\tau_2 = 8$ ps. $\tau_3 = 30$ ps can be assigned to vibrational cooling in the ground state. The fourth time constant which is kept fix at one millisecond is applied for fitting residual signals that occur due to isomerization.

Conclusion

The dynamics of the new azobenzene derivative have been investigated by time-resolved spectroscopy in the mid-IR and visible and the functionality of the photoswitch attached to the peptide could be demonstrated. Thus, the azocollagen appears to be highly promising for exploring early unfolding events of a collagen triple helix, which is the subject of ongoing experiments.

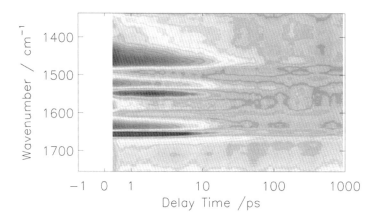

Figure 3: 2D-plot showing the absorption changes after $n\pi^*$-excitation of the *cis*-AB-switch

1   Schrader, T.E., Schreier, W.J., Cordes, T., Koller, F.O., Babitzki, G., Denschlag, R., Renner, C., Löweneck, M., Dong, S.-L., Moroder, L., Tava, P., Zinth, W., PNAS, 104, 15729 (2007)

2   Ihalainene, J.A., Bredenbeck, J., Pfister, R., Helbing, J., Chi, L., von Stockkum, I.H.M., Woolley, G.A. and Hamm, P., PNAS, 104, 5383 (2007)

3   Moroder, L., J. Peptide Sci., 11, 258 (2005)

4   Kusebauch, U., Cadamuro, S.A., Musiol, H.-J., Moroder, L. and Renner, C., Chem. Eur. J., 13, 2966 (2007)

5   Kusebauch, U., Cadamuro, S.A., Musiol, H.-J., Lenz, M. U., Wachtveitl, J., Moroder, L. and Renner, C., Angew. Chem. Int. Ed., 45, 7015 (2006)

6   Huber, R., Köhler, T., Lenz, M.O., Bamberg, E., Kalmbach, R., Engelhard, M. and Wachtveitl,J., Biochemistry 44, 1800 (2005).

7   Staudt, H., Köhler, T., Lorenz, L., Neumann, K., Verhoefen, M.-K. and Wachtveitl, J., Chem. Phys., 347, 462 (2008)

8   Neumann, K., Verhoefen, M.-K., Weber, I., Glaubitz, C. and Wachtveitl, J., Biophys. J.,94, 4796 (2008)

Ultrafast Charge Migration Following Ionization in Oligopeptides

Alexander I. Kuleff, Siegfried Lünnemann, and Lorenz S. Cederbaum

Theoretische Chemie, Physicalisch Chemisches Institut, Universität Heidelberg
Im Neuenheimer Feld 229, 69120 Heidelberg, Germany
E-mail: alexander.kuleff@pci.uni-heidelberg.de

Abstract. Electron correlation can be the driving force for ultrafast charge migration. Using *ab initio* calculations in the present work we demonstrate that the positive charge created by ionization out of an inner- or outer-valence shell of an oligopeptide can migrate throughout the system within just few femtoseconds.

Introduction

Charge transfer is a fundamental phenomenon in nature, playing an essential role in many chemical and biological processes. Traditionally it is assumed that the nuclear dynamics are the mediator of the charge transfer process. However, it was shown [1] that after a localized ionization the created positive charge can migrate throughout the system solely driven by many-electron effects, i.e. by electron correlation and relaxation effects [2-5]. This phenomenon was termed *charge migration*. The charge migration takes place on an ultrafast timescale (usually few femtoseconds) and thus it can be calculated neglecting the nuclear motion as long as one studies the relevant time interval during which this ultrafast migration takes place. Clearly, if we wish to know precisely what happens at later times, nuclear motion must be considered. However, since several or even many electronic states participate, an adequate description of the nuclear motion is rather involved. For a firs attempt to include the nuclear motion quantum mechanically, see Refs [6,7].

In the present work we study the hole charge migration initiated by a sudden ionization of oligopeptides. Apart from the fundamental importance of the charge migration phenomenon an additional motivation for the present study is the series of measurements by Weinkauf and Schlag (see, e.g. [8,9] and references therein), in which, after a localized ionization on a specific site of a peptide chain, a bond breaking on a remote site of the chain occurs. The authors proposed that a fast electronic transfer mechanism is responsible for transporting the positive charge to the latter site of the molecule. However, the underlying mechanism for this charge transport is still unclear.

Methodology

A possible way to trace in time and space the migration of a positive charge is to calculate the so-called *hole density*. The hole density is defined as the difference between the electronic density of the neutral and that of the cation:

$$Q(\vec{r},t) := \langle \Psi_0 | \hat{\rho}(\vec{r},t) | \Psi_0 \rangle - \langle \Phi_i | \hat{\rho}(\vec{r},t) | \Phi_i \rangle = \rho_0(\vec{r}) - \rho_i(\vec{r},t), \tag{1}$$

where $\hat{\rho}$ is the density operator, $|\Psi_0\rangle$ is the ground state of the neutral, and $|\Phi_i\rangle$ is the initially prepared cationic state. The second term in Eq. (1), ρ_i, is time-dependent, since $|\Phi_i\rangle$ is not an eigenstate of the cation. The quantity $Q(\vec{r},t)$ describes the density of the hole at position $\vec{r}$ and time t and by construction is normalized at all times t. For calculating the hole density we use *ab initio* methods only. The cationic Hamiltonian, represented in an effective many-body basis using the Green's function method, is used for directly propagating in the electronic space the initial state via the Lanczos technique [3]. The electronic wave packet thus obtained is then utilized to construct the hole density at each time point via Eq. (1). Theoretical and technical details concerning construction and analysis of the hole density are given in Refs [2-5].

An important feature of the method is that it enables to trace in real time and space the electron dynamics of various de-excitation processes taking into account *all* electrons of the system and their correlations. It should be mentioned also that this methodology is equally suitable for tracing the evolution of the electronic cloud throughout an electronic decay process [10].

Results and Discussion

As an illustrative example we show here our results for the electron dynamics following ionization of the inner-valence orbital $28a'$ of the oligopeptide Gly-Gly-NH-CH$_3$. In Fig. 1 the time evolution of the hole density is shown for the first 6 fs after the ionization. At time zero the positive charge is mainly localized on the "left"-hand side of the system, namely on the CH$_3$, NH, and CO groups. It is clearly seen that as time proceeds the charge migrates to the "right"-hand side of the system and after only 6 fs it is already located almost entirely on the remote NH$_2$, CH$_2$, and CO groups. After that time the process continues in reverse order, till the charge migrates back to its initial position. However, since a large number of ionic states participate in the dynamics, the process is not purely repetitive. Moreover, after, let say, 20 fs the nuclear dynamics could also start to play a role, which will additionally perturb the picture. The movement of the nuclei could even block the reverse process [11,12] and force the charge to remain on the amino group end of the molecule.

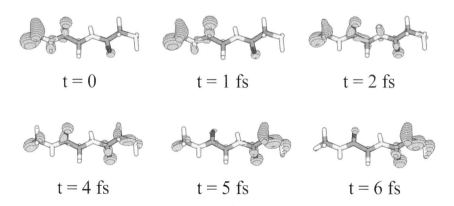

t = 0 t = 1 fs t = 2 fs

t = 4 fs t = 5 fs t = 6 fs

Figure 1: Snapshots of the hole density evolution $Q(\vec{r}, t)$ after ionization out of orbital $28a'$ of Gly-Gly-NH-CH$_3$.

Our calculations on Gly-Gly-NH-CH$_3$ demonstrate that after ionization of an inner-valence electron at the methyl site of the system an ultrafast charge migration to the amine site takes place entirely due to many-electron effects. The charge oscillates between the two sites in only 6 fs. However, the coupling with the nuclear motion at later times could lead to trapping of the charge and thus achieve irreversibility of the process. These findings may give a clue for understanding the results of the measurements of Weinkauf and Schlag, clarifying the underlying mechanism for the observed fast charge transfer.

We hope that our results will inspire further theoretical end experimental investigations on possible charge migration in even larger molecular systems.

Acknowledgements. Financial support by the DFG is gratefully acknowledged.

1   L. S. Cederbaum and J. Zobeley, Chem. Phys. Lett. **307**, 205 (1999).
2   J. Breidbach and L. S. Cederbaum, J. Chem. Phys. **118**, 3983 (2003).
3   A. I. Kuleff, J. Breidbach, and L. S. Cederbaum, J. Chem. Phys. **123**, 044111 (2005).
4   H. Hennig, J. Breidbach, and L. S. Cederbaum, J. Phys. Chem. A **109**, 409 (2005).
5   J. Breidbach and L. S. Cederbaum, J. Chem. Phys. **126**, 034101 (2007).
6   L. S. Cederbaum, J. Chem. Phys. **128**, 124101 (2008).
7   J. Miller, Physics Today (2008), May issue, p. 15.
8   R. Weinkauf, E. W. Schlag, T. J. Martinez and R. D. Levine, J. Phys. Chem. A **101**, 7702 (1997).
9   E. W. Schalg, S.-Y. Sheu, D.-Y. Yang, H. L. Selzle, and S. H. Lin, Angew. Chem. Int. Ed. **46**, 3196 (2007).
10   A. I. Kuleff and L. S. Cederbaum, Phys. Rev. Lett. **98**, 083201 (2007).
11   A. I. Kuleff and L. S. Cederbaum, Chem. Phys. **338**, 320 (2007).
12   S. Lünnemann, A. I. Kuleff, and L. S. Cederbaum, Chem. Phys. Lett. **450**, 232 (2008).

588

Probing Photodynamics of Retinal Protonated Schiff-Base with 7 fs Impulsive Vibrational Spectroscopy

Oshrat Bismuth[1], Amir Wand[1], Noga Friedman[2], Mordechai Sheves[2] and Sanford Ruhman[1]

[1] Institute of Chemistry and Farkas Center for Light Induced Processes, The Hebrew University, Jerusalem 91904, Israel.
  E-mail: sandy@fh.huji.ac.il
[2] Department of Organic Chemistry, the Weitzmann Institute of Science, Rehovot 76100, Israel.

Abstract. Two and three pulses experiments are conducted to record the elusive S_1 vibrational spectrum of Retinal Protonated Schiff-Base in solution. We find a reduction in C=C stretching frequency and other shifted bands in the fluorescent state, whose relevance are discussed.

Introduction

Retinal Protonated Schiff base (RPSB) is the light activated mainspring which energizes biological activity in retinal proteins (RP), such as Rhodopsin and Bacteriorhodopsin (BR). Despite the pivotal role it plays, and years of investigation, the vibrational spectrum of the reactive S_1 state in solvated RPSB, or in the native proteins, has yet to be measured. This information is vital for understanding the debated structural dynamics accompanying internal conversion (IC) and the role of apoprotein in directing reactivity of RPs. Of particular interest are the changes in the frequency of C=C stretches, which are indicative of the alterations in bond order - an essential stage of isomerization.

Remarkably, this frequency has yet to be determined reliably in any RP or in RPSB. In BR, this proves challenging due to the short lifetime of S_1 and to extensive spectral overlap with S_0 bands. Previous studies show that RPSB in solution undergoes IC with bi-exponential kinetics, characterized with fast and slow decay times of 2 and 7 psec respectively [1,2]. While IC kinetics can be assessed from visible transient absorption features, variations in such broad and unstructured electronic bands do not disclose shifts in chemical bonding and molecular geometry. The missing structural information can be obtained by impulsive vibrational spectroscopy which was adopted here. Obtaining the impulsive limit even for the C=C stretching vibrations dictates the use of the sub 10 fsec pulses used in our experiments.

Experimental

In "one pump" experiments a NOPA centered at 550 nm is used to excite and probe RPSB in ethanol. Two pump experiments start with 400nm excitation of S_1, followed by a push/probe sequence with red shifted NOPA pulses - used to improve separation of contributions from ground and excited state. The NOPA was compressed to 7 fs in a prism / deformable mirror shaper. This was verified with X-PG-FROG. Transmitted probe and reference pulses were co-dispersed in an imaging spectrograph / CCD setup

589

to obtain ΔOD(t) spectra. Analysis of the results involved subtraction of a slowly varying background and Fourier analysis of the isolated residuals.

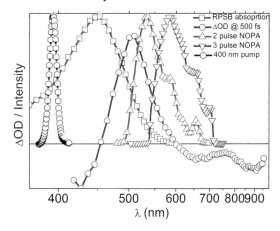

Fig. 1. Normalized absorption spectra of BR and all-trans RPSB in Ethanol together with normalized intensity spectra of typical spectrum of NOPA pulses used in our experiment.

Results

A one dimensional cut at a dispersed probe wavelength of 590 nm, is presented in figure 2A. Inspection of the modulations indicates a prominent mode near 1550 cm^{-1}, which can be assigned to C=C stretching in S_0 or S_1. Decay of observable modulations is considerably shorter than IC which can be assessed from the slowly varying background in the data. This fast decay translates into a large width of the C=C feature measured to be ~40 cm^{-1}. To test the evolution of this modulation a series of DFT spectra in this range were extracted starting at progressively later delays. Results in panel B demonstrate that the low frequency wing recedes more rapidly. Along with the excess width of this band relative to RR data, this suggests red shifted and shorter lived C=C stretching dynamics in S_1.

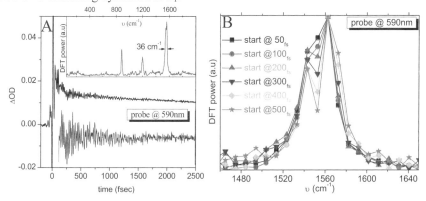

Fig. 2. "One-pump experiment": (A) Transient Absorption (TA) at 590nm (at the center). Isolation of periodical spectral modulations (residuals) done by subtracting a smooth fit (below). DFT analysis of the data is shown in the inset. (B) DFT for TA data in (A), calculated for different starting time, showing the dynamics of the C=C vibration after excitation.

To further investigate this point two pump experiments were initiated which due to the red shifting in the NOPA pulses do not directly excite the RPSB directly, specifically targeting the excited state population. Push-probe data were analyzed as above, with resulting DFTs depicted below (figure 3).

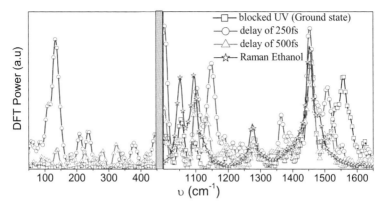

Fig. 3. "Two-pump experiment": DFT analysis of the data at 590nm in different delay between actinic pump and push pulses, together with blocked UV and Raman of pure solvent (Ethanol).

Again, C=C spectral density appears red shifted to ground state, and decays within ~0.5 psec. Additional transient peaks appear – at 1000, 1150, 230 and 130 cm^{-1} - , decaying on a similar timescale. The latter two were previously observed and assigned to vibrational coherences in S_1 of BR and RPSB.

Discussion and Conclusion

1. $\nu(C=C)$ is red-shifted in S_1, in contradiction to quantum chemical predictions, BR Raman experiments [3] and IR of halorhodopsin [4].

2. This conclusion is bolstered by simultaneous appearance of low frequency torsions, previously observed in pump-probe [2], TR fluorescence [5] and by the appearance of features at 1000 and 1150 cm^{-1}, assigned to C-C in IR of halorhodopsin excited state.

3. Susceptibility of nascent S_1 to Raman "push" strangely decays on sub-psec timescale – a phase which exhibits rapid evolution both in RPSB and PRs, and alternately assigned either to IC between excited states [6] or to fast deformation in S_1.

4. Determining correct option in previous paragraph requires more study, but must reflect structural evolution drastically altering $\sigma_{(Raman)}$ in the excited state.

1 P. Hamm, M. Zurek, T. Roschinger, H. Patzelt, D. Oesterhelt and W. Zinth, Chem. Phys. Lett. **263**, 613 (1996);

2 B. Hou, N. Friedman, S. Ruhman, M. Sheves and M. Ottolenghi, JPCB **105**, 7042 (2001).

3 L. Song and M.A. El-Sayed, JACS **120**, 8889 (1998).

4 F. Peters, J. Herbst, J. Tittor, D. Oesterhelt and R. Diller, Chem. Phys. **323**, 109 (2006).

5 G. Zgrablic, S. Haacke and M. Chregui, Chem. Phys. **338**, 168 (2007).

6 O. Bismuth et. al., JPCB **111**, 2327 (2007); G. Zgrablic et. al., Biophys J. **88**, 2779 (2005).

The 2DIR Spectroscopy on C-D Modes of Leucine-d$_{10}$ Side Chain

Sri Ram G. Naraharisetty[1], Valeriy M. Kasyanenko[1], Jörg Zimmermann[2], Megan Thielges[2], Floyd E. Romesberg[2], Igor V. Rubtsov[1]

[1] Department of Chemistry, Tulane University, New Orleans, LA 70118, USA
  E-mail: snaraha@tulane.edu, irubtsov@tulane.edu
[2] The Scripps Research Institute, La Jolla, CA 9203, USA

Abstract. We show that perdeuterated side chain of leucine amino acid and related compounds can serve as a useful structural reporter, suitable for studying proteins using two-dimensional infrared (2DIR) spectroscopy. Strong direct-coupling C-D/C=O and C-D/Am-II cross-peaks were measured by dual-frequency 2DIR in these compounds. The direct-coupling cross peaks were further enhanced up to 5 fold using the relaxation-assisted 2DIR method. The energy transport times (arrival times) and amplification factors were measured and used to assign the C-D absorption peaks and these assignments were compared with the results of anharmonic DFT calculations.

Introduction

The accuracy and specificity of the structural constraints obtained by 2DIR spectroscopy, a new powerful tool for structural measurements,[1] depend strongly on how localized the reporting modes are. Many strong modes such as C=O, O-H, and N-H modes are often delocalized. Exploration of weak IR modes that are spectrally well separated from other modes is a promising solution.[2] A "good" vibrational label should satisfy the following requirements. It should be a localized mode, spectrally resolved from other modes even when placed in large organic molecules. It should not degrade or be involved in dynamic exchanges. It is also important to have a label which can be placed at any desired position in the molecule and its transition(s) would not overlap with strong solvent transitions. C-D stretching modes satisfy most of these requirements; the C-D covalent groups are non invasive and small IR labels offering ultimate labeling flexibility.[2,3]

Results and Discussion

In this work we investigated a set of compounds featuring common Leu-d$_{10}$ residue aiming at understanding properties of the perdeuterated side chain of Leu as structural label for 2DIR measurements. We also used dual-frequency relaxation-assisted 2DIR

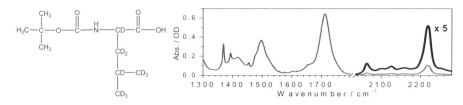

Fig. 1. Structure of Leu-d$_{10}$-Boc (*left*) and its linear spectrum of 150 mM concentration in chloroform solvent (*right*).

method (RA 2DIR) to measure connectivity relations for these compounds.[4] Figure 1 shows the structure of Leu-d_{10}-Boc and its linear spectrum. Several CD transitions are in the range from 2060 to 2220 cm^{-1}. The C=O and amide I (Am-I) modes of Leu-d_{10}-Boc are unresolved in the linear spectrum with the peak centered at 1712 cm^{-1}. The single amide II (Am-II) mode at 1496 cm^{-1} is a convenient mode for studying the energy transport *via* RA 2DIR. The 2DIR spectra were obtained using a dual-frequency three-pulse scheme with heterodyned detection where the k_1 and k_2 IR pulses were centered at 2100 cm^{-1} while the k_3 and local oscillator pulses were centered at 1490 cm^{-1}. Figure 2 shows 2DIR spectra of Leu-d_{10}-Boc amino acid in chloroform measured at three waiting times, 0.2, 2, and 3.0 ps, which is the time delay between the second and third IR pulses. The cross peaks seen in Fig. 2 *left* originate from the direct coupling of various C-D modes of Leu-d_{10} with the Am-II mode. The amplitudes of the C-D / Am-II cross peaks vary significantly as a function of the waiting time, T (Fig. 2).

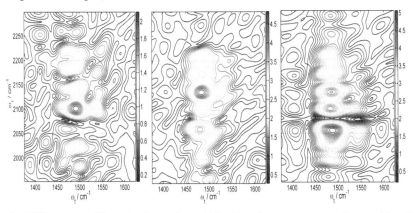

Fig. 2. 2DIR spectra of Leu-d_{10}-Boc amino acid in chloroform measured at three waiting times, T: 0.2, 2, and 3.0 ps (from left to right). The C-D / Am-II cross peak amplitudes show different T-time dependences, demonstrating the ability of preferentially enhancing a particular cross peak in the 2DIR spectrum. The dominant cross peaks at 0.2, 2, and 3 ps delays are the $C_\beta D_2$, $C_\gamma D$, and $(C_\delta D_3)_2$ transitions, respectively.

Figure 3A shows the amplitudes of two cross peaks plotted as a function of waiting time, T. The cross peaks reach maxima at 3.0 and 3.8 ps, respectively; these times represent the energy travel time (arrival time) from the respective CD mode to the vicinity of the Am-II mode. In our previous work, we showed the correlation of the energy arrival times and amplification of the cross peaks with the distance between the modes.[5] The arrival times for different cross peaks detected for Leu-d_{10}-Boc are presented in Table 1. There is a large difference in their values indicating that the side chain behaves as a number of weakly interacting transitions, rather than as a single label comprising strongly coupled modes delocalized over the whole side chain. The DFT anharmonic calculations performed with B3LYP functional using Gaussian 03 software (Gaussian Inc.) confirm this statement. The modeling also suggests that the CD modes of Leu-d_{10} can be divided into several largely independent groups $((C_\delta D_3)_2$, $C_\beta D_2$, and $C_\gamma D/C_\alpha D)$, featuring transitions that are mostly delocalized among the atoms included in the group. The RA 2DIR experimental data agree well with the computational results indicating the correlation of the arrival times with the distance in these compounds for the tested modes. The two strongest transitions in the linear spectrum come from the group that includes two $C_\delta D_3$ moieties; the peaks at 2060 and

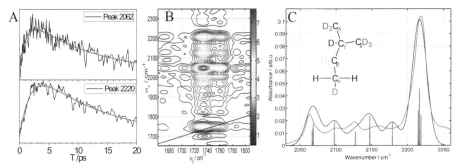

Fig. 3. A. T-delay dependence of the (Am-II, 2062 cm^{-1}) and (Am-II, 2220 cm^{-1}) cross peaks measured for Leu-d$_{10}$-methyl ester. **B.** RA 2DIR spectrum of Leu-d$_{10}$-methyl ester hydrochloride measured at $T = 3$ ps. The C-D/C=O cross peaks are seen. **C.** FTIR spectrum of Leu-d$_{10}$-Boc (red) and DFT calculated spectrum of the fragment shown in the inset (sharp peaks and thin line).

2218 cm^{-1} encompass, respectively, two symmetric and four antisymmetric stretching transitions (Fig. 3C). The longest arrival times are observed for this group. The C$_\beta$D$_2$ moiety forms another group with just two transitions in it: the symmetric stretch at 2098 cm^{-1} and antisymmetric stretch at 2179 cm^{-1}. The C$_\gamma$D stretching transition was computed at 2148 cm^{-1}. While the computed frequency of the C$_\alpha$D transition for the fragment shown in Fig. 3C *inset* is ca. 2130 cm^{-1}, its value in the complete Leu-d$_{10}$ amino acid is less certain.

Table 1. Arrival times, T_{max}, for different C-D / Am-II cross peaks measured for Leu-d$_{10}$-Boc.

Frequency cm^{-1}	T_{max}/ ps
2062	3.0±0.7
2100	1.3±0.6
2125	2.5±0.4
2180	2.5±0.7
2220	3.8±0.5

In conclusion, we have demonstrated that the deuterated side chain of leucine amino acid has a strong potential of becoming a useful structural label suitable for 2DIR measurements. We have demonstrated five-fold amplification of the cross peak for (C$_\delta$D$_3$)$_2$ at 2220 cm^{-1} and Am-II mode *via* RA 2DIR spectroscopy. The basic properties of this label have been determined and the applicability of RA 2DIR method for unraveling connectivity patterns in peptides has been demonstrated.

Acknowledgements. I.V.R. gratefully acknowledges support from the NSF (award number 0750415) and the Louisiana Board of Regents (RCS grant). S.R.G.N. thanks Tulane University and UP 08 organizers for the travel support.

1   P. Hamm, M. Lim, and R. M. Hochstrasser, J. Phys. Chem. B **102,** 6123, 1998.

2   S. G. Naraharisetty, D. V. Kurochkin, and I. V. Rubtsov, Chem. Phys. Lett. **437,** 262, 2007.

3   J. K. Chin, R. Jimenez, and F. E. Romesberg, J. Am. Chem. Soc. USA **123,** 2426, 2001.

4   D. V. Kurochkin, S. G. Naraharisetty, and I. V. Rubtsov, Proc. Natl. Acad. Sci. U.S.A. **104,** 14209, 2007.

5   S. G. Naraharisetty, V. Kasyanenko, and I. V. Rubtsov, J. Chem. Phys. **128,** 104502, 2008.

A Time-resolved Vibrational Spectroscopy Study on Adenine/Thymine Based Nucleic Acid Systems

Susan Quinn[1], Gerard W. Doorley[1], David A. McGovern[1], Anthony W. Parker[2], Kate L. Ronayne[2], Michael Towrie[2] and John M. Kelly[1]

[1] School of Chemistry and Centre for Chemical Synthesis and Chemical Biology, Trinity College, Dublin 2, Ireland.
Email: quinnsu@tcd.ie
[2] Central Laser Facility, Science and Technology Research Council, Rutherford Appleton Laboratory, Chilton, Didcot, Oxfordshire. OX11 OQX, UK

Abstract. The excited state properties of adenine and thymine nucleotides, dinucleotides and polynucleotides (double-stranded) are probed using ultrafast transient infra-red spectroscopy. The differing deactivation processes and the involvement of excimers/exciplexes are considered.

Introduction

In recent years our groups have worked to demonstrate the potential of time-resolved infrared spectroscopy in elucidating the photodynamics of DNA. [1-3] To date the results from this work have served to complement and extend the insights that can be gained from the more traditional ultrafast transient techniques of UV/visible transient absorption and fluorescence spectroscopy. [4,8] Thus while it has been known for many years that individual nucleic acid bases and nucleotides are characterised by ultrafast (< 1 ps) relaxation processes [4,6], our first ps-TRIR study identified clearly an additional slower (2-4 ps) process ascribed to vibrational cooling of the ground state [1]. Subsequently we have used ps-TRIR to observe and assign for the first time a long-lived $n_N \pi^*$ dark state for dCMP and to observe different transient/ground state recovery lifetimes for ring and carbonyl modes in GMP [2,3]. However in higher order structures such as polymeric strands and aggregated species additional longer lived components are observed [4-6]. These are suspected to promote genetic mutations through DNA damage. The proposed mechanism for the extended lifetimes continues to be a subject of debate being attributed to a number of phenomena including base-stacking interactions, excimer/exciplex formation and hydrogen bonding [4,5]. Understanding such mechanisms and their sensitivity to sequence and structural conformation is of great importance and our team has worked to develop time-resolved infrared spectroscopy as a probe to probe such factors.

Results and Discussion

Elucidation of the vibrational spectra of complex biomolecules such as DNA is very challenging. To tackle this problem a stepwise approach has been adopted by studying simple analogues such as dinucleotides which can be considered a basic model for processes within duplex DNA and in this paper ps-TRIR is used to probe the excited state dynamics of adenine and thymine DNA systems: the double-stranded poly(dA-dT).poly(dA-dT) and two dinucleotides namely 2'-deoxyadenyl-(3'-5')thymidine (dApdT), and thymidylyl-(3'-5')-2'-deoxyadenosine (dTpdA).

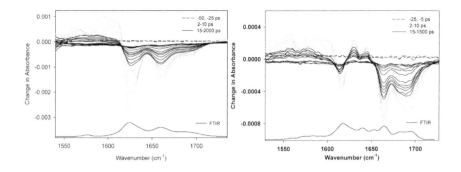

Fig. 1. The ps-TRIR spectra obtained following UV excitation (300 fs, 267 nm) of (a) 10 mM dApT, (b) 10 mM Poly(dA-dT).Poly(dA-dT) in 50 mM potassium phosphate D_2O buffer at pH 7. Short delay times (2-20 ps) are shown in grey with the longer delays (35-1500 ps) shown in black.

The sequence isomer dinucleotides dApT and TpdA are the basic units of the polynucleotide poly(dA-dT).poly(dA-dT), with the two monomer bases being attached via a phosphodiester bond. The bases can participate in some of the same binding interactions that occur in the larger polymeric DNA, especially π-π stacking interactions.

The ps-TRIR spectra of dApT, and poly(dA-dT).poly(dA-dT) obtained following UV excitation (300 fs, 267 nm) are shown in Figure 1. Short delay times (2-20ps) are shown in red with the longer delays (35-2000ps) shown in black. The spectra exhibit common features, with broad transients (1530-1610 cm^{-1}) and strong bleaching at higher wavenumbers. Shifting of the broad transient to higher wavenumbers with time is behaviour that is consistent with vibrational relaxation of the vibrationally hot electronic ground state. This is similar to the behaviour observed for the mononucleotides dAMP and TMP where we have previously demonstrated that, while the excited states of the mononucleotides are not detectable using ps-TRIR, the cooling of their vibrationally hot ground states occur on the order of <5 ps [1].

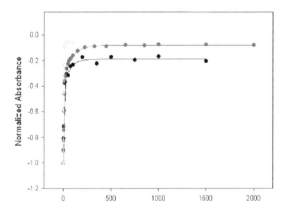

Fig. 2. The kinetic profile at the carbonyl bleach 1623 cm^{-1} for (light triangle) 10 mM dAMP, (gray dot) 10 mM dApT, and (black dot) 10 mM Poly(dA-dT).Poly(dA-dT) in 50 mM potassium phosphate D_2O buffer at pH 7.

In contrast to what is found for the mononucleotides, a persistent long-lived transient and a slower component in the bleaching recovery is seen for the cases of dApT and poly(dA-dT).poly(dA-dT), see Figure 2. This is also observed, but to a smaller extent, for TpdA data not shown. Bleaching centred at 1620 cm^{-1} is predominantly due to adenine with a smaller contribution from the thymine residue. The two remaining bleaches are thymine-based.

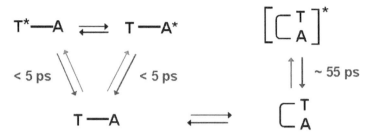

Scheme 1. Proposed interactions in dinucleotide ground state and excited state.

The long-lived species observed is attributed to an excimer/exciplex species, Scheme 1. This is formed to a greater extent in dApT than in TpdA, consistent with greater overlap in the ground state. The dApT excimer also deactivates more slowly than its TpdA analogue. These results represent a further insight into the factors contributing to excited state relaxation processes in mixed sequence DNA and will be discussed in context of our previous, and continuing work on guanine-cytosine systems [1-3].

1   M. K. Kuimova, J. Dyer, M. W. George, D. C. Grills, J. M. Kelly, P. Matousek, A. W. Parker, X. Z. Sun, M. Towrie and A. M. Whelan, Chem. Comm., 1182, 2005.
2   S. Quinn, G. W. Doorley, G. W. Watson, A. J. Cowan, M. W. George, A. W. Parker, K. L. Ronayne, M. Towrie and J. M. Kelly, Chem. Comm., 2130, 2007.
3   D.A. McGovern, S. Quinn, G. W. Doorley, A.M. Whelan, K. L. Ronayne, M. Towrie, A. W. Parker and J. M. Kelly, Chem. Comm., 5158, 2007.
4   D. Markovitsi, A. Sharonov, D. Onidas and T. Gustavsson, ChemPhysChem, **4**, 303, 2003.
5   J. Peon and A. H. Zewail, Chem Phys Lett, **348**, 225, 2001.
6   C. E. Crespo-Hernndez, B. Cohen, P. M. Hare and B. Kohler, Chem. Rev., **104**, 1977, 2004.
7   C. E. Crespo-Hernandez, B. Cohen and B. Kohler, Nature, **436**, 1142, 2005.
8   D. Markovitsi, F. Talbot, T. Gustavsson, D. Onidas, E. Lazzarotto and S. Marguet, Nature, **441**, E7, 2006.
9   M. Daniels, and W. Hauswirth, *Science*, **171**, 675, 1971.

Mapping Parallel Pathways of Energy Flow in LHCII with Broadband 2D Electronic Spectroscopy

Gabriela S. Schlau-Cohen[1,2], Tessa R. Calhoun[1,2], Gregory S. Engel[1,2,3], Elizabeth L. Read[1,2], Naomi S. Ginsberg[1,2], Donatas Zigmantas[1,2,4], Roberto Bassi[5], Graham R. Fleming[1,2]

[1]Dept of Chemistry, University of California, Berkeley, CA 94720-1460 USA
[2]Physical Biosciences Division, Lawrence Berkeley National Laboratory, Berkeley, CA 94720 USA
[3]Current address: Dept of Chemistry, University of Chicago, Chicago, IL 60637, USA
[4]Current address: Dept of Chemical Physics, Lund University PO Box 124 SE-22100, Lund Sweden
[5]Department of Science and Technology, University of Verona, 37134 Verona, Italy
E-mail: GRFleming@lbl.gov

Abstract: Two-dimensional femtosecond broadband electronic spectroscopy was used to simultaneously probe parallel pathways of energy transfer in the major light harvesting complex of Photosystem II from plants. Sub-100 femtosecond relaxation between delocalized excitonic states on highly coupled clusters of chlorophylls and several hundred femtosecond to picosecond components of relaxation between clusters were observed.

Introduction

Photosynthesis, nature's process for conversion of light energy to chemical energy, begins when sunlight is absorbed and the resulting excitation is transferred to the reaction center, where charge-separation occurs with near unity quantum efficiency. In plants, the major light harvesting complex of Photosystem II (LHCII) is the site of the light absorption and of directional energy transfer processes. [1] Two types of chlorophyll (Chl-a and Chl-b) comprise the primary absorbers in the complex. LHCII is trimeric, with three identical monomers each containing fourteen chlorophylls grouped into highly coupled clusters of two or three molecules each. Mapping the dynamics of energy transfer within LHCII and relating these dynamics to the spatial arrangement of chlorophylls within the complex has the potential to show how evolution has given rise to such remarkably efficient energetic dynamics. Due to the two types of chlorophyll as well as to differences in how each chromophore couples to its neighbors, the transitions of the first electronic excited state cover a broad energy range that is fully encompassed by using broad bandwidth laser pulses. Although previous ultrafast experiments and theoretical studies have identified elements of energy transfer [2], broadband 2D electronic spectroscopy generates a map of multiple simultaneous energy transfer and coupling mechanisms in the Q_y region and thus a more comprehensive understanding of LHCII's light-harvesting functionality.

Experimental Methods

Our 2D Fourier-transform technique has been described previously in detail. [3] Briefly, an ultrafast pulse is generated from a non-collinear optical parametric amplifier (NOPA) pumped by a home-built 3.4 kHz Ti:sapphire regenerative amplifier. The 18 fs NOPA output pulses are centered at 640 nm with a 90 nm FWHM. The NOPA output is split into four pulses arranged in a box geometry.

Pulses 1, 2 and 3 interact with the sample to generate the third-order optical response in the phase-matched direction, collinear with beam 4, the local oscillator. The signal is thus heterodyne detected in the frequency domain. The measured electric field is Fourier-transformed with respect to the delay between pulses 1 and 2 to generate a 2D spectrum for each population time, T, the delay between pulses 2 and 3. The full 2D spectrum includes the photon-echo (rephasing) and free polarization decay (non-rephasing) contributions. The non-rephasing component, experimentally separated by switching the time order of pulses 1 and 2, can be used to more clearly isolate cross-peaks. LHCII extracted from *Arabidopsis thaliana* was dissolved in Hepes buffer (pH of 7.6) and mixed with glycerol at 30:70 (v/v). The sample was held in a 200 μm quartz cell and cooled to 77 K. 2D spectra were measured for population times in 10 fs increments from 0 to 500 fs and 1 ps increments from 1 to 15 ps.

Results and Discussion

Absolute value two-dimensional spectra of LHCII and their corresponding non-rephasing components for representative population times are shown in Fig. 1. The absorption of Chl-*b* and Chl-*a* appear along the diagonal in our 2D spectra, around $\omega=15385$ and 14925 cm^{-1}, respectively. The vibronic tail, to at higher energies than the Chl-*b* peak, displays rapid relaxation to the vibrational ground state.

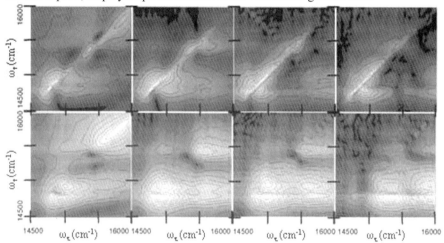

Fig. 1: Absolute value normalized 2D spectra of LHCII at 77 K for (left to right) T=30, 260, 500 fs and 2 ps. The non-rephasing contribution is separated and displayed below the corresponding correlation (total) spectrum.

Overall, many of the energy transfer processes correspond to a model developed by Novoderezhkin, et. al. [2] The chlorophylls are arranged in highly coupled clusters (J=10-100 cm^{-1}), each of which supports energetically separated but spatially overlapping delocalized excitonic wavefunctions. This overlap facilitates sub-100 fs relaxation processes between the delocalized states within both the Chl-a and Chl-b bands, as seen particularly in the cross-peaks emerging within the Chl-*a* band at 30 fs. The T=260 and T=500 fs spectra show energy transferring between clusters within the Chl-*a* manifold on a several hundred femtosecond timescale. The cross-peaks showing energy transfer between the Chl-*b* and Chl-*a* manifolds contain multiple timescale contributions. At T=30 fs, the emergence of two cross-peaks ($\omega_\tau=15300$ cm^{-1}, $\omega_t=14800$ cm^{-1}; $\omega_t=14900$ cm^{-1}) indicates rapid energy transfer between these two regions, perhaps seen here, as with the intra-cluster relaxation, but not in [2] because of the shorter pulse duration than in previous experimentation. The intensity of the cross-peak with the higher ω_t frequency (ω_τ=15300 cm^{-1}, ω_t=14900 cm^{-1}) relative to the cross-peak with the lower ω_t frequency (ω_τ=15300 cm^{-1}, ω_t=14800 cm^{-1}) is stronger in the T=260 fs spectra than in the T=500 fs spectra, showing that some relaxation occurs on a several hundred femtosecond timescale. The cross-peak still exists in the T=2 ps spectra, which shows a picosecond contribution to the cross-peak, potentially corroborating a long-lived intermediate state proposed in [2]. The chlorophylls in LHCII are arranged in two disordered layers, stromal and lumenal, [4] with Chl-*b* to Chl-*a* energy transfer processes occurring on both layers. The Chl-*b* cluster on the stromal level appears to be strongly coupled to neighboring Chl-*a* molecules and might thus show faster relaxation between bands, while the lumenal layer would contribute predominantly to the picosecond Chl-*b* to Chl-*a* energy transfer processes. The increased relative intensity of the cross-peaks along the lowest energy state (ω_t=14750 cm^{-1}), seen in the T=260 fs spectrum and more clearly in the T=500 fs spectrum, shows that the broad range of initial excitations relaxes to the lowest energy state. [5] The longer-time dynamics, notably Chl-*b* transfer to two distinct Chl-*a* levels and Chl-*a* to Chl-*a* transfer, are reproduced with theoretical modeling, as shown in Figure 2.

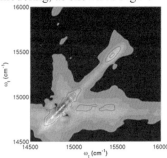 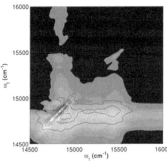

Fig. 2: 77 K normalized to unity simulated LHCII absolute value correlation 2D spectra for T=350 fs and 2 ps (left to right).

Conclusions

2D electronic spectra display an electronic coupling and energy transfer map for the entire Q_y region of LHCII. The spatial overlap of delocalized excitonic wavefunctions facilitates rapid relaxation over large energetic gaps. Overall, the spectra demonstrate multiple simultaneous pathways by which the broad range of initial excitations relaxes to the lowest energy excitonic state. There are two clear pathways of Chl-*b* to Chl-*a* energy transfer through the LHCII complex which could be related to the bilayer design of the pigment-protein complex. Combined with knowledge of the molecular structure, mapping the electronic structure has enabled us to understand more fully the interactions between chromophores and the role specific molecules play in guiding directional energy transfer.

Acknowledgements. Chemical Sciences, Geosciences and Biosciences Division, U.S. Dept. of Energy grant No DE-AC03-76SF000098. The authors thank Rienk van Grondelle for helpful discussion. N.S.G. is supported by the Glenn T. Seaborg postdoctoral fellowship from LBNL.

1 R. E. Blankenship, *Molecular Mechanisms of Photosynthesis*, (Blackwell Science, Oxford/Malden, 2002).
2 R. van Grondelle and V.I. Novoderezhkin, PCCP, **8**, 793 (2006).
3 T. Brixner, et. al., J Chem. Phys. **121**, 4221 (2004).
4 Z.F. Liu, et. al., Nature **428**, 287 (2004).
5 G. S. Schlau-Cohen, et. al., paper in preparation

Dissecting Exciton Dynamics Pathways in Electronic Multidimensional Spectroscopy by Pulse Polarizations

Darius Abramavicius[1], Dmitri V. Voronine[1,2], and Shaul Mukamel[1]

[1] Chemistry Department, University of California Irvine, California, USA
  E-mail: smukamel@uci.edu
[2] Institut für Physikalische Chemie, Universität Würzburg, Am Hubland, Würzburg, 97074, Germany

Abstract. A simulation study shows how coherent exciton dynamics in photosynthetic complexes may be revealed in two-dimensional photon-echo signals by specific laser pulse polarization configurations. Dynamics of single-exciton density matrix cohenrences shows strong signatures of excitonic coherences prior to energy relaxation.

Introduction

Multidimensional correlation spectroscopies are valuable probes of dynamical processes in molecules, which provide detailed dynamical information on complex structures: proteins, excitons, and semiconductors [1]. These techniques are performed by applying four well-separated chronologically-ordered ultrashort laser pulses as shown in Fig. 1.

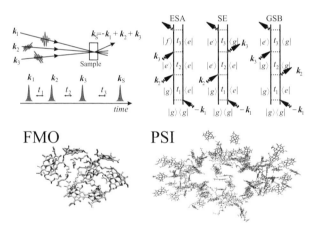

Fig. 1. Top row: Scheme of the coherent third order photon echo technique (left) and the Feynman diagrams of the contributing Liouville space pathways (right). Bottom row: Configurations of BChls in photosynthetic complexes FMO (left) and PSI (right); scales are different.

We consider the photon-echo signal generated in the phase-matching direction $-k_1 + k_2 + k_3$, where the delay times between pulses, t_1, t_2 and t_3 serve as control parameters. A double Fourier transform with respect to $t_1 \rightarrow \Omega_1$ and $t_3 \rightarrow \Omega_3$ at fixed delay t_2 is used to display the two-dimensional coherent spectra (2D CS). Diagonal ($-\Omega_1 = \Omega_3$) peaks carry similar information as linear absorption: peak positions correspond to excitation energies. However, unlike linear techniques, the homogeneous and inhomogeneous broadenings in 2DCS show up in anti-diagonal and diagonal directions, respectively, and can be separated. Crosspeaks $(-\Omega_1 \neq \Omega_3)$ carry novel information about couplings and correlations of different states [2]. We show how symmetry properties of these signals with respect to pulse polarization configurations (PPC) may be used to probe the system's density matrix coherences [3].

Signatures of density matrix coherences

In broad-band impulsive optical techniques, when the pulses are resonant with interband transitions, the signal is proportional to the third order response function $S^{(3)}$ at the photon echo phase-matching direction [5]. The response function is given by a sum over three Liouville space pathways (LSP): excited state emission (ESE), ground state bleaching (GSB) and excited state absorption (ESA) (Fig. 1). We classify the LSPs as follows [4]: coherence (population) pathways, when *bra* and *ket* are different (same) during t_2.

The following set of transitions is characteristic to the population pathways: the population is created from the ground state by two transitions to the same initial state i, it relaxes to a final f state during t_2, and f is deexcited to any other state. The population LSP contains the product $\langle \boldsymbol{\mu}_f^{\nu_4} \boldsymbol{\mu}_f^{\nu_3} \boldsymbol{\mu}_i^{\nu_2} \boldsymbol{\mu}_i^{\nu_1} \rangle$; here $\boldsymbol{\mu}_i$ ($\boldsymbol{\mu}_f$) is the excitation (deexcitation) transition dipole. Angular brackets denote orientational averaging and $\nu_4 \nu_3 \nu_2 \nu_1$ ($\nu = x, y, z$) denote the laser pulse PPC. For three basic tensor components of the response function we have [3]:

$$\langle \boldsymbol{\mu}_f^x \boldsymbol{\mu}_f^x \boldsymbol{\mu}_i^y \boldsymbol{\mu}_i^y \rangle = 15^{-1}(2\boldsymbol{\mu}_f^2 \boldsymbol{\mu}_i^2 - (\boldsymbol{\mu}_f \cdot \boldsymbol{\mu}_i)^2), \tag{1}$$

$$\langle \boldsymbol{\mu}_f^x \boldsymbol{\mu}_f^y \boldsymbol{\mu}_i^x \boldsymbol{\mu}_i^y \rangle = \langle \boldsymbol{\mu}_f^x \boldsymbol{\mu}_f^y \boldsymbol{\mu}_i^y \boldsymbol{\mu}_i^x \rangle = 30^{-1}(-\boldsymbol{\mu}_f^2 \boldsymbol{\mu}_i^2 + 3(\boldsymbol{\mu}_f \cdot \boldsymbol{\mu}_i)^2). \tag{2}$$

We found that the following combination $B \equiv S_{xyxy}^{(3)} - S_{xyyx}^{(3)}$ cancels for all population LSPs [3]. For localized excitons we would have $\boldsymbol{\mu}_i \equiv \boldsymbol{\mu}_f$ and therefore B also vanishes. The B signal therefore solely shows the coherence LSPs of delocalized excitons.

Results and Discussion

We study coherent exciton dynamics in two photosynthetic complexes: the Fenna-Matthews-Olson (FMO) complex of a green sulfur bacteria and the Photosystem I (PSI) photosynthetic complex of a cyanobacteria *Thermosynechococcus elongatus* [4]. Simulations were performed using the Frenkel exciton model of coupled two-level molecules as described in previous publications [3,6]. The FMO complex is one of the most extensively studied photosynthetic pigment-protein complexes. It is a trimer of small noninteracting identical subunits, each consisting of seven bacteriochlorophyll (BChls) molecules (see Fig. 1). The broad absorption spectrum extends from 12000 cm^{-1} to 13000 cm^{-1} (see Fig. 2). The B signal at different t_2 delay times is shown in Fig. 2 (top row). We see well-resolved offdiagonal peaks, which oscillate with t_2. Peak positions can be correlated with the excitons and their wavefunctions. Only highly delocalized excitons contribute to the signal, while the lowest-energy peak (localized exciton) is not observed.

The PSI complex is a larger energy-conversion apparatus appearing in trimeric and monomeric forms. The absorption band of the PSI monomer with 96 chlorophylls extends between 13500 – 15500 cm^{-1}. The B signal of PSI [Fig. 2 (bottom row)] at $t_2 = 0$ contains unresolved features at the bulk antenna region. A distinct pattern of exciton density matrix appears at later delay times. The two well-resolved crosspeaks at 100 fs can be related to the reaction center, which contains very strong couplings between molecules and its excitons are delocalized.

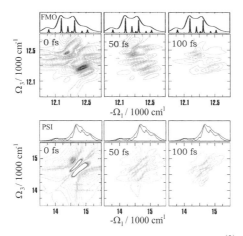

Fig. 2. Absorption and two dimensional photon echo technique $B \equiv S_{xyxy}^{(3)} - S_{xyyx}^{(3)}$ for two photosynthetic complexes: FMO and PSI at various t_2 delay times.

In summary, the single-exciton density matrix can be directly probed by two dimensional signals through crosspeaks in 2D CS using PPCs. The oscillatory pattern of the signal with the delay time t_2 follows propagation of density matrix coherences. It implies exciton delocalization since localized excitons are filtered out by the B signal.

Acknowledgements. The research was supported by the National Science Foundation (CHE-0745892) and the National Institute of Health (GM59230).

1   S. Mukamel, Annu. Rev. Phys. Chem. v 51, 691, 2000.
2   M. T. Zanni, N.-H. Ge,and Y. S. Kim, R. M. Hochstrasser, Proc. Nat. Acad. Sci. USA., v 98, 11265, 2001; T. Brixner, J. Stenger, H. M. Vaswani, M. Cho, R. E. Blankenship, G. R. Fleming, Nature, v 434, 625, 2005.
3   D. Abramavicius, D. V. Voronine, S. Mukamel, Biophys. J., v 94, 3613, 2008.
4   H. van Amerogen, L. Valkunas, R. van Grondelle, Photosynthetic Excitons, World Scientific, Singapore, 2000.
5   S. Mukamel, Principles of Nonlinear Optical Spectrscopy, Oxford University Press, New York, 1995.
6   S. Vaitekonis, G. Trinkunas, L. Valkunas, Photosynth. Res., v 86, 185, 2005; B. Brüggemann, K. Sznee, V. Novoderezhkin, R. van Grondelle, V. May, J. Phys. Chem. B, v 108, 13536, 2004.

Photoselection Polarization Experiments Reveal Ultrafast Electron Hopping Between Distinct Aromatic Residues in the Flavoprotein DNA Photolyase

Andras Lukacs[1], André P. M. Eker[2], Martin Byrdin[3], Klaus Brettel[3], Marten H. Vos[1]

[1] Laboratory for Optical Biosciences, Ecole Polytechnique, 91128 Palaiseau cédex, France
   E-mail: andras.lukacs@polytechnique.edu
[2] Department of Cell Biology and Genetics, Medical Genetics Centre, Erasmus University
   Medical Centre, PO Box 1738, 3000 DR Rotterdam, The Netherlands
[3] Service de Bioénergétique, CEA Saclay, 91191 Gif-sur-Yvette Cedex, France

Abstract. Flavin-excitation initiated electron transfer along three tryptophan amino acids in DNA photolyase was studied. Combining ultrafast polarization and side directed mutagenesis approaches the chain was shown to allow very efficient long-range transprotein electron-transfer in picoseconds.

Introduction

DNA photolyase is one of the very few enzymes that display photocatalytic behavior. Two distinct light driven processes take place in this flavin-containing protein. The photorepair process allows the enzyme to repair UV-induced lesions in the DNA substrate, by electron transfer (ET) initiated by population of the excited state of the reduced flavin $FADH^-$.

Generation of the catalytically active state $FADH^-$ from the stable neutral radical form $FADH^\bullet$ (a form that absorbs visible light) of the isolated protein occurs via a second light activated process. This process – photoreduction – is initiated by the formation of the excited state $FADH^{\bullet*}$, followed by electron extraction from reductants in the solvent via a triad of tryptophans (W382-W359-W306) bridging the flavin cofactor and the aqueous solvent [1]. The distance between the flavin and W306 is ~15Å, the distances between the reactant macrocycles are 4-5 Å. Discriminating between the different tryptophans is challenging, because they are spectroscopically indistinguishable.

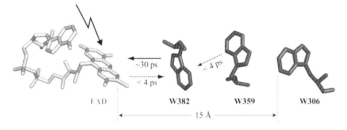

Fig. 1. Scheme of the electron transfer process during photoreduction.

Previously, the above electron transfer scheme has been established by ultrafast spectroscopy of wild type (WT) and mutant proteins where W382 and W359 were replaced by redox-inactive phenylalanines (F) [1-3]. Very briefly, the decay kinetics of FADH$^{\bullet*}$ and the lack of sizeable formation of the FADH$^-$Trp$^{\bullet+}$ state in both these mutants led to a scheme where ET from W382 to FADH$^{\bullet*}$ occurs in 30 ps, followed by much faster stabilization of the charge pair by ET from W359, in competition with recombination. A picture emerges where the Trp chain may act as an ultrafast wire. Here we investigate the speed of ET between the middle and the solvent-exposed Trps, by comparing WT and W306F photolyase. The finding that a stable photoproduct is formed in the mutant, and the fact that the macrocycles of W359 and W306 are near-perpendicularly oriented, allows exploiting polarization photoselection effects to assess the rate of hopping in WT.

Experimental Methods

Wild type and W306F DNA photolyase was overexpressed in *E. coli*, and prepared to ~300 μM in a 1-mm optical path length cell. Multicolor pump-probe spectroscopy was performed with a 1-kHz regenerative amplifier; pump pulses centered at 620 nm were generated using a home-built NOPA and reduced to an energy of ~ 400 nJ. One main challenge of this work is that the protein is relatively instable and some sample scattering is unavoidable. In order to reduce the aggregation of the enzyme the sample was kept at 10°C.

Results and Discussion

Transient absorption experiments were performed with pump pulses centered at 620 nm that solely excite the lowest lying absorption band of FADH$^{\bullet}$ and white light continuum probe pulses. The kinetics and the shape of the transient spectra under isotropic conditions are very similar as in WT. The quantum yield of FADH$^-$Trp$^{\bullet+}$ is also similar.

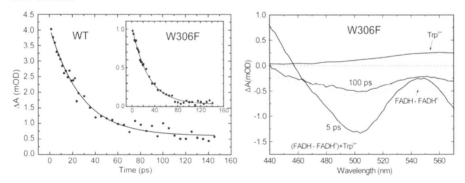

Fig. 2. Kinetics observed in WT and in W306F mutant measured at 440 nm (left) Transient spectra observed in the W306F mutant under isotropic (magic angle) conditions at 5 ps and 100 ps. The latter spectrum can be well modeled by the sum of the flavin difference spectrum and the oxidized Trp spectrum (right).

As in WT [2] the final spectrum is described well by a sum of the (FADH⁻-FADH•) and the Trp•⁺ spectrum (neutral Trp does not absorb in the visible), and this state must reflect FADH⁻W359•⁺ (Fig.1.), implying that this state can be populated to a sizeable extent (the same does not hold for the state FADH⁻W382•⁺ in the W359F mutant photolyase [3]). In WT, it is therefore possible that on the same timescale the positive charge is still on W359, or, if ET from W306 to W359•⁺ is fast, on W306.

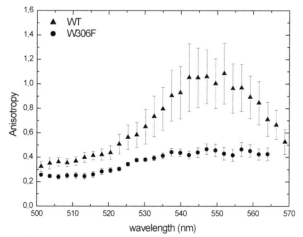

Fig. 3. Anistropy of the asymptotic transient spectra. $r = (\Delta A_{par} - \Delta A_{perp})/(\Delta A_{par} + 2\Delta A_{perp})$

In view of the different orientations of the Trps, the polarization of the Trp signal should be different in both cases. Whereas the Trp signal is relatively small at most wavelengths (Fig. 2), and the dominating polarization signal of the flavin should be identical in both cases, the polarization effects are expected to be strongest near 540 nm, where both isotropic contributions are closest to compensating. Fig. 3 shows the result of photoselection polarization experiments for W306 and WT, clearly indicating strong polarization differences. A detailed spectral analysis, taking into account polarization effects of two flavin absorption bands [4], indicate that the angle between the lowest flavin transition and the probed Trp•⁺ transition is 25-30° and 50-70° for W306 and WT, respectively. Inspection of the crystal structure indicates that these values are consistent with the positive charge being localized on W359 and W306 respectively.

Conclusions

Overall our results imply that the Trp chain acts as an extremely efficient intraprotein "nanowire" that transfers an electron within 30 ps across 15Å. Whereas we have here exploited the photo-activatibility of this chain in photolyase, aromate ET chains may function more widespreadedly in protein systems.

1   C. Aubert, M.H. Vos, P. Mathis, A.P.M. Eker and K. Brettel, Nature, 405, 586, 2000
2   M. Byrdin, A.P.M. Eker, M.H. Vos, K. Brettel, PNAS, 100, 8676, 2003
3   A. Lukacs, A.P.M. Eker, M. Byrdin, S. Villette, J. Pan, K. Brettel and M.H. Vos, J. Phys. Chem. B, 110, 15654, 2006
4   M. Byrdin, S. Vilette, A. Espagne, A.P.M Eker, K. Brettel, J. Phys. Chem.B.,112, 2008

Quantum Coherence Accelerating Photosynthetic Energy Transfer

Hohjai Lee, Yuan-Chung Cheng, and Graham R. Fleming

Department of Chemistry, University of California, Berkeley and Physical Bioscience Division, Lawrence Berkeley National Laboratory, Berkeley, CA 94720
E-mail: GRFleming@lbl.gov

Abstract: We show how long-lasting coherence enhances energy transfer rate in a photosynthetic complex based on an analysis of data collected using a newly developed two-color electronic coherence photon echo technique and theoretical simulations.

Introduction

Long-lasting quantum beating observed in the FMO complex by two-dimensional (2D) spectroscopy suggests that energy transfer in a photosynthetic complex occurs in a wavelike coherent fashion instead of by incoherent hopping [1]. The contribution of the long-lasting quantum coherence in promoting energy transfer efficiency in photosynthetic complexes is an intriguing question. Even though transient absorption and 2D spectroscopic studies have provided evidences of electronic coherence in photosynthetic systems, a quantitative measurement that can unambiguously reveal quantum coherence dynamics is still unavailable. Here, we present a visualization of quantum coherence dynamics obtained by a new technique, two-color electronic coherence photon echo (2CECPE), that exclusively explores the coherence dynamics between two excitonically coupled states. We also show how the quantum coherences can promote energy transfer rate in photosynthesis by theoretical simulations.

The new method was applied to the reaction center (RC) of photosynthetic bacterium, *Rhodobacter Sphaeroides*. The RC is where the solar energy is transformed to chemical energy after being collected by antenna complexes. It is a pigment-protein complex where proteins closely pack six chromophores: a pair of bacteriochlorophylls (P), another bacteriochlorophyll flanking P on each side (B_A and B_B), and a bacteriopheophytin next to each B (H_A and H_B) [2]. The complex has distinct bands at 870, 800 and 750nm which are assigned to the P, B, and H bands, respectively. Energy transfer occurs from H to B in ~100fs and from B to P in ~150fs in isolated RCs. Once P is excited, electron transfer occurs from P to B to H in ~4ps.

Experimental Methods

A schematic diagram of the 2CECPE experiment is illustrated in Fig. 1a. 2CECPE is a four-wave mixing technique where three pulses (ω_1, ω_2, ω_3; numbers indicate pulse sequence) are focused on a sample in equilateral triangle geometry. A similar scheme was used to perform two-color three-pulse photon echo peak shift (2C3PEPS) experiment with the condition $\omega_1=\omega_2\neq\omega_3$ [3]. Conventional 2C3PEPS signals are from third order responses from population states (either excited or ground state) generated by the interaction of the sample with the first two pulses. Therefore 2C3PEPS signals respond to the population evolution. In contrast, the 2CECPE setup requires $\omega_1=\omega_3\neq\omega_2$, which generates coherence between two excitonic states during t_2 (see below). This pulse sequence allows only signals related to the coherence of interest to be detected and avoids signals from the evolution of population. The relevant double-sided Feynman diagrams are shown in Fig. 1b. The time-integrated photon echo signal in the phase matched direction ($k_S = -k_1 + k_2 + k_3$, where k represents a wave vector) is collected by a PMT as a function of the delays between the pulses, t_1 (delay between the 1st and 2nd pulses) and t_2 (delay between the 2nd and 3rd pulses).

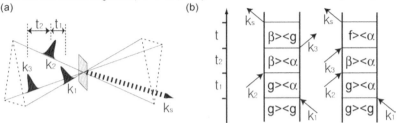

Fig. 1. (a) A diagram of the 2CECPE experiment. (b) The Liouville pathways that contribute to 2CECPE signal. (g: ground state, α and β: one-exciton states, f: two-exciton state $|\alpha\rangle \otimes |\beta\rangle$)

Results and discussion

As the first demonstration of 2CECPE, we studied coherence dynamics between B and H in RC with P oxidized by $K_3Fe(CN)_6$ [4]. The oxidation avoids complexity involved with electron transfer process, but does not affect the energy transfer process. We used 800nm (ω_2) and 750nm (ω_1 and ω_3) pulses.

Fig. 2a shows the results of the 2CECPE experiment on the oxidized RC at 77K. Note that the t_1 and t_2 axes correspond to the evolutions of the $|g\rangle\langle H|$ and $|B\rangle\langle H|$ coherences, respectively (Fig. 1b). It is clear that the $|B\rangle\langle H|$ coherence along t_2 lasts for a surprisingly long time (~440fs of dephasing time) compared to $|g\rangle\langle H|$ coherence along t_1. We simulated the results using an impulsive limit third order response function formalism based on a coupled heterodimer model (Fig. 2b). Our simulation shows that such a long-lasting coherence can only be explained by strong correlation between the protein-induced fluctuations in the transition energy of the B and H chromophores. This result implies that the coherence is protected by the protein environment, which likely promotes efficient energy transfer.

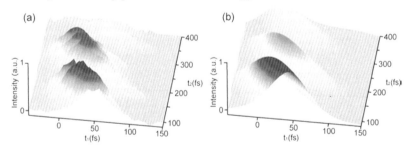

Fig. 2. Experimental (a) and simulation (b) results of the 2CECPE for the oxidized RC at 77K. a.u., arbitrary units. The oscillatory pattern results from a vibrational wavepacket in H.

The surprisingly long-lasting $|B\rangle\langle H|$ coherence also implies that coherences between donors and acceptors should be properly incorporated in description of excitation transfer in the RC although this is not usually done. In Fig. 3a, we show the simulated population dynamics of a coupled two-level system in the site-representation (|b> and |h>) by taking into account the non-zero off-diagonal terms of the density matrix to visualize how excitation moves in space. The coherent picture exhibits oscillations of population on each site, that is, wavelike reversible motion between two sites. The result of a more physical simulation in which an efficient energy trap next to B (e.g. P) is added is depicted in fig. 3b. In this case, the h-site population decays rapidly (~1/150fs) even though the intrinsic $k_{h \to b}$ (=1/500fs) is slow because the ultrafast (1/30fs) trapping process "catches the peaks" of the oscillatory behavior of b-site population.

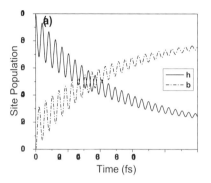 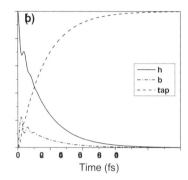

Fig. 3. Simulated population dynamics in the site-representation (a) without a trap site (b) with a trap site. The dynamics is obtained by solving the Bloch equations. A coupling constant $J=220cm^{-1}$, energy gap $\omega_{hb}=680cm^{-1}$, and |B><H| dephasing rate $\gamma=1/440fs$ were used. In addition, the intrinsic h→b rate constant, $k_{h \to b}$, is assumed to be 1/500fs. In (b), the trap site is directly coupled to b, with a trapping rate $k_{b \to trap}=1/30fs$. The rapid $k_{b \to trap}$ effectively couples h directly to the trap, as seen in (b).

Conclusions

2CECPE is a new method ideal for the interrogation of the coherence dynamics between two excitonically coupled states. The first application of 2CECPE on the RC shows that the protein protects the quantum coherence to accelerate the energy transfer process in the photosynthetic pigment-protein complex.

Further work on coherence dynamics of the |B><P| and |H><P| coherences in the RC is in progress. 2CECPE is also being used in our group to study a coherence transfer between difference coherences by frequency resolved detection.

Acknowledgments. This work was supported by U.S. DOE/BES (contract DE-AC03-76SF000098).

References
1. G. S. Engel, T. R. Calhoun, E. L. Read, T.-K. Ahn, T. Mančal, Y. - C. Cheng, R. E. Blankenship and G. R. Fleming, Nature **446**, 782, 2007.
2. U. Ermler, G. Fritzsh, S. K. Buchana and H. Mechel, Structure 2 **10**, 925, 1994.
3. R. Agarwal, B. S. Brall, A. H. Rizvi, M. Yang, and G. R. Fleming, J. Chem. Phys. **116**, 6243, 2002.
4. H. Lee, Y. - C. Cheng, and G. R. Fleming, Science **316**, 1462, 2007.

Ultrafast dynamics of light-harvesting function of β-carotene in carbon nanotube

Masayuki Yoshizawa[1,2], Kenta Abe[1], Daisuke Kosumi[1], Kazuhiro Yanagi[2,3], Yasumitsu Miyata[2,3], and Hiromichi Kataura[2,3]

[1] Department of Physics, Graduate School of Science, Tohoku University, Sendai 980-8578, Japan
E-mail: yoshi@laser.phys.tohoku.ac.jp
[2] JST, CREST, Kawaguchi, Saitama, 332-0012, Japan
[3] Nanotechnology Institute, National Institute of Advanced Industrial Science and Technology, Tsukuba, 305-8562, Japan

Abstract. Ultrafast dynamics of β-carotene encapsulated in single-walled carbon nanotubes (SWCNTs) was investigated by femtosecond absorption spectroscopy. Energy transfer from the excited states of β-carotene to SWCNTs (light-harvesting function) has been observed.

Introduction

Single-walled carbon nanotubes (SWCNTs) attract great interests because of their high aspect ratio, strength, resonant optical absorption, high carrier mobility, and maximum current [1]. However, the spectral range is limited by the specific density of states. The control of the spectral range can be achieved by combining SWCNTs with an appropriate molecule [2]. Recently, β-carotene has been encapsulated inside SWCNTs [3]. It has been well-established that carotenoids have a light-harvesting function in bacterial photosynthesis [4,5]. The light energy absorbed by carotenoids transfers to bacteriochlorophyll from the $1^1B_u^+$ and $2^1A_g^-$ excited states of carotenoids. The light-harvesting function of β-carotene in SWCNTS has been observed by photoluminescence (PL) [6] and femtosecond absorption spectroscopy [7]. In this study, ultrafast optical responses of SWCNT films with and without β-carotene (referred as SWCNT and Car@SWCNT, respectively) have been investigated and the energy transfer between β-carotene and SWCNTs is discussed..

Experimental Methods

We used SWCNTs manufactured by laser vaporization and encapsulated β-carotene [3,6]. The encapsulation was performed as follows. SWCNTs and β-carotene were dissolved in hexane. The mixture was refluxed for about several hours in an N_2 atmosphere. Then the solution was filtered and washed with tetrahydrofuran (THF) several times to remove nonencapsulated β-carotene molecules. In order to prepare proper nonencapsulated reference samples, the same procedure has been repeated without β-carotene.

Photoinduced absorbance change induced by a pump pulse (2.46 eV, 150 fs) was measured using a spectrometer coupled with NMOS and InGaAs image sensors (Hamamatsu S3903-1024Q and G9211-256). The standard error of the obtained absorbance change is smaller than 10^{-4} in visible and near-infrared region (0.9-2.7 eV).

Results and Discussion

A dashed line in Fig.1(a) shows stationary absorption of SWCNT. Two broad bands at 1.2-1.4 eV (S2) and 2.2-2.5 eV (S3) are the optical transitions from the valence to the conduction bands of the semiconducting SWCNTs. Another broad band at 1.6-1.9 eV (M1) is assigned to the metallic SWCNTs. A dash-dotted line shows stationary absorption of Car@SWCNT. The absorption around 2.2-2.7 eV increases with the encapsulation of β-carotene. The difference absorption spectrum (Car@SWCNT-SWCNT, a solid line) is assigned to the $1^1B_u^+$ state of β-carotene [3,6].

Dashed lines in Fig.1(b) show the photoinduced absorbance changes of SWCNT (without β-carotene) following the 2.46 eV pump pulse. Signals at delay times of 0.05 and 0.4 ps are assigned to bleaching of the S2, S3, and M1 bands and transient absorption due to excitons in the semiconducting SWCNTs (1.4-1.6 eV and 2.0-2.1 eV) [8-11]. The signals due to the semiconducting SWCNTs disappears with time constants of 0.18 and 0.88 ps. M1 bleaching decays with time constants of 0.08 and 1.1 ps. The long-lived signal observed at 20 ps is assigned to red-shift of the stationary absorption due to temperature increase.

Solid lines in Fig.1(b) show the difference of photoinduced absorbance changes between Car@SWCNT and SWCNT. A positive peak observed at 1.25 eV decays with a time constant of 100 fs. Comparing to the photoinduced signals of β-carotene in cyclohexane [12], it is assigned to the $1^1B_u^+$ state of β-carotene. Another positive signal at 2.1 eV is assigned to the $2^1A_g^-$ states of β-carotene. The lifetime in SWCNTs is 1.0 ps. It is much shorter than the lifetime of 8.5 ps in cyclohexane solution. The shortening of the lifetime means that the energy transfer occurs from the $2^1A_g^-$ state to SWCNTs. Evidence of the energy transfer from the $1^1B_u^+$ state is observed as relatively smaller intensity of the $2^1A_g^-$ signal in SWCNTs than in cyclohexane. The energy transfer rates from the $1^1B_u^+$ and $2^1A_g^-$ states to SWCNTs are estimated to be $(150\ fs)^{-1}$ and $(1.1\ ps)^{-1}$, respectively.

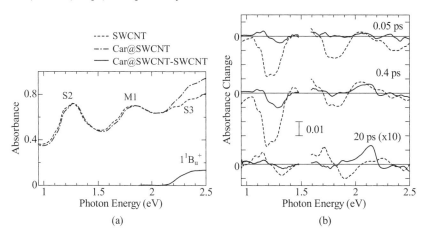

Fig. 1. (a) Stationary absorption spectra of SWCNT (dashed line) and Car@SWCNT (dash-dotted line) and the difference absorption spectrum between Car@SWCNT and SWCNT (solid line). (b) Time-resolved photoinduced absorbance change of SWCNT (dashed line) and the difference between Car@SWCNT and SWCNT (solid line) at 0.05, 0.4, and 20 ps.

A negative difference signal at 1.3 eV means that the S2 bleaching increases because of the energy transfer from β-carotene to the semiconducting SWCNTs. On the other hand, the M1 bleaching (1.7-1.9 eV) does not show clear increase. The light-harvesting function in the metallic SWCNTs is less effective than that in the semiconducting SWCNTs. The difference may be due to the large energy gap between the M1 band and the $1^1B_u^+$ and $2^1A_g^-$ states of β-carotene.

The difference signal at 20 ps has a positive peak at 2.14 ps. It remains much longer than 100 ps. It is assigned to the triplet state of β-carotene. The triplet state is efficiently generated by inter-system crossing from highly excited singlet states induced by multi-photon excitation [13]. However, the 2.46 eV pump is one-photon resonant and the two-photon excitation is negligible. Additional energy transfer from SWCNTs to β-carotene in the $1^1B_u^+$ or $2^1A_g^-$ excited state may generate the highly excited states.

Conclusions

The ultrafast optical response of the SWCNT films with and without encapsulating β-carotene has been investigated. The decay kinetics of β-carotene becomes faster in SWCNTs and the signal due to the semiconducting SWCNTs increases by the encapsulation. The light-harvesting function of β-carotene occurs in Car@SWCNT.

Acknowledgements. This work was partly supported by Grants-in-aid from JSPS (Nos. 18005013 and 19340076).

1 T. Durkop, B. M. Kim, and M. S. Fuhrer, J. Phys.: Condens. Matter **16**, R553 (2004).

2 T. Takenobu, T. Takano, M. Shiraishi, Y. Murakami, M. Ata, H. Kataura, Y. Achiba, and Y. Iwasa, Nat. Mater. **2**, 683 (2003).

3 K. Yanagi, Y. Miyata, and H. Kataura, Adv. Mater. **18**, 437 (2006).

4 N. J. Fraser, H. Hashimoto, and R. J. Cogdell, Photosynth. Res. **70**, 249 (2001).

5 T. Polívka and V. Sundström, Chem. Rev. **104**, 2021 (2004).

6 K. Yanagi, K. Iakoubovskii, S. Kazaoui, N. Minami, Y. Maniwa, Y. Miyata, and H. Kataura, Phys. Rev. **B74**, 155420 (2006).

7 K. Abe, D. Kosumi, K. Yanagi, Y. Miyata, H. Kataura, and M. Yoshizawa, Phys. Rev. **B77**, 165436 (2008).

8 M. Ichida, Y. Hamanaka , H. Kataura, Y. Achida, A. Nakamura, J. Phys. Soc. Jpn. **73**, 3479 (2004).

9 C. Manzoni, A. Gambetta, E. Menna, M. Meneghetti, G. Lanzani, and G. Cerullo, Phys. Rev. Lett. **94**, 207401 (2005).

10 O. J. Korovyanko, C.-X. Sheng, Z. V. Vardeny, A. B. Dalton, and R. H. Baughman, Phys. Rev. Lett. **92**, 017403 (2004).

11 A. Maeda, S. Matsumoto, H. Kishida, T. Takenobu, Y. Iwasa, H. Shimoda, O. Zhou, M. Shiraishi, and H. Okamoto, J. Phys. Soc. Jpn. **75**, 043709 (2006).

12 D. Kosumi, K. Yanagi, N. Nishio, H. Hashimoto, and M. Yoshizawa, Chem. Phys. Lett. **408**, 89 (2005).

13 T. Buckup, T. Lebold, A. Weigel, W. Wohlleben, and M. Motzkus, J.Photochem. Photoboil. A**180**, 314 (2006).

Direct Femtosecond Observation of Tight and Loose Ion Pairs upon Photoinduced Bimolecular Electron Transfer

Omar F. Mohammed[1], Katrin Adamczyk[2], Natalie Banerji[1], Jens Dreyer[2], Bernhard Lang[1], Erik T. J. Nibbering[2], and Eric Vauthey[1]

[1] Department of Physical Chemistry, University of Geneva, 30 Quai Ernest-Ansermet, CH-1211 Geneva 4, Switzerland
E-mail: Eric.Vauthey@chiphy.unige.ch
[2] Max Born Institut für Nichtlineare Optik und Kurzzeitspektroskopie, Max-Born-Strasse 2 A, D-12489 Berlin, Germany
E-mail: nibberin@mbi-berlin.de

Abstract. We observe tight and loose ion pairs in bimolecular electron transfer with ultrafast infrared spectroscopy. For large exergonicity tight donor-acceptor pairs do not rearrange into loose complexes before the reaction proceeds, contrasting generally accepted models.

Photoinduced electron transfer (ET) between donor-acceptor pairs, often considered as the simplest of chemical reactions, plays an important role in many areas of chemistry and biology ranging from harpooning, light harvesting, solar energy conversion, and photoconduction. Marcus theory has been successfully applied to rationalize the kinetics of ET reactions. In its classical formulation Marcus theory predicts, for reactants at a fixed reaction distance, a Gaussian dependence of the ET rate constant on the driving force. Starting from a weak exergonic case, the ET reaction rate increases with increasing energy gap between reactant and product (normal regime), until a crossover occurs, and for large exergonic cases the ET reaction rate diminishes while the energy gap continues to increase (inverted regime).

Since the pioneering work of Weller and coworkers [1, 2] bimolecular photoinduced ET, where donor and acceptor molecules can diffuse freely in liquid solution, has been intensively investigated. While the Marcus inverted regime has been reported for many types of ET reactions, it has never been observed in photoinduced bimolecular charge separation (CS) [1, 3]. Here, after an initial increase of the rate constant with driving force, the process appears to be diffusion-limited regardless of the magnitude of exergonicity. To explain this strong discrepancy usually a driving force dependent reaction distance is invoked. Because the solvation energy of the resulting ion pair increases with increasing interionic distance, the inverted region is shifted to higher exergonicity if ET occurs at longer distances. This also implies that reactants for which bimolecular ET should occur in the inverted regime when at close (tight) contact, will increase their mutual distance by diffusion to reach the crossover region with higher reaction rates. Thus, according to this hypothesis, highly exergonic CS results in the formation of loose ion pairs (LIPs; also dubbed solvent-separated ion pairs), whereas weakly exergonic CS requires contact between the reactants and yields tight ion pairs (TIPs; also named contact ion pairs).

To test whether highly exergonic CS occurs only after rearrangement of tight donor-acceptor complexes into loose reaction pairs we have investigated the photoinduced CS dynamics, and subsequent charge recombination (CR) between 3-methylperylene (MePe) donor and tetracyanoethene (TCNE) acceptor in acetonitrile solution, for which the driving force for CS ΔG_{CS} is large upon local excitation S_1-state of MePe at 400 nm ($\Delta G_{CS} = -2.1$ eV). Until now, ultrafast electronic spectroscopy has been used to study bimolecular CS and CR dynamics. A major experimental problem, however, is that disentanglement of contributions from TIPs and LIPs in transient electronic absorption spectra is hampered due to the large bandwidths and overlapping components. We instead use polarization-sensitive femtosecond infrared spectroscopy [4] to follow the distinct different dynamics of CS and charge recombination of TIPs and LIPs. We find that large exergonic electron transfer reactions predominantly occur in tight donor-acceptor pairs. In addition, tight ion pairs are found to be highly anisotropic, revealing the importance of mutual orientation of the reactants, thus demanding refinement of theoretical models relying on spherical reaction species that solely involve reaction distances.

We record transient infrared absorption spectra by probing with 150 fs infrared pulses (Figs. 1A-B). We have performed two experiments: a) local excitation (LE) of MePe at 400 nm (Fig 1A), or b) charge transfer (CT) excitation of tight MePe-TCNE contact complexes at 800 nm (Fig. 1B). In the case of LE MePe molecules in all different configurations contribute to the observed signals, i.e. tight MePe-TCNE donor-acceptor complexes, loose MePe…TCNE donor-acceptor pairs and MePe fully solvent separated from TCNE. By CT excitation of MePe-TCNE contact complexes we only probe signals of tight contact complexes. Two transient bands are located around 2150 and 2190 cm^{-1}, the frequency positions of the IR-active C≡N stretching vibrations of TCNE$^{\bullet-}$. With LE of MePe and 0.2 M TCNE (Fig. 1A) the spectral shape remains constant during the first 20 ps after local excitation, with a magnitude decay by about 30%. During the next 30 ps both bands undergo a 4-5 cm^{-1} blue shift

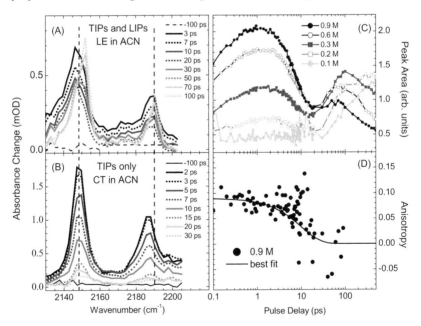

Fig. 1. Transient IR spectra showing contributions of TIPs and LIPs (A) in the case of local excitation (LE), and only TIPs (B) in the case of charge transfer (CT) excitation. The time-dependent integrated peak area as function of quencher concentration shows two time-dependent components (C), where the decay of the early time component occurs on the same time scale as the measured anisotropy decay (D).

and a 8-10 cm^{-1} narrowing. From about 50 ps onwards the shape remains essentially the same but the amplitude now grows in up to a time delay of about 200 ps, after which it remains essentially constant up to 1 ns (the maximum pulse delay in the experiment). This result shows that the spectral changes originate from two distinct spectral components, rather than from a continuous variation of the band shape. Comparison of the transient spectra recorded for different quencher concentrations (0.05 – 0.9 M TCNE) shows that these two components are always present, with the spectral component at early pulse delay dominating at high quencher concentration, and vice versa for low quencher concentration.

Fig. 1C shows the time evolution – depicted on a logarithmic scale – of the integrated area of the early C≡N stretching spectral component at 2150 cm^{-1} estimated by fitting a Lorentzian lineshape to the experimental data, measured at various TCNE concentrations. The amplitude of the spectral component at early times scales with quencher concentration, whereas the dynamics remains essentially the same. The early time component appears to originate from the fraction of donor-acceptor pairs at close range, that do not evolve by molecular diffusion. The dynamics of the frequency upshifted spectral component, however, strongly varies with TCNE concentration, hinting at a pronounced influence of diffusion on this signal component.

More insight into the origin of the two spectral components in the MePe + TCNE in ACN measurements is obtained when the results are compared with those obtained for CT excitation of MePe + TCNE in ACN (Fig. 1B). Here only the red-shifted spectral component is observed. Since in the case of CT excitation only TIPs are expected to play a dominant role, it is evident that TIPs and LIPs can be distinguished by their respective frequency positions of the C≡N stretching bands for TCNE$^{\bullet-}$, that is the component on the red side originates from TIPs and the component on the blue side from LIPs. Quantum chemical calculations suggest that in the case of TIPs the vibrational red-shift points to a stronger coupling of TCNE$^{\bullet-}$ with MePe$^{\bullet+}$, with however, solvent induced frequency shifts being of similar magnitude.

The electronic transition dipole moment for the local excitation transition is directed along the in-plane long axis of MePe. The IR transition dipole moments are aligned in the plane of TCNE, either parallel or perpendicular to the ethylenic bond, for the bands at 2190 and 2150 cm^{-1}, respectively. We have measured the anisotropy for both transient bands at early pulse delay times to be 0.09 ± 0.03 (Fig. 1D). Such a value for both bands means that TCNE$^{\bullet-}$ is oriented with its molecular plane parallel to the LE dipole of MePe. This finding is fully consistent with a sandwich co-facial arrangement.

Based on our experimental observations we conclude that two types of ion pairs exist, TIPs with face-to-face alignment of the ions with strong couplings leading to fast formation and decay rates on subpicosecond time scales, and LIPs where the CS and CR reaction rates are typically lying in the picosecond range. Considering these distinctly different orders of magnitude for the time scales for CS and CR in the case of tight and loose reaction pathways, as well as the fact that diffusional motions are slow, we conclude that rearrangement of tight reaction pairs into loose reaction pairs does not occur before charge separation proceeds, as such the predictions by the standard model are not met.

1   D. Rehm, and A. Weller, Isr. J. Chem. **8**, 259, 1970.

2   A. Weller, Pure & Appl. Chem. **54** 1885, 1982.

3   E. Vauthey, J. Photochem. Photobiol. A **179**, 1, 2006.

4   E. T. J. Nibbering, H. Fidder, and E. Pines, Annu. Rev. Phys. Chem. **56**, 337, 2005.

Ultrafast Dynamics of Dansylated POPAM Dendrimers and Energy Transfer in their Dye Complexes

J. Aumanen[1], T. Kesti[2], V. Sundström[2], F. Vögtle[3], J. Korppi-Tommola[1]

[1] Department of Chemistry, Nanoscience Center, P.O. Box 35, FIN-40014 University of Jyväskylä, Finland
[2] Department of Chemical Physics, Lund University, Chemical Center, Box 124, SE-22100 Lund, Sweden
[3] Kekulé-Institut für Organische Chemie und Biochemie der Rheinischen Friedrich-Wilhelms-Universität Bonn, Gerhard-Domagk Strasse 1, 53121 Bonn, Germany
E-mail: jukka.aumanen@jyu.fi

Abstract: We have studied internal dynamics of dansylated poly(propyleneamine) dendrimers of different generations in solution and excitation energy transfer from dansyl chromophores to xanthene dyes that form van der Waals complexes with the dendrimers.

Introduction

Multi-branched, highly symmetric molecules, dendrimers, have attracted a lot of scientific attention in recent years. We have studied ultrafast dynamics of four dansylated poly(propyleneamine) dendrimer (POPAM) generations (G1-G4) in chloroform solution and excitation energy transfer (EET) from dansyl to the dye complexed with the dendrimer. Figure 1 shows structural formulae of G1 and G3 dendrimers, and N-propyldansylamine used as a monodansyl reference compound.

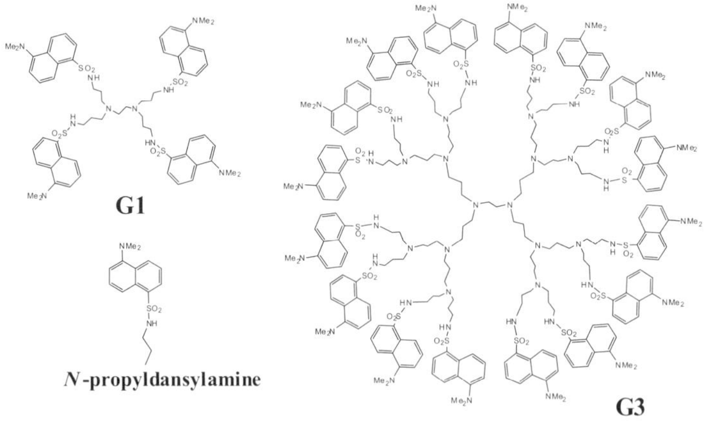

Fig. 1. Structures of G1 and G3 dansylated POPAM dendrimers and monodansyl reference compound

616

Experimental

Time-resolved fluorescence spectra were recorded with a streak camera (Hamamatsu C 4742-95). Excitation pulses (400 nm, 100 fs, 0.1 nJ, 4 MHz) were produced by frequency doubling the output of a Ti:Sapphire oscillator (Spectra Physics Tsunami). Fluorescence polarized both parallel and perpendicular to the excitation light was recorded.

Results and Discussion

In all dendrimer generations sub-nanosecond decay of fluorescence was wavelength dependent; decay in the blue end of the fluorescence band changed to a rise in the red end. This observation together with 8 nm spectral shift towards red in time indicates 15-30 ps solvent relaxation. Similar solvation rates obtained for all dendrimer generations refer to high solvent density and mobility also inside larger dendrimers.

Anisotropy decay of dansyl fluorescence turned out to depend on the dendrimer generation. Two decay time constants were resolved, one in the range of 130-63 ps and the other 320-930 ps, the former decreasing and the latter increasing as the size of dendrimer grows. Table 1 presents the time constants and the relative amplitudes of anisotropy decays. It was concluded that the shorter decay component represents the motions of the individual dansyl chromophore and the longer one overall rotation of the dendrimer. Shorter decay could, in principle, be due to EET between dansyls, but almost nonexistent spectral overlap between absorption and fluorescence bands of dansyl makes EET inefficient, suggesting local motions of the dansyls to be the main reason for the observed fast anisotropy decay. Mono-exponential anisotropy decay of the reference compound gives an upper rate limit of 51 ps for a nearly freely rotating dansyl in the dendrimer.

Table 1: Time-constants and relative amplitudes of the fluorescence anisotropy decays

Sample	τ_1 (ps)	τ_2 (ps)	V_h (nm^3)
N-propyldansylamine	51 (100 %)		
G1	140 (22 %)	310 (78 %)	2.3
G2	100 (53 %)	730 (47 %)	5.3
G3	80 (41 %)	1100 (59 %)	8.1
G4	62 (49 %)	1300 (51 %)	9.2

The longer anisotropy component was assigned to overall rotation of the dendrimers. We estimated hydrodynamic volumes of the dendrimers by using the Stokes-Einstein-Debye model to describe the overall rotation and by assuming spherical shape (Table 1). Relatively slow increase in hydrodynamic volume as generation grows suggests that in solution dendrimer branches are considerably back-folded, which results in unexpectedly small radii of rotation of the dendrimers in solution. This observation is in full agreement with previously published theoretical and experimental studies [1].

Besides pure dansylated POPAM dendrimers, their complexes that may be formed with several xanthene dyes [2], have been studied. In POPAM-eosin complex selective excitation of dansyls leads to a fast energy transfer to the complexed eosin dye [3], as illustrated in Figure 2. Fluorescence intensity of the eosin dye (circle) rises concurrent with a decay of the dansyl fluorescence (square), indicating EET from dansyl to eosin. Such fast decay component was not observed for pure G3 POPAM (triangle).

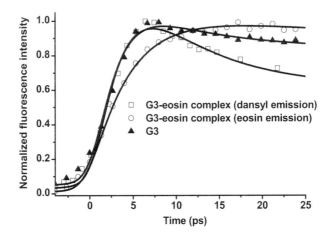

Fig. 2: Magic angle fluorescence rise and decay traces of dansylated G3 POPAM dendrimer and its eosin complex.

In the complexes the overlap integral of dansyl emission and eosin absorption is about 10000 times larger than that of dansyl-dansyl energy transfer. We estimate that the measured energy transfer rates (from 1 to 6 ps) suggest excitation energy to be transferred from the most distant dansyl of G4 POPAM to eosin(s) without help from dansyl-dansyl hopping type energy transfer.

Conclusions

Fluorescence lifetimes and spectral shifts are practically independent of dendrimer generation suggesting only very weak interactions between dansyl groups and high mobility of the solvent molecules inside dendrimers.

Generation dependent fluorescence anisotropy decays are assigned to local motions of the dansyls and the overall rotation of the dendrimer. Hydrodynamic volumes calculated from anisotropy decays suggest considerable back-folding of the dendriric arms in higher generation dendrimers.

In dendrimer-dye complexes excitation energy is transferred effectively from excited dansyl to acceptor dye without help from dansyl-dansyl hopping type EET.

1 M. Ballauff and C. Likos, Angew. Chem. Int. Ed. 43, 2998, 2004

2 V. Balzani, P. Ceroni, S. Gestermann, M. Gorka, C. Kauffmann and F. Vögtle, Tetrahedron 58, 629, 2002.

3 J. Aumanen, V. Lehtovuori, N. Werner, G. Richardt, J. van Heyst, F. Vögtle and J. Korppi-Tommola, Chem. Phys. Lett. 433, 75, 2006.

Electron Transfer in a Donor/Acceptor System Coupled to the Surface of Metal Oxide Nano-porous Films: Direct vs. Surface Confined Electron Transfer

Victor V. Matylitsky, Lars Dworak and Josef Wachtveitl

Institute for Physical and Theoretical Chemistry, J. W. Goethe-University Frankfurt Max-von-Laue-Strasse 7, D-60438 Frankfurt am Main, Germany
E-mail: matylitsky@chemie.uni-frankfurt.de

Abstract. Photophysics of a molecular donor/acceptor pair coupled to the surface of metal oxide nanoporous films was studied via transient absorbance spectroscopy. Competition between interfacial and intermolecular electron transfer was studied for the donor/acceptor pair coupled to a reactive TiO_2 surface. The subsequent transfer of the conduction band electron to the electron acceptor was analyzed by monitoring the free charge carrier absorption in the mid-IR (~5μm) spectral region.

Introduction

Interfacial electron transfer (ET) between molecular adsorbates and semiconductor nanomaterials has attracted much interest during the past years. In many applications, such as water purification [1], solar energy conversion [2], and molecular electronics [3], interfacial ET plays a key role. A detailed description of this process is complicated by the presence of surface states and structural inhomogeneities of the systems [4-6].

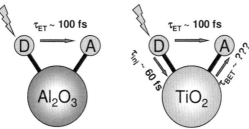

Fig. 1. Schematic representation and dynamics of the photophysical processes for a molecular donor/acceptor pair coupled to the surface of semiconductor nanoparticles.

In this work, an approach employing a molecular donor/acceptor pair coupled to the surface of nanoporous films is used. This composite model system allows to reveal the contribution of surface trap states and their influence on the interfacial ET as well as charge trapping and spatial diffusion processes in the semiconductor nanoporous films. A similar system has been previously established for the construction of a new type of photochromic device, which is based on an ET process involving a donor/acceptor pair coadsorbed onto the surface of nanocrystalline TiO_2 [7]. As a sensitizer we used the dye alizarin, which is well known as suitable electron donor [4-6] and a phosphonated derivative of viologen (PV^{2+}) [8] can serve as an electron acceptor.

Experimental Methods

TiO$_2$, and Al$_2$O$_3$ nanoporous films were prepared from colloidal nanoparticles in acidic solution (5-10 nm in diameter). As an electron donor we used alizarin (purity 97%, Aldrich). The electron acceptor PV^{2+} has been synthesized in the laboratory of Dr. J. E. Moser (Ecole Polytechnique Fédérale Lausanne, Switzerland). The experimental setup used for the transient absorbance measurements was previously described [6, 9]. The desired excitation pulses (480 nm, 20 fs) were obtained from a home built nonlinear optical parametric amplifier. The samples were probed in the visible (347–675 nm) and mid-IR (5.1–5.5 µm) spectral regions.

Results and Discussion

In order to study the intrinsic ET in the donor/acceptor couple, the Al$_2$O$_3$ film was used as a bearing substrate. The coupling of the dyes should be similar in Al$_2$O$_3$ and TiO$_2$, but in Al$_2$O$_3$ any electron injection into the semiconductor is prevented due to its much higher band gap ($\sim 8 - 9$ eV). A complete set of the transient absorbance data obtained for the alizarin sensitized Al$_2$O$_3$ film in the visible spectral range is shown in Figure 2a. The three dimensional plot of the alizarin/Al$_2$O$_3$ system consists of three dominant characteristics: the excited state absorption (ESA) in the spectral range from 350 to 580 nm, the stimulated emission (SE) for probe wavelengths longer than 580 nm and the ground state bleach (GSB) with a maximum at $\lambda_{probe} \approx 480$ nm, that is seen as a gap in the ESA band. Finally, the alizarin/Al$_2$O$_3$ system reflects the simple relaxation of the excited state of coupled alizarin.

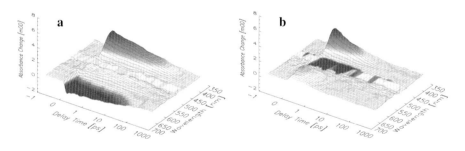

Fig. 2. Transient absorption data of alizarin sensitized Al$_2$O$_3$ film a.) without electron acceptor PV^{2+} and b.) with PV^{2+}. The time scale is plotted linear from -1 to 1 ps and logarithmic thereafter.

Additional coupling of the electron acceptor PV^{2+} to the alizarin sensitized Al$_2$O$_3$ film leads to significant changes in the photoinduced dynamics, which become evident from the comparison of the respective 3D plots in Figures 2a and 2b. First of all, the negative absorbance difference in the red spectral region, which was assigned as the SE of alizarin, disappears completely. Instead of SE, a positive difference signal is developed in the first 200 fs. The absence of the stimulated emission is in agreement with a complete quenching of the alizarin fluorescence observed upon addition of PV^{2+}. Moreover, at the same time two absorption bands with maxima at 400 and 600 nm (maxima of the absorption bands of monoreduced PV$^{\bullet+}$ radical) were formed.

The effects of the fluorescence quenching observed by steady state spectroscopy and absence of SE in time-resolved experiments as well as appearance of the new absorption bands unambiguously indicate that intrinsic ET ($S^* + PV^{2+} \rightarrow S^+ + PV^{\bullet+}$) in the donor/acceptor couple proceeds on the ultrafast time scale. A global fit analysis of the experimental data results in the intrinsic ET time of ~100 fs.

In the case of TiO_2 semiconductor as the bearing substrate, the electron injection from photoexcited alizarin to the TiO_2 conduction band ($S^* \rightarrow S^+ + e^-_{CB}$) has to be taken into account. A careful analysis of the time resolved experimental data of the alizarin/PV^{2+} pair adsorbed onto a TiO_2 film shows that the initially generated photoexcited state of alizarin relaxes during the first 50 fs. During this time formation of the monoreduced $PV^{\bullet+}$ radical and injection of the electron to the conduction band of TiO_2 were observed. Moreover, additional transient absorbance experiments in the mid-IR spectral region were done in order to unambiguously observe the possible reduction of the electron acceptor PV^{2+} by the electron in the conduction band ($PV^{2+} + e^-_{CB} \rightarrow PV^{\bullet+}$). These measurements show no hints of ET from the conduction band of TiO_2 to PV^{2+} for $t_{delay} > 300$ fs, where 300 fs is the resolution of our experimental setup in the mid-IR spectral range. These results are in some disagreement with the experimental results for PV^{2+}/TiO_2, where efficient and ultrafast ET of the photoexcited electron from the conduction band of TiO_2 to PV^{2+} was observed.

Conclusions

A detailed analysis of the photophysical processes in the photoexcited donor/acceptor pair coupled to the surface of different metal oxide mesopourous films was performed. From a comparison of the time resolved data for alizarin/TiO_2 and alizarin/TiO_2/PV^{2+} coupled systems we can conclude that ET between the molecular donor and acceptor coadsorbed onto a TiO_2 surface proceeds rather directly through "space" than through the surface of the bearing semiconductor.

Acknowledgements. This work is supported by the Deutsche Forschungsgemeinschaft (DFG, WA 1850/2-2). We gratefully acknowledge K. Neumann for his help in the measurements of the time resolved spectra in the mid-IR spectral range. We thank Dr. R. Eichberger and U. Michalczik (Hahn-Meitner-Institute, Berlin, Germany) for their help in preparation of the nanostructured films and Dr. J. E. Moser (Ecole Polytechnique Fédérale de Lausanne, Switzerland) for providing us the chemical samples.

1   T.X. Wu, G.M. Liu, J.C. Zhao, H. Hidaka and N. Serpone, J. Phys. Chem. B **103**, 4862, 1999.

2   B. O'Regan and M. Grätzel, Nature **353**, 737, 1991.

3   A. Nitzan and M.A. Ratner, Science **300**, 1384, 2003.

4   R. Huber, S. Spörlein, J.E. Moser, M. Grätzel and J. Wachtveitl, J. Phys. Chem. B **104**, 8995, 2000.

5   R. Huber, J.E. Moser, M. Grätzel and J. Wachtveitl, J. Phys. Chem. B **106**, 6494, 2002.

6   V.V. Matylitsky, M.O. Lenz and J. Wachtveitl, J. Phys. Chem. B **110**, 8372, 2006.

7   M. Biancardo, R. Argazzi and C.A. Bignozzi, Inorg. Chem. **44**, 9619, 2005.

8   T. Watanabe and K. Honda, J. Phys. Chem. **86**, 2617, 1982.

9   K. Neumann, M. Verhoefen, I. Weber, C. Glaubitz and J. Wachtveitl, Biophys J. **94**, 4796, 2008.

Aqueous Proton Transfer Pathways in Bimolecular Acid-Base Neutralization

Omar F. Mohammed[1], Katrin Adamczyk[1], Dina Pines[2], Ehud Pines[2], and Erik T. J. Nibbering[1]

[1] Max Born Institut für Nichtlineare Optik und Kurzzeitspektroskopie, Max-Born-Strasse 2 A, D-12489 Berlin, Germany
E-mail: nibberin@mbi-berlin.de
[2] Department of Chemistry, Ben Gurion University of the Negev, P.O. Box 653, Beer-Sheva 84125, Israel
E-mail: epines@bgu.ac.il

Abstract. We expand the classic Eigen-Weller reaction model with solvent-switch pathways, mediating proton transfer between acids and bases, having one or several water molecules, activated by the solvent and controlled by the base strength.

Modern discussions of acid-base reactions have evolved from the seminal studies of Eigen and Weller [1,2]. The general kinetic approach for acid-base reactions in aqueous solutions consists of three reaction branches [2]: (a) direct proton exchange between acid and base, (b) acid dissociation to solvent (protolysis) followed by proton scavenging by the base, and (c) water hydrolysis by the base followed by the neutralization reaction of the acid by the hydroxyl anion. Judging by the magnitude of the reaction radius in typical (diffusion-controlled) acid-base reactions it has been estimated that up to 2-3 water molecules separate when acid and base exchange a proton through pathway (a) [1]. In reality, however, this value is likely to be an average for several encounter complexes each ultimately leading to proton transfer by a mechanism which in general involves n water molecules and k rearrangement steps. We present a detailed femtosecond infrared study of the aqueous neutralization of a photoacid, pyranine, by a family of carboxylate bases ($^-$OOCCH$_{3-x}$Cl$_x$) ($x = 0\text{-}3$) with a broad range of reactivity. Photoacids can be used as a means to follow proton transfer dynamics to a neutralising base in real time by photoinitiation. Monitoring vibrational marker modes provide unprecedented insight into the elementary reaction stages of this complex bimolecular solution phase reaction. From our modeling using reversible proton transfer reaction steps we conclude that protons are channeled through several pathways in the reactive encounter complexes, consisting of $n = 1$ or $n \geq 2$ water molecules serving as a solvent bridge between acid and base.

The outcome of the observed dynamics, i.e. the reaction rates and yields depend on the relative strengths and concentrations of acid and base. Whereas with the stronger bases acetate and monochloroacetate ($x = 0\text{-}1$) the proton transfer proceed up to full reaction yield, with the weakest base trichloroacetate ($x = 3$) only a fraction of the reactants will show a net transfer to the products. Thus we are able to dictate the outcome of the reaction dynamics by tuning the number of chlorine atoms x.

We have identified two innermost types of encounter complexes (tight, $n = 0$, and loose, $n = 1$), resulting in a sub-150 fs proton dissociation lifetime of the photoacid [3-6]. The step-wise, von-Grotthuss type [4-6], proton transfer in loose complexes involves a first fast step leading to a H$_3$O$^+$ like cation resembling the proton solvation core in the Eigen cation, H$_9$O$_4^+$, and a second and final transfer to the base on much slower picosecond time scales. The transient appearance of the hydrated proton in the loose ($n = 1$) complex increases with decreasing reactivity (basicity) of the carboxylate base as measured in bulk water. Within the smallest reactive complexes ($n = 0,1$) we observe the first reactive step to be always proton dissociation from the acid either directly to the proton acceptor or to the water molecule spacing between acid and base occurring within experimental time resolution (150 fs) [3-5]. This is irrespective of the acidity of the carboxylate acid used in the range of $0 < pK_a < 5$ (Fig. 1a). The rise of the solvated proton band always occurs within time resolution. This is even true for the relatively slow proton transfer reaction from HPTS to $^-$OOCCCl$_3$ ($x = 3$). This points to a base-assisted solvent switch mechanism within the reactive ($n = 1$) complexes that we tentatively ascribe to Coulomb interactions between the charged acid and bases.

We have applied a unified reaction dynamics mechanism where we have modeled all possible configurations between acid and base-by tight ($n = 0$), loose ($n = 1$) and solvent switch ($n \geq 2$) complexes, as well as by acid and base fully separated by the solvent (protolysis pathway). Whereas the fully separated acid and base first have to diffuse before interacting, all other complexes are reactive and are interchanging through reversible proton transfer steps. The issue arising when investigating the weaker di- and trichloroacetate ($x = 2,3$) bases, is the increasing importance of reversibility of proton transfer with decreasing base strength. A more appropriate modeling of the observed reaction kinetics includes forward and backward proton transfer obeying detailed balancing for every reaction step. We thus solve the rate equations using time-independent (reversible) rate constants, that could be diffusion-limited for a fraction of the possible reaction steps. We avoid the inclusion of time-dependent reaction rates , such as in the Debye-Smoluchowski model using Collins-Kimball on-contact reaction probabilities, that have diminished importance in slower on-contact reactions and making the inclusion of detailed balancing a nontrivial procedure. The slower the on-contact proton transfer proceeds, the more useful steady-state rate constants become. We have recently applied this approach to the case HPTS + mono-, di- and trichloroacetate ($x = 1$-3) [5, 6].

We report on the comparative fitting result for $^-$OOCCH$_{(3-x)}$Cl$_x$ ($x = 1$-3). From our fitting results we conclude that the main channel for proton transfer for most bases ($x = 0$-2) is through the loose ($n = 1$) complexes. Step-wise proton shuttling through water provides a route for proton transfer which circumvents further desolvating the acidic and basic groups necessary for direct transfer. Only for the weakest base, trichloroacetate ($x = 3$) the net proton transfer becomes slow enough for other reaction pathways with larger water bridges ($n \geq 2$) to compete significantly. As such only satisfactory fit results (solid lines) are obtained when including an additional water bridge with more than one water molecule. Proton transfer through the loose pathway only leads to a strong deviation (dashed) of the hydrated deuteron signal (Fig. 1b).

The final transfer step of the aqueous proton in the loose and solvent switch complexes arriving at the base is activated and controlled by the basicity of the proton acceptor: the lower the basicity the slower the proton transfer rate. The key role played by the mediating water is clear when comparing the reaction rates found here with those of previously investigated proton transfer of photoacid to water and to accepting bases. The isotope effect on the decay kinetics of the hydrated proton/deuteron intermediate (Fig. 1c) exhibits features indicating that the observed kinetics are not caused by a single elementary reactive step such as the unimolecular dissociation of an O-H bond. These features are: (a) non-exponential decay of both the H$_3$O$^+$ and D$_3$O$^+$ kinetics that can only be satisfactorily fitted with at least two exponential decay curves, (b) an apparent isotope effect that changes with reaction time, and (c) an isotope effect that depends on the relative reactivity of the acid-base pair. The apparent magnitude of the isotope effect changes between 1.9 and 3 getting larger at long decay times. The maximum largest isotope effect is found for $^-OOCCHCl_2$ where the change in free energy of the proton transfer between donor and acceptor is about zero. Here the forward and backward rate constants of the proton intermediate – base system have been found to be almost equal, ie, $(75\ ps)^{-1}$ and $(85\ ps)^{-1}$. We conclude that the measured time-dependent isotope effect in the observed kinetics may be decomposed into several elementary contributions. We suggest that besides the actual proton transfer solvent shell rearrangements are actually involved in facilitating the proton transfer reaction.

In harmony with our reaction modeling, only three different populations of acid-base pairs have been directly identified by our experimental method ($n = 0, 1, n$). Only when other marker bands are identified of the larger solvent switches new structural information may be obtained. Observation of a broad featureless absorption throughout the mid-infrared spectral region may suggest access to this. However, we have found that the pH-dependence of this signal exhibits a time-dependent behaviour opposite to the net dissociation dynamics of pyranine (Fig 1d). Even more so, for the methoxy-derivative of pyranine this broad featureless absorption is also observed, excluding the involvement of the proton dissociation pathways to this signal.

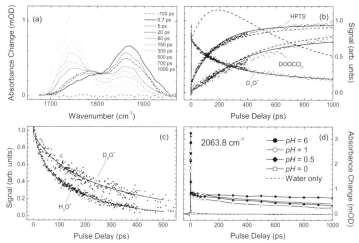

Fig. 1. Transient IR spectra showing contributions of the hydrated deuteron D_3O^+ at 1850 cm^{-1} and the carboxylate acid at 1730 cm^{-1} for 2 M trichloroacetate (a). Appropriate fits of the marker modes signals (dots) are not possible for the 2 M trichloroacetate data when only the $n = 0,1$ pathways are included (b). The decay of the hydrated deuteron D_3O^+ and hydrated proton (H_3O^+) show a clear isotope effect with 1 M dichloroacetate (c). The broad featureless absorption shows pH-dependent dynamics (d).

1   A. Weller, Progr. React. Kinet. **1**, 187, 1961.

2   M. Eigen, Angew. Chem. Int. Ed. **3**, 1, 1964.

3   M. Rini, B.-Z. Magnes, E. Pines and E.T.J. Nibbering, Science **301**, 349, 2003.

4   O. F. Mohammed, D. Pines, J. Dreyer, E. Pines and E. T. J. Nibbering, Science **310**, 83, 2005.

5   O. F. Mohammed, D. Pines, E. T. J. Nibbering, E. Pines, Angew Chem. Intl. Ed. **46**, 1458, 2007.

6   O. F. Mohammed, D. Pines, E. Pines, E. T. J. Nibbering, Chem. Phys. **341**, 240 (2007).

The solvated electron dynamics in aqueous bromide studied by three-pulse-spectroscopy

Martin K. Fischer, Hristo Iglev, and Alfred Laubereau

Physikdepartment E11, Technische Universität München,
James-Franck-Straße, 85748 Garching, Germany
E-mail: mfischer@ph.tum.de

Abstract. We demonstrate manipulation of the ultrafast electron detachment and recombination dynamics using femtosecond pump-repump-probe spectroscopy. The repump-induced kinetics provide convincing evidence for formation of a bromide-electron pair with bonding energy of 0.15 ± 0.02 eV and effective lifetime of 19 ± 2 ps at room temperature.

Introduction

We report on the electron photodetachment dynamics of aqueous bromide. The process is studied by femtosecond pump-probe and pump-repump-probe (PREP) spectroscopy. In latter technique an additional 810-nm pulse is used for secondary excitation of the intermediate species involved in the photodetachment process. Three-pulse spectroscopy was applied by Barbara and coworkers [1] in their investigation on the equilibrated solvated electron. The technique was further developed by Schwartz et al. [2] demonstrating an optical manipulation of electron transfer reactions. Using a modification of the technique convincing evidence for the formation of bromine-electron pair will be presented.

Experimental Methods

The Ti:Sapphire laser system used in the present study is described in details elsewhere [3]. The investigated sample is 15 mM NaBr in aqueous solutions in a free-flow sample jet of 100 µm thickness. The solution temperature is varied from 5 to 65 °C. The photodetachment of aqueous bromide is reached by the forth harmonic at 202 nm and pulse energy of 1 µJ. The changes of optical density ΔOD induced in the samples are monitored near to the absorption maximum of the solvated electron at 700 nm. The data are measured with polarization resolution and time resolution of about 150 fs. The secondary excitation of the excess electron is utilized by additional, third pulse at 810 nm, called repump pulse. The PREP difference-signal is calculated by subtracting the three and two-pulse data. The pulse sequence and the definition of the delay times used in our experiment are schematically illustrated in Fig. 1d.

Results and Discussion

Using femtosecond pump-probe spectroscopy we studied the electron photodetachment dynamics after one-photon excitation of aqueous bromide. CTTS transition produces a state that is bound only by the polarization of the surrounding water molecules [4]. The first intermediate shows a broad absorption spectra in the near IR and a lifetime of about 150 fs. Subsequent separation of the excess electron from the parent atom and assembling of an atom-electron pair is suggested [5,6]. The non-equilibrated atom-electron complex is characterized by an absorption spectrum centered at 810 nm. The well-known absorption band of the hydrated electron is reached after 1 ps. It should be noted, that the spectral signatures of the quasi-equilibrated atom-electron pair and the solvated electron are very similar. The subsequent slower dynamics is governed by the competition of pair dissociation and geminate recombination. This physical picture is consistent with the MD simulation of the halide CTTS system [5,6].

In order to verify the existence of the predicted atom-electron pair in the halide CTTS systems PREP spectroscopy was applied. Examples for the time-resolved data measured in aqueous bromide after excitation at 202 nm and re-excitation at 810 nm are shown in Fig. 1. The pulse sequence and the definition of the used delay times are schematically shown in Fig. 1d. The time evolution of the absorbance changes measured at 700 nm is depicted as a contour plot in Fig. 1c. Data measured for fixed repump time t_{12}= 0.5 ps and 170 ps are presented in Figs. 1f and g, respectively. The repump-induced bleaching is seen to peak for short probe delay at $t_{23} \sim 0.15$ ps. The feature is assigned to cooling of the electron solvent cavity and occurs with time constant of 700±100 fs. For long probe delay t_{23} a significant absorption increase is noted in Fig. 1f. The feature does not occur for late repump pulses (Fig. 1g). The finding clearly indicates a rise of the number of surviving solvated electrons for properly delayed repump pulses.

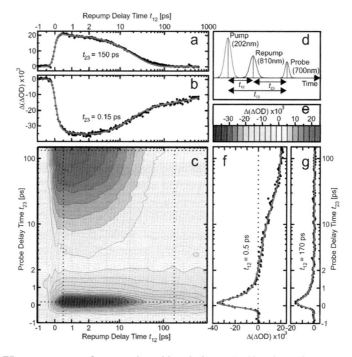

Fig. 1. PREP spectroscopy of aqueous bromide solution. **a, b,** Signal transients measured in the bleaching maximum at t_{23} = 0.15 ps (b) and in the absorption maximum at t_{23} = 150 ps (a) **c,** Two dimensional contour plot of the absorption changes as a function of both delay times. **d,** Pulse sequence, the used wavelengths and definition of the delay times used in the pump-repump-probe experiment. **f, g,** Signal transients measured at t_{12} = 0.5 and 170ps.

Of special interest are the data on the temporal evolution of both characteristic features of bleaching and recombination suppression. Figs. 1a and b show data for fixed t_{23} = 150 ps and 0.15 ps, respectively. As mentioned before, the bleaching (Fig. 1b) is proportional to the number of the excess electrons, except the small overshoot at shorter delay time (< 1 ps) due to the stronger absorption of the non-equilibrated bromine-electron pair. The transient signal measured 150 ps after the 810 nm pulse shows a significant variation in the relaxation dynamics. The rapid initial signal increase is followed by an exponential decay with characteristic time constant of 19±2 ps (see Fig. 1a). The dynamics is obviously due to suppression of the recombination process in the bromine-electron pair.

The observation is the first direct evidence for the existence of a contact pair in the electron photodetachment of aqueous halides. In order to determine the free energy barrier for pair dissociation ΔG we measured the effective lifetime of the contact pair τ_{pair} at various temperatures. The observed pair lifetime decrease from 25±2 to 9±2 ps increasing the solution temperature from 278 to 335 K, respectively. The temperature dependence of the pair lifetime is summarized in Fig. 2 via Arrhenius plot: $ln(\tau_{pair})$ = A + $\Delta G/k_B T$. The data show a free energy barrier for dissociation of the bromine-electron pair of ΔG = 0.15±0.02 eV. The number is in qualitative agreement with the values reported for iodide (0.08 eV [7] and 0.11±0.01 eV [8]) and for chloride (0.08 eV [5]).

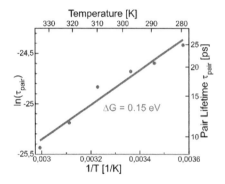

Fig. 2. Arrhenius-plot of the lifetime of the contact pair in respect to the temperature. Increasing the temperature from 278 to 335 K reduces the lifetime from 25 to 9 ps.

References

1   J. Alfano, P. Walhout, Y. Kimura, and P. Barbara, J. Chem. Phys. 98 (1993), 5996
2   I. Martini, E. Barthel, B. Schwartz, Science Vol 293 (2001) 462-465
3   A. Thaller, R. Laenen, and A. Laubereau, J. Chem. Phys. 124 (2006), 024515
4   S. Bradforth, P. Jungwirth, J. Pys. Chem. A106 (2002) 1286-1298
5   A. Staib, D. Borgis, J. Chem. Phys. 104 (1996) 9027
6   W. Sheu, P. Rossky, J. Phys. Chem. 100 (1996) 1297-1302
7   J. Kloepfer, Vilchiz, Lenchenkov, Chen, Bradforth, J. Chem. Phys. 117 (2002) 766-778
8   H. Iglev, A. Trifonov, A. Laubereau, et al., Chem. Phys. Lett. 403 (2005) 198-204

Naphthalene Bisimides: on the Way to Ultrafast Opto-electronic Devices

I. Pugliesi[1], P. Krok[1], A. Błaszczyk[2], M. Mayor[2,3], and E. Riedle[1]

[1]LS für BioMolekulare Optik, LMU München, Oettingenstrasse 67, D-80538 Munich, Germany
[2]Institute for Nanotechnology, Forschungszentrum Karlsruhe GmbH, P.O. Box 3640, D-76021 Karlsruhe, Germany
[3]Department für Chemie, Universität Basel, St. Johanns-Ring 19, CH-4056 Basel, Switzerland
E-mail: igor.pugliesi@physik.uni-muenchen.de

Abstract: For core-substituted naphthalene bisimides and their dimers we observe ultrafast charge transfer and Resonance Energy Transfer processes that change their conduction properties. This makes them suitable candidates for optoelectronic switches with terahertz response times.

Naphthalene bisimides as candidates for opto-electronics

Understanding the excited state energy dissipation mechanisms of core-substituted naphthalene bisimides (NBI) plays an important role in their design as ultrafast opto-electronic switches in nano-electronic devices. We envision single NBIs clamped between two electrodes at the N-termini. A sub-picosecond excitation should then control the conductivity through the molecules. A bi-chromophore should allow an additional control by a preceding pulse of different color. From our spectroscopic investigations of differently substituted NBIs we want to derive design rules for optimized properties and characterize the dynamic response.

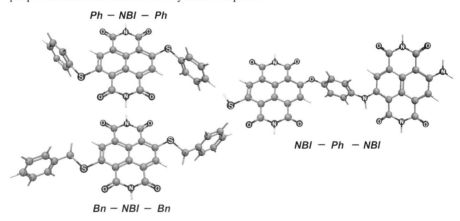

Fig. 1. Naphthalene bisimides (NBI) investigated by time resolved pump probe spectroscopy (Ph = Phenyl, Bn = Benzyl)

So far most spectroscopic studies on this class of molecules have concentrated on determining how steady state absorption and emission properties are affected by core substitution [1,2]. Most notably it is found that the linker atom in a core-substitution allows for the tuning of the lowest energy absorption band. Much less is known about their excited state dynamics [3]. Here we present a time-resolved study on a class of benzyl-alkyl-thio core-substituted NBIs, and a related dimer (see figure 1) combined with ab initio calculations of their excited states. The fluorescence quenching in NBIs where the phenyl group is directly attached to the sulfur core substituents (as is the case for Ph-NBI-Ph) is identified as ultrafast electron transfer. We show that the complex excited state dynamics is not just a simple charge transfer process from the phenyl substituent to the NBI chromophore but that it is accompanied by large conformational changes. In the dimer NBI-Ph-NBI, excitation of the blue absorbing (oxygen-sulfur core substituted) chromophore (see figure 1) does also lead to a complex decay pattern. However, the first step is unexpectedly not a charge transfer from the phenyl to the chromophore but rather an ultrafast Resonance Energy Transfer (RET) of about 150 fs from the blue-absorbing to the red-absorbing (nitrogen-nitrogen core-substituted) chromophore. Such fast RET rates have so far only been observed in oligo(p-phenylenevinylene)-fullerene dyads [4].

Ultrafast processes in the naphthalene bisimides

We have performed time-resolved measurements in a broadband spectrometer setup using the output of a noncollinear parametric amplifier (NOPA) as pump pulse and white light generated in a CaF_2 plate as probe pulse. With a duration of 30 fs for the pump pulses and the predominantly linear chirp of the probe pulse the temporal resolution was about 100 fs. The excitation wavelength was adjusted to 520 nm for all sulfur core substituted NBIs, while for the dimer the oxygen-sulfur core-substituted chromophore was excited with 490 nm and the nitrogen-nitrogen core-substituted chromophore with 608 nm. The probe light was used from 450 to 720 nm. All substances were dissolved in chloroform

The transient spectra and time traces recorded for Ph-NBI-Ph (see figure 2) show a pronounced reshaping in the picosecond time scale and complex multiexponential time behavior. This is in contrast to Bn-NBI-Bn where a uniform decay ($\tau = 41$ ps) of all signals is observed. To interpret the time traces a rate model with two intermediate states has to be assumed resulting in a fast step with $\tau = 1$ ps and two equal decay components of 6 ps. In conjunction with ab initio calculations, this excited state behavior can be explained by photo-induced charge transfer from the directly bound donor to the chromophore upon an asymmetric conformational change: after electronic excitation the molecule accesses the first intermediate state with 1 ps by twisting the phenyl ring into the plane of the chromophore. In this quasi-planar conformation the benzene localized π orbital can overlap with the π system of the NBI chromophore and donates an electron to the initially depopulated NBI-π orbital. From this partial-CT state the molecule accesses the second full-CT state within 6 ps by twisting the whole phenyl-thio substituent out of the plane of the NBI chromophore and thereby lowers its energy by at least 1 eV. With a further 6 ps time constant the molecule returns to the ground state by electron back transfer.

Time traces for the di-chromophore NBI-Ph-NBI obtained by excitation of the blue absorbing (oxygen-sulfur substituted) chromophore also show complex excited state dynamics and could best be fitted with a triexponential function that had a fast step of 150 fs and two slower decay times of 1.4 and 23 ps. Excitation of the red-absorbing (nitrogen-nitrogen substituted) chromophore leads to a biexponential decay whose time constants are within the experimental error margins equal to the slow decay times observed when the blue absorbing chromophore is excited. This indicates that the fast component is a RET from the blue absorbing to the red absorbing chromophore, while the slower time constants can again be attributed to charge transfer from the phenyl linker to the red absorbing chromophore and subsequent relaxation into the ground state.

The photoinduced electron transfer observed in Ph-NBI-Ph can drastically change the conduction properties of the dye since it generates a Coulomb blockade on the chromophore. The transfer leads to a double occupancy of the HOMO and a single electron in the LUMO. Since the LUMO corresponds to the conduction band of a n-type single molecule conductor, the optical investigation renders information on the potential electrical behavior of the naphthalene bisimide dyes. The observed picosecond charge transfer time means that also the conductivity is stabilized in such a short time. As a consequence, the investigated molecules might become first examples of opto-electronic switches with response times approaching the THz range.

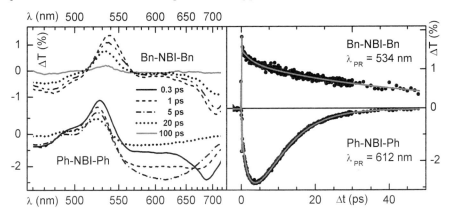

Fig. 2. Transient absorption spectra and time traces of Bn-NBI-Bn and Ph-NBI-Ph in chloroform.

Similarly, the dimer NBI-Ph-NBI would be a THz fluorescence switch. This could be achieved by using a two-color excitation scheme. If the nitrogen-nitrogen substituted NBI is excited before the oxygen-sulfur core-substituted NBI, then the phenyl group bound between the two chromophores will donate its available electron to the former, while the latter can fluoresce after optical excitation from a second delayed pulse. If the nitrogen-nitrogen substituted NBI is not excited, then fluorescence from the oxygen-sulfur core-substituted NBI is quenched by Resonance Energy Transfer to the nitrogen-nitrogen substituted NBI.

In the present examples of NBIs the rather fast back transfer limits the period the molecules could be useful opto-electronic devices. It has already been shown that alkylamino-substituted NBIs have a fluorescence quantum yield of 60% [2] and we can therefore expect that modification in this direction will lead to nanosecond times and solve the issue of the back transfer. By suitable chemical bridges the large amplitude motions of the core-substituents could be influenced and thereby the speed and the efficiency of the initial charge transfer.

1   A. Blaszczyk, M. Fischer, C. von Hänisch, and M. Mayor, Helvetica Chimica Acta **89**, 1986, 2006.

2   Ch. Thalacker, C. Röger, and F. Würthner, J. Org. Chem. **71**, 8098, 2006.

3   F. Chaignon, M. Falkenström, S. Karlsson, E. Blart, F. Odobel and L. Hammarström, Chem. Commun. 64, 2007.

4   P. A. van Hal, R. A. J. Janssen, G. Lanzani, G. Cerullo, M. Zavelani-Rossi, and S. De Silvestri, Phys. Rev. B **64**, 075206, 2001.

Ultrafast Charge Photogeneration in MEH-PPV Charge-Transfer Complexes

Artem A. Bakulin[1], Dmitry Yu. Paraschuk[2], Maxim S. Pshenichnikov[1], Paul H.M. van Loosdrecht[1]

[1] Zernike Institute for Advanced Materials, University of Groningen, Nijenborgh 4, 9747 AG Groningen, The Netherlands
E-mail: a.a.bakulin@rug.nl
[2] Faculty of Physics and International Laser Center, Lomonosov Moscow State University, Leninskie Gory, 119991 Moscow, Russia

Abstract. Visible-pump – IR-probe spectroscopy is used to study the ultrafast charge dynamics in MEH-PPV based charge-transfer complexes and donor-acceptor blends. Transient anisotropy of the polymer polaron band provides invaluable insights into excitation localisation and charge-transfer pathways.

Introduction

The process of charge photogeneration in conjugated polymer based materials is of key importance for organic solar cells and photodetectors [1]. It is well-known that exciton dissociation into charges is extremely fast (<100 fs) and efficient [2] in polymer-fullerene blends which are usually considered as a simple mixture of two compounds without noticeable charge-transfer interaction in the electronic ground state. At the same time, the formation of a weak ground-state charge-transfer complex (CTC) has recently been identified in a number of blends of conjugated polymers with low molecular weight organic acceptors [3]. These donor-acceptor CTCs are interesting for an extended photosensitivity in the red and near-IR ranges due to the formation of a CTC absorption band in the optical gap of the polymer. In such capacity, polymer-based CTCs constitute a promising way of developing low-bandgap materials for plastic solar cells and photodiodes.

Recently, it has been demonstrated that visible pump – IR probe spectroscopy is one of the most efficient ways of studying the initial dynamics of photogenerated charges in conjugated polymers [1]. In such photoinduced absorption (PIA) experiment, the concentration of (positive) charges on the polymer chains is monitored via the time evolution of characteristic IR bands associated with the photoinduced charges. Amongst those bands [4], the so-called low-energy (LE) polaron transition around 3 μm is the most suitable for IR optical detection (Fig.1a). In particular, it allows polarization-sensitive PIA experiments that can provide information on charge mobility [5] and energy transfer among conjugated segments.

Results and Discussion

Here we report on the early-time photophysics of polymer-acceptor blends with a pronounced ground-state CTC formation. The CTCs formed between a MEH-PPV polymer and two organic acceptors (TNF, DNAQ) with different electron affinities were used and a blend of MEH-PPV with a fullerene derivative (PCBM) was used as a reference. Figure 1b compares the time evolution of the LE PIA band in MEH-PPV/TNF, MEH-PPV/DNAQ CTCs, and in a blend of MEH-PPV with PCBM probed

631

at 3300 cm-1. The CTCs were excited at 650 nm (i.e. at the red flank of the CTC absorption band) while the MEH-PPV/PCBM blend was excited near the polymer absorption maximum at 540 nm. The LE band in the MEH-PPV/PCBM blend displays a decay with time constants of 3.5 and 30 ps including 50% constant offset; therefore approximately half the photoexcited charges can be considered long-lived. In contrast, the PIA decay in the CTCs has a pronounced fast component at multiple time scales up to 30 ps. Although the weights and time constants of the fast components are slightly different for both CTCs, the decrease of PIA during the first 100 ps is much more prominent than in the MEH-PPV/PCBM blend. Most likely, PIA decrease due to back electron transfer to the donor accompanied by CTC relaxation to the ground state. The observed recombination results in a long-time survival probability of the separated charges in CTCs of around 5% which is approximately a factor of 10 lower than in MEH-PPV/PCBM blend.

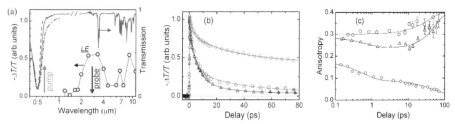

Fig. 1. a) Curves present the transmission of different samples: MEH-PPV/DNAQ (solid), MEH-PPV/TNF (dash-dotted) CTCs, and MEH-PPV/PCBM blend (dashed). Open circles show the PIA spectrum of MEH-PPV/DNAQ CTC excited at 650 nm at 1 ps delay time. b) Isotropic PIA transients of the LE band in MEH-PPV/DNAQ (circles) and MEH-PPV/TNF (triangles) CTCs excited at 650 nm, and MEH-PPV/PCBM blend excited at 540 nm (diamonds). c) Transient anisotropies for the same samples as in (b).

Figure 1c presents transient anisotropies observed for the LE band in the MEH-PPV/PCBM excited at 540 nm and in the CTCs excited at 650 nm. In MEH-PPV/PCBM, the initial anisotropy value is merely 0.18. This is consistent with the loss of the polarization memory during exciton diffusion before the charge separation. The following anisotropy decay can be assigned to the hole transport in the polymer. The anisotropy in the CTCs is persistently higher than in the MEH-PPV/PCBM blend. The initial anisotropy value of 0.3 indicates that energy migration does not play an important role in CTCs like it does in MEH-PPV/PCBM blend and, therefore, the charge separation is confined within the initially-excited CTC pair. The delayed anisotropy rise cannot be explained within a single CTC ensemble model as it would imply that the memory for the initial polarization is recovered after having been lost. This leads us to the conclusion that there are at least two sub-ensembles amongst the excited CTCs. While most of the charge-induced LE transitions have the initial anisotropy of 0.3 and recombine within 50 ps, some of them (about 5%) live longer than 1 ns and have a constant anisotropy value of 0.4.

A possible way to diminish geminate recombination in CTCs is to provide a consecutive electron transfer to another acceptor with a higher electron affinity and mobility. We put this idea under scrutiny test by performing the PIA experiments on the blends of CTCs with C60 fullerene. Figures 2a,b compare the isotropic PIA decays in CTCs with those in ternary blends MEH-PPV/DNAQ/C60 and MEH-PPV/TNF/C60. In the MEH-PPV/DNAQ/C60 blend (Fig.2a), the amplitude of the slow component has increased by the factor of 3 compared to the MEH-PPV/DNAQ

CTC implying that the supplementary acceptor enhances generation of long-lived charges. In contrast, in the TNF-containing materials there is no difference between the simple CTC and the ternary blend (Fig.2b). We assign this to a higher electron affinity of TNF which makes the CTC-fullerene electron transfer unfavorable. Therefore, our data suggest that the observed enhancement is a consequence of the serial (as opposed to parallel) electron transfer from the CTC acceptor to the fullerene.

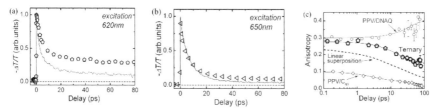

Fig. 2. Isotropic PIA transients of the LE band in ternary blends MEH-PPV/DNAQ/C60 (a, pentagons) and MEH-PPV/TNF/C60 (b, triangles) excited at the CTC band. Curves represent the respective transients for the neat CTCs. (c): Transient anisotropy of the LE band in ternary MEH-PPV/DNAQ/C60 blend (pentagons) after 620 nm excitation. The dashed curve displays the prediction for the superposition of responses calculated from the transient PIA and anisotropies of MEH-PPV/DNAQ CTC (circles) and MEH-PPV/C60 (diamonds).

The anisotropy measurements on MEH-PPV/DNAQ/C60 (Fig. 2c) also confirm that the CTC participates in the charge photogeneration in the ternary blend. The initial value of transient anisotropy of ~0.3 is similar to the one observed in MEH-PPV/DNAQ and is a factor of 3 higher than in respective MEH-PPV/C60 blend. This indicates that excitation in the ternary blend occurs locally as it is the case in pure CTC. Furthermore, the transient anisotropy cannot be explained by considering two parallel processes of charge separation (fig.2c, dashed curve). Therefore, we conclude that generation of long-lived charges in the ternary blend is mediated by the CTC but does not occur though direct charge separation on the polymer-fullerene interface.

Conclusions

After excitation of the CTCs in the visible, an immediate (<100 fs) formation of the positive charges at the polymer was observed, followed by an efficient and fast (<30 ps) geminate recombination. According to the polarization-sensitive experiments, photogeneration and recombination of charges in the CTCs occur locally (i.e. within a single donor acceptor pair) in contrast to the MEH-PPV/PCBM blend where energy and charge transfer takes place. The long-lived charges generation in the polymer CTCs can be enhanced by adding to blend a supplementary fullerene acceptor.

1. D. Moses, A. Dogariu, A.J. Heeger, Chem. Phys. Lett. 316, 356 (2000).
2. C.J. Brabec, G. Zerza, G. Cerullo, S. De Silvestri, S. Luzzati, J. C. Hummelen, S. Sariciftci, Chem. Phys. Lett. 340, 232 (2001).
3. A.A. Bakulin, S.G. Elizarov, A.N. Khodarev, D.S. Martyanov, I.V. Golovnin, D.Yu. Paraschuk, M.M. Triebel, I.V. Tolstov, E.L. Frankevich, S.A. Arnautov, E.M. Nechvolodova, Synthetic Metals 147, 221 (2004).
4. X. Wei, Z. V. Vardeny, N. S. Sariciftci, A. J. Heeger, Phys. Rev. B 53, 2187 (1996).
5. J.G. Müller, J.M. Lupton, J. Feldmann, U. Lemmer, M.C. Scharber, N.S. Sariciftci, C.J. Brabec, U. Scherf, Phys. Rev B 72, 195208 (2005).

Photomodulation of Interfacial Electron Transfer by Optical Switches

Lars Dworak, Victor V. Matylitsky, Josef Wachtveitl

Institute for Physical and Theoretical Chemistry, Johann Wolfgang Goethe-University Frankfurt, Max-von Laue Str. 7, 60438 Frankfurt am Main, Germany wveitl@chemie.uni-frankfurt.de

Abstract. This The dynamics of 4-(phenylazo)benzoic acid coupled to Al_2O_3 and TiO_2 films is described. The drastically altered photochemistry of the optical switch upon absorption to TiO_2 films reflects the competition between electron transfer and intramolecular relaxation.

Introduction

The investigation of the molecular reaction mechanisms of interfacial ET is an active field of research, motivated by applications in solar energy conversion [1] or molecular electronics [2]. For the latter, we try to incorporate photoswitchable elements, that allow to control of electronic properties on the molecular level, in donor-bridge-acceptor systems. These donor-bridge-acceptor systems should perform a light-controllable interfacial electron transfer (ET) from the photoexcited state of a molecular donor to the conduction band of the semiconductor TiO_2. The photoactive bridge acts as an optical on-off switch for this ET. Azobenzene and its derivates can photoisomerize between the cis and the trans isomer and is thus a promising candidate for the incorporation into donor-bridge-acceptor systems. Because of its unique photochemistry azobenzene and its derivates have been frequently studied in different application fields such as light-triggered switches [3,4], optical data storage devices [2] and light driven molecular machines [5]. The molecular donor in our system can be realized by an organic dye. The composition of a light harvesting organic dye and nanocrystalline TiO_2 films is commonly used in dye sensitized solar cells (DSSC) [1].

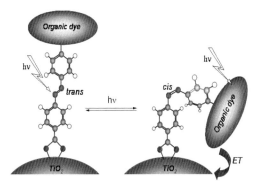

Fig. 1. Photomodulation of FET in a donor-bridge-acceptor system by reversible photoswitching of the bridge. Optical control of ET is achieved by conformation dependent ET-efficiency.

As a first step towards a complete characterisation of our system we started studies on the fast photodynamics of the bridge [4-(penylazo)benzoic acid (PAB)] coupled to the surface of Al_2O_3 and TiO_2 mesoporous films with transient absorbance measurements. We concluded that PAB-Al_2O_3 can be seen as a non-reactive reference system. For the PAB-TiO_2 coupled system we observed a competition between a forward electron transfer (FET) to the conduction band of the TiO_2 and an intramolecular relaxation of PAB excited state.

Experimental Methods

TiO_2 and Al_2O_3 mesoporous films were prepared from colloidal nanoparticles in acidic solution. Preparation of the colloidal nanoparticles followed a published procedure [6] and particles with size of 5-10 nm in diameter were obtained. Mesoporous films were prepared on AF 15 glass (Schott) with polyethylene glycol as surfactant (Fluka, Carbowax 20000). The photoswitch PAB (purity 98%, Sigma-Aldrich) is bound to the TiO2 surface with the carboxyl group. The time-resolved measurements have been carried out by a classical pump/probe setup using a CLARK CPA 2001 laser/amplifier system operating at a repetition rate of 1 kHz with a central wavelength of 775 nm. Pump pulses at the desired excitation wavelength (480 nm) were generated with a noncolliniar optical parametric amplifier (NOPA) [7]. The excitation energies of the pump pulses were below 120 nJ/pulse. A Single-filament white-light continuum of probe pulses is generated by focusing 775 nm light into a CaF2 window. After the correction for the coherent signal and the group velocity dispersion a Marquardt downhill algorithm was used to fit the experimental data by n exponential decays for all wavelengths simultaneously. The fit amplitudes for each kinetic component are wavelength dependent fitting parameters [8].

Results and Discussion

The reaction dynamics of the trans → cis isomerisation for PAB coupled to an Al_2O_3 and a TiO_2 mesoporous film were recorded after nπ* excitation of PAB at 480 nm. The results for the PAB-Al_2O_3 coupled system exhibit an initial strong induced absorption, which is attributed to the electronically excited S_1 state (nπ*) of the trans isomer. The decay of the signal can be fitted biexponentially with the time constants 0.47 and 7.6 ps. Similar biphasic dynamics of azobenzene in solution was previously observed and explained by different relaxation pathways [9].

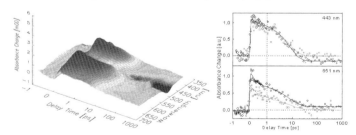

Fig. 2. a.) Transient absorbance changes for the trans → cis isomerisation of PAB on a TiO_2 film and b.) transient absorbance changes at selected probing wavelengths for PAB/Al_2O_3 (gray squares) and PAB/TiO_2 (black circles).

The transient spectrum for the PAB-TiO$_2$ (Fig. 2a.) coupled system also shows an initial positive absorbance change over the whole investigated spectral range. Interestingly, the transient at λ_{pr} = 443 nm in Fig. 2b. shows, that after 10 ps the positive signal, which is the result the intramolecular relaxation of the PAB excited state, converts into a small long-living negative absorbance change in the case of the PAB/TiO$_2$ system. Moreover, the positive signal at λ_{pr} = 651 nm is more pronounced than in the PAB-Al$_2$O$_3$ coupled system and persists until the end of the temporal range (1000 ps). The formation of a negative signal at short wavelengths in the spectral region of the PAB nπ* transition can be assigned to the bleached ground state of the coupled PAB. We ascribe the long living positive absorbance change (550-700 nm) to electrons which are injected to the conduction band of the TiO$_2$. These spectral characteristics are the consequence of a FET from the photoexcited state of PAB to the conduction band of TiO$_2$. The dynamics of the PAB-TiO$_2$ coupled system could be best fitted with the time constants 0.34 and 7.8 ps. They can be assigned to the competition between intramolecular relaxation and the FET. An offset is necessary to account for the slow recombination of the charge separated state.

Conclusions

Time-resolved absorbance measurements for PAB coupled to an Al$_2$O$_3$ film showed a complete and biphasic decay of the excited state, indicating that it can serve as a stable reference system. In contrast, a long living charge separation in the PAB-TiO$_2$ system was detected, which leads to a characteristic difference spectrum at long delay times. This we explained by a competition between the reversible photoisomerization of the PAB and the FET to the TiO$_2$ acceptor. For an unambiguous conformation dependent investigation of the ET dynamics from the donor to the TiO$_2$ acceptor there has to be a clear spectral separation between the excitation of the bridge and the molecular donor.

Acknowledgements.

1   M. K. Nazeeruddin, A. K. Rodicio, R. Humphry-Baker, E. Müller, P. Liska, N. Vlachopoulos and M. Grätzel, in J. Am. Chem. Soc., Vol. 115, 6382 , 1993

2   Z. F. Liu, K. Hashimoto and A. Fujishima, in Nature, Vol. 347, 658 , 1990

3   J. Bredenbeck, J. Helbing, A. Sieg, T. Schrader, W. Zinth, C. Renner, R. Behrendt, L. Moroder, J. Wachtveitl, and P. Hamm, in PNAS Vol. 100, 6452 , 2003

4   S. Spörlein , H. Carstens, Helmut Satzger, C. Renner, R. Behrendt, L. Moroder, P. Tavan, W. Zinth and J. Wachtveitl in PNAS, Vol. 99, 7998 , 2002

5   T. Hugel, N. B. Holland, A. Cattani, L. Moroder, M. Seitz and H. E. Gaub, in Science, Vol. 296, 1103 , 2002

6   J. Moser and M. Grätzel, in J. Am. Chem. Soc., Vol. 105, 6547 , 1983

7   T. Wilhelm, J. Piel and E. Riedle, in Opt. Lett. Vol. 22, 1494 , 1997

8   R.Huber, S. Spörlein, J. E. Moser, M. Grätzel and J. Wachtveitl, in J. Phys. Chem. B, Vol. 104, 8995-9003 (2000)

9   T. Nägele, R. Hoche, W. Zinth and J. Wachtveitl, in Chem. Phys. Lett., Vol. 272, 1997 489-495

Two-color two-dimensional Fourier transform spectroscopy of energy transfer

Kristin L. M. Lewis[1], Jeffrey A. Myers[1], Patrick F. Tekavec[1], and Jennifer P. Ogilvie[1]

[1] Department of Physics and Biophysics, University of Michigan, Ann Arbor MI 48109, USA
   E-mail: jogilvie@umich.edu

Abstract. We report two-color 2D electronic spectra obtained using a diffractive-optics based approach. We employ the two color method to study a simple system consisting of a donor/acceptor pair exhibiting fluorescence resonance energy transfer.

Introduction

Two-dimensional electronic spectroscopy (2DES) is a powerful tool for studying energy transfer in biological systems. In analogy to 2D NMR, 2DES reveals the coupling between electronic transitions, which appear as cross-peaks in the 2DES spectrum. The underlying lineshapes are also revealed, free from inhomogeneous broadening [1]. These unique features of 2DES make it an ideal spectroscopy for mapping the flow of energy through multichromophoric systems: traditional nonlinear spectroscopies are a subset of this measurement. Here we apply two-color 2DES to the study of a simple donor/acceptor system exhibiting fluorescence resonance energy transfer (FRET). The use of separately tunable pump and probe pulses is motivated by the need to follow energy transfer pathways over a broad range of frequencies in natural photosynthesis.

FRET is a ubiquitous energy transfer mechanism exploited by many natural light-harvesting systems. It is also frequently employed as a spectroscopic ruler in single molecule spectroscopy. For a donor/acceptor separation of ~0.5–10 nm, FRET is the dominant energy transfer mechanism, and FRET efficiency is highly sensitive to small changes in separation. The Förster theory of resonant energy transfer, however, breaks down in systems where the chromophores are no longer small compared to their separation. A simple system for observing FRET is a DNA construct consisting of donor and acceptor chromophores covalently bound to the 3' and 5' end of a short piece of DNA. This provides a FRET pair with a rigid, well-defined separation that can be readily varied by changing the number of DNA base pairs.

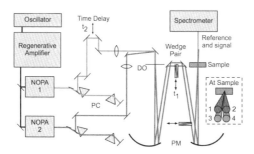

Fig. 1. Diffractive-optics-based two-color 2DES setup. NOPA: noncollinear optical parametric amplifier, PM: parabolic mirror, PC: prism compressor.

Experimental Setup. Our experimental apparatus for two color 2DES is shown in Fig.1. A Titanium sapphire oscillator (Femtolasers Synergy) seeds a 1 kHz regenerative amplifier (Spectra Physics Spitfire Pro) to produce 40 fs pulses with an energy of 1 mJ at 800 nm. To generate tunable visible-near IR light, we divide the output from the regenerative amplifier to pump two noncollinear optical parametric amplifiers (NOPA). These NOPAs produce ~10 μJ, 20 fs pulses over the broad range from ~475–1000 nm [2]. Having independently tunable pump and probe pulses adds flexibility to previous implementations of 2DES by providing access to electronic transitions spanning the visible to near-IR. We generate the excitation pulse sequence for the experiment by sending the pump and probe beams into a passively phase-stabilized diffractive optics (DO) arrangement, as in our previous work [3, 4]. Although active phase-stabilization schemes have been shown to work [5], they are considerably more difficult to implement. The passive phase-stability of the DO approach is derived from the fact that the first order diffracted beams from the DO provide the two pulse pairs for the experiment, and pass through almost identical optical pathways between the DO and the sample. We have previously demonstrated a high degree of phase stability ($\lambda/90$) between the relevant pulse pairs, leading to a high degree of phase stability in the signal [4]. To scan the t_1 delay while maintaining this passive phase stability, we use a refractive delay line as previously demonstrated [4], employing pairs of wedges [6]. Two pairs of wedges allow us to obtain the signals necessary to separate absorptive and dispersive components of the 2D spectra. We use a standard delay line for the t_2 delay, where phase stability is unimportant. We control the delay lines with motorized micrometers (Newport LT-HA) controlled by a digital signal processor (Motorola DSP56F807) to provide position data at a 1 kHz repetition rate synchronously with our signal detection. We spectrally disperse our heterodyned signal in a spectrometer (Jobin Yvon HR320) and detect it with a CCD camera (Princeton Instruments Pixis 100F) via spectral interferometry. Before the sample, a neutral density filter attenuates the reference pulse so as to maximize interference with the signal. Our ability to detect the signal at 1 kHz allows rapid 2D spectra acquisition (~10 s per 2D spectrum).

The two NOPA setup allows independent tuning of the pump and probe pulses, allowing us to examine regions of the 2D spectrum far from the diagonal. Fig. 2 shows the 2D spectrum from our DNA construct sample, consisting of an eight base-pair piece of DNA labeled with donor (Cy3) and acceptor (Cy5) dyes at either end (purchased from IDT technologies). Here we excited the donor (500

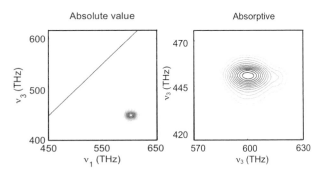

Fig. 2: Two-color 2D spectra exhibiting energy transfer between donor and acceptor on the DNA construct. Left: absolute value spectrum, right: absorptive spectrum.

nm) and probed the acceptor emission (675 nm) at t_2 = 71 ps. The cross peak present in the 2D spectrum is indicative of energy transfer during t_2. Fluorescence studies of the DNA construct show an energy transfer efficiency of 25% and an energy transfer time of ~1 ns, comparable to the lifetime of the acceptor. Thus direct growth of the cross-peak is not expected to be readily observed in this particular construct. In addition, the pump pulse may directly excite a small fraction of acceptor molecules. To confirm that the cross peak contains information about energy transfer and not simply emission from the directly excited acceptor we performed an additional control experiment where we studied a sample containing the single-stranded, donor-labeled DNA. By pumping and probing the donor with and without the acceptor attached, we found the decay of the cross-peak of the absolute value 2D spectrum to be 45% faster in the presence of the acceptor, indicating that energy transfer was occurring to the acceptor.

In summary we have demonstrated two-color two-dimensional electronic spectroscopy on a simple energy transfer system. The method will permit energy transfer studies between widely separated electronic transitions that occur in natural photosynthetic complexes.

1   D. M. Jonas, "Two-Dimensional Femtosecond Spectroscopy," Annual Review of Physical Chemistry **54**, 425-463 (2003).

2   T. Wilhelm, J. Piel, and E. Riedle, "Sub-20-fs Pulses Tunable across the Visible from a Blue-Pumped Single-Pass Noncollinear Parametric Converter," Opt. Lett. **22**, 1494-1496 (1997).

3   J. P. Ogilvie, M. L. Cowan, M. Armstrong, A. M. Nagy, and R. J. D. Miller, "Diffractive Optics-Based Heterodyne Detected Three-Pulse Photon Echo," in *Ultrafast Phenomena XIII*, R. J. D. Miller, M. M. Murnane, N. F. Scherer, and A. M. Weiner, eds. (Springer, Vancouver, 2002), pp. 571-573.

4   M. L. Cowan, J. P. Ogilvie, and R. J. D. Miller, "Two-Dimensional Spectroscopy Using Diffractive Optics Based Phased-Locked Photon Echoes," Chemical Physics Letters **386**, 184-189 (2004).

5   X. Li, T. Zhang, C. N. Borca, and S. T. Cundiff, "Many-Body Interactions in Semiconductors Probed by Optical Two-Dimensional Fourier Transform Spectroscopy," Phys. Rev. Lett. **96**, 057406 (2006).

6   T. Brixner, I. V. Stiopkin, and G. R. Fleming, "Tunable Two-Dimensional Femtosecond Spectroscopy," Optics Letters **29**, 884-886 (2004).

7   L. Lepetit, and M. Joffre, "Two-Dimensional Nonlinear Optics Using Fourier-Transform Spectral Interferometry," Optics Letters **21**, 564-566 (1996).

Electron Injection Dynamics of Perylene Derivatives into ZnO and TiO$_2$ Particle Films

J. Szarko, A. Neubauer, L. Socaciu-Siebert, A. Bartelt, F. Birkner, K. Schwarzburg, and R. Eichberger

Dynamics of Interfacial Reactions–SE 4, Helmholtz Centre Berlin, Glienicker Strasse 100, 14109 Berlin, Germany
E-mail: eichberger@helmholtz-berlin.de

Abstract. The electron injection dynamics of two perylene dyes bound to ZnO and TiO$_2$ nanoparticles was investigated with femtosecond transient absorption simultaneously monitoring the rise of the cationic and the decay of the excited state. Electron transfer from the chromophores was slower when attached to ZnO compared to TiO$_2$. The excited state decay and the cationic state rise showed very good agreement at early times, indicating direct electron injection into the conduction band for both semiconductors in absence of intermediate states.

Introduction

Heterogeneous electron transfer at metal oxide interfaces is the initial process in dye-sensitized solar cells (DSSCs) [1,2] based on TiO$_2$ and ZnO electrodes. In order to drive these devices towards higher efficiency and better stability, the dynamics of the interfacial charge separation needs to be well understood. Numerous attempts using a variety of different molecular absorbers and preparation techniques were taken in the past to elucidate the main mechanisms involved in the electron injection process at metal oxides, but the experimental reports do not at all point in the same direction. In particular, the electron transfer dynamics for dye-sensitized ZnO samples reported in the literature are considerably slower than for TiO$_2$ [3] and vary on a large scale from < 300 fs to 450 ps [4-6] using Ruthenium bipyridyl complexes as the dye. Aggregates due to the formation of Zn^{2+}-dye complexes and exciplexes [7-9] at ZnO surfaces have been proposed to be the origin for the observed slower electron injection rates as compared to the chemically more stable TiO$_2$ counterpart.

Here we compare the transient spectral signatures of photo-induced electron transfer and early recombination for different chromophore units anchored on ZnO and TiO$_2$ nanoparticle films. Two perylene derivatives were studied: 2,5-Di-*tert*-butyl-perylene-9-yl-propionic acid and 2,5-Di-*tert*-butyl-perylene-9-yl-acrylic acid, which differ only in a single bond. The additional *tert*-butyl groups prevented excimer formation of the molecule at the surface while the acid groups chemically bound the molecules to the surface. Perylene is an ideal chromophore for studying heterogeneous electron transfer because the absorption bands of the excited state and cationic state of perylene are energetically well separated, so the two states can be experimentally distinguished and compared when measured in parallel by transient absorption. In a systematic approach the decay of the excited state was measured simultaneously with the rise of the cationic state of the perylene derivatives employing 15 fs pump pulses and a time-zero corrected fs white light continuum (WLC) probe spectrally covering both states. For high temporal resolution the spectrum of the WLC was narrowed with razor blades for better chirp compensation. In contrast to most previous reports, all the measurements were performed in ultrahigh vaccum (UHV).

Results and Discussion

The measured rise times for the cationic state of perylene bound to ZnO via the propionic and the acrylic acid were 260 ± 60 fs and 200 ± 50 fs. The fast decay components of the excited states for the same systems were 260 ± 60 fs and 190 ± 60 fs, respectively. The difference in the observed lifetimes is ascribed to the different chemical bridges between the chromophore and the semiconductor surface, with the saturated and unsaturated species representing the electronically decoupled and coupled case. The fits of the cationic rise time for the ZnO samples are in good agreement with the fast decay times of the excited state signals for both molecules studied in these experiments. This is exemplarily shown in Figure 1. for the propionic acid species.

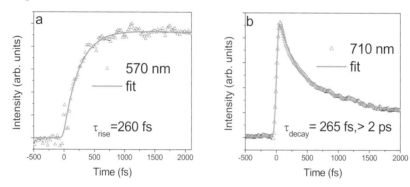

Fig. 1. Rise of the cationic state (left) of the perylene derivative attached to ZnO nanoparticles via a propionic acid group, and the decay of the excited state (right) for the same experimental system

The results indicate that the decay of the excited state did not form any intermediate species on the surface of the ZnO colloids, and the rise of cation absorption signal was a measure for direct electron transfer to the semiconductor corresponding to the depletion time of the excited state.

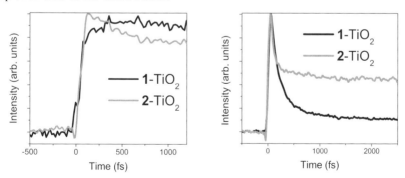

Fig. 2. Left: Rise of the cationic state of the perylene derivatives attached to TiO_2 nanoparticles via a propionic acid group ($1\text{-}TiO_2$) and via a acrylic acid group ($2\text{-}TiO_2$); Right: Decay of the excited state for the same experimental systems

A similar trend was observed for the perylene derivatives attached to TiO_2 (Figure 2.), but the electron injection times were much faster. The measurements were in agreement with previously reported experimental results [10] of 60 fs and 10 fs for the same electronically decoupled and coupled molecular absorber, respectively. For the latter system a contribution of a direct optical charge transfer state is believed to be responsible for the extremely fast electron transfer time. A strong ground state bleaching observed in the broadband transients of the electronically coupled TiO_2 species (not shown here) supports this assumption.

The approximate 200 fs injection time for the perylene acrylic acid coupled to the ZnO electrode was the shortest overall injection time on any dye-sensitized ZnO colloid system currently recorded. This injection time agrees well with theoretical predictions [11] concerning the electron transfer at semiconductor electrodes. The slower electron transfer at ZnO electrodes is best explained by the lower number of conduction band acceptor states compared to TiO_2 [12]. The presence of an isosbestic point in the transient WLC spectra between the cationic and excited state absorption bands was observed for all studied samples. This is a further indication that only two electronic species and no intermediate states are involved in the electron transfer process. In addition the cationic signals of the ZnO systems also fit well to a monoexponential rise substantiating the latter statement.

Conclusions

For all experimental systems investigated here, the decay time of the perylene excited state was remarkably similar to the rise time of the cationic state, which implies that the electron is injected directly into the ZnO conduction band like it is known for TiO_2. The simultaneous temporal and spectral measurements showed no presence of additional states, such as exciplex states or aggregates affecting the injection rate.

Acknowledgements. We would like to thank Sven Kubala for his UHV expertise and Ursula Michalczik for the metal oxide sample preparation.

1   M. Grätzel, Nature 338, 414, 2001.

2   K. Keis, E. Magnusson, H. Lindstrom, S.E. Lindquist, and A. Hagfeldt, Sol. Energy Mater. Sol. Cells 73, 51, 2002.

3   P. Persson, M. J. Lundqvist, R. Ernstorfer, W. A. Goddard, and F. Willig, J. Chem. Theory Comput. 2, 441, 2006.

4   J. B. Asbury, Y. Q. Wang, and T. Q. Lian, J. Phys. Chem. B103, 6643, 1999.

5   W. Beek, M. Wienk, M. Kemerink, X.. Yang, and R. Janssen, J. Phys. Chem. B109, 9505, 2005.

6   C. Bauer, G. Boschloo, E. Mukhtar, and A. Hagfeldt, J. Phys. Chem. B105, 5585, 2001.

7   K. Keis, J. Lindgren, S. E. Lindquist, and A. Hagfeldt, Langmuir 16, 4688, 2000.

8   A. Furube, R. Katoh, K. Hara, S. Murata, H. Arakawa, and M. Tachiya, J. Phys. Chem. B107, 4162, 2003.

9   A. Furube, R. Katoh, T. Yoshihara, K. Hara, S. Murata, H. Arakawa, and M.Tachiya, J. Phys. Chem. B108, 12583, 2004.

10   R. Ernstorfer, PhD thesis, Free University, Berlin, 2004.

11   P. Persson, S. Lunell, and L. Ojamae, Chem. Phys. Lett. 364, 469, 2002.

12   J. Szarko, A. Neubauer, A. Bartelt, L. Socaciu-Siebert, F. Birkner, K. Schwarzberg, T. Hannappel, and R. Eichberger, J. Phys. Chem. C., in print

Part VIII

THz Science and Technology, Nano-Optics and Plasmonics

Dynamic Metamaterials at Terahertz Frequencies

H. T. Chen[1], Abul K. Azad[1], J. F. O'Hara[1], A. J. Taylor[1], W. J. Padilla[2],
R. D. Averitt[3]

[1]MPA-CINT, MS K771, Los Alamos National Laboratory, Los Alamos, New Mexico 87545
[2]Department of Physics, Boston College, Chestnut Hill, Massachusetts 02467
[3] Department of Physics and Photonics Center, Boston University, Boston, Massachusetts 02215
E-mail: raveritt@physics.bu.edu

Abstract. Metamaterials fabricated for operation at terahertz frequencies are presented. Optical excitation enables control of the metamaterial resonance amplitude and frequency.

Introduction

Electromagnetic metamaterials have experienced enormous excitement and growth recently due to the demonstration of exotic effects such as invisibility cloaking [1,2], negative refractive index [3,4], and perfect focusing [5]. In fact, metamaterials offer an electromagnetic response which has been shown to be impossible to achieve in naturally occurring materials. Metamaterials are artificial structures which are fabricated to yield a designed resonant response to electromagnetic radiation. Typically, these materials consist of sub-wavelength metallic inclusions embedded within or on top of a dielectric matrix or substrate material. A great deal of research into metamaterials has used microwave radiation due in part to the ease of fabrication of sub-wavelength structures at these frequencies. For example demonstrations of negative refractive index media composed of negative permittivity (ε) and negative permeability (μ) metamaterial elements was first demonstrated at microwave frequencies [3]. This has led to intense efforts to extend metamaterial response to terahertz ($1\text{THz}=10^{12}\text{Hz}$) [6], near-infrared, and visible frequencies [7-9].

Although most effort in metamaterials research is advancement to optical frequencies, some work has concentrated on THz frequencies [6,10-15]. There are many potential uses for metamaterials specifically within this regime, as response from naturally occurring materials is somewhat rare and there is a noticeable lack of

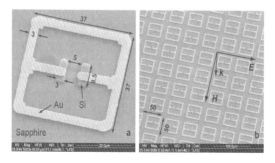

Fig. 1: (a) split ring resonator on SOS. The SOS has been etched away except for the bars in the capacitive gap region. (b) Expanded view of the planar array. The numbers in the figure are dimensions in microns.

high power sources, efficient detectors, and other standard device components.

The choice of substrate confers additional functionality to metamaterials and in our previous work we have shown that it is possible to create optical and voltage controlled terahertz modulators using metamaterials fabricated on semiconductor substrates [11,15]. In the present work, fabrication of metamaterials on ErAs:GaAs nanoisland superlattices enables ultrafast optical switching of the resonant response while photoexcitation of   metamaterials fabricated on silicon-on-sapphire (SOS) enables optical tuning of the resonance frequency.

Results and Discussion

Figure 1a shows a scanning electron image of the unit cell of a frequency tunable THz metamaterial fabricated on SOS. It is a variant of previously demonstrated split ring resonators [12,13] that exhibit a Lorentz-like resonant response described by a complex permittivity $\varepsilon(\omega) = \varepsilon_1 + i\varepsilon_2$. The split gap at the center of the unit cell can be thought of as a capacitor where charges accumulate at resonance. By constructing the capacitor plates out of a semiconductor (silicon) we can control the conductivity upon the photoexcitation of free charge carriers, thereby altering the effective size of this capacitor and tuning the capacitance C. Since the resonance frequency is strongly dependent on the capacitance, i.e. $\omega_0 \sim (LC)^{1/2}$, where L is the effective inductance of the SRRs, then ω_0 should shift monotonically to lower frequencies as the photoexcitation power increases.

Experimental results for the frequency tunable metamaterial are shown in Fig. 2a. The transmitted THz electric field amplitude with no photoexcitation is shown as the black curve. A transmission minimum $t(\omega) = 19\%$ occurs near 1.06 THz, whereas the off-resonance transmission is approximately $t(\omega) \sim 90\%$. As the optical pump power is increased (10-20 mW) the resonance initially only weakens and broadens, showing little perceptible shift in frequency. Upon further increase (50-100 mW) the

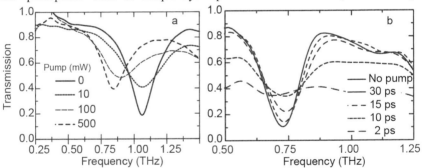

Fig.2: (a) Optical tuning results for metamaterials fabricated on SOS. (b) Dynamic switching for metamaterials on ErAs:GaAs.

resonance continues to weaken and saturates at $t(\omega) = 49\%$, but now shifts to significantly lower frequencies. Increasing the optical pump power beyond 100 mW causes continued shifting to lower frequencies and also re-establishes the resonance strength and narrows the linewidth. At an optical pump power of 500 mW we achieved $t(\omega) = 40\%$ at 850 GHz (red curve). This represents a tuning range of 20 percent in the resonance frequency and is in good agreement with electromagnetic simulations (not shown).

Figure 2b shows the experimental results for Au split ring resonators (with similar dimensions to the SRR in Fig. 1a) fabricated on ErAs:GaAs nanoisland superlattices which are engineered to have a carrier recombination time of ~10 ps [14]. In this case the ErAs:GaAs makes up the entire substrate without selective etching. Therefore, excitation with above bandgap 800 nm 50-fs pulses generates carriers in the substrate which short the capacitive gap thereby turning off the SRR resonance. As the carrier recombine, the resonance is restored. Figure 2(b) shows the transmission as a function of frequency at various times following excitation with a fluence of 2 $\mu J/cm^2$. The resonance is switched off within 2ps and fully recovers within 15 ps consistent with the carrier lifetime. These results show that it is possible to create ultrafast terahertz modulators using planar metamaterial arrays opening possibilities for high bit rate communication at THz frequencies.

1. Pendry, J. B.; Schurig, D.; Smith, D. R., Controlling electromagnetic fields. *Science* 2006, 312, 1780-1782.

2. Schurig, D.; Mock, J. J.; Justice, J. B.; Cummer, S. A.; Pendry, J. B.; Starr, A. F.; Smith, D. R., Metamaterial electromagnetic cloak at microwave frequencies. *Science* 314, 977, 2006.

3. Shelby, R. A.; Smith, D. R.; Schultz, S., Experimental verification of a negative index of refraction. *Science* 292, 77, 2006.

4. Smith, D. R.; Padilla, W. J.; Vier, D. C.; Nemat-Nasser, S. C.; Schultz, S., Composite medium with simultaneously negative permeability and permittivity. *Phys. Rev. Lett. 84,* 4184, 2000.

5. Fang, N.; et al, Sub-diffraction-limited optical imaging with a silver superlens. *Science* 308, 534, 2005.

6. Yen, T. J.; Padilla, W. J.; Fang, N.; Vier, D. C.; Smith, D. C.; Pendry, J. B.; Basov, D. N.; Zhang, X., Terahertz Magnetic Response from Artificial Materials. *Science* 303, 1491, 2004.

7. Dolling, G.; Enkrich, C.; Wegener, M.; Soukoulis, C. M.; Linden, S., Simultaneous negative phase and group velocity of light in a metamaterial. *Science* 312, 892, 2006

8. Yuan, H. K.; Chettiar, U. K.; Wenshan, C.; Kildishev, A. V.; Boltasseva, A.; Drachev, V. P.; Shalaev, V. M., A negative permeability material at red light. *Opt. Express* 15, 7872, 2007.

9. Zhang, S.; Fan, W.; Panoiu, N. C.; Malloy, K. J.; Osgood, R. M.; Brueck, S. R., Optical negative-index bulk metamaterials consisting of 2D perforated metal-dielectric stacks. *Opt. Express* 14, 6778, 2006.

10. Azad, A. K.; Zhang, W., *Opt. Lett.* 30, 2945, 2005.

11. Padilla, W. J.; Taylor, A. J.; Highstrete, C.; Lee, M.; Averitt, R. D., Dynamical electric and magnetic metamaterial response at terahertz frequencies. *Phys. Rev. Lett.* 96, 107401, 2006.

12. Padilla, W. J.; Aronsson, M. T.; Highstrete, C.; Lee, M.; Taylor, A. J.; Averitt, R. D., Novel Electrically Resonant Terahertz Metamaterials. *Phys. Rev. B, Rapid* 75, 041102, 2007.

13. Chen, H. T.; O'Hara, J. F.; Taylor, A. J.; Averitt, R. D.; Highstrete, C.; Lee, M.; Padilla, W. J., Complementary planar terahertz metamaterials. *Opt. Express* 15, 1084, 2007.

14. Chen, H. T.; Padilla, W. J.; Zide, J. M. O.; Bank, S. R.; Gossard, A. C.; Taylor, A. J.; Averitt, R. D., Ultrafast optical switching of terahertz metamaterials fabricated on ErAs/GaAs Nanoisland Superlattices. *Opt. Lett.* 32, 1620, 2007.

15. Chen, H. T.; Padilla, W. J.; Zide, J. M. O.; Gossard, A. C.; Taylor, A. J.; Averitt, R. D., Active terahertz metamaterial devices. *Nature* 44, 597, 2006.

Effect of Spin-Polarized Electrons on THz Emission from Photoexcited GaAs(111)

James M. Schleicher, Shayne M. Harrel, and Charles A. Schmuttenmaer

Yale University, Dept. of Chemistry, 225 Prospect St., New Haven, CT 06520-8107 USA
E-mail: charles.schmuttenmaer@yale.edu

Abstract. We report the dependence of optical rectification and shift currents in unbiased GaAs(111) on the excitation beam polarization using THz emission spectroscopy. When exciting slightly above bandgap with elliptically polarized light, the emission is strongly influenced by spin-polarized electrons.

Background and Introduction

We report the dependence of optical rectification and shift currents in unbiased GaAs(111) on the polarization of the excitation beam. The emitted THz waveform is used as the probe. The dependence of the response on both linear and elliptical excitation polarizations is studied. We find that the dependence of the emission on the polarization angle of linearly polarized excitation is well described by known theory for both above and below bandgap excitation. The dependence on elliptical polarization also agrees well when exciting below bandgap and far above the bandgap. However, THz emission when exciting slightly above the bandgap is strongly influenced by spin-polarized electrons. We propose that the magnetic field generated by these spin-polarized electrons is responsible for altering the direction of their own trajectories via the Hall effect.

Terahertz Emission Spectroscopy (TES) is a method which has been shown to characterize dynamical processes in physical and chemical systems [1]. THz emission is the difference-frequency analog of second harmonic generation, and is based on the second order nonlinear susceptibility, $\chi^{(2)}$, of the material. For semiconductors, the exact form of $\chi^{(2)}$ is inherently complicated as numerous processes can lead to difference-frequency generation. In particular, optical rectification and shift current both contribute to $\chi^{(2)}$ in GaAs [2-3]. For pulses well below bandgap, optical rectification dominates. When the photon energy is tuned above bandgap, carriers in the semiconductor are promoted from the valence band to the conduction band. Due to the spatial localization of the electrons in the conduction and valence bands in GaAs (the electrons around the Γ–point of the conduction band reside primarily on the Ga atoms, while those around the Γ-point of the valence band are predominantly located on the As atoms), the photoexcitation of carriers can generate a net shift in the center of charge within the unit cell, which is known as a shift current.

When exciting GaAs with light that is either elliptically or circularly polarized, and also has photon energy 0.34 eV or less above the bandgap (800 nm light has 1.55 eV photon energy, which is 0.13 eV above the bandgap), spin-polarized electrons are generated [4]. The net spin polarization orients itself parallel or anti-parallel to the propagation direction of the excitation beam. This range of photon energies allows transitions to the conduction band from the light-hole and heavy-hole bands but not from the split-off band, and the origin of these spin-polarized electrons has been described in great detail elsewhere [4]. Briefly, the selection rules in GaAs are such that when photoexciting with right-handed, circularly-polarized light, σ^+, a 50% excess of spin "backward" electrons is produced. The opposite holds true for σ^- light wherein a 50% excess of spin "forward" electrons is produced. More importantly for these studies is the fact that the spin-polarized electron population will vary sinusoidally as a function of the angle between the quarter-wave plate and the input beam's linear polarization, i.e., no spin-polarized electrons when it is linear, and a maximum amount when it is circular.

Experimental Methods

The experimental setup has been described elsewhere [1]. In brief, a regeneratively amplified Ti:Sapphire laser produces ~ 100 fs pulses centered at a wavelength of 800 nm with a repetition rate of 1 kHz. The beam is split into two parts, pump and readout, where the pump photoexcites the sample and the readout beam detects the emitted THz radiation in the far-field via FSEOS in a 0.5 mm thick ZnTe(110) crystal [5]. A wire-grid polarizer that rejects vertically polarized THz radiation is placed in front of the detector crystal to ensure maximum sensitivity towards horizontally polarized THz light.

GaAs(111) wafers, purchased from Litton, are placed in an optical cryostat under low vacuum (30 mtorr). No bias voltage is applied. The sample is excited at 800 nm (~500 mW average power) at room temperature for the above bandgap studies, and at 835 nm (300 mW average power) at 77 K for below bandgap measurements. The dependence of the THz emission on the polarization of the excitation field was performed by placing a half- and quarter-wave plate in the excitation beam path. The dependence on linear polarization was obtained by rotating the half-wave plate in steps of 5° and for the elliptical polarization the quarter-wave plate was also rotated in steps of 5°.

Results and Discussion

THz emission for above and below bandgap excitation with linear polarization is in excellent agreement with theory and will not be discussed further. The data sets with elliptical polarization taken below bandgap also agree with theory: The horizontal THz emission varies with wave plate angle as $-\sin(4\theta)$, as shown in Fig. 1(a).

Similarly, when the excitation beam is frequency doubled to 400 nm, the THz emission is essentially identical to the calculated behavior, as shown in Fig. 1(b). This wavelength corresponds to a photon energy of 3.2 eV, which far above the bandgap for GaAs, and does not produce spin-polarized electrons.

However, when exciting above bandgap with 800 nm excitation, the deviation from theory is striking. The observed signal shown in Fig. 1(c) does not vary as $-\sin(4\theta)$. In fact, it has the opposite sign as expected from the calculations (which are shown with the solid line) between 0° and 30° and between 150° and 180°. Therefore, we conclude that spin-polarized electrons are the source of the "anomalous" results observed when exciting at 800 nm with elliptically polarized light.

A simple phenomenological model has been developed to capture the salient features when exciting GaAs(111) with elliptically polarized 800 nm light. The population of spin-polarized electrons will be at a maximum when the light is circularly polarized (wave plate angles of 45° and 135°) and will be zero when linearly polarized (0°, 90°, and 180°). Additionally, light of one handedness will provide an excess of spin-polarized electrons of one polarization while the light of the opposite handedness will provide an excess of the opposite spin. On the other hand,

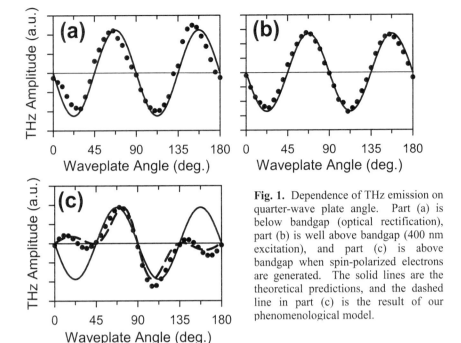

Fig. 1. Dependence of THz emission on quarter-wave plate angle. Part (a) is below bandgap (optical rectification), part (b) is well above bandgap (400 nm excitation), and part (c) is above bandgap when spin-polarized electrons are generated. The solid lines are the theoretical predictions, and the dashed line in part (c) is the result of our phenomenological model.

the shift current will be maximum when the light is linearly polarized, and zero when it is circularly polarized. It is a combination of these effects which leads to the observed behavior.

Spin-polarized electrons that are generated within the sample are polarized normal to the surface and thus generate a magnetic field that is also normal to the surface. This field will alter the trajectory of the carriers contributing to the shift current via the Hall Effect, and thereby affect the polarization of the THz emission. The net result is shown with the dashed line in Fig. 1(c).

These simulated results resemble the data quite well, and capture most of the anomalous features in GaAs(111) that the generic theory of optical rectification and shift currents are unable to account for. This observation is unique in that the electrons themselves are creating the magnetic field that is affecting their own trajectory. These results are reminiscent of observation of the (elliptical) polarization dependence of photogalvanic current in SiGe(113) quantum wells [6] (which is forbidden in bulk zinc-blende materials due to symmetry [6]), and to our knowledge, this is the first observation of such behavior in bulk GaAs.

Acknowledgements. The National Science Foundation (CHE-0616875) is gratefully acknowledged for partial support of this work.

1   C. A. Schmuttenmaer, Chem. Rev. **104**, 1759 (2004).

2   P. Kral, J. Phys. Cond. Matt. **12**, 4851 (2000).

3   J. E. Sipe and A. I. Shkrebtii, Phys. Rev. B **61**, 5337 (2000).

4   D. T. Pierce and F. Meier, Phys. Rev. B **13**, 5484 (1976).

5   Q. Wu and X. C. Zhang, Appl. Phys. Lett. **67**, 3523 (1995).

6   S. D. Ganichev, U. Rossler, W. Prettl, E. L. Ivchenko, V. V. Bel'kov, R. Neumann, K. Brunner and G. Abstreiter, Phys. Rev. B **66**, 075328 (2002).

Nonlinear Lattice Response Observed Through Terahertz SPM

János Hebling[1,2], Matthias C. Hoffmann[1], Ka-Lo Yeh, György Tóth[2], and Keith A. Nelson[1]

[1] Department of Chemistry, Massachusetts Institute of Technology, Cambridge, Massachusetts 02139, USA
 E-mail: hebling@mit.edu
[2] Department of Experimental Physics, University of Pécs, 7624 Hungary
 E-mail: hebling@fizika.ttk.pte.hu

Abstract. Self-phase-modulation of ultrashort THz pulses was observed in lithium niobate at the 100 MW/cm^2 intensity level. The effect, observed in time and frequency domains, suggests 1000x larger nonlinear index n_2 than at visible wavelengths.

Introduction

Self-phase-modulation (SPM) is a well known phenomenon in the visible and near infrared ranges. The reason for it is an intensity dependent index of refraction. This dependence is described by the nonlinear index n_2, which is proportional to the third order nonlinear susceptibility $\chi^{(3)}$. For the $LiNbO_3$ (LN) nonlinear optical crystal, n_2 = 5.3×10^{-15} cm^2/W in the visible [1]. This means that SPM can be observed usually for laser intensity on the order of 100 GW/cm^2. Such high intensity THz sources do not exist. It is known [2,3], however, that the second order susceptibility contains an ionic part, and this can be significantly larger than the electronic part in the far infrared or THz spectral range close to lattice vibrational resonances. Similar enhancement is expected for $\chi^{(3)}$, too [4]. Here we describe observation of THz SPM in LN. The observation indicates a strong lattice anharmonicity that is driven by our THz pulses. Using tilted pulse front excitation [5] we are able to routinely produce near-single-cycle or few-cycle THz pulses with intensities up to 500 MW/cm^2 [6].

Experimental Methods

High intensity THz pulses were generated by optical rectification of 100 fs duration Ti:sapphire laser pulses in LN using the tilted pulse front set-up [6]. In some experiments the LN crystal was inserted into a cryostat and cooled down to 10 K or 80 K. The THz waveform emitted directly by the LN THz source or after transmission through a separate 2-mm thick LN crystal positioned near the output of the generation crystal was measured by electro-optic sampling in ZnTe. The optic axes of the LN crystals were parallel to the polarization of the THz radiation. In other experiments a tilted pulse front pumped LN THz pulse source was used at room temperature and the generated THz pulses were focused into a sample positioned at the focus of an off-axis parabolic mirror set-up. The intensity of the THz pulses at the focus (sample) position reached a value of 100 MW/cm^2. The transmitted THz radiation was collimated and lightly focused into a sandwiched ZnTe crystal with 1 mm total thickness and 0.1 mm sensitive ([110] surface) thickness. The THz waveforms without and with the sample were measured.

Results and Discussion

Changing the pump intensity resulted in strong changes in the temporal and spectral profiles of the THz waveforms generated at 10 K [7], as shown in Figure 1. In the leading part of the waveforms the oscillation frequency was reduced, and on the tailing part it was increased, at high pump intensities. Accordingly, the spectra measured at high pump intensities were broader than for low intensities. This behavior is typical for SPM caused by a positive Kerr type nonlinear refractive index.

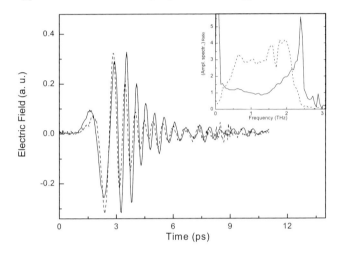

Fig. 1. Temporal shape of THz pulses generated at 10 K in a LN crystal pumped by 1.5 mJ (solid) and 0.8 mJ (dashed) pump pulses. The curve for lower energy pumping is multiplied by 2.3 in order to facilitate comparison of the two curves. The inset shows the amplitude spectrum (dashed) at high pumping and the ratio of the spectra (solid) obtained for high and low pumping.

We also examined the effect in an external LN crystal at 80 K, using a near-single-cycle THz pulse. Figure 2 depicts the measured temporal shapes of the THz pulse with and without the external LN crystal in its path. It can be seen that the distance between the first positive and the first negative lobes increases, while the distance between the two negative peaks decreases, with the LN crystal in place. This is again an indication of SPM caused by positive n_2. Theoretical modeling in a manner similar to that described in [8] yielded far better agreement with the transmitted waveform that we observed if the effect of the nonlinear refractive index was taken into account than if only linear absorption and dispersion were considered. A value of $n_2 = 5.4 \times 10^{-12}$ cm^2/W resulted the best fit. This value of n_2 is 1000 times larger then n_2 of LN in the visible range.

The reason for this extremely large nonlinear index is that at THz frequencies, lattice vibrational anharmonicity may result in a dramatic enhancement of nonlinear responses. The enhancement is expected to be particularly large for soft phonon modes of ferroelectric crystals, which are strongly anharmonic. Systematic investigation of THz SPM can provide insight into the anharmonic lattice potentials and nonlinear lattice dynamics of these materials.

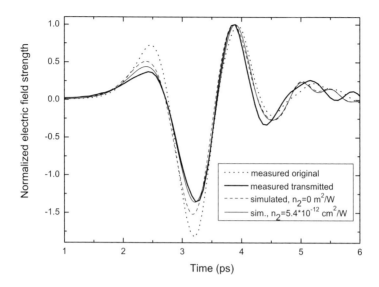

Fig. 2. Measured THz waveform with (bold) and without (dotted) a 2 mm thick LN crystal in the pulse path and the simulated transmitted waveform supposing nonlinear refractive index of $n_2 = 0$ (dashed) and $n_2 = 5.4 \times 10^{-12}$ cm^2/W (solid).

Conclusions

SPM of ultrashort THz pulses was observed for the first time. It was observed not only in LN at low temperature, but also at room temperature for several dielectrics (LiTaO$_3$, BaTiO$_3$, BGO). Comparison of the measured change of the transmitted THz pulse shape with simulation shows a 1000 times larger nonlinear index for LN in the THz range than in the visible. The lattice vibrational anharmonicity may be the cause of this dramatic enhancement of the nonlinear response.

Acknowledgements. This work was supported in part by ONR grant no. N00014-06-1-0463.

1   L. Heping et al., Appl. Phys. B **64**, 659, 1997.
2   G. D. Boyd, and M. A. Pollack, Phys. Rev. B **7**, 5345, 1973.
3   T. Dekorsy et al., Phys. Rev. Lett. **90**, 0555081, 2003.
4   Chr. Flytzanis, and N. Bloembergen, Prog. Quant. Electr. **4**, 271, 1976.
5   J. Hebling, G. Almási, I. Z. Kozma, and J. Kuhl, Opt. Express **10**, 1161 2002.
6   K.-L. Yeh et al., Optics Commun. **281**, 3567, 2008.
7   J. Hebling, K.-L. Yeh, M. C. Hoffmann, and K. A. Nelson, J. Sel. Top. Quant. Electr. **14**, 345, 2008.
8   J. Seres, and J. Hebling, J. Opt. Soc. Am. B **17**, 741, 2000.

Ultrafast Electron Cascades Driven by Intense Femtosecond THz Pulses

Haidan Wen[1], M. Wiczer[2], and Aaron Lindenberg[1,3]

[1] PULSE Institute, Stanford Linear Accelerator Center, Menlo Park, CA 94025
  E-mail: hwen@slac.stanford.edu
[2] Department of Physics, University of Illinois, Urbana, Champaign, Urbana, IL 61801
  E-mail: mwiczer2@uiuc.edu
[3] Department of Materials Science and Engineering, Stanford University, Stanford, CA 94305
  E-mail: aaronl@stanford.edu

Abstract. A table-top THz source has been employed to study the nonlinear response of semiconductors to near-half-cycle intense femtosecond pulses. We report nonlinear field-induced changes in the far infrared absorption coefficient, associated with THz-induced impact ionization processes occurring within the THz pulse.

Introduction

Terahertz (THz) radiation has wide applications in the study of atoms, molecules, and materials, in which THz pulses are mainly used as a probe of sample properties, e.g. transport phenomena. The interaction of high field THz pulses with matter, in contrast, is just beginning to be explored, as new sources for ultrashort high field pulses are developed in both the laboratory and at accelerator-based sources [1-4]. These can be used to initiate dynamics and control materials properties and form the basis for the developing field of nonlinear THz spectroscopy [4-7]. In this work, we report the observation of nonlinear field-induced changes in the far infrared absorption coefficient induced by near-half-cycle THz pulses. We show that this can be associated with free carrier generation by far-below-band-gap photons by the leading edge of the THz pulse, and subsequent absorption by the trailing edge of the pulse, through an impact-ionization process.

Experimental Methods

THz pulses with ~ 50 kV/cm peak field and 300fs duration were generated by focusing bichromatic 400 nm and 800 nm ultrafast pulses in the air [3]. Through a nonlinear process mediated by the generated plasma, intense THz pulses are emitted, which can be collected, collimated and refocused by a pair of parabolic mirrors onto a sample. Semiconductor samples were irradiated normally by the focused THz radiation and the transmitted THz intensity was recorded by a liquid-helium-cooled Si bolometer or through electro-optic sampling while the sample was translated along the THz propagation axis (z-axis) near the focus. In this way, the THz field was continuously varied in a z-scan technique, so that the field-dependent absorption cross-section could be measured.

Results and Discussion

Fig. 1 shows measurements of the THz transmission in both GaAs and InSb samples. It is observed that in InSb, a dramatic decrease in the transmitted THz radiation is

654

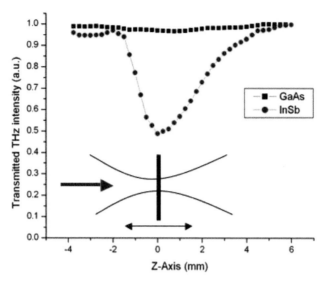

Fig. 1. Transmitted THz intensity as a function of sample position along the THz propagation direction for both InSb and GaAs, normalized to the intensity when the sample is far from the focus. Inset figure shows experimental setup for z-scan measurements.

observed at high fields i.e. near the THz focus. By measuring the field-dependent reflectivity, we have verified that these effects are dominantly due to intensity-dependent changes in the THz absorption coefficient. We interpret this nonlinear absorption as due to free carrier excitation by the leading edge of the intense THz field and subsequent free carrier absorption by the trailing edge of the pulse, acting as a probe of the nonlinear ionization dynamics. It is observed that the effect is significantly smaller at low temperatures (80K), for which the thermally-excited free carrier population is reduced. In addition, by creating a second time-delayed THz pulse, we observe that the effect lasts for at least 30 ps. It is detected in InSb (0.17eV band gap) but not in GaAs (Eg=1.41eV). Because this effect depends on the existing free carrier population and is observed only in small band-gap, small effective mass semiconductors, we hypothesize it to be a result of impact ionization, in which existing free carriers are accelerated in the quasi-half-cycle THz field, leading to avalanche processes and subsequent THz absorption on ultrafast time-scales.

In order to check this further, we have used the transmission data to calculate the change in carrier concentration in a Drude model as a function of peak THz field, calibrated as a function of z using both EO sampling and spot-size measurements. Fig. 2 shows experimentally measured values at a sample temperature of 220 K. Using known theoretical models for the impact ionization rate as a function of electron energy [8], and modeling the conduction band electrons as being classically accelerated in the applied THz field, we have also numerically calculated this dependence using the experimentally measured THz pulse shape. Excellent agreement between experiment and theory is obtained.

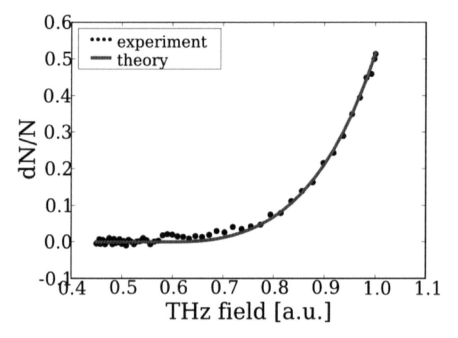

Fig. 2. Measured change in free carrier density as a function of applied peak THz field, shown in comparison to theoretical model (solid line).

Conclusions

In conclusion, we report the observation of nonlinear THz absorption induced by high intensity quasi-half-cycle field pulses. Future experiments based on this work may enable the development of a new class of avalanche photodiodes in which ultrafast THz fields are used as a bias field in order to produce femtosecond temporal resolution, high quantum efficiency detectors.

1   T. Bartel, P. Gaal, K. Reimann, M. Woerner, T. Elsaesser et. al., Optics Lett., Vol 30, 2805 (2005).
2   X. Xie, J. Dai, X.-C. Zhang, Phys. Rev. Lett., Vol. 96, 075005 (2006).
3   K.-L. Yeh, M.C. Hoffmann, J. Hebling, K.A. Nelson, Appl. Phys. Lett., Vol. 90, 171121 (2007).
4   Y. Shen et al., Phys. Rev. Lett., Vol. 99, 043901 (2007).
5   P. Gaal, et. al., Phys. Rev. Lett., Vol. 96, 187402 (2006)
6   T. Hornung, K.-L. Yeh, K.A. Nelson, in Ultrafast Phenomena XV, (Spring-Verlag, Berlin 2007).
7   P. Gaal et al., Nature Vol. 450, 1210 (2007).
8   J.T. Devreese, R.G. Welzenis, R.P. Evrard, Appl. Phys. A Vol. 29, 125 (1982).

Rabi Oscillations in a Shallow Donor System Driven by Intense THz Radiation

P. Gaal[1], W. Kuehn[1], K. Reimann[1], M. Woerner[1], T. Elsaesser[1], R. Hey[2]

[1] Max-Born-Institut für Nichtlineare Optik und Kurzzeitspektroskopie, 12489 Berlin, Germany
E-mail: woerner@mbi-berlin.de
[2] Paul-Drude-Institut für Festkörperelektronik, 10117 Berlin, Germany

Abstract. Carrier-wave Rabi oscillations between bound impurity levels are demonstrated by ultrafast THz propagation experiments. Modelling with an ensemble of two-level systems yields good agreement up to a driving field of 5 kV/cm.

Introduction

Shallow donors are an interesting model system for coherent light-matter interaction. They are promising candidates for quantum information processing [1]. In n-type GaAs with very low doping concentrations ($N_D \leq 10^{14}$ cm^{-3}) a transition energy of 4.1 meV and a dephasing time of several picoseconds has been observed [2]. For higher doping concentrations one expects a faster dephasing due to increased scattering and a shift to higher energies of the optical transition due to wavefunction overlap between the donor levels [3]. Radiative coupling, however, can overcome the decoherence rate of individual transitions [4], thus leading to Rabi oscillations between bound impurity states. At the given transition energies, already moderate THz driving fields lead to a Rabi frequency of the same magnitude as the transition frequency. Such carrier-wave Rabi oscillations exhibit new features like the generation of higher harmonics [5]. In this contribution we report on carrier-wave Rabi oscillations between bound states of shallow donors in GaAs with a donor concentration of $N_D = 2 \times 10^{16}$ cm^{-3}. A comparison with the Bloch equations for radiatively coupled two-level systems [4] shows that the Rabi oscillation picture holds for driving fields up to 5 kV/cm, while for higher fields the two-level approach breaks down.

Experiment

In our experiment, a few-cycle THz pulse with a center frequency of 2 THz excites a semiconductor sample at a temperature of 100 K placed in the focus of a pair of parabolic mirrors. The electric field of the transmitted THz pulse is measured via electro-optic sampling in a thin ZnTe crystal [3]. The entire optical path of the THz beam is placed in vacuum to avoid spurious absorption from water molecules. The sample investigated was grown by molecular beam epitaxy and consists of a 500 nm thin layer of Si-doped GaAs clad between two 300 nm thin Al$_{0.4}$Ga$_{0.6}$As barrier layers. In particular, the sample is much shorter than the THz wavelength of 150 μm and can therefore be considered as two dimensional.

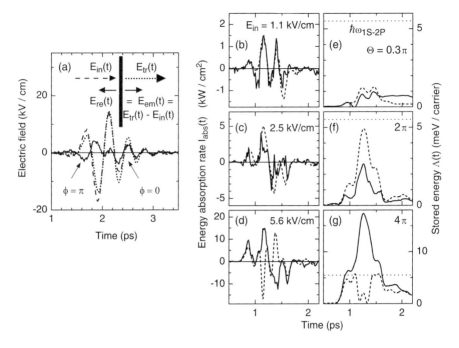

Fig. 1. (a) Time dependence of the incident, $E_{in}(t)$ (dashed line), of the transmitted, $E_{tr}(t)$ (dotted line), and of the emitted electric fields $E_{em}(t) = E_{tr}(t) - E_{in}(t)$ (solid line). The phase between $E_{tr}(t)$ and $E_{em}(t)$ determines periods with absorption ($\phi = \pi$) or with gain ($\phi = 0$). (b) – (d) $I_{abs}(t)$ [Eq. (1)] (solid line: experiment, dashed line: theory) for different electric field amplitudes. (e) – (g) $\Lambda(t)$ [Eq. (2)] for (b) – (d).

Results and Discussion

In Fig. 1(a) the electric field of the incident THz pulse $E_{in}(t)$ (dashed line) and of the transmitted pulse $E_{tr}(t)$ (dotted line) are plotted as a function of time. Due to the two-dimensional sample geometry all donor atoms experience the same local THz field $E_{loc}(t)$, which is equal to $E_{tr}(t)$ [4]. The sample coherently emits the field $E_{em}(t) = E_{tr}(t) - E_{in}(t)$ (solid line). For times $t < 2$ ps, $|E_{tr}(t)| < |E_{in}(t)|$, i.e., the sample absorbs part of the incident pulse. In contrast, for $t > 2$ ps $|E_{tr}(t)| > |E_{in}(t)|$, i.e., the sample shows stimulated emission. Absorption is characterized by a phase of $\phi = \pi$ between the emitted field $E_{em}(t)$ and the local driving field $E_{tr}(t)$, whereas for gain we find $\phi = 0$. The time dependent energy transfer rate between the THz field and the sample is given by:

$$I_{abs}(t) = \varepsilon_0 c \left[E_{in}(t)^2 - E_{re}(t)^2 - E_{tr}(t)^2 \right] = 2\varepsilon_0 c E_{em}(t) E_{tr}(t). \tag{1}$$

The energy that is transiently absorbed per electron is the time integral of $I_{abs}(t)$:

$$\Lambda(t) = \frac{1}{N_D d} \int_{-\infty}^{t} I_{abs}(t') \, dt'. \tag{2}$$

$I_{abs}(t)$ and $\Lambda(t)$ are shown in Figs. 1(b) – (g) as the solid lines for different electric field amplitudes together with the results from a theoretical calculation (dashed lines).

In the linear case [Figs. 1(b) and (e)] $I_{\mathrm{abs}}(t)$ oscillates around zero due to the slightly off-resonant center frequency of the driving pulse (2 THz) compared to the 1S – 2P transition frequency (1.5 THz). Hence, the transiently absorbed energy $\Lambda(t)$ increases with strong oscillations as a function of time. For higher field amplitudes one finds a change between absorption and stimulated emission [Figs. 1(b) and (g)], a typical signature of Rabi oscillations.

We model our result with the Maxwell-Bloch equations for a two-level system [4,6]. For strong radiative coupling which results from the two-dimensional sample geometry and the high doping, the local driving field acting on the sample is given by $E_{\mathrm{loc}}(t) = E_{\mathrm{tr}}(t) = E_{\mathrm{in}}(t) + E_{\mathrm{em}}(t)$. The strong radiative coupling leads to a collective response of the ensemble of impurity transitions and therefore to a much longer decoherence time compared to the individual two-level system. The result of the calculation is shown as dashed lines in Figs. 1(b) – (g). Since in the case of strong radiative coupling the energy and phase relaxation times are of minor relevance, the model does not contain any fitting parameters. The quantitative agreement between experiment and theory is good for field amplitudes up to 5 kV/cm. For higher amplitudes [Figs. 1(d) and (g)] the two-level approximation breaks down. This is seen in Fig. 1(g) where at $t = 1.25$ ps the energy per carrier $\Lambda(t)$ is much larger than the maximum energy $\hbar\omega_{2P-1S} = 5.5$ meV possible in the two-level model [shown as the dotted line in Figs. 1(e) – (g)], pointing to a significant population of other levels during the driving pulse.

Conclusion

We studied the THz response of a thin n-type GaAs layer at low temperatures. The emitted THz radiation is directly measured in phase-resolved nonlinear propagation experiments and demonstrates carrier-wave Rabi oscillations between bound levels of shallow impurities. The Rabi oscillation picture holds for driving fields up to 5 kV/cm. For higher driving fields the two-level approach breaks down.

1   B. E. Cole, J. B. Williams, B. T. King, M. S. Sherwin, and C. R. Stanley, Nature **410**, 60, 2001.
2   M. F. Doty, B. T. King, M. S. Sherwin, and C. R. Stanley, Physical Review B **71**, 201201(R), 2005.
3   P. Gaal, K. Reimann, M. Woerner, T. Elsaesser, R. Hey, and K. H. Ploog, Physical Review Letters **96**, 187402, 2006.
4   T. Stroucken, A. Knorr, P. Thomas, and S. W. Koch, Physical Review B **53**, 2026, 1996.
5   S. Hughes, Physical Review Letters **81**, 3363, 1998.
6   C. W. Luo, K. Reimann, M. Woerner, T. Elsaesser, R. Hey, and K. H. Ploog, Physical Review Letters **92**, 047402, 2004.

Nonlinear optical effects in germanium in the THz range: THz-pump - THz-probe measurement of carrier dynamics

János Hebling[1,2], Matthias C. Hoffmann[1], Harold Y. Hwang[1], Ka-Lo Yeh[1], Keith A. Nelson[1]

[1] Dept. of Chemistry, Massachusetts Institute of Technology, Cambridge, MA 02139, USA
 E-mail: hebling@mit.edu
[2] Dept. of Experimental Physics, University of Pécs, 7624 Hungary
 E-mail: hebling@fizika.ttk.pte.hu

Abstract. Nonlinear high intensity THz transmission measurements were made on Ge, showing nonlinear optical effects in the THz range including self-phase modulation and absorption saturation. THz-pump - THz-probe measurements were performed on n-type Ge, GaAs, and Si to follow the ultrafast free carrier dynamics that underlie the nonlinear effects seen in the transmission measurements. Complex relaxation behavior was observed, arising from inter- and intra-valley scattering in the conduction bands of these semiconductors.

Introduction

Although there has been extensive investigation of nonlinear terahertz effects in semiconductors with long pulses or continuous wave radiation [1], time-resolved studies on the ps timescale using ultrashort THz pulses are rare [2-4]. These recent investigations explore bound systems like excitons [2], impurity states [3] or polarons [4]. Nonlinear optical effects of free carriers in semiconductors were observed earlier in the THz range using a far-infrared molecular laser [5]. Although, in accordance with theory, a relatively strong third-order nonlinearity was observed, a detailed analysis of the nonlinear optical phenomena requires high intensity THz sources with temporal resolution. Using tilted pulse front excitation [6] we are able to routinely produce near-single-cycle or few-cycle THz pulses with intensities up to 300 MW/cm² [7,8]. Here we describe the use of these pulses for observation of self-phase modulation (SPM) and absorption saturation dynamics in n-type Ge. THz pump-probe studies on n-type Si and GaAs as well as Ge are presented.

Experimental Methods

For nonlinear THz transmission measurements, high intensity THz pulses were generated by optical rectification of 100 fs duration 5.6 mJ pulses from a 1 kHz Ti:sapphire laser system (Coherent LEGEND) in lithium niobate using the tilted pulse front (TPF) setup [7]. THz pulses with up to 3 μJ energy were generated in this way. These pulses were focused into the sample by two parabolic mirrors. The THz intensity could be attenuated continuously by rotating one of two wiregrid polarizers inserted before the sample. The transmitted THz radiation was collimated and lightly focused into a sandwiched ZnTe crystal with 1 mm total thickness and 0.1 mm sensitive ([110] surface) thickness. The THz field with and without a sample in the THz focus was recorded for different THz intensities by electro-optic sampling. The maximum THz intensity (for 2.6 μJ THz pulse energy at the sample position) was about 250 MW/cm². For comparison, a linear transmission measurement was carried out using a THz-TD spectrometer based on photoconductive switches with THz pulse energies of only several femto-Joules.

THz-pump - THz-probe measurements were performed in a collinear geometry with a similar setup. The optical pump for THz generation was split into pump and probe arms using a beamsplitter, then recombined for TPF excitation in the LN generation crystal. A delay stage in the pump arm allowed for arbitrary separation of the THz pump and probe pulses in time.

Results and Discussion

Free carrier absorption in the THz range is readily observed in bulk doped semiconductors, and can be described by the Drude model [5]. Linear THz-TDS measurements for Ge (Figure 1b) show standard Drude behavior as expected. At high THz pulse energies, THz absorption by free carriers saturates, corresponding to a dramatic phase shift in the THz field temporal profile, indicating self-phase modulation in Ge. As the THz intensity is attenuated, absorption returns to linear Drude behavior

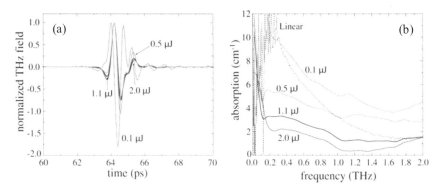

Fig. 1. (a) Electro-optically measured temporal profile of transmitted THz pulse in nonlinear transmission measurements. (b) Corresponding absorption spectra for different THz pulse energies. Linear response is plotted for comparison.

Time and frequency resolved THz-pump - THz-probe measurements were employed to follow the dynamics of free carrier absorption saturation in Ge (Figure 2a). At zero probe delay, absorption saturation occurs as expected from the high intensity transmission measurements. At positive probe delays, absorption recovery on a several ps timescale is observed (Figure 2b). The complex behavior in absorption is related to the mobility of carriers in the conduction band. The pump THz pulse excites the free carriers to states of higher energy in the initial conduction band valley from which they may scatter into side valleys. For Ge, electrons may be scattered from the L valley to the Γ and X valleys. Initial scattering into side valleys causes absorption saturation due to the lower mobility and higher effective mass of the electrons in the X side valley. Different relaxation times are expected for inter- and intra-valley relaxation. Half of the initially lost absorption is recovered after 2.5 ps in Ge, 1.5 ps in Si, and 2 ps in GaAs. All three semiconductors exhibit similar absorption saturation and recovery. The details of the absorption recovery are intimately related to the conduction band structure of each sample.

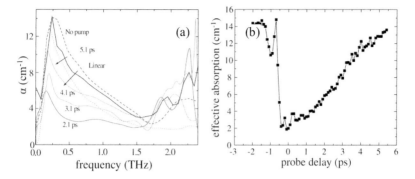

Fig. 2. (a) Nonlinear absorption spectra in Ge for different probe delay times. Linear THz-TDS measurement and measurement without pump pulse are shown for comparison. (b) Spectrally integrated absorption coefficient of Ge from THz-pump - THz-probe measurements.

Conclusions

Self-phase modulation and absorption saturation in the THz range were observed in Ge from THz transmission measurements. The origin of these effects has been studied with time and frequency resolved THz-pump - THz-probe spectroscopy. High intensity THz pulses excite free carriers that may scatter into higher energy valleys in the conduction band. This leads to a change in absorption and refractive index, underlying the self-phase modulation mechanism. The absorption saturation originates from carrier scattering into regions of lower mobility. Subsequent relaxation back to the bottom of the initial conduction band valley occurs on the timescale of a few ps. Multiple relaxation pathways are expected, including electron-phonon and electron-electron scattering from different regions of the conduction band. Measurements have been carried out on n-type Si and GaAs yielding similar results.

Acknowledgements. This work was supported in part by Office of Naval Research Grant **N00014-06-1-0459**

1 S. D. Ganichev, and W. Prettl, in Intense Terahertz Excitation of Semiconductors, Oxford University Press Inc., New York, 2006.

2 R. Huber et al., Phys. Rev. Lett. **96**, 017402, 2006.

3 P. Gaal et al., Phys. Rev. Lett. **96**, 187402, 2006.

4 P. Gaal, et al., Nature **450**, 1210, 2007.

5 Mayer A., and F. Keilmann, Phys. Rev. B **33**, 6962, 1986.

6 J. Hebling, G. Almási, I. Z. Kozma, and J. Kuhl, Opt. Express **10**, 1161, 2002.

7 K.-L. Yeh, J. Hebling, M. C. Hoffmann, and K. A. Nelson, Optics Commun. **281**, 3567, 2008.

8 K.-L. Yeh, M. C. Hoffmann, J. Hebling, and K. A. Nelson, Appl. Phys. Lett. **90**, 171121, 2007.

Terahertz Nonlinear Response and Coherent Population Control of Dark Excitons in Cu_2O

T. Kampfrath[1], S. Leinß[2], K. v. Volkmann[1], M. Wolf[1], D. Fröhlich[3], A. Leitenstorfer[2], and R. Huber[2]

[1] Fachbereich Physik, Freie Universität Berlin, Arnimallee 14, 14195 Berlin, Germany
[2] Fachbereich Physik, Universität Konstanz, Universitätsstrasse 10, 78464 Konstanz, Germany
  Email: rupert.huber@uni-konstanz.de
[3] Fachbereich Physik, Universität Dortmund, 44221 Dortmund, Germany

Abstract. An optically dark, dense, and cold 1s para exciton gas is prepared by two-photon generation of electron-hole pairs and subsequent phonon cooling. Intense multi-terahertz fields of order MV/cm coherently promote 70% of the quasiparticles from the 1s to the 2p state via a partial internal Rabi oscillation. Electro-optic sampling monitors the Larmor precession of the Bloch vector in real time.

Introduction

Sophisticated quantum optical protocols had been a prerequisite for the first observation of Bose-Einstein condensation (BEC) of atomic gases [1]. Excitons – bosonic Coulomb pairs of one electron and one hole – have been envisaged as another potential candidate for BEC [2]. Yet optical control of promising excitons for BEC by intense light fields has been a challenge since relevant systems often exhibit weak if any radiative interband coupling. 1s para-excitons in Cu_2O are a prime example [3]. Terahertz (THz) pulses, in contrast, couple resonantly to the internal hydrogen-like fine structure, irrespective of interband selection rules. This concept has provided novel insight into formation dynamics, fine structure, density, and temperature of excitons [4-6]. The observation of stimulated THz emission from intra-excitonic transitions [7] has raised the hope for future coherent manipulation of dark exciton ensembles similar to atomic quantum optics. Here, we exploit intense THz fields of the order of MV/cm to systematically study the nonlinear response of optically dark, dense, and cold 1s-para excitons in Cu_2O. A partial Rabi flop of the 1s-2p transition allows us to control the internal quantum state on a sub-ps time scale.

Formation and cooling dynamics of 1s para excitons in Cu_2O

The sample is a 334-μm thick, naturally grown single crystal of Cu_2O kept at a lattice temperature T_L = 5 K. Near-infrared femtosecond pulses centered at an energy of 1.5 eV are absorbed via two-photon transitions to generate unbound electron-hole pairs with homogeneous density throughout the crystal length. We first trace the ultrafast formation and cooling dynamics of 1s excitons via time-delayed multi-THz transients probing the internal 1s-2p absorption. Strong exchange interaction splits the n = 1 state into a triplet ortho-exciton and a lower-lying, optically inaccessible singlet para-exciton. The respective 1s-2p Lyman absorption lines are located at photon energies of 116 meV and 129 meV [6]. Fig. 1(a) depicts the pump-induced absorption changes in this frequency window for various delay times t_D after generation of e-h pairs. While the spectrally broad THz absorption at $t_D = 100$ fs indicates unbound e-h pairs, the hallmark 1s-2p lines are already discernable at $t_D = 11$ ps. Within 100 ps, the lines narrow and shift slightly towards lower frequencies while the ratio of para- versus ortho-exciton absorption increases. The subsequent decay of exciton populations follows a complex non-exponential dynamics (not shown). The strength of THz absorption is an absolute measure of the densities $N_{1s,para}$ and $N_{1s,ortho}$ [5]. Owing to the vastly different effective masses of 1s and 2p excitons [3] a detailed analysis of the THz line shape reflects the temperature T_{1s} of the ensemble [8]. For $t_D = 100$ ps [dots in Fig. 1(a)], we find best agreement with the experiment for a 1s-para exciton density $N_{1s,para} = 2 \times 10^{16}$ cm^{-3} and a temperature $T_{1s} = 8 \pm 2$ K, close to the phonon bath. To our knowledge, this density is among the highest directly measured for a cold exciton gas in Cu_2O.

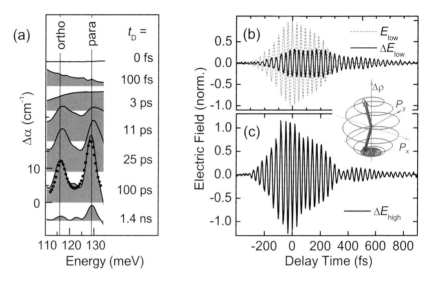

Fig. 1. (a) Pump-induced changes $\Delta\alpha$ of the mid-infrared absorption for various delay times t_D after two-photon absorption of 12-fs near-infrared pulses; vertical lines: energies of the 1s-2p resonances in the ortho and para system at vanishing momentum. Circles: Numerical line fit (see text). (b) Field-resolved polarization response (black curve) of 1s-para excitons to an external THz field (gray dashed curve, peak electric field: 0.1 MV/cm) resonant to the 1s-2p transition ($t_D = 1$ ns). (c) Reemitted field for peak values of the driving field of 0.5 MV/cm. Inset: schematic of the Bloch sphere of the 1s-2p two-level system.

Terahertz nonlinear control of the internal quantum state of excitons

We next demonstrate how to control the internal quantum state of the dark, dense, and cold para-exciton gas formed at $t_D = 1$ ns. Intense THz transients with peak fields as high as 0.5 MV/cm are generated via optical rectification of the output of a high-power Ti:sapphire laser amplifier. An acousto-optic phase and amplitude modulator is used to prepare optimally shaped 0.1-mJ pulses for efficient phase-matched rectification in a GaSe emitter of a thickness of 200 μm.

Fig. 1(b) depicts a THz pulse of moderate intensity (E_{peak} = 0.1 MV/cm) on resonance with the two-level system consisting of the 1s- and 2p-para states. The oscillating electric field of the reemitted light is directly measured in real time via ultrabroadband electro-optic sampling, on a femtosecond scale (black curve). The response is out of phase with the driving field by 180° characteristic of an absorptive transition in a resonant two-level system. The free induction decay following the external THz pulse exhibits a decoherence time of 0.8 ps. Fig. 1(c) compares this response with the polarization induced by a THz pulse of the same temporal profile, but a peak electric field scaled by a factor of 5. Surprisingly, the system polarization is not a linearly scaled version of the low-field response. Rather, the re-emitted field initially rises rapidly within the first seven cycles, reaches its maximum amplitude, and decreases within the decoherence time of the internal transition.

This dynamics is a manifestation of a partial Rabi oscillation explained in a Bloch picture of the two-level system (Fig. 1, inset). The diagonal (population difference ρ_{22}-ρ_{11}) and off-diagonal (polarizations P_x and P_y) elements of the density matrix are depicted as the vertical and the two horizontal coordinates. For low driving fields, the Bloch vector performs a Larmor precession in the vicinity of the south pole of the Bloch sphere. Experimentally, the projection of this trajectory onto the polarization axis is directly mapped out in real time [Fig.1(b)]. With increasing THz field, the state vector may be driven towards the north pole inducing strong population inversion. During this partial Rabi cycle, the projection onto the polarization axis reaches a maximum at the equator and decreases from there on. The real-time data of the polarization [Fig.1(c)] directly reflect this dynamics and allow us to reconstruct the actual motion of the Bloch vector. From a comparison of the data with a numerical solution of the Maxwell-Bloch equations, we find that up to 70% of the optically dark states are promoted from the 1s into the 2p orbital. For yet larger driving pulse areas, the reemitted field exhibits an oscillatory dependence on increasing field strength, indicative of up to 1.5 Rabi cycles (not shown). Quantitative modelling of the nonlinear THz data with state-of-the-art microscopic semiconductor theory, accounting for ponderomotive contributions as well as highly excited and ionized exciton states is under way [9, 10].

In conclusion, our results point out a novel route towards ultrafast nonlinear control of optically dark exciton states. Advanced protocols known from atomic systems may now open new perspectives for preparing ultracold and dense exciton gases via high-field THz transients.

1 M. H. Anderson, et al., Science **269**, 198 (1995); K. B. Davis et al., Phys. Rev. Lett. **75**, 3969 (1995).
2 J. Kasprzak et al., Nature **443**, 409 (2006).
3 J. Brandt et al., Phys. Rev. Lett. **99**, 217403 (2007) and references therein.
4 R. A. Kaindl et al., Nature **423**, 734 (2003).
5 R. Huber et al., Phys. Rev. B Rapid **72**, 161314(R) (2005).
6 M. Kubouchi et al., Phys. Rev. Lett. **94**, 016403 (2005).
7 R. Huber et al., Phys. Rev. Lett. **96**, 017402 (2006).
8 K. Johnsen et al., Phys. Rev. Lett. **86**, 858 (2001).
9 J.R. Danielson et al., Phys. Rev. Lett. **99**, 237401 (2007).
10 M. Kira et al. in Progress in Quantum Electronics **30**, 155 – 296 (2006), G. Eden ed. (Elsevier, Amsterdam, 2006).

Impact Ionization in InSb studied by THz-Pump-THz-probe spectroscopy

Matthias C. Hoffmann[1,*], János Hebling[1,2], Harold Y. Hwang[1], Ka-Lo Yeh[1] and Keith A. Nelson[1]

[1] Department of Chemistry, Massachusetts Institute of Technology, Cambridge, Massachusetts 02139, USA
*E-mail: mch@mit.edu
[2] Department of Experimental Physics, University of Pécs, 7624 Hungary

Abstract. We observe impact ionization and saturation of free carrier absorption in indium antimonide at 200K and 80K. We employ a novel THz-pump-THz-probe scheme with pump fields up to 100 kV/cm amplitude and 100 MW/cm^2 intensity.

Introduction

Indium antimonide has the highest electron mobility and saturation velocity of all known semiconductors. This makes transistors with extremely high switching speed possible [1]. The carrier dynamics on the ultrashort timescale are hence of great technological as well as fundamental interest. The material is a direct semiconductor with a band gap of 170 meV at room temperature, making it well suited for applications in infrared sensors covering the wavelength range from 1 – 5 μm[2].

Impact ionization by high electric fields is a well known phenomenon in InSb [3,4]. Strong THz fields can directly achieve impact ionization on the picosecond time scale, avoiding additional experimental complications by using photon energies well below the band gap. An intensity dependent THz pump experiment on InSb has been reported recently by Lindenberg [5].

Due to their strong interaction with free carriers, THz pulses can also be used as a very sensitive probing tool to monitor the subsequent carrier dynamics after THz excitation. In this article we demonstrate THz-pump-THz-probe spectroscopy and apply it, with frequency as well as time resolution, to the study of impact ionization and hot carrier effects caused by THz fields greater than 100 kV/cm in InSb.

Experimental Methods

Our THz pump-probe method was based on the use of intense THz pulses generated by tilted pulse front excitation in LiNbO$_3$ [6]. A 6-mJ optical pulse from a kHz repetition rate Ti:sapphire amplifier was split into two parts using a 10:90 beam splitter and the more intense pulse was variably delayed. The optical pulse fronts were tilted with a grating-lens combination and directed to a common region of a LiNbO$_3$ crystal to generate collinear THz pump and probe pulses that could be variably delayed. The single-cycle THz pulses were focused to the sample inside a cryostat using a set of off-axis parabolic mirrors resulting in a focus size of 1 mm, as verified by a razor blade scan. The THz pulse energy at the sample spot was 2 μJ. A second pair of parabolic mirrors was used to relay-image the THz field onto an electro-optic (EO) sampling setup. A ZnTe crystal with 0.1 mm active and 1 mm total thickness was used to keep the EO signal in the linear range. EO traces of the THz pulses were recorded with and without the sample in place at every pump step covering a time window of 45 ps. Chopping only the laser beam used to generate the probe THz pulse ensured efficient suppression of the pump THz field. The sample was n-type InSb:Te (110) with a thickness of 450 um and a carrier concentration of 2.0×10^{15} cm^{-3} at 77 K, and a manufacturer-specified mobility of 2.5×10^5 cm^2/Vs at room temperature.

Results and Discussion

From the sample and reference data we can calculate a frequency averaged effective absorption coefficient

$$\alpha_{eff} = -\frac{1}{d} \ln \left(\frac{\int dt E_{sam}^2(t)}{\int dt E_{ref}^2(t)} \right)$$

where d is the sample thickness and $E(t)$ is the electro-optically measured probe field temporal profile. Figure 1 shows our pump-probe results at 200K and 80K. At 200 K, the initial effect of the pump pulse is to accelerate the preexisting carriers, providing access to higher-energy states (including those in neighboring valleys) with lower mobility and higher effective mass. This results in reduced THz probe pulse absorption, consistent with our observations in THz pump-probe measurements of Ge, Si, and GaAs as well as observations of THz absorption saturation in nonlinear transmission measurements of these materials and InSb (reported elsewhere in this volume). The THz probe absorption then increases and reaches levels far higher than its initial value, indicating the presence of additional carriers produced through impact ionization. This behavior was not observed in the other semiconductors studied, whose bandgaps are too wide for impact ionization to occur under these conditions. Carrier cooling during the next 10 ps leads to continued increase in THz probe absorption. Partial reflection of the THz pump pulse within the sample leads to a second weaker response starting at around 10 ps that adds to the rising signal response already under way. At 80 K there are fewer preexisting carriers and their initially reduced absorption appears to be outweighed by the increase in carrier population at all probe delay times. We observe an overall sevenfold increase in the absorption after 30 ps due to impact ionization. Assuming proportionality between the effective absorption and the carrier concentration, we estimate that 1.4×10^{16} cm^{-3} additional carriers were produced, using the ratio of the absorption coefficient before and after the THz pump and the known carrier concentration at 80K.

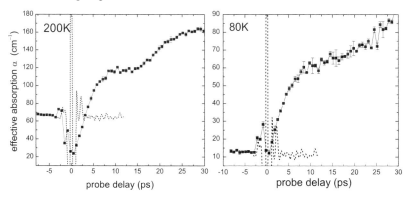

Fig. 1. The effective absorption coefficient observed before and after excitation with a 2 μJ THz pulse at sample temperatures of 200K and 80K. The dashed lines show the shape of the pump THz pulse.

Figure 2 shows the frequency resolved absorption spectra at 200K and 80K for selected time points. At negative time delays (i.e. before THz excitation) the absorption follows essentially a Drude model. At 200 K the absorption at positive delays also can be described by a Drude model. At 80K the behaviour is remarkably different. In the absorption spectrum we observe a broad feature around 1.1 THz that might be related to interaction with phonon modes [7]. We did not observe this absorption peak in nonlinear transmission measurements at any temperature or THz field strength.

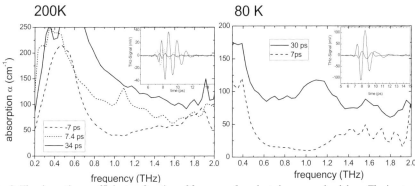

Fig. 2. The absorption coefficient as function of frequency for selected pump-probe delays. The insets show the corresponding transmitted THz probe fields.

Conclusions

The newly developed THz pump/THz probe technique is a sensitive method that monitors carrier dynamics in semiconductors on ultrafast timescales. In InSb at 200K, we observed an initial drop in THz probe absorption due to electron heating, then an increase in absorption beyond its initial level due the generation of new carriers through impact ionization. At 80 K the number of carriers increases by a factor of 7 and the absorption spectrum shows a long-lived non-Drude peak at 1.1 THz.

Acknowledgements. This work was supported in part by ONR grant no. N00014-06-1-0463.

1   T. Ashley et al,. Proc. 7th Intl. Solid State and Integrated Circuits Tech. Conf., Beijing, China, pp. 2253-2256, (2004).

2   D. G. Avery, D. W. Goodwin and Miss A. E. Rennie , J. Sci. Instr. **34**, 394 (1957).

3   C. L. Dick. and B. Ancker-Johnson, Phys. Rev. B. **5**, 526 (1972).

4   A. Lobad and L. A. Schlie, J. Appl. Phys. **95**, 97 (2004).

5   A.M Lindenberg, H. Wen, E. Szilagyi, CLEO 2008.

6   K. Yeh, J. Hebling, M. C. Hoffmann and Keith. A. Nelson, Opt. Commun. **281**, 3567 (2008).

7   D.L. Price and J.M. Rowe, Phys. Rev. B **3**, 1268 (1971).

Single Shot Linear Detection of THz Electromagnetic Fields on the Fs to Ps Scale

Uli Schmidhammer, Vincent De Waele, and Mehran Mostafavi

Laboratoire de Chimie Physique - ELYSE, UMR8000 CNRS-Université Paris Sud
91405 Orsay, France
E-mail: uli.schmidhammer@lcp.u-psud.fr

Abstract. We present single shot electro-optic sampling based on spectral decoding with a chirped supercontinuum: The wavelength dependent polarization state of the probe is analyzed in polychromatic balanced detection. The frequency bandwidth of over 300 THz and the linear, normalizing detection allow widely tunable and broadband diagnostic of THz waveforms without signal distortion. The technique is compatible to state of the art fs to ps laser.

Electro-Optic Sampling for THz Spectroscopy and e⁻ Bunch Monitoring

While the last decade the diagnostic of THz electromagnetic fields via electro-optic (EO) sampling techniques has been a field of greatly increasing interest. One motivation is the variety of applications in THz spectroscopy and imaging with the possibility of non-destructive measuring. On the other hand there is the development of ultrafast electron accelerators used for the generation of new brilliant radiation sources. The nearly transverse Coulomb field of such relativistic electron pulses is correlated to their longitudinal shape allowing for non-invasive characterisation in a nearby placed electro-optic crystal. Here the need for synchronisation of the bunch with accelerating fields or external experiments is inherently linked to the single shot capability of the diagnostic. The fastest possible diagnostic is also desired for the THz characterisation of moving objects, dynamic processes or experiments with low repetition rate.

In general there are two ways to substitute the delay line for conventional, repetitive EO scanning: the birefringence induced in the EO crystal by the THz electric field is encoded as polarization state modulation either spatially to the wavefront of the optical probe pulse or to its temporally dispersed spectrum. The existing techniques are compared in a more exhaustive manner in [1]. The often called spatial encoding techniques are well suited for the sub-ps regime but usually restricted to some ps. One dimensional imaging of the THz waveform in the EO-material is not possible without lateral averaging. In contrast, the spectral encoding can be applied in a stable and tunable manner on the ps scale [2]. Temporal resolution and detection window are in this case connected in analogy to the energy-time uncertainty of the Heisenberg principle via the bandwidth of the probe pulse [3]. This effect limits the applicability especially towards the short time scale in the way that only for the state of the art sources delivering ultrabroad spectra a reasonable operating range can be chosen. Moreover, the single shot EO techniques presented in the literature use a detection scheme based on crossed polarizer operated at (near) optical bias. In contrast to the state of the art high sensitive delay scan methods this configuration needs recording of only one polarization state. However its signal response is nonlinear with different terms that depend on the intensity distributions of the probe, the optical bias, scattering contributions and the amplitude of the electromagnetic field to be detected. Signal distortions and possible artefacts are the consequence [4,5].

669

EO Sampling by Supercontinuum Encoding with Balanced Detection

These restrictions of single shot EO sampling have been overcome to a large extent in the following way: The temporal distribution of the electric field in the EO crystal is encoded to a supercontinuum (SC) whose wavelength dependent polarisation state is then analyzed with balanced detection (see Fig. 1). The generation of the SC not only delivers an ultrabroad bandwidth of the probe whose Fourier Tranform limit is < 5 fs. This strongly nonlinear process allows decoupling the characteristics of the diagnostic to a high degree from the initial laser source. So the required intensity to start SC generation can easily be achieved at different wavelengths, pulse durations and energies just by focusing an appropriate amount of energy smoothly into a sufficiently non-linear medium. In the presented setup we focus about 1 μJ of a Ti: Sapphire laser with a duration of about 200 fs into a Sapphire plate and obtain a highly stable single filament SC covering the visible spectral range.

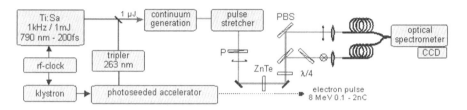

Fig. 1. Scheme of the electro-optic single shot diagnostic based on supercontinuum decoding in polychromatic balanced detection. The configuration for the non-invasive monitoring of the longitudinale electron pulse shape at the laser triggered ps radiolysis facility ELYSE is shown.

In the following the spatial and temporal dispersion of the ultrabroad probe must be controlled. The inherent dispersion of the setup is kept on the sub-ps level up to the EO material. From this minimum detection window suited for highest temporal resolution the time scale can be adapted just by adding glass as pulse stretcher. The remaining dispersive optics must be chosen to be of high achromatic quality in order to avoid significant wavelength dependent beam deviations. So the used polarizer (P) and polarizing beam splitters (PBS) are based on the optical wiregrid technology. To optimize the balance of the two perpendicular polarization components the effective pathways from the separating PBS are realized symmetrically.

Kerr Effect with an Optical Pulse as Fs Reference

To test the reliability and sensitivity on the fs scale we used the electro-optic Kerr effect with the Ti:Sapphire laser source as electric field. The interacting medium was 1 mm quartz. This experiment provides well defined conditions concerning the electromagnetic field to be studied: free space THz pulses in contrast have usually complicated waveforms that are difficult to analyze in an independent and reliable single shot experiment. The calculated correlation time to wavelength was verified experimentally by varying the optical delay (see inset figure 2a). Moreover, the influence of the stability of the laser source generating the SC on the diagnostic can be elucidated. One clearly observes the shot to shot jitter of the laser (see figure 2a). On the other hand the base line of the EO-measurements does not exhibit significant instabilities, i.e. the single shot EO sampling is to a high degree laser fluctuation free.

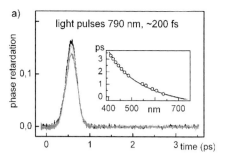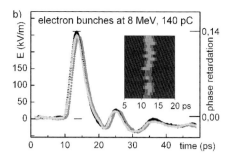

Fig. 2. Three consecutive single shot EO measurements: (a) On the fs scale with the Ti:Sapphire laser pulse as optical Kerr gate in a fused silica plate. The inset shows the calculated and measured wavelength dependence of the time scale. (b) On the ps scale for e⁻-bunches whose electric field was probed in 0.5 mm ZnTe. The arrival time of a series of 20 e⁻-bunches is shown in the inset exhibiting a jitter of ~ 1.5 ps.

Monitoring of Electron Bunches on the Ps Scale

Adapting the dispersion by adding glass the identical set-up can be used on the large ps scale. Passing through 20 cm SF57 results in a 60 ps time window over the spectrum transmitted by the EO crystal, i.e. 0.5 mm ZnTe, and a resolution <1 ps [3]. These conditions are unprecedented for single shot EO sampling on the ps scale and allow for the characterisation of the longitudinal electron bunch distribution at the ELYSE laser triggered ps pulse radiolysis facility [6] (see Fig. 2b). Electron bunch monitoring at the 100 pC charge level is possible even for the low beam energy and brightness of the accelerator at ELYSE. The measurements reveal oscillations caused by reflections of electric field contributions inside the ZnTe crystal. These can be inherent to the EO sampling and may be remodelled. Due to the non-linear response of the crossed polarizer detection they may not clearly be resolved in the state of the art single shot measurements. The balanced detection allows for the first time in a single shot experiment the direct determination of the phase retardation and so to deduce the electric field amplitude probed in the EO crystal.

Conclusion

Due to the combination of the SC encoding and the polychromatic balanced detection the field amplitude within the EO crystal can be determined in an undistorted manner with a detection window several times longer than the THz pulse. For a time window easily tunable between 0.7 to 100 ps the possible time resolution is 90 fs and 1 ps, respectively. The diagnostic is to a high degree laser fluctuation free and its sensitivity shot noise limited to a phase retardation below 0.005 radians. These optimized features can be achieved with ultrafast laser systems operating in the UV-Vis to NIR.

1   K. Y. Kim, B. Yellampalle, A. J. Taylor, G. Rodriguez, and J. H. Glownia, in Optics Express 32, 1968, 2007.
2   Z. Jiang and X.-C. Zhang, in Applied Physics Letters 72, 1945, 1998.
3   J. R. Fletcher, Optics Express 10, 1425, 2002.
4   Z. Jiang, F. G. Sun, and Q. Chen, in Applied Physics Letters 74, 1191, 1999.
5   Y. Li, in Applied Physics Letters 88, 251108, 2006.
6   J. Belloni, *et al.* in Nuclear Instruments and Methods Research Section A 539, 527, 2005.

Intense THz Pulses and 11-fs Electro-optic Sampling with a Multi-Branch Er:fiber/Ti:sapphire Hybrid Amplifier

Alexander Sell, Rüdiger Scheu, Alfred Leitenstorfer, and Rupert Huber

Department of Physics and Center for Applied Photonics, University of Konstanz,
Universitätsstraße 10, 78464 Konstanz, Germany
E-mail: alexander.sell@uni-konstanz.de

Abstract. We combine a four-branch Er:fiber laser with a high-power Ti:sapphire amplifier for high-field THz generation and electro-optic detection with 11-fs pulses. Frequency mixing of phase-correlated fiber branches generates multi-THz seed spectra up to 100 THz.

Introduction

Recent progress of sources and detectors in the elusive terahertz (THz) region of the electromagnetic spectrum has opened new avenues in basic science and technology [1]. The possibility to synthesize precisely defined wave forms and record their oscillating electric field directly in the time domain has been extended throughout the far and mid infrared [2]. While carrier envelope phase (CEP) stabilization of optical pulses has employed complex electronic schemes [3], THz pulses are inherently phase stable by generation.

Novel applications, such as coherent THz control of condensed matter systems [4] or high-harmonic generation [5], call for intense THz radiation – a domain which has long been reserved to large-scale facilities like synchrotrons or free-electron lasers. First table-top sources exploiting optical frequency conversion of amplified millijoule laser pulses for broadband THz generation have relied entirely on Ti:sapphire technology [6], suffering from limitations of long-term stability, efficiency, and bandwidth. Here, we present a novel, flexible approach combining the superior stability and tunability of femtosecond Er:fiber systems with the high power level of a Ti:sapphire amplifier. The four-branch Er:laser generates intense THz fields while at the same time providing 11 fs near-infrared pulses and CEP-stable mid-infrared seed transients for a future multi-THz power amplifier.

Multi-branch Er:fiber – Ti:sapphire hybrid laser

The concept of our laser is schematically depicted in Fig. 1(a). We start with a femtosecond Er:fiber master oscillator (repetition rate: 49 MHz) [7] seeding four parallel low-noise amplifiers, each one featuring an erbium-doped gain fiber of a length of 1.75 m [8]. An input power of less than 1 mW is sufficient to drive each amplifier into saturation. Pulses of an energy of 6 nJ per stage are coupled into free space and subsequently recompressed by silicon prisms to a duration of 70 fs. We have recently demonstrated that all branches are synchronized with high precision: The relative rms timing jitter integrated from 1 Hz to the Nyquist frequency amounts to only 11 attoseconds corresponding to less than one percent of an optical cycle. The mutual phaselock is preserved even after frequency conversion in bulk fibers and

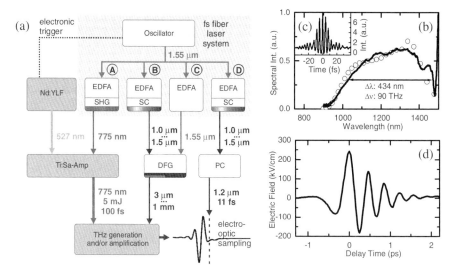

Fig. 1. (a) Schematic of the four-branch femtosecond Er:fiber laser with frequency conversion and power amplification; Nd:YLF: Q-switched pump lasers; EDFA: Er:fiber amplifier (energy: 4 nJ, pulse duration: 70 fs); SHG: second harmonic generation in PPLN; SC: supercontinuum generation in highly nonlinear bulk fiber; Ti:Sa-Amp: Ti:sapphire regenerative amplifier with booster stage; DFG: tunable difference frequency generation in PPLN; PC: prism compressor; (b) Spectrum of an optimized supercontinuum from a bulk fiber (curve: experiment; open circles: parameter-free modeling); (c) Autocorrelation trace demonstrating a compressed pulse duration of 11 fs; (d) THz field from optical rectification of 4.5 mJ pulses from the Ti:sapphire amplifier, electro-optically sampled by the 11 fs pulses in a 10-μm-thick ZnTe crystal.

free-space optics [9]. This principle allows us to design a femtosecond laser system with unprecedented flexibility:

A first output arm [labeled D in Fig. 1(a)] is designated for generation of ultrashort gate pulses for electro-optic sampling. For this purpose, we exploit massive temporal compression of the 1.55 μm pulses in a few centimeters of a tailor-cut highly nonlinear bulk fiber followed by an adaptive SF10 prism compressor. The optimum fiber parameters and input pulse characteristics are determined by a quantitative numerical simulation of the pulse propagation via a modified nonlinear Schrödinger equation. Both instantaneous nonlinearities and Raman scattering are accounted for [10]. No fit parameters enter the calculation. For an optimized scenario, our theory predicts pulses of a duration as low as 9 fs and an unstructured ultrabroad spectrum [open circles in Fig. 1(b)], which is impressively validated by the experiment [black curve in Fig. 1(b)]. The continuum is centered at 1250 nm and exhibits a width of 430 nm (FWHM). The center wavelength may be tuned to some extent by variation of the prechirp. An interferometric autocorrelation yields a pulse duration of 11 fs [Fig. 1(c)]. To the best of our knowledge, these pulses are the shortest reported of an all-fiber source. They constitute an ideal gate for ultrabroadband electro-optic detection up to a potential bandwidth of 100 THz [1,2].

The second Er:fiber arm [Fig. 1(a), label A] is amplified up to the mJ level. Since nonlinearities in fiber-based amplifiers preclude such energies, we continue in a free-space scheme: We first frequency-double the fiber output using a custom-designed periodically poled crystal of MgO:LiNbO$_3$ (PPLN). The resulting 1-nJ pulses exhibiting a nearly Gaussian spectrum of a width of 12 nm (center wavelength: 775 nm) are seeded into a commercial Ti:sapphire regenerative amplifier with a double-pass power booster (repetition rate: 1 kHz). We obtain amplified pulses of an energy of 5 mJ and a duration of 100 fs. Pulse-to-pulse power fluctuations are found to be less than 0.2% (rms) integrated over a time window of 30 s. No amplified spontaneous emission is discernible.

In a first demonstration, we exploit these pulses to generate intense THz radiation directly via optical rectification in a large-area ZnTe crystal (thickness: 500 µm, diameter: 20 mm) [1]. The resulting few-cycle field transient is fully characterized by electro-optic detection in a 10-µm-thick ZnTe element, gated by the 11-fs output of branch D [Fig. 1(d)]. The peak amplitude approaching 300 kV/cm already renders this pulse an interesting source for coherent control of condensed matter systems. In addition, alternative rectification schemes discussed in the literature [6] may be used to potentially achieve yet higher intensities.

A particularly promising aspect of the multi-branch system is the possibility to generate tunable phase-locked mid-infrared pulses by frequency mixing of the remaining two amplifier arms [B and C in Fig. 1(a)]. To this end, we generate 20-fs light pulses widely tunable from 0.95 µm to 1.45 µm in a tailored highly nonlinear fiber incorporated into the third arm [11]. Difference frequency mixing of this output with the fundamental wave emitted by the fourth branch [12] is used to generate up to 1.4 mW of mid-infrared pulses tunable from 2.8 µm to 5.6 µm. Since both branches maintain a stable relative phase [8], the carrier envelope offset frequency should be eliminated upon difference frequency generation. We expect the resulting transients to be inherently CEP stable, representing an attractive seed source for attosecond pulse [5] and high-power terahertz generation [6]. The complete fiber laser system including the difference frequency stage and all prism compressors is mounted on a breadboard of a footprint of only 1 square meter, improving the compactness and stability of the overall system.

In conclusion, the novel multiple-branch hybrid laser unites the benefits of fiber lasers with the high power levels achievable in Ti:sapphire amplifiers for the generation and detection of intense terahertz pulses. Phase-preserving amplification of the widely tunable phase-locked mid-infrared pulses promises an extremely attractive source for high harmonic generation, attoscience, and coherent control.

This work was supported by the Konstanz Center for Applied Photonics (CAP) and the Emmy Noether Program of the Deutsche Forschungsgemeinschaft (DFG).

1   B. Ferguson et al., Nature Materials 1, 26 (2002) and references therein.
2   C. Kübler et al., Appl. Phys. Lett. 85, 3360 (2004); T. Zentgraf et al., Opt. Express 15, 5775 (2007).
3   D. J. Jones et al., Science 288, 635 (2000).
4   S. Ganichev et al., Intense Terahertz Excitation of Semiconductors, (Oxford University Press, Oxford, 2005).
5   Henry Kapteyn et al., Science 317, 775 (2007).
6   A. G. Stepanov et al., Opt. Express 13, 5762 (2005); T. Löffler et al., Semicond. Sci. Technol. 20, 134 (2005); X. Xie et al., Appl. Phys. Lett. 90, 141104 (2007); T. Bartel et al., Opt. Lett. 30, 2805 (2005).
7   K. Tamura et al., Opt. Lett. 18, 1080 (1993).
8   F. Tauser et al., Opt. Express 11, 594 (2003).
9   F. Adler et al., Opt. Lett. 32, 3504 (2007).
10 P. V. Mamyshew et al., Opt. Lett. 15, 1076 (1990).
11 F. Tauser et al., Opt. Lett. 29, 516 (2004).
12 C. Erny et al., Opt. Lett. 32, 1138 (2007).

Frequency selective surface sensor for terahertz bio-sensing applications

M. Nagel[1], G. Klatt[2], M. Awad[1], H. Kurz[1], A. Bartels[2] and T. Dekorsy[2]

[1] Institute of Semiconductor Electronics, RWTH Aachen University, 52074 Aachen, Germany
[2] Department of Physics and Center for Applied Photonics, University of Konstanz, D-78457 Konstanz, Germany
  E-mail: nagel@iht.rwth-aachen.de

Abstract. Using high-speed asynchronous optical sampling, read-out of novel terahertz surface sensors directed at bio-sensing applications is presented. The surface sensor is based on periodically arranged metallic THz split ring resonators on a 27-μm-thin polymer membrane.

Introduction

Frequency selective surfaces (FSS) are based on metallic resonator array structures and show application specific tailored transmission and reflection properties. Throughout more than 4 decades they are widely used for applications in the microwave domain, for example as frequency selective reflectors in antenna applications or as radar absorbers on stealth fighter airplanes [1]. Although, the realization of the first terahertz FSS has been reported very early in the 1960′s [2], the application range and diversity of THz FSS up to now has been comparatively limited. One reason for this situation might be the larger technological effort (such as free-standing microstructures or membrane-based structures [3,4]) necessary in comparison with microwave FSS to make a "good" terahertz FSS – mainly resulting from the wavelength dependent smaller dimensions and higher material losses at THz frequencies. Very recent studies show, that a novel promising range of application for THz FSS may be found in the field of sensing of minute amounts of (bio-)material [5,6]. The approach is based on the fundamental property of a resonant structure to locally store excitation energy for a finite time, thereby increasing the interaction between the probe beam and an attached material dramatically. The senor response is given by a frequency shift of the undisturbed resonant frequency due to the dielectric loading by the sample material. Waveguide-coupled terahertz resonators have already been applied successfully to demonstrate this basic principle for base sequence selective label-free DNA detection [7]. FSS which are read out via free-space radiation are an attractive solution to parallelize the read out of multiple arrayed sensors. In this work, a new and relatively simple fabrication method is presented enabling the realization of terahertz FSS optimized for sensing applications. A terahertz time-domain spectroscopy system based on high-speed asynchronous optical sampling (ASOPS) is used to characterize the fabricated FSS. By avoiding opto-mechanical time-delay components the high-speed ASOPS system provides considerably decreased read-out times in comparison to standard systems.

Surface sensor design and fabrication

The geometric design of the fabricated FSS is depicted in Fig. 1. It consists of asymmetric double-split metallic rings arranged on a quadratic grid with a pitch of p = 197 μm. The rings are placed on top of a 27-μm-thick polymer substrate made of benzocyclobutene (BCB) with an assumed permittivity $\varepsilon_r = 2.6$ and dielectric loss factor tanδ = 0.0001. The FSS is fabricated by spin-on deposition of the BCB layer on a silicon host substrate including thermal curing at 210°C for 2h. Next the ring structures are patterned on top of the BCB using a standard photolithography process, e-beam metal evaporation (Cr/Au with 10nm/200nm thickness) and photo-resist lift-off. In a last step the FSS is lifted off the silicon substrate in KOH solution.

Experimental set-up

The high-speed ASOPS terahertz spectrometer is described in detail elsewhere [8]. Two femtosecond lasers with 1 GHz repetition rate and 50 fs pulse duration (Gigajet TWIN, Gigaoptics GmbH, Germany) are asynchronously linked with a 10 kHz repetition rate offset in a master-slave configuration. One laser serves as pump laser, the other as probe laser. The time-delay between pairs of pump- and probe-pulses is automatically ramped between zero and 1 ns at a rate of 10 kHz. The system enables us to achieve a high signal-to-noise ratio of approximately 60 dB in a data acquisition time of 60 s. The pump laser (800 mW, 830 nm) drives a high-efficiency large area GaAs-based terahertz emitter which emits linearly polarized terahertz radiation [9]. The generated terahertz pulses are steered by four off-axis parabolic mirrors and focused onto a ZnTe crystal for electro-optical sampling. The probe pulses (820 nm) are focused on the detection crystal via a hole in the fourth off-axis parabolic mirror. The polarization rotation induced by the THz radiation in the ZnTe crystal is analyzed and detected using appropriate polarization optics and a single fast photoreceiver. The transient photoreceiver signal directly yields the terahertz electric field in amplitude and phase after applying the appropriate calibration factor for the time axis [8]. Spectral data are obtained by Fourier transforming the electric field transient. For the Fourier transformation we use 200 ps of the total measured 1 ns time delay. This yields a spectral resolution of 5 GHz. Although the spectral resolution is about 5 GHz, the first resonance at 661 GHz can be determined with much higher accuracy. This is due to the fact that we fit the Fourier transformation to an asymmetric double sigmoidal function including 54 FFT samples. Within a measurement time of 2.6 s we can determine the resonance frequency to 661.13 ± 0.350 GHz.

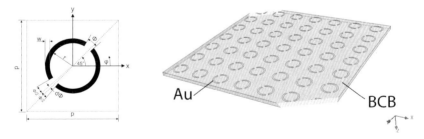

Fig. 1. Lateral unit cell structure of the applied asymmetric double split ring resonators (left) and three dimensional picture of the applied array arrangement on a 27-μm-thick BCB substrate. The resonators are dimensioned with $w = 10$ μm, $\Phi = 22°$, $d\Phi = 5°$, $p = 197$ μm and $r = 49$ μm.

Results and discussion

In a recent numerical study it is postulated, that FSSs based on asymmetric split ring resonators can be sensitive enough to detect locally applied organic films with a thickness of only 10 nm [5]. The high sensitivity is reasoned to originate from an extra-ordinary strong field enhancement in the gap regions of the ring resonators and an interference-like resonance phenomenon used for sensing. Since the ring resonators are split asymmetrically they act as a pair of coupled resonators with different main resonance frequencies. Such a configuration ideally shows a null in the reflection spectrum and accordingly a perfect transmission at the frequency where the summation of the interfering resonator currents exactly vanishes. The transmission spectrum of our FSS demonstrates this effect very well as can be seen from the data in Fig. 2 where the local transmission maximum is marked with an arrow. The FSS further shows a strong dependence on polarization, which needs to be taken into account for future sensor application. The excellent agreement between the measured and numerically simulated data confirms the high precision and accuracy of the applied fabrication and measurement technologies. A further important property of our FSS in this regard is given by low permittivity and thickness of the applied BCB substrate generating only a small dielectric (pre)-loading of the resonators and therefore increasing the sensitivity to dielectric loading by sample material. Further, detailed investigations will be presented at the conference.

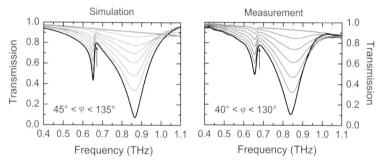

Fig. 2. Simulated (left) and measured (right) transmission amplitude in the frequency range of 0.4–1.1 THz for polarization angles of $40° < \varphi < 135°$ (black lines representing the smallest angle and light gray lines highest angles).

Conclusion

In this work, the fabrication and characterization of terahertz FSS designed towards sensing applications is successfully demonstrated using an optimized fabrication method and high-speed ASOPS measurement equipment.

Acknowledgements. We thank D. Debey for assistance in structure fabrication.

1   B. A. Munk, *Frequency Selective Surfaces: Theory and Design*, (Wiley & Sons, New York, 2000).
2   K. F. Renk and L. Genzel, Appl. Opt. **1**, 643-648 (1962).
3   R. Dickie, et al., IEEE Trans. Ant. Propag. **53**, 6, 1904-1911 (2005).
4   M. E. MacDonald, et al., IEEE Trans. Microw. Theor. **48**, 4, 712-718 (2000).
5   C. Debus and P. H. Bolivar, Appl. Phys. Lett. **91**, 184102 (2007).
6   H. Yoshida et al., Appl. Phys. Lett. **91**, 253901 (2007).
7   M. Nagel, F. Richter, P. H. Bolivar and H. Kurz Phys. Med. Biol. **48**, 3625–36 (2003).
8   A. Bartels, et al., Rev. Sc. Instr. Vol. **78**, 035107 (2007).
9   A. Dreyhaupt, S. Winnerl, T. Dekorsy, M. Helm, Appl. Phys. Lett. **86**, 121114 (2005).

Single cycle THz pulses in 1D and 2D photonic crystal structures

Peter Peier[1], Soenke Pilz[1], Hannes Merbold[1], Vladimir Pashinin [2], Taras Kononenko [2], Sergei Pimenov [2], and Thomas Feurer [1]

[1] Institute for Applied Physics, University of Bern, Sidlerstrasse 5, 3012 Bern, Switzerland
E-mail: peter.peier@iap.unibe.ch
[2] General Physics Institute, Russian Academy of Science, Vavilov-Str. 38, 119991 Moscow, Russia

Abstract. We present coherent time-resolved near-field imaging of single-cycle THz pulses in 1D and 2D photonic crystals. The results agree well with simulations and reveal the bandgaps and the dispersive properties of the photonic structures.

Introduction

The interaction of electromagnetic waves with randomly or periodically structured media has gained considerable interest, especially when the structure sizes become comparable to the wavelength. For example, the propagation of an electromagnetic wave within such structures is restricted and depending on the wavelength, the propagation direction, and the polarization, wave propagation may not be allowed. A great variety of powerful numerical techniques have been developed to simulate wave propagation through those media [1,2], however, few experimental techniques are appropriate to measure the electromagnetic field distribution with sufficient spatial and temporal resolution, for example near-field microscopy combined with interferometry [3,4]. Ideally, the spatial resolution should be a fraction of a wavelength and the temporal resolution a fraction of an oscillation period.

Suitable subjects for electromagnetic field measurements with sufficient spatial and temporal resolution are THz waves in ferroelectric crystals [5]. They are generated with femtosecond laser pulses through optical rectification [6] and are spectrally broadband and close to single-cycle. Their wavelengths are between tens and hundreds of microns and their oscillation periods are roughly a picosecond. They propagate through the crystal with almost light-like speeds and their electric field induces a change of the crystal's refractive index, whose sign and magnitude reflects the electric field amplitude. That is, snapshots of their electric field distribution can be measured [7] at any given time after excitation by a time-delayed probe pulse which illuminates the whole crystal and which experiences a phase modulation proportional to the electric field distribution of the THz wave. Since the probe pulses are at visible wavelengths and have a duration of 100fs, the spatial and temporal resolution are sufficiently high when compared to THz wavelengths and periods.

Experiment and Simulation

Photonic crystals require an accurate arrangement of at least two dielectric materials with different refractive indices. Here, we used the high index material MgO:LiNbO3 (LN), which has a refractive index of $n_{LN} = 5.11$, the medium index material silicon with $n_{Si} = 2.1$, and the low index material air with $n_{Air} = 1$. One dimensional photonic crystal structures were realized by stacking thin crystals of appropriate thickness in between two bulk LN crystals. The 2D photonic structures were fabricated through laser machining of $50\mu m$ thin crystals [8]. Experiments were performed using an amplified femtosecond laser system delivering 100fs, 0.3mJ pulses at 1kHz repetition rate. While the pump beam was focused to LN crystal by a cylindrical lens, the probe beam illuminated the whole crystal including the photonic structure. Therefore, we were able to record the incident, the reflected and the transmitted THz wave and also the field dynamics within the photonic structure. The time delay between the pump and the probe pulse was scanned from 0 to 40ps and 2D electric field images were recorded for every time step. All structures were designed to have at least one bandgap within the frequency range of interest, i.e. between 0.1 and 1.5 THz.

We used the commercial software BandSolve (RSoft) to calculate the band diagrams in two dimensions and a Matlab-based FDTD code to simulate the THz single-cycle waves propagating through the different photonic structures.

Results and Discussion

A snapshot of a typical THz waveform is shown in figure 1a). It was taken at a time delay of 4.5ps and shows only the THz waveform traveling to the right. The electric field is almost uniform along the vertical dimension, y, and shows a single oscillations only along the horizontal dimension, x, which coincides with the direction of propagation. The images taken at different time delays were integrated along the y-axis yielding 1D arrays representing the electric field distribution as a function of x at the given delay. Then, these arrays were stacked on top of each other yielding a space-time-plot as depicted in figure 1b). In a first series of experiments, different 1D photonic crystals, i.e. dielectric or Bragg mirrors, were assembled and analyzed. The reflected and the transmitted THz waveforms show complementary information and allow extracting all bandgaps within the THz spectrum. Experimental results and simulations agree perfectly well. We demonstrated high-reflection and anti-reflection coatings and frequency filters.

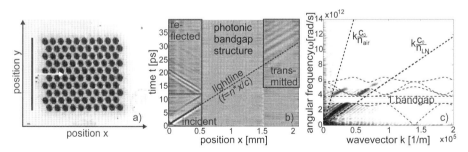

Fig. 1. a) 2D photonic crystal structure with incident THz phonon-polariton. b) Intensity plot of the space-time evolution of the THz electric field showing the incident the reflected and the transmitted wavepackets. The propagation within the structure is faster than the speed of light in the bulk crystal (light-line). The transmitted field is strongly chirped due to the dispersion of the photonic structure. c) Dispersion relation of a hexagonal photonic crystal structure. The lines depict the results calculated with BandSolve.

In a second series of experiments, 2D photonic crystal structures were investigated. A typical measurement is shown in figure 1b). The THz wavepacket was generated at $t = 0$ and $x = 0$ and as time goes on it propagated to the right towards the bandgap structure. At the bulk-structure interface one part of the incident THz wavepacket was reflected ($t > 12$ps and $x < 0.55$mm) and the rest was transmitted trough the structure ($t > 20$ps and $x > 1.5$mm). The velocity of the transmitted THz wavepacket was higher than the bulk group velocity due to the lower effective index of refraction of the photonic structure. The strong temporal chirp of the transmitted signal indicates a strong dispersion. For a more quantitative analysis we performed a two dimensional Fourier transformation of the data shown in figure 1b) (transmitted part only, i.e. $x > 1.5$mm) with the result shown in figure 1c). The intensity plot shows the content of the THz wavepacket in k_x-ω space which can be directly compared to the calculated band structure. Ideally, the maxima of the 2D Fourier transform should coincide with the dispersion functions of the different modes. The bandgap simulation were made in two dimensions, whereas the measured system is 2.5 dimensional. Nevertheless, the position of the band gaps should be predicted correctly as can be seen in figure 1c). All frequency components within the first bandgap are clearly missing in the transmitted waveform.

Conclusions

We analyzed the propagation of THz phonon-polaritons in and close to 1D and 2D photonic crystal structures. The experimental results of the 1D systems agree well with simulations based on dielectric multilayer systems and with FDTD simulations. For a variety of 2D photonic structures the spectrum of the transmitted THz wavepacket reveals the locations and the widths of the lowest order band gaps and the results are in good agreement with the simulations.

Acknowledgements. We would like to acknowledge the fruitful collaboration with E.R. Statz and K.A. Nelson and the Swiss National Science Foundation (project number 200021-111693) for funding.

1   S.C. Hagness and A. Taflove, Computational Electrodynamics, Artech House, 2000.
2   J.N. Winn, J.D. Joannopoulos and R.D. Meade, Photonic Crystals: Molding the Flow of Light, Princeton University Press, 1995.
3   A.M. Otter, J.P. Korterik, L. Kuipers, N.F. van Hulst, E. Flueck, and M. Hammer, in J. Lightwave Technology 21(1), 1, 2003.
4   H. Gersen, T.J. Karle, R.J.P. Engelen, W. Bogaerts, J.P. Korterik, N.F. van Hulst, T.F. Krauss, and L. Kuipers, in Phys. Rev. Lett. 94(7), 073903-1-073903-4, 2005.
5   T. Feurer, N.S. Stoyanov, D.W. Ward, J.C. Vaughan, E.R. Statz, and K.A. Nelson, in Annu. Rev. Mater. Res. 37, 317, 2007.
6   A. Kleinman, D.H. Auston, in IEEE J. QE 20, 964, 1984.
7   P. Peier, S. Pilz, F. Müller, K. A. Nelson, and T. Feurer, in JOSA B, Vol. 25, Issue 7, pp. B70-B75, 2008.
8   D.W. Ward, E.R. Statz, and K.A. Nelson, in Appl. Phys. A 86, 49, 2007.

Terahertz wave from coherent LO phonon in a GaAs/AlAs multiple quantum well under an electric field

K. Mizoguchi[1], Y. Kanzawa[2], M. Nakayama[2], S. Saito[3], K. Sakai[3]

[1] Department of Physical Science, Osaka Prefecture University, Gakuen, Naka-ku, Sakai, 599-8531, Japan
E-mail: k.mizoguchi@p.s.osakafu-u.ac.jp
[2] Department of Applied Physics, Osaka City University, Sugimoto, Sumiyoshi-ku, Osaka, 558-8585, Japan
[3] KARC, National Institute of Information and Communications Technology, Iwaoka, Nishi-ku, Kobe, 651-2492, Japan

Abstract. We report the terahertz wave from the coherent LO phonon in a GaAs/AlAs multiple quantum well by applying an electric field. It is found that the intensity of the THz wave from the coherent GaAs-like LO phonon is resonantly enhanced under the condition that the intersubband energy is tuned to the energy of the GaAs-like LO phonon.

Introduction

The coherent generation and detection of terahertz (THz) electromagnetic waves generated with ultrashort pulse laser have been widely investigated for applications to spectroscopy, imaging, non-destructive sensing and communication [1]. Recently, the generation of the intense THz waves emitted from coherent GaAs-like longitudinal optical (LO) phonons in GaAs-well layers of GaAs/AlAs multiple quantum wells (MQWs) was reported [2,3]. It is well known that the GaAs-like LO phonon is confined in the GaAs layer and its symmetry is the infrared-active B_2 mode. The LO-phonon confinement corresponds to the occurrence of translational symmetry breaking at each GaAs/AlAs interface. When the polarizations due to the coherent LO phonons in respective GaAs layers oscillate in phase, the THz waves generated in respective GaAs layers are constructively superimposed. These are the key points for the generation of the intense THz waves from the coherent LO phonons in GaAs/AlAs MQWs. On the other hand, the enhancement of the coherent LO phonon in GaAs/AlAs MQWs with use of intersubband energy tuning by application of an electric field has been investigated by using a reflection-type pump-probe technique [4]. In the case that the intersubband energy in GaAs/AlAs MQWs is tuned to the LO phonon energy of GaAs (E_{LO}) by the electric field owing to a quantum confined Stark effect (QCSE), the amplitude of the coherent LO phonon is intensively enhanced in comparison with the amplitude in a low electric field regime. We can expect the enhancement of the THz waves from the coherent LO phonons in GaAs/AlAs MQWs by tuning the intersubband energy to E_{LO}. In this work, we have investigated the THz wave from the coherent LO phonon in a GaAs/AlAs MQW under an electric field.

Experimental Methods

The sample used was an undoped GaAs/AlAs MQW embedded in a p-i-n structure

681

grown on a (001) n-GaAs substrate by molecular beam epitaxy, where the p and n layers consist of $Al_{0.5}Ga_{0.5}As$ layers. The GaAs/AlAs MQW has periodic heterostructures of $(GaAs)_{44}/(AlAs)_{16}$ with 20 periods, where the subscript denotes the number of the monolayers in the constituent layers (thickness of one monolayer = 0.283 nm). We performed photocurrent (PC) measurements at 10 K in order to estimate the transition energies including various higher subbands. The measurements of THz waves were carried out at 10 K by using Ti:sapphire laser pulses with a duration of 40 fs. The pump energy was tuned to 1.55 eV, which was located around the center energy between the E1HH1 and E1HH2 excitons at 110 kV/cm. Here, $EiHHj$ indicates the transition between the electron subband with the quantum number $m = i$ and the heavy-hole subband with $m = j$. The energy spacing between the E1HH1 and E1HH2 excitons around 110 kV/cm of the electric field becomes close to E_{LO} owing to QCSE. The pump-power density was kept at approximately 0.4 $\mu J/cm^2$. The samples were excited under 45 degree incidence by pump pulses. The THz waves emitted from the samples were collected with a pair of parabolic mirrors and detected by a photoconductive dipole antenna fabricated on a low-temperature-growth GaAs film, where the gap width is 5 μm. The time-resolved waveforms of the THz waves were obtained by measuring the photocurrent in the dipole antenna and varying the time delay between the pump and gating pulses.

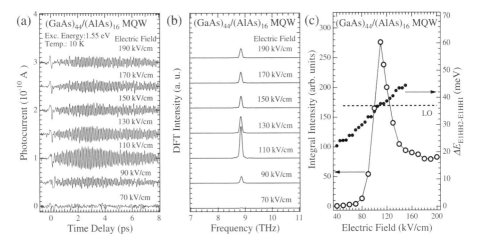

Fig. 1. (a) Temporal THz waveforms from the $(GaAs)_{44}/(AlAs)_{16}$ MQW at various electric fields. (b) Fourier transformed spectra of THz waves emitted from the coherent LO phonon at various electric fields. (c) Electric-field dependence of the integral intensity of THz signals from the coherent LO phonon (open circle). Closed circles indicate the energy spacing between E1HH1 and E1HH2 excitons in the $(GaAs)_{44}/(AlAs)_{16}$ MQW plotted as a function of electric field. The broken line indicates the LO phonon energy.

Results and Discussion

Figure 1(a) shows the temporal THz waveforms emitted from the $(GaAs)_{44}/(AlAs)_{16}$ MQW, and Fig. 1(b) exhibits the Fourier-transformed spectra of the THz waveforms in the whole time region. The THz signal around the time delay of 0 ps is due to the transient photocurrent. The oscillatory THz signals after 0 ps is assigned to the THz wave from the coherent GaAs-like LO phonon, because the observed frequency of the oscillatory THz signal corresponds to E_{LO}. The amplitude and decay rate of the oscillatory THz wave of the coherent LO phonon clearly change with the electric field. Figure 1(b) also indicates that the intensity of the coherent LO phonon in the MQW is obviously enhanced around 110 kV/cm.

In order to clarify the electric-field dependence of the THz wave from the coherent LO phonon, the integral intensity of the coherent LO phonon in the frequency region from 8 to 10 THz is plotted as a function of electric field in Fig. 1(c). The closed circles show the variation of the energy spacing between the E1HH1 and E1HH2 excitons ($\Delta E_{E1HH2-E1HH1}$) estimated from the PC spectra as a function of electric field. The intensity of the THz wave from the coherent LO phonon dramatically changes with the electric field and is resonantly enhanced around 110 kV/cm under the condition that $\Delta E_{E1HH2-E1HH1}$ is tuned to E_{LO} owing to the QCSE. The enhancement factor is over 200 as compared with the intensity of the THz wave in the low electric field region. If this enhancement will originate from the coupling of the coherent LO phonon to the impulsive excitonic interference between the E1HH1 and E1HH2 excitons, the pump-energy dependence of the THz wave from the coherent LO phonon should have a peak around the center energy between the E1HH1 and E1HH2 excitons [4]. However, the pump-energy dependence indicated that the intensity of the THz wave from the coherent LO phonon have a peak at the E1HH2 exciton energy (not shown here). The enhancement of the THz wave from the coherent LO phonon will originate from the double resonance in Raman scattering process[5]. When the incoming resonance at the E1HH2 exciton energy and the outgoing resonance at the E1HH1 exciton energy occur simultaneously, the intensity of the coherent LO phonon will be dramatically enhanced under the condition of $\Delta E_{E1HH2-E1HH1} \sim E_{LO}$.

Conclusions

We have investigated the enhancement of the THz wave from coherent GaAs-like LO phonon in a GaAs/AlAs MQW under an electric field. It is found that the intensity of the THz wave from the coherent GaAs-like LO phonon is intensively enhanced by a factor of 200 in comparison with that in a low electric field regime, under the condition that the intersubband energy is tuned to the energy of the GaAs-like LO phonon.

Acknowledgements. This work was supported by the Ministry of Public Management, Home Affairs, Posts, and Telecommunications, Japan, and by Grant-in-Aid for the Scientific Research from Japan Society for the Promotion of Science.

1   K. Sakai, Terahertz Optoelectronics, (Springer-Verlag, Berlin Heidelberg 2005).
2   K. Mizoguchi, T. Furuichi, O. Kojima, M. Nakayama, S. Saito, A. Syouji, and K. Sakai, Appl. Phys. Lett., 87, 093102 (2005).
3   K. Mizoguchi, A. Mizumoto, M. Nakayama, S. Saito, A. Syouji, K. Sakai, N. Yamamoto, and K. Akahane, J. Appl. Phys., 100, 103527 (2006).
4   O. Kojima, K. Mizoguchi, and M. Nakayama, Phys. Rev. B70, 233306 (2004).
5   B. Jusserand and M. Cardona, in Light Scattering in Solids V, Edited by M. Cardona and G. Güntherodt, (Springer-Verlag, Berlin Heidelberg, 1989), Chap. 3.

Improved Fast Scanning TeraHz Pulse System

Bernhard Heinemann[1], Colleen J. Fox[2], Hermann Harde[1]

[1] Helmut-Schmidt-Universitaet, Holstenhofweg 85, 22043 Hamburg, Germany
E-mail: harde@hsu-hh.de
[2] Thayer School of Engineering, Dartmouth College, 8000 Cummings Hall, Hanover, New Hampshire 03755-8001
E-mail: colleen.j.fox@dartmouth.edu

Abstract. We demonstrate the operation of a fast scanning laser system that was modified to improve and to increase the time resolution as well as spectral width for femtosecond time-resolved optical pump-probe or THz time-domain spectroscopy.

Introduction

Both time-resolved optical pump-probe spectroscopy as well as THz time-domain spectroscopy are typically using pump and probe pulses originating from a single laser. One pulse is employed to excite the sample or to generate the THz pulse, while the second pulse is delayed to probe the sample or the THz signal. Commonly the time delay is realized with an optical delay line consisting of a retro-reflector mounted on a mechanical translation stage or a vibrating shaker. These scanning mechanisms, however, have some significant limitations in their scan rates or the achievable time delay. Additional disadvantages may result from vibrations, some lateral shift of the beam or any spot size variation with increasing delay time.

These problems are eliminated by asynchronous optical sampling (ASOPS), where two mode-locked lasers serve as pump and probe lasers and are linked to each other at a fixed repetition rate difference Δf [1-4]. We have applied this ASOPS technique for fast scanning and data acquisition of femtosecond THz pulses.

Experimental Method and Results

The central component of the experimental set-up, shown in Fig.1, is a commercial dual-laser system (Gigajet TWIN, Gigaoptics, Germany) which consists of two Ti:sapphire femtosecond oscillators acting at repetition rates f_1 and f_2 of approximately 1 GHz [2 - 4]. Each oscillator delivers an average output power of about 750 mW with pulses of 70 - 80 fs duration. While laser 2 drives a large area optoelectronic GaAs transmitting antenna to generate the THz pulses, laser 1 is utilized for electrooptic sampling of the THz signal.

Both oscillators are linked to each other by an active feedback loop, which stabilizes the repetition rate difference $\Delta f = f_2 - f_1$ to 10 kHz. To accomplish this, the pulse rate of laser 1, which acts as the master laser, is detected by a fast photodiode and the measured frequency up-shifted by 10 kHz via a frequency shifter. Simultaneously the pulse rate of laser 2 (slave) is registered by a second photodiode and phase locked to the frequency shifted signal of the master laser. As phase detector a double-balanced mixer (DBM) is used and the output of the mixer fed back to a piezoelectric transducer, by which the cavity length of laser 2 and therefore its repetition rate is controlled.

684

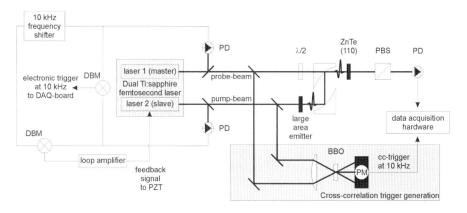

Fig.1. Experimental set-up of the fast scanning THz pulse System

The difference rate Δf determines the scan rate and thus the time $T_S = 1/\Delta f = 100$ µs which is required for a single scan over the temporal measurement window of 1 ns. Generally the time delay τ between the two pulse combs can be expressed by the relation $\tau(t) = \Delta f/f_2 \cdot t$, where t represents the real-time scale. Therefore, any transients on a femtosecond scale are transformed to a nanosecond time scale by the factor $f_2/\Delta f = 10^5$, and the other way round any bandwidth limited signals in real-time are converted to the time-delay scale by $\Delta f/f_2 = 10^{-5}$. Also the minimum time shift from pulse to pulse and therefore the absolute time resolution $\Delta \tau = \Delta f/(f_1 \cdot f_2) = 10$ fs is derived from this upper relation when choosing t as the pulse period $T_1 = 1/f_1$.

The transmitted intensity of the probe pulses, which sample the THz signal in an electrooptic crystal, is monitored by a 125 MHz bandwidth photoreceiver and then stored by a fast data acquisition unit as a function of the delay time. Due to the limited bandwidth the receiver integrates over approximately 8 pulses and therefore causes an uncertainty on the time-delay scale of about 80 fs, quite similar to the pulse width of the lasers. The data acquisition unit works with 12,500 channels, so one channel also corresponds to a 80 fs time-delay width.

In order to start the recording of data a trigger is generated from a second DBM, which uses the signals of the two fast photo diodes as input. This trigger signal appears periodically with the difference rate Δf and, therefore, is used to integrate over many scans with the scan rate Δf.

A typical measurement derived with this set-up and averaged over 2^{20} scans (in 400 s) is shown in Fig. 2a, which for clarity only displays the first 20 ps of the THz pulses which are propagating through dry air.

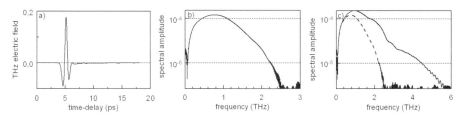

Fig. 2. a) Measured THz pulse, b) Fourier transform spectrum of pulses, measured with the standard set-up, c) Fourier transform spectrum obtained with the modified system (solid line) in comparison to the standard set-up (dashed line)

The respective Fourier transform spectrum depicted in Fig. 2b is significantly narrower than expected from the THz pulses and the time resolution of the measurement system. To characterize the system more accurately and to determine its true time resolution, we measured the cross-correlation signal between the two lasers by applying non-collinear sum-frequency generation in a barium-β-borate crystal.

The cross-correlation signal is detected with a 200 MHz bandwidth photomultiplier and recorded by a digital sampling oscilloscope. The measured pulses are displayed in Fig. 3a on an expanded scale with the solid line representing a single scan and the dashed graph as the superposition of 512 scans. Obviously is the width of the single cross-correlation scan with 124 fs only determined by the bandwidth of the photomultiplier and the pulse durations, while averaging over many scans significantly broadens the overall pulse structure up to 590 fs.

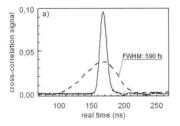 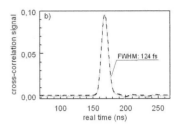

Fig. 3. a) Cross-correlation signals between laser 1 and 2 shot with electronic trigger. b) Cross-correlation signal recorded with optical trigger. Solid lines: single scan, dashed graphs: average over 512 scans. The real time scale is converted to delay-time scale by the factor 10^{-5}

Analysis of these measurements immediately makes clear that the quality of the feedback electronics (DBM 1) and the trigger generator (DBM 2) strongly determine the true time delay between the pulses and their appearance with respect to the next scan. So, any jitter or temporal offset from the desired pulse rate causes deviations in the relative time position of the pulse combs, and phase changes of the electronic trigger produce a shift of the whole time scale from scan to scan.

To overcome both these effects we replace the electronic trigger by an optical trigger which itself is derived from the cross-correlation signal. This trigger comes up with the right periodicity Δf and inherently defines the origin of the time-delay scale. Therefore, any shifts of the pulses in their relative and absolute position to each other no longer affect the correct storage for subsequent scans. A cross-correlation signal measured with the optical trigger and averaged over 512 scans (dashed line) is shown in Fig. 3b. Compared to a single scan almost no additional broadening is observed. Using such optical trigger for a THz measurement, (see Fig.1 - lower part) we observe under otherwise similar conditions a distinctly broadened spectrum as represented by Fig. 2c.

So, we have demonstrated the operation of a modified laser system that can be used for femtosecond time-resolved optical pump-probe and THz spectroscopy and that allows to take scans over one nanosecond time delay at 10 kHz scan rate with an improved time resolution and increased spectral width.

1   P. A. Elzinga et al., in Appl. Opt., Vol. 26, 4303, 1987.

2   C. Janke et al., in Opt. Lett., Vol. 30, 1405, 2005.

3   A. Bartels et al., in Appl. Phys. Lett., Vol. 88, 041117, 2006.

4   A. Bartels et al., Optics Express, Vol. 430, 2006.

Ultrafast photoemission electron microscopy: imaging light with electrons on femto-nano scale

Hrvoje Petek[1,2] and Atsushi Kubo[1,3]

[1]*Department of Physics and Astronomy, University of Pittsburgh, Pittsburgh, PA 15260 USA*
[2]*Donostia International Physics Center, Donostia-San Sebastian 20018 Spain*
[3]*PRESTO, Japan Science and Technology Agency, 4-1-8 Honcho Kawaguchi, Saitama, Japan*
E-mail: petek@pitt.edu

Abstract. Attosecond movies (330 as/frame) of surface plasmon polariton dynamics at a nanostructured silver/vacuum interface are recorded with a photoelectron emission microscope employing phase-locked pulse pair excitation. Examples of simple surface plasmon optical elements are given.

Introduction

Multidimensional spectroscopic imaging combining laser excitation and electron detection makes it possible to study femtosecond dynamics on the nanometer spatial scale [1-3]. By combining 10 fs laser interferometric pump-probe pulse excitation with photoemission electron microscopy we achieve 330 as/frame, <40 nm resolution imaging of coherent PHz electromagnetic fields at an Ag/vacuum interface. We describe ultrafast microscopic studies of the spatio-temporal evolution of surface plasmon polariton fields coupled through nanofabricated plasmon optics [4, 5].

Experimental method

We study the evolution of surface plasmon polaritons coupled with simple geometric structures lithographically defined in thin (~50 nm) Ag films on Si substrates. The samples held at the focal plane of a photoelectron emission microscope (PEEM) are excited by an identical pair of pump and probe laser pulses (10 fs duration, ~400 nm wavelength) with interferometrically defined mutual delay. The spatial distribution of photoelectrons emitted from the Ag surface through the nonlinear two-photon photoemission (2PP) is imaged with <40 nm resolution using the PEEM electron-optical column. Photoelectrons can be detected either with a MCP/CCD detector to provide spatial images of the energy integrated photoelectron yield, or with a crossed-wire x, y, t detector, which simultaneously measures the position and energy - through a time-of-flight measurement – of each photoelectron. By interferometrically scanning the pump probe delay in ¼ cycle (330 as) increments [1, 6], we can record interferometric time-resolved PEEM (ITR-PEEM) movies of light-driven electron polarization or population dynamics on the nanometer scale.

Results

We studied the surface plasmon polariton (SPP) dynamics for a variety of coupling structures defined in atomically smooth Ag films on Si substrates. Following lithographic forming of the coupling structures in Si, the Ag films were deposited by e-beam evaporated Ag vapor source. Because of the momentum mismatch between the external 400 nm excitation light incident at 65° from the surface normal, the SPP mode cannot be excited at a continuous vacuum/metal interface. By engineering

687

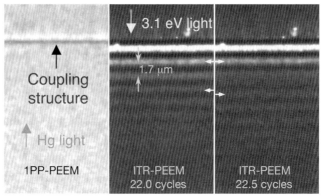

Fig. 1. One-photon photoemission (1PP) PEEM image of a grove defined in an Ag film taken with Hg lamp (4.9 eV) light, and two frames of an ITR-PEEM movie excited with 3.1 eV light. The pump-probe delays are 22.0 and 22.5 optical cycles. Advancing the delay by ½ optical cycle, or 0.66 fs, advances the pump-probe interference by 0.85 μm, or ½ period of the polarization grating. Arrows indicate the direction of light incidence.

abrupt structures with dimensions that are comparable to the wavelength of light, however, the efficient coupling into the SPP mode can be realized. The SPP mode has both the transverse and longitudinal components, and the coupling at a discontinuity primarily occurs into the longitudinal component. The incoming wavevector of the external field and the shape of the coupling structure define the spatio-temporal phase of the excited SPP wave packet, and hence its subsequent evolution. We have investigated structures for the coherent addition, interference, and focusing of SPP waves on the femto/nano scales.

Figure 1 shows a single-photon photoemission (1PP) image of a linear coupling structure (grove) defined in an Ag film, and two frames from an ITR-PEEM movie for 20.0 and 20.5 fs pump-probe optical cycle delays (the period of oscillation of 400 nm light is 1.33 fs). The 1PP image was recorded with 4.9 eV light from an Hg lamp. The image shows the topographic contrast of the Ag slit. Light incoming from the lower part of the image casts a shadow at the downward sloping edge, making it dark, and illuminates the rising slope of the coupling grove, making it bright. The grainy contrast over the flat surface is given by the shot noise in the image acquisition. ITR-PEEM images also show the slit, but in addition there is an interference pattern with a spatial period of 1.7 μm, and other random bright features. The random features represent defects or debris of the lithographic process, which can cause local roughness. The wavelength scale discontinuity at the slit makes it possible to couple the excitation light into the localized plasmon modes. The locally enhanced fields result in the modulation of the two-photon photoemission yield.

The interference pattern has the same wavevector as the momentum mismatch between the external excitation and the SPP fields [4]. The SPP field is generated locally at the coupling structure and it propagates along the vacuum/Ag interface with the corresponding complex wavevector. The SPP wave packet is a nonlocal source of polarization propagating along the surface at the local speed of light. The transverse component of this nonlocal field creates a polarization grating in the Ag film through the interference with the local polarization created by the external excitation field. The two-photon photoemission signal has a quartic dependence on the total polarization impressed in the Ag film. The fringe contrast depends on the local and nonlocal polarization fields having nearly the same amplitude. The mechanism for decay of the collective SPP field into single particle excitations is not yet fully understood. Nevertheless, the interference between the two fields points to the coherent two-photon absorption exciting electron-hole pairs with energy above the 4.3 eV work function of the Ag film.

The two frames of the ITR-PEEM movie show how the polarization grating evolves as a function of the pump-probe delay. The grating shows two types of fringes: 1) Delay independent fringes fixed to the coupling structure; and 2) delay dependent fringes that dominate far from the coupling structure at longer delays. The former can be observed even with single pulse excitation, and arise from the interference between the external pulse and its own SPP wave packet. The latter represent the interference between the pump induced SPP wave packet and the external field of the probe pulse. This pump-probe interference component measures the SPP wave packet dispersive propagation and dephasing. The stationary self-interference is seen to be independent of delay or the phase for in-phase (22.0 cycles) and out-of-phase (22.5 cycles) excitation. By contrast, changing the delay by 0.5 cycle (0.66 fs) causes a π-phase shift between the fringes for the pump-probe interference grating. The decay length of the propagating SPP wave packet is ~3-4 μm is determined by the complex dielectric constant of the Ag/vacuum interface.

Summary

ITR-PEEM movies, such as in Fig. 1, provide information on the evolution of coherent electromagnetic fields at a solid/vacuum interface. Simple structures such as the grove define the SPP wave phase front excited in the sample. More complex structures with different phase fronts can be used as different SPP optical elements, and the subsequent evolution of SPP fields monitored with femtosecond time and <40 nm spatial resolution. The ITR-PEEM technique should also be applicable to electronic excitations on clean and adsorbate covered metal and semiconductor surfaces. The phase sensitivity makes ITR-PEEM uniquely capable of imaging quantum dynamics on the nanometer spatial and femtosecond temporal scales.

Acknowledgements. We thank the support from NSF [CHE-0507147, ECS-0403865, and DMR-0116034], PRESTO JST, and the Basque Science Foundation, Ikerbasque.

[1] A. Kubo, K. Onda, H. Petek, Z. Sun, Y.-S. Jung and H.-K. Kim, "Femtosecond Imaging of Surface Plasmon Dynamics in a Nanostructured Silver Film" Nano Lett. **5**, 1123 (2005).

[2] A. H. Zewail, "4D Ultrafast electron diffraction, crystallography, and microscopy" Annu. Rev. Phys. Chem. **57**, 65 (2006).

[3] G. Schönhense, H. J. Elmers, S. A. Nepijko and C. M. Schneider, "Time-Resolved Photoemission Electron Microscopy" Advances in Imaging and Electron Physics **142**, 159 (2006).

[4] A. Kubo, N. Pontius and H. Petek, "Femtosecond Microscopy of Surface Plasmon Polariton Wave Packet Evolution at the Silver/Vacuum Interface" Nano Lett. **7**, 470 (2007).

[5] A. Kubo, Y. S. Jung, H. K. Kim and H. Petek, "Femtosecond microscopy of localized and propagating surface plasmons in silver gratings" J. Phys. B: **40**, S259 (2007).

[6] H. Petek, A. P. Heberle, W. Nessler, H. Nagano, S. Kubota, S. Matsunami, N. Moriya and S. Ogawa, "Optical phase control of coherent electron dynamics in metals" Phys. Rev. Lett. **79**, 4649 (1997).

Ultrafast Electron Dynamics in Quantum Well States of Pb/Si(111) Investigated by Two-Photon Photoemission

Patrick S. Kirchmann and Uwe Bovensiepen

Fachbereich Physik, Freie Universität Berlin, Arnimallee 14, Berlin, Germany
  Email: patrick.kirchmann@physik.fu-berlin.de

Abstract. We investigated ultrafast energy relaxation in unoccupied quantum well states of Pb/Si(111) with femtosecond time-resolved two-photon photoemission. Decay rates of $6p_z$ quantum well states are compatible with Fermi liquid theory if inter-subband scattering is considered.

Scattering of excited electrons occurs in metals on femtosecond (fs) imescales due to (i) the large phase space for electron-electron scattering and (ii) efficient screening of the hole related to the electron by the underlying elementary excitation of an electron-hole pair. Since in sp- and noble metals the electron-ion interaction is weak the electron scattering rates Γ can reasonably well be approximated by the free electron model [1] for metals. Γ is dominated by electron-electron (e-e) scattering and follows approximately an quadratic energy scaling with respect to the Fermi level E_F, $\Gamma_{\text{e-e}} \sim (E - E_F)^2$, as predicted by Fermi-liquid theory [2].

Time-resolved two-photon photoemission (2PPE) is used to monitor the decay of an excited electron population directly in the time domain. Two time-delayed femtosecond laser pulses first excite hot electrons into bound intermediate states. Subsequently electrons are photoemitted into vacuum states. Ultrathin films of simple metals like Al and Pb grown epitaxially in ultrahigh vacuum on semiconducting substrates, e.g. Si(111) [3], present interesting systems to study ultrafast electron dynamics since electron confinement in the film leads to formation of well-defined occupied and unoccupied quantum well states (QWS) in two dimensions.

In Pb/Si(111) the quantization of the nearly-free-electron $6p_z$ Pb band along the [111] direction gives rise to a series of (un)occupied QWS [3]. In Fig. 1 a) 2PPE spectra of Pb/Si(111) are shown as a function of intermediate state energy $E - E_F$ for different film thickness d ranging from 8-19 monolayer (ML). The spectra exhibit a series of four peaks, which are identified as unoccupied QWS. The QWS binding energies are analyzed by a fit of four Lorentzian peaks and an exponentially decreasing secondary electron background, which is convoluted with a Gaussian instrument function. Fig. 1 b) depicts the energies of unoccupied QWS extracted from 2PPE spectra as function of d in combination with results (+) of a density functional calculation of a bare Pb slab [4]. The reasonable agreement of experimental data and theoretical calculations suggest that Pb/Si(111) serves as a well characterized two-dimensional model system for the study of the ultrafast electron dynamics in metallic QWS.

To study the energy relaxation dynamics UV pulses are time-delayed with respect to visible laser pulses and 2PPE spectra are measured as a function of pump probe delay t, as shown in Fig. 2 a). We prepare a 15 ML film where the first unoccupied state (QWS+1) is situated in the Si bandgap, whereas the second unoccupied QWS (QWS+2) is degenerate with Si bands. Here, the electron dynamics are particularly interesting as electrons excited into QWS+1 are confined to the Pb film, contrary to QWS+2. In the

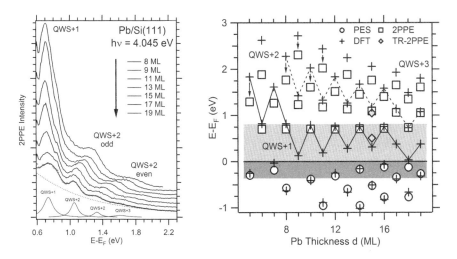

Fig. 1. a) Monochromatic 2PPE spectra for Pb/Si(111) ultrathin films with varying film thickness. The QWS binding energies are analyzed by a fit of four Lorentzian peaks and an exponential background, convoluted with a Gaussian instrument function. b) Thickness dependence of QWS energies for occupied and unoccupied states. Open symbols depict energies determined by laser-induced 1PPE (○), by monochromatic (□) and time-resolved (◇) 2PPE. Crosses indicate results from density functional theory taken from [4]. The shaded area indicates the indirect Si bandgap.

central panel of Fig. 2 a) the 2PPE intensity is shown as a function of pump probe delay and intermediate state energy. The right part of the figure shows a 2PPE spectrum integrated over t from −65 to 190 fs. The bottom panel of Fig. 2 a) shows cross-correlation (XC) traces which have been integrated over the energy intervals, which are indicated by bars on the left. The XC traces of QWS+1 and QWS+2 show a complex behavior. Its explanation requires a number of decay channels. The relaxation of QWS+1 is characterized by a delayed rise and a single exponential decay, whereas in QWS+2 a biexponential decay featuring a fast and a slow component is encountered.

To evaluate the decay dynamics of QWS+1 and QWS+2 consistently, we use a rate equation model which takes inter-subband scattering from the higher lying QWS+2 into QWS+1 into account as sketched in Fig. 2 b). For QWS+1 the fit results in: $\tau_1 = 142(5)$ fs for the decay and $\tau_{12} = 54(5)$ fs for the delayed rise. The analysis of the biexponential decay of QWS+2 results in 30(5) fs for the fast and 130(10) fs for the slow decay component. For the origin of the delayed rise we consider the optical excitation of QWS+2 and explain the delayed rise in QWS+1 by inter-subband scattering QWS+2→QWS+1 at a rate $1/\tau_{12}$. The population build up due to this inter-subband scattering is described quantitatively by the coupled rate equations in Fig. 2, which are fitted to the transient 2PPE data. The population decay of QWS+2 and the simultaneous build up in QWS+1 at the same rate $1/\tau_{12}$ is the direct consequence. We also determine $\tau_2 = 103(10)$ fs and interpret this as scattering from QWS+2 into unoccupied states of the Si substrate. Both decay channels of inter-subband scattering QWS+2→QWS+1 and scattering into Si states account for the fast decay of QWS+2 of 30(5) fs. The slow decay, which is not described by the rate equation model, might represent a further decay channel involving the Si substrate.

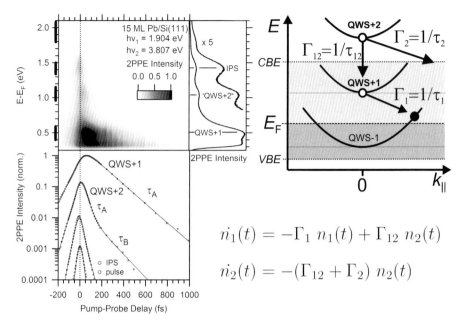

Fig. 2. a). Center panel: False color representation of the 2PPE intensity as a function of pump-probe delay and intermediate state energy $E - E_F$. Vertical bars indicate energy windows in which the intensity was integrated to monitor the time dependence of four spectral features (bottom panel). They are attributed to QWS+1, QWS+2, an image potential state (IPS), and high-energy electrons. Solid lines are fits following a coupled rate equation model. These spectral signatures are also seen in the 2PPE spectrum given in the right panel. b) Sketch of the dispersion of the (un)occupied QWS as function of parallel momentum $k_{\|}$ and the inter-subband scattering QWS+2$\rightarrow$QWS+1, which accounts for the delayed rise of intensity in QWS+1 at a rate Γ_{12}. The decay rates Γ_1 and Γ_2 describe the decay of electrons into unoccupied states of the metal film or the Si substrate. The shaded area visualizes the Si bandgap. Coupled rate equations for the transient populations in QWS+1 and QWS+2 to describe the population decay are given, where $n_i(t)$ denotes the transient population in the ith QWS+i.

In summary, we have investigated ultrafast electron dynamics in QWS of Pb/Si(111) by fs-time-resolved 2PPE. Generally, the decay times of the two lowest QWS are compatible with Fermi liquid theory. A closer investigation makes it necessary to explicitly take inter-subband scattering from the higher lying QWS into the lower lying QWS into account as well as scattering from photoexcited carriers of the Si substrate.

1   D. Pines and P. Nozieres, The Theory of Quantum Liquids, (Benjamin, New York, 1966)
2   E. V. Chulkov et al., Chem. Rev. 106, 4160 (2006)
3   P. S. Kirchmann et al., Phys. Rev. B 76, 075406 (2007)
4   C. M. Wei and Y. M. Chou, Phys. Rev. B 66, 233408 (2002)

Direct Visualization of Electron Emission during Femtosecond Laser Ablation

Christoph T. Hebeisen, Germán Sciaini, Maher Harb, Ralph Ernstorfer, Sergei G. Kruglik and R. J. Dwayne Miller

Institute for Optical Sciences and Departments of Physics and Chemistry, University of Toronto, 80 St. George St., Toronto, ON, M5S 3H6 Canada
Email: dmiller@LPhys.chem.utoronto.ca

Abstract. We use femtosecond electron pulses to probe the charge distributions and transient electric fields in the early stages of femtosecond laser ablation of a silicon (100) surface.

While femtosecond laser ablation has a wide range of applications including laser micromachining [1], laser surgery [2] and matrix assisted laser desorption [3], the various physical processes that lead to material removal are not all well-understood.

Usually, laser ablation is studied by mass spectrometry of the ejecta [4-7]. These methods detect the particles far from their origin and hence, they carry information integrated over their entire trajectory. Insight into the ultrafast ablation dynamics has to be retrieved by indirect means. Time-resolved shadowgraphy [8,9] can record ablation plume dynamics but this technique is not sensitive to electric fields, which play an important role in femtosecond laser ablation. Okano and coworkers [10] used electron pulses to directly probe the electric fields during ablation. The duration of the electron pulses limited their study to dynamics in the tens of picosecond range.

When a solid surface is illuminated by an intense femtosecond laser pulse, large numbers of electrons are emitted. The amount of charge initially emitted can be much larger than the charge on the sample detected in the long time limit since recapture of emitted electrons and emission of ions can significantly reduce the charge of the sample [11]. The charge separation produces electric fields and hence modifies the trajectories of charged particles produced during ablation. In the present study, we use femtosecond electron pulses as a probe of these transient electric fields. This technique allows us to observe the electric fields with subpicosecond resolution, providing unprecedented insight into the earliest times of the ablation process.

The experiments were performed in a modified femtosecond electron diffraction setup [12] with the laser pump pulse at $90°$ w.r.t. the electron beam. The electron pulses were 300 fs long. The sample, a 1 mm wide silicon strip, was placed inside in the electron beam path at the intersection of the electron beam and the excitation laser. As the electron pulse passes the excited part of the sample, the electrons are deflected by the electric field of the charge distribution produced in the ablation process. The distribution of the deflected probe electrons is then detected by a microchannel plate / phosphor screen detector. A series of probe electron maps for different delays between the laser excitation and electron probe pulses are shown in Fig. 1. A fresh spot on the sample was used for each delay. Initially, the probe electrons are deflected away from the sample surface by the electric field of the emitted electrons that form a dense cloud near the sample surface. At later times, when the cloud of emitted electrons has spread further due to self-acceleration [11], the attraction towards the countercharge on the sample dominates.

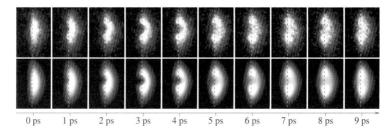

0 ps   1 ps   2 ps   3 ps   4 ps   5 ps   6 ps   7 ps   8 ps   9 ps

Fig. 1. Top row: probe electron density maps on the detector; the sample is on the left side, the laser is incident from the right. The dashed line indicates the approximate position of the edge of the sample "shadow". Bottom row: fit results for the corresponding images in the top row, produced by using the charge distribution model.

Since the recorded probe electron maps contain electrons that passed the excited sample at different distances and "heights", these maps cannot be directly inverted to obtain a map of the electric field at the sample position. Instead, we use a model to describe the charge distribution generated during ablation. This model consists of a positive charge on the sample surface and a cloud of emitted electrons above the surface (see Fig. 2). Both parts of the charge distributions are radially symmetric and Gaussian in the plane parallel to the sample surface and the electron cloud falls off exponentially with the distance from the surface along the surface normal. Assuming this charge distribution and a probe electron distribution as the probe pulse arrives at the sample, we can determine the probe electron density at the detector by calculating the deflection of individual probe electrons as a result of the model charge distribution. We fit the calculated maps to the experimentally recorded ones by varying the model parameters using a genetic search algorithm (similar to [13]).

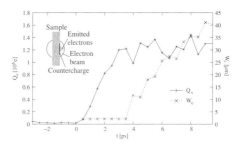

Fig. 2. Countercharge Q_s and width of the cloud of emitted charge W_c vs. time. Inset: the sample is placed vertically inside the electron beam so that the probe electrons pass the sample parallel to the sample surface. The laser beam is incident from the right along the surface normal.

The fits were performed independently for each time delay. The fit parameters were the amount of positive charge on the sample surface Q_s, the radial size of that charge distribution (FWHM) H_s, the width ($1/e$) of the electron cloud in front of the sample surface W_c and the radial size (FWHM) of the electron cloud H_c. As expected, the total charge $Q_s + Q_c$ was consistent with 0 when the amount of charge in the electron cloud Q_c was treated as an independent parameter. Hence, we set $Q = Q_s = -Q_c$ in order to reduce the number of parameters to make the fits more stable and less computationally expensive. The model works well for time delays of less than 10 ps. After that, the restrictions of the model such as the functional forms of the charge distribution and the absence of escaping charges in the model lead to poor fit results. A more complex model that introduces an additional electron cloud escaping from the surface produced smaller errors but proved to be much more computationally expensive because of the additional parameters while producing unstable results because of the weak dependencies of the results on some combinations of parameters. The graph in Fig. 2 shows the time series of Q and W_c. Q plateaus within 3 ps after the laser pulse at 1.2×10^6 electrons. At that time W_c starts to grow rapidly. A linear fit to W_c yields a speed of expansion of the width of the electron cloud of about 2% of the speed of light corresponding to a kinetic energy of 100 eV. Our implementation of the above charge distribution model requires a non-zero minimum for W_c, potentially leading to underestimation of Q at early times.

We have observed the transient charge distributions produced during femtosecond laser ablation from a Si (100) surface on the subpicosecond timescale. Through fits to a simple charge distribution model, we determined that 1.2×10^6 electrons (5.3×10^{11} electrons/cm^2) are emitted within the first 3 ps after excitation. This method could be used to verify dynamic models for the behaviour of charged particles during ultrafast ablation processes. Such experiments combined with a suitable model could e.g. determine if Coulomb explosion exists as a pathway for ablation in semiconductors and metals, which has been disputed [14,15]. Since Coulomb explosion is driven by electric fields, the technique we demontrate in this paper is particularly suitable to differentiate Coulomb explosion from other ablation processes.

Acknowledgements. Funding for this project was provided by the Natural Science and Engineering Research Council of Canada.

1   C. Y. Chien and M. C. Gupta, Appl. Phys. A 81, 1257 (2005).
2   A. Vogel, J. Noack, G. Hüttman, and G. Paltauf, Appl. Phys. B 81, 1015 (2005).
3   R. Knochenmuss, Analyst 131, 966 (2006).
4   A. Cavalleri, K. Sokolowski-Tinten, J. Bialkowski, M. Schreiner, and D. von der Linde, J. Appl. Phys. 85, 3301 (1999).
5   R. Stoian, D. Ashkenasi, A. Rosenfeld, and E. E. B. Campbell, Phys. Rev. B 62, 13167 (2000).
6   W. G. Roeterdink, L. F. Juurlink, O. P. H. Vaughan, J. D. Diez, M. Bonn, and A. W. Kleyn, Appl. Phys. Lett. 82, 4190 (2003).
7   H. Dachraoui, W. Husinsky, and G. Betz, Appl. Phys. A 83, 333 (2006).
8   N. Zhang, X. Zhu, J. Yang, X. Wang, and M. Wang, Phys. Rev. Lett. 99, 167602 (2007).
9   J. König, S. Nolte, and A. Tünnermann, Opt. Express 13, 10597 (2005).
10 Y. Okano, Y. Hironaka, K. Kondo, and K. G. Nakamura, Appl. Phys. Lett. 86, 141501 (2005).
11 T. L. Gilton, J. P. Cowin, G. D. Kubiak, and A. V. Hamza, J. Appl. Phys 68, 4802 (1990).
12 J. R. Dwyer, C. T. Hebeisen, R. Ernstorfer, M. Harb, V. B. Deyirmenjian, R. E. Jordan, and R. J. D. Miller, Phil. Trans. Roy. Soc. A 364, 741 (2006).
13 D. Zeidler, S. Frey, K.-L. Kompa, and M. Motzkus, Phys. Rev. A 64, 023420 (2001).
14 N. M. Bulgakova, R. Stoian, A. Rosenfeld, I. V. Hertel, and E. E. B. Campbell, Phys. Rev. B 69, 054102 (2004).
15 R. Stoian, A. Rosenfeld, I. V. Hertel, N. M. Bulgakova, and E. E. B. Campbell, Appl. Phys. Lett. 82, 4190 (2003).

Attosecond Nanoplasmonic Field Microscope

M. I. Stockman[1,2], M. F. Kling[2], Ulf Kleineberg[3], and F. Krausz[2,3]

[1] Department of Physics and Astronomy, Georgia State University, Atlanta, Georgia 30303, USA
E-mail address:mstockman@gsu.edu
[2] Max-Planck-Institut für Quantenoptik, Hans-Kopfermann-Straße 1, D-85748 Garching, Germany
[3] Ludwig-Maximilians-Universität München, Department für Physik, Am Coulombwall 1, D-85748 Garching, Germany

Abstract. We propose an approach that will provide direct, non-invasive access to the nanoplasmonic collective dynamics, with nanometer-scale spatial resolution and ~100 attoseconds temporal resolution. It combines techniques of photoelectron emission microscopy and attosecond streaking spectroscopy.

1. Introduction

Nanoplasmonics deals with collective electronic dynamics on the surface of metal nanostructures, which arise as a result of excitations called surface plasmons. This field has recently undergone rapid growth, with understanding that applications such as computing and information storage on the nanoscale, the ultrasensitive detection and spectroscopy of physical, chemical, and biological nanosized objects, and the development of optoelectronic devices would benefit. Due to their broad spectral bandwidth, surface plasmons undergo ultrafast dynamics with timescales as short as a few hundred attoseconds. So far, the spatio-temporal dynamics of optical fields localized on the nanoscale has been hidden from direct access in the real space and time domain. Here, we propose an approach that will for the first time provide direct, non-invasive access to the nanoplasmonic collective dynamics, with nanometre-scale spatial resolution and on the order of 100 attosecond temporal resolution [1]. The method, which combines two modern techniques of photoelectron emission microscopy and attosecond streaking spectroscopy, offers a valuable new way of probing nanolocalized optical fields that will be interesting both from a fundamental point of view and in light of the existing and potential applications of nanoplasmonics.

The proposed attosecond nanoplasmonic field microscope, illustrated in Fig. 1, combines two modern techniques: photoelectron emission microscopy (PEEM) and attosecond streaking spectroscopy [2]. A plasmonic nanostructure is excited by an intense, waveform-controlled field in the optical spectral range that drives collective electron oscillations (the quantum of which is called a surface plasmon). This generates optical fields localized on the nanometre scale, which henceforth will be known as nanolocalized optical fields. An attosecond extreme ultraviolet (XUV) pulse, which is produced from and synchronized with the driving optical pulse, is then sent to the system. This XUV pulse produces photoelectrons that, owing to their large energy and short emission time (determined by the attosecond pulse duration), escape from the nanometre-sized regions of local electric fields enhanced by plasmon resonances within a fraction of the oscillation period of the driven plasmonic field. The ultrashort escape time implies a final energy change of the emitted photoelectrons that is proportional to the local electric potential at the surface at the instant of electron release. This is in sharp contrast to previous attosecond streaking experiments performed in macroscopic volumes of gas-phase media, where the electron escape times are longer than the optical period; consequently, the change in electron energy probes the vector potential of the optical field [2]. For nanoplasmonic systems, the imaging of the emitted XUV-induced photoelectrons by an energy-resolving PEEM probes the electric field at the surface as a function of the XUV pulse incidence time and the position at the surface with an attosecond temporal and nanometre-scale spatial resolution.

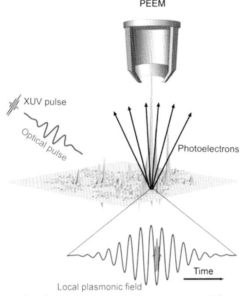

Fig. 1. Schematic of the system and photoprocesses. The nanosystem is shown in a plane denoted by a gray rectangle. Instantaneous local fields, which are excited by the optical pulse, are shown as a 3D plot. The local optical field at the point of the maximum field ("hottest spot") is shown as a function of time by a bold waveform, enhanced with respect to the excitation field. The application of an XUV pulse is shown by a thin waveform that is temporally delayed with respect to the excitation field. The XUV excitation causes the emission of photoelectrons (shown by the arrows), which are accelerated by the local plasmonic potential. They are detected with spatial and energy resolution by PEEM.

2. Calculations and Results

The predicted femtosecond dynamics of the local plasmonic fields as recorded by the attosecond nanoplasmonic field microscope is illustrated in Fig. 2 for two instants separated by approximately 1 fs. The calculations have been carried out for a model of a nanostructured silver layer of 65×65 nm^2 area (in the xz plane) and 4 nm thickness with the minimum-scale features of 2 nm. The plasmonic oscillations are induced by a near-infrared 5.5-fs, waveform-controlled [3], pulse with the central frequency of 1.55 eV (corresponding to 800 nm vacuum wavelength). The maximum near-IR pulse intensity at the target is kept at a moderate level of $I = 10$ GW/cm^2 to assure non-damaging conditions of the excitation. We assume an XUV pulse with 91 eV photon energy. We adopt the values of the XUV photon flux for the state-of-the-art XUV pulses [4] $\sim 10^9$ 1/s (the flux averaged over time). The XUV pulses can be focused [5] to a spot with a radius $\sim 1~\mu$m. This yields an XUV photon fluence $\sim 10^{16}$ cm^{-2}s^{-1}. We compute the energy shift of an electron emitted by such an XUV pulse, which is due to the acceleration in the local plasmonic fields.

Panels (a) and (b) of Fig. 2 show the electron energy distributions within a half period of the driving field. The distributions are shown as three-dimensional maps. Even for the moderate excitation intensities used, the energy shift ΔE_{XUV} in hot spots of the plasmonic potential is rather large (~ 10 eV) and, consequently, relatively easily measurable. There is pronounced nanometer-attosecond kinetics of the electron energy observed in these distributions with sharp hot spots indicative of those of the local fields.

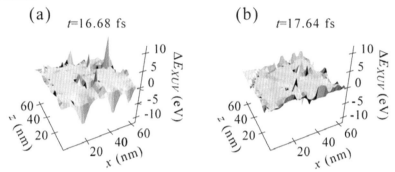

Fig. 2. Spatio-temporal kinetics of the local field potential as detected by attosecond plasmonic field microscope. (a)-(b) Distributions of energy shift ΔE_{XUV} of electrons emitted by XUV pulse in the plane of this nanostructure shown for different moments t_{XUV} (as indicated in the panels) of the XUV pulse incidence within the half cycle of local field oscillations.

3. Conclusions

The proposed attosecond nanoplasmonic field microscope will open up unique possibilities to study directly and control ultrafast photoprocesses in surface-plasmonic nanosystems and circuits. It images the local nanoplasmonic field in real space with nanometer spatial resolution and in real time with ~ 100 as temporal resolution. The perspective applications include nanoplasmonic systems with open metal/vacuum interfaces that enable electron emission. This method will be especially important in ultrafast nanoplasmonics where very tight nanolocalization of optical fields takes place.

References

[1] M. I. Stockman, M. F. Kling, U. Kleineberg, and F. Krausz, Nature Photonics **1**, 539 (2007).

[2] P. B. Corkum and F. Krausz, Nature Physics **3**, 381 (2007).

[3] A. Baltuska, et al., Nature **421**, 611 (2003).

[4] M. Schultze, E. Goulielmakis, M. Uiberacker, M. Hofstetter, J. Kim, D. Kim, F. Krausz, and U. Kleineberg, New J. Phys. **9**, 243 (2007).

[5] M. Schnurer, Z. Cheng, S. Sartania, M. Hentschel, G. Tempea, T. Brabec, and F. Krausz, Appl. Phys. B **67**, 263 (1998).

Coherent Control of Surface Plasmon Propagation Directions

S. B. Choi[1], D. J. Park[1], Y. K. Jung[1], J. H. Kang[2], Q-Han Park[2], C. C. Byeon[3], M. S. Jeong[3] and D. S. Kim[1*]

[1]Department of Physics and Astronomy, Seoul National University, Seoul 151-747, Korea,
 *E-mail: dsk@phya.snu.ac.kr
[2]Department of Physics, Korea University, Seoul 136-701, Korea
[3]Advanced Photonics Research Institute, GIST, Gwangju 500-712, Korea.

Abstract. We have demonstrated the directional control of surface plasmon polariton(SPP) waves through propagating in an asymmetric plasmonic Bragg resonator using femtosecond temporal-phase control via the resonant coupling of SPPs and the interference of SPPs. The near-field images display significant temporal-phase dependence, switching between left and right propagation after the Bragg resonator. Our results would be a key step toward the control of surface plasmon propagation direction in nano-scaled plasmonic applications.

Introduction

A Surface plasmon polaritons (SPP) are surface-bound waves at the metal-dielectric interface, owing their existence to the surface charge density of the conductor. The surface-bound nature of SPP gives rise to the exciting possibility of overcoming the diffraction limit of light. For example, SPP in nano-optic systems enable one to build subwavelength scale optic circuits, which include components for propagation, diffraction, and reflection, etc[1]. With integration of these nano-optic components, the issue arises how to transfer energy to a desired spatial position with subwavelength precision.[2] Here, ultra-broad band femtosecond lasers offer the important additional degree of freedom to coherently control the spatial distribution of the excitation, since the time delay between two pulses, i.e., their relative phase, chirp and polarization state all directly afftect the nano-localized electromagnetic field.

We designed and studied a model-nano optic system for femtosecond coherent control. We use this model under femtosecond laser excitation condition. We experimentally demonstrate the additional degree of control of the SPP propagation direction by control of temporal-phase of femtosecond pulses. Specifically, we clarify the important role played by the spectral dependence of the nano-optical patterns: the temporal-phase controlled SPP direction directly reflects the sensitive change of the two-pulse spectrum with phase between two pulses.

Experimental Methods

The coherently controlled laser beam excites a slit-groove structure fabricated by focused ion beam (FIB) milling on 300 nm thick gold film (Fig. 1(a)). Figure 1(b) shows that in Michelson interferometer (MI), a fs-laser pulse (depicted as solid line) is split into two pulses and these two pulses suffer different time delay thus the phase delay. A continuous-wave (CW) beam from a He-Ne laser (depicted as dashed line) is

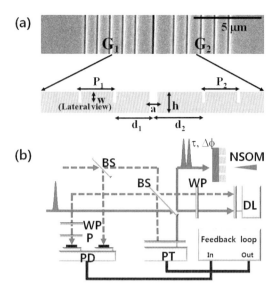

Figure 1: (a) SEM image of slit-grooves sample and experimental parameter. P1=780 nm, P2=830 nm, a=100 nm, h=300 nm, w=160 nm, d1=1170nm, d2=1245 nm. (b) Schematic diagram of actively feedback Michelson interferometer.

used to stabilize the τ and $\Delta\phi$. In this experiment time delay τ as fixed 13.3 fs and phase delay $\Delta\phi$ was changed as 0 and $-\pi$. After passing though actively feedbacked MI, the phase-locked TM polarized (E-field is perpendicular to the slit direction) pulses are then normally incident on the slit-grooves structure from the substrate side. Near-field images are obtained using an apertured metal coated fiber tip. The collected optical signal is analyzed by spectrometer.

Results and Discussion

The Figure 2 shows the experimental data from slit-grooves structure. The near-field images display striking phase dependence, switching between right- and left-propagating SPP images by changing the phase between the two pulses (Fig. 2(b) and Fig 2(c)). This striking phase dependence of the near-field image can be understood when we examine the spectral change of the incident light upon between two pulses (Fig. 2(d)): the coherent-controlled spectrum changes sensitively with ϕ, which determines the relative contributions of surface plasmon resonances at each grooves[3].

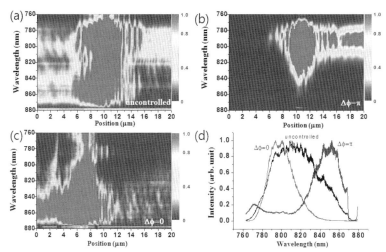

Figure 2: Spectrum resolved near-field images with (a) uncontrolled pulse illumination, (b) $\Delta\phi=\pi$ and (c) $\Delta\phi=0$ phase delayed pulses. (d) Spectral composition changes of coherent controlled pulses. Center solid line stands for uncontrolled spectrum, left and right lines are $\Delta\phi=0$, $\Delta\phi=\pi$ (13.3fs and 14.1 fs) delayed pulses respectively.

In conclusion, we have shown that the spatial propagation of the optical near field can be controlled by the varying the relative phase between a pair of ultrashort illumination pulses. In the investigated one-dimensional nano-optical system, the electromagnetic energy is controlled at the desired direction defined by the in-plane wavevector of the interfering SPP fields. In the present case, the temporal-phase dependent optical nearfield distribution is governed by the spectrally selective coupling to SPP modes at grooves. In general such ultrafast direction selective SPP propagation in a nano-system can also be achieved by chirp- or polarization-shaping of ultrafast light pulses and is expected to find important applications in nano-optical systems, specifically in controlling energy transfer processes on the nano scale.

References

[1] F. Lopez-Tejeira, Sergio G. Rodrigo, L. Martin-Moreno et al., "Efficient unidirectional nanoslit couplers for surface plasmons," Nat. Phys. **3**, 324 (2007).
[2] Mark I. Stockman, David J. Bergman, and Takayoshi Kobayashi, "Coherent control of nanoscale localization of ultrafast optical excitation in nanosystems," Physical Review **69**, 054202 (2004).
[3] F. Lopez-Tejeira, F. J. Garcia-Vidal, and L. Martin-Moreno, "Scattering of surface plasmons by one-dimensional periodic nanoindented surfaces," Physical Review B **72**, 161405 (2005).

Ultrafast Laser-Induced Electron Emission from Field Emission Tips

Catherine Kealhofer[1], Seth M. Foreman[1], Peter Hommelhoff[2], and Mark A. Kasevich[1]

[1] Physics Department, Stanford University, Stanford, CA 94305, USA
E-mail: ckealhof@stanford.edu
[2] Max Planck Institute of Quantum Optics, Garching, Germany

Abstract. We describe a laser-triggered electron source based on a field emission tip. Numerical results indicate that the electron emission times can be sub-femtosecond. We are exploring applications of this source to ultrafast SEM, and have begun characterization of emission mechanisms in an alternate tip material, hafnium carbide. Here we observe a higher efficiency in laser emission and, at high power, non-additive scaling as the number of incident laser pulses per second is doubled.

Introduction

Techniques to characterize nm-scale devices and understand their operation are becoming increasingly important. A version of ultrafast time-resolved scanning electron microscopy (SEM) has already been used to study charge transport in nanostructures [1]. One could imagine also doing time-resolved SEM in a pump-probe mode, where the time resolution is determined by the length of the pump and probe pulses, which might be some combination of electron and laser pulses. So far, the record timing resolution for ultrafast electron microscopy and diffraction experiments has remained longer than 100 fs [1, 2, 3]. Since the characteristic time scale of electronic processes in solids is closer to 1 fs, it is of great interest to obtain faster timing resolution. However, improved time resolution will require both new electron sources and the development of techniques to re-compress electron pulses that have broadened due to their inherent energy spread and possibly space charge.

Laser assisted electron emission from field emission tips can produce electrons that are extremely localized in both space and time [4]. In these sources, the electrons are emitted from sites on the tip that are as small as hundreds of atoms down to a single atom, and the emission process has been demonstrated in tungsten tips to be prompt, following the intensity envelope of the exciting laser pulse or even showing sub-optical cycle emission. As such, this system is a promising candidate source for the development of ultrafast time-resolved SEM.

Experimental Methods

Pulses of 8-fs duration and up to 4-nJ energy derived from a Ti-Sapph oscillator with 150-MHz repetition frequency are focused (spot size ~ 5 μm) onto a sharp metal tip. The tips studied are etched from tungsten or hafnium carbide and have end radii of 50-250 nm. A DC voltage is applied between the tip and the front of a microchannel plate (MCP) placed a few cm from the tip. Due to the small size of the tip, AC field enhancement is expected to occur, leading to peak laser field values on the order of a few GV/m near the surface. Emitted electrons are amplified by the MCP and the laser beam is chopped to allow lock-in detection of the laser-induced current.

Emission Mechanisms

We have characterized the electron emission from tungsten as a function of laser power, polarization, and DC bias voltage. From these measurements, we infer that the electron emission in tungsten is prompt [4, 5]. The qualitative character of the emission changes as a function of laser power. In the low-power limit where the Keldysh parameter [6] is much larger than one, electrons can be emitted due to multi-photon excitation or photo-assisted field emission (see [4]). In the high power limit (Keldysh parameter $<<1$), optical field emission is expected. In this process, electrons tunnel from the tip due to the strong electric field of the laser. As a result, electrons should only be emitted during part of the optical cycle, leading to one or more sub-femtosecond electron pulses for a few-cycle laser pulse at 800 nm.

For the highest laser power that we have studied, the Keldysh parameter is of order one. As a result, there is not a simple analytic model for the process. In order to better understand the temporal features of the emission we developed a numerical model by integrating the Schrödinger equation for a single electron tunneling from a surface state in the presence of the laser field [4]. The model adequately explains the observed dependence of emission non-linearity on bias voltage. Furthermore, it predicts that the emission is confined to only part of an optical cycle, as would be expected for optical field emission. This is consistent with recent experimental and theoretical studies of tunnel ionization in atoms, where the electron emission shows sub-optical cycle features in the intermediate (Keldysh parameter ~1) regime [7, 8].

High Current Field Emitters

Tips made from hafnium carbide and other transition metal carbides have been shown to support current densities greater than 10^8 A/cm^2 in stable DC emission [9]. This is three orders of magnitude larger than the highest current densities that can be obtained with tungsten [10].

We have begun characterizing laser-induced emission from hafnium carbide tips, and have obtained up to 10^4 electrons emitted per exciting laser pulse with 100 mW of average laser power. However, the emission behaviors at average powers below and above approximately 30 mW are qualitatively different. At low power, if we double the number of pulses incident on the tip without changing the pulse energy, the laser-triggered current also doubles as expected for both optical field emission and photo-assisted field emission. However, when the average power is increased above this point, the presence of twice the number of pulses on the tip much more than doubles the measured signal. This effect is illustrated in figure 1, and is independent of the delay between the two pulses for delays between 50 fs and 3.5 ns.

There are also significant differences in the emission dependence on laser polarization between the low power (<30 mW laser power) and high power regimes. In the low power regime, the emission is maximized for laser polarization parallel to the tip, whereas for high power, the emission is maximized for polarization perpendicular to the tip, which is also the polarization expected to most-efficiently heat the tip. From this, it is clear that at least two distinct mechanisms are at play and that the mechanism of high power emission observed in hafnium carbide cannot be modeled as simple cold-cathode optical field emission. Even if heating plays a role in the high-power emission mechanism, this does not require that the electron emission times are long. A thermally-assisted prompt emission model is qualitatively consistent with what we observe: doubling the number of pulses would provide a higher-temperature tip, increasing the probability of electron emission as a highly nonlinear function of power, while maintaining prompt emission timescales. Further investigation of these effects is under way.

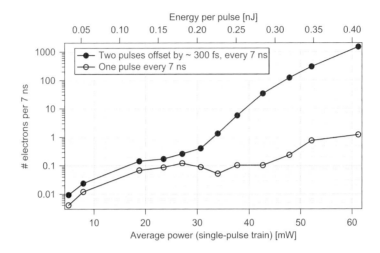

Fig. 1. Current as a function of average laser power for single and double pulse trains.

Conclusion

Photo-assisted field emission and optical field emission have been studied in tungsten and shown to have sub-10-fs down to sub-fs emission times. These short emission times, paired with their nanometer-scale emission areas, make them excellent candidates as sources for time-resolved SEM. As techniques are developed for the recompression of short electron pulses [11, 12], it may become possible to break current timing records. Hafnium carbide holds promise as a new tip material because it allows the possibility of reaching much higher currents than have been achieved with tungsten tips, while reducing the necessary laser intensity. We have recently begun characterizing laser-assisted emission from hafnium carbide.

1   M. Merano et al., Nature, Vol. 438, 479, 2005.

2   V. A. Lobastov, R. Srinivasan, A. H. Zewail, Proc. Nat. Acad. Sci., Vol. 102, 7069, 2005.

3   B. J. Siwick, J. R. Dwyer, R. E. Jordan, and R. J. D. Miller, Science, Vol. 302, 1382, 2003.

4   P. Hommelhoff, C. Kealhofer, and M. A. Kasevich, Phys. Rev. Lett., Vol. 97, 247402, 2006; ibid. in P. Corkum, D. Jonas, R.J.D. Miller, and W. Weiner (eds.), *Ultrafast Phenomena XV*, 746, 2007.

5   P. Hommelhoff, Y. Sortais, A. Aghajani-Talesh, and M. A. Kasevich, Phys. Rev. Lett., Vol. 96, 077401, 2006.

6   T. Brabec and F. Krausz, Rev. Mod. Phys. Vol. 72, 545, 2000.

7   M. Uiberacker et al., Nature, Vol. 446, 627, 2007.

8   G. L. Yudin and M. Y. Ivanov, Phys. Rev. A, Vol. 64, 013409, 2001.

9   W. A. Mackie, R. L. Hartman, and P. R. Davis, Appl. Surf. Sci., Vol. 67, 29, 1993.

10 R. Gomer, Field Emission and Field Ionization, AIP, New York, 1993.

11 L. Veisz et al., New Journal of Physics, Vol. 89, 415, 2007.

12 P. Baum and A. H. Zewail, Proc. Nat. Acad. Sci., Vol. 104, 18409, 2007.

Simultaneous Spatial and Temporal Control of Nanooptical Fields

Martin Aeschlimann[1], Michael Bauer[2], Daniela Bayer[1], Tobias Brixner[3], Stefan Cunovic[4], Frank Dimler[3], Alexander Fischer[1], Walter Pfeiffer[4], Martin Rohmer[1], Christian Schneider[1], Felix Steeb[1], Christian Strüber[4], and Dmitri V. Voronine[3]

[1] Fachbereich Physik, TU Kaiserslautern, Erwin-Schrödinger Str. 46, 67663 Kaiserslautern, Germany
[2] Institut für Experimentelle und Angewandte Physik, Universität Kiel, Leibnizstr. 19, 24118 Kiel, Germany
[3] Institut für Physikalische Chemie, Universität Würzburg, Am Hubland, 97074 Würzburg, Germany
[4] Fakultät für Physik, Universität Bielefeld, Universitätsstr. 25, 33516 Bielefeld, Germany
Corresponding Author e-mail address: pfeiffer@physik.uni-bielefeld.de

Abstract. Using time-resolved two-photon photoemission electron microscopy we demonstrate simultaneous spatial and temporal control of nanooptical fields. Cross correlation measurements reveal the ultrafast spatial switching of the local excitation on a subdiffraction length scale.

Introduction

Improving the spatial resolution in time resolved experiments is of utmost importance for many applications of ultrafast laser spectroscopy. In a theoretical investigation we have proposed to achieve simultaneously ultrahigh spatial and temporal resolution by exciting a nanostructure with polarization-shaped laser pulses [1]. For example, pump-probe sequences can be generated in which pump and probe excitations occur at different locations separated by less than the diffraction limit. This opens a route towards space-time-resolved spectroscopy on nanometer length-scales and femtosecond time-scales with the potential of direct observation of nanoscale energy or electron transport [1]. To experimentally demonstrate this scheme we have recently shown experimentally that adaptive optimization of polarization-shaped laser pulses allows tailoring the spatial excitation with subwavelength resolution [2]. Here we extend this experiment to include not only spatial degrees of freedom but also the temporal evolution of the nanooptical field on a femtosecond timescale. This is achieved by the use of polarization shaped pulses. The spatio-temporal response induced in this process is probed by time-resolved cross correlation measurements between the polarization-shaped pulses and a circularly polarized probe pulse.

Experiment

We combine femtosecond polarization shaping [3] and time-resolved two-photon photoemission electron microscopy (PEEM) [4]. By this the spatial origin of emitted photoelectrons can be determined with approximately 50 nm resolution. The experimental setup is depicted in Figure 1. A polarization-shaped pump pulse (75 MHz, 795 nm, FWHM 24 nm, and maximum flux $2.5 \cdot 10^{-2}$ mJ cm^{-2}) is used to excite a planar silver nanostructure (Figure 2a) that has been manufactured by e-beam lithography on an indium-tin-oxide (ITO)/glass substrate. The work function of the

705

Ag structure has been reduced to just below 3.1 eV by slight Cs dosage such that a two-photon PEEM image at 795 nm excitation can be recorded. The probe beam is split off from the incoming beam and bypasses the polarization pulse shaper. A mechanical delay stage (S1) allows adjusting the temporal delay between the shaped excitation pulse and the circularly polarized probe pulse. To average out interference effects between excitation and probe pulse a piezo-driven delay stage (S2) introduces an oscillating phase jitter between both pulses. Both beams are focussed on the sample inside the vacuum chamber. The polarization-shaped pulses are characterized by spectral interferometry (SI) with a reference pulse that itself is characterized by spectral phase interferometry for direct electric field reconstruction (SPIDER). Since the pulses are shaped not only in their temporal profile but also in their polarization state the SI is performed for two perpendicular polarization components [3].

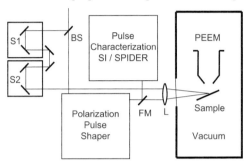

Fig. 1. Experimental setup: BS beam splitter for probe beam, S1 pump-probe delay stage, S2 piezo-stage introducing a phase jitter, FM flip mirror for pulse characterization, L 30cm focusing lens, PEEM with detector.

Results and Discussion

To demonstrate ultrafast spatial switching of the local excitation we use a double pulse sequence to illuminate the nanostructure. The first sub-pulse is almost linearly polarized along -45° with respect to the incidence plane and seen in the direction of the propagating laser beam. The second sub-pulse is delayed by +400 fs and its polarization direction is almost perpendicular to the first sub-pulse (Figure 2e). A slight deviation from linearly polarized sub-pulses is caused by non-diagonal elements in the Jones matrix describing the propagation of pulses from pulse shaper to the experiment. The orientation of the incident polarization influences the spatial distribution of the optical near-field in the vicinity of the nanostructure. Thus, the spatial distribution of energy deposited in the nanostructure also exhibits a spatial variation that can be influenced by the incident polarization. We probe the local excitation delivered by the polarization-shaped excitation pulse using a probe pulse with variable delay. Figure 2b and c show the obtained 2PPE pattern (after subtraction of the 2PPE pattern generated by independent excitation and probe pulses) for temporal overlap of the probe pulse with the first sub-pulse and the second sub-pulse of the excitation pulse, respectively. The emission pattern exhibits a striking variation - the maximum of the emission is "rotated" from the topmost arm of the nanostructure to the next arm on the left side. This corresponds to a lateral spatial shift of the excitation maximum of about 200 nm. In addition also the shape of the entire emission pattern changes. The variation of the emission pattern directly reflects the spatial switching of the local excitation between the two sub-pulses of the excitation pulse. Another way to visualize this switching is shown in Figure 2d. The time-resolved cross correlation traces for different regions of the nanostructure show a clear variation of their relative intensities in the time span of 400fs. This confirms that the polarization-shaped incident laser pulse does indeed switch between two different excitation patterns within a time scale that can be controlled almost freely and is limited only by the spectral bandwidth of the used coherent light source. These results go significantly beyond the nanooptical spatial control reported earlier [2] in the sense that here the spatial *and* temporal degrees of freedom are manipulated and measured.

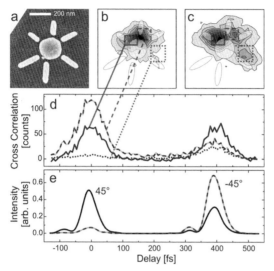

Fig. 2. Time-resolved cross correlation PEEM measurements: a) scanning electron microscope image of a single nanostructure. b),c) 2P-PEEM emission patterns for temporal overlap of the probe pulse with the first and second sub-pulse of the excitation pulse, respectively, from the same nanostructure. The grey solid lines indicate the position of the nanostructure and the rectangles show the regions of interest used for the cross correlation plot shown in part d). d) Time-resolved averaged cross correlation signals for different regions of the nanostructure as indicated by the lines. e) Reconstructed excitation pulse shape represented as projections of the pulse envelope on the 45° (solid line) and -45° (dashed line) polarization direction with respect to the incidence plane and seen in direction of the propagating beam.

Summary

Time-resolved 2-PEEM cross correlation measurements demonstrate for the first time that polarization-shaped laser pulses can be used to control the local excitation of a nanostructure with subwavelength spatial resolution and ultrafast time resolution. This demonstration represents a major step towards the realization of new space-time-resolved spectroscopies.

1   T. Brixner, F. J. García de Abajo, J. Schneider, and W. Pfeiffer, Phys. Rev. Lett. **95**, 093901, 2005.

2   M. Aeschlimann, M. Bauer, D. Bayer, T. Brixner, F. J. García de Abajo, W. Pfeiffer, M. Rohmer, C. Spindler and F. Steeb, Nature **446**, 301, 2007.

3   T. Brixner, G. Krampert, P. Niklaus and G. Gerber, Appl. Phys. B **74**, 133, 2002.

4   O. Schmidt, M. Bauer, C. Wiemann, R. Porath, M. Scharte, O. Andreyev, G. Schönhense, M. Aeschlimann, Appl. Phys. B **74**, 223, 2002.

Nano-Confined Light and Electron Sources Driven by Few-Cycle Optical Pulses

C. C. Neacsu[1,2], C. Ropers[1], T. Elsaesser[1], M. Albrecht[3], M. B. Raschke[2], C. Lienau[1,4]

[1] Max-Born-Institut für Nichtlineare Optik und Kurzzeitspektroskopie, D-12489 Berlin, Germany
E-mail: ropers@mbi-berlin.de
[2] Department of Chemistry, University of Washington, Seattle, Washington 98195-1700, USA
[3] Institut für Kristallzüchtung, D-12489 Berlin, Germany
[4] Institut für Physik, Carl von Ossietzky Universität Oldenburg, D-26129 Oldenburg, Germany

Abstract. Flat and nanostructured metal nano-tips driven by sub-10 fs pulses at an 80-MHz repetition rate serve for nano-confined light and electron generation. We demonstrate control of spatial emission properties and analyze nonlinear generation processes.

Introduction

Ultrafast optics at a sub-wavelength length scale plays a central role for studying the fundamental properties of inorganic and organic nanosystems, including the spatio-temporal dynamics and localization of elementary excitations. In this field, the interaction of light with metallic nanostructures such as periodic thin film gratings, arrays of sub-wavelength apertures and/or sharp conical metal tips has attracted much interest. Light-induced surface plasmon polaritons (SPPs) propagating along metal surfaces and/or interfaces with other materials represent a sensitive electromagnetic field probe of nanometer sized structures [1]. Localized plasmon excitations associated with a strong field enhancement allow for generating and analyzing nonlinear processes in a highly confined spatial range [2]. As a result, SPPs are expected to play a key role for bridging the gap between electronics and photonics in integrated systems [3]. In this paper, we demonstrate novel nano-confined sources that provide photons and electrons on a length scale of tens of nanometers, much shorter than the wavelength of the sub-10 fs optical driving pulses. Grating-coupled excitation of SPPs propagating along a nanostructured metallic tip is applied to implement a nanometer light source with spatially separated excitation and emission areas. The strong field enhancement at the apex of metallic nanotips is exploited for nonlinear frequency conversion and the generation of ultrashort electron pulses with an 80 MHz repetition rate.

Experimental Methods

Gold tips with a radius of curvature at the tip apex of approximately 20 nm and an opening angle of 15 degrees were produced by wet chemical etching. On the shaft of the nanostructured tips, a one-dimensional grating with a 750 nm period was written by focused gallium ion beam sputtering [4]. The grating is located several micrometers away from the tip apex (Figs. 1 a,b). Both unstructured and nanostructured tips were illuminated with 7 fs optical pulses at a center wavelength of 800 nm, generated in a mode-locked Ti:sapphire oscillator (repetition rate 80 MHz). The laser pulses were focused onto the metal tips with a microscope objective

708

(numerical aperture 0.35), resulting in spot sizes between 1 and 5 micrometers. For precise positioning of the laser focus on the nano-tips, either the microscope optics or the tip were positioned with a piezo scanner. The laser pulses are linearly polarized along the tip axis. The light scattered and/or emitted from the tip is collected with a second objective and imaged onto a video camera or the entrance slit of a spectrometer. Electrons emitted from the (unstructured) tips are detected with a microchannel plate and counted with an electronic discriminator [5].

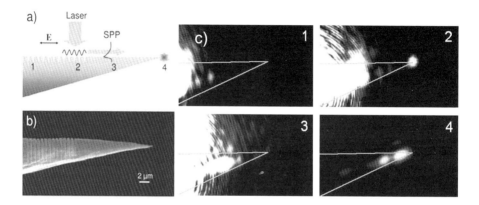

Fig. 1. (a) Scheme of grating-coupled SPP excitation. A 7-fs optical pulse is focused onto the linear grating located on the shaft of a gold tip and generates SPPs traveling to the tip apex. (b) Scanning electron micrograph of the nanostructured gold tip (grating period 750 nm). (c) Series of optical microscope images recorded with excitation at the positions 1 to 4 in Fig. 1 (a). Image 2 shows the nonlocal excitation of the tip apex via illumination of the grating.

Results and Discussion

In Fig. 1 (c), microscope images of light scattered from and/or emitted by a structured tip are summarized for excitation at the positions 1 to 4 (cf. Fig. 1 a). In all cases, SPP generation was in the linear regime and the laser polarization was perpendicular to the grooves of the grating in order to allow for SPP excitation. For illumination on the left side of the grating (1), the scattered light originates mainly from the grating itself. Excitation on the right side of the grating generates an intense emission strongly confined to the tip apex (2). Such emission is caused by SPPs that are generated on the grating, propagate along the tip shaft over a distance of >10 microns, and are reradiated into the far-field at the tip apex. For optimized coupling conditions and minimal surface roughness between the grating and the tip apex, a fraction of 0.1 to 1 percent of the incident laser power is reradiated from the nanometer sized tip apex. For illumination between the grating and the tip apex (3), the localized emission at the apex disappears because of the inefficient excitation of SPPs on the unstructured shaft. Direct excitation of the tip apex (4) results in a spatial scattering pattern with 2 maxima close to the tip apex and on the shaft. Here, the light scattered from the small apex is weaker than light scattered from the larger surface of the shaft.

The spectrum of the light generated with nonlocal excitation (Image 2) depends on the angle of incidence of the laser pulse onto the grating [4]. It can be tuned over

~60 nm by varying the angle over 2 degrees. This behavior is due to the change of the (in-plane) k-vector of the generated SPPs with the angle of incidence which results in a change of emission wavelength according to the SPP dispersion relation. The observed changes of the spectrum are in agreement with model calculations. The time structure of the emission is determined by the SPP lifetime of up to several tens of femtoseconds and the chirp acquired during propagation along the conical tip.

Beyond the regime of linear light-metal interaction, the strong enhancement of the driving field at the tip apex results in a spatially selective higher order response, allowing for optical frequency conversion and the generation of ultrashort electron pulses via multiphoton absorption [2,5-8]. The light emitted from the nonlinearly driven tip displays a broad spectrum extending from 400 to 600 nm. It consists of a strong second harmonic contribution due to the second order optical nonlinearity of the tip and a weaker continuum, predominantly reflecting the third-order nonlinearity. This emission is spatially confined to a small volume at the tip apex. Comparing the intensity of the light emitted from the smooth tip shaft with that from the apex, we estimate a field enhancement factor of 10 to 15.

Optically induced electron emission has been observed in the same regime of driving intensities [5,7,8]. We studied electron generation in a wide range of laser intensities and dc bias voltages on the tip. For an energy of 150 pJ of the 7-fs driving pulses (repetition rate 80 MHz), one generates up to 10^7 electrons/s at zero bias. In the case of gold tips, we identify multiphoton emission of athermal electrons as the main generation process at low bias whereas tunneling of low-energy electrons dominates at high bias voltages. The local field enhancement at the tip apex determines to a large extent the electron yield. This fact allows to implement a new type of scanning probe imaging in which the contrast originates from a modification of the field enhancement by interaction between the tip and the nearby sample. We demonstrate a spatial resolution of few tens of nanometers.

In conclusion, the photon and electron sources presented here combine a spatial confinement in the nanometer range with a temporal resolution of the order of 10 fs. We envisage a broad range of applications in ultrafast nanoscience.

1   W. L. Barnes, A. Dereux, and T. W. Ebbesen, Nature **424**, 824 (2003).

2   A. Bouhelier, M. Beversluis, A. Hartschuh, and L. Novotny , Phys. Rev. Lett. **90**, 013903 (2003).

3   E. Ozbay, Science **311**, 189 (2006).

4   C. Ropers, C. C. Neacsu, T. Elsaesser, M. Albrecht, M. B. Raschke, and C. Lienau, Nano Lett. **7**, 2784 (2007).

5   C. Ropers, D. R. Solli, C. P. Schulz, C. Lienau, and T. Elsaesser, Phys. Rev. Lett. **98**, 043907 (2007).

6   C. C. Neacsu, G. A. Reider, and M. B. Raschke, Phys. Rev. B **71**, 201402 (2005).

7   C. Ropers, T. Elsaesser, G. Cerullo, M. Zavelani-Rossi, and C. Lienau, New J. Phys. **9**, 397 (2007).

8   P. Hommelhoff, C. Kealhofer, and M. A. Kasevich, Phys. Rev. Lett. **97**, 247402 (2006).

Nonlinear Optical Response of Metal Nanoantennas

Barbara Wild, Jörg Merlein, Tobias Hanke, Alfred Leitenstorfer, and Rudolf Bratschitsch

Department of Physics and Center for Applied Photonics, University of Konstanz, 78464 Konstanz, Germany, barbara.wild@uni-konstanz.de

Abstract. We have excited bowtie-shaped metal nanoantennas fabricated via colloidal lithography with ultrashort light pulses. The spectrum emitted by the nanoantennas consists of a broadband continuum overlapped with a narrowband second harmonic signal.

Introduction

Resonant optical nanoantennas hold great promise for applications in physics and chemistry. Examples are nanometer-scale lithography [1], field enhanced spectroscopy [2], and nanoscopic light emitters with tailored absorption and emission characteristics [3-5]. Their operation relies on the ability to concentrate light on spatial scales much smaller than the wavelength. A second fundamental property of these metallic nanostructures is the extreme sensitivity of their plasmon resonances to their surroundings. Hence, metal nanoantennas may be used in biology and medicine as advanced nanometer-sized biochemical sensors. In this paper we report on light emission from bowtie-shaped gold nanoantennas fabricated with a colloidal mask technique, which are excited by picosecond light pulses. The nonlinear spectrum emitted by the nanoantennas consists of a broadband continuum overlapped with a narrowband second harmonic signal.

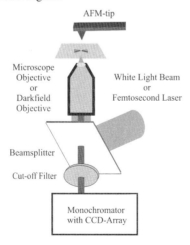

Fig. 1. Schematic drawing of the experimental setup consisting of an atomic force microscope (AFM), an inverted optical microscope for darkfield scattering spectroscopy and optical excitation with a femtosecond laser source.

Experimental Methods

Our experimental setup allows for simultaneous scanning of the sample topography and nanooptic characterization (Fig. 1). It consists of two main parts: (i) an atomic force microscope (AFM) and (ii) an inverted microscope with attached spectrometer and CCD detector array. With the optical setup we can either record the darkfield scattering spectra of the structures or excite them with a modelocked laser.

The studied metal nanoantennas consist of an array of paired gold nanotriangles arranged in tip to tip geometry, which are fabricated with colloidal lithography [6]. A self-assembled monolayer of touching polystyrene spheres of 310 nm diameter on a glass substrate forms the evaporation mask. After deposition of a 40 nm gold film and removal of the nanospheres, a metal nanostructure with triangular shape remains. Figure 2 shows an atomic force microscope (AFM) topograph of the sample. The side length of the equilateral nanotriangles is 147 nm and the antenna feedgap (tip to tip distance of two triangles) amounts to 100 nm.

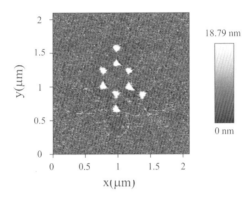

Fig. 2. AFM topograph of the nanoantenna array fabricated with colloidal lithography

Results and Discussion

The antenna array is optically characterized in the linear regime by darkfield scattering spectroscopy. The scattering spectrum from 500 nm to 1000 nm shows a strong resonance in the visible which is assigned to the dipole plasmon oscillation of the gold triangles (Fig. 3a, black line). The resonant behaviour is determined not only by the material but also by the shape and surroundings of the nanostructures [7].

We have excited the nanoantennas with ultrashort light pulses of a repetition rate of 64 MHz with an average power as low as 1 mW focused to a spot diameter of 1 µm. In our present setup, the femtosecond pulse duration becomes stretched to 0.6 ps due to the large dispersion of the microscope objective. The center wavelength was set to 790 nm, due to the peak of the plasmon pole of the antenna array (Fig. 3a, dotted line). The nonlinear emission spectrum shows a broadband continuum overlapped with a narrowband second harmonic signal at $\lambda = 395$ nm (Fig. 3b, marked by an arrow). Note that the data were recorded with a cut-off filter at $\lambda = 680$ nm in order to block the intense fundamental. Previous publications have reported incoherent emission due to spontaneous luminescence from hot carriers excited in the Au nanostructure [8, 10]. In contrast, we see an additional component that seems to originate from the nonresonant Kerr nonlinearity of the glass substrate close to the antenna gaps. Since all materials in the nanoantenna array are centrosymmetric, we expect the second harmonic signal to originate from the symmetry break at the sharp edges of the metal structure. At these positions, the largest resonant field enhancement is located.

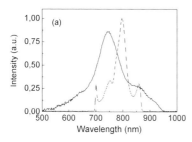 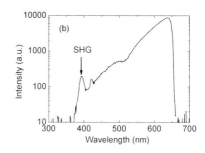

Fig. 3. (a) black line: darkfield scattering spectrum, dashed line: laser spectrum, (b) nonlinear emission of the bowtie-shaped gold nanoantenna array

Outlook

In the next step our experimental setup is adapted in a way that allows us to pump the nanoantennas directly with 10 fs light pulses. In this case, the pulse duration becomes of the same order of magnitude as the dephasing time of the plasmon resonance and optimum field enhancement is predicted. We expect that the thresholds for harmonics generation and continuum emission can be lowered even further under these conditions. It is evident that the possibility of driving second and third order nonlinear frequency conversion at such low pulse energies and average powers will find various applications in ultrafast technology and precision metrology.

1   A. Sundaramurthy, P. J. Schuck, N. R. Conley, D. P. Fromm, G. S. Kino, and W. E. Moerner, Nano Lett. **6**, 355 (2006).

2   J. Aizpurua, G. W. Bryant, L. J. Richter, F. J. Garcia de Abajo, B. K. Kelly, and T. Mallouk, Phys. Rev. B **71**, 235420 (2005).

3   J. N. Farahani, D. W. Pohl, H. J. Eisler, and B. Hecht, Phys. Rev. Lett. **95**, 17402 (2005).

4   E. Cubukcu, E. A. Kort, K. B. Crozier, and F. Capasso, Appl. Phys. Lett. **89**, 93120 (2006).

5   L. Novotny, Phys. Rev. Lett. **88**, 266802 (2007).

6   H. W. Deckman and J. H. Dunsmuir, Appl. Phys. Lett. **41**, 377 (1982).

7   J. Merlein, M. Kahl, A. Zuschlag, A. Sell, A. Halm, J. Boneberg, P.Leiderer, A. Leitenstorfer, and R. Bratschitsch, Nature Photonics **2**, 230 (2008).

8   P. Mühlschlegel, J.-J.Eisler, O.J. F. Martin, B. Hecht, and D.W. Pohl, Science **208**, 1607 (2005).

9   M.R. Beverluis, A. Bouhelier, and L. Novotny, Phys. Rev. B **68**, 115433 (2003).

10 M. Dankwerts and L. Novotny, Phys. Rev. Lett. **98**, 026104 (2007).

Near-Field Imaging of Single-Cycle THz Pulses Transmitted Through Sub-Wavelength Metallic Slit Structures

Hannes Merbold and Thomas Feurer

Institute of Applied Physics, University of Bern
Sidlerstr. 5, CH-3012 Bern, Switzerland
E-mail: hannes.merbold@iap.unibe.ch

Abstract. We experimentally and numerically investigate the spatiotemporal evolution of single-cycle THz pulses transmitted through sub-wavelength metallic slit structures. Employing a polaritonic approach the near field of the THz wave is monitored and compared to simulations.

Introduction

Ebbesen and coworkers observed an extraordinary high transmission of visible light through arrays of sub-wavelength apertures in optically thick metal films [1]. This unexpected observation has sparked considerable interest and subsequently the transmission of light through a variety of plasmonic structures has been investigated [2,3]. The observed phenomena have been linked to surface plasmon polaritons (SPPs) [4], however, the microscopic mechanisms are still discussed controversially. Our experimental approach allows for spatiotemporal imaging of THz pulses transmitted through plasmonic structures with sub-wavelength resolution. Resolving the THz near field in the immediate vicinity of the metallic structures might help to draw conclusions to the underlying mechanisms of the transmission process. Here we present a proof of principle experiment where we visualize THz waveforms transmitted through single sub-wavelength slits and compare the experimental findings to numerical simulations.

Experiment and Simulation

Our approach is based on THz polaritonics enabling us to visualize THz wave propagation in $LiNbO_3$ crystals with very high spatial and temporal resolution. The experimental details are described in detail in Ref. 5. It is possible to sandwich plasmonic structures between a pump and a probe crystal and obtain a real space image of the transmitted THz waveform in the near field regime of the metallic structure. Compared to scanning techniques that have recently been put forward [6] our approach has the advantage that the acquisition time for 2D images is only fractions of a second and that we do not have any detector that might influence the transmitted waveform. We direct our attention to the transmission through 1D single slits with sub-wavelength slit widths. The electric field of the incident THz pulse is polarized perpendicular to the slit's long axis.

We also perform 2D numerical simulations of single cycle THz pulses passing through the slit structures. The usual method of choice for modeling the time dependent propagation of broadband pulses is the finite difference time domain (FDTD) method. In FDTD the simulation domain is divided into a finite amount of points on a uniform rectangular lattice. The interaction lengths in metals ($\sim 100\,nm$) are about three orders of magnitude smaller than the free-space THz wavelength ($\sim 100\,m$). In order to resolve the interaction of THz pulses with metallic structures one would therefore have to refine the grid size which in turn would lead to huge computational demands. We have therefore chosen another approach based on finite element method (FEM) [7]. In FEM the simulation domain is discretized using a triangular mesh with variable mesh size. This allows one to refine the mesh only on the metal-dielectric interfaces yielding an acceptable value for the total number of mesh points. Starting from Maxwell's equations one can obtain a single equation for the vector potential **A** describing the propagation of electromagnetic radiation. In order to simulate the coupling of the radiation to polar lattice vibrations in ionic crystals (phonon-polaritons) and plasma oscillations at metal surfaces (surface plasmon polaritons) the ionic and electronic displacements need to be included in the model. Since we have broadband THz pulses it is furthermore important to correctly take the frequency dependence into account. This is not straightforward since all calculations are performed in the time-domain. We address this issue by expanding the model with two additional differential equations describing the normalized ionic displacement **Q** and current density **J**. Thus we have three coupled differential equations which are solved through iterative solvers using a commercial package.

Results

According to standard waveguide theory the slits have zero cutoff frequency for the given polarization. The incident pulses are therefore transmitted without significant modification of their temporal waveform or spectra. However, the slit structure diffracts the incident plane wavefront into a spherical wave as can be seen in Fig. 1 where the measured and simulated THz waveforms transmitted through single slits with different slit widths are shown.

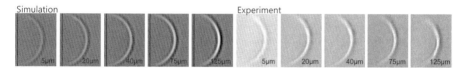

Fig. 1. Experimentally recorded and simulated THz waveforms transmitted through single slits with varying widths in stainless steel

One can see that we obtain good agreement between the simulated and measured waveforms and that we were able to visualize the transmitted THz waveform for slits as small as 5 μm (corresponding to $\lambda/200$). Examples of near field images are shown in Fig. 2.

Fig. 2. Temporal sequence of near field images showing the transmission of a THz pulse through a 40 μm wide slit in stainless steel

Here a temporal sequence showing the transmission of a THz pulse through a 40 μm slit in stainless steel is displayed. One can see how the transmitted pulse emerges and evolves into a propagating spherical wave. Insight into the transmission process can be gained from Fig. 3 where a sequence of images obtained from FEM simulations is shown.

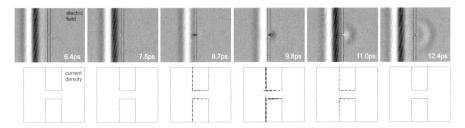

Fig. 3. Simulation showing the transmission of a THz pulse through a metallic slit. A close up on the slit region reveals the coupling to surface currents

A close up on the gap region reveals that the em field couples to surface currents during the transmission process. A comparison with a similar slit in a perfect electric conductor revealed that the surface currents have only little effect on the overall transmitted intensity. This is not surprising since for the given geometry the transmission is explained by waveguide modes. However, we expect surface currents to play an important role in transmission processes for more complex plasmonic structures.

Conclusion and Outlook

In a proof of principle experiment we have shown that we are able to measure and simulate the transmission of THz pulses through metallic structures. In the near future we would like to extent our investigations to more complex plasmonic structures such as hole arrays [1], single holes surrounded by periodic corrugations on the input side [2] or single holes with periodic structures on both sides exhibiting beaming properties [3]. Laser machining can be used to directly cut structures into $LiNbO_3$ that are later filled or coated with metal.

Acknowledgements. We thank the Swiss National Science Foundation (200021-111693) for financial support.

1   T. Ebbesen, H. Lezec, H. Ghaemi, T. Thio, and P. Wolff, Nature **391**, 667 (1998)
2   T. Thio, K.M. Pellerin, R.A. Linke, H.J. Lezec, and T.W. Ebbesen, Opt. Lett. **26**, 1972 (2001)
3   H. Lezec, A. Digeron, E. Devaux, R.A. Linke, L. Martn-Moreno, F.J Garca-Vidal, and T. Ebbesen, Science **297**, 820 (2002)
4   H. Ghaemi, T. Thio, D.E. Grupp, T. Ebbesen, and H. Lezec, Phys. Rev. B **58**, 6779 (1998)
5   T. Feurer, N.S. Stoyanov, D.W. Ward, J.C. Vaughan, E.R. Satz, and K.A. Nelson, Annu. Rev. Mater. Res. **37**, 317-350 (2007)
6   M.A. Seo, A.J.L. Adam, J.H. Kang, J.W. Lee1, S.C. Jeoung, Q.H. Park, P.C.M. Planken, and D.S. Kim, Opt. Exp. **15**, 11781 (2007)
7   J. Jin, The Finite Element Method in Electromagnetics, Wiley-IEEE Press, 2nd edition (2002)

Nanoscale Optical Microscopy in the Vectorial Focusing Regime

K. A. Serrels, E. Ramsay, R. J. Warburton and D. T. Reid

Ultrafast Optics Group, School of Engineering and Physical Sciences, Heriot-Watt University, Edinburgh, EH14 4AS, UK
Email: k.a.serrels@hw.ac.uk

Abstract. By using extreme numerical-aperture solid-immersion microscopy at 1553 nm we demonstrate, under certain circumstances, polarization-sensitive imaging with resolution values approaching 100 nm which substantially surpass the classical scalar diffraction-limit embodied by Sparrow's resolution criterion.

Introduction

The spatial distribution of an optical field at the focus of a high numerical aperture (NA) lens displays an increasing ellipticity as the NA of the system approaches unity. This is because at high NA polarization starts to play a more dominant role in the focal properties of the incident light. In the effort to obtain the smallest focal spot size attainable one must take this into account since polarization-sensitive effects are no longer negligible. The first theoretical analysis of focused linearly polarized light was described by Richards and Wolf in 1959 [1], however it is the analysis of spatially non-uniform polarization distributions – typically radial and azimuthal – which has recently dominated the search for the smallest focused spot [2].

As an introduction to vectorial effects, consider the results of a model calculating the focal-plane time-averaged intensity distribution of linearly polarized light using the theory presented in [1] for low and high NA. The results are shown in Fig.1 and illustrate the elongation of the point-spread function along the polarization direction. In an extension to this model, aperture-function engineering effects were investigated in order to reduce the focal spot size further, shown in Fig.2. The line-sectional plots (circles – x-axis, crosses – y-axis) were compared against a scalar analysis and highlight an additional reduction in spot size along the x-axis.

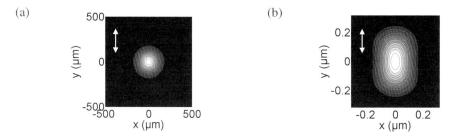

Fig.1. Calculated time-averaged focused intensity distribution of 1530 nm linearly polarized light (arrows) light focused in a silicon substrate at (a) low and (b) high NA.

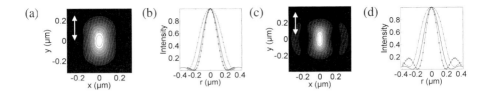

Fig.2. Calculated time-averaged focused intensity distribution of 1530 nm linearly polarized light (arrows) light focused in a silicon substrate under (a) high NA, clear aperture and (c) high NA, annular aperture conditions; (b), (d) are the cross-section plots for (a) and (c) respectively (circles – x-axis, crosses – y-axis), compared with scalar theory (plain solid line).

The elliptical nature of the focal spot, illustrated in the model, is accessible experimentally in high-NA focusing environments such as solid-immersion-lens (SIL) imaging inside silicon. We have studied such polarization effects experimentally and report here ultrahigh-resolution sub-surface imaging of a silicon flip-chip using a SIL [3], obtaining average resolution values from 122 – 240 nm, depending on the polarization state used, and the relative orientation of the features under inspection.

Experiment

In order to circumvent the contradictory requirements of high bulk transmission through the backside silicon substrate and SIL along with sufficient near-IR absorption at the μm-scale device layer, a two-photon optical beam induced current (TOBIC) effect was adopted. In this work, two-photon excitation was achieved by using a 1553 nm modelocked erbium-fiber laser whose linearly-polarized output beam overfilled an objective lens of NA = 0.55. A half-wave plate positioned before the objective lens controlled the incident polarization. The laser generated 160 fs pulses at a repetition frequency of 30 MHz with an average (peak) output power of 75 mW (15.6 kW), and its output was focused by the objective lens through a silicon super-SIL into the device under test. The sample was a 0.35 μm feature size CMOS silicon flip-chip and had an exposed silicon substrate which had been optically polished to a thickness of 100 μm. The sample and SIL were scanned using stepper motors with a minimum physical step of 100 nm which corresponded to a lateral optical step of 8.3 nm. Two-photon photocurrent generation was detected using a digital oscilloscope, and a desktop PC was used to control the scanning and visualize the data.

Results and Discussion

A coarse stepping increment was initially used to provide an image suitable for basic navigation, shown in Fig. 3(a). The area of interest contained three 3×3 matrices of tungsten vias on a finger-like structure of n-doped silicon and is highlighted by the rectangular white box. TOBIC images of this highlighted area were acquired in a raster fashion using different incident polarization directions at maximum resolution (Fig. 3(b)). The linear polarization vectors are indicated by the double-arrows. The imaging improvement is apparent by the narrowing of the gaps in between the finger structure since a beam polarized parallel to an edge will register a sharper transition from a low signal to a high signal due to its smaller PSF diameter along that axis

To quantitatively study this polarization effect we examined individual line-scan data across the horizontal and vertical edges of the fingers. The vertical line-scans through the image data are shown in Fig. 3 (a), (c), the horizontal in (b) and (d), and were fitted to a Gaussian error function to determine the width of the point-spread function for each polarization state. The results show a twofold difference in resolution for data acquired with opposite polarization states when scanning across the gaps between neighbouring fingers (122 nm and 251 nm) and produce a similar performance when observed in the orthogonal scanning direction when scanning into the fingertip (119 nm and 172 nm).

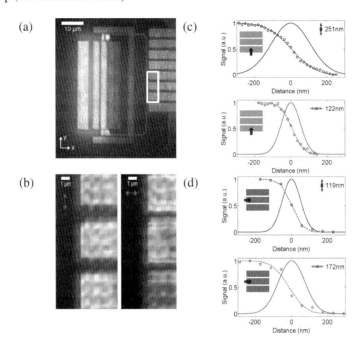

Fig. 3. (a) Wide field-of-view TOBIC image; (b) TOBIC images of the boxed region obtained under the linear polarizations indicated by the double-arrows; (c),(d) Resolution measurements obtained by scanning across the finger structures under different polarization conditions

Conclusions

In conclusion, we have demonstrated the significant role that polarization plays in high NA imaging. We have shown performance surpassing Sparrow's scalar diffraction-limited resolution value of 158 nm for TOBIC imaging of a silicon integrated-circuit with polarization sensitive average resolution values from 122 – 240 nm and minimum resolutions of around 100 nm [4].

1. B. Richards and E. Wolf, "Electromagnetic diffraction in optical systems II. Structure of the image field in an aplanatic system," Proc. R. Soc. London Ser. A **253**, 358 - 379, 1959.
2. G.M. Lerman and U. Levy, "Tight focusing of spatially variant vector optical fields with elliptical symmetry of linear polarization", Optics Letters **32**, 2194-2196, 2007.
3. E. Ramsay et al, "Three-dimensional nanoscale subsurface optical imaging of silicon circuits," Appl. Phys. Lett. **90**, 131101, 2007.
4. K.A. Serrels et al, Nature Photonics **2**, 311 - 314, 2008.

Ultrafast Wide-Field Fluorescence Microscopy

Lars Gundlach, and Piotr Piotrowiak

Department of Chemistry, Rutgers University Newark, 73 Warren St, Newark, NJ 07102, USA.
E-mail: larsg@andromeda.rutgers.edu

Abstract. A Kerr-gated microscope capable of collecting diffraction limited 2D fluorescence images with sub 100 fs time resolution is presented. The concept is based on the insertion of a solid state optical Kerr gate into a wide-field microscope. The ultrafast fluorescence dynamics of gold nanoparticles is presented as an example of the capabilities of the instrument.

Introduction

The static and dynamic properties of nanoscale materials exhibiting a broad range of heterogeneities call for new experimental approaches. Among the dynamic properties photo-driven transport of energy is of particular interest, since it is essential for processes as fundamental as photosynthesis, photo-catalysis and the operation of optoelectronic devices. Charge transport on the nanoscale occurs on an ultrashort time scale. Therefore, monitoring such processes demands techniques that are capable of simultaneously resolving spatial features on the nanometer scale and femtosecond dynamics. Thanks to its high sensitivity and dynamic range fluorescence microscopy is an especially useful tool for the detection of small ensembles of emitters down to the single molecule level [1]. Time-resolved fluorescence microscopy is a direct measure of excited-state lifetimes and decay channels. It is a powerful technique as it allows one to probe the spatial, temporal and spectral inhomogeneities that are not discernible in ensemble measurements yet frequently are crucial in determining the overall behavior of the system of interest [2].

Experimental Method

The core of our approach involves the insertion of a non-linear optical Kerr gate [3,4] into the output pathway of a wide-field microscope. The collected emission light can pass through the Kerr gate only when the gating pulse is incident upon the Kerr medium. The transmitted light is detected by a CCD camera, resulting in a nearly single photon sensitivity of the method. By delaying the gating pulse with respect to the excitation pulse, the time evolution of the imaged object can be followed and the corresponding emission decays can be assembled. In addition to the imaging mode of operation, the Kerr-gated microscope can be arranged in a spectrally dispersed configuration and function as a femtosecond fluorescence spectrometer. The chief objective of developing the instrument was to bridge the temporal resolution gap between bulk femtosecond pump-probe experiments and time-resolved microscopy. The experimental setup is shown in Fig.1. The amplified Ti:sapphire laser system delivered 60 fs pulses with 0.5 mJ pulse energy at 790 nm at 1.25 kHz. One part of the output is used to pump a non-collinear optical parametric amplifier (NOPA) the remaining part is utilized to drive the Kerr gate.

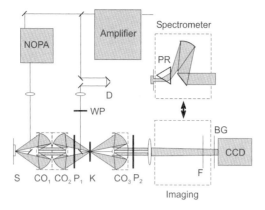

Fig. 1. Setup for Kerr-gated fluorescence microscopy. S: sample, CO_{1-3}:Cassegrain objectives, P_{1-2}: polarizer, K: Kerr medium, D: delay line, F,BG: spectral filter, WP: waveplate, PR: prism.

A detailed description of the setup is given in Ref. 3. Briefly, the microscope consists of three Cassegrain objectives (CO_{1-3}). CO_1 collects the light emitted from the sample. The excitation pulse (tunable from 250 nm to 1100 nm, µJ, 10-50 fs) is directed onto the sample (S) via a mirror in front of CO_1. The collected light is focused into the Kerr medium (K, 0.5 mm fused silica) via CO_2 after passing a polarizer (P_1) and subsequently recollimated via CO_3. The gate pulse (790 nm, 40 µJ, ~ 60 fs) is superimposed with the linearly polarized light from the sample in the Kerr medium (K) via another small mirror. The recollimated light passes through the second polarizer (P_2) before it is either projected onto a CCD or dispersed by a prism and then focused onto a CCD. The spatial resolution of the setup was approximately 1 µm. The design of the microscope allows one to switch easily between collecting fluorescence images and spectrally resolving the light emitted from the same area of the sample by inserting a simple spectrometer consisting of a slit, a prism, a curved mirror and a planar mirror.

Results and Discussion

Kerr-gated time-resolved images of a cluster of gold nano-spheres, as well as the corresponding spectra and ultrashort lifetime of the surface plasmon emission are presented. We used commercially available nanoparticles to demonstrate the single emitter capability of the microscope. Nanoparticles exhibit a high absorption cross section combined with a higher resistance to photobleaching than molecular dyes, although shape transformation of gold nanoparticles have been observed at higher fluence [5].

Absorption and emission spectra of 25 nm Au particles (Nanocs Inc.) in the UV/Vis substrate were measured. The onset of the emission on the red side of the spectrum resembles closely the onset of the absorption as it is expected for plasmon emission. However, the blue part of the spectrum differs slightly. The difference may be due to aggregation effects present in the dried film used for the emission measurements, but not in the aqueous solution used for measuring the absorption.

Next we measured the temporal evolution of the emission. The corresponding transient is shown in Fig. 2 (a).

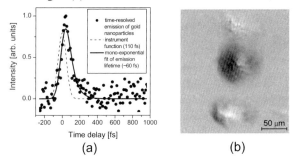

(a)   (b)

Fig. 2. (a) Time-resolved emission of 25 nm Au nanoparticles coated on a fused silica substrate ($\tau \sim 60$ fs). (b) Time resolved emission image of a cluster of gold nanoparticles at t_0. Darker colors indicate a higher signal.

Since there is no signal from the scattered 300 nm excitation pulse in this measurement because of the employed filters, the instrument function was determined separately without the filters. In this case, an exact determination of t_0 is difficult. Precise knowledge of t_0 is crucial in order to extract lifetimes that are shorter than the instrument response function. Thus, the 60 fs, obtained by fitting the data with a mono-exponential rate equation and convolution with the measured instrument response function, remains an upper limit for the emission lifetime. Nevertheless, this value is in good agreement with reported values [6], and with the fact that relaxation processes in metal nanoparticles can be extremely fast. The time resolved emission image of a cluster of gold nanoparticles is shown in Fig. 3(b). Darker colors indicate a higher signal. An image taken at 10 ps negative delay was subtracted from an image taken at t_0 to remove the remaining leakage through the different filter and the Kerr gate. Although the image shown in Fig. 3(b) is not yet a single nanoparticle image, it demonstrates the potential of the new technique.

Conclusions

The setup presented here is capable of recording luminescence images with a time resolution well below 100 fs.

Acknowledgements. We are grateful for the financial support from the DOE, Grant DE-FG02-06ER15828, and the NSF, Grant No. 0342432.

1 R. M. Dickson, A. B. Cubitt, R. Y. Tsien, and W. E. Moerner, in Nature, Vol.388, 355, 1997.

2 W. E. Moerner, in J. Phys. Chem. B, Vol. 106, 910, 2002.

3 L. Gundlach and P. Piotrowiak, in Opt. Lett., Vol. 33, 992, 2008.

4 T. Fujino, T. Fujima, and T. Tahara, in Appl. Phys. Lett. Vol. 87, 131105, 2005.

5 A. Bouhelier, R. Bachelot, G. Lerondel, S. Kostcheev, P. Royer, and G. P. Wiederrecht, in Phys. Rev. Lett., Vol. 95, 267405, 2005.

6 E. Dulkeith, T. Niedereichholz, T. A. Klar, J. Feldmann, G. von Plessen, D. I. Gittins, K. S. Mayya, and F. Caruso, in Phys. Rev. B, Vol 70, 205424, 2004.

Measurement of Dispersion Properties of Silver Nanowires Used as Plasmon Waveguides

Jess M. Gunn[1], Scott H. High[2], and Marcos Dantus[1,2]

[1] Department of Chemistry, Michigan State University, East Lansing, MI 48824, USA
[2] Department of Physics and Astronomy, Michigan State University, East Lansing, MI 48824, USA
E-mail: dantus@msu.edu

Abstract. Surface plasmon waves created by shaped femtosecond pulses are used to control the two-photon induced plasmon emission of silver nanoparticles. A quantitative measurement of the dispersion properties of surface plasmon waveguides is given.

Introduction

In the field of plasmonics, which explores the transport of information via surface plasmons on nanowires, the ability to pre-determine and control the flow of energy is critical. Previous work in our group has shown that exciting a thin film of silver nanoparticles at normal incidence with ultrashort laser pulses results in the transfer of energy as far as 100 μm from the focal spot. This transfer is observed by detecting the two-photon-induced localized surface plasmon resonance emission from individual nanoparticles far from the focal spot (Fig. 1, top, shows schematic. Fig. 1, bottom left, shows experimental results).

The intensity of emission from the discrete regions observed can be controlled by changing the polarization and spectral phase of the excitation laser [1], as illustrated in Fig. 1, bottom right. Our hypothesis is that the observed control is due to the different dispersion characteristics of the surface plasmon waveguide that carries the excitation pulse. We measure the dispersion characteristics by assuming that the maximum two-photon-induced emission will be observed for transform-limited (zero-phase) pulses at the emitting nanoparticle.

Experimental Methods

The synthesis of thin films of silver nanoparticles has been presented previously [1]. To prepare the thin films, silver nanoparticles are synthesized by the citrate reduction of $AgNO_3$. Cluster formation is then induced by the addition of fumaric acid. The clusters are allowed to precipitate onto quartz substrates for three days. The samples were excited at normal incidence by a Ti:sapphire femtosecond laser (800 nm center wavelength, 10 fs pulse duration, 100 nm bandwidth) brought through an inverted microscope and focused on the sample with a 60x/1.45 N.A. microscope objective. Transform-limited pulses were achieved at the focal spot via the MIIPS method [2]. Emitted light from two-photon-induced localized surface plasmon resonance was imaged with an electron-multiplying CCD camera.

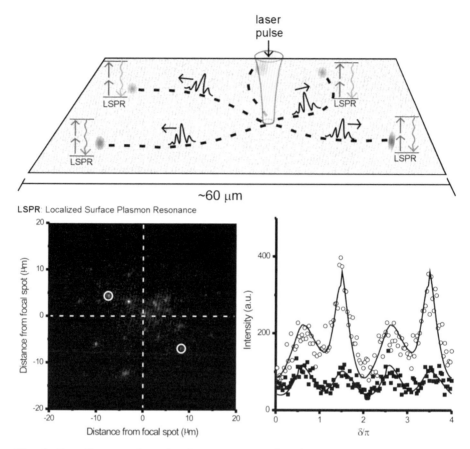

Fig. 1. Top: Cartoon schematic of energy propagation through nanoparticle surface and subsequent two-photon-induced surface plasmon resonance. Bottom, left: Image collected from sample when laser is focused at center of crosshairs. Bottom, right: Intensity of the two circled regions as a function of sinusoidal phase functions of the form $1.5\pi \sin(12\omega - \delta)$. Note that the regions have different dependencies on the phase.

To quantify the amount of second- and third-order dispersion introduced by the sample as the energy propagates to each point, a series of spectral phases, generated to correspond to a list of second- and third-order dispersion values was applied via a pulse shaper with a spatial light modulator (SLM) to the transform-limited laser pulses, and the emission imaged. Each discrete region of emission was analyzed to determine which applied phase resulted in the maximum signal. Because the maximum signal is achieved when the pulse has a zero-phase at the location of emission, we can conclude that the applied phase giving the maximum signal is the negative of the actual phase introduced by the sample.

Results and Discussion

The second- and third-order dispersion values for each of 150 points of discrete emission on a single sample are shown in Fig. 2, left. As a control, the same experiment was run with an additional well-defined phase corresponding to the

724

application of -200 fs^2 and -2000 fs^3. The corresponding shift in retrieved dispersion values (Fig. 2, right) suggests that the assumption that the maximum signal is observed when the pulse is transform-limited at the sample is valid. It should be noted that no radial dependence on average intensity or dispersion values has been observed. Both positive and negative values for second- and third-order-dispersion are observed.

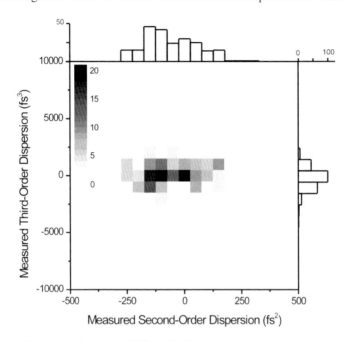

Fig. 2. Histogram of second- and third-order dispersion values measured for more than 150 discrete regions of emission. Note that both positive and negative values of second- and third-order dispersions are measured.

Conclusions and Future Work

We have reported here on the sensitivity of remote regions of emission to the spectral phase of the excitation femtosecond pulse, and on our ability to measure the spectral phase introduced as the energy propagates through the sample by measuring the intensity of two-photon induced emission. This could lead to the development of a 'plasmonic switch,' in which the laser pulse itself carries the information (phase) that would determine along which well-defined path the energy would be transported.

Future work on this topic will include studying the propagation and control of surface plasmons through gold nanoparticle systems, as well as controlling surface plasmon propagation along a single 'wire.'

Acknowledgements. The authors would like to thank Dr. Vadim Lozovoy for the insight provided through valuable discussions.

[1]. J. M. Gunn, M. Ewald, and M. Dantus, Nano Letters **6**, 2804, 2006.
[2] B.W. Xu, et al., JOSA B. **23**, 750, 2006.

Part IX

Novel Pulsed Sources: oscillators, amplifiers, nonlinear mixing

Pulse energies exceeding 20 µJ directly from a femtosecond Yb:YAG oscillator

Joerg Neuhaus[1,2], Dominik Bauer[1,2], Jochen Kleinbauer[1], Alexander Killi[1], Sascha Weiler[1], Dirk H. Sutter[1], and Thomas Dekorsy[2]

[1] TRUMPF-Laser GmbH + Co. KG, Aichhalder Str. 39, 78713 Schramberg, Germany
[2] Department of Physics, University of Konstanz, 78465 Konstanz, Germany and
 Center for Applied Photonics, University of Konstanz, 78465 Konstanz, Germany
 E-mail: joerg.neuhaus@uni-konstanz.de

Abstract. We demonstrate the generation of the highest pulse energies ever reported directly from a mode-locked oscillator: Twenty-five microjoules from a thin-disk oscillator operating in air, corresponding to an average output power of seventy-three watts.

Introduction

Ultrashort laser pulses in the microjoule regime are of prime importance for many applications, including high-speed micromachining, pumping of optical parametric oscillators, as well as basic research, e.g. in high-field physics and for the generation of attosecond electron pulses [1-4]. To some extend the pulse energies from an oscillator can be increased by using extended resonator cavities [4-6] or by cavity dumping [7]. Previous record pulse energies obtained directly from an oscillator *in ambient atmosphere* were below 2 µJ. These pulse energies were limited by the strong self-phase modulation (SPM) of air [8]. Higher pulse energies were only obtained *in a He-flooded cavity* with the highest pulse energies for sub-ps pulses obtained so far of 5.1 µJ [9] and 11 µJ [4].

One way to decrease the SPM is to use larger output coupling rates in combination with a high-gain medium. In order to also reach high average powers together with high pulse energies, a thin-disk laser crystal is the medium of choice, allowing for true power scalability by increasing the beam size and exploiting the excellent cooling properties of the disk. However, the low gain of a thin-disk laser has to be overcome. Hence, we have increased the roundtrip gain by passing the gain medium successively under different angles within one roundtrip. This approach has previously been used in laser amplifiers with conventional solid state rod geometries [10], as well as in thin-disk amplifiers [11]. Recently we have obtained pulse energies of 13.4 µJ with a novel active multiple-pass cell operated in air, already surpassing the previous energy levels obtained with He-flooded oscillators [12]. Here we report on the generation of record-breaking pulse energies of up to 25.9 µJ in ambient atmosphere, surpassing the results published in [8] by more than an order of magnitude.

Experiment

The experimental setup of the laser is shown in Fig. 1. The pumping chamber provides 20 passes through the gain medium, leading to a reasonable absorption (> 70 %) of the fiber-coupled pump power of up to 236 W at a center wavelength of 940 nm. Passive mode-locking of the laser was started and stabilized with the help of a SESAM [4,8,9]. The SESAM has a modulation depth of approximately 2%.

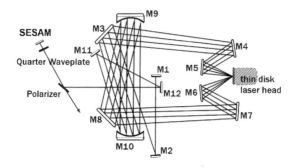

Fig. 1: The inset shows a schematic layout of a Yb:YAG thin-disk laser with angular multiplexing. Highly dispersive mirrors were used for mirrors M2-M10.

Strong SPM balanced by large negative group delay dispersion (GDD) resulted in soliton mode-locking as the predominant pulse shaping mechanism [13]. The estimated B-integral per round-trip was $B = 0.9$ rad, mainly originating from the nonlinear refractive index of air. Dispersive mirrors introduced a total GDD of − 207800 fs^2 per roundtrip. The resonator beam passed the active multiple-pass cell 13 times (corresponding to 52 passes through the gain medium) within one round-trip, making up a total resonator length of 51 m. The drastically increased round-trip gain allowed for output-coupling of up to 80%, which is - to our knowledge - the highest output-coupling rate of any thin-disk laser reported so far. Thereby the ratio of intracavity pulse energy to external energy was greatly reduced, such that He flooding could be avoided. Compared to our first results [12], the maximum pump power was increased by using a larger pump spot (2 mm) on the thin-disk. By using a SESAM incorporating nitrogen in a diluted concentration in the InGaAs quantum well, the onset of double pulsing at higher pump powers was efficiently suppressed.

Results and Conclusion

A maximum average output power of 76 W was obtained at a total pump power of 236 W and output coupling rate of 78%. At a repetition rate of 2.94 MHz this corresponds to pulse energies of 25.9 µJ. In this case a pulse duration of 928 fs, deduced from the autocorrelation assuming an ideal $sech^2$ shape, and a spectral bandwidth of 1.22 nm (FWHM) at a center wavelength of 1030 nm were measured. The resulting time bandwidth product of 0.32 is within 10% of the transform limit of 0.315 for soliton pulses. Even shorter pulses of 811 fs with a spectral bandwidth of 1.48 nm and a TBP of 0.34 were measured for pulse energies of up to 20 µJ (Fig. 2, inset). Soliton mode-locking is confirmed by the reciprocal dependence of the pulse width on the pulse energy (Fig. 2). In contrast to [12], the CW-background was completely suppressed. Single-pass frequency doubling has been accomplished with up to 60 % conversion efficiency in a BBO crystal with maximum green average powers of 28 W.

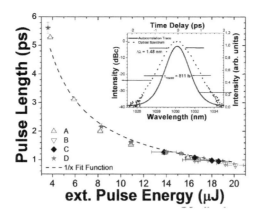

Fig. 2: Pulse length dependence on the external pulse energy for an active multiple-pass cell with 11 passes over the thin-disk gain medium and different SESAMs (A-D). The inset shows the optical spectrum (dashed line) and autocorrelation of the measured output pulses with 20 µJ of energy.

To our knowledge, the reported pulse energies are by far the highest ever obtained directly from an unamplified solid-state laser oscillator. This laser architecture appears to be the least complex for generating high average power ultrafast pulses. This work was partly funded by the German Federal Ministry of Education and Research, contract 13N8580.

1 F. Brunner, E. Innerhofer, S. V. Marchese, T. Südmeyer, R. Paschotta, T. Usami, H. Ito, S. Kurimura, K. Kitamura, G. Arisholm, and U. Keller, in Optics Letters, Vol. 29, 1921, 2004.

2 L. Shah, M. E. Fermann, J. W. Dawson, and C. P. J. Barty, in Optics Express, Vol. 14, 12546, 2006.

3 P. Baum and A. H. Zewail, in PNAS, Vol. 104, 18409, 2007.

4 S. V. Marchese, C. R. Baer, A. G. Engqvist, S. Hashimoto, D. J. Maas, M. Golling, T. Südmeyer and U. Keller, in Optics Express, Vol. 16, 6397, 2008.

5 S. Naumov, A. Fernandez, R. Graf, P. Dombi, F. Krausz, and A. Apolonski, in New Journal of Physics, Vol. 7, 216, 2005.

6 S. Dewald, M. Siegel, T. Lang, C.D. Schröter, R. Moshammer, J. Ullrich, and U. Morgner, in Optics Letters, Vol. 31, 2072, 2006.

7 G. Palmer, M. Siegel, A. Steinmann, and U. Morgner, in Optics Letters, Vol. 32, 1593, 2007.

8 E. Innerhofer, T. Südmeyer, F. Brunner, R. Häring, A. Aschwanden, R. Paschotta, C. Hönninger, M. Kumkar, and U. Keller, in Optics Letters, Vol. 28, 367, 2003.

9 S. V. Marchese, T. Südmeyer, M. Golling, R. Grange, and U. Keller, in Optics Letters, Vol. 31, 2728, 2006.

10 A. M. Scott, G. Cook, and A. P. G. Davies, in Applied Optics, Vol. 40, 2461, 2001.

11 M. Kumkar: U.S. Pat. No. 6,765,947, 2006

12 J. Neuhaus, J. Kleinbauer, A. Killi, S. Weiler, D. Sutter and T. Dekorsy, in Optics Letters, Vol. 33, 726, 2008.

13 F. X. Kärtner and U. Keller, in Optics Letters, Vol. 20, 16, 1995.

Fundamentally Mode-locked 3 GHz Femtosecond Erbium Fiber Laser

Jian Chen[1], Jason W. Sickler[1], Hyunil Byun[1], Erich P. Ippen[1], Shibin Jiang[2], and Franz X. Kärtner[1]

[1] Department of Electrical Engineering and Computer Science, Massachusetts Institute of Technology, 77 Massachusetts Avenue, Cambridge, Massachusetts 02139, USA
E-mail: Jchen9@mit.edu
[2] NP Photonics Inc., UA Science & Technology Park, 9030 S. Rita Rd. - Suite 120, Tucson, AZ 8574, USA
E-mail: Jiang@npphotonics.com

Abstract. Generation of low jitter 500-fs pulse trains at a fundamental repetition rate of 3.2 GHz, from an Er-fiber laser passively mode-locked with a saturable Bragg reflector. This constitutes an increase in repetition rate of femtosecond fiber lasers by about one order of magnitude. Laser performance and potential applications are discussed.

Introduction

Femtosecond sources with multi-gigahertz repetition rates at optical communications wavelengths are the critical building blocks for numerous applications such as femtosecond laser frequency comb generation for frequency metrology, optical arbitrary waveform generation, high speed optical sampling, and the calibration of astrophysical spectrographs [1,2]. Currently, only a few approaches [3,4] meet the stringent requirements in terms of pulse duration, repetition rate, operating wavelength, and noise performance simultaneously. However, these current approaches are bulky, expensive, and of limited robustness, since they employ Fabry-Perot filters locked to the mode comb for multiplying the repetition rate to the multi-gigahertz range, either inside or outside of the low fundamental-repetition-rate laser cavities. A more simple, compact, robust and cost efficient approach is desirable. Lasers with fundamental repetition rates in the multi-gigahertz range are ideal candidates for some of the intended applications.

With the constraints of achieving femtosecond pulse duration, and low timing jitter, passive mode-locking is the only path to reach multi-gigahertz fundamental repetition rate. Record high repetition rates of a few hundred MHz have been reported in previous attempts [3,5] using polarization additive pulse mode-locking (P-APM). Also passively mode-locked fiber lasers using saturable Bragg reflectors (SBRs) with up to 2 GHz repetition rate, with setups very similar to the one presented here, have been reported, however, those attempts failed to produce femtosecond pulses [6], which is of key importance for low jitter and for frequency metrology applications.

In this paper, we demonstrate, for the first time, a fundamentally mode-locked femtosecond erbium-doped fiber laser (EDFL) with a repetition rate of 3.2 GHz and a timing jitter of 19 fs [10kHz-40MHz]. This result exceeds previous attempts to increase the repetition rate of fiber lasers by about ten fold, and sets a clear pathway to achieving low timing jitter, mode-locked femtosecond sources at 1550 nm with up to ten GHz of repetition rate.

Design Considerations

Since the cavity length of a multi-gigahertz repetition rate laser is only several centimeters, the P-APM mechanism is too weak; mode-locking by a SBR is however possible. SBR mode-locking only results in femtosecond pulses when soliton pulse shaping [7], governed by Eq. (1), can be employed. The average power and, therefore, pulse energy is limited by the available pump power and doping concentration of the gain fiber. In order to generate the shortest pulse duration τ with limited pulse energy W, the cavity dispersion D needs to be minimized and the cavity nonlinearity   needs to be maximized. Note that D needs to be kept negative for soliton formation.

$$W = \frac{4|D|}{\delta \cdot \tau} \tag{1}$$

Fortunately, unlike traditional fiber lasers, dispersion of the fiber section no longer dominates the total cavity dispersion for multi-gigahertz lasers. Other components such as the SBR play important roles for dispersion compensation. It is also critical to keep the finesse of the laser cavity high to increase pulse energy.

Experimental Setup

The experimental setup is shown in Fig. 1 (a). Guided by simulation based on the above considerations, we choose a 3 cm-long highly doped erbium-ytterbium fiber (NP Photonics) with a group-velocity dispersion (GVD) of -60 fs2/mm and a nonlinear coefficient of ~3 W-1km-1 as the gain medium. One end of the gain fiber is butt-coupled to an SBR, and the other to a 2% dielectric output coupler. We carefully choose the SBR so that the dispersion of the SBR is slightly positive with 1000 fs2 to compensate about half of the negative dispersion from the gain fiber. The SBR also has a 6% modulation depth, a 2 ps recovery time, and a saturation fluence of 25 μJ/cm2. Pump light is provided by two polarization combined 500 mW 980-nm-laser diodes, free-space coupled via an aspherical lens into the gain fiber through the output coupler. The output signal is separated from the pump light by a dichroic mirror.

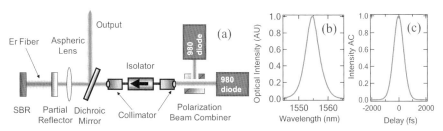

Fig. 1. (a) Experimental setup. (b) Optical spectrum of the pulse train. (c) Intensity autocorrelation measurement

The setup is simple, compact, and self-starting. This laser can be made reliable in a turn-key configuration. When the laser is pumped with ~700 mW of 980 nm pump power the mode-locked optical spectrum is 5.3 nm wide shown in Fig. 1(b).

Fig. 1 (c) shows the intensity autocorrelation of the pulse train after amplification by an erbium-doped fiber amplifier resulting in a pulse duration of 600 fs, which was limited by dispersion compensation of the amplifier. Further compression to its transform-limited pulse duration of 500 fs is in progress. The average output power from the laser cavity is 2 mW and was used to seed a subsequent erbium doped fiber amplifier for applications demanding higher average power. The intracavity average power is calculated to be 100 mW yielding a pulse energy of 33 pJ. These numbers agree relatively well with soliton theory and our simulations.

Figure 2(a) shows the measured RF signal using a 10 GHz photodetector; Figure 2(b) shows one of the RF comb lines with a 3 kHz resolution bandwidth indicating clean mode-locked operation with a noise suppression better than 80 dB. Measurements of the timing jitter were carried out using an HP5052a signal source analyzer. Figure 2 (c) shows the single sideband phase noise measurement and the timing jitter progressively integrated from 40 MHz down to 1 kHz. The timing jitter integrated from 10 kHz to 40 MHz is 19 fs.

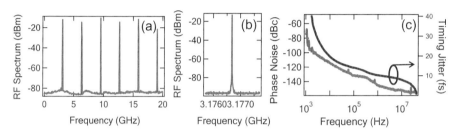

Fig. 2. (a) RF spectrum, span=20 GHz, (b) RF spectrum, span =2 MHz, RBW=3kHz, (c) Phase noise measurement, and integrated timing jitter starting from 40 MHz down to 1 kHz

Conclusions

We have demonstrated the first compact, multi-gigahertz, fundamentally mode-locked EDFL producing femtosecond-duration pulses. The output pulse train exhibits a low timing jitter of 19 fs [10kHz-40MHz] as necessary for a range of applications from frequency metrology, arbitrary optical waveform generation, high speed sampling and calibration of astrophysical spectrographs. The laser design sets a clear pathway to achieve even shorter pulses and higher fundamental repetition rates beyond 10 GHz.

1 S. T. Cundiff, Nature, vol. 450, pp. 1175-1176 (2007)

2 C-H. Li, A. Benedick, P. Fendel, A. G. Glenday, F. X. Kärtner, D. Phillips, D. Sasselov, A. Szentgyorgyi and R. Walsworth, Nature 452, 610-612 (2008)

3 J. Chen, J. Sickler, E.P.Ippen, and F.X. Kartner, Optics Letters, Vol. 33 (9), pp.959-961 (2008)

4 F. Quinlan, S. Gee, S. Ozharar, and P. J. Delfyett, Optics Letters, Vol. 31 Issue 19, pp.2870-2872 (2006)

5 T. Wilken, T.W. Hänsch, R. Holzwarth, P. Adel, M. Mei, in Conference on Laser and Electro-Optics. OSA, 2007.

6 J. J. McFerran, L. Nenadovi´c, W. C. Swann, J. B. Schlager and N. R. Newbury, Optics Express, 15, 13155, (2007).

7 F. X. Kärtner and U. Keller, Opt. Lett. 20, 16-18, 1995..

Ultrabroadband Er:fiber Systems and Applications

A. Leitenstorfer[1], A. Sell[1], D. Träutlein[1], F. Adler[1], K. Moutzouris[1], F. Sotier[1], M. Kahl[1], R. Bratschitsch[1], R. Huber[1], and E. Ferrando-May[2]

[1] Department of Physics and Center for Applied Photonics, University of Konstanz, D-78457 Konstanz, Germany
E-mail: aleitens@uni-konstanz.de
[2] Department of Biology and Center for Applied Photonics, University of Konstanz, D-78457 Konstanz, Germany

Abstract. Compact and low-noise Er:fiber lasers allow efficient frequency conversion from the near ultraviolet into the mid infrared. These widely tunable sources enable multi-color experiments in applications ranging from precision metrology via bioimaging to pump-probe measurements on single electron systems.

Introduction

Since the implementation of the first femtosecond Er:fiber laser [1], these systems have represented a key enabling technology of compact, robust and cost effective sources of ultrashort pulses. The invention of the stretched pulse oscillator [2] capable of generating pulse durations below 100 fs at repetition rates around 50 MHz and average powers up to 100 mW promoted output levels into a regime that allowed first real-world applications. Efficient frequency doubling resulted in a hands-off seed source for Ti:sapphire regenerative amplifiers [3]. More recently, Yb:fiber amplifiers took off to generate high-power femtosecond pulse trains suitable for materials processing [4,5]. Due to the advent of Yb:oscillators employing microstructured fibers for dispersion compensation [6], all-fiber implementations became possible also at 1050 nm. Nevertheless, the emission band of Er^{3+} at 1.55 μm brings about a series of benefits including compactness, excellent noise performance and highly versatile options for frequency conversion. As we shall see, these advantages are fundamentally connected to the large repertoire of standard telecom components available in this wavelength regime such as highly nonlinear bulk fibers with dispersion profiles that may be easily tailored varying the Ge content in the core.

Systems Performance and Applications

Our first contribution to this field was the design of a single-pass nonlinear Er:fiber amplifier [7]. This device overcompensates gain narrowing via self-phase modulation whose influence is precisely controlled by setting the correct prechirp for the seed pulse. These systems generate relatively clean pulses with duration below 60 fs at repetition rates between 30 MHz and 150 MHz. The highest average output achieved so far is 500 mW. Due to the enhanced peak intensities it became possible to generate octave spanning supercontinua with an all-fiber system. This result led to the first demonstration of a self-referenced frequency comb from a fiber laser [7]. Due to their inherent long-term stability and robustness, these systems represent promising tools for practical applications of high-precision frequency metrology [8,9].

The continuum generation in highly nonlinear germanosilicate fibers may be tailored precisely by varying the prechirp on the 1.55 μm pump pulse via a simple silicon prism compressor [10]. Based on realistic simulations of the nonlinear pulse propagation inside these fibers, we are able to obtain dispersion profiles that allow tunability of the emission peaks from 970 nm to 2.4 μm, see Fig. 1(a). Due to the excellent coherence properties and low phase noise, the broadband spectra may be compressed down to pulse durations as short as 13 fs employing a sequence of SF10 prisms with negligible third-order dispersion in the near infrared [11].

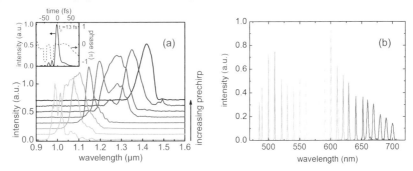

Fig. 1. (a) Short-wavelength part of the supercontinua from highly nonlinear fibers tuned via the prechirp of the 1.55 μm input pulse. A FROG trace of a recompressed 13 fs pulse centered at 1.25 μm is shown in the inset. (b) Visible spectra obtained via frequency doubling of the near-infrared output from the highly nonlinear fibers.

For many applications, laser radiation in the visible range is highly desirable. It turns out that both second harmonic generation and sum frequency mixing within the broadband near infrared spectra may be phase matched in fan-out poled $MgO:LiNbO_3$ crystals. In this way, pulse trains tunable between 490 nm and 700 nm are generated with average powers up to 10 mW [12], see Fig. 1(b). It should be mentioned that the relative amplitude noise of this radiation amounts to less than 10^{-6} $Hz^{-1/2}$ for frequencies above 20 Hz [13], although it is generated in a conversion process that is effectively of 6th order. This measurement demonstrates the excellent stability of the entire generation chain. The visible spectrum is completed via highly efficient 2nd, 3rd and 4th harmonic generation [14] in addition to sum frequency mixing from two-branch systems for the deep blue [15].

An important aspect of the nonlinear single-pass Er:amplifiers is that an input power of less than 1 mW is sufficient to drive the system into saturation. Therefore, setups with several parallel amplifiers which may be custom designed can be seeded synchronously from the same oscillator. Application examples for such multi-branch systems are two-color pump-probe spectroscopy and frequency comb generators where one amplifier provides an octave spanning continuum for self-referencing and the other branches produce optimized output power in the spectral regimes of interest [8]. While it is known that supercontinua generated in parallel setups pumped by the same source can exhibit an extremely high degree of mutual coherence [16], the question is whether similar findings also apply for two branches of our systems where several meters of prestretch, amplifier and highly nonlinear fiber are involved. We have therefore measured the relative phase noise between two parallel amplifier branches seeded by the same Er:fiber oscillator and found a pleasant surprise: The rms timing jitter integrated in an interval from 1 Hz up to the Nyquist frequency is as low as 11 attoseconds directly from the Er:amplifiers. After additional frequency conversion in a highly nonlinear fiber including some free space optics, this value is increased to 43 as, i.e. still approximately two orders of magnitude less than an optical cycle [11]. It was demonstrated that up to 1 mW of tunable mid-infrared pulses may be generated via difference frequency mixing from a two-branch system [17]. Due to the extremely low relative phase jitter, we expect these transients to exhibit a high degree of phase stability, resulting in an attractive seed source for attosecond pulse and high-power terahertz generation.

Owing to their high degree of compactness and wideband tunability, the femtosecond Er:fiber lasers are promising sources for applications in femtosecond spectroscopy and confocal microscopy. Very recently, we were able to carry out two-color femtosecond absorption measurements of a single electron in a semiconductor quantum dot. We have also constructed a setup for linear and nonlinear imaging techniques in the life sciences [13]. This system is used for linear fluorescence excitation with a tunable source and two-photon imaging in the optimum transparency window of live tissue around 1.1 μm. It is even powerful enough to induce local photodamage of DNA via three-photon absorption.

Conclusions

In summary, the versatility of femtosecond Er:fiber systems renders them promising tools for a broad spectrum of possible applications. Salient features of these lasers are compactness, robustness, excellent noise performance and the possibility of broadband tunable frequency conversion via highly nonlinear bulk fibers. We predict that such sources will play a key role for the implementation of future femtosecond quantum optics based on single-electron systems in solid-state nanostructures.

1   I. N. Duling, Electron. Lett. **27**, 544 (1991).

2   K. Tamura, E.P. Ippen, H. A. Haus, and L. E. Nelson, Opt. Lett. **18**, 1080 (1993).

3   L. E. Nelson, S. B. Fleischer, G. Lenz, and E. P. Ippen, Opt. Lett. **21**, 1759 (1996).

4   A. Malinovski, A. Piper, J. H. V. Price, K. Furusawa, Y. Jeong, J. Nilsson, and D. J. Richardson, Opt. Lett. **29**, 2073 (2004).

5   F. Röser, D. Schimpf, O. Schmidt, B. Ortac, K. Rademaker, J. Limpert, and A. Tünnermann, Opt. Lett. **32**, 2230 (2007).

6   H. Lim, A. Chong, and F. W. Wise, Opt. Express **13**, 3460 (2005).

7   F. Tauser, A. Leitenstorfer, and W. Zinth, Opt. Express **11**, 594 (2003).

8   F. Adler, K. Moutzouris, A. Leitenstorfer, H. Schnatz, B. Lipphardt, G. Grosche, and F. Tauser, Opt. Express **12**, 5872 (2004).

9   P. Kubina, P. Adel, F. Adler, G. Grosche, T. W. Hänsch, R. Holzwarth, A. Leitenstorfer, B. Lipphardt, and H. Schnatz, Opt. Express **13**, 904 (2005).

10 F. Tauser, F. Adler, and A. Leitenstorfer, Opt. Lett. **29**, 516 (2004).

11 F. Adler, A. Sell, F. Sotier, R. Huber, and A. Leitenstorfer, Opt. Lett. **32**, 3504 (2007).

12 K. Moutzouris, F. Adler, F. Sotier, D. Träutlein, and A. Leitenstorfer, Opt. Lett. **31**, 1148 (2006).

13 D. Träutlein, F. Adler, K. Moutzouris, A. Jeromin, A. Leitenstorfer, and E. Ferrando-May, J. Biophoton. **1**, 53 (2008)

14 K. Moutzouris, F. Sotier, F. Adler, and A. Leitenstorfer, Opt. Express **14**, 1905 (2006).

15 K. Moutzouris, F. Sotier, F. Adler, and A. Leitenstorfer, Opt. Comm. **274**, 417 (2007).

16 P. Baum, S. Lochbrunner, J. Piel, and E. Riedle, Opt. Lett. **28**, 185 (2003).

17 C. Erny, K. Moutzouris, J. Biegert, D. Kühlke, F. Adler, A. Leitenstorfer, and U. Keller, Opt. Lett. **32**, 1138 (2007).

Compact high Power Ytterbium based fs-Oscillator-Amplifier System

P. Rußbüldt[1], T. Mans[2], D. Hoffmann[1], A.-L. Calendron[3], M.J. Lederer[3], R. Poprawe[1,2]

[1]Fraunhofer Institute for Laser Technology, Steinbachstr. 15, 52074 Aachen, Germany
E-mail: Peter.Rusbueldt@ilt.fraunhofer.de
[2]Chair for Laser Technology RWTH Aachen, Steinbachstr. 15, 52074 Aachen, Germany
[3]High Q Laser Production GmbH, Kaiser-Franz-Josef-Strasse 61, 6845 Hohenems, Austria

Abstract: A compact diode-pumped Yb:YAG Innoslab fs-oscillator-amplifier system, scalable to several 100W, was realized. Nearly transform and diffraction limited 786fs pulses at 77W average output power and 63.2MHz repetition rate are achieved so far.

Introduction

Ultrafast photonics in atomic, plasma and nuclear physics like XUV-generation or wakefield acceleration are evolving from a few scientific facilities to tabletop tools. To transfer femtosecond technology to industrial applications, laser sources of high power and high repetition rate are essential. These lasers have to combine sub picosecond pulse duration of ultrashort lasers and mean output power of high power lasers at nearly diffraction limited beam quality. Mainly due to commercial available laser diodes of continually increasing power new solid state laser concepts like disc, fiber and partial pumped slab lasers (Innoslab) established beside rod lasers.

An Innoslab laser or amplifier comprises a longitudinal partially pumped slab crystal. The grinded mounting surfaces of the slab suppress parasitic oscillations of the line-shaped, homogenously pumped cross section inside the crystal. One dimensional heat conduction out of the pumped gain volume to the large cooled mounting surfaces of the thin slab allow for efficient heat removal and a homogenous cylindrical thermal lens. A confocal arrangement of two cylindrical mirrors folds the laser radiation several times through the gain volume [1]. The beam is expanding each passage throughout the laser crystal by a constant factor in the plane of the pumped volume. At each passage a new section of the gain volume is evenly saturated resulting in an efficient exploitation of the stored energy. Amplification factors of 10-1000 per slab are possible.

Setup

For amplification of sub picosecond laser pulses the utilized gain medium have to support several nanometer gain bandwidth. Yb:YAG with a gain bandwidth $\Delta\lambda\approx5$nm was employed in an Innoslab amplifier for the first time. For efficient amplification a quasi three-level system like Yb:YAG has to be pumped in the scale of the saturation intensity $I_{sat}(\lambda_p)=28$kWcm^{-2} at the wavelength $\lambda_p=940$nm of the used laser diodes. By recent enhancement of brilliance commercially available laser diode bars provide these intensities now.

The employed horizontal laser diodes stack comprises 4 laser diode bars (JOLD-408-HSC-4L) and provides up to 480W in a 46x0.8mm^2 aperture with 150x4mrad^2 divergence. The radiation is focused in a planar waveguide for homogenization in

738

slow axis and imaged into the 1x10x10mm³ Yb:YAG crystal of 2% doping level. Up to 430W of the pump radiation are available at the laser crystal, of those 320W are absorbed in a single pass. The confocal cavity of magnification M=1.2 consists of two cylindrical mirrors and two plane dichroic folding mirrors. For stable longtime operation the whole setup is temperature stabilized by water cooling.

The amplifier was seeded by a newly developed High Q Laser ultra-compact saturable absorber mirror (SESAM) mode-locked [2] short pulse oscillator (High Q Laser UC-1030-200 fs). The UC-platform allows the integration of various laser media for ps- and fs-pulse generation around 1µm wavelength, requiring only minimal design variations. Thanks to the compact dimensions (200x140x100mm³), excellent pointing stability over temperature and shock resilience (50g-tested), these oscillators are ideal for any fs/ps seeder applications, giving potential additional features such as air cooling and SHG. The seed oscillator used in the experiments was fitted with an Yb-glass active medium, which enables fs pulse generation between 1025nm and 1065nm. The circular and non stretched seed radiation passes the gain volume of the Innoslab amplifier 9 times before extraction.

Results

For optimization of the cavity and doping level the amplifier was numerically simulated. Considering the rate equations of the quasi three level system the static three dimensional gain distribution was calculated. The pump radiation was propagated by ray tracing and laser radiation by physical optical propagation of the electromagnetic field or by mode expansion. Diffraction at the saturated gain volume and the in- and output mirrors are considered, likewise temperature distribution and thermal lens. Nonlinear effects are calculated to be irrelevant.

In a first experiment the amplifier was seeded with a Yb:YAG oscillator operated at λ_s=1030nm in continuous wave. Up to P=111W output power at P_p=409W pump- and P_s=6W seed power is achieved. Above a threshold pump power of 200W a slope efficiency of about 55% is attainable. In the future, folding back the transmitted pump radiation will increase efficiency further. Output power is only limited by the wavelength shift of the laser diodes away from peak absorption at 940nm wavelength, decreasing absorbed power at P_p>409W. Apart from this the measured amplification is in very good agreement with the calculated lines in Fig. 1a.

For the short pulse experiments we seeded the amplifier with the Yb:glass fs-oscillator without isolator, stretcher or cylindrical beam transformation.

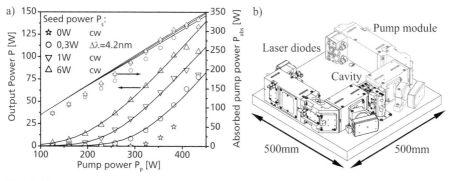

Fig. 1. (a) Measured (symbols) and calculated (lines) output and absorbed power of amplifier
(b) Amplifier setup (measurements with one pump module)

The wavelength was tuned to 1030nm, where the output power was 300mW. The oscillator had a repetition rate of 63.2MHz, a pulse duration of 280fs (inferred from the intensity autocorrelation) and a corresponding spectral width of $\Delta\lambda$=4.5nm, resulting in nearly transform limited pulses (Fig. 2).

Up to 77W of fs-radiation was extracted from the amplifier (Fig. 1a). To increase absorbed pump power we lowered water temperature of the amplifier laser diodes by 5°C. Due to the short 9x10mm passages inside the laser crystal the pulses of τ=786fs duration ($sech^2$) are nearly transform limited without compression (Fig. 2). Gain narrowing reduces the spectral bandwidth $\Delta\lambda_s$=4.5nm of the oscillator to $\Delta\lambda$=1.62nm.

After amplification a cylinder telescope transforms radiation into a round beam. Beam quality was measured to M_x^2=1.27 and M_y^2=1.17 (Spiricon M²-200/v.4.2) at full output power. The RF spectrum measured by a photodiode is noise free down to 10Hz (70dB). There is no sign of background radiation or nonlinear spectral broadening. Lasing of the amplifier like in Fig. 1a is completely suppressed by the seed radiation. Inside amplification bandwidth the amplifier is seeded by ~120mW translating to an amplification factor exceeding G=600.

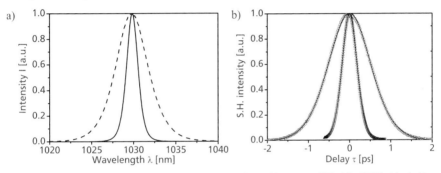

Fig. 2. (a) Spectrum oscillator (dashed) P_s=0,3W, $\Delta\lambda_s$=4,28nm, amplified P=77W, $\Delta\lambda$=1,62nm
(b) Autocorrelation oscillator (triangles) τ=280fs, amplified (dots) τ=786fs, $\tau\Delta\nu$=0,36

Scaling the output power to multi 100W by pumping the laser crystal double pass from both sides is in progress. Without chirped pulse amplification (CPA) some 10µJ pulse energy at repetition rates >10MHz are attainable. With CPA some 10mJ at >10kHz repetition rate are expected to be feasible, taking nonlinear effects and damage thresholds into account. Lower repetition rates or pulsed pumping will not gain technical advantages, as heat removal is not limiting the performance. Due to the larger beam cross section and shorter passes Innoslab amplifiers can sustain two to four orders of magnitude higher peak powers than fiber amplifiers. At high average power, Innoslab amplifiers bridge the gap between fiber lasers with high amplification but high nonlinearities, small cross section and damage threshold and disc lasers with large cross sections but low amplification.

Acknowledgements. This work was supported by the BMBF (FKZ 13N8720)

1. B. Luther-Davies, V.Z. Kolev, M.J. Lederer, N.R. Madsen, A.V. Rode, J. Giesekus, K.M. Du, M. Duering „Table-top 50-W laser system for ultra-fast laser ablation" Appl. Phys. A **00**, 1–5 (2004)
2. U. Keller, K. J. Weingarten, F. X. Kärtner, D. Kopf, B. Braun, I. D. Jung, R. Fluck, C. Hönninger, N. Matuschek, J. Aus der Au, "Semiconductor saturable absorber mirrors (SESAMs) for femtosecond to nanosecond pulse generation in solid-state lasers" IEEE J. Selected Topics in Quantum Electronics (JSTQE) **2**, 435-453 (1996)

Fiber laser pumped high average power single-cycle THz pulse source

Matthias C. Hoffmann[1,*], Ka-Lo Yeh[1], Harold Y. Hwang[1], Tom Sosnowski[2], János Hebling[1,3], and Keith A. Nelson[1]

[1] Department of Chemistry, Massachusetts Institute of Technology, Cambridge, Massachusetts 02139, USA
*E-mail: mch@mit.edu
[2] Clark-MXR, Inc., 7300 West Huron River Drive, Dexter, Michigan 48130, USA
[3] Department of Experimental Physics, University of Pécs, 7624 Hungary

Abstract. Near single-cycle THz pulse generation from lithium niobate crystal by optical rectification of pulses from an Yb fiber laser/amplifier system was investigated. The pump laser supplied 250 fs long pulses with 14 μJ energy at up to 1 MHz repetition rate. The THz average power exceeded 0.25 mW.

Introduction

THz technology is being used in an increasingly wide field of applications including non-destructive evaluation, quality control in the pharmaceutical industry, biological and medical sciences and high-speed communication systems [1]. However the lack of compact high average power THz sources has limited the practical application of THz radiation in these fields. Optical rectification of femtosecond laser pulses is one of the standard methods for generating single cycle THz pulses [2]. Lithium niobate (LN) has a high nonlinearity and a large bandgap, but its high dielectric constant prevents collinear velocity matching between optical and THz generation. Noncollinear velocity matching has been demonstrated by various methods [3-5], most effectively by tilting the pump pulse intensity front with a grating-lens combination [6-9].

In this article, we demonstrate near-single-cycle THz pulse generation in LN pumped by an Yb-doped fiber laser system to yield 0.25 mW average THz output power. The development of such high average power, high repetition rate THz devices will have direct applications in real-time THz spectroscopic imaging, high throughput screening and signal processing.

Experimental Methods

An Yb-doped fiber oscillator amplifier system (CLARK Impulse) producing 250 fs duration pulses at 1.035 μm, with up to 14 μJ energy at repetition rates up to 1 MHz, was used as the pump source. The intensity front of the pulses was tilted by a 1800 lines/mm grating and imaged with a 60 mm lens, with a demagnification of 1.1, onto the input surface of a 0.6% MgO doped stoichiometric $LiNbO_3$ (sLN) crystal. A beam telescope with a negative first lens was used to focus the pump beam onto the surface of the grating. The spot size at the crystal was about 0.3 (horizontal) x 0.2 (vertical) mm. The generated THz radiation was collected by an off-axis parabolic reflector or a polymer lens. A calibrated pyroelectric detector (Microtech Instruments) was used to measure the average THz output power at different pump intensities and repetition rates. Electro-optic sampling using a 2 mm thick <110> cut GaP crystal was used to characterize the THz pulse shape and spectral content.

Results and Discussion

The temporal profile of the generated THz pulses at 1 MHz repetition rate is depicted in Fig. 1. The near-single-cycle pulse has a duration of about 1 ps with a spectral maximum at 1.0 THz and frequency components extending above 2 THz. The effective sensitivity of the pyroelectric detector was calculated by weighting the detector sensitivity curve with the measured spectrum of the generated THz radiation.

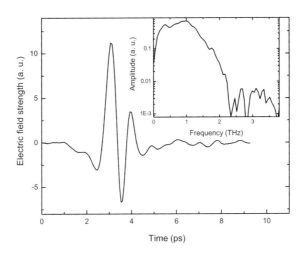

Fig. 1. Temporal shape of the THz pulse generated at 1 MHz repetition rate and (inset) corresponding amplitude spectrum.

The THz pulse energy as a function of the pump laser pulse energy for three different repetition rates (200 kHz, 500 kHz, and 1 MHz) is shown in Fig. 2. Neither the dependence of the THz energy on the pump pulse energy nor its dependence on the repetition rate indicates any heating effect even at high average pump powers above 10 W. The fact that for a given pump energy the generated THz energy is higher for higher repetition rate is probably due to the shortening of the pump pulse duration at higher repetition rate in the fiber laser. The pump pulse durations were about 450fs, 350 fs, and 250 fs at 200 kHz, 500 kHz, and 1 MHz repetition rates, respectively.

To the best of our knowledge, this is the highest output ever achieved in the generation of single-cycle THz pulses using fiber laser pumping. In fact, it is only a factor of four smaller than the highest average power obtained from multicycle THz pulses [10] and more than two times higher than what was obtained very recently using a fiber laser pumped GaP waveguide [11]. The quadratic fitting curve displayed in Fig. 2 matches the measured data quite well, up to more than 10 W, indicating the absence of saturation in the conversion efficiency in this pumping range. Further developments in fiber laser technologies will make the current method of terahertz generation even more attractive.

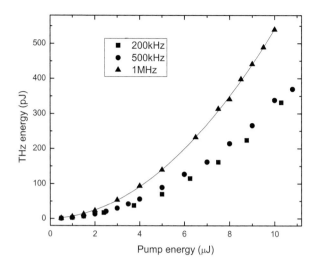

Fig. 2. Dependence of THz pulse energy on the pump pulse energy for 200 kHz (squares), 500 kHz (dots), and 1 MHz (triangles) pump pulse repetition rate, respectively. The solid line is a quadratic fit to the data.

Conclusions

We have demonstrated the generation of 0.25 mW of terahertz average power by tilting the intensity front of pulses from a 1 MHz fiber amplifier laser at 1.035 μm using LN as the nonlinear optical crystal. The growing availability of fiber amplifier lasers will pave the way for development of compact high power THz sources that are well suited for incorporation into spectroscopic and imaging devices.

This work was supported in part by the MIT Deshpande Center for Technological Innovation.

1   M. Tonouchi, Nature Photonics 1, 97 (2007).
2   B. B. Hu, X.-C. Zhang, D. H. Auston, and P. R. Smith, Appl. Phys. Lett. 56, 506 (1990).
3   R.M. Koehl and K.A. Nelson, *Chem. Phys.* **267**, 151-159 (2001).
4   T. Feurer, J.C. Vaughan, and K.A. Nelson, *Science* **299**, 374-377 (2003).
5   T. Feurer, N.S. Stoyanov, D.W. Ward, J.C. Vaughan, E.R. Statz, and K.A. Nelson, *Annu. Rev. Mater. Res.* **37**, 317-350 (2007).
6   J. Hebling, G. Almási, I. Z. Kozma, and J. Kuhl, Opt. Express 10, 1161 (2002).
7   A. G. Stepanov, J. Hebling, and J. Kuhl, Appl. Phys. Lett. 83, 3000 (2003).
8   M. C. Hoffmann, K.-L. Yeh, J. Hebling, and K. A. Nelson, Opt. Express 15, 11706 (2007).
9   K.-L. Yeh, M. C. Hoffmann, J. Hebling, and K. A. Nelson, Appl. Phys. Lett. 90, 171121 (2007).
10   J. E. Schaar, K. L. Vodopyanov, and M. M. Fejer, Opt. Lett. 32, 1284 (2007).
11   G. Chang et al., Opt. Express 15, 16308 (2007).

Millijoule Pulse Energy High Repetition Rate Femtosecond Fiber CPA System: Results, Micromachining Application and Scaling Potential

F. Röser[1], J. Rothhardt[1], T. Eidam[1], O. Schmidt[1], D.N. Schimpf[1], A. Ancona[1], S. Nolte[1], J. Limpert[1], and A. Tünnermann[1,2]

[1] Institute of Applied Physics, Friedrich-Schiller-University Jena, Albert-Einstein-Str. 15, D-07745 Jena, Germany
[2] Fraunhofer Institute for Applied Optics and Precision Engineering, Jena, Germany
E-mail: fabian.roeser@uni-jena.de

Abstract. We report on an ytterbium-doped fiber CPA system delivering millijoule energy 800 fs pulses at high repetition rates and average powers exceeding 100 W. A micromachining application and average power scaling potential are also presented.

Introduction

Fiber lasers and amplifiers have the reputation to be immune against thermo-optical distortions due to the fiber geometry itself and the diffraction-less confined propagation of the laser radiation. Hence, they are considered as an average power scalable solid-state laser concept. However, the extraction of ultra-short laser pulses possessing high peak powers is significantly more challenging than in bulk laser systems due to the high intensities in the fiber core sustained over considerable interaction length, which evoke nonlinear pulse distortions, mainly by self-phase modulation (SPM). Fiber CPA systems with reduced nonlinearity applying ytterbium-doped large-mode-area step-index fibers as main amplifier with pulse energies of a few 100 μJ have been presented years ago [1]. Those systems operated at low repetition rates, abandoning the advantages of fiber lasers in terms of average power and were pushed to surprisingly high B-integrals. Here we report on a two-stage fiber CPA system comprising an 80 μm ytterbium-doped core photonic crystal fiber as main amplifier delivering compressed pulse energies up to 1 mJ and average powers up to 100 W at repetition rates up to 200 kHz. The setup consists of a passively mode-locked Yb:KGW oscillator, a dielectric grating stretcher-compressor unit, an acousto-optical modulator as pulse selector and two ytterbium-doped photonic crystal fibers both used in single-pass configuration as amplification stages.

Experimental Setup and Results

The long-cavity Yb:KGW oscillator delivers transform-limited 400 fs pulses at a repetition rate of 9.7 MHz and an average power of 1.6 W at 1030 nm center wavelength. The stretcher-compressor-unit employs two 1740 lines/mm dielectric diffraction gratings and stretches the 3.3 nm bandwidth pulses to 2 ns. A quartz based acousto-optical modulator is used to reduce the pulse repetition rate and possesses diffraction efficiency as high as 75 %. The pre-amplifier comprises a 1.2 m long 40 μm core single-polarization air-clad photonic crystal fiber pumped by a fiber coupled diode laser emitting at 976 nm. This stage is capable of delivering a single-pass gain as high as 35 dB and average powers up to 6 W. However, we have operated the preamplifier just up to a few μJ of pulse energy to avoid excessive accumulation of nonlinear phase in this stage. The main amplifier is constructed using a low-nonlinearity air-cladding photonic crystal rod-type fiber similar to the fiber presented in [2]. The inner cladding has a diameter of 200 μm (NA = 0.58), the active core is as large as 80 μm. This structure possesses a small signal pump light absorption at 976 nm of 30 dB/m. The fiber has no polymer coating and is embedded in a water cooled aluminum body. The core supports very few transverse modes; however, stable excitation of the fundamental mode only is achieved by seed mode matching. A power independent beam quality characterized by a M2-value of less than 1.2 and a mode-field diameter as large as 71 μm, corresponding to an effective mode-field area of 4000 μm2, have been measured. The fiber length used in this experiment is 1.2 m. The degree of polarization of the fiber amplifier output is 98% allowing for an efficient recompression of the pulses. The throughput efficiency of the compressor is 70%. At 200 kHz repetition rate a compressed output power of more than 100W could be reached, corresponding to pulse energy of 500 μJ. At 50 kHz and 70 mW seed power we achieved 71 W of average power with a pump power of 180 W, corresponding to 1.45 mJ energy. leading to a compressed average power of 50 W equal to pulse energy of 1 mJ.

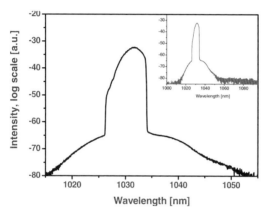

Fig. 1. Spectral characteristics of the output at 1.45 mJ pulse energy; inset: larger wavelength scan

The output spectrum of the high performance fiber CPA system at the highest extracted pulse energy is shown in figure 1. Amplified spontaneous emission and intermediated (non-selected) pulses are suppressed better than 35 dB. This inset of figure 1 shows a larger wavelength scan of the spectrum excluding the onset of stimulated Raman scattering.

The autocorrelation trace of the compressed pulses at low energy and at 1 mJ is shown in Fig. 2. This reveals a wing structure growing with pulse energy, which can be attributed to the imposed nonlinear phase. The total B-integral at the highest energy is calculated to 7. The compressed pulses exhibit an autocorrelation width of 1.23 ps (equivalent to 800 fs pulse duration). The corresponding pulse peak power is approximately 1 GW.

With this novel femtosecond fiber CPA laser system operating in the parameter regime relevant for industrial applications we conducted laser drilling experiments on copper and stainless steel sheets. The influences of repetition rate (up to 1 MHz) and average power (up to 70 Watt) on the processing time and on the resulting hole quality were investigated [3].

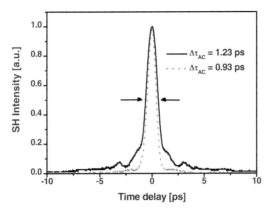

Fig. 2. Measured autocorrelation traces; dotted: low pulse energy, solid: 50 kHz and 1 mJ

To show the average power scaling potential of this particular fiber we built a simple cw laser cavity pumping this fiber from both ends by fiber-coupled diode lasers emitting at 976 nm. A maximum output power of more than 710W could be reached. This value corresponds to an extracted power per unit length of 570 W/m. Even at this high extracted power per unit length the laser operated stable, no thermal degradation occurred.

Conclusions

We have demonstrated the extraction of millijoule pulse energy femtosecond laser pulses from a fiber based CPA system at high average powers. To our knowledge, these are the highest pulse energies ever reported from a fiber based ultra-fast laser system; together with the high repetition rates this constitutes a unique performance. Key element of the system is an 80 μm core photonic crystal fiber with the ability of stable fundamental mode propagation, possessing low nonlinearity and allowing for the extraction of high average power from a short fiber length. The fiber demonstrated the capability of producing more than 710 W output power in cw operation. This shows the feasibility of a fiber based ultrashort pulse system with pulse energies above 1 mJ and average powers in the kW level. Ultrashort pulse laser drilling experiments have been performed on metals at high repetition rates up to the MHz regime. Processing times have been reduced to a few ms, demonstrating great potential for industrial applications.

This work is partly funded by the Bundesministerium für Bildung und Forschung (BMBF) under contract 13 N 8579.

1   A. Galvanauskas, G. C. Cho, A. Hariharan, M. E. Fermann, and D. Harter, Opt. Lett. 26, 935-937 (2001).

2   J. Limpert, O. Schmidt, J. Rothhardt, F. Röser, T. Schreiber, A. Tünnermann, S. Ermeneux, P. Yvernault and F. Salin, Opt. Express 14, 2715–2720 (2006).

3   A. Ancona, F. Röser, K. Rademaker, J. Limpert, S. Nolte, and A. Tünnermann, Opt. Express 16, 8958-8968 (2008)

Femtosecond thin disk lasers with >10 µJ pulse energy for high field physics at multi-megahertz repetition rates

T. Südmeyer[1], S. V. Marchese[1], C. R. E. Baer[1], S. Hashimoto[1], M. Golling[1], A. G. Engqvist[1], D. J. H. C. Maas[1], G. Lépine[2], G. Gingras[2], B. Witzel[2], and U. Keller[1]

[1] Department of Physics, Institute of Quantum Electronics, ETH Zurich, 8093 Zurich, Switzerland
    E-mail: sudmeyer@phys.ethz.ch
[2] Centre d'optique, photonique et laser, Université Laval, Pav. d'optique-photonique Québec G1V 0A6, Canada

Abstract. We present a modelocked femtosecond thin disk laser that generates pulse energies beyond the 10-µJ limit. We discuss the first photoelectron imaging spectroscopy measurements at multi-megahertz repetition rate, which is advantageous due to high signal-to-noise ratio and reduced measurement time. A maximum peak intensity of $6 \cdot 10^{13}$ W/cm^2 was achieved at 14 MHz repetition rate.

Introduction and motivation

Numerous applications in science and industry require high energy pulses with durations in the femtosecond regime. The direct generation of such pulses from a multi-megahertz solid-state laser oscillator has significant advantages over more complex amplifier systems, which are typically operating at kilohertz repetition rates with average powers of only a few watts. We present an Yb:YAG thin disk laser generating pulses with a duration of 791 fs and an energy of 11 µJ. By performing photoelectron imaging spectroscopy (PEIS) in xenon and argon, we demonstrate the first application of a thin disk laser in high field science. Driving PEIS with a multi-megahertz repetition rate laser results in high signal-to-noise ratio, short measurement time, and high accuracy.

Femtosecond thin disk laser

The Yb:YAG thin disk laser head used in this experiment is described in detail in reference 1. Passive mode locking is started and stabilized using a semiconductor saturable absorber mirror (SESAM)[2] with a saturation fluence of 115 µJ/cm^2 and a modulation depth of $\approx 0.6\%$. A set of GTI-type dispersive mirrors lead to a total negative dispersion of $\approx$ -19900 fs^2 per cavity roundtrip, and a glass plate with a thickness of 1 mm inserted at Brewster's angle ensures linear polarization of the laser output. Operation in helium atmosphere was used to eliminate the large contribution of the air to the nonlinearity inside the laser cavity.[3] The increase in pulse energy presented in this paper was achieved by significantly extending the cavity length using a 4f-extension and a MPC.[4] The former is realized with two equally curved mirrors, separated by their radius of curvature ($R = 2f = 5$ m), whilst the latter consists of two flat mirrors and one curved mirror (ROC = 10 m). The total beam path in the MPC is 23.4 m. To inject and extract the beam of the laser cavity into the MPC we use a single flat mirror in front of a flat MPC mirror (Fig. 1b).

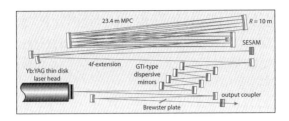

Fig. 1. Schematic of the 11-μJ laser setup including a multiple-pass cavity (MPC).

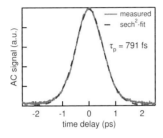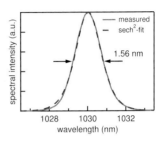

Fig. 2. Autocorrelation (left) and optical spectrum (right) of the output pulses. The dashed curves represent a sech²-fit with a duration of 791 fs and an optical bandwidth of 1.56 nm. The time bandwidth product is 0.35 (ideal: 0.315).

All these measures to increase the cavity length resulted in a 4-MHz resonator with an average output power of 45 W corresponding to a pulse energy of 11 μJ.[5] The sech²-shaped pulses had a FWHM-duration of 791 fs and a spectral bandwidth of 1.56 nm (Fig. 2), resulting in a time-bandwidth product of 0.35 (Fourier limit: 0.315). A peak power of 12.5 MW was achieved. The beam quality was nearly diffraction limited with an M^2 value of 1.1 (measured at 9.4 μJ).

Photoelectron imaging spectroscopy (PEIS)

Photoelectron Imaging Spectroscopy (PEIS) is a sensitive method to determine the photoelectron momentum distribution after a multi-photon ionization process.[6,7] By focusing a linearly polarized laser pulse with sufficient peak power into a gas target, atoms or molecules can be ionized. PEIS yields a detailed image pattern that gives insight into the excitation path of a photoelectron. Driving PEIS with a multi-megahertz repetition rate laser results in a high signal-to-noise ratio, short measurement time, and high accuracy. Increasing the repetition rate by up to 4 orders of magnitude compared with amplified kilohertz systems allows a large number of photoelectrons to be accumulated, while operating at a low number of ionization events per pulse, which avoids space charge effects and results in a close to single atom response.

For the PEIS experiments, the output pulses of a passively modelocked thin disk laser running at 14 MHz repetition rate were temporally compressed using a microstructured large mode area (LMA) fiber and a single prism pair as described in reference 8. The laser was operated in air atmosphere with 17 W of average power. Two different compressor setups were used to generate pulse durations of 35 fs and 79 fs respectively, with which we measured photoelectron images from ionization of xenon and argon. An image is obtained by accumulating the electron signals from more than 10^9 laser pulses.

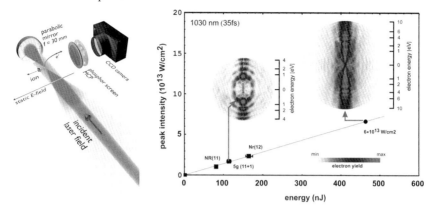

Fig. 3. Photoelectron imaging spectroscopy setup, and calibration of the peak intensity.

At low laser intensities the ionization is dominated by non-resonant multi-photon ionization (NRMPI) as well as ionization involving the population of excited electronic states of the atom.[9] Due to the electric field in the laser focus, the linear ponderomotive shift of high lying Rydberg states enable enhanced multi-photon ionization (REMPI) at specific laser intensities.[10] Their appearance in the measured spectra allows a calibration of the peak intensity in the laser focus as a function of the incident pulse energy. To perform this calibration, the appearance of the resonant (11+1)-photon ionization via the $5g$ state was used, as well as two calibration points obtained from channel closing of the non-resonant 11- and 12-photon ionization, and the zero-energy level.[11] A maximum peak intensity of $6 \cdot 10^{13}$ W/cm^2 was achieved using the compressed 35-fs pulses.

Summary and conclusions

We have demonstrated the first passively modelocked Yb:YAG thin disk laser generating pulse energies beyond the 10-µJ level. These lasers can strongly improve signal-to-noise ratio and reduce measurement time in high field physics applications, as we illustrated by the first PEIS experiments performed at multi-megahertz repetition rate.

1   Marchese, S. V. et al. Opt. Lett. 31 (18), 2728-2730 (2006).
2   Keller, U. et al. IEEE J. Sel. Top. Quant. 2 (3), 435-453 (1996).
3   Nibbering, E. T. J. et al. J. Opt. Soc. Am. B 14 (3), 650-660 (1997).
4   Herriott, D. et al. Appl. Opt. 3 (4), 523-526 (1964).
5   Marchese, S. V. et al. Opt. Express 16 (9), 6397-6407 (2008).
6   Helm, H. et al. Phys. Rev. Lett. 70, 3221 (1993).
7   Wiehle, R. et al. Phys. Rev. Lett. 89 (22), 223002 (2002).
8   Südmeyer, T. et al. Opt. Lett. 28, 1951-1953 (2003).
9   Helm, Hanspeter et al. Phys. Rev. A 49 (4), 2726-2733 (1994).
10 Wiehle, R. et al. Phys. Rev. A 67, 063405 (2003).
11 Schyja, V. et al. Phys. Rev. A 57 (5), 3692-3697 (1998).

Ultra-high intensity-High Contrast 300-TW laser at 0.1 Hz repetition rate.

V. Yanovsky[1], V. Chvykov[1], G. Kalinchenko[1], P. Rousseau[1], T. Planchon[1*], T. Matsuoka[1], A. Maksimchuk[1], J. Nees[1], G. Cheriaux[2], G. Mourou[2], and K. Krushelnick[1]

[1]FOCUS Center and Center for Ultrafast Optical Science, University of Michigan, Ann Arbor, Michigan 48109

[2]Laboratoire d'Optique Appliqu´ee, UMR 7639 ENSTA,-CNRS-Ecole Polytechnique, F91761, Palaiseau Cedex, France

*Current address Colorado School of Mines, Golden, CO 80401

Abstract: We demonstrate the highest intensity - 300 TW laser   by developing booster amplifying stage to the 50-TW-Ti:sapphire laser (HERCULES). To our knowledge this is the first Petawatt-scale laser at 0.1 Hz repetition rate.

1. Introduction

Recently, we demonstrated intensity as high as 10^{22} W/cm^2 [1] by focusing a 50 TW-laser (HERCULES [2]) into a wavelength-limited spot. Later, the nanosecond-contrast of the laser was improved by 3 orders of magnitude to 10^{11} [2] . Significant increase of the intensity above 10^{22} W/cm^2 can only be accomplished by increasing the laser energy as there is not much prospect in shortening high-power-pulse duration below 10 fs.

Here we report the upgrade of the HERCULES laser to 300 TW output power at 0.1 Hz repetition rate. To our knowledge, this is the first Petawatt-scale laser at high repetition rate. By using adaptive optics and f/1 parabola we focused the output   beam into a 1.3 μ focal spot corresponding to unprecedented intensity of ~ 2 10^{22} W/cm^2 .

2. Laser design

HERCULES laser design is based on chirped-pulse amplification with cleaning of amplified spontaneous emission (ASE) noise after the first amlifier. Output pulse of the short pulse oscillator of the HERCULES laser is preamlified in the two-pass preamlifier to the microjoule energy level. ASE added by the two-pass amplifier is removed by the cleaner based on cross-polarized-wave generation [2]. High-energy regenerative amplifier [3] and cryogenically cooled 4-pass amplifier bring the pulse energy to a joule-energy level with nearly diffraction-limited beam quality.

Two sequential 2-pass-Ti:sapphire amplifiers of 1' and 2" beam diameter respectively raise the output energy to a value approaching 20 J. We designed our own frequency-doubled Nd:glass pump laser [4] for pumping of the final two amplifiers of the HERCULES laser. The pump laser has two stages of amplification. The frequency-doubled output of the first stage is used for pumping of 1"-diameter T:sapphire amplifier, while the unconverted infrared light is injected into the second stage of the pump laser for further amplification. The frequency- doubled output of the second stage is used for pumping of the booster (2"-diameter) amplifier of the HERCULES laser. The pump laser has a quasi-flat-top beam profile that was achieved at 0.1 Hz repetition rate through relay imaging and thermally-introduced birefringence compensation. The booster two-pass amplifier uses 11-cm-diameter Ti:sapphire crystal. Only a portion of this crystal is used to amplify 2" - diameter output beam of the HERCULES laser. In order to suppress parasitic oscillations the side surface of the crystal is covered with a thin layer of index-matching thermoplastic coating (Cargille Laboratories, Inc.) doped with organic dye absorbing at 800 nm.

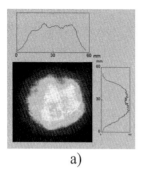
a)

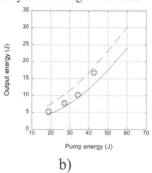
b)

Fig.1. Measurements at the output of the booster amplifier. a) Output beam profile of the HERCULES laser; b) Output energy of the booster amplifier in dependence on the pump energy, crosses- experimental data, solid line-Frantz-Nodvic calculations for 1.7 J input energy, dashed line- for 3J input energy

3. Experimental results

Output beam profile (Fig. 1(a)) is quasi-flat-top as a result of using flat-top pump beams and of the image relaying of the amplified beam through the whole laser chain. Output energy of 17 J corresponding to 300 TW power after compression has been reached so far (Fig. 1(b)). The pump energy for the booster Ti:sapphire -amplifier (2"-diameter) is controlled by changing the pumping level of the oscillator of the pump laser. Because the same oscillator provides seeding pulse for both stages of the pump laser, changing the oscillator energy changes the pump energy for the last two Ti:sapphire amplifiers. It means that not only the pumping energy of the booster amplifier changes but the input energy changes as well if the oscillator energy is changed. As a result, in predicting the performance of the booster amplifier we calculate several Frantz-Nodvic curves, corresponding to varying input energy. Two of them, corresponding to input energy 1.7 J and 3 J are shown in Fig. 1(b).

The output pulse is compressed in a 4-grating compressor to ~30 fs. Because the beam size in the compressor is rather large (6"-diameter) achromatic lenses are used in the final relays to prevent spatially varying group delay across the beam. The autocorrelator that is sensitive to spatial variation of the group delay (autocorrelator with inversion [5]) is used to control this effect. The pulse width is measured at full energy using beam leak through a mirror. The results of the measurements are shown in Fig. 2 (a,b) . The experimental spectrum profile (Fig. 2 (b)) is closely fitted by the Gaussian shape curve of 37 nm FWHM. The pulse-width-bandwidth product (~0.5) is close to the value of 0.44 for a transform-limited ~30 fs Gaussian pulse of 37 nm bandwidth. The final amplifier added no more than 0.2 to the estimated B-integral value of the laser chain. This value is too low to influence the

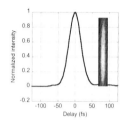 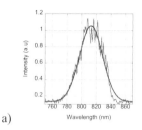 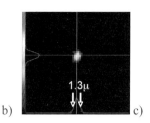

a) b) c)

Fig.2. Compressed-pulse measurements. a) Autocorrelation of 300 TW pulse, the experimental autocorrelation picture (insert) demonstrates that there is no amplitude front tilt or other spatial variations of the pulse arrival time; b) Output spectrum and Gaussian fit (FWHM=37nm); c) Focal spot focused by f/1 parabolic mirror

compression that is further evidenced by excellent quality of the compressed pulse (Fig. 2(a)). Although final amplifiers are only water-cooled, the thermal effects in them are minimal as the average absorbed pump power for them is quite modest (<5W, they work at 0.1Hz repetition rate).The beam is down-collimated by the all-reflective telescope after the compressor to 4"-diameter and is send to the interaction chamber where it is focused by the parabolic mirror.

A deformable mirror (4"-diameter, 177 actuators, made by Xinetics), controlled by the signal from a wavefront sensor located after the f/1 parabolic mirror (at low energy) compensates the aberrations of the parabolic mirror, astigmatism of the telescope and the residual aberrations of the laser beam. The focal distribution (FWHM= 1.3 μ) shown in Fig. 2(c) is characterized by using the method that we developed in [1].

4. Conclusion

In conclusion, by upgrading HERCULES's laser power to 300 TW we demonstrated the highest intensity ($\sim2 \times 10^{22}$ W/cm^2) laser. This intensity can be raised to 0.5×10^{23} W/cm^2 by using f/0.6 paraboloid (as we did in [1]) opening the radiation dominated regime of electron –light interaction for experimental studies. To our knowledge this is also the first Petawatt-scale laser at high repetition rate.

Acknowledgements

This study was supported by the National Science Foundation through the Frontiers in Optical and Coherent Ultrafast Science Center at the University of Michigan.

1.   S.-W. Bahk, P. Rousseau, T. Planchon, V. Chvykov, G. Kalintchenko, A. Maksimchuk, G. Mourou,V. Yanovsky, Opt. Lett. **29**, 2837-2839 (2004).
2.   V. Chvykov, P. Rousseau, S. Reed, G. Kalinchenko, and V. Yanovsky, Opt. Lett. **31**, 1456-1458 (2006).
3.   V. Yanovsky, C. Felix , and G. Mourou, "High-energy Broadband Regenerative Amplifier for Chirped-pulse Amplification" IEEE J. Sel. Top. Quant. Electr." **7**, 539-541 (2001).
4.   V. Yanovsky, G. Kalinchenko, P. Rousseau, V. Chvykov, G. Mourou, and K. Krushelnick, Appl. Opt. **47**, 1968-1972 (2008)
5.   Z. Sacks, G. Mourou, Danielius, Opt. Lett., **26**, 462-464 (2003).

Highly Efficient, Low-Cost Diode-Pumped Femtosecond Cr^{3+}:LiCAF Lasers

Umit Demirbas[1], Alphan Sennaroglu[1,2], Franz X. Kärtner[1], and James G. Fujimoto[1]

[1]Department of Electrical Engineering and Computer Science and Research Laboratory of Electronics, Massachusetts Institute of Technology, Cambridge, MA 02139, USA
[2]Laser Research Laboratory, Department of Physics, Koç University, Rumelifeneri, Sariyer, 34450 Istanbul, Turkey
E-mail: jgfuji@mit.edu

Abstract. Low-cost, single-mode diode pumping of Cr^{3+}:LiCAF generates 72-fs pulses with 178 mW power, at a record 54% slope efficiency. Single-mode and multimode diode pumped Cr^{3+}:LiCAF are compared as an alternative to femtosecond Ti:Sapphire technology.

Introduction

Cr^{3+}-doped colquiriite gain media (Cr^{3+}:LiSAF, Cr^{3+}:LiSGaF, Cr^{3+}:LiSCAF, and Cr^{3+}:LiCAF) can be directly diode-pumped and can enable the development of low-cost femtosecond laser technology that is an alternative to Ti:Sapphire lasers [1-5]. Low-cost, diode-pumped lasers include broad-stripe single-emitter diodes and single transverse-mode laser diodes. Currently, single transverse-mode laser diodes provide output powers of ~150 mW at costs of only $100 to $200. Broad-stripe, multimode diodes can produce an order of magnitude more power and cost ~$1000, but they suffer from poor beam quality with typical M^2 values of ~10 along the junction axis.

In this work, we describe and compare the results of continuous-wave (cw) and cw mode-locked operations of a Cr^{3+}:LiCAF laser, pumped either by multimode broad-area or single-mode laser diodes. In cw laser experiments with multimode diode pumping, up to 590 mW of cw output power and a slope efficiency of 22% were obtained. On the other hand, single-mode diode pumping yielded slope efficiencies up to 54%, with an output power as high as 280 mW. We believe this slope efficiency in cw operation is the highest obtained to date from any diode-pumped Cr^{3+}:colquiriite laser.

Stable, self-starting mode-locked operation was obtained in both pumping configurations using a semiconductor saturable absorber mirror (SESAM); also known as a saturable Bragg reflector (SBR). For multimode diode pumping, 97-fs pulses were obtained with 390 mW of average power at a repetition rate of 140 MHz (~2.8-nJ pulse energy); corresponding to an electrical-to-optical conversion efficiency of 4.2%. For single-mode diode pumping, 72-fs, 1.4-nJ pulses were obtained with 178 mW of average power at a pulse repetition rate of 127 MHz. An electrical to optical conversion efficiency of up to 7.8% was demonstrated. To the best of our knowledge, these are the highest average mode-locked powers demonstrated by using a diode-pumped Cr^{3+}:LiCAF laser system, and the electrical-to-optical conversion efficiencies for the single-mode diode-pumped cavity are the highest obtained conversion efficiencies for a femtosecond laser to date.

Experimental Setup

A schematic of the multimode diode-pumped Cr^{3+}:LiCAF laser is shown in Fig. 1. Two linearly polarized (TM) 1.6-W, 150-μm wide, single-emitter diodes (DM1-DM2) were used as the pumps. Diode output was diffraction limited along the fast axis and multimode along the slow axis with M^2 ~10. The laser resonator was an astigmatically compensated, x-folded cavity consisting of two curved pump mirrors (M1 and M2, R=75 mm); a flat, highly reflecting end mirror (M3); and a flat output coupler (OC). A slit near the OC was used to control the transverse mode structure of the laser beam with multimode diode pumping. The gain medium was a 2-mm-long, Brewster-cut, 10% Cr-doped Cr^{3+}:LiCAF crystal. For mode-locking experiments, a double-chirped mirror (DCM) was used to provide negative dispersion. A low-loss SESAM/SBR was used to initiate and sustain mode locking. For the single-mode diode-pumped cavity, multimode diodes were replaced with four linearly polarized, ~660-nm, single-mode diodes. Each single-mode diode provided up to ~160 mW of pump power, and the total absorbed pump power was ~580 mW.

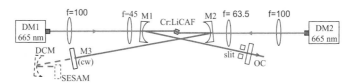

Fig. 1. Schematic of the multimode diode-pumped Cr^{3+}:LiCAF laser. Dashed lines indicate the mode-locked cavity. For single-mode diode pumping, four diodes are used, two on each side.

Results and Discussion

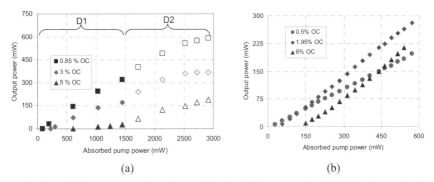

(a)   (b)

Fig. 2. Cw efficiency curves for (a) multimode diode-pumped and (b) single-mode diode-pumped Cr:LiCAF lasers. High slope and conversion efficiencies are achieved.

Figures 2 (a) and (b) show cw efficiency of the laser using different output couplers for multimode and single-mode pumping, respectively. Multimode diode pumping enabled higher cw output powers (590 mW with a 0.85% output coupler), however, the slope efficiencies were low (22%). On the other hand, record slope efficiency (54%) with medium power levels (280 mW with a 1.95% output coupler) was obtained from the single-mode diode-pumped cavity. Also, in the case of multimode diode pumping, mode mismatch between the pump and the laser beams created strong thermal loading due to upconversion. This causes the output power saturation observed at high pumping levels (Fig. 2 (a)). Furthermore, thermal effects became stronger and slope efficiency decreased with increasing output coupler transmission in multimode pumping due to the increased role of upconversion. Without the slit, using the 1.4% output coupler, the multimode diode-pumped system produced up to 880 mW of cw output power with 30% slope efficiency, however the laser output was multimode.

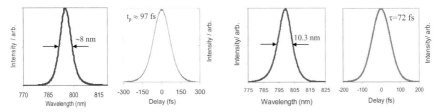

Fig. 3. Measured spectra and background-free intensity autocorrelation trace for (left) multimode and (right) single-mode, diode-pumped cw mode-locked laser cavities.

Figure 3 summarizes the cw mode-locking results for both diode pumping configurations. 97-fs, 2.8-nJ pulses at 140-MHz repetition rate with 390 mW of average power were obtained with multimode diode pumping. In comparison, single-mode diode pumping gave 72-fs, 1.4-nJ pulses at 127 MHz with 178 mW of average power. Although the pulse energies were lower for single-mode pumping, the calculated electrical-to-optical conversion efficiency was 7.8%, which is almost twice that of multimode diode pumping (~4.2%).

Conclusions

In conclusion, we report cw as well as cw mode-locked performance of a Cr^{3+}:LiCAF laser pumped by multimode and single-mode diodes near 660 nm. Single-mode diode pumping yielded record slope efficiencies up to 54% with pulses as short as 72-fs and 178 mW of average power. These results suggest that diode-pumped Cr^{3+}:LiCAF is emerging as a low-cost, alternative technology to femtosecond Ti:Sapphire lasers.

Acknowledgements. We thank Andrew Benedick and Aleem Siddiqui for help during the initial experiments and acknowledge support by the National Science Foundation (ECS-0456928 and ECS-0501478), Air Force Office of Scientific Research (FA9550-07-1-0014 and FA9550-07-1-0101) and the Scientific and Technical Research Council of Turkey (Tubitak, 104T247).

1   S. A. Payne, L. L. Chase, H. W. Newkirk, L. K. Smith, and W. F. Krupke, IEEE Journal of Quantum Electronics **24**, 2243-2252, 1988.

2   A. Miller, P. LikamWa, B. H. T. Chai, and E. W. Van Stryland, Optics Letters **17**, 195-197, 1992.

3   G. J. Valentine, J. M. Hopkins, P. Loza-Alvarez, G. T. Kennedy, W. Sibbett, D. Burns, and A. Valster, Optics Letters **22**, 1639-1641, 1997.

4   J-M. Hopkins, G. J. Valentine, B. Agate, A. J. Kemp, U. Keller, and W. Sibbett , IEEE Journal of Quantum Electronics **38**, 360-368, 2002.

5   U. Demirbas, A. Sennaroglu, F. X. Kärtner, and J. G. Fujimoto, Optics Letters **33**, 590-592, 2008.

Environmentally stable 200-fs Yb-doped fiber laser with dispersion compensation by photonic crystal fiber

Samuli Kivistö[1], Robert Herda[1], Alexey F. Kosolapov[2], Andrei E. Levchenko[2], Sergei L. Semjonov[2], Evgueni M. Dianov[2], and Oleg. G. Okhotnikov[1]

[1]Optoelectronics Research Centre, Tampere University of Technology, FIN-33101 Tampere, Finland
 E-mail: samuli.kivisto@tut.fi
[2]Fiber Optics Research Centre, Moscow, 119333, Russia
 E-mail: kaf@fo.gpi.ru

Abstract. We report environmentally stable mode-locked Yb-doped fiber laser with dispersion compensation by index-guided solid-core photonic crystal fiber. The photonic crystal fiber and Faraday rotator in the cavity allow for robust 200-fs operation at 1 µm.

1. Introduction

Practical femtosecond fiber lasers require all-fiber low-loss means for group-velocity dispersion compensation at 1 µm wavelength region. Among the different types of photonic crystal fibers (PCFs), photonic bandgap (PBG) fibers - both solid-core and hollow-core - can generate a large amount of anomalous dispersion and may have large mode size [1, 2]. Studies have shown; however, that PBG fibers exhibit a large amount of high-order dispersion that has a notable effect on the pulse formation [3]. Another constraint expected with PBG fibers is that the presence of a bandgap structure eventually limits the shortest pulsewidths achievable with such fibers. With these arguments in mind, index-guided PCF may provide a competitive alternative to PBG fibers because of the absence of bandgap restrictions and low higher-order dispersion.

2. Photonic crystal fiber

The preform from which the photonic crystal fiber was drawn was fabricated by mechanical drilling. The drilled preform was then etched, resized, and jacked. During drawing and jacketing, a noble gas at excess pressure was introduced into the internal holes to compensate for surface tension forces tending to collapse the holes. Thermal processing was performed in the flame of a hydrogen-oxygen burner. The cross-section and the measured group-velocity dispersion of the PCF are shown in Fig. 1. As can be seen, the air filling factors are different in the first and second ring of holes. The air filling factor for the first ring of holes is $k_1=d/\Lambda=0.75$, and for the second ring of holes $k_2=0.79$. The core diameter of the fiber is 2.9 µm and the measured zero dispersion wavelength is 874 nm. The loss at 1060 nm is 7 dB/km. The calculated effective mode area A_{eff} is 4.86 µm², and the calculated nonlinear coefficient γ is 26 $W^{-1}\cdot km^{-1}$.

(a)   (b)

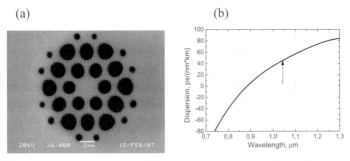

Fig. 1. (a) The cross-section image, and (b) the measured group-velocity dispersion of the PCF. The arrow shows the operation wavelength of the laser.

3. Environmentally stable mode-locked laser

The schematic of the fiber laser used in the experiments is shown in Fig. 2. The active material is an 80-cm long ytterbium doped fiber pumped by a 300 mW diode laser. The light is coupled out via a 25 % tap coupler. The total length of the passive HI-1060 fiber in the cavity is 120 cm. The high birefringence of the cavity caused by the PCF [4] is compensated for by a Faraday rotator mirror acting as a cavity end reflector. The self-starting passive mode-locking is initiated by an In-GaAs-GaAs quantum well semiconductor saturable absorber mirror (SESAM) with a modulation depth of 10% and a saturation fluence of 7 $\mu J/cm^2$.

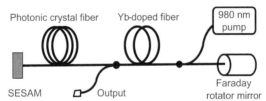

Fig. 2. Schematic of the mode-locked laser cavity with dispersion compensation by PCF.

The photonic crystal fiber with a length of 3.4 m and a total dispersion of 0.09 ps^2 was spliced to a single mode fiber using a standard fusion splicer. Reflections from the PCF to silica fiber interface and losses due to the mode field mismatch were minimized using repeated arc discharges. This allows for smooth collapse of the holes and causes an adiabatic mode field transformation in the PCF resulting in optimized mode matching [5]. The splice loss was measured to be 1.5 dB. The net cavity dispersion was calculated to be -0.04 ps^2 assuming a dispersion of 25 ps^2/km for the ytterbium and the HI 1060 fiber.

The PCF dispersion compensation results in soliton operation with transform limited pulses. Fig. 3 shows the optical spectrum and the autocorrelation of the mode-locked laser. The pulse duration of 199 fs was derived using a Gaussian fit resulting in a time-bandwidth product of 0.42. The pulse energy was 460 pJ at a fundamental repetition rate of 18 MHz.

Remarkable feature of the laser performance was the obvious tendency to operation with a single pulse in the cavity. There was no evidence of multiple pulse mode-locking at the available pumping power. Without the use of the Faraday rotator, the pulse operation start-up and mode-locking performance were very sensitive to changes in the polarization state. A polarization controller was then essentially needed to optimize the pulse duration and quality. Due to environmental changes the polarization controller has to be frequently realigned to maintain operation state. In contrast, with the Faraday rotator, the self-starting and the steady-state operation was independent of the fiber bending up to few centimeters bending radius. Stable mode-locking was observed for several hours without the need for any readjustment and there was no need for readjustment when restarting the laser after few days.

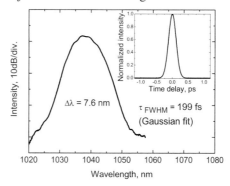

Fig. 3. Optical spectrum and autocorrelation (inset) of the mode-locked pulse.

4. Conclusions

In conclusion, we have demonstrated an environmentally stable femtosecond ytterbium-doped fiber laser using index-guided PCF for dispersion compensation at 1 μm and a Faraday rotator mirror for the compensation of the birefringence induced by the PCF. This approach may constitute an important step towards highly practical ultrafast fiber oscillators since index-guided solid-core PCFs allow for better control of dispersion and are free from severe bandwidth limitation and large higher-order dispersion typical for crystal fibers using bandgap guiding mechanism.

Acknowledgements. The authors acknowledge the support of the graduate school of Tampere University of Technology, the Finnish Foundation for Economic and Technology Sciences, the Jenny and Antti Wihuri Foundation, the Nokia Foundation, and the Emil Aaltonen foundation.

1   A. Isomäki and O. G. Okhotnikov, "Al-fiber ytterbium soliton mode-locked laser with dispersion control by solid-core photonic bandgap fiber" Opt. Express 14, 4368-4373, 2006.

2   A. Isomäki and O. G. Okhotnikov, "Femtosecond soliton mode-locked laser based on ytterbium-doped photonic bandgap fiber", Opt. Express 14, 9238-9243, 2006.

3   R. Herda, A. Isomäki and O. G. Okhotnikov, "Soliton sidebands in photonic bandgap fibre lasers," Electron. Lett., 42, 19-20, 2006.

4   H. Lim, F. Ilday, and F. Wise, "Femtosecond ytterbium fiber laser with photonic crystal fiber for dispersion control," Opt. Express 10, 1497-1502, 2002.

5   L. Xiao, W. Jin, and M. S. Demokan, "Fusion splicing small-core photonic crystal fibers and single-mode fibers by repeated arc discharges," Opt. Lett. 32, 115-117, 2007.

Noncollinear Optical Parametric Amplification Pumped by the Third Harmonics of a Ti:sapphire Laser

Takashi Tanigawa, Keisaku Yamane, Taro Sekikawa, and Mikio Yamashita

Department of Applied Physics, Hokkaido University, and Core Research for Evolutional Science and Technology, Japan Science and Technology Agency, Kita-13 Nishi-8, Kita-ku, Sapporo, 060-8628, Japan
E-mail: tanigawa@hokudai.ac.jp

Abstract. Broardband amplification in the 380-490 nm region was achieved by ultraviolet (UV) pumped noncollinear optical parametric amplification. This result leads to 6 fs UV pulse generation and can be utilized to amplify monocycle pulses.

Introduction

Noncollinear optical parametric amplification (NOPA) of laser pulses in the ultraviolet region is desired from various points of view. Firstly, amplified ultrashort UV pulses are demanded in the field of time-resolved spectroscopy. Secondly, NOPA of the UV region is utilized to generate intense monocycle optical pulses.

Generation of optical pulses in the monocycle region has been achieved using a gas-filled hollow fiber which broadens the spectrum via self-phase modulation (SPM) and a $4f$-SLM (spatial light modulator) phase compensator [1], where it is necessary for the diameter of the hollow fiber to be small enough to cause strong SPM. Therefore it is difficult to input a intense pulse to the gas-filled hollow fiber to avoid ionization of the gas. Consequently, monocycle pulses generated by this technique have too small pulse energy of the order of 10 μJ for various applications such as high harmonic generation which requires the order of 100 μJ pulses. We have been aiming to generate more intense monocycle pulses by amplifying over-octave bandwidth pulses from the hollow fiber in the whole range of the spectrum (400-1100 nm) [1]. Our approach is as follows: the shorter wavelength part (shorter than 500nm) in the entire pulse spectrum is amplified by UV pumped NOPA [2, 3] and the longer wavelength part (longer than 500 nm) is amplified by angularly dispersed NOPA [4, 5]. We have been studying both UV pumped and angularly dispersed NOPAs.

In this paper we will discuss the generation of ultrashort UV pulses based on NOPA for the present aim. And we intend to expand this system to the one that can generate monocycle pulses in the near future.

Experiments

We used β-barium borate (BBO) crystal for UV pumped NOPA. And we used third harmonics of output pulses from a Ti:sapphire laser amplifier as a pump source for NOPA. Figure 1 shows type I phase matching curves of BBO crystals pumped at 267 nm. From the simple calculation, provided that the pump beam is monochromatic and both seed and pump beams are CW, the most broad gain spectrum is obtained under the condition that the noncollinear angle between seed and pump beams is about 6.2° and the crystal angle is about 52°. Under this condition, the gain spectrum is in the range of

370-440 nm.

Figure 2 shows the schematic of the experiment. We used a Ti:sapphire laser amplifier system (KMLabs Red Dragon) which generates 800 nm, 5 mJ, 30fs and 1kHz optical pulses as a source.

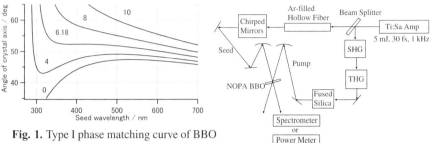

Fig. 1. Type I phase matching curve of BBO crystal pumped by 267 nm. Values in the figure indicate noncollinear angles in degree.

Fig. 2. Schematic of the experimental setup

Fundamental pulses from this laser system was split into two fractions. The 10% of the beam was sent through an argon gas filled hollow fiber where the fiber core diameter was 140 μm and the gas pressure was 2 atm. SPM in the hollow fiber broadened the spectrum to 350-1000 nm which is sufficient for generating monocycle pulses. Those ultra-broad pulses were used as the seed of NOPA. This seed pulse has a better spatial profile compared to the case where white light continuum from CaF_2 is used as the seed. The rest fraction of the beam was used to generate 260 nm pump pulses via SHG and THG BBO crystals. Group velocity mismatching between 260 nm pump pulses and 400 nm centered seed pulses in NOPA BBO crystal is about 300 fs/mm. Therefore, pump pulses were stretched to about 1 ps by group velocity dispersion in 8 cm fused silica glass. Energy of the pump pulse incident on NOPA BBO crystal is 40 μJ at the most. This low energy is mainly due to absorption losses in the fused silica glass. Seed pulses were compressed to some degree by a pair of chirped mirrors. Although those chirped mirrors have a high reflectivity only in 300-500nm, this experiment was intended to examine amplification only in this wavelength region and this band-pass filtering doesn't matter. Then, pump and seed pulses were focused into a 3 mm thick BBO crystal (cut at 52°) by concave mirrors with focal lengths of 25 cm and 30 cm, respectively. The noncollinear angle between seed and pump pulses was set to about 6.2° for amplification with the maximum bandwidth. After passing BBO crystal, the intensity spectrum of seed pulses was measured by a fiber coupled monochromator and its power was also measured by a power meter.

Results and Discussion

Figure 3 shows a observed gain spectrum. The corresponding temporal width of transform-limited pulse is 5.8 fs. Seed pulses were amplified over the very broadband range from 380 to 490 nm that almost matches the region demanded to UV pumped NOPA. This broadband amplification range is more than the range supposed from the simple calculation. We surmise that it is owing to the fact that pump pulses are non-monochromatic (3 nm bandwidth) and angularly dispersion accompanying beam focusing. But the amplification gain is relatively low in the present circumstances. This is due to several reasons. Main two reasons are low pump energy and pump depletion in BBO crystal caused by two photon absorption (TPA). Low pump pulse energy can be improved by replacing the fused silica glass with more adequate materials such as CaF_2. In addition, pump depletion in BBO crystal due to TPA is a problem accompanying the UV pumped NOPA. We are now investigating theoretically as well as experimentally the optimum condition that maximizes the amplification gain under the existence of TPA in BBO crystal. Moreover, we are constructing a $4f$-SLM pulse shaper to compress UV pulses down to transform-limited time widths. For this pulse shaper, we have fabricated a new liquid crystal SLM that works for the UV region in addition to the visible region [6, 7].

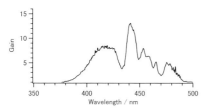

Fig. 3. Observed gain spectrum of third harmonics pumped NOPA

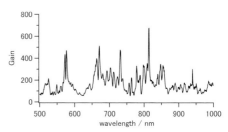

Fig. 4. Gain spectrum of angularly dispersed NOPA

Conclusion

We showed the broadband amplification of NOPA pumped by the third harmonics of a Ti:sapphire laser. Together with the broadband gain spectrum of angularly dispersed NOPA(Figure 4) observed by us [5], Third harmonics pumped NOPA can be utilized to amplify the monocyle pulses.

1   E. Matsubara, K. Yamane, T. Sekikawa, and M. Yamashita, J. Opt. Soc. Am. B **24**, 985, 2007.
2   P. Tzankov, T. Fiebig, and I. Buchvarov, Appl. Phys. Lett. **82**, 517, 2003.
3   G. Kurdi, K. Osvay, M. Csatari, I. N. Ross, and J. Klebniczki, IEEE J. Sel. Top. in Quantum. Electron. **10**, 1259, 2004.
4   G. Arisholm, J. Biegert, P. Schlup, C. Hauri, and U. Keller, Opt. Express **12**, 518, 2004.
5   K. Yamane, T. Tanigawa, T. Sekikawa, and M. Yamashita, Technical Digest of Conference on Lasers and Electo-Optics (CLEO), CTuEE4, 2008.
6   K. Hazu, T. Sekikawa, and M. Yamashita, Opt. Lett. **32**, 3318, 2007.
7   K. Hazu, N. Nakagawa, S. Fang, Y. Nakano, T. Sekikawa, and M. Yamashita, Technical Digest of Conference on Lasers and Electo-Optics (CLEO), CThDD7, 2008.

Sub-10 fs Pulse Generation in Vacuum Ultraviolet Using Chirped Four Wave Mixing in Hollow Fibers

I. Babushkin, F. Noack, J. Herrmann.

Max Born Institute for Nonlinear Optics and Short Pulse Spectroscopy, Max-Born-Str. 2a, D-12489 Berlin, Germany
E-mail: jherrman@mbi-berlin.de

Abstract. We investigate the potential of four-wave mixing for VUV pulse generation in hollow waveguides with unprecedented short pulse durations (up to 2.5fs) at 160nm using broadband chirped 800nm idler pulses.

Introduction

Ultrashort pulse sources in the UV and VUV spectral range are essential tools for time-domain investigations of the fundamental dynamics of atoms and molecules in physics, chemistry, biology and material sciences. So far, sub-10 fs UV pulses in the region of 275-335 nm were generated by non-collinear optical parametric amplification in nonlinear crystals [1]. An alternative method is the use of four wave mixing (FWM) in hollow waveguides filled with a noble gas, which is transparent in the UV and VUV region. Generation of 8 fs pulses with 1 µJ at 270 nm were reported with this method [2].

Recently in our group it was demonstrated that fs pulse generation by FWM in gas-filled hollow waveguides can be extended to reach the VUV range at 160 nm using the fundamental and third harmonic of the Ti:sapphire laser as idler and pump, respectively: $\omega_{signal} = \omega_{pump} + \omega_{pump} - \omega_{idler}$ with $\omega_{pump} = 3\omega_{idler}$ and $\lambda_{idler} = 800$ nm [3]. The efficiency was greatly improved by coupling to higher-order transverse modes, as well by coating the inner surface of the waveguide with aluminum, which allows an increase of the energy of the 160 nm pulse up to 600 nJ with a pulse duration of 160 fs.

In the first part of the present paper we study the potential of FWM in hollow waveguides filled with argon for the generation of sub-5 fs VUV pulses by using bandwidth-limited intense 10 fs pulses at 800 nm and longer and weaker third harmonic pulses at 267 nm. As will be shown, the VUV signal pulses are split into two parts with different velocities, one propagating with the linear group velocity at 150 nm while the other is locked to the pulse at 800 nm. The shortest pulses with 2.3 fs duration are achieved if both input pulses are coupled into the fundamental mode only.

In the second part of the paper, we consider the generation of VUV pulses by chirped FWM with broad-band positively chirped idler pulses and pump pulses with large pulse durations. Assuming these input pulses, we obtain negatively chirped signal pulses at 160nm with a pulse energy of about 1 µJ which can be compressed to 6 fs by a layer with normal dispersion. The pulse energy is scalable to higher values up to 100 µJ by corresponding input pulses with higher energies.

Results and Discussion

For the numerical simulations, we used a model based on a generalized version for the forward Maxwell equation [4] with inclusion of the transverse higher-order mode structure and without the assumption of the slowly varying envelope approximation. Our model incorporates generation of all possible frequencies from the input fields, multimode light propagation including nonlinear coupling between the transverse modes, influence of plasma and photoionization, self- and cross-phase modulations, and other higher-order nonlinear effects for all spectral components and transverse modes of the waveguide [5].

We consider hollow waveguides with an inner wall made from SiO_2 and a diameter of 100 μm. Because of the waveguide contribution to dispersion different transverse modes of the waveguide exhibit varying dispersion and therefore unequal phase-matching pressures. The excitation of higher-order transverse modes EH_{1n}, $n > 1$ can be easily controlled by the waist diameter w of the input Gaussian beam.

In Fig. 1 an example for sub-5 fs pulse generation by bandwidth-limited pulses is presented for fundamental transverse mode excitation only ($w = 0.64d$). The spectral maximum of the signal pulse is shifted to 153 nm (Fig. 1a) and the spectrum is extended from 142-174 nm (solid curve) with an almost linear increasing phase (dotted curve) in the main part of the spectrum. In Fig. 1b the pulse shapes of all three pulses are presented (note the different scale for the signal pulse). As seen the signal pulse (solid curves) splits into two parts with different velocities. The pulse moving with the linear group velocity at 153 nm has a duration of 11 fs. The other part is locked to the idler at 800 nm and is almost bandwidth-limited with a duration of only 2.5 fs. The spectrum of the free moving pulse is extended from 150 to 157 nm while the locked pulse has a spectrum from 150 to 175 nm. The later is spectrally broadened by cross-phase modulation from the idler pulse.

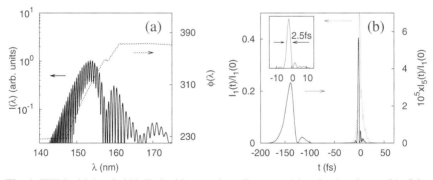

Fig. 1. FWM with bandwidth-limited input pulses. Spectrum (a) and pulse shapes (b) of the signal pulse for a 10 fs 8 μJ input idler at 800 nm, and 50 fs 2.4 μJ pump pulses at 267 nm with a fiber length $z = 10$ cm, a diameter $d = 100$ μm, and pressure $p = 700$ Torr. In (a) the spectrum is shown by the solid line and the phase by the dotted line. In (b), the intensity $I(t)$ for the idler (dotted line) and signal (solid line) is presented. The insert shows the details of the VUV pulse in the vicinity of $t = 0$.

Let us consider now the case of chirped FWM with broadband streched idler input pulses at 800nm. In particular, we assumed for the idler broadband positively chirped pulses with a pulse energy of 0.1mJ stretched from 3fs to 300 fs pulse duration. For the pump at 273nm unchirped 300 fs pulses with 0.1 mJ energy were assumed. In Fig. 2 one can see the results for VUV pulse generation by chirped FWM with the above described parameters after $z = 10$ cm propagation. The pulse energy now is 1.4 µJ and the spectrum in Fig.2a (solid curve) reaches from 147nm to 175nm, the spectral phase (dotted curve) shows a negative chirp opposite to the chirp of the idler. It allows compression of the output signal pulse (solid curve in Fig.2b) by a layer of normal-dispersion material, here assuming a 1 mm layer made from MgF_2. The dotted curve in Fig. 2b represents the compressed pulse with a pulse duration of 6.0 fs.This method allows a further significant increase of the VUV pulse energy up to 100 µJ by using input idler pulses with about 1 mJ energy.

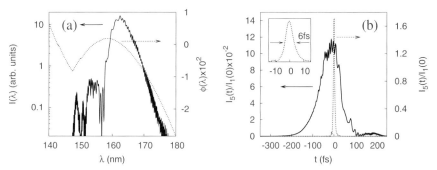

Fig. 2. High energy VUV pulse generation by chirped FWM. Spectrum (a) and pulse shape (b) of the VUV pulse after propagation in a fiber with length $z = 10$ cm and a diameter $d = 100$ µm at 1.3 atm for 300 fs 0.1 mJ input idler (800 nm) and pump (267 nm) pulses. The spectral width of the idler pulse corresponds to 3 fs duration. In (a) the spectral phase is shown by the dotted line and the intensity is shown by the solid line. In (b) the temporal intensity distribution of the chirped signal pulse is shown by the solid line and the compressed pulse by the dotted line. The insert shows the details of the compressed 6.0 fs VUV pulse.

Conclusions

In conclusion, we studied the potential of four-wave mixing for VUV pulse generation in hollow waveguides with unprecedented short pulse durations. By using bandwidth-limited 10fs input idler pulses at 800nm the generation of 2.5fs VUV nJ pulses at 160 nm are predicted. The pulse energy can be significantly increased by using broadband chirped input idlers. As an example, we predict the generation of VUV pulses with 1.4 µJ energy and 6 fs duration for 300 fs, 0.1 mJ idler pulses. The VUV pulse energy can be increased up to the range of 100 µJ, the pulse duration in this case is about 8 fs.

1 P. Baum, S. Lochbrunner, and E. Riedle, "Tunable sub-10-fs ultraviolet pulses generated by achromatic frequency doubling," Opt. Lett. **29**, 1686 (2004).
2 C. G. Durfee III *et al.*, "Intense 8-fs pulse generation in the deep ultraviolet," Opt. Lett. **24** 697 (1999).
3 P. Tzankov *et al.*, "High-power fifth-harmonic generation of femtosecond pulses in the vacuum ultraviolet using a Ti:sapphire lasers," Opt. Expr. **15**, 6389 (2007).
4 A. V. Husakou and J. Herrmann, "Supercontinuum Generation of Higher-Order Solitons by Fission in Photonic Crystal Fibers." Phys. Rev. Lett. **87**, 203901 (2001).
5 I. Babushkin, A. Husakou, J. Herrmann, (to be published)

Generation of High Energy Pulses from a Fiber-based Femtosecond Oscillator

Jungkwuen An[1], Dongeon Kim[1], Jay W. Dawson[2], Michael J. Messerly[2] and
Christopher P. J. Barty[2]

[1] Department of Physics, Pohang University of Science and Technology, Pohang 790-784,
South Korea
E-mail: kimd@postech.ac.kr
[2] Photon Science and Applications Program, Lawrence Livermore National Laboratory,
Livermore, California 94550, USA

Abstract. The high energy pulse can be achieved by exploiting self-similar
prapagation regime. In this regime, mode-lock pulse can be generated without
dispersive optics such as gratings or prisms in the cavity.

Introduction

Generating a single train of high energy pulses in a fiber-based oscillator is a
significant experimental challenge because wave-breaking is more likely to happen
than the bulk solid-state oscillator due to the inherent high nonlinearity of a fiber. In
the case of using nonlinear polarization evolution as mode-locking mechanism, the
angular alignment tolerance of the cavity's wave plates tends to be inversely
proportional to the product of the pulse energy and the fiber length. We estimate that
for a 25 nJ, 100 fs pulse propagating through a 10 m section of fiber having a modal
effective area of 30 $\mu m2$, a change in the polarization orientation of the order of a few
hundredths of a degree is sufficient to alter the nonlinear phase shift by 2π, allowing
an additional pulse stream to circulate.

A simple solution to overcome the severe alignment tolerance, is to minimize the
length of fiber in the cavity. Note that by reducing the fiber length from 10 to 1 m we
increase the alignment tolerance by an order of magnitude. In addition, the GDD of 1
m of fiber is roughly the amount required to form 10 nJ pulses under self-similar
propagation regime [1] and thus the cavity should no longer require a grating pair to
trim its dispersion.

Experimental Methods

The oscillator depicted in Figure. 1 is a unidirectional ring cavity. The core of the Yb-
doped, double-clad gain fiber has a diameter of 7 μm and a numerical aperture (NA)
of 0.12 (Nufern SM-YDF-7/210). We varied the fiber length from 1 to 2 m. The gain
fiber is pumped by a diode laser array having a center wavelength of 976 nm and a
maximum output power of 60W; it is coupled to a fiber bundle having a diameter of
400 _m and an NA of 0.22 (LIMO 60-F400-DL976). A pair of dichroic mirrors,
which transmit the pump and reflect the 1053 nm oscillator signal, couples the pump
into the cavity. The half-wave plate between the isolator and polarizing beam splitter
adjusts the fraction of the power that exits the cavity; it was typically set so that 97%
exits.

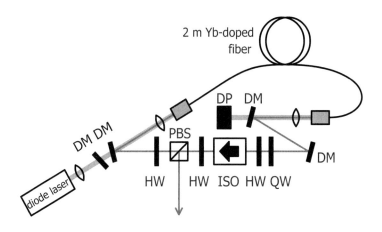

Fig. 1. Schematic diagram of a grating-less fiber oscillator: HW, half-wave plate; QW, quarter-wave plate; ISO, isolator; PBS, polarizing beam splitter; DP, beam dumper; DM, dichroic mirror which is HT for 978 nm and HR for 1053 nm

Results and Discussion

We obtained higher pulse energies with longer lengths of gain fiber: 25 nJ for a 2.0 m length and 20 nJ for a 1.2 m length. In both cases higher pump power was available, but if applied resulted in multi-pulsing. Net dispersions of fiber are 0.004 ps^2 for 1.2 m length and 0.006 ps^2 for 2 m length. Corresponding maximum pulse energies are about 20 and 25 nJ for each, according to the numerical simulation plotted in Fig. 3(a) of [1].

We verified the single-pulse operation by monitoring the long range autocorrelation (600 ps range) and a fast photodiode signal (0.3 ns resolution). We also monitored the stability of the pulse train with a radio-frequency (RF) spectrum analyzer. The RF analyzer always showed a signal-to-noise ratio better than 80 dB RF (40 dB optical) and a linewidth less than 8 kHz which is shown in Figure. 2(b).

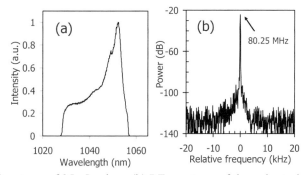

Fig. 2. (a) Spectrum of 25 nJ pulses. (b) RF spectrum of the pulse trains.

Conclusions

We have demonstrated a novel design for a high energy fiber laser that does not contain a grating pair. This design exploits the energy scaling properties of self-similar pulses to eliminate the need for dispersion compensation inside the cavity, resulting in a simpler design whose polarization evolution can be readily controlled and maintained. With the grating-less Yb-doped fiber oscillator, 25 nJ pulse energy has been achieved at a repetition rate of 80 MHz, producing 2 W average output power. The pulses were compressible to 150 fs. Further experiments to improve the pulse energy and analysis on the current propagation regime are in progress. Details of this work were described in Ref .2

Acknowledgements. This work was performed under the auspices of the U.S. Department of Energy by the University of California, Lawrence Livermore National Laboratory under Contract No. W-7405-Eng-48. J. An and D. Kim also acknowledge the support of the National Research Laboratory project (No. M10500000066-06J0000-06610) funded by Korean Science and Engineering Foundation (KOSEF), Brain Korea 21 project funded by Korean Research Foundation (KRF), and Core Technology Development Program funded by the Ministry of Commerce, Industry and Energy of Korea.

1   F. O. Ilday, J. R. Buckley, W. G. Clark, and F. W. Wise in Physical Review Letters, Vol. 92, 213902, 2004.
2   J. An, D. Kim, J. W. Dawson, M. J. Messerly, and C. P. J. Barty, in Optics Letters, Vol. 7, 2010, 2007.

Femtosecond passively mode-locked fiber lasers using saturable Bragg reflectors

Hyunil Byun, Jason Sickler, Jonathan Morse, Jeff Chen, Dominik Pudo, Erich P. Ippen, and Franz X. Kärtner

Department of Electrical Engineering and Computer Science, and Research Laboratory of Electronics, Massachusetts Institute of Technology, 77 Massachusetts Avenue, Cambridge, Massachusetts 02139, USA
Email: hibyun@mit.edu

Abstract. We demonstrate a soliton fiber laser with 280-fs pulses at 408-MHz repetition rate, and a stretched-pulse regime fiber laser with 102-fs pulses at 234-MHz repetition rate. Both use saturable Bragg reflectors for mode-locking and/or self-starting.

Introduction

Compact sources of femtosecond laser pulses are an attractive and versatile technology for a variety of applications, such as frequency metrology [1] and ultrafast sampling [2]. Passive mode-locking enables low jitter femtosecond pulses and alleviates the need for an external microwave oscillator. In the past, both polarization additive-pulse mode-locking (P-APM) and/or saturable Bragg reflector (SBR) mode-locking [3,4] were used. The former has been successfully used in high repetition rate fiber lasers in the soliton [5] and stretched-pulse [6] regimes. The latter can lead to a more compact cavity with fewer components required. Used in combination with P-APM, SBRs can enable self-starting and increase stability when fibers are too short for P-APM alone to self-start, allowing scaling up the repetition rate, while allowing for ultrashort pulse durations.

In this paper, we demonstrate two high repetition-rate, self-starting, passively mode-locked femtosecond erbium-doped fiber (EDF) lasers using commercial SBRs. The first is a simple, soliton-regime linear cavity modulated solely by an SBR, and the second is a compact, stretched-pulse-regime sigma cavity modulated by a combination of P-APM and an SBR. The linear cavity soliton source generates a pulse train at up to 408 MHz, with a corresponding full-width half maximum (FWHM) inferred pulse width of 280 fs, while the stretched-pulse sigma cavity results in a pulse train at 234 MHz repetition rate, consisting of 242 pJ, 102 fs pulses.

Experimental Results

Linear soliton laser

The experimental setup is depicted in Fig. 1. The laser cavity consists of a 25 cm section of EDF with a group-velocity dispersion (GVD) of -20 fs^2/mm. One end of the cavity is butt-coupled to an SBR, and the other to a dielectric mirror, which acts as the output coupler. The SBR is a commercial unit with 14% modulation depth, a 2 ps recovery time, and a saturation fluence of 25 µJ/cm^2. Pump is provided by a 980 nm laser diode, free-space coupled through a dichroic beamsplitter, and focused by a collimating lens through the output coupler and into the EDF. The output signal follows the same path in reverse, and is separated from the pump by the dichroic beamsplitter. The output pulses are then amplified using an EDFA (980 nm pump, 200mA), detected using a 10 GHz photodiode, and measured with a 500 MHz sampling scope and a signal source analyzer (Agilent E5052).

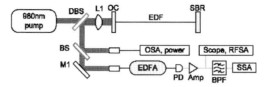

Fig. 1. Experimental setups of the linear soliton laser.

The 25 cm section of EDF yields a 408 MHz pulse repetition rate. Fig. 2 depicts a sampling scope trace, an optical spectrum, a 2 GHz bandwidth RF spectrum, and phase noise/timing jitter traces, respectively. Integrating the phase noise from 1 kHz to 10 MHz yields a root-mean-squared timing jitter of 196 fs.

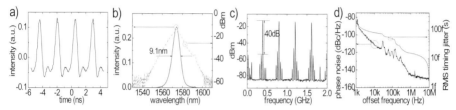

Fig. 2. Measurement traces at 400 MHz: a) sampling scope trace, b) optical spectrum, c) RF spectrum, and d) phase noise.

The 40 dB RF side-mode suppression ratio indicates an energy stability to better than 1%. The 9.1 nm FWHM optical bandwidth implies 280 fs duration transform-limited pulses. An autocorrelation measurement is in progress. All measurements were done with 130 mW of cavity-coupled pump power; the intracavity signal power was measured to be 136 mW, resulting in 330 pJ intracavity pulse energies. The laser was self-starting; as the pump power increased, the laser first operated in an unstable Q-switching state, changing to a continuous-wave soliton mode-locked state at pump powers of 115 mW. For pump powers greater than 160 mW, multiple pulsing occurred.

We subsequently increased the EDF length to 50 cm, resulting in a 197.8 MHz repetition rate. With 58 mW of cavity-coupled pump power, the intracavity signal power was measured to be 22.4 mW, yielding 113 pJ intracavity pulse energies. The optical signal exhibited a 6 nm FWHM bandwidth, corresponding to a 420 fs transform-limited pulse duration. Here, the continuous-wave soliton mode-locking, and multiple-pulsing threshold pump powers were 50 mW and 60 mW, respectively. The smaller threshold powers (as compared to the 408 MHz setup) come from the fact that a lower repetition rate at the same intracavity power correspondingly increases the pulse energy, and along with it, the nonlinear phase shift. Phase noise measurements yielded a timing jitter of 502 fs from 1 kHz to 10 MHz.

Stretched-pulse laser

The stretched-pulse laser is shown in Fig. 3a. The cavity is in a sigma configuration to provide a point of reflection for the SBR, and includes an isolator for unidirectional operation, a polarizing beam splitter as the P-APM analyzer and output coupler, and various waveplates to control the polarization evolution. A silicon slab is included in the cavity to prevent residual pump power from reaching the SBR, and to provide normal GVD that, together with the anomalous GVD of the 60 cm of gain fiber, leads to stretched-pulse operation.

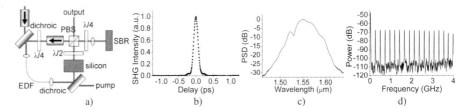

Fig. 3. The stretched pulse schematic is shown in a). Laser performance is demonstrated with b) the autocorrelation trace ($\tau_{\text{gaussian}} = 102$ fs), c) optical spectrum ($\Delta\omega_{\text{FWHM}}$=35.8 nm), and d) RF spectrum of the detected pulse train.

Fig. 3b-d shows the laser performance. The laser is free-space pumped with 980 nm light, resulting in an average laser output power of 56.7 mW. At the repetition rate of 234 MHz, this corresponds to 242 pJ pulses. The output pulses are normally chirped, and compress to 102 fs using 71.7 cm of single-mode fiber. The optical spectrum FWHM is 35.8 nm, which corresponds to 73 fs transform-limited pulses, indicating the presence of some residual chirp. The clean RF-spectrum indicates single pulse operation as was the case for the linear cavity laser.

Discussion and conclusion

We demonstrated two stable, passively mode-locked lasers using SBRs. The first is a soliton laser generating 280 fs pulses at 408 MHz, using 25 cm of EDF as the cavity. Such a design provides a simple, fiber-compatible low-jitter femtosecond source without the need for driving electronics for active modulation. The laser provides good stability in a simple and potentially scalable design. No polarization control or active stabilization is required, and the laser self-starts. Still higher repetition rates can be achieved by further reducing the cavity length while optimizing the pumping scheme.

The second is a stretched-pulse laser that produces 242 pJ pulses compressible to at least 102 fs at 234 MHz repetition rate. The SBR enables self-starting, and P-APM provides a strong saturable absorber mechanism, leading to very stable, and most importantly, shorter pulses. The laser offers a higher pulse energy alternative to the linear cavity, and with improved cavity and pump optimization, such a system should also be scalable to higher repetition rates.

1   S. T. Cundiff, in Nature, Vol. 450, 1175, 2007.
2   A. Bartels, R. Cerna, C. Kistner, A. Thoma, F. Hudert, C. Janke, and T. Dekorsy, in Review of Scientific Instruments, Vol. 78, 035107, 2007.
3   H. A. Haus, in Journal of Applied Physics, Vol. 46, 3049, 1975.
4   R. Paschotta, U. Keller, in Applied Physics B, Vol. 73, 653, 2001.
5   J. Chen, J.W. Sickler, E.P. Ippen, and F.X. Kartner, in *Conference on Laser and Electro-Optics*. OSA, 2007.
6   T. Wilken, T.W. Hansch, R. Holzwarth, P. Adel, M. Mei, in *Conference on Laser and Electro-Optics*. OSA, 2007.

Noncollinear optical parametric amplification of cw light, continua and vacuum fluctuations

Markus Breuer, Christian Homann, and Eberhard Riedle

LS für BioMolekulare Optik, Ludwig-Maximilians-Universität München, Oettingenstraße 67, 80538 München, Germany
E-mail: markus.breuer@physik.uni-muenchen.de

Abstract: Seed sources for NOPAs are compared. Single-mode cw light renders Fourier-limited femtosecond and fully tunable picosecond μJ output pulses, OPG leads to random spectral fluctuations and a sapphire continuum delivers identical pulses on every shot.

Influence of the seed light on the output of parametric amplifiers

Optical parametric amplifiers (OPAs) seeded by a continuum or by parametric generation (OPG) are the prime source of tunable radiation for ultrafast spectroscopy. In chirped pulse OPAs (OPCPAs) unprecedented levels of peak power and pulse shortness are reached and envisioned [1]. White light seeded OPAs are known to deliver reproducible pulses, however, for ultrabroadband pulses the spectrum starts to deviate from a Gaussian shape and weak satellites are found in the temporal structure. Even more severely, OPGs are known for large intensity and spectral fluctuations. In the nanosecond regime superior pulses are generated by amplification of monomode cw seed light in dye amplifiers and optical parametric oscillators [2,3].

For the ultrafast regime the principle has so far not been exploited, likely due to the small number of seed photons contained within the temporal amplification interval. Yet, the ever increasing availability of low cost diode sources would make it an attractive method. We report both a picosecond and femtosecond noncollinear OPA seeded with cw light that display smooth output spectra with Fourier-limited width. From the observed bandwidth for the femtosecond NOPA we conclude that the spectral modulations and pulse-to-pulse variations in an OPG seeded NOPA are evidence that the OPG starts from vacuum fluctuations. In contrast, the output pulses of a continuum seeded NOPA are identical for each shot.

Amplification of cw light in femtosecond and picosecond pumped NOPAs

For the evaluation of the different seeding sources under comparable conditions, we have used a BBO-based two-stage OPA with a noncollinear amplifier geometry [4]. Noncollinear phase matching is used in many blue or green pumped systems for its extremely broad amplification bandwidth. In addition, it provides high small signal gain and efficiency and avoids the need for dichroic optics.

The first NOPA was pumped by 130 μJ of the 532 nm frequency doubled output of a 10 ps, 5 kHz Nd:YVO$_4$ system. As seed laser we used a fiber coupled single mode cw diode laser system with automated tunability from 1260 to 1630 nm. Due to the low damage threshold of the BBO amplifier crystals for ps pumping, weak focusing and a moderate intensity around 10 GW/cm^2 had to be used. The resulting low gain coefficient was compensated by the use of 8 mm BBO crystals, in accord with previous work on OPG/OPA systems [5]. The resulting ps NOPA was tunable over the full tuning range of the seed laser (Fig. 1a) with an output energy of up to 4 μJ. Compared to the mW seeding power, this is a total amplification of 10^9. Only a slight readjustment of the phase matching angle of the amplifier crystal was needed. Due to the continuous seed light no adjustment of the seed pump delay is needed. Figure 1b) shows the amplified spectra, which result by electronic tuning of the cw-seed laser only, without any adjustment of the phase matching. The spectral bandwidth was only 0.63 nm (at 1.30 µm) with a clean Gaussian distribution very close to the Fourier limit of the 5.3 ps duration. The background due to spontaneous (non-seeded) emission was on the order of only 1 %.

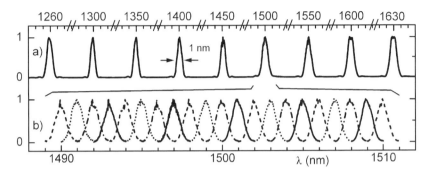

Fig. 1. Typical output spectra of the ps-pumped cw-seeded noncollinear optical parametric amplifier, showing a) wide range tunability and b) fine tunability by electronic adjustment only

Motivated by the success of the cw seeded ps NOPA we also investigated the possibility of cw seeding a femtosecond unit. The 150 fs duration of frequency-doubled Ti:sapphire pump pulses corresponds to just 500 seed photons of the low power 532 nm laser. Nonetheless we were able to observe a µJ output with 88 fs duration and 6 nm bandwidth, corresponding to a time bandwidth product of 0.55 without external pulse compression. The bandwidth of the output pulses is considerably broader than the bandwidth of the seed-light. The additional spectral components originate from the 100 fs amplification. When the 150 fs pump pulse was stretched to 300 fs by the dispersion of a fused silica slab, the bandwidth decreased further.

The present implementation of the cw seeded fs NOPA produces about an equal amount of amplified output and amplified parametric superfluorescence. The ratio depends critically on the alignment, in particular the spatial overlap of the beams. In the ps NOPA we also find that the imaging of the fiber output into the amplifier BBO with a high quality aspherical lens is essential for the optimum performance. As it has been recognized that the spontaneous emission of the first amplifier in an OPCPA is a crucial issue [6] and the proper imaging has been shown to improve the contrast ratio [7], we believe that our investigations of low power cw seeding can contribute important fundamental understanding toward the proper design and alignment of the OPCPAs. Indeed, the peak power level of a highly chirped ultrabroadband Ti:sapphire oscillator is not much higher than the seeding level used by us.

Comparison of cw-, continuum- and OPG-seeded NOPAs in the fs-regime

When the single longitudinal mode cw laser is used as seed in the fs NOPA, all individual pulses show the same 6 nm wide spectrum (middle row in Fig. 2b). The output pulse energy does, however, fluctuate due to the low number of seed photons and some mode beating of the cw laser. These fluctuations are not found for the ps system since 100 times more seed photons are available over the longer pump pulse duration. A white light seeded NOPA was investigated for comparison and we find fluctuations of only 1 % and again identical, yet slightly structured and non-Gaussian spectra for each shot (top row of Fig. 2b). The width of these spectra can be chosen by varying the chirp of the continuum generated in a sapphire plate. Typically the pulses can be compressed close to the Fourier limit.

When neither the cw seed nor the continuum is used, the second amplifier stage is seeded by the parametric superfluorescence generated in the first stage. This is the typical OPG/OPA configuration. The single shot spectral analysis shows that the spectrum is highly structured and varies from shot to shot dramatically (bottom row in Fig. 2b). About 500 shots averaging is needed to converge to a stable and smooth distribution. The width of each individual spectral structure is identical to the one found in cw seeding but the height varies largely (Fig. 2a). We conclude that we amplify individual photons out of the vacuum fluctuations [8] and each amplification process starts at a different depth in the first amplifier crystal.

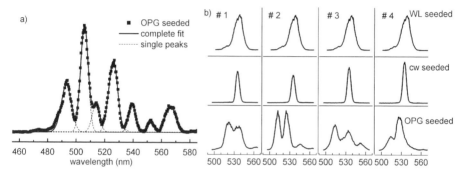

Fig. 2. a) dots: single shot spectrum of OPG-seeded NOPA; dashed line: Gaussian fits with nearly same width; solid line: fit to amplified OPG-seeded spectrum. b) shot-to-shot comparison of amplified spectra of a fs-NOPA with different seed sources

The OPG/OPA setup makes it possible to directly visualize the vacuum fluctuations. By suitable imaging of the parametric superfluorescence ring from the first noncollinear amplifier, a spatial and spectral blinking is observed on a white card by the bare eye. It is the simultaneous parametric generation and amplification in the BBO crystal at well chosen pump intensity that allows for this startling effect. A detailed analysis will allow the evaluation of the statistics in the OPG and OPA process.

1 R. Butkus, R. Danielius, A. Dubieitis, A. Piskarskas, and A. Stabinis, Appl. Phys. B **79**, 693 (2004).

2 M. M. Salour, Opt. Commun. **22**, 202 (1977).

3 O. Votava, J. R. Fair, D. F. Plusquellic, E. Riedle, and D. J. Nesbitt, J. Che. Phys. **107**, 8854 (1997).

4 E. Riedle, M. Beutter, S. Lochbrunner, J. Piel, S. Schenkl, S. Spörlein, and W. Zinth, Appl. Phys. B **71**, 457 (2000).

5 X. D. Zhu and L. Deng, Appl. Phys. Lett. **61**, 1490 (1992).

6 F. Tavella, A. Marcinkevicius, and F. Krausz, Opt. Expr. **14**, 12822 (2006).

7 S. Witte, R. Th. Zinkstok, A. L. Wolf, W. Hogervorst, W. Ubachs, and K. S. E. Eikema, Opt. Express **14**, 8168 (2006).

8 R. Glauber and F. Kaake, Phys. Lett. **68A**, 29 (1978).

Modeling of Octave-Spanning Sub-Two-Cycle Titanium:Sapphire Lasers: Simulation and Experiment

Michelle Y. Sander, Helder M. Crespo, Jonathan R. Birge, and Franz X. Kärtner

Department of Electrical Engineering and Computer Science and Research Laboratory of Electronics, Massachusetts Institute of Technology, Cambridge, Massachusetts, 02139, USA
E-mail: sanderm@mit.edu

Abstract. It is shown that a one-dimensional temporal laser model under optimized intracavity dispersion settings can quantitatively predict the spectral output and temporal pulse shape of octave-spanning, sub-two-cycle Ti:sapphire lasers.

Introduction

Various models have been derived to describe the pulse generation in mode-locked lasers [1]. For femtosecond lasers, the dispersion managed mode-locking (DMM) theory [2] takes into account pulse shaping effects of broadening and recompression during each cavity round-trip, based on the sign of group delay dispersion. Simulations based on DMM have been used to accurately describe the behaviour of standard non-octave-spanning mode-locked lasers [3]. To our knowledge, this approach has not yet been demonstrated with sub-two-cycle octave-spanning lasers.

In this work, we present an extensive one-dimensional temporal numerical analysis of an actual sub-two-cycle octave-spanning laser. This model incorporates high-order material dispersion and real mirror data from measurements.

Our model can capture the octave-spanning output spectrum and generated sub-two-cycle pulse in great detail, allowing for a direct and precise comparison with experimental measurements. Thus, the laser dynamics, stability and steady-state operation output based on realistic parameters can be predicted and optimized.

Laser Model

During one cavity round-trip, the pulse shaping is determined by nonlinear propagation through the Ti:sapphire crystal and linear evolution through the air path, mirrors and wedges. The pulse propagation in the crystal is modelled by applying the Split Step Fourier Method to the Nonlinear Schrödinger Equation [4]. The Kerr-Lens mode-locking mechanism is approximated with a fast saturable absorber and the pulse experiences gain saturation with Lorentzian filtering. Properly balancing self-phase modulation, filtering and dispersion (second and higher order material dispersion from the crystal, double-chirped mirrors (DCMs) and wedges), the system evolves to a steady-state solution. Therefore, this approach allows us to accurately follow the temporal and spectral breathing of the pulse within the laser cavity for any general laser configuration.

To accurately reproduce the experimental conditions, the DCM pairs are described by the measured reflectivity and group delay data shown in Figs. 1(a) and (b). The DCMs were designed and optimized to provide exact compensation of the material dispersion in the cavity while maintaining a smooth phase over a broad wavelength range [5]. Important features of the mirrors are the pump window around 532 nm (mirrors M1, M3) and a 50% transmission around 1160 nm and 580 nm, corresponding to the 1f to 2f frequency components that are used for carrier-envelope offset (CEO) phase stabilization of the laser. The output coupler with a 2% reflectivity for the main part of the spectrum supports the generation of a broad intracavity spectrum while a reflectivity higher than 50% below 590 nm and above 1040 nm effectively enhances and couples out the spectral wings.

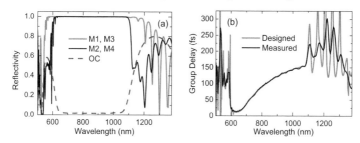

Fig. 1. (a) Reflectivity curves for the double-chirped mirrors and output coupler; (b) Group delay of the designed and measured DCM pair; measured data was determined by white-light interferometry for a normal incidence angle

Experimental Setup

The simulated laser system is a Kerr lens mode-locked Ti:sapphire ring laser with a 2-mm-long crystal and a repetition rate of 500 MHz. The crystal is placed near the focus between two concave DCMs with a 5 cm radius of curvature and the cavity consists of two other flat DCMs (see Fig. 2). A broadband output coupler is coated on a fused silica wedge, which is used in combination with a BaF_2 wedge to fine-tune the intracavity dispersion. This configuration simultaneously produces a main carrier-envelope phase stabilized pulse from the reflective output coupler and a 1f-2f output coupled out through the flat mirror M3, used for detection of the CEO frequency in the 1f-2f interferometer [6].

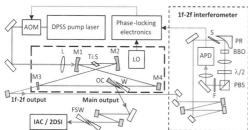

Fig. 2. Octave-spanning sub-two-cycle, carrier envelope phase stabilized Ti:Sapphire ring laser system [6]

Pulse Dynamics in the Laser

The temporal numerical analysis demonstrates the capability of capturing all of the significant laser characteristics: the output spectrum in Figs. 3(a) and (b) shows excellent agreement with the measured data and accurately reproduces the main features in the spectral wings. In addition, we can directly predict the optimized power levels achievable for the 1f and 2f output (with 1160 nm and 580 nm as 1f and 2f frequencies as seen in Fig. 3(b)). As the pronounced ripples in the measured main output spectrum are assumed to mostly have been caused by etalon filtering effects in the experimental setup, the numerical analysis reproduces the envelope of the measured output spectrum but does not exhibit the same strong oscillations.

The importance of optimized dispersion compensation becomes obvious when adjusting the inserted material dispersion by a few fs^2. The wings in the spectrum are notably decreased (around -18 dB at 1160 nm for changes in material insertion corresponding to a group delay dispersion of ~7 fs^2 at 800 nm), in excellent agreement with the required experimental fine-tuning of the wedges. With external compression of the main output pulse (using four additional octave-spanning DCMs, a BaF_2 plate, and two thin FS wedges for fine dispersion compensation), the numerically achievable minimum pulse duration is 4.9 fs. This result exactly reproduces the retrieved pulse from two-dimensional spectral shearing interferometry (2DSI) measurements [7], as shown in Fig. 3(c). Therefore, the numerical analysis clearly establishes that sub-two-cycle pulses can be generated in Ti:sapphire oscillators.

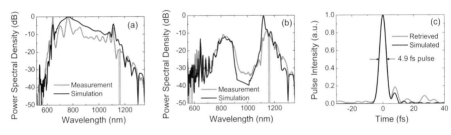

Fig. 3. (a) Simulated and measured main output spectrum; (b) simulated and measured 1f-2f output spectrum; (c) retrieved pulse intensity from 2DSI measurements and simulated output pulse after external compression

Conclusions

We successfully demonstrated that under optimized dispersion settings, a realistic one-dimensional numerical analysis can accurately describe the laser dynamics of octave-spanning lasers with sub-two-cycle pulses. By optimizing the dispersion compensation we can predict the best achievable performance in terms of output spectrum, frequency enhancement of the wings, and achievable pulse shape and duration in an actual experimental setup, providing a powerful tool for analysis and laser design.

Acknowledgements. This work was supported in part by NSF grant ECS-0501478, DARPA grant HR0011-05-C-0155 and AFOSR grant FA9550-07-1-0014.

1   H. A. Haus, IEEE, J. Sel. Top. Quantum Electronics **6**, 1173, 2000.
2   Y. Chen, F. X. Kärtner, U. Morgner, S. H. Cho, H. A. Haus, E. P. Ippen, and J. G. Fujimoto, J. Opt. Soc. Am. B. **16**, 1999, 1999.
3   M. V. Tognetti, M. N. Miranda, and H. M. Crespo, Phys. Rev. A **74**, 033809, 2006.
4   G. P. Agrawal, Nonlinear Fiber Optics, Academic Press, Boston, 2001.
5   J. R. Birge and F. X. Kärtner, Appl. Opt. **46**, 2656, 2007.
6   H. M. Crespo, J. R. Birge, E. L. Falcão-Filho, M. Y. Sander, A. Benedick, and F. X. Kärtner, Opt. Lett, **33**, 833, 2008.
7   J. R. Birge, R. Ell, and F. X. Kärtner, Opt. Lett. **31**, 2063, 2006.

Ultra-Broadband Infrared Pulses from a Potassium-Titanyl Phosphate Optical Parametric Amplifier for VIS-IR-SFG Spectroscopy

Oleksandr Isaienko and Eric Borguet

Chemistry Department, Temple University, 1901 N. 13th Street, Philadelphia, Pennsylvania, 19122, USA
E-mail: eborguet@temple.edu

Abstract. A non-collinear KTP-OPA to provide ultra-broadband mid-infrared pulses was designed and characterized. With proper pulse-front and phase correction, the system has a potential for high-time resolution vibrational VIS-IR-SFG spectroscopy.

Many vibrational spectral features, e.g. OH stretching modes, extend over several hundreds of wavenumbers. However, because of phase matching conditions, the bandwidth of IR pulses from most available optical parametric amplifiers is limited to ~150 cm^{-1}. For this reason, the term "broadband Sum-Frequency Generation (SFG) spectroscopy" is applied to the studies in which the infrared pulses have bandwidth of ~150 cm^{-1} (see, e.g., [1]), with only few accounts of SFG-acquisition over ~500-600 cm^{-1} bandwidth [2]. Rather than acquiring the spectrum in a single shot, SFG spectra from such systems as, e.g., the air-water interface, are normally obtained by tuning the output of an IR source over the broad spectral feature. Moreover, with ~150 cm^{-1} pulses the time-resolution of surface vibrational dynamics measurements is limited to ~100 fs, so information about ultrafast processes (e.g., charge transfer, vibrational dephasing) is inaccessible.

Recently, there has been a growing interest in the generation of ultra-broadband ultrashort near- and mid-IR pulses via non-collinear optical parametric amplification (NOPA) in nonlinear optical crystals [3, 4]. We applied the concept of NOPA to bulk potassium-titanyl phosphate (KTP) [5]. Calculations of phase matching curves for type-II OPA (o-pump $\rightarrow$ e-signal + o-idler) in XZ-plane of KTP with 800-nm pump, suggested that the combination of a non-collinear signal-pump geometry and the use of a divergent white-light (WL)-seed (rather than collimated) can phase match the near-IR signal in a broad wavelength range [5]. Generation of broadband signal pulses covering simultaneously ~1.05-1.45 μm range was achieved experimentally by stretching the pump pulses to compensate for any possible chirp in the WL-seed [5].

The optical setup for generation of broadband near- and mid-IR pulses is shown in Fig. 1, together with the setup for acquisition of idler+800-nm SFG-spectra. In order to obtain the mid-IR idler pulses with sufficient energies, we added a second KTP-NOPA stage to amplify the broadband signal generated in the first NOPA-stage, further called "pre-amp seed". In order to stretch the pump pulses, two equilateral SF18 prisms with face size 2.5 cm were used at optimal geometry [5]. At KTP-1, the internal phase matching angle for pump was set at ~48-49°, the internal non-collinear signal-pump angle was adjusted to ~3.5-4.0°; the angular divergence of the WL-seed was created by slightly displacing the spherical mirror collecting the WL from the sapphire plate. The geometry of the pre-amp seed and pump at KTP-2 was as close as possible to that of the WL-seed and pump at KTP-1.

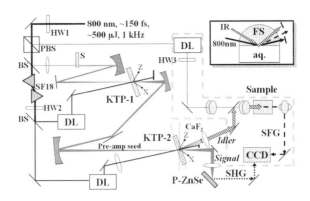

Fig. 1. Optical setup for generation of broadband infrared pulses and (inside the dashed contour) acquisition of the broadband SFG-spectra from silica-water interface: KTP - 2-mm thick crystals ($\theta=42°$, $\varphi=0°$); P-ZnSe – 2-mm thick polycrystalline ZnSe crystal; S – 2-mm thick sapphire plate; PBS – polarizer-beam splitter; HW – half-wave plates; SF18 – equilateral prisms; BS – BK7 beam splitter; DL – delay line; CaF_2 – 50-mm lens; all other optics - BK7-based lenses, flat and spherical mirrors. Orientation of the z- and x-axes is shown. SFG – idler-800nm sum-frequency generation from reference or samples (see text). SHG – second-harmonic generation from the signal pulses. Inset: side-view of the beam geometry at the sample interface; "FS" – IR-grade fused-silica hemicylinder prism; "aq." – aqueous solution

The spectra of the pre-amp seed and amplified signal were acquired by measuring their SHG-spectra in reflection off a polycrystalline (P-) ZnSe crystal [6], as shown in Fig. 1. This material ensures high-efficiency conversion and has insignificant phase-matching restrictions on the converted bandwidth. The spectra of idler from KTP-2 were measured via SFG with 800-nm pulses (Fig. 1) with P-ZnSe placed instead of the sample.

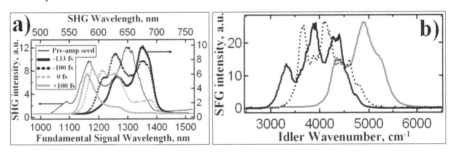

Fig. 2. (a) SHG-spectra of the pre-amp seed (dashed line) and amplified signal pulses at different pump – pre-amp seed relative delays at KTP-2 (thick lines). (b) Spectra of the idler pulses from KTP-2

A spectrum of the pre-amp seed after KTP-1 is shown as a dashed line in Fig. 2(a), along with spectra of the signal after OPA in KTP-2 at different relative time delays between the pre-amp seed and the pump pulses. Although the pump pulses are stretched (from ~180 fs to >500 fs [5]), apparently, it was not enough to compensate for the chirp of pre-amp pulses. Tuning of the signal SHG-spectra excluded 1- or 2-photon fluorescence from ZnSe as a source. Idler spectra were derived from idler+800nm SFG on P-ZnSe (Fig. 2(b)). As the entire bandwidth of the pre-amp seed pulses could not be amplified at once, the idler spectra obtained were narrower than one would expect based on the pre-amp seed spectrum. Additionally, to get idler frequencies in higher wavenumber region (>5000 cm^{-1}), we also needed to slightly adjust the pump phase-matching angle within ~1–2°. The energy of the near-IR pre-amp seed pulse is ~2-3 μJ, and that of the signal and idler pulses from KTP-2 is on the order of 11-14 μJ and 4-5 μJ, respectively, indicating ~3-4% energy conversion.

As the idler pulses are obtained in NOPA, they inevitably have a large angular dispersion [7]. Despite this fact, we decided to perform proof-of-principle experiments to check the feasibility of ultra-broadband VIS-IR-SFG spectroscopy on the fused silica - water interface, a system that has vibrational features spanning over ~1000 cm^{-1}. For this, the idler pulses were nearly collimated with a 50-mm CaF$_2$ lens (Fig. 1), and focused tightly at the sample interface with another (not shown).

In conclusion, we have demonstrated a KTP-NOPA that generates ultra-broadband near- and mid-IR pulses in ~1.0-1.4 μm and ~1.9 – 3 μm regions, respectively. Preliminary results on the vis-IR-SFG spectroscopy from the silica-water interface show that ultra-broadband SFG-vibrational spectroscopy of features extending over >1000 cm^{-1} is feasible. As suggested from the idler bandwidth, generation of sub-20 fs mid-IR pulses should be possible once the pulse-front of the idler pulses is corrected with subsequent compression. Use of a grating -telescope (or prism-telescope) setup ([7]) will provide for compensation of the signal and idler angular dispersion. Additionally, introduction of a larger chirp into the pump pulses should enable production of ~2000 cm^{-1} broad idler pulses.

The NOPA described here will allow the generation of broadband mid-IR pulses for a wide range of applications. Our initial focus will be studies of vibrational dynamics of processes such as heat/energy transfer between different broad modes and/or chemical species with sub-20 fs time resolution. Another interesting perspective is a realization of time-domain ultra-broadband IR spectroscopy with pulses covering simultaneously vibrational transitions of multiple species of interest, in analogy to pulsed NMR spectroscopy.

Acknowledgements. The authors acknowledge the support of the US Department of Energy – Office of Basic Energy Sciences.

1. Bonn, M., Ueba, H., and Wolf, M., *Theory of sum-frequency generation spectroscopy of adsorbed molecules using the density matrix method - broadband vibrational sum-frequency generation and applications.* J. Phys.: Condens. Matter **17**(8), S201-S220 (2005)

2. Hommel, E.L., Ma, G., and Allen, H.C., *Broadband vibrational sum frequency generation spectroscopy of a liquid surface.* Anal. Sci. **17**(11), 1325-1329 (2001)

3. Cirmi, G., Brida, D., Manzoni, C., Marangoni, M., De Silvestri, S., and Cerullo, G., *Few-optical-cycle pulses in the near-infrared from a noncollinear optical parametric amplifier.* Opt. Lett. **32**(16), 2396-2398 (2007)

4. Brida, D., Manzoni, C., Cirmi, G., Marangoni, M., De Silvestri, S., and Cerullo, G., *Generation of broadband mid-infrared pulses from an optical parametric amplifier.* Opt. Express **15**(23), 15035-15040 (2007)

5. Isaienko, O. and Borguet, E., *Generation of ultra-broadband pulses in the near-IR by non-collinear optical parametric amplification in potassium titanyl phosphate.* Opt. Express **16**(6), 3949-3954 (2008)

6. Chinh, T.D., Seibt, W., and Siegbahn, K., *Dot patterns from second-harmonic and sum-frequency generation in polycrystalline ZnSe.* J. Appl. Phys. **90**(5), 2612-2614 (2001)

7. Shirakawa, A., Sakane, I., and Kobayashi, T., *Pulse-front-matched optical parametric amplification for sub-10-fs pulse generation tunable in the visible and near infrared.* Opt. Lett. **23**(16), 1292-1294 (1998)

Chirped-pulse Raman amplification for two-color high-intensity laser experiments

Peng Dong[1], Franklin Grigsby[1], and Mike Downer[1]

[1]FOCUS Center, University of Texas at Austin, Department of Physics, Austin, TX 78712, USA
  E-mail: Donwer@physics.utexas.edu

Abstract. We report generation and compression of millijoule-level first Stokes sideband (873nm) of 800nm TW pulses by inserting a multi-stage barium nitrate Raman shifter-amplifier into a conventional Ti:sapphire chirped pulse amplification system.

Introduction

In many high-field experiments it is desirable to accompany the main terawatt (TW) pulse with a moderately powerful (~0.1 TW), temporally synchronized ultrashort pulse at a slightly shifted (~100 nm) center wavelength outside the bandwidth of the main pulse. Zhavoronkov et al. [1] demonstrated the generation of 80 μJ, 190fs, sub-GW, 870 nm pulses by stimulated Raman scattering (SRS) of chirped 1.5 mJ, 800 nm pulses in barium nitrate. Because it uses chirped pulses to avoid damage and self-phase modulation in the Raman-active medium, this technique is well suited to TW-scale laser systems based on chirped pulse amplification (CPA), but so far has been demonstrated only on a much lower power system.

In this paper, we report scaling the chirped-pulse Raman amplification (CPRA) technique to a TW CPA laser system, resulting in 873 nm pulses up to 3 mJ, 0.03 TW synchronized with 800 nm, 5 TW pulses. This energy was achieved purely by CPRA, without adding any pump lasers to the CPA system and without compromising the amplified energy of the main 800 nm pulses. Further amplification of the 873 nm output by conventional methods e.g. a multi-pass Ti:sapphire amplifier with an additional pump laser prior to compression, appears straightforward. Fig. 1 shows the setup of Raman shifter and amplifier.

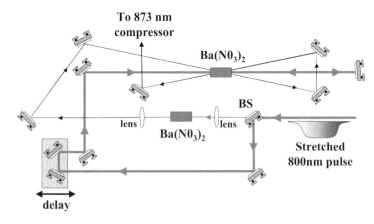

Fig. 1. Details of stimulated Raman shifter and two-pass Raman amplifier.

Results and Discussion

The two main hurdles encountered in scaling CPRA from microjoule [1] to millijoule output energy are: 1) onset of self-focusing and filamentation; and 2) competition between first Stokes and higher-order Raman amplification.

The first issue limits output energy of the first (SRS) stage to $<\sim 0.1$ mJ. Below this threshold, a high quality Gaussian mode can be achieved. Attempts to extract higher energy result in a filamented beam of limited usefulness, as shown in Fig. 2(a). Twenty percent conversion (i.e. ~ 1 mJ) to 1st Stokes is possible, but only with a pump that exceeds the critical power for self-focusing. In normal operation, we restrict first stage output to $\sim 20 \mu J$, in order to completely eliminate filamentation. Higher first Stokes energy is then achieved by seeding a second Raman amplification stage with the high quality output mode of the first stage. This reinitiates Raman amplification with the non-filamented second-stage pump. The mode quality of the first stage output is then preserved through amplification to millijoule energies in the second stage, as shown in Fig. 2(b) for energy ~ 1 mJ. At this energy, Raman beam "cleanup" [2] occurs - i.e. the amplified Stokes beam assumes the pump beam profile with smoothed irregularities. For example, the feature in the upper left corner of the profile in Fig. 2(b) was transferred from the pump, where it was much more prominent. To maintain good beam quality, overall gain in the seeded second stage amplifier was kept $<\sim 300$. Because a seed was injected, pump intensity could be kept below the level (~ 1 GW/cm2) required for single-pass Raman amplification to the same energy, thus preventing filamentation.

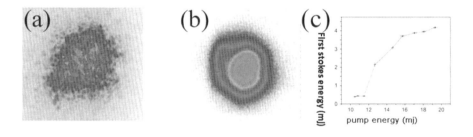

Fig. 2. (a,b) Transverse profiles of first Stokes beam: (a) produced by single pass stimulated Raman amplification to ~ 1 mJ energy, showing filamentation; (b) similar energy produced by seeded Raman amplification in second stage, showing nearly Gaussian mode. Pump intensity is much lower in the seeded amplifier. (c) Stokes beam energy out of the seeded second-stage, two-pass Raman amplifier as a function of second-stage pump energy.

The second issue sets the primary limit on Raman gain in the second stage. Higher order Raman processes (e.g. four-wave mixing) become important as the energy of the growing Stokes pulse approaches $\sim 10\%$ of pump energy. These competing processes deplete both first Stokes and pump energy, causing first Stokes gain to saturate well below the theoretical quantum efficiency (91.6%) of SRS. In order to manage these saturation effects, we divided the second stage amplifier into two passes. After the first pass, the first Stokes component of the incipient Raman cascade is spectrally and spatially filtered, then preferentially seeds second-pass Raman gain. In this manner, first Stokes gain saturates at a higher level than for a single pass through a crystal twice

781

as long. Figure. 2(c) shows typical Raman gain saturation for the complete two-stage Raman amplifier. Second stage gain is observed above a threshold pump energy $\sim$12 mJ. Saturation becomes significant above $\sim$16 mJ pump energy, indicating that the amplifier is approaching optimum efficiency. Uncompressed Stokes pulse energy of 6 mJ is routinely achieved with pump energy of 25 to 30 mJ, sufficient for multi-GW compressed power and 10^{16} W/cm^2 focused intensity.

Even though the CPRA pulse acquires the chirp of the main pulse, it is poorly compressed by 1200 line/mm gratings matching those in the main pulse compressor and stretcher [1] because the Stokes shift breaks the symmetry between stretching and compression mechanisms. Optimal compression of the CPRA pulse requires gratings mismatched to those in the main pulse compressor and stretcher. Based on ZEMAX simulation, we constructed a CPRA compressor using 830.8 line/mm gratings. A single-shot autocorrelator measured compressed pulse durations to be 115fs with the Ti:S and CPRA systems operating at approximately 10 nm bandwidth. See Fig. 3(a). Even at 10 nm bandwidth, this compression is significantly better than previous results with matched gratings in the 800 nm and 873 nm systems [1].

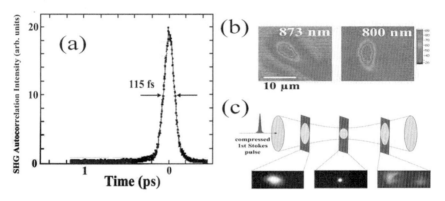

Fig. 3. (a) Measured autocorrelation trace of 1st Stokes pulse. (b) Intensity profiles of 1st Stokes and fundamental pulses at focus of f/6 parabola. (c) Schematic, top, and digital photographs, bottom, showing the incident focused Stokes beam fluorescencing on a screen (left), the laser produced spark at focus (center), and the ionization defocused beam after the focus (right). Ionization blue shifted light from the IR Stokes beam is visible.

Other notable features of the Raman-amplified output are its high contrast compared to the main pulse, making it potentially useful as a pump in high-intensity solid target experiments, and the simultaneous production of a cascade of chirped anti-Stokes (738, 685 and 639 nm) and second-Stokes (961 nm) pulses as by-products of the chirped-pulse Raman amplifier, each compressible in principle with a suitably designed grating pair, and potentially useful for synthesizing few and sub-femtosecond optical pulses.

1   N. Zhavoronkov, F. Noack, V. Petrov, V. P. Kalosha, and J. Herrmann, Opt.Lett. 26, 47-49 (2001).
2   J. T. Murray, W. L. Austin and R. C. Powell, Opt. Mat. 11, 353-371 (1999).
3   F. B. Grigsby, P. Dong and M. C. Downer, J. Opt. Soc. Am. B. 25, 346-350 (2008).

Generation of Broadband mid-infrared Pulses from an Optical Parametric Amplifier

C. Manzoni[1], D. Brida[1], G. Cirmi[1], M. Marangoni[1], S. De Silvestri[1], and G. Cerullo[1]

[1] National Laboratory for Ultrafast and Ultraintense Optical Science, INFM-CNR,
Dipartimento di Fisica, Politecnico di Milano, Piazza Leonardo da Vinci 32, 20133 Milano,
Italy.
E-mail: cristian.manzoni@polimi.it

Abstract. We generate broadband mid-IR pulses from an 800-nm-driven optical parametric amplifier in LiIO$_3$. Exploiting its broad phase-matching bandwidth around 1 μm, we produced 2-μJ idler pulses in the 3-4 μm range supporting 30-fs transform-limited duration.

The generation of few-optical-cycle light pulses in the mid-IR (3-6 μm) wavelength range is crucial for many applications ranging from time-resolved spectroscopy to high-field science. Such wavelength range overlaps several important vibrational transitions in molecules [1], as well as electronic transitions in highly correlated materials and intersubband transitions in quantum confined semiconductors [2]. In addition mid-IR pulses are expected to increase the cut-off energy for high harmonic generation. Traditionally, mid-IR pulses are produced by inter-or intra-pulse difference-frequency generation (DFG) [3,4]; mid-IR pulses have also been directly generated as idler pulses of an OPA, by tuning the signal as close as possible to the pump wavelength [5]. Up to now, due to the quasi-monochromatic seeding, mid-IR OPAs have generated only relatively narrowband pulses, with transform-limited (TL) pulsewidths longer than 50 fs.

In this work we show that several crystals with extended IR transparency, when pumped at 800 nm, display a broad phase-matching bandwidth around 1 μm, allowing the direct generation of broadband idler pulses spanning the 3-5 μm wavelength range and with duration down to two optical cycles. We experimentally demonstrate this configuration with LiIO$_3$ producing up to 2-μJ energy pulses with spectra supporting 30-fs TL duration, corresponding to approximately three optical cycles. Further advantages of our approach are energy scalability and the generation of pulses with a stable carrier-envelope phase (CEP) [6]. The basic idea underlying our configuration is to achieve broadband amplification of signal pulses at a carrier wavelength as close as possible to the pump wavelength, thus pushing the carrier wavelength of the resulting broadband idler as far as possible to the infrared.

In an OPA, the phase matching bandwidth is given to the first order by $\Delta\omega \propto 1/|\delta_{si}|$ where $\delta_{si} = 1/v_{gs} - 1/v_{gi}$ is the group velocity mismatch (GVM) between signal and idler. Figure 1(a) plots δ_{si} as a function of signal wavelength for several nonlinear optical crystals with broad mid-IR transparency range, high refractive index and zero dispersion wavelength (corresponding to a maximum of the group velocity) around 1.9-2 μm. Apart from degeneracy, these crystals exhibit group-velocity matching also between 950 and 1050 nm. It is therefore possible to amplify a broad signal bandwidth around 1 μm and produce a correspondingly broad idler in the mid-IR [7]. This prediction is confirmed by calculating the frequency-dependent parametric gain assuming monochromatic and undepleted pump (Fig. 1(b,c)): this analysis indicates that LiIO$_3$ is suitable for the generation of broadband idler pulses. In addition to broad phase-matching bandwidth, it also exhibits extended mid-IR tunability and low pump-signal GVM; this crystal has therefore been chosen for the experiments.

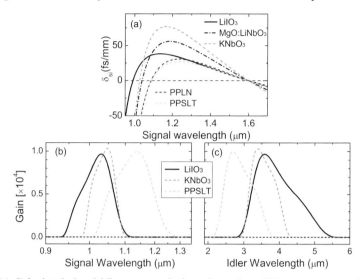

Fig. 1. (a) Calculated signal-idler group-velocity mismatch at 800 nm pump wavelength for several nonlinear optical crystals with extended mid-IR transparency. (b,c) Calculated frequency dependence of the parametric gain for 2-mm-thick 800-nm pumped LiIO₃, KNbO₃ and PPSLT crystals, assuming phase-matching at zero GVM point.

The experimental setup of the mid-IR OPA is shown in Fig. 2(a). It is pumped by 300 μJ, 50 fs, 800 nm, 1 kHz pulses from a regeneratively amplified Ti:sapphire laser. A 3-μJ fraction of the energy is focused in a 2-mm-thick sapphire plate to generate a single-filament WLC, used as a seed. The spectral portion of the WLC around 1 μm is first amplified in a non-collinear OPA (seed generation stage) and then fed into the mid-IR OPA, which generates the 3-5 μm idler in a collinear geometry. For the seed generation we used 400-nm pumped 1-mm-thick type I BBO crystal ($\theta = 29°$), which provided enough energy (1 μJ) to seed the mid-IR OPA. The mid-IR OPA used a 2-mm-thick type I LiIO₃ crystal ($\theta = 21.5°$) pumped by 200-μJ 800-nm pulses, and produced more than 8-μJ signal energy around 1-1.1 μm. After the OPA the idler beam was collimated by a germanium lens, that also filtered the pump and signal beams. The idler pulse energy was 2 μJ, in good agreement with the Manley-Rowe relations. The mid-IR spectra were measured with a grating monochromator and a lithium tantalate pyroelectric detector.

Figure 2(b) shows a sequence of idler spectra obtained by changing the pump-signal delay and slightly tuning the phase-matching angle. We obtained pulses tunable from 3 to 4.5 μm with an unprecedented combination of high energy (μJ-level) and broad bandwidth (≈30 fs TL duration). We note that the dip observed around 4.2 μm is due to atmospheric CO_2 absorption, so that even broader bandwidth could be achieved by purging the system. We could tune the system further to the IR up to 5 μm but with reduced bandwidth. The idler bandwidth is currently limited by the chirp introduced on the 1-μm seed by the refractive optics, which prevents temporal overlap of all the spectral components with the short pump pulse. Reducing this chirp by the use of reflective optics and/or stretching the pump pulses, we should

784

be able to improve the pump-seed temporal overlap and amplify broader bandwidths, so as to approach the predicted ≈ 20 fs limit.

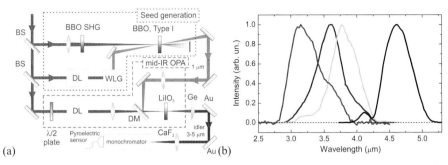

Fig. 2. (a) Experimental setup; BS, beam splitter; DL, delay line; SHG, second-harmonic generation; WLG, white light generation; DM, dichroic mirror; Au, gold mirror. (b) Idler spectra from the second stage

These pulses should be compressible to the TL by propagation in a properly chosen thickness of materials such as silicon, germanium or zinc selenide. At the highest pump intensity levels, we observed optical damage of the $LiIO_3$ crystal in the second OPA stage; we expect that more robust operation and further energy scaling should be achievable by crystals with slightly less favourable phase matching characteristics but larger nonlinear optical coefficient and higher damage threshold, such as $KNbO_3$ and PPSLT (see Fig. 1).

In conclusion, we have proposed and experimentally demonstrated the possibility of generating broadband high-energy mid-IR pulses from an OPA employing some crystals with extended IR transparency. In a $LiIO_3$ crystal we have obtained 2-μJ pulses with bandwidths supporting three-optical-cycle durations. Moreover these pulses, being generated by a DFG process between two pulses derived from the same source [6], are expected to display stable CEP.

References

1   M. L. Cowan, B. D. Bruner, N. Huse, J. R. Dwyer, B. Chugh, E. T. J. Nibbering, T. Elsaesser, R. J. D. Miller, Nature **434**, 199-202 (2005).

2   C. W. Luo, K. Reimann, M. Woerner, T. Elsaesser, R. Hey, and K. H. Ploog, Phys. Rev. Lett. **92**, 047402 (2004).

3   R. A. Kaindl, M. Wurm, K. Reimann, P. Hamm, A. M. Weiner, M. Woerner, J. Opt. Soc. Am. B **17**, 2086-2094 (2000).

4   T. Zentgraf, R. Huber, N. C. Nielsen, D. S. Chemla, and R. A. Kaindl, Opt. Express **15**, 5775-5781 (2007).

5   V. Petrov, F. Noack, and R. Stolzenberger, Appl. Opt. **36**, 1164-1172 (1997).

6   C. Manzoni, D. Polli, G. Cirmi, D. Brida, S. De Silvestri, and G. Cerullo, Appl. Phys. Lett. **90**, 171111 (2007).

7   D. Brida, C. Manzoni, G. Cirmi, M. Marangoni, S. De Silvestri, and G. Cerullo, Opt. Express **15**, 15035-15040 (2007).

Optimized 2-micron Optical Parametric Chirped Pulse Amplifier for High Harmonic Generation

Jeffrey Moses, Oliver D. Mücke, Shu-Wei Huang, Andrew Benedick, Edilson L. Falcão-Filho, Kyung Han Hong, Aleem M. Siddiqui, Jonathan R. Birge, F. Ömer Ilday and Franz X. Kärtner

Department of Electrical Engineering and Computer Science and Research Laboratory of Electronics, Massachusetts Institute of Technology, Cambridge, MA 02139, USA
E-mail: j_moses@mit.edu

Abstract. An optical parametric chirped pulse amplification system producing high-energy, few-cycle pulses at 2.0-μm wavelength for high harmonic generation is demonstrated. Simultaneous optimization of conversion efficiency, bandwidth and signal-to-noise ratio is obtained.

Since the prediction of high-yield keV-energy photon generation through high harmonic generation (HHG) with long-wavelength driver pulses [1], the development of high-energy, few-cycle, carrier-envelope phase- (CEP-) stabilized light pulse sources in the infrared has been of great importance. A 2-μm-wavelength driver system is particularly suitable because it may be feasibly built by means of optical parametric chirped pulse amplification (OPCPA) [2]. There are several advantages of a 2-μm OPCPA system: it is degenerate with the most efficient pump laser sources available to date, Yb:YAG; the amplifier and pump may be directly seeded by a single octave spanning Ti:sapphire (Ti:S) laser (the amplifier by difference frequency generation (DFG) of the laser, and the pump by the Ti:S output itself), and the pump is then automatically synchronized; and the resulting pulse is automatically CEP-stabilized [2]. Previously, groups pursued such an OPCPA system at 2 μm, but found that due to the limited seed power from the DFG process, the parasitic amplification of spontaneous parametric generation, or superfluorescence, overcomes the signal pulse at the 100-μJ level [3], below the energy at which efficient, phase-matched gas-jet HHG experiments can be performed.

In this paper we demonstrate a 2-μm OPCPA system designed carefully for superfluorescence suppression. A thorough study of amplifier performance with variation of signal pulse duration and placement of an acousto-optic programmable dispersive filter (AOPDF) before the power amplification stage has allowed us to simultaneously optimize conversion efficiency and signal bandwidth while suppressing superfluorescence to suitable levels, allowing the generation of high-energy, few-cycle, CEP-stabilized pulses at 2 μm that are suitable for HHG.

A schematic of our OPCPA system is shown in Fig. 1. An octave-spanning Ti:S laser generates passively CEP-stabilized broadband 2-μm seed pulses by DFG in a MgO-doped periodically-poled lithium niobate (MgO:PPLN) crystal (2 mm; poling period, Λ, is 13.1 μm). A Nd:YLF regenerative amplifier is injection-seeded by the 1047-nm spectral component of the oscillator after pre-amplification in a fiber amplifier, and provides 9-ps, 800-μJ pump pulses for the OPCPA stages (70 μJ, first stage; 730 μJ, second.) The seed pulses are stretched in 30-mm of bulk silicon to 5.0-ps length and preamplified in OPA1 (3-mm MgO:PPLN, Λ = 31.0 μm). The pulses are then amplified to 70 μJ in OPA2 (3-mm MgO:PPSLT, Λ = 31.4 μm) and compressed in 30 cm of near-lossless Suprasil300 glass. The infrared AOPDF

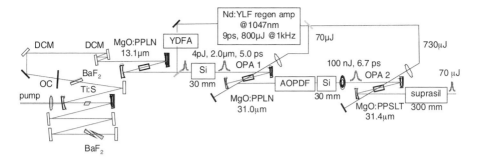

Fig. 1. Schematic of the OPCPA system

(Fastlite), placed between amplification stages, adds programmable dispersion to allow the pulse duration to reach near transform limit after the compressor. An additional 30-mm block of silicon is placed between OPA1 and OPA2 in order to set the AOPDF chirp such that its diffraction efficiency is maximized at the same time that the chirp of the seed pulse in OPA2 (6.7 ps) is optimized for amplification. An aperture is used prior to OPA2 in order to eliminate spatial chirp caused by the AOPDF and to eliminate superfluorescence in high order spatial modes.

We implement four design features to suppress superfluorescence. First, since the ratio of amplified signal to superfluorescence noise in OPCPA depends strongly on the initial seed energy [4], our stretcher-compressor scheme maximizes 2-μm seed energy at OPA1 by avoiding lossy elements in the pulse stretcher. The stretcher consists of a single block of silicon. The AOPDF is placed between pre- and power-amplification stages so its 90% transmission losses do not affect the seed energy of OPA1 and can be compensated by gain in the second amplification stage. Second, we implement two apertures after OPA1 to eliminate superfluorescence-dominated high-order spatial modes. The first is a hard aperture; the second is an effective aperture in OPA2 created by setting the pump beam width narrower than the signal beam width. Third, we use a 2-μm seed with a spectrum spanning 1.6-2.5 μm, which fully covers the amplifier gain bandwidths. Fourth, we have carefully tailored the signal pulse durations in each amplification stage to fully seed the temporal gain profiles of the amplifiers while maintaining high conversion efficiency and wide signal bandwidth. These last two design features suppress superfluorescence by maintaining high signal-to-noise contrast in the seed for all transverse coordinates of the beam for which there is significant gain. We have obtained a signal to superfluorescence level of >80:1, a worst-case estimate determined by blocking the seed of the OPCPA and measuring the noise power. This will allow us to further amplify the signal to the millijoule level with a higher-energy pump laser while maintaining suitable superfluorescence suppression and allowing efficient HHG in gases.

The results of our investigation of optimal seed pulse duration appear in Fig. 2. In order to suppress superfluorescence, the full temporal gain profile of the amplifier must be seeded to allow quenching of the gain by signal amplification before the noise can significantly deplete the pump. This is especially important at the wings of the pulse, where noise gain is higher than signal gain [5]. The result is a small loss in amplifier bandwidth, but high conversion efficiency with little degradation of signal to noise. Since the AOPDF was placed before OPA2, we were able to carefully determine the optimal power-amplifier chirp (Fig. 2). We varied the power-amplifier seed pulse linear chirp from 10,000 to 50,000 fs^2, thus varying the duration from 2 to

11 ps. The amplified bandwidth drops as chirp is increased due to the reduced gain of the high- and low- frequency wings of the signal. However, the conversion efficiency increases since the gain profile becomes more evenly seeded. The efficiency-bandwidth product curve maximum, at ~7 ps, corresponds to a duration where the balance of efficiency, bandwidth and signal to noise ratio is optimized. We set our compressor path length to 30 cm in order to dechirp the necessary 30,000 fs^2 of chirp for a 6.7-ps seed pulse in OPA2. Note, since optimum seed duration depends on the gain [5], the OPA1 (5.0 ps) and OPA2 (6.7 ps) durations have been set accordingly.

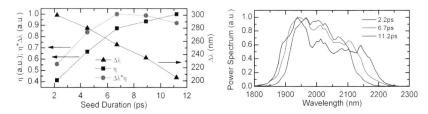

Fig. 2. (a) Conversion efficiency η, signal bandwidth Δλ, and efficiency-bandwidth product η*Δλ versus power-amplifier seed duration. (b) Amplified signal spectra corresponding to several values of seed duration in (a).

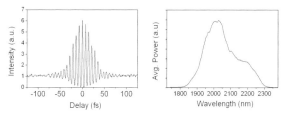

Fig. 3. (a) IAC of amplified signal pulse after compression, measuring 5 cycles (35 fs). (b) Corresponding spectrum with calculated transform limit of 27 fs.

Fig. 3(a) shows a 2nd-order interferometric autocorrelation (IAC) of the compressed pulse using two-photon absorption photocurrent in InGaAs. The linear absorptive response of the detector is still significant at >1800 nm, and there is some linear response distorting the 1:8 ratio of the IAC. The trace denotes a ~5-cycle (35-fs) compressed pulse, nearly compressed to its transform limit (27 fs). We are currently pursuing alternative diagnostics capable of measuring a 2-cycle pulse [6] and optimizing the OPCPA for a larger signal bandwidth. Previous results have already verified the CEP stability of the OPCPA process with self-CEP-stabilized seed pulses generated by DFG [3]. CEP measurements on the presented OPCPA system are also in progress.

1   A. Gordon and F. X. Kärtner, Opt. Express **13**, 2941, 2005.
2   F. X. Kärtner, F. Ö. Ilday and W. Graves, MIT Workshop on Seeded FELs, June 17th, 2004, Cambridge, MA.
3   T. Fuji, N. Ishii, C. Y. Teisset, X. Gu, T. Metzger, A. Baltuška, N. Forget, D. Kaplan, A. Galvanauskas, and F. Krausz, Opt. Lett. **31**, 1103, 2006.
4   F. Tavella, A. Marcinkevicius, and F. Krausz, New J. of Phys. **8**, 219, 2006.
5   J. Moses, C. Manzoni, S.-W. Huang, G. Cerullo, and F. X. Kärtner, Ultrafast Phenomena, Stresa, Italy, 2008.
6   J. R. Birge, R. Ell, F. X. Kärtner, Opt. Lett. **31**, 2063, 2006.

Generation of sub-20-fs, two-color deep-ultraviolet pulses by four-wave mixing through filamentation in gases

Takao Fuji, Takuya Horio, and Toshinori Suzuki

Chemical Dynamics Laboratory, RIKEN, Hirosawa 2–1, Wako, Saitama, 351–0198, Japan
  E-mail: takaofuji@riken.jp

Abstract. Generation of ultrashort pulses at 260 nm and 200 nm by four-wave mixing through filamentation in neon gas was demonstrated. Fundamental (ω_1) and second-harmonic (ω_2) pulses of 25 fs Ti:sapphire laser output were focused into neon gas, and 260 nm (ω_3) pulses were produced by a four-wave mixing process, $\omega_2 + \omega_2 - \omega_1 \rightarrow \omega_3$, through an ~15 cm filament. At the same time, 200 nm (ω_4) pulses were also genereted by a cascaded process, $\omega_3 + \omega_2 - \omega_1 \rightarrow \omega_4$, and/or by a sum frequency process, $\omega_2 + \omega_1 + \omega_1 \rightarrow \omega_4$. The both pulses were simultaneously compressed by a grating-based compressor, and characterized by a trangient grating frequency-resolved optical gating. The estimated pulse widths of the 260 nm and 200 nm pulses were 14 fs and 16.5 fs, respectively.

For many experimental studies in fundamental chemical dynamics, ultrashort pulses (~10 fs) in deep-ultraviolet (DUV, 200~300 nm) region are required since a number of small molecules have resonances in the wavelength region and the motion of important atoms (O, C, etc.) during chemical reaction is 10 fs time scale. In particular, a light source which generates synchronized twin femtosecond pulses at different wavelengths in the DUV region is very useful for pump-probe photoelectron spectroscopy [1].

One interesting method to generate ultrashort DUV pulses with different colors at the same time is to use third order cascaded processes in a gas-filled hollow waveguide [2]. By mixing the second harmonic (2ω) and the fundamental (ω) of powerful Ti:sapphire laser pulses in noble gas, one can generate 3ω, 4ω, and 5ω pulses all together with the cascaded third order nonlinear processes. However, the compression was successful only for the 3ω (267 nm) component. The compression for shorter wavelength components was too difficult because of the large nonlinear dispersion of air in the wavelength region and the power of the generated pulses (<1 μJ at 200 nm) was probably too low for precise femtosecond pulse characterization.

Recently, we proposed the use of the self-guiding effect, so called filamentation [3], of intense ultrashort pulses instead of the hollow waveguide to allow higher output energy of DUV ultrashort pulses [4]. We succeeded in generating ultrashort pulses at 260 nm and 200 nm simultaneously with more than two times larger pulse energy than that by the hollow waveguide scheme. In this contribution, we report the improvement of the system for two-color pump-probe experiment. We have succeeded in compressing the 260 nm and 200 nm pulses at the same time with a single grating compressor. Although the grating compressor had large loss (~10% throughput), enough energy for precise pulse characterization remained thanks to the large energy pulses from the filamentation system.

The experimental setup is shown in Fig. 1. The light source was based on a cryogenically cooled Ti:sapphire amplifier system which delivered 25-fs, 2.0-mJ, 775-nm pulses at 1-kHz repetition rate. The second harmonic (ω_2, 0.5 mJ) and the fundamental

(ω_1, 0.5 mJ) of the light source were gently focused by concave mirrors ($r = 2$ m) independently, then they were spatially overlapped and introduced into a stainless steel gas cell 1.5 m in length through a 1 mm thick calcium fluoride (CaF_2) brewster window. Neon was filled in the gas cell at $\sim$0.1 MPa. Bright orange-colored plasma column with $\sim$15 cm length (as is shown in the picture in Fig. 1) was seen when the delay time between the fundamental and the second harmonic pulses was optimized for generating DUV pulses.

The generated DUV spectra are shown in Fig. 2(a) and (b). The 260 nm component was generated by four-wave mixing process, $\omega_2 + \omega_2 - \omega_1 \rightarrow \omega_3$. The 200-nm component could result from $\omega_2 + \omega_1 + \omega_1 \rightarrow \omega_4$ and/or the cascaded four-wave mixing process, $\omega_3 + \omega_2 - \omega_1 \rightarrow \omega_4$. By adjusting the beam overlap carefully with monitoring the power of the 200 nm components, pulse energy of the 200 nm component reached more than 4 μJ, which is at least 4 times more than that by the hollow fiber scheme. The pulse-to-pulse energy stability (rms) of the 200 nm pulses was 2.5%.

We built a grating based compressor to compress the 260 nm and 200 nm pulses simultaneously as is shown in left-hand side of Fig. 1. We can avoid large third-order dispersion and nonlinear effect of the material by choosing a grating compressor instread of a prism compressor. The compressor basically consisted of one grating (2400 lines/mm, 250 nm blaze), two cylindrical concave mirrors ($r = 500$ mm), and two retroreflectors. In this way, we can compress the both components independently. In order to minimize the losses of the compressor, we use single-pass configuration. Although some spatial dispersion cannot be avoided with the configuration, the estimated dispersion is only 0.5 mm shift between 190 nm and 210 nm components, which is negligible with the current beam diameter, 5 mm. The total throughput of the compressor was $\sim$10%, then the energies of the compressed 260 nm and 200 nm pulses were $\sim$2.5 μJ and 0.5 μJ, respectively.

These energies were large enough for pulse charaterization with dispersion-free transient-grating frequency-resolved optical gating (TG-FROG) [5]. A 0.2-mm thick CaF_2 crystal was used for the nonlinear medium. From the measured FROG traces shown in Fig. 2(c) and (d), the pulse widths of 260 nm and 200 nm pulses were calculated to be 14 fs and 16.5 fs, respectively. The both FROG errors for the matrix size 256×256 were $\sim$0.005. To the best of our knowledge, the 16.5 fs pulse is the shortest at 200 nm. Significant higher order dispersion still remained for 200 nm pulses. It comes from nonlinear dispersion of air and the exit window on the gas cell.

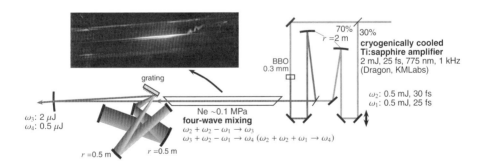

Figure 1: Schematic of the experiment.

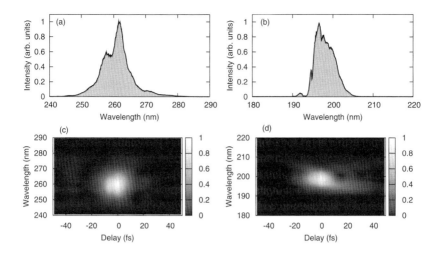

Figure 2: The spectra of the generated 260 nm (a) and 200 nm (b) pulses through filamentation in neon and TG-FROG traces of the compressed 260 nm (c) and 200 nm (d) pulses.

In summary, we have demonstrated the efficient generation of 260 nm and 200 nm ultrashort pulses by four-wave mixing processes through filamentation in neon gas. The energy of the 200 nm pulse reaches 4 μJ. The 260 nm and 200 nm pulses were, at the same time, compressed down to 14 fs and 16.5 fs, respectively, by a grating compressor. The pulses were characterized by TG-FROG. Even after the lossy grating compressor, we can expect half micro joule pulse energy at 200 nm which is sufficient for gas phase spectroscopy. The two-color DUV ultrashort pulse generation method can be a unique light source for ultra-high time-resolution photoelectron spectroscopy.

Acknowledgements. This study was supported by JSPS through a Grant-in-Aid for Young Scientists (A).

1   T. Suzuki, "Femtosecond time-resolved photoelectron imaging," Ann. Rev. Phys. Chem. **57**, 555–592 (2006).

2   L. Misoguti, S. Backus, C. G. Durfee, R. Bartels, M. M. Murnane, and H. C. Kapteyn, "Generation of broadband VUV light using third-order cascaded processes," Phys. Rev. Lett. **87**, 013,601 (2001).

3   A. Braum, G. Korn, X. Liu, D. Du, J. Squier, and G. Mourou, "Self-channeling of high-peak-power femtosecond laser pulses in air," Opt. Lett. **20**, 73–75 (1995).

4   T. Fuji, T. Horio, and T. Suzuki, "Generation of 12-fs deep-ultraviolet pulses by four-wave mixing through filamentation in neon gas," Opt. Lett. **32**, 2481–2483 (2007).

5   M. Li, J. P. Nibarger, C. Guo, and G. N. Gibson, "Dispersion-free transient-grating frequency-resolved optical gating," Appl. Opt. **38**, 5250–5253 (1999).

Efficient ultrafast four-wave optical parametric amplification in condensed bulk media

A. Dubietis, H. Valtna, G. Tamošauskas, J. Darginavičius and A. Piskarskas

Department of Quantum Electronics, Vilnius University, Vilnius, Lithuania
  E-mail: audrius.dubietis@ff.vu.lt

Abstract. Efficient four-wave optical parametric amplification in bulk fused silica is demonstrated under mJ-pumping in the ultrashort pulse regime by means of cylindrical beam focusing and non-collinear phase matching geometry without onset of filamentation and optical damage. We experimentally show that the developed four-wave optical parametric amplification scheme is able to support sub-10 fs pulses in the near infrared and in the ultraviolet spectral region.

Introduction

The discovery of the four-wave-mixing-driven optical parametric amplification dates back to the dawn of nonlinear optics [1]. However, real progress in the field by employing transparent isotropic bulk amplifying media has made a sensible breakthrough only recently. A widely used approach for the four-wave parametric amplification in the ultrashort pulse regime exploits self-guided propagation of intense pump beam in noble gas-filled capillary waveguide [2]. More recently, four-wave optical parametric amplification was demonstrated in the filamentation regime, where the signal was amplified along the intense femtosecond filament in noble gas or air [3] and has led to amplification of optical pulses with duration down to few optical-cycles [4]. These techniques combine all the essential prerequisites for efficient four-wave parametric amplification to take place – phase-matching, long interaction length maintaining high pump intensity, and broadband amplification supported by low dispersion of the gaseous media, and typically deliver pulses of $\sim 10~\mu$J energy with pumping at the mJ energy level.

In condensed bulk media, due to high material dispersion, non-degenerate four-wave interactions could be phase-matched only noncollinearly, which thus impose substantial reduction of the interaction length in tight focusing geometry. Moreover, high intensity of the applied laser field gives rise to a series of competing nonlinear effects – self- and cross-phase modulation, self-focusing, beam break-up and filamentation, and eventually optical breakdown, which occur almost at the same intensity threshold [5]. So far, typical energy of the pulses amplified via parametric four-wave processes in bulk solid state media were in the order of $\sim 1~\mu$J in the visible [6] and few hundreds of nJ in the mid-infrared [7] spectral range.

Experimental Methodology

A new route in practical implementation of efficient four-wave optical parametric amplification in condensed bulk medium (water) was suggested by the use of cylindrical focusing geometry, which allowed to fulfil noncollinear phase matching condition without reducing the interaction length [8]. Interestingly, in this configuration, catastrophic self-focusing of high-intensity elliptical pump beam is quenched by the injection of a weak seed signal. As a result, the nonlinearly coupled pump, signal and idler beams simultaneously reshape into stable 1-dimensional spatial solitons, where noncollinearly propagating signal and idler beams act as an efficient damping mechanism, which suppresses transverse instabilities. Under this setting, the setup operates reasonably below the optical damage threshold with energy conversion from pump to parametric radiation as high as 25%. It is important to note, that 1-dimensional spatial solitons are not limited in their wider transverse dimension, and therefore can be made to carry a large amount of energy without altering the propagation dynamics.

In this Contribution, we demonstrate that the four-wave optical parametric amplifier in bulk fused silica safely and efficiently operates under milijoule pumping and has broad amplification bandwidth in the near infrared and ultraviolet spectral range, which might support sub-10 fs pulses.

Results and Discussion

Figure 1 plots the phase matching curves for fused silica pumped by $\lambda_p = 1054$ nm (a) and $\lambda_p = 527$ nm (b) pulses, which correspond to fundamental and second harmonics of the amplified Nd:glass laser system. Despite different phase-matching angle range in the graphs, both curves exhibit quite similar characteristics, which suggest broadband amplification in the near infrared, around 730 nm (with $\lambda_p = 1054$ nm) and in the ultraviolet, around 340 nm (with $\lambda_p = 527$ nm).

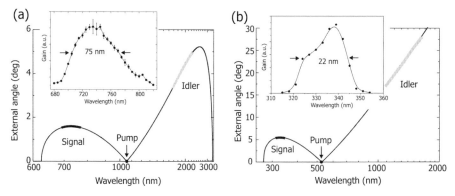

Fig.1. Phase matching curves in fused silica pumped by (a) $\lambda_p = 1054$ nm and (b) $\lambda_p = 527$ nm pulses. Expected wavelength ranges for broadband amplification of the signal are highlighted by the bold curves. Insets depict the experimentally measured gain profiles.

In the experiment, the UV-grade fused silica sample was pumped by 1 ps pulses either at fundamental ($\lambda_p = 1054$ nm) or second harmonic ($\lambda_p = 527$ nm) frequency pulse of the amplified Nd:glass laser. The tunable, narrow-band seed signal was generated by the commercial OPA system TOPAS (Light Conversion Ltd.). The pump and seed signal beams were matched by suitable telescopes and focused by $f_y = +750$ mm, $f_x = \infty$ cylindrical lens onto the input face of the fused silica sample. Keeping the external beam crossing (phase-matching) angle θ_{ext} fixed, we have varied the seed pulse wavelength in 5 nm steps. The insets of Fig. 1 depict the obtained gain profiles, as measured in the gain saturation regime: (a) $E_p = 1.8$ mJ, $E_{seed} = 10$ μJ, $\theta_{ext} = 1.59°$ with $\lambda_p = 1054$ nm and (b) $E_p = 0.8$ mJ, $E_{seed} = 10$ μJ, $\theta_{ext} = 5.51°$ with $\lambda_p = 527$.

The scanned FWHM gain bandwidths are as broad as 75 nm and 22 nm, respectively, which are able of supporting the amplification of sub-10 fs pulses.

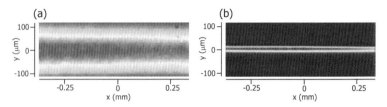

Fig. 2. Beam profiles of (a) input seed signal and (b) amplified signal at the output.

In the near infrared, the largest amplified signal ($\lambda_s = 730$ nm) energy of 180 μJ with $E_p = 1.8$ mJ in 12-mm-long fused silica sample was measured, pointing to pump-to-signal energy conversion efficiency of 9.2% [9]. The amplified signal energy, to the best of our knowledge, outreaches that produced in gaseous media so far by almost 20 times. Fig. 2 shows the central portion of the intensity profiles of the seed and amplified signal, demonstrating how it reshapes into a smooth light stripe having $\sim$ 17 μm beam width imposed by the propagation dynamics of the intense pump beam via cross-phase modulation. In the ultraviolet, several limiting factors, e.g. nonlinear absorption and group velocity mismatch come into play, nevertheless a still appreciable amplified signal ($\lambda_s = 340$ nm) energy of 50 μJ with $E_p = 0.9$ mJ pump in 5-mm-long fused silica sample was measured. We note that in our experiments the highest pump intensity inside the sample was at least 1.5 times below the optical damage threshold.

Conclusions

In conclusion, we have demonstrated an operating condition for highly efficient broad-band four-wave optical parametric amplification in isotropic transparent condensed bulk medium with Kerr nonlinearity (fused silica) in the near infrared and ultraviolet.

Acknowledgements. Authors acknowledge financial support from the Lithuanian State Science and Studies Foundation (project No. B-29/2008). H.V. acknowledges support from the Sixth EU Framework Programme, contract MEST-CF-2004-008048 (ATLAS).

1　R. L. Carman, R. Y. Chiao, and P. L. Kelley, Phys. Rev. Lett. 17, 1281, 1966.
2　C. G. Durfee III, S. Backus, M. M. Murnane, and H. C. Kapteyn, Opt. Lett. 22, 1565, 1997.
3　F. Théberge, N. Aközbek, W. Liu, A. Becker, and S. L. Chin, Phys. Rev. Lett. 97, 023904, 2006.
4　T. Fuji and T. Suzuki, Opt. Lett. 32, 3330, 2007.
5　A. Penzkofer and H. J. Lehmeier, Opt. Quantum Electron. 25, 815, 1993.
6　H. Crespo, J. T. Mendonça, and A. Dos Santos, Opt. Lett. 25, 829, 2000.
7　H.-K. Nienhuys, P. C. M. Planken, R. A. van Santen, and H. J. Bakker, Opt. Lett. 26, 1350, 2001.
8　A. Dubietis, G. Tamošauskas, P. Polesana, G. Valiulis, H. Valtna, D. Faccio, P. Di Trapani, and A. Piskarskas, Opt. Express 15, 11126, 2007.
9　H. Valtna, A. Dubietis, G. Tamošauskas, and A. Piskarskas, Opt. Lett. 33, 971, 2008.

Cascaded four-wave mixing technique for high-power few-cycle pulse generation

Helder Crespo[1] and Rosa Weigand[2]

[1] CLOQ/IFIMUP, Departamento de Física, Faculdade de Ciências, Universidade do Porto, 4169-007 Porto, Portugal
E-mail: hcrespo@fc.up.pt
[2] Departamento de Óptica, Facultad de Ciencias Físicas, Universidade Complutense de Madrid, 28040 Madrid, Spain
E-mail: weigand@fis.ucm.es

Abstract. We describe a new technique for synthesizing high-power few-cycle pulses by coherent superposition of multiple highly-nondegenerate cascaded four-wave mixing pulses. The cascaded process was characterized with broadband cross-correlation frequency-resolved optical gating. The resulting octave-spanning spectrum allows for the generation of near-single-cycle 2.2 fs visible-UV pulses without the need of complex phase control.

Introduction

Achieving efficient and reproducible generation of carrier-envelope phase stabilized near-single-cycle optical pulses in the visible and ultraviolet ranges is expected to have a strong impact on a wide number of fields, such as the study of phase-sensitive nonlinear optical and physical phenomena and the generation of isolated attosecond pulses [1]. Different techniques have been proposed for single-cycle pulse generation in the near infrared and visible ranges, from adaptive compression of a white-light supercontinuum [2] to temporal superresolution of an octave-spanning Ti:sapphire laser oscillator [3]. To our knowledge, the only experimental demonstration of single-cycle optical pulse synthesis in the visible range [4] relies on laser-driven molecular modulation in Deuterium where phase optimization of the resulting multi-line spectrum resulted in a train of 10^6 single-cycle 1.6 fs pulses separated by 11 fs (imposed by the fixed Raman sideband spacing), although it is difficult to isolate a single pulse from the train and the energy per pulse is very low.

The first demonstration of highly-nondegenerate cascaded four-wave mixing (CFWM) of femtosecond pulses in bulk media [5] showed that high-power multi-band coherent spectra can be obtained when two ultrashort laser pulses with different frequencies ω_0 and ω_1 propagate through $\chi^{(3)}$ nonlinear media, such as common glass, at an optimized crossing angle. Efficient CFWM has also been demonstrated in other spectral ranges (see, e.g., [6]), and can occur for quite different conditions and materials, provided that the nonlinear medium is transparent to both pump and generated wavelengths, and the necessary phase-matching conditions are met over a sufficiently broad range. The frequencies of the cascaded beams are given by $\omega_n = \omega_0 + n\Delta\omega$, where n is the beam order, $\Delta\omega = \omega_1 - \omega_0$ is the beat frequency, and $n > 1$ ($n < 0$) denotes frequency upconverted (downconverted) pulses. The central frequency of the total spectrum and the separation between sidebands can be adjusted by tuning the pump frequencies. The analytical solution for CFWM assuming perfect phase-matching and constant coupling coefficients [7] is formally equivalent to that obtained for molecular modulation, and it can be shown that the sign of chirp in the total CFWM field depends on the initial phase difference between pumps.

Experimental setup and results

The experimental setup is shown in Fig. 1(a) and includes generation of the CFWM beams, pulse synthesis by beam recombination, and diagnostics. In this proof-of-principle experiment, like in previous work [5], two synchronized visible laser pulses from a dual-wavelength (615 and 563 nm) 10 Hz amplifier were used as pumps. The noncollinear horizontally-polarized orange and green pulses (90 fs, 32 μJ and 45 fs, 38 μJ, respectively) are focused in a 150-μm-thick fused silica slide FS1, producing 2 frequency downconverted pulses and a fan of 21 upconverted pulses (with wavelengths down to 200 nm) as shown in Fig. 1(b). The total energy in the cascaded beams is ~ 6 μJ (10% efficiency). The corresponding 2-octave spectrum measured with a calibrated fiber-coupled spectrometer (250-1100 nm) is shown in Fig. 1(c).

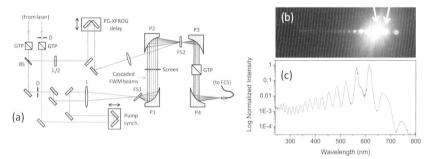

Fig. 1. (a) Experimental setup: GTP, Glan-Thomson Polarizers; D, diaphragms; BS, ultrafast beamsplitter; P1-P4: Al-coated off-axis parabolic mirrors; FS1-FS2: fused-silica slides; FCS, fiber-coupled spectrometer. (b) Image of the generated cascade projected onto a white screen and (c) corresponding measured spectrum. The white arrows denote the pump beams.

Two Al-coated off-axis parabolic mirrors P1 and P2 collimate and focus the angularly separated beams in a second slide FS2; the residual pumps are attenuated with an apertured screen. Simultaneous phase-matching and ensures that the CFWM pulses are phase-locked; the short air path between the two slides provides a small amount of second-order dispersion while minimizing higher-order dispersion, and the λ/8 mirror quality minimizes phase distortions. Hence a train of white-light ultrashort pulses is expected to be synthesized at the focal plane of P2. To characterize the pulses, we used polarization gating induced in slide FS2 by a portion of the orange pump. A second set of parabolic mirrors P3 and P4 sends the gated pulses through a Glan-Thompson analyzer and into the fiber-coupled spectrometer. Polarization gating cross-correlation frequency resolved optical gating (PG XFROG) traces were obtained by averaging 10 gated spectra for each time delay [Fig. 2(a)].

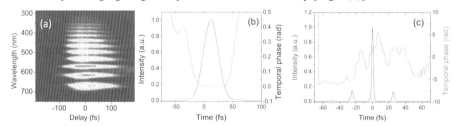

Fig. 2. (a) Measured PG XFROG trace. (b) Retrieved first frequency upshifted pulse (30.6 fs FWHM duration; 531 nm center wavelength). (c) Synthesized pulse in the time domain.

The measured traces reveal multiply synchronized CFWM pulses. Due to the practically instantaneous nature of the nonlinear response and the low dispersion of the experimental setup, each CFWM order is also near-transform-limited. Since the relative phase between the pumps is not stabilized, the synthesized pulses will still be randomly time-shifted and their duration will change from shot to shot. This, together with the use of a long gate pulse, results in smearing of fine temporal structure in the measured traces. The complete electric field of each of the 15 gated cascaded orders can nevertheless be retrieved unambiguously, as illustrated in Fig. 2(b) for the first frequency upconverted order. By assuming zero relative phase between each of the generated CFWM pulses (equivalent to introducing small, fixed time delays between pulses) we obtain a synthesized field composed of a main 2.2 fs pulse - less than 1.3 optical cycles at the center wavelength of 513 nm - and 2 smaller side pulses at a distance of 23 fs (the pump beat period), as shown in Fig. 2(c). The energy in the main pulse is ~ 10 μJ, corresponding to an instantaneous power of ~ 4 GW.

Conclusions

A technique to Fourier-synthesize high-power near-single-cycle pulses from multi-band coherent spectra in the visible-UV generated through cascaded four-wave mixing was presented. Broadband PG XFROG measurements show the possibility of synthesizing 2.2 fs 1.3-cycle pulses with multi-gigawatt powers without significant manipulation of the intermediate cascaded beams. The use of more stable pump pulses and shorter (~ 25 fs) gate pulses provided by a Ti:sapphire laser amplifier coupled to an optical parametric amplifier or hollow-fiber compressor is expected to significantly improve the generation and measurement of near-single-cycle pulse trains and isolated single-cycle pulses using this technique.

Acknowledgements. This work was partly supported by the Access to Research Infrastructures activity in the Sixth Framework Programme of the EU (contract RII3-CT-2003-506350, Laserlab Europe). The authors gratefully acknowledge J. Etchepare and G. Mourou for fruitful discussions, and A. dos Santos for technical support.

1   G. Sansone, E. Benedetti, F. Calegari, C. Vozzi, L. Avaldi, R. Flammini, L. Poletto, P. Villoresi, C. Altucci, R. Velotta, S. Stagira, S. De Silvestri, M. Nisoli, "Isolated single-cycle attosecond pulses," Science 314, 443 (2006).

2   E. Matsubara, K. Yamane, T. Sekikawa, and M Yamashita, "Generation of 2.6 fs optical pulses using induced-phase modulation in a gas-filled hollow fiber," J. Opt. Soc. Am. B 24, 985 (2007).

3   T. Binhammer, E. Rittweger, U. Morgner, R. Ell, and F. X. Kärtner, "Spectral phase control and temporal superresolution toward the single-cycle pulse," Opt. Lett. 31, 1552 (2006).

4   M. Y. Schverdin, D. R. Walker, D. D. Yavuz, G. Y. Yin, and S. E. Harris, "Generation of a single-cycle optical pulse," Phys. Rev. Lett. 94, 033904 (2005).

5   H. Crespo, J. T. Mendonça, and A. Dos Santos, "Cascaded highly nondegenerate four-wave mixing phenomenon in transparent isotropic condensed media," Opt. Lett. 25, 829 (2000).

6   L. Misoguti, S. Backus, C. G. Durfee, R. Bartels, M. M. Murnane, and H. C. Kapteyn, "Generation of broadband VUV light using third-order cascaded processes," Phys. Rev. Lett. 87, 013601 (2001).

7   J. T. Mendonça, H. Crespo, and A. Guerreiro, "A new method for high harmonic generation by cascaded four-wave mixing," Opt. Comm. 188, 383 (2001).

2 MHz repetition rate - 15 fs fiber amplifier pumped optical parametric amplifier

Steffen Hädrich[1], Jan Rothhardt[1], Fabian Röser[1], Damian N. Schimpf[1], Jens Limpert[1], and Andreas Tünnermann[1,2]

[1] Friedrich Schiller Universitt Jena, Institute of Applied Physics, Albert-Einstein-Str. 15, 07745 Jena, Germany

[2] Fraunhofer Institute for Applied Optics and Precision Engineering, Albert-Einstein-Str. 7, 07745 Jena, Germany
E-mail: haedrich@iap.uni-jena.de

Abstract. An optical parametric amplifier pumped by a fiber amplifier producing ultrashort pulses with durations of 15.6 fs at 2 MHz repetition rate is presented together with scaling considerations to tens of μJ pulse energy.

Introduction

The availability of ultrashort and high peak power optical pulses has driven breathtaking advances in applications ranging from industrial to fundamental science. Unfortunately, a number of ultrafast processes initiated by such pulses are characterized by a low probability or a low conversion efficiency. Hence, detection systems comprise very sophisticated and sensitive apparatus precluding these from real world applications. An increase of few orders of magnitude in the repetition rate could dramatically decrease the necessary measurement times. It is well known that fiber laser systems are average power scalable due to the fiber design itself, and femtosecond fiber amplification systems have been demonstrated with average powers above 100 W and pulse energies up to 1 mJ [1]. Though, their gain bandwidth does not support pulse durations of few 10 fs [2]. On the other hand optical parametric amplifiers (OPA) offer an enormous amplification bandwidth [3], up to 200 THz in a non-collinear configuration (NOPA), and are inherently immune against thermo-optical problems due to the fulfilled energy conservation during the nonlinear amplification. Furthermore, a high gain can be achieved in just few millimeter long crystals, therefore, the B-integral (accumulated nonlinear phase) is negligible. Based on these facts, OPA is a promising way in order to generate high peak-power pulses at high repetition rates. Our approach involves the transfer of the high pulse energy, at high average power, of a 1 μm laser source to pulses of significantly shorter pulse durations via parametric amplification.

Experiment and Results

The experimental setup is shown schematically in fig. 1. A cavity dumped Ti:Sapphire oscillator providing 25 fs , 15 nJ pulses at 810 nm wavelength is used to seed the NOPA and the fiber based amplification stage as well. To generate a signal for the fiber amplification stage pulses with an energy of 1 nJ are coupled into a photonic crystal fiber with a zero dispersion wavelength of 975 nm providing soliton generation. The coupled pulse energy is carefully aligned to generate a single soliton at 1030 nm center wavelength and a few pJ of pulse energy and a spectral bandwidth of more than 30 nm. These pulses are pre-amplified in a double pass 6 μm core and a single pass 10 μm core Yb doped fiber amplifier to 360 mW average power. Approximately 40 m of delay fiber are used within the first double-pass amplifier to ensure temporal overlap with the next signal pulse in the NOPA (at 2 MHz repetition rate).

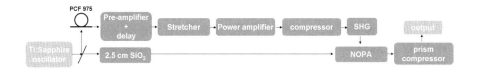

Fig. 1. Experimental setup of the fiber amplifier pumped NOPA.

Before further amplification, the pulses are stretched to about 320 ps and spectrally cut to 6 nm bandwidth using a grating-stretcher (1740 lines/mm). The power amplifier is operated in a single-pass configuration using a polarizing-large-mode-area double-clad photonic crystal fiber. After amplification the pulses are compressed by a grating pair (1740 lines/mm) to a pulse duration of 640 fs with a pulse energy of 10 μJ. The infrared pulses are frequency doubled in a 2 mm critical phase matched LBO crystal with 41 % conversion efficiency, resulting in a 4.1 μJ, 515 nm pump source for the NOPA. With a pump-signal angle of 2.6° broadband phase matching and therefore broadband amplification can be obtained. The Ti:Sapphire laser pulses are parametrically amplified either directly or after additional spectral broadening. Direct amplification leads to 500 nJ pulses which are compressed in a simple fused silica prism compressor (efficiency 90 %) to high quality pulse with 20.1 fs width and a spectral bandwidth of 59 nm resulting in a peak power of 20 MW.

Additional spectral broadening in a 1.5 cm photonic crystal fiber results in an increased bandwidth of 124 nm (fig. 2). Parametric amplification of these pulses results in 300 nJ pulse energy and a pulses duration of 15.6 fs (fig. 2), which is only slightly above the transform limit (13.2 fs).

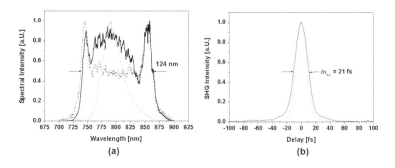

Fig. 2. (a) Normalized spectrum of the Ti:Sapphire oscillator (gray) , the spectrally broadened pulse (dotted) and amplified spectrum (black). (b) Autocorrelation trace of the compressed pulses (black) and fourier transformation of the measured spectrum (dotted).

Scaling Considerations

Recent demonstration of a mJ fiber-chirped-pulse amplification system [1] has proven the capability of scaling the output of this NOPA experiment to much higher pulse energies at moderate repetition rates. First proof of principle experiments revealed that the necessary pump pulse generation for the parametric amplification can be achieved by frequency doubling the output of the fiber CPA system at repetition rates as high as 200 kHz with a pulse energy of 205 μJ (41 W) and a pulse duration of 720 fs (fig. 3) .

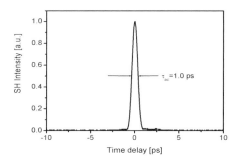

Fig. 3. Autocorrelation trace of the SHG signal at 205 μJ pulse energy and 200 kHz repetition rate.

Using the same experimental configuration as shown in fig.1 combined with the high energy fiber CPA system offers the possibility to generate ultrashort pulses (20 fs) and pulse energies of tens of μJ at a repetition rate of 200 kHz.

Conclusions

A new approach for parametric amplification that combines the advantages of a broadband Ti:Sapphire oscillator and a high average power Yb-doped-fiber-amplifier system is presented. Using the frequency doubled output pulses of a fiber amplifier with up to 4.1 μJ pulse energy, efficient parametric amplification of the Ti:Sapphire oscillator pulses is possible. Output energies up to 500 nJ were achieved with a 5 mm long BBO amplifier crystal. The shortest pulse duration of 15.6 fs was obtained with additional spectral broadening in a photonic crystal fiber. The scalability of this approach to pump pulse energies of 205 μJ at 200 kHz repetition rate has been proven with the help of a state-of-the-art fiber-chirped-pulse amplification system and is currently under investigation experimentally.

Acknowledgements. This work has been partly funded by the German Federal Ministry of Education and Research (BMBF) with project 03ZIK455 'onCOOPtics'.

1 F. Röser, T. Eidam, J. Rothhardt, O. Schmidt, D.N. Schimpf, J. Limpert, and A. Tünnermann, Opt. Lett. 32, 3495-3497, 2007.
2 R. Paschotta, J. Nilsson, A. Tropper, and D. Hanna, IEEE J. Quantum Electron . 33, 1049-1056, 1997.
3 G. Cerullo and S. Silvestri, Review of Scientific Instruments 74, 1, 2003.

Octave-wide tunable NOPA pulses at up to 2 MHz repetition rate

Christian Homann, Christian Schriever, Peter Baum, and Eberhard Riedle

LS für BioMolekulare Optik, Ludwig-Maximilians-Universität München, Oettingenstraße 67, 80538 München, Germany
E-mail: christian.homann@physik.uni-muenchen.de

Abstract: Based on noncollinear parametric amplification, we demonstrate frequency conversion of the 230 fs pulses of a high repetition rate ytterbium-doped fiber amplifier system to octave wide tunable femtosecond pulses with down to 20 fs duration.

Complete spectral coverage for ultrafast spectroscopy

For numerous applications in ultrafast spectroscopy and nonlinear microscopy, spectrally tunable sources of intense femtosecond laser pulses are needed. Fiber amplifiers offer high peak power together with a high repetition rate, high energy stability and excellent beam quality. However, they so far only provide pulse durations above 100 fs and lack flexibility in output wavelength. Therefore a device is desirable that can convert the fiber amplifier pulses to an as wide as possible wavelength range, ideally in combination with temporal shortening to the 20 fs regime.

Here we present a noncollinear optical parametric amplifier (NOPA [1]) that converts the 1035 nm, 230 fs output pulses of an ytterbium-doped fiber amplifier laser system to pulses with continuous tunability from 440 to 990 nm [2]. Pulse durations of down to 19.8 fs and pulse-to-pulse energy stability of 1.3% (rms) at up to 2 MHz repetition rate are demonstrated. The NOPA is pumped with UV pulses from the third harmonic of the fiber amplifier at 345 nm and is seeded by a spectrally smooth supercontinuum generated in sapphire. The leftover second harmonic light from the frequency tripling process is used to pump an additional independently tunable NOPA with a tuning range of 600 to 970 nm. Together the two NOPAs provide powerful sources for two-color pump-probe spectroscopy at MHz repetition rates.

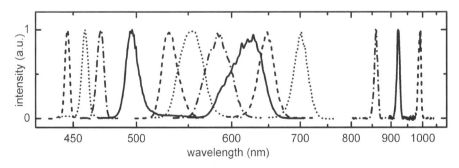

Fig. 1. Typical output spectra of the 345 nm pumped noncollinear optical parametric amplifier, showing the more than octave wide continuous tuning range

Interference experiments show that the two NOPAs have a precisely locked relative phase, despite of being pumped by different harmonics with random phase jitter. This directly proves that parametric amplification preserves the phase of the seed light.

Octave-wide tunability with femtosecond UV pumping

As primary pump source we use a commercially available ytterbium-doped fiber oscillator/amplifier system (IMPULSE; Clark-MXR, Inc.) that delivers 10 µJ output pulses with 230 fs duration at 1035 nm and at a variable repetition rate between 200 kHz and 2 MHz. Pulses with an energy of 1.5 µJ are split off and focused into a 4 mm thick rotating sapphire disc for supercontinuum generation. The rotation is needed to prevent accumulative damage at high repetition rates. The remaining part of the output pulses is used to generate the NOPA pump pulses. We use type I frequency doubling and subsequent type II sum-frequency mixing of the fundamental and second harmonic pulses in specially selected BBO crystals to generate the third harmonic at 345 nm by collinear propagation only. Due to the group velocity mismatch, the second harmonic pulses leave the first type-I BBO crystal later than the transmitted fundamental pulses. With the particular choice of type-II sum frequency mixing in the second BBO crystal, the sign of the group velocity mismatch is reversed and the two pulses restore perfect temporal overlap inside the crystal for efficient conversion (total 15%).

The third harmonic and the remaining second harmonic are each separated from the left over fundamental with a dichroic mirror and used to pump two independently tunable NOPA units. Fig. 1 shows output spectra from the 345 nm pumped NOPA unit. The output pulses are continuously tunable over more than one optical octave (440 to 990 nm). Autocorrelation measurements yield pulse durations of about 20-30 fs between approximately 500 and 700 nm. The smooth Gaussian shaped spectra allow for pulse shapes without temporal satellites, as observed in the autocorrelation traces.

Towards the short and long wavelength side of the tuning range, the Fourier-Limits of the measured spectra increase. For short wavelengths, this is the result of the strong dispersion of the seed light and can be overcome in future experiments by precompression of the seed light. A calculation of the phase matching bandwidth lets us expect 13 fs blue pulses. On the long wavelength side, above the NOPA's degeneracy point at 690 nm, the group velocity mismatch between signal and idler cannot be compensated for by noncollinearity. However, the secondary NOPA unit operates with visible pump pulses and renders pulse durations of ~10-30 fs all the way up to 970 nm [3]. The output pulse energies are in the range of 150 nJ for the 345 nm pumped NOPA, best at the center of its tuning range. Simultaneously, the second harmonic pumped NOPA delivers pulses with up to 250 nJ energy. These pulse energies are well sufficient for frequency doubling and thereby extending the available spectral range to the ultraviolet region. The demonstrated octave-wide tuning range together with a single additional nonlinear conversion will lead to a gapless coverage from below 250 nm to nearly 1 µm. The high output stability of our setup is evident in pulse-to-pulse energy fluctuations of less than 1.3% rms, measured by recording sets of 100.000 subsequent single pulse energies.

Investigation of phase dependencies in optical parametric amplification

The overlapping amplification regions of the two differently pumped NOPA units allow for an instructive interference experiment that renders direct information about the phase dependencies in the parametric amplification process. The two NOPA units, seeded by the same supercontinuum, are tuned to the same center wavelength of 720 nm and brought to interference with a small angle on a distant screen (see Fig. 2a). In earlier experiments it was shown that such an arrangement leads to stable interference fringes, when the NOPAs were seeded with the same supercontinuum and when pump pulses with similar phase fluctuations were applied [4]. In contrast, the presented experiment involves pump pulses that are derived from different harmonics of the primary fiber laser system, which is not phase stabilized. The second and third harmonic pulses therefore have carrier-envelope phases with twice or threefold the original phase fluctuations, which makes their relative phase jitter at random.

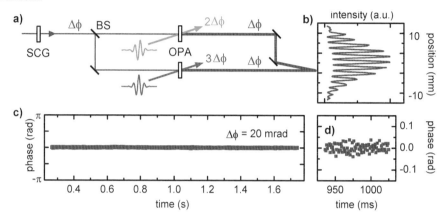

Fig. 2. Interference experiment to investigate phase dependencies in optical parametric amplification (OPA). a) Experimental setup. SCG: supercontinuum generation, BS: beam splitter. b) Measured spatial interference pattern. c) and d) Phase of the interference pattern over time

Nevertheless we observe a very stable interference pattern of the NOPA outputs (Fig. 2b), which, as can be seen in Fig. 2c) and d), shows very low residual phase fluctuations of less than 20 mrad rms (0.1-1000 Hz). This demonstrates that optical parametric amplification, despite possible saturation effects, preserves the phase of the seed light independently of phase fluctuations of the pump pulses to an extreme precision. In combination with the ultrashort pulse durations, the octave-wide tunability, and the high repetition rate, the presented system has potential to significantly advance nonlinear microscopy and ultrafast spectroscopic applications.

1   T. Wilhelm, J. Piel, and E. Riedle, "Sub-20-fs pulses tunable across the visible from a blue pumped single pass noncollinear parametric converter", Opt. Lett. **22**, 1494, 1997.

2   C. Homann, C. Schriever, P. Baum, and E. Riedle, "Octave wide tunable UV-pumped NOPA: pulses down to 20 fs at 0.5 MHz repetition rate", Opt. Express **16**, 5746, 2008.

3   C. Schriever, S. Lochbrunner, P. Krok, and E. Riedle, "Tunable pulses from below 300 to 970 nm with durations down to 14 fs based on a 2 MHz ytterbium-doped fiber system", Opt. Lett. **33**, 1, 2008.

4   P. Baum, S. Lochbrunner, J. Piel, and E. Riedle, "Phase-coherent generation of tunable visible femtosecond pulses", Opt. Lett. **28**, 185, 2003.

Asymptotic pulse shapes and pulse self-compression in femtosecond filaments

Carsten Krüger[1,2], Ayhan Demircan[1], Stefan Skupin[3], Gero Stibenz[2], Nickolai Zhavoronkov[2], and Günter Steinmeyer[2]

[1] Weierstraß-Institut für Angewandte Analysis und Stochastik, Mohrenstr. 39, 10117 Berlin,Germany, e-mail: krueger@wias-berlin.de
[2] Max-Born-Institut, Max-Born-Straße 2a, 12489 Berlin, Germany, e-mail: steinmey@mbi-berlin.de
[3] Max-Planck-Institut für Physik komplexer Systeme, Nöthnitzer Str. 38, 01187 Dresden, Germany, e-mail: skupin@pks.mpg.de

Abstract. The balance of Kerr-type and plasma-mediated self-amplitude modulations can give rise to self-stabilizing asymptotic pulse shapes in filament propagation. These soliton-like solutions resemble experimental data and constitute the major mechanism for self-compression in femtosecond filaments.

Self-guided propagation of femtosecond pulses in filaments has found novel applications, including detection of atmospheric pollution, terahertz generation [1], and few-femtosecond pulse generation in the VUV [2]. One of the most surprising effects, however, is the self-compression of millijoule pulses in the filamentary channel [3]. Using input pulses of 2.6 mJ input energy and 45 fs duration, a sixfold compression down to 7.3 fs duration has been achieved in argon, without any need for external dispersion compensation schemes, see Fig. 1. Such a source is highly interesting for applications in high-field physics, because it also removes energy constraints imposed by guiding fibers as they are used in hollow fiber compressors. Consequently, filament self-compression allows for the tightest concentration of optical energy at kHz repetition rates. Elaborate numerical models have been developed for an analysis of the self-compression scenario [4], and these models reproduce most experimental findings very well. However, the exact origin of stable self-compressing light bullets is still not completely understood.

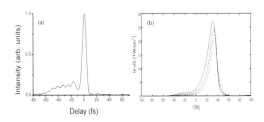

Fig. 1. (a) Experimentally measured pulse shapes generated by self-compression in an argon filament. (b) Numerically simulated sequence of pulse shapes illustrating the pulse shaping mechanism during propagation in the filament.

It is rather undisputed that a balance between Kerr-type self-focusing and plasma-induced self-defocusing causes self-stabilization of the beam profile, giving rise to a stable beam profile with a typical diameter of 200 microns, the longitudinal extension of which can easily exceed the confocal parameter by a factor 5 to 10. Here we introduce an unconditionally stable pulse shape arising from a similar balance along the direction of propagation. The resulting pulses display the same characteristic asymmetry as observed in the experiments, with a slowly rising edge and a steep trailing edge. These temporal pulse profiles are also observed in numerical simulations, see Fig. 1(b). The balance of nonlinear effects can compensate for small perturbations, making these pulse shapes an asymptotic solution in filamentary propagation. Similar to solitons in nonlinear fiber optics, such pulse shapes automatically appear once a certain threshold is reached. Excess energy is stripped off into the spatial reservoir surrounding the filament, similar to the transfer into the temporal continuum in fiber soliton propagation. It is important to understand that the formation of a stable filament is a condition that requires a balance of focusing and defocusing effects in every temporal point of a propagating pulse. Kerr-type self-focusing is an instantaneous effect in noble gases. Plasma-induced defocusing, however, will monotonically increase over the pulse as recombination of electrons occurs on much longer time scales than the pulse duration. The balancing condition

$$n_2 I(t) = \eta_0 \frac{q_e^2}{2 m_e \varepsilon_0 \omega_0^2} \int_{-\infty}^{t} w(I(t'))dt' \qquad (1)$$

therefore gives rise to characteristically asymmetric pulse shapes. Here n_2 is the nonlinear refractive index, η_0 the number density of atoms, ε_0 the dielectric constant, ω_0 the angular frequency of the light, and w the ionization rate, e.g., as calculated by the Ammosov-Delone-Krainov (ADK) theory. q_e and m_e denote electron charge and mass, respectively. Typical solutions of Eq. (1) are shown in Fig. 2. Solution of Eq. (1) can be simplified by replacing the ADK ionization rate by a simple power law $w = \sigma_{N^*} I^{N^*}$ with an effective nonlinearity N^* ranging from 8 to 9 and a Keldysh parameter σ_{N^*}. With these simplifications, Eq. (1) can be solved analytically, yielding solutions of the type $I(t) \propto (-t)^{1/(1-N^*)}$. Quite clearly, these root-like pulse shapes reflect the balance between an instantaneous and a non-instantaneous nonlinear optical effect, with a slowly varying pedestal-like rising edge and an indefinitely steep falling edge, see Fig. 2. Other than Schrödinger solitons in nonlinear fiber optics, stable pulse shapes in filaments are not localized, may exhibit a pole, and show diverging energy.

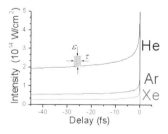

Fig. 2. Characteristically asymmetric asymptotic pulse shapes calculated from Eq. (1) using ADK-theory. Shown are the cases of helium, argon, and xenon. The gray shaded area illustrates the definitions of τ and ε, describing the perturbation used for the stability analysis. The balance of nonlinear optical effects has to transfer the energy in the gray shaded area into the reservoir for any duration lasting longer than the optical cycle.

Despite their awkward mathematical properties, however, the root-like solutions of Eq. (1) are locally stable. Phenomenologically, this is quite easy to understand, as any positive deviation from the balancing solution $I(t)$ will increase the number of electrons generated, cf. Fig. 2. As the ionization rate w is highly nonlinear with I, the resulting additional defocusing will easily outweigh the increased self-focusing effect, which causes an effective transfer of radiation into the reservoir and restores the balancing condition. The same argument holds for a negative deviation as the threshold-like intensity dependence of plasma generation causes stalling of defocusing effects. With strongly reduced plasma defocusing, energy in the filament will reconcentrate, i.e., compensate the original perturbation. Mathematically, this self-restoration requires

$$\frac{2n_2 m_e \varepsilon_0 \omega_0^2}{\eta_0 q_e^2} < \tau \frac{\partial w}{\partial I} \approx N^* \sigma_{N^*} I^{N^* - 1} \tag{2}$$

where τ is the duration of the perturbation. Demanding that any distortion of $I(t)$ be restored within the duration of an optical cycle $2\pi/\omega_0$, one finds that this condition is met for effective nonlinearities $N^* < 9$, i.e., for all noble gases at 800 nm. This means that Ti:sapphire amplifiers can unconditionally benefit from self-compression, if only the critical power for self-focusing is exceeded within a significant portion of the temporal profile and provided that filamentary propagation lasts long enough for shaping of these pulses. This explains why early experiments did not see self-compression as the observed filament lengths did not exceed the confocal range by much [5]. Our considerations indicate that Eq. (2) is also fulfilled for wavelengths in the ultraviolet and farther in the infrared, which may explain the remarkably short self-compressed 9.7 fs pulses observed in a filament at 290 nm [2] and similar observations at 2 μm [6]. Pulse self-compression in a filament is therefore not restricted to Ti:sapphire wavelengths and the use of argon gas as a medium. As pulse compression is otherwise very difficult in the ultraviolet, filament self-compression may be the method of choice for producing sub-10-fs pulses in otherwise difficult to access spectral regions. In conclusion, we introduced asymptotic pulse shapes that explain the self-compression in the leading edge of a pulse propagating through a filament above the critical power for self-focusing. Combined with the action of self-steepening and plasma defocusing in the trailing part, stable pulses with characteristic asymmetry can originate from this scenario. Of course, the nonlinear optical effects inside the filament are involved, and other effects may modify the resulting pulse shapes, such that no exact correspondence to the predicted root-like shapes is anticipated. Still, the resemblance between experimental data and the solutions of Eq. (1) is very close. Moreover, filament self-compression appears to be an extremely interesting route for the generation of few-cycle pulses in the blue and ultraviolet, opening a completely new perspective for a region where conventional compression schemes are hard-pressed.

1   C. D'Amico, A. Houard, M. Franco, B. Prade, A. Mysyrowicz, Opt. Express **15**, 15274 (2007).
2   S. A. Trushin, K. Kosma, W. Fuß, W. E. Schmid, Opt. Lett. **32**, 2432 (2007).
3   G. Stibenz, N. Zhavoronkov, G. Steinmeyer, Opt. Lett. **31**, 274 (2006).
4   S. Skupin, G. Stibenz, L. Bergé, F. Lederer, T. Sokollik, M. Schnürer, N. Zhavoronkov, G. Steinmeyer, Phys. Rev. E **74**, 056604 (2006).
5   C. P. Hauri, W. Kornelis, F. W. Helbing, A. Heinrich, A. Couairon, A. Mysyrowicz, J. Biegert, U. Keller, Appl. Phys. B **79**, 673 (2004).
6   C. P. Hauri, R. B. Lopez-Martens, C. I. Blaga, K. D. Schultz, J. Cryan, R. Chirla, P. Colosimo, G. Doumy, A. M. March, C. Roedig, E. Sistrunk, J. Tate, J. Wheeler, L. F. DiMauro, and E. P. Power, Opt. Lett. **32**, 868 (2007).

Efficient and Highly Coherent Extreme-Ultraviolet High-Harmonic Source

Sven Teichmann[1], Bo Chen[2], Jeffrey Davis[1], Peter Hannaford[1] and Lap Van Dao[1]

[1]ARC Centre of Excellence for Coherent X-Ray Science and Centre for Atom Optics and Ultrafast Spectroscopy, Swinburne University of Technology, Melbourne, Australia 3122
E-mail: steichmann@swin.edu.au
[2]ARC Centre of Excellence for Coherent X-Ray Science and School of Physics, The University of Melbourne, Parkville, Victoria 3052, Australia

Abstract. We report on a femtosecond laser-based high-harmonic generation argon-cell source that efficiently delivers two to seven highly coherent and Gaussian beam-like beam in wavelength range from 26 to 43 nm.

Introduction

Table-top femtosecond-laser high-harmonic sources complement large installations such as synchrotrons or x-ray free-electron lasers by providing spatially and temporally coherent ultrashort pulses of extreme ultraviolet radiation and soft x-rays [1]. Although increasing the brightness of high-harmonic sources remains a major challenge, these sources are highly flexible and allow tailoring of the resulting characteristic output for technological and biophysical research as well as applications in atomic and molecular spectroscopy, condensed matter physics, imaging on a (sub-) nanometer-scale, and plasma physics.

Considerable progress in the development of high-harmonic sources and on-going fundamental research has recently led to rapid developments in applications of high-harmonic radiation. Depending on the application, the effects of the experimental parameters which have a critical impact on the harmonic yield have to be known and, eventually, controlled. We have carried out systematic studies of the conversion efficiency, the spatial beam profile, the degree of spatial coherence and the spectral characteristics of femtosecond laser-based high-harmonic generation of a small band of harmonics, ranging from 26 to 43 nm, in an argon gas-cell geometry.

Experimental Methods

Our studies were performed by varying the key parameters for microscopic and macroscopic phase-matching: (i) aperturing of the driving laser beam, (ii) adjustment of the pressure inside the gas cell, and (iii) adjustment of the position of the laser focus with respect to the exit plane of the semi-infinite gas cell. These studies follow on from the research and preliminary results presented in [2]. (i) When aperturing of the driver beam is performed, effects of a strong interplay of many physical quantities leading to phase-matched radiation have to be taken into account regarding the generation process itself as well as propagation in the generating medium [3]. (ii) By properly adjusting the pressure of the argon gas cell and adjusting the focussed beam intensity by altering the diameter of the driving beam phase-matching can be achieved; i.e., the positive generating material dispersion is balanced by the negative plasma dispersion. If phase-matching cannot be achieved, the harmonic output can then be optimized by tailoring the single atom response under macroscopic phase-matching conditions [4]. (iii) By positioning the laser focus carefully with respect to the generating medium, the Gouy-phase shift due to focussing can be exploited to generate harmonics with almost the same phase over an extended propagation length [5]. All three parameters strongly affect the harmonic beam with regard to its flux, its spatial, spectral and temporal domain, and its degree of spatial coherence.

In our experimental setup high-harmonic generation is driven by the output of a 30 fs, 1 kHz Ti:sapphire multi-stage-amplification laser system operating at a centre wavelength of 810 nm and providing energies of up to 6.5 mJ per pulse. The 12 or 17 mm diameter ($1/e^2$) laser beam is focussed by a 50 cm focal length lens into a semi-infinite argon-gas cell that is normally operated at a pressure between 5 and 50 Torr. The harmonic beam is separated from the fundamental beam by at least one 250 nm thick aluminium filter. The flexible setup allows the detection and usage of the harmonic beam directly, after application of appropriate spatial and spectral filters, or after insertion of a grazing incidence spectrometer of spectral range 4.4 - 85 nm into the beam path.

Results and Discussion

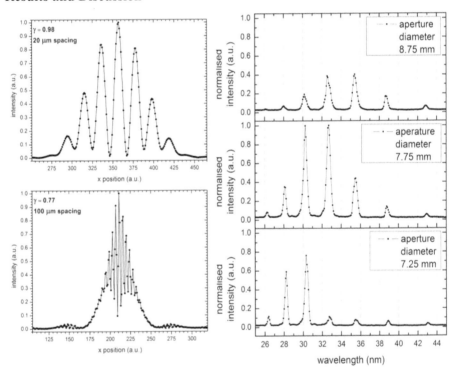

Fig. 1.. The spatial coherence of the multi-harmonic beam is obtained by using different Young double slits (left figures). The multi-harmonic beam covers a spectral range from 26 to 43 nm (right figures) and, depending on the phase-matching conditions, consists of two or more dominating harmonic orders and some smaller satellite harmonics.

By properly adjusting the phase-matching conditions we are able to generate the harmonics H19 to H31 with a net average flux of up to 10^{10} photons/(s cm^2) per harmonic order. By avoiding the usage of the spectrometer, which has poor transmission in the extreme ultraviolet, a net average flux of up to 10^{13} photons/(s cm^2) for the multi-harmonic beam can be detected. A 3D-Gaussian fit yields a somewhat larger $1/e^2$-diameter of 1.3/2.5 mm (x/y) of the Gaussian-like multi-harmonic beam at a distance of about 125 cm from the harmonic source. Several pairs of Young double slits, mounted on precision translation stages and driven by motors with a resolution better than 30 nm, allow us to precisely centre the Young slit pairs with respect to the centre of the whole or spatially filtered multi-harmonic beam in order to determine the degree of spatial coherence about 90 cm after the source. The degree of spatial coherence is determined as almost unity when the Young slits are separated by 20 µm (Fig. 1, top left) and as 0.77 when the slits are separated by 100 µm (Fig. 2, bottom left). We pay particular attention to the spectral properties of the harmonic radiation depending on the experimental parameters. In general, when shifting the focal point of the laser beam from the exit plane towards the centre of the gas cell we observe an overall spectral shift towards higher wavelengths as well as an increased flux until absorption counterbalances any further gain. Changes within the spectrum regarding relative intensities and number of harmonics are also evident. Similar behaviour is found when the pressure is increased inside the gas cell. For a particular pressure and focal position the reduction of the diameter of the fundamental driver beam by means of an aperture leads to a shift towards higher-order harmonics (Fig. 1, top to bottom right). Particular settings yield a symmetrically distributed spectrum, an over-all high-flux multi-harmonic beam or even an harmonic beam consisting of two dominating harmonic orders and very small satellites (Fig. 1., bottom right).

Conclusions

Our systematic studies exemplify how the characteristic output of harmonic sources can be easily tailored. Thus, the requirements of special applications can be met by adjusting the main experimental parameters such as aperturing the laser beam and by adjusting the pressure inside the gas cell as well as the focal position of the laser beam inside the cell. In particular, the output of this compact, narrow-bandwidth, high-flux, highly spatially coherent, ultrafast harmonic source of extreme ultraviolet radiation can be used directly to carry out studies of coherent diffraction imaging.

References

1 J. Zhou, J. Peatross, M. M. Murnane, H. C. Kapteyn, and I. P. Christov, in Physical Review Letters, Vol. 76, 752, 1996. J. J. Macklin, J. D. Kmetec, and C. L. Gordon, in Physical Review Letters, Vol. 70, 766, 1993.

2 L. Van Dao, S. Teichmann, J. Davis, and P. Hannaford, in Journal of Applied Physics (in press).

3 S. Kazamias, F. Weihe, D. Douillet, C. Valentin, T. Planchon, S. Sebban, G. Grillon, F. Augé, D. Hulin, and Ph. Balcou, in The European Physical Journal D, Vol. 21, 353, 2002.

4 J. E. Constant, D. Garzella, P. Breger, E. Mével, Ch. Dorrer, C. Le Blanc, F. Salin, and P. Agostini, in Physical Review Letters, Vol. 82, 1668, 1999. Y. Tamaki, J. Itatani, Y. Nagata, M. Obara, and K. Midorokawa, in Physical Review Letters, Vol. 82, 1422, 1999. K. Midorikawa, Y. Tamaka, J. Itatani, Y. Nagata, and M. Obara, in IEEE Journal of Quantum Electronics, Vol. 5, 1475, 1999..

5 P. Salières, A. L'Huillier, and M. Lewenstein, in Physical Review Letters, Vol. 74, 3776, 1995.

Single-stage Pulse Compression and High-Energy Supercontinuum generation from a Chirped-pulse oscillator

Alexander Fuerbach[1,2], Christopher Miese[1] and Wolfgang Koehler[2]

[1] Centre for Ultrahigh bandwidth Devices for Optical Systems (CUDOS) and MQPhotonics
Research Centre, Macquarie University, Sydney, NSW 2109, Australia
E-mail: fuerbach@ics.mq.edu.au
[2] Femtolasers Produktions GmbH, Fernkorngasse 10, 1100 Wien, Austria
E-mail: wolfgang.koehler@femtolasers.com

Abstract. We demonstrate the generation of high-energy supercontinuum pulses by coupling the uncompressed pulses of a Ti:sapphire Chirped-pulse oscillator into a microstructure fibre which features a highly anomalous dispersion at the centre wavelength of the laser.

Introduction

The so-called chirped-pulse oscillator (CPO) concept has been proven to be one of the most successful techniques to generate high-energy femtosecond laser pulses at MHz repetition rates [1]. The basic idea is to insert a Herriot-type multipass cell [2] into a Ti:sapphire oscillator, which acts as an intracavity optical delay line, thus reducing the repetition rate and, at constant average output power, increasing the energy per laser pulse. In order to avoid pulse splitting due to excessive nonlinearities in the gain material, the oscillator is operated at net-positive intracavity dispersion. The output pulses are therefore strongly chirped and have to be compressed by a suitable extra-cavity delay line, typically a prism pair.

Supercontinuum generation is an extremely active research topic, driven by the wide range of potential applications in various fields like frequency metrology, optical coherence tomography or pump-probe spectroscopy [3]. The introduction of microstructured optical fibres [4] opened the way to generate octave spanning super-continua by pumping them with femtosecond laser pulses from standard low-energy oscillators. However, due to the typically very small core-diameter of the fibre, this approach is limited the onset of damage on the front facet of the fibre when pulses with several tens of Nanojoules come into play. One solution which has been suggested in the past was to negatively pre-chirp the oscillator pulses, using an additional prism pair, and then launching them into a standard single mode fibre, where they are recompressed and subsequently spectrally broadened [5]. While the damage problem can be circumvented with this method, the setup is rather complex and the spectral broadening that has been achieved was relatively weak. Photonic crystal fibres (PCFs) with two zero dispersion wavelengths on the other hand, have been shown to generate ultrabroad supercontinuum spectra with high spectral density and coherence [6].

In this paper we show how these two approaches can be combined. We demonstrate the use of a Photonic crystal fibre with two zero dispersion wavelength to compress the strongly chirped pulses emitted by a Chirped-pulse oscillator followed by the generation of an extremely broad white-light supercontinuum in a single fibre. As the input pulses are almost two nanoseconds in duration, damage of the front facet can be avoided up to input energies in excess of 150nJ. In addition, the fairly flat and anomalous dispersion of the fibre at 800 nm makes any external prism compressor obsolete.

Experimental setup

The laser used in our experiments was a commercially available chirped-pulse oscilla-tor (Femtosource scientific XL 200, Femtolasers GmbH), delivering pulses with an energy of >200 nJ and a pulse duration <50 fs at a 5 MHz repetition rate. We have extracted the pulses before the internal prism compressor and have launched the heavily positively chirped laser pulses (Fig. 1) into the PCF (a SEM image of which is shown in Fig. 2a). Using multipole simulations [7] we have calculated the disper-sion of this fibre. As can be seen in Fig. 2b, at the centre wavelength of the CPO, the dispersion is anomalous and, moreover, fairly flat, ideal for compressing the output pulses.

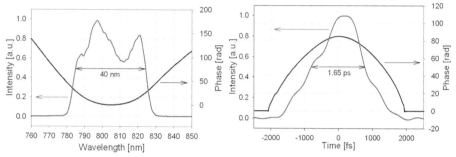

Fig. 1. The output pulses of the CPO measured by Frequency-Resolved Optical Gating (FROG).

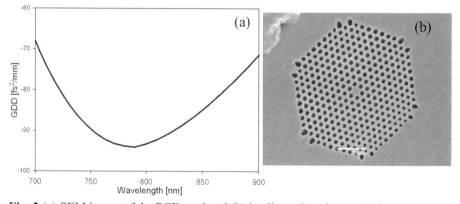

Fig. 2 (a) SEM image of the PCF used and (b) its dispersion characteristics

Results and Discussion

Numerical solutions of the nonlinear Schrödinger equation (NLSE) by the split-step Fourier method [8] show that for low input energy levels, the laser pulses undergo linear compression in the fibre due to its anomalous dispersion. At higher input ener-gies, the pulses also initially compress. However, once the peak intensity becomes sufficiently high, nonlinear processes begin to dominate the pulse propagation. The pulses become spectrally broadened by self-phase modulation (SPM) and then degenerate and non-degenerate four-wave mixing as well as higher-order soliton fission lead to the formation of an extremely broadband super-continuum. Depending on the initial peak power of the lunched pulses, these nonlinear processes kick in at a certain propagation distance z. The higher the input power, the earlier the supercontinuum is generated. However, after a long enough propagation distance, i.e. at the output of a 40 cm long piece of PCF, the generated spectrum, ranging from below 350 nm up to 1600 nm is remarkably independent on the input energy. Therefore, the initial pulse compression phase can be regarded as a mechanism to adjust the peak intensity of the pulses to the threshold level for efficient supercontinuum generation.

As can be seen in Fig. 3, the supercontinuum spectra are virtually indistinguishable for output energies ranging from 5 nJ up to 20 nJ, corresponding to an input energy of 20 nJ to 120 nJ. Damage of the front facet of the PCF did occur only for input energies of > 150 nJ. With this scheme, a maximum energy of the white-light pulses of 22 nJ could be achieved.

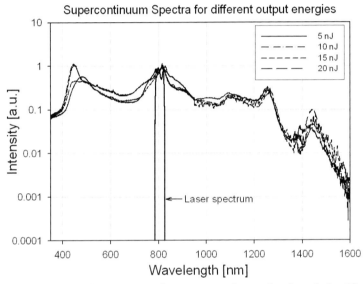

Fig. 3. Generated supercontinuum spectra for output energies ranging from 5 nJ to 20 nJ

Acknowledgements. This work was funded by the Australian Research Council under the Centres of Excellence scheme

1   A. Fernandez, T. Fuji, A. Poppe, A. Fuerbach, F. Krausz, and A. Apolonski in Optics Letters **29**, 1366, 2004.
2   D.Herriott, H. Kogelnik, R. Kompfner in Applied Optics **3**, 523, 1964.
3   R.R. Alfano, The Supercontinuum Laser Source, 2nd ed., Springer, 2006.
4   P.S.J. Russell in Science **299**, 358, 2003.
5   P. Dombi, P. Antal, J. Fekete, R. Szipoecs, Z. Varallyay in Appl. Phys. B **88**, 379, 2007.
6   K.M. Hilligsøe, T.V. Andersen, H.N. Paulsen, C.K. Nielsen, K. Mølmer, S. Keiding, R. Kristiansen, K.P. Hansen, J.L. Larsen in Opt. Express **12**, 1045, 2004.
7   T.P. White, B.T. Kuhlmey, R.C. McPhedran, D. Maystre, G. Renversez, C.M. de Sterke, L.C.Botten in J. Opt. Soc. Amer. B **19**, 2322, 2002.
8   G. P. Agrawal, Nonlinear Fiber Optics, 2nd ed., Academic Press, San Diego, 1995.

An All-Optical Synchrotron Light Source

H. Schwoerer[1,2], H.P. Schlenvoigt[2], K. Haupt[1], A. Debus[2], E. Rohwer[1], J. Gallacher[3], R. Shanks[3], D. Jaroszynski[3]

[1] Laser Research Institute, Physics Department, Stellenbosch University, Private Bag X1, Matieland 7602, South Africa
E-mail: heso@sun.ac.za
[2] Institut für Optik und Quantenelektronik, Universität Jena, Max-Wien-Platz, 07743 Jena, Germany
[3] Department of Physics, Scottish Universities Physics Alliance, University of Strathclyde, Glasgow G4 0NG, United Kingdom

Abstract. We report on the generation of synchrotron radiation from laser accelerated relativistic electrons propagating through an undulator. We indicate that this provides exciting novel opportunities in ultrafast spectroscopy.

Introduction and Motivation

Ultrashort coherent light pulses are an invaluable tool to study the microscopic dynamics of matter. Femtosecond lasers in combination with optical nonlinear devices deliver pulses with wavelengths between the UV and the NIR spectral region and are therefore in the range of relevant electronic excitations of molecules and solids. Shorter wavelengths down to a few nm allow a direct view onto the molecular structure through diffraction. They can be generated by synchrotron radiation using electron storage rings or linear accelerators equipped with undulators. In particular if the undulator, operated in the free electron laser mode (FEL), extremely brilliant, ultrashort, polarized and coherent light pulses are produced. Tunability is achieved by varying the electron energy, coherence is obtained by an intrinsically generated modulation of the electron pulses during the interaction with the self-generated light field. This process is called self-amplified spontaneous emission (SASE), which is the basis of all present-day FEL at short wavelengths. The short duration of the emission arises from the temporal structure of the SASE electron pulse itself. Currently, constructed free electron lasers promise a wide applicability spanning from atomic and cluster physics, through temporally resolved structural analysis of complex molecules to plasma physics and even quantum electrodynamics in high external fields [1]. However, nowadays FEL require kilometer long electron accelerators due to the maximum energy gain per length in the order of tens of MeV/m which is set by the material breakdown of the radio frequency cavity.

An alternative electron acceleration mechanism is accomplished by the fs lasers themselves: if the output of conventional fs lasers is extended by powerful short pulse laser amplifiers, laser pulses with peak powers of tens of terawatts and peak intensities in the laser beam focus of more than $10^{20}\,W/cm^2$ can be achieved [2]. If these laser pulses correctly interact with a self- or externally generated plasma, electrons can be accelerated to energies up to a GeV with a few percent bandwidth and within a well collimated beam [3]. The underlying acceleration mechanism is called forced wake field or bubble acceleration and relies on the generation of a spatially confined plasma wave, which captures electrons and accelerates them within a few millimeters to several hundreds of MeV. The energy gain per length is significantly larger than in radio frequency accelerators because the acceleration is based on acceleration in a plasma, which has to be avoided in the conventional approach. The electron pulse duration has been measured to be not longer than the laser pulse duration ($<$ 50 fs), simulations suggest that it might be even shorter. We report here on a first successful generation of synchrotron radiation from laser accelerated relativistic electrons [4]

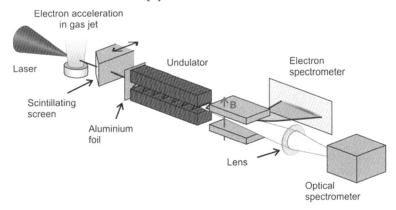

Fig. 1. The experimental setup of the all optical synchrotron light source consists of a laser wakefield accelerator as source of a relativistic electron beam, electron beam diagnostics, an undulator, an electron spectrometer and an optical spectrometer. All parts are aligned along a line and are located inside vacuum chambers, except the optical spectrometer. An essential feature of the experiment is that the acceleration region, the electron spectrum and the undulator spectrum were simultaneously recorded for each individual shot

Results and Prospects

Ultrashort monochromatic electron pulses with energies around 60 MeV are produced by the interaction of a high intensity laser (Jena high-intensity Titanium:Sapphire laser JETI, 80 fs, 430 mJ on target) with a He gas jet. These electrons propagate through a static undulator of 2 cm period length, where they undergo oscillations perpendicular to the magnetic field and the propagation direction and therefore emit polarized radiation. The wavelength λ of the emitted light is mainly determined by the undulator period λ_u and the electron energy $E_e = \gamma m_0 c^2$ (γ being the Lorentz factor, m_0 the electron's rest mass) and to second order by the undulator parameter K, determining the amplitude and therefore the anharmonicity of the oscillatory motion. Furthermore the energy of the emitted photons is peaked in the forward direction. For $K < 1$ the emitted wavelength in forward direction is approximately $\lambda \cong \lambda_u / 2\gamma^2$. For electron energies around 60 to 70 MeV the wavelength emitted by our undulator $\left(\lambda_u = 2 \text{ cm} \right)$ is in the visible spectral range, see Fig. 2.

In the presented simple approach of laser electron acceleration, the generation of the plasma with its required complex density profile, and the acceleration of the electrons therein, have to be accomplished by the same intense laser pulse. Since several nonlinear processes are involved, this scenario is not perfectly predictable and repeatable, resulting in shot to shot variations of energy, spectral width, yield and direction. However, different plasma target designs as colliding pulses or capillary discharges have proven that more stable conditions can be achieved [5,3]. Due to the favourable $1/\gamma^2$ scaling of the undulator radiation, ultrashort, incoherent light pulses in the UV and soft x-ray spectral range can be produced with today's table-top high intensity lasers by the method described here. The generation of coherent undulator radiation is more challenging since it requires electron pulses shorter than the emitted wavelength. With today's laser accelerated electrons this might be possible in the infrared spectral range, using undulators with a long period. In the UV or even x-ray region one has to rely on the SASE process. An interesting aspect of a purely laser driven synchrotron source is the temporal coupling of the laser light and the undulator radiation: for each time resolved experiment at least two pulses are required, one to start the fast process to be observed and one to probe it after a well defined temporal delay. One of the two pulses typically is a femtosecond laser pulse, which can be generated by the same laser source which drives the electron acceleration and is by that perfectly synchronized to the electron pulse and the undulator radiation respectively. In conclusion, the combination of laser electron acceleration and undulator technology opens exciting novel opportunities for ultrafast spectroscopy with short wavelength brilliant light pulses.

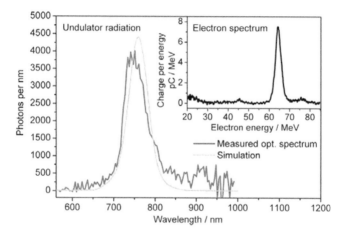

Fig. 2. Electron spectrum (inset) and undulator radiation spectrum (dark grey line). Both spectra were recorded for the same laser shot. The electron spectrum is peaked at 64 MeV, has a width of 3.4 MeV (FWHM) and contains a charge of 10 pC. The undulator radiation is peaked at 740 nm with a bandwidth of 55 nm, containing 250,000 photons. Expected undulator radiation was simulated (light grey line) from the measured electron spectrum (inset) taking into account undulator parameters and solid angle of radiation detection. There is an excellent agreement of the spectral position and width between measured and expected undulator radiation and a reasonable agreement in yield [4]

1   H. Chapman et al., in Nature Physics, Vol. 2, 839, 2006.
2   M. Pittman, S. Ferré, J. Rousseau, L. Notebaert, J. Chambaret, G. Chériaux, in Applied Physics B, Vol. 74, 529, 2002.
3   W. Leemans, B. Nagler, A. Gonsalves, Cs. Toth, K. Nakamura, C. Geddes, E. Esarey, C. Schroeder, S. Hooker, in Nature Physics, Vol. 2, 696, 2006.
4   H.-P. Schlenvoigt, K. Haupt, A. Debus, F. Budde, O. Jäckel, S. Pfotenhauer, H. Schwoerer, E. Rohwer, J. Gallacher, E. Brunetti, R. Shanks, S. Wiggins, D. Jaroszynski, in Nature Physics, Vol. 4, 130, 2008.
5   J. Faure, C. Rechatin, A. Norlin, A. Lifschitz, Y Glinec, V. Malka, in Nature, Vol. 444, 737, 2006.

Compression of an Ultraviolet Pulse by Molecular Phase Modulation and Self-Phase Modulation

Yuichiro Kida[1], Shin-ichi Zaitsu[1], and Totaro Imasaka[1,2]

[1] Department of Applied Chemistry, Graduate School of Engineering, Kyushu University, 744, Motooka, Nishi-ku, Fukuoka 819-0395, Japan
E-mail: y-kida@cstf.kyushu-u.ac.jp
[2] Division of Translational Research, Center for Future Chemistry, Kyushu University, 744, Motooka, Nishi-ku, Fukuoka 819-0395, Japan
E-mail: imasaka@cstf.kyushu-u.ac.jp

Abstract. A compression scheme for an ultraviolet pulse to sub-15 fs is reported. Frequency modulation of an ultraviolet pulse by molecular rotations and by self-phase modulation results in a compressed pulse with small intensities of sub-pulses.

Introduction

An ultrashort ultraviolet (UV) pulse is a useful mean for the investigation of the ultrafast phenomena of an organic compound. For generation of the pulse, techniques based on self-phase modulation (SPM), difference-frequency mixing (DFM), non-collinear parametric amplification (NOPA), and Molecular Phase Modulation (MPM) have been reported. In the technique based on MPM, the frequency of a UV pulse (probe pulse) is modulated by a coherent molecular motion induced by an intense pulse (pump pulse). In this case, the energy of the modulated pulse is determined by the energy of the original probe pulse [1]. A high-energy ultrashort UV pulse is, therefore, generated by use of a high energy probe pulse whose width is long enough to prevent the ionization of the molecules which leads to undesired phenomena such as deterioration of beam profile. The use of the long probe pulse, however, leads to generation of sub-pulses in the temporal profile [2].

To reduce the intensities of the sub-pulses, SPM induced by the probe pulse is considered here. When the energy of the probe pulse is high enough for inducing SPM, the frequency of the probe pulse is simultaneously modulated by a coherent molecular motion and by SPM. The spectrum of the modulated probe pulse has a broad width arising from the molecular phase modulation, and also has a dense structure arising from SPM.

Experimental Methods

A near-infrared pulse provided from a Ti:sapphire chirped pulse amplifier (784 nm, 1.2 mJ, 110 fs) was passed through a LiB_3O_5 crystal to generate a UV pulse (392 nm) that was used as a probe pulse. The near-infrared pulse (pump pulse) and the probe pulse emitted from the crystal were separated from each other with a dichloic mirror for controlling the time delay of the probe pulse with respect to the pump pulse. The two pulses were then recombined and were focused into a Raman cell filled with pressurized hydrogen gas (10 atm). The probe beam transmitted from the Raman cell was collimated with a concave mirror before compensation for the GDD and TOD of the probe pulse by use of a prism compressor and a grating compressor. The compressed probe pulse in the compressors was, then, propagated to a self-diffraction

(SD) autocorrelator. The time delay of the probe pulse was adjusted to 0.9 ps, and the energies of the input pump and probe pulses were 260 μJ and 80 μJ, respectively.

Results and Discussion

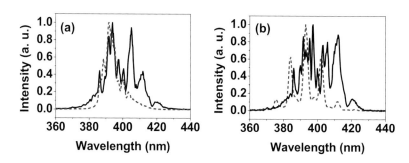

Fig. 1. (a) The spectra of the probe pulse emitted from the Raman cell (solid line). The spectrum of the probe pulse measured in the case that only the probe pulse was focused into the cell (broken line). (b) The spectrum of the probe pulse measured in front of the autocorrelator (solid line). A typical spectrum measured under the situation that SPM is not induced (Ref. 2) is shown in the figure with broken line.

The spectrum of the probe pulse emitted from the Raman cell is shown in Fig. 1 (a). The frequency of the probe pulse was modulated by the coherent rotation of *ortho*-hydrogen and by SPM. The spectral width (FWHM) was broader than that expanded only by SPM [broken line in Fig. 1 (a)], and the spectral structure was denser than that modulated without SPM [broken line in Fig. 1 (b)] [2]. After the dispersion compensation for the probe pulse, the relative intensity of each spectral component was changed to that shown with solid line in Fig. 1 (b). This was due to the fact that the output energy from the compressors was different in each spectral component, and was also due to the fact that the frequency modulation was induced efficiently in the center part of the beam. The center part was extracted using an aperture before the measurements of the spectrum [Fig. 1 (b)] and the autocorrelation trace described below. The spectrum shown in Fig. 1 (b) is broad enough for generation of a 10-fs UV pulse as indicated by the inverse Fourier-transform [IFT, Fig. 2 (a)] of the spectrum [solid line in Fig. 1 (b)]. The waveform of the IFT contains no sub-pulses, while the measured SD-autocorrelation trace shown in Fig. 2 (b) consists of a main peak structure and small sub-pulses in the vicinity of the structure. In principle, the spectral phase of a probe pulse modulated by the phase modulations shows a complicated structure and it can not be compensated only by GDD and TOD compensations. The residual phase distortion leads to the generation of the sub-pulses even when perfect compensation for GDD and TOD are demonstrated. Though complete removal of the sub-pulses from the temporal profile was not demonstrated, the intensities of the sub-pulses were lower than those in the case where SPM was not induced [2]. Hence the generation of SPM is useful for suppressing the generation of the sub-pulses in the temporal profile. The FWHM of the main peak in the autocorrelation trace was 17 fs, corresponding to the FWHM of the SD-autocorrelation trace of a 14-fs Gaussian pulse [solid line in Fig. 2 (b)]. The

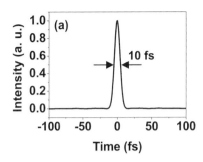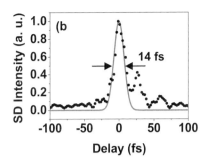

Time (fs)   Delay (fs)

Fig. 2. (a) Inverse Fourier transform of the spectrum (Fig. 1 (b)). (b) The measured (solid circles) and calculated (solid line) SD-autocorrelation traces.

good correspondence between the main structure of the measured trace and the trace of the Gaussian pulse suggests that the FWHM of the main structure of the compressed probe pulse was ca. 14 fs. The width of 14 fs is slightly longer than that of the width of the IFT (10 fs), which would be due to imperfect compensation for the TOD. Though a frequency-resolved SD-autocorrelation trace (spectrogram) was measured to estimate the amount of the dispersion, it could not be easily estimated since the spectrogram is not appreciably distorted by a small amount of TOD [3]. The width of 14 fs is, however, much shorter than the width of the original probe pulse (ca. 100 fs). Furthermore, the energy of the compressed pulse was not so small (4.5 µJ) despite the much energy loss in the two compressors.

Conclusions

As reported here, the spectrum of a 100-fs UV pulse is modulated by the coherent rotation of hydrogen molecules and by SPM for generation of a 10-fs UV pulse. The resultant spectral width is wide enough for generation of a 10-fs pulse, and the pulse width of the compressed pulse is sub-15 fs which would be determined mainly by the precision in the dispersion compensation. The temporal profile of the compressed pulse has the sub-pulses whose intensities are small compared to those in the case where SPM is not induced. Since the GDD and TOD are compensated with a prism and grating compressors, this approach can be applied to compression of a deep-UV (DUV) pulse by only replacing the probe pulse with a DUV pulse.

Acknowledgements. This research was supported by Research Fellowships of the Japan Society for the Promotion of Science (JSPS) for Young Scientists, Grants-in-Aid for Scientific Research, and a Grant-in-Aid for the Global COE program, "Science for Future Molecular Systems", from the Ministry of Education, Culture, Science, Sports and Technology of Japan.

1   N. Zhavoronkov and G. Korn, Phys. Rev. Lett. **88**, 203901, 2002.

2   Y. Kida, T. Nagahara, S. Zaitsu, M. Matsuse, and T. Imasaka, Opt. Exp. **14**, 3038, 2006.

3   K. W. DeLong, R. Trebino, and D. J. Kane, J. Opt. Soc. Am. B **11**, 1595, 1994.

Temporal Optimization of Ultrabroadband Optical Parametric Chirped Pulse Amplification

Jeffrey Moses[1], Cristian Manzoni[2], Shu-Wei Huang[1], Giulio Cerullo[2], and Franz X. Kärtner[1]

[1] Department of Electrical Engineering and Computer Science and Research Laboratory of Electronics, Massachusetts Institute of Technology, Cambridge, MA 02139, USA
E-mail: j_moses@mit.edu
[2] ULTRAS-INFM-CNR Dipartimento di Fisica, Politecnico, Piazza L. da Vinci 32, 20133 Milano, Italy
E-mail: cristian.manzoni@polimi.it

Abstract. Critical optimization considerations are presented for ultrabroadband, high-power optical parametric chirped-pulse amplifiers, where simultaneous suppression of superfluorescence and maximization of both conversion efficiency and bandwidth is required. Numerical simulations verify theory.

Today's demands on light sources for high-intensity ultrafast optics research are stringent: peak power must be maximized by scaling both to high energy and near-single-cycle duration, signal to noise contrast must be high, and often pulses at nontraditional wavelengths must be generated. These requirements have led to the rapid development of ultrabroadband optical parametric chirped pulse amplification (OPCPA) pumped by powerful picosecond pulses, in which gain bandwidth is stretched to near-octave breadths by group-velocity matching between signal and idler. In recent years, several problems in the construction of these amplifiers have become relevant. The coupling of temporal gain narrowing and spectral narrowing results in a trade-off between conversion efficiency and bandwidth. Additionally, the amplifier seed energy is often low while total gain is high, resulting in high levels of parametric superfluorescence and poor signal to noise ratio [1, 2]. While the effect of temporal gain narrowing on ultrabroadband OPCPA has been investigated [3], a study of simultaneous optimization of conversion efficiency, signal bandwidth and signal-to-superfluorescence ratio has not yet been presented, and several details of the temporal optimization problem have been neglected.

In this paper we study the simultaneous optimization of conversion efficiency, signal bandwidth and noise suppression in ultrabroadband OPCPA as a function of the ratio of pump and seed pulse durations, their relative energy and the total parametric gain set by the pump intensity. The optimal pump-seed pulse duration ratios for maximization of the efficiency-bandwidth product and signal-to-noise ratio are found to depend on the total gain.

The basic compromise between conversion efficiency and amplified signal bandwidth in OPCPA is conceptually well understood: for a chirped seed pulse, frequency is mapped to time, and thus when conversion efficiency is maximized by stretching the seed pulse to cover the full temporal gain profile of the amplifier, the short- and long-wavelength wings of the seed pulse experience significantly lower gain than the peak. Hence, temporal gain narrowing results in spectral narrowing. The difference between chirped pulse and non-chirped pulse parametric amplification is depicted in Fig. 1 for a Gaussian pump pulse and a peak gain of 100. If pump beam depletion can be neglected, parametric gain is $G = 1 + (\Gamma^2/g^2)\sinh^2(gL)$, where g

$= [\Gamma^2 - (\Delta k/2)^2]^{1/2}$, Δk is the wavevector mismatch, and Γ^2 is the nonlinear drive, proportional to the pump intensity I_p. Given a pump profile $I_p(t)$, with perfect phase matching and at high gain we may find the temporal region of overlap ($|t| < t_g$) of the seed with the pump pulse where gain is at least e^{-a} times the gain at $t = 0$ by setting

$$\sqrt{\frac{I_p(t_g)}{I_p(t=0)}} = \frac{g(t=0) - a}{g(t=0)}, \tag{1}$$

where $a = 1$. Assuming the pump has a Gaussian temporal profile, we can find t_g by setting $I_p(t_g)/I_p(0) = \exp(-t_g^2/\tau_0^2)$. We find

$$t_q = \tau_0\sqrt{-2ln\,[1 - a(ln[4G(t=0)])^{-1}]}. \tag{2}$$

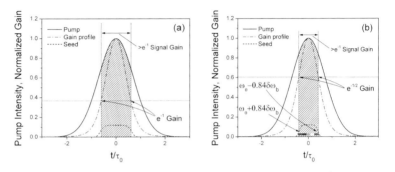

Fig. 1. Schematics of high-conversion-efficiency (a) optical parametric amplification, (b) OPCPA in which amplified signal bandwidth is maximized.

The case for an unchirped seed is depicted in Fig. 1(a). In order to maximize conversion efficiency and suppress superfluorescence noise, the temporal extent of the seed pulse should exactly fill the region $|t| < t_g$ [shaded region in Fig. 1(a)]. If the pulse is too short, a portion of the available pump energy remains unconverted to signal after amplification and regions where there is significant gain remain unseeded, resulting in noise amplification without quenching of the gain by the signal. If the pulse is too long, additional gain is needed to saturate the amplifier. An important feature of Eq. 2 is the dependence of the gain temporal width on the peak gain. As a consequence, the optimal seed duration varies from stage to stage in a multiple-stage amplifier. Note that in the pump-depletion regime the region of significant gain becomes slightly wider, since the gain at the temporal and spectral wings increases with respect to the peak gain.

In the case of a chirped seed pulse [Fig. 1(b)], the spectral and temporal gain profiles simultaneously affect the width of the significant gain region. In order to employ the full gain bandwidth $2\delta\omega_b$ of the amplifier [defined as $G(\omega_0 \pm \delta\omega_b) = e^{-1}G(\omega_0)$] the gain temporal width is significantly narrower than in the unchirped seed case: at a certain temporal coordinate, both the drop in pump intensity and the instantaneous wavevector mismatch combine to reduce the gain by e^{-1}. An appropriate definition of the new gain temporal width, $|t| < t'_g$, and the amplification bandwidth, $2\delta\omega_b'$, is where each effect reduces the gain by $e^{-1/2}$, or when $a = 1/2$ in Eq. 2. In this case the gain bandwidth is reduced by 30% in a regular OPCPA and by only 16% in the case of broadband group-velocity-matched amplification.

820

When suppression of superfluorescence is necessary, however, the optimization problem is more complex: the seed is temporally chirped but the incoherent noise is not. In Fig. 1, the signal and noise fields have different gain widths, with ratio ~0.7:1. As a result, the average noise gain across the pump pulse is higher than the average signal gain. Moreover, if the seed is stretched such that it fills the region $|t| < t'_g$ [to maximize amplifier bandwidth, as in Fig. 1(b)], the low seed energy in the region $t'_g < |t| < t_g$, will allow significant conversion of pump to noise. Thus, we conclude that the seed should be chirped to at least cover the region $|t| < t_g$ (Eq. 2, $a = 1$), resulting in some reduction in bandwidth but maximizing conversion efficiency and preventing strong degradation of signal to noise ratio.

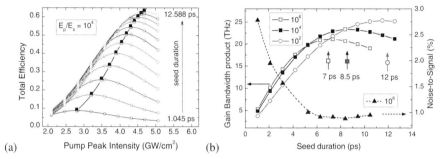

(a)   (b)

Fig. 2. (a) Simulation results for conversion efficiency (signal + idler) for several values of seed pulse duration and $G = 10^4$. The squares denote where maximum pump depletion occurs. (b) Efficiency-bandwidth product (at maximum possible conversion efficiency) versus seed pulse duration for $G = 10^2$, 10^4, and 10^6. Triangles denote noise-to- signal ratio for $G = 10^6$.

To verify predictions of the analytical model, we numerically simulated an OPCPA by solving the nonlinear coupled equations for the case of collinear interaction and plane-waves. We considered a 3-mm thick PPLN crystal, pumped by 9-ps Gaussian pulses at 1.047 μm for broadband degenerate amplification at 2 μm. The super-Gaussian seed had a bandwidth corresponding to 12-fs transform limit, larger than the acceptance of the crystal. By varying both the pump intensity and the seed second order chirp, we explored the parameter space as shown in Fig. 2(a) for $G = 10^4$. Maximum conversion efficiency (squares) increases strongly with increasing seed duration until the significant gain region of the pump is filled. At this point the growth in conversion efficiency relaxes, while at the same time bandwidth begins to strongly decrease. Using these datapoints from Fig. 2(a), the efficiency-bandwidth product [panel (b), filled squares] is maximized when the seed duration is close to the gain region width predicted by Eq. 2. The same numerical analysis was conducted for the cases of a pump to signal energy ratio of 10^2 (circles) and 10^6 (open squares): a comparison between the three amplification regimes shows that the optimal seed to pump duration ratio decreases with increasing pump-to-signal energy ratio, as predicted by Eq. 2. In other words, the optimal seed duration decreases for increasing gains. Finally, a minimum in noise to signal ratio is found close to the maximum gain-bandwidth product (Fig. 2(a), triangles [$G = 10^6$]).

1   F. Tavella, A. Marcinkevicius, and F. Krausz, New J. of Phys. **8**, 219, 2006.

2   T. Fuji, N. Ishii, C. Y. Teisset, X. Gu, T. Metzger, A. Baltuška, N. Forget, D. Kaplan, A. Galvanauskas, and F. Krausz, Opt. Lett. **31**, 1103, 2006.

3   S. Witte, R. T. Zinkstok, W. Hogervorst, K. S. E. Eikema, Appl. Phys. B **87**, 677, 2007.

Third Harmonic X-waves Generation by Filamentation of Infrared Femtosecond Laser Pulses in Air

Han Xu[1], Hui Xiong[1], See Leang Chin[2], Ya Cheng[1] and Zhizhan Xu[1]

[1] State Key Laboratory of High Field Laser Physics, Shanghai Institute of Optics and Fine
  Mechanics, Chinese Academy of Sciences
  P.O. Box 800-211, Shanghai 201800, China
[2] Centre d'Optique, Photonique et Laser (COPL) and Département de physique, de génie
  physique et d'optique, Université Laval, Québec, Québec G1K 7P4, Canada
E-mail: ycheng-45277@hotmail.com, zzxu@mail.shcnc.ac.cn

Abstract. We report the first measurement of the hyperbolic featured angularly resolved spectra of the X-shaped third harmonic generated in infrared femtosecond pulses pumped filament in air. We show that at low pump intensity, phase matching between the fundamental and third harmonic waves dominates the nonlinear optical effect and induces a ring structure of the third harmonic beam; whereas at high pump intensity, the dispersion properties of air begins to affect the angular spectrum, leading to the formation of nonlinear X-wave at third harmonic.

Introduction

In propagation of intense laser pulse in Kerr nonlinear medium, stable optical filament could be generated, and efficient third harmonic (TH) generation [1-5] could arise due to the very high intensity achieved in filamentation. Due to the potential application in laser frequency conversion to shorter wavelength, TH generation through optical filaments in air has been intensively studied. Spatiotemporal coupling during nonlinear propagation would naturally result in complex spatiotemporal structures in both the fundamental and TH pulses. The traditional spectrum detection technique, however, is not sufficient for the diagnosis of the complex wavepacket, while a simple but powerful diagnostic method, namely, the measurement of angularly resolved spectra, has been developed [6] for this purpose. Angularly resolved spectra could provide surprisingly detailed portrait of complex wavepacket. In this Letter, we report on the experimental measurement of angularly resolved spectrum of third harmonic generated by intense infrared pulse after its filamentation in air. The X featured angularly resolved spectrum of TH indicates that the TH wavepacket is transformed into nonlinear X wave after filamentation [6], and the intensity-dependent spectrum gives a clear evidence of the strong nonlinear phase locking between the FW and the TH during their co-propagation [1-3].

Experimental Methods

The carrier wavelength from our OPA is tuned to 1270nm with single pulse energy of up to 487nm and pulse duration of about 20fs-25fs. After being separated from the idler using a broadband reflection mirror (M1, high reflection at 1100nm-1300nm), the infrared pump pulse is then tightly focused by a gold coated concave mirror with a focal length of 250mm to generate a single light filament in air near the geometrical focal point. After filamentation, an angularly resolved spectrometer was employed to portrait the angular resolved spectrum of the output TH pulse; the far-field spatial pattern as well as the filament is thus captured on a digital camera (Coolpix995, Nikon, Japan). The angular-resolved (k_y-ω) spectrum is recorded by angularly resolved spectrometer consists of a positive lens (L, focal length: 300mm) and a grating spectrometer (SpectraPro 300i, Acton)

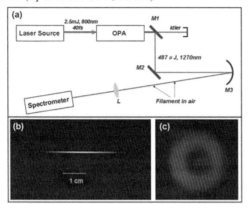

Fig.1. (a) The experimental setup, and the digial camera captured images of (b) the side view of the filament in air, and (c) the far-field spatial pattern of the third harmonic.

Results and Discussion

Figure 2 shows the evolution of the angular-resolved spectrum (θ-λ) of TH with the increase of the pumping pulse power from 30 µJ to 487 µJ. We note that there are mainly two mechanisms contributing to the process of third-order harmonic generation. When the IR pulse energy is between ~200 µ J and ~400 µ J, the mechanism of longitude phase matching between the generated TH and the fundamental wave dominates the TH generation process, which requires:

$$k_z(3\omega)=3k(\omega). \tag{1}$$

The conversion efficiencies of both the ring and axial components increase with the increasing pump energy, and the ring TH is much stronger due to the longitude phase matching condition could be satisfied on the ring. When the IR pulse energy is further increased to > 400 µJ, two hyperbolic structured tails (tail2, solid curve and tail3, dash-dot curve) emerge and grow rapidly. TH frequency components growing along these tails obey the mechanism of group velocity matching between the generated TH and the pump wave, which requires [7]:

$$k_z(\omega)= k(\omega_s)+(\omega- \omega_s)/V_g \tag{2}$$

,where ω_s is the frequency of the scattered TH wave, and V_g is group velocity of the pump wave. By fitting the measured angularly resolved spectra with equation 2, the group velocity of tail 2 and tail 3 could be obtained, which reveals that both the tails propagating at the same group velocity and locked to the same peak of the pump pulse.

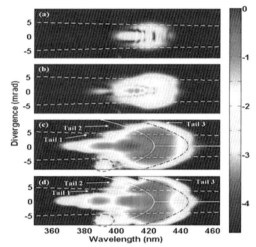

Fig. 2. Normalized angular resolved spectrum of TH in air with FW pulse energy at (a) 135μJ , (b) 240μJ , (c) 410μJ , and (d) 481μJ (in logarithmic scale and normalized by maximum intensity of TH).

Conclusions

We systematically investigate the evolution of angular spectrum of TH wave generated by IR ultrashort pulse filamentation in air. We speculate that there exist two physical mechanisms governing the evolution of angular spectrum of TH wave, implying different origins of conical ring and X-wave.

Acknowledgements. This work was supported by the National Basic Research Program of China (Grant No. 2006CB806000), Shanghai Commission of Science and Technology (Grant No. 07JC14055), and National Natural Science Foundation of China (Grant No. 10523003). S. L. C acknowledges the support of the Canada Research Chairs program..

1   N. Aköbek, A. Iwasaki, A. Becker, M. Scalora, S.L. Chin, and C.M. Bowden, Phys. Rev. Lett. Vol. 89, 143901, 2002
2   F. Théberge, N. Aközbek, W. Liu, J.–F. Gravel, and S. L. Chin, Optics Communications. Vol. 245, 399, 2005
3   F. Théberge, N. Aközbek, W. Liu, J. Filion, and S. L. Chin, Optics Communications. Vol. 276, 298, 2007.
4   H. Xu, H. Xiong, R. Li, Y. Cheng, Z. Z. Xu, and S. L. Chin, Appl. Phys. Lett., Vol. 92, 011111, 2008
5   H. Xiong, H. Xu, Y. Fu, Y. Cheng, Z. Z. Xu, and S. L. Chin, Phys. Rev. A, Vol. 77, 043802, 2008
6   D. Faccio, P. Di Trapani, S. Minardi, A. Bramati, F. Bragheri, C. Liberale, V. Degiorgio, A. Dubietis, A. Matijosius, J. Opt. Soc. Am. B, Vol. 22, 862, 2005
7   D. Faccio, et al., Opt. Express, Vol. 15, 13077, 2007

Generation and control of coherent conical pulses in seeded optical parametric amplification

Ottavia Jedrkiewicz[1], Matteo Clerici[1], Daniele Faccio[1] and Paolo Di Trapani[1,2]

[1] Cnism and Dipartimento di Fisica e Matematica, Università dell'Insubria, Via Valleggio 11, 22100 Como (Italy)
 E-mail: ottavia.jedrkiewicz@uninsubria.it
[2] Department of Quantum Electronics, Vilnius University, Sauletekio 9, LT-10222 Vilnius (Lithuania)

Abstract. We propose a new technique for high-energy conical pulse generation based on continuum seeded parametric amplification process in quadratic nonlinear media. We show that by using an appropriate broadband week seed we are able to lock the many spatiotemporal modes of the parametric radiation to obtain a single, coherent conical pulse.

Introduction

Conical waves are peculiar wave-packets, in which the energy flow is not directed along the beam axis, as in conventional waves. In contrast here, the energy arrives laterally, i.e. from a cone-shaped surface, leading to the appearance of a very intense and localized interference peak at the cone vertex, as in Bessel (or Durnin) beams, in the continuous wave limit, and in the so-called X-waves, in the ultra-short pulse regime. The conical nature of these waves allow them to carry angular dispersion, i.e. a controlled dependence of temporal frequencies on angles. As a consequence, they can propagate stationary in linear as well as in nonlinear media, in spite of diffraction and material dispersion, and exhibit tunable "effective" phase and group velocities, which result in unique dispersion-management features. Recent experiments in phase-mismatched second harmonic generation [1] have highlighted the existence of interlinked spatial and temporal processes, which cannot be considered separately within the nonlinear dynamics, as consequence of the presence of angular dispersion. In quadratic processes, this space and time coupling has also been observed in parametric down-conversion where the generated superfluorescence could be interpreted as a stochastic "gas" of quasi-stationary X-type modes characterized in the spatiotemporal domain by a skewed correlation surrounding a very sharp peak [2,3]. The work presented here extends that previous study with the goal of generating and controlling single coherent conical pulses in seeded Optical Parametric Amplification (OPA).

Here we show that by using an appropriate broadband week seed in OPA we are able to lock the many incoherent spatiotemporal modes of the parametric down-converted radiation to obtain a single, coherent, conical pulse.

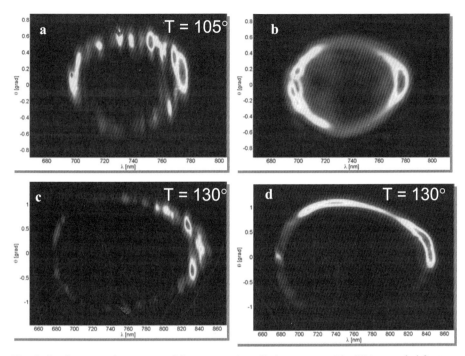

Fig. 1. Spatiotemporal spectrum of the parametric radiation generated in OPA recorded for two different crystal temperatures (two different phase-matching configurations) in non seeded (a and c) and continuum seeded (b and d) configurations respectively.

Experimental Methods

The OPA was seeded by a broadband coherent week radiation, which was obtained by spectral windowing a filament induced by focusing in a 5cm BK7 bulk sample a 1ps laser pulse at 1055nm. The continuum seed generated over a smooth bandwidth in the range of 600-800nm (with energy in the μJ range) was suitably focused and injected collinearly with the pump inside the LBO crystal. The temporal overlap of the pump pulse and seed pulse inside the crystal was controlled by means of a delay line. The statistical and coherence properties of the radiation generated in the OPA process can be studied by recording the far-field spatiotemporal (θ,λ) spectrum, in analogy to [2,3]. To this end the radiation at the output of the LBO crystal was collected by means of a telescope and the far-field diagnostic was based on an imaging spectrometer coupled with a 16 bit high efficiency CCD camera (Andor) for the visible, and a InGAs CCD (Xenics) for the infrared region.

Results and Discussion

The single shot spatiotemporal spectra of the signal recorded in non seeded and seeded configuration respectively are presented in Fig.1 for two different crystal temperatures, both corresponding to emission out of degeneracy (for best superposition range with spectral characteristics of continuum).

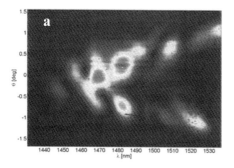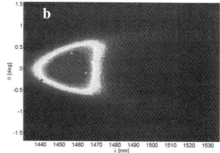

Fig. 2. Portion of single shot spatiotemporal spectrum of the generated idler field recorded by an InGAs CCD.

The parametric down converted radiation is generated along the so-called phase-matching curve of the quadratic nonlinear process, satifying the energy and momentum conservation laws. Thus when the broadband coherent continuum seed is injected inside the crystal and the temporal superposition is guaranteed, the amplification process occurs over the same broadband region dictated by the phase-matching curves. In addition we can clearly observe the spatiotemporal mode locking process, which leads to smoothed spatiotemporal spectral curves (in single shot) of the amplified signal. Note that because of the phase-conjugation property of amplified signal/idler radiation in OPA, the single shot idler spectrum in infrared region turns out to be similarly smoothed. Experimentally, because of the huge spatial and spectral bandwidth of the idler radiation, only a portion of the expected ring type (similarly to the signal) spectrum of the idler could be recorded, as shown in Fig.2.
The spectral distribution of the obtained OPA radiation is characteristics of the so-called O-waves, which are stationary modes in materials with anomalous dispersion ($\lambda > 1 \mu m$) [4]. Moreover in analogy to what done in [2,3], the evaluation of the space-time coherence function of the generated radiation leads to an "onion" type structure in the (r,t) space, characterized by a central core of about 10μm and 10fs respectively.

Conclusions

The results presented here permit to evidence the possibility of generating in seeded OPA single ultrashort localized conical wave modes in space and time, with peak dimensions of the order of few μm and fs respectively, in analogy with previous work [2,3]. Further work on pulse reshaping and control in OPA, in particular around degeneracy, with the aim of generating X-type or Bessel like pulses is in progress, together with the study of stationarity of the single wave modes.

1   P. Di Trapani, G. Valiulis, A. Piskarskas, O. Jedrkiewicz, J. Trull, C. Conti, S. Trillo, Phys. Rev. Lett. **91**, 093904 (2003).
2   O. Jedrkiewicz, A. Picozzi, M. Clerici, D. Faccio, P. Di Trapani, Phys. Rev. Lett. **97**, 243903 (2006).
3   O. Jedrkiewicz, M. Clerici, A. Picozzi, D. Faccio, and P. Di Trapani, Phys. Rev. A **76**, 033823 (2007).
4   M. A. Porras and P. Di Trapani, Phys. Rev. E **69**, 066606 (2004).

Generation of Ultrashort Optical Pulses Using Multiple Coherent Anti-Stokes Raman Scattering Signals in a Crystal and Observation of the Raman Phase

Eiichi Matsubara, Taro Sekikawa, and Mikio Yamashita

Department of Applied Physics, Hokkaido University, and Core Research for Evolutional Science and Technology (CREST), Japan Science and Technology Agency, Kita-13, Nishi-8, Kita-ku, Sapporo, 060-8628, Japan
E-mail: matsu-e@eng.hokudai.ac.jp

Abstract. We demonstrate Fourier synthesis of the multiple coherent anti-Stokes Raman-scattering signals in $LiNbO_3$. Both angle and temporal dispersions of the signals are compensated for by a modified $4f$ configuration. As a result, isolated pulses with 25-fs duration (640–780 nm) are generated and discrete phase shifts due to Raman coherence are observed.

Introduction

Generation of one-optical-cycle pulses is attractive for many applications. One of the promising methods to achieve it is the Fourier synthesis of multiple Raman sidebands. So far, by using a cooled D_2 gas as a Raman medium, a train of 1.6-fs pulses have already been generated [1]. However, its repetition period is too short (11 fs) because spectra of the adjacent Raman sidebands do not overlap at all. Recently, it has been found that multiple coherent anti-Stokes Raman scattering (CARS) signals with broad bandwidths can be generated even in crystals [2] by the excitation with two-color femtosecond laser pulses at room temperature, in which all the spectra of adjacent CARS peaks overlap so that the entire spectrum is continuous. This property gives two advantages over the cases of a gas. One is the potential of the generation of "isolated" pulses. The other is that the spectral phase can be directly measured by spectral phase interferometry for direct electric field reconstruction (SPIDER). This is important not only from the viewpoint of the ultrafast optical pulse technology but also from the viewpoint of physics in quantum optics because whether or not discrete phase shifts of multiple CARS signals due to Raman coherence formation is an interesting issue.

Experimental Setup and Results

The experimental set up is shown in Fig. 1. The YZ-surface of a $LiNbO_3$ crystal with a 0.5-mm thickness (LN) was simultaneously irradiated by two slightly chirped (+100 to 200 fs^2) fundamental laser pulses from a multi-pass Ti:sapphire laser amplification system (center wavelength: 810 nm, duration: 30 fs, repetition rate: 1 kHz). The input pulse energies were 6 and 11 μJ. The relative angle between the two beams was $1.9°$ in the air, because in the case of crystals refractive index dispersions are large so that two collinear input beams cannot generate CARS signals efficiently. The angle dispersion of the signals was compensated into one white-continuum beam by a modified $4f$ configuration which consists of a spherical mirror with a 100-mm focal length (SM), a rectangular mirror (RM), a grating with a groove density of 1200 lines/mm and a blaze wavelength at 500 nm (G). To tune the magnification of f_1/f_2 arbitrarily, only one spherical mirror was used. The collimated beam was picked up by a small square mirror (SQM), and collimated both vertically and horizontally by two cylindrical mirrors (CM1 and CM2: focal lengths were 100 and 250 mm, respectively). Figure 2 shows the intensity spectrum and the spectral phase profiles of the collimated CARS beam measured by the modified SPIDER. It was found that the sign of the dispersion could be both positive (upper dotted curve) and negative (lower dotted curve). By appropriately adjusting the rotation angle and the position of the grating, the spectral phase became almost flat, within a 10-rad variation in the frequency range from 13000 to 15600 cm^{-1} (middle dotted curve). Here, we can see many step-like changes in the spectral phase. They originate from the Raman phase which is expected to slip discretely with increasing order of CARS signals. The temporal intensity profile of the generated pulse (black solid curve) and that of the Fourier-transform-limited (TL) one (black dotted curve) are shown in Fig. 3. The gray solid curve shows the temporal phase profile of the generated pulse, which is almost flat during the pulse duration. An isolated pulse with a duration of 25 fs is observed, while that of the TL one is 11 fs. This difference seems to come from the incompletely flattened spectral phase.

Discussion

Now we discuss the present results. First, we explain how the compensations of both the angle and the temporal dispersion of the multiple CARS signals are achieved in our experimental setup. The position and the angle of the grating (G) play the most important role. It is known that angle dispersive elements, such as prisms and gratings in a 4f configuration, introduce group delay dispersion (GDD: $d^2\Psi/d\omega^2$) in such a manner as [3],

$$\frac{d^2\Psi}{d\omega^2} = -\frac{1}{c}(z'M^2 + z)\left\{\left(2\frac{d\theta}{d\omega} + \omega\frac{d^2\theta}{d\omega^2}\right)\sin\theta + \omega\left(\frac{d\theta}{d\omega}\right)^2\cos\theta\right\} \quad (1)$$

Here, θ is the diffracted angle of a ray with an angular frequency ω, and is measured from the direction of the center-frequency component. z is the distance between the crystal and the first focal point, and z' is the distance between the second focal point and the grating. Because $d\theta/d\omega$ is a function of the angle of the grating, the GDD depends on both the angle and the position of the grating. Using parameters of $z=0$, $z'=\pm 300$ μm, and $M=0.74$, the applied GDD is ∓ 200 fs^2 at 700 nm. If we rotate the grating by $1°$, the GDD changes by 5 fs^2. Thus we can understand that the temporal dispersion of the collimated beam is adjustable by slightly moving the grating. Next we discuss why the spectral phase was not flattened perfectly. Although it is almost flat within 1-rad fluctuation in the frequency range from 13200 to 14500 cm^{-1}, it increases a little steeply to 11 rad in the range from 14600 to 15600 cm^{-1}. We think this is due to the nonnegligible higher-order dispersion applied by the modified 4f configuration. It will be certain that, by using a programmable multi-channel spatial light modulator [4] or well-designed chirped mirrors, we can generate TL pulses with this scheme. Finally we discuss the discrete phase shift due to Raman phase. Figure 4 shows the expanded spectral phase profile in Fig. 2 (the middle one). As eye-guided by dotted and dashed vertical lines, ripples with spacings of 155 and 369 cm^{-1} are seen, which correspond to the frequencies of TO (E) phonons. On the other hand, the spectral resolution of SPIDER determined by the spectral shear (170 cm^{-1}) is 340 cm^{-1} according to the Nyquist limit, so that one might think that such phonon frequencies are not observable. However, a simple simulation shows that some information on periodicity can still be retrieved even if the sampling step is nearly equal to the frequency of interest.

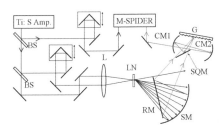

Fig. 1. Experimental setup for the generation of ultrashort optical pulses using multiple CARS signals in LiNbO$_3$.

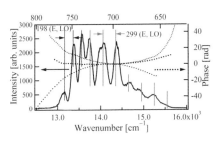

Fig. 2. Intensity spectrum (black solid curve) and spectral phase profiles (black dotted curves) of the generated pulse.

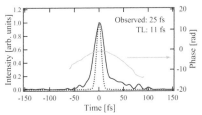

Fig. 3. Temporal-intensity profiles of the generated pulse (solid curve) and Fourier-transform limited one (dotted curve). Gray curve shows the temporal phase profile of the generated pulse.

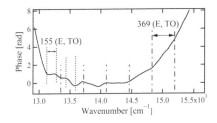

Fig. 4. Spectral-phase profile shown in Fig. 2. As eye-guided by dotted and dashed lines, ripples with spacings of 155 and 369 cm^{-1} are seen.

Conclusions

We demonstrated Fourier synthesis of the multiple CARS signals from a LiNbO$_3$ single crystal. Isolated pulses with 25-fs duration were generated. In addition, discrete phase shifts of multiple CARS signals due to Raman coherence were observed.

References

1 M. Y. Shverdin, D. R. Walker, D. D. Yavuz, G. Y. Yin, and S. E. Harris: Phys. Rev. Lett. **94**, 033904 (2005).

2 M. Zhi and A. V. Sokolov, New J. Phys. **10**, 025032 (2008), and references therein.

3 J-C. Diels and W. Rudolph: "*Ultrashort Laser Pulse Phenomena*", pp75-78, Academic press, San Diego (1996).

4 M. Yamashita, K. Yamane, and R. Morita: IEEE J. Sel. Top. Quantum Electron. **12**, 213 (2006).

Generation of High-power Visible and UV/VUV Supercontinua and Self-compressed Single-cycle Pulses in Metal-dielectric Hollow Waveguides

J. Herrmann, A. Husakou

Max Born Institute for Nonlinear Optics and Short Pulse Spectroscopy,
Max-Born-Str. 2a, D-12489 Berlin, Germany
email: jherrman@mbi-berlin.de

Abstract. We investigate high-power soliton-induced supercontinuum generation in visible and UV/VUV based on argon-filled metal-dielectric hollow waveguides. We predict the generation of MW/nm spectral power densities with ∼0.1 mJ energy and self-compressed isolated 1.7-fs pulses.

The discovery of supercontinua (SC) in microstructure fibers [1] using fs pulses with nJ energy has attracted widespread interest, it has been studied extensively in recent years and found several fascinating applications. The dramatic spectral broadening for low pulse energies is related to the crucial role of soliton dynamics and soliton emission [2], requiring anomalous dispersion at the input wavelength and thus small fiber radii. However, simultaneously small radii severely restrict the possible total pulse energies to few nJ and spectral power densities to tens of W/nm.

In this contribution, we predict that using specifically designed metal-dielectric hollow waveguides one can generate two-octave-broad SC with five order of magnitude higher power densities in the range of MW/nm and 0.1 mJ total energy. Besides, for optimized waveguide parameters self-compressed isolated sub-cycle pulses with a duration of 1.7 fs are predicted. In addition, we show that by pumping with the third harmonic of Ti:sapphire laser VUV supercontinua in the wavelength range of 150-500 nm can be generated.

Hollow dielectric waveguides are a key element in modern ultrafast nonlinear optics, e. g. for pulse compression and attosecond pulse generation. Unfortunately, these waveguides provide tolerable loss only for relatively large radii in the range of 50 μm, thus inhibiting the control over group-velocity dispersion (GVD). In this contribution we predict that metallic hollow waveguides coated with a nm-scale dielectric layer have moderate loss even for small radii in the range of 10 to 25 μm, with a broad range of anomalous dispersion even for high gas pressures.

We consider cylindrical straight hollow waveguides with metallic walls coated on the inner surface by a layer of dielectric to improve the guiding properties. The calculation of the waveguide loss and group velocity dispersion is made in the formalism of transfer matrices[3], including roughness loss modelled by pointlike scatterers. The simulation of the nonlinear pulse propagation is performed using the Forward Maxwell equations[2] including dispersion to all orders, Kerr nonlinearity and higher-order nonlinear effects, energy transfer to higher-order modes, as well as photoionization and plasma effects. In Fig. 1(a) the loss (red curve) and the group velocity dispersion (green curve) are presented for a 40-μm-radius silver waveguide coated with 45 nm of SiO_2. One can see that the waveguide loss remains in the range of 0.01-0.1 dB/m over the broad spectral range including the whole visible range, in contrast to dielectric hollow

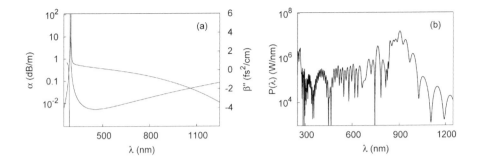

Fig. 1. Characteristics of the waveguide (a) and peak spectral power density (b) for a silver-wall waveguide with a radius of 40 μm coated by a 45-nm layer of SiO_2 with a average scatterer diameter of 100 nm and argon gas filling at 1 atm. In (b), we assume a 100-fs, 100 TW/cm^2 input pulse at 800 nm; the propagation length is 50 cm. The output pulse duration was used for the calculation of the *peak* spectral power density.

waveguides which have much higher loss for this radius. The group velocity dispersion is anomalous for all wavelengths above 600 nm for argon filling at 1 atm. The output peak spectral power density for a 100-fs, 100-TW/cm^2 input pulse is presented in Fig. 1(b). One can see that a spectrum reaching from 300 nm to 1500 nm is achieved, with peak power spectral density in the range of MW/nm and energy of 0.11 mJ, which are

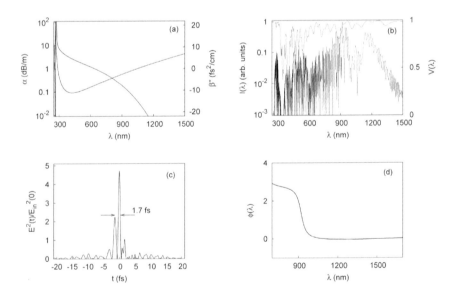

Fig. 2. Waveguide characteristics (a), spectrum (b), temporal shape (c) and the phase of the compressed pulse (d) for the a silver-wall waveguide with radius of 15μm, argon-filled at 13 atm, coating layer of 40 nm, an average scatterer diameter of 100 nm, and length of 71.8 cm. The input pulse has a duration of 50 fs, peak intensity of 20 TW/cm^2, and central wavelength of 800 nm.

4 to 5 orders of magnitude higher than in microstructure fiber. Such dramatic increase of the spectral power density is brought about both by the larger radius of the waveguide and by the increased allowable input intensity related to the high ionization threshold of argon. The mechanism of supercontinuum generation is due to soliton fission and soliton emission, similar to microstructured fibers.

To study SC generation for a higher nonlinerity, we consider a higher gas pressure which requires a lower waveguide radius. The loss [red curve in Fig. 2(a)] in the case of a 15-μm waveguide remains in the range of 1-10 dB/m. The GVD [green curve in Fig. 2(a)] is anomalous in the broad spectral range, despite the high argon pressure of 13 atm. In Fig. 2(b) the output spectrum is presented, exhibiting a spectral width of roughly two octaves. For propagation lengths in the range from 71 to 73 cm an ultrashort pulse is formed, as illustrated in Fig. 2(c) with smooth phase [Fig. 2(d)]. Its FWHM duration is $\sim$1.7 fs, and it has only a weak pedestal. Thus, we predict the generation of a coherent isolated sub-cycle pulse in a hollow waveguide without any external chirp compensation. Further, we have modeled the influence of quantum noise and calculated the first-order coherence [or visibility V(λ)] of the generated spectra. As illustrated by the green curve in Fig. 2(b) the whole spectrum is highly coherent with an average coherence of 0.98. Finally we studied the possibility to shift the SC to the VUV range using aluminium walls coated with SiO_2 and pumping with the third harmonic of Ti:sapphire lasers at 266 nm. We have predicted that VUV supercontinua in the wavelength range of 150-500 nm can be generated in this way (not shown).

1   J. K. Ranka, R. S. Windeler, and A. J. Stentz, "Visible continuum generation in air-silica microstructure optical fibers with anomalous dispersion at 800 nm", Opt. Lett. **25**, 2527 (2000).
2   J. Herrmann et al., "Experimental evidence for supercontinuum generation by fission of higher-order solitons in photonic fibers", Phys. Rev. Lett. **88**, 173901 (2002).
3   P. Yeh, A. Yariv, and E. Marom, "Theory of Bragg fiber", J. Opt. Soc. Am. **68**, 11961201 (1978).

Part X

Frequency Combs and Waveform Synthesis

CEO-Phase Stabilized Few-Cycle Field Synthesizer

Stefan Rausch[1], Thomas Binhammer[2], Anne Harth[1], Franz X. Kärtner[3] and Uwe Morgner [1,4]

[1] Institute of Quantum Optics, Leibniz University of Hannover, Welfengarten 1, 30167 Hannover, Germany
E-mail: rausch@iqo.uni-hannover.de
[2] VENTEON Femtosecond Laser Technologies by Nanolayers, Maarweg 30, 53619 Rheinbreitbach, Germany
[3] Research Laboratory of Electronics, Massachusetts Institute of Technology (MIT), Cambridge, MA, USA
[4] Laser Zentrum Hannover (LZH), Hollerithallee 8, 30419 Hannover, Germany

Abstract. We present an optical field synthesizer consisting of a CEO-phase stabilized octave-spanning Ti:sapphire laser oscillator and prism-based pulse shaper allowing for full control of the electric field on a sub-femtosecond time-scale.

Introduction

The electric field of few-cycle femtosecond laser pulses can be controlled on a sub-femtosecond time-scale by manipulating the spectral phase and amplitude together with the carrier envelope offset (CEO) phase of the pulse train. An octave-spanning Ti:sapphire oscillator provides the spectral width required for direct CEO-phase stabilization and few-cycle shaping experiments. This field synthesizer is a versatile tool for coherent control and CEO-phase sensitive experiments.

Field Synthesizer

The light source is a prism-less Ti:sapphire oscillator with specially designed output coupling mirror and double chirped mirror pairs [1], delivering an average output power of about 100 mW at a pulse repetition rate of 80 MHz. The octave-spanning output spectrum, shown in Fig. 1, supports a Fourier-limited pulse duration as short as 3.7 fs.

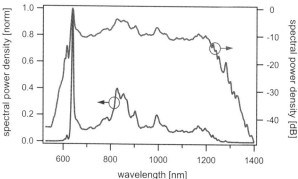

Fig. 1. Octave-spanning output spectrum supporting Fourier-limited pulses as short as 3.7 fs, shown on a linear (left axis) and logarithmic scale (right axis).

To obtain full control over the electric field of the few-cycle laser pulses the oscillator has to be CEO-phase stabilized. Due to the octave-spanning output spectrum no additional spectral broadening is necessary for the f-2f measurement approach [2]. The beat signal reveals a signal-to-noise ratio greater than 30 dB @ 100 kHz resolution bandwidth and is sufficient to stabilize the laser using a PLL locking electronic to control the pump power. This is schematically illustrated in Fig. 2 (upper right corner).

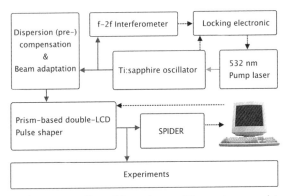

Fig. 2. Field synthesizer - overall working-scheme; The system consists of a CEO-phase stabilized Ti:sapphire oscillator, dispersion compensation and beam adaptation, prism-based double-LCD pulse shaper, SPIDER measurement system and personal computer.

The phase-stabilized pulses propagate through a prism-sequence to compensate for the highly dispersive prisms used in the 4-f geometry in which the double-LCD spatial light modulator is positioned in the Fourier-plane to independently manipulate the spectral phase and amplitude of the input femtosecond pulses. This shaper allows for highly efficient pulse manipulation covering the whole spectral octave. In the next step the pulses are analyzed with SPIDER. The computer-based analysis allows for a direct control of the shaper. If the required shape is generated and verified, the pulses can be guided towards the desired experiment.

This system allows for flexible manipulation of the CEO stabilized few-cycle input pulses, and full control over the electric field of these pulses on the sub-cycle scale is achieved.

Results and Discussion

The systems capability with respect to spectral phase and amplitude shaping is demonstrated in Fig. 3. The upper row shows a simple example: A shaped pulse with rectangular spectral amplitude and flat spectral phase. The calculated and measured time-domain sinc-shaped pulses match nearly perfectly.

Various spectral shapes can be realized, such as Gaussian or triangular. Also sharp and flexible wavelength filtering, e.g. edge or band pass, can be achieved using this technique.

The bottom part for Fig. 3 illustrates simulation results showing a double pulse sequence generated by simultaneously modulating the spectral phase and amplitude resulting in two clean sub-two-cycle double pulses in the time domain. The duration and shape of the pulses contained within the sequence is thereby similar to the input pulse, whereas the time delay is variable depending on the modulation frequency.

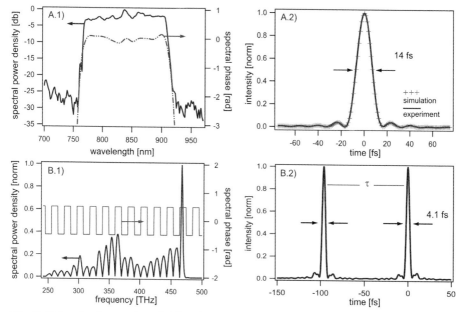

Fig. 3. Pulse shaping results; A) Sinc-shaped pulse formed by phase and amplitude shaping. B) Pulse sequence generated by simultaneous modulation of the spectral phase and amplitude (simulation results).

By using super-resolution techniques as presented in [3] pulse durations even below 3.7 fs are possible by accepting stronger wings in the pulse intensity profile. Such extreme pulse durations with controlled CEO-phase allow for novel experiments with sub-cycle scale resolution where the position and control of the electric field becomes crucial.

Conclusions

In conclusion, we demonstrated a phase-stabilized few-cycle field synthesizer build from an octave-spanning laser oscillator, a prism-based double-LCD pulse shaper for independent phase and amplitude shaping, and SPIDER measurement system. This unique combination is capable of forming versatile pulse shapes and sequences in time and frequency domain down to the single cycle.

Acknowledgements. The author thanks VENTEON Femtosecond Laser Technologies by Nanolayers for the close cooperation.

1   T.R. Schibli, O. Kuzucu, J.W. Kim, E.P. Ippen, J.G. Fujimoto, F.X. Kärtner, V. Scheuer, G. Angelow, in IEEE J. of Selected Topics in Quantum El., Vol. 9 (4), 990-1001, 2003.

2   U. Morgner, R. Ell, G. Metzler, T. R. Schibli, F. X. Kärtner, J. G. Fujimoto, H. A. Haus, and E. P. Ippen, in Phys. Rev. Lett. 86, No. 24, 5462-5465, 2001.

3   T. Binhammer, E. Rittweger, R. Ell, F.X. Kärtner and U. Morgner, in Opt. Lett. 31, 1552-1554, 2006.

High-power, mHz linewidth Yb:fiber optical frequency comb for high harmonic generation

T. R. Schibli[1], I. Hartl[2], D. C. Yost[1], M. J. Martin[1], A. Marcinkevičius[2], M. E. Fermann[2], and J. Ye[1]

[1] JILA, National Institute of Standards and Technology and University of Colorado, Boulder, CO 80309, USA
E-mail: trs@colorado.edu
[2] IMRA America, Inc., 1044 Woodridge Ave., Ann Arbor, MI 48105, USA
E-mail: ihartl@imra.com

Abstract. We present a fully phase-stabilized, high-power Yb:fiber frequency comb with record-low sub-mHz relative linewidths. Utilizing coherent pulse-addition inside a passive optical cavity, we achieve >3 kW average power and 100 fs pulse duration for high harmonic generation at >100 MHz pulse repetition rate. On a phosphor screen we visually observe up to the 21st harmonic generated in Xe at 136 MHz pulse repetition rate.

Introduction

Growing demands for high average and peak powers in extreme nonlinear optics, attosecond pulse, and XUV-comb generation experiments can find a powerful solution in mode-locked fiber lasers. Fiber lasers have demonstrated the capability to reliably produce sub-100 fs pulse trains with unprecedented average powers. In this paper we report on a fully phase-stabilized Yb:fiber laser system capable of producing an ultra-coherent fs-comb with sub-mHz relative linewidth at >10 W average powers and more than 3 kW with cavity-enhancement, empowering HHG at 136 MHz repetition rate.

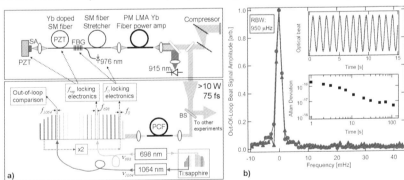

Fig. 1. a) Setup of the fiber comb and its precise characterizations: SA: Saturable absorber; PZT: Piezo actuator; FBG: Fiber Bragg grating; PM LMA: Polarization-maintaining large-mode-area fiber; PCF: Photonic crystal fiber; BS: Beam splitter; 698 nm: sub-Hz linewidth cw-laser for repetition rate locking; 1064 nm: cw-Nd:YAG laser locked to a Ti:sapphire frequency comb for the out-of-loop comparison. **b)** Two independent records of the out-of-loop heterodyne beat note between the stabilized 1064 nm laser and the Yb:fiber comb, showing an measurement limited linewidth of 950 µHz. Top inset: 15 s record of the out-of-loop beat note (dots) with a sinusoidal fit (line). Lower inset: Relative stability between the two combs.

Experimental Methods and Results

Figure 1a) shows the experimental setup for the high-power fiber comb. A Fabry-Perot type Yb-similariton oscillator [1,2] modelocked with a sub-picosecond lifetime saturable absorber is used as a seed-source. The dispersion of the optical fibers inside the oscillator is compensated by a chirped fiber Bragg grating to a net-normal dispersion value. The oscillator generates ~40 nm of bandwidth centered at 1065 nm with >100 mW average power. This 136 MHz pulse train is amplified in a chirped-pulse amplifier (CPA) in which the pulses are first stretched in an anomalous third-order dispersion single-mode fiber to ~70 ps to avoid nonlinear phase shifts in the subsequent power amplifier. After amplification, the pulses are recompressed using two fused-silica transmission gratings, yielding 75 fs pulses with >10 W of average power. The spectrum is then broadened to more than one frequency octave (675 nm - 1450 nm) in a 15 cm long photonic crystal fiber (PCF). After the PCF, a fraction of the continuum is sent into a nonlinear f-$2f$ interferometer, which yields the carrier-envelope offset frequency f_0. A (for fiber lasers) record low linewidth for the free-running f_0 signal of less than 10 kHz (inset of Fig. 2a) is routinely observed. Since the CPA is operating in a linear regime, the signal-to-noise ratio and the linewidth of f_0 do not degrade at higher amplifier output powers (Fig. 2a). To phase-stabilize the frequency comb, we lock f_0 to a Cs-clock using two loop filters, one of which controls the pump power of the fiber oscillator and the other the temperature of the fiber Bragg grating in the oscillator. When locked, the f_0 beat note contains 90% of the RF power within a coherent, mHz line-width carrier. The repetition rate of the oscillator is stabilized by phase-locking the heterodyne beat note between one of the comb teeth around 698 nm and a sub-Hz linewidth external cavity laser diode that is stabilized to a high finesse optical cavity ($F \sim 250,000$) [3]. This sub-Hz cw-laser provides better short-term stability than any commercial microwave source available to date. The heterodyne beat note was locked to the same Cs-reference using two loop filters, one controlling a fast and the other a slow PZT actuator inside the fiber oscillator. The locked heterodyne beat note again contains ~90% of the RF power within the coherent carrier.

To evaluate the comb's performance we conduct a thorough comparison between this novel, high-power fiber comb and an octave-spanning Ti:sapphire laser-based frequency comb. Fig. 1a) shows the experimental setup for this comparison: The Ti:sapphire laser and the Yb:fiber laser are both locked to the same sub-Hz linewidth reference laser at 698 nm. f_0 of each of the combs is independently stabilized using f-$2f$ setups. Utilizing the 698 nm cw-laser as a common optical reference allows us to compare the two combs without being limited by the stability of the reference laser. However, since the 698 nm reference and the two combs are each separated by tens of meters it is crucial to actively stabilize the fiber links between the three setups to a small fraction of an optical wavelength [4]. Finally, we use a Nd:YAG non-planar ring oscillator (NPRO) as a transfer laser by locking it to one of the comb teeth of the stabilized Ti:sapphire comb at 1064 nm. The light of this transfer laser is then delivered through a third noise-cancelled fiber link to the fiber-comb setup. The heterodyne beat note between the stabilized Nd:YAG laser and the fiber comb is recorded with an FFT analyzer and a frequency counter. From the FFT analyzer we find measurement-limited linewidths of less than 1 mHz (Fig. 1b). The top inset of Fig. 1b) shows a 15 s record of the out-of-loop beat note (dots) with a sinusoidal fit (line). The lower inset shows the Allan deviation of the out-of-loop beat.

By coherent addition of the fs-pulses produced by this laser system, we have obtained more than 3 kW average power and 100 fs pulse duration inside a passive optical cavity [5]. Inside the passive enhancement cavity we reach intracavity intensities as high as $3 \cdot 10^{14}$ W/cm^2. To confirm these record-high levels of intensity at >100 MHz repetition rates, we measure electric currents through the laser induced plasma for several noble gases (see Fig. 2b). A theoretical calculation for Kr (green, dotted line; calculation by Robin Santra, Argonne National Laboratory, IL) shows an excellent agreement with the measured data. In addition we observed strong high-harmonic radiation produced in Xe at 136 MHz repetition rate (inset Fig. 2b).

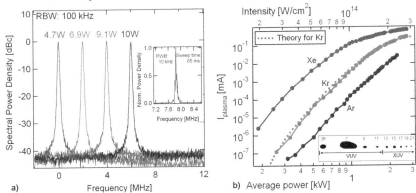

Fig. 2a) Free running f_0 beat notes obtained from a nonlinear f-$2f$ interferometer at four different levels of average output powers (4.7W, 6.9W 9.1W and 10W). Each trace shows an average of 25 single scans at 100 kHz RBW (20 s accumulation per trace.) The four traces are offset in x-direction for clarity. Inset: Free running beat note on a linear scale with 10 kHz RBW (sweep time: 65 ms.) **b)** Electrical currents through the laser-induced plasma for various noble gases at 750 mTorr under 10 V/mm bias as a function of laser power and peak intensity. The perfect agreement with a theoretical calculation (dotted line) and the clear onset of saturation confirm the $3 \cdot 10^{14}$ W/cm^2 peak intensity levels. The inset shows the fluorescence of high-harmonic radiation produced in Xe at 136 MHz repetition rate on a sodium salicylate screen after separation on a VUV grating. The power per harmonic across the VUV range (5th – 11th harmonic) exceeds 1 µW while the average power for the 13th – 19th harmonic is ~100 nW/harmonic. The 21st harmonic marks the beginning of the cut-off region.

Conclusions

We have demonstrated that Yb:fiber lasers in conjunction with chirped pulse Yb:fiber amplifiers are not only perfectly suited to produce very high average and peak powers, but are also capable of producing ultra-precise optical frequency combs. In conjunction with coherent pulse addition inside a passive optical cavity we observed strong HHG at 136 MHz, yielding all odd-harmonics up to the 21st. This work certainly marks an important milestone towards coherent as-comb generation in the XUV domain as well as highly-nonlinear light-matter interaction experiments that require precise control over the optical phase.

1   M. E. Fermann, "Ultrafast fiber oscillators', in Ultrafast lasers:technology and applications," (eds. M. E. Fermann, A. Galvanauskas, G. Suchaeds) Marcel Dekker, New York, 2003.

2   F. Ő. Ilday, et al., Phys. Rev. Lett., **92,** 213902, 2004.

3   A. D. Ludlow, et al., Opt. Lett., **32,** 641-643, 2007.

4   S. M. Foreman, et al., Rev. Sci. Instrum. **78,** 021101/1-25, 2007.

5   I. Hartl, et al., Opt. Lett, **32,** 2870-2872, 2007.

High Harmonic Frequency Combs for High Resolution Spectroscopy

A. Ozawa[1], J. Rauschenberger[1,2], Ch. Gohle[1], M. Herrmann[1], D. R. Walker[1]
V. Pervak[2], A. Fernandez[1], A. Apolonski[2], R. Holzwarth[1], F. Krausz[1,2]
T. W. Hänsch[1,2] and Th. Udem[1]

[1] Max-Planck-Institut für Quantenoptik, Hans-Kopfermann-Strasse 1, 85748 Garching, Germany
   E-mail: akira.ozawa@mpq.mpg.de
[2] Department für Physik der Ludwig-Maximilians-Universität München, Am Coulombwall 1, 85748 Garching, Germany

Abstract. Intracavity high harmonic generation is demonstrated in an external cavity, seeded by a Ti:sapphire mode-locked laser at a repetition rate of 10.8MHz. Harmonics up to 19th order at 43 nm were observed with plateau harmonics at the µW power level.

Introduction

The extreme ultraviolet (XUV) frequency comb technique is expected to play an important role in extending high-resolution laser spectroscopy to the XUV region where many interesting transitions are located. Conventional frequency combs generated at infrared wavelengths can be frequency converted to XUV wavelengths with high harmonic generation (HHG). However, until recently, a low repetition rate in the kHz range was required to generate sufficient pulse energy to produce high harmonics. The resulting frequency comb is far too dense to be useful for high-resolution spectroscopy. Direct frequency comb spectroscopy requires the separation of the modes, i.e., the repetition rate, to be much larger than the measured linewidth. This problem has been solved by employing intracavity HHG, achieving the required intensity by resonantly enhancing the pulse train from a mode-locked laser without compromising on the repetition rate [1-4]. Unfortunately, the powers generated in the XUV with this technique have been far too low for a reasonable excitation rate of a narrowband XUV transition. We report a dramatic enhancement of the XUV output power by almost 5 orders of magnitude. This is achieved by an elaborate dispersion compensating scheme and by reducing the repetition rate to 10.8MHz, which is still sufficient to resolve narrowband atomic transitions.

Experimental Methods

Our mode-locked laser has a long cavity and runs in the positive GDD regime. Its linear cavity has a length of 13.9m which is achieved by using a Herriott-type multi-path delay line. The output of about 1.5W of average power is compressed by a pair of LaK16 (SCHOTT) prisms (Fig. 1). The enhancement cavity is located in a vacuum setup and consists of 10 mirrors including 8 quarter-wave-stack mirrors, 4 of which are curved, 2 chirped mirrors, and one input coupler. Two homemade chirped mirrors are used to compensate the dispersion of the Brewster XUV output coupler made of sapphire and the contribution from the remaining cavity mirrors. Firstly, we design one chirped mirror by using the estimated total dispersion of the cavity. Then the residual dispersion is measured by analyzing the spectral cavity enhancement as described in [5]. Using this information, we generate several different coating designs with the correct dispersion properties and choose the design which is the least sensitive to the coating errors. With this dispersion compensating scheme, we obtain broad enhancement over 40nm. The circulating power is determined to be 100W by measuring the residual transmission through one of the highly reflecting cavity mirrors. This corresponds to an average power enhancement of 100. Intracavity pulsewidth is measured to be 57fs. Asymmetric focusing with two differently curved mirrors (radii of curvature, 0.1m and 0.24m) is used to focus to a waist size of $w_0 = 13$ μm into the gas target emerging from a nozzle. At the focus, a peak intensity of > 5 10^{13} W/cm^2 is obtained. When injecting Xe, Ar, or air through the gas nozzle, a fluorescing plasma can be observed. The gas flow is estimated to be $1\ 10^{-2}$ mbar l/s. The XUV output is used to illuminate the entrance slit of a scanning grating spectrometer (Jobin-Yvon, LHT30) that has an estimated resolution of 1.4 nm and is equipped with a channeltron detector (Burle, CEM4839). The pulse compressor, gas flow rate, nozzle position, and carrier-envelope offset frequency are optimized to maximize the XUV signal. The measured XUV spectrum is divided by the specified wavelength dependent grating diffraction efficiency. In order to independently calibrate the XUV power, the (spectrally unresolved) total power is measured with a calibrated Si photodiode (IRD, AXUV20HS1) placed directly after the Brewster XUV output coupler. A 150μm thickness Al filter (Lebow) is used to remove the residual reflection of the fundamental laser beam. The transmission of the Al filter is then measured by comparing the XUV spectra with and without it. Knowing the undistorted XUV spectrum, the total power and the spectrally resolved filter transmission, we determine the absolute spectral power density of outcoupled XUV beam.

Results and Discussion

The obtained spectral power density of XUV beam is shown in Fig. 2. A μW power level is obtained at plateau harmonics. Compared to previously reported powers [1], this represents an improvement of 10^4-10^5. High harmonics up to the 19th order (41 nm) are clearly observed, which agrees with the calculated cutoff located between the 13th and 15th harmonics. In addition to the odd harmonics, a broad peak at 104 nm is observed. Its origin is not yet understood but appears to be related to the occupancy of excited bound states of the Xe atom [1]. With the power level obtained here, precision spectroscopy in the XUV comes into reach for the first time. An example could provide the 1S-2S transition in He$^+$. In this system, the ground-state ions may be excited to the 2S state by two-photon absorption by using the 13th harmonic around 60.8 nm. A third photon can further ionize to He^{2+} which can be accumulated in the ion trap and detected with unity efficiency. If the full 0.84μW generated so far could be focused on the He ion with a waist size of 0.5 μm, an ionization rate close to 1Hz (0.88Hz) could be obtained. This is a typical rate for a high precision spectroscopy experiment on trapped ions. Since there is virtually no background for this type of detection, we believe precision spectroscopy is realistic even with count rates below 1Hz.

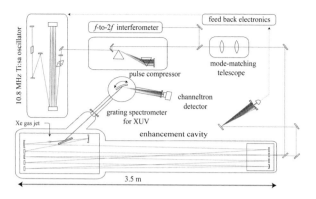

Fig. 1. Experimental Setup

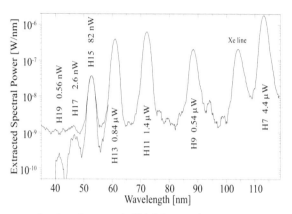

Fig. 2. Spectrum power density of outcoupled high harmonics

Conclusions

Generation of high power XUV frequency combs with µW power level is demonstrated. An elaborate dispersion compensation scheme and the use of a moderate repetition rate allow for this significant improvement in output power. With this power level and repetition rate, high-resolution spectroscopy in the extreme ultraviolet (XUV) region becomes conceivable. Our improved XUV source demonstrates that generating high harmonics in external enhancement cavities has now moved beyond proof-of-principle experiments to become a tool capable of performing spectroscopy at previously inaccessible wavelengths.

1   Ch. Gohle, Th. Udem, M. Herrmann, J. Rauschenberger, R. Holzwarth, H.A. Schuessler, F. Krausz, and T. W. Hänsch, in Nature, 436, 234, 2005.

2   R. J. Jones, K. D. Moll, M. J. Thorpe, and Jun Ye, in Phys. Rev. Lett., 94, 193201, 2005.

3   I. Hartl, T. R. Schibli, A. Marcinkevicius, D. C. Yost, D. D. Hudson, M. E. Fermann, and Jun Ye, in Opt. Lett., 32, 2870, 2007.

4   D. C. Yost, T. R. Schibli, and Jun Ye, in Opt. Lett. 33, 1099, 2008.

5   A. Schliesser, Ch. Gohle, Th. Udem and T. W. Hänsch, in Opt. Express, 14, 5975, 2006.

Ultrafast double pulse parametric amplification for precision Ramsey metrology

D.Z. Kandula, A. Renault, C. Gohle, A.L. Wolf, S. Witte, W. Hogervorst, W. Ubachs, and K.S.E. Eikema

Laser Centre, Vrije Universiteit Amsterdam, De Boelelaan 1081, 1081HV Amsterdam, Netherlands
E-mail: kjeld@nat.vu.nl

Abstract. An optical parametric chirped pulse amplifier system for pulse pairs is presented. The differential phase stability of the pulse pairs is 20 mrad, giving good prospects for high resolution Ramsey spectroscopy in the extreme ultraviolet.

Introduction

The convergence of ultrafast optics and high precision optical frequency metrology has led to fascinating new possibilities. The advent of femtosecond laser optical frequency combs (OFC) [1] turned the determination of arbitrary optical frequencies into a routine task. This enabled new high accuracy test of quantum electrodynamics (QED) [2] and the determination of tight lower bounds on the current time dependence of fundamental physical constants [3] as well as it opened the possibility of creating optical frequeny standards which have the potential to give 18 digits of accuracy. Simultaneously OFC opened the door to attosecond physics [4].

In recent years a new branch of high-precision physics is emerging from the crossfertilization between laser frequency metrology and ultrafast technology. Ultrashort pulses provide an extremely high peak power suitable for relatively efficient nonlinear frequency conversion of the available power into frequency ranges where no laser sources exist. If these pulses originate from a phase stable OFC, they interfere to create a frequency comb inside the THz bandwidth of a single femtosecond pulse. Each of the modes of this frequency comb can be extremely narrow and are suitable as a probe for high resolution spectroscopy experiments. As the nonlinear conversion can preserve this comb structure, high-precision experiments in the extreme ultraviolet spectral range come into reach, and first proof of principle experiments have been perfomed to demonstrate this idea [5,6].

Quite a few interesting atomic transitions exist in the extreme ultraviolet (XUV) wavelength range below 100 nm. As an example, the $1s^2$ $^1S_0 - 1s4p$ $^1P_1$ transition in atomic helium at 52 nm wavelength could be used to improve the value of the ground state Lamb shift in this atom by at least an order of magnitude. One specifically intriguing possibility is to determine the frequency of the 1s-2s transition in He II, a 2-photon transition at 60 nm. He II is a Hydrogen like system which is specifically simple to model in QED, and comparisons between theory and experiment could in principle be carried out at an extreme level of accuracy. Compared to Hydrogen, the ground state energy in He II is four-fold lower which leads to stronger relativistic effects. Specifically, the ground state Lamb shift (i.e. the QED correction to the Dirac energy level structure) is 16 fold stronger than in Hydrogen.

Currently, there exist two approaches to coherent XUV generation for high resolution spectroscopy. The first possibility is to enhance the entire bandwidth of a MHz repetition rate femtosecond frequency comb laser in a passive optical resonator in order to achieve the intensities required for high order harmonic generation (HHG) [7]. On the other hand it can be shown that optical Ramsey type spectroscopy, where only two identical pulses with a variable but precisely known delay and carrier envelope phase are used to excite the atom, is essentially equivalent. Such pulse pairs can be obtained by pulsed amplification of an OFC seed source. While the former is a continuous wave technique which avoids many systematic effects due to transients in the optically active materials in the setup, we prefer the latter as it can rely on cutting edge non collinear optical parametric chirped pulse amplifier (NOPCPA) technology [8]. This offers more than three orders of magnitude larger peak power, so that the nonlinear conversion becomes a lot more efficient, facilitating the actual spectroscopy experiment.

Phase-stable double pulse NOPCPA

The double-pulse NOPCPA is based on the single-pulse system presented previously [9]. To achieve double-pulse amplification, the pump pulses are split, delayed with respect to each other and superimposed in a symmetrized relay imaged Mach-Zehnder
interferometer. In this way two identical (in spatial profile) pump pulses with a delay of 6.6 ns are generated. This delay matches the time between two pulses from our 151 MHz OFC seed oscillator. In this way we can amplify a pair of subsequent pulses from our OFC to the millijoule level with a pulse duration down to 10 fs. This pair can be upconverted into the XUV spectral and used for a Ramsey experiment with a spacing of the Ramsey fringes of 151 MHz.

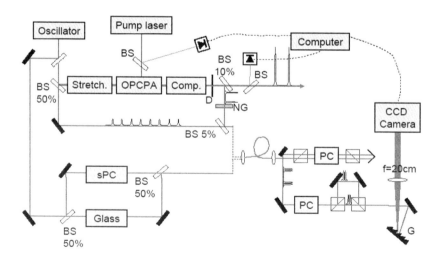

Fig. 1. NOPCPA and Mach-Zehnder interferometer for measurement of the differential phase shift accumulated during the amplification in a NOPCPA, and the setup used to test the reliability of the measurement (dotted beam line). BS: Beam splitter, PC: Pockels cell, sPC – slow Pockels cell, NG: neutral grey filter, Comp.: compressor, G: 1200 l/mm grating

If the wavefront and the intensity of the two pulses is not identical, the two amplified pulses can acquire different phase shifts during the amplification process. This would lead to a shift of the Ramsey fringes with respect to the original comb mode positions. Such a shift is multiplied up by the upconversion process, which means that it needs to be accurately measured and monitored.

Figure 1 shows a schematic of the setup used to monitor differential phase shift induced by the NOPCPA system on the amplified pulse pair. It consists of a Mach-Zehnder interferometer including the NOPCPA, the stretcher and compressor in one arm. The output of this interferometer is analysed using spectral interferometry. In order to maintain a good contrast in the interference signal, the output of the interferometer is spatially filtered using a single mode optical fiber and temporally gated to remove the pulses from the seed oscillator which have no amplified counterpart using a double passed Pockels cell (PC1). Like this we obtain a fringe contrast of almost unity with a background from leaking oscillator pulses of less than 10^{-4}. The two remaining pulses are spatially split using a second Pockels cell so that two interferograms can be recorded, one for each pulse. The information on the differential phase shift is now contained in the relative position of the two spectral interferograms. As only differential phase shifts are of importance for the spectroscopy the absolute position of the two interferograms is not important and the interferometer needs only to be stable on a 10 ns time scale (the separation between the pulses). This is passively guaranteed even for the almost 10 meter arm length of the interferometer.

Results

The accuracy of the phase measurement scheme was found to be better than 10 mrad. The single shot stability of the interferometer output is on the same order of magnitude, so that shot to shot fluctuations can be accurately measured and taken into account. The rms phase stability of the NOPCPA system was found to be 20 mrad. This is already sufficient for the observation of clear Ramsey fringes even at the 15th harmonic of the fundamental signal frequency at 800 nm, that is required for the $1s^2$ $^1S_0 - 1s4p\ ^1P_1$ transition in He I. The achievable accuracy for the lamb shift due to the amplifier phase shift is therefore at least 15 MHz which would be a threefold improvement over previous results.

1   T. Udem, R. Holzwarth and T.W. Hänsch, Nature. **416**, 233, 2002.

2   M. Niering, R. Holzwarth et al., Phys. Rev. Lett., **84**, 5496, 2000.

3   M. Fischer, N. Kolachevsky et al., Phys. Rev. Lett., **92**, 230802, 2004.

4   A. Baltuska, Th Udem et al., Nature, **421**, 611, 2003.

5   S. Witte, R.T. Zinkstok, W. Ubachs, W. Hogervorst and K.S.E. Eikema, Science, **307**, 400, 2005.

6   R.T. Zinkstock, S. Witte, W. Ubachs, W. Hogervorst and K.S.E. Eikema, Phys. Rev. A, **73**, 061801, 2006.

7   C. Gohle, Th. Udem, M. Herrmann, J Rauschenberger, R. Holzwarth, H. A. Schuessler, F. Krausz and T.W. Hänsch, Nature, **436**, 234, 2005.

8   S. Witte, R. T. Zinkstok, A.L. Wolf, W. Hogervorst, W. Ubachs and K.S.E Eikema, Opt. Express, **14**, 8168, 2006

9   A. Renault, D.Z. Kandula, S. Witte, A.L. Wolf, R.T. Zinkstok, W. Hogervorst and K.S.E Eikema, Opt. Lett., **32**, 2363, 2007

Towards Versatile Coherent Pulse Synthesis using a Femtosecond Laser and Synchronously Pumped Optical Parametric Oscillator

Barry J. S. Gale, Jinghua Sun, and Derryck T. Reid

Ultrafast Optics Group, School of Engineering and Physical Sciences,
Heriot-Watt University, Edinburgh EH14 4AS
E-mail: d.t.reid@hw.ac.uk

Abstract. Pulses from a femtosecond optical parametric oscillator and its Ti:sapphire pump laser were phase-locked as a prerequisite to coherent synthesis from different wavelengths. Mutual coherence was demonstrated using spectral interferometry and cross-correlation.

Introduction

Pulse compression and shaping are limited by the available spectral components of a single laser. Micro-structured fibres are capable of broadening a laser spectrum to a supercontinuum sufficient to support a pulse of 2.6 femtosecond duration [1], but they require high pulse power and have little flexibility to obtain variable spectral shapes. Coherent synthesis of the outputs from multiple laser sources offers a direct and intuitive way to achieve the desired spectral components [2-4], but it requires not only carrier-envelope phase slip (CEPS) locking, but also precise synchronization of pulse repetition frequencies (F_{rep}) to ensure the coherence of the combined pulses over significant time periods. An alternative route uses multi-colour pulses from an optical parametric oscillator (OPO). This requires active control of only the carrier-envelope phase slip (CEPS) frequencies, because of the inherent passive synchronization between the pump and all OPO outputs, which makes coherent synthesis easier and more robust. We previously generated coherent outputs from a femtosecond OPO centered at 1240 nm and 1330 nm [5]. Here we report the achievement of mutual coherence between the pump and frequency-doubled signal pulses, both at 780 nm, as a prerequisite for coherent synthesis from different wavelengths.

Experiment

The 1.25 W output from a self-mode-locked Ti:sapphire laser was split by mirror M1 (see Fig. 1), with 1 W and 0.25 W used to pump a quasi-phase-matched MgO:PPLN OPO and directed to an f-to-2f non-linear interferometer respectively. The latter employed photonic crystal fiber to generate a super-continuum reference spectrum against which the pump and signal CEPS frequencies were measured.

The OPO was operated close to degeneracy with a signal (idler) wavelength of 1560 (1640) nm. The non-phase-matched sum-frequency mixing (SFM) between pump and signal (529 nm), which was reflected from the PPLN crystal rear surface, was collected through mirror M4 and heterodyned with the pump super-continuum in a second interferometer to obtain the signal CEPS frequency. Signal second-harmonic (SH) light with an average power of 15 mW at 780 nm was generated by an intra-cavity frequency-doubling BBO crystal, and collected through mirror M7. Pump pulses at 780 nm with average power 10 mW were selected with an interference filter (IF) from the residual pump leaking through OPO mirror M5, and after suitable delay combined with signal SH pulses in polarizing beam splitter PBS3.

The pump and signal CEPS frequencies were locked to 50 MHz ($F_{rep}/4$) and 25 MHz ($F_{rep}/8$) respectively, so that the signal SH CEPS frequency was also 50 MHz. The locking bandwidths were 1.2 kHz and 2.7 kHz respectively. The CEPS frequencies of the pump and signal were controlled using a travelling-wave acousto-optic modulator (not shown) to fine-tune the power from the Verdi laser, and a piezoelectric translator (PZT) mirror M6 to tune the OPO cavity length.

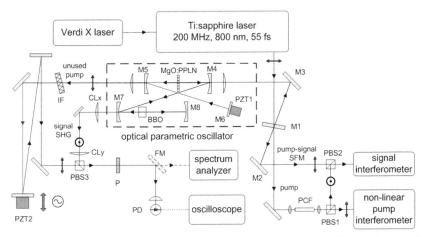

Fig. 1 Optical layout. BBO, β barium borate crystal. FM, flipper mirror. IF, interference filter. P, polariser. PBS1-3, polarising beam splitters. PZT1,2, piezoelectric transducers.

Results

When the CEPS frequencies of the pump and signal SH were locked to the same value (50 MHz), spectral interference between the two outputs at 780 nm showed deep fringes (Figure 2a) on the screen of the optical spectrum analyzer (see Figure 1) during a scan time of 4 ms, indicating strong coherence between the sources. Fringe visibility was limited by imperfect spectral overlap, spectrometer resolution (0.3 nm), and small differences in the beam size and divergence. Scanning PZT2 (see Fig. 1) gave an interferometric second-order cross-correlation trace with high-contrast fringes (Figure 2c) in a sweep time of 20 ms, which was stable for several seconds. The 6:1 contrast ratio (as compared with 8:1 for autocorrelation) does not indicate lack of coherence, only that the input powers were not exactly balanced. Fourier transformation of the in-loop phase-locking errors showed that the pulses from both the Ti:sapphire laser and the OPO had sub-ms coherence times individually, but because the OPO inherited the pump laser noise, the mutual coherence time could be much longer. A phase-noise power spectral density analysis revealed that both oscillators possessed significant acoustic noise at 220 Hz.

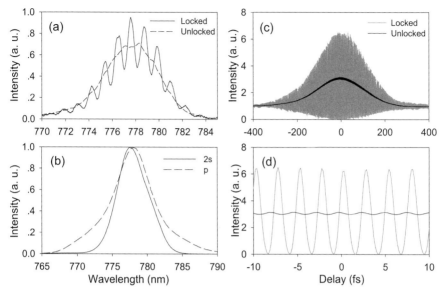

Fig. 2. (a) Spectral interference observed in a 4 ms scan between SH signal and pump pulses with CEPS frequencies locked to 50 MHz (solid line) and unlocked (dashed line). (b) Typical constituent spectra, recorded separately from measurement (a). 2s, SH of the signal. p, the pump. (c) Second-order cross-correlation between SH signal and pump pulses when their CEPS frequencies were locked to 50 MHz (lighter lines) or unlocked (darker lines). (d) Expanded detail of fringes around zero delay.

Conclusions

We have generated two coherent pulse sequences by locking the CEPS frequencies of pulses from a femtosecond OPO and its pump laser to sub-harmonics of their common repetition rate. Spectral interferometry and cross-correlation indicated that coherence between the pulses was maintained for at least 20 ms. Such a level of coherence has been shown sufficient for synthesizing waveforms [3].

Acknowledgements. We gratefully acknowledge support for this research from Coherent, Inc, and the Engineering and Physical Sciences Research Council, UK.

1   E. Matsubara, K. Yamane, T. Sekikawa, and M. Yamashita, J. Opt. Soc. Am. B **24**, 985 (2007).
2   T. W. Hänsch, Opt. Commun. **80**, 71 (1990)
3   R. K. Shelton, L. S. Ma, H. C. Kapteyn, M. M. Murnane, J. L. Hall and J. Ye, Science **293**, 1286 (2001)
4   Y. Kobayashi, H. Takada, M. Kakehata and K. Torizuka, Appl. Phys. Lett. **83**, 839 (2003)
5   J. H. Sun, B. J. S. Gale and D. T. Reid, Opt. Lett. **32**, 1396 (2007)

Frequency comb spectroscopy on calcium ions in a linear Paul trap

A.L. Wolf[1], S.A. v.d. Berg[2], C. Gohle[1], E.J. Salumbides[1], W. Ubachs[1], K.S.E. Eikema[1]

[1] Laser Centre Vrije Universiteit, De Boelelaan 1081, 1081 HV Amsterdam, The Netherlands
[2] NMi van Swinden Laboratorium BV, Thijsseweg 11, 2629 JA Delft, The Netherlands

Abstract. To add to the debate on a possible variation of the fine structure constant, we perform frequency comb spectroscopy on laser cooled (calcium) ions in a linear Paul trap.

Introduction

In 1999, Webb et al. claimed a possible variation of the fine structure constant α of $\Delta\alpha/\alpha = 0.57(10) \times 10^{-5}$ over cosmological timescales [1]. These results are based on a comparison of the wavelengths of atomic resonances between absorption lines in quasar spectra observed at high redshift, and the current laboratory values. Since spectral lines have a different dependence on a change in α, such an analysis can be used to find a non-zero value for $\Delta\alpha/\alpha$ over time spans of many billion years. Currently, many ionic lines that are interesting for this analysis are only known to a precision of a few tens of MHz. By using cooled and trapped ions, the lines can be measured to much higher precision.

An interesting ion for comparison to quasar spectra is Ca^+. We have developed a setup for trapping and laser cooling calcium ions to Coulomb crystallization, to obtain a much improved transition frequency for the $4s\,^2S_{1/2} - 4p\,^2P_{1/2}$ transition.

Experimental methods

Trapping and laser cooling of calcium ions. Calcium ions are trapped and cooled in a linear Paul trap. Atomic calcium is first ionized in the trapping region using a frequency doubled Ti:Sapphire laser (422 nm) and a frequency tripled Nd:YAG laser (355 nm). A 3.3 MHz radio-frequency potential for trapping the ions is supplied by a waveform generator, resonantly enhanced by a helical resonator. A grating stabilized diode laser at 397 nm is used for laser cooling on the $4s\,^2S_{1/2} - 4p\,^2P_{1/2}$ transition, while an additional diode laser at 866 nm is used for repumping of the ions that leak into the $3d\,^2D_{3/2}$ state. Fluorescence from the trapping region is imaged onto a photomultiplier for signal detection.

On scanning the cooling laser over the transition, an asymmetric fluorescence signal is detected (see Figure 1): at energies below the transition the ions are cooled, above the transition ions are heated, and hence blown out of the laser focus, leading to a loss of fluorescence. The phase transition to a Coulomb crystal is clearly visible as a sudden decrease of the fluorescence on the low frequency side of the transition [2].

To make the spectroscopy independent of the coolig dynamics, the cooling laser frequency is fixed, while an additional weak probe laser is used for the spectroscopy.

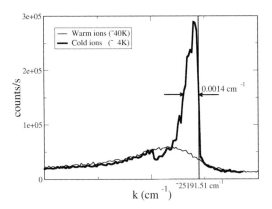

Fig. 1. Measured fluorescence by scanning the 397 nm laser cooling diode laser over the Ca$^+$ $4s\,^2S_{1/2} - 4p\,^2P_{1/2}$ transition

Spectroscopy on calcium ions The $4s\,^2S_{1/2} - 4p\,^2P_{1/2}$ transition is measured on the laser cooled calcium ions using a frequency-doubled diode laser. This beam necessarily has a low intensity ($\sim 1\mu$W) to prevent heating of the ion cloud by the spectroscopy laser. To keep the ions cold during the measurement, the ions are alternately cooled and probed. The probe laser is calibrated by referencing it to a frequency comb laser. An interference beat note is generated between the spectroscopy laser and the comb laser modes, by overlapping the near-infrared fundamental output of the spectroscopy laser with the output of the frequency comb laser.

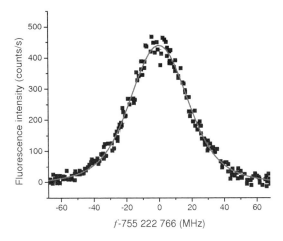

Fig. 2. An example of the measured fluorescence spectrum for the $4s\,^2S_{1/2} - 4p\,^2P_{1/2}$ transition in Ca$^+$ and the corresponding Voigt fit

Results and Discussion

An example of a calibrated scan over the $4s\,^2S_{1/2} - 4p\,^2P_{1/2}$ transition in Ca^+, after subtraction of the background and correction for ion loss during the scan, is shown in figure 2. The cooling is insufficient to reach the natural linewidth of the transition, hence a Voigt profile is fitted. The width of the Lorentzian part is fixed to the natural linewidth of the transition (22.4 MHz), while the width of the Gaussian part is fitted. The Gaussian FWHM of the line varies from scan to scan (depending on the cooling conditions), typically between 28 and 43 MHz, corresponding to an average temperature of $T \approx 0.2K$.

Effects that can introduce systematic and statistical errors have been investigated (see table 1). The $^{40}Ca^+$ $4s\,^2S_{1/2} - 4p\,^2P_{1/2}$ transition follows from the statistical average of the measurements, corrected for the measured shifts, which in total adds up to 755 222 766.2(1.7) MHz. The present result is consistent with the most accurate previously reported value of $f = 755\,222\,740(60)$ MHz [3].

Table 1. Measured systematic shifts and uncertainty budget (1σ). All values in MHz.

Effect	Shift(MHz)	1σ Uncertainty (MHz)
Zeeman	0.0	0.0
AC Stark repumper	-0.4	0.6
AC Stark spectroscopy laser	-0.4	0.8
RF Stark effect	0.0	1.2
Comb calibration	0.0	0.2
Statistics		0.6
Total	-0.8	1.7

Conclusions

We have measured the $4s\,^2S_{1/2} - 4p\,^2P_{1/2}$ transition in $^{40}Ca^+$ to be at 755 222 766.2(1.7) MHz, in a laser cooled ion crystal. The level of accuracy, at $\Delta\lambda/\lambda = 2 \times 10^{-9}$, is such that for comparison with state-of-the-art astrophysical data, the laboratory value can be considered exact. The technique employed in this work will be used for spectroscopy on other transitions and ions in the near future. We intend to measure the $4s\,^2S_{1/2} - 4p\,^2P_{3/2}$ and the $4s\,^2S_{1/2} - 5p\,^2S_{1/2}$ transitions in Ca^+, the latter using direct two-photon frequency comb spectroscopy [4,5]

Acknowledgements. This project is financially supported by the Foundation for Fundamental Research on Matter (FOM), and the Nederlands Meetinstituut (NMi). These contributions are gratefully acknowledged.

1   J.K. Webb, V.V. Flambaum, C.W. Churchill, M.J. Drinkwater, and J.D. Barrow, Phys. Rev. Lett. 82, 884, 1999.
2   F. Diedrich, E. Peik, J.M. Chen, W. Quint, and H. Walther, Phys. Rev. Lett. 415, L7, 2004
3   U. Litzén, Private communication, 2008
4   A. Marian, M.C. Stowe, D. Felinto, and J. Ye, Phys. Rev. Lett. 95, 023001, 2005
5   S. Witte, R.Th. Zinkstok, W. Ubachs, W. Hogervorst, K.S.E. Eikema, Science 307, 400, 2005

Generation of octave-spanning Raman comb stabilized to an optical frequency standard

M. Katsuragawa[1, 2], F. L. Hong[3, 4], M. Arakawa[1], and T. Suzuki[1, 2]

[1] Department of Applied Physics and Chemistry, University of Electro-Communications, 1-5-1 Chofugaoka, Chofu, Tokyo 182-8585, Japan
[2]PRESTO, JST, 4-1-8 Honcho Kawaguchi, Saitama, Japan
[3]National Institute of Advanced Industrial Science and Technology, 1-1-1, Umezono, Tsukuba 305-8563, Ibaraki, Japan
[4]CREST, JST, 4-1-8 Honcho Kawaguchi, Saitama, Japan
E-mail: katsura@pc.uec.ac.jp

Abstract. We show a novel octave-spanning comb generation having precise frequency-spacing of a Raman transition. It is shown that the carrier-envelope-offset of the Raman comb is precisely controlled by stabilizing the driving lasers to an optical-frequency-standard.

Introduction

A novel approach using an adiabatic Raman technique to generate ultrashort pulses, has been extensively discussed. The technique relies on the production of maximal coherence through the adiabatic Raman process and the resultant collinear broad Raman generation [1-3]. The generated sidebands are mutually phase coherent and have a wide, equidistant frequency-spacing. Recently, the generation of an ultrahigh-repetition-rate train of monocycle pulses has been demonstrated by Fourier-synthesizing such Raman sidebands [4]. It has also been shown based on well-established techniques of evaluating a temporal waveform, an intensity autocorrelation [5] and a frequency-resolved optical-gating [6], that high quality ultrashort pulses with potential as an actual light source can be constituted from such Raman sidebands.

Here, we demonstrate generation of octave-spanning Raman sidebands with accurate control of the carrier envelope offset (ceo) by stabilizing the two-wavelength driving-laser radiations to an optical frequency standard. The realized broad Raman sidebands have potential to produce monocycle ultrashort pulses with an absolute-phase control.

Experimental Methods

The conceptual scheme and the experimental setup are illustrated in Fig. 1. The central part of the driving laser system is a dual-wavelength, injection-locked, pulsed Ti:sapphire laser [7]. The key performance of this laser is to emit two-wavelength, transform-limited, nanosecond-pulses from a single laser resonator, enabling the perfect overlap of the two nanosecond-pulsed-outputs (typically 6 ns full-width, half-maximum) in both time and space. This performance is realized by simultaneously injecting the two-wavelength continuous wave laser radiations as seeds, which are generated from the two independent external-cavity controlled diode-lasers,

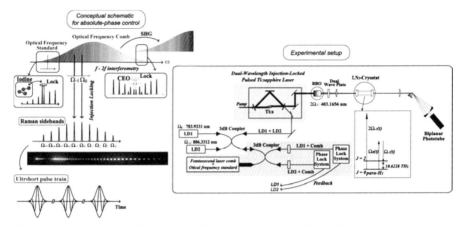

Fig. 1. Conceptual scheme for absolute-phase control and experimental setup.

respectively. In the present driving laser system, these two diode lasers were further phase-locked with each other through the femtosecond-laser optical-frequency-comb. This femtosecond-laser optical-frequency-comb had an absolute frequency stability equivalent to an optical frequency standard by employing both of the ceo control with an well-known f-2f self-referencing technique and the phase-locking of the comb to the optical frequency standard. As for the optical frequency standard, we employed the iodine stabilized Nd:YAG laser, having a stability better than 2×10^{-14} for 60-s average (absolute frequency uncertainty: 8×10^{-13}) [8]. These absolute and relative frequency stabilities were transferred to the two independent diode lasers via the phase-locking procedures.

The Raman sidebands were generated by adiabatically driving the pure rotational transition ($v = 0$, $J = 2 \leftarrow v = 0$, $J = 0$) in *para*-hydrogen. The wavelengths of the two driving-lasers, Ω_0, Ω_{-1}, were set to 783.9331 and 806.3312 nm, respectively, slightly detuned on the positive side (as shown in Fig. 1) by 700 MHz from the Raman-resonance (10.6235 THz) to satisfy the adiabatic condition. In order to realize an octave-spanning Raman sideband generation, in addition to these two driving-lasers, we further introduced the second harmonic, $2\Omega_{-1}$ of the driving laser Ω_{-1}. The Raman sidebands are generated from the driving lasers, Ω_{-1}, Ω_0 and simultaneously from the second harmonic, $2\Omega_{-1}$. It should be noted that this sideband generation scheme also provides us the ceo information through an overlap of the both sidebands, similarly to the f-2f self-referencing technique in a femtosecond laser comb.

Results and Discussion

Figure 2 shows stability of the phase locking for the diode laser, Ω_0, measured with a frequency counter. The stability was better than 4 mHz in 500 s. It was also confirmed that the other diode laser, Ω_{-1} had nearly same frequency stability. Keeping this stability of the two diode lasers as seeds, we carried out the Raman sideband generation. The broad sidebands over an octave-spanning, $371 \sim 941$ nm, were clearly seen with high beam quality. We picked up the sidebands at 487 nm and introduced them into a high-speed detection system having a frequency response broader than 10 GHz. The beat signal corresponding to the ceo of the Raman

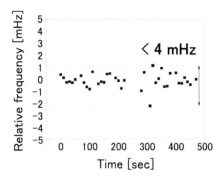

Fig. 2. Frequency stability of diode laser, Ω_0.

sidebands was observed directly in time domain as shown in the inset of Fig. 3. The observed clear beat signal reveals that the sidebands generated from the fundamental driving lasers and the second harmonic, were phase-coherent with each other, across the whole beam cross section. Figure 3 plots the ceo estimated from the observed beats as a function of the relative frequency of the seed diode laser, Ω_0, which was controlled with a synthesizer set in the phase-locking loop. The red line represents the theoretically predicted line in this ceo control, proportional to 36 times of the tuning frequency of Ω_0. It is clearly shown that the ceo is accurately controlled along the theoretical red line. The error bars indicate the standard deviation for 100 measurements. When we further tuned the absolute frequency of Ω_0, close to the zero ceo condition, we observed the temporal waveform with a smooth envelope as expected. This implies that the sidebands from fundamental and that from SHG were overlapped in phase at least in this time scale.

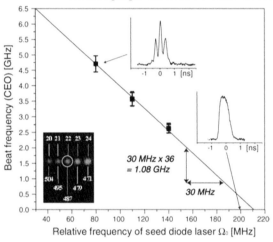

Fig. 3. Beat frequency corresponding to ceo of Raman sidebands vs the relative frequency tuning of seed diode laser, Ω_0.

1   S. E. Harris and A. V. Sokolov, Phys. Rev. A **55**, R4019-4022 (1997).

2   A. V. Sokolov, D. R. Walker, D. D. Yavuz, G. Y. Yin, and S. E. Harris, Phys. Rev. Lett. **85**, 562-565 (2000).

3   J. Q. Liang, M. Katsuragawa, F. Le Kien, and K. Hakuta, Phys. Rev. Lett. **85**, 2474-2477 (2000).; M. Katsuragawa, J. Q. Liang, J.Z. Li, M. Suzuki and K. Hakuta, CLEO/QELS '99, Technical Digest, QthE2, pp.195-196, Baltimore, USA, May 23-28 (1999).

4   M. Y. Shverdin, D. R. Walker, D. D. Yavuz, G. Y. Yin, and S. E. Harris, Phys. Rev. Lett. **94**, 033904-033907 (2005).

5   M. Katsuragawa, K. Yokoyama, T. Onose, and K. Misawa, Optics Express. Vol. **13**, No.15, 5628-5634 (2005); M. Katsuragawa, T. Onose, K. Yokoyama, and K. Misawa, CLEO/QELS 2006, QELS Technical Digest, QFE1, California, USA, May. 21-26 (2006).

6   M. Katsuragawa, T. Onose, T. Suzuki, and K. Misawa, Nonlinear Optics 2007, MB2, Kona, Hawaii, USA, 30 July – 3 August (2007).

7   M. Katsuragawa and T. Onose, Opt. Lett. **30**, 2421‐2423 (2005).

8   F. L. Hong, J. Ishikawa, Y. Zhang, R. Guo, A. Onae, and H. Matsumoto, Opt. Comm. **235**, 377‐385 (2004).

Tunable, octave-spannning supercontinuum driven by X-Waves formation in condensed Kerr media.

Alessandro Averchi[1,5], Daniele Faccio[1,5], Antonio Lotti[1,5], Miroslav Kolesik[2], Jerome V. Moloney[2], Arnaud Couairon[3], and Paolo Di Trapani[1,4,5]

[1] CNISM and Department of Physics and Mathematics, University of Insubria, Via Valleggio 11, 22100 Como, Italy
  E-mail: alessandro.averchi@uninsubria.it
[2] ACMS and Optical Sciences Center, University of Arizona, Tucson, AZ 85721, USA
[3] Centre de Physique Théorique, CNRS, École Polytechnique, F-91128, Palaiseau, France
[4] Department of Quantum Electronics, Vilnius University,
  Sauletekio Ave. 9, bldg. 3, LT-10222, Vilnius, Lithuania
[5] Virtual Institute for Nonlinear Optics, Centro di Cultura Scientifica Alessandro Volta, Villa Olmo, Via Simone Cantioni 1, 22100 Como, Italy

Abstract. We generate an enhanced blue-shifted continuum in bulk Kerr media in ultrashort laser pulse filamentaiton at 1055 nm. At threshold, a spectrally isolated blue peak appears, while at higher energies the continuum expands from the blue peak and spans more than an octave in the spectrum. The central wavelength of the peak can be tuned in a 150 nm range. The effect is explained in terms of X-waves generation.

Supercontinuum (SC) generation, i.e. the generation of an ultra-broadband spectrum starting from a laser pulse is attracting much interest due to the potential applications in a wide range of areas, such as LIDAR, few-cycle pulses generation and ultrafast spectroscopy [1]. In particular, filamentation in Kerr media has been proven to be an efficient way to generate SC: the key mechanism for spectral broadening is self phase modulation (SPM) along with shock fronts formations. Even if the fundamental physical processes behind the SC generation are well known, a number of features have been observed in filamentation whose real nature remains unclear. In recent years it has been demonstrated that the filamentation process is inextricably linked with the reshaping of the input pulse into conical waves (X-waves in the case of normal dispersion) [2].

In this work we study the formation of strongly blue-shifted, spectrally isolated radiation associated to filamentation in bulk fused silica in the normal dispersion regime. Similar observations have been previously reported in anomalous dispersion [3] and with chirped input pulses [4]. With our experiment and simulation we show that this blue spectral peak is the blue-shifted tail of the same X-wave in which the pulse is reshaping into during the filamentation process. Increasing the input pulse energy we obtain a SC whose banwidth spans more than an octave. Notably, unlike usual SC generation, the spectral broadening starts from the blue and not from the pump. We also demonstrate that the spectral position of the blue tail depends on the X-Wave group velocity which can be tuned continuously by controlling the input pulse parameters. Experiments were performed with a 1 ps duration (FWHM) 1055 nm laser pulse delivered by a 10 Hz amplified Nd:glass laser (Twinkle, Light Conversion Ltd., Vilnius, Lithuania). The pulse had a diameter of 5 mm (FWHM) and was focused into a 1 cm long sample of fused silica using a 51 cm focal length lens. The input energy was

858

adjusted using a first order half-wave plate and a polarizer.

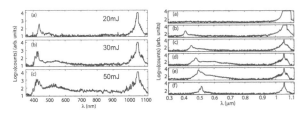

Fig. 1. (Left) Experimental spectra with increasing input energy: (a) 20 μJ, (b) 30 μJ, (c) 50 μJ. (Right) Spectra measured for different positions of the sample with respect to the focusing lens. (a) 49 cm, (b) 50 cm, (c) 51 cm, (d) 52.5 cm, (e) 53 cm and (f) 54 cm.

We performed a first series of measurements fixing the position of the glass sample at z = 52.5 cm and gradually increasing the energy of the input pulse. Figure 1. (Left) shows the spectra recorded using a 14-bit fiber-coupled spectrometer (Ocean Optics) with increasing input energy. At threshold energy we observe a peak in the spectrum at 450 nm. The peak is clearly spectrally isolated, with more than an octave shift from the pump wavelength, and with an energy roughly around 100 nJ, corresponding to a conversion efficency of 0.5%. When increasing the input energy a SC starts to develop; the most notable feature of the bandwidth increase is that, rather than extending from the pump spectrum, as is commonly observed and as would be expected in a SPM-related process, the SC grows starting from the blue peak.

In a second series of measurements we placed the sample at different positions around the focus of the input lens. Here we kept the energy around 20 μJ, slightly adjusting it in each case so as to be just above the blue peak generation threshold. Figure 1. (Right) from (a) to (f) shows the spectra taken respectively at different positions cm from the input lens. By putting the output facet of the sample close to the focus we observed an isolated peak at 405 nm which may be continuously tuned up to 550 nm when shifting the sample further away. To investigate more in detail the process we measured the angularly resolved spectra at the output of the sample using a commercial imaging spectrometer (lot-Oriel MS260i) and recorded it with a modified digital Nikon D70 camera (Figure 2.). In the figure we note the onset of conical emisison around the pump and, most importantly, the angular dispersion of the blue peak. To proof that X-wave reshaping of the pulse is occurring, we fit the conical emission at the pump wavlength using the X-wave relation $k_\perp = \sqrt{k^2 - k_z^2}$ with $k = (\omega/c)$ and

$$k_z(\omega) = k(\omega_{laser}) + \frac{\omega - \omega_{laser}}{v_x} \qquad (1)$$

where v_x is the group velocity of the X-wave, determined directly from the experimental angular spectrum with a method described in [5] as $v_x = 2.035 \times 10^8$ m/s. As may be seen the curve reproduces the conical emission at the pump wavelength and in the blue part of the spectrum it fits very closely the position and the angular dispersion of the blue peak. It is important to note that the position of the blue peak in the spectrum is determined by the precise value of v_x: larger group velocities lead to larger wavelength gaps between the pump and the blue-shifted X tails.

859

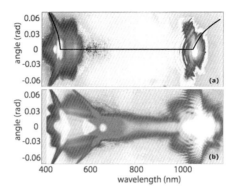

Fig. 2. Experimental angularly resolved spectra for input energies of (a) 20 μJ (b) 40 μJ. The solid black line shows a plot of Equation 1. with $v_x = 2.035 \times 10^8$ m/s.

This explains the tunability of the blue peak we observed: indeed, it has been shown that the group velocity of the X-waves forming at the beginning of filamentation depends on the peak intensity reached by the pulse during the initial collapse stage [2]. By changing the position of the sample respect to the focus of the lens, we are modifiying the input pulse condition and in turn the collapse dynamic, so that the group velocity of the resulting X-wave is varied. From Equation 1. this determines also the central wavelength of the blue shifted peak. In the experiment we noticed that also keeping the position of the sample fixed but changing the aperture of the beam with an iris allows also the tunability of the spectrum. To further proof the validity of our interpretation we performed a series of numerical simulations using the Unidirectional Pulse Propagation Equation solver [6]. To investigate the influence of the input pulse condition on the group velocity of the X-wave and the position of the blue peak we simply changed the beam diameter before the focusing lens from 5 mm to 3 mm. This produced a corresponding variation in the X-wave forming during the filamentation, as expected, and in turn a shift of 50 nm in the blue peak central wavelength, confirming our understanding (*data not shown*).

In conclusion our measurements show the formation of an isolated blue peak, tunable in wavelength by changing the input pulse conditions and highlight a new mechanism for SC generation. The process can be described as due to the X-waves formation which is occurring during filamentation.

1   R. R. Alfano, The Supercontinuum Light Source, Springer-Verlag, New York, 1989.
2   D. Faccio, M. A. Porras, A. Dubietis, F. Bragheri, A. Couairon, and P. Di Trapani in Physics Review Letters, Vol. 96, 193901, 2006.
3   J. Liu, R. Li, and X. Xu in Physics Review A, Vol. 74, 0143801, 2006.
4   V. Kartazaev, and R. R. Alfano in Optics Letters, Vol. 32, 3293, 2007.
5   D. Faccio, A. Averchi, A. Couairon, M. Kolesik, J. V. Moloney, A. Dubietis, G. Tamosauskas, P. Polesana, A. Piskarskas, and P. Di Trapani in Optics Express, Vol. 15, 13077, 2007.
6   M. Kolesik, J. V. Moloney, and M. Mlejnek in Physics Revew Letters, Vol. 89, 283902, 2002.

Toward Ultrafast Optical Waveform Synthesis with a Stabilized Ti:Sapphire Frequency Comb

Matthew S. Kirchner[1], Tara M. Fortier[1], Danielle Braje[1], Andy M. Weiner[2], Leo Hollberg[1], Scott A. Diddams[1]

[1] National Institute of Standards and Technology, Boulder, Colorado 80305, USA
  E-mail: matthew.kirchner@boulder.nist.gov
[2] Electrical and Computer Engineering, Purdue University, West Lafayette, Indiana 47907, USA

Abstract. We have developed a system for line-by-line control of a stabilized Ti:Sapphire optical frequency comb. We show individually-addressed 20 GHz comb modes around 960 nm and apply simple masks to demonstrate individual mode control.

Introduction

Pulse-shaping of ultrafast laser pulses is a valuable tool in the areas of spectroscopy, X-ray generation, quantum control, and communications [1-4]. Recent experiments have shown the ability to perform this pulse shaping in a line-by-line manner by addressing single frequency components of fiber frequency combs or modulated cw lasers[1,5,6]. Such line-by-line shaping enables the generation of waveforms with time durations greater than the repetition period of the source and is a route to arbitrary optical waveform generation [5,6].

   We build on these experiments by demonstrating the ability to individually address and manipulate single frequency components of a Ti:Sapphire frequency comb that can be stabilized in both repetition rate and carrier envelope offset frequency. The absolute stabilization of the frequency modes can provide femtosecond timing jitter in the generated waveforms as well as precise control of the carrier phase within the pulse envelope. These features will expand the capabilities of traditional pulse shaping and should enable new applications in secure communication and data transfer. A unique advantage of using this octave-spanning Ti:Sapphire comb is the opportunity to perform this line-by-line manipulation over a broad range of wavelengths from 650 nm to 1050 nm while retaining the low timing jitter provided by locking the comb to a stable optical frequency reference.

Setup

We employ a 1 GHz repetition rate octave-spanning Ti:Sapphire frequency comb with spectral coverage from 550-1100 nm, as shown in Fig. 1a [7]. Using dichroic mirrors, we pick off light at 550 and 1100 nm to detect the carrier envelope offset frequency with a SNR of 35 dB in 300 kHz resolution bandwidth. The optical setup allows light at 657 nm to be spectrally separated and compressed to 25 fs for use as a reference pulse in cross-correlations as well as for locking the laser to a calcium optical clock. The rest of the light is double passed through a 20 GHz Fabry-Perot filter cavity (F ~ 300) with mirrors centered at 910 nm [8]. The cavity is designed for low dispersion and has a single-pass suppression of off-resonant modes of 27 dB with an acceptance bandwidth of around 100 nm when centered at the peak of the input laser spectrum (970 nm). The double pass configuration allows for an off-resonant mode suppression of 50 dB while maintaining the high acceptance bandwidth of our mirrors. The output spectrum from the filter cavity is shown in Fig. 1b.

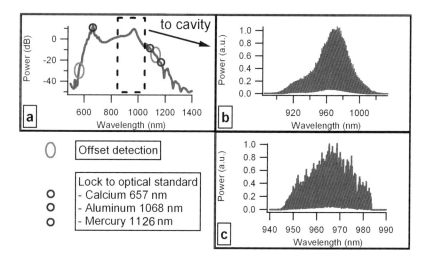

Fig. 1. a) Octave-spanning laser spectrum shown on a log scale. Light at 550 and 1100 nm is used to detect and stabilize the carrier envelope offset frequency. Any one of the cw optical frequency references listed can be used to stabilize the absolute frequency of all the comb modes, leaving a broad swath of spectrum from 650-1050 nm available for pulse shaping. In this case we send light from 850-1050 nm to a 20 GHz filter cavity. b) 20 GHz output of the filter cavity showing good coupling over 100 nm. c) Output of pulse shaper showing a hard edge at each end of the aperture of the SLM. Over 600 modes are captured in the aperture of the SLM.

After the comb is filtered to 20 GHz, it is spatially expanded and sent to a pulse shaper, consisting of a 1200 grooves/mm gold grating, a 1 m focal length lens and a 640 pixel liquid crystal spatial light modulator (SLM) in reflection mode that is capable of both amplitude and phase control. The optics are arranged so that the 20 GHz modes nominally match the 100 μm/pixel pitch of the SLM. The modulated light is retroreflected and is picked off by an optical isolator. Its spectrum is shown in Fig. 1c. The reflected light shows a bandwidth of 40 nm centered around 965 nm which should allow for pulses shorter than 50 fs. The total shaper output power is 200 μW. We have amplified the output to 5-10 mW using a semiconductor optical amplifier.

In principle, we can manipulate this entire bandwidth; however, in the present configuration the physical separation between modes varies across the aperture of the SLM (due to the non-constant angular dispersion of the grating). We achieve good overlap between optical modes and SLM pixels over a subset of the full aperture before the modes walk off of a SLM pixel. Currently we achieve a bandwidth of about 7.5 nm (120 modes) around 959 nm. Approaches to reduce the pixel walk-off limitations include using a spacing of 2 SLM pixels per comb mode or using a grating plus a prism (grism) to minimize dispersion in comb mode separation and provide 1 pixel per comb mode across the entire aperture of the SLM. Eventually, we hope to have full control over 320 comb modes (in the two pixel per comb mode case) or 640 comb modes (in the grism case).

Results

To demonstrate the individual addressing of many comb modes, we performed basic amplitude masking on the 120 comb modes from 958 to 965 nm. In this preliminary demonstration, we turned off every other mode to double the repetition rate as shown in Fig. 2a. The extinction was greater than 10 dB, but should be improved by adjusting the mode size at the SLM. We amplified the light from the shaper with a semiconductor optical amplifier and examined the time domain signal with a fast photodiode and oscilloscope as shown in Fig. 2b.

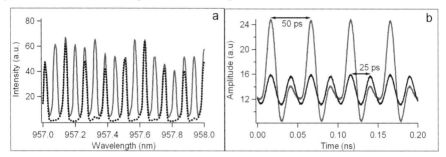

Fig. 2. a) Zoomed in view of all the pixels on (red trace) and every other pixel turned off (black dashed trace). b) Fast oscilloscope trace of the 120 amplified comb teeth with all teeth on showing a 20 GHz repetition rate (red trace) and every other tooth turned off showing a 40 GHz repetition rate (black trace). The photodetector bandwidth is 45 GHz.

Conclusion

We have shown individual control over 120 comb modes around 960 nm from a filtered Ti:Sapphire frequency comb. The optical frequency of all comb teeth can be stabilized to better than one part in 10^{15} by locking the comb to one of the optical references, ultimately enabling novel waveform generation with femtosecond timing jitter.

References

1 A.M.Weiner, Rev. Sci. Instr. **71**, 1929–1960 (2000).
2 R. Bartels, S. Backus, E. Zeek, L. Misoguti, G. Vdovin, I.P. Christov, M.M. Murnane, H.C. Kapteyn, Nature **406**, (164–166 (2000).
3 A.M. Weiner, D.E. Leaird, G.P. Wiederrecht, K.A. Nelson, Science **247**, 1317–1319 (1990).
4 N. Dudovich, D Oron, Y. Silberberg, Nature **418**, 512–514 (2002).
5 Z. Jiang, D. S. Seo, D. E. Leaird, and A. M. Weiner, Opt. Lett. **30**, 1557-1559 (2005).
6 Z Jiang, C. Huang, D.E. Leaird, A.M. Weiner, Nat. Photon. **1**, 463-467 (2007).
7 T. M. Fortier, A. Bartels, and S.A. Diddams, Opt. Lett. **31**, 1011 (2006).
8 K. Yiannopoulos, K. Vyrsokinos, E. Kehayas, N. Pleros, K. Vlachos, H. Avramopoulos, G. Guekos, IEEE Photon. Tech. Lett. **15,** no. 9, 1294-1296 (2003).

Multimillijoule Optically Synchronized and Carrier-Envelope-Phase-Stable Chirped Parametric Amplification at 1.5 μm

O.D. Mücke[1], D. Sidorov[1], P. Dombi[1], A. Pugžlys[1], S. Ališauskas[2], N. Forget[3], J. Pocius[4], L. Giniūnas[4], R. Danielius[4], and A. Baltuška[1]

[1] Photonics Institute, Vienna University of Technology, Gusshausstrasse 27-387, A-1040, Vienna, Austria
  E-mail: oliver.muecke@.tuwien.ac.at
[2] Laser Research Center, Vilnius University, Saulėtekio av. 10, LT-10223 Vilnius, Lithuania
[3] Fastlite, Bâtiment 403, Ecole Polytechnique, 91128 Palaiseau, France
[4] Light Conversion Ltd., P/O Box 1485, Saulėtekio av. 10, LT-10223 Vilnius, Lithuania

Abstract. Efficient infrared 35-THz-wide parametric amplification at 1.5 μm with the energy of ~10 mJ is obtained in a 4-stage OPCPA using a combination of a 1030-nm 200-fs Yb- and a 1064-nm 60-ps Nd amplifier seeded with a common Yb oscillator.

Optical Parametric Chirped Pulse Amplification (OPCPA) [1] has attracted a lot of attention as a promising route toward intensity scaling of few-cycle laser pulses. Intense phase-stable few-cycle laser pulses have numerous intriguing applications in attosecond science and high-field science including attosecond XUV/soft-X-ray pulse generation by high-harmonic generation (HHG), tomographic imaging of molecular orbitals, and laser-induced electron diffraction. A major challenge for using HHG in studies of time-resolved tomography of molecular dissociative states is the low ionization potential I_p of excited molecular states. The resulting competition between state depletion and HHG prevents generation of broad HHG spectra necessary for tomographic reconstruction. One solution are laser sources with high ponderomotive energy $U_p \propto \lambda^2 I$ at moderate intensity level, i.e., infrared phase-stable few-cycle high-power laser systems. High-U_p-sources [2,3] also open the door to experimental investigations of the λ-scaling laws of strong-field physics (Keldysh parameter $\propto \lambda^{-1}$, electron energies $\propto \lambda^2$, HHG cutoff $\propto \lambda^2$, HHG efficiency $\propto \lambda^{-5.5}$, minimum attosecond pulse duration $\propto \lambda^{-1/2}$ [4]), and they would benefit laser-induced electron diffraction because of the shorter de Broglie electron wavelength and consequently higher spatial resolution [5]. The main objective of our work is to generate IR pulses with ~40-fs duration that fully satisfy the requirements for external spectral broadening in gas [6]. In addition, with an IR pulse we expect to surpass the energy limitation (4-5 mJ at 0.8 μm) for gas broadening schemes because the critical power of self-focusing also scales as λ^2.

Using mJ pulses from Ti:sapphire amplifiers at 0.8 μm, coherent X-rays in the keV photon energy range were generated by HHG in helium [7]. A technological problem hindering further scaling of the pulse energy beyond several mJ is gas ionization in the gas-filled hollow-fiber compressors required to achieve few-cycle pulse duration at mJ pulse energies. More fundamentally, ionization in helium saturates when the intensity of a few-cycle pulse at 0.8 μm exceeds ~1 PW/cm^2, thus the HHG cutoff and photon flux is limited by ground-state depletion in helium in these experiments.

Here, we report on the development of a multi-mJ all-optically synchronized and phase-stable OPCPA at 1.5 μm (see Fig. 1). As opposed to our OPCPA systems developed previously, in this work we modify our approach: (1) with the advent of a mature 200-fs Yb MOPA system it became possible to abandon the Ti:sapphire front-end; (2) we avoid working close to the signal-idler wavelength degeneracy and reduce the quantum defect for the signal wave; (3) we employ (nearly) collinear Type II phase matching that, as opposed to Type I, supports a much narrower bandwidth but is free of parasitic self-diffraction. Following the pioneering work of Miller and coworkers [8], we employ Type II KTP/KTA (1030/1064 nm pump, ~1500 nm signal, ~3500 nm idler) because these crystals are transparent for the mid-IR idler wavelength and exhibit a relatively broad bandwidth around 1500 nm. The repetition rate of the Yb:KGW DPSS MOPA (Pharos, Light Conversion, Ltd.), tunable in the range of 1–100 kHz, was set at 10 kHz as a 500-th harmonic of the flash-lamp pumped Nd:YAG amplifier (Ekspla Ltd.) operating at 20 Hz. In our scheme (Fig. 1), both Yb and Nd RA are simultaneously seeded from a single master oscillator that has a modest FWHM bandwidth of 30 nm. To seed the Nd RA, we pick up the 0th-order diffraction beam behind a transmission grating in the pulse stretcher.

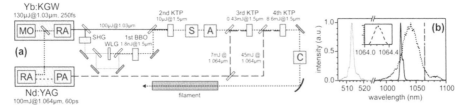

Fig. 1. (a) Scheme of the IR OPCPA setup. MO, master oscillator; RA, regenerative amplifier; PA, double-pass post amplifier; S/C grating-based stretcher/compressor; A, acousto-optic programmable dispersive filter (DAZZLER); WLG, white-light generator in a 4-mm-thick sapphire plate; the CEP-stable idler wave from stage 1 becomes the signal wave in stage 2. Stage 1 (BBO, Type I) is pumped at 515 nm, stage 2 (KTP, Type II) at 1030 nm, stages 3 and 4 (KTP, Type II) at 1064 nm. (b) Spectra of Kerr-lens mode-locked Yb:KGW oscillator (dotted), Yb:KGW RA (solid), SHG of Yb:KGW (grey). The Nd:YAG RA (dashed) contains an intracavity 2-mm-thick etalon to narrow the ps amplifier bandwidth.

The output of the Yb:KGW CPA system is used to pump the first two OPA stages. The frequency-doubled output at 515 nm is used to generate white light continuum in sapphire and as a pump of the 1st stage (collinear Type I BBO). This configuration produces a carrier-envelope phase (CEP) stable idler [9] at 1.5 µm that we further use as seed (signal wave) in the subsequent OPA stages (see Fig. 2(a)). CEP stability of the 2nd stage output was verified by means of f-to-2f interferometry in the wavelength range from 690-830 nm (Fig. 2(b)).

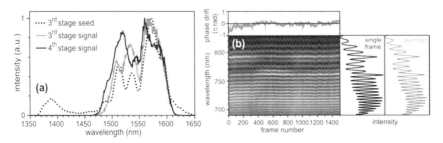

Fig. 2. (a) Spectral properties of the final OPA stages. (b) f-to-2f interferogram reflecting CEP stability measured after the 2nd stage OPA.

The output of the 2ⁿᵈ stage is stretched in a grating stretcher to ~40 ps and a tunable higher-order phase correction is introduced by DAZZLER (Fig.1). The temporally stretched seed is amplified in two final OPA stages (3ʳᵈ and 4ᵗʰ) using a 50-mJ picosecond pump pulse from the Nd:YAG system. The maximum pulse energy at 1.5 μm before the 60% efficient grating pulse compressor is ~10 mJ, as measured through a bandpass filter that blocks off the 3.6-μm idler wave. μJ-level 10-kHz-repetition-rate pulses after the 2ⁿᵈ OPA stage, pulse stretcher, and DAZZLER were compressed to ~50-fs with the grating compressor and measured with SHG FROG, as shown in Fig.3. Work is now in progress to compress and characterize the multi-mJ 20-Hz pulses at the output of the 4ᵗʰ OPA, the bandwidth of which supports virtually the same pulse duration as the 2ⁿᵈ stage. The output of the multi-mJ IR OPCPA system will be broadened in a noble gas, where we expect to reach up to 4 times higher pulse energies in comparison with a filament/hollow-fiber pumped at λ=0.8 μm.

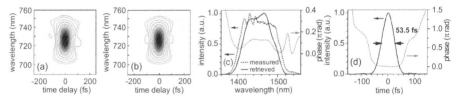

Fig. 3. a) SHG FROG characterization of the stretched and recompressed 1.5 μm pulses: (a) Measured and (b) retrieved SHG FROG trace. (c) Measured spectrum (black dashed), retrieved spectral intensity (black solid) and phase (grey dashed). (d) Retrieved temporal intensity (black solid) and phase (grey dashed) profile indicating a FWHM 53.5 fs pulse duration.

In conclusion, we have developed a prototype CEP-stable IR OPCPA for high field applications which draws on a straightforwardly scalable picosecond pump at the Nd/Yb fundamental wavelength and uses a femtosecond Yb front-end.

Acknowledgements. This work is supported by the Austrian Science Fund (FWF), grants U33-N16 and F1619-N08.

1   A. Dubietis et al., J. Sel. Top. Quantum Electron. **12**, 163 (2006), and references therein.
2   T. Fuji et al., Opt. Lett. **31**, 1103 (2006).
3   C. Vozzi et at., Opt. Express **14**, 10109 (2006); C. Vozzi et at., Opt. Lett. 32, 2957 (2007).
4   P. Colosimo et al., Nature Phys. **4**, 386 (2008); J. Tate et al., Phys. Rev. Lett. **98**, 013910 (2007); K. Schiessl et al., Phys. Rev. Lett. **99**, 253903 (2007); A. Gordon et al., Opt. Express **13**, 2941 (2005); B. Shan et al., Phys. Rev. A **65**, 011804(R) (2001).
5   M. Meckel et al., Science **320**, 1478 (2008); S. N. Yurchenko et al., Phys. Rev. Lett. **93**, 223003 (2004); M. Spanner et al., J. Phys. B **37**, L243 (2004).
6   C. P. Hauri et al., Appl. Phys B **79**, 673 (2004); C. P. Hauri et al., Opt. Lett. **32**, 868 (2006).
7   J. Seres et al., Nature **98**, 433 (2005).
8   D. Kraemer et al., Opt. Lett. **31**, 981 (2006); D. Kraemer et al., JOSA B **24**, 813 (2007).
9   A. Baltuška et al., Phys. Rev. Lett. **88**, 133901 (2002).

5-fs multi-mJ CEP-locked parametric chirped-pulse amplifier at 1 kHz

S. Adachi[1, 3], N. Ishii[1, 3], H. Ishii[1, 3], T. Kanai[1, 3], A. Kosuge[1, 3], Y. Kobayashi[2, 3], D.Yoshitomi[2, 3], K. Torizuka[2, 3], and S. Watanabe[1, 3]

[1] Institute for Solid State Physics, University of Tokyo, Kashiwanoha 5-1-5, Kashiwa Chiba 277-8581, Japan
[2] National Institute of Advanced Industrial Science and Technology (AIST), 1-1-1 Umezono, Tsukuba 305-8568, Japan
[3] CREST, Japan Science and Technology Agency, Sanbancho 5, Chiyoda-ku, Tokyo 102-0075, Japan
adachi@issp.u-tokyo.ac.jp

Abstract. We report an optical parametric chirped-pulse amplifier with 5.5-fs pulse duration, 2.7-mJ pulse energy at a 1-kHz repetition rate, pumped by a 450-nm pulse from a frequency-doubled Ti:sapphire laser.

Introduction

The concept of optical parametric chirped-pulse amplification [1] (OPCPA) has been recognized to be promising for the generation of high-intensity few-cycle laser pulses. So far, sub-10 fs, multi-mJ to multi-tens-of-mJ OPCPA systems have been reported [2-4]. Recently we reported sub-7-fs, 1.5 mJ OPCPA system pumped with a 400-nm pulse from a frequency-doubled Ti:sapphire laser. However, the shorter pulse width (~ 5 fs) is essentially important to obtain an isolated attosecond pulse by a high-harmonic generation process as discussed in Ref [5].

Gain spectra with several pump wavelengths

Even the generation of sub-5-fs pulses has been already demonstrated from noncollinear optical parametric amplifiers (NOPA), but with rather low pulse energies [6-8]. Figure 1 shows the calculated parametric gain bandwidths of a type-I BBO crystal (widely used for OPA) for several pump wavelengths. In the case of 400-nm pump wavelength, the signal spectral range of 520-750 nm [(a)] is the best for broadest amplification [6-8], where the phase mismatch $?k$ of OPA is widely diminished around the center wavelength. In contrast, the spectral range (a) has poor overlap with the spectrum from a typical broadband Ti:sapphire oscillator [(b)], which is used as a seed source in our OPCPA system. Therefore, in the previous report [9], we were obliged to modify the OPA configuration to match these two wavelength ranges, at the expense of broadest OPA gain bandwidth [(d)]. Meanwhile, there is another solution for this problem: Shifting the pump wavelength itself. From the calculations, the pump wavelength of 450-500 nm [(e), (f)] is desirable to amplify the seed from the Ti:sapphire oscillator [(b)]. Though we could choose any pump wavelength within 450-500 nm since Ti:sapphire has a very broad gain bandwidth (typically 650-1050 nm, corresponding frequency-doubled range of 325-525 nm), we set the pump wavelength at 450 nm (corresponding fundamental wavelength of 900 nm) in order to obtain a moderate Ti:sapphire gain.

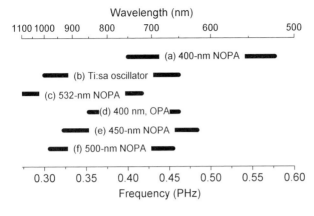

Wavelength (nm)

1100 1000 900 800 700 600 500

(a) 400-nm NOPA
(b) Ti:sa oscillator
(c) 532-nm NOPA
(d) 400 nm, OPA
(e) 450-nm NOPA
(f) 500-nm NOPA

0.30 0.35 0.40 0.45 0.50 0.55 0.60
Frequency (PHz)

Figure 1. Calculated parametric amplification ranges of a BBO crystal for several pump wavelengths. Noncollinear and phase-matching angles for the calculations were optimized for each pump wavelength.

Experiment

The overview of our OPCPA system is described elsewhere [9]. A master oscillator (Venteon OS version, Nanolayers) produced octave-spanning spectrum [10]. The spectral components at 570 and 1140 nm were utilized for the carrier-envelope phase (CEP) stabilization with an f-to-$2f$ setup, while 600-1100 nm spectral component was sent to a stretcher as a seed pulse of OPCPA. We observed a CEP beat note with a SNR of ~ 35 dB in a 100-kHz resolution bandwidth, and the CEP of the oscillator was verified to be stabilized for > 1 hour by a home-built phase-lock loop.

All of the optical elements (mirrors, polarizers, *etc*) in the Ti:sapphire pump laser system were replaced to be optimized at 900 nm. 5-mJ and 20-mJ fundamental radiations at 900 nm were generated from the pump system, and after the frequency-doubling 4-mJ (for pre-OPA) and 15-mJ (for power-OPA) pump pulses at 450 nm were obtained. The stretcher of the OPCPA consisted of a grating pair and a prism pair to give a negative chirp, which was compensated with a bulk material compressor (SF57, 25 cm) after parametric amplification. Its configuration was slightly modified from that in Ref [9] to compensate for broader spectral range (650-930 nm). The duration of the stretched seed pulse was calculated to be ~ 50 ps.

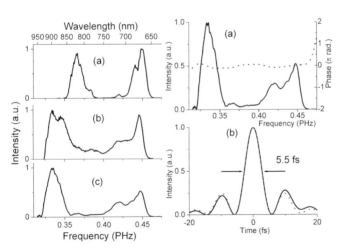

Figure 2. (left) Measured OPCPA spectra pumped with (a) 400-nm pump pulse and (b) (c) 450-nm pump pulse.

Figure 3. (right) Spectral and temporal profiles of recompressed OPCPA output pulse. (a) Solid curve, spectral intensity; dotted curve, spectral phase. (b) Solid curve, temporal intensity; dashed curve, transform-limited intensity.

868

Figure 2 displays the OPCPA spectra pumped at 450 nm for slightly different phase-matching angles [(b), (c)], as well as the previous result obtained with 400-nm pump pulse [(a)]. Now the spectral width became substantially broader enough to support a 5-fs pulse generation. Figure 3 shows spectral and temporal profiles of the recompressed OPCPA output pulse obtained with SPIDER measurement. The spectral phase was almost flat over the whole OPCPA spectral range of > 130 THz after the huge compression ratio of 10^4 (50 ps : 5 fs), enabled by an acousto-optic programmable dispersion filter (Dazzler, Fastlite) with an adaptive phase control. Consequently, the reconstructed pulse width was 5.5 fs, which is almost equal to the transform-limited pulse width [Fig. 3(b)]. The compressed output energy was 2.7 mJ with 15-mJ pump pulse (~ 17 % conversion efficiency). The evaluation of the CEP drift induced by the OPCPA was implemented with the second f-to-$2f$ interferometer placed after the compressor. The CEP of the OPCPA output was stable over 30 sec except for a slow drift, which can be easily compensated by providing feedback to the relative delay in the first f-to-$2f$ interferometer.

Conclusions

We have demonstrated the OPCPA system with 5.5-fs pulse duration, 2.7-mJ pulse energy at a 1-kHz repetition rate, pumped by a 450-nm pulse from a frequency-doubled Ti:sapphire laser. This CEP-locked TW-class few-cycle laser system will offer an ideal light source for the experiments of ultrafast spectroscopy in the soft X-ray in the attosecond timescale.

1   A. Dubietis, G. Jonusauskas, and A. Piskarskas, Opt. Commun. **88**(4-6), 437-440 (1992).

2   N. Ishii, *et al.*, Opt. Lett. **30**(5), 567-569 (2005).

3   S. Witte, *et al.*, Opt. Express **14**(18), 8168-8177 (2006).

4   F. Tavella, A. Marcinkevicius, and F. Krausz, Opt. Express **14**(26), 12822-12827 (2006).

5   R. Kienberger, *et al.*, Nature **427**(6977), 817-821 (2004).

6   A. Shirakawa, I. Sakane, M. Takasaka, and T. Kobayashi, Appl. Phys. Lett. **74**(16), 2268-2270 (1999).

7   M. Zavelani-Rossi, *et al.*, Opt. Lett. **26**(15), 1155-1157 (2001).

8   A. Baltuska, T. Fuji, and T. Kobayashi, Opt. Lett. **27**(5), 306-308 (2002).

9   S. Adachi, *et al.*, Opt. Lett. **32**(17), 2487-2489 (2007).

10   O. Mucke, *et al.*, Opt. Express **13**(13), 5163-5169 (2005).

Sub-two-cycle pulses at 1.6 μm from an optical parametric amplifier

D. Brida[1], G. Cirmi[1], C. Manzoni[1], M. Marangoni[1], S. Bonora[1,2], P. Villoresi[2], S. De Silvestri[1], and G. Cerullo[1*]

[1] National Laboratory for Ultrafast and Ultraintense Optical Science – INFM-CNR,
   Dipartimento di Fisica, Politecnico di Milano, Piazza L. da Vinci 32, 20133 Milano, Italy.
[2] LUXOR - Laboratory for UV and X ray Optical Research – CNR-INFM, D.E.I. - Università
   di Padova, Italy.
[*] E-mail: giulio.cerullo@fisi.polimi.it

Abstract. We demonstrate two optical parametric amplifier schemes, based on β-barium-borate and periodically-poled lithium tantalate respectively, generating ultrabroadband pulses in the 1-2 μm range. Using a deformable mirror compressor we obtain 8.5-fs pulses at 1.6 μm.

Light pulses with duration of just few optical cycles are important for a number of applications, ranging from time-resolved optical spectroscopy to high field science. Such pulses have been generated by a variety of techniques: directly from a laser oscillator, by spectral broadening in a fiber or by ultrabroadband optical parametric amplifiers (OPAs). OPAs seeded by a white-light continuum (WLC) support under suitable conditions ultrabroad gain bandwidths [1, 2]. In fact the phase matching bandwidth in an OPA depends on the group velocity (GV) mismatch between signal and idler, so that broadband gain requires matching the GVs of signal and idler. This condition occurs either for type I phase matching around the degeneracy point or in a non-collinear OPA (NOPA), in which the signal GV is matched to the projection of the idler GV along the signal direction. The broad gain bandwidths of OPAs have been exploited for few-cycle pulse generation in the visible [1] and around 800 nm [2], but so far the 1-2 μm spectral range has not been explored and the shortest pulses in this region have ≈15 fs duration.

Here we report on two different schemes for the generation of ultrabroadband near-IR pulses: a degenerate OPA in β-barium borate (BBO) and a NOPA in Periodically Poled Stoichiometric Lithium Tantalate (PPSLT). Both schemes, when pumped at 800 nm, produce, in a simple single stage setup, pulses with μJ-level energy and spectra spanning the 1-1.7 μm range (for the NOPA) and the 1.2-2.1 μm range (for the degenerate OPA). Using an adaptive pulse shaper employing a Deformable Mirror (DM), we produce nearly transform-limited (TL) pulses with 8.5 fs duration at 1.6 μm, corresponding to less than two optical cycles. These are to our knowledge the shortest light pulses generated in this wavelength range.

Fig. 1 shows the experimental setup of the broadband near-IR OPAs. The pump pulses are derived from a Ti:sapphire laser (80 μJ, 150 fs at 800 nm and 1 kHz). A ≈2 μJ fraction of the pump is focused in a 3-mm-thick sapphire plate to generate a WLC seed; the remaining energy pumps the nonlinear crystal. For the degenerate OPA we use a 3-mm-thick type I BBO crystal (θ=21°) in a nearly collinear configuration. The amplified signal has an energy of 2÷3 μJ and a spectrum covering the 1200-2100 nm wavelength range (dashed line in Fig. 2(a)). One can shift the amplified bandwidth to

870

the blue by using non-collinear phase matching in PPSLT [3], for which, in contrast to BBO, the GV of the idler is larger than that of the signal. For the IR NOPA we use a 1.2-mm-thick PPSLT crystal, with poling period Λ=20.9 μm and non-collinear angle $\alpha = 2°$. We obtain again μJ-level pulses with a spectrum shown as solid line in Fig. 2(a), which is blue-shifted by $\approx$ 150 nm with respect to the degenerate OPA one.

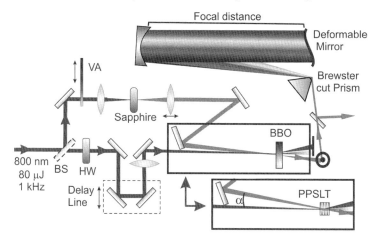

Fig. 1. Experimental setup of the broadband near-IR OPAs based on BBO and PPSLT. It is possible to switch from one configuration to the other by just replacing the nonlinear crystal and varying the angle between pump and signal beams. VA, variable attenuator; BS, beam splitter, HW half-wave plate.

Both OPA schemes generate ultrabroadband near-IR pulses supporting sub-10-fs duration, corresponding to about two optical cycles in this wavelength range. Their spectral phase shows a strong high-order dispersion contribution, which is impossible to correct by using prism or grating pairs. To achieve dispersion compensation to all orders, we implemented an adaptive system based on a DM. The DM is placed in the Fourier plane of a 4f zero-dispersion pulse shaper [4] consisting of a Brewster-cut SF56 prism and a spherical gold mirror. The DM is a rectangular silver-coated membrane activated by 30 linear electrodes. We chose SF56 as prism material because it shows good angular wavelength dispersion, enabling to fill the DM nearly completely, while adding a low contribution to the pulse spectral phase. Furthermore, the use of a prism instead of a grating reduces the losses. The mirror response was calibrated in order to define the influence function matrix describing the effect of each electrode on the membrane shape. The mirror shape for optimum pulse compression was obtained by first measuring the pulse spectral phase with Second Harmonic Generation Frequency Resolved Optical Gating (SHG-FROG) and then introducing the additional phase required to compensate it. Such process was iteratively repeated acting on the residual phase after the previous correction. In such approach it is crucial to accurately map each wavelength on the mirror by measuring with a spectrometer the wavelengths corresponding to 5 different positions and interpolating the results with the angular dispersion function of the prism. This procedure also enables to reproduce the compression results on a day to day basis.

Fig. 2(b) shows the SHG-FROG trace, measured with a 10-μm-thick BBO crystal, of the compressed pulses from the degenerate OPA, while Fig. 2(c) reports the retrieved temporal intensity and phase profile. The measured 8.5 fs pulsewidth is very close to the TL duration (8.2 fs) and corresponds to less than two cycles of the 1.6 μm carrier frequency (5.3 fs period). Similar results are expected for the PPSLT-based NOPA.

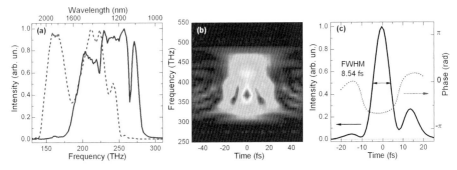

Fig. 2. (a) Spectra of the degenerate OPA in BBO (dashed line) and the NOPA in PPSLT (solid line). (b) SHG-FROG field trace (128×128 pixels) for the degenerate OPA pulse compressed by the DM system. (c) Retrieved temporal intensity and phase profile of the compressed pulse (reconstruction error = 0.0107).

In conclusion, we have demonstrated two OPA configurations, based on BBO and PPSLT respectively, capable of generating ultrabroadband pulses in the 1-2 μm wavelength range, and achieved nearly TL 8.5 fs at 1.6 μm using a DM compressor [5]. The μJ-level pulses produced by our simple single-stage system are already suitable for time-resolved spectroscopy in the near-IR with unprecedented resolution. It should be straightforward, by adding one or two similar OPA stages, to scale the output energy by 2÷3 orders of magnitude, enabling the application of the energetic sub-two-cycle pulses to high harmonic generation. For such applications, it will also be possible to stabilize the Carrier-Envelope Phase (CEP) of the pulses by seeding the OPA with a CEP stable broadband pulse produced either by intrapulse difference frequency generation [6] or by WLC generation from the idler of an OPA in which pump and signal are derived from the same source [7].

References

1   A. Baltuška, T. Fuji, and T. Kobayashi, Opt. Lett. **27**, 306-308 (2002).
2   S. Witte, R. T. Zinkstok, A. L. Wolf, W. Hogervorst, W. Ubachs, and K. S. E. Eikema, Opt. Express **14**, 8168-8177 (2006).
3   G. Cirmi, D. Brida, C. Manzoni, M. Marangoni, S. De Silvestri, and G. Cerullo, Opt. Lett. **32**, 2396-2398 (2007).
4   E. Zeek, K. Maginnis, S. Backus, U. Russek, M. Murnane, G. Mourou, H. Kapteyn, and G. Vdovin, Opt. Lett. **24**, 493-495 (1999).
5   D. Brida, G. Cirmi, C. Manzoni, S. Bonora, P. Villoresi, S. De Silvestri, and G. Cerullo, Opt. Lett. **33**, 741-743 (2008).
6   C. Vozzi, G. Cirmi, C. Manzoni, E. Benedetti, F. Calegari, G. Sansone, S. Stagira, O. Svelto, S. De Silvestri, M. Nisoli, and G. Cerullo, Opt. Express 14, 10109-10116 (2006).
7   C. Manzoni, D. Polli, G. Cirmi, D. Brida, S. De Silvestri, and G. Cerullo, Appl. Phys. Lett. **90**, 171111 (2007).

Carrier envelope offset control of broad Raman sidebands by locking two pump laser frequencies to a single optical cavity

T. Suzuki[1,2], M. Hirai[1], R. Tanaka[1], and M. Katsuragawa[1,2]

[1] Department of Applied Physics and Chemistry, University of Electro-Communications, 1-5-1 Chofugaoka, Chofu, Tokyo 182-8585, Japan
[2] PRESTO, JST, 4-1-8 Honcho Kawaguchi, Saitama, Japan
   E-mail: t-suzuki@pc.uec.ac.jp

Abstract. We generate broad Raman sidebands with zero carrier-envelope-offset by frequency-locking the pump lasers to a single optical cavity. It is shown in both spectral and temporal domains that the carrier-envelope-offset is controlled to discrete values.

Introduction

In the last decade, there has been significant progress in generating high coherence in molecular vibrational and/or rotational transitions by means of adiabatic Raman excitation [1,2]. A molecular ensemble with such high coherence, in turn, strongly modulates the pump laser fields and generates high order Raman sidebands collinear to the pump beams. Since the generated sidebands are mutually coherent, ultrashort pulses can be constructed by synthesizing them [3]. The main feature of this ultrashort pulse generation scheme is that the ultrashort pulses are produced from two single-frequency lasers. Here, we demonstrate that the carrier envelope offset (CEO) of such Raman sidebands can be controlled to discrete values by locking the two pump laser frequencies to a single optical cavity, leading to the generation of ultrashort pulses with a constant carrier envelope phase (CEP).

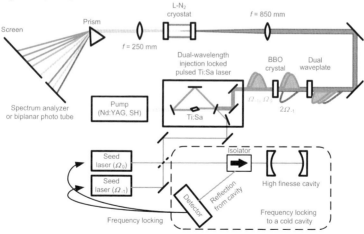

Fig. 1. Schematic diagram of the experimental setup

873

Experimental Methods

The Raman sidebands with zero CEO can be generated when the two pump laser frequencies (Ω_{-1}, Ω_0) are set to $\Omega_0 = n\Delta\Omega$ and $\Omega_{-1} = (n-1)\Delta\Omega$ where $\Delta\Omega$ is the frequency difference between the two pump lasers and n is an integer. We realize this critical condition by locking the two pump laser frequencies to a single optical cavity and then choosing the appropriate longitudinal modes.

The schematic of the experimental setup is illustrated in Fig. 1. The CEO control with the single optical cavity was examined for the following two cases. One is the case using the cavity of the pump laser itself and the other is the case using an external high-finesse cold cavity. The two-frequency pump pulses to drive the high Raman coherence are generated from the dual-wavelength injection-locked Ti:Sa laser, which consists of a single cavity. The two pump-laser frequencies, Ω_{-1}, Ω_0, are thereby necessarily restricted to integer multiples of the free-spectral-range (FSR) of the laser cavity. We set the two frequencies, Ω_{-1}, Ω_0, around the zero CEO condition by choosing the appropriate longitudinal cavity modes.

The Ti:Sa laser cavity has small dispersion due to the Ti:Sa crystal and thus the FSR is different for each frequency components. Therefore the ratio of two longitudinal cavity mode numbers which gives the smallest CEO is not equal to the simple integer ratio of $n-1$ to n. However, the CEO can be controlled by steps of the FSR, because the deviation of the FSR is small compared with the FSR itself. By using a cold cavity to lock the frequencies instead of the Ti:Sa cavity, the small internal dispersion of the cavity can be removed and zero CEO is realized with the ideal mode ratio of $n-1$ to n. Once we lock two frequencies to the ideal modes of the cavity, the CEO remains zero, in principle, irrespective any drift in cavity length. Moreover, because our cold cavity has a finesse of ~ 200 which is one order higher than the Ti:Sa cavity, it is expected that the accuracy of the frequency locking will be improved.

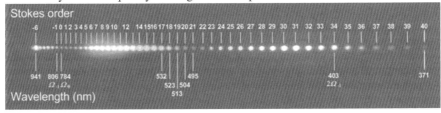

Fig. 2. Photograph of broad Raman sidebands generated by two pump pulses Ω_0 and Ω_{-1}, and the second harmonic $2\Omega_{-1}$.

In order to evaluate the CEO, we further introduced the second harmonic of one of the pump lasers, $2\Omega_{-1}$, in addition to the pump lasers, Ω_{-1}, Ω_0. The Raman sidebands are generated from the pump pulses, Ω_{-1}, Ω_0, and simultaneously from the second harmonic, $2\Omega_{-1}$, giving the CEO information through an overlap of both the sidebands, which is well-known as the $f - 2f$ technique in a femtosecond laser comb. It should be noted that this scheme also provides us with an octave-spanning Raman sideband.

Results and Discussion

Figure 2 shows a photograph of the generated Raman sidebands dispersed with a prism. It is clearly seen that the sidebands originating from the fundamental pump lasers, Ω_{-1},

Ω_0, and those from the second harmonic, $2\Omega_{-1}$, are overlapped with each other. We picked up the sidebands at 513 nm, and introduced them into an optical spectrum analyzer. Figure 3a shows CEOs obtained from the frequency deviation between the pair of sidebands versus the selected longitudinal modes of the pump lasers. It is clearly shown that the CEO was controlled with steps of the FSR (1.24 GHz) and could be set to zero by choosing the appropriate longitudinal modes. We also carried out the same CEO measurement, but in the temporal domain, by employing a fast detection system with a response time better than 10 GHz. In the temporal domain, the CEO is observed as a beat between a pair of sidebands. Figure 3b shows the results obtained with the same pair of sidebands as in Fig. 3a. It is confirmed that the CEO is controlled consistently to that found in the frequency domain.

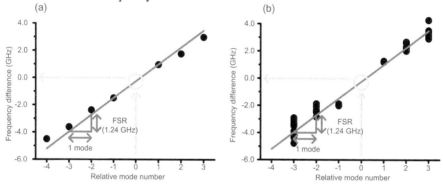

Fig. 3. CEOs measured in (a) spectral and (b) temporal domains. Ti:Sa pump laser cavity has a FSR of 1.24 GHz. The CEOs can be controlled with the discrete FSR steps.

We also examined the case of the cold cavity frequency locking. As shown in fig. 1, a high-finesse cold cavity was used for frequency locking. In the same way as before, we obtained in the temporal domain that the CEOs were controlled with steps of the FSR of the cold cavity (1.94 GHz). The difference of the FSR for the two frequency components is three orders less than that of the Ti:Sa cavity, and the controllability of the CEO is estimated to be as precise as 10^8 Hz. It means that the CEP is maintained at least for a few nanoseconds, which is enough to cover a train of ultrashort pulses over nanosecond pulsed envelope.

1   S. H. Harris and A. V. Sokolov, Phys. Rev. **A55**, R4019, 1997.
2   J. Q. Liang, M. Katsuragawa, F. Le Kien, and K. Hakuta, Phys. Rev. Lett. **85**, 2474, 2000.
3   M. Katsuragawa, K. Yokoyama, T. Onose, and K. Misawa, Opt. Express **13**, 5628, 2005.

Cancellation of the coherent accumulation in rubidium atoms excited by a train of femtosecond pulses

T. Ban, D. Aumiler, H. Skenderovic, S. Vdovic, N. Vujicic and G. Pichler

Institute of Physics, Bijenicka 46, HR-10 000 Zagreb, Croatia
E-mail: ticijana@ifs,

Abstract. In present experiments gradual change from the frequency comb excitation to pulse by pulse excitation of Rb atoms is observed. Shown results could lead to the development of a new method for system coherence monitoring.

Introduction

Our recent work in single photon spectroscopy of rubidium [1,2] combines fixed comb lines, broad absorption (rubidium atoms at room temperature) and an additional cw scanning probe laser. Resonant excitation of the rubidium atoms by discrete frequency comb optical spectrum results in the comb-like velocity distribution of the excited state hyperfine level populations and velocity-selective population transfer between the Rb ground state hyperfine levels. A modified DFCS was developed which uses the fixed frequency comb for the Rb $5^2S_{1/2} \rightarrow 5^2P_{1/2,3/2}$ excitation and the weak cw scanning probe laser for ground levels population monitoring. Observed modulations in the probe absorption are a direct consequence of the velocity-selective optical pumping (VSOP) induced by the frequency comb excitation. The fs pulse train excitation of Rb four $5^2S_{1/2} \rightarrow 5^2P_{1/2}$ and six $5^2S_{1/2} \rightarrow 5^2P_{3/2}$ level Doppler-broadened system was investigated theoretically in the context of the density matrix formalism. The analogous effect has also been observed at theoretically treated in the case of cesium atoms [3].

This work [4] improves the sensitivity of the detection by introducing the lock-in technique. This technique eliminates the Doppler background from the signal and results in direct monitoring of the modulation of the probe laser transmission. The structure and depth of the observed modulations are unique for each of the four Doppler broadened absorption lines and they reflect the structure of hyperfine levels and values of corresponding transition dipole moments. The enhanced sensitivity (over ten times) enables us to study various effects in frequency comb spectroscopy of Doppler broadened lines more thoroughly. In this work, we investigate how the frequency comb induced VSOP is influenced by the cw probe laser intensity. It turns out that in the strong probe regime the frequency comb excitation leads to the increase of the probe absorption and to the disappearance of the modulation structure. The disappearance of the modulation structure implies the disappearance of the VSOP. This imposes the conclusion that in the strong probe case, the atomic excitation is not driven by pulse train, but rather by pulse by pulse excitation. The physical mechanism behind this effect is the effective shortening of the atomic coherence relaxation time due to the strong probe laser. The numerical calculations of the density matrix evolution for this system, already developed in [1,2], are amended by introducing the probe laser electric field. Calculated probe laser absorption supports the experimental findings.

Experimental Methods

A Tsunami (Spectra Physics) mode-locked Ti:sapphire laser with pulse duration of about 100 fs generates 5.8 THz broad optical frequency comb. The comb frequencies were not tunable and the frequency comb was kept constant during the measurements. The laser repetition rate f_r, measured with a fast photodiode amounts 80 MHz. The fs laser beam, chopped with a SR540 mechanical chopper at 3.5 kHz repetition rate, was weakly focused onto the center of the glass cell containing rubidium vapor at room temperature. The ^{85,87}Rb $5^2S_{1/2}$ hyperfine ground state populations were monitored with a cw diode laser (Toptica DL100, ECDL at 780 nm), which propagated anti-collinearly with the fs laser, intersecting it under a small angle in the center of the cell. The probe frequency was slowly scanned across the Doppler-broadened ^{85,87}Rb $5^2S_{1/2} \rightarrow 5^2P_{3/2}$ hyperfine transitions at 3 GHz/s scanning rate. The probe laser transmission was simultaneously detected with two Hamamatsu Si photodiodes and fed into a digital oscilloscope (Tektronix TDS5140).

Results and Discussion

Transmission across all four Doppler broadened $5^2S_{1/2} \rightarrow 5^2P_{3/2}$ absorption lines at 780 nm for the fs laser tuned to $5^2S_{1/2} \rightarrow 5^2P_{1/2}$ transition at 795 nm is shown in Fig.1. Two outer absorption lines result from the ^{87}Rb absorption, whereas the inner two come from ^{85}Rb absorption. The excited state $5^2P_{3/2}$ hyperfine levels are not resolved due to the Doppler broadening. The modulations of the absorption profiles are the result of the fs pulse train excitation of Rb atoms. The signal from the second photodiode PD2, Fig. 1, was fed to the lock-in amplifier referenced to the mechanical chopper on the fs laser beam. The lock-in signal represents the change of the probe laser transmission induced by fs laser. The lock-in output is monitored on the digital oscilloscope. The advantage of the lock-in detection technique is obvious, since the broad Doppler background is eliminated. Additionally, the signal to noise ratio is greatly enhanced providing insight to finer details of the modulations.

We investigated the change in the probe transmission due to the fs laser excitation, for different probe laser powers. By increasing the probe laser power the structure and depth of the modulations change significantly. First, the fs laser induced increase of the absorption for all velocity groups is observed in the strong probe case. Second, as the probe laser power is increased the modulation depth decreases and a broad background with negative change in trasmission for all velocity groups appears. The velocity selection due to the frequency comb excitation is therefore lost, and the atoms interact with individual pulses rather than with the pulse train. This result is supported by the theoretical calculation where in the strong probe case no accumulation of population and coherence is obtained.

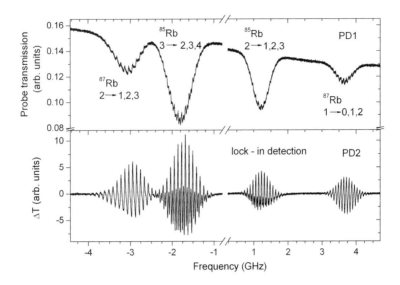

Fig. 1. Measured probe transmission for the Rb vapor at room temperature in the case when the fs laser is tuned to $5^2S_{1/2} \to 5^2P_{1/2}$ transition at 795 nm. The probe laser frequency is continuously scanned across all four Doppler broadened $5^2S_{1/2} \to 5^2P_{3/2}$ absorption lines at 780 nm.

Conclusions

We have presented enhanced sensitivity measurements of the velocity selective optical pumping (VSOP) of the Rb ground state hyperfine levels induced by the pulse train excitation. Using this approach we were able to directly measure the change of the probe laser transmission as a result of the resonant fs laser excitation of the Rb atoms. The most interesting result has been obtained in the case of the strong probe laser field. In this case, the velocity selection due to the frequency comb excitation is lost, corresponding to the interaction of atoms with the individual pulses rather than with the pulse train.

Exploitation of quantum memory for information storage and data processing is one of the fundamental aspects in the experimental physics. In the experiments where the coherence of the system has to be controlled and maintained for as long a time as possible despite higher field intensities, our method could be applied for system coherence monitoring.

1 D. Aumiler, T. Ban, H. Skenderovic and G. Pichler, Phys. Rev. Lett. 95, 233001 (2005).
2 T. Ban, D. Aumiler, H. Skenderovic and G. Pichler,Phys. Rev. A 73, 043407 (2006).
3 N. Vujicic, S. Vdovic, D. Aumiler, T. Ban, H. Skenderovic and G. Pichler, Eur. Phys. J. D 41, 447 (2007).
4 T. Ban, D. Aumiler, H. Skenderovic, S. Vdovic, N. Vujicic and G. Pichler, Phys. Rev. A 76, 043410 (2007).

Part XI

Optics, Optoelectronics, Measurement, Diagnostics and Instrumentation

Sub-10-fs XUV Tunable Pulses at the Output of a Time-Delay-Compensated Monochromator

L. Poletto[1], P. Villoresi[1], E. Benedetti[2], F. Ferrari[2], S. Stagira[2], G. Sansone[2], M. Nisoli[2]

[1] CNR-INFM - D.E.I. - Università di Padova, Padova, Italy
[2] CNR-INFM - Dipartimento di Fisica, Politecnico di Milano, Piazza L. da Vinci 32, 20133 Milano, Italy
E-mail: mauro.nisoli@fisi.polimi.it

Abstract. Extreme-ultraviolet pulses, produced by high-order-harmonic generation, have been spectrally selected by a time-delay-compensated monochromator. Temporal characterization has been obtained using cross-correlation method: pulses as short as 8 fs, with high photon flux, have been measured.

Introduction

Extreme ultraviolet (XUV) radiation produced by high-order harmonic generation (HHG) is attracting a large and rising interest for the ample range of possible applications. High brightness, femtosecond XUV pulses are important in various research fields, ranging from time-resolved spectroscopy to holography, microscopy and free-electron laser injection [1]. Various techniques have been implemented to select spectral portions of the XUV spectrum such as metallic filters and XUV dielectric multilayers; however, both of them are limited to a fixed spectral range and do not present any tunability. On the other hand, the use of a single diffraction grating causes severe broadening of pulse duration and significant attenuation.

In this work we have measured the duration of the harmonic pulses and the corresponding photon flux after spectral selection by a time-delay-compensated monochromator (TCM) based on the off-plane diffraction mount, where the incident and diffracted wave vectors are almost parallel to the grating grooves. Such off-plane mount allows one to achieve a remarkable improvement in terms of tunability and throughput with respect to the classical diffraction mount [2]. We demonstrate that the developed monochromator allows one to preserve the XUV pulse duration after spectral filtering. Indeed, we have generated broadly tunable coherent XUV pulses, with duration as short as 8 fs and high photon fluxes at the output of the TCM [3]. Such source lends itself as an important tool for a number of applications of femtosecond XUV pulses ranging from atomic and molecular spectroscopy to solid-state physics.

Experimental Results

The TCM is characterized by two equal sections, with two toroidal mirrors and a plane grating, as shown in Fig. 1. Since the grating has to be operated in parallel light, the first mirror of each section acts as the collimator and the second mirror as the focusing element. The first section generates a spectrally dispersed image of the harmonic source on the intermediate plane, where a slit is used for spectral selection of the harmonics. The selected spectral portion, composed by a single harmonic or a set of few harmonics, propagates toward the second section, which compensates for both the temporal broadening and the spectral dispersion and generates a stigmatic image on its focal plane. The four mirrors are operated at equal grazing angle and unity magnification to minimize the aberrations. Wavelength scanning is achieved by rotating the gratings around an axis tangent to their vertices and parallel to the grooves.

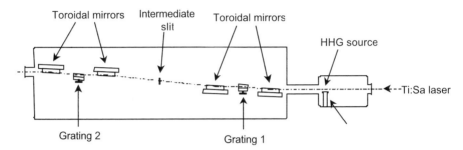

Fig. 1. Optical setup of the time-delay-compensated monochromator.

In this experiment, the laser pulses (25-fs duration, 800-nm central wavelength and 1-kHz repetition rate) are split in two parts using a drilled mirror. The inner part is focused on an Argon cell with static pressure for HHG. The XUV radiation propagates inside the TCM for spectral selection of single harmonics by the intermediate slit. An Argon jet is located at the output focal point of the monochromator. The outer annular part of the infrared (IR) beam is focused onto the same Argon jet, for the cross-correlation measurement with collinear geometry. The delay between the two pulses is controlled by a piezoelectric translator. The photoelectrons generated by single-photon absorption of the XUV pulses are collected by a time-of-flight spectrometer. The duration of the spectrally selected harmonic pulses is obtained by measuring the cross-correlation between the XUV and the 25-fs IR pulses. The harmonic XUV pulse ionizes a gas (Argon) in the presence of the IR field. When the two pulses overlap in time and space on the gas jet, sidebands appear in the photoelectron spectrum, spectrally shifted by the IR photon energy, determined by the absorption of one harmonic photon plus the absorption or the emission of one IR photon. The sideband amplitude as a function of the delay, between the XUV and IR pulses provides the cross-correlation signal [4].

Figures 2(a) and 2(b) show (dots) the temporal evolution of the amplitude of the first sideband vs delay in the case of XUV pulses obtained by selecting the 19th and 23rd harmonic, respectively. In order to obtain the XUV pulse durations, we have first calculated the evolution of the sideband amplitude vs delay assuming the measured IR intensity and duration, for different values of the XUV pulse duration. As shown in Fig. 2(a)-(b), the measured cross-correlation traces can be well fitted assuming an XUV pulse duration $\tau = 13 \pm 0.5$ fs (FWHM) in the case of the 19th harmonic and $\tau = 8 \pm 1$ fs (FWHM) in the case of the 23rd harmonic. The relative durations of the XUV and generating pulses turn out to be in good agreement with numerical simulations based on the nonadiabatic saddle-point method [5]. The experimental results clearly demonstrate that spectral selection of the XUV pulses obtained by using the TCM has been achieved preserving the harmonic pulse duration. We have then characterized the output of the TCM in terms of total photon yield, by placing an absolutely calibrated XUV photodiode at the TCM output. The TCM throughput efficiency is higher than 10% in the 20-40 nm spectral region with a peak of 18% around 30 nm. Using a 230-μJ pump pulse, we have measured a photon flux at the output of 6.5×10^8 ph/s and 1.3×10^9 ph/s, in the case of the 19th and 23rd harmonic, respectively.

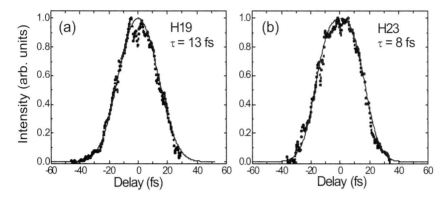

Fig. 2. Amplitude of the first sideband in the case of the 19th (a) and 23rd (b) harmonics as a function of the delay between the XUV and IR pulses. The dots are the experimental results; the solid lines are calculated as explained in the text.

Conclusions

In conclusion, coherent XUV pulses produced by HHG have been spectrally selected by a time-delay-compensated monochromator. Pulses as short as 8 fs have been measured at the output of the monochromator with high photon flux. Such high flux, combined with the very short duration, enables a number of novel and intriguing applications of femtosecond XUV pulses.

1   A. L'Huillier, D. Descamps, A. Johansson, J. Norin, J. Mauritsson, and C. –G. Wahlström, Eur. Phys. J. D **26**, 91, 2003.
2   L. Poletto, Appl. Phys. B **78**, 1013, 2004.
3   L. Poletto, P. Villoresi, E. Benedetti, F. Ferrari, S. Stagira, G. Sansone, M. Nisoli, Opt. Lett. **32**, 2897, 2007.
4   T. E. Glover, R. W. Schoenlein, A. H. Chin, and C. V. Shank, Phys. Rev. Lett. **76**, 2468, 1996.
5   G. Sansone, C. Vozzi, S. Stagira, and M. Nisoli, Phys. Rev. A **70**, 013411, 2004.

First Step Towards a Femtosecond VUV Microscope: Zone Plate Optics as Monochromator for High-Order Harmonics.

Jérôme Gaudin[1,2], Stefan Rehbein[1], Peter Guttmann[1], Sophie Godé[1], Gerd Schneider[1], Philippe Wernet[1] and Wolfgang Eberhardt[1]

[1] Berliner Elektronenspeicherring Gesellschaft für Synchrotronstrahlung BESSY, Albert-Einstein-Strasse 15, D-12489 Berlin , Germany
E-mail: wernet@bessy.de
[2] European X-Ray Free Electron Laser XFEL - Deutsches Elektronen-Synchrotron, Notkestrasse 85, D-22607 Hamburg, Germany
E-mail: jerome.gaudin@desy.de

Abstract. We report the use of zone plate optics as a monochromator for the spectral selection of a single high-order harmonic of a femtosecond laser generated in a rare gas medium in the photon energy range from 30 up to 70 eV while keeping the pulse duration in the femtosecond range. This is our first step towards a VUV microscope with sub-micrometer spatial resolution and femtosecond time resolution.

Introduction

Nowadays microscopy using soft X-Rays has became a routine method for high resolution imaging. By combining this technique with a light source delivering ultrashort pulses one would be able to perform time resolved imaging experiments with both high-spatial and high temporal resolution. Such light sources in the VUV domain are now available: High-order Harmonics (HHs) of femtosecond lasers. These type of sources can be optimized to routinely deliver high photon flux and ultrashort pulses which make them suitable for imaging experiments and first results have been recently published [1,2]. But so far the use of zone plates (ZPs) with HH-based sources has been limited to objective lenses with HHs of around 100 eV. The main limitation is that the transmission of the standard material for the ZP substrate, silicon nitride, is very low for energies below about 90 eV. To overcome this transmission problem we used a Si foil as substrate. We show that ZP optics provide a sufficient monochromaticity to select only a particular harmonic and that the transmission properties enable to record CCD images in a comparably short acquisition time for photon-energies of a few ten eV.

Experimental Set-up

The planed microscope is shown in Fig.1 . The results presented in this article concern only the condenser ZP (parts numbered in red in the Fig.1, see also [3] for a more detailed description). HH generation is driven by an infra-red laser (1.5 mJ, 40 fs per pulse at 1kHz, photon energy = 1.57 eV) which is focused inside a glass capillary (2 cm long, 300 μm inner diameter). The IR beam is then blocked by two 150 nm thick aluminum filters (not shown in Fig.1). At a distance of 1455 mm from the capillary, the condenser ZP is mounted on a two-axis motorized stage. The ZP substrate is a 180 nm thin Si foil. The ZP pattern is made of a polymer with a thickness of 120 nm.

884

This polymer has been chosen as it is known to be stable under VUV irradiation. The total transmission can to be estimated in the order of 5% to 10% in the energy range considered. The ZP consists of N = 2500 zones with a diameter D = 2 mm diameter. The maximum temporal broadening is given by $t = N\lambda/2c$ (for the harmonic 33 (51.8 eV) $t \approx 100$ fs). A pinhole of $d=10$ μm diameter is used as an order sorting aperture (OSA). The OSA is mounted on a 3-axis linear stage allowing for a variation of the distance to the ZP with an accuracy of 1 μm. All the measurements are performed with a back illuminated soft X-Ray CCD camera.

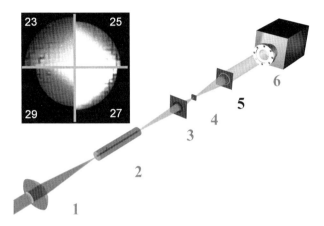

Fig. 1. The full field VUV microscope, only elements numbered in red have been implemented yet. Top left insert: Image of the 4 main harmonics in Argon. Numbers in the corners indicate the corresponding harmonic order.1: focusing lens - 2: glass capillary for HH generation - 3: condenser ZP - 4: order sorting aperture -5: micro-zoneplate - 6: X-Ray CCD

Results

In order to select one harmonic we take advantage of the fact that ZPs are chromatic optics. Hence the different harmonics are focused at different points. Since f increases with decreasing wavelength (i.e. increasing harmonic order) the diameter of the corresponding circle on the CCD is smaller the higher the harmonic order. Adding an OSA allows us to block the -1 order. Moreover, if it is placed exactly at the focus of one harmonic only this harmonic is transmitted. This provides the monochromatization and we obtain CCD images showing only one circle. Images taken for the 4 most intense harmonics as generated in Argon are presented in Fig.1.

If one now varies the distance between the ZP and the pinhole, and records an image for each step one obtains a full emission spectrum. Spectra measured for the generation gases Argon and Neon are shown in Fig.2. This allowed us to test the ZP in the photon energy range from 30 to 70 eV. Using the zone plate formula $f(\lambda)=D \cdot dr_n/m \cdot \lambda$ where m stands for the diffraction order, together with the thin lens formula we convert the ZP/OSA distance to photon energy. The graph also shows the centre energies as calculated according to the fundamental laser energy of 1.57 eV. The calculated values are in a good agreement with experimental peak maxima for low harmonics. The mismatch for the highest harmonics can be explained by the blue shift of the fundamental laser while propagating in an ionized medium [4]. This effect is more important in Neon as we focused the IR beam more tightly by using a shorter focal lens, enhancing HH generation efficiency but also enhancing ionization. The typical energy resolution amounts to $E/\Delta E = 42$ (FWHM). Finally, the short acquisition time of 10 s (60 s) for Argon (Neon) should be noted.

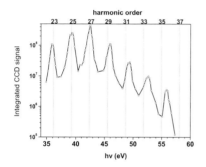 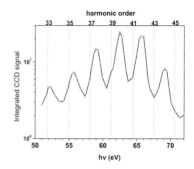

Fig. 2. HH spectrum obtained in Argon (left) and in Neon (right)

Conclusions

In conclusion, we have demonstrated that ZPs can be used as monochromating optics for HH in the VUV range. Moreover this first successful step opens the way to the development of a full field VUV microscope. In fact by adding a zone plate (element number 5 in Fig. 1) to the existing experiment we will be able to set-up a full field microscope. This so called micro ZP will be the microscope objective. It will image the focal spot of the condenser ZP. The temporal broadening induced will be negligible. The spatial resolution expected is in the order of dr_n of the condenser ZP. Such a table-top microscope should allow to perform time resolved imaging experiments with a sub-micrometer/sub-picosecond resolution.

1   I.R. Früke, J. Kutzner, T. Witting, H. Zacharias and Th. Wilhein, Eur. Phys. Lett. 72, 915, 2005
2   R.L. Sandberg, A. Paul, D.A. Raymondson, S. Härich, D.M. Gaudiosi, J. Holtsnider, R. I. Tobey, O. Cohen, M.M. Murnane and H.C. Kapteyn, Phys. Rev. Lett. 99, 098103, 2007
3   J. Gaudin, S. Rehbein, P. Guttmann, S. Godé, G. Schneider, Ph. Wernet and W. Eberhardt, J. Appl. Phys. 104, 033112, 2008.
4   C.G. Wahlström, J. Larsson, A. Persson, T. Starczemski, S. Svanberg, P .Salières, Ph. Balcou and A. L'Huillier, Phys. Rev. A 48, 4709, 1993.

Measurement of Electron Pulse Duration by Attosecond Streaking

P. Reckenthaeler[1,2]*, M. Centurion[1], V.S. Yakovlev[2], M. Lezius[1], F. Krausz[1,2] and E.E. Fill[1]

[1]Max-Planck-Institut für Quantenoptik, Hans-Kopfermann-Strasse 1, D-85748 Garching, Germany
[2]Department für Physik der Ludwig-Maximilians-Universität München, Am Coulombwall 1, D-85748 Garching, Germany
* E-mail: peter.reckenthaeler@mpq.mpg.de

Abstract: We propose a new method to measure the duration of ultrashort electron pulses using the principle of laser assisted Auger-decay.

Introduction

The quest for ever shorter electron pulses is motivated by a number of exciting applications, including ultrafast electron diffraction, crystallography and microscopy [1-3], electron imaging [4, 5], the generation of ultrashort X-ray pulses [6, 7] and pumping of X-ray lasers [8].

A key problem in most applications is to measure the duration of ultrashort electron pulses. Several methods have been investigated and applied for this purpose including streak cameras [9], interferometry of coherent transition radiation [9, 10], radio-frequency zero-phasing [11], terahertz radiation diagnostics [12], electro-optic encoding [13] and ponderomotive interaction of a laser with the electron pulse [14]. None of these methods, however, has proved to be applicable in a wide range of parameters: Streak cameras and radio-frequency zero-phasing cannot be used for pulses shorter then a few hundred fs, the coherent transition radiation method as well as the terahertz radiation and electro-optic methods require large number of electrons per pulse to provide sufficient signal and the ponderomotive interaction requires high laser intensities and is limited by the duration of intense laser pulses.

Proposed Experiment

We present a new method for measuring the duration of electron pulses, which holds promise for being applicable to an unprecedentedly broad range of parameters: from femtoseconds to attoseconds, from electronvolts to megaelectronvolts, and from millions of electrons per bunch to single-electron pulses [15]. It is similar to that known as the "attosecond-streak camera" [16,17], an ingenious tool for investigating ultrafast processes. So far the latter has been used for measurement of soft x-ray pulses, characterizing laser fields and investigating ultrafast processes in atoms [18].

The underlying idea is based on the fact that and electron pulse impinging on a solid target will generate, through impact ionization, a pulse of Auger electrons with duration equal to that of the incident pulse convolved with the duration of the Auger decay. The Auger decay can be much faster than the duration of the pulses and is easily accounted for. The energy of an Auger electron created in the presence of a

887

laser field is altered by the field. The spectrum is dependent on the relative time delay between the laser and the electron pulse. This can be used to establish cross-correlation between electron and laser pulses and thus measure the duration of the electron pulse. Calculation of the spectra following a quantum-mechanical approach yields side bands spaced by the laser photon energy. Such spectra have been seen in the so-called laser-assisted Auger decay [19].

Numerical Results

In the following, we present results of the calculations for two different cases, viz. relatively long laser and electron pulses (of order 100 fs), and very short pulses, viz. a few-cycle laser pulse with an as-electron pulse. The "streaked" spectra, generated by plotting the modified Auger spectra as functions of delay between laser and electron pulse, are shown in Fig. 1. The very different parameter ranges illustrate the broad applicability of the method. In the long-pulse case (electron pulse duration longer than the laser cycle), the electron spectrum is broadened, since the electron energies are swept through a large number of laser cycles. In the short-pulse regime (electron pulse duration shorter than laser cycle), the electron spectrum at any particular delay will be broadened and shifted. An Auger line with an appropriate decay time must be chosen for optimum results in the two regimes.

The following parameters were used for the calculation of the long-pulse case: Electron pulse duration 80 fs, laser pulse duration 50 fs. Both pulses were assumed to be Gaussian in time. For the Auger transition, the Oxygen KLL line with energy of 500 eV is chosen. The natural width of this line is 0.15 eV corresponding to a decay time of 4.4 fs. The laser intensity is assumed to be 1.6×10^{12} W/cm^2 at a wavelength of 800 nm. The maximum energy shift is calculated to ±20 eV. The long pulse case is shown in Fig. 1a. As the overlap between the two pulses increases, more and more side bands appear on both sides of the main Auger line. From such a correlation measurement, the duration of the electron pulse can easily be retrieved if the laser pulse envelope is known.

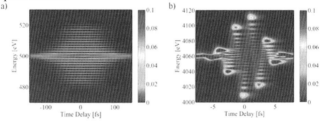

Fig. 1: Calculated streaked electron spectra. Time zero is defined as the temporal delay at which both pulse maxima coincide. The colour scale is normalized to the intensity of the original Auger line. a) Spectra for long-pulse conditions: Electron pulse duration 100 fs; laser pulse duration 80 fs. Auger line applied oxygen KLL at 500 eV. b) Spectra for short-pulse conditions: Electron pulse duration 800 as; laser pulse duration 5 fs. Auger line Ti KLL at 4.06 keV.

The short-pulse case (Fig. 1b) is calculated with the parameters: electron pulse duration 800 as, a few-cycle cosine-laser pulse with duration of 4 fs at a wavelength of 800 nm. A laser intensity of 1.6×10^{12} W/cm^2, was assumed. Note that the laser pulse must be carrier-envelope stabilized, otherwise the temporal resolution would only be given by its duration. For the Auger transition, the titanium KLL Auger line at

4.06 keV with a natural width of 0.94 eV and a corresponeing decay time of 702 as is chosen. Here, the maximum energy shift of the electrons is ± 57 eV. Features quite different from the long-pulse case are now observed: The main change in the spectrum is an energy shift which occurs in synchronism with the laser vector potential. Slight broadening results from the combined smearing of electron pulse duration and Auger decay time over the laser cycle.

For successful realization of the experiment, the number of Auger electrons generated must be high enough to produce a spectrum in an appreciable amount of time. We calculated the number of Auger electrons per pulse electron taking into account the cross-section for the generation of a K-hole, the fluorescence yields for oxygen and titanium, the number density of the solid and the escape depth of the Auger electrons from the solid. The result is, that with an electron gun running at 1,000 electrons per pulse at a kHz repetition rate or with 1 electron per pulse at a MHz repetition rate, and an electron spectrometer with an acceptance angle of about 1 sterad, a complete spectrum can be recorded in less than an hour.

Conclusion

In conclusion, we present a new method of electron pulse duration measurement. By drawing on laser assisted Auger electron emission induced by impact ionization, the method can be used for determining electron pulse durations in a broad range of parameters of electron energy and pulse duration. Electron energies may range from a few keV up to highly relativistic ones in the GeV range and pulse durations from a few 100 fs to sub fs. The method also works independently of the number of electrons per pulse.

1   A. H. Zewail, Ann. Rev. Phys. Chem. 57, 65 (2006).
2   J. R. Dwyer, C. T. Hebeisen, R. Ernstorfer, M. Harb, V. B. Deyirmenjian, R. E. Jordan, and R. J. Dwayne Miller, Phil. Trans. R. Soc. A 364, 741 (2006).
3   V. A. Lobastov, R. Srinivasan, and A. H. Zewail, PNAS 102, 7069 (2005).
4   Y. Okano, Y. Hironaka, K. G. Nakamura, and K. Kondo, Appl. Phys. Lett. 83, 1536 (2003).
5   C. G. Serbanescu and R. Fedosejevs, Appl.Phys. B 83, 521 (2006).
6   H. Schwoerer, B. Liesfeld, H.-P. Schlenvoigt, K.-U. Amthor, and R. Sauerbrey, Phys. Rev. Lett. 96, 014802 (2006).
7   R. Schoenlein, W. Leemans, A. Chin, P. Volfbeyn, T. Glover, P. Balling, M. Zolotorev, K. Kim, S. Chattopadhyay, and C. Shank, Science 274, 236 (1996).
8   D. Kim, C. Toth, and C. P. J. Barty, Phys. Rev. A 59, R4129 (1999).
9   T. Watanabe, M. Uesaka, J. Sugahara, T. Ueda, K. Yoshii, Y. Shibata, F. Sakai, S. Kondo, M. Kando, H. Kotaki, and K. Nakajima, Nucl. Instrum. Meth. in Phys. Res. A 437, 1 (1999).
10 H. C. Lihn, P. Kung, C. Settakorn, H. Wiedemann, and D. Bocek, Phys. Rev. E 53, 6413 (1996).
11 D. X. Wang, G. A. Krafft, and C. K. Sinclair, Phys. Rev. E 57, 2283 (1998).
12 J. van Tilborg, C. B. Schroeder, C. V. Filip, C. Toth, C. G. R. Geddes, G. Fubiani, E. Esarey, and W. P. Leemans, Phys. Pasmas 13, 056704 (2006).
13 I. Wilke, A. M. MacLeod, W. A. Gillespie, G. Berden, G. M. H. Knippels, and A. F. G. van der Meer, Phys. Rev. Lett. 88, 124801 (2002).
14 C. T. Hebeisen, R. Ernstorfer, M. Harb, T. Dartigalongue, R. E. Jordan, and R. J. Dwayne Miller, Opt. Lett. 31, 3517 (2006).
15 P. Reckenthaeler, M. Centurion, V. S. Yakovlev, M. Lezius, F. Krausz and E. E. Fill, Phys. Rev. A 77, 042902 (2008).
16 J. Itatani, F. Quere, G. L. Yudin, M. Y. Ivanov, F. Krausz, and P. B. Corkum, Phys. Rev. Lett. 88, 173903 (2002).
17 M. Kitzler, N. Milosevic, A. Scrinzi, F. Krausz, and T. Brabec, Phys. Rev. Lett. 88, 173904 (2002).
18 R. Kienberger, et al., Nature 427, 817 (2004).
19 J. M. Schins, P. Breger, P. Agostini, R. C. Constantinescu, H. G. Muller, G. Grillon, A. Antonetti, and A. Mysyrowicz, Phys. Rev. Lett.73, 2180 (1994).

Nanoscale Spatial Effects of Pulse Shaping

Daan Brinks[1], Fernando D. Stefani[1], and Niek F. van Hulst[1,2]

[1] ICFO – Institut de Ciencies Fotoniques, Mediterranean Technology Park, 08860 Castelldefels (Barcelona), Spain.
E-mail: daan.brinks@icfo.es
[2] ICREA - Institució Catalana de Recerca i Estudis Avançats, 08015 Barcelona, Spain
E-mail: Niek.vanHulst@ICFO.es

Abstract. Commonly used pulse-shaping techniques create a coupling between the spatial and temporal characteristics of the shaped femtosecond laser pulse. Consequently, measured apparent responses to shaped pulses that might seem produced by time domain molecular dynamics could in fact be due to spatial changes in the electric field.

Introduction

Shaping ultrashort laser pulses has opened the possibility of influencing molecular dynamics through interaction with light. To steer molecular dynamics one requires to shape fs pulses in the time-frequency domain. Most commonly used pulse shapers are based on spatial light modulators (SLM)[1] and Acousto-Optic Programmable Dispersive Filters (AOPDF)[2]. In both techniques, the spatial and temporal characteristics of the shaped pulse are coupled. Since detection of the effect of shaped pulses necessarily takes place in a limited spatial region, i.e. a focus or the overlap volume of a pump and a probe beam, this coupling can influence the outcome of experiments. It is therefore important to know the magnitude of this spatio-temporal coupling.

Spatio-temporal coupling

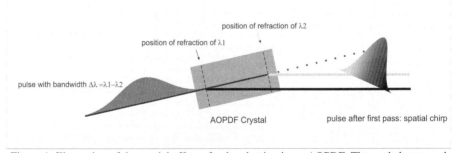

Figure 1: Illustration of the spatial effect of pulse shaping in an AOPDF. The angle between the input and output beam and the different positions of refraction in the AOPDF crystal create a chirped output beam.

In an AOPDF, the transit time of light through a birefringent crystal is controlled by polarization rotation through acousto-optic interaction at a certain position inside the crystal. Different travelling times for different frequency components cause a phase modulation of the pulse; the efficiency of the acousto-optic interaction for different frequency components causes an amplitude modulation[3].

Since the Poynting vectors of the optical and the acoustic waves are aligned and not their k-vectors[4], the acousto-optic interaction causes a slight change in direction; i.e. each frequency component that undergoes the acousto-optic interaction exits the crystal at a different position on the output facet. This effect is depicted in figure 1. This phenomenon complicates the interpretation of experiments in several ways:

1. Different positions across the beam profile will be dominated by different frequencies, which can influence the outcome of multi-photon processes.
2. Different frequency components will have different k-vectors in a focus, influencing the efficiency of interaction with (transition) dipoles depending on their respective orientation.
3. Spectral amplitude modulation causes the transverse beam profile to change. As a result every temporal pulse shape will also have a different transverse spatial profile.

Spatial shaping effects in a focus

In order to quantify the magnitude of the spatio-temporal coupling and its possible effects on experiments, we perform simulations of shaping operations as they are typically used in investigations of molecular dynamics. We simulate common shaping configurations and calculate the spatial effect of compensating chirp by phase-shaping and creating phase-locked pulse pairs and multipulse-sequences by amplitude modulation.

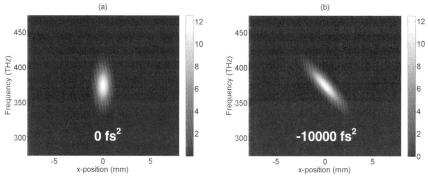

Figure 2: Frequency-space beam profile showing the frequency content versus the spatial dimension in which the frequencies have been dispersed for shaping (x-axis). Shown are (a) an unshaped pulse and (b) a pulse with -10000 fs^2 chirp added. The pulse chirped in time is clearly also chirped in space.

In figure 2 the coupling between temporal and spatial chirp after shaping is illustrated. The plots display the intensity of the frequency components of the laser pulse as a function of the position along the shaping coordinate x. Figure 2a shows a Gaussian laser pulse of 10 fs centered at 800 nm and with 10000 fs^2 chirp (as can be acquired in the optics of a typical optical setup). Figure 2b shows the same pulse after adding -10000 fs^2 chirp compensation in the shaper. Clearly, the shaped pulse presents different spectral components at different positions across the beam.

In Coherent Control and Multidimensional Spectroscopy experiments, multi-pulse sequences are the basic constituents of the pulse shapes used to probe and control molecular dynamics [1,5]. Therefore, we took the pulse of figure 2b and simulated the shaping into a multi-pulse sequence and the subsequent imaging in a focus.

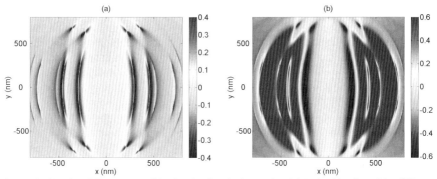

Figure 3: Spatial difference profiles in the focal plane of a 1.4 NA objective. (a): difference between the focal profile of a phase locked double pulse and a single chirped 10 fs pulse. (b): difference between the focal profile of a phase locked quadruple pulse a single chirped 10 fs pulse.

In figure 3 we compare the intensity distribution of a focused multi-pulse sequence to a focused single pulse, all in the focal plane of a 1.4 NA objective. The normalized differences (i.e. $(I_{multi-pulse} - I_{single\,pulse})/(I_{multi-pulse} + I_{single\,pulse})$) are plotted as a function of position in the diffraction limited focus. The single pulse is in all cases a 10 fs pulse, as in figure 2b. In figure 3a the multi-pulse consists of two such pulses with a 10 fs delay and phase locked at 0 rad. In figure 3b the multi-pulse consists of four pulses with an interpulse delay of 10 fs and again phase locked at 0 rad.

Clearly changing the shape of the double pulse creates local changes in the intensity of up to 40%. This effect grows to 60% in the case of the quadruple pulse. Any molecule exposed to a field varying locally like this will respond accordingly. It is worth noting that the variations are not only very strong, but depending on the position in the focus can be very local: positions in the focus that are 10 nanometer apart already experience field intensities developing very differently with varying pulse shapes.

Conclusions

To our best knowledge, current pulse shaping techniques, used in for instance coherent control, create a coupling between the temporal and spatial characteristics of the shaped pulses. Manifest effects include spatial chirp and different spatial intensity profiles for different temporal pulse shapes. The spatio-temporal coupling will influence the outcome of experiments, as not all molecules exposed to the shaped pulses experience the same changes in field strength for varying pulse-shapes. Interestingly, the development of the local field as a function of temporal pulse shape can differ profoundly for positions in a tight focus that are separated by no more than 10 nm.

1   J. Herek et al., Nature **417**, 533, 2002.
2   V.I. Prokhorenko et al., Science **313**, 1257, 2006.
3   F. Verluise et al., Opt. Lett **25,** 575, 2000.
4   F. Verluise et al., J. Opt. Soc. Am. B, **17,** 138, 2000.
5   N.F. Scherer et al., J. Chem Phys. **95**, 1487, 1991

Designer Femtosecond Pulse Shaping Using Grating-Engineered Quasi-Phasematching in Lithium Niobate

Łukasz Kornaszewski[1], Markus Kohler[1], Usman K. Sapaev[2], Derryck T. Reid[1]

[1] Ultrafast Optics Group, School of Engineering and Physical Sciences, Heriot-Watt University, Edinburgh, EH14 4AS, UK
[2] Laser-Matter Interaction Laboratory, NPO Akadempribor, Academy of Sciences of Uzbekistan, Tashkent 700125, Uzbekistan
E-mail: lwk1@hw.ac.uk

Abstract. The generation of tailored femtosecond pulses with fully engineered intensity and phase profiles is demonstrated using second-harmonic generation of an Er:fibre laser in an aperiodically-poled lithium niobate crystal in the undepleted pump regime. Second harmonic pulse-shapes including Gaussian, stepped, square and multiple pulses have been characterised using cross-correlation frequency-resolved optical gating and shown to agree well with theory.

Introduction

Designer pulse shaping requires the independent manipulation of the spectral intensity and phase of an optical waveform. Aperiodically-poled quasi-phasematched (QPM) crystals have already been used to compress pulses produced by second-harmonic generation (SHG) [1], and to create sub-ps pulse sequences and shaped multi-ps pulses [2]. This concept can be extended to shape individual fs pulses by controlling the local duty cycle and the period in a QPM grating, as we previously showed theoretically [3, 4]. We now present the first experimental validation of this approach.

Our earlier work [3, 4] described a theoretical model containing a crystal transfer function that took account of each domain size and its position in the crystal, evaluated using the analytical formula [3, 4]:

$$E_{CRYS}(\omega) = \frac{-\kappa \, d_{ijk}}{\Delta k(\omega)} \left\{ 1 - (-1)^n \exp\left[i \, \Delta k(\omega) Q_n\right] + \sum_{m=1}^{n-1} 2(-1)^m \exp\left[i \, \Delta k(\omega) Q_m\right] \right\} \quad (1)$$

where: κ is ω_{SHG}/cn_{SHG}, d_{ijk} is the absolute value of the nonlinear coefficient; $\Delta k(\omega)$ is the magnitude of the wavevector mismatch in the process, and Q_m is the end position of domain m in the grating which contains a total of n domains, as described in [4]. Using the crystal transfer function it is easy to calculate the SH pulse, $E_{OUT}(t)$, generated from an input pulse, $E_{IN}(t)$, by a grating characterised by $E_{CRYS}(\omega)$ [5]:

$$E_{OUT}(t) = F^{-1}\left\{ F\left[E_{IN}^2(t)\right] E_{CRYS}(\omega) \right\} \quad (2)$$

where F and F^{-1} are direct and inverse Fourier-transform operators, respectively.

Crystal design and experiment

Following the procedure outlined in [4] we used a simulated annealing algorithm to find the appropriate grating designs for 9 different target SH pulses. The design assumed an unchirped 150 fs Gaussian fundamental pulse centred at 1530 nm which was similar, but not identical to the actual pulses used in the experiment. The 4 mm long APPLN crystal was phasematched for an input wavelength of 1530 nm and had 9 gratings of differing lengths. The design (inset, Fig. 1) was optimised for the outputs: short / long Gaussian pulses (1, 2), stepped / square pulses (3, 4), triangular pulses (5), identical but oppositely chirped pulses (6, 7), and multiple (8, 9) pulses.

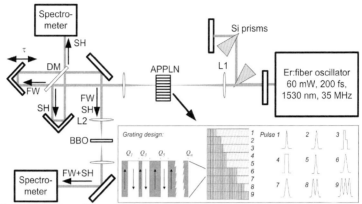

Fig. 1. Experimental configuration. Pulses from the Er:fibre oscillator are compressed and then focused into the APPLN crystal. After collimation, the second-harmonic (SH) and the fundamental wave (FW) light enter a Michelson interferometer. A dichroic mirror (DM) acts as a beamsplitter, and the nonlinear mixing occurs in a 100 μm-thick BBO crystal. FW+SH denotes the sum-frequency beam resulting from nonlinear mixing in the BBO crystal. L1 and L2 are lenses with focal lengths of 100 mm and 15 mm respectively. Inset: schematic of the APPLN crystal, and the target SH profiles designed assuming 150 fs Gaussian input pulses. The first few domains are shown schematically for grating 6.

Fig. 1 shows the experimental arrangement. The compressed output pulses from an Er:fibre oscillator were focused into the SHG crystal, after which the fundamental and SH pulses were mixed in a BBO crystal and the XFROG traces recorded (Fig. 2).

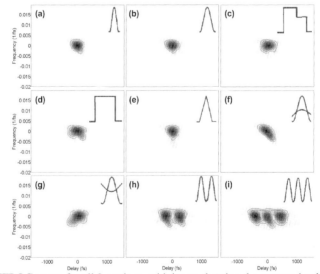

Fig. 2. XFROG traces for all 9 gratings, with insets showing the expected pulse shapes.

Results and discussion

Retrieved results are presented in Fig. 3. The measurements indicated qualitative agreement between experiment and theory, but to allow a quantitative comparison we used equations (1) and (2), along with the complex field amplitude of the Er:fibre pulses determined from the XFROG measurements, to evaluate the expected SH pulse shapes, also shown in Fig. 3.

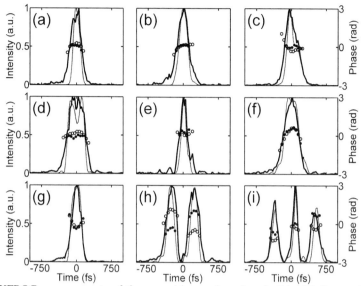

Fig. 3. XFROG measurements of the second-harmonic pulses (solid thick line — intensity; open circles — phase), compared with the shapes calculated using the fundamental pulses (dotted thin line – intensity; filled circles – phase).

The agreement between the experimental and calculated SHG pulses is generally good, in terms of both their amplitude and phase. However, for shorter grating designs there is a significant difference probably due to a relatively long region of non-poled material which does not contribute to the pulse shaping but rather only adds unwanted dispersion to already created pulse.

Conclusions

In conclusion, the use of fully grating-engineered crystals for femtosecond pulse shaping is a simple and robust alternative to an adaptive optics system, and our results show the strong potential of this technique. Extension of the technique to the high-depletion regime will widen its applicability.

1 M. A. Arbore, A. Galvanauskas, D. Harter, M. H. Chou and M. M. Fejer, Opt. Lett. 22, 1341 (1997).
2 G. Imeshev, A. Galvanauskas, D. Harter, M. A. Arbore, M. Proctor, and M. M. Fejer , Opt. Lett. 23, 864 (1998).
3 D. T. Reid, J. Opt. A 5, S97 (2003).
4 U. K. Sapaev and D. T. Reid, Opt. Expr. 13, 3264 (2005).
5 M. A. Arbore, O. Marco and M. M. Fejer, Opt. Lett. 22, 865 (1997).

Direct Measurement of Spectral Phase for Ultrashort Laser Pulses Based on Intrapulse Interference

Bingwei Xu, Vadim V. Lozovoy, Yves Coello, and Marcos Dantus

Michigan State University, Department of Chemistry. East Lansing, Michigan 48824, USA
E-mail: dantus@msu.edu

Abstract. We present a method for the direct spectral phase measurement of ultrafast laser pulses. The second-derivative of the unknown spectral phase is revealed by the experimental 2D-contour plot and can be measured without mathematical manipulation.

Introduction

Pulse characterization and compression are of critical importance in ultrafast laser science and technology. Among the available spectral phase characterization techniques, the early development of Yamada and coworkers [1], and the development of FROG [2] and SPIDER [3] represent milestones in the field. There are also optimization algorithm approaches to achieve pulse compression with or without spectral phase measurements [4, 5]. Ideally, a spectral phase measurement should be simple, direct and relatively insensitive to noise. Here, we report on such a method, based on a simple chirp scan.

It has long been known that nonlinear optical (NLO) processes are sensitive to the second derivative of the phase. If we introduce a reference function $f''(\omega)$ to measure the second derivative of the unknown phase $\phi''(\omega)$, NLO processes are maximized at frequency ω when the equation $\phi''(\omega)-f''(\omega)=0$ is satisfied [6, 7]. Since $f''(\omega)$ is known, the value of $\phi''(\omega)$ can be easily obtained. Based on this observation, our group has been using the multiphoton intrapulse interference phase scan (MIIPS) method, which typically uses a sinusoidal function for $f(\omega)$ [6, 7]. In this contribution, we successively impose a set of quadratic phases (chirp) instead of sinusoidal phases on the ultrashort pulses and record the corresponding nonlinear spectra. The second-derivative of the unknown spectral phase is directly revealed by the experimental 2D-contour plot resulting from the single chirp scan. An accurate measurement can be extracted from the experimental data without any mathematical treatment or approximation [8]. Given that chirp can be introduced using standard passive optics such as a prism, grating or grism-pair arrangement, this method can be conveniently implemented without the need of an adaptive pulse shaper.

Experimental Methods

For the experiments carried out with an adaptive pulse shaper we used a folded all-reflective grating-based system containing a grating, a long focal length spherical mirror, and a 640-pixel dual-mask spatial light modulator (SLM-640, CRi Inc.). After the pulse shaper, the laser was focused onto a thin KDP crystal and the second harmonic generation (SHG) signal was directed to a spectrometer. The detailed setup of the experiments described here can be found in publications [8, 9].

896

Results and Discussion

Measurements with an adaptive pulse shaper

Transform-limited pulses were first obtained by measuring and compensating the spectral phase of the system using the sinusoidal MIIPS method [6, 7]. A 3000 fs^3 cubic spectral phase function was then introduced to the pulses via the pulse shaper, and measured with a simple chirp scan. The resulting 2D contour plot is shown in Fig. 1(a). Quantitative results are shown in Fig 1(b) together with the spectrum of the laser. A sinusoidal spectral phase function was also introduced and measured. The experimental trace is shown in fig. 1(c). Fig. 1(d) shows the introduced (black curve) and measured second derivative of the phase (dots). Note that in both cases the experimental traces directly reveal the measured $\phi''(\omega)$. The accuracy of this method, especially for complex phases, improves with an iterative measurement-compensation routine [8, 9].

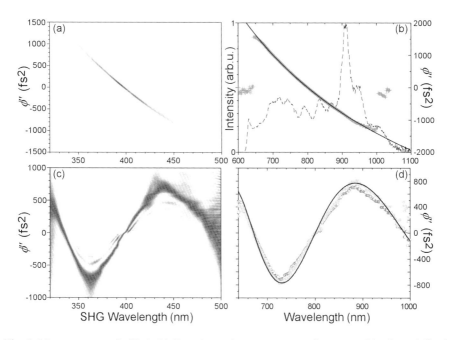

Fig. 1. Measurements of $\phi''(\omega)$. (a) Experimental trace corresponding to a cubic phase defined by $\phi(\omega)=1/6*3000\text{fs}^3*(\omega-\omega_0)^3$. (b) shows the introduced (black curve) and measured functions (crosses), the dashed curve corresponds to the spectrum of the laser. (c) Experimental trace corresponding to a sinusoidal phase defined by $\phi(\omega)=5\pi \sin[7\text{fs}*(\omega-\omega_0)]$. (b) The introduced function is shown in black. Black and grey dots correspond to the functions measured after a single scan and after one iteration, respectively.

Once $\phi''(\omega)$ is obtained, as demonstrated before, double integration can be used to calculate $\phi(\omega)$. Fig. 2(a) shows complex spectral phase measurements. For these experiments, two independent pulse shapers were used. One pulse shaper introduced the desired spectral phase, while the other was used to measure it using MIIPS. The agreement between the introduced and measured phases illustrates the performance of the method for the case of complex spectral phases.

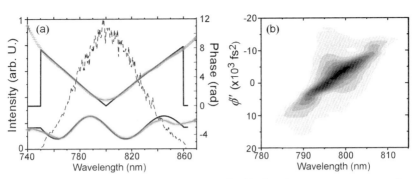

Fig. 2. (a) Complex spectral phase measurements. The black and grey curves correspond to the introduced and measured phases. The dotted curve corresponds to the spectrum of the pulses (b) Phase measurement without an adaptive pulse shaper. The figure shows the experimental trace obtained after a chirp scan with a grating compressor. The linear feature corresponds to a cubic spectral phase.

Measurements without an adaptive pulse shaper

Here, we demonstrate spectral phase measurements without the use of an adaptive pulse shaper. Instead, different amounts of linear chirp were introduced to amplified pulses using the built-in compressor in the regenerative amplifier by varying the spacing between the grating pair. Fig. 2(b) shows the experimental trace obtained after the chirp scan. The linear $\phi''(\omega)$ dependence indicates the presence of a cubic phase distortion, also known as third-order dispersion (TOD). No effort was made here to eliminate the measured TOD.

Conclusions

In conclusion, a new MIIPS implementation based on a simple chirp scan was presented. The corresponding experimental trace directly yields the second derivative of the unknown spectral phase, without any mathematical treatment.

Acknowledgements. We gratefully acknowledge funding for this research from the National Science Foundation, Major Research Instrumentation grant CHE-0421047.

1   K. Naganuma, K. Mogi, and H. Yamada, IEEE J. Quantum Elect. **25**, 1225, 1989.

2   R. Trebino, and D. J. Kane, J. Opt. Soc. Am. A. **10**, 1101, 1993.

3   C. Iaconis, and I. A. Walmsley, Opt. Lett. **23**, 792, 1998.

4   T. Baumert, T. Brixner, V. Seyfried, M. Strehle, and G. Gerber, Appl. Phys. B-Lasers O. **65**, 779, 1997.

5   D. Yelin, D. Meshulach, and Y. Silberberg, Opt. Lett. **22**, 1793, 1997.

6   V. V. Lozovoy, I. Pastirk, and M. Dantus, Opt. Lett. **29**, 775, 2004.

7   B. Xu, J. M. Gunn, J. M. Dela Cruz, V. V. Lozovoy, and M. Dantus, J. Opt. Soc. Am. B **23**, 750, 2006.

8   V. V. Lozovoy, B. Xu, Y. Coello, and M. Dantus, Opt. Express **16**, 592, 2008.

9   Y. Coello, V. V. Lozovoy, T. C. Gunaratne, B. Xu, I. Borukhovich, C. Tseng, T. Weinacht, and M. Dantus, J. Opt. Soc. Am. B **25**, 140, 2008.

Two Dimension Spatial Light Modulator with an Over-Two-Octave Bandwidth for High-Powered Monocycle Optical Pulses

K. Hazu, T. Tanigawa, N. Nakagawa, Y. Sakakibara, Sh. b. Fang, T. Sekikawa, and M. Yamashita

Department of Applied Physics, Hokkaido University, and Core Research for Evolutional Science and Technology, Japan Science and Technology Agency, Kita-13, Nishi-8, Kita-ku, Sapporo, 060-8628 Japan
E-mail: mikio@eng.hokudai.ac.jp

Abstract. We carried out feedback chirp compensation using a two-dimension spatial light modulator operating in a wavelength range from 260 to 1100 nm, which is useful for the application to ultrabroadband and high-powered optical pulses.

1. Introduction

A liquid crystal spatial light modulator (LC-SLM) is a very powerful tool for chirp compensation and shaping of ultrashort optical pulses. LC-SLMs have enabled phase control over the octave broad spectrum to generate monocycle pulses by the feedback (FB) phase control technique using a 4-f phase compensator [1-3]. On the other hand, pulse shaping has been reported for quantum coherent control, high-speed optical information processing and material characterization. Furthermore, the interest in the application for coherent control in high-field physics is increasing and the technique to generate a few mJ pulses with a duration in the 5 fs region has been proposed. However, LC-SLMs have a bandwidth limitation in the UV region because of UV absorption by the liquid crystal and low damage threshold.

2. Structure of 2D-SLM

The fabricated UV-LC-SLM has vertically two-channels (a large height of 9.8 mm for two dimension (2D) with 200 μm channel gaps: we also fabricated a one-channel 2D·UV-LC-SLM with the same height) and horizontally 648-pixels (98 μm width: 5 μm pixel gap). A new nematic LC of a mixture of cyclohexane derivatives with fluorine substituents (20 μm thickness) operating in the range from 260 to 1100 nm was sandwiched between two fused-silica glasses (0.5 mm thickness), deposited indium tin oxides (100 nm thickness) and parallel-oriented organic films [4]. The maximum values of phase modulation are 31.7, 13.1 and 7.0 rad at 305, 600 and 1100 nm, respectively [4], which are all larger than 2π rad.

3. Optical damage of 4-f phase compensator with 2D·UV-LC-SLM

We investigated optical damage characteristics of the 2D·UV-LC-SLM using an amplified Ti:sapphire laser system (30 fs, 1 mJ, 800 nm, 1 kHz). The result showed that the UV-LC-SLM has no damage under the direct irradiation of the laser pulse with photon density of 29 GW/cm^2 for longer than 10 hours.

We calculated the photon density of the 1-mJ, 30-fs input pulse in a conventional 4-f phase compensator. For the 4-f compensator consisting of a pair of concave mirrors (f=300 mm) and gratings (300 grooves/mm), the photon density at one pixel of the UV-LC-SLM was 378 GW/cm^2. This is much higher than the damage threshold (29 GW/cm^2) of the UV-LC-SLM. To solve this damage problem, we introduced cylindrical mirrors (f=1000 mm) instead of concave mirrors, and a fused silica prism instead of gratings into the 4-f folded phase compensator with the 2D·UV-LC-SLM [Fig. 1]. The prism has more transmission efficiency than gratings over the broadband, and is able to avoid spatial overlapping of the second-order diffracted light. The photon density in the new cylindrical 4-f phase compensator was 22 GW/cm^2 at one pixel of the 2D·UV-LC-SLM under the same condition (30 fs, 1 mJ), which is smaller than the damage threshold. Ignoring losses of optical components, the maximum input pulse energy to be allowed is 1.3 mJ for the new 4-f phase compensator.

4. Feedback chirp compensation experiment in the near-infrared (NIR) region

A feedback (FB) chirp compensation experiment was carried out using an amplified Ti:sapphire laser system (40 fs, 220 μJ, 780 nm, 1 kHz). A 23-mm-long fused silica glass was used as a dispersion medium yielding a strong chirp. The spectral phase was characterized by the modified spectral phase interferometry for direct electric-field reconstruction (M-SPIDER) [2]. The delay time, the spectral shear and the spectral shift were 990 fs, 23.6 rad·THz and 2.35 rad·PHz, respectively. Figure 2(a) shows the spectral phase before and after FB compensations. The chirped pulse before FB compensation has a spectral phase variation over 30 rad in the spectral range from 720 to 860 nm and a duration of 118 fs [Fig. 2(b)]. The group delay dispersion (GDD) obtained by curve fitting was 846 fs^2 at 800 nm. Those agree with the GDD of the dispersion glass and other optical components. After FB compensation, the spectral phase is almost flat within the smaller GDD than 20 fs^2. The corresponding temporal intensity profile shows a duration of 24 fs, which is very close to the transform-limited pulse duration of 18 fs.

5. Feedback chirp compensation experiment in the UV region

We carried out a FB chirp compensation experiment in the UV region. We used a nonlinear optical crystal BBO (0.1 mm, type I) to generate UV light (24 μJ, 392 nm, 1 kHz). The spectral phase was characterized by the same method above. The delay time, the spectral shear and the spectral shift were 557 fs, 15.7 rad·THz, 2.41 rad·PHz, respectively. Figure 3(a) shows the spectral phase before and after FB compensations. The chirped pulse before FB compensation has a duration of 270 fs [Fig. 2(b)]. The GDD was 2418 fs^2 at 392 nm. Those agree with the GDD of the prism material in the new 4-f phase compensator and other optical components. After FB compensation, the fitted GDD was 730 fs^2. The corresponding temporal intensity profile shows a duration of 34 fs, which is close to the transform-limited pulse duration of 21 fs.

6. Conclusions

We have investigated optical damage characteristics of the 4-f phase compensator with the UV-LC-SLM. The allowed maximum input pulse energy of 1.3 mJ has been obtained. We also carried out FB chirp compensations in the NIR region and the UV region using the same 2D·UV-LC-SLM. In the NIR region, after FB compensation, the spectral phase becomes almost flat and its temporal duration is 24 fs (TL: 18 fs). In the UV region, GDD approaches zero and its temporal duration is 34 fs (TL: 21 fs). These results suggest that the 2D·UV-LC-SLM enables us to generate sub-femtosecond optical pulses by applying it to induced phase and self-phase modulated output, and to generate high-powered optical pulses by applying it to output from an angularly-dispersed, non-colinear optical parametric amplifier.

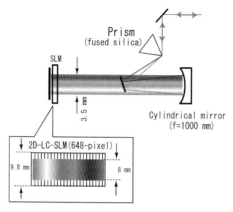

Fig. 1 The 4-*f* folded configuration consists of a cylindrical mirror, a prism and a 2D·UV-LC-SLM.

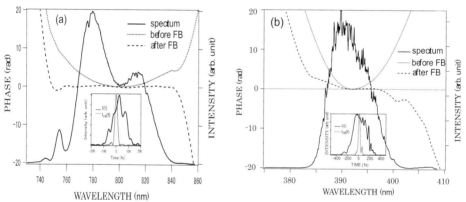

Fig. 2 **NIR region**: (a) The intensity spectrum and the spectral phase before and after FB compensations. Inset shows the temporal intensity profiles before and after FB compensations. **UV region**: (b) The intensity spectrum and the spectral phase before and after FB compensations. Inset shows the temporal intensity profiles before and after FB compensations.

References

1 M. Yamashita, H. Shigekawa and R. Morita, " *Mono-cycle photonics and optical scanning tunnelling microscopy*", M. Yamashita et al. eds. (Springer Verlag, Berlin, 2005).
2 M. Hirasawa, N. Nakagawa, K. Yamamoto, R. Morita, H. Shigekawa, and M. Yamashita, Appl. Phys. B **74**, S225 (2002).
3 M. Yamashita, K. Yamane, and R. Morita, IEEE J. Sel. Top. Quantum Electron. **12**, 213 (2006).
4 K. Hazu, T. Sekikawa, and M. Yamashita, Opt. Lett. **32**, 3318 (2007).

Vector Pulse Shaper Assisted Short Pulse Characterization

Andreas Galler[1] and Thomas Feurer[1]

[1] Institute of Applied Physics, University of Bern, CH-3012 Bern, Siwtzerland
 E-mail: andreas.galler@iap.unibe.ch

Abstract. We demonstrate that shaper-assisted pulse characterization is able to imitate most standard pulse characterization methods. If a polarization shaper is used even more complex schemes, such as SPIDER, can be realized.

Introduction

In most pulse shaping experiments the shaping of linearly polarized light pulses and their characterization is performed in two separate optical arrangements. Only a handful of experiments have been reported where the pulse shaper was used as an integral part of the diagnostic setup. Those are the multi-photon intra-pulse interference phase scan method [1], the shaper-assisted collinear SPIDER [2], and time-domain interferometry with an acousto-optic modulator [3]. Here, we show that a pulse shaper may be used to imitate most standard short-pulse characterization arrangements. Thus, shaped optical waveforms can be characterized exactly where it is needed since the corresponding nonlinear element can be mounted instead of the sample of interest. In addition, the unique capability of the pulse shaper to control the carrier envelope phase can be used to reduce the number of required sample points in interferometric measurements. Our main objective is to replace a standard short pulse characterization setup by a single nonlinear element and to use the pulse shaping apparatus, first, to create the desired shaped waveform and, second, to produce two replica of the shaped waveform or modified versions thereof and to scan the delay between them. Both operations are linear and can be performed simultaneously by the same device. We show results for scanning second order autocorrelation measurements, for FROG, for STRUT, and for triple correlation measurements. While pulse shaping techniques for linearly polarized light are well established, only recently attention has been devoted to the development of programmable vector pulse shaping techniques [4]. Here, we demonstrate a novel design with full phase and amplitude control of both polarization components of the shaped vector waveform. The design is common path and requires only a single spatial light modulator. Additionally, we demonstrate that such a vector pulse shaper may be used to imitate a SPIDER measurement.

Experiment and Simulation

All experiments were performed with a Ti:Sapphire oscillator (KML) delivering pulses with an energy of approximately 1nJ, a bandwidth of 80 nm at a centre wavelength of 820 nm. Linearly polarized laser pulses were phase- and amplitude modulated in a standard pulse shaping apparatus consisting of a double display spatial light modulator (Jenoptik SLM640-d) in the symmetry plane of a 4f zero-dispersion compressor. They were then focused to a suitable nonlinear crystal by an additional lens. Alternatively, the pulses were sent to a polarization shaper. It consisted of three main sections; a polarization beam splitter, a two lens telescope to adjust the angles of incidence on the grating, and a conventional folded 4f spectral filtering arrangement. A schematic of the complete setup is shown in Fig. 1. The polarization of the incoming beam was linear and tilted by 45 deg with respect to the reference frame. After passing through the specially designed Wollaston prism the two orthogonal polarizations were angularly separated. The magnification of the following two lens telescope was adjusted such that the angle between the two beams impinging on the grating was about 3 deg. Then, both spectra in the Fourier plane of the actual pulse shaper covered about one half of the modulator's active area. This setup allows reaching virtually every point on the Poincaré sphere.

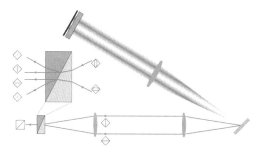

Fig. 1.) The polarization shaper consists of a Wollaston prism, a two-lens telescope, and a folded zero dispersion compressor. Amplitude modulation of a given frequency component is realized by, first, the SLM changing its state of polarization and, second, the Wollaston prism directing its undesired polarization component into a different direction.[5]

Results and Discussion

In order to imitate a specific characterization scheme it is necessary to identify the transfer function which produces the required pulse sequence, for example two replica with a given time delay between them. The two replica are inherently parallel, which under normal circumstances would lead to an interferometric-type measurement. With the shaper we are able to delay the slowly varying envelope only, which is realized by selecting $\gamma = 0$ in the transfer function

$$M(\omega) = \frac{1}{2}\left[e^{i(\omega-(1-\gamma)\omega_c)\tau/2} + e^{-i(\omega-(1-\gamma)\omega_c)\tau/2} \right].$$ (1)

Varying τ within reasonable limits then results in an intensity-like autocorrelation. When the nonlinear response to the waveform produced by eq.(1) with $\gamma = 0$ is recorded an interferometric autocorrelation is obtained. If the nonlinear response is spectrally resolved an interferometric FROG is obtained [6]. This scheme can easily be extended to three pulses with two delay times to adjust, which together with third harmonic generation would yield a so-called triple correlation [7].The following transfer function

$$M(\omega) = \begin{cases} A & \omega < \omega_0 - \frac{\Delta\omega}{2} \vee \omega > \omega_0 + \frac{\Delta\omega}{2} \\ A + (1-A)e^{i\omega\tau} & \omega_0 - \frac{\Delta\omega}{2} \leq \omega \leq \omega_0 + \frac{\Delta\omega}{2} \end{cases},$$ (2)

produces a waveform which results in an interferometric version of a STRUT measurement [8,9] and an example of such a measurement is shown in Fig. 2.

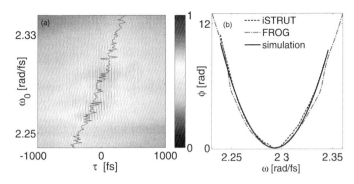

Fig. 2.) (a) Shaper assisted STRUT measurement of a laser pulse with a GVD of 3000 fs². The position of the maximum correlation signal is indicated by the black line. (b) Comparison between the extracted phase and the phase retrieved from a reference FROG.

By using the vector pulse shaper one can realize pulse sequences where the sub-pulses have orthogonal polarizations. This is a necessary prerequisite to imitate a SPIDER measurement [10]. We use a type II BBO for the second harmonic generation and mix the two orthogonally polarized waveforms. The ordinary beam consists of two replica of the pulse to be characterized and the extraordinary beam is a chirped pulse. Thus, the transfer function has two parts,

$$M_o(\omega) = \tfrac{1}{2}\left[e^{i\omega\tau/2} + e^{-i\omega\tau/2}\right] \quad \text{and} \quad M_e(\omega) = e^{i\varphi^{(2)}} \quad , \tag{3}$$

with the inter pulse delay τ and the quadratic phase $\varphi^{(2)}$. The first replica of the ordinary waveform is mixed with a different frequency than the second replica and upon recording the generated second harmonic signal a standard SPIDER trace is recovered.

1 V. V. Lozovoy, I. Pastirk, and M. Dantus, "Multiphoton intrapulse interference 4; Characterization and compensation of the spectral phase of ultrashort laser pulses," Opt. Lett. 29, 775 (2004).
2 B. von Vacano, T. Buckup, M. Motzkus, "In situ broadband pulse compression for multiphoton microscopy using a shaper-assisted collinear SPIDER," Opt. Lett. 31, 1154 (2006).
3 A. Monmayrant, M. Joffre, T. Oksenhendler, R. Herzog, D. Kaplan, and P. Tournois, "Time-domain interferometry for direct electric-field reconstruction by use of an acousto-optic programmable filter and a two-photon detector," Opt.Lett. 28, 278 (2003).
4 T. Brixner, G. Gerber, "Femtosecond polarization pulse shaping," Opt. Lett. 26, 557 (2001).
5 M.Ninck, A.Galler, T.Feurer, T.Brixner, "Programmable common-path vector field synthesizer for femtosecond pulses," Opt.Lett. 32, 3379 (2007).
6 G. Stibenz, G. Steinmeyer, "Inteferomtric frequency-resolved optical gating," Opt. Express 13, 2617 (2005).
7 T. Feurer, S. Niedermeier, R. Sauerbrey, "Measuring the temporal intensity of ultrashort laser pulses by triple correlation," Appl. Phys. B 66, 163 (1998).
8 J.L.A. Chilla, O.E. Martinez, "Direct determination of the amplitude and the phase of femtosecond light pulses," Opt. Lett. 16, 39 (1991).
9 A. Galler, T. Feurer, "Pulse shaper assisted short laser pulse characterization," Appl. Phys. B 90, 427 (2008).
10 C. Iaconis, I.A. Walmsley, "Spectral phase interferometry for direct electric-field reconstruction of ultrashort optical pulses," Opt. Lett. 23, 792 (1998).

Femtosecond Spectral Interferometry with Attosecond Accuracy by Correction for Spectrometer Resolution Asymmetry

Michael K. Yetzbacher, Trevor L. Courtney, William K. Peters and David M. Jonas[1]

[1] Department of Chemistry and Biochemistry, University of Colorado, Boulder, Colorado 80309-0215, USA
E-mail: david.jonas@colorado.edu

Abstract. Asymmetry in the line spread function of the spectrometer causes delay dependent phase shifts. Fourier deconvolution with the complex-valued optical transfer function allows accurate spectral phase recovery.

Introduction

Spectral Interferometry [1] is unique in its ability to measure time delays and constant spectral phase differences between ultrafast pulses [2]. Such measurements are essential to distinguish, for example, a photon echo from a nonlinear free induction decay. Spectral interferomtery is widely used in 2D spectroscopy, SPIDER [3] and OCT. Accuracy of the algorithms for spectral interferometry has been established for dispersion measurements, but has not, to our knowledge, been experimentally tested for linear or constant phase terms. The line-spread function (LSF) of the measuring instrument (including spectrometer and pixellated detector) distorts spectra. The measured spectral interferogram dispersed by a grating spectrometer is given by discrete sampling of

$$I(\lambda) = \{| E_1(\lambda) |^2 + | E_2(\lambda) |^2$$
$$+ | E_1(\lambda)E_2(\lambda) | \cos[\phi_1(\lambda) - \phi_2(\lambda)]\} \otimes LSF(\lambda) \tag{1}$$

The convolution is written in the wavelength domain as the pixel spacing is approximately constant in λ. The optical transfer function (OTF(ξ)) is the Fourier transform of the LSF. Dorrer et al. [1] developed an algorithm to correct for the influence of the modulation transfer function MTF=|LSF|. When a typical spectrometer is used at other than the (coma-free) design wavelength, it displays an asymmetric LSF [4] which will shift the fringes to one side and generate a complex valued OTF [5].

Experiment

Pulses from a Ti:sapphire laser with a ~40 nm wide spectrum were sent into a Mach-Zehnder interferometer which was PZT stabilized using feedback from the difference between the two outputs for a CW laser (632.8 nm, optical period T). Stabilization, measured out of loop with an additional CW laser, was $\lambda/250$ at 594.5 nm over 0.1 s. Pulse-pairs with relative delays nT (integer n) were coupled through single-mode fiber into a 0.34m Czerny-Turner spectrograph with a 300 grooves/mm grating. The LSF is measured by adapting an algorithm used to measure the point-spread function (PSF) of the Hubble Space Telescope [6]. Figure 1 shows the measured LSF for the CCD spectrograph, formed by interleaving dithered atomic line spectra.

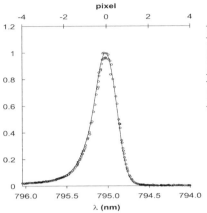

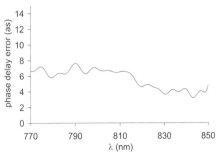

Fig. 1. The instrument response (LSF) determined with an atomic line at 795 nm. The low wings are asymmetric.

Fig. 2. The phase delay error $e(\omega) = [\delta\Delta\phi(\omega)/\omega] - nT$ recovered from a difference of two spectral phases. Inflection points are spaced with a period of 15 pixels, the step-and-repeat CCD fabrication interval.

Analysis

Although there is no unique way to determine the center for an asymmetric function, this poses no difficulty if the same center is used for calibrating and deconvolution. Line centers from another atomic emission lamp are retrieved to within 1/71 of a pixel. Spectra are deconvolved with the LSF by complex division in the wavelength conjugate (ξ) domain. After deconvolution and removal of the first two terms of eq. (1), the data has the form $|E_1(\lambda)E_2(\lambda)|\cos[\phi_1(\lambda) - \phi_2(\lambda)]$. Spectral interferograms are processed according to the algorithm in refs. [1,2]. The recovered spectral phase, $\Delta\phi(\omega) = \phi_1(\omega) - \phi_2(\omega)$, is corrected for interferometer dispersion by subtracting a reference spectral phase, $\Delta\phi_r(\omega)$, determined in the same interferometer. This difference, $\delta\Delta\phi(\omega) = \Delta\phi(\omega) - \Delta\phi_r(\omega)$, is then least-squares fit to a line $\delta\Delta\phi(\omega) = \phi_0 + \omega\tau$ to retrieve the ω=0 intercept phase shift, ϕ_0, and the slope delay, τ.

Results and Conclusions

The magnitude of the Fourier transform of the measured LSF was compared with the fringe amplitude for several interferograms. These two measurements of the MTF agree past the Nyquist limit of 8 ps. Figure 2 shows the phase delay error over the central 80 nm. The phase delay error compares favourably with the ±35 as error reported for 2D spectral shearing interferometry [7]. Figure 3 shows intercept phase shifts with and without LSF deconvolution. Compared to a systematic intercept phase shift error of 1 mrad per fs delay in prior work[2], errors are 20x smaller after deconvolution. These errors can be reduced 3x more by pseudo-deconvolution using the wavelength dependent LSF (the width varies by 10% across the spectrograph).

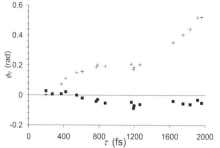
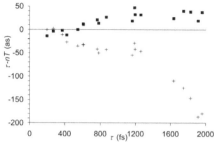

Fig. 3. Delay dependence of the intercept phase shift retrieved with LSF deconvolution (squares) and without (crosses).

Fig. 4. Slope delay error (τ-nT) at several delays without accounting for the LSF (crosses) and after deconvolution of the LSF (squares).

Without correcting for the LSF, slope delay and intercept phase errors are strongly correlated. Figure 4 shows the slope delay error in attoseconds. Note the mirror image similarity to Fig. 3. Interferograms with large time delays are more strongly affected by the LSF. Surprisingly, correction for the MTF alone removes less than half the slope delay error for delays greater than 1.5 ps. Pseudo-deconvolution (using the LSF for each wavelength to deconvolve the entire spectrum for restoration of the undistorted spectrum at that wavelength) reduces the phase delay error significantly, but still leaves the systematically spaced inflection points that likely arise from random CCD step-and-repeat fabrication errors of 100 nm after every 15 pixels (pixel widths are 25 microns). A fine wavelength re-calibration procedure to remove the linear time delay dependence of the phase delay at each pixel can reduce these errors, leading to rms phase delay errors of ±2.4 as, consistent with the measured interferometer stabilization error over 0.1 s and the 1s interferogram acquisition time. With this fine re-calibration, the maximum slope delay error is 20 as. Using SPIDER, delay errors of 10 as can lead to pulse duration errors of 1 fs for single cycle pulses [7]. Thus, our data shows that SPIDER measurements of pulse duration can be significantly affected by the phase of the complex-valued OTF. Further, asymmetry in the instrument response can appear in both dimensions for an imaging spectrometer. Therefore, spatially encoded arrangements [8,9] may be subject to similar distortions by the two-dimensional complex-valued OTF.

1 C. Dorrer, N. Belabas, J. Likforman, and M. Joffre, J. Opt. Soc. Am. B, **17**, 1795, 2000.

2 A. W. Albrecht, J.D. Hybl, S.M. Gallagher Faeder, and D.M. Jonas, J. Chem. Phys. **111**, 10934, 1999.

3 P. Baum, and E. Riedle, J. Opt. Soc. Am. B, **22**, 1875, 2005.

4 J. Reader, J. Opt. Soc. Am., **59**, 1189, 1969.

5 E. Hecht, Optics, 2nd ed., (Addison-Wesley, Reading, 1990).

6 J. Anderson, and I.R. King, Pub. Astronomical Soc. Pacific, **112**, 1360, 2000.

7 J.R. Birge, R. Ell, and F. Kartner, in Ultrafast Phenomena XV, P. Corkum, D. Jonas, R.J.D. Miller, A.M. Weiner, eds. Springer, Berlin, 2007.

8 E.M. Kosik, A.S. Radunsky, I.A. Walmsley, and C. Dorrer, Opt. Lett., **30**, 326, 2005.

9 P. Bowlan, P. Gabolde, A. Shreenath, K McGresham, R. Trebino, and S. Akturk, Optics Express **14**, 11892, 2006.

Spatial phase control and applications of high-order harmonics

C. Valentin[1], J. Gautier[1], E. Papalazarou[1], Ch. Hauri[1], G. Rey[1], Ph. Zeitoun[1], S. Sebban[1], V. Hajkova[2], J. Chalupsky[2], L. Vysin[3] and L. Juha[2]

[1] Laboratoire d'Optique Appliquée, ENSTA-Ecole Polytechbique-CNRS UMR 7639, Chemin de la Hunière, F 91761 Palaiseau Cedex, France
E-mail: constance.valentin@ensta.fr
[2] Institute of Physics, Na Slovance 2, Cz 18221 Prague, Czech Republic
[3] Faculty of Biomediacal Engineering, Zikova 4, Cz 16636 Prague, Czech Republic

Abstract. We present experimental results of control of high-order harmonic wave-fronts. We have reached a spatial phase with average distortions of $\lambda/7$ at 32 nm when controlling the fundamental laser beam wave front. We apply our results to experiments requiring tight focusing conditions.

Introduction

The tremendous progress of gas high-order harmonic generation (HHG) based ultra-short light source developed in the Extreme Ultra-Violet (EUV) and soft x-rays, during the past few years, opened up a whole range of applications. Assets such as the short pulse duration, the high spatial and temporal coherences as well as its compactness make HHG a handy laboratory tool with respect to other EUV sources (i.e. synchrotron radiation, EUV-Free Electron Lasers).

In this proceeding, we will show how to control EUV beam wave-front using a deformable mirror on the path of the fundamental laser beam. The aim of this experiment is to improve the EUV beam focusing, thus to achieve high intensities. We will then show application experiments of such our harmonic source as digital in-line holography and EUV-induced ablation on PMMA, in attosecond time scale.

EUV spatial phase shaping

In order to shape the EUV spatial phase, we need to measure the wave-fronts of our harmonic source. For this, we have developed an EUV Hartman sensor in collaboration with Imagine Optic [1]. We have first calibrated this sensor using high-order harmonics generated in argon (λ =32nm) with $\lambda/50$ accuracy ($\lambda/120$ at 13 nm [2]). Moreover, we have previously demonstrated that the root-mean-square (r.m.s.) values for EUV wave-front distortions do not depend dramatically on the generation parameters (as pressure in the gas, iris diameter) [3].

In standard generation conditions, IR and EUV beams present a strong astigmatism with r.m.s. distortion values of $\lambda/5$ for the IR beam and $\lambda/4$ for the EUV beam (at λ = 32 nm). For this experiment, we have studied the correlations between the IR and EUV spatial distortions, measuring wave-fronts by a Shack-Hartmann (SH) sensor (HASO, Imagine Optic) for the IR beam and the Hartmann sensor for the EUV beam. We placed a deformable mirror (BIM31, CILAS) on the IR beam before the generation setup. The IR beam is provided by a Ti:Sapphire laser system (810nm, 6 mJ, 35 fs, 1 kHz,). The beam diameter was measured to be 36 mm at $1/e^2$. The deformable mirror (DM) is used to control the IR beam wave-front: We could either flatten the spatial phase, either impose a specific aberration, thanks to the servo-loop between the SH sensor and the DM. All the measurements were performed when the iris was fully open. Then we generated harmonics around H25 in argon ($\lambda = 32$ nm) and we measured the EUV wave front in the propagation axis. For these measurements, the iris was closed to a diameter of 15 mm to ensure best phase-matching. We present only the results obtained with astigmatism at 0°, corresponding to the 5th Zernike polynomial (cf figure 1).

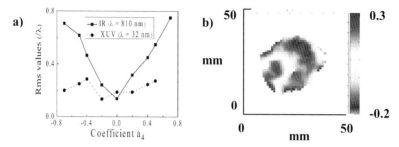

Fig. 1. a) Correlations between IR and EUV r.m.s. spatial distortions measured with respect to wavelength , b) EUV wave-front distortions measured by the Hartmann sensor.

Figure 1a shows that IR and EUV r.m.s. distortions measured with respect to wavelength are correlated. The harmonic beam appeared to have less spatial phase distortions than the measured IR ones, because of an apertured at 15 mm IR beam. Moreover, harmonic flux and wave-front depend on IR wave-front but in a non-trivial manner. We measured the flattest EUV wave-front to be $\lambda/7$ r.m.s. distortions (figure 1b) at a wavelength of 32 nm, leading to twice the diffraction limit according to Marechal's criterion. Very tight focusing is then possible with 40% of the encircled energy in the focal spot.

Applications of high-order harmonic based source

We have performed two application experiments using our high-order harmonic beam-line: Digital in-line holography and ablation of PMMA. For both applications, we needed to focus efficiently the harmonic beam composed of 5 successive odd orders around 25 (H21-H29). For this purpose, we used a multilayer off-axis parabolic mirror (f=65mm, θ=5°). Multilayer structure has been calculated in order to keep the attosecond structure of the pulse train [4].

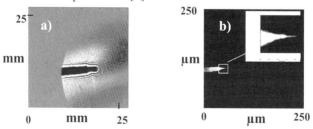

Fig. 2. a) Raw hologram recorded in 10 seconds, b) reconstructed image of the tip from the hologram, the insert is a zoom on the tip showing a resolution of 740 nm.

The first proof-of-principle holography experiment has demonstrated a resolution of 620 nm with a harmonic beam composed of one order (H25) [5]. Using all the harmonics generated in the Argon cell (H21-H29), we have measured the dimensions of a tip placed after the focusing parabola with a resolution of 740 nm, keeping the attosecond structure (cf figure 2).

The second experiment, performed in collaboration with L. Juha's group demonstrates that ablation of poly (methyl methacrylate) PMMA with high-order harmonic beam is possible. The figure 3 shows the AFM image of the ablation hole with dimensions 2 μm x 3 μm obtained with $6 \cdot 10^4$ laser shots. The total focused harmonic flux is then over the damage threshold of PMMA at 32 nm (2 mJ/cm^2) [6]. We measured simultaneously the harmonic beam wave-front with r.m.s. distortions of $\lambda/6$ after the parabola (figure 3b). The shape of the crater corresponding to the focal spot can be compared to the point Spread Function (PSF) calculated by the Hartmann sensor software (figure 3c). The two measurements are well correlated.

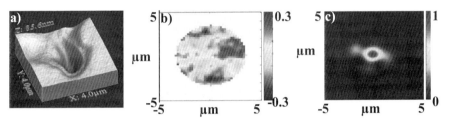

Fig. 3. a) AFM (Atomic force microscopy) image of ablation hole in PMMA performed by an EUV beam around 32 nm, b) wave-front measurement after the off-axis parabolic mirror showing r.m.s. distortions of $\lambda/6$ at 32 nm, c) calculated PSF given by the Hartmann sensor software.

Conclusions

In summary, strong correlations between IR beam and generated harmonic beam have been highlighted. Controlling the EUV wave-front using a deformable mirror for IR beam is then possible.

Improvements are foreseen for both application experiments: Getting more flux and controlling the spatial phase will allow single shot experiments. They pave the way to studies of bio-molecules with sub-micron and attosecond resolutions.

1   Imagine Optic : www.imagine-optic.com.
2   P. Mercère *et al.*, Opt. Lett. **28**, 1534, 2003..
3   J. Gautier *et al.*, Euro. Phys. J. D, DOI: 10.1140/epjd/e2008-00123-2, 2008.
4   A.-S. Morlens *et al.*, Opt. Lett.**31**, 1558, 2006.
5   A.-S. Morlens *et al.*, Opt. Lett.**31**, 3095, 2006.
6   J. Chalupsky *et al.*, Opt. Exp. **15**, 6036, 2007.

A New Generalized Projections Algorithm Geared Towards Sub-100 Attosecond Pulse Characterization

Justin Gagnon[1], Vladislav S. Yakovlev[1,2], Eleftherios Goulielmakis[1], Martin Schultze[1] and Ferenc Krausz[1,2]

[1] Max-Planck-Institut für Quantenoptik, D-85748 Garching, Germany
[2] Department für Physik, Ludwig-Maximilians-Universität München, D-85748 Garching, Germany
  Email: justin.gagnon@mpq.mpg.de

Abstract: We developed a new algorithm for characterizing attosecond pulses from streaked spectra. We compare our algorithm to the current one used for attosecond characterization, and show that it is better suited for sub-100 attosecond pulses.

Introduction

Since their first experimental demonstration, extreme-ultraviolet (XUV) attosecond pulses have become increasingly shorter due to improvements in technology. This progress calls for advances in methods used to characterize these pulses.

Currently, the standard characterization technique is attosecond streaking, first proposed by J. Itatani et al. [1]. When electron spectra are "streaked" by an infrared (IR) laser field, and are measured for several delays between the IR and XUV pulses, they contain all the information about the XUV and IR fields.

Characterization techniques previously developed for femtosecond light pulses [2,3] were adapted for the need of attosecond metrology [4,5]. One of the most general and versatile retrieval algorithms is FROG CRAB [4]. While this algorithm was demonstrated for ~130 as pulses [6], we found that several underlying approximations and assumptions need to be revised in order to accurately characterize ever shorter and broadband attosecond XUV pulses. In this paper, we propose an improved implementation of the attosecond FROG retrieval that is free from several important drawbacks of the original algorithm. We show that our improvements are necessary in order to rely on the retrieved IR and XUV fields, and we establish the range of reliability of the attosecond FROG technique

Electron spectra recorded at different delays τ form a streaking spectrogram. The streaking spectrogram can be described, in atomic units, by [7]

$$S(p,\tau) = \left| \int_{-\infty}^{\infty} E_X(t)\, d(p + A(t+\tau)) \exp\left[-i \int_{t+\tau}^{\infty} \left(p\, A(t') + \frac{1}{2} A^2(t') \right) dt' \right] e^{i \frac{p^2}{2} t}\, dt \right|^2 , \qquad (1)$$

where $E_X(t)$ is the electric field of an XUV pulse and $A(t)$ is the vector potential that describes the infrared (IR) streaking pulse. In order to process the spectrogram $S(p,\tau)$ with a FROG algorithm, there cannot be terms inside the integral that depend both on momentum and time. An inspection of (1) reveals two such terms: $d(p + A(t))$ and $pA(t)$. We remove the momentum dependence of these terms by making the susbstitution $p \to p_0$, where p_0 is the central momentum of the unstreaked electrons.

Methods

The Principal Components Generalized Projections Algorithm (PCGPA) [8] has long been the standard used for the inversion of FROG traces and has recently been applied to attosecond streaking spectrograms [4]. However, this algorithm is not optimized for attosecond streaking. First of all, by its very construction the PCGPA requires the spectrogram to satisfy periodic boundary conditions with respect to the delay. Although this condition holds true for

911

conventional FROG applications, the expression for $S(p,\tau)$ shows that the spectra do not necessarily satisfy periodic boundary conditions.

Furthermore, the PCGPA requires the spectrogram to be interpolated so that its energy and delay steps, $\delta\varepsilon$ and $\delta\tau$, obey the sampling relation $\delta\varepsilon\delta\tau=2\pi/N_\varepsilon$ (in atomic units). For a typical attosecond streaking range of $\Delta\varepsilon = 70$ eV, this implies a nominal delay step of $\delta\tau \sim 60$ as, which may be difficult to achieve experimentally. In practice, delay steps of ~100 as or greater are used in order to speed up the acquisition time, but most importantly to minimize the drift of experimental conditions. Our findings demonstrate that interpolating the spectrogram along the delay axis dramatically affects the quality of the retrieval.

To alleviate these shortcomings, we have developed a new version of generalized projections which retains the overall robustness of other algorithms, but obviates the need to interpolate the spectrogram along the delay axis, and does not assume periodic boundary conditions for the gate. Generalized projections algorithms employ the strategy of enforcing alternating constraints between the time and frequency domains [8] in order to find a pulse and gate pair that reproduce the measured spectrogram. A signal matrix is initially computed from time-sampled guesses for the pulse P and gate G. In our algorithm, the signal matrix S is obtained by shifting the pulse elements with respect to the gate elements by a time interval equal to the delay step, according to the prescription $S_{j,i}=P_jG_{j+L(i-1)}$. The inclusion of the integer "L", which is the number of time samples contained in a delay step, makes it possible to avoid interpolating the spectrogram along the delay axis during the preprocessing stage.

Results

To compare our algorithm to the PCGPA, a streaking spectrogram was simulated using a 750 nm laser field with a peak intensity of 2.4×10^{13} W/cm^2, and a FWHM duration of 4 fs. The spectrogram's 65 spectra were calculated at delay intervals of 100 as. The attosecond XUV pulse consisted of a sequence of two pulses, which is a typical feature of attosecond pulses obtained through spectral filtering of high harmonics [9]. The FWHM duration of the stronger pulse was 85 as, and weaker one had a duration of 77 as.

Figure 1 shows a comparison between the pulses retrieved by (A) our algorithm, and (B) the PCGPA. The same parameters were used for both algorithms. However, due to the coarse delay step of 100 as, the spectrogram that was fed to the PCGPA had to be interpolated to 256 delays points to satisfy the sampling constraint $\Delta\varepsilon\Delta\tau=2\pi$, whereas our algorithm only made use of the 65 known spectra. The PCGPA was unable to correctly retrieve the pulses and the vector potential because it was given three interpolated spectra for every know spectrum.

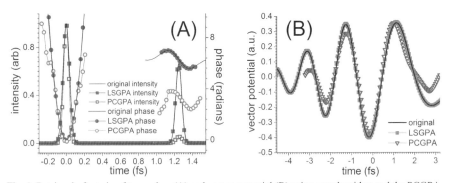

Fig. 1. Retrieval of a train of two pulses (A) and vector potential (B) using our algorithm and the PCGPA

Although our algorithm avoids the interpolation of the spectrogram, another source of error arises due to the substitution $p \to p_0$ in (1). To investigate the effect this approximation has on the retrieved XUV and IR fields, we gave our algorithm a synthetic spectrogram calculated without making this substitution. To reproduce realistic experimental conditions, we again

912

chose a double XUV pulse structure, separated by a half-cycle of the IR field. The XUV pulses were given phases with higher-order terms. The most intense one had a duration of 90 as, whereas the weaker one was more extended in time, with a temporal intensity profile modulated on the attosecond time scale. The theoretical dipole transition matrix element of Neon was used[1], and the IR streaking field was modeled with a 3.5 fs Gaussian pulse with a peak intensity of 2.4×10^{13} W/cm2, and a central wavelength of 750 nm. The spectrogram was again calculated at delay intervals of 100 as. The results of the retrieval are shown in Fig. 2.

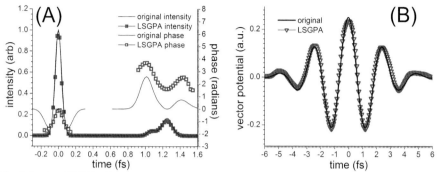

Fig. 2. Retrieved XUV pulse (A) and vector potential (B) from a spectrogram calculated without the approximation $p \rightarrow p_0$, and using the dipole transition matrix element of Neon.

As a consequence of replacing the momentum p with the central electron momentum p_0, the retrieved attosecond XUV pulses and IR vector potential exhibit slight temporal distortions, but most importantly the relative phase between the attosecond pulses is off by approximately 90°. This error is due to the substitution $pA(t) \rightarrow p_0A(t)$ in expression (1), which entails an error in the phase acquired by the electrons during their interaction with the laser field [10].

Conclusion

We have shown that our implementation of the attosecond FROG retrieval technique is a reliable and accurate method for characterizing ever shorter attosecond pulses. Constant improvements in technology will allow for temporally resolving physical phenomena on scales approaching the atomic unit of time (~ 24 as).

Acknowledgements. This work was supported by the DFG Cluster of Excellence: Munich-Centre for Advanced Photonics. The authors are grateful to X. Gu and F. Krausz.

1   J. Itatani et al., Phys. Rev. Lett., Vol. 88, 173903, 2002.

2   R. Trebino and D. J. Kane, J. Opt. Soc. Amer. A, Vol. 10, 1101, 1993.

3   C. Iaconis and I. A. Walmsley, IEEE J. Quantum Electron., Vol. 35, 501, 1999.

4   Y.Mairesse and F. Quéré, Phys. Rev. A, Vol. 71, 0011401(R), 2005.

5   F. Quéré et al., Phys. Rev. Lett., Vol. 90, 073902, 2003.

6   G. Sansone et al., Science, Vol. 314, 443, 2006.

7   Markus Kitzler et al., Phys. Rev. Lett., Vol. 88, 173904, 2002.

8   D. J. Kane, IEEE J. Quantum Electron., Vol. 35, 421, 1999.

9   M Schultze, et al., New J. Phys., Vol 9, 243, 2007.

10   J Gagnon, E Goulielmakis and V. S. Yakovlev, Appl. Phys. B, Vol. 92, 25, 2008.

1The dipole transition matrix element from the ground state of neon was calculated in the Hartree-Fock (HF) approximation. The continuum states were modeled as a linear combination of the s and d waves forming a frozen-core HF solution propagating in the direction of observation (courtesy of Y. Komninos).

Characterization of Mid-Infrared Pulses by Time-Encoded Arrangement

Kevin F. Lee[1,2], Adeline Bonvalet[1,2], and Manuel Joffre[1,2]

[1] Laboratoire d'Optique et Biosciences, Ecole Polytechnique, Centre National de la Recherche Scientifique, 91128 Palaiseau, France
[2] Institut National de la Santé et de la Recherche Médicale, U696, 91128 Palaiseau, France
E-mail: kevin.lee@polytechnique.edu

Abstract. We characterize mid-infrared pulses using upconversion to the visible regime by mixing with two collinear time-delayed replicas of an 800 nm chirped pulse. The phase is encoded as a function of the time-delay.

Introduction

Mid-infrared laser light is an important tool for studying molecules. To characterize the relative phases of the frequency components within a laser pulse, there are techniques such as modified-SPIDER (spectral phase interferometry for direct electric-field reconstruction) [1] and Zero-Additional Phase (ZAP) SPIDER [2], which have previously been adapted to the mid-infrared [3] by upconverting to the visible region, where standard charged-coupled device (CCD) detectors can be used.

Here, we adapt two-dimensional spectral shearing interferometry (2DSI) [4] to the mid-infrared with upconversion to the visible. To differentiate from related 2D techniques such as spatially encoded arrangement for SPIDER (SEA-SPIDER) [5,6], we will call our method time-encoded arrangement SPIDER (TEA SPIDER) [7]. TEA SPIDER has two key advantages over ZAP SPIDER, the ability to calibrate the frequencies involved, and a simpler collinear optical arrangement.

TEA SPIDER

In this experiment, we want to characterize a 3.9 μm pulse, which was generated by taking the difference frequency of the signal and idler from an optical parametric amplifier driven by an 800 nm regenerative amplifier. We normally make measurements in the infrared using chirped-pulse upconversion (CPU). This involves focusing the mid-infrared light in a nonlinear crystal along with a strongly chirped 800 nm pulse which, in our case, is uncompressed output from the regenerative amplifier. Sum-frequency generation gives a visible pulse near 664 nm, the spectrum of which we measure for each laser shot using a spectrometer and a CCD.

To perform a TEA SPIDER measurement, we add a Michelson interferometer to the 800 nm beam, giving us two collinear replicas at 800 nm, with a time delay controlled by a motorized delay stage in one arm of the interferometer. If either of the interferometer arms is blocked, the experiment returns to being a chirped pulse upconversion detection scheme. With both arms unblocked, the two pulses will cause spectral interference that is a function of the time delay. The desired phase information is carried in this interference pattern.

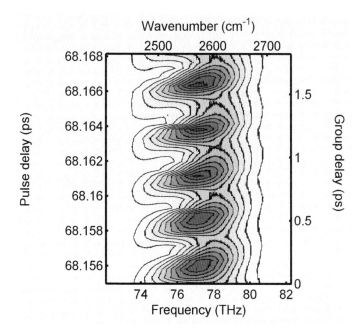

Fig. 1. TEA SPIDER trace of a chirped mid-infrared pulse recorded by the spectrometer as the pulse delay between the 800 nm pulses is varied. The tilt of the fringes away from the horizontal corresponds to the group delay of the pulse at that frequency, as shown by the approximate scale on the right. A large pulse delay corresponding to a 2.8 THz shear frequency was chosen to improve the visibility of the fringe tilt.

An example of a raw TEA SPIDER trace is shown in Figure 1. If there was no dispersion in the mid-infrared pulse, the fringes would be horizontal. The pulse measured in Figure 1 was purposely chirped by passing the beam through a 2 cm piece of CaF_2, resulting in tilted fringes. The two-dimensional phase $\Phi(\omega,\tau)$ of a TEA SPIDER trace is written as [4]:

$$\Phi(\omega,\tau) = \Omega_1\tau + \varphi(\omega - \Omega_1) - \varphi(\omega - \Omega_2) \approx \Omega_1\tau + \Omega \, d\varphi/d\omega \qquad (1)$$

where Ω_1 and Ω_2 are the frequencies of the quasi-monochromatic part of the 800 nm fields with which the mid-infrared pulse is mixing, with Ω_1 associated with the moving arm. The difference of these two frequencies is the shear frequency $\Omega = \Omega_2 - \Omega_1$. φ is the phase of the mid-infrared pulse, and τ is the time delay between the 800 nm pulses. Noting that a fringe corresponds to a constant value of Φ, the group delay can be found by multiplying the delay τ by Ω_1/Ω, giving the group delay axis on Figure 1.

Looking more carefully at the TEA SPIDER trace, we can convert this to the relative phase of the different frequency components. The analytic signal can be retrieved from the raw data by doing a Fourier transform with respect to time, zeroing components at negative frequencies, and transforming back. The result is the phase derivative which can be integrated to retrieve the desired phase. This phase retrieval and the measured upconverted spectrum, was used to create the reconstructed pulse shown in Figure 2.

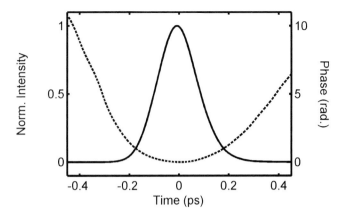

Fig. 2. The pulse intensity (—) and phase (---) of a chirped mid-infrared pulse as reconstructed by using TEA SPIDER. The data used here was an average of 1000 laser shots, with a shear frequency of 0.5 THz.

The TEA SPIDER trace also provides a measure of the frequency component of the adjustable pulse with which the mid-infrared pulse is mixing. The fringe spacing is $\Omega_1\tau$, so it gives directly the frequency Ω_1, assuming a good knowledge of the translation stage delay. For very good precision, one might add a beam from a helium-neon laser to the interferometer, and use the resulting beating to continuously measure the translation stage motion, effectively using the helium-neon wavelength as a reference for the 800 nm beam, and subsequently the mid-infrared spectrum.

Conclusions

When using mid-infrared pulses with chirped pulse upconversion, TEA SPIDER is the most convenient way to characterize the mid-infrared pulse. With the addition of an interferometer to the chirped pulse, one can calibrate the frequencies involved, and measure the relative phases in the mid-infrared beam.

1   M. Hirasawa, N. Nakagawa, K. Yamamoto, R. Morita, H. Shigekawa, and M. Yamashita, Appl. Phys. B. **74**, S225, 2002.
2   P. Baum and E. Riedle, J. Opt. Soc. Am. B. **22**, 1875, 2005.
3   K.J. Kubarych, M. Joffre, A. Moore, N. Belabas, and D.M. Jonas, Opt. Lett. **30**, 1228, 2005.
4   J.R. Birge, R. Ell, F.X. Kartner, Opt. Lett. **31**, 2063, 2006.
5   E.M. Kosik, A.S. Radunsky, I.A. Walmsley, C. Dorrer, Opt. Lett. **30**, 326, 2005.
6   A.S. Wyatt, I.A. Walmsley, G. Stibenz, G. Steinmeyer, Opt. Lett. **31**, 1914, 2006.
7   K.F. Lee, K.J. Kubarych, A. Bonvalet, M. Joffre, J. Opt. Soc. Am. B. **25**, A54, 2008.

Intensity and phase measurements of the spatiotemporal electric field of focusing ultrashort pulses

Pamela Bowlan,[1] **Ulrike Fuchs,**[2] **Pablo Gabolde,**[1] **Rick Trebino**[1] **and Uwe D. Zeitner**[2]

[1]Georgia Institute of Technology, School of Physics, 837 State St NW, Atlanta, GA 30332, USA
[2]Fraunhofer-Institut für Angewandte Optik und Feinmechanik, Albert-Einstein-Str. 7, 07745 Jena, Germany
Email: PamBowlan@gatech.edu

Abstract: We demonstrate a spectral interferometer with NSOM probes for measuring focusing ultrashort pulses with high spatial and spectral resolution. We measure a 0.44 NA focus and, for the first time to our knowledge, we observe the forerunner pulse.

Introduction

Nearly all ultrashort pulses are used at a focus, where their intensity is highest. And because the quality of many experiments depend on the quality of the focus, it is important to be able to measure the pulse at the focus. But this has remained a difficult challenge since the origin of the field of ultrafast optics. Focused pulses can easily have extremely complex spatiotemporal structure when lens aberrations are present[1, 2]. As a result, simply making one measurement vs. time (or frequency) and another vs. space is not a sufficient characterization of the pulse; a complete *spatiotemporal* measurement must be made to characterize a focused pulse. And because pulses are routinely focused to spot sizes less than a few microns, a technique with sub-micron spatial and high temporal resolution is needed.

Recently we demonstrated a technique for measuring the spatiotemporal field of a focusing ultrashort pulses, which we call SEA TAPDOLE[3]. SEA TADPOLE is a high spectral resolution and experimentally simple version of spectral interferometry. It uses fiber optics to introduce the unknown and reference beams into the device. The fiber can also be used to spatially resolve the unknown focusing beam, yielding spatial resolution equal to the fiber mode size. This is done by scanning the fiber that collects the unknown beam so that an interferogram is measured at every x, y, and z along the focusing beam's cross section. Then $E(t)$ can be found for each fiber position which gives us the spatiotemporal field of the focusing pulse or $E(x,y,z,t)$.

In our original setup, we used single-mode optical fibers with a mode size of 5.6 μm, which limited our measurements to foci with NAs less than 0.12. Here we replace the fiber that samples the focusing beam with single-mode fiber with a near-field scanning optical microscopy (NSOM) probe at its tip[4]. The NSOM probe is essentially an aperture on the end of the fiber, which can be as small as 30 nm in diameter. Using NSOM fibers, SEA TAPDOLE can measure even tightly focused pulses, so that previously unmeasured complex effects, such as the fore-runner pulse, can now be observed for the first time.

Experimental Results

We built our SEA TADPOLE using an NSOM fiber probe that was purchased from Nanonics and it had an aperture diameter of 500nm. The fiber used in this probe was the same as the fiber that we use to collect the reference beam. Fig. 1 shows the experimental setup that we used.

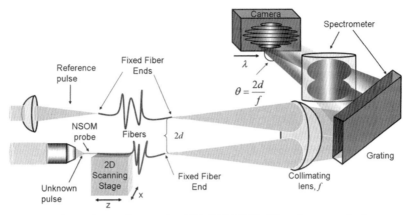

Fig. 1. Experimental setup for scanning SEA TADPOLE: The reference and unknown pulses enter the device via equal-length, single-mode optical fibers. The unknown pulse is collected with an NSOM fiber which has an aperture diameter that is smaller than the focused spot size. The NSOM probe is scanned in x (y) and z directions, so that multiple measurements of the field can be made at different positions in space. Once inside SEA TADPOLE, in the horizontal dimension, the light is collimated and then spectrally resolved at the camera using the grating and the cylindrical lens. In the vertical dimension, the light emerging from the two fibers crosses at a small angle and makes horizontal spatial fringes at the CCD camera.

We focused our Ti:Sa oscillator pulse with a bandwidth of 26 nm (FWHM) and a spot size of 4 mm (FWHM) using an aspheric lens purchased from New Focus with a focal length of 8 mm, yielding a NA of 0.44. A 4X telescope was used to increase the spot size of the beam before the aspheric lens. We placed a 25µm pin hole at the focus of the telescope to remove any aberrations that may have been introduced by the first lens (the higer NA one) in the telescope. The second lens in the telescope had an NA less than 0.01 and therefore its aberrations were negligible. We put a beam splitter after the telescope to obtain a reference pulse and loosely focused it into a single-mode optical fiber. The tightly focused pulse was sampled with an identical fiber, which had the NSOM probe on its end. The fiber was scanned in x and z, so that $E(x,\omega)$ was measured at nine different values of z.

To check our experimental results, we performed a simulation using the method described in reference [5]. Because we did not focus through a microscope side, as this lens is designed to do, there is a little spherical aberration present in this focus, though the main distortion is chromatic aberration. To determine the aberration parameters of this lens for our simulation, we did ray tracing using the program OSLO and the lens coefficients that were supplied by New Focus. Figure 2 shows the results of our simulations and experiments.

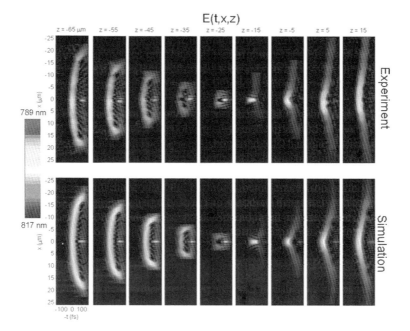

Fig. 2. Experimental (bottom) and simulation (top) results: Each box shows E(x,t) as a function of x and t at a distance z from the geometric focus where the color indicates the instantaneous frequency. A pulse preceding the main pulse is observed.

The data shows E(x,t) where the color indicates the instantaneous frequency and shows that the pulse is chirped which is due to the glass in the lens. The color also varies with transverse position (x) due to the chromatic aberrations. The spot size of the redder colors is bigger than that for the bluer colors before the focus, because blue focuses before red. Due to the combination of chromatic aberration and overfilling a lens, there is an additional pulse that is before the main pulse that appears before the focus[1]. This pulse is often referred to as the "forerunner pulse". The spot size of the (at one t) additional pulse is smaller than 1μm in some places which illustrates our sub-micron spatial resolution.

Acknowledgements. Pamela Bowlan acknowledges support from the NSF fellowship IGERT-0221600 and would like to thank the OSA for permission to use content from several of its publications which are listed below.

1   Z. Bor and Z. L. Horvath, "Distortion of femtosecond pulses in lenses. Wave optical description," Opt. Commun. **94**, 249-258 (1992).
2   U. Fuchs, U. D. Zeitner and A. Tuennermann, "Ultra-short pulse propagation in complex optical systems," Opt. Express **13**, 3852-3861 (2005).
3   P. Bowlan, P. Gabolde and R. Trebino, "Directly measuring the spatio-temporal electric field of focusing ultrashort pulses," Opt. Express **15**, 10219-10230 (2007).
4   P. Bowlan, U. Fuchs, R. Trebino and U. D. Zeitner, "Measuring the spatiotemporal electric field of tightly focused ultrashort pulses with sub-micron spatial resolution," Opt. Express **16**, 13663-13675 (2008).
5   M. Kempe and W. Rudolph, "Femtosecond pulses in the focal region of lenses," Phys. Rev. A **48**, 4721-4729 (1993).

Polarization, ionization and spatial gates in single attosecond pulse generation

Valer Tosa[1], Carlo Altucci[2], and Raffaele Velotta[2]

[1] National Institute for R&D Isotopic and Molecular Technologies, Donath 71-103, 400293 Cluj-Napoca, Romania,
E-mail: tosa@itim-cj.ro
[2] CNISM-Dipartimento di Scienze Fisiche, Università di Napoli "Federico II", via Cintia,26 - 80126, Napoli, Italia
E-mail: caltucci@unina.it, rvelotta@unina.it

Abstract. We show that in polarization-gating technique, ionization dynamics and three-dimensional propagation effects act as additional filters in single attosecond pulse generation. We propose a novel laser field configuration generating single harmonic bursts from multi-cycle laser pulses, allowing the use of laser pulses up to 25 fs to yield a single XUV pulse.

Introduction

High Harmonic Generation (HHG) in gas can be understood by the so-called three-step model [1]. In the first step the electron tunnels through the Coulomb potential barrier lowered by the slowly varying laser electric field, then gains kinetic energy moving in the laser field and finally returns to the vicinity of the ionic core where recombines and emits a burst of light shorter than one femtosecond. From this description it is clear that a single attosecond pulse can only be achieved if one is able to select the HHG emission within half an optical cycle (o.c.) of the laser field. Using multi-cycle pulses to obtain a single attosecond pulse would represent a great step forward, due to their typical high energy per pulse and their commercial availability even with stabilised CEP.

In this paper we propose a new approach to achieve isolated attosecond pulses from multi-cycles driving laser sources delivered by commercial laser systems, which combines a polarisation gating generated on the leading edge of the driving pulse with a subsequent ionisation gating whose role is to prevent the emission of additional attosecond pulses from the rest of the pulse. Such a method proves to be effective with pulse duration up to 25 fs and can be used with intensity as high as $\approx 10^{15}$

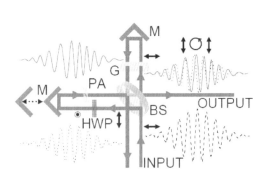

Fig. 1 Scheme of the proposed method to generate single attosecond pulse from multi-cycle laser pulses.

W·cm^{-2} or potentially higher. A scheme of principle of our method is reported in

Figure 1, where two copies of a linearly polarised, multi-cycles laser pulse are differently chirped and superimposed with crossed polarisation in a single Michelson interferometer. The frequency modulation (chirping) of one of the two beams (continuous line) is achieved when the beam crosses a piece of suitable material (G), whereas the adjustable delay between the two arms is realised by the translational stage in the left arm. Eventually, the output of the interferometer is a superposition of two crossed linearly polarised beams, one of which is chirped and delayed. This leads to an electric field with a time varying polarisation; as sketched in Figure 1 (output arm).

Results and Discussion

Appropriate values of the experimental parameters can realise linear polarisation gates which can be correlated with the evolution of the ionization. This is shown in the Figure 2(a) where the ellipticity $\varepsilon=E_y(t)/E_x(t)$ is reported for a laser pulse with $\tau_p=20$ fs. The two pulses are in phase ($\varepsilon=0$) only for approximately half a cycle thus realising the single emission, which is the main requirement for the generation of an attosecond pulse.

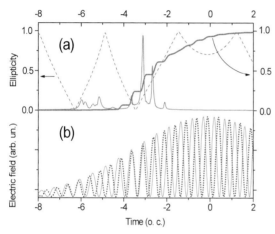

Fig. 2. (a) Ellipticity (dashed line), electron fraction (solid thick line), and single dipole emission (solid thin line), as function of time for a pulse of 20 fs and CEP=0.4π rad, other conditions as described in text. (b) Chirped (E_x, solid line) and chirp-free (E_y, dotted line) squared electric fields.

The main recollision events leading to single attosecond pulse occur in the time window centred at $t\approx-3.5$ o.c., since in the previous window centred at $t\approx-6.3$ o.c. the pulse intensity is still too low for an effective HHG. Moreover, in the earlier window the two fields are in phase with their maximum rather than with their minimum, the second condition being more effective in HHG [2]. The central part of the pulse is intense enough to fully ionise the medium, while the delay between the pulses assures that the ellipticity is quite high (≈0.7) at the peak of the combined pulse. The above two factors impede further contributions to the atomic polarization from the rest of the pulse. Such a description is supported by the single dipole emission, (see Figure 2(a)) calculated within the Strong Field Approximation [3,4] generalized for a field with time-dependent ellipticity [5]. Non-adiabatic, three-dimensional propagation of fundamental and harmonic fields through to the gas target has been accounted for by extending the model [6] for linear polarization to the case of time-dependent polarization. Medium ionization has been calculated by using the non-adiabatic Ammosov-Delone-Krainov model [7]. Throughout our simulations we have assumed a pulse peak intensity at focus of 8×10^{14} W·cm^{-2}, a 1-mm long Ar jet of 3.3×10^3 Pa local pressure placed at 1.7 mm in a diverging beam of 3.5 mm confocal parameter.

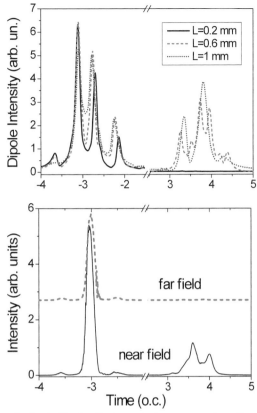

The filtering action of field propagation, which is essential to the formation of a single, short and well-shaped attosecond pulse, is illustrated in Figure 3 where on-axis, single dipole emission is shown versus time for three different medium lengths. These data elucidate the physical mechanism leading to the formation of a single attosecond pulse at the end of the medium, strongly guided by transient phase-matching of all the emitted elemental HHG contributions. The single dipole emission is characterised by a complicated multi-peak structure in time, varying both along the propagation and the radial directions, but *only* the first main peak always occurs at the same time, i.e. $t \approx -3.5$ o.c., thus implying that *only* this harmonic field contribution experiences the same phase at each medium slab.

Fig. 3. (top) Single dipole emission versus time for three different Ar medium lengths. (bottom) Harmonic near and far field illustrating the effect given by a 0.7 mm iris placed at 0.5 m distance. For clarity, far field is shifted on intensity axis.

In conclusion, in our method a polarization gating still induces a train of bursts in the single atom response. A strong ionization suppress bursts developed during the second half of the pulse, while the propagation acts as an additional, spatial filter both in axial and in radial direction, giving rise to a strong single attosecond pulse. Our calculations show that the scheme is robust against carrier to envelope as well as intensity fluctuations, and that pulses up to 25 fs can be used to generate a single isolated attosecond pulse. The experimental scheme is easy to implement in a typical ultrafast laser facility, and therefore represents a vital step to enable the access to attoscience to a wide scientific community.

1   P.B. Corkum, F. Krausz, Nature Physics 3, 381, 2007.

2   C. Altucci, V. Tosa, R. Velotta, Phys. Rev A 75, 061401(R), 2007.

2   P.B. Corkum, Phys. Rev. Lett. 71, 1994, 1993.

3   M. Lewenstein, P. Balcou, M. Yu. Ivanov, A. L'Huillier, P.B. Corkum, Phys. Rev. A 49, 2117, 1994.

4   P. Antoine, A. L'Huillier, M. Lewenstein, Phys. Rev. A 53, 1725, 1996.

5   V. Tosa, V., H.T. Kim, I.J. Kim, C.H. Nam, Phys. Rev. A 71, 063807 2005.

6   M.V. Ammosov, N.B. Delone, V.P. Krainov, Sov. Phys. JETP 64, 1191, 1986.

All dispersive mirrors compressor for femtosecond lasers

V. Pervak[1], C. Teisset[2], A. Sugita[2], F. Krausz[1,2], A. Apolonski[1,2]

[1] Ludwig-Maximilians-Universitaet Muenchen, Am Coulombwall 1, D-85748 Garching,
Germany
E-mail: vladimir.pervak@physik.uni-muenchen.de
[2] Max-Planck-Institut für Quantenoptik, Hans-Kopfermann-Str. 1, D-85748 Garching,
Germany

Abstract. We report on the development of highly dispersive mirrors for chirped-pulse amplifiers (CPA). The designed mirrors are potentially capable of replacing the prisms in the existing CPA compressors making them more compact and stable.

Introduction

High-energy femtosecond laser systems deal with dispersion of the materials in pulse compressors, in spectral broadening stages, either inside the laser oscillator cavity or outside. In all cases, the amount of the (absolute) dispersion to be compensated usually grows with the pulse energy. In the case of negative-dispersion oscillators, or chirped-pulse Ti:Sa oscillator CPO, such a monotonic dependence had already been proven theoretically and experimentally [1-5] providing in such a way a stable soliton-like intracavity pulse. In kHz systems the compressor size and its intrinsic dispersion grows with energy because of both the size of the dispersive components and the propagation distance through the components to be used. In high-energy femtosecond Ti.Sa oscillator-amplifier systems, usually prism or grating compressors are in use because they allow compensating large material dispersion of the order of $(2-5) \times 10^4$ fs^2 in a wavelength bandwidth of interests (~40 nm), resulting in sub-60 fs pulses. For a broader spectral range, uncompensated third-order dispersion (TOD) becomes so large that the pulse can not be compressed down to the targeted duration (usually 20-30 fs) and additional high-dispersive (hereafter: chirped) mirrors must be added. An alternative to this approach can be an all-chirped-mirror (CM) compressor. "Standard" CMs with the group delay dispersion (GDD) of the order of -50-100 fs^2 cannot be used in compressors of such type or/and in high-energy (of μJ-level) oscillators due to a large number of bounces required. To make the problem clear, let us make a rough estimate of the throughput of the compressor equipped with "standard" CMs. For the dispersion to be compensated of the order of 2×10^4 fs^2, the number of necessary bounces must be as high as 200. For a typical CM reflectance of 0.995, it leads to a throughput of only $0.995^{200} = 0.37$. After 500 bounces which one needs to compensate a material dispersion of 5×10^4 fs^2 with 100 fs^2-CMs, the throughput of less than 10% becomes completely unacceptable. A second, even more important obstacle, is that the initial pulse will be completely destroyed after such amount of bounces due to accumulated deviations of the realized GDD curve from the targeted one. Meantime, based on the progress in the CM development [6-13], one can expect CMs with dispersion values of $\sim 10^3$ fs^2 and reflectivity of some 0.995 for a spectral range of at least 50 nm around a central wavelength of 800 nm. Such mirrors, being realized, could replace prisms and gratings in the compressors giving thus a compact cheap device with a stable output beam free of residual TOD. The absolute value of the CM dispersion can be even higher as the spectrum becomes narrower, as it happens in a case of Yb:YAG disk oscillator [14], where mirrors with the dispersion around 10^3 fs^2 per bounce were successfully demonstrated for the spectral width of several nm. Due to the fact that CMs have higher losses and lower damage threshold in comparison to Bragg (= high) reflectors, CMs with GDD of the order of up to 10^4 fs^2 are desirable for that spectral range in order to decrease the total amount of chirped mirrors in the oscillator cavity.

High dispersive chirped mirrors

As a first step in direction of highly dispersive mirrors (HDCM) for CPA, we demonstrate the usability of highly dispersive mirrors for high-energy femtosecond oscillators, namely for i) CPO [3] and ii) an Yb:YAG disk oscillator. In both these oscillators GDD to be compensated is around $2x10^4$ fs^2, of the order of the nominal dispersion present in the CPA compressor. By definition, HDCM is characterized by a high group delay of different spectral components. Because the delay is proportional to the optical thickness of the layers involved, HDCM has thick layers and a big total multilayer structure thickness. From that we can formulate the conditions of manufacturing HDCMs: we need a very stable deposition process allowing us to deposit a thick multilayer structure with high accuracy. The total amount of GDD of a HDCM compressor needed for obtaining chirp-free high-energy pulses out of a Ti.Sa CPO is of the order of $2.5x10^4$ fs^2 at 800 nm and this value is achievable with only 20 bounces of the HDCM shown in Figure 1 (a).

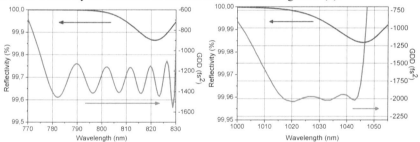

Fig. 1 (a,b). The calculated GDD and reflectivity of HDCMs. Left: HDCM for Ti:sapphire CPO, right: for Yb:YAG oscillator.

We have now to prove that such amount of bounces will not deteriorate the incident test chirp-free pulse in terms of its duration and energy. In the analysis, the main part of the dispersion was taken away and only residual GDD ripples were included. For virtual compression experiment, we used an incident 60-fs pulse realized in Ti:Sa CPO [3]. The reflected pulse does not become longer when the ripples are absent or small. Calculations show that after 20 bounces i) the exiting pulse preserves its incident duration and ii) the main pulse contains >95% of the energy from the initial value. Based on the analysis above, we hope for efficient compression of highly-chirped pulses exiting the CPO. In a Yb:YAG oscillator, HDCM will provide enough negative dispersion for keeping high-energy soliton pulse inside the oscillator cavity, Figure 1 (b). A HDCM compressor was successfully applied for compressing 2 ps chirped pulse out of Ti:Sa CPO down to 65 fs pulse at 5.3 MHz repetition rate. The group delay dispersion of the HDCM is -1300 fs^2 per reflection @800 nm that represents the highest negative dispersion value in the 40 nm wavelength bandwidth, realized so far. The HDCM at 1030nm has a nominal dispersion of ~ -2000 fs^2.Three HDCMs inside the Yb:YAG disk oscillator allowed us to generate stable 6 μJ 800 fs pulses. The low amount of HDCMs allowed us to minimize the losses in the optical part of the oscillator.

Conclusions

We demonstrate that the required CPA dispersion of the order of 10^4 - 10^5 fs^2 can possibly be introduced by a set of high-dispersive chirped multilayer dielectric mirrors offering several advantages including simplicity, alignment-insensitivy, and the potential for increased efficiency. As a first step toward an all-HDCM CPA compressor, we have shown 2 sets of HDCMs with both bandwidth and the main value of the dispersion comparable to what one expect in CPA lasers. The mirrors were manufactured and successfully tested in μJ-level laser oscillators.

Acknowledgements. This work was supported by the DFG Cluster of Excellence "Munich Centre for Advanced Photonics" (www.munich-photonics.de).

1   F. Krausz, M.E.Fermann, Th.Brabec, P.F.Curley, M.Hofer, M.H.Ober, Ch.Spielmann, E.Wintner, A.J.Schmidt, IEEE J.Quant. Electron. **28**, 2097-2122, 1992.

2   A. Fernandez, T. Fuji, A. Poppe, A. Fürbach, F. Krausz, and A. Apolonski, Opt. Lett. **29**, 1366-1368, 2004.

3   S. Naumov, A. Fernandez, R. Graf, P. Dombi, F.Krausz, and A. Apolonski, New J. Phys. **7**, 216, 2005.

4   V. L. Kalashnikov, E. Podivilov, A.Chernykh, S.Naumov, A.Fernandez, R.Graf, A.Apolonski, New J. Phys. **7**, 217, 2005.

5   A.Fernandez, A.Verhoef, V.Pervak, G.Lermann, F.Krausz, A.Apolonski, Appl.Phys. B 87 395-398, 2007.

6   R. Szipöcs, K. Ferencz, C. Spielmann, and F. Krausz, Opt. Lett. 19, 201–203, 1994.

7   V. Pervak, S. Naumov, G. Tempea, V. Yakovlev, F. Krausz, A. Apolonski, Proc. SPIE Vol. 5963, pp. 490-499, 2005.

8   V. Pervak, A. V. Tikhonravov, M. K. Trubetskov, S. Naumov, F. Krausz, A. Apolonski, Appl. Phys. B. **87**, 5-12, 2007.

9   G. Steinmeyer, G. Stibenz, Appl. Phys. B **82**, 175-181, 2006.

10 G. Steinmeyer, Appl. Opt. **45**, 1484–1490, 2006.

11 N. Matuschek, L. Gallmann, D. H. Sutter, G. Steinmeyer, U. Keller, Appl. Phys. B **71**, 509-522, 2000.

12 F. X. Kärtner, N. Matuschek, T. Schibli, U. Keller, H. A. Haus, C. Heine, R. Morf, V. Scheuer, M. Tilsch, and T. Tschudi, Opt. Lett. **22**, 831–833, 1997.

13 G. F. Tempea, B. Považay, A. Assion, A. Isemann, W. Pervak, M. Kempe, A. Stingl, and W. Drexler, in Biomedical Optics, Technical Digest (CD) (Optical Society of America, 2006), paper WF2.

14 S. V. Marchese, T. Südmeyer, M. Golling, R. Grange, and U. Keller, Opt. Lett. **31**, 2728-2730, 2006.

Optical Mapping of Attosecond Ionization Dynamics by Few-Cycle Light Pulses

A.J. Verhoef[1], A. Mitrofanov[1], E.E. Serebryannikov[2], D. Kartashov[1], A.M. Zheltikov[2], and A. Baltuška[1]

[1] Photonics Institute, Vienna University of Technology, Gusshausstrasse 27-387, A-1040, Vienna, Austria
E-mail: averhoef@mail.tuwien.ac.at
[2] Physics Department, International Laser Center, M.V. Lomonosov Moscow State University, Moscow, Russia

Abstract. Few-cycle light pulses are used to map ultrafast ionization dynamics in the time and frequency domain by all-optical means. Tunneling ionization encodes an attosecond phase mask, suggesting a promising method for attosecond shaping of high-intensity optical fields.

Propagation of high-intensity ultrashort light pulses through ionizing gas-phase media involves a complicated ultrafast dynamics of the laser field in space, time, and frequency domain. Filament formation [1] and compression of laser pulses to few-cycle pulse widths [2] are among the most interesting scenarios of such ultrafast pulse-propagation dynamics. Ionization induced by few-cycle laser fields is, on the other hand, an intriguing ultrafast physical process, whose significance for ultrafast science and technologies is still to be fully realized. In the presence of a high-intensity optical field, electrons are released from atoms on an attosecond time scale. Moreover, in the tunneling regime, this process displays a strong sensitivity to the carrier–envelope phase (CEP) of a few-cycle light pulse [3]. In a recent experiment [4], attosecond steps in the ion yield, measured by time-of-flight spectrometry, have been resolved by using an isolated XUV attosecond pump pulse and a few-cycle optical probe pulse. In this work, we focus on the possibility of all-optical mapping of attosecond ionization dynamics by few-cycle light pulses and show that this dynamics encodes an attosecond phase mask that projects itself a) as a frequency sweep and b) as a gradual phase velocity increase of the traveling optical pump field. This information can only be probed in the optical frequency range and cannot be accessed with XUV pulses because of the plasma dispersion $n_p = \left(1 - \omega_p^2 / \omega^2\right)^{1/2}$, where n_p is the plasma refractive index, ω_p is the plasma frequency, and ω is the radiation frequency. For potential attosecond spectroscopy applications, another advantage of gaining an attosecond response directly with an optical pulse rather than with an XUV pulse is related to the ω^{-3} dependence of transition cross-sections and the very low yield $\sim 10^{-7}$–10^{-8} of isolated attosecond XUV pulses.

In this work we propose to use the ratio of spectral intensities of low-order harmonics as a measure of tunneling ionization rate. The relation between a step-wise change of the free electron plasma density (Fig. 1a), resulting from a twice-per-optical-cycle tunnel ionization (TI), and the appearance of a low-order harmonic spectrum was first introduced by Brunel [5] and explained in terms of tranverse plasma current. This harmonic radiation emission mechanism is distinctly different from the Corkum model for higher-order harmonic generation [6] that comprises a sequence of TI, quasi-free electron acceleration, and re-collision with the parent ion. Since the Brunel mechanism in gas does not rely on a re-collision, this type of harmonic emission, in contrast with the higher-order harmonics [6], is a direct frequency-domain reflection of the TI dynamics. The Brunel type of harmonic emission has been tackled experimentally and theoretically [7, 8], but the contribution of these harmonics has never been disentangled experimentally from other harmonic generation mechanisms, i.e. the Corkum type and the Kerr-nonlinearity type. In this work we employ crossed-beam geometry with a strong linearly polarized few-cycle pump pulse and a weak cross-polarized chirped probe pulse for a background-free detection of the Brunel type of emission (Fig. 1b). This emission process can be readily understood in terms of cross-phase modulation impinged on the probe pulse by a step-like temporal phase (=refractive index change) mask. Therefore, the intensity of the emitted Brunel harmonics scales linearly with the probe pulse intensity and does not depend on the polarization state of the probe pulse.

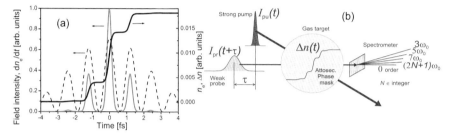

Fig. 1. (a) Field intensity (dashed line), electron density $n_e(t)$ and refractive index change (bold line), and time derivative of the electron density (thin line) calculated as functions of time for Keldysh parameter $\gamma=1$. (b) Schematic of an all-optical measurement of the attosecond phase mask. The observed orders of Brunel harmonics are odd harmonics of the pump and probe frequency ω_0.

To assess whether or not practical information on TI dynamics can be extracted from a cross-correlation measurement of Brunel harmonics (Fig. 1b), we developed a full 3-D code to study propagation effects. Fig. 2a shows the dependence between the extent of the harmonic spectrum and the rise time of the phase steps, confirming the intuitively clear conclusion that a faster tunneling dynamics translates into a wider spectrum by causing steeper phase steps.

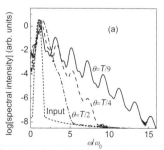 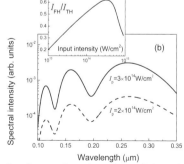

Fig. 2. Numerical simulations. (a) Spectrum of an input ~2-cycle pulse and Brunel harmonic spectra for various phase step rise times θ given as a fraction of the cycle duration T. (b) Spectra of optical harmonics from a 5-fs Gaussian-shape pulse in neon at 300 mbar after 1 mm propagation. Inset shows the spectral intensity ratio of the 5th vs. the 3rd harmonic.

Fig. 2b shows the effect of propagation suggesting that for propagation distances in gas on the order of 1 mm the phase mask can survive without being significantly washed out, that would lead to a structure-less spectrum. Our numerical simulations also show that for our experimental conditions the contributions of "direct" cross-phase modulation harmonics due to $\chi^{<3>}$, $\chi^{<5>}$, $\chi^{<7>}$,... and cascaded $\chi^{<3>}$ nonlinearities can be neglected.

Measured and calculated spectro-temporal cross-correlation maps for the 5th and 7th Brunel harmonics in krypton are presented in Fig. 3. Corresponding frequency-unresolved cross-correlation traces obtained by integrating the spectro-temporal maps along the frequency axis for three harmonic orders are shown in Fig. 3e. The intensity ratios of individual harmonics are still to be determined because in the present setup different detectors were used to measure each of the harmonic orders.

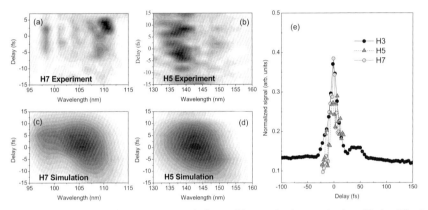

Fig. 3. Measured and simulated cross-correlations of low-order harmonics in a Kr jet. The jet thickness was 2 mm and the krypton pressure 400 mbar. The intensity of the 5-fs 760-nm 180-μJ pump pulse in the interaction region was $I_p = 5 \times 10^{14}$ W/cm^2. A 1-μJ pulse replica was stretched to ~20 fs and used as a probe. The pump–probe noncollinearity angle was 20 mrad.

In conclusion, we present the first to our knowledge cross-correlation detection of Brunel harmonic radiation in gas which provides direct information on tunneling ionization dynamics. This all-optical technique is much simpler than photo-electron and ion detection and is particularly promising for studying the TI dynamics in larger molecules (unsuitable for cold targets) and possibly bulk solids (unsuitable for photo-electron spectroscopy).

Acknowledgements. This work is supported by the Austrian Science Fund (FWF), grants U33-N16 and F1619-N08. We are grateful to Misha Ivanov, Olga Smirnova, and Paul Corkum for stimulating discussions.

1   A. Braun et al., Opt. Lett. **20**, 73 (1995); J. Kasparian and J.-P. Wolf, Opt. Express **16**, 466 (2008).
2   C.P. Hauri et al., Appl. Phys. B **79**, 673 (2004); Opt. Express **13**, 7541 (2005).
3   G. L. Yudin and M.Yu. Ivanov, Phys. Rev. A **64**, 013409 (2001).
4   M. Uiberacker et al.Nature **446**, 627(2007).
5   F. Brunel, J. Opt. Soc. Am. B **7**, 521 (1990).
6   P.B. Corkum, Phys. Rev. Lett. **71**, 1994 (1993).
7   N. Burnett, C. Kan, P.B. Corkum, Phys Rev A **51**, R3418 (1995).
8   C.W. Siders, et al., PRL **87**, 263002 (2001).

Polarization, Phase and Amplitude Control and Characterization of Ultrafast Laser Pulses

Philip Schlup[1], Omid Masihzadeh[1], Lina Xu[2], Rick Trebino[2], and Randy A. Bartels[1]

[1] Colorado State University, Department of Electrical and Computer Engineering,
Fort Collins CO 80523, USA
E-Mail: philip.schlup@colostate.edu
[2] School of Physics, Georgia Institute of Technology, Atlanta GA 30332, USA
E-Mail:rick.trebino@physics.gatech.edu

Abstract. We demonstrate complete control over the polarization, phase and amplitude state of an ultrafast laser pulse using a single, linear spatial light modulator, and introduce a self-referenced method for characterization the polarization state.

The shaping of ultrafast pulses using programmable pulse shapers [1] has become ubiquitous and indispensable for experiments in diverse fields including physics and biochemistry. Since in many physical systems the full vector field, rather than just the spectral phase, plays a critical role, control over amplitude and phase [2, 3], as well as polarization [4–7], have been independently demonstrated. Simultaneous control over all aspects of the field has recently become feasible [8–10]. By contrast, there have been few techniques reported for the characterization of such polarization-shaped fields. The POLLIWOG method [11] requires a well-characterized reference pulse that may be cumbersome to obtain in practice.

We here demonstrate and characterize a polarization-amplitude-phase pulse shaper [10] that offers benefits over published alternatives as it operates in a near-common path geometry, minimizing relative phase instability between the polarization components, while using only a single, linear liquid crystal spatial light modulator (SLM) element. In addition, we introduce a new, self-referenced method for fully characterizing the polarization state of an arbitrary ultrashort pulse [12]. It can be used with any well-established measurement technique that yields intensity and phase information for single polarization components, and involves combining three or more such measurements at different angles of an analyzing polarizer.

The shaper is configured as a folded 4-f Martinez stretcher (Fig. 1) using a Si prism as dispersive element. An imaging telescope maps the angular separation between the two orthogonal polarization components passing through a Wollaston polarizer onto the prism so that the components are spatially separated at the SLM (Boulder Nonlinear Systems, Lafayette, CO). The resulting near-common-path, common-optic design has a high relative phase stability. Over-sampling of the spatial modes at the Fourier plane allows us to control the amplitude individual frequency components via spatial diffraction by applying a rapidly-oscillating phase grating, independent of a slowly-varying phase mask [13]. We insert a $\lambda/2$ plate to obtain parallel p polarizations within the shaper to maximize the prism transmission and match the LC-SLM response. The resulting relative group delay yields spectral-interferometry fringes after an analyzing polarizer at the output of the shaper, with which we calibrate the wavelength distribution and pixel phase responses for each polarization [10]. A compensating plate (C in Fig. 1) is inserted to restore close temporal overlap between the shaped pulses. Coarse

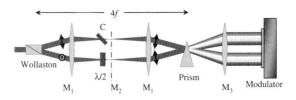

Figure 1: Schematic layout. M_1, M_3, concave, M_2 flat mirrors; C compensating plate.

adjustments are possible by observing the interference fringes between the two pulses after a polarizer at 45°. We determined the remaining delay from a type-II SHG cross-FROG measurement, applying linear phases to each polarization to effect the requisite variable relative delay [12].

To characterize the polarization-shaped pulses, we write the arbitrary field in the frequency domain as $\tilde{\mathbf{E}}(\Omega) = \tilde{E}_x(\Omega)\hat{\mathbf{x}} + r\tilde{E}_y(\Omega)e^{-\mathrm{i}(\Omega\tau+\theta)}\hat{\mathbf{y}}$ where r, τ and θ are the relative amplitude, delay, and phase offset between the $\tilde{E}_x$ and $\tilde{E}_y$ components. Power measurements of the $\tilde{E}_x$ and $\tilde{E}_y$ components and normalizing the reconstructed fields such that $\int |\tilde{E}(\Omega)|^2 \mathrm{d}\Omega = 1$ determines the amplitude ratio $r = (P_y/P_x)^{1/2}$. The establishment of τ and θ, to which self-referenced single measurements are insensitive, requires an additional measurement at angle η relative to the x-axis that takes the form $r_\eta \tilde{E}_\eta(\Omega)e^{-\mathrm{i}(\Omega\tau_\eta+\theta_\eta)} = \cos\eta\, \tilde{E}_x(\Omega) + r\sin\eta\, \tilde{E}_y(\Omega)e^{-\mathrm{i}(\Omega\tau+\theta)}$. Solving for the relevant variables yields

$$-\mathrm{i}(\Omega\tau+\theta) = \ln\left[\frac{r_\eta \tilde{E}_\eta(\Omega)e^{-\mathrm{i}(\Omega\tau_\eta+\theta_\eta)} - \cos\eta\, \tilde{E}_x(\Omega)}{r\sin\eta\, \tilde{E}_y(\Omega)}\right]. \tag{1}$$

The imaginary part represents a straight line over frequency Ω, from which the slope τ and intercept θ may be extracted. Manipulating the real part of Eq. (1), we write the fields in terms of amplitudes and spectral phases as $\tilde{E}(\Omega) = \tilde{A}(\Omega)e^{-\mathrm{i}\phi(\Omega)}$ and find another straight-line expression for τ_η and θ_η: $\Omega\tau_\eta + \theta_\eta = \phi_x(\Omega) - \phi_\eta(\Omega) - \cos^{-1}[\Gamma(\Omega)]$. Here $\Gamma = \left[r_\eta^2 \tilde{A}_\eta^2 + \cos^2\eta\, \tilde{A}_x^2 - r^2\sin^2\eta\, \tilde{A}_y^2\right]/2r_\eta\cos\eta\, \tilde{A}_\eta \tilde{A}_x^2$. The complete polarization characterization is thus in principle possible from just three measurements using existing methods. We term this approach tomographic ultrafast retrieval of transverse linear E-fields, or TURTLE.

In experiments, we find the additional redundant information contained in FROG traces leads to more robust reconstructions using a simple fitting procedure. An example is shown in Fig. 2. The orthogonal components $\tilde{E}_x$ and $\tilde{E}_y$ [Fig. 2(a), (b)] were measured and reconstructed using the standard algorithm. FROG traces at angles $\eta = \pm45°$ [Fig. 2(c), (d)] were measured and then curve-fit, using the reconstructed $\tilde{E}_x$ and $\tilde{E}_y$, with respect to the values of τ and θ. A simplex algorithm minimized the least-squares difference between simulated and measured traces. In the "phantom" traces of Figs. 2(a)-(d), the left half plane shows the measured FROG trace compared to the reconstruction in the right half. Figure 2(e) shows the three-dimensional reconstruction of the field. We found good agreement between the retrieved delay ($\tau = 55$ fs) and that posted to the SLM mask (50 fs) over a wide range of delays between 5 and 200 fs, as shown in Fig. 2(f). Although independent power measurements for the determination of r are indispensable in some instances (e.g., amplitude-only shaping), simulations

930

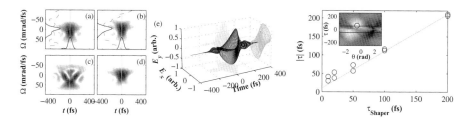

Figure 2: TURTLE retrieval of arbitrary shaped pulse. (a) $\tilde{E}_x$ and (b) $\tilde{E}_y$ reconstructed FROG traces. (c), (d) Fitted traces to find parameters in Eq. (1). (e) Retrieved temporal field. (f) Measured τ as function of delay applied to shaper; inset, error surface.

show that highly-structured FROGs arising from complex amplitude-and-phase shaped pulses can obviate the need for these measurements. We note that the lack of spurious structure in the 45°-projection FROGs of Fig. 2(c), (d) indicates good relative phase stability ($\ll \pi$ rad) between the polarization components over our $\sim$15-min acquisition time of high-resolution FROG traces.

In summary, a single, high-resolution linear LC-SLM was used to achieve complete control over the phase, amplitude, and polarization state of an ultrafast pulse. The individual polarization components are shaped on different sections of the shaper. The near-common-path, common-optic geometry ensures a stable relative delay and phase between the polarization components. We introduced a self-referenced technique to determine the polarization state of an arbitrary ultrafast pulse and used it to characterize the shaped pulses. The technique uses as few as three measurements by existing methods that measure the intensity and phase of a single pulse polarization

Acknowledgements. The authors wish to thank Carmen Menoni and Cameron Moore for the loan of the OSA. The authors gratefully acknowledge support from NSF CAREER Award ECS-0348068, ONR Young Investigator Award, the Beckman Young Investigator Award, and Sloan Research Fellowship support for R.A.B.

1   A. M. Weiner, Review of Scientific Instruments Vol. 71, 1929, 2000.
2   J. C. Vaughan, T. Hornung, T. Feurer, and K. A. Nelson, Optics Letters Vol. 30, 323, 2005.
3   J. W. Wilson, P. Schlup, and R. A. Bartels, Optics Express Vol. 15, 8979, 2007.
4   T. Brixner, and G. Gerber, Optics Letters Vol. 26, 557, 2001.
5   L. Polachek, D. Oron, and Y. Silberberg, Optics Letters Vol. 31, 631, 2006.
6   C. G. Slater, D. E. Leaird, and A. M. Weiner, Applied Optics Vol. 45, 4858, 2006.
7   T. Suzuki, S. Minemoto, T. Kanai, and H. Sakai, Physical Review Letters Vol. 92, 133005, 2004.
8   S. M. Weber, F. Weise, M. Plewicki, and A. Lindinger, Applied Optics Vol. 46, 5987, 2007.
9   M. Ninck, A. Galler, T. Feurer, and T. Brixner, Optics Letters 32, 3379, 2007.
10   O. Masihzadeh, P. Schlup, and R. A. Bartels, Optics Express Vol. 15, 18025, 2007.
11   W. J. Walecki, D. N. Fittinghoff, A. L. Smirl, and R. Trebino, Optics Letters Vol. 22, 81, 1997.
12   P. Schlup, O. Masihzadeh, L. Xu, R. Trebino, and R. A. Bartels, Optics Letters Vol. 33, 267, 2008.
13   J. W. Wilson, P. Schlup, and R. A. Bartels, Optics Express Vol. 15, 8979, 2007.

Silicon-Chip-Based Single-Shot Ultrafast Optical Oscilloscope

Mark A. Foster[1], Reza Salem[1], David F. Geraghty[1], Amy C. Turner[2], Michal Lipson[2], and Alexander L. Gaeta[1]

[1] School of Applied and Engineering Physics, Cornell University, Ithaca, NY 14853
[2] School of Electrical and Computer Engineering, Cornell University, Ithaca, NY 14853
E-mail: a.gaeta@cornell.edu

Abstract. We demonstrate a single-shot ultrafast optical oscilloscope using a four-wave-mixingbased parametric temporal lens integrated on a CMOS-compatible silicon photonic chip. Experimentally, we demonstrate waveform measurement with a 100-ps record length and sub-750-fs resolution.

Introduction

Measurement of optical waveforms on the sub-picosecond time scale is of great interest for next generation highspeed optical communications and for studies of ultrafast chemical and physical phenomena [1]. Current electronic oscilloscopes are capable of single-shot measurements of waveforms with slightly better than 100-ps resolution. Optical techniques such as frequency resolved optical gating [2] and spectral phase interferometry for direct electric-field reconstruction [3] can provide single-shot measurements of optical waveforms with fewfemtosecond resolution but with record lengths limited to a few picoseconds. To fill this gap in measurement capabilities, researchers have explored techniques using the space-time duality of electromagnetic waves [4-10]. To date the implementations of these techniques for sub-picosecond resolution typically require second-order nonlinear materials and wavelength conversion to non-standard wavelength bands. Furthermore, simultaneous 100-ps record lengths and sub-ps resolution has yet to be demonstrated. Here we demonstrate a single-shot ultrafast optical oscilloscope (UFO) with a 100-ps record length and a resolution considerably better than 750-fs. The implementation relies upon our newly developed parametric time-lens based on the third-order nonlinear process of four-wave mixing (FWM). This parametric process yields an output that is generated at a wavelength near those of the pump and input waves. This enables the frequencies of all the interacting waves to be in the S-, C-, and L-telecommunication bands, which allows for manipulation of all the waves using the well-established instrumentation and components available for these bands. Furthermore since all materials posses a third-order nonlinear moment, the FWM based time-lens can be implemented in any material platform including the CMOS-compatible SOI photonic platform used here.

Time-to-Frequency Conversion

The temporal processing technique we employ is based on the well-known feature of spatial imaging systems that an object positioned at the front focal plane of a lens will produce the two-dimensional Fourier transform of the object at the back focal plane. Extending the spatial Fourier transform processor to the temporal domain in which diffraction and the spatial lens are replaced by dispersive propagation and the application of a quadratic temporal phase generated here using FWM, respectively, yields a device that converts the spectral (temporal) profile of the input to the temporal (spectral) profile of the output. Since in the temporal domain the Fourier transform of the waveform is its optical spectrum, a measurement of the spectrum at the Fourier plane will provide the temporal profile of the incident waveform and this process is termed time-to-frequency conversion [4]. Since this spectrum does not change during propagation from the lens to the second focal plane, propagation through the second dispersive path is unnecessary and can be removed in a time-to-frequency converter.

Silicon-Chip-Based Ultrafast Optical Oscilloscope

Here, we demonstrate a silicon-based single-shot UFO in which the time-to-frequency conversion is implemented using a FWM-based parametric time-lens in a 1-cm-long CMOS-compatible embedded SOI nanowaveguide with a cross-sectional size of 300 nm by 750 nm. The strong optical confinement of these silicon structures allows for highly efficient nonlinear processes and for engineerable group-velocity dispersion (GVD) that can yield conversion bandwidths greater than 150 nm [11]. We have experimentally observed conversion bandwidths approaching 500 nm. The pump and input waves are generated from an optical parametric oscillator (OPO) that produces 150-fs pulses at a 76-MHz repetition rate. The pulses from the OPO are spectrally separated into a 280-fs pump pulse with 15-nm of bandwidth centered at 1545 nm and a variable bandwidth signal pulse centered at 1575 nm. The input waveform is passed through a dispersive element consisting of a 240-m length of single-mode optical fiber. For this length to correspond to the focal length of the FWM time-lens, the input wave is mixed with a pump wave, which has been passed through twice the dispersive length of the input signal. The pump wave is amplified using an erbium-doped fiber amplifier (EDFA) and FWM is carried out. The resulting FWM generated spectrum is measured with an optical spectrum analyzer (OSA) for multi-shot measurements or with a monochromator and IR-camera for single-shot measurements.

We experimentally characterize the record length and temporal resolution of our system by sending in input waveforms of various complexity. First we inject a two narrowband 900-fs pulses and vary their temporal separation to observe the record length of the device. As shown in Fig. 1(b), we are able to measure the two pulses separated by a record length of 100 ps. To characterize the resolution of the FWM-based UFO, we decrease the signal pulsewidth to 400 fs and then reduce the separation of the pulses. As shown in Fig. 1(c), we are able to clearly differentiate the two 400-fs pulses when separated by 750 fs.

To demonstrate the ability of the FWM-based UFO to measure complex pulses, we amplify the short input pulse in an EDFA and induce nonlinear spectral broadening in the amplifier and subsequently pass this pulse through 20-m of optical fiber. The resulting pulses are then measured using the UFO and using cross-correlation with the 280-fs pump pulse. The measurements of three different pulses and the respective cross-correlations are shown in Fig. 2(a)-(c). To further demonstrate the utility of the FWM-based UFO, we reduce the amount of fiber in the dispersive paths to 75 m and 150 m. This new system has a smaller record length but a better temporal resolution due to the reduced effect of TOD. Using this system, we measure the beating between two spectrally separated pulses that are temporally overlapped as shown in Fig. 2(d). To demonstrate the single-shot capability of the device, we replace the OSA used to measure the output spectrum of the UFO with a single-shot spectrometer and we measure a single-shot optical waveform composed of two pulses separated by 3.5 ps as shown in Fig. 2(e).

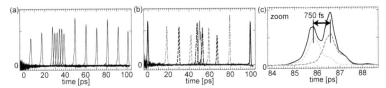

Fig. 1. (a) Collective measurements of a single 900-fs pulse scanned across the 100-ps record length of the system. (b) Measurements of two 900-fs pulses of variable separation. (c) Experimental characterization of the temporal resolution.

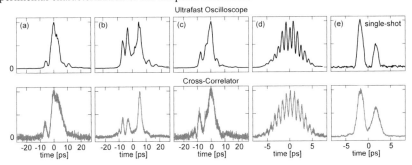

Fig. 2. (a)-(d) Comparison of a variety of complex pulses measured using the ultrafast optical oscilloscope and a crosscorrelator. (e) Single-shot measurement of a waveform using the ultrafast oscilloscope and comparison with a multishot cross-correlation.

1 C. Dorrer, "High-speed measurements for optical telecommunication systems," IEEE Select. Topics Quantum Electron. 12, 843-858 (2006).

2 D. J. Kane and R. Trebino, "Single-shot measurement of the intensity and phase of an arbitrary ultrashort pulse by using frequencyresolved optical gating." Opt. Lett. 18, 823-825 (1993).

3 C. Dorrer, B. de Beauvoir, C. Le Blanc, S. Ranc, J. P. Rousseau, P. Rousseau, J. P. Chambaret, and F. Salin, "Single-shot real-time characterization of chirped-pulse amplification systems by spectral phase interferometry for direct electric-field reconstruction." Opt. Lett. 24, 1644-1646 (1999).

4 M. T. Kauffman, W. C. Banyal, A. A. Godil, and D. M. Bloom, "Time-to-frequency converter for measuring picosecond optical pulses," Appl. Phys. Lett. 64, 270-272 (1994).

5 C. V. Bennett, R. P. Scott, and B. H. Kolner, "Temporal magnification and reversal of 100 Gb/s optical data with an upconversion timemicroscope," Appl. Phys. Lett. 65, 2513-2515 (1994).

6 L. K. Mouradian, F. Louradour, V. Messager, A. Barthelemy, and C. Froehly, "Spectro-temporal imaging of femtosecond events," IEEE J. Quantum Electron. 36, 795-801 (2000).

7 C. V. Bennett, B. D. Moran, C. Langrock, M. M. Fejer, and M. Ibsen, "Guided-wave temporal imaging based ultrafast recorders," Conference on Lasers and Electro-Optics, OSA Technical Digest Series (CD) (Optical Society of America, 2007), paper CFF1.

8 B. H. Kolner, "Space-time duality and the theory of temporal imaging," IEEE J. Quantum Electron. 30, 1951-1963 (1994).

9 J. Van Howe and C. Xu, "Ultrafast optical signal processing based upon space-time dualities," J. Lightwave Technol. 24, 2649-2662 (2006).

10 C. V. Bennett and B. H. Kolner, "Principles of parametric temporal imaging—Part I: System configurations," IEEE J. Quantum Electron. 36, 430-437 (2000).

11 M. A. Foster, A. C. Turner, R. Salem, M. Lipson, and A. L. Gaeta, "Broad-band continuous-wave parametric wavelength conversion in silicon nanowaveguides," Opt. Express 15, 12949-12958 (2007).

934

Time-resolved off-axis digital holography for characterization of ultrafast phenomena in water

T. Balciunas[1], A. Melninkaitis[1], G. Tamosauskas[2], and V. Sirutkaitis[1]

[1] Laser Research Centre, Vilnius University, Vilnius LT-10223, Lithuania
  E-mail: tadas.balciunas@ff.vu.lt
[2] Department of Quantum Electronics, Vilnius University, Vilnius LT-10222, Lithuania

Abstract. We present the application of time-resolved off-axis digital holography for the investigation of refractive index/ transmission properties of laser-induced plasma filaments in water. Time evolution of both amplitude and phase contrast images of the self-focused beam in water was characterized with temporal resolution better than 50 fs. The images reveal the picture of the early dynamics of plasma.

Introduction

The ability to excite matter with ultrashort light pulses and probe its subsequent evolution on the femtosecond time scale has opened up complete new realms of science. However, up to now the in situ characterization of laser-induced refractive index and transmission change with spatial and temporal resolution was a challenging task.

We attempt to combine the merits of a "traditional" time-resolved pump–probe technique and holographic phase-contrast imaging to explore light-induced nonlinear changes in the transparent media and to monitor their time evolution [1]. Several other techniques such as time-resolved shadowgraphy [2], scattering-based imaging [3], inline digital holography [4], or conventional holography [5] have been applied for similar purposes. Despite their merits, they all have serious limitations. The advantages of digital holography, compared with conventional holography, are that (a) no wet processing of the data medium is necessary, (b) retrieving the holographic information is possible practically in real time, and (c) phase measurement is quantitative. We employed off-axis experimental design that is superior to in-line holography [4] due to the fact that it does not impose size limitations on the objects to be observed and works well with highly absorptive as well as transparent samples. To our knowledge, this study is the first attempt to implement off-axis digital holography with femtosecond time resolution.

Method

Digital holography employs a digital image sensor to record the interferogram, and, subsequently, numerical algorithms described in [5] are used for the reconstruction of the original hologram. The recorded interferogram is usually referred to as a "digital hologram". The optical layout of the experiment is virtually that of a Mach–Zehnder interferometer. The transparent sample to be characterized is the source of amplitude and phase modulation of the transmitted light, resulting in a so-called object wave O. At the exit of the interferometer the interference between the object wave O and the reference (unperturbed) wave R creates the hologram which is later reconstructed numerically using a computer.

The general layout of the experimental setup is shown in Fig. 1. Femtosecond pulses from an amplified Ti:Sapphire laser system were split into two parts: the first was focused in the water cell to generate light filaments, whereas the second one was used to pump the NOPA. The duration of the initial pulses was τ =130 fs. Pump pulses were focused into the water cell using a lens. The cell containing water was placed in one of the interferometer arms for probing.. These pulses were further used for probing in a Mach–Zehnder interferometer. The area around the waist of the pump beam in the water cell was selected for imaging. A delay line was used for changing the delay of the probe pulse.

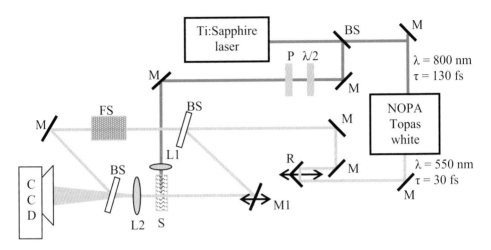

Fig. 1. Experimental setup of ultrafast digital holography. P - polarizer, λ/2 - half-wave plate, M - mirror, BS – beam splitter, R - delay line, L1 and L2 - lenses, S - water cell, CCD - digital camera, FS - dispersive element.

Results and Discussion

Ultrashort pulse propagation in water was observed by registering the interferogram of 30 fs probe pulse imaged on the CCD camera. Phase-contrast and amplitude-contrast images reconstructed numerically from the interferograms are shown in Fig. 2. Very early pulse propagation dynamics can be clearly seen in the phase-contrast images. During the light filamentation, two competing phenomena the optical Kerr effect and plasma generation occur. Their respective contributions carry opposite signs [4]. The total effective refractive index can then be expressed as:

$$n = n_0 + n_2 \times I + \Delta n_p \tag{2}$$

where the first term corresponds to the intrinsic refractive index of water, the second term characterizes the influence of the optical Kerr effect (n_2 denotes the Kerr coefficient), and the third term is the refractive index change due to the generation of plasma. Due to the high temporal resolution: $\tau_{probe} << \tau_{pump}$ intensity distribution within the pump pulse can be observed through the Kerr effect. This is an exclusive property of UDH compared with other [2–4] methods where the duration of the pump and probe pulses is equal.

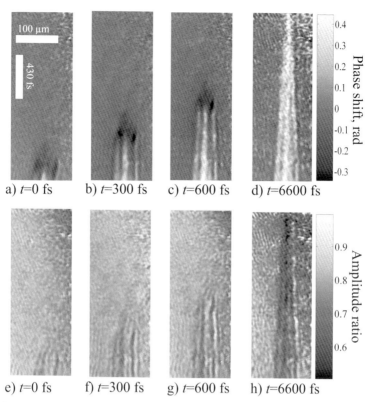

Fig. 2. Phase contrast (a)–(d) and amplitude (e)–(h) contrast images for various time delays.

Conclusions

In conclusion, the Time-resolved Digital Holography technique presented here was used for the characterization of femtosecond pulse filamentation and plasma generation in water. Quantitative phase-contrast and amplitude-contrast images were recorded with temporal resolution better than 50 fs. The images reveal the picture of the early dynamics of plasma with hitherto unprecedented details.

Acknowledgements. Authors are thankful to S. Minardi's group for valuable discussions, R. C. Eckardt, M. Vengris, project Laserlab-Europe and Lithuanian State Science and Studies Foundation for support.

1   T. Balciunas, A. Melninkaitis, G. Tamosauskas, and V. Sirutkaitis, Opt. Lett. 33, 58-60 (2008).
2   A. Gopal, S. Minardi, and M. Tatarakis, Opt. Lett. 32, 1238–1240 (2007).
3   C. Schaffer, N. Nishimura, E. Glezer, A. Kim, and E. Mazur, Optics Express 10, 196–203 (2002).
4   M. Centurion, Y. Pu, Z. Liu, D. Psaltis, and T. H¨ansch, Opt. Lett. 29, 772–774 (2004).
5   E. Cuche, P. Marquet, and C. Depeursinge, Appl. Opt 38, 6994–7001 (1999).

3 GHz RF Streak Camera for Diagnosis of sub-100 fs, 100 keV Electron Bunches

T. van Oudheusden[1], J.R. Nohlmans[1], W.S.C. Roelofs[1], W.P.E.M. Op 't Root[1], O.J. Luiten[1]

[1] Department of Applied Physics, Eindhoven University of Technology, P.O. Box 513, 5600 MB Eindhoven, The Netherlands
E-mail: t.v.oudheusden@tue.nl

Abstract. We have designed and built a 3 GHz radio-frequency cavity for use as an ultrafast streak camera to measure with 10 fs resolution the duration of electron bunches that are suitable for single-shot ultrafast electron diffraction experiments. We present our first measurements of 10 ps electron bunches and explain how we will measure sub-100 fs bunch durations.

Among other techniques ultrafast electron diffraction (UED) is a promising approach for complete structural characterization on the timescale of atomic motion. Typically dynamics are initiated with an ultrashort light pulse (pump) and then –at various delay times- the sample is probed with an ultrashort electron bunch (probe pulse). By recording diffraction patterns as a function of the pump-probe delay it is possible to follow various aspects of the real-space atomic configuration of the sample as it evolves. Time resolution is fundamentally limited by the electron bunch duration. To study samples that do not return to their initial (pre-pump) state the diffraction pattern at each pump-probe delay has to be recorded within a single shot. We have proposed a new electron source concept for single-shot sub-100 fs electron diffraction in the 100 keV energy range [1], which is based on compression of bunches that have a linear velocity-position correlation. This setup is currently being commissioned in our lab, which involves mainly characterization of the electron bunches. In this paper we focus on the bunch length diagnostics.

To measure sub-100 fs electron bunch durations we have designed and built a 3 GHz radio frequency (RF) cavity, operating in the TM_{110} mode, for use as an ultrafast streak camera [2]. An incoming electron bunch is deflected transversely by the rapidly varying magnetic field in such a way that the front of the bunch is deflected more then the rear, or vice versa. This way the bunch is swept over a phosphorous screen, generating a streak of which the length is a direct measure of the bunch duration. In the ideal case of a mono-energetic bunch moving on-axis the streak length X_{str} is given by

$$X_{str} = \frac{4eB_0}{m\omega} L \sin\left(\frac{\omega}{2}t_b\right)\cos(\omega t_0 + \varphi_0)\sin\left(\frac{\omega}{2}t_{tr}\right), \qquad (1)$$

with e and m the electron charge and mass respectively, ω the frequency and B_0 the amplitude of the time-dependent magnetic field, t_b the bunch duration, t_{tr} the transit time of an electron, t_0 the time at which the center of the bunch is at the center of the cavity, φ_0 a phase offset, and L the distance between the cavity and the phosphorous screen. We have optimized the cavity length and the phase offset, so that, using $\omega t_b/2$ $\ll 1$, the streak length is given by

$$X_{str} \approx \frac{2eB_0 t_b}{m} L \cdot \tag{2}$$

From this equation it follows that a time resolution of 10 fs can be obtained if $B_0 =$ 10 mT, and if the streak is imaged onto a CCD camera with a pixel size of 5 μm, which is positioned at $L = 500$ mm. By increasing this distance the resolution can in principle be increased indefinitely, but in practice it will be limited by other factors like off-axis behaviour, energy spread, and angular spread. We have investigated these aberrations both by analytical calculations and by detailed particle tracking simulations with the General Particle Tracer (GPT) code [3].

In the analytical calculations the well-known closed expressions for the electro-magnetic fields inside a standard cylindrical cavity are used. In addition GPT simulations have been performed for the this cavity. However, to be able to use a compact solid state 1 kW RF amplifier we have designed a more efficient cavity, which is show in figure 1 (left panel). It needs only 860 W to produce the desired magnetic field strength [2]. The electro-magnetic field inside this cavity has been calculated numerically with CST Microwave Studio [4]. The on-axis magnetic field of the actual cavity has been measured and agrees to a high degree of precision with the simulation. The 3-dimensional field map created by CST Microwave Studio is included in GPT to do particle tracking simulations with a realistic field.

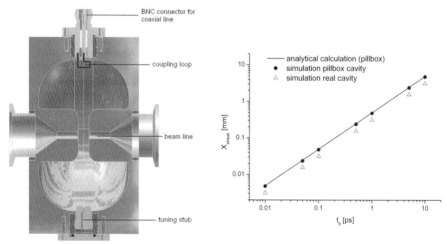

Fig. 1. Left panel: Cross-section of the power-efficient cavity. Right panel: Calculated and simulated streak length as a function of bunch duration.

In the right panel of figure 1 we show the results of the analytical calculations and the GPT simulations. The streak length X_{str} is plotted as a function of bunch duration for the case of an ideal, 1-dimensional (1D) bunch that has zero width and bunch duration t_b, traveling on-axis. The screen is placed at $L = 100$ mm and the amplitude of the magnetic field is 16.7 mT. We find that the streak length X_{str} depends linearly on the bunch duration t_b, in agreement with equation (2), and that the analytical calculations and the GPT simulations fully agree for the pillbox cavity, in particular if small relativistic corrections are taken into account. The simulations for the realistic field of the newly designed cavity result in the same linear dependence with slightly smaller streak lengths. Deviations from the case of an ideal, zero-energy-spread, zero-divergence, 1D bunch have also been investigated with GPT simulations [2]. The main limiting factor turns out to be the finite transverse bunch size. We find however that the resulting decrease in temporal resolution can be completely counteracted by using a 10 μm pinhole to limit the transverse bunch size just before entering the cavity.

In figure 2 we show one of the first streaks of 90 keV electron bunches that we have measured with our RF streaking camera. The picture is taken in a single shot, which was possible because the bunch had a charge of 0.3 pC. The line profile is the convolution of the transverse and longitudinal particle density profile of the bunch, which are both Gaussian. The standard deviation X_{streak} of the streak profile is therefore given by

$$ X_{streak} = \sqrt{\sigma_R^2 + \left(\frac{2e}{m} B_0 L t_b\right)^2}, \tag{3} $$

where σ_R is the RMS bunch radius, and t_b the RMS bunch duration. Calculating the bunch durations with equation (3) for the measurements with different magnetic field strengths and averaging leads to a bunch duration of $t_b = (11.8 \pm 0.6)$ ps. This is consistent with our detailed particle tracking simulations yielding a bunch duration of 11.4 ps.

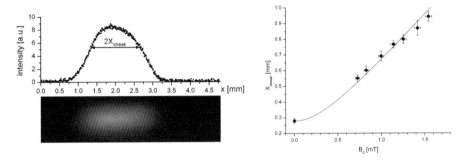

Fig. 2. Left panel: Typical streak of a 10 ps electron bunch and its intensity profile. Right panel: Streak length as a function of the magnetic field strength. The line is governed by equation (3), where the measured spotsize and duration of the bunch, and the actual distance from the cavity to the screen are used.

In summary, we have developed and built a power efficient 3 GHz RF cavity that can be used as a streak camera. We have done single-shot bunch length measurements of 90 keV electron bunches that have a duration of 11.8 ps. The resolution of the streak camera can be increased to 10 fs by placing a 10 μm pinhole just before the cavity.

1   T. van Oudheusden, B.J. Siwick, E.F. de Jong, S.B. van der Geer, W.P.E.M. Op 't Root, and O.J. Luiten, in Journal of Applied Physics, Vol. 102, 093501, 2007.

2   J.R. Nohlmans, *"RF cavities for ultrafast electron microscopy"*, Master thesis, Eindhoven University of Technology, 2007.

3   http://www.pulsar.nl/gpt.

4   http://www.cst.com.

Simulations of Frequency-Resolved Optical Gating for measuring very complex pulses

Lina Xu, Erik Zeek and Rick Trebino

School of Physics, Georgia Institute of Technology, Atlanta, 30332
E-mail: *gth665y@mail.gatech.edu*

Abstract. Frequency-Resolved Optical Gating (FROG) and its variations are the only techniques available for measuring complex pulses without a well-characterized reference pulse. We study the performance of the FROG generalized-projections (GP) algorithm for retrieving the intensity and phase of very complex ultrashort laser pulses (with time-bandwidth products of up to 100) in the presence of noise. We compare the performance of three versions of FROG: second-harmonic-generation (SHG) FROG, polarization-gate (PG) FROG, and cross-correlation FROG (XFROG).

Introduction

The shaping of ultrashort laser pulses into complex intensities and phases vs. time is finding many applications, including coherent control, telecommunications, micro-machining, and multi-photon imaging. In addition, commercial pulse shapers have become available and can generate pulses with time-bandwidth products (TBP's) up to ~ 100. Unfortunately, methods for independently checking the actual shapes (intensity and phase vs. time) of such shaped complex pulses have not received much attention. Currently, the most commonly used (and the simplest) method for measuring shaped pulses is second-harmonic-generation (SHG) frequency-resolved optical gating (FROG) (see Fig 1) [1].

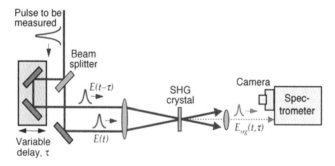

Fig. 1. Schematic of an SHG FROG apparatus. A pulse is split into two, one pulse gates the other in an SHG crystal, and the relative delay between the two pulses is varied. The nonlinear-optical signal pulse spectrum is then measured vs. delay. In PG FROG, the nonlinearity is polarization-gating in glass, and crossed polarizers are used. In XFROG, an independent (previously measured) reference pulse is used instead of one of the unknown pulses.

But other FROG variations could also work well for this application. So in this work, we study three versions of FROG: SHG FROG, PG FROG, and XFROG. While the first two are self-referencing, the last uses the unshaped pulse to measure the shaped one, which is generally fine since it is readily available. The general equation for all FROG versions is:

941

$$I_{FROG}(\omega,\tau) = \left| \int_{-\infty}^{\infty} E(t) E_g(t-\tau) \exp(-i\omega t) dt \right|^2$$

where $E(t)$ is the unknown input-pulse electric field that we are trying to measure. The various versions of FROG are distinguished by their gate pulses: in SHG FROG, $E_g(t) = E(t)$; in PG FROG $E_g(t) = |E(t)|^2$; and in XFROG, $E_g(t)$ is an independently measured reference pulse (the unshaped pulse in pulse-shaping applications).

The algorithm used in these FROG techniques is the Generalized Projection (GP) algorithm. It involves iteratively projecting onto two constraint sets, one involving the measured trace and the other involving the mathematical expression for the signal field in terms of the pulse field. The intersection of these two constraints—the final solution—can be reached using random noise as the initial guess. The GP algorithm performs very well for retrieving simple pulses and achieves convergence quickly (in well under 1 s). In order to test the performance of different FROG techniques based on the GP algorithm for measuring complex pulses, we generated very complex pulses with time-bandwidth products of up to ~100 by beginning with noise and multiplying both the time- and frequency-domain fields by Gaussian temporal and spectral shapes. The resulting pulses are complex in both the time and frequency domains. To simulate experimental measurements, we added 1% Poisson noise to all the resulting traces. Typical XFROG, PG FROG, and SHG FROG traces and their complex pulses are shown in Figs. 2-4.

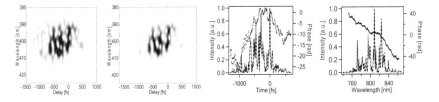

Fig. 2. Example of XFROG for measuring complex pulses (here a pulse with TBP = 66). From the left: generated FROG trace with noise, retrieved trace (with less noise), generated (solid line) and retrieved (dashed line) spectral intensity and phase, generated and retrieved temporal intensity and phase. Note that excellent convergence has occurred.

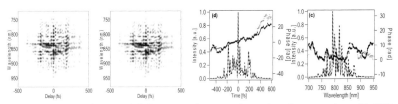

Fig. 3. Example of PG FROG for measuring complex pulses (here a pulse with TBP = 15.5).

942

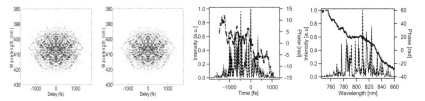

Fig. 4. Example of SHG FROG for measuring complex pulses (here a pulse with TBP = 94.3).

We performed a statistical analysis of the performance of the FROG GP algorithm for retrieving pulses with rms TBP's from 1 to 100. We define convergence as a FROG error of less than 1%, which, we find, corresponds to visually nearly perfect retrieved intensities and phases. We also find that, even if the pulse is extremely complicated, the intensity and phase of the pulse can almost always be retrieved even in the presence of such noise—provided that there is sufficient resolution and the FROG trace is contained in the data window, that is, is not cropped. We find that the XFROG algorithm converges with 100% reliability on the first initial guess, in agreement with an existing proof that spectrogram (XFROG) inversion should always succeed [2]. On the other hand, we find, as expected, that PG and SHG FROG are not 100% reliable, but are not far from it. PG FROG achieves more than 95% for complex pulses. And the SHG FROG algorithm converged for over 80% of even the most complex pulses we tried if multiple initial guesses were allowed. The results are shown in Fig 5. Fortunately, when the algorithm does not converge, the retrieved trace visibly disagrees with the measured one, so, in measurements, it is clear that another initial guess is required; there is no danger of an incorrect measurement. Also, in this work, we found no new ambiguities.

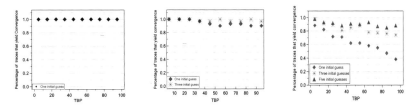

Fig 5. Statistical analysis of the performance of the XFROG, PG FROG, and SHG FROG pulse-retrieval algorithms. From the left: the percentage of traces that yielded convergence using the XFROG, PG FROG, and SHG FROG algorithms.

Conclusions

The XFROG, PG FROG and SHG FROG techniques are able to measure complex ultrashort pulses with TBP of 100 in the presence of noise. XFROG is the best of the three, but it requires a reference pulse. PG FROG and SHG FROG also work reasonable well if a reference pulse is not available.

References:

1. R. Trebino, "Frequency-Resolved Optical Gating: The Measurement of Ultrashort Laser Pulses, " Kluwer Academic Publishers (2002).
2. Richard A. Altes, "Detection, estimation, and classification with spectrograms," J. Acoust. Soc. Am. 67(4), 1232-46, 1980.

Electron density gradient measurement for laser wakefield accelerator

Jaehoon Kim[1], Hyojae Jang[2], Min Sup Hur[1], and Jong-Uk Kim[1]

[1] Korea Electrotechnology Research Institute, 1271-19, Sa-dong, Sangnok-gu, Ansan-city, Gyeonggi, 426-170, Rep. Korea
E-mail: jkim@keri.re.kr
[2] Pohang University of Science and Technology, San 31, Namgu, Pohang, Kyungbuk, 790-784, Rep. Korea

Abstract. The electron density structure is studied for the electron injection in the laser wakefield acceleration. A shock structure of plasma channel was generated to make a sharp downward electron density structure and was measured by interferometer. A total regularization method was used for Abel inversion. The measured electron density structure indicates that just after laser pulse we could generate very sharp electron density transition.

Introduction

A laser wakefield accelerator is a new tool to generate a femto-second electron bunch. Due to the development of a ultrahigh intensity laser, an electron beam generation by the laser wakefield accelerator is already demonstrated. The main issue for the stable operation of the laser wakefield accelerator is the injection problem. Many injection schemes were proposed. One of the injection schemes is using a sharp electron density transition [1, 2]. For this scheme, a very sharp electron density transition (density change scale length less than the plasma wavelength, $\lambda_p = 2\pi c/\omega_{pe} \sim 3.3 \times 10^{10} \, n_e^{-1/2}$ μm, where ω_{pe} is the plasma frequency) structure is needed. If the laser pulse duration, τ, is 35 fs, we need plasma wavelength around 10 μm. In this case, the electron density change scale length must be shorter than 10 μm. A shock structure of the plasma channel is a feasible candidate for such as a density transition [3, 4]. The shock structure of the plasma was studied with an interferometer. The change of the electron density transition structure was measured at the different time after laser pulse.

Experimental Setup

In this experiment, plasmas were generated by 200 ps laser pulse which was focused at the gas target and the electron densities were measured using biprism interferometer. A schematic diagram of the experiment is shown in Fig. 1. A femtosecond 20 TW laser system was used in this experiment. Before a compressor, the laser beam was split in two parts. The uncompressed laser beam was used to generate plasmas. The laser energy was 300 mJ and the pulse duration was 200 ps. The other laser beam was compressed and the small parts of the compressed laser beam was converted into second harmonic wavelength by BBO crystal and used as a probe beam for the interferometer. The pulse duration of the compressed laser pulse was 40 fs. The uncompressed laser beam was focused on the top of the nozzle by a lens with a focal length 300 mm. The nozzle size is 4 mm long and 1 mm wide. The neutral gas density was controlled by the back pressure of the nozzle and measured by the interferometer. The measured neutral gas density was 1.5×10^{19} cm^{-3} at the interaction region with a back pressure 80 bar. A biprism interferometer was used to

944

measure the electron density structure. The fast Fourier transform was used to recover the phase change from the measured interferogram. A total variation regularization method was used for the sharp edge reserving Abel inversion

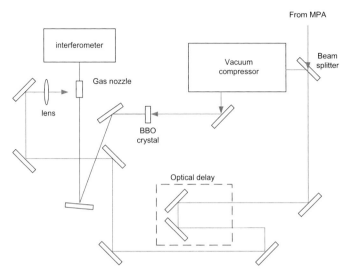

Fig. 1. Experimental setup. Red line shows the laser pulse path which was used to generate plasmas. Blue line shows the probe beam for the interferometer direction.

Results and Discussion

Figure 2 shows the measured electron density shape. Figure 2(a) shows the interferogram and Fig. 2(b) shows the electron density structure recovered by total variation regularization Abel inversion. The sharp electron density structure is well reserved. Figure 2(c) shows the cross section of electron density at the centre of the plasma with different time delay. We can generate sharp electron density structure just after the laser pulse. Figure 3 shows the peak electron density and the distance, Δr, from the peak to zero plasma density at the edge with difference time delay.

The electron density structure has a very sharp density transition. Out of the density transition, the neutral gas can be ionized by the radiation from the plasma and electron collision [5]. Due to the ionization of the gas and diffusion of the electron to the neutral gas region, the scale length, Δr, increases with time as shown in Fig. 3. For a laser wakefield accelerator, the scale length shorter than the plasma wavelength is needed[2]. If the plasma density is 1×10^{19} cm^{-3} than we need scale length shorter than 10 μm. The measured plasma density indicates that the time delay between the laser beam which is used to generate the electron density transition and the laser beam which is used for the laser wakefield accelerator must be shorter than 2 ns for our electron density range. This parameter will be used for the laser wakefield accelerator.

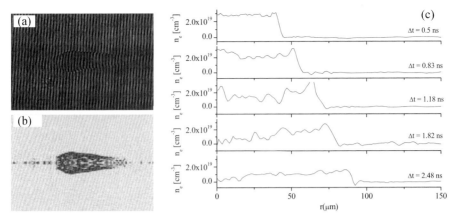

Fig. 2. (a) measured interferogram, (b) electron density structure recovered by Abel inversion. The interferogram was taken at time delay 0.5 ns. (c) the electron density at the center of the plasma with different time.

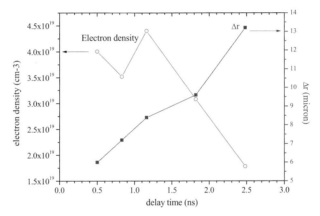

Fig. 3. Measured peak electron density(open circle) and density change scale(closed square) at different time after laser pulse.

1   H. Suk, N. Barov, and J. B. Rosenzweig, "Plasma electron trapping and acceleration in a plasma wake field using a density transition," Phys. Rev. Lett. 86, 1011 (2001).

2   Hyyong Suk, Hae June Lee, and In Soo Ko, "Generation of high-energy electrons by a femtosecond terawatt laser propagating through a sharp downward density transition," J. Opt. Soc. Am. B 21. 1391 (2004).

3   J. U. Kim, N. Hafz, and H. Suk, "Electron trapping and acceleration across a parabolic plasma density profile," Phys. Rev. E, 69, 026409 (2004)

4   J. Kim, H. Jang, S. Yoo, M. Hur, I. Hwang, J. Lim, V. Kulagin, H. Suk, I. W. Choi, N. Hafz, H. T. Kim, K.-H. Hong, T. J. Yu, J. H. Sung, T. M. Jeong, Y.-C. Noh, D.-K. Ko, and J. Lee, "Quasi-Monoenergetic Electron-Beam Generation Using a Laser Accelerator for Ultra-Short X-ray Sources," J. Korean Phys. Soc. 51 397-401 (2007).

5   C. G. Durfee III, J. Lynch, and H. M. Milchberg, "Development of a plasma waveguide for high-intensity laser pulses," Phys. Rev. E, 51, 2368 (1995)

10-femtosecond Precision, Long-term Stable Timing Distribution Over Multiple Fiber Links

Jonathan A. Cox, Jungwon Kim, and Franz X. Kärtner

Research Laboratory of Electronics and Dept. of Electrical Engineering and Computer Science, Massachusetts Institute of Technology, 77 Massachusetts Ave., Cambridge MA, USA
E-mail: joncox@mit.edu

Abstract. The distribution of an ultrafast pulse train over two, 300-meter fiber links with less than 1 femtosecond of timing jitter [0.1Hz, 100kHz] and 6.4 femtoseconds rms of drift over 72 hours of continuous operation [4µHz, 0.5Hz] is demonstrated.

Ultra-low noise timing distribution over the ~1 km range is critical to the implementation of next-generation timing/phase sensitive systems, such as x-ray free electron lasers (XFEL) and phased antenna arrays for radio astronomy [1]. The low timing jitter, <10 fs rms [10 kHz, 10 MHz], of commercially available ultrafast, modelocked fiber lasers and the reduced temperature sensitivity of an all-optical timing distribution system make timing distribution over optical fiber an attractive technique [2]. In addition, by directly distributing a ~200 Mhz ultrafast pulse train, as opposed to a ~200 THz optical carrier, expensive and fragile optical frequency division techniques at each remote end point are not necessary to extract a useable timing signal [3]. Here, we demonstrate the distribution of an ultrafast, optical pulse train over two, independent, 300 meter long, stabilized fiber links with less than 1 femtosecond (fs) rms of timing jitter [0.1 Hz, 100 kHz], and 6.4 fs rms of drift over 72 hours of continuous, unaided operation [4 µHz, 0.5Hz]. This corresponds to a long-term fractional timing stability of $2 \cdot 10^{-20}$ between the outputs of both links. Coupling the ultrafast timing distribution system with the low-noise, low-drift (<6.8 fs rms [27µHz, 1 MHz]) optoelectronic microwave signal regeneration system makes robust, cost-effective and long-term, sub-10-femtosecond timing distribution possible [4].

The system operates by stabilizing the total group delay, or time-of-flight, of a dispersion compensated single mode fiber link with a motorized free space delay and a piezoelectric fiber stretcher, as shown in Figure 1. In this way, a ~100 fs pulse train from a 200 MHz erbium fiber laser is delivered across 300 meters of fiber. Measurement of the optical delay is made by performing an optical cross-correlation between a new pulse entering the link from the laser and an old pulse that was reflected back from the end of the link by the 50 percent faraday rotating mirror (FRM). This cross-correlation is performed with a periodically poled KTP (PPKTP) balanced cross-correlator, providing a measurement of the delay between the two pulses [1]. This error signal, in turn, drives the loop filter controlling the fiber stretcher, to remove the majority of fast fiber length fluctuations. The closed-loop bandwidth of this system is approximately 1 kHz, which is sufficient to eliminate the vast majority of thermal and acoustic fiber fluctuations. In addition, a motor controller monitors the loop filter output and adjusts the free space delay to keep the loop filter in range. To overcome the link loss, an in-loop EDFA is also implemented.

Since the thermal expansion of aluminum is about 770 as/°C/cm, a standard optics layout on the order of one meter in length will introduce significant timing uncertainty. Thermal drift of the optical path length of the pulse splitting optics, including the reference delay arms and the pulse recombination optics, (Figure 1) directly effect timing distribution and measurement accuracy. Therefore, it was necessary to construct these optics on a temperature stabilized Invar breadboard that was held to 0.1 °C, providing a thermal stability of about 4 as/cm. To further suppress unwanted thermal drifts in the optical components, the optics of both links were carefully arranged as symmetric, mirror images. As a result, it is now possible to accurately characterize the timing drift of one or more timing links over many hours.

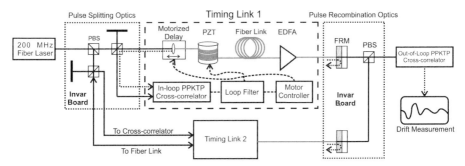

Fig. 1. A schematic of the dual timing link system with path length sensitive optics mounted in a symmetric fashion on temperature stabilized Invar breadboards.

Measurement of the timing stability between the outputs of the two links is performed with the out-of-loop PPKTP cross-correlator, as shown in Figure 1. The response of the cross-correlator is calibrated against the time delay provided by the piezo fiber stretcher. Over a frequency interval from 0.1 Hz to 100 kHz, we observed a timing jitter below 1 fs rms, as shown in Figure 2. In addition, from 35 µHz to 100 kHz, as computed from the average of nine, eight-hour intervals, we observed a timing jitter of 3.3 fs rms. Furthermore, over a 72 hour interval of continuous, unaided operation, shown in Figure 3, we observed only 6.4 fs rms drift from 4 µHz to 0.5 Hz. Moreover, the timing signal remained within 16.4 fs pk-to-pk over the full 72 hours.

Since the optics are now constructed on a highly temperature invariant breadboard, as described above, we believe the performance is fundamentally limited by the polarization mode dispersion (PMD) of SMF-28e fiber. With an upper limit on PMD (PMD_Q) of 60 fs/√km, one expects no more than 33 fs PMD per link, depending on the stress applied to the fiber. In fact, by rewinding the fiber spool for the second link in a looser fashion, we were able to reduce the PMD of link 2 from ~700 fs/√km, where two separate ~500 fs pulses are clearly observable at the link output, to a level where only a single pulse was discernable. In addition, the looser winding reduced the fiber drift of link 2 by a factor of three, as revealed by the free space delay imparted by the motors (Figure 3). Consequently, we expect the performance of the system with standard single mode fiber (SMF) to be, in large part, limited by the PMD of the SMF and the physical stresses exerted upon it.

In conclusion, we have demonstrated the first timing distribution system capable of operating with 10 fs absolute, or $2·10^{-20}$ relative, precision over multiple days. Strategies for overcoming PMD effects can be implemented to reach long term stable sub-femtosecond performance. When combined with optoelectronic microwave signal regeneration [4], this technology represents a complete set of tools for distributing and extracting a 10 fs timing signal in a robust and cost-effective fashion.

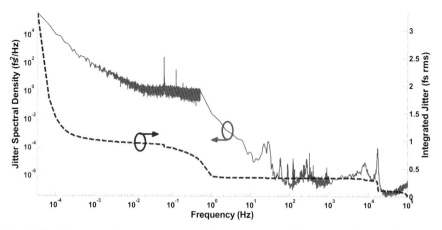

Fig. 2. The jitter spectral density measured between the two link outputs. The total jitter is 3.3 fs rms [35 μHz, 0.5 Hz] and below 1 fs rms [0.1 Hz, 100 kHz].

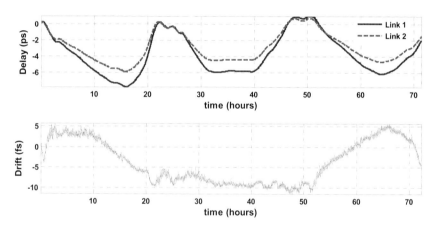

Fig. 3. The long-term timing drift measured over 72 hours of operation. The upper plot shows the free space delay from the motors. The lower plot shows the timing drift between the outputs of both links, which was 6.4 fs rms [4 μHz, 0.5 Hz] or 16.5 fs pk-to-pk.

Acknowledgements. We thank EuroFEL, Univ. of Wisconsin, the O.N.R. (N00014-02-1-0717), and the A.F.S.O.R. (FA9550-06-1-0468) for their generous support.

1   J. Kim, J. Chen, Z. Zhang, F. Wong, F. Kaertner. "Long-term femtosecond timing link stabilization using a single-crystal balanced cross correlator," *Opt. Lett.* **32**, 1044 (2007).

2   J. Kim, J. Chen, J. Cox, F.X. Kaertner, "Attosecond-resolution timing jitter characterization of freerunning mode-locked lasers using balanced optical crosscorrelation," Opt. Lett. **32**, **3519** (2007).

3   N. Newbury, P. Williams, W. Swann. "Coherent transfer of an optical carrier over 251 km," *Opt. Lett.* **32**, 3056 (2007).

4   J. Kim, F.X. Kaertner. "Microwave Signal Regeneration from Mode-Locked Lasers with 1.9×10^{-19} Stability," *Lasers and Electro-Optics, 2008. (CLEO), Conference on.*

Two-dimensional pulse shapers capable of more than phase & amplitude modulation

Ge.Wang[1], Hiroki Yazawa[1], Yoshihiro Esumi[1], Tomoaki Abe[1], and Fumihiko Kannari[1]

[1] Department of Electronics and Electrical Engineering, Keio University
3-14-1, Hiyoshi, Kohoku-ku, Yokohama, 223-8522, Japan
E-mail: kannari@elec.keio.ac.jp

Abstract. We hereby prensent two nova pulse shaping applications of two-dimensional spatial light modulators. We firstly performed polarization control as well as phase & amplitude modulation using a 2D LC-SLM, and then proposed a scheme of pulse shaping on $\omega + 2\omega$ two-color laser with a 2D MEMS-MMA SLM.

Introduction

Femtosecond pulse shaping technology has been holding an essential position in the development of ultrafast science and found widespread use in spectroscopy, microscopy, molecular coherent control and many other research fields. Nowadays, it is routine to generate user-defined waveforms through automated control and the invention of two-dimensioanl liquid-crystal spatial light modulators (2D LC-SLMs) faciliates spatio-temporal pulse shaping[1]. Besides, 2D SLMs play a pivotal role in the simultaneous phase & amplitude modulation of femtosecond laser pulses as well. Nelson group initially proposed the diffraction-based method of simultaneous shaping of both phase and amplitude using a 2D LC-SLM[2]. Recently, Silberberg analyzed the simultaneous femtosecond phase & amplitude shaping using 2D SLM more in detail and proposed a new zero-order approach for simultaneous phase and amplitude modulation[3].

As a matter of fact, 2D SLMs are capable of more than simultaneous phase & amplitude shaping. We hereby present two nova applications of 2D SLMs. Firstly, we performed polarization control together with phase & amplitude modulation using a 2D optically-addressed LC-SLM. We also proposed a scheme to gain full control over $\omega + 2\omega$ two-color laser pulse with a 2D micro-electro-mechanical-system micromirror-array(MEMS-MMA) SLM[4].

Polarization control along with phase & amplitude modulation using multi-pass 2D LC-SLM

Femtosecond pulse laser proves to be an extremely effective tool in coherent control of quantum systems. Optical electric field is defined with three paprameters: spectral phase, spectral amplitude and spectral polarization. Conventional pulse shaping techniques, however, deal with only scalar properties (phase and amplitude) of ultrafast laser pulses, while in fact the laser electric field is a vectorial quantity. Actually, the left behind parameter, polarization, is able to offer access to a further level of control over quantum systems[5]. Silberberg reported full control over spectral polarization of ultrashort pulses using four LC-SLMs[6].The results seem very attractive, whereas the systems would be space-consuming and quite sophisticated due to the cascaded 4-f optical configuration. Besides, several other groups also brought out their original designs for polarization shaping as well, while they also confront some silimar difficulties.

We would like to present you with a newly-designed , much easier-to-build pulse shaper for perfect shaping of time-dependent polarization with only one single 2D LC-SLM. And the experimental setup is showed as follows.

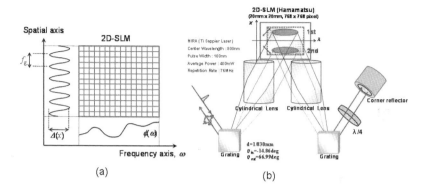

Fig. 1: (a) Spectral mask pattern along the spatial & frequency axis on 2D-SLM. (b) Setup of the two-time reflective pulse shaper using a 2D-SLM

The femtosecond laser pulse with wavelength centered 800 nm and 40 nm wide spectrum is guided to the first grating with 45° linear polarization and its Fourier-plane (FP) is imaged on the 2D-SLM by a cylindrical lens. After modulation on FP, the polarization of laser pulse is adjusted to -45° by passing through a quarter-wave plate twice employing a corner reflector. Then, the pulse is imaged on FP of the 2D-SLM for modulation once again, afterwards the frequency components are then transformed back to the time domain by the incident grating-lens pair. Finally time-dependent polarization shaped pulses are obtained. We adopted the zero-order approach for the phase and amplitude modulation of each polarization of the pulse each[3]. For evaluation of the time-dependent polarization, we utilized the dual-channel spectral interferometry technique.

Full control over optical e-field superposition of $\omega + 2\omega$ pulses

Recently, quite inspiring experiments on molecular orientation has been performed using the superposition of two-color $\omega + 2\omega$ laser pulses[7]. Nevertheless, the interferometric two-color pulse shaping scheme used in the abovementioned experiments is realtively unstable and lacks the abilty to respectively modulate the fundamental pulse and the second harmonic generation(SHG) pulse. Owing to the superior pulse shaping features of the newly-developed 2D MEMS-MMA SLM, in particular its broad operating range from near-infrared(NIR) to deep ultraviolet(DUV)[8], the unprecedent full pulse shaping control over two-color 400 nm & 800 nm pulses now becomes possible.

As shown in the schematic experimental setup, Ti-Sapphire pulse laser amplified by chirped pulse amplifier(CPA) with central wavelength of 800 nm, spectral bandwidth of 25 nm, and temporal width of approximately 40 fs (FWHM) is chosen as the light source. A 0.5 mm thick BBO crystal functions to generate frequency-doubled 400 nm pulse with spectral width of about 4 nm. A dual-waveplate is used here to adjust the polarization of the two pulses to the same direction. Then the dual pulse is lead to

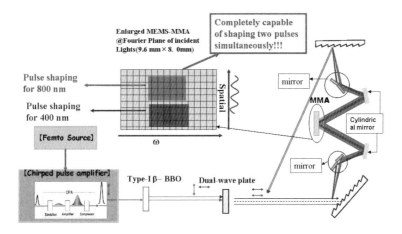

Fig. 2: Simplified schematic diagram of the experimental setup for dual pulse shaping

the first diffraction grating of the pulse shaper.High reflectivity mirrors are placed after the first grating and before the cylindrical lens. After modulated at the Fourier plane, the pulses are recombined again at the second grating in order to obtain full control over dual pulse independently. In a similar way with the polarization shaping scheme, we apply sinusoidal modulation pattern along vertical(spatial) directionas well as phase modulation function along horizontal(frequency) direction.As shown in Fig.2, the spectral components are modulated in different locations on the MEMS-MMA.We plan to investigate two-color laser superposition through the Coulomb explosion experiment of small organic molecules such as ethanol.

1   J. Vaughan, T. Feurer, and K. Nelson,J. Opt. Soc. Am. B 19, 2489(2002).
2   J. Vaughan, T. Hornung, T. Feurer, and K. Nelson, Opt. Lett. 30, 323(2005).
3   E. Frumker and Y. Silberberg, J. Opt. Soc. Am. B 24, 2940(2007).
4   M. Hacker, G. Stobrawa, R. Sauerbrey, T.Buckup, M. Motzkus, M. Wildenhain, and A. Gehner, Appl. Phys. B 76, 711(2003).
5   T. Brixner, G. Krampert, T. Pfeifer, R. Selle, G. Gerber, M.Wollenhaupt, O. Graefe, C. Horn, D. Liese, and T. Baumert, Phys. Rev. Lett. 92, 208301 (2004).
6   L. Polachek, D. Oron, and Y. Silberberg, Opt. Lett. 31, 631(2006).
7   H. Ohmura and T. Nakanaga, J. Chem. Phys. 120, 5176(2004).
8   K. Stone, M. Milder, and K. Nelson, in UltrafastPhenomena XV,(Spring-Verlag, 2007).

Adaptive Phase Shaping in a Fiber Chirped Pulse Amplification System

Nikita K. Daga, Fei He, Hazel S.S. Hung, N. Naz, Jerry Prawiharjo, David C. Hanna, David J. Richardson, David P. Shepherd

Optoelectronics Research Centre, University of Southampton, Southampton SO17 1BJ, UK
E-mail: nid@orc.soton.ac.uk

Abstract. We demonstrate adaptive spectral phase shaping in a fiber chirped pulse amplification system. The adaptive process, controlled by a simulated annealing algorithm, resulted in three times improvement in the autocorrelation peak intensity of 65μJ pulses.

Introduction

Fiber lasers offer high efficiency and high beam quality, while also dissipating heat efficiently. In order to achieve high-power femtosecond pulses in fiber laser systems, chirped pulse amplification (CPA) is employed. Power scaling of the fiber CPA system is, however, non-trivial. The tight confinement of the fiber geometry leads to nonlinear effects, especially self-phase-modulation (SPM), while gain shaping in the fiber amplifiers limits the spectral bandwidth and alters the spectral profile. The long interaction length of the fiber introduces significant dispersion, which can not always be fully compensated by the compressor. The combined effect of these factors results in significant degradation of the pulse. In order to improve the pulse quality, a programmable pulse shaper can be implemented.

The use of phase modulators to correct for SPM-induced phase distortions up to 2π has been demonstrated in fiber CPA systems, [1], and adaptive pulse shaping has been used to control the propagation of ultrashort pulses in optical fibers [2]. Recent work combining adaptive pulse shaping with fiber-based CPAs has been demonstrated with amplitude-only shaping [3]. This technique can be used to offset the gain shaping and SPM phase contribution, but cannot compensate for the residual dispersion in the system.

In this paper we demonstrate the use of spectral phase shaping in an adaptive fiber CPA system. The adaptive process was controlled by a simulated annealing algorithm and was used to optimize the coefficients of a Taylor expansion of the spectral phase profile. Three times improvement in the autocorrelation peak intensity was demonstrated, with close to transform-limited pulse durations of 800fs at pulse energies of 65μJ.

Experimental Details

A schematic of the CPA system is shown in Fig 1. The pump source is a mode-locked 1.053μm Nd:glass laser delivering 600fs pulses at a repetition rate of 80MHz. These pulses are shaped with a phase-only pulse shaper before being stretched to ~1ns with a grating stretcher. An electro-optic modulator then reduces the repetition rate to 100kHz before amplification to 0.1W using the core-pumped Yb-fiber pre-amplifiers. High pulse energies were achieved with two cladding-pumped amplifiers, each of which used a 2 m length of double-clad Yb-doped PCF from Crystal Fiber (core diameter 40 μm, NA 0.03; inner cladding diameter 170 μm, NA 0.6) and fiberised 975 nm pump laser diodes (diode powers were 6 W and 20 W respectively). Acousto-optic modulators prevented ASE build-up, and reduced the final repetition rate to 16.67 kHz. A dielectric grating compressor, with 65% overall transmission efficiency, recompressed the pulses. The maximum-to-minimum contrast of the autocorrelation trace within the measuring window (30ps) was used as the feedback parameter for the adaptive loop, which modified the phase profile applied at the shaper.

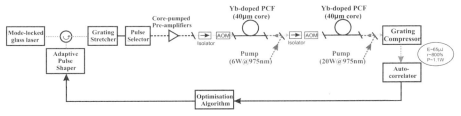

Fig. 1. Schematic of the CPA system

The pulse shaper is arranged in a 4-f configuration [4]. It consists of a diffraction grating with a groove density of 1100 lines/mm, cylindrical lens (f=200mm), spatial light modulator (SLM) and a flat mirror to retro-reflect the input beam. The SLM is a phase-only liquid crystal array of 128 pixels covering a distance of 12.8mm (CRI SLM-128-MIR). The 10nm bandwidth input beam was expanded to a collimated beam radius of 5.9mm and was incident upon the diffraction grating at an angle of 10 degrees, with a corresponding diffracted beam angle of 80 degrees. The spectral resolution as defined in ref. [4] was calculated to be 0.04nm. The maximum complexity, η, is limited by the number of pixels to 128 such that the finest achievable spectral feature, δλ, is 0.078nm and the time window, T, is 20ps.

In this work, we use an optimisation algorithm known as the Generalised Simulated Annealing (GSA) algorithm [5]. The main parameters that affect the convergence are the initial starting temperature, T_0, the acceptance parameter, q_a, and the visiting parameter, q_v. Typically, the algorithm ran for 150 iterations with T_0 set to be twice the expected value of the optimized feedback parameter (i.e. twice the maximum minus the minimum voltage of the autocorrelation), $q_a=1$, and $q_v=2.5$.

The phase was described by a Taylor series expansion so that the 128 pixels could all be addressed with only a few optimisation parameters and thus achieve fast convergence. The adaptive CPA system modified the β_2 to β_6 dispersion terms of the spectral phase. Figure 2 below shows a set of typical convergence data for the feedback parameter and the first three optimisation parameters. It can be seen from this data that the algorithm converges quickly within the first 100 iterations, which took ~2 minutes.

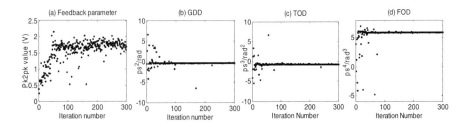

Fig. 2. Convergence data for (a) the feedback parameter and the (b) second, (c) third, and (d) fourth order dispersion parameters of the GSA algorithm.

To confirm the calibration of the pulse shaper, low energy pulses suffering from minimal SPM were adaptively compressed at various grating compressor roof mirror positions. The second order phase parameters obtained from these experiments agreed well with the expected change in dispersion from the compressor.

Results

The quality of the pulses without shaping is shown in Fig.3(a). The CPA system yields high quality pulses at low energy but a large pedestal remains at high pulse energies of 65µJ which have a calculated B-integral of ~2.5π. With the shaper in the system, Fig.3(b) shows that at 65µJ, the pulse quality is significantly improved with the pedestal almost completely depleted and with a three times improvement in the autocorrelation peak. The autocorrelation is compressed from 1.4ps to 1.1ps, close to the flat phase limit calculated from the Fourier transform of the measured spectral data. The spectra after the shaper with and without shaping are shown in Fig.3(c) showing minimal spectral distortion due to the applied phase.

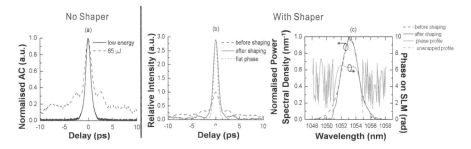

Fig. 3. Autocorrelations for (a) the unshaped pulses of the low energy and 65µJ energy pulses and (b) the high energy shaped pulses. (c) The spectra after the shaper with and without shaping, and with the applied phase profile are also shown.

Conclusions

We have demonstrated high quality 65µJ pulses from an adaptive fiber CPA system by controlling the spectral phase pre-compensating for SPM effects induced in the amplifier stages. A GSA algorithm was used to optimize the second to sixth order dispersion parameters of the phase resulting in pulse compression near to the flat phase limit within 150 iterations. A three-fold increase in the autocorrelation peak was also observed.

Further work into phase and amplitude shaping in fiber CPA systems, investigating the compensation of SPM at even higher energies in addition to alternative optimisation techniques, is anticipated.

1   F.G.Omenetto, D.H.Reitze, B.P.Luce, M.D.Moores, and A.J.Taylor, IEEE J. Select. Top. Quantum Electron. **8**, 690, 2002.
2   G. Zhu, J. Edinberg, and C. Xu, Opt. Express **15**, 2530-2534, 2007.

Two-dimensional Fourier transform electronic spectroscopy with a pulse-shaper

J. A. Myers[1], K. L. M. Lewis[1], P. F. Tekavec[1], and J. P. Ogilvie[1]

[1] Department of Physics and Biophysics, University of Michigan, Ann Arbor MI 48109, USA
E-mail: jogilvie@umich.edu

Abstract. We report 2D electronic spectra obtained using a pulse-shaper in a pump-probe geometery. We demonstrate the method at visible wavelengths on a dye system and discuss the benefits of this approach compared to other implementations.

Introduction

Two-dimensional electronic spectroscopy (2DES) has recently emerged as a powerful technique for studying energy transfer in photosynthetic and solid-state systems. Despite the rich chemical information available from 2DES, the relative difficulty of implementing the experiment has limited the degree to which this method has been utilized. This is particularly true at visible wavelengths where the demand for high phase-stability generally requires diffractive-optics-based [1, 2] or actively phase stabilized approaches [3]. Alternately, collinear experiments combined with phase-cycling [4], phase modulation [5], or noncollinear, pulse-shaping-based methods [6] have been pursued. It was suggested some time ago that absorptive 2D spectra could be obtained using a combination of a collinear pump pulse pair and noncollinear probe [7]. Recently, 2D spectra have been collected in this "pump-probe" geometry at infrared [8, 9] and near-infrared wavelengths [10], greatly simplifying the experimental approach. Here we implement a method that employs an acousto-optic pulse-shaper used in a pump-probe geometry to obtain absorptive spectra in a simple dye system. This approach takes advantage of the high degree of precision with which a pulse-shaper can create phase-locked pulse pairs and automatically retrieves absorptive 2D spectra.

The real (imaginary) part of a 2D correlation spectrum can be roughly interpreted as the change in absorption (dispersion) at the detection frequency (v_3) due to excitation at frequency v_1. To obtain spectra that have properly separated real and imaginary parts, rephasing and non-rephasing contributions to the third-order response must be combined with the proper phase relationship. This separation of absorptive and dispersive spectra is desirable to increase the resolution and information content of the spectra and to reveal homogeneous lineshapes that are masked by inhomogeneous broadening [7]. In the commonly-used box car geometry, the rephasing and non-rephasing signals are emitted in different phase-matched directions. To obtain absorptive spectra, the two contributions must be measured separately and "phased" by fitting them to a separate pump-probe measurement. This procedure can often be difficult. In addition, in order to make the necessary interferometric measurements, the phase stability between the pulses must be much better than an optical wavelength. The pulse-shaper-based pump-probe geometry approach removes these difficulties and makes the recording of 2D absorptive spectra a straight-forward task.

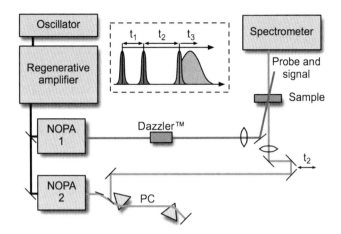

Fig. 1. Experimental setup. The probe pulse is compressed with a prism compressor (PC) and the pump pulses are compressed with the DAZZLER. Inset shows the pulse sequence at the sample.

Our experimental setup is shown in Fig. 1. A Ti:Sapph oscillator (Synergy, Femtolasers) seeds a regenerative amplifier (Spitfire Pro, Spectra Physics), producing 800nm, 1mJ pulses at 1kHz. The amplified beam is split and fed into two home-built non-collinear optical parametric amplifiers (NOPAs). One NOPA creates the pump beam, and the second NOPA creates the independently tunable probe beam. The pump beam is sent into an acousto-optic pulse-shaper (Dazzler, Fastlite), which creates the first two excitation pulses with a variable delay, t_1. These pulses serve as the "pump" in the pump-probe geometry, with the Dazzler also acting to chop the excitation at 500Hz. A time t_2 later, the sample interacts with the probe pulse. The resulting signal/probe beam is sent into a spectrometer (Horiba Jobin Yvon iHR320) and recorded on a CCD camera (Pixis 100F), providing the v_3 axis of the 2D spectrum. In the pump-probe configuration of Fig 1., both the rephasing and nonrephasing contributions to the third order polarization are emitted in the phase-matched direction of the probe pulse. Thus both signals can be collected simultaneously without timing errors that can lead to difficulty in extracting the absorptive component. In addition, since the signals are emitted in the same direction as the probe beam, the probe serves as an intrinsically phase stable local oscillator for heterodyne detection. After scanning the t_1 delay and performing the Fourier transform, the second axis of the 2D spectrum (v_1) is obtained, yielding a properly-phased 2D absorptive spectrum. We note that the non-oscillatory pump-probe component to the signal appears near zero frequency in 2D spectrum. Generating the t_1 delay with the Dazzler also removes any uncertainty as to the location of $t_1 = 0$.

To demonstrate the method we have performed 2DES on cresyl violet in ethanol. Fig. 2 shows absolute value, absorptive (real) and imaginary components of the signal at $t_2 = 1.2$ ps. The central peak is Stokes shifted and broadened as memory of the initial excitation frequency is lost, consistent with 2D measurements of similar dye systems.[11] The vertical asymmetry of the lineshape is attributed to unequal bandwidths of the pump and probe pulses.

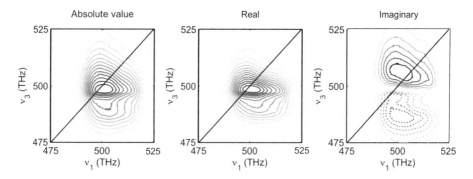

Fig. 2. 2DES spectra for cresyl-violet in ethanol at $t_2 = 1.2$ ps.

In summary we have demonstrated a pulse-shaping-based approach for obtaining absorptive 2D spectra in a noncollinear pump-probe geometry at visible frequencies. Future work will compare the pros and cons of this approach with previously applied diffractive-optics-based methods and will apply 2DES to studying energy transfer in biological systems.

1 M. L. Cowan, J. P. Ogilvie, and R. J. D. Miller, "Two-Dimensional Spectroscopy Using Diffractive Optics Based Phased-Locked Photon Echoes," Chemical Physics Letters **386**, 184-189 (2004).

2 T. Brixner, I. V. Stiopkin, and G. R. Fleming, "Tunable Two-Dimensional Femtosecond Spectroscopy," Optics Letters **29**, 884-886 (2004).

3 X. Q. Li, T. H. Zhang, C. N. Borca, and S. T. Cundiff, "Many-Body Interactions in Semiconductors Probed by Optical Two-Dimensional Fourier Transform Spectroscopy," Physical Review Letters **96** (2006).

4 P. F. Tian, D. Keusters, Y. Suzaki, and W. S. Warren, "Femtosecond Phase-Coherent Two-Dimensional Spectroscopy," Science **300**, 1553-1555 (2003).

5 P. F. Tekavec, G. A. Lott, and A. H. Marcus, "Fluorescence-Detected Two-Dimensional Electronic Coherence Spectroscopy by Acousto-Optic Phase Modulation," Journal of Chemical Physics **127** (2007).

6 J. C. Vaughan, T. Hornung, K. W. Stone, and K. A. Nelson, "Coherently Controlled Ultrafast Four-Wave Mixing Spectroscopy," Journal of Physical Chemistry A **111**, 4873-4883 (2007).

7 S. M. G. Faeder, and D. M. Jonas, "Two-Dimensional Electronic Correlation and Relaxation Spectra: Theory and Model Calculations," Journal Of Physical Chemistry A **103**, 10489-10505 (1999).

8 S. H. Shim, D. B. Strasfeld, Y. L. Ling, and M. T. Zanni, "Automated 2D IR Spectroscopy Using a Mid-IR Pulse-shaper and Application of This Technology to the Human Islet Amyloid Polypeptide," Proceedings of the National Academy of Sciences of the United States of America **104**, 14197-14202 (2007).

9 L. P. DeFlores, R. A. Nicodemus, and A. Tokmakoff, "Two Dimensonal Fourier Transform Spectroscopy in the Pump-Probe Geometry," Optics Letters **32**, 2966-2968 (2007).

10 E. M. Grumstrup, S.-H. Shim, M. A. Montgomery, N. H. Damrauer, and M. T. Zanni, "Facile Collection of Two-Dimensional Electronic Spectra Using Femtosecond Pulse-Shaping Technology," Optics Express **15**, 16681-16689 (2007).

11 D. M. Jonas, "Two-Dimensional Femtosecond Fourier Transform Spectroscopy," Ann. Rev. Phys. Chem. **54**, 425-463 (2003).

Probing Anomalous Spectral Diffusion and Exciton Fluctuations by Coherent Multidimensional Spectroscopy

František Šanda[1], and Shaul Mukamel[2]

[1] Charles University, Faculty of Mathematics and Physics, 121 16 Praha, Czech Republic
 E-mail: sanda@karlov.mff.cuni.cz
[2] University of California, Department of Chemistry, Irvine, CA, USA

Abstract. Novel signatures of anomalous algebraic spectral relaxation, non Gaussian fluctuations and bath-induced transition dipole moments in two dimensional optical lineshapes of excitonic aggregates are predicted using stochastic models with long algebraic relaxation tails.

1. Introduction

Two dimensional infrared and optical spectroscopy probes exciton dynamics through cross (correlation) peaks extending ideas proven in NMR to the femtosecond timescale.

Multidimensional lineshapes are obtained by the coherent nonlinear response to three laser pulses where three time intervals t_1, t_2 and t_3 are controlled. Frequency-frequency correlation plots in Ω_1, Ω_3, the Fourier conjugates to t_1, t_3, give valuable insights into the pathways and time profiles of bath spectral diffusion dynamics, as observed in the contour shapes of 2D peaks and in the cross peaks dynamics respectively. Virtually all modelling of these signals in molecular aggregates is limited to white noise fluctuations of exciton couplings (Redfield equations) and Gaussian fluctuations of frequencies. These describe multiexponential relaxation decay of correlations. Long algebraic tails of anomalous relaxation are common in many complex systems (Glasses, proteins, quantum dots) [1]. These require different dynamical models such as continuous time random walks (CTRW) characterized by the probability distribution function (WTDF) $\psi(t)$ of waiting times for successive jumps between bath states.

We assume an excitonic Hamiltonian $H = \sum_i \varepsilon_i |i\rangle\langle i| + \sum_{ij} J_{ij} |i\rangle\langle i|$ [2] whose parameters ε_i, J_{ij} depend on stochastic bath variables $\sigma(t)$. All physical observables are related to the weighted average over the ensemble of stochastic paths $\sigma(t)$ of the bath

$$\rho(t) = \left\langle \operatorname{T}\exp-i\int_0^t \Lambda(\sigma(t'))dt' \right\rangle \qquad \Lambda \equiv \frac{1}{\hbar}[H, \ldots] \qquad (1)$$

2. Anomalous relaxation and 2D lineshapes

Stochastic quantum dynamics can be computed by a direct generation of random bath paths $\sigma(t)$. More efficient methods are at hand for specific bath models. The key point of the CTRW model, which makes it tractable, is that all memory is erased at the time of jump and the WTDF for the next jump is independent of the history. This is known as renewal. We developed a general algorithm for calculating nonlinear response functions for this model [3,4]. When the asymptotic decay of $\psi(t)$ is fast and all moments of $\psi(t)$ exist the 2D lineshapes show a typical Markovian relaxation pattern (represented by exponential WTDF). A qualitatively different behaviour is observed when the first or the second moments of WTDF diverge $\psi(t) \approx 1/t^{\alpha+1}$.

2.1. Stationary ensembles ($1 < \alpha < 2$).

When the first moment κ_1 of $\psi(t)$ is finite the process is stationary. Stationary ensembles are defined by prescribing special WTDF $\psi'(t) = \kappa_1^{-1}\int_t^\infty \psi(t')dt'$ for the first jump (it may differ from $\psi(t)$ depending on the initial preparation) [3]. In Fig 1 2D absorptive lineshapes are plotted for of a single two level chromophore whose transition frequency $\varepsilon_1 = \pm\omega_0$ is modulated by a two state jump CTRW in slow limit ($\kappa_1\omega_0 \gg 1$). Novel signatures of algebraic WTDF seen in Fig 1 are: divergencies $\sim \Omega^{\alpha-3}$ at peaks (-1,-1), (1,1), discontinuity of the first derivatives at fundamental frequencies $\Omega_1 = \pm 1$ and $\Omega_3 = \pm 1$ and algebraic growth $\sim t_2^{1-\alpha}$ of cross peaks.

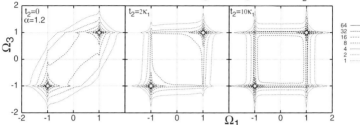

Fig. 1. 2D absorptive lineshape of a single chromophore coupled to two state anomalous spectral CTRW ($\alpha = 1.2, \omega_0 = 1$). Diagonal peaks correspond fundamental frequencies of the bath states; cross-peaks volume gives the fraction of paths that are in different states during intervals t_1, t_3.

2.2. Aging ; spectral diffusion with ($0 < \alpha < 1$).

Random walks with a diverging first moment of $\psi(t)$ retain their memory and never equilibrate. They represent nonstationary ensembles depending on choice of $\psi'(t)$. Usually it is assumed that all allowed bath paths have a jump at some fixed time, where the random walk is started (and $\psi'(t) = \psi(t)$). The mobility of particles will vanish at long times since more random paths $\sigma(t)$ will be trapped due to the long tails of $\psi(t)$.

The nonlinear optical response shows the dependence on the time t_0 elapsed from the start of the random walks to the first laser pulse. (For Markovian relaxation this memory is lost at long t_0.) The 2D lineshapes displayed in Fig 2 interpolate between motional narrowing limit of fast fluctuations at short t_0 (this is similar to Markovian case). However, the static peaks at (-1,-1) and (1,1) which grow with t_0 represent the "trapped" paths (Fig 2). Both types of peaks coexist for anomalous diffusion, what may not be described by master equations with time-dependent rates [4].

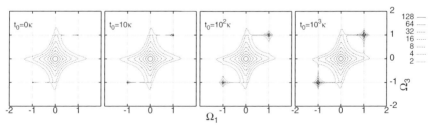

Fig. 2. Aging in 2D absorptive lineshape for nonstationary CTRW spectral diffusion $\psi(t) \sim (\kappa/t)^{\alpha-1}$ ($\alpha = 0.98, t_2 = 0$) on single chromophore. Peaks at -1 and 1 show up as the static fraction of particles trapped in long tails of WTDF is developed with t_0. The central motional narrowing peak is consequence of fast jumps and (very) slowly decays with t_0.

3. Peaks induced by slow bath fluctuations

For Markovian random walks the full information is contained in the density matrix in the joint space of the system and the bath. The evolution of the joint density matrix may be described by the Stochastic Liouville equations, which combine the Liouville-von Neumann equation with the master equation for Markovian stochastic variable σ (described by L^σ).

$$\frac{d\rho(\sigma,t)}{dt} = -i\Lambda(\sigma,t)\rho(t) + L^\sigma\rho(\sigma,t) \tag{2}$$

The third order response functions now factorize into products of Green's functions for the free evolution between pulses and the dipole moment elements that describe the action of the laser pulses. Fast $\sigma(t)$ modulation agrees with the Redfield equations, since all memory effects of the fast bath variables are erased.

New effects which cannot be described by the Redfield equations (where the bath is formally eliminated) are expected for slow or intermediate bath timescales. We consider a homo-dimer with parallel dipole moments. One exciton state carries all oscillator strength and the other is dark. Due to slow Gaussian-Markovian fluctuations of the site frequencies, the σ dependent dark state acquires a transient dipole moment, and additional diagonal peak at (-1,-1) is observed in Fig 3 for the delay times during bath equilibration, but not longer.

The cross peaks (-1,1) and (1,-1) appear at all delay times, because these are induced by the transient dipole moments at either t_1 or t_3 intervals.

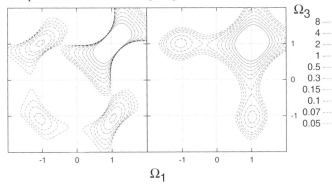

Fig. 3. 2D absorptive lineshape of two coupled chromophores ($J_{12} = 1$) with slow Gaussian-Markovian spectral difusion (of ε_1 and ε_2) with various delay times t_2, shorter and larger then the bath autocorrelation time. The lower transient peak correspond to antibonding orbital, which is dark at equilibrium, but carries transient dipolemoment induced by slow spectral diffusion.

Acknowledgements. F. Š. acknowledges the support of the MŠMT ČR (MSM 0021620835) and the GAČR (202/07/P245). S.M. acknowledges the support of the NSF (CHE-0745892) and NIRT (EEC-0303389) and NIH(GM 59230).

1 J. Klafter and I.M. Sokolov , Physics World 18, 29 , 2005.
2 F. Šanda, S. Mukamel, "Stochastic Liouville Equations for Multidimensional Coherent Spectroscopy of Excitons," submitted for J. Phys. Chem , 2008.
3 F. Šanda, S. Mukamel, Phys. Rev. Lett. 98, 080603, 2007.
4 F. Šanda, S. Mukamel, J. Chem. Phys. 127, 154107, 2007.

Noninterferometric Two-Dimensional Fourier Transform Spectroscopy

J. A. Davis[1], L. V. Dao[1], P. Hannaford[1], K. A. Nugent[2], and H. M. Quiney[2]

[1] Centre for Atom Optics and Ultrafast Spectroscopy, Swinburne University of Technology, Melbourne, 3122 Australia
E-mail: JDavis@swin.edu.au
[2] School of Physics, University of Melbourne, Victoria 3010, Australia

Abstract. We demonstrate a technique that determines the phase of the photon echo emission from spectrally resolved intensity data, without requiring phase stabilised input pulses. The validity is demonstrated using simulated data based on a model three level system with two-colour excitation.

Introduction

Two-dimensional Fourier transform spectroscopy (2DFTS) has become an important tool for studying electronic and vibrational transitions and dynamics in a diverse range of condensed phase systems [1-6]. Traditionally, the intensity of a four wave mixing (FWM) signal is measured. However, because the phase information is lost in detection, details of many body effects and coupling between states, is limited. The technique of 2DFTS measures the phase of the emission in addition to intensity, by using heterodyne detection. Spectrally resolved measurements are taken as a function of delay τ, between the first two pulses (or T, between second and third pulses). Then, by performing a Fourier transform with respect to τ (or T), two-dimensional correlation spectra can be obtained, with one axis representing the frequency of the polarization in the sample ω_τ (ω_T), in the period τ (T) ,and the other axis the frequency of the third-order emission, ω_t. This representation allows straightforward identification of states that are coupled, and the dynamics that govern these processes. Further dissection of the signal into rephasing and non-rephasing, and real and imaginary components reveals even greater details of the interactions in the sample. In order to get this complete picture, phase stability is required between all three input pulses, plus the local oscillator pulse. This may be achieved provided that all three pulses are of the same wavelength and polarization. However, this is not always desirable and some important information can only be obtained when this is not the case. This information then remains largely hidden, as it becomes virtually impossible to maintain phase stability and employ heterodyne detection to measure the phase.

Experimental Methods

We have developed a method that extends 2DFTS to such cases, without the need to employ heterodyne detection, and which does not require phase stability of the pulses. The signal is instead spectrally resolved and averaged over all phase differences, so that the intensity of $E_{sig}(\tau, \omega_t)$, the detected signal, is measured. The phase distribution, which is lost in the intensity measurement, is recovered using an

962

iterative phase-retrieval algorithm that utilizes *a priori* information about the effective domains of $s_T(\tau, t)$ and $S_T(\omega_\tau, \omega_t)$, which are related to $E_{sig}(\tau, \omega_t)$ by 1D Fourier transforms with respect to ω_t and τ, respectively [7]. This technique is drawn from a combination of other well-known phase retrieval methods used in frequency-resolved optical gating [8] and coherent diffractive imaging (CDI) [9].

The key to this, and all phase retrieval techniques, is that the sampling rate is sufficient. In the present case, this means that the delay and spectral step sizes are sufficiently small, and the ranges sufficiently large. The major difficulty with all phase retrieval techniques is their propensity to fall into solutions that represent local minima, but which do not accurately represent the phase. As a result, the phase retrieval algorithms often need to be run many times with different initial guesses to ensure the correct solution has been identified. Significant effort over the years has been devoted to finding ways to prevent such algorithms stalling in local minima. We are able to adapt these techniques to enhance the performance of our algorithm.

We have implemented a technique equivalent to the shrink-wrap algorithm used in CDI [10], whereby the support constraints applied to the data in time and frequency are adjusted as the algorithm proceeds. We found that this enhanced the repeatability, but did not remove all the problems with additional signals [7].

An additional advance that is not applicable to CDI, where the object being imaged can be completely arbitrary, comes from the fact that the laws of physics must be obeyed in non-linear spectroscopy. Specifically, we know that the signal must be smooth and continuous. We therefore force the signal to decay exponentially outside the supports rather than setting it abruptly to zero. This greatly enhanced both the repeatability of finding the correct solution (by nearly an order of magnitude) and the quality of the solution obtained for the simulated data presented in Fig 1.

The simulated data were calculated for a *V-type* three-level system by solving the semiconductor Bloch equations using the method described in [11]. The two excited states were separated by 6 meV, coupled solely through the shared ground state. Inhomogeneous broadening of 3meV was introduced, and lifetimes of 6 ps and coherence times of 3 ps were used for both excited states. We used two-colour excitation, with pulse 1 energy of 1.406 eV, resonant with the |1> to |3> transition, and pulses 2 and 3 energy of 1.400 eV, resonant with the |1> to |2> transition.

Results and Discussion

Figure 1(a) shows the output from the simulation $E_{sig}(\tau, \omega_t)$ which is used as the pseudo-experimental data. The signal at 1.394 eV is due to diffraction from the second order coherence established between states |2> and |3> and the pure FWM signal. The extended signal at 1.400 eV is due to coherent energy transfer from state |3> to state |2> and then photon echo emission from the |2> to |1> transition, with quantum beats due to the quantum coupling with state |3>.

The correlation spectra obtained directly from the simulation and from the spectral intensity data following phase retrieval are shown in Fig. 1(b) and (c), respectively. The agreement between the actual and calculated correlation spectra in this case is very good. In the basic algorithm problems in the reconstruction such as aliasing due to the abrupt edges imposed by the support constraints were identified [7]. These have almost all been removed by implementing the shrinkwrap algorithm and exponential decay outside the support region.

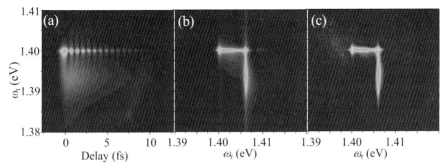

Fig. 1. (a) The simulated spectrally resolved intensity data as a function of delay, τ. The correlation spectra obtained directly from simulation, (b), and from the spectral intensity data via the phase retrieval algorithm, (c). In two-colour experiments, the diagonal in correlation spectra represents the line where the emission frequency, ω_t, is expected for a given ω_τ.

A closer look at the correlation spectra reveals additional details not identifiable in $|E_{sig}(\tau, \omega_t)|^2$. The signal at $(\omega_\tau, \omega_t) = (1.406, 1.394)$ corresponds to diffraction from the second order coherence established between states $|2>$ and $|3>$, and the expected FWM signal from the input laser energies. The signal at $(1.406, 1.400)$ is an off-diagonal signal, and identifies coherent energy transfer from state $|3>$ $(1.406$ eV$)$ to state $|2>$ $(1.400$ eV$)$ as expected. The signal at $(1.400, 1.400)$ was not predicted and is due to the $|1>$ to $|2>$ transition being driven off-resonantly by pulse 1.

Conclusion

In conclusion, we present a method that allows the recovery of the full temporal evolution of the polarization for photon echo experiments. We have demonstrated its validity using simulated data, establishing its ability to retrieve the phase accurately from intensity measurements. The great advantage of this technique compared to direct experimental determination of the phase is that phase stability is not required allowing it to be applied to multi-colour/multi-polarization experiments, hence revealing information not previously accessible. This method also maintains the generality of heterodye techniques, with the ability to separate rephasing and non-rephasing signals, and their real and imaginary components preserved

Acknowledgements. We would like to thank the ARC for financial support.

1   H. Lee, Y-C Cheng, and G. R. Fleming, Science 316, 1462 (2007).

2   G. S. Engel, et al., Nature 446, 782 (2007).

3   T. Brixner, et al., Nature 434, 625 (2005).

4   J. Zheng, K. Kwak, and M. D. Fayer, Acc. Chem. Res. 40, 75 (2007).

5   D. M. Jonas, Annu.,Rev. Phys. Chem. 54, 425 (2003).

6   T. Zhang, et al., Proc. Natl Acad. Sci. USA 104, 14227 (2007)

7   J. A. Davis, et al., Phys Rev. Lett. 100, 227401.

8   R. Trebino, and D. J. Kane, J. Opt. Soc. Am. A 10, 1101 (1993).

9   J. R. Fienup, Appl. Opt. 21, 2758 (1982).

10 S. Marchesini, et al., Phys. Rev. B 68, 140101(R) (2003).

11 J. A. Davis, J. J. Wathen, V. Blanchet and R. T. Phillips, Phys. Rev. B 75, 35317 (2007).

Part XII

Applications of Ultrashort Pulses

Filament-induced electric events in thunderstorms

J. Kasparian[1,2], R. Ackermann[1], Y.-B. André[3], G. Méchain[3], G. Méjean[1], B. Prade[3], P. Rohwetter[4], E. Salmon[1], L. A. Schlie[5], K. Stelmaszczyk[4], J. Yu[1], A. Mysyrowicz[3], R. Sauerbrey[6], L. Wöste[4], J.-P. Wolf[1,2]

[1] Teramobile, Université Lyon 1; CNRS; LASIM UMR 5579, 43 Bd du 11 novembre 1918, F-69622 Villeurbanne Cedex, France
[2] GAP, Université de Genève, 20 rue de l'École de Médecine, CH-1211 Genève 4, Switzerland E-mail: jerome.kasparian@physics.unige.ch
[3] Teramobile, LOA, UMR CNRS 7639, ENSTA—Ecole Polytechnique, Chemin de la Hunière, F-91761 Palaiseau Cedex, France
[4] Teramobile, Institut für Experimentalphysik, Freie Universität Berlin, Arnimallee 14, D-14195 Berlin, Germany
[5] Directed Energy Directorate (AFRL/DELS), Air Force Research Laboratory, 3550 Aberdeen Blvd, SE, Kirtland AFB, NM 87117, USA
[6] Teramobile, Institut für Optik und Quantenelektronik, Friedrich Schiller Universität, Max-Wien-Platz 1, D-07743 Jena, Germany

Abstract. Under conditions of high electric field during two thunderstorms, we observed a statistically significant number of electric events synchronized with the ionized filaments generated by ultrashort laser pulses in the atmosphere. This observation suggests that corona discharges have been triggered by filaments.

Introduction

Lightning has always been considered as fascinating but hazardous. Extensive work has been dedicated to the quantitative understanding of this natural phenomenon. Its randomness raises the need to trigger lightning strikes. Rockets pulling wires into thunderclouds can achieve such triggering [1, 2]. To overcome the limited number of rockets available and the delay of several seconds that elapse between the launch of a rocket and the triggered lightning strike, lasers has been considered since the 1970's although without success. [3,4]. The advent of high-power, ultrashort (~100 fs) lasers renewed this perspective. Such lasers generate multiple ionized filaments in the atmosphere [5], with a length up to hundreds of meters. They can be generated away from the laser source [6], propagate in clouds or turbulent atmospheres [7-9], and control high-voltage discharges even in rain [5,10].

Experimental methods

We investigated the effect of femtosecond plasma channels on thunderclouds during a field campaign at the Langmuir Laboratory (New Mexico, USA, See Fig. 1). The *Teramobile* femtosecond-terawatt laser [12] was fired during two thunderstorms (labeled T1 and T2, respectively) on September 24 and 25, 2004, from South Baldy Peak, 3209 m above sea level, at a repetition rate $f = 10$ Hz. It generated multiple filamentation with significant ionization over a typical length of 100 m, a few hundreds of meters above ground. We analyzed the raw data from five receivers located within 1 km distance from the laser (See Fig. 1), which detected and located pulses at 63 MHz generated by the electric activity in the atmosphere. We looked for such RF pulses synchronized with the laser pulses.

Results and discussion

Figure 1 displays the result for T1. At the location of the laser filaments (arrow head), 3 pulses out of 7 are synchronized with the laser repetition rate (Fig. 1(a)). A statistical analysis (Fig. 1(b)) shows a high confidence level (0.987) only at the location of the laser filaments, suggesting that the laser did induce the observed electric events. A similar effect was observed during T2. No effect was observed in low or negative (downward pointing) electric field. Therefore, a small fraction (0.24 %, *i.e.* ~1 event/minute) of the filaments appear to have initiated electric events in a strong positive electric field. [12]

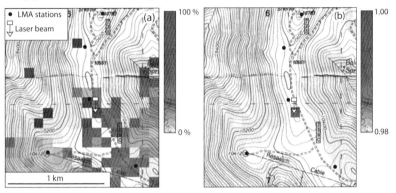

Fig. 1. Statistical analysis of the electric events detected during thunderstorm T1. (a) Rate of pulses synchronized with the laser repetition rate; (b) Corresponding statistical confidence level. Arrow head: laser-induced plasma channel. Topographic background courtesy of USGS.

The very limited effect of the plasma channel left behind by the filaments, which triggered no full lightning strike, is due to their limited lifetime of ~1 μs [13,14], corresponding to an effective length on the meter-scale for a leader propagating at a speed of a few 10^6 m/s [15]. Since wires of a few tens of meters are sufficient for rocket triggering of lightning, the plasma lifetime has to be enhanced up to several tens of μs. This could be achieved by a subsequent nanosecond, multi-Joule laser pulse [16,17] or sequences of ultrashort pulses at 800 nm [18].

Conclusion

Our results suggest that plasma filaments generated in the atmosphere by ultrashort laser pulses can trigger electric events in thunderclouds under high positive electric field, although the efficiency achieved in our experiment is low. The triggering of actual lightning strikes requires further development, in particular to enhance the plasma lifetime and density within the filament, *e.g.* by using sequences of laser pulses.

Acknowledgements. This work has been performed in the framework of the *Teramobile* collaboration (www.teramobile.org), funded jointly by the CNRS, DFG, French and German ministries of Foreign affairs, the French ANR (grant # NT05-1_43175), Swiss FNS (grants #200021-111688/1 and 200021-116198/1), and the Swiss SER (COST P18 action C06.0114). The authors also acknowledge AFOSR and

AFRL/DE for their financial support. K. S. acknowledges financial support by the Alexander von Humboldt Foundation.

1. R. Fieux, C. Gary, and P. Hubert, Nature **257**, 212-214 (1975).

2. V. A. Rakov, M. A. Uman, and K. J. Rambo, Atmos. Res. **76**, 503-517 (2005).

3. M. Miki, Y. Aihara, and T. Shindo, J. Phys. D: Appl. Phus **26**, 1244-1252 (1993).

4. S. Uchida, Y. Shimada, H. Yasuda, S. Motokoshi, C. Yamanaka, T. Yamanaka, Z. Kawasaki, and K. Tsubakimoto, J. Opt. Technol. **66**, 199-202 (1999).

5. J. Kasparian and J.-P. Wolf, Opt. Express **16**, 466-493 (2008).

6. M. Rodriguez, R. Bourayou, G. Méjean, J. Kasparian, J. Yu, E. Salmon, A. Scholz, B. Stecklum, J. Eislöffel, U. Laux, A. P. Hatzes, R. Sauerbrey, L. Wöste, and J.-P. Wolf, Phys. Rev. E **69**, 036607 (2004).

7. G. Méjean, J. Kasparian, J. Yu, E. Salmon, S. Frey, J.-P. Wolf, S. Skupin, A. Vinçotte, R. Nuter, S. Champeaux, and L. Bergé, Phys. Rev. E **72**, 026611 (2005).

8. G. Méchain, G. Méjean, R. Ackermann, P. Rohwetter, Y.-B. André, J. Kasparian, B. Prade, K. Stelmaszczyk, J. Yu, E. Salmon, W. Winn, L. A. V. Schlie, A. Mysyrowicz, R. Sauerbrey, L. Wöste, and J.-P. Wolf, Appl. Phys. B **80**, 785-789 (2005).

9. R. Salamé, N. Lascoux, E. Salmon, J. Kasparian, and J. P. Wolf, Appl. Phys. Lett. **91**, 171106 (2007).

10. J. Kasparian, M. Rodriguez, G. Méjean, J. Yu, E. Salmon, H. Wille, R. Bourayou, S. Frey, Y.-B. André, A. Mysyrowicz, R. Sauerbrey, J.-P. Wolf, and L. Wöste, Science **301**, 61-64 (2003).

11. H. Wille, M. Rodriguez, J. Kasparian, D. Mondelain, J. Yu, A. Mysyrowicz, R. Sauerbrey, J.-P. Wolf, and L. Wöste, Eur. Phys. J. - Appl. Phys. **20**, 183-190 (2002).

12. J. Kasparian, R. Ackermann, Y.-B. André, G. Méchain, G. Méjean, B. Prade, P. Rohwetter, E. Salmon, K. Stelmaszczyk, J. Yu, A. Mysyrowicz, R. Sauerbrey, L. Wöste, and J.-P. Wolf, Optics Express **16**, 5757-5763 (2008).

13. B. La Fontaine, F. Vidal, D. Comtois, C.-Y. Chien, A. Deparois, T. W. Johnston, J.-C. Kieffer, H. P. Mercure, H. Pépin, and F. A. M. Rizk, IEEE Trans. Plasma Sci. **27**, 688-700 (1999).

14. S. Tzortzakis, B. Prade, M. Franco, and A. Mysyrowicz, Opt. Commun. **181**, 123-127 (2000).

15. M. Rodriguez, R. Sauerbrey, H. Wille, L. Wöste, T. Fujii, Y.-B. André, A. Mysyrowicz, L. Klingbeil, K. Rethmeier, W. Kalkner, J. Kasparian, E. Salmon, J. Yu, and J.-P. Wolf, Opt. Lett. **27**, 772-774 (2002).

16. X. M. Zhao, J.-C. Diels, C. Y. Wang, and J. M. Elizondo, IEEE J. Quantum Electron. **31**, 599-612 (1995).

17. P. Rambo, J. Schwarz, and J.-C. Diels, J. of Opt. A: Pure Appl. Opt. **3**, 146-158 (2001).

18. Z. Hao, J. Zhang, Y. T. Li, X. Lu, X. H. Yuan, Z. Y. Zheng, Z. H. Wang, W. J. Ling, and Z. Y. Wei, Appl. Phys. B **80**, 627-630 (2005).

Optimizing laser-induced refractive index changes in "thermal" glasses

Razvan Stoian[1], Alexandre Mermillod-Blondin[1], Cyril Mauclair[1], Nicolas Huot[1], Eric Audouard[1], Igor M. Burakov[2], Nadezhda M. Bulgakova[2], Yuryi P. Meschcheryakov[3], Arkadi Rosenfeld[4], and Ingolf V. Hertel[4]

[1] Laboratoire Hubert Curien CNRS UMR 5516, Université Jean Monnet, 42000 Saint Etienne, France
[2] Institute of Thermophysics SB RAS, 630090 Novosibirsk Russia
[3] Design and Technology Branch, Lavrentyev Institute of Hydrodynamics SB RAS, 630090 Novosibirsk, Russia
[4] Max-Born-Institut für Nichtlineare Optik und Kurzzeitspektroskopie, 12489 Berlin, Germany

Abstract. Ultrafast laser radiation usually induces rarefaction and negative refractive index changes in glasses with high thermal expansion. Acting on the laser-induced heat source, tailored intensity envelopes can generate in turn thermo-mechanical compaction. This leads to positive refractive index changes and to potential creation of light guiding structures. Using additional spatial wavefront corrections, the compaction mechanisms can be preserved at arbitrary depths.

Introduction

Ultrafast lasers emerged as promising tools to structure dielectric materials for optical applications [1]. The nonlinear energy localization determines refractive index changes required to design compact optical devices; notably index increase for embedded waveguide structures. However, in certain glasses (e.g. borosilicate crown BK7) characterized by slow relaxation and high expansion under thermal load, the refractive index change induced by standard ultrafast radiation is mainly negative, detrimental for waveguide writing. The reason relates to the formation of hot regions, where, due to rapid expansion and mechanical rarefaction, the material is quenched in a low density phase. With the purpose to reverse the rarefaction tendency and to induce positive index changes we have modulated the spatio-temporal characteristics of the laser pulse in order to initiate and preserve the compaction behavior.

Experimental Methods

The irradiation tool includes a high repetition rate (100 kHz) ultrafast laser system equipped with two light control devices; a spectral phase modulator for temporal pulse shaping and an optical valve for spatial phase design. Radiation is focused inside the glass materials using a NA 0.45 microscope objective. To guide and evaluate the laser action results, a feedback loop driven by an adaptive search strategy connects an in-situ phase contrast microscope (PCM) with the pulse control units.

Results and Discussion

To counteract rarefaction, adaptive programmable pulse temporal tailoring [2] was used to initiate compaction. The optimally tailored radiation triggers a transition from a behavior dominated by thermal expansion to a regime of inelastic flow. Material compaction occurs, leading to a positive refractive index change. The index flip in BK7 is shown in Figure 1. (a,b) and results in guiding structures [Fig. 1. (c,d)]. The inversion occurs for relatively high repetition rates (100 kHz) where mechanical relaxation cannot be achieved completely between consecutive pulses.

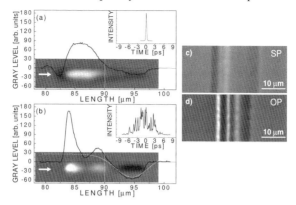

Fig. 1. PCM images of refractive index flip in BK7 under short (a) and optimally tailored (b) irradiation in a multipulse (10^5) accumulation regime at 0.17 μJ pulse energy. White colors represent negative refractive index changes, black colors represent positive refractive index changes. The onset of an axial compression zone is noticeable under optimal conditions. Inset: corresponding laser pulse intensity envelopes. Transverse traces written by laser pulses in standard (c) and optimized (d) conditions at 1.1 μJ energy/pulse and 50 μm/s scan velocity.

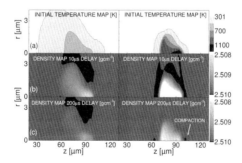

Fig. 2. Single pulse excitation maps (absorbed energy-top) and subsequent material density maps (thermoelastoplastic calculations) under standard (left) and optimized (right) excitation at 0.5 μJ. Axial compaction becomes visible for optimal pulse shapes upon cooling.

The material response is simulated using an approach that couples together nonlinear pulse propagation and the thermoelastic material response [3], namely the evolution of heat and strain waves in the glass material. A description of the laser-induced heat sources and subsequent density maps is given in Figure 2. The relaxed density maps after laser excitation with different pulse durations and repetition rates show that axial densification becomes possible for ps pulse envelopes at repetition rates on the timescale of mechanical relaxation. In this case narrow energy distribution for optimized ps pulses determines shock-induced plastic deformations accompanied by partial healing of the lateral strain due to preferential heat flow. The momentum relaxation generates directional on-axis material compaction.

971

The material behavior indicates that compaction can be triggered if the energy density stays above a certain threshold. This can be achieved in two particular ways, plasma regulation and spatial confinement. Minimizing plasma defocusing by pulse temporal regulation allows preserving the energy concentration. Furthermore, when focusing deeper into the bulk, wavefront distortions are generated at the air-dielectric interface and the laser photoinscription process degrades due to focal elongation. Using adaptive spatial tailoring of ultrashort laser pulses, spherical aberrations can be dynamically compensated in synchronization with the writing procedure. Aberration-free structures can thus be induced at different depths [4], showing higher flexibility for 3D processing [Fig. 3. (a)]. This enables optimal writing of homogeneous longitudinal waveguides over significant lengths [Fig. 3. (b)].

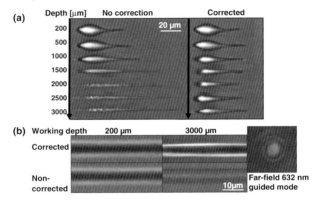

Fig. 3. (a) Non-corrected (left) and spatially-corrected (right) static structures induced at different depths. The structures are generated by 10^5 pulses of 150 fs duration and 1.25 μJ energy at 100 kHz. (b) Longitudinal scan in BK7 (10 μm/s). Guiding is observed on longer distances when wavefront corrections are used.

Conclusions

Spatio-temporal design of ultrafast heat sources enforces key control factors for refractive index change; geometry and temperature level. Creating positive refractive index changes in materials which do not allow it in standard irradiation conditions is beneficial for applications in photonic technologies, enabling as well a fundamental perspective into the mechanisms of material changes on microscales.

Acknowledgements. We acknowledge the support of PICS and ANR programs.

1   K. M. Davis, K. Miura, N. Sugimoto, and K. Hirao, Opt. Lett. **21**, 1729, 1996.

2   A. M. Weiner, Rev. Sci. Instrum. **71**, 1929, 2000.

3   A. Mermillod-Blondin, I. M. Burakov, Y. P. Meshcheryakov, N. M. Bulgakova, E. Audouard, A. Rosenfeld, A. Husakou, I. V. Hertel, and R. Stoian, Phys. Rev. B **77**, 104205, 2008.

4   C. Mauclair, A. Mermillod-Blondin, N. Huot, E. Audouard, and R. Stoian, Opt. Express, **16**, 5841, 2008.

Femtosecond laser fabrication for the integration of optical sensors in microfluidic lab-on-chip devices

R. Osellame[1], R. Martinez Vazquez[1], C. Dongre[2], R. Dekker[2], H.J.W.M. Hoekstra[2], M. Pollnau[2], R. Ramponi[1], and G. Cerullo[1]

[1] Istituto di Fotonica e Nanotecnologie del CNR – Dipartimento di Fisica del Politecnico di Milano, P.zza L. da Vinci 32, 20133 Milano, Italy
E-mail: Roberto.osellame@polimi.it
[2] Integrated Optical MicroSystems, MESA+ Institute for Nanotechnology, University of Twente, PO Box 217, 7500 AE Enschede, The Netherlands
E-mail: M.Pollnau@ewi.utwente.nl

Abstract. Femtosecond lasers enable the fabrication of both optical waveguides and buried microfluidic channels on a glass substrate. The waveguides are used to integrate optical detection in a commercial microfluidic lab-on-chip for capillary electrophoresis.

A lab-on-chip (LOC) is a device that squeezes onto a single glass substrate the functionalities of a biological laboratory, by incorporating a network of microfluidic channels, reservoirs, valves, pumps and micro-sensors [1]. It offers the capabilities of preparation, transport, reaction and analysis of very small volumes (nano- to picoliters) of biological samples. Its main advantages are high sensitivity, speed of analysis, low sample and reagent consumption and the possibility of measurement automation and standardization. The LOC concept has a huge application potential in many fields, ranging from basic science (genomics and proteomics), to chemical synthesis and drug development, point-of-care medical analysis and environmental monitoring. The next technological challenge of LOCs is direct on-chip integration of photonic functionalities, by manufacturing optical waveguides for sensing of biomolecules flowing in the microchannels. Such integrated approach has many advantages over traditional free space optical sensing, including compactness, sensitivity, enhanced device portability and the possibility of multipoint excitation. However, standard waveguide fabrication methods are planar multistep processes, which considerably complicate the LOC device production.

Femtosecond-laser induced refractive index modification is a powerful technique enabling direct, maskless three-dimensional fabrication of optical waveguides in glass [2], and appears to be particularly suited for their integration into LOCs. It allows to position optical waveguides (or more complex photonic devices such as splitters and interferometers) inside a pre-existing LOC without affecting the manufacturing procedure of the microfluidic part of the device, thus greatly simplifying the production process. In addition, femtosecond-laser irradiation of fused silica followed by chemical etching in HF solution allows the manufacturing of directly buried microfluidic channels [3], due to the enhanced (by up to two orders of magnitude) etching rate of the irradiated material with respect to the pristine one. This opens the possibility of using a single femtosecond laser system for the production and the integration of microfluidic channels and optical waveguides [see Fig. 1(a)].

In this work we demonstrate the fabrication, by femtosecond laser irradiation, of both high-quality optical waveguides and microfluidic channels on the same fused silica substrate. We also use the femtosecond laser to inscribe optical waveguides on a

973

commercial LOC for capillary electrophoresis (CE), for the realization of compact, highly sensitive integrated optical sensors.

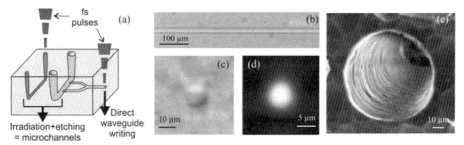

Fig. 1. (a) Conceptual scheme of the use of the femtosecond laser for the fabrication of optical waveguides and microfluidic channels; (b,c) top and end view of a femtosecond written waveguide;(d) waveguide mode at 532 nm;(e) SEM image of a microfluidic channel obtained by femtosecond laser irradiation followed by chemical etching.

The femtosecond laser fabrication process employs a regeneratively amplified Ti:sapphire laser generating 150-fs, 500-μJ, 800-nm pulses at 1 kHz. A fraction of the beam is focused, by either a 20× (NA = 0.3) or a 50× (NA = 0.6) objective, at a depth ranging from 100 μm to 500 μm below the surface of the fused silica sample, which is moved perpendicularly to the beam propagation direction by a precision translation stage at a speed of ≈20 μm/s. In order to produce modifications with a circular cross-section, the beam is astigmatically shaped by passing it through a cylindrical telescope [4]. The optical waveguides, fabricated using the 20× objective and 4 μJ pulse energy, are very uniform, have a nearly circular cross section with 10 μm diameter and are single mode in the visible [see Figs. 1(b)-(d)]. The peak refractive index change is 1×10^{-3}. Propagation losses in the visible are 0.5÷1 dB/cm, which is generally a lower value than those obtained for waveguides integrated on LOCs by photolithographic techniques. The microfluidic channels are manufactured by higher intensity laser irradiation (pulse energy of 4 μJ through a 50× objective) and subsequent etching for 3 hours in an ultrasonic bath with a 20% HF solution in water. With double side etching, channel lengths up to 3 mm with 100 μm cross section are obtained [5]. The channels are directly buried and can be positioned at any depth with respect to the sample surface, in a three-dimensional fashion. Increased channel lengths, required for the LOC applications, can be obtained by a combination of iterative etching, concentration gradient and non-uniform sample irradiation. The quality of the microchannels is high, as shown in the SEM image of Fig.1e.

To test the applicability of our fabrication technique to real-world devices, we inscribed optical waveguides in a commercial LOC for CE (LioniX bv, the Netherlands), a schematic layout of which is shown in Fig.2(a). This chip has two crossing microchannels, that are responsible for the sample injection (channel going from reservoir 1 to 3) and for the electrophoretic separation (channel going from reservoir 2 to 4). Several optical waveguides have been inscribed perpendicular to the separation channel towards its end, to provide highly localized excitation for laser-induced fluorescence (LIF). A precise alignment procedure has been developed in order to have the 10 μm-diameter waveguides exactly crossing the 12-μm-high microfluidic channel. A low power green laser was coupled to a waveguide by a single mode optical fiber and the separation channel was filled with a solution of Rhodamine 6G. Figure 2(b) shows the yellow fluorescence excited by the green light coupled into the waveguide. The excitation is selective in space (10 μm, as the waveguide diameter), indicating low light leakage out of the waveguide. In addition, the fluorescence covers the whole width (50 μm) of the channel due to the low divergence of light coming from a waveguide with a numerical aperture below 0.1

As shown in Fig.2(a), the LIF signal is collected by an ultrahigh numerical aperture optical fiber (NA = 0.5) glued to the chip in correspondence to the excited portion of the microchannel in a 90° geometry with respect to the exciting waveguide, thus strongly suppressing the excitation light background. The signal is detected by a photon counting photomultiplier after notch and interference filters, used to further reject the excitation light and chip autofluorescence. Measurements indicate a limit of detection of 40 pM, comparable to the best results obtained with standard free-space optics, with the clear advance of enhanced compactness and portability.

In conclusion, the femtosecond laser is a powerful tool for the fabrication of optical waveguides and microfluidic channels and their combination in optofluidic devices. It can be used to integrate optical excitation/detection in commercial LOCs without affecting their manufacturing process, strongly increasing device compactness and portability.

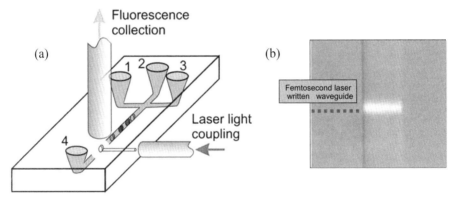

Fig. 2. (a) Schematic diagram of the commercial microfluidic chip with femtosecond laser inscribed optical waveguides; (b) fluorescence induced in the microchannel by the optical waveguide.

Acknowledgements. This work was funded by the European Commission, 6th FP STREP Project Contract No. IST-2005-034562 [Hybrid Integrated Biophotonic Sensors Created by Ultrafast laser Systems (HIBISCUS)].

1   G.M.Whitesides, Nature **442**, 368-373 (2006).

2   K. M. Davis, K. Miura, N. Sugimoto and K. Hirao, Opt. Lett. **21**, 1729-1731 (1996).

3   A. Marcinkevicius, S. Juodkazis, M. Watanabe, M. Miwa, S. Matsuo, and H. Misawa, Opt. Lett. **26**, 277-279 (2001).

4   R. Osellame, S. Taccheo, M. Marangoni, R. Ramponi, P. Laporta, D. Polli, S. De Silvestri, and G. Cerullo, J. Opt. Soc. Am. B **20**, 1559-1567 (2003).

5   R. Osellame, V. Maselli, R. Martinez Vazquez, R. Ramponi, and G. Cerullo, Appl. Phys. Lett. **90**, 231118, 2007.

Tailored Femtosecond Pulses for Nanoscale Laser Processing of Dielectrics

Lars Englert[1], Matthias Wollenhaupt[1], Lars Haag[1], Cristian Sarpe-Tudoran[1], Bärbel Rethfeld[2] and Thomas Baumert[1]

[1]Institut fuer Physik und CINSaT, Universitaet Kassel, Heinrich-Plett-Str. 40,
 D - 34132 Kassel, Germany
 E-mail: baumert@physik.uni-kassel.de
[2] Fachbereich Physik, Technische Universitaet Kaiserslautern, Erwin-Schroedinger-Strasse 46,
 D - 67663 Kaiserslautern Germany

Abstract. Laser control of two basic ionization processes in dielectrics on intrinsic time and intensity scales with temporally asymmetric pulse trains is investigated. We create robust structures one order below the diffraction limit.

Introduction

Lasers delivering ultrashort pulses have emerged as a promising tool for processing wide band gap materials for a variety of applications ranging from precision micromachining on and below the wavelength of light to medical surgery [1] [2]. Within this context it is the transient free-electron density in the conduction band of the dielectric that plays a fundamental role in addition to various propagation and relaxation mechanisms. A large number of experiments makes use of the threshold of observed damage as experimental evidence for exceeding a certain critical electron density after the laser interaction. These involve pulse duration measurements [3] [4] [5] and recent pulse-train experiments [6] all showing a strong dependence of the damage threshold on pulse duration and on pulse separation. Direct studies of transient electron densities range from intensities below [7] [8] up to well above the breakdown threshold [9] [10]. The temporal evolution of the free-electron density and the role of the fundamental ionization processes are strongly depending on time and intensity as well as on the instantaneous frequency [11] [12] [13]. Two main processes for generating free electrons are multi photon ionization (MPI) and Avalanche-Ionization (AI). MPI requires no free initial free electrons and has highest efficiency for shortest pulses AI on the other hand needs initial free electrons and needs time to establish. In our work we make use of temporally asymmetric femtosecond pulses in order to control MPI and AI. Control leads to different final electron densities (and energies) as the direct temporal profile and the time inverted profile address the two ionization processes in a different fashion. This results in observed different thresholds for material modification in fused silica as well as in reproducible lateral structures being an order of magnitude below the diffraction limit.

Experiment

In our experiment we combine femtosecond pulse shaping techniques [14] with a microscope setup for material processing [15] [16]. Linear polarized laser pulses with 35 fs full width at half maximum (FWHM) pulse duration and a central wavelength of 790 nm are provided by an amplified Ti:Sapphire laser system. After

976

passing a calibrated home built spectral phase modulator [17] the pulses are focused via a 50 x 0.5 NA objective to a spot diameter of 1.4 μm (1/e² value of intensity profile). The pulse shaper is operated in a parameter regime far away from severe space time coupling effects [18] as can be seen for ex. from quantum optical measurements with the same set up [19]. Shaped pulses are characterized in the interaction region. The sample is translated by a 3-axis piezo table to a new position for each shot. A typical measurement pattern consists of an array of points where we vary the pulse shapes, energy and focal z-position. After laser processing the samples are analyzed via scanning electron microscopy (SEM).

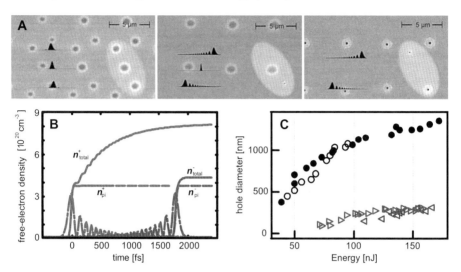

Fig. 1: A) SEM micrographs of a measurement pattern on fused silica: For an applied energy E and focal position a triplet of applied laser pulses is highlighted by the ellipse. Negative, zero and positive TOD were used. Normalized temporal intensity profiles are sketched for comparison between different TODs. *Left:* low TOD (statistic pulse duration of 2 σ = 50 fs; E = 77 nJ) results in negligible differences between created structures. *Middle:* high positive TOD (statistic pulse duration of 2 σ = 960 fs; E = 71 nJ) results in a change of structure size and threshold energy. *Right:* the threshold energy for ablation with high negative TOD is reached with E = 110 nJ. Here the unshaped pulse is suppressed in order not to mask structures with TOD. **B)** Transient free-electron density n_{total} (solid lines) as calculated with help of the MRE, together with the density of electrons provided by photoionization n_{pi} (dashed lines) and the corresponding transient intensities (dashed-dotted lines) of the pulse with positive TOD (index +) and negative TOD (index -), respectively. **C)** Diamaters of ablation structures as a function of pulse energy for unshaped pulses (circles), for (+) shaped pulses (triangles pointing right) and for (-) shaped pulses (triangles pointing left). Without changing the focus spot diameter structures below 300 nm are obtained over a large energy range thus providing a large process window for creation of nanostructures. The smallest structures are about 100 nm in diameter.

Results and Discussion

Systematic studies with phase shaped laser pulses based on third order dispersion (TOD) leading to asymmetric temporally shaped laser pulses revealed a change in the threshold depending on whether the direct pulse shape or the time inverted profile was used (see Fig. 1A). Theoretical simulations based on a multiple rate equation (MRE) model described in [12] show, that it is the timing of an intense photoionizing sub-pulse that can turn on or off AI [16] (see Fig. 1B). The observed nanoscale structures are an order of magnitude below the diffraction limit and remarkably stable with respect to variations in laser fluence (see Fig. 1C) [20]. We propose that it is the interplay of MPI creating free electrons in a spatially very confined region followed by AI that restricts the area of reaching the critical electron density (and energy) that may eventually lead to the nanoscale structures seen for positive and negative TOD pulses.

Conclusions

We conclude that control of ionization processes with tailored femtosecond pulses is an important prerequisite for robust control of laser processing of high band gap materials and that our strategy opens the route to develop tailored pulse shapes for controlled nanoscale material processing of dielectrics.

1   H. Misawa and S. Juodkazis, in 3 D Laser Microfabrication, (WILEY 2006).

2   A. Vogel, J. Noack, G. Hüttman, and G. Paltauf, in Appl Phys B, Vol. 81, 1015, 2005.

3   B. C. Stuart, M. D. Feit, A. M. Rubenchik, B. W. Shore, and M. D. Perry, in Phys Rev Lett, Vol. 74, 2248, 1995.

4   A. C. Tien, S. Backus, H. C. Kapteyn, M. M. Murnane, and G. Mourou, in Phys Rev Lett, Vol. 82, 3883, 1999.

5   M. Lenzner, J. Krüger, S. Sartania, Z. Cheng, Ch. Spielmann, G. Mourou, W. Kautek, and F. Krausz, in Phys Rev Lett, Vol. 80, 4076, 1998.

6   R. Stoian, M. Boyle, A. Thoss, A. Rosenfeld, G. Korn, and I. V. Hertel, in Appl Phys A, Vol. 77, 265, 2003.

7   S. S. Mao, F. Quéré, S. Guizard, X. Mao, R. E. Russo, G. Petite, and P. Martin, in Appl Phys A, Vol. 79, 1695, 2004.

8   V. V. Temnov, K. Sokolowski-Tinten, P. Zhou, A. El-Khamhawy, and D. von der Linde, in Phys Rev Lett, Vol. 97, 237403-4, 2006.

9   I. H. Chowdhury, X. Xu, and A. M. Weiner, in Appl Phys Lett, Vol. 86, 151110-3, 2005.

10 C. Sarpe-Tudoran, A. Assion, M. Wollenhaupt, M. Winter, and T. Baumert, in Appl Phys Lett, Vol. 88, 261109-3, 2006.

11 A. Kaiser, B. Rethfeld, M. Vicanek, and G. Simon, in Phys Rev B, Vol. 61, 11437, 2000.

12 B. Rethfeld, in Phys Rev Lett, Vol. 92, 187401-4, 2004.

13 E. Louzon, Z. Henis, S. Pecker, Y. Ehrlich, D. Fisher, M. Fraenkel, and A. Zigler, in Appl Phys Lett, Vol. 87, 241903-3, 2005.

14 A. M. Weiner, in Rev Sci Instr, Vol. 71, 1929, 2000.

15 A. Assion, M. Wollenhaupt, L. Haag, F. Mayorov, C. Sarpe-Tudoran, M. Winter, U. Kutschera, and T. Baumert, in Appl Phys B, Vol. 77, 391, 2003.

16 L. Englert, B. Rethfeld, L. Haag, M. Wollenhaupt, C. Sarpe-Tudoran, and T. Baumert, in Optics Express, Vol. 15, 17855, 2007.

17 A. Präkelt, M. Wollenhaupt, A. Assion, C. Horn, C. Sarpe-Tudoran, M. Winter, and T. Baumert, in Rev Sci Instr, Vol. 74, 4950, 2003.

18 B. J. Sussman, R. Lautsen, and A. Stolow, in Phys Rev A, Vol. 77, 043416-11, 2008.

19 T. Bayer, M. Wollenhaupt, and T. Baumert, in J Phys B, Vol. 41, 074007-13, 2008.

20 L. Englert, M. Wollenhaupt, L. Haag, C. Sarpe-Tudoran, B. Rethfeld, and T. Baumert, in Appl Phys A in print, 2008.

Electric Field Detection of Near-Infrared Light Using Photoconductive Sampling

M. Ashida[1,2], R. Akai[1], H. Shimosato[1], I. Katayama[3], K. Miyamoto[4] and H. Ito[4,5]

[1] Graduate School of Engineering Science, Osaka University, 1-3 Machikaneyama-cho, Toyonaka, 560-8531, Japan
[2] PRESTO, Japan Science and Technology Agency, 4-1-8, Honcho, Kawaguchi, Saitama 332-0012, Japan
[3] Interdisciplinary Research Center, Yokohama National University, 79-7 Tokiwadai, Hodogaya, Yokohama, 240-8501, Japan
[4] RIKEN, 519-1399 Aramaki-Aoba, Aoba, Sendai 980-0845, Japan
[5] Graduate School of Engineering, Tohoku University, 6-6-04 Aramaki-Aoba, Aoba, Sendai 980-8579, Japan
E-mail: ashida@mp.es.osaka-u.ac.jp

Abstract. We demonstrate electric field detection of light using photoconductive sampling with a combination of ultrashort laser pulses of 5 fs duration and a DAST crystal. We successfully observed near-infrared component beyond 170 THz.

Introduction

Direct measurement of electric field with ultrashort laser pulses is widely used in THz spectroscopy. This method, "time-domain spectroscopy" (TDS), is very powerful for infrared spectroscopy, because it provides us many advantages, e.g. simultaneous determination of real and imaginary parts of the dielectric function of materials [1]. However, the detection frequency is usually still limited up to several THz and its extension is highly desirable. There are two detection methods for the TDS: electro-optic (EO) sampling and photoconductive (PC) sampling. The former utilizes phase distortion of ultrashort pulses by EO effect in a nonlinear optical (NLO) crystal such as ZnTe or GaSe. Kübler *et al.* reported the EO sampling of infrared light in the frequency beyond 120 THz [2]. However, this method does not suit for the detection in the Restrahlen band due to optical phonons of the crystals (between 5 and 10 THz for typical NLO crystals). On the other hand, the PC sampling with a PC switch or PC antenna, which detects photocurrent induced by incident electric field gated with ultrashort pulses, shows smooth spectral sensitivity without insensitive regions [3]. Recently we demonstrated ultra-high frequency detection up to 100 THz with the PC switch using a combination of 10 fs pulses and a thin GaSe crystal, and also suggested the detection limit can be extended to 130 THz with an advanced setup [4]. In the present work, we have demonstrated ultra-high frequency detection using PC sampling with a combination of ultrashort pulses with a duration of 5 fs and a DAST (4-dimethylamino-N-methyl-4-stilbazollium tosylate) crystal.

Experimental Methods

For the generation of near-infrared pulses, we utilized optical rectification by using a DAST crystal with a thickness of 0.4 mm. For the detection of the infrared field, we used a PC switch of gold with a 5 μm gap fabricated on a low-temperature grown

GaAs substrate. We used a mode-locked Ti: Sapphire laser of which pulse duration is approximately 5 fs at a repetition rate of 80 MHz and center wavelength is 800 nm with an average power of 200 mW for the generation and detection. The chirping of the laser pulses was controlled to obtain the highest frequency component generated by the DAST crystal. The time-integrated spectrum was measured with a combination of a monochromator and an InAs detector. The sensitivity of this system was calibrated with a standard lamp.

Results and Discussion

Figure 1 shows a time-domain waveform of ultra-broadband infrared electric field from the DAST crystal detected with the PC switch. Though the waveform is very noisy because of many absorption lines of the DAST in mid-infrared region and strong chirping, it includes an ultra-high frequency component of which period is less than 10 fs shown in the inset, a closeup of the circled region. Figure 2 shows the Fourier-transformed amplitude spectrum of Fig. 1 in logarithmic scale. The ultra-broadband spectrum of the infrared light pulse extends beyond 170 THz. The high frequency limit is determined by the absorption of a Ge filter, which is used to cut the tail of the laser spectrum down to 200 THz. If the sharper low-pass filter is used, the high frequency limit may extend to 200 THz, because the emission spectrum from the DAST crystal measured with the monochromator was observed to extend beyond 200 THz. The large nonlinearity and high transparency of the DAST crystal in near-infrared region gives rise to this ultra-high frequency generation. On the other hand, the noisy structures in the lower frequency part below 100 THz are mainly caused by the absorption of phonons of the DAST.

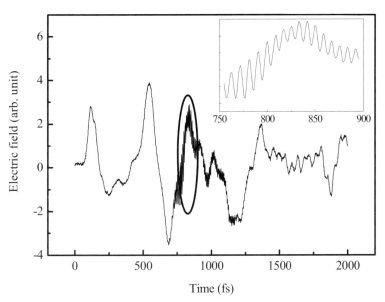

Fig. 1. A waveform of infrared electric field detected with a PC switch. The region surrounded by the circle is enlarged in the inset.

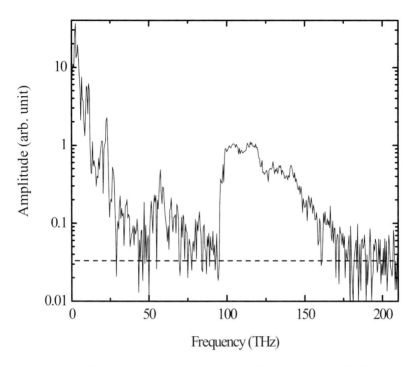

Fig. 2. Fourier-transformed amplitude spectrum of Fig. 1.

The spectral sensitivity of the PC sampling was estimated by the comparison between the obtained data shown in Fig. 2 and the calibrated emission spectrum measured with the time-integrated detection system. The result can be reproduced by a simple model based on the Drude theory [4]. This model predicts the feasibility of the detection beyond 200 THz using ultrashort pulses with 5 fs duration.

Conclusions

With a combination of 5 fs pulses and a DAST crystal, we successfully observed near-infrared component beyond 170 THz using a photoconductive switch. We found that the Drude theory is still valid in such an ultra-high frequency region. A simple model predicts that the optical communication band around 200 THz can be covered by the photoconductive sampling.

Acknowledgements. The authors would like to thank Dr. S. Saito, Prof. K. Sakai, and Prof. T. Itoh for stimulating discussions.

1   M. Tonouchi, Nature Photonics **1**, 97 2007.
2   C. Kübler, R. Huber, S. Tübel, A. Leitenstorfer, Appl. Phys. Lett. **85,** 3360, 2004.
3   S. Kono, M. Tani and K. Sakai, IEE Proc. Optoelectronics **149**, 105, 2002.
4   H. Shimosato, M. Ashida, T. Itoh, S. Saito, and K. Sakai, in Ultrafast Optics, Edited by S. Watanabe and K. Midorikawa Vol. 5, 317, 2007.

Fluorescence-Detected Two-Dimensional Electronic Coherence Spectroscopy by Acousto-Optic Phase Modulation

Patrick F. Tekavec[1], Geoffrey A. Lott[1], and Andrew H. Marcus[2]

[1] Department of Physics, University of Oregon, Eugene, OR 97403, USA
[2] Department of Chemistry, University of Oregon, Oregon Center for Optics, Eugene, OR 97403, USA
E-mail: ahmarcus@uoregon.edu

Abstract. We present a robust and high signal-to-noise strategy for phase-selective ultrafast electronic coherence spectroscopy. We demonstrate our approach using atomic Rb, isolating specific non-linear signal contributions to the excited state population.

Introduction

Two-dimensional electronic coherence spectroscopy (ECS) is an important method to study the coupling between distinct optical modes of a material system [1-3]. Such studies often involve excitation using a sequence of phased ultrashort laser pulses. In conventional approaches, the delays between pulse temporal envelopes must be precisely monitored or maintained. Here we introduce a new experimental scheme for phase-selective non-linear ECS, which combines acousto-optic phase modulation with ultrashort laser excitation to produce intensity modulated non-linear fluorescence signals. We isolate specific non-linear signal contributions by synchronous detection, with respect to appropriately constructed references. Our method effectively decouples the relative temporal phases from the pulse envelopes of a collinear train of four sequential pulses. We thus achieve a robust and high signal-to-noise scheme for phase-selective ECS to investigate the resonant non-linear optical response of photo-luminescent systems. We demonstrate the validity of our method using a model quantum three-level system – atomic Rb vapor. Moreover, we show how our measurements determine the resonant complex-valued third-order susceptibility.

Experimental Methods

In our approach [4, 5], we use continuously driven acousto-optic (AO) Bragg cells to impart time-varying phase shifts to consecutive incident pulses of a laser pulse train. By placing two such AO cells, each in separate arms of a Mach-Zehnder interferometer and operating at well-defined and distinct frequencies (Ω_1 and Ω_2), we create a train of collinear pulse pairs with relative phase modulated at the AO difference frequency ($\Omega_{21} = \Omega_2 - \Omega_1$). Based on the above technique, we construct a train of four collinear pulses consisting of two independently phase modulated pulse-pairs ($\Omega_{21}/2\pi \sim 5$ kHz and $\Omega_{43}/2\pi \sim 8$ kHz; see figure 1).

When this train excites a quantum system, such as the D-line transitions of atomic Rb vapor, the excited state population depends on both linear and non-linear coherences, which contribute to the fluorescence signal. While the linear signal terms oscillate at the fundamental frequencies Ω_{21} and Ω_{43}, the non-linear terms oscillate at the combination (side-band) frequencies $\left(\Omega_{43} - \Omega_{21}\right)/2\pi \sim 3$ kHz and $\left(\Omega_{43} + \Omega_{21}\right)/2\pi \sim 13$ kHz.

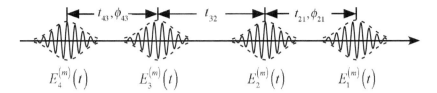

Fig. 1. Train of four sequential phase-modulated pulses. Each pulse sequence is labeled by the superscript m; individual pulses are labeled by the subscripts 1 – 4. A pulse sequence is characterized by the interpulse delays t_{21}, t_{32}, and t_{43}, and the relative temporal phases ϕ_{21} and ϕ_{43}.

We synchronously detect the fluorescence signal using "sum" and "difference" frequency reference waveforms, which we generate from the pulse spectral densities. This approach, combined with a consistent phase calibration method, ensures that the signal phase is measured relative to that of the excitation pulses. We use the resulting de-modulated signals to construct complex-valued interferograms, proportional to overlaps between third-order wave packets excited by three of the four pulses, and first-order wave packets excited by the remaining pulse. These fluorescence-detected "difference" and "sum" frequency signals are, respectively, analogous to the rephasing and non-rephasing third-order polarizations, characterized in four-wave mixing experiments and excluding the effects of non-resonant interactions.

Results and Discussion

In figures 2 and 3, we compare our measurements of the rephasing and non-rephasing

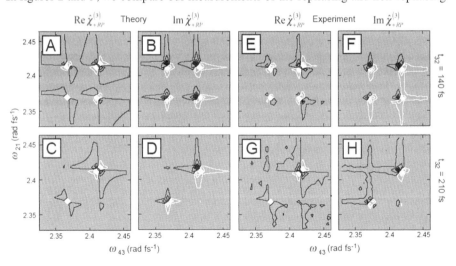

Fig. 2. Comparison between theoretical calculations [panels (a) – (d)] and experimental data [panels (e) – (h)] for the rephasing susceptibility obtained by phase-modulation ECS.

third-order susceptibilities to theoretical calculations, which are based on the known three-level Hamiltonian for the Rb system. We determine the susceptibilities by Fourier transforming the "sum" and "difference" frequency interferograms.

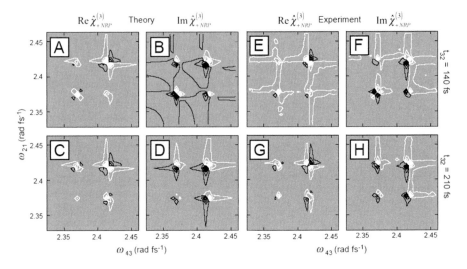

Fig. 3. Comparison between theoretical calculations [panels (a) – (d)] and experimental data [panels (e) – (h)] for the non-rephasing susceptibility obtained by phase-modulation ECS.

Both rephasing and non-rephasing susceptibilities agree quantitatively with our calculations. We emphasize that both the real and imaginary parts of the spectra are accurately described, confirming that the data are properly phased.

Conclusions

We have developed a phase-selective technique for two-dimensional femtosecond spectroscopy, called phase-modulation electronic coherence spectroscopy (PM-ECS). PM-ECS provides a straightforward route to measure the complex-valued resonant third-order susceptibility. The establishment of this method is an important step toward future experiments on photo-luminescent molecular systems, in which the optical nonlinearity results from direct coupling between optical transitions.

Acknowledgements. This work is supported by the National Science Foundation.

1   R. M. Hockstrasser, Proc. Nat. Acad. Sci. **104**, 14190, 2007.

2   J. C. Vaughan, T. Hornung, K. W. Stone, and K. A. Nelson, J. Phys. Chem. A **111**, 4873, 2007.

3   D. M. Jonas, Annu. Rev. Phys. Chem. **54**, 425, 2003.

4   P. F. Tekavec, T. R. Dyke, and A. H. Marcus, J. Chem. Phys. **125**, 194303, 2006.

5   P. F. Tekavec, G. A. Lott, and A. H. Marcus, J. Chem. Phys. **127**, 214307, 2007.

Single-pulse standoff nonlinear Raman spectroscopy using shaped femtosecond pulses

Adi Natan[1], Ori Katz[1], Salman Rosenwaks[2], and Yaron Silberberg[1]

[1] Dep. of Physics of Complex Systems, Weizmann Institute of Science, Rehovot 76100, Israel.
E-mail: adi.natan@weizmann.ac.il
[2] Dep. of Physics, Ben Gurion University of the Negev, Beer Sheva 84105, Israel

Abstract. We demonstrate fast standoff (>10 m) single-pulse coherent anti-Stokes Raman spectroscopy (CARS), from trace amounts of solids, under ambient light conditions, using phase or amplitude shaped femtosecond pulses.

Introduction

Standoff spectroscopy and detection schemes span almost every instrumental method available, including diverse optical spectroscopy methods, such as Coherent anti-Stokes Raman scattering (CARS). In resonant CARS, a pump (ω_p) a Stokes (ω_s) and a probe (ω_{pr}) photon mix to excite a signal photon $\omega_{CARS} = \omega_R - \omega_s + \omega_{pr}$. One of the main problems in nonlinear spectroscopy and in CARS in particular, is the strong non-resonant background signal that often masks the resonant one (Fig 1a).

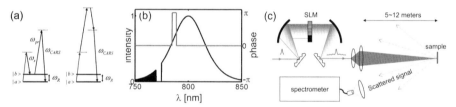

Fig. 1. (a) Energy level diagrams of resonant (left) and non-resonant (right) CARS processes. (b) Schematic drawing of the spectral intensity (black line) and phase (gray line) of a spectral π phase-gated pulse. Also shown is the spectral region (dark area) where the CARS signal is measured. (c) Outline of the experimental setup.

Much of the experimental effort is therefore directed at the reduction of this background term to a minimum. Several approaches to this issue used tailoring of the probe-pulse width and temporal delay [1, 2], as well as polarization shaping [3]. While these techniques are beneficial for applications such as chemical imaging in CARS microscopy they are not optimal for standoff spectroscopy, where the collected signal scatters off the material. For example, polarization shaping for standoff detection has been recently studied [4], but it is sensitive to depolarization processes caused by the multiple random scattering inside a sample. Other weak resonant signal detection schemes [2, 5] focus on the reduction of the strong coherent non-resonant background signal and naturally do not benefit from its existence.

Recently, our group introduced a novel approach [6], where the non-resonant background serves as a local oscillator for phase-sensitive detection of the resonant signal. This is especially effective when the non-resonant background is stronger than the resonant signal, as often occurs for short pulses. By π shifting the spectral phase of a narrow frequency bandwidth of the excitation pulse, the resonant signal is π/2 phase shifted relative to the non-resonant background with which it was initially in quadrature. The small intensity variation due to the resonant signal is amplified by homodyne detection with the strong non-resonant background. Consequently, high-resolution spectroscopy is achieved although the non-resonant signal level remains practically unaffected. Alternatively, we show that applying an amplitude gate will have a similar effect, resulting in a different interference signature. The use of such approach opens the possibility to considerably simplify the experimental setup, by replacing the pulse shaper apparatus with custom made notch filters.

Results and Discussion

In order to demonstrate the ability to measure CARS spectra from a distance, we measured the CARS power spectrum both from a transform limited and from either a π phase gate shaped pulse (4-pixel wide, corresponding to 18 cm^{-1}), or a similar amplitude gate shaped pulse. The application of the phase gated probe has two effects. First, it induces two peaks at the resonant spectrum, which are located at the phase probe boundaries shifted by the Raman energy. Second, the relative phase between the resonant and the non-resonant components fluctuate rapidly across the phase gate. As a result, destructive interference with the non-resonant background is generated at the red-wavelength edge of the phase gate, whereas constructive interference is generated at the blue-wavelength edge. Accordingly, a dip and a peak appear in the overall signal due to the interference of the two effects, corresponding to the two respective edges of the phase gate probe. The combined effect exhibit a narrow spectral interference structure on the measured spectrum originating from the existence of a resonant level. A notable spectral structure also appears for the amplitude gated pulse shape, since it is equivalent to the coherent addition of flat phase and π/2 phase gate amplitudes that similarly interfere between the resonant and the non-resonant components across the amplitude gate.

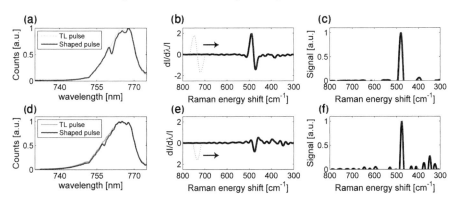

Fig. 2. The measured CARS signal of Sulfur. (a) The transform limited signal and the phase gated probe signal. The characteristic spectral interference structure around 761 nm is produced from a phase gated probe located around 790 nm. (b) The normalized intensity derivative difference of the phase gated probe spectrum from the transform limited pulse is used to separate the spectral structure from the non-resonant background. (c) Convolving this structure with a predetermined matched filter function resolves the vibrational levels spectrum and reveals the strong 473 cm^{-1} line of Sulfur. (d)-(f) Following the same procedure for an amplitude gated pulse. In this case the probe was located around 786, manifesting in an interference spectral structure around 757 nm.

Thus, by measuring the signal obtained from a substance such as Sulfur, that has a resonant Raman level at 473 cm^{-1}, a distinct maximum followed by a minimum is exhibited for the phase gated shaped pulse (Fig. 2a). For the case of the amplitude shaped pulse, only a minimum is exhibited (Fig. 2d). The level structure is then extracted by normalized intensity derivative difference of the measured CARS that is convolved with a predetermined matched filter function (Fig. 2b-2c and Fig. 2e-2f).

The experiment was realized (Fig 1c) using a Ti:S multipass amplifier system (Femtolasers GmbH) producing 1 KHz of 30 fs pulses. Following that, the laser beam was directed into a 4-f pulse shaper setup with a programmable liquid crystal spatial light modulator (Jenoptik Phase SLM-640) located in the Fourier plane. The SLM was used to implement spectral phase manipulations of the pulses, as well as for filtering out undesired short wavelength spectral components (Fig. 1b). The amplitude shaping was done by blocking a spectral component in the Fourier plane of the pulse shaper with a thin wire. We further spectrally filtered the beam (Omega Optical), and focused it to our sample from distances of 5 to 12 meters. In order to prevent any possibility of collecting a reflection of the forward CARS signal, in all measurement we placed an absorptive background behind the sample. The focused beam diameter on the sample was ~1 mm, resulting in peak powers of ~$3 \cdot 10^{11}$ W/cm^2. The CARS signal was gathered from the same distance by a 7.5' lens onto a fiber bundle configuration coupled to an imaging spectrometer, with liquid nitrogen cooled CCD (Jobin Yvon). All measurements were performed under ambient lab light conditions, and the entire CARS spectra were obtained with integration times of 100-3000 milliseconds. We further experimented with minute amounts of substances (<1000 μg) of other materials such as, Urea, KNO$_3$, RDX, and obtained similar results [7].

In summary, we demonstrated fast standoff CARS spectroscopy from trace amounts of solids, under ambient lab light conditions. We believe that the schemes presented here can be applied to improve current standoff detection tools, in particular of hazardous materials such as explosives, and chemical agents.

Acknowledgements. We would like to thank Dan Oron and Nirit Dudovich for helpful discussions.

1   D. Pestov, R. K. Murawski, G. O. Ariunbold, X. Wang, M. Zhi, A. V. Sokolov, V. A. Sautenkov, Y. V. Rostovtsev, A. Dogariu, Y. Huang, and M. O. Scully, "Optimizing the Laser-Pulse Configuration for Coherent Raman Spectroscopy", Science, **316**, 265 (2007).

2   D. Oron, N. Dudovich, D. Yelin and Y. Silberberg, "Narrow-Band Coherent Anti-Stokes Raman Signals from Broad-Band Pulses", Phys. Rev. Lett. **88**, 63004 (2002).

3   D. Oron, N. Dudovich and Y. Silberberg "Femtosecond Phase-and-Polarization Control for Background-Free Coherent anti-Stokes Raman Spectroscopy", Phy. Rev. Lett. **90**, 213902 (2003).

4   H. Li, A. D. Harris, B. Xu, P. Wrzesisnki, V. V. Lozovoy, M. Dantus "Detection of chemicals at a standoff >10 m distance based on single-beam coherent anti-Stokes Raman scattering," Optics Express **16**, 5499-5504 (2008).

5   N. Dudovich, D. Oron, and Y. Silberberg, "Single-pulse Coherently Controlled Nonlinear Raman Spectroscopy and Microscopy", Nature **418**, 512 (2002).

6   D. Oron, N. Dudovich, and Y. Silberberg, "Single-pulse Phase-Contrast Nonlinear Raman Spectroscopy", Phys. Rev. Lett. **89**, 273001 (2002).

7   O. Katz, A. Natan, S. Rosenwaks, and Y. Silberberg, "Standoff detection of trace amounts of solids by nonlinear Raman spectroscopy using shaped femtosecond pulses", Appl. Phys. Lett. **92**, 171116 (2008).

Multiphoton Microscopy by Multiexcitonic Ladder Climbing in Colloidal Quantum Dots

Nir Rubin Ben-Haim and Dan Oron

Dept. of Physics of Complex Systems, Weizmann Institute of Science, Rehovot 76100, Israel
E-mail: dan.oron@weizmann.ac.il

Abstract. Depth resolved multiphoton microscopy is performed by collecting the fluorescent emission of two-exciton states in colloidal quantum dots. This process involves two consecutive resonant absorption events, thus requiring unprecedented low excitation energy and peak power.

Introduction

The growing interest in multiphoton microscopy is mostly due to its ability to perform three-dimensional imaging in scattering media, by virtue of the localization of the excitation of a nonlinear process [1]. Microscopy based on a variety of multiphoton processes has already been demonstrated, most notably two- [2] and three-photon [3] fluorescence but also coherent processes such as second- [4] and third-harmonic [5] generation as well as coherent Raman processes [6]. All the above processes share a common feature – they are nearly instantaneous, and thus require excitation by ultrashort pulses, and peak intensities approaching the ionization thresholds of the probed media. This results in a need to precompensate dispersion of the excitation pulses, and may lead to significant sample photodamage. One attempt to partially overcome these difficulties is by the design of probes with a high nonlinear cross section, in particular for two-photon excitation fluorescence.

Here we describe a different approach to multiphoton microscopy, which relies on sequential resonant absorption in colloidal quantum dots (QDs). Since this is a doubly resonant process, the required excitation energies are much smaller than for nonresonant excitation. Moreover, the efficiency of this process remains unchanged as long as the excitation pulse is shorter than the lifetime of the two-exciton state, which is typically of the order of hundreds of picoseconds. This further reduces the requirements on peak power, resulting, in turn, in significantly reduced photodamage. The main difficulty of this method is to differentiate between the emission of the singly excited state (exciton) and that of the doubly excited state (biexciton), which we achieve by observing the transient emission dynamics.

Background

Semiconductor QDs can support multiple optical excitations. While for a well-passivated QD the single exciton state decays radiatively from the band edge (with a typical radiative lifetime of the order of tens of nanoseconds), there are two possible decay routes for the biexciton. The radiative lifetime of the biexciton is roughly half of that of the exciton, due to the twofold degeneracy. In strongly confined QDS, however, where the two excitons are forced to spatially overlap, a nonradiative decay channel, whereby a recombination event is followed by nonradiative energy transfer to

988

a spectator charge carrier, termed Auger recombination is enabled [7]. The Auger recombination lifetime is proportional to the QD volume and can, for small QDs, be as short as several tens of picoseconds. Thus, the intrinsic quantum yield of biexcitons is relatively low, of the order of 1% [7].

Typical molar absorption coefficients of QDs are in the range of 10^5-10^6 (M cm)$^{-1}$ at the exciton peak [8]. Thus, absorption of a single photon (on average) requires an energy density of the order 1-10pJ/μm^2. For pulsed excitation at or slightly below this level, biexciton creation is a second order nonlinear process, due to the Poissonian distribution of the number of absorbed photons. If the excitation pulse is shorter than the biexciton Auger recombination lifetime, the emission transient should then be a biexponential function. The ratio of the amplitudes of the fast (biexciton) component to the slow (exciton) component is an accurate measure of the energy density, and therefore a nonlinear signal enabling us to perform multiphoton microscopy.

Unlike conventional multiphoton microscopy, the nonlinear signal is independent of the excitation pulse duration, as long as it is shorter than the biexciton Auger lifetime. This enables to induce this nonlinear signal with peak powers as low as 10^5W/cm^2, and at a laser average power (at 100MHz) of much less than 1mW.

Experimental demonstration

We demonstrate this principle by showing depth-resolved detection of biexciton emission from core/shell/shell ZnSe/CdS/ZnS QDs emitting at 588nm. A frequency doubled Ti:Sapphire laser at 405nm serves as the excitation pulse. Emission is detected by a fiber-coupled avalanche photodiode (id100, id quantique) and analyzed using time-correlated single photon counting (Picoharp 300, Picoquant). In the experiments, a thin (<1μm) layer of QDs in PMMA is spin coated on a cover glass and excited through a NA=1.3 oil immersion microscope objective. Epi-fluorescence time traces are collected as a function of the layer height. To evaluate the depth resolution afforded by this method, we perform a z-scan of our QD sample with a relatively low fluence of approximately 0.25ph/dot/pulse at the focal spot, a regime in which the generation of BX is clearly quadratic in the excitation intensity. Four typical time traces, as the sample is driven out of focus, are shown in Fig. 1(a). As can be seen, while the sample is far from focus (3μm - bottom), only a slow (exciton) component is apparent. As the sample is driven into focus (top curve), a significant fast transient (BX) component arises. The decomposition to the fast and slow components as a function of the sample position is shown in Fig. 1(b), for fitted lifetimes of τ_X^{rad}=14ns and τ_{BX}^{Aug} =150ps, which gives a relative quantum yield of approximately 2% for the BX.

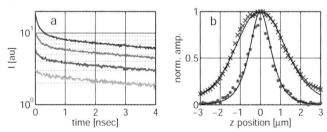

Fig. 1. (a) Measured histograms at 0.25ph/dot/pulse for z positions of 0, 0.75, 1.5 and 3μm (top to bottom). (b) Decomposition of the time traces to slow (crosses) and fast (dots) normalized components accompanied by model calculation (solid line).

Clearly, the depth resolution of 1.4μm afforded by the fast component, corresponding to the second-order nonlinear signal, is significantly better than that exhibited by the slow component (2.8μm). The latter is due to the effective confocal pinhole induced by the 100μm multimode fiber through which the signal is coupled to the detector.

A simple quantitative evaluation of the performance of this method, taking into account both the excitation geometry and the Poissonian distribution describing the probability of photons absorption by the QDs is proposed. The results of this calculation (solid lines in Fig. 1(b)) were found to reproduce well the z-scan profiles.

Conclusions

While the above presented results are representative of the strength of the doubly resonant multiphoton microscopy technique, they are far from being an optimal realization. The two most important parameters are the quantum yield of the biexciton state and the peak energy density required for multiphoton excitation. For single component QDs, the Auger recombination rate is inversely proportional to the volume [7, 9] while the radiative lifetime is nearly volume independent [10]. Thus, by increasing the QD volume, the BX quantum yield can be easily increased to the 10-20% region. This would also be accompanied by an increase in the absorption cross section, and a resulting decrease of the peak intensity required for excitation of biexcitons. In order to further reduce the required pulse energy, one can increase the exciton radiative lifetime, so that absorption of two photons can occur during more consecutive pulses. This can be accomplished, for example, by utilizing type-II quantum dot heterostructures in which charge separation occurs within the QD itself [11]. In this case, the reduction of the required peak power will, however, inevitably involve a reduction in the emission rate. We believe this holds great promise for excitation deep inside a scattering tissue, particularly since, at the required low excitation levels, photodamage to the observed tissue is completely eliminated. To summarize, considering the dramatic improvement in controlled synthesis of complex quantum dot structures in recent years, the doubly resonant multiphoton microscopy scheme outlined here holds great promise for simple and practical three-dimensional fluorescence imaging.

1 W.R. Zipfel, R.M. Williams, W.W. Webb, in Nature Biotech. 21, 1369 (2003).

2 W. Denk, J.H. Strickler, W.W. Webb, in Science 248, 73 (1990).

3 S. Maiti, J.B. Shear, R.M Williams, W.R. Zipfel, W.W., Webb, in Science 275, 530 (1997).

4 J. Gannaway, C.J.R. Sheppard, in Opt. Quant. Electron. 10, 435, (1978).

5 D. Yelin, Y. Silberberg, in Opt. Express 5, 169 (1999).

6 A. Zumbusch, G.R. Holtom, X.S. Xie, in Phys. Rev. Lett. 82, 4142 (1999).

7 V.I. Klimov, A.A. Mikhailovsky, D.W. McBranch, C.A. Leatherdale, M.G. Bawendi, in Science 287, 1011 (2000)

8 W.W. Yu, L. Qu, W. Guo, X. Peng, in Chem. Mater. 15, 2854 (2003).

9 B. Fisher, J.-M. Caruge, Y.-T. Chan, J. Halpert, M. G. Bawendi, in Chem. Phys. 318, 71 (2005).

10 A. F. van Driel, G. Allan, C. Delerue, P. Lodahl, W. L. Vos and D. Vanmaekelbergh, in Phys. Rev. lett 95, 236804 (2005).

11 S. Kim, B. Fisher, H.-J. Eisler, M. Bawendi, in J. Am. Chem. Soc. 125, 11466 (2003); D. Oron, M. Kazes, U. Banin, in Phys. Rev. B 75, 035330 (2007).

Real-time wave-packet engineering using a sensitive wave-packet spectrometer and a pulse-shaper

Kazuhiko Misawa[1,2] and Kengo Horikoshi[1,2]

[1] Department of Applied Physics, Tokyo University of Agriculture and Technology,
Naka-cho, Koganei 184-8588, Japan
E-mail: kmisawa@cc.tuat.ac.jp
[2] CREST, JST, 5 Sanban-cho, Chiyoda-ku, Tokyo 102-0075, Japan

Abstract. Real-time wave-packet engineering was demonstrated by full capture of the phase-controlled wave-packet motions. Optimal pulses were obtained by adaptive pulse-shaping for excitation of coherently controlled wave-packet motions. Selective excitation of twisting and bending modes was successful in a cyanine dye molecule by using a chirped pulse sequence.

Introduction

Adaptive control using tailored femtosecond pulses is one of the promising methods for controlling photo-induced structural changes in molecules or condensed matters [1]. At the early stage of such photo-chemical processes, direct observation and control of wave-packet dynamics are essentially important. In the previous experiments, iterative optimization of the excitation pulses has been realized mainly by a learning algorithm based on the feedback from the amount of final photo-product [2]. Wave-packet dynamics has not been very much concerned in the learning-algorithm experiments, because of the relatively long acquisition time for capturing dynamical wave-packet motions.

We developed a highly sensitive wave-packet spectrometer, which enabled us to measure the time-resolved difference transmission spectra at 20 frame-per-second [3]. Owing to the advantage of our rapid-scanning wave-packet spectrometer, one minute is sufficient for capturing full information of wave-packet motions. In the present paper, we demonstrate controlled excitation and real-time observation of vibrational wave-packets using a pulse shaper. We chose chirped pulse sequence with a controlled sub-pulse interval [4]. Vibrational wave-packet associated with a twisting and bending motions was selectively excited in cyanine dye molecules.

Experimental Methods

The experimental setup was composed of a mode-locked Ti:S oscillator, a pulse shaper, and a rapid-scanning wave-packet spectrometer, as shown in Fig. 1. The duration, pulse energy, and repetition rate of femtosecond pulses (FemtoSource, FemtoLasers) were of 13 fs, 1 nJ, and 78 MHz, respectively. The sensitive wave-packet spectrometer is a pump-probe setup with a rapid-scanning optical delay stage (ScanDelay 15, APE) [3]. The probe light was spectrally dispersed in a polychromator equipped with a linear photo-diode array, each pixel of which was connected to an electrical high-pass filter for extracting the pump-dependent signal. This combination of the rapid-scanning stage and the electric high-pass filters is essential for improvement of the sensitivity and the data acquisition time.

An optical phase modulator based on a spacial light modulater (SLM-256,CRI) was

991

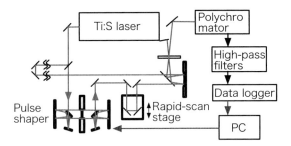

Figure 1: Experimental setup composed of a mode-locked Ti:sapphire oscillator, a pulse shaper, and a rapid-scanning wave-packet spectrometer.

inserted in the pump path of the sensitive wave-packet spectrometer. Adaptive modulation of pulse profiles and measurement of wave-packet dynamics were automatically performed. A multiple pulse sequence was generated with a sum of quadratic and sinusoidal phase shifts. The adjustable parameters were the second-order dispersion and the interval of the chirped pulses. The incident pulse was divided into three subpulses which share almost equal amplitude.

Results

Figure 2 shows the contour plot of the time-resolved difference transmission spectra ((a), (b)) and their Fourier transforms ((c), (d)) in case of two optimized dynamics. The sample molecule was diethylthiatricarbocyanine iodide (DTTCI) dissolved in ethanol. Because each time-resolved spectrum could be obtained within 3 minutes, parameter scans were also possible for optimization of the excitation pulses.

Oscillatory structure with 230- or 67-fs round-trip time of the wave-packet motion is evident in the transient spectra. These two components correspond to 4.5- and 15-THz vibrational mode frequencies, which are strongly related to the twisting and bending motions of DTTCI molecules, respectively [5].

The optimized pulses favorably excited the twisting ((a), (c)) or bending ((b), (d)) motion. A single positively chirped pulse with $+300$-fs^2 GDD efficiently induced the 4.5-THz twisting mode, whereas a pulse sequence of negatively chirped pulses with -300-fs^2 GDD induced the 15-THz bending mode. Exclusive generation of 15-THz mode could not be achieved by a single chirped pulse. By adjusting the interval of the sequence to 67fs, the triplet pulse generated only 15-THz mode suppressing the lower frequency modes. This mode selectivity was 1.5-times enhanced by providing negative GDD to make negative-chirped triplet pulse.

Discussion and Conclusion

The important result here is that the optimization of the spectral phase could be realized by full capture of wave-packet motion. The transient profiles due to the wave-packet dynamics were controlled only by the spectral phase of the excitation pulses.

Full capture of wave-packet motion is important, because the full information tells us how to discriminate between the excited and ground-state wave-packets [6]. The maximal intensity in the Fourier transforms ((c), (d)) corresponds to the turning position

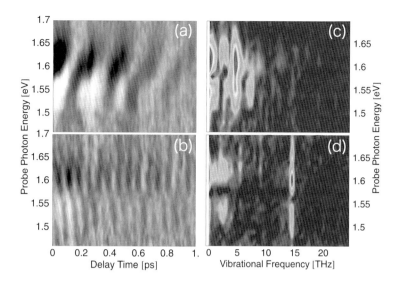

Figure 2: Contour plot of the time-resolved difference transmission spectra ((a), (b)) and their Fourier transforms ((c), (d)) for selective excitation of the 4.5-THz twisting mode ((a), (c)) and 15-THz bending mode ((b), (d)).

of the round-trip motion, and the minimal intensity between the two maxima represents the equilibrium position on the potential surface. The inner turning position and midpoint of the round-trip motion should be observed at the Franck-Condon transition at the stationary absorption and fluorescence peaks, respectively, in case of excited-state wave-packets. In the present result, the inner turning position at 1.62 eV and the midpoint at 1.56 eV were actually located at the stationary absorption and fluorescence peaks. The present result unambiguously suggests that the controlled wave-packet is on the excited-state potential surface. This result is essentially significant, because the photo-induced chemical reaction is effective via excited-state potential surfaces.

In conclusion, we developed an experimental system composed of a sensitive wave-packet spectrometer and a pulse-shaper for real-time wave-packet engineering. Owing to the rapid-scanning spectrometer, full capture of wave-packet motion greatly improved the optimization procedure of adaptive pulse shaping. We achieved almost arbitrary vibrational-mode control by using single and multiple chirped pulses, which were generated by combining parabolic and sinusoidal phase modulation. The twisting and bending modes were individually driven in DTTCI molecules by using a chirped pulse sequence.

1 R. S. Judson and H. Rabitz, Phys. Rev. Lett. **68**, 4, 1500 (1992).
2 A. Assion, T. Baumert, M. Bergt, T. Brixner, B. Kiefer, V. Seyfried, M. Strehle, and G. Gerber, Science, **282**, 919 (1998).
3 K. Horikoshi, K. Misawa, R. Lang, Ultrafast Phenomena XV, 175 (2007).
4 J. L. Herek, W. Wohlleben, R. J. Cogdell, D. Zeidler, M. Motzkus, Nature **417**, 533 (2002).
5 K. Ishida, F. Aiga, and K. Misawa, J. Chem. Phys. **127**, 194304 (2007).
6 K. Horikoshi, K. Misawa, R. Lang, J. Chem. Phys. **127**, 054104 (2007).

Grating Enhanced Ponderomotive Scattering for Characterization of Femtosecond Electron Pulses

Christoph T. Hebeisen, German Sciaini, Maher Harb, Ralph Ernstorfer, Thibault Dartigalongue, Sergei G. Kruglik, and R. J. Dwayne Miller

Institute for Optical Sciences and Departments of Physics and Chemistry, University of Toronto, 80 St. George St., Toronto, ON, M5S 3H6 Canada
Email: dmiller@LPhys.chem.utoronto.ca

Abstract. We demonstrate a method for measuring the duration of femtosecond electron pulses capable of 10 fs accuracy, using the ponderomotive force of the intensity grating produced by counterpropagating laser pulses in the microjoule energy range.

Femtosecond Electron Diffraction (FED) has made enormous progress towards directly observing structural dynamics at the atomic level in the past few years [1,2]. The atomic motions in question happen on the 100 fs timescale, the time it takes an atom to move the distance of a chemical bond at the speed of sound. Laser pulses shorter than 100 fs are readily available but the generation of intense femtosecond electron pulses remains a technical challenge. To obtain the desired time resolution, one must not only be able to produce such pulses but also to characterize their duration as well as their exact time of arrival at the sample [3].

In the past, electron pulse durations have been determined through streak cameras [4]. While streak camera technology has reached the sub-100 fs regime, the pulse parameters of the electron pulses used in FED make this technique unsuitable. Due to large electron densities, the pulse shape and duration change while the electron pulse traverses the streaking field, rendering the duration an ill-defined quantity. Also, due to the extended nature of the streaking plates, the pulse duration at the sample position cannot be probed directly. To avoid these problems, a strongly time-dependent interaction has to take place over a short section of the propagation path at the sample position. We demonstrated a method that meets these conditions based on the ponderomotive force of a very intense focused femtosecond laser pulse [5]. However, this method is not practical for most researchers because it requires pulse energies in excess of 10 mJ. In this paper, we describe a new measurement, which reduces the pulse energy requirement by two orders of magnitude.

The ponderomotive force is proportional to the gradient of the light intensity [6]. In the case of a single laser pulse, as used in our previous experiments, the intensity gradients - and therefore the ponderomotive force - arise due to the spatial and temporal envelope of the laser pulse. The length scale over which the intensity varies is given by the $c\tau_l$ (where τ_l is the duration of the laser pulse) in the direction of propagation of the laser pulse and by the beam waist in the lateral directions. These two characteristic length scales were 30 μm and 14 μm, respectively in our previous experiments. While shorter pulses and tighter focusing increase the ponderomotive force, both lead to reductions in the observed scattering. In the present study we used two counterpropagating laser pulses. When these two pulses overlap, they form a transient intensity grating, in which the distance between maxima and minima is $\lambda/4$ i.e. 194 nm for a 775 nm Titanium Sapphire laser. Hence, the intensity gradients, which produce the ponderomotive force, are significantly enhanced. The use of standing light waves to scatter and diffract electrons was proposed by Kapitza and Dirac [7] in 1933 and has been realized with relatively low energy electrons [8].

While the ponderomotive force of a single laser pulse acts in all directions as dictated by the spatial and temporal envelope of the laser pulse, the transient grating only enhances the force along the propagation direction. For a pulse duration of at least a few cycles and a laser beam waist significantly larger than the laser wavelength, the force in this direction dominates.

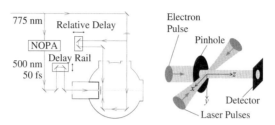

Fig. 1. Left side: the optical setup. A 775 nm, 200 fs laser pulse is split into two arms. One arm is used to drive a NOPA which generates a 500 nm, 50 fs pulse used to create the electron pulse after a delay stage. The other arm is focused by a lens and the converging beam is split evenly into two arms, whose foci are overlapped to create the transient intensity grating at their crossing with the electron beam. Right side: detailed view of the laser-electron interaction region. The diameter of the electron beam is reduced by a pinhole to match the laser focus size. After interaction with the intensity grating, the electrons are detected by a microchannel plate / phosphor screen detector.

The setup of the pulse duration measurement is shown in Fig. 1. Due to Coulomb interaction between the electrons, a high charge density electron pulse undergoes accelerated spreading while it propagates. To keep the electron pulse as short as possible by the time it reaches the sample in FED experiments, we use a very compact electron gun so that the pulse propagation distance is less than 30 mm [1].

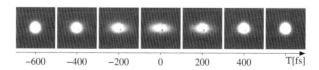

Fig. 2. Electron beam profiles at different laser pulse-electron pulse delay times. The images at ±600 fs are virtually identical to the electron beam profile without laser interaction. $10,000$ electrons per pulse, 135 μJ pulse energy.

The two laser pulses and the electron pulse all have to overlap in time and in space for the electrons to get scattered. The multi-dimensional search for this condition is facilitated by first using a rectangular pinhole at $45°$ between the electron and laser pulse propagation directions and aligning all beam paths through this pinhole. The approximate temporal overlap is then determined separately between each of the laser pulses and the electron pulse using a laser generated plasma, a method commonly used in FED to determine the $t = 0$ position [1]. The final adjustments of spatial and temporal overlap are made by maximizing the deflected electron probe signal itself. We characterize the temporal profile of the electron beam by scanning the delay to scatter different temporal sections of the electron pulse. The resulting series of images (Fig. 2) is then analyzed to yield a linear cross-correlation trace (Fig. 3) of the profiles of the laser pulse and the electron pulse. By deconvolving the laser pulse duration and a geometrical contribution from this trace, the electron pulse duration and the temporal overlap of the electron and laser pulses can be recovered. We have performed this measurement for a large range of charge densities in the electron pulse. The results are in good agreement with numerical simulations of femtosecond electron bunch behaviour. This result demonstrates the accuracy of both the measurement and the theoretical models for high density electron pulse propagation used for the design of next generation electron guns [9]. This work has led to pulses as short as 200 fs with sufficient electrons for near single shot structure determinations.

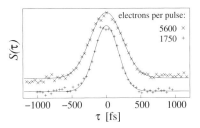

Fig. 3. Electron pulse-laser pulse cross-correlation traces. The Gaussian fits have FWHM widths of 546 fs and 404 fs, respectively.

We have demonstrated a method for the complete temporal characterization of femtosecond electron pulses. As opposed to an earlier technique, this new method can be used with common tabletop laser systems and represents a practical solution that can now be routinely implemented.

Acknowledgements. Funding for this project was provided by the Natural Science and Engineering Research Council of Canada.

1   J. R. Dwyer, C. T. Hebeisen, R. Ernstorfer, M. Harb, V. B. Deyirmenjian, R. E. Jordan, and R. J. D. Miller, Phil. Trans. Roy. Soc. A 364, 741, 2006.
2   A. H. Zewail, Annu. Rev. Phys. Chem. 57, 65, 2006.
3   W. E. King, G. H. Campbell, A. Frank, B. Reed, J. F. Schmerge, B. J. Siwick, B. C. Stuart, and P. M. Weber, J. Appl. Phys. 97, 111101, 2005.
4   B. J. Siwick, J. R. Dwyer, R. E. Jordan, and R. J. D. Miller, J. Appl. Phys. 92, 1643, 2002.
5   C. T. Hebeisen, R. Ernstorfer, M. Harb, T. Dartigalongue, R. E. Jordan, and R. J. D. Miller, Opt. Lett. 31, 3517, 2006.
6   T. W. B. Kibble, Phys. Rev. Lett. 16, 1054, 1966.
7   P. L. Kapitza, and P. A. M. Dirac, P. Camb. Philos. Soc. 29, 297, 1933.
8   P. H. Bucksbaum, D. W. Schumacher, and M. Bashkansky, Phys. Rev. Lett. 61, 1182, 1988.
9   C. T. Hebeisen, G. Sciaini, M. Harb, R. Ernstorfer, T. Dartigalongue, S. G. Kruglik, and R. J. D. Miller, Opt. Express 16, 3334, 2008.

CARS Microspectrometer with a Suppressed Nonresonant Background

R. Zadoyan, T. Baldacchini, M. Karavitis, J. Carter

Technology and Applications Center, Newport Corp. 1791 Deere Ave. Irvine, CA 92606 USA
E-mail: ruben.zadoyan@newport.com

Abstract: We describe a multiplex CARS microspectrometer utilizing a photonic crystal fiber. Enhanced contrast is achieved by subtracting the nonresonant signal in real time. The approach is demonstrated on images of polystyrene beads.

Introduction

CARS has become a powerful imaging technique with submicron resolution providing unique or/and complimentary information to traditional imaging methods [1, 2]. Multiplex CARS based on supercontinuum in photonic crystal fiber offers a cost effective and relatively simple solution for CARS microscopy and spectroscopy [3]. Similar to other CARS techniques, the multiplex CARS also suffers from the contrast reducing nonresonant background. Several rather nontrivial approaches have been used to reduce the nonresonant background [4, 5]. We describe a multiplex CARS microspectrometer that employs a simple nonresonant signal subtraction technique based on differential image acquisition.

Experimental Methods

The block diagram of the setup is shown in Fig. 1. It is based on Spectra-Physics Mai Tai® or Tsunami® femtosecond oscillator and SCG-800 photonic crystal fiber from Newport. A portion of the 800 nm, 100 fs, 80 MHz output of the laser is used to generate supercontinuum in the SCG-800. The infrared part of the supercontinuum (longer than 800 nm) is used as a Stokes pulse. The remaining part of the 800 nm beam is spectrally narrowed to 3 nm by a bandpass filter and is used as a pump beam. Both pump and Stokes beams are overlapped in time and space using delay lines, routing mirrors and 800 nm Razor Edge Long Pass filter (Semrock). The colinearly propagating beams are directed into a XY galvo scanner positioned close to the 20x focusing objective, FO. When acquiring spectra the galvos are centered and are not scanned. The microscope slide with a single layer of 10 μm polystyrene beads is attached to the XYZ stage for the sample positioning. The forward generated CARS signal is collimated by a 10x objective. After filtering with several shortpass filters the CARS spectrum is recorded using a fiber spectrometer (OSM-series, Newport).

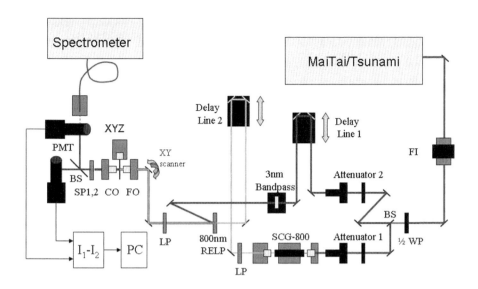

Fig 1. Block diagram of the experimental setup. FI-Faraday isolator, 1/2 WP – half wave plate, BS – beam splitter, RELP – razor edge long pass filter, LP – long pass filter, SCG-800 – supercontinuum generation kit, FO – focusing objective, CO – collimating objective, SP – combination of short pass filters, BS – polarizing beam splitter cube with ½ wave plate, PMT – photomultiplier.

Results and Discussion

The experimental CARS spectrum of the polysterene beads is shown in Fig. 2. It exhibits a well known strong peak around 3000cm^{-1} corresponding to C-H stretch, and a broad background.

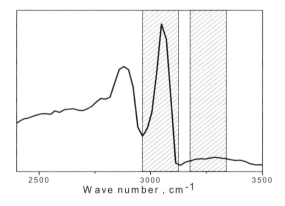

Fig. 2. CARS spectrum of polystyrene beads. The shaded area represents the FWHM transmission window of the bandpass filters in front of the two photo-multipliers.

When acquiring images of the beads, the CARS beam is directed into a polarizing cube, BS, combined with a ½ wave plate. Rotation of the wave plate allows splitting the signal into two orthogonal polarization components with variable intensity. Two identical PMTs are placed in front of two signal beams. One PMT contains a 10 nm bandpass filter centered at 650 nm corresponding to the CARS signal from C-H stretching mode. Similarly, the second PMT contains a bandpass filter centered 630 nm for the broadband background signal detection. The background signal from the photomultiplier with 630 nm bandpass filter is subtracted from the signal from the PMT with 650 nm bandpass at each position of the XY scanner. The differential signal is used to generate the image on the screen. By adjusting the splitting ratio in the beam splitting cube the best quality image can be achieved. Figure 3a. represents images of 10 µm polystyrene beads aquired at 650 nm corresponding to C-H stetching mode of polystyrene. Figure 3b is the same image at 630 nm and represents background. The differential image is shown in Fig. 3c. The white lines are intensity profiles of one bead in the field of view. Clearly, the background signal from the substrate is significnatly suppressed. The image acquired at 630 nm contains some signal from polystyrene, and the weak image of the beads is observable. This is due to the fact that transmission of the 630 nm bandpass filter partially overlaps with the C-H peak. The the presence of the residual background on the differential image can be explained by slightly different alignment of the CARS beam in each channel. It can be minimized by scanning the sample instead of the laser beams.

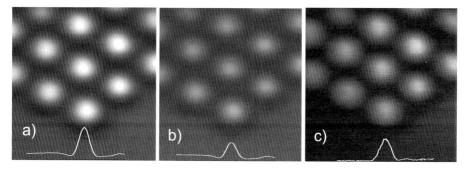

Fig. 3. Images of 10 µm polystyrene beads taken at 650nm (a), 630nm (b), and the differential image (c). The white lines are the intensity profiles of the lowest bead in the field of view.

Conclusions

We decribed a simple multiplex CARS microspectrometer based on supercontinuum generated in photonic crystal fiber. A new differential technique for background subtraction employing two channel acquisition of the resonant and nonresonant signal is proposed. The approach is demonstrated on CARS images of polystyrene beads on a glass substrate. The background signal from the substrate was reduced and better contrast was achieved.

1   J. X. Cheng and X. S. Xie, J. Phys Chem. B. **108**, 827, 2004
2   J. X.Cheng J, Applied. Spectroscopy, **61**, 197, 2007
3   H. Kano, H. Hamaguchi, Appl. Phys. Lett. **86**, 121113, 2005
4   S. H. Lim, A. G. Caster, O. Nicolet, S. R. Leone, J. Phys. Chem. B, **110** (11), 5196, 2006
5   Y. S .Yoo, D. H. Lee, H. Cho, Opt. Lett, **32**, 3254, 2007

Resonant and Nonresonant Stimulated Parametric Fluorescence Microscopy

Xuejun Liu, James Thomas, Wolfgang Rudolph

Department of Physics and Astronomy, University of New Mexico, 800 Yale Blvd NE, Albuquerque NM 87131, USA. E-mail: jthomas@unm.edu

Abstract. A femtosecond four-wave mixing microscopy with polarized detection has been applied to selectively image dyes while suppressing signals from host materials. The image signal persists even after photobleaching, making this technique attractive for biological microscopy.

Introduction

Microscopies using nonlinear optical imaging signals that are produced with ultrashort light pulses have gained interest because of their 3D imaging and novel contrast capabilities. Four-wave mixing using ultrashort pulse lasers at two different frequencies (ω_1 and ω_2) has been applied to generate coherent anti-Stokes Raman scattering (CARS) signals that allow one to image the distribution of molecular vibrational resonances [1].

Four-wave mixing may also be enhanced through a two-photon absorption of either laser (Fig. 1a); with near-IR input frequencies, such an absorption is typically of electronic nature. Signals of frequency $2\omega_2-\omega_1$ are generated by parametric stimulated emission, and can be used to image the distribution of these electronic resonances [2]. This process, also referred to as 'stimulated parametric fluorescence' (SPF) cannot readily be distinguished from possible CARS signals if $\omega_2 > \omega_1$; for that reason, we have focused our attention on the IR SPF signal obtained when $\omega_2 < \omega_1$, for which no significant CARS contribution is possible.

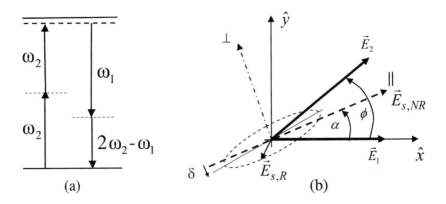

Fig. 1. (a) Schematic diagram of the generation of IR stimulated parametric fluorescence (IR SPF). The dashed lines indicate virtual levels that may be near in energy to a real resonance (solid line.) (b) Polarization geometry of the input fields E_1 and E_2 and the signal E_s for resonant (R) and nonresonant (NR) nonlinearity.

Experimental Methods

An SPF microscope was constructed using a mode-locked Ti-sapphire laser ($\lambda_1=790$ nm), which is used to pump an OPO ($\lambda_2=1035$ nm). Output pulses from both lasers (100 MHz, ~70 fs and 100 fs respectively) overlap temporally and spatially at the sample, which is positioned in the focal plane of a microscope objective (EC-Plan-NEOFLUOR 20x/0.5 Zeiss). Either the beams or the sample may be scanned, using galvano-mirrors or a piezostage. SPF at 1502 nm is filtered and detected in the forward direction after a second microscope objective (EDSCORP 20x/0.4 Edmund Scientific). In some cases fluorescent molecules were incorporated into the sample and two-photon fluorescence (TPF) was recorded in a second channel. To reduce the non-resonant contribution (background) to the signal arising from solvent molecules or host materials, we exploited a polarization technique pioneered for CARS [3,4]. With input beams linearly polarized at an angle of ϕ with respect to each other the nonresonant four-wave mixing signal is linearly polarized. The resonantly enhanced signals are elliptically polarized (Fig. 1b), where the ellipse major axis is rotated with respect to direction ∥. An analyzer at an angle β along direction ⊥ can thus block the background signal and allow detection of a component of the resonant SPF signal. In practice, with some of the nonresonant signal leaking into the detection channel, the best signal to background ratio is obtained for $\phi=72°$ [3]. The polarization of the input beams was controlled by half wave plates, and quarter wave plates were used to compensate birefringence of optical components in the beam paths. The polarization of both incident fields have extinction ratios of ~600:1 or better, measured both in front of and after the objectives. Cross-correlation IR SPF measurements, with signal

$$S(\tau) \propto \int \left| E_2^2(t) E_1^*(t-\tau) \right|^2 dt\,,$$

indicated that IR-SPF is instantaneous on a 20-fs time scale, which can be expected from the broad electronic resonances.

Results and Discussion

With collinearly polarized input beams the SPF signal changes quadratically with the concentration of rhodamine 6G dye in methanol, where the dye contributes a signal that is largely out of phase with the background.

Photobleaching is a persistent and ubiquitous problem in biological microscopy, limiting the achievable resolution, contrast, and sensitivity. Thus, the possibility of imaging fluorescent tags without photobleaching is of great interest. To study the effect of photobleaching on the resonant SPF signal, we used a PMMA film doped with rhodamine 6G molecules. Figure 2 shows the TPF signals and the IR SPF signals, measured with two different analyzer

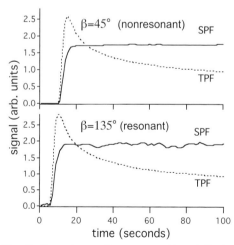

Fig. 2. Two-photon fluorescence (dotted line) and IR-SPF (solid line) signals from a rhodamine-doped PMMA film.

1001

orientations, as a function of time. Remarkably, although the dye is rapidly photobleached by the excitation pulse train, the IR SPF signal, *including the resonant signal observed along direction* ⊥, is essentially undiminished. This result implies that the features of $\chi^{(3)}$ providing the resonant IR SPF enhancement are retained even upon photobleaching.

Microscopic TPF and IR SPF images of undoped polystyrene beads (diameter ≅ 6 μm) and beads doped with a rhodamine dye (diameter ≅ 7 μm, Duke Scientific) are shown in Fig. 3. Using two-channel data collection, each IR SPF image (top row) was taken simultaneously with a TPF image (lower row.) The top left panel shows a nonresonant IR SPF image (analyzer oriented along ∥). Both undoped and doped beads are visible in the IR SPF image, at nearly equal intensities; only the doped beads are visible in the TPF image. A resonant IR SPF image taken with the analyzer oriented along ⊥ shows strong signals from doped beads with a small background leakage from undoped beads, center panel. The field was then intentionally photobleached by repeated scanning; this strongly reduced the TPF, but the resonant IR SPF signals were essentially unchanged, right panels.

The visible SPF signal from PMMA films doped with rhodamine 6G or regatta blue (a green dye) also showed persistence in the presence of photobleaching.

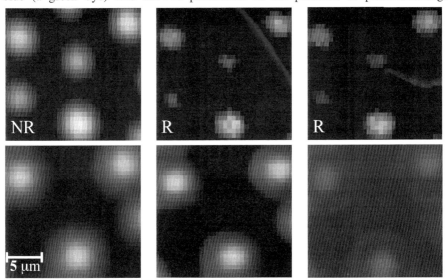

Fig. 3. IR-SPF (top) and TPF images (bottom) of mixed dyed and blank beads, taken as simultaneous pairs. After the second pair, the beads were intentionally photobleached. The fluorescence was greatly diminished, but the resonant SPF (R) was unchanged. (Note that undyed beads give a small signal in the resonant polarization, likely owing to scattering depolarization.)

Acknowledgements. We are grateful to the Army Research Office (ARO W911NF-05-1-0464) for support of this work.

1   A. Zumbusch, G.R. Holtom, and X.S. Xie, Phys. Rev. Lett. , Vol. 82, 4142, 1999.

2   K. Isobe, M. Kataoka, R. Murase, W. Watanabe, T. Higashi, S. Kawakami, S. Matsunaga, K. Fukui and K. Itoh, Opt. Exp., Vol. 14, 786, 2006.

3   J-L. Oudar, R.W. Smith, and Y.R. Shen. Appl. Phys. Lett., Vol. 34, 758, 1979.

4   J.X. Cheng, L.D. Book, X.S. Xie. Opt. Lett., Vol. 26, 1341, 2001.

Femtosecond pump-probe spectroscopy as an instrument for nanostructured materials investigation.

Sergey V. Chekalin
Institute of Spectroscopy RAS, 142190 Troitsk, Moscow Region, Russia;
E-mail:chekalin@isan.troisk.ru

Abstract: The femtosecond pump-probe technique was used to investigate the difference spectra dynamics in heterophase fullerene-metal nanostructures. The relaxation at the same metal-to-fullerene ratio strongly depends on the mutual distribution of nanocomposite components.

Information about the spatial distribution of nanostructure components can be gained by investigating the relaxation material properties. Among the most powerful instruments suited to this purpose are nondestructive optical techniques, which permit retaining all functional properties of the material under study during investigation. Clearly, the spatial distribution of nanostructure components, as well as the mutual packing of the subunits in the nanostructured material, are critical to the properties of the material as a whole. In turn, information about it can be gained by investigating the relaxation material properties. The nanometer sizes of the subunits suggest that the shortest excitation relaxation times range into the femtoseconds. That is why femtosecond spectroscopy may turn out to be one of the most adequate techniques for investigating and certifying nanostructured materials. It should be noted that the dependence of the relaxation character on the structure of the object has being observing for a rather long time in investigation of primary photoinduced processes, for example, in various bacteriochlorophyll(BChl)-containing light-harvesting complexes [1]. The main absorption bands of the complexes are due to absorption by similar BChl molecules in ring structures with a characteristic dimension of the order of several dozen nanometers, which vary mainly in the number and mutual orientation of the BChl. Despite a strong band overlap and a fast excitation energy transfer between the components of the complex, the obtained kinetic curves show the dependence on packing of components in the complex. Although the main instrument for investigation of spatial structure of ordered complexes is the X-ray analysis but it is not the case for disordered ones. Recently a spatial structure of disordered light harvesting system (associates of tightly bounded LH2 (B800-850) and LH3 (B800-820) complexes from *Thiorhodospira sibirica*) has been investigated by femtosecond pump-probe spectroscopy [2]. This experiment revealed a negative value of differential absorption polarization for the spectral range of 860-900 nm which is in contrast with similar measurements for other purple bacteria and can be indicative of larger angle between the B850 and B820 dipoles in *Trs sibirica* associates.

The femtosecond pump-probe technique was used to investigate the dynamics of transmission and reflection difference spectra in heterophase C60/Sn nanostructures, fabricated as 50-150nm films on thin quartz substrates by vacuum deposition from two different sources for the metal and the fullerene. The diagnostics of the samples thus fabricated, which were performed with the aid of X-ray structural analysis, electron microscopy, and optical and Raman spectra, showed that varying the deposition mode allowed obtaining films consisting of fullerene polymers, 10nm tin nanocrystals covered with fullerene anions (CT samples), oriented submicron-sized tin crystallites or amorphous layered structures containing metal fractals [3-5]. The kinetic curves obtained for the four above structures exhibit an extremely high sensitivity of the dynamics of ultrafast photo-induced processes in composite media to the quantitative ratio between the components and their spatial packing. The substantial difference between the relaxation processes observed for samples with various geometries of the nanocomposites (Fig. 1) arises from the distinctions in charge carrier generation and in charge transfer between the metal and the fullerene and back.

Relaxation in tin-doped polymeric samples probed in the IR range differs from that in the pure C60 films in that the relaxation rate in the initial stage increases with the metal content. This effect can be related to increased efficiency of charge carrier recombination in the polymerized fullerene. A significantly faster decay of the IR absorption bands as compared to the 500-nm band studied in [3–6] casts doubt on the interpretation that this bend belongs to the excited anion. The possible assignment of longlived photoproducts being absorbed at 500 nm to the triplet state requires additional experimental verification.

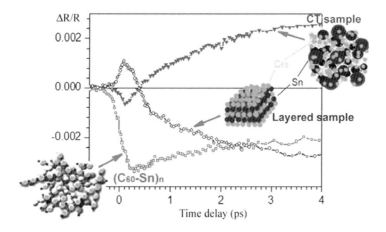

Fig. 1. Dynamics of the differential reflectance (in optical density units) measured at a wavelength of 1250 nm for various nanostructured samples with the same ratio of fullerene/metal.

The more complicated relaxation dynamics observed in the layered samples is related to the generation of nonequilibrium electrons in the metal and the formation of excited species in fullerene layers at the moment of excitation. The over-barrier transfer of nonequilibrium excited electrons from the metal to adjacent fullerene molecules leads to the formation of anions at the metal surface. The degree of excitation of these ions depends on the kinetic energies of electrons. The maximum electron energy rapidly decreases due to the electron–electron and electron–phonon relaxation. As a result, the reverse process begins whereby electrons are transferred from excited anions to the metal. At the same time, neutral excited molecules formed in the fullerene layer at the moment of excitation diffuse toward the metal, where they are effectively "quenched" by fullerene anions with the formation of additional excited anions, after which the electron transfer from excited anions to the metal is repeated. In the CT samples, ions and positively charged metal crystallites are initially present. The samples with sufficiently large metal particles are practically identical to purely metal films and exhibit relaxation behavior typical of the metals. In the samples with nanodimensional metal crystallites, the main contribution to the observed relaxation dynamics is due a submonolayer of anions adsorbed on the metal surface. The oscillating character of relaxation in these samples is related to the excitation of anions, the subsequent charge transfer to the metal with the formation of excited neutral molecules, and the excitation energy transfer to anions in the ground state, which is followed by repeated charge transfer from excited anions to the metal. It should be noted that oscillations in the signals of difference reflection observed in the films containing 10-nm crystallites could not be related to coherent acoustic waves excited in tin nanocrystals by femtosecond pulses, since in that case the amplitude and period of observed oscillations would increase with the size of particles [7], which is not observed in the experiments described above. Such nanostructures with separated charges, in which the exciting laser pulse initiates the CT from metal to fullerene and back over a period of time on the order of a few picoseconds (accompanied by the appearance and disappearance of local electric fields) can be used as ultrafast optical nanoswitches and as nanodimensional sources of ultrashort pulses in the terahertz range.

This study demonstrated the potentialities of the method of femtoseconds pump–probe spectroscopy for characterizing nanostructural materials. The main problem to be solved remains an appropriate modeling of the ultrafast processes in photoexcited nanostructures under investigation. One more improvement which is in progress now is the investigation of single complexes instead of groups.

Acknowledgements. I would like to thank Dr's A.P. Razjivin, N.F.Starodubtzev, V.O.Kompanets and A.P.Yartzev for helpful cooperation. This work was supported in part by grants from RFBR 06-04-49-072 and 06-02-16186.

1   V. Sundstrom, T. Pullerits, R. Van Grondelle, in Journal of Physical Chemistry B, Vol. 103, 2327, 1999.
2   A.P. Razjivin, I.A. Stepanenko, V.S. Kozlovsky, V.O. Kompanets, S.V. Chekalin, Z.K. Makhneva, A.A. Moskalenko, A.V. Tikhonov, V.O. Popov, T.V. Tikhonova, in Biochemistry (Moscow) Supplement Series A: Membrane and Cell Biology, Pleiades Publishing, Ltd., Vol. 1, 336, 2007.
3   S.V.Chekalin, in Journal of Experimental and Theoretical Physics, Vol. 103, 756, 2006.
4   S.V.Chekalin, V.O.Kompanets, M.S.Kurdoglyan, A.N. Oraevsky, N.F. Starodubtsev, V. Sundstrom, A.P. Yartsev, in Synthetic Metals, Vol. 139, 799, 2003.
5   S.V. Chekalin, V.O. Kompanets, N.F. Starodubtsev, V. Sundstrom, A.P. Yartsev, in Femtochemistry and Femtobilology, Ultrafast Events in Molecular Science, Edited by M.Martin, J.T.Hynes, Elsevier, 553, 2004.
6   S.V. Chekalin, A.P. Yartsev, V. Sundstrom, in Recent Advances in Ultrafast Spectroscopy, p.37, Edited by S.Califano, P.Foggi, R.Righini, Firenze, 2003.
7   M. Nisoli, S. De Silvestri, A. Cavalleri, A. M. Malvezzi, A. Stella, G. Lanzani, P. Cheyssac, and R. Kofman, in Physical Review B, Vol. 55, 13424, 1997.

Selective excitation in nonlinear microscopy by using an ultra-broadband pulse

Keisuke Isobe[1], Akira Suda[1], Masahiro Tanaka[2], Fumihiko Kannari[2],
Hiroyuki Kawano[3], Hideaki Mizuno[3], Atsushi Miyawaki[3], and Katsumi Midorikawa[1]

[1] Laser Technology Laboratory, RIKEN, 2-1 Hirosawa, Wako, Saitama 351-0198, Japan
E-mail: kisobe@riken.jp
[2] Department of Electronics and Electrical Engineering, Keio University, 3-14-1 Hiyoshi,
Kohoku-ku, Yokohama 223-8522, Japan
[3] Laboratory for Cell Function Dynamics, Brain Science Institute, RIKEN, 2-1 Hirosawa,
Wako, Saitama 351-0198, Japan

Abstract. We show that the selective excitation of vibrational mode is achieved by a single broadband pulse to focus its bandwidth into a narrow spectral region. The spectral focusing is performed by controlling the spectral phase.

Introduction

Four-wave mixing (FWM) microscopy has become an important tool for imaging structures and dynamic interactions in biological samples. By taking advantage of electronic and/or Raman spectroscopic information, specified molecules can be selectively imaged. The spectral bandwidth of the electronic resonance is much broader than that of the Raman resonance. If the spectral bandwidth of an excitation light is matched to the electronic bandwidth, spectroscopic information is generally dominated by the electronic contribution [1]. Thus, it is difficult to obtain both the electronic and Raman spectroscopic information. However, even though a 20-fs broadband pulse is used as an excitation light source, the selective excitation of the Raman mode at Ω_R is achieved by a periodic modulation of the spectral phase in the pulse [2]. In the periodic modulation, the excitation of Raman mode with a wavenumber of $N\Omega_R$, where N is integer, is also accompanied. In this paper, we propose a novel pulse shaping technique for the selective excitation of a single Raman mode. This technique is similar to the spectral focusing technique based on the chirping excitation pulses to focus their bandwidth into a narrow spectral region [3]. It is a different point that the spectral focusing is realized not by using a pair of equally linear-chirped pulses but by controlling the spectral phase of a single broadband pulse.

Principle of selective excitation of single vibrational mode

We propose the spectral phase for selective excitation of single vibrational mode in FWM process induced by a single broadband pulse. In FWM process with the Raman resonance, a pump and Stokes photons with frequencies of ω_p and ω_s excite a coherent vibrational motion with energy Ω_R. A probe photon with ω_{pr} interacts with the coherent motion to emit a signal photon with $\omega_p+\omega_{pr}-\omega_s$. We assume that FWM process is induced by a single broadband pulse. In a broadband pulse, many vibrational coherence are possibly excited by the combination of ω_p and ω_s. In order to excite a single vibrational motion Ω_R, it is necessary to focus the difference-frequency spectrum of a broadband pulse into a narrow spectral region. The difference-frequency spectrum is shaped by modulating the spectral phase in the following manner. The quadratic phase $\phi''(\omega-\omega_0)^2/2$ is applied in the entire spectral region and then the linear phase $\phi''\Omega_R(\omega-\omega_{max}+\Omega_R)$ is given in the spectral region where frequency is smaller than $\omega_{max}-\Omega_R$. Here, ϕ'', ω, ω_{max} and ω_0 are the group delay dispersion (GDD), the frequency in the pulse, the maximum frequency in the pulse and the central frequency in the pulse. The typical time-spectral distribution of the shaped pulse is shown in Fig. 1(a). As shown in Fig. 1(a), the frequency-difference between ω_p and ω_s is constant over entire pulse duration. Therefore, only single vibrational motion can be excited by the spectral focusing.

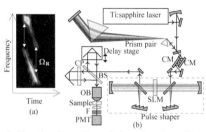

Fig. 1. (a) Time-spectral distribution of modulated pulse. (b) Experimental setup.

Experimental

A schematic of the experimental setup is shown in Fig. 1(b). As a broadband light source, we employed a Ti:sapphire laser operating at a repetition rate of 80 MHz. The laser spectrum ranged from 670 nm to 1100 nm. In order to compensate for the second-order dispersion of all optical components, the laser pulse passed through a fused silica prism pair and chirp mirrors (CMs). A grating-pair-formed pulse shaper with a liquid-crystal spatial light modulator (SLM) took on the task of compensating for the higher-order dispersion and additional shaping for the selective excitation by modulating the spectral phase. The modulated pulse was focused into a sample by an objective lens (OB) with a numerical aperture of 0.9. The resultant signal was detected by a photomultiplier tube (PMT).

To confirm the spectral focusing, we measured difference-frequency spectrum of the shaped pulse to selectively excite a single vibrational motion. Without an IR spectrometer, which is used for measuring directly difference-frequency generation, the Fourier-transform spectroscopic technique [4] was applied to obtain the difference-frequency spectrum. An interferometric autocorrelation (IAC) signal by nonresonant second-harmonic generation process includes fringe components at frequencies of around 0, ω_p, ω_s, $\omega_p-\omega_s$, $2\omega_p$, $2\omega_s$, and $\omega_p+\omega_s$. By focusing on a frequency component of $\omega_p-\omega_s$, we obtained difference-frequency spectrum. A 10-μm-thick β-barium borate (BBO) crystal was used as a nonresonant sample. Figure 2(a) shows the GDD dependence of difference-frequency spectrum. We confirmed that the bandwidth of difference-frequency spectrum narrowed with increasing the GDD. Figure 2(b) shows the difference-frequency spectra obtained by the shaped pulses to selectively excite different vibrational motions. From this result, we found that the arbitrary vibrational motion in the laser spectrum could be selectively excited.

Next, we demonstrated the selective excitation of the vibrational mode. Sample was ethanol sandwiched between a hole-slide glass and a cover glass. We obtained FWM images near the interface between ethanol and cover glass, where selective excited wavenumbers were set to 2800 cm^{-1} and 3400 cm^{-1}. Figure 3(a) shows FWM image at a selective excited wavenumber of 2800 cm^{-1}, which is identified by the CH stretching mode. FWM signal in ethanol is much higher than that in the cover glass region. Figure 3(b) shows FWM image at a selective excited wavenumber of 3200 cm^{-1}, which is a nonresonant frequency. FWM signal in ethanol is much lower than that in the cover glass region. We successfully demonstrated the selective excitation of CH stretching mode. Finally, we demonstrate the practical applicability to biological samples. Unstained HeLa cell was used as a biological sample. Figures 3(c) and (d) show the FWM images at selective excited wavenumbers of 2800 cm^{-1} and 3200 cm^{-1}, respectively. In Fig. 3(c), FWM signal was enhanced at mitochondrial membrane, while in Fig. 3(d), FWM signal was enhanced at water around cell. A wavenumber of 3200 cm^{-1} is identified by the OH stretching mode. We successfully demonstrated the multi-spectral imaging by the selective excitation.

Fig. 2. (a)GDD dependence of difference-frequency spectrum. (b)Focusing

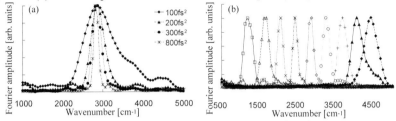

difference-frequency spectrum into various narrow spectral regions.

Fig. 3. FWM images near the interface between ethanol and cover glass at 2800 cm^{-1} (a) and 3400 cm^{-1} (b). Selective Raman-resonant FWM images of HeLa cell at 2800 cm^{-1} (c) and 3200 cm^{-1} (d).

Conclusions

We proposed that the spectral focusing into a narrow spectral region can be achieved by controlling the spectral phase of a single broadband pulse. The spectral focusing technique is powerful tool for the selective excitation of a single vibrational mode. Only by the phase modulation, the excitation mode can be switched from the electronic excitation to the Raman excitation.

Acknowledgements. We thank T. Kitajima, I. Ishikawa, T. Kogure, T. Kitaguchi, and Y. Wada in RIKEN for providing the biological sample. This research was supported by the Special Postdoctoral Researchers Program of RIKEN. This work was partly supported by a Grand-in-Aid for Scientific Research (No. 18656023) from the Ministry of Education, Culture, Sports, Science and Technology, Japan.

1   K. Isobe *et al.*, Opt. Express Vol. 14, 785, 2006.

2   N. Dudovich *et al.*, Nature Vol. 418, 512, 2002.

3   T. Hellerer *et al.*, Appl Phys. Lett. Vol. 85, 25, 2004.

4   K. Isobe *et al.*, Phys. Rev. A 2008 (in press).

Interferometrically Detected Femtosecond CARS in a Single Beam of Shaped Femtosecond Pulses

Bernhard von Vacano, Tiago Buckup, Jean Rehbinder and Marcus Motzkus

Physikalische Chemie, Philipps Universität Marburg, D-35043 Marburg, Germany.
E-mail: motzkus@staff.uni-marburg.de

Abstract. Photonic integration of functions such as excitation, probing and interferometry in shaped broadband pulses allows huge simplification of coherent anti-Stokes Raman scattering (CARS) for microspectroscopy, paving the way to cost-efficient implementations, e. g. all-fibre solutions.

Introduction

Nonlinear Raman scattering techniques, such as coherent Anti-Stokes Raman scattering (CARS) allow rapid three-dimensional microscopic imaging and spectroscopy of untreated samples. Image contrast is based on the intrinsic vibrational quantum level structure of the sample, and thus gives maps of the chemical composition [1]. As no labelling is needed and fast imaging can be achieved, CARS is a very promising tool for the study of living biological cells or complex composite materials. Technical challenges for the implementation of CARS are the necessity of high peak intensities, the availability of different coherent laser colours, presence of nonresonant background and the usually low sensitivity due to its quadratic dependence on the concentration of scattering molecules. All these issues can elegantly be addressed employing broadband femtosecond laser sources in combination with pulse shaping: The broad bandwidth of the fs-pulses contains all necessary frequency components to drive the CARS process, pulse shaping ensures spectral resolution despite the broadband excitation [2-4], and use of the phase-modulated blue spectral wing of the excitation for interferometry with the CARS signal provides highly sensitive detection [5,6]. Using the pulse shaper, additional quantum control can be exerted to increase selectivity [7]. The simplicity of the scheme makes it very attractive for coupling to an unamplified fibre-laser source to perform chemical microanalytics, which is an important step towards low-cost systems for real-life applications in microscopy, analytics or threat detection.

Single-Beam Heterodyne CARS

Using an auxiliary optical field as local oscillator with intensity I_{LO} brought to interference with the CARS signal I_{CARS}, linearization and amplification for sensitive detection can be achieved:

$$I_{Det} = I_{CARS} + 2\sqrt{I_{CARS}}\sqrt{I_{LO}} \times \cos[\Delta\phi] + I_{LO} \qquad (1)$$

with the characteristic interference term being proportional to the square roots of I_{CARS} and I_{LO} and dependent on the relative phase $\Delta\phi$. As $I_{CARS} \ll I_{LO}$, the interference term and hence the heterodyne CARS signal can be amplified by several orders of magnitude [5,6]. But most importantly, the sample concentration only enters

1009

linearly into the heterodyne signal, pushing CARS microspectroscopy towards analytical applications.

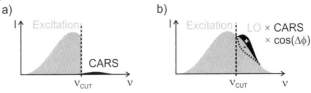

Fig. 1. a) "Conventional" single-beam CARS, where the blue wing of the excitation spectrum is suppressed in the pulse shaper in order to only detect the weaker CARS signal in this spectral position. b) Single-beam heterodyne CARS approach. The blue wing of the excitation spectrum is not blocked, but used as a local oscillator (LO) field for interferometric detection of the CARS signal. The LO phase difference $\Delta\phi$ can be controlled precisely by the pulse shaper, in order to maximize the interference signal.

Heterodyne or interferometric detection has been implemented in CARS microscopy [8-10], also in order to suppress nonresonant background. Common to all interferometric approaches is a relatively complex experimental setup. In single-beam CARS it is much easier to introduce interferometric detection. In a frequency-domain scheme, the nonresonant background generated in the sample has deliberately been used as LO [3]. One problem is however the dependence of the LO on the sample composition, and hence difficult quantitative analysis in mixtures and a limited signal amplification. Recently, we have demonstrated a novel, simplified approach to single-beam heterodyne CARS [5]. Instead of blocking the blue part of the broadband excitation spectrum for frequencies $\nu > \nu_{cut}$ in the pulse shaper (Fig. 1a), we use it as strong and fully phase-controlled LO (Fig. 1b). Hence, interferometric CARS detection is realized with intrinsic stability. Due to the interference with the strong external LO, signal amplification of a factor larger than 5000 was achieved. Fig. 2 shows the linear concentration dependence resulting from interferometric detection and consequently the ability to measure CARS spectra of a diluted sample with concentrations in the $\sim$ 100mM range. For this data, single-beam heterodyne CARS microspectroscopy has been performed on $CHBr_3$ diluted in Ethanol for different concentrations in molar fractions ranging from 1 to 0.01. Even at the lowest concentration of $x(CHBr_3) = 0.01$ used in this study, corresponding to estimated 14 attomole (8×10^6 molecules) in the excitation volume of approx. 80 attolitres, the chemically selective CARS signal was detected.

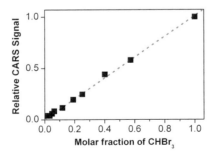

Fig. 2. Linear concentration dependence achieved with single-beam heterodyne CARS. The sample used was CHBr3 diluted in Ethanol, at different molar fractions ranging from 1 to 0.01. The indicated linear fit (slope 1.01 ± 0.01) through the origin confirms excellent agreement with the expected slope of 1.

All-Fibre-Implementation

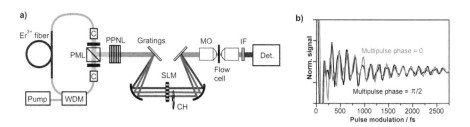

Fig. 3. a) Experimental setup. Laser source is a Er^{3+}-fs-fibre laser, frequency-doubled with a periodically poled lithium niobate crystal (PPNL). In a $4f$-pulse shaper, the red part of the spectrum is chopped (CH) for lock-in detection. CARS spectroscopy is performed in a flow cell; the signal is detected after an interference bandpass filter (IF) and a photodiode. b) Femtosecond transient measured for different phases controlled by the pulse shaper. Double quadrature spectral interferometry (DQSI) robustly retrieves the interferometric signal.

Recently, laser research has seen rapid advances of fibre technology, leading to powerful amplified systems and also to compact and robust ultrafast fibre lasers. Especially unamplified diode-pumped fs-fibre laser oscillators are now available with low operating costs and high reliability. Fibre lasers can thus pave the way for applications of ultrafast nonlinear spectroscopy outside the laser lab. In this contribution, we demonstrate for the first time single-beam heterodyne CARS spectroscopy using a fs-fibre laser (Fig. 3a). For this source, we introduce an extended measurement scheme (Fig. 3b) using a specifically designed set of phases created by the pulse shaper to always maximize signal contrast and ensure maximum constructive interference with the LO. This is necessary, as the measured signal amplitude crucially depends on the relative phase between signal and LO, which cannot fully be disentangled otherwise. The simplicity and the increased sensitivity of the scheme should prove very useful in developing single-beam CARS for microanalytics, as demonstrated in a first application by monitoring a chemical flow in a microfluidic cell.

1   J. X. Cheng, X. S. Xie, in Journal of Physical Chemistry B, Vol. 108, 827, 2004.

2   N. Dudovich, D. Oron, Y. Silberberg, in Nature, Vol. 418, 512, 2002.

3   S. H. Lim, A. G. Caster, S. R. Leone, in Physical Review A, Vol. 72, 41803, 2005.

4   B. von Vacano, W. Wohlleben, M. Motzkus, in Optics Letters, Vol. 31, 413, 2006.

5   B. von Vacano, T. Buckup, M. Motzkus, in Optics Letters, Vol. 31, 2495, 2006.

6   M. Cui, M. Joffre, J. Skodack, J. P. Ogilvie, in Optics Express, Vol. 14, 8448, 2006.

7   B. von Vacano, M. Motzkus, in Journal of Chemical Physics, Vol. 127, 144514, 2007.

8   C. L. Evans, E. O. Potma, X. S. N. Xie, in Optics Letters, Vol. 29, 2923, 2004.

9   D. L. Marks, C. Vinegoni, J. S. Bredfeldt, S. A. Boppart, in Applied Physics Letters, Vol. 85, 5787, 2004.

10 M. Greve, B. Bodermann, H. R. Telle, P. Baum, E. Riedle, in Appl. Phys. B, Vol. 81, 875, 2005.

Advantages of Two-photon Microscopy with Ultrashort Pulses

Yair Andegeko, Kyle E. Sprague, Dmitry Pestov, Vadim V. Lozovoy, Marcos Dantus

Michigan State University, Department of Chemistry, East Lansing, Michigan 48824, USA
E-mail: dantus@msu.edu

Abstract. Ultrashort pulses are expected to be beneficial for multiphoton microscopy. To utilize their advantages, however, chromatic dispersion must be compensated. We use multiphoton intrapulse interference phase scan (MIIPS) to measure and then eliminate phase distortions of pulses with a FWHM spectral bandwidth greater than 100 nm. Once compensated, the transform limited pulses (<15 fs) deliver higher two-photon excited fluorescence signal intensity, which translates into deeper optical penetration depth.

Introduction

Since the introduction of two-photon microscopy (TPM) by Webb and Denk [1] it has been known that the two-photon excited fluorescence (TPEF) signal intensity should be linearly proportional to the inverse of the pulse duration. Despite this advantage, pulses shorter than 100 fs are rarely used for multiphoton imaging, the primary reason being chromatic dispersion introduced by high numerical objectives. We utilize a method called multiphoton intrapulse interference phase scan (MIIPS) [2] to measure and compensate chromatic dispersion to obtain transform limited (TL) pulses at the focal plane. Here we demonstrate significant advantages to imaging with dispersion free ultrashort sub-15 fs pulses.

Experimental Methods and Results

Schematics of the laser system and microscope is shown in Figure 1(a). The experiments were carried out with a broad-bandwidth (over 100 nm) Ti:Sapphire oscillator with a repetition rate of 86 MHz (KMLabs, Boulder, CO). Second order dispersion (SOD), also known as group delay dispersion (GDD), was compensated by a prism pair. High-order dispersion (HOD) was measured and compensated for by MIIPS, using a 4f pulse shaper with a liquid-crystal spatial light modulator (SLM).

In order to demonstrate the dependence of TPEF on pulse duration, we measured intensities of TPEF as a function of the spectral bandwidth for GDD-compensated and TL pulses, Fig. 1(b). The experiments were performed with the same average power at the sample (2.3 mW) for all bandwidths, using a red fluorescence standard slide (Chroma Technologies). Our results show that in the case of TL pulses, obtained using MIIPS compensation, the magnitude of TPEF increases linearly almost over the whole studied range. It is enhanced by a factor of 6 as the laser bandwidth is tuned from 10 to 80 nm. In contrast, the laser pulses compensated only for SOD demonstrate a modest effect on the signal gain. Once the bandwidth reaches 30 nm, no further increase in TPEF is observed. Therefore, we conclude that above 30-nm bandwidth HOD, introduced by the prisms and the microscope objective, becomes dominant in the degradation of the signal intensity in TPM. Direct demonstration that we can deliver ultrashort TL pulses at the focus of a high-NA objective via MIIPS

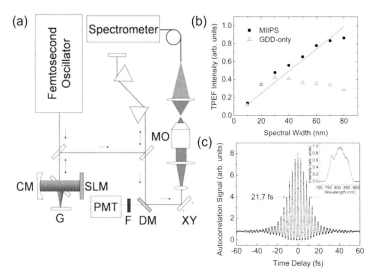

Fig. 1. (a) Schematics of the experimental setup. G, grating; CM, curved mirror; SLM, spatial light modulator (CRi SLM-640-P or SLM-640-D), DM, dichroic mirror; XY, galvanic beam scanner (QuantumDrive-1500, Nutfield Technology, Inc.); MO, microscope objective (Zeiss LD C-APOCHROMAT 40x/1.1 NA); F, emission filter; PMT, photomultiplier tube. (b) Dependence of the TPEF signal on the laser spectral bandwidth for TL (filled squares) and SOD-compensated (open triangles) pulses. The average power at the sample is 2.3 mW for all bandwidths. The sample is a red fluorescence standard slide (Chroma Technologies) [The figure is adapted from [3] with permission.]. (c) Interferometric autocorrelation of the MIIPS-corrected pulse at the focus of the objective. Phase-amplitude shaping is used to split the laser pulse into two attenuated replicas with adjustable time delay. The integrated SHG signal from a 100 μm KDP crystal at the objective focus is recorded as a function of the pulse timing, controlled by the pulse shaper.

compensation is given in Figure 1(c). We show an interferometric autocorrelation profile obtained by focusing sub-15 fs pulses on a 100-μm KDP crystal (EKSMA). The signal collected is a spectrally integrated SHG signal as a function of time delay between two collinear pulses generated by the phase-amplitude pulse shaper. The spacing between the interference features is determined by the optical cycle of the incident light (~ 2.67 fs).

The results shown above are in agreement with theory. However, it is important to demonstrate that they are applicable towards actual TPM of biological samples. TPM images on a fixed sample of mouse intestine (Molecular Probes, F-24631), given in Figure 2, were obtained with GDD-only compensation and after full MIIPS compensation for HOD. The average increase in the TPEF signal between these two conditions is at least ~5 fold.

In order to study the implications of spectral phase correction for two-photon depth-resolved imaging, we imaged a thick section of mouse brain cross section. Images obtained for TL and GDD compensated pulses are shown in Figure 3. The signal intensity and the signal-to-noise ratio for TL pulses is 5 times greater than for GDD compensated pulses.

TL GDD 6x GDD

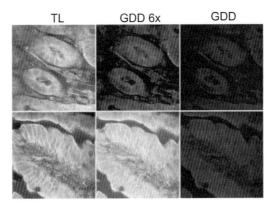

Fig. 2. TPEF imaging with MIIPS compensated pulses (TL) compared to GDD-only compensated pulses on mouse intestine tissue. Cross section of fixed intestine, stained with Alexa Fluor 350 wheat germ agglutinin (goblet cells), phalloidin Alexa Fluor 568 phalloidin (actin) and SYTOX green nucleic acid stain (nuclei). The images display a surface area expansion of villus showing epithelial cells that cover the villi (top) and microvilli (bottom) in the small intestine. The intensity with GDD compensated pulses were amplified 6x to show image detail. Image size 75 µm (top), 100 µm (bottom).

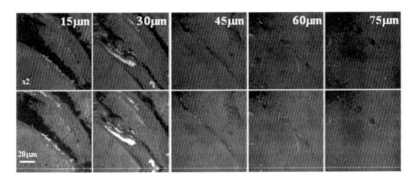

Fig. 3. TPEF images of brain: cerebrum labeled with MitoTracker 488 and Phalloidin 568, obtained at different depths, with TL pulses (top) and GDD compensated pulses (bottom).

Summary

The autocorrelation measurements confirm that sub-15 fs pulses can be delivered at the focus of a high-NA objective. The use of these ultrashort pulses provides several advantages for TPM as demonstrated in this work: increased signal intensity and extended penetration depths. The greater efficiency allows imaging with lower laser intensities, thereby reducing photobleaching and photodamage [4].

1. W. Denk, J.H. Strickler, and W.W. Webb, Science **248**, 73, 1990.
2. B. W. Xu, J. M. Gunn, J. M. Dela Cruz, V. V. Lozovoy and M. Dantus, J. Opt. Soc. Am. B **23**, 750, 2006.
3. P. Xi, Y. Andegeko, L.R. Weisel, V. V. Lozovoy, M. Dantus, Opt. Com. **281**, 1841, 2008.
4. Dela Cruz, J.M, Lozovoy, V.V and M. Dantus. J. Photochem. Photobio. A **180**, 307, 2006.

Development of laser-based imaging systems for medical diagnostics

S. Witte[1], M. Salumbides[1], E.J.G. Peterman[1], R. Brakenhoff[2], G. van Dongen[2], R. Toonen[3], H.D. Mansvelder[3] and M.L. Groot[1]

[1] Laser Centre Vrije Universiteit, De Boelelaan 1081, 1081 HV Amsterdam, The Netherlands
   E-mail: switte@few.vu.nl
[2] Otolaryngology/Head-Neck Surgery, Vrije Universiteit Medical Centre, De Boelelaan 1117, 1081 HV Amsterdam, The Netherlands
[3] Center for Neurogenomics and Cognitive Research, Vrije Universiteit, De Boelelaan 1085, 1081 HV Amsterdam, The Netherlands

Abstract. We present a laser system with high wavelength flexibility, suitable for nonlinear microscopy and optical coherence tomography, for visualization of disease-related morphological changes in vivo. A single-shot 2D OCT system is demonstrated.

Introduction

Visualization of tissue samples at the sub-micrometer level is an attractive tool for medical diagnostics, since various disease-related changes in cell or tissue function are manifested through changes in cell morphology. Laser-based imaging systems hold promise as versatile, non-invasive diagnostic tools to identify such morphological changes in vivo. For the past century, the standard for detection and diagnosis of such diseases has been histopathology, which is an invasive technique that requires tissue biopsies and ex-vivo microscopic examination.

Especially ultrafast lasers have potential for sub-cellular imaging: the high peak intensity facilitates multiphoton microscopy, while the broad bandwidth and good coherence are highly useful for optical coherence tomography. As different imaging techniques can complement each other, we are developing a system that can be used for various types of multiphoton microscopy and OCT simultaneously. With our setup, we are aiming for in-vivo visualization of neuronal networks, and of possible changes in such networks that are caused by various diseases such as Parkinson and Alzheimer [1]. In addition, we are developing a technique for early detection of cancer in the oral cavity.

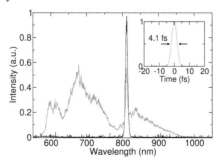

Fig. 1. Spectrum of the Ti:Sapphire laser (black trace) and the supercontinuum generated in a photonic crystal fiber (grey trace). The inset shows the temporal coherence function corresponding to the supercontinuum spectrum.

Laser system with wavelength flexibility

The heart of our setup is a modelocked Ti:Sapphire laser, which is used to pump an optical parametric oscillator (OPO) with intracavity frequency-doubling. The combination of Ti:Sapphire and OPO provides gap-free wavelength tunability between 500 and 1600 nm, as well as the OPO idler output which is tunable between 1600 and 3000 nm. With this system, various types of nonlinear interactions such as SHG, THG, sum-frequency generation and CARS can be used for imaging. Especially the near-infrared (1000-1600 nm) is promising for high-depth imaging, due to reduced light scattering by tissue in this wavelength range.

Part of the Ti:Sapphire light is split off from the OPO pump beam, and sent into a photonic crystal fiber with a zero dispersion wavelength of 790 nm. The specific properties of supercontinuum generation in this fiber lead to a relatively smooth, ultrabroadband spectrum that spans from about 600 to 1000 nm (Fig. 1), with excellent beam profile and coherence properties. This broadband light source is used in the development of an ultrahigh-resolution OCT system [2]. The width of the spectrum corresponds to a coherence function with a FWHM of 4.1 fs, leading to an achievable OCT depth resolution of 1.2 micron in air.

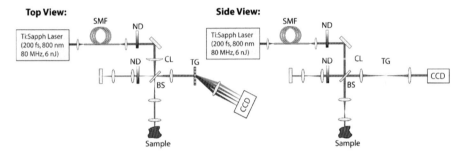

Fig. 2. Schematic layout of the constructed 2D-OCT setup. In one dimension, the light is focused onto the sample, and the scattered light is sent into an imaging spectrometer. In the other dimension, the light is collimated when reaching the sample, and this line focus is imaged onto the CCD camera. SMF: single-mode fiber, CL: cylindrical lens, BS: 50% beamsplitter, ND: variable neutral-density filter, TG: transmission grating.

Single-shot optical coherence tomography

Two major issues in OCT are the achievable resolution and the imaging speed. Especially for in-vivo applications such as medical diagnostics, the speed should be high enough for the patient to remain interferometrically still during data acquisition, and to provide the physician with real-time data. The resolution of such an image should be sufficient for visualization of all the relevant fine details.

The ideal situation for OCT is that a complete image can be measured without the need for moving parts in the setup. Spectral domain OCT has provided a breakthrough in this respect, by retrieving a complete depth scan from a single spectral interferogram [3]. With a two-dimensional CCD camera as the detector, it should in principle be possible to obtain two-dimensional images in a single shot as well. We have developed such a system, using cylindrical focusing optics to provide simultaneous depth- and transverse-direction imaging. A schematic of the optical layout of this setup is shown in Fig. 2. Initial tests using a 100 l/mm grating as the imaged object confirm a transverse resolution of 3 μm.

A drawback of the depth scans provided by spectral-domain OCT is that they contain a mirror image of the measured sample, which is caused by the presence of the complex conjugate of the signal in the spectral interferogram. While several groups have shown that this complex-conjugate ambiguity can be resolved using an additional scanning mirror [4], we use an alternative approach that does not require any moving parts in the setup and is suitable for our 2D-OCT setup. Our approach exploits the dispersion sensitivity of ultra-broadband light pulses: by producing a dispersion imbalance between the reference and sample arm of the interferometer, an additional frequency-dependent phase shift is imprinted onto the interferogram (see Fig. 3). Since the real signal and its complex conjugate have phases of opposite sign, this added phase shift allows us to determine whether a peak in the retrieved depth scan belongs to the actual signal or to its complex conjugate. Numerical phase compensation then allows the removal of the unwanted mirror images. The removal of these mirror images effectively increases the usable depth range by a factor of 2.

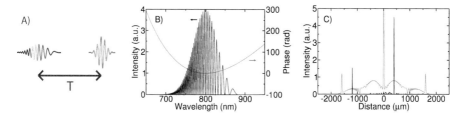

Fig. 3. A) Schematic of two pulses with a spectral phase difference, separated by a time delay T. B) The resulting interference pattern induced by these pulses in a spectrometer (black), and the spectral phase retrieved from this interferogram (grey). C) Simulation of the signal from a sample with two reflective surfaces at z = +400 μm and z = -1200 μm (with different reflection coefficients) before (grey) and after (black) numerical dispersion compensation.

Results and prospects

We have developed an OCT system capable of taking 2D images in a single shot, with a resolution of a few μm. In the near future, we plan to use this system to characterize tissue cultures with and without squamous cell carcinoma precursor lesions that occur in the oral cavity. In addition, experiments on visualization of neuronal networks in mouse brain slices, using both OCT and two-photon microscopy, are under way.

1   B.J. Bacskai and B.T. Hyman, Neuroscientist **8**, 386, 2002.
2   Y. Wang, Y. Zhao, J.S. Nelson, Z. Chen, and R.S. Windeler, Opt. Lett **28**, 182, 2003.
3   N.A. Nassif, B. Cense, M.C. Pierce, S.H. Yun, B.R. White, B.E. Bouma, G.E. Tearney, and J.F. de Boer, Opt. Express **12**, 367, 2004.
4   M. Wojtkowski, A. Kowalczyk, R. Leitgeb, and A.F. Fercher, Opt. Lett. **27**, 1415, 2002.

Frequency shifts at the fiber-optical event horizon

Christopher Kuklewicz[1], Thomas Philbin[1,2], Scott Robertson[1], Stephen Hill[1], Friedrich König[1], and Ulf Leonhardt[1]

[1] School of Physics and Astronomy, University of St Andrews,North Haugh, St Andrews, Fife, KY16 9SS, UK
E-mail: chrisk@alum.mit.edu
[2] Max Planck Research Group of Optics, Information and Photonics, Günther-Scharowsky-Str. 1, Bau 24, D-91058 Erlangen, Germany

Abstract. Event horizons can be simulated by waves in a moving medium. Using ultrashort pulses in microstructured optical fibers, we have performed the first experimental demonstration of an artificial event horizon in optics.

Introduction

In general relativity, an event horizon is either a barrier that prevents light from returning (a black-hole horizon), or from entering (a white-hole horizon). The situation is analogous to waves propagating in a medium with non-uniform speed; the locations where the velocity of the medium equals the wave's group velocity act as event horizons: points of no return or no entry [1].

We create such a moving medium by using an ultrashort pulse in a microstructured fiber. The $\chi^{(3)}$ nonlinearity of the glass causes the effective index of refraction, n, to change underneath the pulse. This changing n causes light to change speed as if the medium it is traversing were changing speed. Some nearly group-velocity-matched wavelengths of light can catch up to this pulse, but due to the cross phase modulation these wavelengths are not free to propagate past the trailing edge of the pulse. Thus this system possesses a white-hole horizon. In general relativity, the wavelength of light approaching such a horizon shrinks without limit until the unknown physics at the Planck scale is reached. In our fiber-optical analogue the behavior is contained in the wavelength-dependent refractive index. The wavelength shift stops when the group-velocity dispersion (GVD) of the fiber has sufficiently changed the light's speed relative to the pulse. For the normal GVD region, the wavelength shortened light is shifted to higher energy and this slows it down and it falls behind the pulse.

Experiment

We use the experimental setup shown in Fig. 1 to produce and measure this blue-shifting effect of the white-hole horizon. The Titanium Sapphire laser produces 70 fs-long pulses at 803 nm which are non-dispersing solitons in the 1.5-m long microstructured fiber, which has anomalous dispersion between the zero GVD points near 740 nm and 1235 nm. The fiber was selected such that the group velocity at the pulse's wavelength matches that of light near 1500 nm. Our source of probe light is a continuous-wave grating-stabilized diode laser that can be tuned from 1460 nm to 1540 nm. Due to the high birefringence ($\delta n \approx 6 \times 10^{-4}$) of our fiber, the exact group velocity-matched probe wavelength, λ_m, depends on the choice of pulse and probe polarizations; for pulse and probe co-polarized along the slow axis this matching occurs for $\lambda_m = 1499.5$nm.

Measurements were made by recording a spectrum of the probe light without the pulses and a spectrum with the pulses present and then taking the difference. Close to the probe wavelength the difference in the two spectra is too noisy from probe fluctuations, so there is a small gap around the probe wavelength in the plotted spectrum.

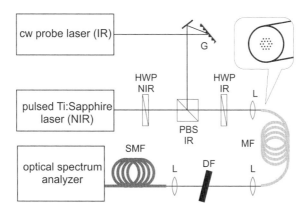

Fig. 1. Experimental setup. The pulse is combined with the cw probe inside the microstructured fiber (MF). The pulse is removed by the dichroic filter (DF) and the spectrum around the probe is recorded on the OSA. Also shown are the half-wave plates (HWP), polarizing beamsplitter (PBS), and single-mode fiber (SMF).

Results

The probe wavelength, $\lambda_p = \lambda_m + \delta\lambda$, was tuned to be slightly longer than λ_m. Since $\delta\lambda > 0$, the probe photons travel faster and part of the probe light catches up to the pulse. At the pulse, the interacting probe photons see a horizon and can no longer freely propagate forward, which opens a gap in the probe light. This gap in the cw light broadens the probe bandwidth, and the spectrum of this broadening is seen in Fig. 2 as the peak around the probe wavelength to the right of λ_m. Larger $|\delta\lambda|$ opens a longer gap which creates a narrower spectral peak. The light that interacts with the horizon is predicted to have its wavelengths shortened to $\lambda_p' = \lambda_m - \delta\lambda < \lambda_p$. This blue-shifted probe light is showing up as the peak to the left of λ_m on the spectra in Fig. 2. Also, these peaks are narrower for larger $|\delta\lambda|$. The total conversion efficiency for interacting probe light is less than 100%, due to our pulses being short enough to allow part of the probe light to tunnel through the pulses.

Further experiments have been performed with different polarization combinations of the pulses and probe light and experiments with probes with $\delta\lambda < 0$. In the latter case, probe photons are caught up to by the leading edge of the pulse, which acts like a black-hole horizon. These photons undergo the reverse of the blue-shifting process described above and are red-shifted.

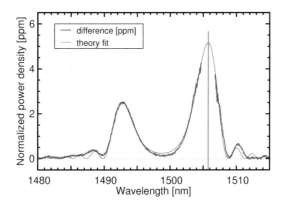

Fig. 2. Spectrum for a particular probe wavelength, which is marked with the vertical line. The pulse causes the probe light to be blue shifted to the opposite side of group velocity matching wavelength, $\lambda_m = 1499.5$nm.

Conclusion

These observations are the first demonstration of the ability to create an analogue model of an event horizon in optics. Future experiments will look for Hawking radiation created at such horizons.

1   *Artificial black holes*, edited by M. Novello, M. Visser, and G. E. Volovik (World Scientific, Singapore, 2002).

Index of Contributors